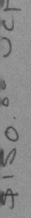

Time–Saver Standards for Architectural Lighting

Time–Saver Standards for Architectural Lighting

Gary Steffy, LC, FIALD

Gary Steffy Lighting Design Inc.

McGraw–Hill

New York   San Francisco   Washington, D.C.   Auckland   Bogotá
Caracas   Lisbon   London   Madrid   Mexico City   Milan
Montreal   New Delhi   San Juan   Singapore
Sidney   Tokyo   Toronto

Library of Congress Cataloging-in-Publication Data

Steffy, Gary R.
    Time-saver standards for architectural lighting / Gary Steffy.
        p.   cm
    Includes index.
    ISBN 0-07-061046-0
    1. Electric lighting–Standards–Handbooks, manuals, etc. 2. Lighting,
Architectural and decorative–Standards–Handbooks, manuals, etc. 3. Electric
lamps–Standards–Handbooks, manuals, etc. I. Title.
    TK4188 .S753 2000
    621.32–dc21   00-036148

McGraw-Hill
A Division of The McGraw-Hill Companies

1 2 3 4 5 6 7 8 9 0   KGP/KGP   0 6 5 4 3 2 1 0

ISBN 0-07-061046-0

*The sponsoring editor for this book was Wendy Lochner, the editing
supervisor was Frank Kotowski, Jr. and the production supervisor was
Sherri Souffrance.*

This book was set in Folio Light by Gary Steffy.

Printed and bound by Quebcor / Kingsport.

McGraw-Hill books are available at special quantity discounts to use as
premiums and sales promotions, or for use in corporate training programs. For
more information, please write the Director of Special Sales, McGraw-Hill,
Two Penn Plaza, New York, NY 10121-2298. Or contact your local bookstore.

This book is printed on recycled, acid-free paper containing
a minimum of 50% recycled, de-inked fiber.

Contents

Contents

Contents

Contents

Contents

Contents

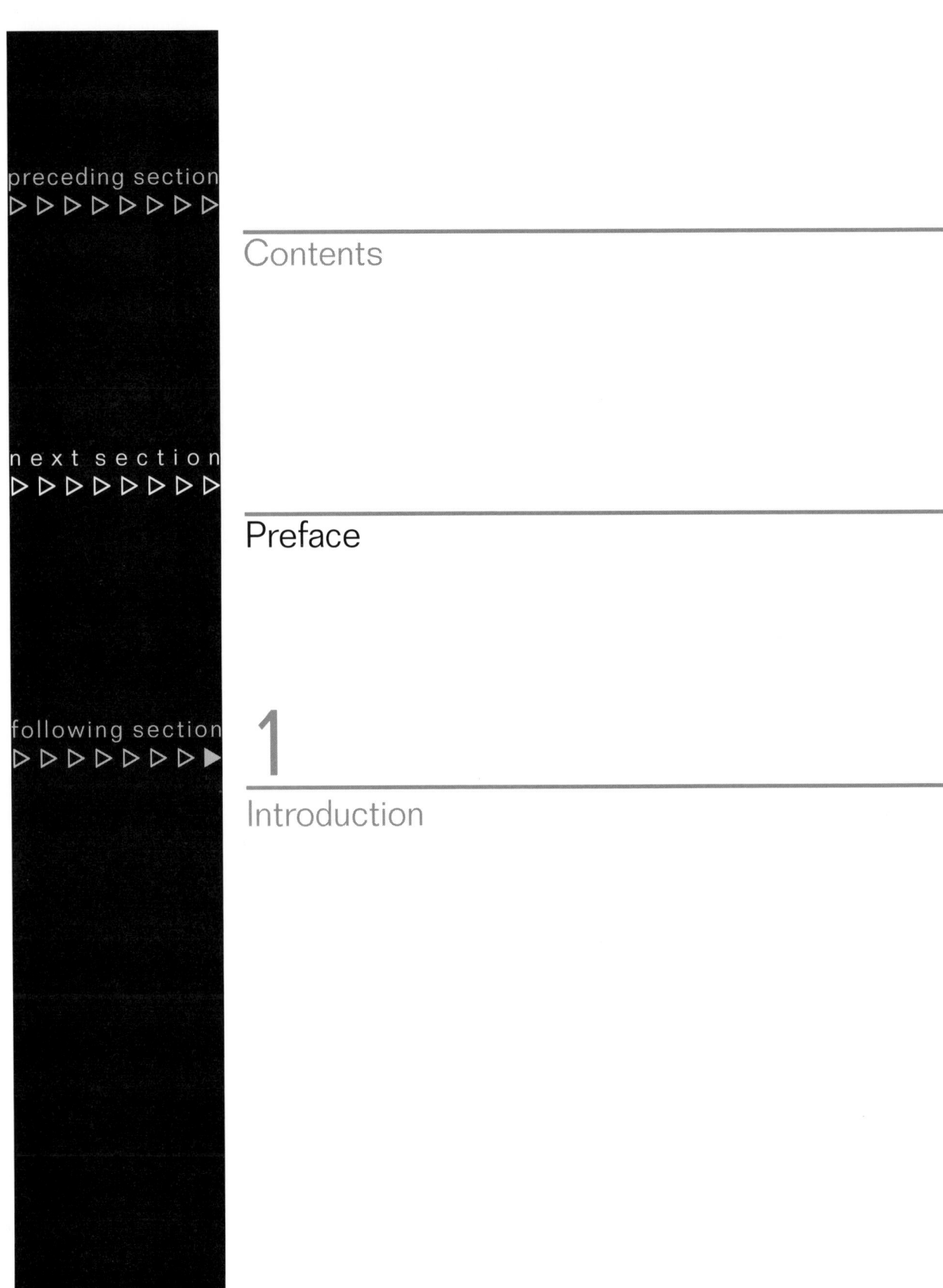

preceding section
▷ ▷ ▷ ▷ ▷ ▷ ▷ ▷

next section
▷ ▷ ▷ ▷ ▷ ▷ ▷ ▷

following section
▷ ▷ ▷ ▷ ▷ ▷ ▷ ▶

Contents

Preface

1

Introduction

Preface

Wow. The minutiae of specialties.

The seed for this book was *Architectural Lighting Graphics* by Flynn and Mills in 1962 (published by Reinhold and long out of print). The impetus came from our editor at McGraw–Hill, Wendy Lochner, who wrote in 1995 about updating a text by the late Lee Watson. This Time–Saver Standards reference combines specific application needs, lighting technologies and computed data with a graphically rich presentation style to give the designer and engineer a resource reference for developing lighting concepts.

Why you want this book

Back to the very first sentence above. One could take the tack that with a few basic lighting principles, anyone in the design and engineering fields could go off and design lighting. There are more than a few basic principles. To learn them sufficiently well to apply their premises on each and every project would, indeed, make you a lighting designer. If you're not a lighting designer, you want some store of knowledge and ideas (that just can't fit in your head with all the other stuff) that is applicable—to today's work. And that consists of more than formulae, geometric diagrams, and table upon table of multipliers and coefficients (all of which are totally meaningless unless you understand those darned principles, in which case you'd be a lighting designer and not in this position of needing a good clue, right now, about the lighting parameters of a cove, for example—like, how deep should the cove be; how high does the ceiling need to be; what kind of light (well, here you mean luminaire and lamp) should be used in the cove; and, well, how much light (here you mean illuminance) will the cove generate). This book is the store of such information.

Preface

So, the philosophy here is to present you with lighting technologies, techniques, and concepts. Stuff that has, through practice, worked out pretty well for many applications. If you want fundamental knowledge about light (well, actually you may need to look under "visible radiation") or you want to know about those lighting principles (things like radiative transfer, vision, psychological aspects), then please check out the good number of lighting texts currently available (a search on Amazon.com is a good start). Another fine fundamental reference is the Illuminating Engineering Society's Ninth Edition handbook, *The IESNA Handbook/Reference and Application*.

Acknowledgments

Many folks helped with this effort and I wish to acknowledge a few here. First, I could not have agreed to undertake this effort on behalf of Gary Steffy Lighting Design Inc. had I not had a supportive family and staff. Once again (they're getting used to this book stuff), my wife Laura and daughter Heather have been very understanding as I missed movies, lunches and/or dinners, or stayed up too late, or as I asked for household quiet (much of the writing and editing took place out of the office to avoid the distractions of all of the other work). Thank you and I love you both.

My staff helped by developing original artwork and performing computer simulations and calculations and editing my editing and maintaining the normal workload of a small consultancy. Thank you to Damon Grimes and Gary Woodall, LC, IALD. Independent contractors Mark de la Fuente (while studying at University of Kansas) and Emily Koonce (while studying at University of Colorado) were very helpful in compiling data and results for Sections 9 and 10.

Second, thanks to McGraw–Hill for the opportunity, support and consideration (on those deadlines). Robin Gardner and Wendy Lochner were very supportive and helpful throughout the process. Tom Kowalczyk was most helpful on the technicalities of computer publishing (this book was done in PageMaker 6.52 and output electronically to the printer—that's right, if you don't like the layout, fonts, graphic style, etc. it's my fault). Steve Musser at the print house was also a great resource and helped me work through the electronic files for printing. Frank Kotowski, Jr. and a freelance copyeditor helped me tow the line. Thank you all.

Preface

Third, trite as it may sound, thank you to the wonderful taskmasters known as the AE Department at The Pennsylvania State University in the 1970's. These folks helped me understand buildings and, in particular, lighting. However, they were instrumental in setting the tone for a professional career—with the well–rounded exercises which included some form of graphic and/or verbal presentation. I feel this contributed to, if not originated my ability to communicate to others in a form such as this book. Thanks to a special college professor, the late John Flynn, for piquing my interest in lighting as well as providing strong direction on both lighting design and graphic presentation skills. And thanks to his co–author Samuel Mills for illustrating *Architectural Lighting Graphics* in a fashion that is visually pleasing and easy to understand.

Photography

Without visiting an actual installation, photos help immensely. Here they are used to illustrate applications of specific lamps and luminaires to achieve specific design results. Most of these are in black and white, but a 32–page color section allowed us to also show some of these in color. The photographers' work used here is very much appreciated—thanks to them for both their talent (to show installations with little or no photo fill light so that what you see is what you actually get in situ) and their very gracious permission to allow the use of these images here. Indeed, I was so pleased with their work and the enthusiastic response from each of these photographers that they deserve specific citation here:

Robert Eovaldi
376 Mayfair Street
Holland, MI 49424

Stephen Graham
Stephen Graham Photography
1120 West Stadium Boulevard
Suite 2
Ann Arbor, MI 48103–5854

Preface

Balthazar Korab
Balthazar Korab Ltd. Photography
P.O. Box 895
Troy, MI 48099–0895

Christian Korab
4104 42nd Avenue South
Minneapolis, MN 55406

Gary Quesada
Hedrich Blessing Photographers
11 West Illinois Street
Chicago, IL 60610

Laszlo Regos
Spectrum Photo
3127 West Twelve Mile Road
Berkley, MI 48072

Vance Roth
VRA Photography
P.O. Box 786
Mentor, OH 44061–0786

Projects and clients

Appreciation of all clients' projects needs stating. For nearly twenty years, clients have allowed Gary Steffy Lighting Design Inc. the privilege of lighting their facilities—commercial, residential, hospitality, institutional, religious, performing arts, educational, gallery—inside and outside. Thanks to them, their architects and engineers for the experience and, in those cases where photographic record was desired, for the permission to photograph. In particular, thanks here to The Saginaw Chippewa Indian Tribe of Michigan and to Steelcase Inc.

Late News

As this book went to press, lamp manufacturers announced new products in the infrared halogen category, the ceramic metal halide category, and the triphosphor fluorescent category. This is a clear sign that these lamp categories offer great promise in meeting a host of users' needs while minimizing energy use and maximizing lamp life. GE announced planned introductions of various ceramic metal halide products and T5 triphosphor fluorescent products. Osram Sylvania announced PAR halogen infrared and ceramic metal halide product introductions. Philips announced the introduction of its low voltage MR halogen infrared products and an expansion of its ceramic metal halide product line. Use the website addresses provided throughout for these lamp manufacturers to learn about their latest products.

LED lamps are quite new to the architectural lighting scene. Much development work is reportedly under way and these lamps may soon be appropriate for more useful applications (as opposed to the very decorative nature they now serve). The significant benefits of these lamps are low energy use and very long life expectancy.

Book layout

You will see a few blank pages throughout the text. You'll also see that some segments within each section begin on the even page (typically, new segments begin on odd pages). Blank pages occur where the presentation of information for a given lamp or detail took two pages. To keep the two pages together for more convenient reference, there were times that an odd page or "filler" was used to force the layout.

Feedback

Your feedback is very much encouraged and will be greatly appreciated. Please tell me if you find errors—citation of page number and the mistake/correction will be very helpful. If you see a particular need not addressed here, please forward a note outlining the need—and, if you would be so kind, citing specifics about the kinds of data, graphics and/or imagery that would be helpful in meeting the need. E–mail me at gsteffy@tssal.com. Errata will be posted at www.tssal.com on a periodic basis.

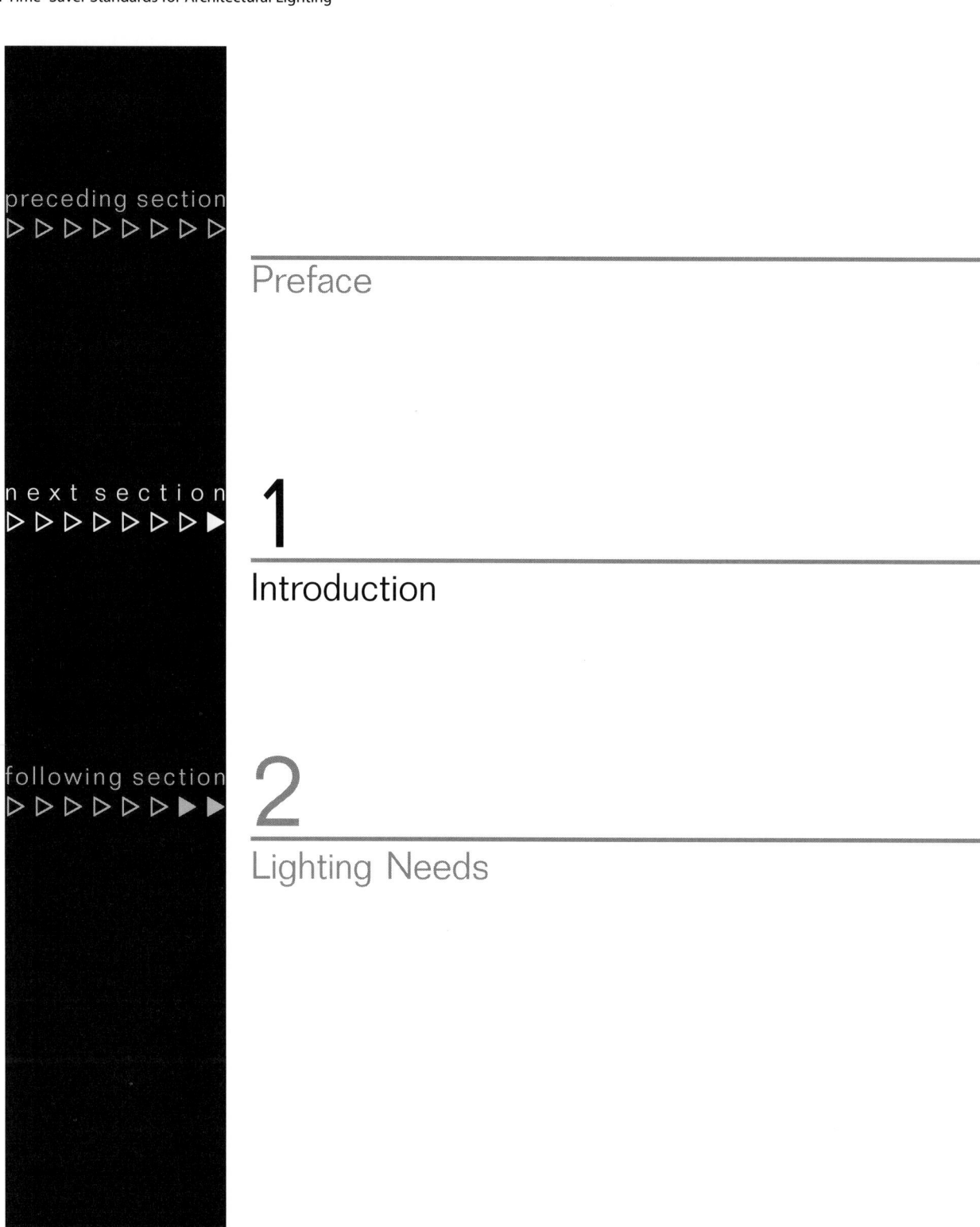

Introduction
The Lighting Reference

1

Welcome to this reference. Here is a repository of lighting design information. Not the usual volume of formulae, geometric diagrams and related multipliers. Nor a compilation of "initial" data without regard for in situ conditions. Nor the reference text of outdated "industry statistics." *Time–Saver Standards for Architectural Lighting* offers visual images (Performance Sketches™) for easy assessment of lamp performance for accent and art lighting. Concept Starters™ help with lighting layouts. Details are illustrated with Perspective Sketches™ and Performance Sketches™ of the anticipated brightness patterning and intensities. This introduction will outline what is presented, why it is presented and highlights these handy features introduced for ease of reference.

Introduction
The Lighting Reference

This is a reference. Intended to be a handy, convenient (well, once you get the "lay of the land") reference to developing lighting solutions for most any kind of project. It may be useful for simply pulling together accent lighting for a residential project. Or it may be used to develop a conceptual plan for retail highlighting. It may also be useful for pulling together a rather complete lighting concept. For an office project, ambient, task and accent lighting layers can be generated. Guessing is minimized if not eliminated. And as a result, more energy efficient lighting solutions are expected to prevail.

Indeed, one goal of this reference is to simply leave behind old, inefficient, poor quality, and just plain poor lighting. Only efficient (efficacious) light sources are presented. No old standard filament incandescent lamps. No poor–color, T12 (large diameter) halophosphate (cool white and warm white) fluorescent lamps. No standard lensed luminaires. In their day, these products brought efficient and quality lighting to the masses. This stuff just doesn't cut it today and isn't referenced.

The book is divided into sections. Each section is a stand–alone segment about that particular topic—so there is some repetition in introductory portions of sections which are related.

Lamps

Lamps are the engines of any lighting design. So, they are given lots of attention. While lamp data from manufacturers changes from time to time, such changes are typically minor (less than 10 or 15 percent) unless new lamp

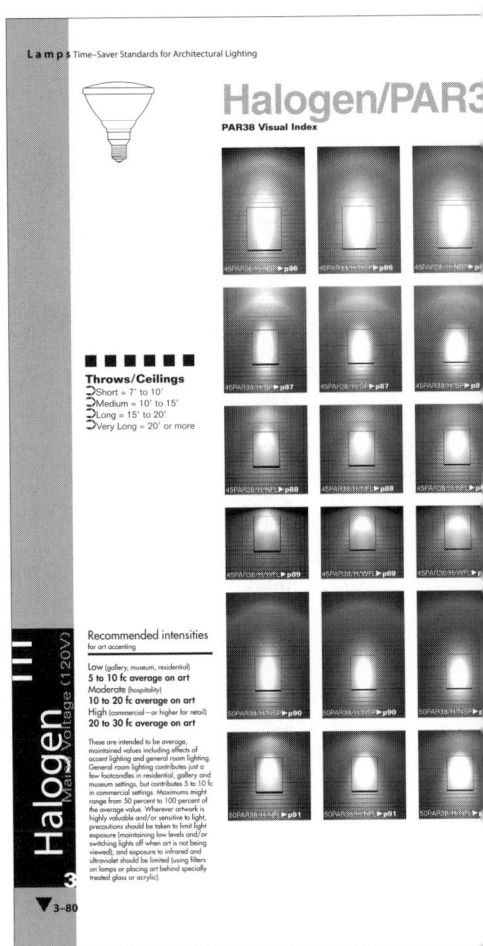

Introduction
The Lighting Reference

technologies are introduced—in which case a new family or category of lamp is created. Some typical applications for lamps at some typical mounting heights and aiming angles (defined here by distance from walls) are used to illustrate actual light spread and intensity.

Thumb to any of the lamp sections noted in the contents. Shaded margins mark lamp subcategory sections, and each of these is indexed with thumbnails and lamp performance highlights (an excerpted layout of one visual index from Section 3 is shown below).

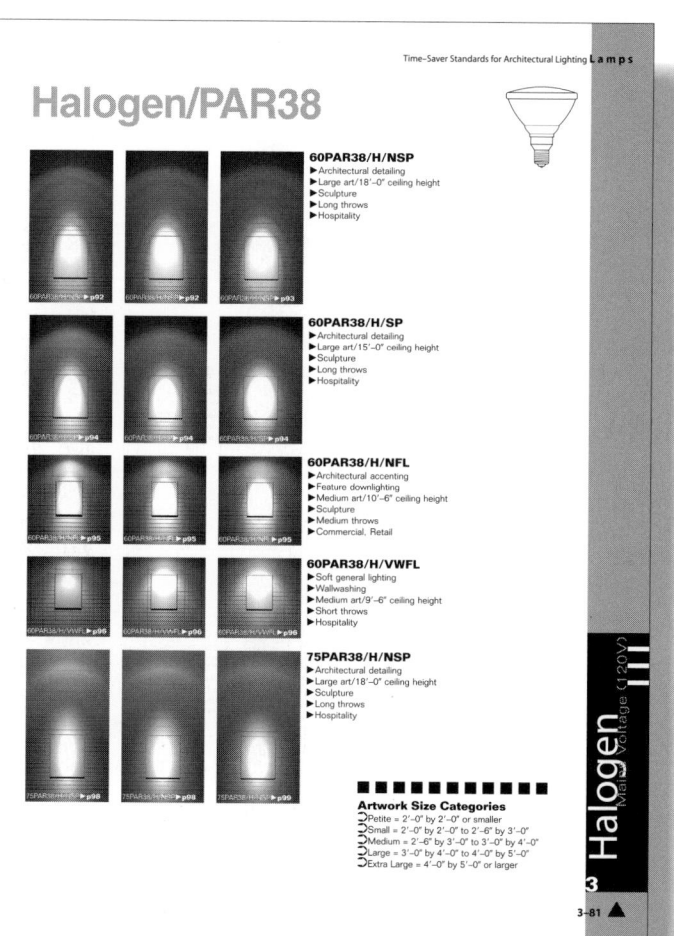

Visual Indexes
Thumbnail performance renderings show comparative ceiling heights and spread of light over typical artwork sizes. Referencing specific pages cited in this visual index leads to more detailed lamp information, including aiming and design application tips.

Concept Starters™
Concept layouts are offered for various luminaires meeting
prescribed criteria.

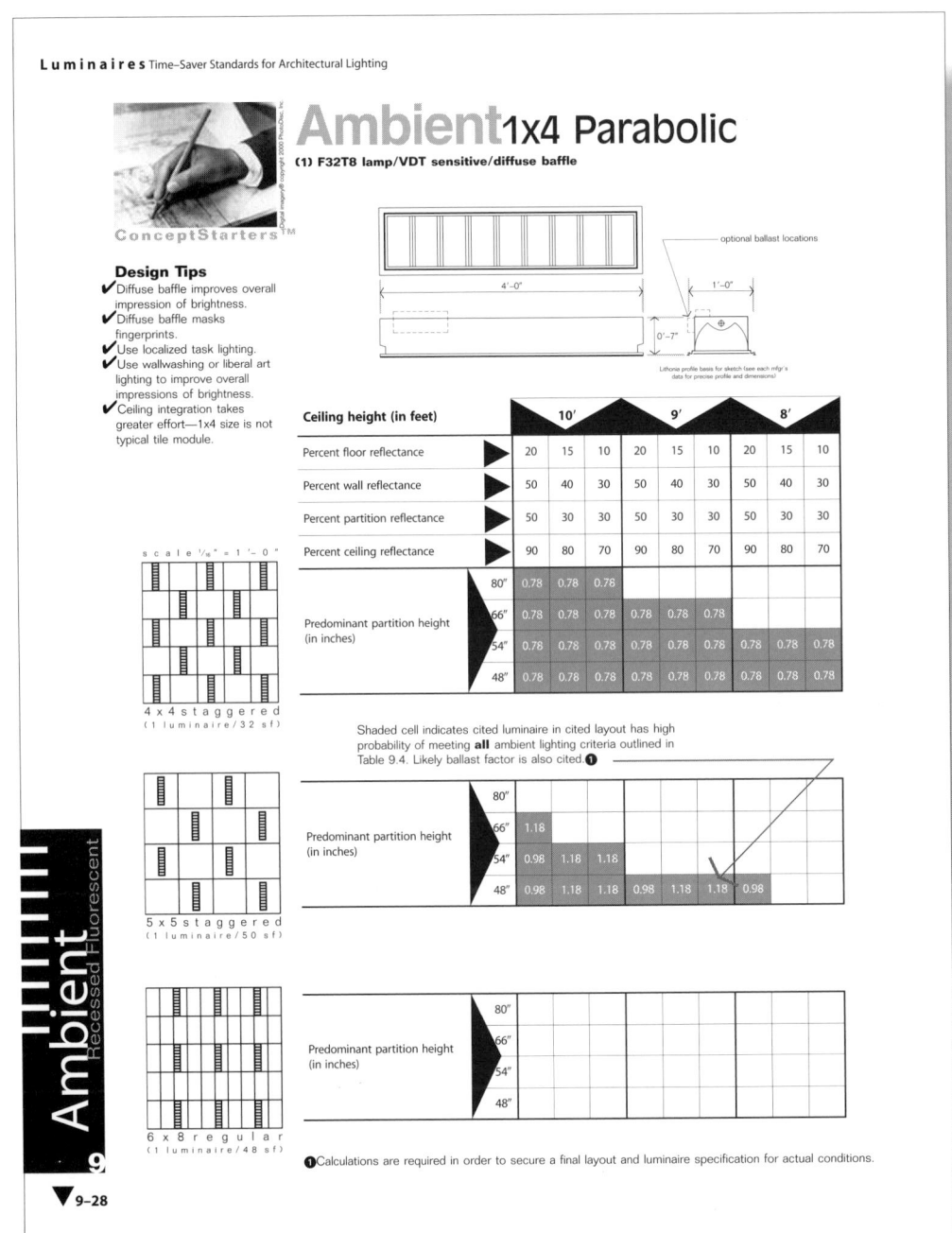

Luminaires Time–Saver Standards for Architectural Lighting

Ambient 1x4 Parabolic

(1) F32T8 lamp/VDT sensitive/diffuse baffle

optional ballast locations

4'–0"   1'–0"   0'–7"

Lithonia profile basis for sketch (see each mfg's
data for precise profile and dimensions)

Design Tips
✔ Diffuse baffle improves overall impression of brightness.
✔ Diffuse baffle masks fingerprints.
✔ Use localized task lighting.
✔ Use wallwashing or liberal art lighting to improve overall impressions of brightness.
✔ Ceiling integration takes greater effort—1x4 size is not typical tile module.

Ceiling height (in feet)		10'			9'			8'		
Percent floor reflectance		20	15	10	20	15	10	20	15	10
Percent wall reflectance		50	40	30	50	40	30	50	40	30
Percent partition reflectance		50	30	30	50	30	30	50	30	30
Percent ceiling reflectance		90	80	70	90	80	70	90	80	70

scale ¹/₁₆" = 1'–0"

4 x 4 staggered
(1 luminaire / 32 sf)

Predominant partition height (in inches)	80"	0.78	0.78	0.78						
	66"	0.78	0.78	0.78	0.78	0.78	0.78			
	54"	0.78	0.78	0.78	0.78	0.78	0.78	0.78	0.78	0.78
	48"	0.78	0.78	0.78	0.78	0.78	0.78	0.78	0.78	0.78

Shaded cell indicates cited luminaire in cited layout has high
probability of meeting **all** ambient lighting criteria outlined in
Table 9.4. Likely ballast factor is also cited. ❶

5 x 5 staggered
(1 luminaire / 50 sf)

Predominant partition height (in inches)	80"									
	66"	1.18								
	54"	0.98	1.18	1.18						
	48"	0.98	1.18	1.18	0.98	1.18	1.18	0.98		

6 x 8 regular
(1 luminaire / 48 sf)

Predominant partition height (in inches)	80"									
	66"									
	54"									
	48"									

❶ Calculations are required in order to secure a final layout and luminaire specification for actual conditions.

Ambient Recessed Fluorescent

9

Introduction
The Lighting Reference

For each lamp, three renderings or Performance Sketches™ are shown. Some appear to be just right, while others look a bit out of whack by comparison. Go for the better look, but recognize that during construction, as questions arise in the field, this reference can help determine if lights should move or if consideration might be given to reconfiguration of conflicting elements.

Quick lighting studies of art walls and accent effects can be done by tracing light patterns onto onion–skin overlays or by photocopying key Performance Sketches™. Alternatively, onion–skin sketches of various artwork sizes can be overlaid on several Performance Sketches™ to assess the best lamp for a particular situation. Lamps can be used in situations other than those cited herein, but calculations are encouraged for a better understanding of expected performance.

Throughout the reference, citations are made to lamp manufacturers' websites where more up–to–date information can be viewed along with announcements on new lamps. Look for the connect–for–more cursor. Many manufacturers now allow users to download cutsheets and technical data for use in presentations and calculations.

Luminaires

Luminaires are, in some instances, decoration and therefore are subject to design movements and fashion trends; particularly so on track lights and monopoints. On recessed adjustable accent luminaires, the reflector cone helps minimize direct view of the lamp and thus shields viewers from direct glare. However, the lamp itself performs the bulk of the optical spread and intensity of light. So, for accent lighting, this reference does not provide guidance on the selection of luminaires. Rather, here you'll find guidance on the lamps needed to achieve cited effects. For ambient lighting systems in commercial applications, luminaires are less decorative and more optically functional. Sections 9 and 10 address ceiling recessed and pendent ambient lighting respectively. Concept Starters™ are offered—concept layouts that meet prescribed lighting criteria (see excerpt to the left). Cost magnitude data is supplied for each lumi-

Introduction
The Lighting Reference

naire type reviewed. Throughout the reference Internet connection addresses are cited where specific luminaire models can be reviewed. Since the Internet is a new concept for luminaire manufacturers, not many prominent manufacturers are on the web. Check into a search engine from time to time to find more recent lighting additions.

Lighting details

The section on lighting details covers various lighting techniques which are traditionally achieved through architectural detailing. Here lighting effects are illustrated along with the requisite detail dimensions and/or lighting equipment. Wallwash lighting techniques are included with cove and wall slot techniques.

Key reference features

Certain information is presented in consistent format throughout the reference—here's a rundown:

Energy saver sidebars.

Energy saver sidebars

Energy savers are highlighted in sidebar form with an energy bolt. Sidebars without the energy bolt highlight key technologies with efficiency ramifications.

Informational tips

Sidebars with the bold dashed overhead bar offer information on hardware hazards, hardware applications, or design tips.

Informational tips

Connect for more information

Sidebars with the connect–for–more cursor are citations of online addresses of relevant suppliers, vendors or additional references. As of publication date, these addresses connected directly to primary vendors—no intermediary resource centers are used, although these do exist and can be found through many of the Internet search engines. Only vendors with addresses providing some level of product information and detail are cited (many others

Introduction
The Lighting Reference

had a single page website offering little in the way of technical information or product support). Remember, only vendors with useful websites are listed. Many, many more vendors of quality products can be contacted through snail mail and voicemail. Consider checking search engines or www.iesna.org for directories. Most lighting magazines have an annual directory of products—Architectural Lighting, Lighting Design & Application, Lighting Dimensions.

CONNECT FOR MORE

Color material

Application photographs are interspersed throughout. However, light is as much about color as it is about brightness contrast. A 32–page color spread is included with color images of various projects on which many of the lamps, luminaires and details cited herein have been used successfully.

Conventions

In the States, the lighting industry (along with many others) has yet to make much progress toward the Systeme Internationale (SI) (although some new lamps are, finally, metric). An initial effort to make this reference entirely metric soon foundered as lamping nomenclature (related to inches) resulted in odd hard conversions—not to mention the text–bloat and ever–increasing cross–referencing problems. Here is a table to ease conversion for those needing data in metric form.

Measure	U.S. Customary	Systeme Internationale (SI)
Length	1 in 1 ft 3 ft–3 in	25.4 mm 305 mm 1 m
Illuminance	1 footcandle (fc)	10.76 lux (lx)
Luminance	1 footlambert (fL)	3.4 candelas per square meter (cd/m²)

Introduction
The Lighting Reference

Manufacturers cited

Nothing is more irritating than the presentation of an idea, particularly in this world of instant planning, specification, purchasing and construction, which has the earmarks of a theory without real–life existence. So, here you'll find lots of references to manufacturers' data and products in both written and graphic form. Beware that this material changes all the time. That's why we've included addresses to websites of manufacturers' literature. We included those which appeared to have a reasonably robust web presence—you could, most likely, pull off catalog cutsheets and/or view image shots of installations for review by your team. New sites are added daily, so a global search for lighting manufacturers or a specific manufacturer's name is encouraged from time to time. In any event, our manufacturer citations are not all–inclusive nor are our website address listings.

Cost magnitudes

We tried to be practical here. You want pricing (well, at least your client does). We'd love to give you pricing. Unfortunately, the electrical industry has some extraordinarily byzantine pricing and quoting practices. So, we refer to costs as magnitudes—they should be something in this order, but they may fluctuate as much as 50 percent—lower or higher! That's right, depending on the factory, your lighting rep (in the location where the project is specified *as well as* in the location where the project is being constructed), the electrical distributor, the contractor, so–called market conditions, the astrological signs and other vagaries, you'll find pricing varies a lot. Quite unfortunate, since this allows for the now infamous value engineering technique to put lots of value in places and pockets other than the clients' and end users' environment. Ah, but we digress from the real reason for this particular book.

Technical data and computer output

Data presented here were presumed accurate at time of manuscript preparation. We found that not all data were available for all products—so we made some estimates of performance and noted such.

Introduction
The Lighting Reference

Manufacturers' current data should always be checked prior to your design and specification efforts. Calculation results herein are estimated to have an error range of plus/minus 15 percent. Some of this is due to generalizing of data—where several manufacturers have good competing products and where data were available, we averaged the results. Your specific application situations (such as final selection of surface materials and their resulting reflectances) and which manufacturer's equipment you select will affect this error range. Additional deviations can result depending on actual project voltage conditions, environmental dirt conditions and the like.

Lighting software has included some rendering or imaging techniques for some time. None of these are perfect (or even near–perfect). Be advised—these programs take an awful amount of time for the learning curve. Input/output practices are horrendously laborious (clearly this stuff has been written by computer code experts and not folks familiar with the needs of the design community). Computer time is excrutiatingly long. Results are, at best, marginal. Still, this is better than building it, only to find out it isn't close to what you had in mind. Some of these image inaccuracies are likely due to how lights are tested and how this test data is then used. Still, other inaccuracies are due to computer run–time limits. The more problematic images herein are the cove images in Section 11. The nuances of striplights with their socket shadows (which are visiible in many real–world situations) or the much better asymmetric cove luminaires and their smoother light distribution (which result in uniformly lighted coves in many real–world situations) could not be illustrated in these images—even though each image took about 3 hours of computer time!

Using this material

Other than the photos and images copyrighted by other individuals and entities cited, any of the material herein may be copied for exploring lighting concepts on projects. No material may be copied for use in other publications or on the Internet without the written consent of McGraw–Hill. Substantive portions of this material may not be copied as handouts to other team members, employees, comrades, partners, friends, students, etc.

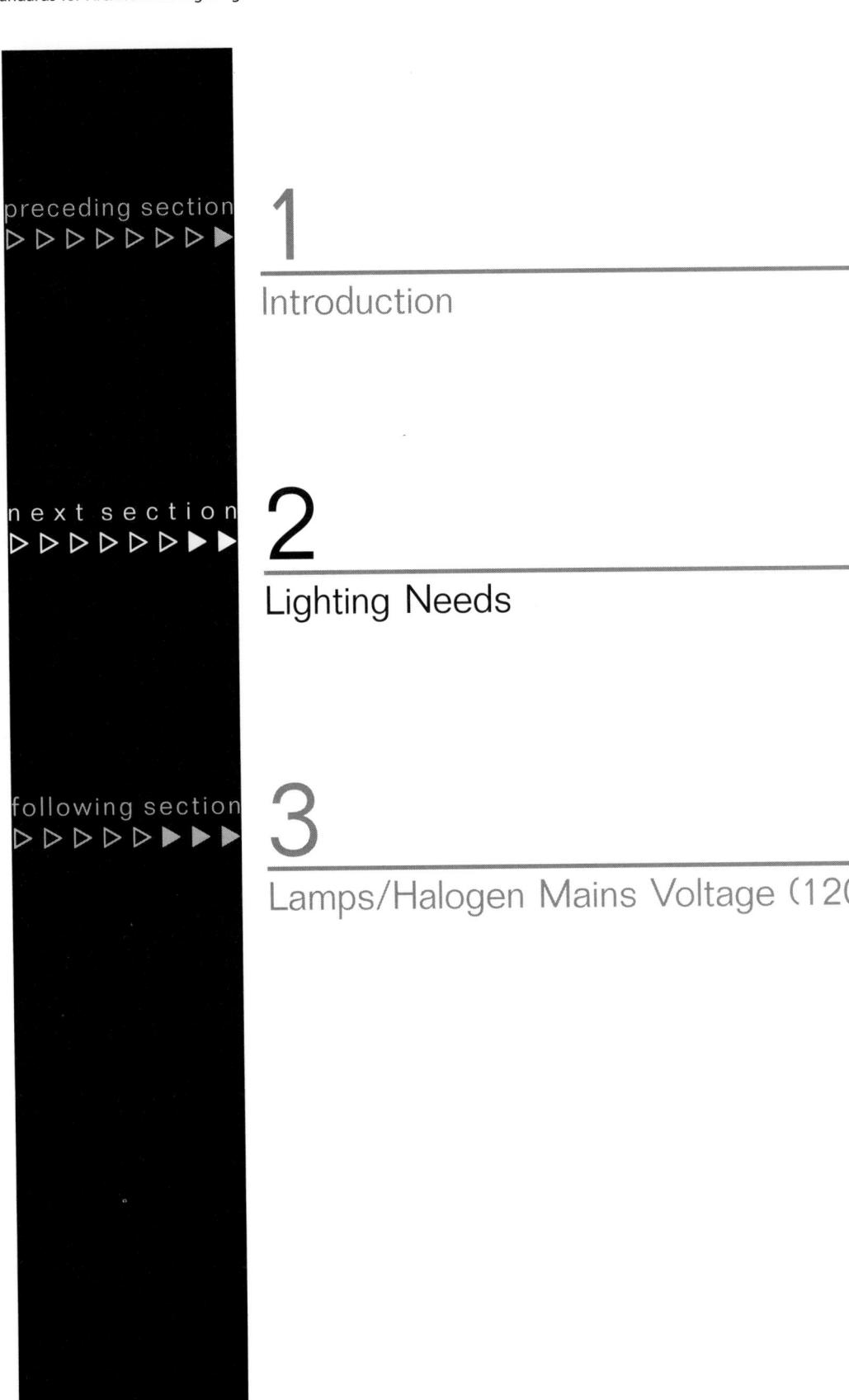

Lighting Needs
Criteria and Priorities

2

This section is a quick guide to resources on establishing lighting needs and priorities. Throughout the rest of *Time–Saver Standards for Architectural Lighting*, relevant criteria are cited—based on experience and critieria references noted here in Section 2.

Lighting Needs
Criteria and Priorities

Lighting Needs
Criteria and Priorities

2

The California Energy Commission's Title 24 and the ASHRAE/ IESNA's 90.1/1999 are model guidelines for lighting power budget criteria.

Criteria and communication—two reasons why projects succeed or fail. Establishing criteria is a must. Knowing which criteria are correct and which are higher on the priority list than others takes communication. If the communication chain is not with the end–user or even the client, then guesses must be made on criteria priorities. If nothing else, establish criteria, record it, and share it with the team. If guesses need be made on priorities, then use previous experience, other professionals' opinions and, most importantly, published industry guidance. Lighting criteria can be categoried in terms of quantity and quality issues and code issues.

Lighting quantity and quality criteria

The Illuminating Engineering Society of North America (IESNA) has a cadre of publications available—from the industry–tome Ninth Edition Handbook and Reference Volume to technical memorandums, design guides, and recommended practices. While these guidelines are not code requirements, they do outline the multitude of criteria that are appropriate for many different space types/ functions. Many European countries, Japan, Australia/New Zealand and others have simlar organizations espousing similar guidelines.

Throughout *Time–Saver Standards for Architectural Lighting*, references are made to quantity and quality criteria as they relate to specific topics (e.g., accent lighting for artwork in Sections 3, 4, 5, 6 and 7; ambient lighting for commercial offices in Sections 9 and 10). These criteria have been culled from the references discussed here and/or cited later as well as from experience of the authoring entity.

Lighting codes

Lighting codes depend on governmental jurisdiction and can be classified as life safety or energy. Reviews of project–specific municipality, state/region, and/or country codes are necessary to establish life safety lighting criteria. Web searches can be helpful in this regard.

In the United States, energy codes are subject to the 1992 EPAct (Energy Policy Act). Most states have adopted the ASHRAE/IESNA Energy Standard for Buildings Except Low–Rise Residential Build-

Lighting Needs
Criteria and Priorities

ings (ASHRAE/IESNA 90.1/1999) to conform with the EPAct. Present requirements necessitate careful design practice (e.g., overlighting is no longer an option), use of controls and daylight integration as well as use of low wattage light sources. References to this code standard are made in Sections 9 and 10 as it relates to commercial office lighting.

A model lighting energy code is California's Energy Efficiency Standards for Residential and Nonresidential Buildings (commonly referenced as Title 24/Section 6; revised in 1999). Indeed, the ASHRAE/IESNA 90.1/1999 standard closely parallels the California Title 24 lighting document.

Environmental issues

Energy codes were initially established to respond to the oil embargos of the 1970s. Now, they are intended to both limit oil dependence and environmental destruction, including the greenhouse effect. Other more recent environmental issues include sustainability—care in the selection of materials and how resources are managed to create products and how/if such products are disposed/recycled. Simply, the idea is to use little energy to make products which themselves require little energy for operation during their life and which can be recycled or, if absolutely necessary, disposed without harm to the environment.

Table 2.1 Lighting Issues/References

Issues Category	References	Website Address
Quality/Quantity	• Illuminating Engineering Society of North America (IESNA) • Chartered Institution of Building Services Engineers (CIBSE)	http://www.iesna.org/ http://www.cibse.org/
Product Safety	• Underwriters Laboratories (UL)	http://www.ul.com/
Energy	• American Society of Heating, Refrigerating and Air Conditioning Engineers (ASHRAE) • California Energy Commission (CEC)	http://www.ashrae.org/ http://www.energy.ca.gov/title24/index.html
Life Safety	• Building Officials and Code Administrators (BOCA)	http://www.bocai.org/boca_codes.htm
Installation	• National Electrical Code (NEC) • National Electrical Contractors Association (NECA)	http://www.nfpa.org/Codes/index.html http://www.necanet.org/
Document Searches	• National Standards Systems Network (NSSN) • Global Engineering Documents	http://www.nssn.org/ http://global.ihs.com/

Lighting Needs
Criteria and Priorities

Table 2.2 Lighting Criteria Overview

Some Criteria	Meaning	Rationale
Horizontal illuminances	Lighting intensities on horizontal surfaces: • worksurfaces • laps • floors	• Task visibility
Vertical illuminances	Lighting intensities on vertical surfaces: • merchandise • walls • objects on walls • faces • computer screens	• Task visibility (note: in some cases less is better, such as computer screens) • Psychological impressions (improve senses of overall brightness and spaciousness)
Luminaire luminances	Light fixture brightnesses	• Direct glare • Reflected glare
Surface luminances	Surface brightnesses	• Adaptation • Transient adaptation effects
Power budget	Also known as connected load —quantity of watts per unit area necessary to operate the lighting system	• Limit resources expended on lighting
Cost budget	Design fees and initial hardware and installation costs	• Limit capital expended on lighting
Productivity/Satisfaction	User pleasure, comfort and/or productivity over the life of the installation	• Was it worth doing in the long run?
Life cycle	Issues and costs associated with manufacture, procurement, operation and disposal	• Limit resources expended on lighting
Maintenance	• Relamping frequency • Reballasting frequency • Difficulty of either	• Limit capital and resources expended on lighting

Lighting Needs
Criteria and Priorities

Here's the point. Connected lighting loads are now legislated at quite low levels. Generally, 1.1 to 2.0 w/sf depending on the space function—commercial facilities at the lower end and retail facilities at the higher end. Low wattage (with relatively high output) lamps, electronic ballasts and/or transformers and well–designed luminaire optics will become the norm. Using fewer components is desirable (this minimizes natural resource use, initial cost and solid waste). Lighting layers—accent, architectural and ambient/task—will only succeed if the designer is extremely careful in how each layer is achieved. Using daylight to supplant or supplement ambient lighting is a necessity near windows and/or at skylights. Lighting controls interfacing daylight with electric light are also necessary.

■ ■ ■ ■ ■ ■ ■ ■ ■ ■ ■ ■ ■ ■

Criteria and Priorities
❶ Communicate/understand needs
❶ Review/establish criteria
❶ Prioritize critera/most important to least important
❶ Properly assess cost and maintenance—these aspects should not drive the solution
❶ Communicate/priorities and design suggestions

Lamp, ballast and luminaire technologies outlined herein will have to be used. Halogen infrared (HIR) mains voltage and low voltage lamps or ceramic metal halide PAR lamps (for higher ceilings and greater intensities) will be necessary where accent lighting is desired. Triphosphor Triple Tube Compact and Tubular Linear T8 or T5 fluorescent lamps operating on electronic ballasts will be necessary for most ambient lighting. General lighting intensities will be designed to moderate–to–low levels and local task lighting will be used to provide sufficient intensities on work areas or areas of interest. Task lighting will use low wattage triphosphor fluorescent lamps with multilevel user controls. Value will have to be placed on lighting as a medium to an end—to visual comfort, user pleasure and satisfaction, and productivity over the long term.

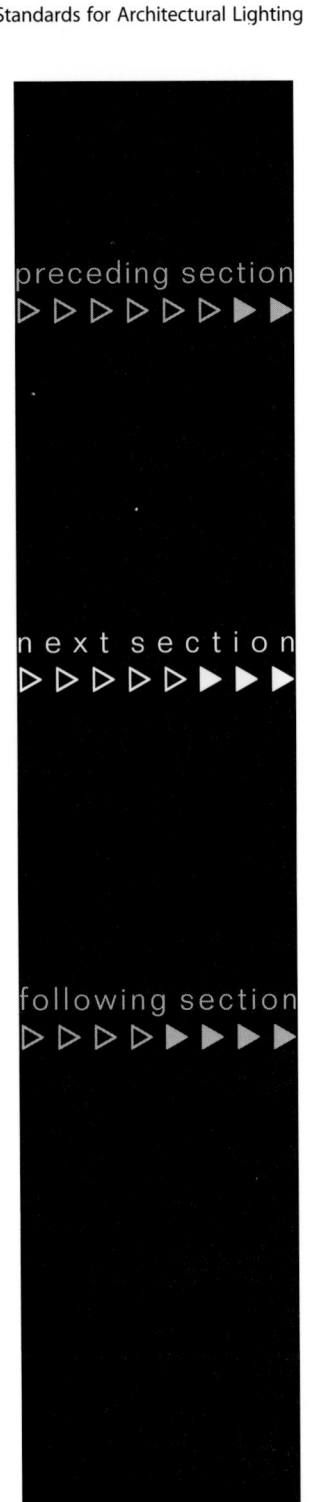

Halogen Lamps
Mains Voltage (120V)

3

This section addresses 120–volt (mains voltage) standard halogen incandescent lamps. For purposes of this reference book, these halogen lamps are considered the basic or common incandescent lamp of the day. Standard tungsten filament (non–halogen) incandescent lamps which are still widely available, particularly to the consumer through the hardware and do–it–yourself home centers, are inefficient and short–lived (but with a very low initial cost). Halogen lamps' improved efficacy and longer life make them a greener alternative. There are also low voltage (12–volt) halogen lamps (see *Halogen Lamps—Low Voltage (12V)* in Section 4) and infrared halogen lamps (see *HIR Lamps— Mains Voltage (120V)* in Section 5 and *HIR Lamps—Low Voltage (12V)* in Section 6). Low voltage lamps and halogen infrared lamps are more efficacious and typically longer lived than halogen lamps. As such, where incandescent lighting is determined as necessary, then low voltage halogen and halogen infrared lamps offer even less costly operation and are more environmentally appropriate than the standard halogen mains voltage lamps outlined in this section.

Halogen Lamps
Mains Voltage (120V)

Halogen lamps are the efficient incandescent alternative to standard or "energy–saver" tungsten filament incandescent lamps.

Halogen lamps are an extension of the incandescent lamp family. Because of the efficacy and life advantages of halogen lamps, they have replaced standard tungsten filament incandescent lamps in most residential and hospitality applications where incandescent lamps are desired. With energy costs, cooling load costs and the pollution concerns associated with electricity production, it is difficult to justify the use of standard tungsten filament incandescent lamps. Halogen lamps should be considered the common incandescent lamp—the norm. Halogen lamps come in a variety of shapes for different hardware and design applications (see Table 3.1). For even greater efficacy, compact fluorescent lamps offer significant energy and life cycle cost advantages over halogen. For more efficient accenting, halogen infrared PAR lamps and ceramic metal halide PAR lamps are recommended over halogen PAR lamps.

Table 3.1 Halogen Mains Voltage (120V) Lamps

Halogen Lamp	Use/Replaces	Lamp Profile
TB19	New or Retrofit • Standard Incandescent A19	
PAR16	New or Retrofit (consider PAR20) • Interchangeable with PAR20 • Few dedicated luminaires available	
PAR20	New or Retrofit • Standard Incandescent R20 • Standard Incandescent R30	
PAR30S	New • Standard Incandescent R30 • Standard Incandescent R40	
PAR30L	Retrofit • Standard Incandescent R30	
PAR38	New or Retrofit • Standard Incandescent PAR38 • Standard Incandescent R40	

PAR16 lamp image courtesy of and copyright by Philips 2000. All other lamp images courtesy of and copyright by GE 2000.

Halogen Mains Voltage (120V)

3

Halogen Lamps
Mains Voltage (120V)

Halogen lamp color characteristics

Color of light (color temperature), color rendering (how colors look under the light in question) and easy, low–cost dimming are the advantages of halogen lamps. Color temperature of most halogen lamps is "crisp white." Good for contemporary interiors. Where halogen lamps are used for accent lighting, this whiteness adds to the sense of brightness and attraction.

Halogen lamps have excellent color rendering—perceived by most folks as rendering all colors "true." As such, halogen PAR lamps are best for highlighting merchandise or art.

Halogen dimming

Dimming halogen lamps can increase lamp life and also will warm the color of light somewhat. This helps to "take the edge" off the light and makes halogen an efficient alternative in more traditional residential and hospitality settings. Since most halogen lamps are surrounded by cast glass enclosures and use small filaments which themselves are enclosed in capsules, there is little hum from dimming that is more common with standard tungsten filament incandescent lamps.

The halogen cycle results in improved efficacy and long life in halogen lamps. This same cycle, however, can be "short circuited" with continual dimming. Hence, where dimming is employed, occasionally the lamps must be operated at full bright to allow the halogen cycle to operate.

■ ■ ■ ■ ■ ■ ■ ■ ■ ■ ■ ■ ■

Halogen dimming
❶Increases lamp life—doubling it, but only if lamps are brightened to full output occasionally.
❶Yellows color of light—more cozy, intimate appearance.
❶Results in little or no hum (which is common in standard tungsten filament incandescent).

Halogen hazards

The hazards of halogen lamps have been highly publicized in recent years. The very reason a halogen lamp operates so efficiently is the same reason it can be a hazard—the high temperature operation of the filament. Early halogen lamps were actually comprised of quartz (literally clear stone) envelopes—this stone was the only substance at the time that could withstand the high temperatures of the halo-

Cost magnitude^{2000}
for halogen and HIR lamps

Low
US$7.^{50} to US$12.^{00}
Moderate
US$12.^{00} to US$17.^{00}
High
US$17.^{00}–plus

Costs vary based on quantities, distributor and contractor markups, and market conditions, manufacturing situations, and annual inflation. These values are for preliminary, magnitude budgeting and do not represent quotes nor actual final pricing.

Halogen
Mains Voltage (120V)

3

3–3 ▲

Halogen Lamps
Mains Voltage (120V)

gen gas and filament. These first lamps were linear, typically several inches in length and of relatively high wattage. This linear version of the halogen lamp is still available today and is usually found in floor torchieres. If the hot quartz or hot surrounding metal components come in contact with flammable materials (e.g., drapes or sheers), fire can certainly result. There have been cases where the quartz envelope fails violently—in this situation shards of hot quartz and filament bits explode onto surrounding surfaces. Again, flammability is an issue with room/furniture finishes. If the torchiere falls while the lamps are energized, the breaking halogen lamp is likely to cause fire. Because of their very portability, torchieres can be unstable and/or are easily placed too close to ceilings, walls, drapes and the like.

■ ■ ■ ■ ■ ■ ■ ■ ■ ■ ■ ■ ■

Halogen hazards

⊘ Linear halogen (aka quartz) lamps should only be used in luminaires that are UL–listed and labeled and fitted with tempered glass lenses.

⊘ Portable halogen luminaires should be well–balanced and heavily weighted to prevent tipping.

⊘ Off–brand halogen A– and PAR–lamps may not offer the same protective quality of cast glass enclosure as the brand–name lamps.

⊘ Luminaires with halogen lamps typically are hot to touch.

⊘ Ultraviolet light from halogen lamps can be damaging to humans and artwork unless UV filters are used.

⊘ Keep flammable materials/surfaces sufficient distance from halogen lamps and luminaires.

This flammability hazard can be reduced if linear quartz lamps are used in well–constructed, well–balanced luminaires fitted with tempered glass lenses; the luminaire is located away from circulation paths; and clear, safe distances (at least 1 foot from any part of the torchiere) are maintained between the torchiere and any materials of any kind. If halogen versions of A lamps (called "TB" lamps) and/or "PAR" lamps are used from any of the name–brand manufacturers, then the lamp failure hazard has been addressed in manufacturing with thick, cast or pressed glass enclosures surrounding a halogen capsule. Nevertheless, lamps should be used with care. Maintain sufficient distance from lamp to temperature–sensitive surfaces or materials.

Luminaires should never be overlamped. Label(s) on the luminaire detail the kind of lamp (e.g., "PAR20" or "PAR38") and the wattage limit. Overlamping a luminaire with a higher wattage lamp than is permitted on the label OR than was originally anticipated by the design team and construction team can lead to thermal protectors cutting off (lights blink on and off frequently), circuit breakers popping or, worse, to fire. Increasing lamp bulb size (from PAR20 to PAR30, for example) is also a dangerous proposition—hot lamp

Halogen
Mains Voltage (120V)

3

Halogen Lamps
Mains Voltage (120V)

components may now be placed in direct contact with luminaire housing materials and/or architectural materials. This could result in overheating and ultimately to fire.

Another hazard with halogen lamps is ultraviolet radiation. Ultraviolet output is higher in halogen lamps than in standard tungsten filament incandescent lamps. Artwork requires protection from this UV, otherwise degradation of the pigments, colors and/or base media (e.g., paper) may be significant and irreparable over a fairly short period of time. Similarly, people require protection from this UV. Tempered glass lenses help reduce UV light output from halogen lamps. For sensitive art, however, UV reduction lenses or filters should be used with halogen lamps.

When to use halogen
Halogen lamps are appropriate on projects where construction budgets are low, life cycle budgets are unimportant or intended use is not prolonged. Typically, residential projects fit this profile. Further, where fine, low–level dimming is a critical criterion or where the budget cannot support fluorescent dimming, halogen lamps are justified. Religious facilities and some hospitality facilities (e.g., banquet facilities, meeting facilities) fit this profile. Finally, where historic restoration is required and where lamping is behind frosted or otherwise non–image–preserving glass or faux alabaster type media, halogen lamps may be appropriate— offering better efficacy than standard or "energy–saver" tungsten filament incandescent lamps and offering improved maintenance cycles.

Environments where temperature extremes exist or where the temperature is likely to remain at or below 45°F are candidates for halogen lamps. Halogen lamps are particularly useful in cold settings if dimming and/or instant "bright" or "full–on" conditions are a requirement of the lighting operation.

Where flashing of lights is a requirement, either for effect or for signaling purposes, 120V halogen lamps deserve consideration.

When not to use halogen
Most commercial, institutional and retail projects deserve compact fluorescent and/or ceramic metal halide lamp solutions. Standard

■ ■ ■ ■ ■ ■ ■ ■ ■ ■ ■ ■ ■ ■

Halogen cycle
❶ Heated filament metal boils away into vapor (as happens in standard tungsten filament incandescent lamps).
❷ Vaporized filament metal is redirected by halogen gas back onto filament—the filament is regenerated!
❸ Yields longer life lamp or more efficacious lamp or a lamp with some increase in life and some increase in light output.

Halogen
Mains Voltage (120V)

Halogen Lamps
Mains Voltage (120V)

Table 3.2 Efficacious Alternatives to Halogen Lamps

Halogen Lamp	Efficacious Alternative	Hardware Application
TB19	• SLS15–SLS25/Screw–in (compact fluorescent)❶	• Decorative Luminaires❷ • Portable Luminaires❸
	• 13CFTriple (compact fluorescent)	• Decorative Luminaires❷ • Portable Luminaires❸
	• 18CFTriple (compact fluorescent) • 26CFTriple (compact fluorescent) • 32CFTriple (compact fluorescent)	• Decorative Luminaires❷ • Downlight Luminaires
PAR16	• Halogen Low Voltage MR16 • Halogen PAR20	• Accent Luminaires • Decorative Luminaires❺ • Downlight Luminaires
PAR20	• Halogen Low Voltage MR16 • Halogen PAR30S • Halogen PAR30L • Halogen Infrared PAR30	• Accent Luminaires • Decorative Luminaires❺ • Downlight Luminaires
PAR30S	• Halogen Infrared PAR30 • CMH PAR20 (ceramic metal halide)❹	• Accent Luminaires • Downlight Luminaires
PAR30L	• Halogen Infrared PAR30 • Halogen Infrared PAR38 • CMH PAR20 (ceramic metal halide)❹ • CMH PAR30 (ceramic metal halide)❹	• Accent Luminaires • Downlight Luminaires
PAR38	• Halogen Infrared PAR38 • CMH PAR20 (ceramic metal halide)❹ • CMH PAR30 (ceramic metal halide)❹ • CMH PAR38 (ceramic metal halide)❹	• Accent Luminaires • Downlight Luminaires • Wallwash Luminaires

❶Only a select few of these lamps are dimmable—confirm with manufacturer.
❷Decorative Luminaires include pendents and sconces.
❸Portable Luminaires include table lights, floor lights and task lights.
❹This alternative lamp is not dimmable inexpensively, if at all.
❺Decorative Luminaires include pendents.

Triple–tube compact fluorescent lamps (Section 8) are the efficient alternative to halogen lamps for general lighting. For efficient accent lighting, consider halogen infrared mains voltage (HIR) PAR lamps (Section 5), HIR low voltage MR16 lamps (Section 6) or ceramic metal halide (CMH) PAR lamps (Section 7).

Net Addresses/Exterior Accent
http://www.bega-us.com/home/home.html
http://www.hadcolighting.com/html/bronzelite.htm
http://www.hydrel.com
http://www.kimlighting.com/ingrd1.html

CONNECT FOR MORE

Halogen Mains Voltage (120V)

Halogen Lamps
Mains Voltage (120V)

halogen lamps do not offer the low energy use and long life that these facilities require to be cost–competitive and energy efficient. If accent lighting is required in these kinds of facilities, then at a minimum halogen infrared lamps should be considered, with ceramic metal halide lamps offering much more accent capability in smaller lamp sizes and wattages. Halogen lamps should not be used where very close proximity to people for prolonged periods is expected (e.g., task lighting)—given their high operating temperatures (resulting in higher ambient temperatures of nearby air and yielding potential for skin burns) and ultraviolet radiation. Table 3.2 outlines more efficacious alternatives to halogen lamps.

Halogen/TB19

Stats <small>Data varies manufacturer to manufacturer and changes from time to time</small>

Stats Guide
⊃Longer life is better
⊃Higher efficacy is better
⊃Lower wattage is better
⊃Lower color of light is warmer
⊃130 volt versions last longer, but are less bright and less efficient
⊃Very low cost range for halogen is less than US$7.⁵⁰

Life/2000 and 3500 hours
Different versions from different manufacturers. A typical office environment might be occupied about 2600 hours each year.

Efficacy/14 LPW

Wattages/42, 50, 52, 60, 72, 75, 90, 100 <small>(see Light Output Table opposite page)</small>

Color of Light/2800°K

Beam spreads/Not Applicable
This is a general service lamp intended for table lights, floor lights, sconces and downlights where an all around glow of light is desired. Do not use where distinct accent light or focused light is desired—for accent use Halogen PAR lamps (p3–14), HIR PAR lamps (Section 5) or CMH PAR lamps (Section 7).

Size/4⁷⁄₁₆″ L by 2³⁄₈″ Ø

Voltage/120V and 130V

Cost Magnitude/Very Low⁽²⁰⁰⁰⁾

Manufacturers
General Electric
Osram Sylvania
Philips

Net Addresses
http://www.ge.com/lighting/business/index.htm
http://ecom.sylvania.com/osicatalog/
http://www.lighting.philips.com/

CONNECT FOR MORE

Advantages
Crisp white light
Standard voltage (no transformers)
Easily dimmed
Low initial cost
Much longer life than typical A lamps
Better alternative to less efficient standard filament A lamps
Withstand extreme temperature range
Very little lamp lumen depreciation (light loss) over life

Disadvantages
High operating costs
◆Inefficiency and short life may make this cost prohibitive in typical commercial applications, depending on ease of changing lamps and maintenance staffing.
Hot to touch
Hot-shock sensitive
◆Lamp may fail if sharp vibration occurs while lamp is energized.
Water sensitive
◆Lamp may fail violently if placed in contact with water (not to be used exposed in outdoor lighting situations).

For more efficient accent lighting, use HIR low voltage lamps (Section 6); HIR PAR lamps (Section 5) or CMH PAR lamps (Section 7).

Halogen Mains Voltage (120V)

Halogen/TB19

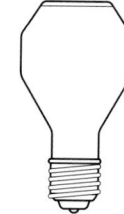

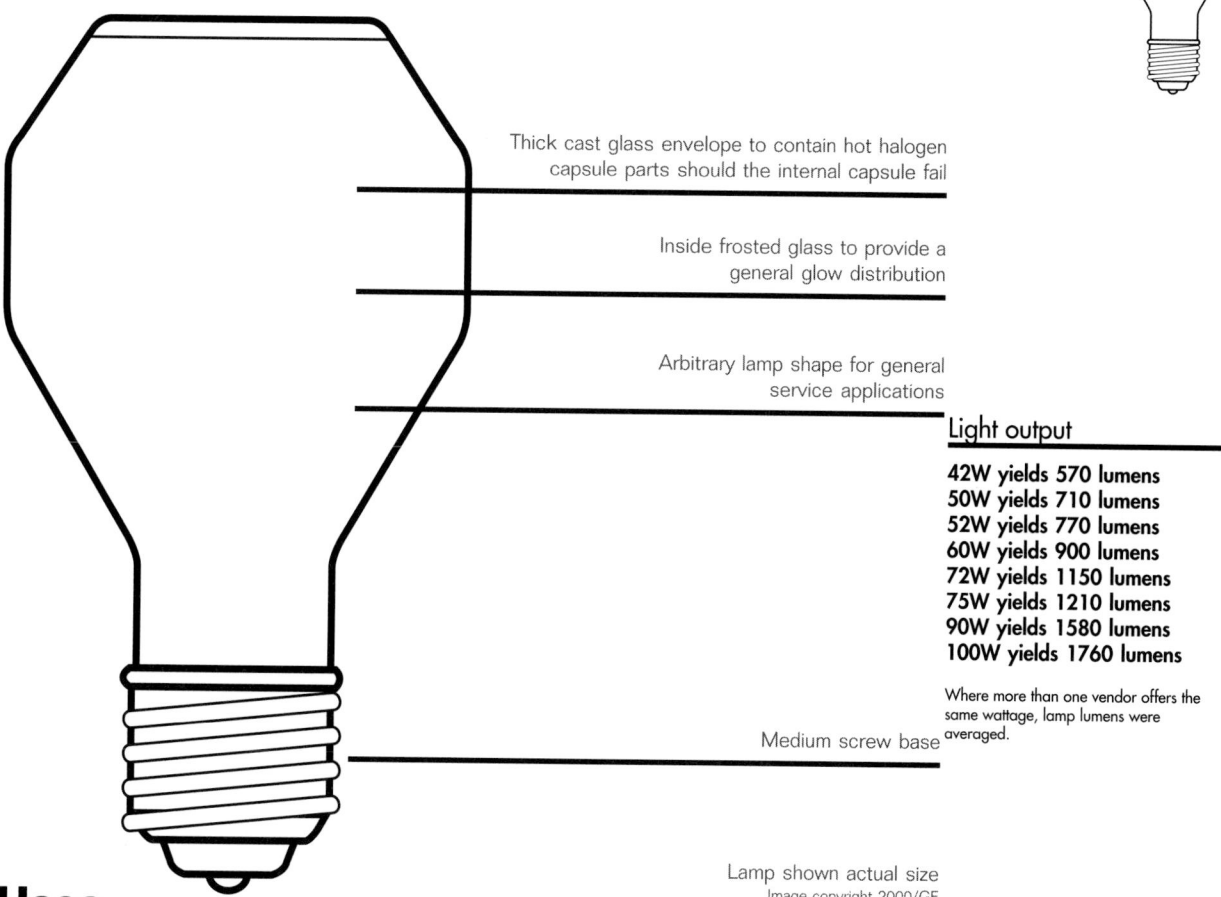

Thick cast glass envelope to contain hot halogen capsule parts should the internal capsule fail

Inside frosted glass to provide a general glow distribution

Arbitrary lamp shape for general service applications

Medium screw base

Light output

42W yields 570 lumens
50W yields 710 lumens
52W yields 770 lumens
60W yields 900 lumens
72W yields 1150 lumens
75W yields 1210 lumens
90W yields 1580 lumens
100W yields 1760 lumens

Where more than one vendor offers the same wattage, lamp lumens were averaged.

Lamp shown actual size
Image copyright 2000/GE

Uses

Decorative

Hospitality
- ◆Floor lights
- ◆Table lights
- ◆Sconces

Residential
- ◆Floor lights
- ◆Table lights
- ◆Sconces

Downlighting

Circulation
Conference rooms
Hospitality
Residential

■ ■ ■ ■ ■ ■ ■ ■ ■ ■ ■ ■ ■ ■ ■

TB19 Lighting Design Tips
✔ Best in lower ceiling spaces (10'–0" or less).
✔ Dimming increases life (typically doubling life), but shifts color warmer.
✔ Beats standard incandescent lamping on maintenance cycles and energy use.

 CONNECT FOR MORE

For general downlighting in low–to–moderate ceilings, consider compact fluorescent triple tube lamps (see Section 8).

Halogen Mains Voltage (120V)

3

3–9 ▲

Halogen/TB19

Figures 3.1 and 3.2 (opposite)

Here the table lights and custom wall sconces are considered to be a significant component in setting the stage for an inviting lobby/lounge. Light sources in close proximity to visitors and guests were specified as incandescent for warmth and dimmability. Dimming enables the lobby to change character and mood during the 24–hour operation; and also increases lamp life. Halogen lamps do require operation at or very near full output for some period of time (an hour or two each day in this 24–hour operation) in order to maintain the halogen cycle for best efficacy and life. TB19 lamps are used in the wall sconces and table lights. A closer view of the custom sconces is shown in Figure 3.2 on the opposing page. The general service halogen lamps provide a soft, diffuse glow through translucent materials—in this case shades of Dupont Nomex™ brand fiber on the table lights and Sterling Products Natural Horn faux alabaster acrylic in the sconces. The sconces are mounted so that the bottom is above 6 feet/8 inches, thereby meeting ADA requirements (i.e., projection is 0 feet/4 inches or less or alternatively, bottom of luminaire is mounted at 6 feet/8 inches or higher AFF). Table lights by Martin. Sconces by Baldinger Custom Architectural Lighting. Pendent by Baldinger Custom Architectural Lighting. Downlights and accents by Lightolier.

Halogen/TB19

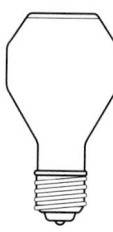

Halogen Main Voltage (120V)

3

Halogen/TB19

Large-scale rawhide wall sconces using TB19 lamps are located between the column detail "supporting" each arch. These sconces are important toward achieving a perception of warmth and invitation. The hide material is backlit with the TB19 lamps. See Figure 3.4 for more detail. Sconces are Baldinger Architectural Lighting/Custom.

Figure 3.3

Halogen
Main Voltage (120V)

Halogen/TB19

Figure 3.4

In this large–scale rawhide wall sconce seven TB19 lamps provide diffuse, consistent light. 130–volt lamps result in a doubling of lamp life for an improved maintenance cycle in a hospitality facility. Sconce measures 6 feet in height and 2 feet in width with a projection of 2 feet at the top. With such a hand–wrought luminaire, the juncture of the sconce and wall would not allow for a neat, continuous, light–sealed edge. Hence, the luminaire sits away from the wall about an inch. Sconces are Baldinger Architectural Lighting/Custom.

Halogen Mains Voltage (120V)

3

Halogen/PAR

PAR lamps—PAR is an acronym for parabolic aluminized reflector—have been effective accent lamps since the 1950's. With the halogen technology, the PAR lamp provides more light at lower wattage with a longer life. A small halogen capsule is enclosed by a relatively thick, cast/pressed glass envelope. The reflector portion of this envelope is coated with specular (polished) aluminum. The front of the lamp has a specially designed glass lens which refracts the light concentrated by the reflector into a pattern—from a narrow spot (NSP) to a wide flood (WFL). Some manufacturers offer watertight versions for direct exposure to weather. Tables 3.3 through 3.10 are lamp guides—outlining halogen PAR lamps that might be considered for various tasks and effects in several applications and ceiling heights. More detailed information about these specific lamps and information about their intensities and coverages can be found on the pages cited.

■■■■■■■■■■■■■

Efficient Incandescent Lighting

❶ Don't use R lamps.
❶ Don't use BR or Krypton–filled lamps.
❶ Consider Halogen/PAR lamps, or better yet...
❶ Consider Halogen Infrared/PAR (HIR/PAR) lamps.

Less efficient reflector or R lamps were inexpensive and therefore popular prior to the 1992 Energy Policy Act (EPAct), at which point the lamps were essentially legislated out of existence. Standard incandescent R lamps cannot meet the legislated efficacy requirements of the EPAct. Today, some variations on R lamps remain available—typically known as BR lamps or Krypton–filled R lamps. For better efficacy in accenting, however, halogen and halogen infrared PAR lamps should be used. Halogen PAR lamps are now the common replacement for R lamps, with PAR20s replacing R20s; PAR30Ss and PAR30Ls replacing R30s; and PAR38s replacing R40s. The figures on the facing page illustrate various R and BR lamps with notes regarding the better–advised PAR lamp alternatives. PAR lamp replacements are ghosted background images to illustrate relative size. The PAR lamps offer smaller bulb envelopes and therefore can be fit into smaller housings, or in retrofit situations will eliminate the glary and unsightly exposed bulb envelope of the R lamp bulging from the luminaire.

Halogen Mains Voltage (120V)

3

Halogen/PAR

R20 Lamp/Replaced by Halogen/PAR20

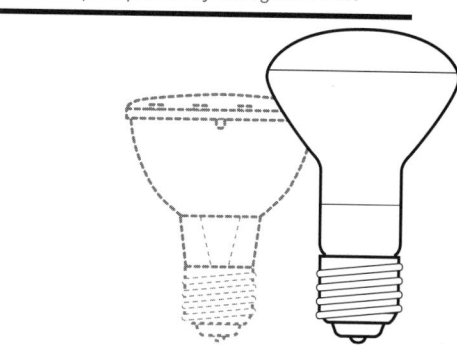

R40 Lamp/Replaced by Halogen/PAR38

R30 Lamp/Replaced by Halogen/PAR30S and PAR30L

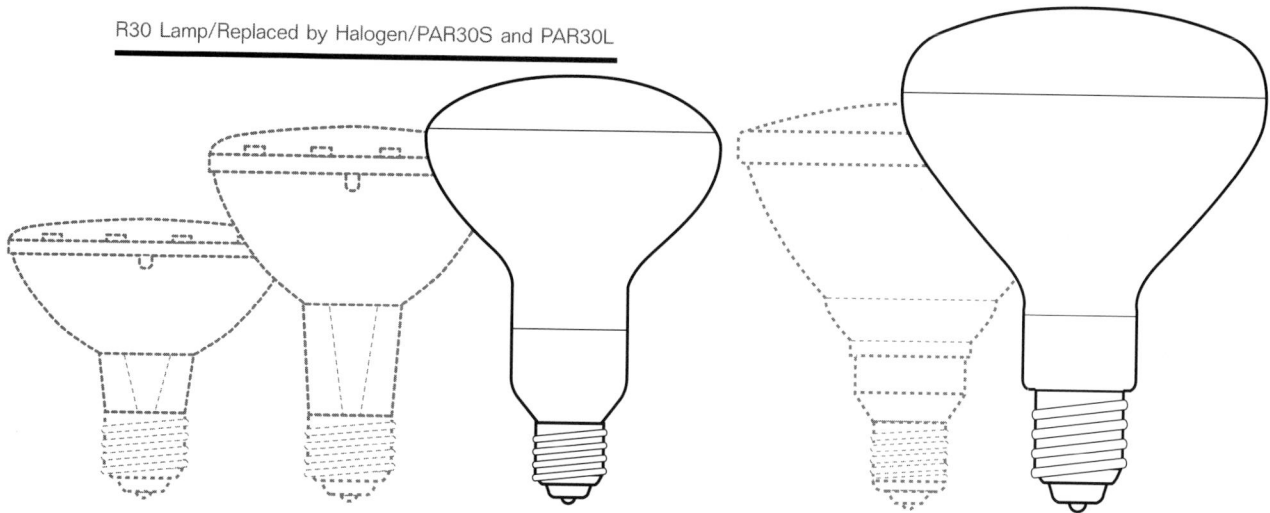

BR40 Lamp/Consider Halogen/PAR38

BR30 Lamp/Consider Halogen/PAR30

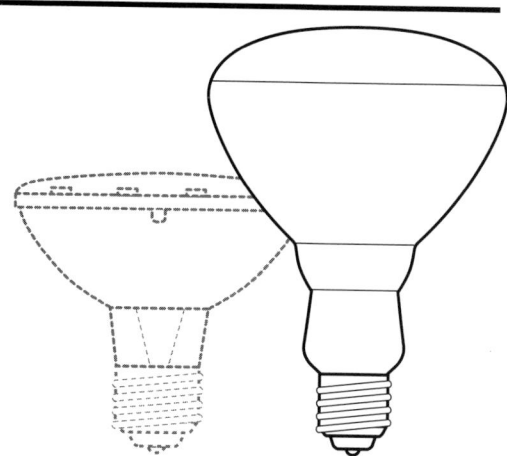

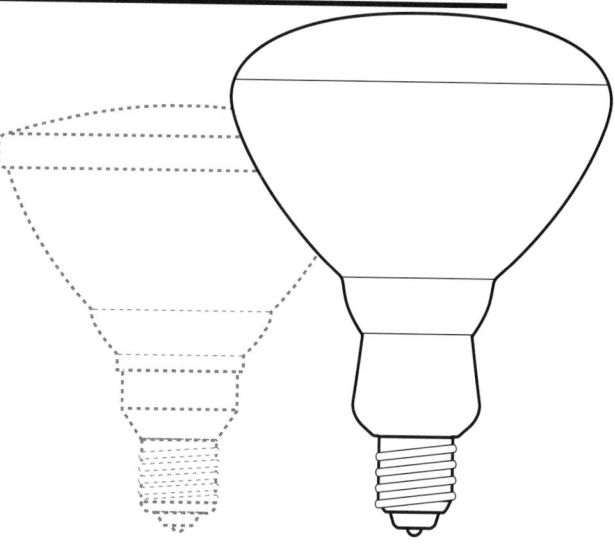

▶The halogen PAR lamps improved optics along with the smaller "point source" created by the halogen capsule allow for shallower lamps.
▶Old inefficient R and slightly improved BR lamps shown in solid outline for convenient reference when looking to retrofit existing luminaires.
▶Halogen PAR lamp replacements shown as dashed background.
▶All lamps shown to rough relative scale.

Images copyright 2000/GE

Halogen Mains Voltage (120V)

3

Halogen/PAR

Halogen PAR Lamp Designation

50PAR30S/H/NSP

Last set of letters identifies beam spread

FLOODS
- 24° to 32° for narrow flood (NFL)
- 33° to 44° for flood (FL)
- 45° to 54° for wide flood (WFL)
- 55° or greater for very wide flood (VWFL)

SPOTS
- 7° or less for very narrow spot (VNSP)
- 8° to 10° for narrow spot (NSP)
- 11° to 14° for spot (SP)
- 15° to 18° for wide spot (WSP)
- 19° to 23° for very wide spot (VWSP)

Ranges of beam spreads used for convenient/consistent reference in this text and do not necessarily correspond to each manufacturer's nor any ANSI or NEMA definitions.

This letter indicates "Halogen"

"S" for short neck lamps and "L" for long neck (only applies to PAR30 lamps)

These two digits represent lamp diameter in eighths inches (e.g., 30/8 or 3–3/4 inches)

These letters indicate this is a parabolic aluminized reflector (PAR) lamp

First two or three digits represent lamp wattage

[This designation used for convenient reference throughout this *Time–Saver Standards for Architectural Lighting*—but not necessarily used by all lamp manufacturers (although it would be convenient if manufacturers would come to agreement on standard designations.]

Halogen/PAR lamp designation

Halogen PAR lamps are typically identified with an alphanumeric designation outlined above. Since each manufacturer has slightly different means for lamp designation, the designation method presented above is solely for convenient reference throughout the text. Actual lamp manufacturers' designations will need to be used in final specifications. Perhaps the lamp manufacturers will standardize lamp designations in the near future in an effort to ease lamp cross referencing and specifications.

Halogen/PAR lamps for accenting

PAR lamps are great for art accenting. With so many choices on the spread of the light pattern (beam spread), on the intensity of the pattern (candlepower) and on wattage, accenting can be achieved from a wide range of ceiling heights and lateral distances from the objects to be highlighted. For the PAR20, PAR30 and PAR38 lamps, a series of Performance Sketches™ are presented which offers quick visual assessment of what lamp wattage and beam spread might be appropriate for given ceiling heights and artwork sizes.

Recommended intensities
for art accenting

Low (gallery, museum, residential)
5 to 10 fc average on art
Moderate (hospitality)
10 to 20 fc average on art
High (commercial—or higher for retail)
20 to 30 fc average on art

These are intended to be average, maintained values including effects of accent lighting and general room lighting. General room lighting contributes just a few footcandles in residential, gallery and museum settings, but contributes 5 to 10 fc in commercial settings. Maximums might range from 50 percent to 100 percent of the average value. Wherever artwork is highly valuable and/or sensitive to light, precautions should be taken to limit light exposure (maintaining low levels and/or switching lights off when art is not being viewed); and exposure to infrared and ultraviolet should be limited (using filters on lamps or placing art behind specially treated glass or acrylic).

Halogen Mains Voltage (120V)

Halogen/PAR

Performance Sketches™ are shown at a scale of ¼ inch equals 1 foot. Tracing for evaluation of various elevations is encouraged. Copying is permitted (according to the conventions outlined in "Licensing Agreement" in the Introduction of this book). Each Sketch includes an outline of artwork of the size category which may be lighted by the cited lamp. This is to assist in judgement of area of coverage. Location or distance from wall is listed with each Sketch as are center beam aiming angle and average light level over the area of the artwork outline. Maximum maintained intensity on artwork outline is also listed. For museum quality artwork where conservation of art is a priority, intensities should be 5 to 10 footcandles average, maintained, with maximums of perhaps 10 to 15 fc. In other applications, average maintained intensities on art will depend on the general or ambient lighting within the space(s). In typical hospitality applications, artwork accenting might exhibit average maintained intensities of 10 to 20 fc, with maximum intensities of 20 to 40 fc. In commercial applications, average, maintained intensities on artwork might range from 20 to 30 fc, with maximums of 40 fc and greater. These Performance Sketches™ are based on the cited manufacturer's data at time of manuscript preparation. Where two or more manufacturers offer nearly identical products, only one manufacturer's data (boldfaced) has been used for development of the Performance Sketches™. Data cited in the sidebar "Specs" is averaged from all available manufacturers' data. Where any manufacturer's data did not fall within 15 percent of this averaged data, the respective manufacturer's lamp is cited separately along with separate Performance Sketches™. Manufacturers' catalog data may be of one date and for one set of lamps while photometric data available from manufacturers may be based on another set of lamps of another date (this may have occurred with some data in this reference). Performance Sketches™ were generated with this data in Radiance (a Unix–based lighting simulation and rendering program available free of charge at www.radsite.lbl.gov/radiance/refer/long.html).

These Performance Sketches™ illustrate a specific condition. As ceiling heights change and focal objects' sizes change, the cited lamp may

■ ■ ■ ■ ■ ■ ■ ■ ■ ■ ■ ■ ■

Performance Sketches™ and Information
❶ Typical application is shown
❶ Other applications may also be acceptable
❶ Other applications may depend on ceiling heights
❶ Maintained values based on 0.9 maintenance factor
❶ Cited data is for a single lamp—two or more lamps may be necessary for other applications

Radiance (lighting rendering software)
http://www.radsite.lbl.gov/radiance/refer/long.html

◄ **CONNECT FOR MORE**

Halogen Mains Voltage (120V)

3

Halogen/PAR

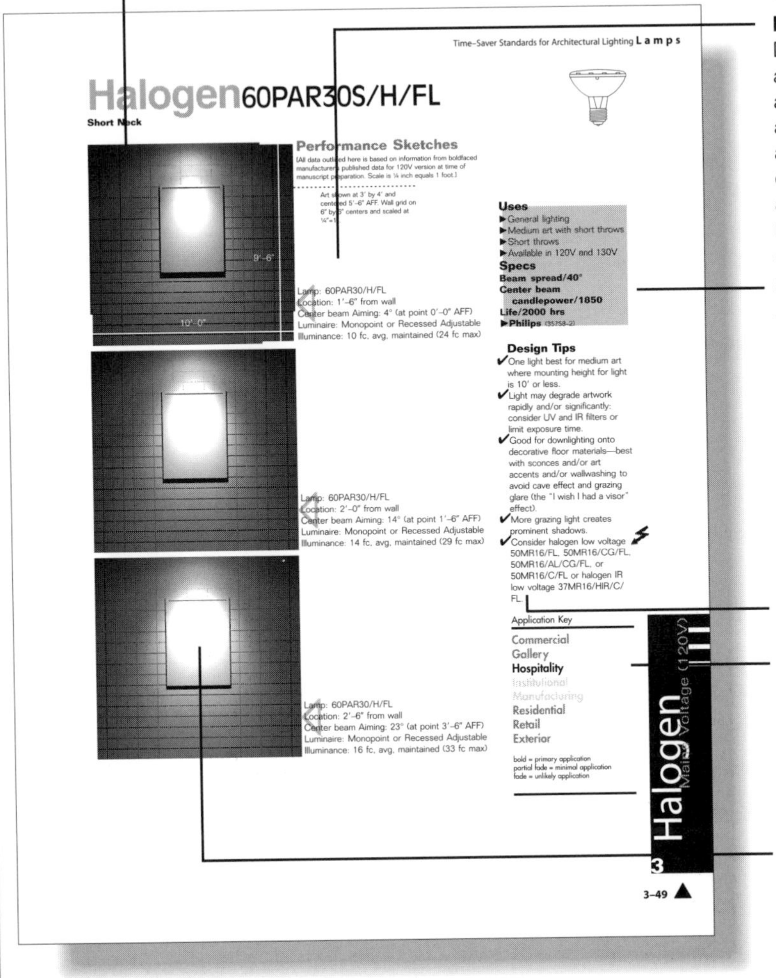

Performance Sketch™

A quarter–inch scale rendering illustrates lighting effect for a typical situation. Note how the area of coverage, intensity and frame shadow change with beam spread, ceiling height and distance from wall. Higher or lower ceilings and longer or shorter distances from walls will impact intensities, light patterns and application(s).

Performance Details

Details such as lamp distance from wall, aiming angle, point above floor at which lamp is aimed, along with maintained average and maximum illuminance are reported. The aiming angle should be used to select a luminaire capable of achieving such an angle. Maximum illuminances are calculated maintained over time and are reported to help assess likelihood of artwork degradation. Average illuminance is calculated maintained over time just for the size of the artwork shown.

Suggested Uses/Lamp Specs

A bullet list outlines uses for which the lamp seems best suited at cited ceiling height. Lamp specifications are outlined for quick reference and to assist in specification writing. Where two or more offer nearly identical products, only one manufacturer's data (boldfaced) has been used for development of the Performance Sketches™. An SKU– number or product code is included parenthetically. Data cited in the sidebar "Specs" is averaged from all available manufacturers' data.

Energy Tip

The energy bolt ⚡ is used to denote an energy saving option.

Design Tips

Key considerations are offered for the specific lamp.

Application Key

Those applications for the cited ceiling height where the specific lamp may be most useful are highlighted in bold. Applications where the lamp may be of lesser use are faded—the softer the fade, the less appropriate is the lamp for that application in cited ceiling height. Application appropriateness will change with ceiling heights.

Artwork Outline

An outline of two–dimensional artwork of size that might be acceptably lighted by lamp cited.

Inset sample page content:

Time–Saver Standards for Architectural Lighting **Lamps**

Halogen 60PAR30S/H/FL

Short Neck

Performance Sketches

[All data outlined here is based on information from boldfaced manufacturer's published data for 120V version at time of manuscript preparation. Scale is ¼ inch equals 1 foot.]

Art shown at 3' by 4' and centered 5'–6" AFF. Wall grid on 6' by 6" centers and scaled at ¼"×¼".

9'–6"

10'–0"

Lamp: 60PAR30/H/FL
Location: 1'–6" from wall
Center beam Aiming: 4° (at point 0'–0" AFF)
Luminaire: Monopoint or Recessed Adjustable
Illuminance: 10 fc, avg. maintained (24 fc max)

Lamp: 60PAR30/H/FL
Location: 2'–0" from wall
Center beam Aiming: 14° (at point 1'–6" AFF)
Luminaire: Monopoint or Recessed Adjustable
Illuminance: 14 fc, avg. maintained (29 fc max)

Lamp: 60PAR30/H/FL
Location: 2'–6" from wall
Center beam Aiming: 23° (at point 3'–6" AFF)
Luminaire: Monopoint or Recessed Adjustable
Illuminance: 16 fc, avg. maintained (33 fc max)

Uses
► General lighting
► Medium art with short throws
► Short throws
► Available in 120V and 130V
Specs
Beam spread/40°
Center beam
 candlepower/1850
Life/2000 hrs
► Philips (35758-2)

Design Tips
✔ One light best for medium art where mounting height for light is 10' or less.
✔ Light may degrade artwork rapidly and/or significantly: consider UV and IR filters or limit exposure time.
✔ Good for downlighting onto decorative floor materials—best with sconces and/or art accents and/or wallwashing to avoid cave effect and grazing glare (the "I wish I had a visor" effect).
✔ More grazing light creates prominent shadows.
✔ ⚡ Consider halogen low voltage 50MR16/FL, 50MR16/CG/FL, 50MR16/AL/CG/FL, or 50MR16/C/FL or halogen IR low voltage 37MR/HIR/C/FL.

Application Key

Commercial
Gallery
Hospitality
Institutional
Manufacturing
Residential
Retail
Exterior

bold = primary application
partial fade = minimal application
fade = unlikely application

Halogen Main Voltage (120V)

3

3–49 ▲

■ ■ ■ ■ ■ ■ ■ ■ ■ ■

Artwork Size Categories

⊃ Petite = 2'–0" by 2'–0" or smaller
⊃ Small = 2'–0" by 2'–0" to 2'–6" by 3'–0"
⊃ Medium = 2'–6" by 3'–0" to 3'–0" by 4'–0"
⊃ Large = 3'–0" by 4'–0" to 4'–0" by 5'–0"
⊃ Extra Large = 4'–0" by 5'–0" or larger

Halogen Main Voltage (120V)

Halogen/PAR

be quite appropriate for other applications. Generally, in lower ceiling settings, light intensities will be greater and area of coverage will be smaller. In higher ceiling settings, light intensities will be less and area of coverage will be larger. Lamp data changes from time to time. As manufacturing techniques change, as materials' technologies change, and as the market changes, lamp manufacturers will revise wattages, beam spreads and intensities. Manufacturers' data should be checked periodically for any possible updates.

■ ■ ■ ■ ■ ■ ■ ■ ■ ■

Manufacturers' Update Data:
❶ Periodically due to materials' technology
❶ Periodically due to manufacturing techniques
❶ Periodically due to governmental regulations
❶ Performance changes may result

Performance Sketch™ thumbnails

At the beginning of each PAR lamp section (PAR20, PAR30S, PAR30L and PAR38) there are thumbnails of the lighting images to further help provide quick visual assessment of various lamp beam spreads and intensities. Thumbnails serve as an index to the lamps in that category. The pages containing thumbnails are highlighted with gray margin bleeds.

Uses

For convenient reference, a shaded box entitled "Uses" appears in the upper corner of each page detailing the more likely uses for which the lamp seems suited. Artwork sizes are based on the rough dimensions as outlined in the tip box on the facing page (lower left). Lamp "Specs" are also included in the shaded box. These specs are averages of the listed manufacturers' data at time of manuscript preparation. Since products are constantly upgraded or deleted, a check of the current status of the listed lamp is recommended before including it in specifications.

Lighting characteristics

In addition to beam spread and intensity, these images illustrate beam striations and cutoff patterns to some extent. When lamps are being considered for a specific project, actual samples should be obtained and mocked up for visual inspection of quality of light patterning. Beam striations and aberrations may appear more pronounced in reality since Performance Sketches™ herein are based on finite candlepower reports (typically at 2.5° increments—a bright spot or dead spot might occur at an increment not reported on the photometry but which could be seen visually on mockup). Color of light cannot be illustrated in this black and white format. These

Halogen/PAR

Performance Sketches™ will help narrow selections so that appropriate lamp mockups can be made to assess actual beam striations and color of light.

Luminaires

PAR lamps are used in recessed downlights, recessed adjustable accent lights and track and monopoint adjustable accents. Since the PAR lamp is a self–contained optical package, track and monopint luminaires do not need to offer much more than a lamp socket and perhaps a decorative shroud to hide the bare PAR lamp. With downlights and adjustable accents, the reflector cone in the luminaire can help to control some light spill and/or minimize direct glare. Nevertheless, the lamp's optics provide the bulk of the lighting performance. Hence, the Performance Sketches™ shown here illustrate expected results from PAR lamps in track and monopoint adjustable accents as well as recessed adjustable accents.

Selection guides

Selection guides are offered on the next several pages for various uses in various applications. These guides are intended to help limit the designer's, engineer's or facility engineer's search and offer a good starting point from which design or alternative analyses can progress. Lamps which have similar beam spreads and intensities may be suggested for different ceiling heights and art sizes simply to show the variety of situations in which lamps might be used. Uses for Halogen/PAR lamps include:

▶pinspot accent
  An intense, well–controlled, focused lighting effect.
▶accent/medium art
  Relatively intense lighting effect on medium piece of art.
▶accent/large art
  Relatively intense lighting effect on large piece of art.
▶wallwashing/matte materials
  Relatively uniform wash of light on wall materials exhibiting little or no sheen, polish or specularity.
▶wallwashing/polished materials
  Relatively uniform wash of light on polished or specular wall materials.
▶feature downlighting
  Somewhat intense lighting effect on flooring material details.
▶general downlighting
  Relatively uniform wash of light at lap height, table height or on floor.

Halogen Mains Voltage (120V)

3

Halogen/PAR

While there are exceptions, Halogen PAR lamps are appropriate in the following applications:

▶commercial /retail

▶gallery/residential

▶hospitality

For low ceiling conditions—less than 10 feet—see Tables 3.3 (Commercial/Retail), 3.4 (Gallery/Residential), and 3.5 (Hospitality). For high ceiling conditions—10 feet or greater in height—see Tables 3.6 (Commercial/Retail), 3.7 (Gallery/Residential) and 3.8 (Hospitality).

Extrapolating information

There will be situations where perhaps twice as much light is desired (to provide a significant focal point) or where half as much light is desired (very sensitive artworks—paper-based pieces or media which fades easily). Find a ceiling height condition and artwork size which matches the planned situation. Then, to almost double light intensities, consider using two lights instead of a single light. Lights should be spaced on center at about the same distance they are spaced from the wall. So, if a single light spaced 2 feet/6 inches from the wall provides 17 fc average on a piece of art that is 3 feet wide by 4 feet long, then two such lights spaced 2 feet/6 inches from the wall and each spaced 2 feet/6 inches on center from the other centered on the art will provide about 30 fc average on the piece of art.

Another way to double the light is to find a lamp with similar beam spread (within 5 degrees) but twice the center beam candlepower. A search over several lamp families is suggested (e.g., perhaps a halogen low voltage MR lamp exists with nearly identical beam spread and twice the center beam candlepower as a halogen mains voltage PAR lamp). This is generally less costly than adding a second light, both initially and operationally. Most fine art installations, however, will have two lights aimed onto each piece for maximum viewing quality from most any angle—eliminating the harsh veiling reflections (particularly problematic with oils and acrylics) which are evident with single light accents.

To halve the light level, look for lamps of similar beam spreads (within 5 degrees) but with half the center beam candlepower. Light intensities can be reduced with neutral density filters, although this does waste energy and thus should be a last resort.

Halogen Mains Voltage (120V)

Halogen PAR Lamp Selection Guide

Table 3.3 Commercial or Retail Low Ceiling (less than 10 feet)[1][2]

◀Smaller lamp/smaller luminaire

Larger lamp/larger luminaire▶

Lighting Task	PAR20	PAR30S	PAR30L	PAR38
Pinspot Accent	Consider (1) or more depending on desired impact			
• Petite to small art • Feature merchandise • Small merchandise	• 35/NSP p3–34 beam spread wattage			
Key Area Accent	Consider (2) or more depending on desired impact			
• Small to medium art • Feature display • Moderately sized merchandise	• 50/NSP p3–37 • 50/WSP p3–38	• 50/FL p3–46 • 75/FL p3–52	• 35/NSP p3–61	• 60/NFL p3–95 • 120/WFL p3–119
Feature Lighting	Consider (2) or more depending on size of featured area			
• Medium to large art • Floor feature • Destination focus (e.g., cashier)	• 50/NFL p3–39	• 50/FL p3–46 • 75/FL p3–52	• 75/FL p3–74	• 60/NFL p3–95 • 120/WFL p3–119
General Downlighting	Consider symmetric arrangement[3]			
• Overall lighting of lap, table or floor		• 50/FL p3–46 • 60/FL p3–49		• 60/VWFL p3–96 • 90/VWFL p3–110

Typically dimmer▲ / ▼Typically brighter

[1] These lamp guides are intended to direct your attention to lamps which might meet your needs. If daylighting is prevalent, more lamps may be necessary to provide a visual focus to the artwork or merchandise. Design analysis is required to finalize lamp selections for projects.

[2] For retail applications it is reasonable to also use lamps cited in Table 3.6—intensities will be significantly greater in low ceiling spaces.

[3] Spacing between lights should not exceed 0.75 to 1.0 of the distance between the surface being downlighted and the ceiling.

Recommended intensities
for art accenting

Low (gallery, museum, residential)
5 to 10 fc average on art
Moderate (hospitality)
10 to 20 fc average on art
High (commercial—or higher for retail)
20 to 30 fc average on art

These are intended to be average, maintained values including effects of accent lighting and general room lighting. General room lighting contributes just a few footcandles in residential, gallery and museum settings, but contributes 5 to 10 fc in commercial settings. Maximums might range from 50 percent to 100 percent of the average value. Wherever artwork is highly valuable and/or sensitive to light, precautions should be taken to limit light exposure (maintaining low levels and/or switching lights off when art is not being viewed); and exposure to infrared and ultraviolet should be limited (using filters on lamps or placing art behind specially treated glass or acrylic).

Halogen Mains Voltage (120V)

Halogen PAR Lamp Selection Guide

Table 3.4 Gallery or Residential Low Ceiling (less than 10 feet)❶❷

◀Smaller lamp/smaller luminaire   Larger lamp/larger luminaire▶

Lighting Task	PAR20	PAR30S	PAR30L	PAR38
Pinspot Accent	Consider (1)			
• Petite art • Objet d'art • Centerpieces	• 35/NSP p3–34❷			
Accent/Small Artwork	Consider (1)			
• Small art	• 35/NFL p3–35❷ • 35/FL p3–36		• 35/WFL p3–62 beam spread	
Accent/Medium Artwork	Consider (1)			wattage
• Medium art • Medium sculpture • Architectural detail or feature	• 50/NFL p3–39❷	• 50/FL p3–46❷ • 60/FL p3–49❷	• 50/FL p3–67 • 50/WFL p3–68 • 50/VWFL p3–69 • 75/WFL p3–75 • 75/VWFL p3–76❷	• 45/WFL p3–89 • 75/VWFL p3–103 • 90/WFL p3–109❷
Accent/Large Artwork	Consider (1)			
• Large art • Large sculpture • Larger architectural detail or feature		• 75/FL p3–52❷	• 75/FL p3–74❷	
General Downlighting	Consider symmetric arrangement❸			
• Overall lighting of lap, table or floor	• 35/FL p3–36		• 35/WFL p3–62 • 50/WFL p3–68 • 50/VWFL p3–69 • 75/WFL p3–75	• 45/WFL p3–89 • 75/VWFL p3–103 • 90/WFL p3–109

▲Typically dimmer
▼Typically brighter

❶These lamp guides are intended to direct your attention to lamps which might meet your needs. Design analysis is required to finalize lamp selections for projects.

❷Neutral density filter will be required in front of halogen lamp to limit intensity to between 5 and 10 fc for art preservation.

❸Spacing between lights should not exceed 0.75 to 1.0 of the distance between the surface being downlighted and the ceiling.

■ ■ ■ ■ ■ ■ ■ ■ ■ ■

Artwork Size Categories

ᗡPetite = 2'–0" by 2'–0" or smaller
ᗡSmall = 2'–0" by 2'–0" to 2'–6" by 3'–0"
ᗡMedium = 2'–6" by 3'–0" to 3'–0" by 4'–0"
ᗡLarge = 3'–0" by 4'–0" to 4'–0" by 5'–0"
ᗡExtra Large = 4'–0" by 5'–0" or larger

Halogen Mains Voltage (120V)

Halogen PAR Lamp Selection Guide

Table 3.5 Hospitality Low Ceiling (less than 10 feet)❶❷

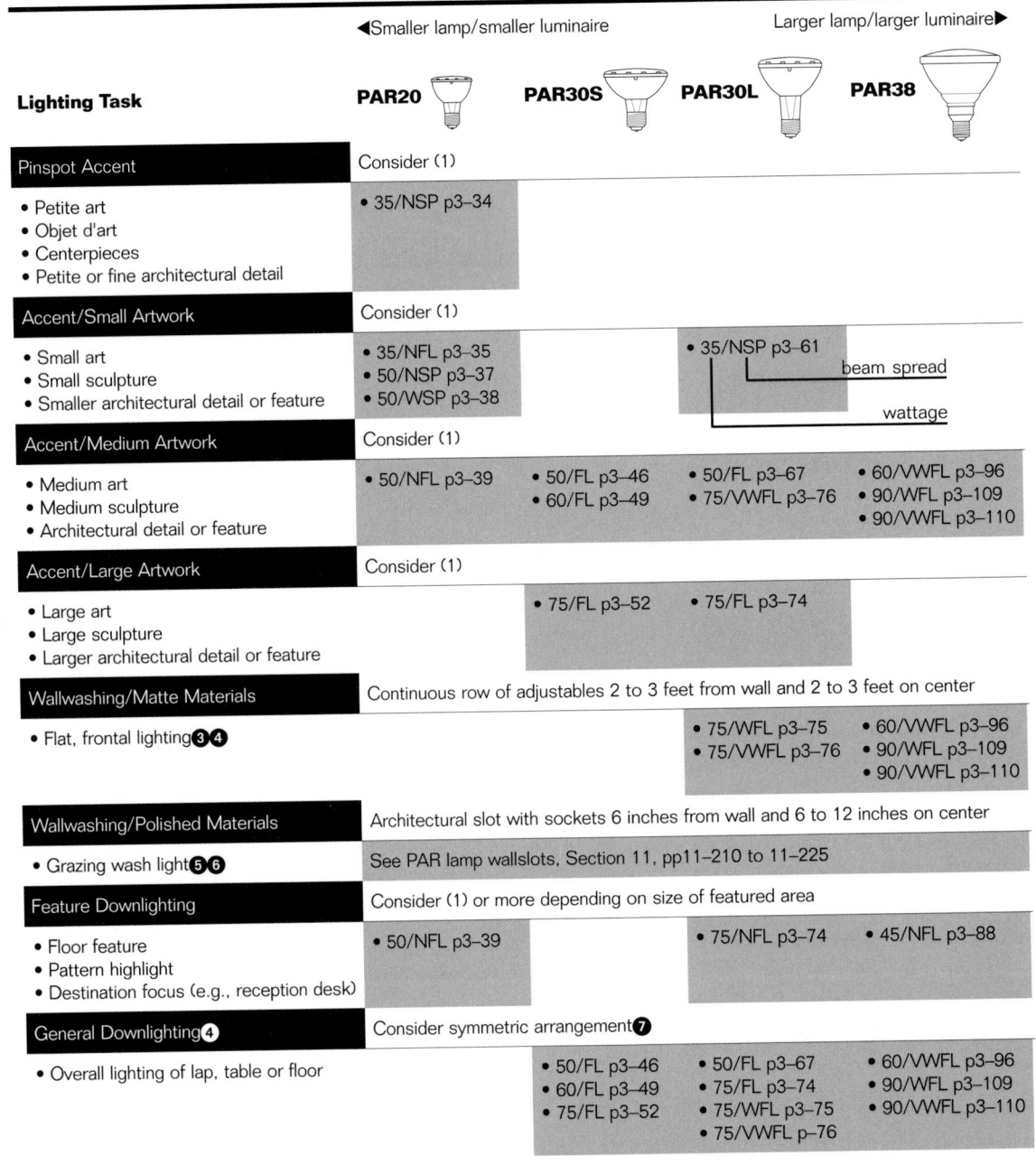

◄Smaller lamp/smaller luminaire   Larger lamp/larger luminaire►

Lighting Task	PAR20	PAR30S	PAR30L	PAR38
Pinspot Accent	Consider (1)			
• Petite art • Objet d'art • Centerpieces • Petite or fine architectural detail	• 35/NSP p3–34			
Accent/Small Artwork	Consider (1)			
• Small art • Small sculpture • Smaller architectural detail or feature	• 35/NFL p3–35 • 50/NSP p3–37 • 50/WSP p3–38		• 35/NSP p3–61	
Accent/Medium Artwork	Consider (1)			
• Medium art • Medium sculpture • Architectural detail or feature	• 50/NFL p3–39	• 50/FL p3–46 • 60/FL p3–49	• 50/FL p3–67 • 75/VWFL p3–76	• 60/VWFL p3–96 • 90/WFL p3–109 • 90/VWFL p3–110
Accent/Large Artwork	Consider (1)			
• Large art • Large sculpture • Larger architectural detail or feature		• 75/FL p3–52	• 75/FL p3–74	
Wallwashing/Matte Materials	Continuous row of adjustables 2 to 3 feet from wall and 2 to 3 feet on center			
• Flat, frontal lighting❸❹			• 75/WFL p3–75 • 75/VWFL p3–76	• 60/VWFL p3–96 • 90/WFL p3–109 • 90/VWFL p3–110
Wallwashing/Polished Materials	Architectural slot with sockets 6 inches from wall and 6 to 12 inches on center			
• Grazing wash light❺❻	See PAR lamp wallslots, Section 11, pp11–210 to 11–225			
Feature Downlighting	Consider (1) or more depending on size of featured area			
• Floor feature • Pattern highlight • Destination focus (e.g., reception desk)	• 50/NFL p3–39		• 75/NFL p3–74	• 45/NFL p3–88
General Downlighting❹	Consider symmetric arrangement❼			
• Overall lighting of lap, table or floor	• 50/FL p3–46 • 60/FL p3–49 • 75/FL p3–52	• 50/FL p3–67 • 75/FL p3–74 • 75/WFL p3–75 • 75/VWFL p–76		• 60/VWFL p3–96 • 90/WFL p3–109 • 90/VWFL p3–110

35/NSP p3–61
└── beam spread
        wattage

Typically dimmer▲
Typically brighter▼

❶These lamp guides are intended to direct your attention to lamps which might meet your needs. Design analysis is required to finalize lamp selections for projects.
❷Higher light levels on artwork and features are typically expected in hospitality spaces compared to residential and gallery spaces. Where artwork preservation is a priority, neutral density filters will be required in front of these lamps or lamps suggested for Gallery or Residential Applications should be used.
❸Energy intensive—only consider for exclusive zones of small size in highly visible public areas. Also see spreadlens wallwashers, pp11–136 to 11–143
❹High–wattage lamps not listed—for high intensities, more efficacious sources should be considered.
❺Extraordinarily energy intensive and should be reserved for very exclusive zones of small size in highly visible public areas.
❻This approach accentuates any wall imperfections.
❼Spacing between lights should not exceed 0.75 to 1.0 of the distance between the surface being downlighted and the ceiling.

Halogen Mains Voltage (120V)

Halogen PAR Lamp Selection Guide

Table 3.6 Commercial or Retail High Ceiling (10 feet or greater)❶❷

◀Smaller lamp/smaller luminaire   Larger lamp/larger luminaire▶

Lighting Task	PAR20	PAR30S	PAR30L	PAR38
Pinspot Accent	Consider (1) or more depending on ceiling height and desired impact			
• Petite to small art❸ • Feature merchandise • Small merchandise		• 50/NSP p3–44	• 50/NSP p3–63	• 90/NSP p3–104 • 120/NSP p3–116
Key Area Accent	Consider (2) or more depending on ceiling height and desired impact			
• Small to medium art❸ • Feature display • Moderately sized merchandise		• 60/NSP p3–47 • 60/NFL p3–48	• 75/NSP p3–70 • 75/WSP p3–72 beam spread wattage	• 60/NFL p3–95 • 90/SP p3–106 • 120/NFL p3–116 • 250/SP p3–120❹ • 250/NFL p3–122❹
Feature Lighting	Consider (2) or more depending on size of featured area and ceiling height			
• Medium to larger art❸ • Floor feature • Destination focus (e.g., cashier)		• 50/NFL p3–45 • 60/NFL p3–48 • 75/NFL p3–51	• 75/SP p3–71 • 75/WSP p3–72	• 90/NFL p3–108 • 100/NFL p3–114 • 120/NFL p3–118 • 250/SP p3–120 • 250/NFL p3–122
General Downlighting	Consider symmetric arrangement❺			
• Overall lighting of table or floor		• 75/FL p3–52		• 120/WFL p3–119

Typically brighter▲ / Typically dimmer▼

❶These lamp guides are intended to direct your attention to lamps which might meet your needs. If daylighting is prevalent, more lamps may be necessary to provide a visual focus to the artwork or merchandise. Design analysis is required to finalize lamp selections for projects.

❷Some lamps' performance sketches which are referenced here are not optimized for retail applications. Lower ceiling heights will result in higher intensities on merchandise.

❸This is the accenting typically found with the smaller–sized, lower wattage lamps listed. Larger lamps and higher wattages will accent larger art pieces.

❹These are very powerful lamps typically intended for high ceilings and/or where very strong intensities are required.

❺Spacing between lights should not exceed 0.75 to 1.0 of the distance between the surface being downlighted and the ceiling.

Artwork Size Categories
⊃Petite = 2'–0" by 2'–0" or smaller
⊃Small = 2'–0" by 2'–0" to 2'–6" by 3'–0"
⊃Medium = 2'–6" by 3'–0" to 3'–0" by 4'–0"
⊃Large = 3'–0" by 4'–0" to 4'–0" by 5'–0"
⊃Extra Large = 4'–0" by 5'–0" or larger

Recommended intensities
for art accenting

Low (gallery, museum, residential)
5 to 10 fc average on art
Moderate (hospitality)
10 to 20 fc average on art
High (commercial—or higher for retail)
20 to 30 fc average on art

These are intended to be average, maintained values including effects of accent lighting and general room lighting. General room lighting contributes just a few footcandles in residential, gallery and museum settings, but contributes 5 to 10 fc in commercial settings. Maximums might range from 50 percent to 100 percent of the average value. Wherever artwork is highly valuable and/or sensitive to light, precautions should be taken to limit light exposure (maintaining low levels and/or switching lights off when art is not being viewed); and exposure to infrared and ultraviolet should be limited (using filters on lamps or placing art behind specially treated glass or acrylic).

Halogen Mains Voltage (120V)

Halogen PAR Lamp Selection Guide

Table 3.7 Gallery or Residential High Ceiling (10 feet or greater)❶❷

◀Smaller lamp/smaller luminaire   Larger lamp/larger luminaire▶

Lighting Task	PAR20	PAR30S	PAR30L	PAR38
Pinspot Accent	Consider more efficient spot lamps❸			
• Petite art • Objet d'art • Centerpieces	• Halogen low voltage MR16 VNSP • Halogen low voltage PAR36 VNSP			
Accent/Small Artwork	Consider more efficient spot lamps❸			
• Small art	• Halogen low voltage MR16 NSP and VNSP • Halogen low voltage PAR36 VNSP		wattage	
Accent/Medium Artwork	Consider (1)		beam spread	
• Medium art • Medium sculpture • Architectural detail or feature			• 50/SP p3–64❹ • 50/WSP p3–65❹	• 45/NFL p3–88❹ • 50/NSP p3–90❹ • 90/VWSP p3–107❹
Accent/Large Artwork	Consider (1)			
• Large art • Large sculpture • Larger architectural detail or feature	• 75/NFL p3–51❹	• 75/NFL p3–73❹	• 60/NSP p3–92❹ • 75/NSP p3–98❹ • 75/NFL p3–101❹ • 75/NFL p3–102❹ • 90/WFL p3–109❹	
General Downlighting	Consider symmetric arrangement❺			
• Overall lighting of lap, table or floor			• 75/FL p3–74	• 60/VWFL p3–96 • 90/WFL p3–109 • 90/VWFL p3–110

▲Typically brighter ▼Typically dimmer

❶These lamp guides are intended to direct your attention to lamps which might meet your needs. Design analysis is required to finalize lamp selections for projects.

❷Gallery (museum or museum–quality display of valued artwork) applications typically use less light on artwork than most hospitality or residential applications. For purposes of conserving/preserving artwork, neutral density filters may be required in front of the Halogen/PAR lamps to limit light intensity to between 5 and 10 fc without disturbing beam spread or color of light.

❸To achieve small, well–controlled beam spreads, VNSP lamps in the halogen low voltage MR16 and PAR36 are recommended. The intensities, however, are significant and therefore should be considered for inert artwork or use of dimmers and/or neutral density filters will be necessary to limit intensities.

❹Neutral density filter required in front of lamp to limit intensity to between 5 and 10 fc if art preservation is paramount.

❺Spacing between lights should not exceed 0.75 to 1.0 of the distance between the surface being downlighted and the ceiling.

Footnotes to Table 3.8 (opposite page)▶

❶These lamp guides are intended to direct your attention to lamps which might meet your needs. Design analysis is required to finalize lamp selections for projects.

❷Higher light levels on artwork and features are expected in hospitality spaces compared to residential and gallery spaces. If art preservation is a priority, neutral density filters will be required in front of these lamps or lamps suggested for Gallery or Residential Applications should be used.

❸To achieve small, well–controlled beam spreads, VNSP lamps in the halogen low voltage MR16 and PAR36 are recommended. The intensities, however, are significant and therefore should be considered for inert artwork or use of dimmers and/or neutral density filters will be necessary to limit intensities.

❹Energy intensive—only consider for exclusive zones of small size in highly visible public areas. Also see spreadlens wallwashers, pp11–136 to 11–143.

❺High–wattage lamps not listed—for high intensities, more efficacious sources should be considered.

❻Extraordinarily energy intensive and should be reserved for very exclusive zones of small size in highly visible public areas.

❼This approach accentuates any wall imperfections.

❽Spacing between lights should not exceed 0.75 to 1.0 of the distance between the surface being downlighted and the ceiling.

Halogen Main Voltage (120V)

3

Halogen PAR Lamp Selection Guide

Table 3.8 Hospitality High Ceiling (10 feet or greater)**❶❷**

◄Smaller lamp/smaller luminaire   Larger lamp/larger luminaire►

Lighting Task	PAR20	PAR30S	PAR30L	PAR38
Pinspot Accent	Consider more efficient spot lamps**❸**			
• Petite art • Objet d'art	• Halogen low voltage MR16 VNSP • Halogen low voltage PAR36 VNSP			
Accent/Small Artwork	Consider more efficient spot lamps**❸**			
• Small art • Small sculpture	• Halogen low voltage MR16 VNSP • Halogen low voltage PAR36 VNSP			
Accent/Medium Artwork	Consider (1)			
• Medium art • Medium sculpture • Architectural detail or feature		• 50/NSP p3–44 • 50/NFL p3–45	• 50/NSP p3–63 • 50/SP p3–64 • 50/WSP p3–65 • 50/NFL p3–66	• 45/SP p3–87 • 45/NFL p3–88 • 50/NSP p3–90 • 50/NFL p3–91 • 75/SP p3–100 • 90/VWSP p3–107
Accent/Large Artwork	Consider (1)			
• Large art • Large sculpture • Larger architectural detail or feature		• 75/NSP p3–50 • 75/NFL p3–51	• 75/NSP p3–70 • 75/SP p3–71 • 75/WSP p3–72 • 75/NFL p3–73	• 45/NSP p3–86 • 50/NSP p3–90 • 60/NSP p3–92 • 60/SP p3–94 • 75/NSP p3–98 • 75/NFL p3–101 • 75/NFL p3–102 • 90/NSP p3–104 • 90/SP p3–106 • 90/NFL p3–108 • 100/SP p3–112 • 100/NFL p3–114 • 120/NSP p3–116 • 120/NFL p3–118 • 250/NFL p3–122
Wallwashing/Matte Materials	Continuous row of adjustables 2 to 3 feet from wall and 2 to 3 feet on center			
• Flat, frontal lighting**❹❺**				• 90/VWFL p3–110 • 120/WFL p3–119
Wallwashing/Polished Materials	Architectural slot with sockets 6 inches from wall and 6 to 12 inches on center			
• Grazing wash light**❻❼**	See PAR lamp wallslots, Section 11, pp11–210 to 11–225			
Feature Downlighting	Consider (1) or more depending on size of featured area			
• Floor feature • Pattern highlight • Destination focus (e.g., reception desk)		• 60/NFL p3–48 • 75/NFL p3–51 beam spread wattage	• 50/NFL p3–66 • 75/NFL p3–73	• 45/NFL p3–88 • 50/NFL p3–91 • 60/NFL p3–95 • 75/NFL p3–101 • 75/NFL p3–102 • 90/NFL p3–108 • 100/NFL p3–114
General Downlighting❺	Consider symmetric arrangement**❽**			
• Overall lighting of lap, table or floor		• 60/FL p3–49 • 75/FL p3–52	• 75/FL p3–74 • 75/WFL p3–75	• 90/WFL p3–109 • 90/VWFL p3–110 • 120/WFL p3–119

Typically dimmer▲
▼Typically brighter

Halogen Mains Voltage (120V)

3

Halogen/PAR20

Stats
Data varies manufacturer to manufacturer and changes from time to time

Life/2000 and 2500 hours
Varies manufacturer to manufacturer. A typical office environment might be occupied about 2600 hours each year.

Efficacy/11 LPW
Good beam intensities and distributions available.

Wattages/35 and 50

Color of Light/2800°K

Beam spreads/NSP, WSP, NFL and FL
Narrow spot, spot, narrow flood and wide flood distributions are available.

Size/3⅛" L by 2½" Ø

Voltage/120V and 130V

Cost Magnitude/Low (2000)

Manufacturers
General Electric
Osram Sylvania
Philips

Net Addresses
http://www.ge.com/lighting/business/index.htm
http://ecom.sylvania.com/osicatalog/
http://www.lighting.philips.com/

CONNECT FOR MORE

Advantages
Crisp white light
Good intensities available
Small size
Standard voltage (no transformers)
Easily dimmed
Low initial cost
Alternative to less efficient R lamps
Withstand extreme temperature range
Very little lamp lumen depreciation (light loss) over life

Disadvantages
Beam aberrations
◆Striations
◆Inconsistent beam shape
High operating costs
◆Inefficiency and short life may make this cost prohibitive in typical commercial applications, depending on ease of changing lamps and maintenance staffing.
Hot to touch
Hot-shock sensitive
◆Lamp may fail if rough aiming–adjustment or sharp vibration occurs while lamp is energized.
Water sensitive
◆Lamp may fail violently if placed in contact with water (not to be used exposed in outdoor lighting situations).

Halogen Mains Voltage (120V)

For more efficient accent lighting, use HIR low voltage lamps (Section 6); HIR PAR lamps (Section 5) or CMH PAR lamps (Section 7).

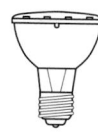

Halogen/PAR20

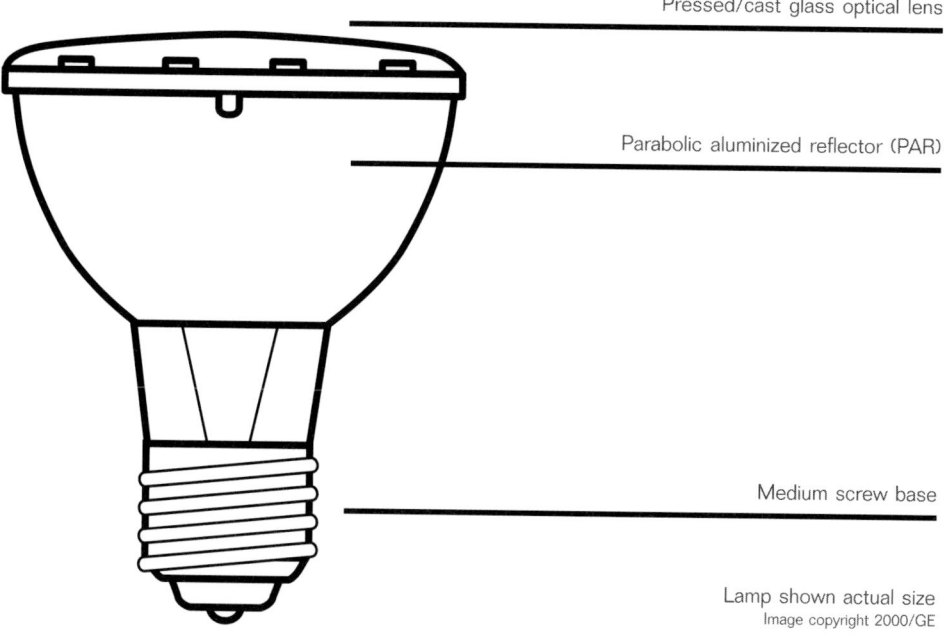

Pressed/cast glass optical lens

Parabolic aluminized reflector (PAR)

Medium screw base

Lamp shown actual size
Image copyright 2000/GE

Uses
Architectural Accenting
Art Accenting
Gallery
Hospitality
Residential
Downlighting
Circulation
Conference rooms
Hospitality
Residential
Retail
Merchandise Accenting
Surface Grazing
Specular/semispecular surfaces
◆Polished/honed stone
◆Polished/satin wood
◆Fritted/etched glass
◆Cut glass

■ ■ ■ ■ ■ ■ ■ ■ ■ ■ ■ ■ ■ ■ ■ ■ ■

PAR20 Lighting Design Tips
✔Best at lower mounting heights (less than 10′–0″).
✔Dimming increases life (typically doubling life), but shifts color warmer.
✔Glary when used without much other general lighting or when aimed across long distances where ceilings are low.
✔Beam striations are most apparent on light–colored, nontextured flat surfaces and may be objectionable.
✔PAR20 Downlighting alone is harsh—combine several techniques for best results.
✔Use smaller scale (e.g., 4–inch diameter) downlights and adjustable accents for attractive, more human scale approach—excellent in residential applications.

Net Addresses/PAR20 Track, Monopoints, Recessed and Exterior
See pages 3–6, 3–32 and 3–33

CONNECT FOR MORE

Halogen Mains Voltage (120V)

3

Halogen/PAR20

Figure 3.5

50PAR20/H/NFL lamps are used in small downlights in the bulkhead along the wood panel wall. Downlights were used to avoid the maintenance issues associated with lamp aiming and to provide consistent light for artwork or for exposed paneling. An additional benefit to downlighting in this application—the paper–based artwork is relatively delicate, grazing downlight softly lights the pieces, whereas adjustable accents would place more light on the works. Finally, the architectural design intent was to create a relatively thin bulkhead/niche for the art—adjustable accents would only be useful if mounted at least 1 foot/3 inches from the wall. The same lamp was used in downlights over the conference table to provide note taking light during presentations (see Figure 3.7).

Halogen Main Voltage (120V)

Image copyright 2000 Robert Eovaldi

Halogen/PAR20

Figure 3.6 (top)
50PAR20/H/NFL lamps are used in small downlights located in the upper coffered ceiling. These are typically off during meetings, but, as shown in Figure 3.7, can be used during presentations.

Figure 3.7 (bottom)
A preset scene control system (integrated into the control monitor at the "head" seating position) allows users to create scenes with the push of a single virtual "button" on the touch screen monitor. Fluorescent striplights (in cove) by Lithonia. Downlights are Kurt Versen.

Images copyright 2000 Robert Eovaldi

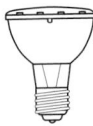

Halogen/PAR20

PAR20 Visual Index

35PAR20/H/NSP
▶ Architectural detailing
▶ Petite art/8'–6" ceiling height
▶ Objet d'art
▶ Short throws
▶ Commercial, Retail

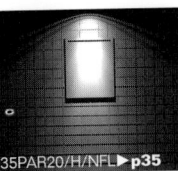  

35PAR20/H/NFL
▶ Feature downlighting
▶ Small art/8'–6" ceiling height
▶ Short throws
▶ Hospitality

  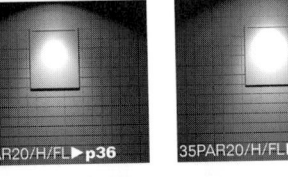 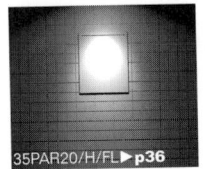

35PAR20/H/FL
▶ General lighting
▶ Wallwashing
▶ Small art/8'–6" ceiling height
▶ Short throws
▶ Gallery, Residential

■ ■ ■ ■ ■ ■

Throws/Ceilings
⟩ Short = 7' to 10'
⟩ Medium = 10' to 15'
⟩ Long = 15' to 20'
⟩ Very Long = 20' or more

 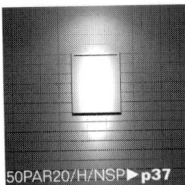 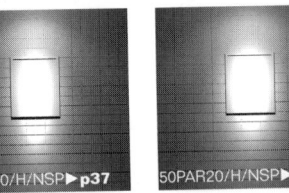

50PAR20/H/NSP
▶ Architectural detailing
▶ Small art/9'–6" ceiling height
▶ Objet d'art
▶ Commercial, Hospitality, Retail

50PAR20/H/WSP
▶ Architectural detailing
▶ Small art/9'–6" ceiling height
▶ Objet d'art
▶ Commercial, Hospitality, Retail

50PAR20/H/NFL
▶ Feature downlighting
▶ Architectural accenting
▶ Medium art/9'–6" ceiling height
▶ Hospitality

Net Addresses/PAR20 Recessed
http://www.thomasltg.com/
http://www.con-techlighting.com/
http://www.cooperlighting.com/
http://www.hubbell-ltg.com/products.htm#Down&Track
http://www.danalite.com/
http://www.kramerlighting.com/
http://www.lightolier.com/
http://www.lithonia.com/
http://www.prescolite.com/

CONNECT FOR MORE

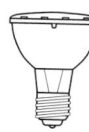

Halogen/PAR20

■ ■ ■ ■ ■ ■ ■ ■ ■ ■

Artwork Size Categories

➲Petite = 2'–0" by 2'–0" or smaller
➲Small = 2'–0" by 2'–0" to 2'–6" by 3'–0"
➲Medium = 2'–6" by 3'–0" to 3'–0" by 4'–0"
➲Large = 3'–0" by 4'–0" to 4'–0" by 5'–0"
➲Extra Large = 4'–0" by 5'–0" or larger

Halogen PAR Lamp Designation

35PAR20/H/NFL

Last set of letters identifies beam spread
FLOODS
• 24° to 32° for narrow flood (NFL)
• 33° to 44° for flood (FL)
• 45° to 54° for wide flood (WFL)
• 55° or greater for very wide flood (VWFL)
SPOTS
• 7° or less for very narrow spot (VNSP)
• 8° to 10° for narrow spot (NSP)
• 11° to 14° for spot (SP)
• 15° to 18° for wide spot (WSP)
• 19° to 23° for very wide spot (VWSP)

Ranges of beam spreads used for convenient/consistent reference in this text and do not necessarily correspond to each manufacturer's nor ANSI's definitions.

This letter indicates "Halogen"

These two digits represent lamp diameter in eighths inches (e.g., 30/8 or 3–3/4 inches)

These letters indicate this is a parabolic aluminized reflector (PAR) lamp

First two or three digits represent lamp wattage

[This designation used for convenient reference throughout this *Time–Saver Standards for Architectural Lighting*—but not necessarily used by all lamp manufacturers (although it would be convenient if manufacturers would come to agreement on standard designations.]

Net Addresses/PAR20 Track and Monopoints

http://www.bartcolighting.com/index2.html
http://www.thomasltg.com/
http://www.con-techlighting.com/
http://www.cooperlighting.com/
http://www.hubbell-ltg.com/products.htm#Down&Track
http://www.danalite.com/
http://www.kramerlighting.com/
http://www.lightingservicesinc.com/
http://www.lightolier.com/
http://www.lithonia.com/
http://www.prescolite.com/
http://www.tslight.com/

CONNECT FOR MORE

Recommended intensities
for art accenting

Low (gallery, museum, residential)
5 to 10 fc average on art
Moderate (hospitality)
10 to 20 fc average on art
High (commercial—or higher for retail)
20 to 30 fc average on art

These are intended to be average, maintained values including effects of accent lighting and general room lighting. General room lighting contributes just a few footcandles in residential, gallery and museum settings, but contributes 5 to 10 fc in commercial settings. Maximums might range from 50 percent to 100 percent of the average value. Wherever artwork is highly valuable and/or sensitive to light, precautions should be taken to limit light exposure (maintaining low levels and/or switching lights off when art is not being viewed); and exposure to infrared and ultraviolet should be limited (using filters on lamps or placing art behind specially treated glass or acrylic).

Halogen Mains Voltage (120V)

3

Halogen 35PAR20/H/NSP

Uses
▶Architectural detailing
▶Petite art
▶Objet d'art
▶Short throws
▶Available in 120V and 130V

Specs
Beam spread/8°
Center beam
candlepower/3000
Life/2500 hrs
▶**Osram Sylvania** (14467)

Design Tips
✔One light best for petite art where mounting height of light is 9' or less.
✔Intense light degrades artwork rapidly and/or significantly— consider UV and IR filters or limit exposure time.
✔Great for glass pieces and floral centerpieces where mounting height of light is 9' or less.
✔Shadows are less severe with flatter light (compare bottom image to top image).
✔Larger protruding frames create harsher shadows.

Performance Sketches

[All data outlined here is based on information from boldfaced manufacturer's published data for 120V version at time of manuscript preparation. Scale is ¼ inch equals 1 foot.]

Art shown at 2' by 2' and centered 5'–6" AFF. Wall grid on 6" by 6" centers and scaled at ¼"=1'.

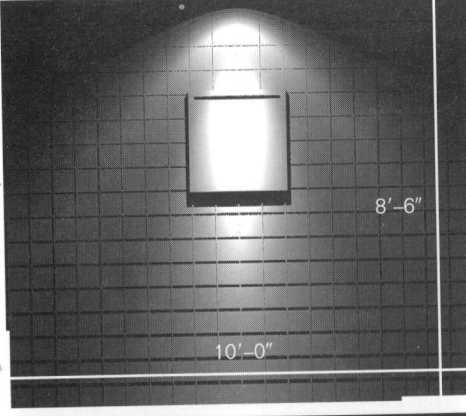

Lamp: 35PAR20/H/NSP
Location: 1'–0" from wall
Center beam Aiming: 14° (at point 4'–6" AFF)
Luminaire: Monopoint or Recessed Adjustable
Illuminance: 18 fc, avg, maintained (55 fc max)

Lamp: 35PAR20/H/NSP
Location: 1'–6" from wall
Center beam Aiming: 23° (at point 5'–0" AFF)
Luminaire: Monopoint or Recessed Adjustable
Illuminance: 23 fc, avg, maintained (84 fc max)

Lamp: 35PAR20/H/NSP
Location: 2'–0" from wall
Center beam Aiming: 30° (at point 5'–0" AFF)
Luminaire: Monopoint or Recessed Adjustable
Illuminance: 24 fc, avg, maintained (89 fc max)

Application Key

Commercial
Gallery
Hospitality
Institutional
Manufacturing
Residential
Retail
Exterior

bold = primary application
partial fade = minimal application
fade = unlikely application

Halogen 35PAR20/H/NFL

Performance Sketches

[All data outlined here is based on information from boldfaced manufacturer's published data for 120V version at time of manuscript preparation. Scale is ¼ inch equals 1 foot.]

Art shown at 2'–6" by 3' and centered 5'–6" AFF. Wall grid on 6" by 6" centers and scaled at ¼"=1'.

Lamp: 35PAR20/H/NFL
Location: 1'–0" from wall
Center beam Aiming: 7° (at point 0'–6" AFF)
Luminaire: Monopoint or Recessed Adjustable
Illuminance: 10 fc, avg, maintained (27 fc max)

Lamp: 35PAR20/H/NFL
Location: 1'–6" from wall
Center beam Aiming: 18° (at point 4'–0" AFF)
Luminaire: Monopoint or Recessed Adjustable
Illuminance: 15 fc, avg, maintained (37 fc max)

Lamp: 35PAR20/H/NSP
Location: 2'–0" from wall
Center beam Aiming: 27° (at point 4'–6" AFF)
Luminaire: Monopoint or Recessed Adjustable
Illuminance: 16 fc, avg, maintained (39 fc max)

Uses
► Feature downlighting
► Small art
► Short throws
► Available in 120V and 130V

Specs
Beam spread/30°
Center beam candlepower/900
Life/2500 hrs
► **Osram Sylvania** (14464)

Design Tips
✔ One light best for small art where mounting height of light is 9' or less.
✔ Striations may occur.
✔ Shadows are less severe with flatter light (compare bottom image to top image).
✔ Larger protruding frames create harsher shadows.

Application Key

Commercial
Gallery
Hospitality
Institutional
Manufacturing
Residential
Retail
Exterior

bold = primary application
partial fade = minimal application
fade = unlikely application

Halogen Mains Voltage (120V)

Halogen 35PAR20/H/FL

Uses
▶ Wallwashing (see 11–111)
▶ Small art
▶ General lighting
▶ Available in 120V and 130V

Specs
Beam spread/40°
Center beam candlepower/600
Life/2500 hrs
▶ **Osram Sylvania** (wide flood)
(14506)

Design Tips
✔ One light best for small art where mounting height of light is 9′ or less.
✔ Multiple units can be used for wallwashing or lighting of larger artwork.
✔ Shadows are less severe with flatter light (compare bottom image to top image).
✔ Larger protruding frames create harsher shadows.
✔ Consider halogen low voltage 20MR16/FL and 20MR16/C/FL.

Performance Sketches

[All data outlined here is based on information from boldfaced manufacturer's published data for 120V version at time of manuscript preparation. Scale is ¼ inch equals 1 foot.]

Art shown at 2′–6″ by 3′ and centered 5′–6″ AFF. Wall grid on 6″ by 6″ centers and scaled at ¼″=1′.

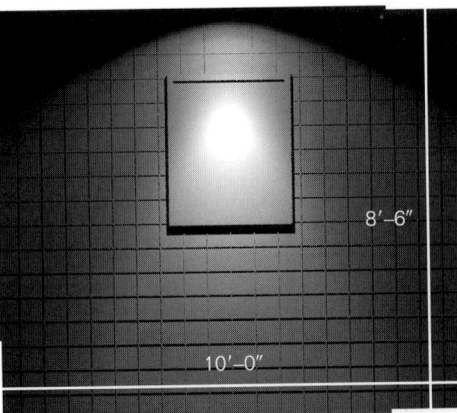

8′–6″

10′–0″

Lamp: 35PAR20/H/FL
Location: 1′–6″ from wall
Center beam Aiming: 11° (at point 1′–0″ AFF)
Luminaire: Monopoint or Recessed Adjustable
Illuminance: 8 fc, avg, maintained (18 fc max)

Lamp: 35PAR20/H/FL
Location: 2′–0″ from wall
Center beam Aiming: 20° (at point 3′–0″ AFF)
Luminaire: Monopoint or Recessed Adjustable
Illuminance: 10 fc, avg, maintained (20 fc max)

Lamp: 35PAR20/H/FL
Location: 2′–6″ from wall
Center beam Aiming: 32° (at point 4′–6″ AFF)
Luminaire: Monopoint or Recessed Adjustable
Illuminance: 13 fc, avg, maintained (25 fc max)

Application Key
Commercial
Gallery
Hospitality
Institutional
Manufacturing
Residential
Retail
Exterior

bold = primary application
partial fade = minimal application
fade = unlikely application

Halogen Main Voltage (120V) III

Halogen 50PAR20/H/NSP

Performance Sketches

[All data outlined here is based on information from boldfaced manufacturer's published data for 120V version at time of manuscript preparation. Scale is ¼ inch equals 1 foot.]

Art shown at 2'–6" by 3' and centered 5'–6" AFF. Wall grid on 6" by 6" centers and scaled at ¼"=1'.

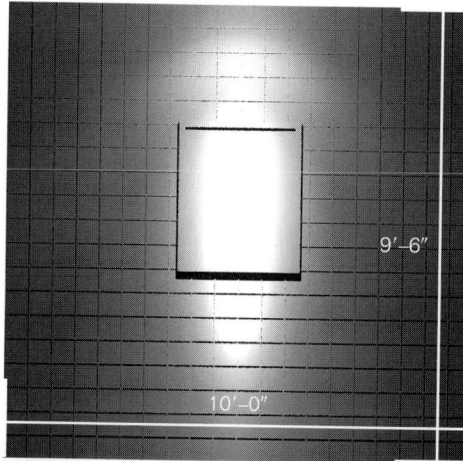

9'–6"

10'–0"

Lamp: 50PAR20/H/NSP
Location: 2'–0" from wall
Center beam Aiming: 18° (at point 3'–6" AFF)
Luminaire: Monopoint or Recessed Adjustable
Illuminance: 17 fc, avg, maintained (47 fc max)

Lamp: 50PAR20/H/NSP
Location: 2'–6" from wall
Center beam Aiming: 24° (at point 4'–0" AFF)
Luminaire: Monopoint or Recessed Adjustable
Illuminance: 20 fc, avg, maintained (63 fc max)

Lamp: 50PAR20/H/NSP
Location: 3'–0" from wall
Center beam Aiming: 31° (at point 4'–6" AFF)
Luminaire: Monopoint or Recessed Adjustable
Illuminance: 23 fc, avg, maintained (83 fc max)

Uses
▶ Architectural detailing
▶ Small art
▶ Objet d'art
▶ Available in 120V and 130V

Specs
Beam spread/10°
Center beam
   candlepower/5600
Life/2150 hrs
▶ **GE** (14927)
▶ Osram Sylvania
▶ Philips

Design Tips
✔ One light best for small art where mounting height of light is 10' or less.
✔ Intense light degrades some artwork rapidly and/or significantly—consider UV and IR filters or limit exposure time.
✔ Great for inert sculptural pieces and floral centerpieces.
✔ Consider halogen low voltage 20MR11/NSP , 20MR11/CG/ NSP or 20MR16/NSP.

Application Key

Commercial
Gallery
Hospitality
Institutional
Manufacturing
Residential
Retail
Exterior

bold = primary application
partial fade = minimal application
fade = unlikely application

Halogen Mains Voltage (120V)

3

Halogen 50PAR20/H/WSP

Uses
▶ Architectural detailing
▶ Small art
▶ Objet d'art
▶ Available in 120V and 130V

Specs
Beam spread/16°
Center beam
   candlepower/3200
Life/2000 hrs
▶ **Philips** (22908–8)

Design Tips
✔ One light best for small art where mounting height of light is 10' or less.
✔ Intense light degrades some artwork rapidly and/or significantly—consider UV and IR filters or limit exposure time.
✔ Great for inert sculptural pieces and floral centerpieces.
✔ Consider halogen low voltage 20MR16/C/WSP and 20MR16/C/CG/WSP.

Performance Sketches
[All data outlined here is based on information from boldfaced manufacturer's published data for 120V version at time of manuscript preparation. Scale is ¼ inch equals 1 foot.]

Art shown at 2'–6" by 3' and centered 5'–6" AFF. Wall grid on 6" by 6" centers and scaled at ¼"=1'.

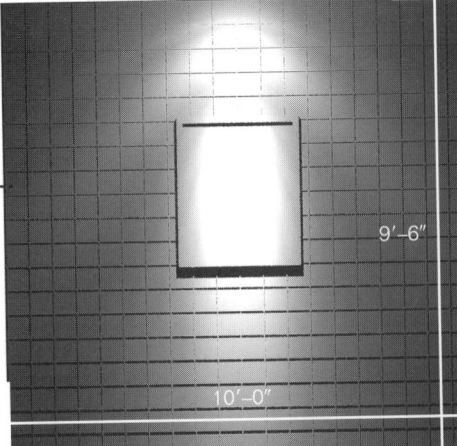

9'–6"

10'–0"

Lamp: 50PAR20/H/WSP
Location: 1'–6" from wall
Center beam Aiming: 14° (at point 3'–6" AFF)
Luminaire: Monopoint or Recessed Adjustable
Illuminance: 17 fc, avg, maintained (40 fc max)

Lamp: 50PAR20/H/WSP
Location: 2'–0" from wall
Center beam Aiming: 20° (at point 4'–0" AFF)
Luminaire: Monopoint or Recessed Adjustable
Illuminance: 20 fc, avg, maintained (45 fc max)

Lamp: 50PAR20/H/WSP
Location: 2'–6" from wall
Center beam Aiming: 27° (at point 4'–6" AFF)
Luminaire: Monopoint or Recessed Adjustable
Illuminance: 23 fc, avg, maintained (52 fc max)

Application Key

Commercial
Gallery
Hospitality
Institutional
Manufacturing
Residential
Retail
Exterior

bold = primary application
partial fade = minimal application
fade = unlikely application

Halogen 50PAR20/H/NFL

Performance Sketches

[All data outlined here is based on information from boldfaced manufacturer's published data for 120V version at time of manuscript preparation. Scale is ¼ inch equals 1 foot.]

Art shown at 3' by 4' and centered 5'–6" AFF. Wall grid on 6" by 6" centers and scaled at ¼"=1'.

Lamp: 50PAR20/H/NFL
Location: 1'–6" from wall
Center beam Aiming: 9° (at point 0'–0" AFF)
Luminaire: Monopoint or Recessed Adjustable
Illuminance: 10 fc, avg, maintained (23 fc max)

Lamp: 50PAR20/H/NFL
Location: 2'–0" from wall
Center beam Aiming: 18° (at point 3'–6" AFF)
Luminaire: Monopoint or Recessed Adjustable
Illuminance: 14 fc, avg, maintained (33 fc max)

Lamp: 50PAR20/H/NFL
Location: 2'–6" from wall
Center beam Aiming: 27° (at point 4'–6" AFF)
Luminaire: Monopoint or Recessed Adjustable
Illuminance: 17 fc, avg, maintained (38 fc max)

Uses
▶ Feature downlighting
▶ Architectural accenting
▶ Medium art
▶ Available in 120V and 130V

Specs
Beam spread/30°
Center beam
  candlepower/1380
Life/2150 hrs
▶ **GE** (14928)
▶ Osram Sylvania
▶ Philips

Design Tips
✔ One light best for medium art where mounting height of light is 10' or less.
✔ More grazing light creates prominent shadows.

Application Key

Commercial
Gallery
Hospitality
Institutional
Manufacturing
Residential
Retail
Exterior

bold = primary application
partial fade = minimal application
fade = unlikely application

Halogen Main Voltage (120V)

Halogen/PAR30S

Short Neck

Stats
_{Data varies manufacturer to manufacturer and changes from time to time}

Life/2000 and 2500 hours
Different versions from different manufacturers. A typical office environment might be occupied about 2600 hours each year.

Efficacy/12 LPW
Good beam intensities and distributions available.

Wattages/ 50, 60 and 75

Color of Light/2800°K

Beam spreads/NSP, NFL and FL
Narrow spot, narrow flood and wide flood distributions are available.

Size/3⅝" L by 3¾" Ø

Voltage/120V and 130V

Cost Magnitude/Low ⁽²⁰⁰⁰⁾

Manufacturers
General Electric
Osram Sylvania
Philips

Net Addresses
http://www.ge.com/lighting/business/index.htm
http://ecom.sylvania.com/osicatalog/
http://www.lighting.philips.com/

CONNECT FOR MORE

Advantages
Crisp white light
Good to excellent intensities available
Relatively small size
Short neck versions fit into shallower/shorter lights
Standard voltage (no transformers)
Easily dimmed
Low initial cost
Alternative to less efficient R lamps
Withstand extreme temperature range
Very little lamp lumen depreciation (light loss) over life

Disadvantages
Beam aberrations
◆Striations
◆Inconsistent beam shape
High operating costs
◆Inefficiency and short life may make this cost prohibitive in typical commercial applications, depending on ease of changing lamps and maintenance staffing.
Hot to touch
Hot-shock sensitive
◆Lamp may fail if rough aiming–adjustment or sharp vibration occurs while lamp is energized.
Water sensitive
◆Lamp may fail violently if placed in contact with water (not to be used exposed in outdoor lighting situations).

Halogen Mains Voltage (120V)

For more efficient accent lighting, use HIR low voltage lamps (Section 6); HIR PAR lamps (Section 5) or CMH PAR lamps (Section 7).

3

Halogen/PAR30S
Short Neck

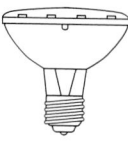

Pressed/cast glass optical lens

Parabolic aluminized reflector (PAR)

Medium screw base

Lamp shown actual size
Image copyright 2000/GE

Uses

Architectural Accenting
Art Accenting
Gallery
Hospitality
Residential
Downlighting
Circulation
Conference rooms
Hospitality
Residential
Retail
Merchandise Accenting
Surface Grazing
Specular/semispecular surfaces
◆Polished/honed stone
◆Polished/satin wood
◆Fritted/etched glass
◆Cut glass

PAR30S Lighting Design Tips
✔Best at low–to–moderate mounting heights (12'–0" or less).
✔Dimming increases life (typically doubling life), but shifts color warmer.
✔Glary when used without much other general lighting or when aimed across long distances where ceilings are low.
✔Beam striations are most apparent on light–colored, nontextured, flat surfaces and may be objectionable.
✔PAR30 Downlighting alone is harsh—combine several techniques for best results.
✔Use smaller scale (e.g., 5–inch diameter) downlights and adjustable accents for attractive, more human scale approach.

Net Addresses/PAR30S Track, Monopoints, Recessed and Exterior
See pages 3–53

CONNECT FOR MORE

Halogen Mains Voltage (120V)

3

3–41 ▲

Halogen/PAR30S

PAR30S Visual Index

50PAR30S/H/NSP
► Architectural detailing
► Medium art/12'–6" ceiling height
► Sculpture
► Medium throws
► Commercial, Hospitality, Retail

50PAR30S/H/NFL
► Architectural accenting
► Feature downlighting
► Medium art/10'–6" ceiling height
► Sculpture
► Medium throws
► Hospitality

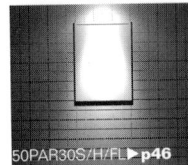  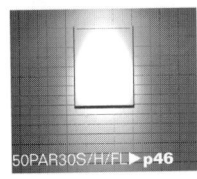

50PAR30S/H/FL
► General lighting
► Medium art/8'–6" ceiling height
► Short throws
► Commercial, Retail

60PAR30S/H/NSP
► Architectural detailing
► Medium art/12'–6" ceiling height
► Sculpture
► Medium throws
► Commercial, Retail

60PAR30S/H/NFL
► Architectural accenting
► Feature downlighting
► Medium art/10'–6" ceiling height
► Sculpture
► Medium throws
► Commercial, Retail

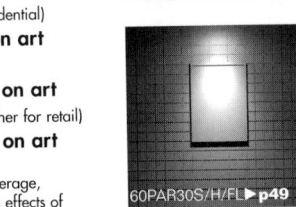    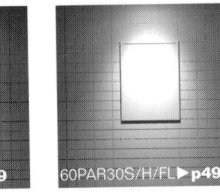

60PAR30S/H/FL
► General lighting
► Medium art/9'–6" ceiling height
► Short throws
► Hospitality

■ ■ ■ ■ ■

Throws/Ceilings
⊃Short = 7' to 10'
⊃Medium = 10' to 15'
⊃Long = 15' to 20'
⊃Very Long = 20' or more

Recommended intensities
for art accenting

Low (gallery, museum, residential)
5 to 10 fc average on art
Moderate (hospitality)
10 to 20 fc average on art
High (commercial—or higher for retail)
20 to 30 fc average on art

These are intended to be average, maintained values including effects of accent lighting and general room lighting. General room lighting contributes just a few footcandles in residential, gallery and museum settings, but contributes 5 to 10 fc in commercial settings. Maximums might range from 50 percent to 100 percent of the average value. Wherever artwork is highly valuable and/or sensitive to light, precautions should be taken to limit light exposure (maintaining low levels and/or switching lights off when art is not being viewed); and exposure to infrared and ultraviolet should be limited (using filters on lamps or placing art behind specially treated glass or acrylic).

Halogen Mains Voltage (120V)

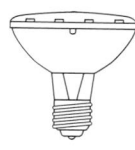

Halogen/PAR30S

Short Neck

75PAR30S/H/NSP
▶Architectural detailing
▶Large art/15'–0" ceiling height
▶Sculpture
▶Long throws
▶Hospitality

75PAR30S/H/NFL
▶Architectural accenting
▶Feature downlighting
▶Large art/12'–6" ceiling height
▶Sculpture
▶Medium throws
▶Hospitality

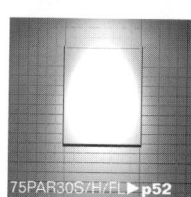

75PAR30S/H/FL
▶General lighting
▶Large art/9'–6" ceiling height
▶Short throws
▶Hospitality

Halogen PAR Lamp Designation

50PAR30S/H/NSP

Last set of letters identifies beam spread

FLOODS
• 24° to 32° for narrow flood (NFL)
• 33° to 44° for flood (FL)
• 45° to 54° for wide flood (WFL)
• 55° or greater for very wide flood (VWFL)

SPOTS
• 7° or less for very narrow spot (VNSP)
• 8° to 10° for narrow spot (NSP)
• 11° to 14° for spot (SP)
• 15° to 18° for wide spot (WSP)
• 19° to 23° for very wide spot (VWSP)

Ranges of beam spreads used for convenient/consistent reference in this text and do not necessarily correspond to each manufacturer's nor ANSI's definitions.

This letter indicates "Halogen"

"S" for short neck lamps and "L" for long neck (only applies to PAR30 lamps)

These two digits represent lamp diameter in eighths inches (e.g., 30/8 or 3-3/4 inches)

These letters indicate this is a parabolic aluminized reflector (PAR) lamp

First two or three digits represent lamp wattage

[This designation used for convenient reference throughout this *Time–Saver Standards for Architectural Lighting*—but not necessarily used by all lamp manufacturers (although it would be convenient if manufacturers would come to agreement on standard designations.]

Artwork Size Categories
➲Petite = 2'–0" by 2'–0" or smaller
➲Small = 2'–0" by 2'–0" to 2'–6" by 3'–0"
➲Medium = 2'–6" by 3'–0" to 3'–0" by 4'–0"
➲Large = 3'–0" by 4'–0" to 4'–0" by 5'–0"
➲Extra Large = 4'–0" by 5'–0" or larger

Halogen Mains Voltage (120V)

3

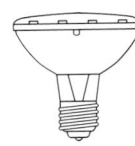

Halogen 50PAR30S/H/NSP

Short Neck

Uses
- ▶ Architectural detailing
- ▶ Medium art with medium throws
- ▶ Sculpture with medium throws
- ▶ Medium throws
- ▶ Available in 120V and 130V

Specs
Beam spread/10°
Center beam candlepower/ 8000
Life/2150 hrs
- ▶ GE (14023)
- ▶ Osram Sylvania
- ▶ Philips

Design Tips
✔ One light best for medium art where mounting height for light is 13′ or less.
✔ Intense light degrades some artwork rapidly and/or significantly—consider UV and IR filters or limit exposure time.
✔ Great for glass pieces and floral centerpieces and where mounting height for light is 13′ or less.
✔ More grazing light creates prominent shadows.
✔ Consider halogen low voltage 35MR11/NSP, 35MR16/AL/ NSP and 35MR16/C/NSP or halogen IR low voltage 37MR16/HIR/C/NSP.

Performance Sketches

[All data outlined here is based on information from boldfaced manufacturer's published data for 120V version at time of manuscript preparation. Scale is ¼ inch equals 1 foot.]

Lamp: 50PAR30S/H/NSP
Location: 3′–0″ from wall
Center beam Aiming: 21° (at point 4′–6″ AFF)
Luminaire: Monopoint or Recessed Adjustable
Illuminance: 18 fc, avg, maintained (53 fc max)

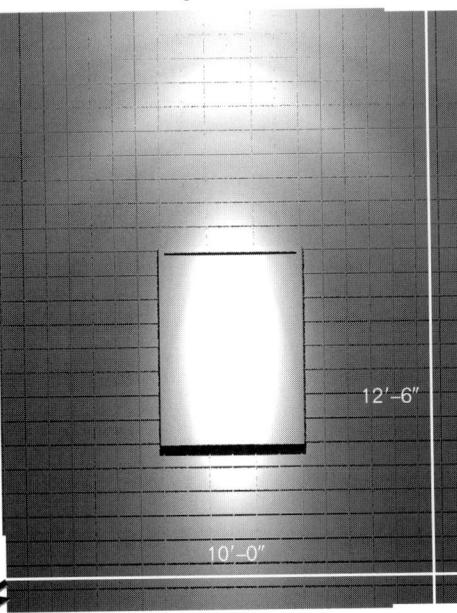

12′–6″

10′–0″

Lamp: 50PAR30S/H/NSP
Location: 4′–0″ from wall
Center beam Aiming: 27° (at point 4′–6″ AFF)
Luminaire: Monopoint or Recessed Adjustable
Illuminance: 19 fc, avg, maintained (59 fc max)

Art shown at 3′ by 4′ and centered 5′–6″ AFF. Wall grid on 6″ by 6″ centers and scaled at ¼″=1′.

Lamp: 50PAR30S/H/NSP
Location: 5′–0″ from wall
Center beam Aiming: 34° (at point 5′–0″ AFF)
Luminaire: Monopoint or Recessed Adjustable
Illuminance: 21 fc, avg, maintained (69 fc max)

Application Key

Commercial
Gallery
Hospitality
Institutional
Manufacturing
Residential
Retail
Exterior

bold = primary application
partial fade = minimal application
fade = unlikely application

Halogen Mains Voltage (120V)

3

Halogen 50PAR30S/H/NFL

Short Neck

Performance Sketches

[All data outlined here is based on information from boldfaced manufacturer's published data for 120V version at time of manuscript preparation. Scale is ¼ inch equals 1 foot.]

Lamp: 50PAR30S/H/NFL
Location: 2'–3" from wall
Center beam Aiming: 19° (at point 4'–0" AFF)
Luminaire: Monopoint or Recessed Adjustable
Illuminance: 15 fc, avg, maintained (38 fc max)

Lamp: 50PAR30S/H/NFL
Location: 2'–9" from wall
Center beam Aiming: 25° (at point 4'–6" AFF)
Luminaire: Monopoint or Recessed Adjustable
Illuminance: 17 fc, avg, maintained (41 fc max)

10'–6"

10'–0"

Art shown at 3' by 4' and centered 5'–6" AFF. Wall grid on 6" by 6" centers and scaled at ¼"=1'.

Lamp: 50PAR30S/H/NFL
Location: 3'–3" from wall
Center beam Aiming: 28° (at point 4'–6" AFF)
Luminaire: Monopoint or Recessed Adjustable
Illuminance: 17 fc, avg, maintained (39 fc max)

Uses

► Architectural accenting
► Feature downlighting
► Medium art with medium throws
► Sculpture with medium throws
► Medium throws
► Available in 120V and 130V

Specs

Beam spread/30°
Center beam candlepower/1900
Life/2150 hrs
►**GE** (17871)
►Osram Sylvania
►Philips

Design Tips

✔ One light best for medium art where mounting height for light is 11' or less.
✔ Great for moderate to larger sculpture where mounting height for light is 11' or less.
✔ Scallops and frame shadows are most severe when light is closer to the wall (compare top left image to bottom image).

Application Key

Commercial
Gallery
Hospitality
Institutional
Manufacturing
Residential
Retail
Exterior

bold = primary application
partial fade = minimal application
fade = unlikely application

Halogen Mains Voltage (120V)

3

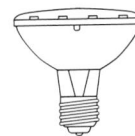

Halogen 50PAR30S/H/FL

Short Neck

Uses
► General lighting
► Medium art with short throws
► Short throws
► Available in 120V and 130V

Specs
Beam spread/40°
Center beam candlepower/1350
Life/2150 hrs
► **GE** (14022)
► Osram Sylvania
► Philips

Design Tips
✔ One light best for medium art where mounting height of light is 9' or less.
✔ Light may degrade artwork rapidly and/or significantly: consider UV and IR filters or limit exposure time.
✔ Good for downlighting onto decorative floor materials— best with sconces and/or art accents and/or wallwashing to avoid cave effect and grazing glare (the "I wish I had a visor" effect).
✔ More grazing light creates prominent shadows.
✔ Consider halogen low voltage 35MR11/FL and 35MR16/FL or halogen IR low voltage 37MR16/HIR/C/FL.

Performance Sketches

[All data outlined here is based on information from boldfaced manufacturer's published data for 120V version at time of manuscript preparation. Scale is ¼ inch equals 1 foot.]

Art shown at 3' by 4' and centered 5'–6" AFF. Wall grid on 6" by 6" centers and scaled at ¼"=1'.

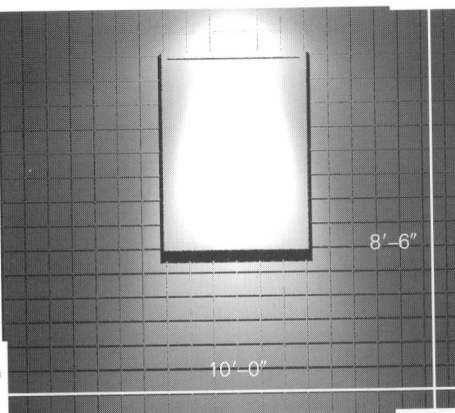

Lamp: 50PAR30S/H/FL
Location: 1'–6" from wall
Center beam Aiming: 14° (at point 2'–6" AFF)
Luminaire: Monopoint or Recessed Adjustable
Illuminance: 18 fc, avg, maintained (53 fc max)

Lamp: 50PAR30S/H/FL
Location: 2'–0" from wall
Center beam Aiming: 24° (at point 4'–0" AFF)
Luminaire: Monopoint or Recessed Adjustable
Illuminance: 22 fc, avg, maintained (61 fc max)

Lamp: 50PAR30S/H/FL
Location: 2'–6" from wall
Center beam Aiming: 32° (at point 4'–6" AFF)
Luminaire: Monopoint or Recessed Adjustable
Illuminance: 23 fc, avg, maintained (60 fc max)

Application Key

Commercial
Gallery
Hospitality
Institutional
Manufacturing
Residential
Retail
Exterior

bold = primary application
partial fade = minimal application
fade = unlikely application

Halogen Mains Voltage (120V)

Halogen 60PAR30S/H/NSP

Short Neck

Performance Sketches

[All data outlined here is based on information from boldfaced manufacturer's published data for 120V version at time of manuscript preparation. Scale is ¼ inch equals 1 foot.]

Lamp: 60PAR30S/H/NSP
Location: 3'–0" from wall
Center beam Aiming: 19° (at point 4'–0" AFF)
Luminaire: Monopoint or Recessed Adjustable
Illuminance: 20 fc, avg, maintained (45 fc max)

Lamp: 60PAR30S/H/NSP
Location: 4'–0" from wall
Center beam Aiming: 28° (at point 5'–0" AFF)
Luminaire: Monopoint or Recessed Adjustable
Illuminance: 25 fc, avg, maintained (65 fc max)

12'–6"

10'–0"

Uses
▶ Architectural detailing
▶ Medium art with medium throws
▶ Sculpture with medium throws
▶ Medium throws
▶ Available in 120V and 130V

Specs
Beam spread/10°
Center beam candlepower/ 10000
Life/2000 hrs
▶ **Philips** (35751–7)

Design Tips
✔ One light best for medium art where mounting height for light is less than 13'.
✔ Light may degrade artwork rapidly and/or significantly: consider UV and IR filters or limit exposure time.
✔ Great for glass pieces and floral centerpieces and where mounting height for light is 13' or less.
✔ These are strong accents and deserve judicious use.
✔ More grazing light creates prominent shadows.
✔ Consider halogen low voltage 50MR16/NSP, 50MR16/CG/ NSP, 50MR16/AL/SP or 50MR16/AL/CG/SP or halogen IR low voltage 37MR16/HIR/C/ NSP.

Art shown at 3' by 4' and centered 5'–6" AFF. Wall grid on 6" by 6" centers and scaled at ¼"=1'.

Lamp: 60PAR30S/H/NSP
Location: 5'–0" from wall
Center beam Aiming: 34° (at point 5'–0" AFF)
Luminaire: Monopoint or Recessed Adjustable
Illuminance: 25 fc, avg, maintained (67 fc max)

Application Key

Commercial
Gallery
Hospitality
Institutional
Manufacturing
Residential
Retail
Exterior

bold = primary application
partial fade = minimal application
fade = unlikely application

Halogen Mains Voltage (120V)

3

Halogen 60PAR30S/H/NFL

Short Neck

Uses

▶ Architectural accenting
▶ Feature downlighting
▶ Medium art with medium throws
▶ Sculpture with medium throws
▶ Medium throws
▶ Available in 120V and 130V

Specs

Beam spread/30°
Center beam candlepower/3300
Life/2000 hrs
▶ **Philips** (35753–3)

Design Tips

✔ One light best for medium art where mounting height for light is 11′ or less.
✔ Light may degrade artwork rapidly and/or significantly: consider UV and IR filters or limit exposure time.
✔ Great for moderate to large sculpture where mounting height for light is 11′ or less.
✔ More grazing light creates prominent shadows.
✔ Consider halogen IR low voltage 37MR16/HIR/C/NFL.

Performance Sketches

[All data outlined here is based on information from boldfaced manufacturer's published data for 120V version at time of manuscript preparation. Scale is ¼ inch equals 1 foot.]

Lamp: 60PAR30S/H/NFL
Location: 2′–3″ from wall
Center beam Aiming: 21° (at point 4′–6″ AFF)
Luminaire: Monopoint or Recessed Adjustable
Illuminance: 20 fc, avg, maintained (51 fc max)

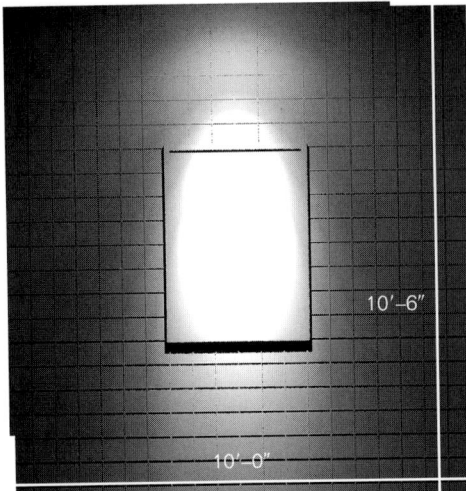

Lamp: 60PAR30S/H/NFL
Location: 2′–9″ from wall
Center beam Aiming: 25° (at point 4′–6″ AFF)
Luminaire: Monopoint or Recessed Adjustable
Illuminance: 21 fc, avg, maintained (48 fc max)

Art shown at 3′ by 4′ and centered 5′–6″ AFF. Wall grid on 6″ by 6″ centers and scaled at ¼″=1′.

Lamp: 60PAR30S/H/NFL
Location: 3′–3″ from wall
Center beam Aiming: 28° (at point 4′–6″ AFF)
Luminaire: Monopoint or Recessed Adjustable
Illuminance: 21 fc, avg, maintained (46 fc max)

Halogen Mains Voltage (120V)

Application Key

Commercial
Gallery
Hospitality
Institutional
Manufacturing
Residential
Retail
Exterior

bold = primary application
partial fade = minimal application
fade = unlikely application

Halogen60PAR30S/H/FL
Short Neck

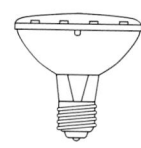

Performance Sketches

[All data outlined here is based on information from boldfaced manufacturer's published data for 120V version at time of manuscript preparation. Scale is ¼ inch equals 1 foot.]

Art shown at 3' by 4' and centered 5'–6" AFF. Wall grid on 6" by 6" centers and scaled at ¼"=1'.

9'–6"

10'–0"

Lamp: 60PAR30/H/FL
Location: 1'–6" from wall
Center beam Aiming: 4° (at point 0'–0" AFF)
Luminaire: Monopoint or Recessed Adjustable
Illuminance: 10 fc, avg, maintained (24 fc max)

Lamp: 60PAR30/H/FL
Location: 2'–0" from wall
Center beam Aiming: 14° (at point 1'–6" AFF)
Luminaire: Monopoint or Recessed Adjustable
Illuminance: 14 fc, avg, maintained (29 fc max)

Lamp: 60PAR30/H/FL
Location: 2'–6" from wall
Center beam Aiming: 23° (at point 3'–6" AFF)
Luminaire: Monopoint or Recessed Adjustable
Illuminance: 16 fc, avg, maintained (33 fc max)

Uses
▶General lighting
▶Medium art with short throws
▶Short throws
▶Available in 120V and 130V

Specs
Beam spread/40°
Center beam
  candlepower/1850
Life/2000 hrs
▶**Philips** (35758–2)

Design Tips
✔One light best for medium art where mounting height for light is 10' or less.
✔Light may degrade artwork rapidly and/or significantly: consider UV and IR filters or limit exposure time.
✔Good for downlighting onto decorative floor materials—best with sconces and/or art accents and/or wallwashing to avoid cave effect and grazing glare (the "I wish I had a visor" effect).
✔More grazing light creates prominent shadows.
✔Consider halogen low voltage 50MR16/FL, 50MR16/CG/FL, 50MR16/AL/CG/FL, or 50MR16/C/FL or halogen IR low voltage 37MR16/HIR/C/FL.

Application Key

Commercial
Gallery
Hospitality
Institutional
Manufacturing
Residential
Retail
Exterior

bold = primary application
partial fade = minimal application
fade = unlikely application

Halogen Mains Voltage (120V)

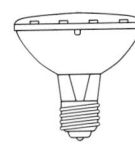

Halogen 75PAR30S/H/NSP

Short Neck

Performance Sketches

[All data outlined here is based on information from boldfaced manufacturer's published data for 120V version at time of manuscript preparation. Scale is ¼ inch equals 1 foot.]

Uses
▶ Architectural detailing
▶ Large art with long throws
▶ Sculpture with long throws
▶ Long throws
▶ Available in 120V and 130V

Specs
Beam spread/10°
Center beam candlepower/12000
Life/2250 hrs
▶ **GE** (14802)
▶ Osram Sylvania
▶ Philips

Design Tips
✔ One light best for large art where mounting height for light is 16' or less.
✔ Light may degrade artwork rapidly and/or significantly: consider UV and IR filters or limit exposure time.
✔ Great for glass pieces and floral centerpieces where mounting height for light is 16' or less.
✔ These are strong accents and deserve judicious use.
✔ More grazing light creates prominent shadows.
✔ Consider halogen low voltage 42MR16/NSP, 50MR16/C/CG/NSP, 65MR16/NSP or 65MR16/C/NSP or halogen IR low voltage 37MR16/HIR/C/NSP.

Lamp: 75PAR30S/H/NSP
Location: 4'–3" from wall
Center beam Aiming: 22° (at point 4'–6" AFF)
Luminaire: Monopoint or Recessed Adjustable
Illuminance: 16 fc, avg, maintained (41 fc max)

15'–0"

10'–0"

Lamp: 75PAR30S/H/NSP
Location: 5'–3" from wall
Center beam Aiming: 27° (at point 4'–6" AFF)
Luminaire: Monopoint or Recessed Adjustable
Illuminance: 17 fc, avg, maintained (43 fc max)

Application Key

Commercial
Gallery
Hospitality
Institutional
Manufacturing
Residential
Retail
Exterior

bold = primary application
partial fade = minimal application
fade = unlikely application

Art shown at 4' by 5' and centered 5'–6" AFF. Wall grid on 6" by 6" centers and scaled at ¼"=1'.

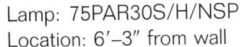

Lamp: 75PAR30S/H/NSP
Location: 6'–3" from wall
Center beam Aiming: 31° (at point 4'–6" AFF)
Luminaire: Monopoint or Recessed Adjustable
Illuminance: 17 fc, avg, maintained (45 fc max)

Halogen Mains Voltage (120V)

3

Halogen 75PAR30S/H/NFL

Short Neck

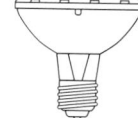

Performance Sketches

[All data outlined here is based on information from boldfaced manufacturer's published data for 120V version at time of manuscript preparation. Scale is ¼ inch equals 1 foot.]

Lamp: 75PAR30S/H/NFL
Location: 3'–0" from wall
Center beam Aiming: 18° (at point 3'–0" AFF)
Luminaire: Monopoint or Recessed Adjustable
Illuminance: 12 fc, avg, maintained (26 fc max)

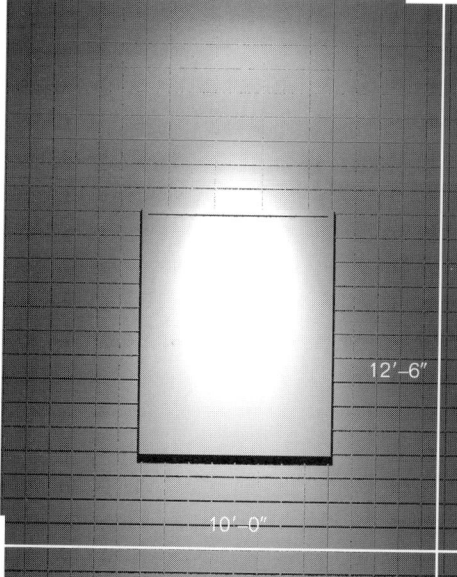

12'–6"

10'–0"

Lamp: 75PAR30S/H/NFL
Location: 4'–0" from wall
Center beam Aiming: 25° (at point 4'–0" AFF)
Luminaire: Monopoint or Recessed Adjustable
Illuminance: 14 fc, avg, maintained (28 fc max)

Uses
▶ Architectural accenting
▶ Feature downlighting
▶ Large art with medium throws
▶ Sculpture with medium throws
▶ Medium throws
▶ Available in 120V and 130V

Specs
Beam spread/30°
Center beam candlepower/ 3000
Life/2000 hrs
▶ **GE** (18057)
▶ Osram Sylvania
▶ Philips

Design Tips
✔ One light best for large art where mounting height for light is 13' or less.
✔ Light may degrade artwork rapidly and/or significantly: consider UV and IR filters or limit exposure time.
✔ Great for medium to large sculpture where mounting height for light is 13' or less.
✔ More grazing light creates prominent shadows.
✔ Consider halogen IR mains voltage 50PAR38/HIR/NSP.

Art shown at 4' by 5' and centered 5'–6" AFF. Wall grid on 6" by 6" centers and scaled at ¼"=1'.

Lamp: 75PAR30S/H/NFL
Location: 5'–0" from wall
Center beam Aiming: 30° (at point 4'–0" AFF)
Luminaire: Monopoint or Recessed Adjustable
Illuminance: 14 fc, avg, maintained (26 fc max)

Application Key

Commercial
Gallery
Hospitality
Institutional
Manufacturing
Residential
Retail
Exterior

bold = primary application
partial fade = minimal application
fade = unlikely application

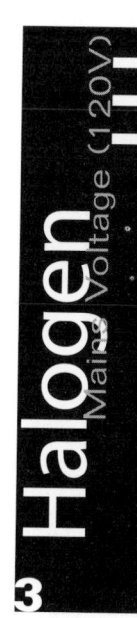

Halogen Mains Voltage (120V)

3

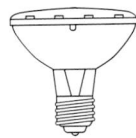

Halogen 75PAR30S/H/FL

Short Neck

Performance Sketches

[All data outlined here is based on information from boldfaced manufacturer's published data for 120V version at time of manuscript preparation. Scale is ¼ inch equals 1 foot.]

Art shown at 4' by 5' and centered 5'–6" AFF. Wall grid on 6" by 6" centers and scaled at ¼"=1'.

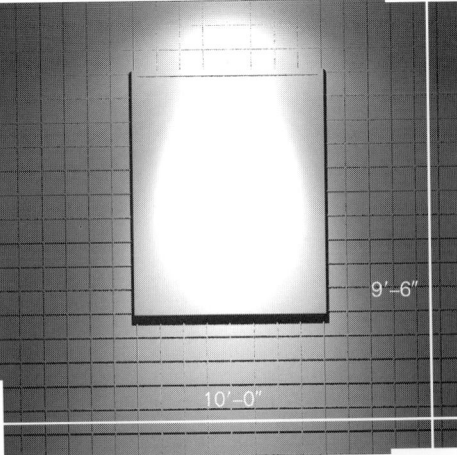

9'–6"

10'–0"

Lamp: 75PAR30/H/FL
Location: 2'–0" from wall
Center beam Aiming: 16° (at point 2'–6" AFF)
Luminaire: Monopoint or Recessed Adjustable
Illuminance: 17 fc, avg, maintained (50 fc max)

Lamp: 75PAR30/H/FL
Location: 2'–6" from wall
Center beam Aiming: 23° (at point 3'–6" AFF)
Luminaire: Monopoint or Recessed Adjustable
Illuminance: 19 fc, avg, maintained (52 fc max)

Lamp: 75PAR30/H/FL
Location: 3'–0" from wall
Center beam Aiming: 31° (at point 4'–6" AFF)
Luminaire: Monopoint or Recessed Adjustable
Illuminance: 22 fc, avg, maintained (60

Uses
▶ General lighting
▶ Large art with short throws
▶ Short throws
▶ Available in 120V and 130V

Specs
Beam spread/40°
Center beam
candlepower/1900
Life/2250 hrs
▶ **GE** (16393)
▶ Osram Sylvania
▶ Philips

Design Tips
✔ One light best for large art where mounting height for light is 10' or less.
✔ Light may degrade artwork rapidly and/or significantly: consider UV and IR filters or limit exposure time.
✔ Good for downlighting onto decorative floor materials— best with sconces and/or art accents and/or wallwashing to avoid cave effect and grazing glare (the "I wish I had a visor" effect).
✔ More grazing light creates prominent shadows.
✔ Consider halogen low voltage 50MR16/FL, 65MR16/FL, 50MR16/C/FL or 65MR16/C/FL or halogen IR low voltage 37MR16/HIR/C/FL.

Application Key

Commercial
Gallery
Hospitality
Institutional
Manufacturing
Residential
Retail
Exterior

bold = primary application
partial fade = minimal application
fade = unlikely application

Halogen
Main Voltage (120V)

3

Halogen/PAR30S

Short Neck

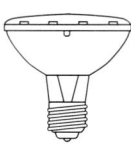

Net Addresses/PAR30S Track and Monopoints

```
http://www.bartcolighting.com/index2.html
http://www.thomasltg.com/
http://www.con-techlighting.com/
http://www.cooperlighting.com/
http://www.hubbell-ltg.com/products.htm#Down&Track
http://www.danalite.com/
http://www.kramerlighting.com/
http://www.lightingservicesinc.com/
http://www.lightolier.com/
http://www.lithonia.com/
http://www.prescolite.com/
http://www.tslight.com/
```

 CONNECT FOR MORE

Net Addresses/PAR30S Recessed

```
http://www.thomasltg.com/
http://www.con-techlighting.com/
http://www.cooperlighting.com/
http://www.hubbell-ltg.com/products.htm#Down&Track
http://www.danalite.com/
http://www.kramerlighting.com/
http://www.lightolier.com/
http://www.lithonia.com/
http://www.prescolite.com/
```

 CONNECT FOR MORE

Net Addresses/Exterior Accent

```
http://www.bega-us.com/home/home.html
http://www.hadcolighting.com/html/bronzelite.htm
http://www.hydrel.com
http://www.kimlighting.com/ingrd1.html
```

 CONNECT FOR MORE

Halogen Mains Voltage (120V)

3

▶Intended for R30 and R40 retrofit applications.

Halogen/PAR30L

Long Neck

Stats
Data varies manufacturer to manufacturer and changes from time to time

Life/2000 and 2500 hours
Different versions from different manufacturers. A typical office environment might be occupied about 2600 hours each year.

Efficacy/13 LPW
Good beam intensities and distributions available.

Wattages/ 50 and 75

Color of Light/2800°K

Beam spreads/NSP, SP, NFL, FL and WFL
Narrow spot, spot, narrow flood, flood and wide flood distributions are available. WARNING—unfortunately, lamp manufacturers have not standardized beam spread descriptors. One manufacturer's "NSP" is another's "SP" and so on. Information reported herein should be cross–verified with manufacturers' current data.

Size/4¾" L by 3¾" Ø

Voltage/120V and 130V

Cost Magnitude/Low (2000)

Manufacturers
General Electric
Osram Sylvania
Philips

Net Addresses
http://www.ge.com/lighting/business/index.htm
http://ecom.sylvania.com/osicatalog/
http://www.lighting.philips.com/

CONNECT FOR MORE

Advantages
Crisp white light
Good to excellent intensities available
Long neck for direct replacement of inefficient R lamps
Standard voltage (no transformers)
Easily dimmed
Low initial cost
Withstand extreme temperature range
Very little lamp lumen depreciation (light loss) over life

Disadvantages
Beam aberrations
♦Striations
♦Inconsistent beam shape
High operating costs
♦Inefficiency and short life may make this cost prohibitive in typical commercial applications, depending on ease of changing lamps and maintenance staffing.
Hot to touch
Hot-shock sensitive
♦Lamp may fail if rough aiming–adjustment or sharp vibration occurs while lamp is energized.
Water sensitive
♦Lamp may fail violently if placed in contact with water (not to be used exposed in outdoor lighting situations).

Halogen Mains Voltage (120V)

For more efficient accent lighting, use HIR low voltage lamps (Section 6)**; HIR PAR lamps** (Section 5) **or CMH PAR lamps** (Section 7)**.**

3

Halogen/PAR30L

Long Neck

►Intended for R30 and R40 retrofit applications

Pressed/cast glass optical lens

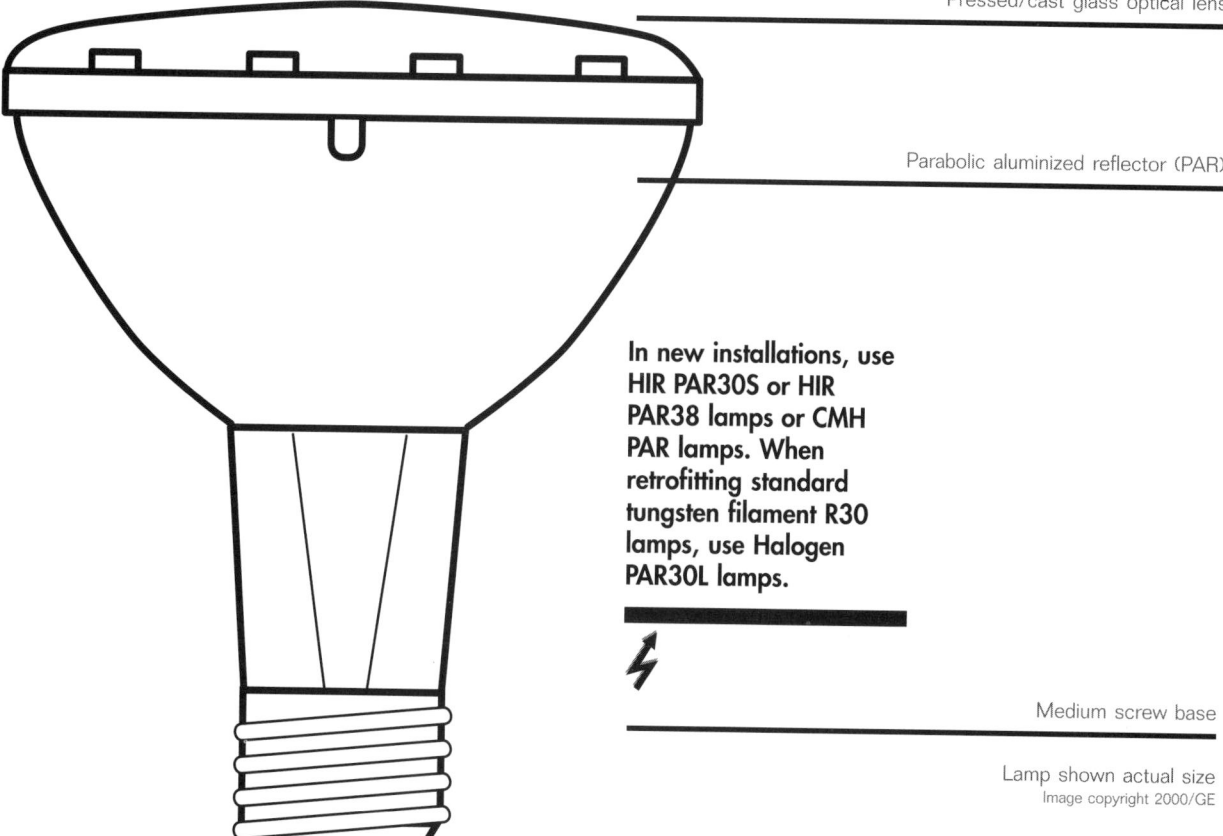

Parabolic aluminized reflector (PAR)

In new installations, use HIR PAR30S or HIR PAR38 lamps or CMH PAR lamps. When retrofitting standard tungsten filament R30 lamps, use Halogen PAR30L lamps.

Medium screw base

Lamp shown actual size
Image copyright 2000/GE

Uses

Architectural Accenting

Art Accenting

Gallery
Hospitality
Religious
Residential

Downlighting

Circulation
Conference rooms
Hospitality
Residential
Retail

Merchandise Accenting

Surface Grazing

Specular/semispecular surfaces
- Polished/honed stone
- Polished/satin wood
- Fritted/etched glass
- Cut glass

PAR30L Lighting Design Tips

✔ Best at moderate mounting heights (10' to 15').
✔ Dimming increases life (typically doubling life), but shifts color warmer.
✔ Glary when used without much other general lighting or when aimed across long distances where ceilings are low.
✔ Beam striations are most apparent on light–colored, nontextured, flat surfaces and may be objectionable.
✔ PAR30L downlighting alone is harsh—combine several techniques for best results (e.g., sconces and art accents in addition to downlighting).
✔ Use medium scale (e.g., 6–inch diameter) downlights and adjustable accents for attractive, more human scale approach.

Net Addresses/PAR30L Track, Monopoints, Recessed and Exterior
These lamps are intended for retrofiitting existing luminaires which use inefficient BR, R and standard filament incandescent PAR lamps.

CONNECT FOR MORE

Halogen Mains Voltage (120V)

3

▶ Intended for R30 and R40 retrofit applications.

Halogen/PAR30L
Long Neck

Image copyright 1992 Balthazar Korab, Ltd.

Figure 3.8
In–floor uplights accent the architectural pilasters and the decorated under–balcony ceilings in the rotunda of the Michigan State Capitol. For size, finish and reasons of integration with the 1870's metal flooring, the uplights needed to do the job were rated for R30, R40 and PAR38 lamps. For efficiency, 75PAR30L/H/NFL lamps were used. Uplights are Lithonia/Hydrel.

Halogen Main Voltage (120V)

Halogen/PAR30L

Long Neck

▶Intended for R30 and R40 retrofit applications

Image copyright 1992 Balthazar Korab. Ltd.

Figure 3.9

Another view showing the in–floor uplights as well as the effect of the uplighting on the pilasters and the decorative balcony ceiling. Without these accent uplights, the architecture would appear drab and dingy. While the uplights do not add appreciably to the light level, they add considerably to the brightness impressions. Note the fluorescent picture lights (suitably lensed with a UV filter and then also louvered). Uplights are Lithonia/Hydrel.

Halogen Main Voltage (120V)

3

▶ Intended for R30 and R40 retrofit applications.

Halogen/PAR30L

PAR30L Visual Index

35PAR30L/H/NSP▶**p61** 35PAR30L/H/NSP▶**p61** 35PAR30L/H/NSP▶**p61**

35PAR30L/H/NSP
▶ Architectural detailing
▶ Small art/9'–6" ceiling height
▶ Objet d'art
▶ Short throws
▶ Commercial, Retail

35PAR30L/H/WFL▶**p62** 35PAR30L/H/WFL▶**p62** 35PAR30L/H/WFL▶**p62**

35PAR30L/H/WFL
▶ General lighting
▶ Wallwashing
▶ Small art/9'–6" ceiling height
▶ Very short throws
▶ Gallery, Residential

Artwork Size Categories
⊃ Petite = 2'–0" by 2'–0" or smaller
⊃ Small = 2'–0" by 2'–0" to 2'–6" by 3'–0"
⊃ Medium = 2'–6" by 3'–0" to 3'–0" by 4'–0"
⊃ Large = 3'–0" by 4'–0" to 4'–0" by 5'–0"
⊃ Extra Large = 4'–0" by 5'–0" or larger

50PAR30L/H/NSP▶**p63** 50PAR30L/H/NSP▶**p63** 50PAR30L/H/NSP▶**p63**

50PAR30L/H/NSP
▶ Architectural detailing
▶ Medium art/15'–0" ceiling height
▶ Sculpture
▶ Long throws
▶ Hospitality

  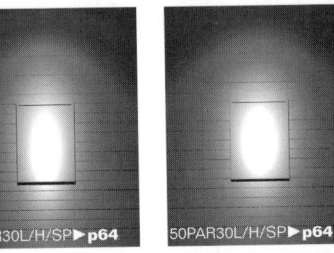

50PAR30L/H/SP▶**p64** 50PAR30L/H/SP▶**p64** 50PAR30L/H/SP▶**p64**

50PAR30L/H/SP
▶ Architectural accenting
▶ Medium art/12'–6" ceiling height
▶ Sculpture
▶ Medium–to–long throws
▶ Hospitality

Recommended intensities
for art accenting

Low (gallery, museum, residential)
5 to 10 fc average on art
Moderate (hospitality)
10 to 20 fc average on art
High (commercial—or higher for retail)
20 to 30 fc average on art

These are intended to be average, maintained values including effects of accent lighting and general room lighting. General room lighting contributes just a few footcandles in residential, gallery and museum settings, but contributes 5 to 10 fc in commercial settings. Maximums might range from 50 percent to 100 percent of the average value. Wherever artwork is highly valuable and/or sensitive to light, precautions should be taken to limit light exposure (maintaining low levels and/or switching lights off when art is not being viewed); and exposure to infrared and ultraviolet should be limited (using filters on lamps or placing art behind specially treated glass or acrylic).

50PAR30L/H/WSP▶**p65** 50PAR30L/H/WSP▶**p65** 50PAR30L/H/WSP▶**p65**

50PAR30L/H/WSP
▶ Architectural accenting
▶ Large art/12'–6" ceiling height
▶ Sculpture
▶ Medium–to–long throws
▶ Hospitality

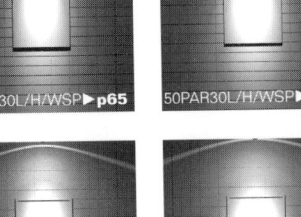  

50PAR30L/H/NFL▶**p66** 50PAR30L/H/NFL▶**p66** 50PAR30L/H/NFL▶**p66**

50PAR30L/H/NFL
▶ Architectural accenting
▶ Feature downlighting
▶ Medium art/10'–6" ceiling height
▶ Sculpture
▶ Medium–to–long throws
▶ Hospitality

 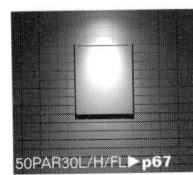 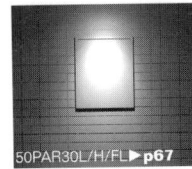

50PAR30L/H/FL▶**p67** 50PAR30L/H/FL▶**p67** 50PAR30L/H/FL▶**p67**

50PAR30L/H/FL
▶ General lighting
▶ Medium art/8'–6" ceiling height
▶ Short throws
▶ Hospitality

Halogen Main Voltage (120V)

Halogen/PAR30L

Long Neck

►Intended for R30 and
R40 retrofit applications

 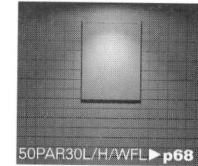 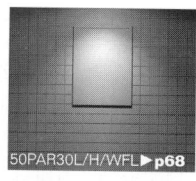

50PAR30L/H/WFL►p68   50PAR30L/H/WFL►p68   50PAR30L/H/WFL►p68

50PAR30L/H/WFL
►General lighting
►Wallwashing
►Medium art/8'–6" ceiling height
►Short throws
►Gallery, Residential

  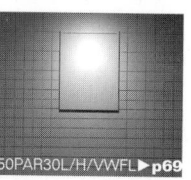

50PAR30L/H/VWFL►p69   50PAR30L/H/VWFL►p69   50PAR30L/H/VWFL►p69

50PAR30L/H/VWFL
►General lighting
►Wallwashing
►Medium art/8'–6" ceiling height
►Short throws
►Gallery, Residential

75PAR30L/H/NSP►p70   75PAR30L/H/NSP►p70   75PAR30L/H/NSP►p70

75PAR30L/H/NSP
►Architectural detailing
►Large art/15'–0" ceiling height
►Sculpture
►Long throws
►Hospitality

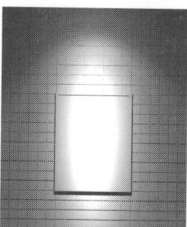  

75PAR30L/H/SP►p71   75PAR30L/H/SP►p71   75PAR30L/H/SP►p71

75PAR30L/H/SP
►Architectural accenting
►Medium art/12'–6" ceiling height
►Sculpture
►Medium–to–long throws
►Hospitality

75PAR30L/H/WSP►p72   75PAR30L/H/WSP►p72   75PAR30L/H/WSP►p72

75PAR30L/H/WSP
►Architectural accenting
►Large art/12'–6" ceiling height
►Sculpture
►Medium–to–long throws
►Commercial, Hospitality, Retail

Throws/Ceilings
⊃Short = 7' to 10'
⊃Medium = 10' to 15'
⊃Long = 15' to 20'
⊃Very Long = 20' or more

75PAR30L/H/NFL►p73   75PAR30L/H/NFL►p73   75PAR30L/H/NFL►p73

75PAR30L/H/NFL
►Architectural detailing
►Feature downlighting
►Large art/15'–0" ceiling height
►Sculpture
►Medium–to–long throws
►Hospitality

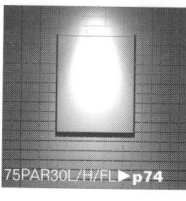 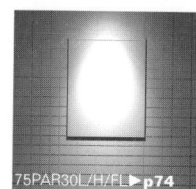 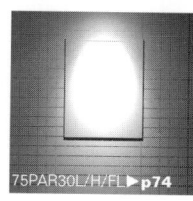

75PAR30L/H/FL►p74   75PAR30L/H/FL►p74   75PAR30L/H/FL►p74

75PAR30L/H/FL
►General lighting
►Large art/9'–6" ceiling height
►Short throws
►Hospitality

Halogen Main Voltage (120V)

3

▶Intended for R30 and R40 retrofit applications.

Halogen/PAR30L

Long Neck

  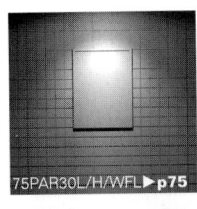

75PAR30L/H/WFL ▶p75 75PAR30L/H/WFL ▶p75 75PAR30L/H/WFL ▶p75

75PAR30L/H/WFL
▶General lighting
▶Wallwashing
▶Medium art/9'–6" ceiling height
▶Short throws
▶Gallery, Residential

75PAR30L/H/VWFL ▶p76 75PAR30L/H/VWFL ▶p76 75PAR30L/H/VWFL ▶p76

75PAR30L/H/VWFL
▶General lighting
▶Wallwashing
▶Medium art/9'–6" ceiling height
▶Short throws
▶Hospitality

Halogen PAR Lamp Designation
50PAR30L/H/NSP

Last set of letters identifies beam spread

FLOODS
• 24° to 32° for narrow flood (NFL)
• 33° to 44° for flood (FL)
• 45° to 54° for wide flood (WFL)
• 55° or greater for very wide flood (VWFL)

SPOTS
• 7° or less for very narrow spot (VNSP)
• 8° to 10° for narrow spot (NSP)
• 11° to 14° for spot (SP)
• 15° to 18° for wide spot (WSP)
• 19° to 23° for very wide spot (VWSP)

Ranges of beam spreads used for convenient/consistent reference in this text and do not necessarily correspond to each manufacturer's nor ANSI's definitions.

This letter indicates "Halogen"

"S" for short neck lamps and "L" for long neck (only applies to PAR30 lamps)

These two digits represent lamp diameter in eighths inches (e.g., 30/8 or 3–3/4 inches)

These letters indicate this is a parabolic aluminized reflector (PAR) lamp

First two or three digits represent lamp wattage

[This designation used for convenient reference throughout this *Time–Saver Standards for Architectural Lighting*—but not necessarily used by all lamp manufacturers (although it would be convenient if manufacturers would come to agreement on standard designations.]

Halogen Main Voltage (120V)

Halogen 35PAR30L/H/NSP

Long Neck

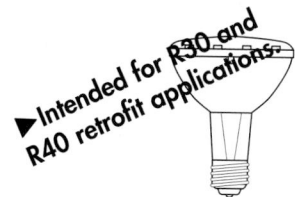

▶Intended for R30 and R40 retrofit applications

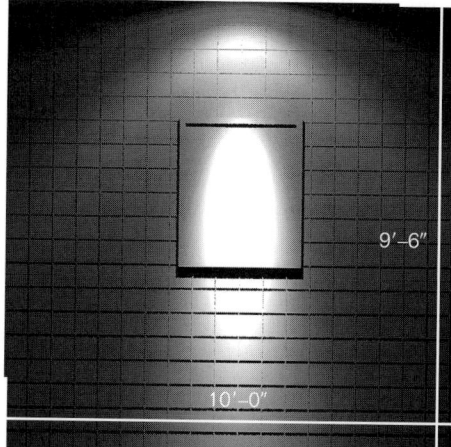

9'–6"

10'–0"

Performance Sketches

[All data outlined here is based on information from boldfaced manufacturer's published data for 120V version at time of manuscript preparation. Scale is ¼ inch equals 1 foot.]

Art shown at 2'–6" by 3' and centered 5'–6" AFF. Wall grid on 6" by 6" centers and scaled at ¼"=1'.

Lamp: 35PAR30L/H/NSP
Location: 1'–6" from wall
Center beam Aiming: 15° (at point 4'–0" AFF)
Luminaire: Monopoint or Recessed Adjustable
Illuminance: 18 fc, avg, maintained (61 fc max)

Lamp: 35PAR30L/H/NSP
Location: 2'–0" from wall
Center beam Aiming: 24° (at point 5'–0" AFF)
Luminaire: Monopoint or Recessed Adjustable
Illuminance: 25 fc, avg, maintained (106 fc max)

Lamp: 35PAR30L/H/NSP
Location: 2'–6" from wall
Center beam Aiming: 32° (at point 5'–6" AFF)
Luminaire: Monopoint or Recessed Adjustable
Illuminance: 28 fc, avg, maintained (138 fc max)

Uses

▶Architectural detailing
▶Small art
▶Objet d'art
▶Available in 120V

Specs

Beam spread/9°
Center beam
 candlepower/6000
Life/2500 hrs
▶**Osram Sylvania** (14759)

Design Tips

✔One light best for small art where mounting height for light is 10' or less.
✔Light may degrade artwork rapidly and/or significantly: consider UV and IR filters or limit exposure time.
✔These are strong accents and deserve judicious use.
✔More grazing light creates prominent shadows.
✔Intended for retrofitting of R30 luminaires.
✔For new construction, consider halogen low voltage 20MR11/NSP, 20MR11/CG/NSP, 20MR16/NSP or 20MR16/C/CG/NSP.

Application Key

Commercial
Gallery
Hospitality
Institutional
Manufacturing
Residential
Retail
Exterior

bold = primary application
partial fade = minimal application
fade = unlikely application

Halogen Main Voltage (120V)

3

▶ Intended for R30 and R40 retrofit applications.

Halogen 35PAR30L/H/WFL

Long Neck

Performance Sketches

[All data outlined here is based on information from boldfaced manufacturer's published data for 120V version at time of manuscript preparation. Scale is ¼ inch equals 1 foot.]

Art shown at 2'–6" by 3' and centered 5'–6" AFF. Wall grid on 6" by 6" centers and scaled at ¼"=1'.

Uses
▶ General lighting
▶ Wallwashing (see 11–115)
▶ Small art
▶ Available in 120V

Specs
Beam spread/50°
Center beam candlepower/350
Life/2500 hrs
▶ **Osram Sylvania** (14764)

Design Tips
✔ One light best for small art where mounting height for light is 10' or less.
✔ Can be used to wash walls without expense of spreadlens wallwashers—units need to be on 1'–6" to 2'–0" centers and placed 1'–3" to 1'–6" from wall being washed.
✔ More grazing light creates prominent shadows.
✔ Intended for retrofitting of R30 luminaires.

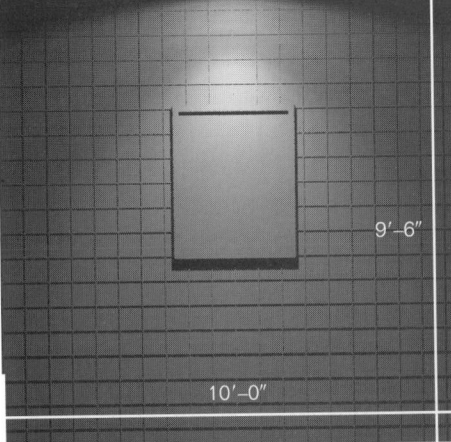

9'–6"

10'–0"

Lamp: 35PAR30L/H/WFL
Location: 1'–6" from wall
Center beam Aiming: 4° (at point 0'–0" AFF)
Luminaire: Monopoint or Recessed Adjustable
Illuminance: 4 fc, avg, maintained (9 fc max)

Lamp: 35PAR30L/H/WFL
Location: 2'–0" from wall
Center beam Aiming: 13° (at point 1'–0" AFF)
Luminaire: Monopoint or Recessed Adjustable
Illuminance: 5 fc, avg, maintained (10 fc max)

Application Key

Commercial
Gallery
Hospitality
Institutional
Manufacturing
Residential
Retail
Exterior

bold = primary application
partial fade = minimal application
fade = unlikely application

Lamp: 35PAR30L/H/WFL
Location: 2'–6" from wall
Center beam Aiming: 18° (at point 2'–0" AFF)
Luminaire: Monopoint or Recessed Adjustable
Illuminance: 5 fc, avg, maintained (9 fc max)

Halogen Mains Voltage (120V)

Halogen 50PAR30L/H/NSP
Long Neck

▶Intended for R30 and R40 retrofit applications

Performance Sketches

[All data outlined here is based on information from boldfaced manufacturer's published data for 120V version at time of manuscript preparation. Scale is ¼ inch equals 1 foot.]

Lamp: 50PAR30L/H/NSP
Location: 3'–0" from wall
Center beam Aiming: 15° (at point 3'–6" AFF)
Luminaire: Monopoint or Recessed Adjustable
Illuminance: 10 fc, avg, maintained (19 fc max)

Lamp: 50PAR30L/H/NSP
Location: 4'–0" from wall
Center beam Aiming: 21° (at point 4'–6" AFF)
Luminaire: Monopoint or Recessed Adjustable
Illuminance: 13 fc, avg, maintained (25 fc max)

15'–0"

10'–0"

Art shown at 3' by 4' and centered 5'–6" AFF. Wall grid on 6" by 6" centers and scaled at ¼"=1'.

Lamp: 50PAR30L/H/NSP
Location: 5'–0" from wall
Center beam Aiming: 25° (at point 4'–6" AFF)
Luminaire: Monopoint or Recessed Adjustable
Illuminance: 14 fc, avg, maintained (26 fc max)

Uses
▶Architectural detailing
▶Medium art with long throws
▶Sculpture with long throws
▶Long throws
▶Available in 120V and 130V

Specs
Beam spread/9°
Center beam candlepower/ 8950
Life/2250 hrs
▶Osram Sylvania
▶**Philips** (22922–9)

Design Tips
✓One light best for medium art where mounting height for light is 16' or less.
✓Light may degrade artwork rapidly and/or significantly: consider UV and IR filters or limit exposure time.
✓Scallops/striations are most severe when light is closer to the wall (compare top left image to bottom image).
✓More grazing light creates prominent shadows.
✓Intended for retrofitting of R30 luminaires.
✓For new construction, consider halogen low voltage 35MR11/ NSP or 35MR11/C/NSP.

Application Key
Commercial
Gallery
Hospitality
Institutional
Manufacturing
Residential
Retail
Exterior

bold = primary application
partial fade = minimal application
fade = unlikely application

Halogen Main Voltage (120V)

3

▶ Intended for R30 and R40 retrofit applications.

Halogen 50PAR30L/H/SP

Long Neck

Performance Sketches

[All data outlined here is based on information from boldfaced manufacturer's published data for 120V version at time of manuscript preparation. Scale is ¼ inch equals 1 foot.]

Uses
▶ Architectural accenting
▶ Medium art with medium to long throws
▶ Sculpture with medium to long throws
▶ Medium to long throws
▶ Available in 120V and 130V

Specs
Beam spread/12°
Center beam candlepower/ 6000
Life/2000 hrs
▶**GE** (14940)

Design Tips
✔ One light best for medium art and where mounting height for light is 13' or less.
✔ Light may degrade artwork rapidly and/or significantly: consider UV and IR filters or limit exposure time.
✔ Scallops/striations are most severe when light is closer to the wall (compare top left image to bottom image).
✔ More grazing light creates prominent shadows.
✔ Intended for retrofitting of R30 luminaires.
✔ For new construction, consider halogen low voltage 35MR16/ C/SP.

Lamp: 50PAR30L/H/SP
Location: 3'–0" from wall
Center beam Aiming: 21° (at point 4'–6" AFF)
Luminaire: Monopoint or Recessed Adjustable
Illuminance: 11 fc, avg, maintained (29 fc max)

Lamp: 50PAR30L/H/SP
Location: 4'–0" from wall
Center beam Aiming: 27° (at point 4'–6" AFF)
Luminaire: Monopoint or Recessed Adjustable
Illuminance: 12 fc, avg, maintained (32 fc max)

12'–6"

10'–0"

Art shown at 3' by 4' and centered 5'–6" AFF. Wall grid on 6" by 6" centers and scaled at ¼"=1'.

Lamp: 50PAR30L/H/SP
Location: 5'–0" from wall
Center beam Aiming: 34° (at point 5'–0" AFF)
Luminaire: Monopoint or Recessed Adjustable
Illuminance: 13 fc, avg, maintained (37 fc max)

Application Key

Commercial
Gallery
Hospitality
Institutional
Manufacturing
Residential
Retail
Exterior

bold = primary application
partial fade = minimal application
fade = unlikely application

Halogen Mains Voltage (120V)

Halogen 50PAR30L/H/WSP

Long Neck

▶Intended for R30 and R40 retrofit applications

Performance Sketches

[All data outlined here is based on information from boldfaced manufacturer's published data for 120V version at time of manuscript preparation. Scale is ¼ inch equals 1 foot.]

Lamp: 50PAR30L/H/WSP
Location: 3'–0" from wall
Center beam Aiming: 21° (at point 4'–6" AFF)
Luminaire: Monopoint or Recessed Adjustable
Illuminance: 12 fc, avg, maintained (27 fc max)

Lamp: 50PAR30L/H/WSP
Location: 4'–0" from wall
Center beam Aiming: 27° (at point 4'–6" AFF)
Luminaire: Monopoint or Recessed Adjustable
Illuminance: 13 fc, avg, maintained (26 fc max)

Art shown at 3' by 4' and centered 5'–6" AFF. Wall grid on 6" by 6" centers and scaled at ¼"=1'.

Lamp: 50PAR30L/H/WSP
Location: 5'–0" from wall
Center beam Aiming: 32° (at point 4'–6" AFF)
Luminaire: Monopoint or Recessed Adjustable
Illuminance: 13 fc, avg, maintained (26 fc max)

Uses

▶Architectural accenting
▶Medium art with medium to long throws
▶Sculpture with medium to long throws
▶Medium to long throws
▶Only available in 120V

Specs

Beam spread/16°
Center beam candlepower/4200
Life/2000 hrs
▶**Philips** (spot) (22923–7)

Design Tips

✓One light best for medium art and where mounting height for light is 13' or less.
✓Light may degrade artwork rapidly and/or significantly: consider UV and IR filters or limit exposure time.
✓More grazing light creates prominent shadows.
✓Intended for retrofitting of R30 luminaires.
✓For new construction, consider halogen low voltage 20MR11/C/WSP.

Application Key

Commercial
Gallery
Hospitality
Institutional
Manufacturing
Residential
Retail
Exterior

bold = primary application
partial fade = minimal application
fade = unlikely application

Halogen Mains Voltage (120V)

▶ Intended for R30 and R40 retrofit applications.

Halogen 50PAR30L/H/NFL

Long Neck

Performance Sketches

[All data outlined here is based on information from boldfaced manufacturer's published data for 120V version at time of manuscript preparation. Scale is ¼ inch equals 1 foot.]

Uses
- ▶ Architectural accenting
- ▶ Medium art with medium throws
- ▶ Sculpture with medium throws
- ▶ Medium to long throws
- ▶ Available in 120V and 130V

Specs
Beam spread/29°
Center beam candlepower/ 1875
Life/2150 hrs
- ▶ GE
- ▶ Osram Sylvania
- ▶ **Philips** (22925–2)

Design Tips
✔ One light best for medium art where mounting height for light is 11' or less.
✔ More grazing light creates prominent shadows.
✔ Intended for retrofitting of R30 luminaires.

Lamp: 50PAR30L/H/NFL
Location: 2'–0" from wall
Center beam Aiming: 13° (at point 2'–0" AFF)
Luminaire: Monopoint or Recessed Adjustable
Illuminance: 13 fc, avg, maintained (28 fc max)

10'–6"

10'–0"

Lamp: 50PAR30L/H/NFL
Location: 2'–6" from wall
Center beam Aiming: 20° (at point 3'–6" AFF)
Luminaire: Monopoint or Recessed Adjustable
Illuminance: 16 fc, avg, maintained (33 fc max)

Art shown at 3' by 4' and centered 5'–6" AFF. Wall grid on 6" by 6" centers and scaled at ¼"=1'.

Lamp: 50PAR30L/H/NFL
Location: 3'–0" from wall
Center beam Aiming: 27° (at point 4'–6" AFF)
Luminaire: Monopoint or Recessed Adjustable
Illuminance: 19 fc, avg, maintained (39 fc max)

Halogen Mains Voltage (120V)

Application Key

Commercial
Gallery
Hospitality
Institutional
Manufacturing
Residential
Retail
Exterior

bold = primary application
partial fade = minimal application
fade = unlikely application

3

Halogen 50PAR30L/H/FL

Long Neck

▶Intended for R30 and R40 retrofit applications

Performance Sketches

[All data outlined here is based on information from boldfaced manufacturer's published data for 120V version at time of manuscript preparation. Scale is ¼ inch equals 1 foot.]

Art shown at 3' by 4' and centered 5'–6" AFF. Wall grid on 6" by 6" centers and scaled at ¼"=1'.

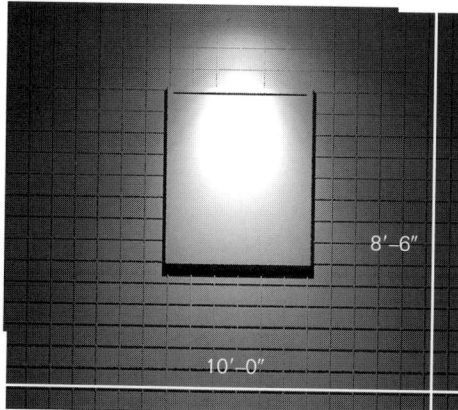

8'–6"
10'–0"

Lamp: 50PAR30L/H/FL
Location: 1'–6" from wall
Center beam Aiming: 11° (at point 1'–6" AFF)
Luminaire: Monopoint or Recessed Adjustable
Illuminance: 10 fc, avg, maintained (27 fc max)

Lamp: 50PAR30L/H/FL
Location: 2'–0" from wall
Center beam Aiming: 17° (at point 3'–0" AFF)
Luminaire: Monopoint or Recessed Adjustable
Illuminance: 12 fc, avg, maintained (26 fc max)

Lamp: 50PAR30L/H/FL
Location: 2'–6" from wall
Center beam Aiming: 23° (at point 3'–6" AFF)
Luminaire: Monopoint or Recessed Adjustable
Illuminance: 12 fc, avg, maintained (25 fc max)

Uses

▶General lighting
▶Medium art with short throws
▶Short throws
▶Available in 120V and 130V

Specs

Beam spread/40°
Center beam candlepower/1125
Life/2000 hrs
▶GE
▶Philips (22927–8)

Design Tips

✔One light best for medium art where mounting height for light is 9' or less.
✔Good for downlighting onto decorative floor materials: best with sconces and/or art accents and/or wallwashing to avoid cave effect.
✔More grazing light creates prominent shadows.
✔Intended for retrofitting of R30 luminaires.
✔For new construction, consider halogen low voltage 35MR11/FL, 35MR16/FL or 35MR16/C/FL.

Application Key

Commercial
Gallery
Hospitality
Institutional
Manufacturing
Residential
Retail
Exterior

bold = primary application
partial fade = minimal application
fade = unlikely application

Halogen Mains Voltage (120V)

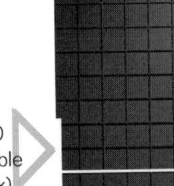

▶ **Intended for R30 and R40 retrofit applications.**

Halogen 50PAR30L/H/WFL

Long Neck

Performance Sketches

[All data outlined here is based on information from boldfaced manufacturer's published data for 120V version at time of manuscript preparation. Scale is ¼ inch equals 1 foot.]

Art shown at 3' by 4' and centered 5'–6" AFF. Wall grid on 6" by 6" centers and scaled at ¼"=1'.

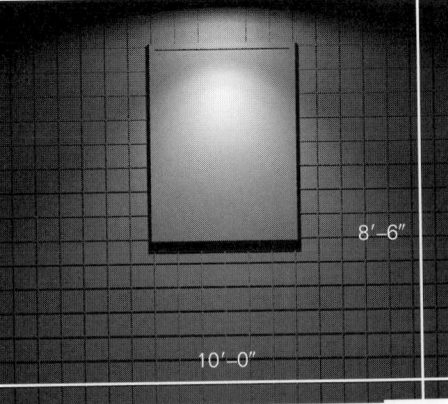

Uses
▶ Soft general lighting
▶ Wallwashing (see 11–113)
▶ Medium art with short throws
▶ Soft accenting
▶ Short throws
▶ Available in 120V and 130V

Specs
Beam spread/50°
Center beam candlepower/ 500
Life/2500 hrs
▶ **Osram Sylvania** (14539)

Design Tips
✔ One light best for medium art where mounting height for light is 9' or less.
✔ Good for soft downlighting: best with sconces and/or art accents and/or wallwashing to avoid cave effect.
✔ More grazing light creates prominent shadows.
✔ Intended for retrofitting of R30 luminaires.

Lamp: 50PAR30L/H/WFL
Location: 1'–6" from wall
Center beam Aiming: 1° (at point 0'–0" AFF)
Luminaire: Monopoint or Recessed Adjustable
Illuminance: 6 fc, avg, maintained (13 fc max)

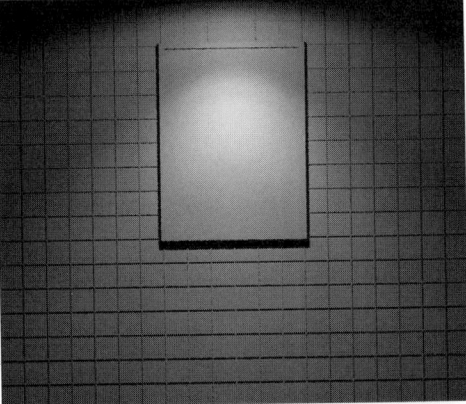

Lamp: 50PAR30L/H/WFL
Location: 2'–0" from wall
Center beam Aiming: 6° (at point 0'–0" AFF)
Luminaire: Monopoint or Recessed Adjustable
Illuminance: 5 fc, avg, maintained (10 fc max)

Application Key

Commercial
Gallery
Hospitality
Institutional
Manufacturing
Residential
Retail
Exterior

bold = primary application
partial fade = minimal application
fade = unlikely application

Lamp: 50PAR30L/H/WFL
Location: 2'–6" from wall
Center beam Aiming: 18° (at point 1'–0" AFF)
Luminaire: Monopoint or Recessed Adjustable
Illuminance: 7 fc, avg, maintained (13 fc max)

Halogen Mains Voltage (120V)

Halogen 50PAR30L/H/VWFL
Long Neck

▶Intended for R30 and R40 retrofit applications

Performance Sketches

[All data outlined here is based on information from boldfaced manufacturer's published data for 120V version at time of manuscript preparation. Scale is ¼ inch equals 1 foot.]

. . . .

Art shown at 3' by 4' and centered 5'–6" AFF. Wall grid on 6" by 6" centers and scaled at ¼"=1'.

Lamp: 50PAR30L/H/VWFL
Location: 1'–6" from wall
Center beam Aiming: 0° (at point 0'–0" AFF)
Luminaire: Monopoint or Recessed Adjustable
Illuminance: 8 fc, avg, maintained (19 fc max)

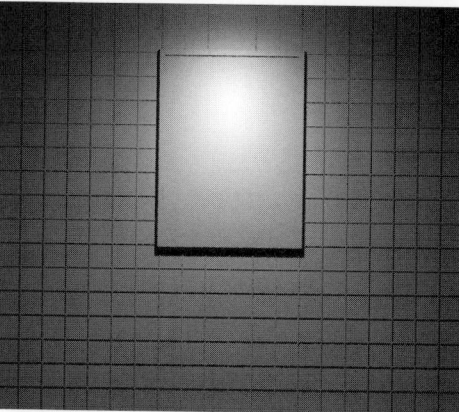

Lamp: 50PAR30L/H/VWFL
Location: 2'–0" from wall
Center beam Aiming: 6° (at point 0'–0" AFF)
Luminaire: Monopoint or Recessed Adjustable
Illuminance: 7 fc, avg, maintained (14 fc max)

Lamp: 50PAR30L/H/VWFL
Location: 2'–6" from wall
Center beam Aiming: 18° (at point 1'–0" AFF)
Luminaire: Monopoint or Recessed Adjustable
Illuminance: 9 fc, avg, maintained (17 fc max)

Uses
▶Soft general lighting
▶Wallwashing (see 11–113)
▶Medium art with short throws
▶Soft accenting
▶Short throws
▶Available in 120V and 130V

Specs
Beam spread/58°
Center beam candlepower/ 575
Life/2000 hrs
▶GE
▶**Philips** (23645–5)

Design Tips
✔One light best for medium art where mounting height for light is 9' or less.
✔Good for soft downlighting: best with sconces and/or art accents and/or wallwashing to avoid cave effect.
✔More grazing light creates prominent shadows.
✔Intended for retrofitting of R30 luminaires.
✔For new construction, consider halogen low voltage 35MR16/ C/VWFL.

Application Key

Commercial
Gallery
Hospitality
Institutional
Manufacturing
Residential
Retail
Exterior

bold = primary application
partial fade = minimal application
fade = unlikely application

Halogen Main Voltage (120V)

▶ Intended for R30 and R40 retrofit applications.

Halogen 75PAR30L/H/NSP

Long Neck

Performance Sketches

[All data outlined here is based on information from boldfaced manufacturer's published data for 120V version at time of manuscript preparation. Scale is ¼ inch equals 1 foot.]

Uses
▶ Architectural detailing
▶ Large art with long throws
▶ Sculpture with long throws
▶ Long throws
▶ Available in 120V and 130V

Specs
Beam spread/9°
Center beam candlepower/14750
Life/2250 hrs
▶ Osram Sylvania
▶ **Philips** (22930–2)

Design Tips
✔ One light best for large art where mounting height for light is 16' or less.
✔ Light may degrade artwork rapidly and/or significantly: consider UV and IR filters or limit exposure.
✔ These are strong accents and deserve judicious use. Limit exposure of sensitive artwork.
✔ Great for glass pieces and floral centerpieces and where mounting height for light is 16' or less.
✔ More grazing light creates prominent shadows.
✔ PAR30L lamps are intended as retrofits for standard R30 lamps.
✔ For new construction, consider halogen low voltage 50MR16/C/CG/NSP, 65MR16/NSP or 65MR16/C/NSP or halogen IR mains voltage 50PAR38/HIR/NSP.

Application Key

Commercial
Gallery
Hospitality
Institutional
Manufacturing
Residential
Retail
Exterior

bold = primary application
partial fade = minimal application
fade = unlikely application

Halogen Mains Voltage (120V) III
3

Lamp: 75PAR30L/H/NSP
Location: 4'–3" from wall
Center beam Aiming: 22° (at point 4'–6" AFF)
Luminaire: Monopoint or Recessed Adjustable
Illuminance: 15 fc, avg, maintained (38 fc max)

15'–0"

10'–0"

Lamp: 75PAR30L/H/NSP
Location: 5'–3" from wall
Center beam Aiming: 27° (at point 4'–6" AFF)
Luminaire: Monopoint or Recessed Adjustable
Illuminance: 16 fc, avg, maintained (41 fc max)

Art shown at 4' by 5' and centered 5'–6" AFF. Wall grid on 6" by 6" centers and scaled at ¼"=1'.

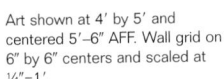
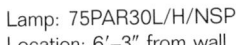

Lamp: 75PAR30L/H/NSP
Location: 6'–3" from wall
Center beam Aiming: 32° (at point 5'–0" AFF)
Luminaire: Monopoint or Recessed Adjustable
Illuminance: 17 fc, avg, maintained (47 fc max)

Halogen 75PAR30L/H/SP

Long Neck

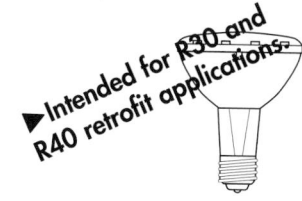

▶ Intended for R30 and R40 retrofit applications

Performance Sketches

[All data outlined here is based on information from boldfaced manufacturer's published data for 120V version at time of manuscript preparation. Scale is ¼ inch equals 1 foot.]

Lamp: 75PAR30L/H/SP
Location: 3'–0" from wall
Center beam Aiming: 21° (at point 4'–6" AFF)
Luminaire: Monopoint or Recessed Adjustable
Illuminance: 17 fc, avg, maintained (48 fc max)

12'–6"

10'–0"

Lamp: 75PAR30L/H/SP
Location: 4'–0" from wall
Center beam Aiming: 27° (at point 4'–6" AFF)
Luminaire: Monopoint or Recessed Adjustable
Illuminance: 18 fc, avg, maintained (52 fc max)

Art shown at 4' by 5' and centered 5'–6" AFF. Wall grid on 6" by 6" centers and scaled at ¼"=1'.

Lamp: 75PAR30L/H/SP
Location: 5'–0" from wall
Center beam Aiming: 32° (at point 4'–6" AFF)
Luminaire: Monopoint or Recessed Adjustable
Illuminance: 18 fc, avg, maintained (53 fc max)

Uses
▶ Architectural accenting
▶ Large art with medium throws
▶ Sculpture with medium throws
▶ Medium throws
▶ Available in 120V and 130V

Specs
Beam spread/12°
Center beam candlepower/ 9000
Life/2000 hrs
▶ **GE** (11124)

Design Tips
✔ One light best for large art where mounting height for light is 13' or less.
✔ Light may degrade artwork rapidly and/or significantly: consider UV and IR filters or limit exposure time.
✔ Great for glass pieces and floral centerpieces and where mounting height for light is 13' or less.
✔ More grazing light creates prominent shadows.
✔ PAR30L lamps are intended as retrofits for standard R30 lamps.
✔ For new construction, consider halogen low voltage 50MR16/ AL/SP or 50MR16/AL/CG/ SP.

Application Key

Commercial
Gallery
Hospitality
Institutional
Manufacturing
Residential
Retail
Exterior

bold = primary application
partial fade = minimal application
fade = unlikely application

Halogen Main Voltage (120V)

▶Intended for R30 and
R40 retrofit applications.

Halogen 75PAR30L/H/WSP

Long Neck

Performance Sketches

[All data outlined here is based on information from boldfaced manufacturer's published data for
120V version at time of manuscript preparation. Scale is ¼ inch equals 1 foot.]

Uses
▶Architectural accenting
▶Large art with medium to
  long throws
▶Sculpture with medium to
  long throws
▶Medium to long throws
▶Only available in 120V

Specs
Beam spread/16°
Center beam
  candlepower/6700
Life/2000 hrs
▶**Philips** (spot) (22934–4)

Design Tips
✔One light best for medium art
  and where mounting height for
  light is 13' or less.
✔Light may degrade artwork
  rapidly and/or significantly:
  consider UV and IR filters or
  limit exposure time.
✔More grazing light creates
  prominent shadows.
✔Intended for retrofitting of R30
  luminaires.
✔For new construction,
  consider halogen low voltage
  50MR16/C/WSP or 50MR16/
  C/CG/WSP.

Lamp: 75PAR30L/H/WSP
Location: 3'–0" from wall
Center beam Aiming: 18° (at point 3'–6" AFF)
Luminaire: Monopoint or Recessed Adjustable
Illuminance: 19 fc, avg, maintained (39 fc max)

12'–6"

10'–0"

Lamp: 75PAR30L/H/WSP
Location: 4'–0" from wall
Center beam Aiming: 24° (at point 3'–6" AFF)
Luminaire: Monopoint or Recessed Adjustable
Illuminance: 20 fc, avg, maintained (39 fc max)

Art shown at 4' by 5' and
centered 5'–6" AFF. Wall grid on
6" by 6" centers and scaled at
¼"=1'.

Lamp: 75PAR30L/H/WSP
Location: 5'–0" from wall
Center beam Aiming: 32° (at point 4'–6" AFF)
Luminaire: Monopoint or Recessed Adjustable
Illuminance: 25 fc, avg, maintained (48 fc max)

Application Key

Commercial
Gallery
Hospitality
Institutional
Manufacturing
Residential
Retail
Exterior

bold = primary application
partial fade = minimal application
fade = unlikely application

Halogen Mains Voltage (120V)

3

Halogen 75PAR30L/H/NFL

Long Neck

▶Intended for R30 and R40 retrofit applications

Performance Sketches

[All data outlined here is based on information from boldfaced manufacturer's published data for 120V version at time of manuscript preparation. Scale is ¼ inch equals 1 foot.]

Lamp: 75PAR30L/H/NFL
Location: 3'–0" from wall
Center beam Aiming: 18° (at point 3'–0" AFF)
Luminaire: Monopoint or Recessed Adjustable
Illuminance: 11 fc, avg, maintained (24 fc max)

Lamp: 75PAR30L/H/NFL
Location: 4'–0" from wall
Center beam Aiming: 25° (at point 4'–0" AFF)
Luminaire: Monopoint or Recessed Adjustable
Illuminance: 13 fc, avg, maintained (26 fc max)

Art shown at 4' by 5' and centered 5'–6" AFF. Wall grid on 6" by 6" centers and scaled at ¼"=1'.

Lamp: 75PAR30L/H/NFL
Location: 5'–0" from wall
Center beam Aiming: 30° (at point 4'–0" AFF)
Luminaire: Monopoint or Recessed Adjustable
Illuminance: 13 fc, avg, maintained (24 fc max)

Uses

▶Feature downlighting
▶Architectural accenting
▶Large art with medium throws
▶Sculpture with medium throws
▶Medium to long throws
▶Available in 120V and 130V

Specs

Beam spread/29°
Center beam candlepower/ 3250
Life/2150 hrs
▶GE (flood)
▶Osram Sylvania
▶**Philips** (22941–9)

Design Tips

✔One light best for medium art where mounting height for light is 13' or less.
✔More grazing light creates prominent shadows.
✔Intended for retrofitting of R30 luminaires.
✔For new construction, consider halogen IR mains voltage 50PAR38/HIR/NFL or 60PAR38/HIR/NFL or halogen IR low voltage 37MR16/HIR/ C/NFL.

Application Key

Commercial
Gallery
Hospitality
Institutional
Manufacturing
Residential
Retail
Exterior

bold = primary application
partial fade = minimal application
fade = unlikely application

Halogen Mains Voltage (120V)

3

▶ Intended for R30 and R40 retrofit applications.

Halogen 75PAR30L/H/FL

Long Neck

Performance Sketches

[All data outlined here is based on information from boldfaced manufacturer's published data for 120V version at time of manuscript preparation. Scale is ¼ inch equals 1 foot.]

Art shown at 4' by 5' and centered 5'–6" AFF. Wall grid on 6" by 6" centers and scaled at ¼"=1'.

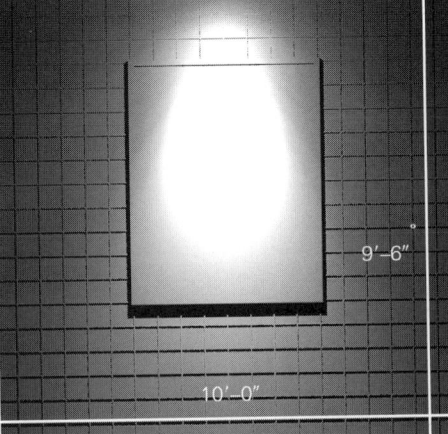

9'–6"

10'–0"

Lamp: 75PAR30L/H/FL
Location: 1'–6" from wall
Center beam Aiming: 11° (at point 2'–0" AFF)
Luminaire: Monopoint or Recessed Adjustable
Illuminance: 14 fc, avg, maintained (53 fc max)

Uses
▶ General lighting
▶ Large art with short throws
▶ Short throws
▶ Available in 120V and 130V

Specs
Beam spread/40°
Center beam candlepower/2200
Life/2000 hrs
▶ **Philips** (22944–3)

Design Tips
✔ One light best for large art where mounting height for light is 10' or less.
✔ Good for downlighting onto decorative floor materials: best with sconces and/or art accents and/or wallwashing to avoid cave effect.
✔ More grazing light creates prominent shadows.
✔ Intended for retrofitting of R30 luminaires.
✔ For new construction, consider halogen low voltage 50MR16/FL, 65MR16/FL, 65MR16/C/FL or 50MR16/C/CG/FL (Philips) or halogen IR mains voltage 60PAR38/HIR/FL or halogen IR low voltage 37MR16/HIR/C/FL.

 Lamp: 75PAR30L/H/FL
Location: 2'–0" from wall
Center beam Aiming: 17° (at point 3'–0" AFF)
Luminaire: Monopoint or Recessed Adjustable
Illuminance: 15 fc, avg, maintained (47 fc max)

Application Key

Commercial
Gallery
Hospitality
Institutional
Manufacturing
Residential
Retail
Exterior

bold = primary application
partial fade = minimal application
fade = unlikely application

Lamp: 75PAR30L/H/FL
Location: 2'–6" from wall
Center beam Aiming: 27° (at point 4'–6" AFF)
Luminaire: Monopoint or Recessed Adjustable
Illuminance: 18 fc, avg, maintained (55 fc max)

Halogen Mains Voltage (120V)

Halogen 75PAR30L/H/WFL

Long Neck

▶Intended for R30 and R40 retrofit applications

Performance Sketches

[All data outlined here is based on information from boldfaced manufacturer's published data for 120V version at time of manuscript preparation. Scale is ¼ inch equals 1 foot.]

. . . .

Art shown at 3' by 4' and centered 5'–6" AFF. Wall grid on 6" by 6" centers and scaled at ¼"=1'.

Lamp: 75PAR30L/H/WFL
Location: 1'–6" from wall
Center beam Aiming: 1° (at point 0'–0" AFF)
Luminaire: Monopoint or Recessed Adjustable
Illuminance: 7 fc, avg, maintained (18 fc max)

Lamp: 75PAR30L/H/WFL
Location: 2'–0" from wall
Center beam Aiming: 6° (at point 0'–0" AFF)
Luminaire: Monopoint or Recessed Adjustable
Illuminance: 7 fc, avg, maintained (14 fc max)

Lamp: 75PAR30L/H/WFL
Location: 2'–6" from wall
Center beam Aiming: 16° (at point 1'–0" AFF)
Luminaire: Monopoint or Recessed Adjustable
Illuminance: 9 fc, avg, maintained (17 fc max)

Uses

▶Soft general lighting
▶Wallwashing (see 11–121)
▶Medium art with short throws
▶Soft accenting
▶Short throws
▶Available in 120V and 130V

Specs

Beam spread/50°
Center beam candlepower/ 750
Life/2500 hrs
▶**Osram Sylvania** (14768)

Design Tips

✔One light best for medium art (3' W by 4' H) where mounting height for light is 10' or less.
✔Good for soft downlighting: best with sconces and/or art accents and/or wallwashing to avoid cave effect.
✔More grazing light creates prominent shadows.
✔Intended for retrofitting of R30 luminaires.

Application Key

Commercial
Gallery
Hospitality
Institutional
Manufacturing
Residential
Retail
Exterior

bold = primary application
partial fade = minimal application
fade = unlikely application

Halogen Main Voltage (120V)

3

▶ Intended for R30 and R40 retrofit applications.

Halogen 75PAR30L/H/VWFL

Long Neck

Performance Sketches

[All data outlined here is based on information from boldfaced manufacturer's published data for 120V version at time of manuscript preparation. Scale is ¼ inch equals 1 foot.]

Art shown at 3' by 4' and centered 5'–6" AFF. Wall grid on 6" by 6" centers and scaled at ¼"=1'.

9'–6"

10'–0"

Uses
▶ General lighting
▶ Wallwashing (see 11–121)
▶ Medium art with short throws
▶ Short throws
▶ Available in 120V and 130V

Specs
Beam spread/60°
Center beam
   candlepower/900
Life/2000 hrs
▶ **GE** (16393)

Design Tips
✔ One light best for medium art where mounting height for light is 10' or less.
✔ Good for downlighting onto decorative floor materials: best with sconces and/or art accents and/or wallwashing to avoid cave effect.
✔ More grazing light creates prominent shadows.
✔ Intended for retrofitting of R30 luminaires.
✔ For new construction, consider halogen low voltage 50MR16/VWFL, 50MR16/C/VWFL, 65MR16/C/VWFL or 50MR16/C/CG/VWFL.

Lamp: 75PAR30L/H/VWFL
Location: 1'–6" from wall
Center beam Aiming: 0° (at point 0'–0" AFF)
Luminaire: Monopoint or Recessed Adjustable
Illuminance: 11 fc, avg, maintained (29 fc max)

Lamp: 75PAR30L/H/VWFL
Location: 2'–0" from wall
Center beam Aiming: 6° (at point 0'–0" AFF)
Luminaire: Monopoint or Recessed Adjustable
Illuminance: 11 fc, avg, maintained (22 fc max)

Lamp: 75PAR30L/H/VWFL
Location: 2'–6" from wall
Center beam Aiming: 16° (at point 1'–0" AFF)
Luminaire: Monopoint or Recessed Adjustable
Illuminance: 14 fc, avg, maintained (26 fc max)

Application Key

Commercial
Gallery
Hospitality
Institutional
Manufacturing
Residential
Retail
Exterior

bold = primary application
partial fade = minimal application
fade = unlikely application

Halogen Mains Voltage (120V)

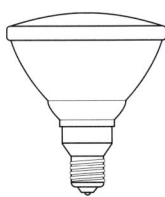

Halogen/PAR38

Guide
⊃Longer life is better
⊃Higher efficacy is better and/or higher intensity from beam spread is better
⊃Lower wattage is better
⊃Lower color of light is yellower
⊃Low cost range for halogen is about US$7.⁵⁰ to US$12.⁰⁰ (except 250PAR38 which is US$30.⁰⁰±)

StatsData varies manufacturer to manufacturer and changes from time to time

Life/2000 and 4200 hours
Different versions from different manufacturers. A typical office environment might be occupied about 2600 hours each year.

Efficacy/12 to 16 LPW
Varies by wattage. Good to excellent beam intensities and distributions available.

Wattages/ 45, 50, 60, 75, 90, 100, 120 and 250
Color of Light/2750°K to 2900°K
Beam spreads/NSP, SP, VWSP, NFL, FL, WFL, VWFL
Narrow spot, spot, very wide spot, narrow flood, flood, wide flood and very wide flood distributions are available.

Size/5⁵⁄₁₆″ L by 4¾″ Ø
Voltage/120V and 130V
Cost Magnitude/Low (2000)
Costs for the high–powered and longer–lived quartz halogen (250PAR38/H) lamps are likely to range from US$30.⁰⁰ to US$40.⁰⁰.

Manufacturers
General Electric
Osram Sylvania
Philips

Net Addresses
http://www.ge.com/lighting/business/index.htm
http://ecom.sylvania.com/osicatalog/
http://www.lighting.philips.com/

CONNECT FOR MORE

Advantages
Crisp white light
Good to excellent intensities available
Standard voltage (no transformers)
Easily dimmed
Low initial cost
Alternative to less efficient R lamps
Withstand extreme temperature range
Withstand moisture contact
◆When used in wet–rated sockets or wet–rated open luminaires.
Very little lamp lumen depreciation (light loss) over life

Disadvantages
Beam aberrations
◆Striations
◆Inconsistent beam shape
High operating costs
◆Inefficiency and short life may make this cost prohibitive in typical commercial applications, depending on ease of changing lamps and maintenance staffing.
Hot to touch
Hot-shock sensitive
◆Lamp may fail if rough aiming–adjustment or sharp vibration occurs while lamp is energized.
Large/bulky
◆PAR30S lamps are smaller; HIRPAR30S lamps are smaller and equally punchy.

Halogen Mains Voltage (120V)

For more efficient accent lighting, use HIR low voltage lamps (Section 6); **HIR PAR lamps** (Section 5) or **CMH PAR lamps** (Section 7).

Halogen/PAR38

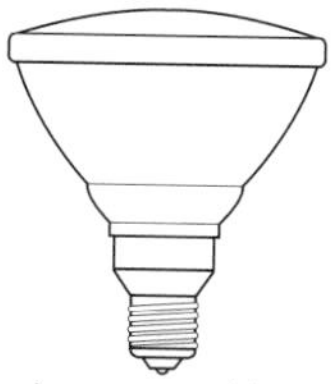

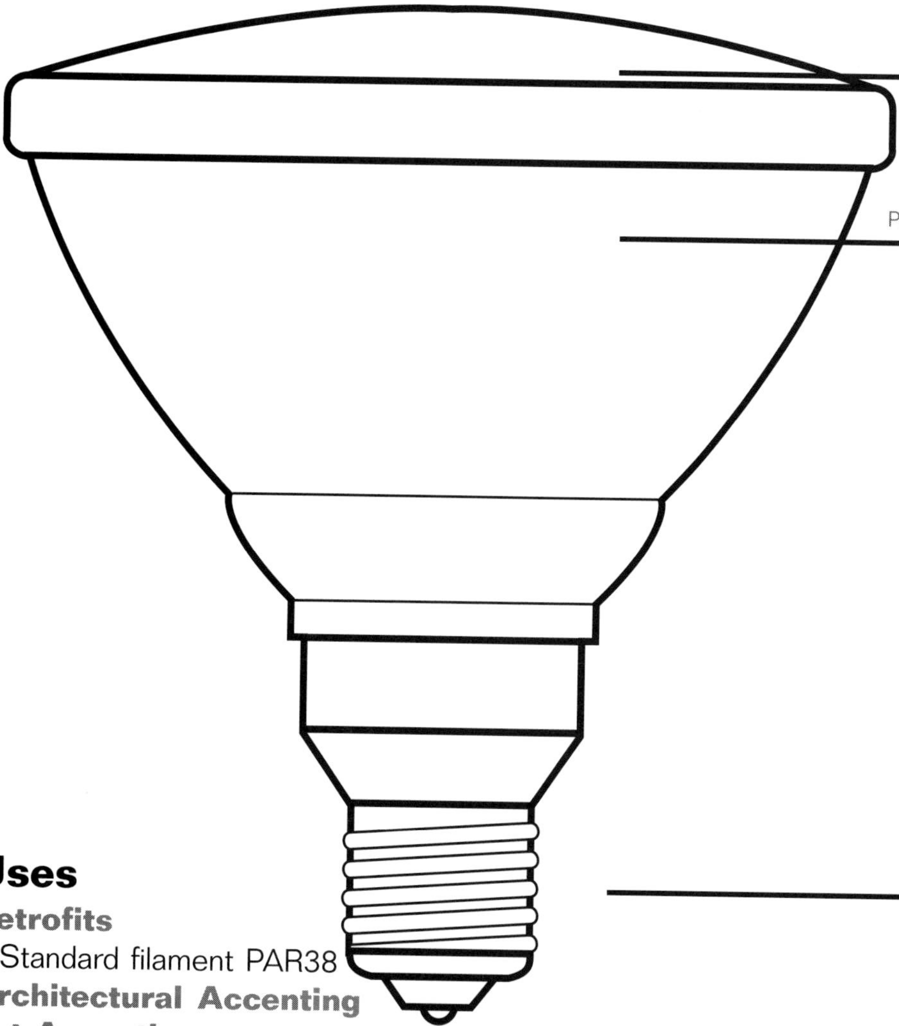

Pressed/cast glass optical lens

Parabolic aluminized reflector (PAR)

In new installations, use HIR PAR30S, HIR PAR38 or HIR low voltage lamps or CMH PAR lamps. In retrofit installations, use HIR PAR38 lamps.

Medium screw base

Lamp shown actual size
Image copyright 2000/GE

Uses

Retrofits
Standard filament PAR38

Architectural Accenting

Art Accenting
Gallery
Hospitality
Residential

Downlighting
Circulation
Conference rooms
Hospitality
Residential
Retail

Merchandise Accenting

Surface Grazing
Specular/semispecular surfaces
◆Polished/honed stone
◆Polished/satin wood
◆Fritted/etched glass
◆Cut glass

PAR38 Lighting Design Tips
✔PAR38 spot versions best at high mounting heights (15'–0" to 20'–0").
✔PAR38 narrow flood versions best at moderate mounting heights (12'–0" to 15'–0").
✔PAR38 flood versions best at low mounting heights (8'–6" to 10'–0").
✔Dimming increases life (typically doubling life), but shifts color warmer.
✔Glary when used without much other general lighting or when aimed across long distances where ceilings are low.
✔Beam striations are most apparent on light–colored, nontextured, flat surfaces and may be objectionable.
✔PAR38 Downlighting alone is harsh—combine several techniques for best results.
✔PAR38 typically rated for exposure to moisture when used in wet–rated sockets or wet–rated open luminaires—confirm with manufacturers' latest datasheets—so can be used in many open exterior utility lights, sign lights and above–grade uplights.

Net Addresses/PAR38 Track, Monopoints, Recessed and Exterior
See page 3–97

CONNECT FOR MORE

Halogen Mains Voltage (120V)

Halogen/PAR38

PAR38 Visual Index

45PAR38/H/NSP
- ▶ Architectural detailing
- ▶ Large art/15'–0" ceiling height
- ▶ Sculpture
- ▶ Long throws
- ▶ Hospitality

45PAR38/H/SP
- ▶ Architectural accenting
- ▶ Medium art/12'–6" ceiling height
- ▶ Sculpture
- ▶ Medium–to–long throws
- ▶ Hospitality

  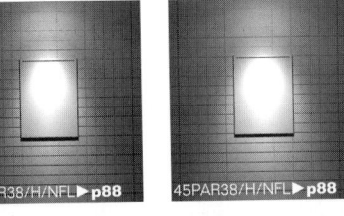

45PAR38/H/NFL
- ▶ General lighting
- ▶ Feature downlighting
- ▶ Medium art/10'–6" ceiling height
- ▶ Short–to–medium throws
- ▶ Gallery, Hospitality, Residential

 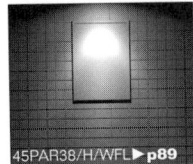 

45PAR38/H/WFL
- ▶ Soft general lighting
- ▶ Wallwashing
- ▶ Medium art/8'–6" ceiling height
- ▶ Short throws
- ▶ Gallery, Residential

  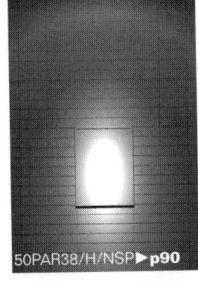

50PAR38/H/NSP
- ▶ Architectural detailing
- ▶ Medium art/15'–0" ceiling height
- ▶ Sculpture
- ▶ Long throws
- ▶ Hospitality

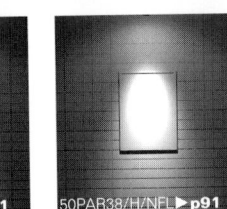     

50PAR38/H/NFL
- ▶ Architectural accenting
- ▶ Feature downlighting
- ▶ Medium art/10'–6" ceiling height
- ▶ Sculpture
- ▶ Medium throws
- ▶ Hospitality

Throws/Ceilings
- ◗ Short = 7' to 10'
- ◗ Medium = 10' to 15'
- ◗ Long = 15' to 20'
- ◗ Very Long = 20' or more

Halogen Mains Voltage (120V)

Recommended intensities
for art accenting

Low (gallery, museum, residential)
5 to 10 fc average on art
Moderate (hospitality)
10 to 20 fc average on art
High (commercial—or higher for retail)
20 to 30 fc average on art

These are intended to be average, maintained values including effects of accent lighting and general room lighting. General room lighting contributes just a few footcandles in residential, gallery and museum settings, but contributes 5 to 10 fc in commercial settings. Maximums might range from 50 percent to 100 percent of the average value. Wherever artwork is highly valuable and/or sensitive to light, precautions should be taken to limit light exposure (maintaining low levels and/or switching lights off when art is not being viewed); and exposure to infrared and ultraviolet should be limited (using filters on lamps or placing art behind specially treated glass or acrylic).

Halogen/PAR38

60PAR38/H/NSP▶**p92** 60PAR38/H/NSP▶**p92** 60PAR38/H/NSP▶**p93**

60PAR38/H/NSP
▶Architectural detailing
▶Large art/18′–0″ ceiling height
▶Sculpture
▶Long throws
▶Hospitality

60PAR38/H/SP▶**p94** 60PAR38/H/SP▶**p94** 60PAR38/H/SP▶**p94**

60PAR38/H/SP
▶Architectural detailing
▶Large art/15′–0″ ceiling height
▶Sculpture
▶Long throws
▶Hospitality

60PAR38/H/NFL▶**p95** 60PAR38/H/NFL▶**p95** 60PAR38/H/NFL▶**p95**

60PAR38/H/NFL
▶Architectural accenting
▶Feature downlighting
▶Medium art/10′–6″ ceiling height
▶Sculpture
▶Medium throws
▶Commercial, Retail

 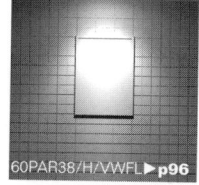 

60PAR38/H/VWFL▶**p96** 60PAR38/H/VWFL▶**p96** 60PAR38/H/VWFL▶**p96**

60PAR38/H/VWFL
▶Soft general lighting
▶Wallwashing
▶Medium art/9′–6″ ceiling height
▶Short throws
▶Hospitality

75PAR38/H/NSP▶**p98** 75PAR38/H/NSP▶**p98** 75PAR38/H/NSP▶**p99**

75PAR38/H/NSP
▶Architectural detailing
▶Large art/18′–0″ ceiling height
▶Sculpture
▶Long throws
▶Hospitality

Artwork Size Categories
⊃Petite = 2′–0″ by 2′–0″ or smaller
⊃Small = 2′–0″ by 2′–0″ to 2′–6″ by 3′–0″
⊃Medium = 2′–6″ by 3′–0″ to 3′–0″ by 4′–0″
⊃Large = 3′–0″ by 4′–0″ to 4′–0″ by 5′–0″
⊃Extra Large = 4′–0″ by 5′–0″ or larger

Halogen Mains Voltage (120V)

3

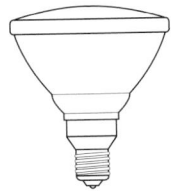

Halogen/PAR38

75PAR38/H/SP
▶ Architectural detailing
▶ Large art/15'–0" ceiling height
▶ Sculpture
▶ Long throws
▶ Hospitality

75PAR38/H/NFL/26°
▶ Architectural accenting
▶ Feature downlighting
▶ Large art/12'–6" ceiling height
▶ Sculpture
▶ Medium throws
▶ Hospitality

75PAR38/H/NFL/30°
▶ Architectural accenting
▶ Feature downlighting
▶ Large art/12'–6" ceiling height
▶ Medium throws
▶ Hospitality

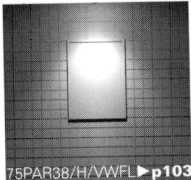 

75PAR38/H/VWFL
▶ Soft general lighting
▶ Wallwashing
▶ Medium art/9'–6" ceiling height
▶ Short throws
▶ Gallery, Residential

■ ■ ■ ■ ■ ■

Throws/Ceilings
◗ Short = 7' to 10'
◗ Medium = 10' to 15'
◗ Long = 15' to 20'
◗ Very Long = 20' or more

90PAR38/H/NSP
▶ Strong architectural detailing
▶ Large art/18'–0" ceiling height
▶ Sculpture
▶ Long throws
▶ Hospitality

Recommended intensities
for art accenting

Low (gallery, museum, residential)
5 to 10 fc average on art
Moderate (hospitality)
10 to 20 fc average on art
High (commercial—or higher for retail)
20 to 30 fc average on art

These are intended to be average, maintained values including effects of accent lighting and general room lighting. General room lighting contributes just a few footcandles in residential, gallery and museum settings, but contributes 5 to 10 fc in commercial settings. Maximums might range from 50 percent to 100 percent of the average value. Wherever artwork is highly valuable and/or sensitive to light, precautions should be taken to limit light exposure (maintaining low levels and/or switching lights off when art is not being viewed); and exposure to infrared and ultraviolet should be limited (using filters on lamps or placing art behind specially treated glass or acrylic).

Halogen
Main Voltage (120V)

3

Halogen/PAR38

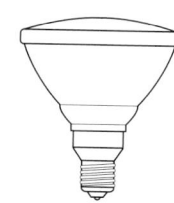

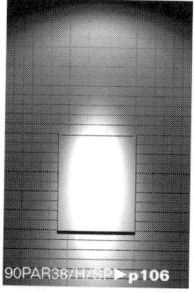  

90PAR38/H/SP
▶Architectural detailing
▶Large art/15'–0" ceiling height
▶Sculpture
▶Long throws
▶Commercial, Hospitality, Retail

90PAR38/H/VWSP
▶Architectural accenting
▶Medium art/12'–6" ceiling height
▶Sculpture
▶Medium throws
▶Hospitality

90PAR38/H/NFL
▶Architectural accenting
▶Feature downlighting
▶Large art/12'–6" ceiling height
▶Sculpture
▶Medium throws
▶Hospitality

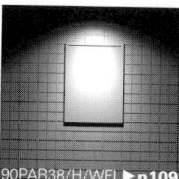  

90PAR38/H/WFL
▶Soft general lighting
▶Wallwashing
▶Medium art/9'–6" ceiling height
▶Short throws
▶Gallery, Hospitality, Residential

  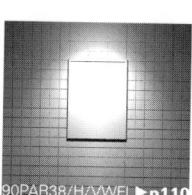

90PAR38/H/VWFL
▶Soft general lighting
▶Wallwashing
▶Medium art/9'–6" ceiling height
▶Short throws
▶Hospitality

Halogen Mains Voltage (120V) **III**

3

Halogen/PAR38

 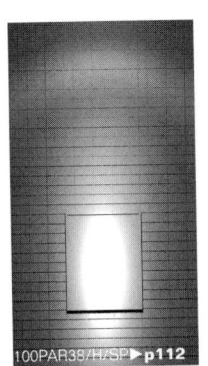 

100PAR38/H/SP
▶ Strong architectural detailing
▶ Large art/18'–0" ceiling height
▶ Sculpture
▶ Long throws
▶ Hospitality

100PAR38/H/NFL
▶ Architectural accenting
▶ Feature downlighting
▶ Large art/12'–6" ceiling height
▶ Sculpture
▶ Medium throws
▶ Hospitality

■ ■ ■ ■ ■ ■

Throws/Ceilings
◗ Short = 7' to 10'
◗ Medium = 10' to 15'
◗ Long = 15' to 20'
◗ Very Long = 20' or more

120PAR38/H/NSP
▶ Strong architectural detailing
▶ Large art/20'–0" ceiling height
▶ Sculpture
▶ Very long throws
▶ Commercial, Hospitality, Retail

120PAR38/H/NFL
▶ Architectural accenting
▶ Feature downlighting
▶ Large art/12'–6" ceiling height
▶ Sculpture
▶ Medium throws
▶ Commercial, Retail

  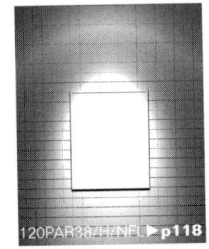

Recommended intensities
for art accenting

Low (gallery, museum, residential)
5 to 10 fc average on art
Moderate (hospitality)
10 to 20 fc average on art
High (commercial—or higher for retail)
20 to 30 fc average on art

These are intended to be average, maintained values including effects of accent lighting and general room lighting. General room lighting contributes just a few footcandles in residential, gallery and museum settings, but contributes 5 to 10 fc in commercial settings. Maximums might range from 50 percent to 100 percent of the average value. Wherever artwork is highly valuable and/or sensitive to light, precautions should be taken to limit light exposure (maintaining low levels and/or switching lights off when art is not being viewed); and exposure to infrared and ultraviolet should be limited (using filters on lamps or placing art behind specially treated glass or acrylic).

120PAR38/H/WFL
▶ General lighting
▶ Wallwashing
▶ Medium art/9'–6" ceiling height
▶ Short throws
▶ Commercial, Retail

Halogen Mains Voltage (120V)

Halogen/PAR38

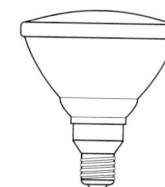

250PAR38/H/SP
▶Very strong architectural detailing
▶Large art/25'–0" ceiling height
▶Sculpture
▶Very long throws
▶Commercial, Retail

250PAR38/H/NFL
▶Architectural accenting
▶Extra large art/18'–0" ceiling height
▶Sculpture
▶Long throws
▶Hospitality

Halogen PAR Lamp Designation

45PAR38/H/NSP

Last set of letters identifies beam spread

FLOODS
• 24° to 32° for narrow flood (NFL)
• 33° to 44° for flood (FL)
• 45° to 54° for wide flood (WFL)
• 55° or greater for very wide flood (VWFL)

SPOTS
• 7° or less for very narrow spot (VNSP)
• 8° to 10° for narrow spot (NSP)
• 11° to 14° for spot (SP)
• 15° to 18° for wide spot (WSP)
• 19° to 23° for very wide spot (VWSP)

Ranges of beam spreads used for convenient/consistent reference in this text and do not necessarily correspond to each manufacturer's nor ANSI's definitions.

This letter indicates "Halogen"

These two digits represent lamp diameter in eighths inches (e.g., 30/8 or 3–3/4 inches)

These letters indicate this is a parabolic aluminized reflector (PAR) lamp

First two or three digits represent lamp wattage

[This designation used for convenient reference throughout this *Time–Saver Standards for Architectural Lighting*—but not necessarily used by all lamp manufacturers (although it would be convenient if manufacturers would come to agreement on standard designations.]

■ ■ ■ ■ ■ ■ ■ ■ ■ ■ ■

Artwork Size Categories
➧Petite = 2'–0" by 2'–0" or smaller
➧Small = 2'–0" by 2'–0" to 2'–6" by 3'–0"
➧Medium = 2'–6" by 3'–0" to 3'–0" by 4'–0"
➧Large = 3'–0" by 4'–0" to 4'–0" by 5'–0"
➧Extra Large = 4'–0" by 5'–0" or larger

Halogen Mains Voltage (120V)

Halogen 45PAR38/H/NSP

Performance Sketches

[All data outlined here is based on information from boldfaced manufacturer's published data for 120V version at time of manuscript preparation. Scale is ¼ inch equals 1 foot.]

Uses
▶ Architectural detailing
▶ Large art with long throws
▶ Sculpture with long throws
▶ Long throws
▶ Available in 120V and 130V versions

Specs
Beam spread/9°
Center beam candlepower/10000
Life/2500 hrs
▶ **Osram Sylvania** (14590)

Design Tips
✔ One light best for large art and where mounting height for light is 16' or less.
✔ These are strong accents and deserve judicious use. Light may degrade artwork rapidly and/or significantly: consider UV and IR filters or limit exposure time.
✔ Great for glass pieces and floral centerpieces and where mounting height for light is 16' or less.
✔ Consider halogen low voltage 35MR16/NSP.

Lamp: 45PAR38/H/NSP
Location: 4'–3" from wall
Center beam Aiming: 22° (at point 4'–6" AFF)
Luminaire: Monopoint or Recessed Adjustable
Illuminance: 14 fc, avg, maintained (33 fc max)

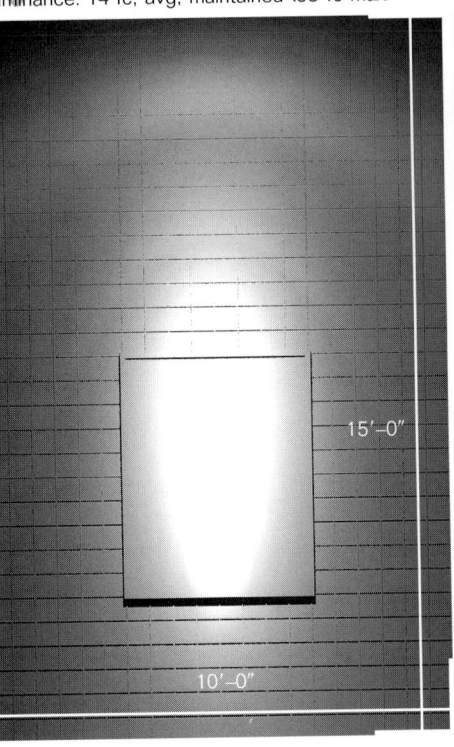

15'–0"

10'–0"

Lamp: 45PAR38/H/NSP
Location: 5'–3" from wall
Center beam Aiming: 27° (at point 4'–6" AFF)
Luminaire: Monopoint or Recessed Adjustable
Illuminance: 15 fc, avg, maintained (35 fc max)

Application Key

Commercial
Gallery
Hospitality
Institutional
Manufacturing
Residential
Retail
Exterior

bold = primary application
partial fade = minimal application
fade = unlikely application

Art shown at 4' by 5' and centered 5'–6" AFF. Wall grid on 6" by 6" centers and scaled at ¼"=1'.

Lamp: 45PAR38/H/NSP
Location: 6'–3" from wall
Center beam Aiming: 32° (at point 5'–0" AFF)
Luminaire: Monopoint or Recessed Adjustable
Illuminance: 15 fc, avg, maintained (38 fc max)

Halogen Mains Voltage (120V)

Halogen 45PAR38/H/SP

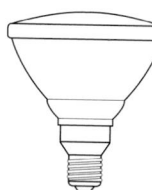

Performance Sketches

[All data outlined here is based on information from boldfaced manufacturer's published data for 120V version at time of manuscript preparation. Scale is ¼ inch equals 1 foot.]

Lamp: 45PAR38/H/SP
Location: 3'–0" from wall
Center beam Aiming: 21° (at point 4'–6" AFF)
Luminaire: Monopoint or Recessed Adjustable
Illuminance: 15 fc, avg, maintained (43 fc max)

12'–6"

10'–0"

Lamp: 45PAR38/H/SP
Location: 4'–0" from wall
Center beam Aiming: 27° (at point 4'–6" AFF)
Luminaire: Monopoint or Recessed Adjustable
Illuminance: 16 fc, avg, maintained (48 fc max)

Art shown at 3' by 4' and centered 5'–6" AFF. Wall grid on 6" by 6" centers and scaled at ¼"=1'.

Lamp: 45PAR38/H/SP
Location: 5'–0" from wall
Center beam Aiming: 32° (at point 4'–6" AFF)
Luminaire: Monopoint or Recessed Adjustable
Illuminance: 16 fc, avg, maintained (49 fc max)

Uses
▶ Architectural accenting
▶ Medium art with medium to long throws
▶ Sculpture with medium to long throws
▶ Medium to long throws
▶ Available in 120V and 130V

Specs
Beam spread/12°
Center beam candlepower/ 6100
Life/2500 hrs
▶ **GE** (17470)
▶ Osram Sylvania
▶ Philips

Design Tips
✔ One light best for medium art and where mounting height for light is 13' or less.
✔ Light may degrade artwork rapidly and/or significantly: consider UV and IR filters or limit exposure time.
✔ Scallops/striations are most severe when light is closer to the wall (compare top left image to bottom image).
✔ More grazing light creates prominent shadows.
✔ Consider halogen low voltage 35MR16/C/NSP.

Application Key

Commercial
Gallery
Hospitality
Institutional
Manufacturing
Residential
Retail
Exterior

bold = primary application
partial fade = minimal application
fade = unlikely application

Halogen Main Voltage (120V)

3

Halogen 45PAR38/H/NFL

Uses
▶ General lighting
▶ Feature downlighting
▶ Medium art with short to medium throws
▶ Short to medium throws
▶ Available in 120V and 130V versions

Specs
Beam spread/29°
Center beam candlepower/1800
Life/2500 hrs
▶ **GE** (flood) (17471)
▶ Osram Sylvania (flood)
▶ Philips (flood)

Design Tips
✔ One light best for medium art where mounting height for light is 11′ or less.
✔ More grazing light creates prominent shadows.

Performance Sketches

[All data outlined here is based on information from boldfaced manufacturer's published data for 120V version at time of manuscript preparation. Scale is ¼ inch equals 1 foot.]

Lamp: 45PAR38/H/NFL
Location: 2′–0″ from wall
Center beam Aiming: 13° (at point 2′–0″ AFF)
Luminaire: Monopoint or Recessed Adjustable
Illuminance: 10 fc, avg, maintained (21 fc max)

Lamp: 45PAR38/H/NFL
Location: 2′–6″ from wall
Center beam Aiming: 18° (at point 3′–0″ AFF)
Luminaire: Monopoint or Recessed Adjustable
Illuminance: 11 fc, avg, maintained (22 fc max)

10′–6″

10′–0″

Art shown at 3′ by 4′ and centered 5′–6″ AFF. Wall grid on 6″ by 6″ centers and scaled at ¼″=1′.

Lamp: 45PAR38/H/NFL
Location: 3′–0″ from wall
Center beam Aiming: 23° (at point 3′–6″ AFF)
Luminaire: Monopoint or Recessed Adjustable
Illuminance: 12 fc, avg, maintained (21 fc max)

Application Key

Commercial
Gallery
Hospitality
Institutional
Manufacturing
Residential
Retail
Exterior

bold = primary application
partial fade = minimal application
fade = unlikely application

Halogen Mains Voltage (120V)

Halogen 45PAR38/H/WFL

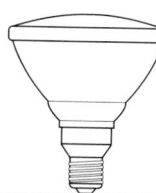

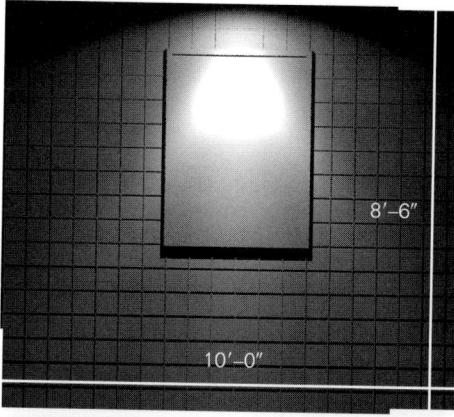

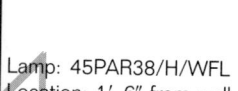

Performance Sketches

[All data outlined here is based on information from boldfaced manufacturer's published data for 130V version at time of manuscript preparation. Data and sketches shown are corrected for 120V operation. Scale is ¼ inch equals 1 foot.]

Art shown at 3' by 4' and centered 5'–6" AFF. Wall grid on 6" by 6" centers and scaled at ¼"=1'.

Lamp: 45PAR38/H/WFL
Location: 1'–6" from wall
Center beam Aiming: 11° (at point 1'–0" AFF)
Luminaire: Monopoint or Recessed Adjustable
Illuminance: 7 fc, avg, maintained (23 fc max)

Lamp: 45PAR38/H/WFL
Location: 2'–0" from wall
Center beam Aiming: 17° (at point 2'–0" AFF)
Luminaire: Monopoint or Recessed Adjustable
Illuminance: 7 fc, avg, maintained (17 fc max)

Lamp: 45PAR38/H/WFL
Location: 2'–6" from wall
Center beam Aiming: 24° (at point 3'–0" AFF)
Luminaire: Monopoint or Recessed Adjustable
Illuminance: 7 fc, avg, maintained (16 fc max)

Uses

▶ Soft general lighting
▶ Wallwashing (see 11–111)
▶ Medium art with short throws
▶ Soft accenting
▶ Short throws
▶ Only available in 130V

Specs

Beam spread/50°
Center beam candlepower/ 460 (at 120V operation)
Life/5000 hrs (at 120V operation)
Wattage/39 (at 120V operation)
▶ **Osram Sylvania** (14010)

Design Tips

✔ One light best for medium art where mounting height for light is 9' or less.
✔ Can be used to wash walls without expense of spreadlens wallwashers—units need to be on 1'–6" to 2'–0" centers and placed 1'–3" to 1'–6" from wall being washed.
✔ More grazing light creates prominent shadows.
✔ Although life is longer, efficacy and intensity are reduced and color temperature is lower (warmer) with these 130V lamps operated at 120V.

Application Key

Commercial
Gallery
Hospitality
Institutional
Manufacturing
Residential
Retail
Exterior

bold = primary application
partial fade = minimal application
fade = unlikely application

Halogen Mains Voltage (120V)

3

Halogen 50PAR38/H/NSP

Performance Sketches

[All data outlined here is based on information from boldfaced manufacturer's published data for 120V version at time of manuscript preparation. Scale is ¼ inch equals 1 foot.]

Uses
▶ Architectural detailing
▶ Medium art with long throws
▶ Sculpture with long throws
▶ Long throws
▶ Available in 120V and 130V versions

Specs
Beam spread/10°
Center beam candlepower/9000
Life/2000 hrs
▶ **GE** (17980)

Design Tips
✔ One light best for medium art and where mounting height for light is 16' or less.
✔ These are strong accents and deserve judicious use. Light may degrade artwork rapidly and/or significantly: consider UV and IR filters or limit exposure time.
✔ Great for glass pieces and floral centerpieces and where mounting height for light is 16' or less.
✔ Consider halogen low voltage 35MR11/NSP or 35MR16/C/NSP or halogen IR low voltage 37MR16/HIR/C/NSP.

Lamp: 50PAR38/H/NSP
Location: 4'–3" from wall
Center beam Aiming: 22° (at point 4'–6" AFF)
Luminaire: Monopoint or Recessed Adjustable
Illuminance: 12 fc, avg, maintained (26 fc max)

15'–0"

10'–0"

Lamp: 50PAR38/H/NSP
Location: 5'–3" from wall
Center beam Aiming: 27° (at point 4'–6" AFF)
Luminaire: Monopoint or Recessed Adjustable
Illuminance: 12 fc, avg, maintained (28 fc max)

Art shown at 3' by 4' and centered 5'–6" AFF. Wall grid on 6" by 6" centers and scaled at ¼"=1'.

Lamp: 50PAR38/H/NSP
Location: 6'–3" from wall
Center beam Aiming: 31° (at point 4'–6" AFF)
Luminaire: Monopoint or Recessed Adjustable
Illuminance: 12 fc, avg, maintained (29 fc max)

Halogen 50PAR38/H/NFL

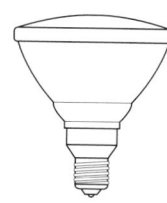

Performance Sketches

[All data outlined here is based on information from boldfaced manufacturer's published data for 120V version at time of manuscript preparation. Scale is ¼ inch equals 1 foot.]

Lamp: 50PAR38/H/NFL
Location: 2'–0" from wall
Center beam Aiming: 15° (at point 3'–0" AFF)
Luminaire: Monopoint or Recessed Adjustable
Illuminance: 12 fc, avg, maintained (27 fc max)

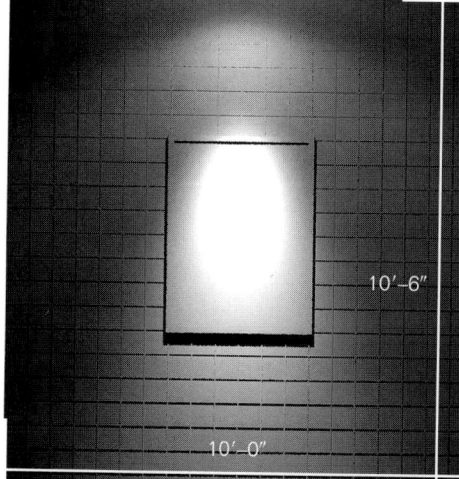

10'–6"

10'–0"

Lamp: 50PAR38/H/NFL
Location: 2'–6" from wall
Center beam Aiming: 21° (at point 4'–0" AFF)
Luminaire: Monopoint or Recessed Adjustable
Illuminance: 14 fc, avg, maintained (31 fc max)

Art shown at 3' by 4' and centered 5'–6" AFF. Wall grid on 6" by 6" centers and scaled at ¼"=1'.

Lamp: 50PAR38/H/NFL
Location: 3'–0" from wall
Center beam Aiming: 27° (at point 4'–6" AFF)
Luminaire: Monopoint or Recessed Adjustable
Illuminance: 16 fc, avg, maintained (34 fc max)

Uses
▶ Architectural accenting
▶ Feature downlighting
▶ Medium art with medium throws
▶ Sculpture with medium throws
▶ Medium throws
▶ Available in 120V and 130V

Specs
Beam spread/25°
Center beam candlepower/ 2200
Life/2000 hrs
▶ **GE** (flood) (17979)

Design Tips
✔ One light best for medium art and where mounting height for light is 11' or less.
✔ Good for downlighting onto decorative floor materials— best with sconces and/or art accents and/or wallwashing to avoid cave effect and "I wish I had a visor" effect.
✔ Consider 37MR16/HIR/C/ NFL.

Application Key

Commercial
Gallery
Hospitality
Institutional
Manufacturing
Residential
Retail
Exterior

bold = primary application
partial fade = minimal application
fade = unlikely application

3

Halogen 60PAR38/H/NSP

Uses
▶Architectural detailing
▶Large art with long throws
▶Sculpture with long throws
▶Long throws
▶Available in 120V and 130V versions

Specs
Beam spread/9°
Center beam
candlepower/16000
Life/2750 hrs
▶Osram Sylvania (spot)
▶**Philips** (23059–9)

Design Tips
✔One light best for large art and where mounting height for light is 19' or less.
✔These are strong accents and deserve judicious use. Light may degrade artwork rapidly and/or significantly: consider UV and IR filters or limit exposure time.
✔Excellent for glass pieces and floral centerpieces and where mounting height for light is 19' or less.
✔Consider halogen low voltage 50MR16/C/CG/NSP or halogen IR mains voltage 50PAR30/HIR/NSP.

Performance Sketches
[All data outlined here is based on information from boldfaced manufacturer's published data for 120V version at time of manuscript preparation. Scale is ¼ inch equals 1 foot.]

Lamp: 60PAR38/H/NSP
Location: 5'–0" from wall
Center beam Aiming: 21° (at point 4'–9" AFF)
Luminaire: Monopoint or Recessed Adjustable
Illuminance: 11 fc, avg, maintained (22 fc max)

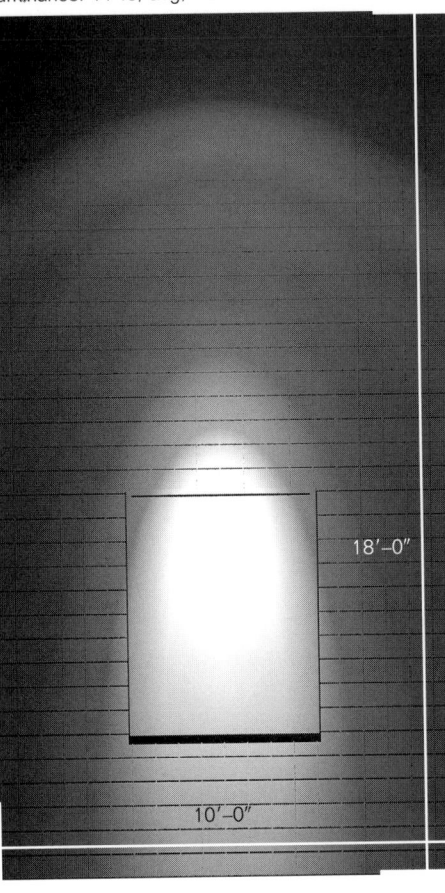

18'–0"

10'–0"

Lamp: 60PAR38/H/NSP
Location: 6'–0" from wall
Center beam Aiming: 25° (at point 5'–3" AFF)
Luminaire: Monopoint or Recessed Adjustable
Illuminance: 12 fc, avg, maintained (24 fc max)

Application Key

Commercial
Gallery
Hospitality
Institutional
Manufacturing
Residential
Retail
Exterior

bold = primary application
partial fade = minimal application
fade = unlikely application

Halogen Mains Voltage (120V)

Halogen 60PAR38/H/NSP

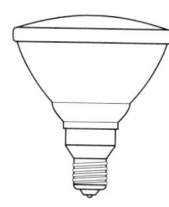

Lamp: 60PAR38/H/NSP
Location: 7'–0" from wall
Center beam Aiming: 29° (at point 5'–3" AFF)
Luminaire: Monopoint or Recessed Adjustable
Illuminance: 13 fc, avg, maintained (24 fc max)

Art shown at 4' by 5' and centered 5'–6" AFF. Wall grid on 6" by 6" centers and scaled at ¼"=1'.

Halogen Main Voltage (120V)

Halogen 60PAR38/H/SP

Uses
▶ Architectural detailing
▶ Large art with long throws
▶ Sculpture with long throws
▶ Long throws
▶ Available in 120V and 130V versions

Specs
Beam spread/12°
Center beam candlepower/ 11750
Life/2750 hrs
▶ Osram Sylvania (listed as wide spot/rated at 10000 cd)
▶ **Philips** (23062–3)

Design Tips
✔ One light best for large art and where mounting height for light is less than 16'.
✔ These are strong accents and deserve judicious use. Light may degrade artwork rapidly and/or significantly: consider UV and IR filters or limit exposure time.
✔ Good for glass pieces and floral centerpieces and where mounting height for light is 16' or less.
✔ Consider halogen low voltage 50MR16/SP.

Performance Sketches
[All data outlined here is based on information from boldfaced manufacturer's published data for 120V version at time of manuscript preparation. Scale is ¼ inch equals 1 foot.]

Lamp: 60PAR38/H/SP
Location: 4'–3" from wall
Center beam Aiming: 21° (at point 4'–0" AFF)
Luminaire: Monopoint or Recessed Adjustable
Illuminance: 13 fc, avg, maintained (29 fc max)

15'–0"

10'–0"

Lamp: 60PAR38/H/SP
Location: 5'–3" from wall
Center beam Aiming: 25° (at point 4'–0" AFF)
Luminaire: Monopoint or Recessed Adjustable
Illuminance: 13 fc, avg, maintained (30 fc max)

Application Key

Commercial
Gallery
Hospitality
Institutional
Manufacturing
Residential
Retail
Exterior

bold = primary application
partial fade = minimal application
fade = unlikely application

Art shown at 4' by 5' and centered 5'–6" AFF. Wall grid on 6" by 6" centers and scaled at ¼"=1'.

Lamp: 60PAR38/H/SP
Location: 6'–3" from wall
Center beam Aiming: 32° (at point 5'–0" AFF)
Luminaire: Monopoint or Recessed Adjustable
Illuminance: 16 fc, avg, maintained (38 fc max)

Halogen Mains Voltage (120V)

Halogen 60PAR38/H/NFL

Performance Sketches

[All data outlined here is based on information from boldfaced manufacturer's published data for 120V version at time of manuscript preparation. Scale is ¼ inch equals 1 foot.]

Lamp: 60PAR38/H/NFL
Location: 2'–0" from wall
Center beam Aiming: 16° (at point 3'–9" AFF)
Luminaire: Monopoint or Recessed Adjustable
Illuminance: 20 fc, avg, maintained (50 fc max)

10'–6"

10'–0"

Lamp: 60PAR38/H/NFL
Location: 2'–6" from wall
Center beam Aiming: 21° (at point 4'–0" AFF)
Luminaire: Monopoint or Recessed Adjustable
Illuminance: 21 fc, avg, maintained (50 fc max)

Art shown at 3' by 4' and centered 5'–6" AFF. Wall grid on 6" by 6" centers and scaled at ¼"=1'.

Lamp: 60PAR38/H/NFL
Location: 3'–0" from wall
Center beam Aiming: 27° (at point 4'–6" AFF)
Luminaire: Monopoint or Recessed Adjustable
Illuminance: 24 fc, avg, maintained (56 fc max)

Uses
► Architectural accenting
► Feature downlighting
► Medium art with medium throws
► Sculpture with medium throws
► Medium throws
► Available in 120V and 130V

Specs
Beam spread/29°
Center beam candlepower/ 3000
Life/2750 hrs
► Osram Sylvania (flood)
► **Philips** (flood) (23065–6)

Design Tips
✔ One light best for medium art where mounting height for light is 11' or less.
✔ Great for moderate to large sculpture where mounting height for light is 11' or less.
✔ More grazing light creates prominent shadows.
✔ Consider halogen IR mains voltage 50PAR38/HIR/NFL or halogen IR low voltage 37MR16/HIR/C/NFL.

Application Key

Commercial
Gallery
Hospitality
Institutional
Manufacturing
Residential
Retail
Exterior

bold = primary application
partial fade = minimal application
fade = unlikely application

Halogen Mains Voltage (120V)

3

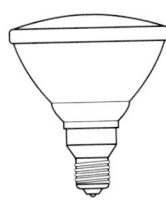

Halogen 60PAR38/H/VWFL

Uses

▶Soft general lighting
▶Wallwashing (see 11–119)
▶Medium art with short throws
▶Short throws
▶Available in 120V

Specs

Beam spread/60°
Center beam
  candlepower/1000
Life/3000 hrs
▶**Philips** (wide flood) (26385–5)

Design Tips

✔One light best for medium art and where mounting height for light is 10' or less.
✔Good for soft downlighting— best with sconces and/or art accents and/or wallwashing to avoid cave effect.
✔Consider halogen low voltage 50MR16/VWFL or 50MR16/ C/VWFL.

Performance Sketches

[All data outlined here is based on information from boldfaced manufacturer's published data for 120V version at time of manuscript preparation. Scale is ¼ inch equals 1 foot.]

Art shown at 3' by 4' and centered 5'–6" AFF. Wall grid on 6" by 6" centers and scaled at ¼"=1'.

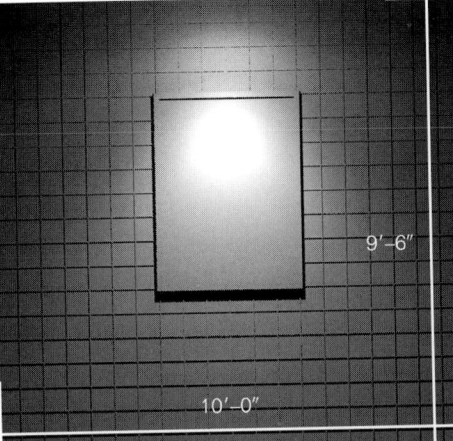

9'–6"

10'–0"

Lamp: 60PAR38/H/VWFL
Location: 2'–0" from wall
Center beam Aiming: 1° (at point 0'–0" AFF)
Luminaire: Monopoint or Recessed Adjustable
Illuminance: 11 fc, avg, maintained (30 fc max)

Lamp: 60PAR38/H/VWFL
Location: 2'–6" from wall
Center beam Aiming: 16° (at point 1'–0" AFF)
Luminaire: Monopoint or Recessed Adjustable
Illuminance: 14 fc, avg, maintained (29 fc max)

Lamp: 60PAR38/H/VWFL
Location: 3'–0" from wall
Center beam Aiming: 22° (at point 2'–0" AFF)
Luminaire: Monopoint or Recessed Adjustable
Illuminance: 15 fc, avg, maintained (26 fc max)

Application Key

Commercial
Gallery
Hospitality
Institutional
Manufacturing
Residential
Retail
Exterior

bold = primary application
partial fade = minimal application
fade = unlikely application

Halogen Main Voltage (120V)

Halogen/PAR38

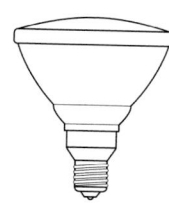

Net Addresses/PAR38 Track and Monopoints

http://www.bartcolighting.com/index2.html
http://www.thomasltg.com/
http://www.con-techlighting.com/
http://www.cooperlighting.com/
http://www.hubbell-ltg.com/products.htm#Down&Track
http://www.danalite.com/
http://www.kramerlighting.com/
http://www.lightingservicesinc.com/
http://www.lightolier.com/
http://www.lithonia.com/
http://www.prescolite.com/
http://www.tslight.com/

CONNECT FOR MORE

Net Addresses/PAR38 Recessed

http://www.thomasltg.com/
http://www.con-techlighting.com/
http://www.cooperlighting.com/
http://www.hubbell-ltg.com/products.htm#Down&Track
http://www.danalite.com/
http://www.kramerlighting.com/
http://www.lightolier.com/
http://www.lithonia.com/
http://www.prescolite.com/

CONNECT FOR MORE

Net Addresses/Exterior Accent

http://www.bega-us.com/home/home.html
http://www.hadcolighting.com/html/bronzelite.htm
http://www.hydrel.com
http://www.kimlighting.com/ingrd1.html

CONNECT FOR MORE

Halogen Mains Voltage (120V)

3

Halogen 75PAR38/H/NSP

Uses
▶ Architectural detailing
▶ Large art with long throws
▶ Sculpture with long throws
▶ Long throws
▶ Available in 120V and 130V versions

Specs
Beam spread/10°
Center beam candlepower/16500
Life/2500 hrs
▶ **GE** (14751)
▶ Osram Sylvania (spot)
▶ Philips (spot)

Design Tips
✔ One light best for large art and where mounting height for light is 19' or less.
✔ These are strong accents and deserve judicious use. Light may degrade artwork rapidly and/or significantly: consider UV and IR filters or limit exposure time.
✔ Excellent for glass pieces and floral centerpieces and where mounting height for light is 19' or less.
✔ Consider halogen low voltage 50MR16/C/CG/NSP or halogen IR mains voltage 50PAR38/HIR/NSP or 60PAR38/HIR/NSP.

Application Key

Commercial
Gallery
Hospitality
Institutional
Manufacturing
Residential
Retail
Exterior

bold = primary application
partial fade = minimal application
fade = unlikely application

Performance Sketches

[All data outlined here is based on information from boldfaced manufacturer's published data for 120V version at time of manuscript preparation. Scale is ¼ inch equals 1 foot.]

Lamp: 75PAR38/H/NSP
Location: 5'–0" from wall
Center beam Aiming: 20° (at point 4'–6" AFF)
Luminaire: Monopoint or Recessed Adjustable
Illuminance: 12 fc, avg, maintained (31 fc max)

Lamp: 75PAR38/H/NSP
Location: 6'–0" from wall
Center beam Aiming: 24° (at point 4'–6" AFF)
Luminaire: Monopoint or Recessed Adjustable
Illuminance: 13 fc, avg, maintained (33 fc max)

18'–0"

10'–0"

Halogen75PAR38/H/NSP

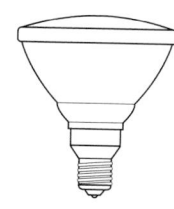

Lamp: 75PAR38/H/NSP
Location: 7'–0" from wall
Center beam Aiming: 27° (at point 4'–6" AFF)
Luminaire: Monopoint or Recessed Adjustable
Illuminance: 13 fc, avg, maintained (34 fc max)

Art shown at 4' by 5' and
centered 5'–6" AFF. Wall grid on
6" by 6" centers and scaled at
¼"=1'.

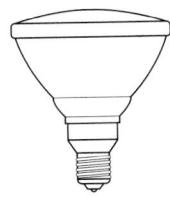

Halogen 75PAR38/H/SP

Uses
▶ Architectural detailing
▶ Medium art with long throws
▶ Sculpture with long throws
▶ Long throws
▶ Available in 120V

Specs
Beam spread/12°
Center beam
  candlepower/11000
Life/2500 hrs
▶ **Osram Sylvania** (wide spot)
  (14510)

Design Tips
✔ One light best for medium art and where mounting height for light is 16' or less.
✔ These are strong accents and deserve judicious use. Light may degrade artwork rapidly and/or significantly: consider UV and IR filters or limit exposure time.
✔ Good for small to moderate sculpture and where mounting height for light is 16' or less.
✔ Consider halogen low voltage 50MR16/SP, 50MR16/AL/SP or 50MR16/AL/CG/SP.

Performance Sketches
[All data outlined here is based on information from boldfaced manufacturer's published data for 120V version at time of manuscript preparation. Scale is ¼ inch equals 1 foot.]

Lamp: 75PAR38/H/SP
Location: 4'–3" from wall
Center beam Aiming: 20° (at point 3'–6" AFF)
Luminaire: Monopoint or Recessed Adjustable
Illuminance: 16 fc, avg, maintained (29 fc max)

15'–0"

10'–0"

Lamp: 75PAR38/H/SP
Location: 5'–3" from wall
Center beam Aiming: 25° (at point 3'–6" AFF)
Luminaire: Monopoint or Recessed Adjustable
Illuminance: 17 fc, avg, maintained (30 fc max)

Application Key

Commercial
Gallery
Hospitality
Institutional
Manufacturing
Residential
Retail
Exterior

bold = primary application
partial fade = minimal application
fade = unlikely application

Art shown at 3' by 4' and centered 5'–6" AFF. Wall grid on 6" by 6" centers and scaled at ¼"=1'.

Lamp: 75PAR38/H/SP
Location: 6'–3" from wall
Center beam Aiming: 31° (at point 4'–6" AFF)
Luminaire: Monopoint or Recessed Adjustable
Illuminance: 20 fc, avg, maintained (38 fc max)

Halogen Mains Voltage (120V)

Halogen 75PAR38/H/NFL26°

Performance Sketches

[All data outlined here is based on information from boldfaced manufacturer's published data for 120V version at time of manuscript preparation. Scale is ¼ inch equals 1 foot.]

Lamp: 75PAR38/H/NFL
Location: 3'–0" from wall
Center beam Aiming: 18° (at point 3'–6" AFF)
Luminaire: Monopoint or Recessed Adjustable
Illuminance: 12 fc, avg, maintained (29 fc max)

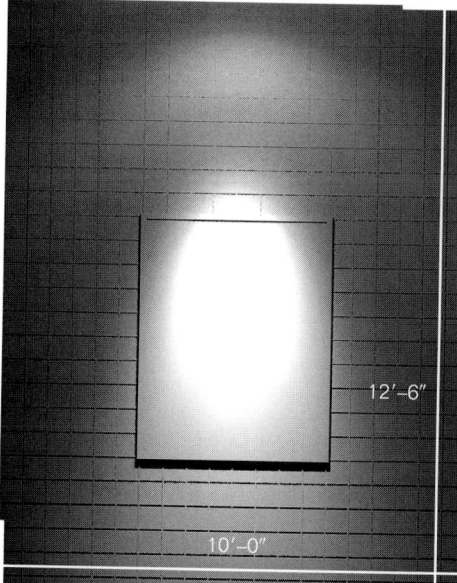

12'–6"

10'–0"

Lamp: 75PAR38/H/NFL
Location: 4'–0" from wall
Center beam Aiming: 25° (at point 4'–0" AFF)
Luminaire: Monopoint or Recessed Adjustable
Illuminance: 14 fc, avg, maintained (28 fc max)

Uses
▶ Architectural accenting
▶ Feature downlighting
▶ Large art with medium throws
▶ Sculpture with medium throws
▶ Medium throws
▶ Available in 120V and 130V

Specs
Beam spread/26°
Center beam candlepower/ 4250
Life/2500 hrs
▶ **GE** (14748)
▶ Philips (flood)

Design Tips
✓ One light best for large art where mounting height for light is 13' or less.
✓ Great for moderate to large sculpture where mounting height for light is 13' or less.
✓ More grazing light creates prominent shadows.
✓ Consider halogen low voltage 65MR16/NFL, 65MR16/C/ NFL or 50MR16/C/CG/NFL or halogen IR mains voltage 60PAR38/HIR/NFL25°.

Art shown at 4' by 5' and centered 5'–6" AFF. Wall grid on 6" by 6" centers and scaled at ¼"=1'.

Lamp: 75PAR38/H/NFL
Location: 5'–0" from wall
Center beam Aiming: 30° (at point 4'–0" AFF)
Luminaire: Monopoint or Recessed Adjustable
Illuminance: 14 fc, avg, maintained (26 fc max)

Application Key

Commercial
Gallery
Hospitality
Institutional
Manufacturing
Residential
Retail
Exterior

bold = primary application
partial fade = minimal application
fade = unlikely application

Halogen Mains Voltage (120V)

3

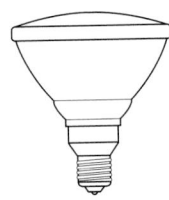

Halogen 75PAR38/H/NFL30°

Performance Sketches

[All data outlined here is based on information from boldfaced manufacturer's published data for 120V version at time of manuscript preparation. Scale is ¼ inch equals 1 foot.]

Uses
▶ Architectural accenting
▶ Feature downlighting
▶ Large art with medium throws
▶ Sculpture with medium throws
▶ Medium throws
▶ Available in 120V and 130V

Specs
Beam spread/30°
Center beam candlepower/ 3150
Life/2500 hrs
▶ **Osram Sylvania** (flood) (14513)

Design Tips
✔ One light best for large art where mounting height for light is 13' or less.
✔ Great for moderate to large sculpture where mounting height for light is 13' or less.
✔ More grazing light creates prominent shadows.
✔ Consider halogen IR mains voltage 60PAR38/HIR/NFL 28°.

Lamp: 75PAR38/H/NFL
Location: 3'–0" from wall
Center beam Aiming: 18° (at point 3'–6" AFF)
Luminaire: Monopoint or Recessed Adjustable
Illuminance: 14 fc, avg, maintained (34 fc max)

12'–6"

10'–0"

Lamp: 75PAR38/H/NFL
Location: 4'–0" from wall
Center beam Aiming: 25° (at point 4'–0" AFF)
Luminaire: Monopoint or Recessed Adjustable
Illuminance: 15 fc, avg, maintained (33 fc max)

Application Key

Commercial
Gallery
Hospitality
Institutional
Manufacturing
Residential
Retail
Exterior

bold = primary application
partial fade = minimal application
fade = unlikely application

Art shown at 4' by 5' and centered 5'–6" AFF. Wall grid on 6" by 6" centers and scaled at ¼"=1'.

Lamp: 75PAR38/H/NFL
Location: 5'–0" from wall
Center beam Aiming: 30° (at point 4'–0" AFF)
Luminaire: Monopoint or Recessed Adjustable
Illuminance: 16 fc, avg, maintained (29 fc max)

Halogen Mains Voltage (120V)

Halogen75PAR38/H/VWFL

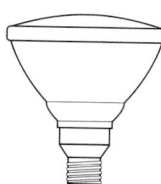

Performance Sketches

[All data outlined here is based on information from boldfaced manufacturer's published data for 130V version at time of manuscript preparation. Data and sketches shown are corrected for 120V operation. Scale is ¼ inch equals 1 foot.]

Art shown at 3' by 4' and centered 5'–6" AFF. Wall grid on 6" by 6" centers and scaled at ¼"=1'.

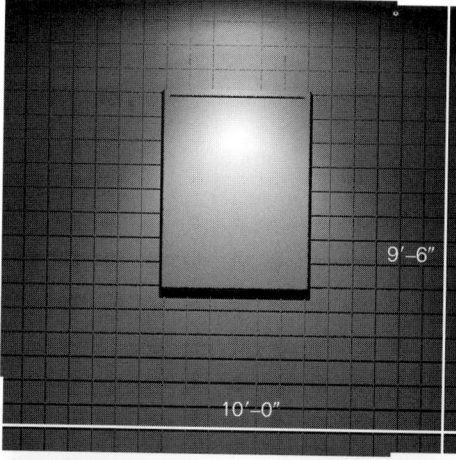
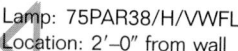

9'–6"

10'–0"

Lamp: 75PAR38/H/VWFL
Location: 2'–0" from wall
Center beam Aiming: 3° (at point 0'–0" AFF)
Luminaire: Monopoint or Recessed Adjustable
Illuminance: 5 fc, avg, maintained (11 fc max)

Lamp: 75PAR38/H/VWFL
Location: 2'–6" from wall
Center beam Aiming: 16° (at point 1'–0" AFF)
Luminaire: Monopoint or Recessed Adjustable
Illuminance: 9 fc, avg, maintained (16 fc max)

Lamp: 75PAR38/H/VWFL
Location: 3'–0" from wall
Center beam Aiming: 22° (at point 2'–0" AFF)
Luminaire: Monopoint or Recessed Adjustable
Illuminance: 9 fc, avg, maintained (15 fc max)

Uses
▶ Soft general lighting
▶ Wallwashing (see 11–121)
▶ Medium art with short throws
▶ Short throws
▶ Only available in 130V

Specs
Beam spread/55°
Center beam candlepower/930 (at 120V operation)
Life/5000 hrs (at 120V operation)
Wattage/66 (at 120V operation)
▶ **Osram Sylvania** (14517)

Design Tips
✔ One light best for medium art and where mounting height for light is 10' or less.
✔ Good for soft downlighting— best with sconces and/or art accents and/or wallwashing to avoid cave effect.
✔ Although life is longer, efficacy and intensity are reduced and color temperature is lower (warmer) with 130V lamps operated at 120V.
✔ Consider halogen low voltage 50MR16/C/CG/VWFL.
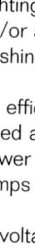

Application Key

Commercial
Gallery
Hospitality
Institutional
Manufacturing
Residential
Retail
Exterior

bold = primary application
partial fade = minimal application
fade = unlikely application

3

3–103 ▲

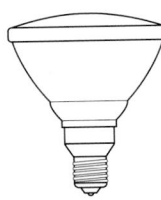

Halogen 90PAR38/H/NSP

Uses
▶Strong architectural detailing
▶Large art with long throws
▶Sculpture with long throws
▶Long throws
▶Available in 120V and 130V versions

Specs
Beam spread/10°
Center beam candlepower/17750
Life/2500 hrs
▶GE
▶**Osram Sylvania** (14586)

Design Tips
✓One light best for large art and where mounting height for light is 19′ or less.
✓These are strong accents and deserve judicious use. Light may degrade artwork rapidly and/or significantly: consider UV and IR filters or limit exposure time.
✓Excellent for glass pieces and floral centerpieces and where mounting height for light is 19′ or less.
✓Good for dimmable downlighting in ceiling heights of 30′ to 40′. [Recognize relamping issues at such heights—lift use or top relamping is necessary]
✓Consider halogen IR mains voltage 60PAR38/HIR/NSP or ceramic metal halide 39PAR20/CMH/3K/NSP.

Application Key

Commercial
Gallery
Hospitality
Institutional
Manufacturing
Residential
Retail
Exterior

bold = primary application
partial fade = minimal application
fade = unlikely application

Performance Sketches

[All data outlined here is based on information from boldfaced manufacturer's published data for 120V version at time of manuscript preparation. Scale is ¼ inch equals 1 foot.]

Lamp: 90PAR38/H/NSP
Location: 5′–0″ from wall
Center beam Aiming: 20° (at point 4′–6″ AFF)
Luminaire: Monopoint or Recessed Adjustable
Illuminance: 17 fc, avg, maintained (35 fc max)

Lamp: 90PAR38/H/NSP
Location: 6′–0″ from wall
Center beam Aiming: 24° (at point 4′–6″ AFF)
Luminaire: Monopoint or Recessed Adjustable
Illuminance: 18 fc, avg, maintained (36 fc max)

Halogen Mains Voltage (120V)

Halogen 90PAR38/H/NSP

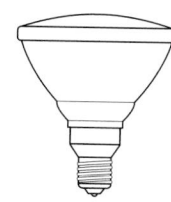

Lamp: 90PAR38/H/NSP
Location: 7'–0" from wall
Center beam Aiming: 27° (at point 4'–6" AFF)
Luminaire: Monopoint or Recessed Adjustable
Illuminance: 18 fc, avg, maintained (38 fc max)

Art shown at 4' by 5' and centered 5'–6" AFF. Wall grid on 6" by 6" centers and scaled at ¼"=1'.

Halogen
Mains Voltage (120V)

3

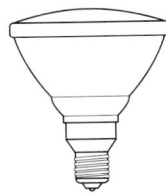

Halogen 90PAR38/H/SP

Uses
▶ Architectural detailing
▶ Large art with long throws
▶ Sculpture with long throws
▶ Long throws
▶ Available in 120V and 130V

Specs
Beam spread/12°
Center beam
   candlepower/13250
Life/2500 hrs
▶ **Osram Sylvania** (14580)
▶ Philips

Design Tips
✔ One light best for large art and where mounting height for light is 16' or less.
✔ These are strong accents and deserve judicious use. Light may degrade artwork rapidly and/or significantly: consider UV and IR filters or limit exposure time.
✔ Good for glass pieces and floral centerpieces and where mounting height for light is 16' or less.
✔ See 75PAR30S/H/NSP for similar results at fewer watts).
✔ Consider halogen low voltage 73MR16/SP.

Performance Sketches
[All data outlined here is based on information from boldfaced manufacturer's published data for 120V version at time of manuscript preparation. Scale is ¼ inch equals 1 foot.]

Lamp: 90PAR38/H/SP
Location: 4'–3" from wall
Center beam Aiming: 21° (at point 4'–0" AFF)
Luminaire: Monopoint or Recessed Adjustable
Illuminance: 17 fc, avg, maintained (38 fc max)

Lamp: 90PAR38/H/SP
Location: 5'–3" from wall
Center beam Aiming: 26° (at point 4'–0" AFF)
Luminaire: Monopoint or Recessed Adjustable
Illuminance: 22 fc, avg, maintained (39 fc max)

15'–0"

10'–0"

Art shown at 4' by 5' and centered 5'–6" AFF. Wall grid on 6" by 6" centers and scaled at ¼"=1'.

Lamp: 90PAR38/H/SP
Location: 6'–3" from wall
Center beam Aiming: 31° (at point 4'–6" AFF)
Luminaire: Monopoint or Recessed Adjustable
Illuminance: 19 fc, avg, maintained (43 fc max)

Halogen
Main Voltage (120V)

Application Key

Commercial
Gallery
Hospitality
Institutional
Manufacturing
Residential
Retail
Exterior

bold = primary application
partial fade = minimal application
fade = unlikely application

3

Halogen 90PAR38/H/VWSP

Performance Sketches

[All data outlined here is based on information from boldfaced manufacturer's published data for 130V version at time of manuscript preparation. Data and sketches shown are corrected for 120V operation. Scale is ¼ inch equals 1 foot.]

Lamp: 90PAR38/H/VWSP
Location: 3'–0" from wall
Center beam Aiming: 18° (at point 3'–0" AFF)
Luminaire: Monopoint or Recessed Adjustable
Illuminance: 10 fc, avg, maintained (20 fc max)

Lamp: 90PAR38/H/VWSP
Location: 4'–0" from wall
Center beam Aiming: 24° (at point 3'–6" AFF)
Luminaire: Monopoint or Recessed Adjustable
Illuminance: 11 fc, avg, maintained (18 fc max)

12'–6"

10'–0"

Uses

►Architectural accenting
►Medium art with medium throws
►Sculpture with medium throws
►Medium throws
►Only available in 130V

Specs

Beam spread/20°
Center beam candlepower/
4685 (at 120V operation)
Life/5000 hrs (at 120V operation)
Wattage/79 (at 120V operation)
►**Osram Sylvania** (narrow flood)
(14601)

Design Tips

✔One light best for medium art and where mounting height for light is 13' or less.
✔Light may degrade artwork rapidly and/or significantly: consider UV and IR filters or limit exposure time.
✔More grazing light creates prominent shadows.
✔Although life is longer, efficacy and intensity are reduced and color temperature is lower (warmer) with 130V lamps operated at 120V.

Art shown at 3' by 4' and centered 5'–6" AFF. Wall grid on 6" by 6" centers and scaled at ¼"=1'.

Lamp: 90PAR38/H/VWSP
Location: 5'–0" from wall
Center beam Aiming: 30° (at point 4'–0" AFF)
Luminaire: Monopoint or Recessed Adjustable
Illuminance: 12 fc, avg, maintained (17 fc max)

Application Key

Commercial
Gallery
Hospitality
Institutional
Manufacturing
Residential
Retail
Exterior

bold = primary application
partial fade = minimal application
fade = unlikely application

Halogen Mains Voltage (120V)

3

Halogen 90PAR38/H/NFL

Uses
▶ Architectural accenting
▶ Feature downlighting
▶ Large art with medium throws
▶ Sculpture with medium throws
▶ Medium throws
▶ Available in 120V and 130V

Specs
Beam spread/29°
Center beam candlepower/ 4000
Life/2500 hrs
▶ GE (flood) (17451)
▶ Osram Sylvania (flood)
▶ Philips (flood)

Design Tips
✔ One light best for large art and where mounting height for light is 13′ or less.
✔ Good for moderate to larger sculpture where mounting height for light is 13′ or less.
✔ This lamp slightly floodier and less intense than lamp cited on previous page.
✔ Consider halogen IR mains voltage 60PAR38/HIR/NFL28° or ceramic metal halide 39PAR20/CMH/3K/NFL.

Performance Sketches

[All data outlined here is based on information from boldfaced manufacturer's published data for 120V version at time of manuscript preparation. Scale is ¼ inch equals 1 foot.]

Lamp: 90PAR38/H/NFL
Location: 3′–0″ from wall
Center beam Aiming: 18° (at point 3′–0″ AFF)
Luminaire: Monopoint or Recessed Adjustable
Illuminance: 15 fc, avg, maintained (34 fc max)

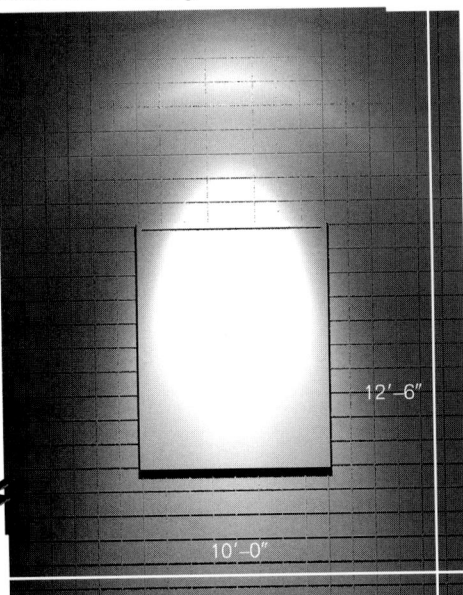

12′–6″

10′–0″

Lamp: 90PAR38/H/NFL
Location: 4′–0″ from wall
Center beam Aiming: 24° (at point 3′–6″ AFF)
Luminaire: Monopoint or Recessed Adjustable
Illuminance: 17 fc, avg, maintained (32 fc max)

Art shown at 4′ by 5′ and centered 5′–6″ AFF. Wall grid on 6″ by 6″ centers and scaled at ¼″=1′.

Lamp: 90PAR38/H/NFL
Location: 5′–0″ from wall
Center beam Aiming: 30° (at point 4′–0″ AFF)
Luminaire: Monopoint or Recessed Adjustable
Illuminance: 18 fc, avg, maintained (31 fc max)

Application Key

Commercial
Gallery
Hospitality
Institutional
Manufacturing
Residential
Retail
Exterior

bold = primary application
partial fade = minimal application
fade = unlikely application

Halogen Mains Voltage (120V)

Halogen 90PAR38/H/WFL

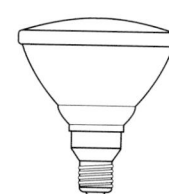

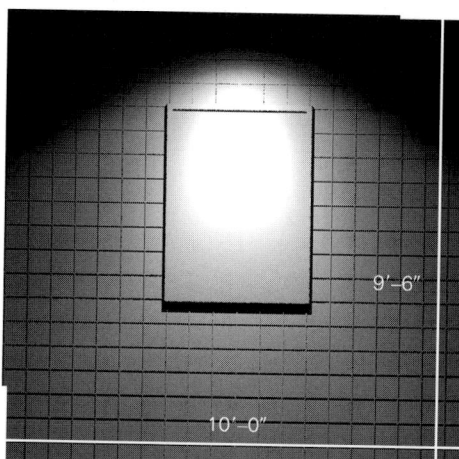

Performance Sketches

[All data outlined here is based on information from boldfaced manufacturer's published data for 130V version at time of manuscript preparation. Data and sketches shown are corrected for 120V operation. Scale is ¼ inch equals 1 foot.]

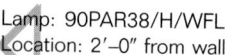

Art shown at 3' by 4' and centered 5'–6" AFF. Wall grid on 6" by 6" centers and scaled at ¼"=1'.

9'–6"

10'–0"

Lamp: 90PAR38/H/WFL
Location: 2'–0" from wall
Center beam Aiming: 3° (at point 0'–0" AFF)
Luminaire: Monopoint or Recessed Adjustable
Illuminance: 9 fc, avg, maintained (17 fc max)

Lamp: 90PAR38/H/WFL
Location: 2'–6" from wall
Center beam Aiming: 16° (at point 1'–0" AFF)
Luminaire: Monopoint or Recessed Adjustable
Illuminance: 14 fc, avg, maintained (26 fc max)

Lamp: 90PAR38/H/WFL
Location: 3'–0" from wall
Center beam Aiming: 22° (at point 2'–0" AFF)
Luminaire: Monopoint or Recessed Adjustable
Illuminance: 14 fc, avg, maintained (24 fc max)

Uses

► Soft general lighting
► Wallwashing (see 11–123)
► Medium art with short throws
► Short throws
► Only available in 130V

Specs

Beam spread/50°
Center beam candlepower/ 1170 (at 120V operation)
Life/5000 hrs (at 120V operation)
Wattage/79 (at 120V operation)
► **Osram Sylvania** (14602)

Design Tips

✔ One light best for medium art and where mounting height for light is 10' or less.
✔ Good for soft downlighting— best with sconces and/or art accents and/or wallwashing to avoid cave effect.
✔ Although life is longer, efficacy and intensity are reduced and color temperature is lower (warmer) with 130V lamps operated at 120V.
✔ Consider halogen IR mains voltage 60PAR38/HIR/WFL.

Application Key

Commercial
Gallery
Hospitality
Institutional
Manufacturing
Residential
Retail
Exterior

bold = primary application
partial fade = minimal application
fade = unlikely application

Halogen Main Voltage (120V)

3

Halogen 90PAR38/H/VWFL

Uses
▶ Soft general lighting
▶ Wallwashing (see 11–123)
▶ Medium art with short throws
▶ Short throws
▶ Available in 120V

Specs
Beam spread/60°
Center beam
 candlepower/1300
Life/2500 hrs
▶ **Philips** (wide flood) (23646–3)

Design Tips
✔ One light best for medium art and where mounting height for light is 10' or less.
✔ Good for soft downlighting—best with sconces and/or art accents and/or wallwashing to avoid cave effect..
✔ Consider halogen low voltage 50MR16/VWFL or halogen IR mains voltage 60PAR38/HIR/WFL.

Performance Sketches

[All data outlined here is based on information from boldfaced manufacturer's published data for 120V version at time of manuscript preparation. Scale is ¼ inch equals 1 foot.]

Art shown at 3' by 4' and centered 5'–6" AFF. Wall grid on 6" by 6" centers and scaled at ¼"=1'.

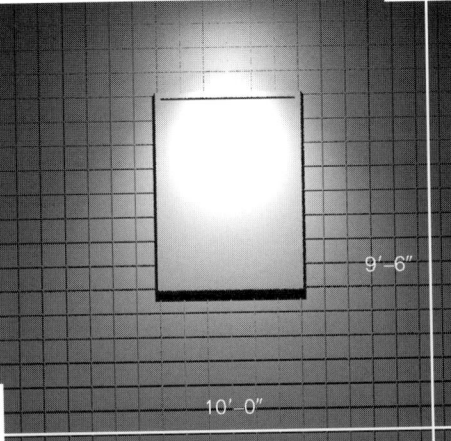

Lamp: 90PAR38/H/WFL
Location: 2'–0" from wall
Center beam Aiming: 3° (at point 0'–0" AFF)
Luminaire: Monopoint or Recessed Adjustable
Illuminance: 13 fc, avg, maintained (25 fc max)

Lamp: 90PAR38/H/WFL
Location: 2'–6" from wall
Center beam Aiming: 16° (at point 1'–0" AFF)
Luminaire: Monopoint or Recessed Adjustable
Illuminance: 18 fc, avg, maintained (35 fc max)

Lamp: 90PAR38/H/WFL
Location: 3'–0" from wall
Center beam Aiming: 22° (at point 2'–0" AFF)
Luminaire: Monopoint or Recessed Adjustable
Illuminance: 18 fc, avg, maintained (31 fc max)

Application Key

Commercial
Gallery
Hospitality
Institutional
Manufacturing
Residential
Retail
Exterior

bold = primary application
partial fade = minimal application
fade = unlikely application

Halogen Mains Voltage (120V)

Halogen 100PAR38/H/SP

Uses
▶Strong architectural detailing
▶Large art with long throws
▶Sculpture with long throws
▶Long throws
▶Available in 120V and 130V versions

Specs
Beam spread/11°
Center beam candlepower/17000
Life/2000 hrs
▶**GE** (17992)

Design Tips
✔One light best for large art and where mounting height for light is 19' or less.
✔These are strong accents and deserve judicious use. Light may degrade artwork rapidly and/or significantly: consider UV and IR filters or limit exposure time.
✔Excellent for glass pieces and floral centerpieces and where mounting height for light is 19' or less.
✔Good for dimmable downlighting in ceiling heights of 30' to 40'. (NOTE: recognize relamping issues at such heights—lift use or top relamping is necessary.)
✔Consider halogen low voltage 71MR16/C/NFL or 71MR16/C/CG/NFL or ceramic metal halide 39PAR20/CMH/3K/NSP.

Application Key

Commercial
Gallery
Hospitality
Institutional
Manufacturing
Residential
Retail
Exterior

bold = primary application
partial fade = minimal application
fade = unlikely application

Performance Sketches

[All data outlined here is based on information from boldfaced manufacturer's published data for 120V version at time of manuscript preparation. Scale is ¼ inch equals 1 foot.]

Lamp: 100PAR38/H/SP
Location: 5'–0" from wall
Center beam Aiming: 18° (at point 3'–0" AFF)
Luminaire: Monopoint or Recessed Adjustable
Illuminance: 12 fc, avg, maintained (21 fc max)

18'–0"

10'–0"

Lamp: 100PAR38/H/SP
Location: 6'–0" from wall
Center beam Aiming: 23° (at point 4'–0" AFF)
Luminaire: Monopoint or Recessed Adjustable
Illuminance: 13 fc, avg, maintained (25 fc max)

Halogen Main Voltage (120V)

Halogen 100PAR38/H/SP

Lamp: 100PAR38/H/SP
Location: 7'–0" from wall
Center beam Aiming: 27° (at point 4'–0" AFF)
Luminaire: Monopoint or Recessed Adjustable
Illuminance: 14 fc, avg, maintained (29 fc max)

Art shown at 4' by 5' and
centered 5'–6" AFF. Wall grid on
6" by 6" centers and scaled at
¼"=1'.

Halogen 100PAR38/H/NFL

Performance Sketches

[All data outlined here is based on information from boldfaced manufacturer's published data for 120V version at time of manuscript preparation. Scale is ¼ inch equals 1 foot.]

Uses
- ▶ Architectural accenting
- ▶ Feature downlighting
- ▶ Large art with medium throws
- ▶ Sculpture with medium throws
- ▶ Medium throws
- ▶ Available in 120V and 130V

Specs
Beam spread/27°
Center beam candlepower/ 4800
Life/2000 hrs
▶ **GE** (flood) (17986)

Design Tips
✔ One light best for large art and where mounting height for light is 13' or less.
✔ Good for moderate to larger sculpture where mounting height for light is 13' or less.
✔ Consider ceramic metal halide 39PAR20/CMH/3K/NFL.

Lamp: 100PAR38/H/NFL
Location: 3'–0" from wall
Center beam Aiming: 18° (at point 3'–0" AFF)
Luminaire: Monopoint or Recessed Adjustable
Illuminance: 17 fc, avg, maintained (35 fc max)

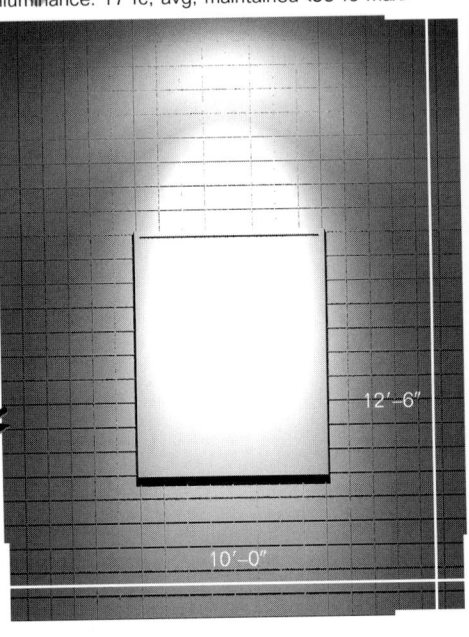

12'–6"
10'–0"

Lamp: 100PAR38/H/NFL
Location: 4'–0" from wall
Center beam Aiming: 24° (at point 3'–6" AFF)
Luminaire: Monopoint or Recessed Adjustable
Illuminance: 19 fc, avg, maintained (37 fc max)

Art shown at 4' by 5' and centered 5'–6" AFF. Wall grid on 6" by 6" centers and scaled at ¼"=1'.

Lamp: 100PAR38/H/NFL
Location: 5'–0" from wall
Center beam Aiming: 30° (at point 4'–0" AFF)
Luminaire: Monopoint or Recessed Adjustable
Illuminance: 19 fc, avg, maintained (33 fc max)

Application Key
Commercial
Gallery
Hospitality
Institutional
Manufacturing
Residential
Retail
Exterior

bold = primary application
partial fade = minimal application
fade = unlikely application

Halogen Mains Voltage (120V)

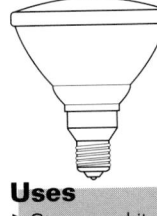

Halogen 120PAR38/H/NSP

Uses
▶ Strong architectural detailing
▶ Large art with very long throws
▶ Sculpture with very long throws
▶ Very long throws
▶ Available in 120V and 130V versions

Specs
Beam spread/9°
Center beam candlepower/25000
Life/3000 hrs
▶ **Osram Sylvania** (spot)
(14856)

Design Tips
✔ One light best for large art and where mounting height for light is 21′ or less.
✔ These are very strong accents and deserve judicious use. Light may degrade artwork rapidly and/or significantly— consider UV and IR filters or limit exposure time.
✔ Excellent for glass pieces and floral centerpieces and where mounting height for light is 21′ or less.
✔ Good for dimmable downlighting in ceiling heights of 30′ to 40′. (NOTE: recognize relamping issues at such heights—lift use or top relamping is necessary.)
✔ Consider ceramic metal halide 39PAR20/CMH/3K/NSP.

Application Key

Commercial
Gallery
Hospitality
Institutional
Manufacturing
Residential
Retail
Exterior

bold = primary application
partial fade = minimal application
fade = unlikely application

Halogen Mains Voltage (120V)

Performance Sketches

[All data outlined here is based on information from boldfaced manufacturer's published data for 120V version at time of manuscript preparation. Scale is ¼ inch equals 1 foot.]

Lamp: 120PAR38/H/NSP
Location: 6′–0″ from wall
Center beam Aiming: 21° (at point 4′–6″ AFF)
Luminaire: Monopoint or Recessed Adjustable
Illuminance: 18 fc, avg, maintained (35 fc max)

Lamp: 120PAR38/H/NSP
Location: 7′–0″ from wall
Center beam Aiming: 24° (at point 4′–6″ AFF)
Luminaire: Monopoint or Recessed Adjustable
Illuminance: 19 fc, avg, maintained (37 fc max)

20′–0″

10′–0″

Halogen 120PAR38/H/NSP

Lamp: 120PAR38/H/NSP
Location: 8′–0″ from wall
Center beam Aiming: 27° (at point 4′–6″ AFF)
Luminaire: Monopoint or Recessed Adjustable
Illuminance: 20 fc, avg, maintained (37 fc max)

Art shown at 4′ by 5′ and centered 5′–6″ AFF. Wall grid on 6″ by 6″ centers and scaled at ¼″=1′.

Halogen Mains Voltage (120V)

3

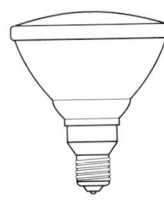

Halogen 120PAR38/H/NFL

Uses
▶ Architectural accenting
▶ Feature downlighting
▶ Large art with medium throws
▶ Sculpture with medium throws
▶ Medium throws
▶ Available in 120V and 130V

Specs
Beam spread/30°
Center beam candlepower/ 5000
Life/3000 hrs
▶ **Osram Sylvania** (flood) (14855)

Design Tips
✔ One light best for large art and where mounting height for light is 13′ or less.
✔ Good for moderate to larger sculpture where mounting height for light is 13′ or less.
✔ Consider ceramic metal halide 39PAR20/CMH/3K/NFL.

Performance Sketches

[All data outlined here is based on information from boldfaced manufacturer's published data for 120V version at time of manuscript preparation. Scale is ¼ inch equals 1 foot. **Note: At press time, actual photometry was unavailable for this lamp. Data were generated using a 90PAR38/H/NFL lamp and rerating candlepower—this is imprecise.**]

Lamp: 120PAR38/H/NFL
Location: 3′–0″ from wall
Center beam Aiming: 21° (at point 4′–0″ AFF)
Luminaire: Monopoint or Recessed Adjustable
Illuminance: 23 fc, avg, maintained (60 fc max)

Lamp: 120PAR38/H/NFL
Location: 4′–0″ from wall
Center beam Aiming: 27° (at point 4′–0″ AFF)
Luminaire: Monopoint or Recessed Adjustable
Illuminance: 24 fc, avg, maintained (52 fc max)

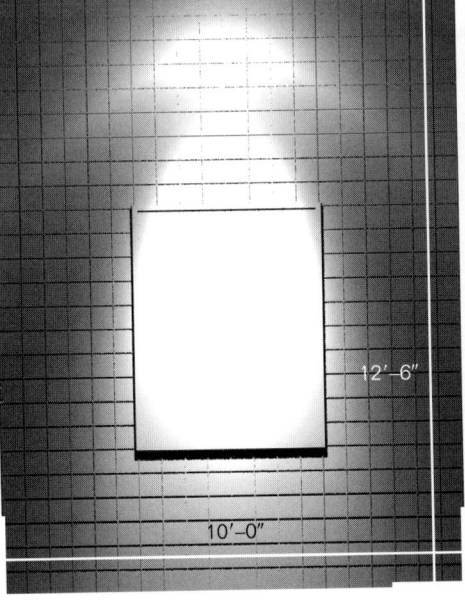

12′–6″

10′–0″

Art shown at 4′ by 5′ and centered 5′–6″ AFF. Wall grid on 6″ by 6″ centers and scaled at ¼″=1′.

Lamp: 120PAR38/H/NFL
Location: 5′–0″ from wall
Center beam Aiming: 32° (at point 4′–0″ AFF)
Luminaire: Monopoint or Recessed Adjustable
Illuminance: 25 fc, avg, maintained (48 fc max)

Application Key

Commercial
Gallery
Hospitality
Institutional
Manufacturing
Residential
Retail
Exterior

bold = primary application
partial fade = minimal application
fade = unlikely application

Halogen Main Voltage (120V)

Halogen 120PAR38/H/WFL

Performance Sketches

[All data outlined here is based on information from boldfaced manufacturer's published data for 120V version at time of manuscript preparation. Scale is ¼ inch equals 1 foot.]

Art shown at 3' by 4' and centered 5'–6" AFF. Wall grid on 6" by 6" centers and scaled at ¼"=1'.

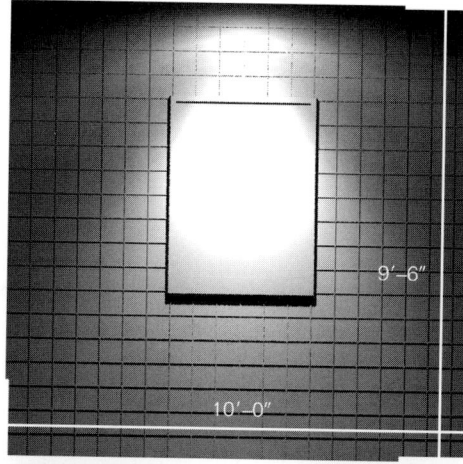

9'–6"

10'–0"

Lamp: 120PAR38/H/WFL
Location: 2'–0" from wall
Center beam Aiming: 6° (at point 0'–0" AFF)
Luminaire: Monopoint or Recessed Adjustable
Illuminance: 19 fc, avg, maintained (83 fc max)

Lamp: 120PAR38/H/WFL
Location: 2'–6" from wall
Center beam Aiming: 16° (at point 1'–0" AFF)
Luminaire: Monopoint or Recessed Adjustable
Illuminance: 25 fc, avg, maintained (84 fc max)

Lamp: 120PAR38/H/WFL
Location: 3'–0" from wall
Center beam Aiming: 22° (at point 2'–0" AFF)
Luminaire: Monopoint or Recessed Adjustable
Illuminance: 26 fc, avg, maintained (64 fc max)

Uses
▶ Soft general lighting
▶ Wallwashing (see 11–125)
▶ Medium art with short throws
▶ Short throws
▶ Available in 120V

Specs
Beam spread/50°
Center beam
  candlepower/2000
Life/3000 hrs
▶ **Osram Sylvania** (14594)

Design Tips
✔ One light best for medium art and where mounting height for light is 10' or less.
✔ Good for soft downlighting— best with sconces and/or art accents and/or wallwashing to avoid cave effect.

Application Key

Commercial
Gallery
Hospitality
Institutional
Manufacturing
Residential
Retail
Exterior

bold = primary application
partial fade = minimal application
fade = unlikely application

Halogen Main Voltage (120V)

3

▶See note about socket.

Halogen 250PAR38/H/SP

Uses
▶Very strong architectural detailing
▶Large art with very long throws
▶Sculpture with very long throws
▶Very long throws
▶Available in 120V

Specs
Beam spread/11°
Center beam candlepower/40000
Life/4200 hrs
▶**GE** (spot) (23719)

Design Tips
✔One light best for large art and where mounting height for light is less than 26'.
✔These are very strong accents and deserve judicious use. Light may degrade artwork rapidly and/or significantly: consider UV and IR filters or limit exposure time.
✔Excellent for glass pieces and floral centerpieces and where mounting height for light is less than 26'.
✔Good for dimmable downlighting in ceiling heights of 30' to 40'. (NOTE: recognize relamping issues at such heights—lifts are a necessity or top relamping is desired.)
✔Consider ceramic metal halide 39PAR30L/CMH/3K/NSP.

Application Key

Commercial
Gallery
Hospitality
Institutional
Manufacturing
Residential
Retail
Exterior

bold = primary application
partial fade = minimal application
fade = unlikely application

Halogen Mains Voltage (120V)

3-120

Performance Sketches

[All data outlined here is based on information from boldfaced manufacturer's published data for 120V version at time of manuscript preparation. Scale is ¼ inch equals 1 foot.]

Lamp: 250PAR38/H/SP
Location: 8'–0" from wall
Center beam Aiming: 21° (at point 4'–6" AFF)
Luminaire: Monopoint or Recessed Adjustable
Illuminance: 25 fc, avg, maintained (39 fc max)

Lamp: 250PAR38/H/SP
Location: 9'–0" from wall
Center beam Aiming: 24° (at point 4'–6" AFF)
Luminaire: Monopoint or Recessed Adjustable
Illuminance: 26 fc, avg, maintained (40 fc max)

25'–0"

10'–0"

▶ **Safety notes: (1)** This lamp operates at very high temperatures and requires a ceramic socket. Verify with luminaire manufacturer that luminaire is fitted with ceramic socket and is rated for this lamp. **(2)** Always check with luminaire manufacturer on allowable luminaire spacing distances for any luminaires using any lamp—from architectural surfaces/materials and from other luminaires. For example, to meet UL listing requirements, luminaires may have to be spaced at least 3 or 4 feet from walls (face of wall to closest face/part of luminaire) and/or at least 3 or 4 feet from other luminaires (outside edge of one luminaire to outside edge of another luminaire).

Halogen250PAR38/H/SP

▶See note about socket

Lamp: 250PAR38/H/SP
Location: 10'–0" from wall
Center beam Aiming: 26° (at point 4'–6" AFF)
Luminaire: Monopoint or Recessed Adjustable
Illuminance: 27 fc, avg, maintained (41 fc max)

Art shown at 4' by 5' and centered 5'–6" AFF. Wall grid on 6" by 6" centers and scaled at ¼"=1'.

Halogen Mains Voltage (120V)

3

▶ See note about socket.

Halogen 250PAR38/H/NFL

Performance Sketches

[All data outlined here is based on information from boldfaced manufacturer's published data for 120V version at time of manuscript preparation. Scale is ¼ inch equals 1 foot.]

Uses
▶ Architectural accenting
▶ Extra large art with long throws
▶ Sculpture with long throws
▶ Long throws
▶ Available in 120V

Specs
Beam spread/32°
Center beam candlepower/ 8000
Life/4200 hrs
▶**GE** (flood) (23718)

Design Tips
✔ One light best for large art and where mounting height for light is 19' or less.
✔ Good for moderate to larger sculpture where mounting height for light is 19' or less.
✔ Good for dimmable downlighting in ceiling heights of 20' to 25'. (NOTE: recognize relamping issues at such heights—lifts are a necessity or top relamping is desired.)
✔ Consider ceramic metal halide 39PAR30L/CMH/3K/NFL.

Lamp: 250PAR38/H/NFL
Location: 5'–0" from wall
Center beam Aiming: 17° (at point 2'–0" AFF)
Luminaire: Monopoint or Recessed Adjustable
Illuminance: 13 fc, avg, maintained (25 fc max)

18'–0"

10'–0"

Lamp: 250PAR38/H/NFL
Location: 6'–0" from wall
Center beam Aiming: 22° (at point 3'–0" AFF)
Luminaire: Monopoint or Recessed Adjustable
Illuminance: 14 fc, avg, maintained (25 fc max)

▶ **Safety notes:** (1) This lamp operates at very high temperatures and requires a ceramic socket. Verify with luminaire manufacturer that luminaire is fitted with ceramic socket and is rated for this lamp. (2) Always check with luminaire manufacturer on allowable luminaire spacing distances for any luminaires using any lamp—from architectural surfaces/materials and from other luminaires. For example, to meet UL listing requirements, luminaires may have to be spaced at least 3 or 4 feet from walls (face of wall to closest face/part of luminaire) and/or at least 3 or 4 feet from other luminaires (outside edge of one luminaire to outside edge of another luminaire).

Application Key

Commercial
Gallery
Hospitality
Institutional
Manufacturing
Residential
Retail
Exterior

bold = primary application
partial fade = minimal application
fade = unlikely application

Halogen Mains Voltage (120V)

3

Halogen250PAR38/H/NFL

▼See note about socket

Lamp: 250PAR38/H/NFL
Location: 7'–0" from wall
Center beam Aiming: 26° (at point 3'–6" AFF)
Luminaire: Monopoint or Recessed Adjustable
Illuminance: 15 fc, avg, maintained (24 fc max)

Art shown at 6' by 8' and
centered 5'–6" AFF. Wall grid on
6" by 6" centers and scaled at
¼"=1'.

Halogen Main Voltage (120V)

3

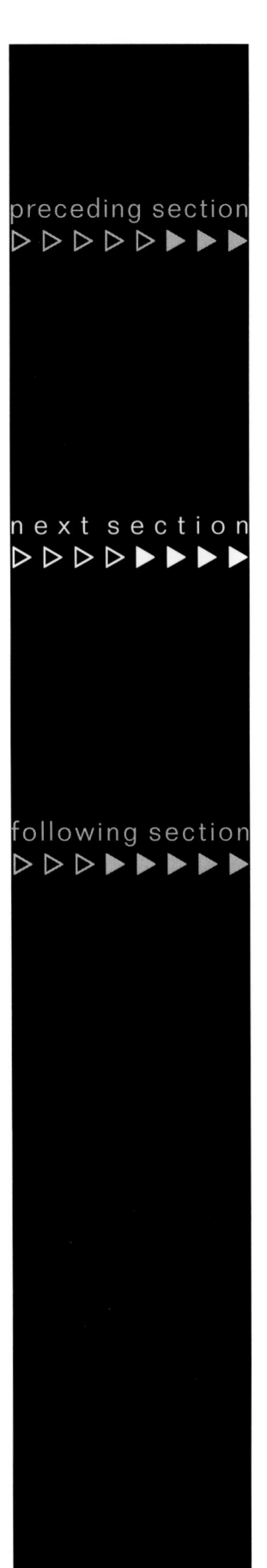

Halogen Lamps
Low Voltage (12V)

4

This section covers 12–volt (low voltage) halogen incandescent lamps. For purposes of this reference book, the 12–volt lamps discussed here are considered as more functional lamps (as opposed to decorative lamps) for architectural lighting applications. Low voltage decorative lamps are addressed in Section 11. Halogen lamps' inherent efficacy combined with the advantages of miniaturization resulting from low voltage operation make for an efficacious, easily dimmed, small, relatively–long life light source. There are also low voltage halogen infrared lamps (see *HIR Lamps—Low Voltage (12V)* in Section 6) which are more efficacious than the standard low voltage halogen lamps outlined here.

Halogen Lamps
Low Voltage (12V)

Low voltage halogen lamps are a result of marrying halogen technology and very low voltages (typically 12V and 24V). The halogen technology, more completely discussed in Section 3, offers excellent efficacy and life advantages over standard tungsten incandescent lamps. Operating lamps at low voltages allow for miniaturization of the filament and halogen capsule, which allows for much better optical control of the light, thereby leading to excellent beam control and very efficient focusing.

Low voltage halogen lamps are used in residential and hospitality projects as accents for artwork or architectural detailing; as general lighting; or as task lighting.

Halogen low voltage lamp color characteristics
Color of light (color temperature) of low voltage halogen lamps is "crisp white" like their mains voltage counterparts. These lamps are very appropriate for contemporary interiors. When used with straw color filters or when dimmed, however, they are also appropriate in more traditional interiors. Indeed, given their typically small–aperture housings, these lamps may be more appropriate for traditional interiors than their mains voltage counterparts.

Color rendering of the low voltage halogen lamps is excellent— typically rated a perfect 100. As such, low voltage MR16 halogen lamps are excellent for highlighting art.

Halogen low voltage dimming
Dimming low voltage halogen lamps can increase lamp life and also will warm the color of light somewhat, although type of dimmer selection is important (see *Low voltage transformers* below). Dimming helps to "take the edge" off the light and makes halogen an efficient alternative in more traditional residential and hospitality settings. Since most low voltage halogen lamps have small filaments which are surrounded by at least a capsule and perhaps even a "cover glass" lens, there is little hum from dimming that is more common with standard tungsten filament incandescent lamps. The sole exception is the low voltage halogen PAR36 lamp which exhibits significant filament singing or hum when dimmed.

The halogen cycle, for which the lamp is named, results in improved efficacy and long life in halogen lamps over standard tungsten incandescent lamps. This same cycle, however, can be "short circuited" with continual dimming. Hence, where dimming is em-

Halogen Lamps
Low Voltage (12V)

Table 4.1 Low Voltage (12V) Halogen Lamps

Low Voltage Halogen Lamp	Use/Replaces	Lamp Profile
T3Q/G4 (20 watts or less)	New • Standard Incandescent 12V R14 • Standard Incandescent 120V R14 • Standard Incandescent R20	
T4Q/GY6.35 (35 watts or more)	New • Standard Incandescent 12V R14 • Standard Incandescent 120V R14 • Standard Incandescent R20 • Standard Incandescent R30 • Standard Incandescent R40	
MR11	New • Standard Incandescent 12V R14 • Standard Incandescent 120V R14 • Standard Incandescent R20	
MR16	New • Standard Incandescent R20 • Standard Incandescent R30 • Standard Incandescent 12V PAR36 • Standard Incandescent R40	
PAR36	New or Retrofit • Standard Incandescent 12V PAR36	

ployed, occasionally the lamps must be operated at full bright to allow the halogen cycle to operate.

Low voltage transformers

For most applications, low voltage lamps are powered by transformers—devices which transform the mains voltage (e.g., 120 volts) to the lamp's operating voltage (e.g., 12 volts). There are two types of transformers—electromagnetic and electronic. Both are essentially "black box" devices which contain wire coils and/or electronic components sealed away from view. Transformers require additional power—typically 3 to 5 watts per lamp.

Electronic transformers are typically smaller than electromagnetic versions. Where dimming is involved, however, care must be taken in specifying a dimmer which is compatible with the transformer—

Where accent lighting is needed, consider low voltage halogen infrared (HIR) MR16 (Section 6), HIR PAR lamps (Section 5) or ceramic metal halide (CMH) PAR lamps (Section 7).

Halogen Lamps
Low Voltage (12V)

and electromagnetic transformers are consistently more reliable than electronic versions when dimming is involved. This aspect cannot be overemphasized. For electromagnetic transformers, use dimmers which are designed specifically to operate electromagnetic transformers. Similarly, for electronic transformers, use only dimmers rated for control of electronic transformers. Otherwise, expect significant system dysfunction—from poor dimming range and non–smooth dimming to humming, short lamp life and/or short transformer life.

Low voltage transformers can be designed to operate a single lamp and therefore are small enough to be integral to the luminaire. For track fittings, however, the transformers are usually larger than the lamp and socket assembly. For these applications, it may be desirable to specify a larger transformer to operate several lamps on a track which is rated for low voltage operation. These larger transformers need to be remotely located, but within proximity of the lamps—otherwise voltage drop (the loss of voltage over length of wire run) is sufficient to operate the lamps in a permanently dimmed (and lower output) state. Further, these remote transformers need to be located in accessible, appropriately ventilated and sound isolated areas. Access is necessary for maintenance and/or replacement. Ventilation is required to limit the amount of ambient heat in the area of the transformer (the higher the ambient temperature, typically the less efficient the transformer operation and the more likely the transformer will fail prematurely) and to prevent overheating of the transformer. Sound isolation is a necessity, as the larger the transformer the more likely humming or buzzing will occur. In any event, installation must conform to manufacturer's instructions and code requirements.

Low voltage halogen dimming

❶Increases lamp life—doubling it, but only if lamps are brightened to full output occasionally.

❶Yellows color of light—more cozy, intimate appearance.

❶Results in little or no filament hum (which is common in standard tungsten filament incandescent) EXCEPT in PAR36/H lamps (which do hum or "sing" when dimmed)

❶Electromagnetic transformers are historically more reliable and typically quieter than electronic transformers counterparts.

❶Only successful when dimmer switch is matched to the transformer type—electromagnetic or electronic.

Low voltage halogen lamp hazards

The hazards of halogen lamps have been highly publicized in recent years. The very reason a halogen lamp operates so efficiently is the

Halogen Lamps
Low Voltage (12V)

same reason it can be a hazard—the high–temperature, pressurized operation of the filament. While halogen low voltage lamps are primarily available in wattages up to 100 watts—much less than the 300– and 500–watt mains voltage linear halogen lamps used in many torchieres—their small size makes them a serious hazard. Low voltage halogen lamps' flammability hazard due to very infrequent "non–passive failure" (exploding) can be reduced with tempered cover glass filters. Additional precaution should be taken by maintaining appropriate distances between lamps and any materials and surfaces.

Tempered cover glass filters help contain any hot filament bits and capsule shards should the lamp fail violently. Common sense should be exercised when placing lamps/luminaires—this follows for any lamp type. Spot lamps of higher wattage act similarly to focusing the sun's rays through a magnifying glass—surface materials within a few inches or even a foot or so of a high wattage spot lamp could ignite depending on the type of material and the ambient conditions. This suggests that millwork–integrated lighting must be limited in wattage and beam spread and that installation must conform to lamp and luminaire manufacturers' instructions and code requirements.

Ultraviolet light output is higher in low voltage halogen lamps than in standard tungsten filament incandescent lamps. Artwork requires protection from this UV, otherwise degradation of the pigments, colors and/or base media (e.g., paper) may be significant and irreparable over a fairly short period of time. Similarly, people require protection from this UV. Tempered glass lenses help reduce UV light output from halogen lamps. For sensitive art, however, UV reduction lenses or filters should be used with halogen lamps.

Low voltage halogen hazards

○ Lamps may fail violently—exploding. Only use luminaires that are UL–listed and labeled and fitted with tempered glass lenses OR use lamps with integral cover glass.
○ Portable low voltage halogen luminaires should be well balanced and heavily weighted to prevent tipping.
○ Portable luminaires should be planned sparingly and users should be warned of fire and burn hazards.
○ Lamps are extremely hot—sufficient distance from people, furnishings and architectural elements is a must.
○ The spottier (more focused) the beam spread, the greater the distance necessary to avoid rapid surface material degradation and/or fire.
○ Ultraviolet light from low voltage halogen lamps can be damaging to humans and artwork—use UV filters.
○ Keep flammable materials/surfaces sufficient distance from low voltage halogen lamps and luminaires.

When to use low voltage halogen

Low voltage halogen lamps are appropriate on a variety of projects. Commercial, hospitality, residential and retail spaces can all be appropriate applications for low voltage halogen lamps. In commer-

Halogen Low Voltage (12V)

4

Halogen Lamps
Low Voltage (12V)

cial spaces, art accenting is very helpful to minimize the "haze" or "fog" of the uniform fluorescent lighting. Here, MR16 lamps of the 5000 and 6000 hour variety in moderate wattages are appropriate. Projects where small, dimmable accent lights are required and/or where construction budgets are moderate, life cycle budgets are of less importance or intended use is not prolonged are candidates for the lower wattage and/or shorter life MR11 and MR16 low voltage halogen. Typically, residential projects fit this profile. Further, where fine, low–level dimming is a critical criterion or where the budget cannot support fluorescent dimming, low voltage halogen MR lamps are justified. Some hospitality facilities (e.g., restaurants, banquet facilities, meeting facilities) and commercial facilities fit this profile.

■ ■ ■ ■ ■ ■ ■ ■ ■ ■ ■

Halogen cycle
✔ Heated filament metal boils away into vapor (as happens in standard tungsten filament incandescent lamps).
✔ Vaporized filament metal is redirected by halogen gas back onto the filament—the filament is regenerated!
✔ Results in either construction of longer life lamp or more efficient lamp or a lamp with some increase in life and some increase in light output.

Executive office suites fitted with motion sensors (to limit operational time) and retail spaces are also candidates for low voltage halogen lighting. Finally, where historic restoration is required and where architectural or fixture detailing accommodates the small low voltage lamp, it can be used as supplemental accent and/or functional lighting without betraying its presence.

When not to use low voltage halogen

Triple–tube compact fluorescent lamps (Section 8) **are the efficient alternative to halogen low voltage lamps for general lighting. For efficient accent lighting, consider halogen infrared mains voltage (HIR) PAR lamps** (Section 5)**, HIR low voltage MR16 lamps** (Section 6) **or ceramic metal halide (CMH) PAR lamps** (Section 7)**.**

Most commercial and institutional projects deserve compact fluorescent and/or ceramic metal halide lamp solutions. Standard low voltage halogen lamps do not offer the low energy use and long life that these facilities require to be cost–competitive and energy efficient. If accent lighting is required in these kinds of facilities, then at a minimum mains or low voltage halogen infrared lamps or, preferably, ceramic metal halide lamps should be considered. Table 4.2 outlines more efficacious alternatives to low voltage halogen lamps. Low voltage halogen lamps should not be used where very close proximity to people for prolonged periods is expected (e.g., task lighting)—given their high operating temperatures (resulting in higher ambient temperatures of nearby air and yielding potential for skin burns) and ultraviolet radiation. Portable task lights and floor lights using halogen lamps should be avoided in children–intensive settings, particularly where adult supervision is unlikely or only periodic.

Halogen Lamps
Low Voltage (12V)

Table 4.2 Efficacious Alternatives to Halogen Low Voltage Lamps

Low Voltage Halogen Lamp	Efficacious Alternative	Hardware Application
T3Q/G4	• F13Triple (compact fluorescent)❶	• Decorative Luminaires❷ • Portable Luminaires❸
	Note: Due to lamp's very small size, no other lamp is a reasonable alternative —larger luminaires will be necessary in order to use alternative.	
T4Q/GY6.35	• Halogen Infrared Low Voltage MR16	• Accent Luminaires
	Note: Due to lamp's very small size, no other lamp is a reasonable alternative —larger luminaires will be necessary in order to use alternative.	
MR11	• Halogen Infrared Low Voltage MR16 • Halogen Infrared PAR30 • Halogen Infrared PAR38	• Accent Luminaires • Decorative Luminaires❹ • Downlight Luminaires
	Note: Due to lamp's very small size, no other lamp is a reasonable alternative —larger luminaires will be necessary in order to use alternative.	
MR16	• Halogen Infrared Low Voltage MR16 • Halogen Infrared PAR30 • Halogen Infrared PAR38	• Accent Luminaires • Decorative Luminaires❹ • Downlight Luminaires
PAR36	• Halogen Infrared Low Voltage MR16 • Halogen Infrared PAR30 • Halogen Infrared PAR38 • CMH PAR20 (ceramic metal halide)❶ • CMH PAR30 (ceramic metal halide)❶	• Accent Luminaires • Downlight Luminaires

❶This alternative lamp is not dimmable inexpensively, if at all.
❷Decorative Luminaires include pendents and sconces.
❸Portable Luminaires include table lights, floor lights and task lights.
❹Decorative Luminaires include pendents.

Cost magnitude[2000]
for halogen and HIR lamps

Low
US$7.⁵⁰ to US$12.⁰⁰
Moderate
US$12.⁰⁰ to US$17.⁰⁰
High
US$17.⁰⁰–plus

Costs vary based on quantities, distributor and contractor markups, and market conditions, manufacturing situations, and annual inflation. These values are for preliminary, magnitude budgeting and do not represent quotes nor actual final pricing.

Halogen Low Voltage (12V)

4

T3/T4 BiPin
Halogen Low Voltage (12V)

Stats Data varies manufacturer to manufacturer and changes from time to time

Life/2000 and 3000 hours
Different versions and various wattages from different manufacturers. A typical office environment might be occupied about 2600 hours each year.

Efficacy/12 to 20 LPW Wattage dependent (higher wattage/higher efficacy) (values include transformer loss)

Wattages/5, 10, 20, 35, 50, 75, 90

Color of Light/2900°K±

Beam spreads/Not Applicable
This is a general service low voltage lamp intended for miniature task lights, sconces, downlights and accents where an all around glow of light is desired; and, with a good reflector, where some accenting is desired. Not suggested where prolonged tasks are performed (this lamp runs hot in a task light—compact fluorescent lamps are much more appropriate for task lighting) nor without a protective tempered glass covering (since the lamp may undergo "non–passive failure"—it may explode accidentally—hot small shards of quartz and filament may result).

Size/1¼" L by ⅜" Ø (T3) and 1¾" L by ½" Ø (T4)

Voltage/12V

Cost Magnitude/Low ⁽²⁰⁰⁰⁾

Manufacturers
General Electric
Osram Sylvania
Philips

Net Addresses
http://www.ge.com/lighting/business/index.htm
http://ecom.sylvania.com/osicatalog/
http://www.lighting.philips.com/

CONNECT FOR MORE

Advantages
Crisp white light
Tiny
Easily dimmed
Relatively long life (for incandescent–type source)
Operates in extreme temperature range
Very little lamp lumen depreciation (light loss) over life

Disadvantages
High operating costs
◆Inefficiency and short life may make this cost prohibitive in typical commercial applications, depending on ease of changing lamps and maintenance staffing
High operating temperatures
◆Hot to touch
◆Fire hazard when too close to fabrics and architectural materials
Hot-shock sensitive
◆Lamp may fail if sharp vibration occurs while lamp is energized
Transformer required
◆Additional cost and visual aspect of transformer must be accommodated
◆Additional wattage requirement of transformer of 3 to 5 watts
◆Potential for audible hum

Where accent lighting is needed, low voltage HIR MR16 (Section 6), HIR PAR lamps (Section 5) or CMH PAR lamps (Section 7) are more efficient than halogen low voltage lamps.

T3/T4 BiPin

Halogen Low Voltage (12V)

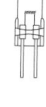

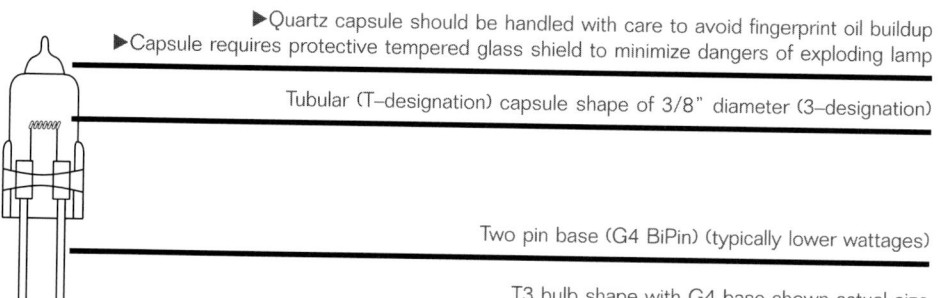

▶Quartz capsule should be handled with care to avoid fingerprint oil buildup
▶Capsule requires protective tempered glass shield to minimize dangers of exploding lamp

Tubular (T–designation) capsule shape of 3/8" diameter (3–designation)

Two pin base (G4 BiPin) (typically lower wattages)

T3 bulb shape with G4 base shown actual size
Image copyright 2000/GE

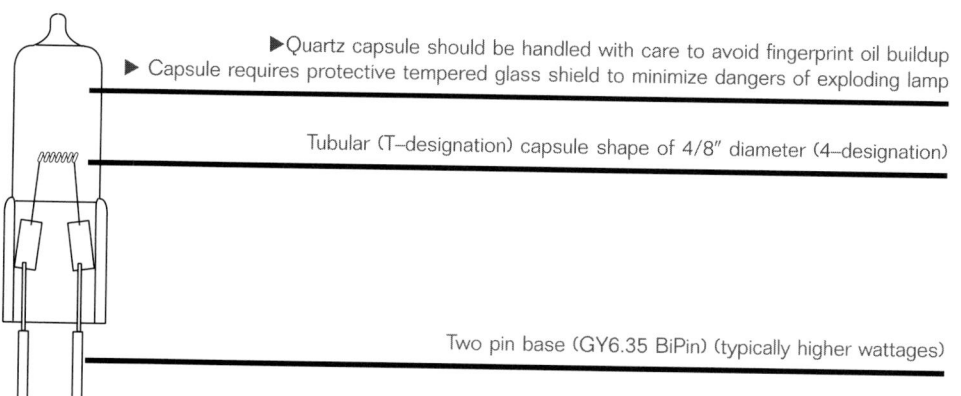

▶Quartz capsule should be handled with care to avoid fingerprint oil buildup
▶ Capsule requires protective tempered glass shield to minimize dangers of exploding lamp

Tubular (T–designation) capsule shape of 4/8" diameter (4–designation)

Two pin base (GY6.35 BiPin) (typically higher wattages)

T4 bulb shape with GY6.35 base shown actual size
Image copyright 2000/GE

For general downlighting in low–to–moderate ceilings, consider triple-tube compact fluorescent lamping (Section 8).

Uses

Decorative/Functional

Hospitality
◆Decorative downlights
◆Steplights

Residential
◆Decorative downlights
◆Decorative pendents
◆Decorative sconces
◆Floor reading lights
◆Steplights
◆Task lights

■ ■ ■ ■ ■ ■ ■ ■ ■ ■ ■ ■ ■ ■ ■ ■ ■ ■ ■ ■

T3/T4 Lighting Design Tips

✔Best in lower light settings.
✔Use decoratively with anticipation of some functional light. Where moderate to high light levels are required use halogen TB19, HIR PAR or compact fluorescent lamps.
✔Dimming increases life (typically doubling life), but shifts color warmer.
✔Match dimmer type with transformer. Electronic transformers require dimmers rated for electronic transformer operation.
✔Handle lamps with care—follow lamp manufacturers' instructions.
✔Beats standard incandescent lamping on maintenance cycles and energy use.
✔Not recommended for task lighting—avoid decorative halogen task lights.

CONNECT FOR MORE

G4 BASE

Halogen Low Voltage (12V)

4

Uses
► Interior low–level steplights
► Exterior low–level steplights
► Low–level decorative downlights
► Residential floor lights
► Residential task lights

Design Tips
✔ Best when used in clusters for effect or to define a setting or path.
✔ Use protective tempered glass covering at all times in open luminaires. Closed luminaires, such as shower lights or steplights, do not require any additional protective covering.
✔ Use in decorative luminaire fittings for most impact.
✔ These lamps are not suitable for general, functional lighting of entire spaces.
✔ Use in low light situations like nightlighting, steps or paths and dining areas.

Application Key

Commercial
Gallery
Hospitality
Institutional
Manufacturing
Residential
Retail
Exterior

bold = primary application
partial fade = minimal application
fade = unlikely application

T3 BiPin/Low Wattage

Halogen Low Voltage (12V)

Performance Data

[All data outlined here is based on information from boldfaced manufacturer's published data for 12V version at time of manuscript preparation.]

▷ **5 watt/Standard/Clear (5T3Q/CL)**
Lumens/60
Life/2000 hrs
Application notes
- Protective tempered glass lens required
- Clear lamp is glary without diffuser or with clear glass lens
- Clear lamp most optically precise

Suggested uses/very low–level steplighting
► **GE**
► **Osram Sylvania**
► **Philips**

Softer light quality

▷ **5 watt/Standard/Frosted (5T3Q/F)**
Lumens/55
Life/2000 hrs
Application notes
- Protective tempered glass lens required
- Frosted lamp is less glary
- Frosted lamp provides softer, but less–precise light distribution

Suggested uses/very low–level steplighting, very soft decorative downlighting, very soft "indicator" lights
► **Osram Sylvania**

Longer life

▷ **5 watt/Long–life/Clear (5T3Q/CL/TR)**
Lumens/60
Life/3000 hrs
Application notes
- Protective tempered glass lens required
- Clear lamp is glary without diffuser or with clear glass lens
- Clear lamp most optically precise

Suggested uses/very low–level steplighting
► **Osram Sylvania**

▷ **10 watt/Standard/Clear (10T3Q/CL)**
Lumens/137
Life/2000 hrs
Application notes
- Protective tempered glass lens required
- Clear lamp is glary without diffuser or with clear glass lens
- Clear lamp most optically precise

Suggested uses/low–level steplighting, soft decorative downlighting
► **GE**
► **Osram Sylvania**
► **Philips**

T3 BiPin/Low Wattage
Halogen Low Voltage (12V)

G4 BASE

Performance Data

[All data outlined here is based on information from boldfaced manufacturer's published data for 12V version at time of manuscript preparation.]

 10 watt/Standard/Frosted (10T3Q/F)  *Softer light quality*
Lumens/120
Life/2000 hrs
Application notes
- Protective tempered glass lens required
- Frosted lamp is less glary
- Frosted lamp provides softer, but less–precise light distribution

Suggested uses/low–level steplighting, soft decorative downlighting
▶**Osram Sylvania**

 10 watt/Long–life/Clear (10T3Q/CL/AX) *Longer life*
Lumens/130
Life/3000 hrs
Application notes
- Protective tempered glass lens required
- Clear lamp is glary without diffuser or with clear glass lens
- Clear lamp most optically precise

Suggested uses/low–level steplighting, soft decorative downlighting
▶**Osram Sylvania**

20 watt/Standard/Clear (20T3Q/CL)
Lumens/330
Life/2000 hrs
Application notes
- Protective tempered glass lens required
- Clear lamp is glary without diffuser or with clear glass lens
- Clear lamp most optically precise

Suggested uses/steplighting, decorative downlighting
▶**GE**
▶**Osram Sylvania**
▶**Philips**

 20 watt/Long–life/Clear (20T3Q/CL/AX)  *Longer life*
Lumens/320
Life/3000 hrs
Application notes
- Protective tempered glass lens required
- Clear lamp is glary without diffuser or with clear glass lens
- Clear lamp most optically precise

Suggested uses/steplighting, decorative downlighting
▶**Osram Sylvania**

Images copyright 2000 Gary Steffy Lighting Design Inc.

Figure 4.1 (top)
Steplights (Bega 1126) using 20T3Q/CL lamps are less than 5″ in diameter and are 4″ in recessed depth.

Figure 4.2 (bottom)
Steplights (Bega 1126) using 20T3Q/CL lamps are used both functionally to light the floor and as indicator lights to accentuate the arcing wall (photo taken while project was under construction).

Application Key

Commercial
Gallery
Hospitality
Institutional
Manufacturing
Residential
Retail
Exterior

bold = primary application
partial fade = minimal application
fade = unlikely application

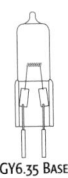

GY6.35 Base

Halogen Low Voltage (12V)

4

Uses
▶Architectural detailing
▶Petite art
▶Objet d'art
▶Downlighting
▶Wallwashing

Design Tips
✔Use protective tempered glass covering at all times in open luminaires. Closed luminaires, such as shower lights, do not require any additional protective covering.

✔Intense light degrades artwork rapidly and/or significantly—consider UV and IR filters or limit exposure time.

✔These lamps are potentially very glary—best when used in well–shielded luminaires and when controlled by dimmers.

✔Use in moderate light situations like kitchens, gallery spaces, living and dining areas.

Application Key

Commercial
Gallery
Hospitality
Institutional
Manufacturing
Residential
Retail
Exterior

bold = primary application
partial fade = minimal application
fade = unlikely application

T4 BiPin/High Wattage
Halogen Low Voltage (12V)

Performance Data
[All data outlined here is based on information from boldfaced manufacturer's published data for 12V version at time of manuscript preparation.]

35 watt/Standard/Clear (35T4Q/CL)
Lumens/575
Life/2000 hrs
Application notes
- Protective tempered glass lens required
- This size lamp at this wattage is very glary
- Clear lamp most optically precise

Suggested uses/downlighting, wallwashing, accenting
▶**Philips**

35 watt/Long–life/Clear (35T4Q/CL/AX)
Lumens/600
Life/3000 hrs
Application notes
- Protective tempered glass lens required
- This size lamp at this wattage is very glary
- Clear lamp most optically precise

Suggested uses/downlighting, wallwashing, accenting
▶**Osram Sylvania**

Longer life

50 watt/Standard/Clear (50T4Q/CL)
Lumens/950
Life/2000 hrs
Application notes
- Protective tempered glass lens required
- This size lamp at this wattage is extremely glary
- Clear lamp most optically precise

Suggested uses/downlighting, wallwashing, accenting
▶**GE**
▶**Osram Sylvania**
▶**Philips**

Softer light quality

50 watt/Standard/Frosted (50T4Q/F)
Lumens/900
Life/2000 hrs
Application notes
- Protective tempered glass lens required
- This size lamp at this wattage is very glary
- Clear lamp most optically precise

Suggested uses/downlighting, wallwashing, accenting
▶**Osram Sylvania**

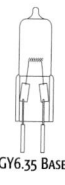

T4 BiPin/High Wattage
Halogen Low Voltage (12V)

GY6.35 BASE

Performance Data

[All data outlined here is based on information from boldfaced manufacturer's published data for 12V version at time of manuscript preparation.]

50 watt/Long–life/Clear (50T4Q/CL/AX)  Longer life
Lumens/950
Life/3000 hrs
Application notes
- Protective tempered glass lens required
- This size lamp at this wattage is extremely glary
- Clear lamp most optically precise

Suggested uses/downlighting, wallwashing, accenting
▶**Osram Sylvania**

75 watt/Standard/Clear (75T4Q/CL)
Lumens/1600
Life/2000 hrs
Application notes
- Protective tempered glass lens required
- This size lamp at this wattage is excruciatingly glary
- Clear lamp most optically precise

Suggested uses/downlighting, wallwashing, accenting
▶**GE**

75 watt/Long–life/Clear (75T4Q/CL/AX) Longer life
Lumens/1450
Life/3000 hrs
Application notes
- Protective tempered glass lens required
- This size lamp at this wattage is excruciatingly glary
- Clear lamp most optically precise

Suggested uses/downlighting, wallwashing, accenting
▶**Osram Sylvania**

90 watt/Long–life/Clear (90T4Q/CL/AX) Longer life
Lumens/1750
Life/3000 hrs
Application notes
- Protective tempered glass lens required
- This size lamp at this wattage is excruciatingly glary
- Clear lamp most optically precise

Suggested uses/downlighting, wallwashing, accenting
▶**Osram Sylvania**

Net Addresses/T3/T4 Track and Monopoints
http://www.cooperlighting.com/
http://www.danalite.com/
http://www.lightolier.com/
http://www.lightproject.com/
http://www.techlighting.com/pro1.html
http://www.tslight.com/

CONNECT FOR MORE

Application Key

Commercial
Gallery
Hospitality
Institutional
Manufacturing
Residential
Retail
Exterior

bold = primary application
partial fade = minimal application
fade = unlikely application

MR
Halogen Low Voltage (12V)

MR lamps—MR is an acronym for multifaceted reflector—are essentially slide projector lamps modified for architectural lighting applications. A very tiny filament possible because of the low voltage (12 volts) operation in a halogen capsule surrounded by a multi-faceted dichroic reflector makes for a tiny, intense light source. The glass reflector has a dichroic coating which selectively transmits infrared radiation (through the back of the lamp) while reflecting visible radiation through the front of the lamp. As such, multifaceted reflectors (MR) built around these small halogen capsules make for potent light sources at relatively low wattages—20 to 73 watts. For architectural lighting and accenting, the MR11 and MR16 low voltage halogen lamps are appropriate. Even more efficient accenting is possible, however, with halogen infrared (HIR) mains voltage lamps (see Section 5), HIR low voltage lamps (see Section 6) and ceramic metal halide lamps (see Section 7).

Low voltage halogen lamps, depending on lamp envelope style, have life ratings ranging from 2000 hours up to 6000 hours. The low voltage halogen MR16 lamps offer the longest life (typically 4000 to 6000 hours)—this is even longer life than that of halogen and HIR mains voltage PAR lamps. The size and shape of the facets establishes the intensity and spread of the light—such as a very narrow spot (NSP), a spot (SP), a narrow flood (NFL) and so forth. There are two sizes (diameters) of MR lamps. MR11 lamps are smallest at 1–3/8 inches (35mm) and MR16 lamps are at 2 inches (51mm).

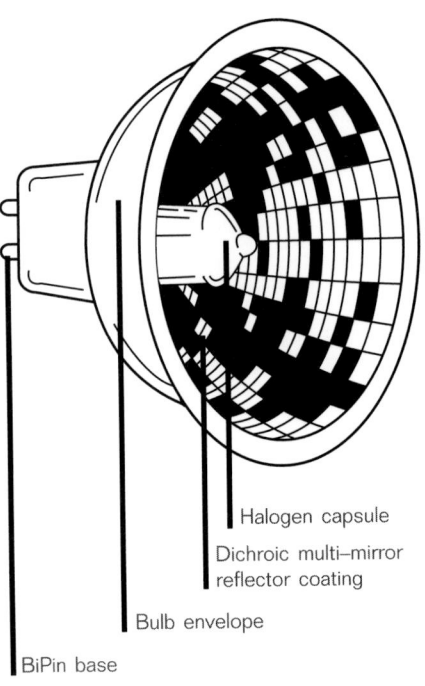

Halogen capsule

Dichroic multi–mirror reflector coating

Bulb envelope

BiPin base

MR16 lamp shown more or less actual size
Image copyright 2000/GE

Figure 4.3
MR16 lamp detail. The MR11 lamp is similar, but smaller.

MR

Halogen Low Voltage (12V)

Table 4.3 Low Voltage (12V) Halogen MR Lamps

MR Lamp Type	Advantages	Disadvantages	Brand Names ❶
Standard MR11 (MR11) **see page 4–30**	• 1⅝ inch diameter • Good beam control	• Color inconsistency lamp to lamp • Significant color change over life • Relatively short life • BiPin base may be unreliable • Requires tempered glass cover • Exhibits back spill light	• GE Standard MR11 • Osram Sylvania Tru–Aim® MR11
Covered MR11 ❷ (MR11/CG) **see page 4–42**	• 1⅝ inch diameter • Good beam control • Integral cover glass	• Color inconsistency lamp to lamp • Significant color change over life • Relatively short life • BiPin base may be unreliable • Exhibits multicolored back spill light • Some efficiency loss	• Philips Continuum™ Pro MRC11
Standard MR16 (MR16) **see page 4–48**	• 2 inch diameter • Better beam control • 4000 hour life (except GE)	• Color inconsistency lamp to lamp • Significant color change over life • BiPin base may be unreliable • Requires tempered glass cover • Exhibits back spill light	• GE Standard MR16 (2000 hour) • Osram Sylvania Tru–Aim® MR16 • Philips Continuum™ MR16
Covered MR16 ❷ (MR16/CG) **see page 4–74**	• 2 inch diameter • Better beam control • Integral cover glass • 4000 hour life	• Color inconsistency lamp to lamp • Significant color change over life • BiPin base may be unreliable • Exhibits multicolored back spill light • Some efficiency loss	• Philips Continuum™ Pro MRC16
Aluminized MR16 (MR16/AL) **see page 4–84**	• 2 inch diameter • Better beam control • 4000 hour life • No back spill light	• Some color inconsistency lamp to lamp • Some color change over life • BiPin base may be unreliable • Requires tempered glass cover	• Osram Sylvania Tru–Aim Brilliant® MR16
Covered Aluminized MR16 ❷ (MR16/AL/CG) **see page 4–98**	• 2 inch diameter • Better beam control • Integral cover glass • 4000 hour life • No back spill light	• Some color inconsistency lamp to lamp • Some color change over life • BiPin base may be unreliable • Some efficiency loss	• Osram Sylvania Tru–Aim Brilliant® MR16 Covered • Philips Continuum™ Pro MRC16 Aluminum Coated
Hard Dichroic MR16 (MR16/C) **see page 4–108**	• 2 inch diameter • Best beam control • 4000 to 6000 hour life • Consistent color of light • Consistent color backlight	• BiPin base may be unreliable • Requires tempered glass cover	• GE ConstantColor® Precise™ MR16 • Osram Sylvania Tru–Aim Titan® MR16
Covered Hard Dichroic MR16 ❷ (MR16/C/CG) **see page 4–138**	• 2 inch diameter • Better beam control • Integral cover glass • 4000 to 6000 hour life • Consistent color of light • Consistent color backlight	• BiPin base may be unreliable • Some efficiency loss	• GE ConstantColor® Precise™ Cover Glass MR16 • Philips Continuum™ Color MRC16

❶Brand names are registered by and/or trademarks of respective manufacturers.
❷Covered lamps are the same as Standard lamps but with integral cover glass over front of lamp which is intended to substitute for a separate tempered glass lens.

Left margin (bottom to top): Most Desirable to Least Desirable

Halogen Low Voltage (12V)

4

Halogen Low Voltage (12V)

MR
Halogen Low Voltage (12V)

There are several versions of MR lamps. For ease of reference in this text, these versions are called Standard, Aluminized Reflector and Hard Dichroic. Each of these versions is available in uncovered and covered options. Uncovered means that the halogen capsule is exposed and open sitting in a reflector compartment. Covered means the lamp is manufactured with a glass cover or lens enclosing the halogen capsule within the reflector compartment. The glass covered style is typically more costly than the open style. While the cover glass may reduce light intensity somewhat, it must be understood that in all applications using the uncovered lamp version a separate tempered glass lens must be placed in front of the lamp to provide a level of safety should the halogen capsule every fail violently. Finally on covered versions, some manufacturers' covered lensing consists of convex (bulging outward) glass which makes the

Halogen Low Voltage MR Lamp Designation
50MR16/C/CG/NFL

Last set of letters identifies beam spread

FLOODS
- 24° to 32° for narrow flood (NFL)
- 33° to 44° for flood (FL)
- 45° to 54° for wide flood (WFL)
- 55° or greater for very wide flood (VWFL)

SPOTS
- 7° or less for very narrow spot (VNSP)
- 8° to 10° for narrow spot (NSP)
- 11° to 14° for spot (SP)
- 15° to 18° for wide spot (WSP)
- 19° to 23° for very wide spot (VWSP)

Ranges of beam spreads used for convenient/consistent reference in this text and do not necessarily correspond to each manufacturer's nor any ANSI or NEMA definitions.

"CG" for built–in cover glass or no designation for no cover glass

Reflector treatment code
- "C" for consistent color (hard dichroic coating) or no letter for inconsistent color
- "AL" for aluminized reflector eliminating back spill light through reflector

These two digits represent lamp diameter in eighths inches (51mm or 2 inches)

These letters indicate this is a multifaceted reflector (MR) lamp

First two digits represent lamp wattage

[This designation used for convenient reference throughout this *Time–Saver Standards for Architectural Lighting*—but not necessarily used by all lamp manufacturers (although it would be convenient if manufacturers would come to agreement on standard designations.]

MR
Halogen Low Voltage (12V)

use of other accessories (e.g., color filters, louvers, tempered glass lenses) difficult or impossible in some luminaires. Check with specific manufacturers for requirements. It is best to view a sample of the lamp prior to specification. This would also be a great opportunity to energize the lamp and assess beam integrity (are there any striations or odd brightness patterns that will conflict with the proposed design application?); assess back spill (does any light leak from the back of the lamp reflector, and, if so, is this an acceptable intensity and/or color of back spill?); assess beam symmetry (does the beam have a nice round pattern or is it oblong or does it exhibit some odd left–to–right or top–to–bottom variation?).

MR lamps, while capable of providing good intensity, are notorious for creating odd striations and scallops on walls, light colored artwork or merchandise. This can be mitigated with the use of tempered soft diffusion lenses.

Efficient Halogen Low Voltage MR Lighting
❶ Don't use Standard MR lamps.
❶ Consider Hard Dichroic lamps, or better yet…
❶ Consider Halogen Infrared/MR (HIR/MR) lamps.

Halogen low voltage MR lamp designation

Halogen low voltage MR lamps are typically identified with an alphanumeric designation outlined on the opposite page. Since each manufacturer has slightly different means for lamp designation, the designation method presented here is solely for convenient reference throughout the text. Actual lamp manufacturers' designations will need to be used in final specifications. Perhaps the lamp manufacturers will standardize lamp designations in the near future in an effort to ease lamp cross referencing and specifications.

Halogen low voltage MR lamps for accenting

MR lamps are great for art accenting. With so many choices on the spread of the light pattern (beam spread), on the intensity of the pattern (candlepower) and on wattage, accenting can be achieved from a reasonable range of ceiling heights and lateral distances from the objects to be highlighted. For MR11 and MR16 lamps in Standard, Aluminized Reflector and Hard Dichroic versions, a series of

MR

Halogen Low Voltage (12V)

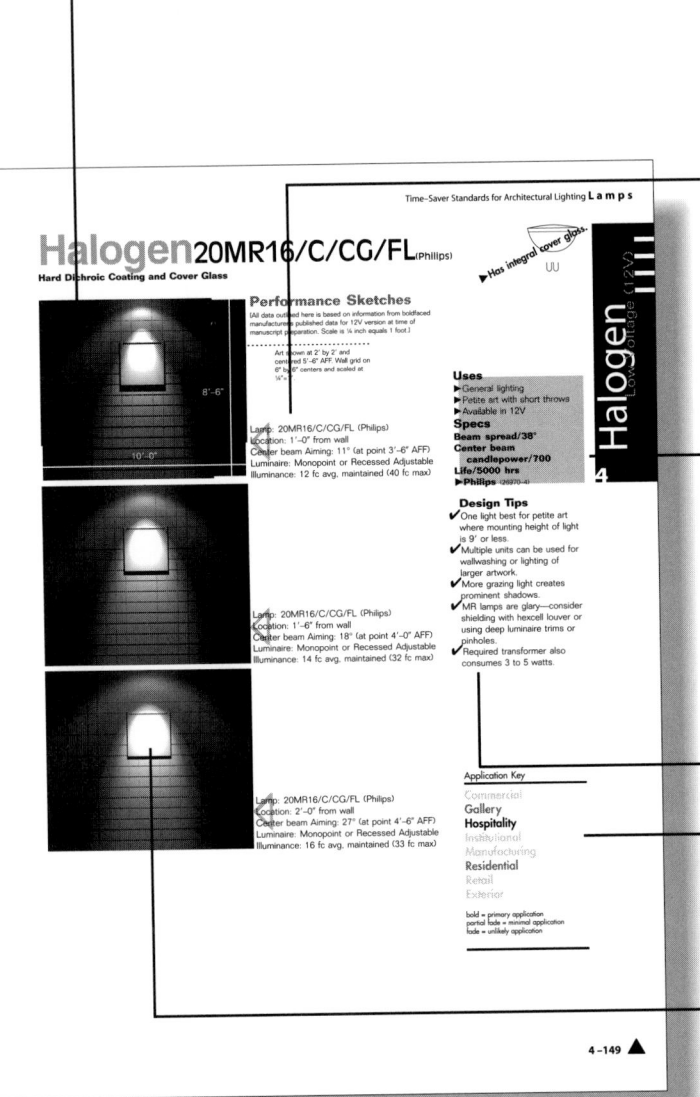

Performance Sketch™
A quarter–inch scale rendering illustrates lighting effect for a typical situation. Note how the area of coverage, intensity and frame shadow change with beam spread, ceiling height and distance from wall. Higher or lower ceilings and longer or shorter distances from walls will impact intensities and light patterns.

Performance Details
Details such as lamp distance from wall, aiming angle, point above floor at which lamp is aimed, and maintained average and maximum illuminance are reported. The aiming angle should be used to select a luminaire capable of achieving such an angle. Maximum illuminances are calculated maintained over time and are reported to help assess likelihood of artwork degradation. Average illuminance is calculated maintained over time just for the size of the artwork shown.

Suggested Uses/Lamp Specs
A bullet list outlines uses for which the lamp seems best suited. Lamp specifications are outlined for quick reference and to assist in specification writing. Where two or more offer nearly identical products, only one manufacturer's data (boldfaced) has been used for development of the Performance Sketches™. An SKU–number or product code is included parenthetically. Data cited in the sidebar "Specs" is averaged from all available manufacturers' data.

Energy Tip
The energy bolt ⚡ is used to denote an energy saving option.

Design Tips
Key considerations are offered for the specific lamp.

Application Key
Those applications where the specific lamp may be most useful are highlighted in bold, while applications where the lamp may be of lesser use are faded—the softer the fade, the less appropriate is the lamp for that application.

Artwork Outline
An outline of two–dimensional artwork of size that might be acceptably lighted by lamp cited.

MR

Halogen Low Voltage (12V)

Performance Sketches™ are presented which offers quick visual assessment of what lamp wattage and beam spread might be appropriate for given ceiling heights and artwork sizes.

See Manufacturers' Updates:
- Periodically due to materials' technology
- Periodically due to manufacturing techniques
- Periodically due to governmental regulations
- May result in performance changes

Performance Sketches™ are shown at a scale of ¼ inch equals 1 foot. Tracing for evaluation of various art sizes and elevations is encouraged. Copying is permitted (according to the conventions outlined in "Licensing Agreement" in the Introduction of this book).

Each Sketch includes an outline of artwork of a size category which might be lighted by the cited lamp. This is to assist in judgement of area of coverage. Location or distance from wall is listed with each Sketch as are center beam aiming angle and average maintained light level over the area of the artwork outline. Maximum maintained light intensity is also listed. For museum quality artwork where conservation of art is a priority, intensities should be 5 to 10 footcandles average, maintained, with maximums of perhaps 10 to 15 fc. In other applications, average maintained intensities on art will depend on the general or ambient lighting within the space(s). In typical hospitality applications, artwork accenting might exhibit average maintained intensities of 10 to 20 fc, with maximum intensities of 20 to 30 fc. In commercial applications, average maintained intensities on artwork might range from 20 to 30 fc, with maximums of 30 to 40 fc.

These Performance Sketches™ are based on the cited manufacturer's data at time of manuscript preparation. Where discrepancies existed between published data and actual photometric tests, the actual photometric test information was used. Where manufacturers' photometric reports were not available, a lamp of similar beam spread was used and data rerated according to published information. Where two or more manufacturers offer nearly identical products, only one manufacturer's data has been used for development of the Perfor-

Radiance (lighting rendering software)
http://www.radsite.lbl.gov/radiance/refer/long.html

CONNECT FOR MORE

Halogen Low Voltage (12V)

4

MR

Halogen Low Voltage (12V)

mance Sketches™. Data cited in the sidebar "Specs" is averaged from all available manufacturers' data. Where any manufacturer's data did not fall within 15 percent of this averaged data, the respective manufacturer's lamp is cited separately along with separate Performance Sketches™. Sketches were generated with this data in Radiance (a Unix–based lighting simulation and rendering program available free of charge at www.radsite.lbl.gov/radiance/refer/long.html).

When specifying lamps of interest, use the beam spread and center beam candlepower as part of the specification along with the generic lamp designation cited in this text. Since data is likely to change from time to time, beam spread and/or candlepower may differ from published results. Further, measurement and calculation techniques may be off by 15 percent or so.

Artwork Size Categories
- Petite = 2'–0" by 2'–0" or smaller
- Small = 2'–0" by 2'–0" to 2'–6" by 3'–0"
- Medium = 2'–6" by 3'–0" to 3'–0" by 4'–0"
- Large = 3'–0" by 4'–0" to 4'–0" by 5'–0"
- Extra Large = 4'–0" by 5'–0" or larger

Performance Sketch™ thumbnails
At the beginning of each MR lamp section (MR11 and MR16) there are thumbnails of the lighting images to further help provide quick visual assessment of various lamp beam spreads and intensities.

Uses
For convenient reference, a shaded box entitled "Uses" appears in the upper corner of each page detailing the more likely uses for which the lamp seems suited. Artwork sizes are based on the rough dimensions as outlined in the tip box on this page (to the left). Lamp "Specs" are also included in the shaded box. These specs are averages of the listed manufacturers' data at time of manuscript preparation. Since products are constantly upgraded or deleted, a check of the current status of the listed lamp is recommended before including it in specifications.

Cost magnitude[2000]
for halogen and HIR lamps

Low
US$7.50 to US$12.00
Moderate
US$12.00 to US$17.00
High
US$17.00–plus

Costs vary based on quantities, distributor and contractor markups, and market conditions, manufacturing situations, and annual inflation. These values are for preliminary, magnitude budgeting and do not represent quotes nor actual final pricing.

MR
Halogen Low Voltage (12V)

Lighting characteristics

In addition to beam spread and intensity, these images also illustrate beam striations and cutoff patterns. Beam striations and aberrations may appear more pronounced in reality since Performance Sketches™ herein are based on finite candlepower reports (typically at 2.5° increments—a bright spot or dead spot might occur at an increment not reported on the photometry but which could be seen visually on mockup). Color of light cannot be illustrated in this black and white format. These Performance Sketches™ will help narrow selections so that appropriate lamp mockups can be made to asess actual beam striations and color of light.

■ ■ ■ ■ ■ ■ ■

MR Lamp Applications
✔ Commercial (secondary)
✔ Gallery (primary)
✔ Hospitality (primary)
✔ Residential (primary)
✔ Retail (primary)

Selection guides

Selection guides are offered on the next several pages for various uses in various applications. These guides are intended to help limit the designer's, engineer's or facility engineer's search and offer a good starting point from which design or alternative analyses can progress. Lamps which have similar beam spreads and intensities may be suggested for different ceiling heights and art sizes simply to show the variety of situations in which lamps might be used. Uses for Halogen/MR lamps include:

► pinspot accent
   An intense, well–controlled, focused lighting effect.
► dramatic accent/medium art
   Relatively intense lighting effect on medium piece of art.
► dramatic accent/large art
   Relatively intense lighting effect on large piece of art.
► soft accent
   Relatively mellow wash of light on an object or area.
► wallwashing/matte materials
   Relatively uniform vertical wash of light on large areas of wall materials exhibiting no sheen, polish or specularity.
► wallwashing/polished materials
   Relatively uniform vertical wash of light on large areas of polished or specular wall materials.

Halogen Low Voltage (12V)

4

MR
Halogen Low Voltage (12V)

▶feature downlighting
Somewhat intense lighting effect on flooring material details.
▶general downlighting
Relatively uniform horizontal wash of light at lap height, table height or on floor.

While there are exceptions, halogen low voltage MR lamps are most appropriate for accent lighting. Small size, relatively low wattage, and long life in the MR16 varieties make these candidates for:

▶commercial
▶gallery
▶hospitality
▶residential
▶retail

However, the higher–efficacy halogen infrared low voltage MR lamps are more appropriate where connected load and energy costs are of great concern.

For low ceiling conditions, Tables 4.4 (Commercial/Retail), 4.5 (Gallery/Residential) and 4.6 (Hospitality) should be referenced. For high ceiling conditions—greater than 10 feet in height, Tables 4.7 (Commercial/Retail), 4.8 (Gallery/Residential) and 4.9 (Hospitality) should be referenced.

Recommended intensities
for art accenting

Low (gallery, museum, residential)
5 to 10 fc average on art
Moderate (hospitality)
10 to 20 fc average on art
High (commercial—or higher for retail)
20 to 30 fc average on art

These are intended to be average, maintained values including effects of accent lighting and general room lighting. General room lighting contributes just a few footcandles in residential, gallery and museum settings, but contributes 5 to 10 fc in commercial settings. Maximums might range from 50 percent to 100 percent of the average value. Wherever artwork is highly valuable and/or sensitive to light, precautions should be taken to limit light exposure (maintaining low levels and/or switching lights off when art is not being viewed); and exposure to infrared and ultraviolet should be limited (using filters on lamps or placing art behind specially treated glass or acrylic).

Halogen Low Voltage MR Lamp Selection Guide

Table 4.4 Commercial or Retail Low Ceiling (less than 10 feet)❶❷

Lighting Task	Standard MR11	Standard MR16	Aluminized MR16	Hard Dichroic MR16
Pinspot Accent❸	Consider (1) or more depending on desired impact			
• Petite to small art • Feature merchandise • Small merchandise	• 20/NSP p4–34 • 20/NFL p4–36	• 20/NSP8 p4–55 • 20/FL p4–57	▶ 20/CG/NSP p4–102	▶ 20/CG/NFL p4–148
Key Area Accent	Consider (2) or more depending on desired impact			
• Small to medium art • Feature display • Moderately sized merchandise		• 35/FL p4–60 　└ beam spread 　　└ wattage		
Feature Lighting	Consider (2) or more depending on size of featured area			
• Medium to large art • Floor feature • Destination focus (e.g., cashier)		• 50/FL p4–64 ▶ 50/CG/FL p4–82 • 65/FL p4–69 • 73/FL p4–72	• 50/FL p4–94 • 65/FL p4–97	• 50/FL p4–128 ▶ 50/CG/FL p4–158 ▶ 50/CG/FL p4–159 • 65/FL p4–133 • 71/FL p4–137 ▶ 71/CG/FL p4–163
General Downlighting	Consider symmetric arrangement❹			
• Overall lighting of lap, table or floor		• 35/FL p4–60 • 50/FL p4–64 • 50/VWFL p4–65	• 50/FL p4–94 ▶ 50/CG/FL p-106	• 50/FL p4–128 • 50/VWFL p4–129

Typically dimmer ▲
▼ Typically brighter

❶These lamp guides are intended to direct your attention to lamps which might meet your needs. If daylighting is prevalent, more lamps may be necessary to provide a visual focus to the artwork or merchandise. Design analysis is required to finalize lamp selections for projects.

❷For retail applications it is reasonable to also use lamps cited in Table 4.7—intensities will be significantly greater in low ceiling spaces.

❸For intense pinspot accenting, also consider higher wattage VNSP and/or NSP lamps, which for purposes of this reference have been suggested for higher ceiling situations and/or other task applications. Lower ceiling will result in smaller area of coverage and more intense light within the smaller area of coverage.

❹Spacing between lights should not exceed 0.75 to 1.0 of the distance between the surface being downlighted and the ceiling. Downlighting references may be to luminaires for spaces of other task functions and/or other ceiling heights. Lamps aimed straight downward will generally produce more light on the horizontal plane (lap, table or floor elevation) than on wall (vertical) surfaces. Calculations should be performed to establish final lamp wattage and spacing for general downlighting.

▶Designates lamp versions with integral cover glass. All other low voltage MR lamps require auxiliary tempered glass lens.

■ ■ ■ ■ ■ ■ ■ ■ ■ ■

Artwork Size Categories
ↄPetite = 2'–0" by 2'–0" or smaller
ↄSmall = 2'–0" by 2'–0" to 2'–6" by 3'–0"
ↄMedium = 2'–6" by 3'–0" to 3'–0" by 4'–0"
ↄLarge = 3'–0" by 4'–0" to 4'–0" by 5'–0"
ↄExtra Large = 4'–0" by 5'–0" or larger

Halogen Low Voltage MR Lamp Selection Guide

Table 4.5 Gallery or Residential Low Ceiling (less than 10 feet)❶

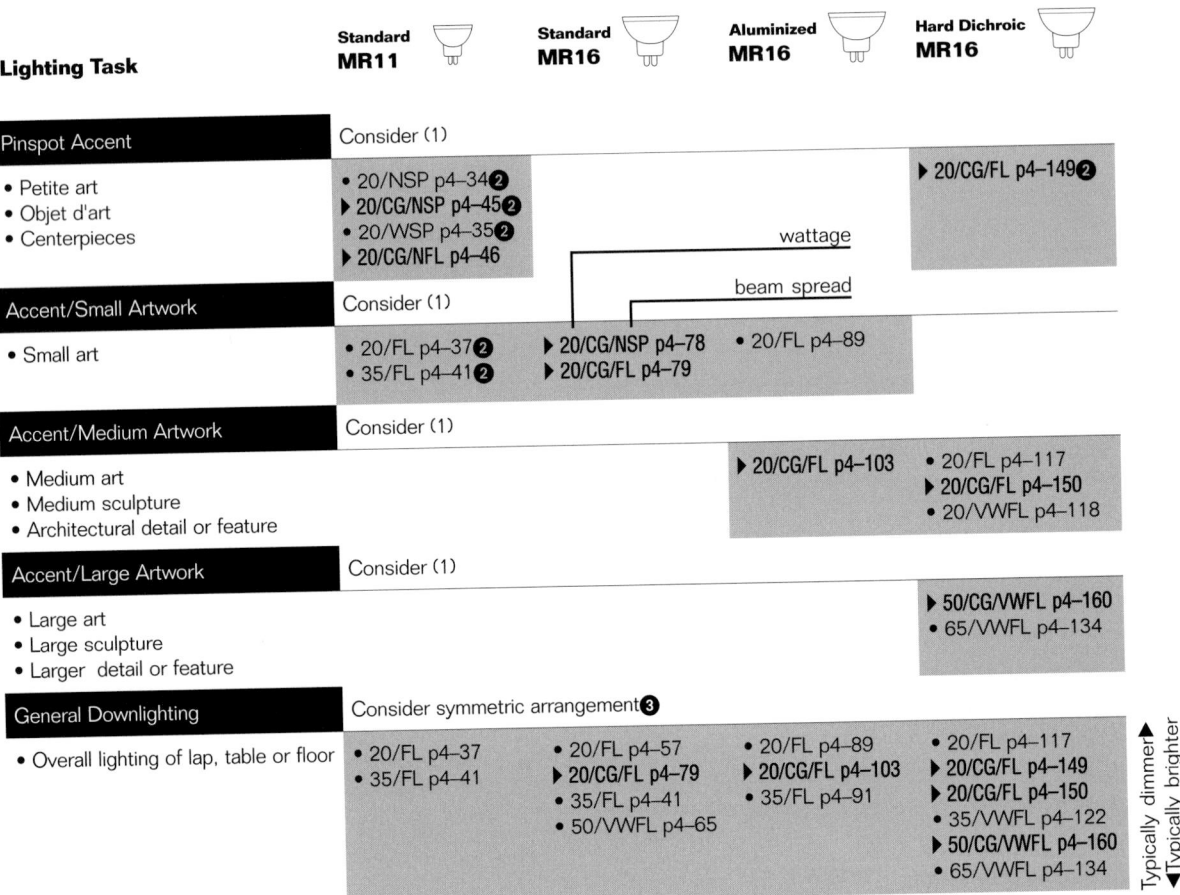

Lighting Task	Standard MR11	Standard MR16	Aluminized MR16	Hard Dichroic MR16
Pinspot Accent	Consider (1)			
• Petite art • Objet d'art • Centerpieces	• 20/NSP p4–34❷ ▶ 20/CG/NSP p4–45❷ • 20/WSP p4–35❷ ▶ 20/CG/NFL p4–46			▶ 20/CG/FL p4–149❷
Accent/Small Artwork	Consider (1)			
• Small art	• 20/FL p4–37❷ • 35/FL p4–41❷	▶ 20/CG/NSP p4–78 ▶ 20/CG/FL p4–79	• 20/FL p4–89	
Accent/Medium Artwork	Consider (1)			
• Medium art • Medium sculpture • Architectural detail or feature			▶ 20/CG/FL p4–103	• 20/FL p4–117 ▶ 20/CG/FL p4–150 • 20/VWFL p4–118
Accent/Large Artwork	Consider (1)			
• Large art • Large sculpture • Larger detail or feature				▶ 50/CG/VWFL p4–160 • 65/VWFL p4–134
General Downlighting	Consider symmetric arrangement❸			
• Overall lighting of lap, table or floor	• 20/FL p4–37 • 35/FL p4–41	• 20/FL p4–57 ▶ 20/CG/FL p4–79 • 35/FL p4–41 • 50/VWFL p4–65	• 20/FL p4–89 ▶ 20/CG/FL p4–103 • 35/FL p4–91	• 20/FL p4–117 ▶ 20/CG/FL p4–149 ▶ 20/CG/FL p4–150 • 35/VWFL p4–122 ▶ 50/CG/VWFL p4–160 • 65/VWFL p4–134

(labels within table: wattage, beam spread)

Typically dimmer ▲
▼ Typically brighter

❶These lamp guides are intended to direct your attention to lamps which might meet your needs. Design analysis is required to finalize lamp selections for projects.

❷Neutral density filter will be required in front of halogen lamp to limit intensity to between 5 and 10 fc if art preservation is paramount.

❸Spacing between lights should not exceed 0.75 to 1.0 of the distance between the surface being downlighted and the ceiling. Downlighting references may be to luminaires for spaces of other task functions and/or other ceiling heights. Lamps aimed straight downward will generally produce more light on the horizontal plane (lap, table or floor elevation) than on wall (vertical) surfaces. Calculations should be performed to establish final lamp wattage and spacing for general downlighting.

▶Designates lamp versions with integral cover glass. All other low voltage MR lamps require auxiliary tempered glass lens.

Halogen Low Voltage MR Lamp Selection Guide

Table 4.6 Hospitality Low Ceiling (less than 10 feet)❶❷

Halogen Low Voltage (12V)

4

Lighting Task	Standard MR11	Standard MR16	Aluminized MR16	Hard Dichroic MR16
Pinspot Accent	Consider (1)			
• Petite art • Objet d'art • Centerpieces • Petite or fine architectural detail	▶ 20/CG/NSP p4–45 • 20/WSP p4–35 • 20/NFL p4–36	• 20/NSP p4–55 • 20/FL p4–57	• 20/NSP p4–88 ▶ 20/CG/NSP p4–102	▶ 20/CG/NFL p4–148 ▶ 20/CG/FL p4–149 beam spread wattage
Accent/Small Artwork	Consider (1)			
• Small art • Small sculpture • Smaller architectural detail or feature		• 20/FL p4–37 • 35/NFL p4–40 • 35/FL p4–41	• 20/NSP p4–56 ▶ 20/CG/NSP p4–78	
Accent/Medium Artwork	Consider (1)			
• Medium art • Medium sculpture • Architectural detail or feature			• 35/FL p4–91 • 50/FL p4–94 ▶ 50/CG/FL p4–106	• 35/FL p4–121 ▶ 35/CG/FL p4–153 • 35/VWFL p4–122
Accent/Large Artwork	Consider (1)			
• Large art • Large sculpture • Larger architectural detail or feature		• 50/VWFL p4–65	• 65/FL p4–97	• 50/VWFL p4–129 • 65/VWFL p4–134 ▶ 71/CG/FL p4–163
Wallwashing/Matte Materials	Continuous row of adjustables 2– to 3–feet from wall and 2– to 3–feet on center			
• Flat, frontal lighting❸		• 50/VWFL p4–65 • 65/FL p4–69		• 35/VWFL p4–122 • 50/VWFL p4–129 • 65/VWFL p4–134 ▶ 71/CG/FL p4–163
Wallwashing/Polished Materials	Architectural slot with sockets 6–inches from wall and 6– to 12–inches on center			
• Grazing wash light❹❺	See PAR lamp wallslots, Section 11, pp11–210 to 11–225			
Feature Downlighting	Consider (1) or more depending on size of featured area			
• Floor feature • Pattern highlight • Destination focus (e.g., reception desk)				
General Downlighting	Consider symmetric arrangement❻			
• Overall lighting of lap, table or floor		• 50/VWFL p4–65		▶ 20/CG/FL p4–149 • 35/FL p4–121 • 35/VWFL p4–122 • 50/VWFL p4–129 • 65/VWFL p4–134 ▶ 71/CG/FL p4–163

▲Typically dimmer ▼Typically brighter

❶These lamp guides are intended to direct your attention to lamps which might meet your needs. Design analysis is required to finalize lamp selections for projects.
❷Higher light levels on artwork and features are typically expected in hospitality spaces compared to residential and gallery spaces. Where artwork preservation is a priority, neutral density filters will be required in front of these lamps or lamps suggested for Gallery or Residential Applications should be used.
❸Energy intensive—only consider for exclusive zones of small size in highly visible public areas. Also see spreadlens wallwashers, pp11–136 to 11–143.
❹Extraordinarily energy intensive and should be reserved for very exclusive zones of small size in highly visible public areas.
❺This approach accentuates any wall imperfections.
❻Spacing between lights should not exceed 0.75 to 1.0 of the distance between the surface being downlighted and the ceiling. Downlighting references may be to luminaires for spaces of other task functions and/or other ceiling heights. Lamps aimed straight downward will generally produce more light on the horizontal plane (lap, table or floor elevation) than on wall (vertical) surfaces. Calculations should be performed to establish final lamp wattage and spacing for general downlighting.
▶Designates lamp versions with integral cover glass. All other low voltage MR lamps require auxiliary tempered glass lens.

Halogen Low Voltage MR Lamp Selection Guide

Table 4.7 Commercial or Retail High Ceiling (10 feet or greater)❶❷

Lighting Task	Standard MR11	Standard MR16	Aluminized MR16	Hard Dichroic MR16
Pinspot Accent❸	Consider (1) or more depending on ceiling height and desired impact			
• Petite to small art • Feature merchandise • Small merchandise				• 20/NSP p4–115 ▶ 20/CG/NSP p4–146 ▶ 20/CG/NFL p4–148 • 20/WSP p4–116
Key Area Accent	Consider (2) or more depending on ceiling height and desired impact			
• Small to medium art • Feature display • Moderately sized merchandise		• 35/NSP p4–58	• 35/NSP p4–90 • 50/NFL p4–93	• 35/VWSP p4–120 • 35/NSP p4–119 • 42/NSP p4–123
Feature Lighting	Consider (2) or more depending on size of featured area and ceiling height			
• Medium to large art • Floor feature • Destination focus (e.g., cashier)		• 50/SP p4–62 ▶ 50/CG/NSP p4–80 • 50/NFL p4–63 _beam spread_ _wattage_	▶ 50/CG/NFL p4–105 • 65/NSP p4–95	• 50/NSP p4–124 • 50/WSP p4–125 • 50/NFL p4–126 • 50/NFL p4–127 ▶ 50/CG/NFL p4–156 ▶ 50/CG/NFL p4–157 • 71/WSP p4–135 • 71/FL p4–137 ▶ 71/CG/NFL p4–162
General Downlighting	Consider symmetric arrangement❹			
• Overall lighting of lap, table or floor		▶ 50/CG/FL p4–82 • 73/FL p4–72		• 35/FL p4–121 • 71/FL p4–137

Typically dimmer ▲
Typically brighter ▼

❶These lamp guides are intended to direct your attention to lamps which might meet your needs. If daylighting is prevalent, more lamps may be necessary to provide a visual focus to the artwork or merchandise. Design analysis is required to finalize lamp selections for projects.

❷Some lamps' performance sketches which are referenced here are not optimized for retail applications. Lower ceiling heights will result in higher intensities on merchandise.

❸For intense pinspot accenting, also consider higher wattage VNSP and/or NSP lamps, which for purposes of this reference have been suggested for other task applications. Lower ceiling will result in smaller area of coverage and more intense light within the smaller area of coverage.

❹Spacing between lights should not exceed 0.75 to 1.0 of the distance between the surface being downlighted and the ceiling. Downlighting references may be to luminaires for spaces of other task functions and/or other ceiling heights. Lamps aimed straight downward will generally produce more light on the horizontal plane (lap, table or floor elevation) than on wall (vertical) surfaces. Calculations should be performed to establish final lamp wattage and spacing for general downlighting.

▶Designates lamp versions with integral cover glass. All other low voltage MR lamps require auxiliary tempered glass lens.

Artwork Size Categories
⊃Petite = 2'–0" by 2'–0" or smaller
⊃Small = 2'–0" by 2'–0" to 2'–6" by 3'–0"
⊃Medium = 2'–6" by 3'–0" to 3'–0" by 4'–0"
⊃Large = 3'–0" by 4'–0" to 4'–0" by 5'–0"
⊃Extra Large = 4'–0" by 5'–0" or larger

Halogen Low Voltage MR Lamp Selection Guide

Table 4.8 Gallery or Residential High Ceiling (10 feet or greater)❶

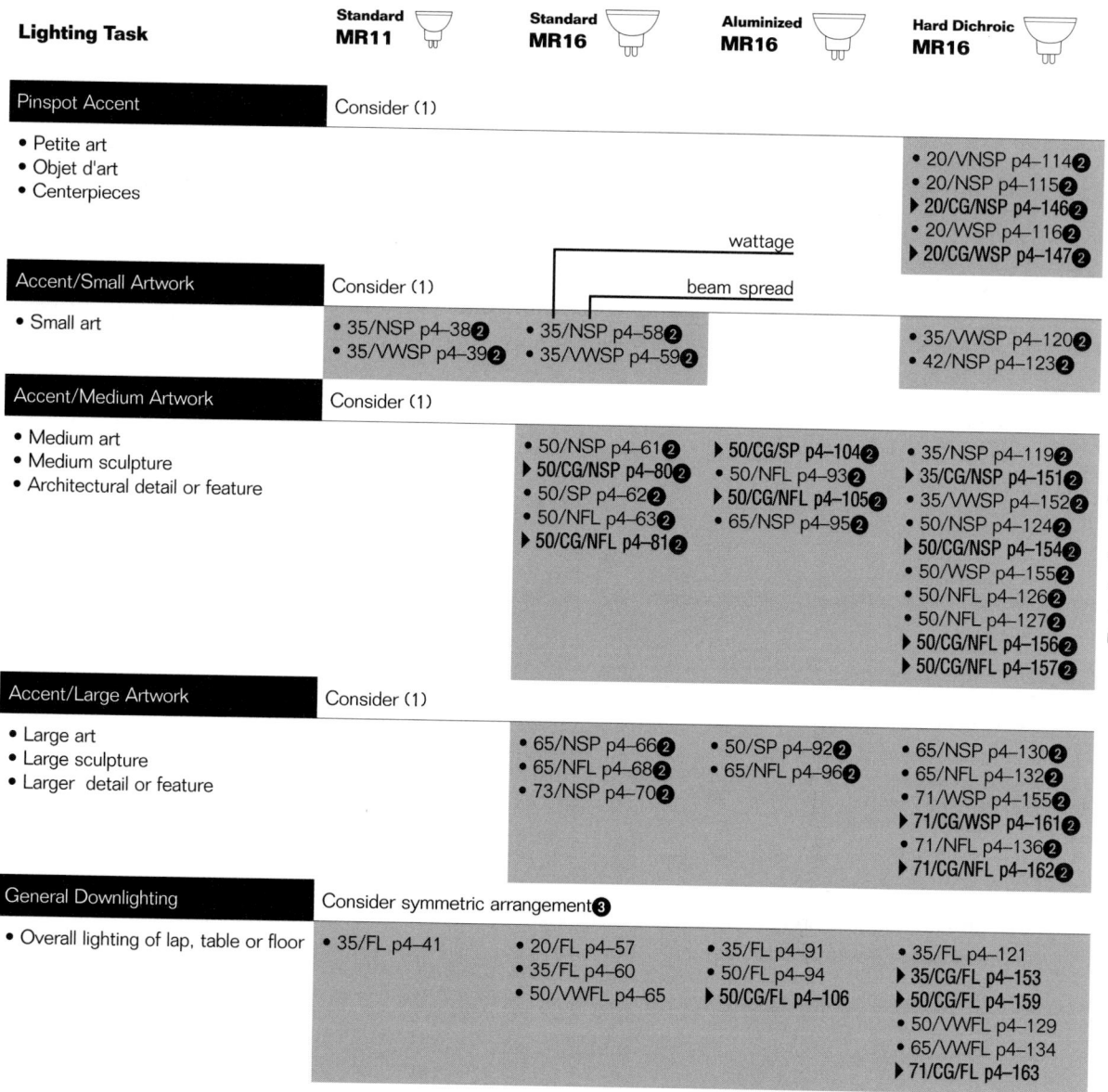

Lighting Task	Standard MR11	Standard MR16	Aluminized MR16	Hard Dichroic MR16
Pinspot Accent	Consider (1)			
• Petite art • Objet d'art • Centerpieces				• 20/VNSP p4–114❷ • 20/NSP p4–115❷ ▶ 20/CG/NSP p4–146❷ • 20/WSP p4–116❷ ▶ 20/CG/WSP p4–147❷
Accent/Small Artwork	Consider (1)			
• Small art	• 35/NSP p4–38❷ • 35/VWSP p4–39❷	• 35/NSP p4–58❷ • 35/VWSP p4–59❷		• 35/VWSP p4–120❷ • 42/NSP p4–123❷
Accent/Medium Artwork	Consider (1)			
• Medium art • Medium sculpture • Architectural detail or feature		• 50/NSP p4–61❷ ▶ 50/CG/NSP p4–80❷ • 50/SP p4–62❷ • 50/NFL p4–63❷ ▶ 50/CG/NFL p4–81❷	▶ 50/CG/SP p4–104❷ • 50/NFL p4–93❷ ▶ 50/CG/NFL p4–105❷ • 65/NSP p4–95❷	• 35/NSP p4–119❷ ▶ 35/CG/NSP p4–151❷ • 35/VWSP p4–152❷ • 50/NSP p4–124❷ ▶ 50/CG/NSP p4–154❷ • 50/WSP p4–155❷ • 50/NFL p4–126❷ • 50/NFL p4–127❷ ▶ 50/CG/NFL p4–156❷ ▶ 50/CG/NFL p4–157❷
Accent/Large Artwork	Consider (1)			
• Large art • Large sculpture • Larger detail or feature		• 65/NSP p4–66❷ • 65/NFL p4–68❷ • 73/NSP p4–70❷	• 50/SP p4–92❷ • 65/NFL p4–96❷	• 65/NSP p4–130❷ • 65/NFL p4–132❷ • 71/WSP p4–155❷ ▶ 71/CG/WSP p4–161❷ • 71/NFL p4–136❷ ▶ 71/CG/NFL p4–162❷
General Downlighting	Consider symmetric arrangement❸			
• Overall lighting of lap, table or floor	• 35/FL p4–41	• 20/FL p4–57 • 35/FL p4–60 • 50/VWFL p4–65	• 35/FL p4–91 • 50/FL p4–94 ▶ 50/CG/FL p4–106	• 35/FL p4–121 ▶ 35/CG/FL p4–153 ▶ 50/CG/FL p4–159 • 50/VWFL p4–129 • 65/VWFL p4–134 ▶ 71/CG/FL p4–163

wattage

beam spread

Typically dimmer▲
▼Typically brighter

❶These lamp guides are intended to direct your attention to lamps which might meet your needs. Design analysis is required to finalize lamp selections for projects.

❷Neutral density filter will be required in front of halogen lamp to limit intensity to between 5 and 10 fc if art preservation is paramount.

❸Spacing between lights should not exceed 0.75 to 1.0 of the distance between the surface being downlighted and the ceiling. Downlighting references may be to luminaires for spaces of other task functions and/or other ceiling heights. Lamps aimed straight downward will generally produce more light on the horizontal plane (lap, table or floor elevation) than on wall (vertical) surfaces. Calculations should be performed to establish final lamp wattage and spacing for general downlighting.

▶Designates lamp versions with integral cover glass. All other low voltage MR lamps require auxiliary tempered glass lens.

Halogen Low Voltage MR Lamp Selection Guide

Table 4.9 Hospitality High Ceiling (10 feet or greater)❶❷

Lighting Task	Standard MR11	Standard MR16	Aluminized MR16	Hard Dichroic MR16
Pinspot Accent	Consider (1)			
• Petite art • Objet d'art • Centerpieces • Petite or fine architectural detail				• 20/VNSP p4–114 • 20/NSP p4–115 ▶ 20/CG/NSP p4–146 • 20/WSP p4–116 ▶ 20/CG/WSP p4–147
Accent/Small Artwork	Consider (1)			
• Small art • Small sculpture • Smaller architectural detail or feature		• 35/NSP p4–38 • 35/VWSP p4–39	• 35/VNSP p4–59	• 42/NSP p4–123
Accent/Medium Artwork	Consider (1)			
• Medium art • Medium sculpture • Architectural detail or feature		• 50/NSP p4–61 ▶ 50/CG/NSP p4–80 • 50/SP p4–62 ▶ 50/CG/NFL p4–81	• 35/NSP p4–90 ▶ 50/CG/SP p4–104 • 50/NFL p4–93 ▶ 50/CG/NFL p4–105 • 65/NSP p4–95	▶ 35/CG/SP p4–151 ▶ 35/CG/VWFL p4–152 ▶ 35/CG/NFL p4–157 • 50/NSP p4–124 ▶ 50/CG/NSP p4–154 • 50/WSP p4–125 • 50/NFL p4–126 • 50/NFL p4–127 ▶ 50/CG/NFL p4–156 ▶ 50/CG/NFL p4–157
Accent/Large Artwork	Consider (1)			
• Large art • Large sculpture • Larger architectural detail or feature		• 65/NSP p4–66 • 65/NFL p4–68 • 73/NSP p4–70	• 50/SP p4–92 • 65/NFL p4–96	▶ 50/CG/WSP p4–155 • 65/NSP p4–130 • 65/NFL p4–132 ▶ 71/CG/WSP p4–161 • 71/NFL p4–136
Wallwashing/Matte Materials	Continuous row of adjustables 2 to 3 feet from wall and 2 to 3 feet on center			
• Flat, frontal lighting❸		• 50/VWFL p4–65 • 65/FL p4–69		• 65/VWFL p4–134 ▶ 71/CG/FL p4–163
Wallwashing/Polished Materials	Architectural slot with sockets 6 inches from wall and 6 to 12 inches on center			
• Grazing wash light❹❺	See PAR lamp wallslots, Section 11, pp11–210 to 11–225			
Feature Downlighting	Consider (1) or more depending on size of featured area			
• Floor feature • Pattern highlight • Destination focus (e.g., reception desk)		▶ 50/CG/NFL p4–81 • 65/NFL p4–68 • 73/NSP p4–70	• 50/NFL p4–93 • 65/NFL p4–96 ▶ 50/CG/NFL p4–105	• 50/NFL p4–126 • 50/NFL p4–127 ▶ 50/CG/NFL p4–156 ▶ 50/CG/NFL p4–157 • 65/NFL p4–132 • 71/NFL p4–136
General Downlighting	Consider symmetric arrangement❻			
• Overall lighting of lap, table or floor		• 65/FL p4–69 • 73/FL p4–72	• 65/FL p4–97 ▶ 50/CG/FL p4–106	• 50/FL p4–128 • 65/FL p4–133 ▶ 71/CG/FL p4–163

wattage

beam spread

Typically dimmer ▲
▼ Typically brighter

Halogen Low Voltage MR Lamp Selection Guide

◀ Footnotes to Table 4.9 (opposite page)

❶These lamp guides are intended to direct your attention to lamps which might meet your needs. Design analysis is required to finalize lamp selections for projects.

❷Higher light levels on artwork and features are typically expected in hospitality spaces compared to residential and gallery spaces. Where artwork preservation is a priority, neutral density filters will be required in front of these lamps or lamps suggested for Gallery or Residential Applications should be used.

❸Energy intensive—only consider for exclusive zones of small size in highly visible public areas. Also see spreadlens wallwashers, pp11–136 to 11–143.

❹Extraordinarily energy intensive and should be reserved for very exclusive zones of small size in highly visible public areas.

❺This approach accentuates any wall imperfections.

❻Spacing between lights should not exceed 0.75 to 1.0 of the distance between the surface being downlighted and the ceiling. Downlighting references may be to luminaires for spaces of other task functions and/or other ceiling heights. Lamps aimed straight downward will generally produce more light on the horizontal plane (lap, table or floor elevation) than on wall (vertical) surfaces. Calculations should be performed to establish final lamp wattage and spacing for general downlighting.

▶Designates lamp versions with integral cover glass. All other low voltage MR lamps require auxiliary tempered glass lens.

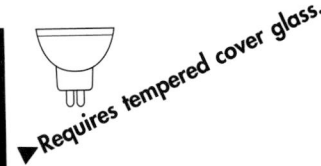

▼ *Requires tempered cover glass.*

Halogen Low Voltage (12V)

4

Halogen/MR11

Standard

Stats
<small>Data varies manufacturer to manufacturer and changes from time to time</small>

Life/3000 and 3500 hours
Varies manufacturer to manufacturer. A typical office environment might be occupied about 2600 hours each year.

Efficacy/Not reported
Excellent beam intensities and distributions available for the wattage.

Wattages/20 and 35
Color of Light/2900°K
Beam spreads/NSP, SP, NFL and FL
Narrow spot, spot, narrow flood and flood distributions are available.

Size/1⅜" L by 1⅜" Ø
Voltage/12V
Cost Magnitude/Moderate ⁽²⁰⁰⁰⁾
Manufacturers
General Electric
Osram Sylvania

Net Addresses
http://www.ge.com/Lighting/business/index.htm
http://ecom.sylvania.com/osicatalog/

CONNECT FOR MORE

Advantages
Crisp white light
Good intensities available
Very small size
Easily dimmed
Low initial cost
Alternative to less efficient R and halogen PAR lamps
Withstand extreme temperature range
Very little lamp lumen depreciation (light loss) over life

Disadvantages
Beam aberrations
◆Striations
◆Inconsistent beam shape
Color inconsistencies
◆Lamp to lamp color variations
◆Over time, color shifts—very evident with exposed–lamp monopoints or trackheads where back of lamp is visible
High operating costs
◆Inefficiency and short life may make this cost prohibitive in typical commercial applications
Hot to touch
Tempered glass lens required
◆Prevents glass and filament shards from falling onto flammable surfaces/materials should a violent lamp failure occur (known as a non–passive failure)
Transformer required
◆Additional cost and visual aspect of transformer must be accommodated
◆Additional wattage requirement of transformer of 3 to 5 watts
◆Potential for audible hum

For more efficient accent lighting, use HIR low voltage lamps (Section 6)**; HIR PAR lamps** (Section 5) **or CMH PAR lamps** (Section 7).

Halogen/MR11
Standard

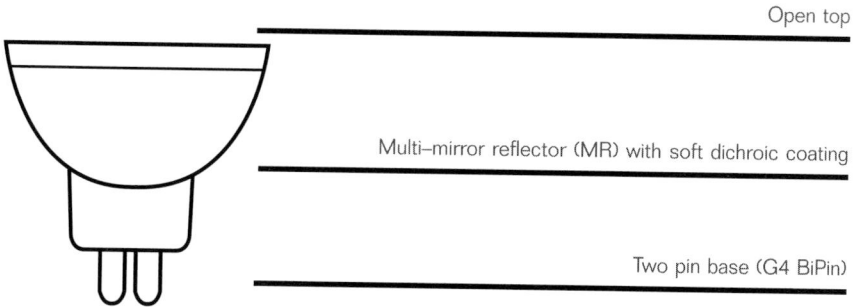

Open top

Multi–mirror reflector (MR) with soft dichroic coating

Two pin base (G4 BiPin)

Lamp shown actual size
Image copyright 2000/GE

Uses
Architectural Accenting
Residential
Art Accenting
Residential
Downlighting
Residential

MR11 Lighting Design Tips
✔Best at lower mounting heights (less than 10′–0″).
✔Dimming increases life (typically doubling life), but shifts color warmer.
✔Very glary when used without much other general lighting or when aimed across long distances where ceilings are low.
✔Beam striations are most apparent on light–colored, nontextured flat surfaces and may be objectionable.
✔MR11downlighting alone is harsh—combine several techniques for best results.
✔Use smaller scale (e.g., 4–inch diameter) downlights and adjustable accents for attractive, more human–scale approach—excellent in residential applications.
✔Use tempered glass lens for safety.
✔Use in recessed luminaires or in details or in enclosed trackheads to avoid seeing color inconsistencies which are most visible from back of lamp.

Net Addresses/MR11 Recessed
http://www.ardeelighting.com/index.htm
http://www.artemide.com/
http://www.bartcolighting.com/index2.html

 CONNECT FOR MORE

Halogen Low Voltage (12V)

▶ Requires tempered cover glass.

Halogen/MR11

Standard MR11 Visual Index

Throws/Ceilings
⊃ Short = 7' to 10'
⊃ Medium = 10' to 15'
⊃ Long = 15' to 20'
⊃ Very Long = 20' or more

Recommended intensities
for art accenting

Low (gallery, museum, residential)
5 to 10 fc average on art
Moderate (hospitality)
10 to 20 fc average on art
High (commercial—or higher for retail)
20 to 30 fc average on art

These are intended to be average, maintained values including effects of accent lighting and general room lighting. General room lighting contributes just a few footcandles in residential, gallery and museum settings, but contributes 5 to 10 fc in commercial settings. Maximums might range from 50 percent to 100 percent of the average value. Wherever artwork is highly valuable and/or sensitive to light, precautions should be taken to limit light exposure (maintaining low levels and/or switching lights off when art is not being viewed); and exposure to infrared and ultraviolet should be limited (using filters on lamps or placing art behind specially treated glass or acrylic).

20MR11/NSP ▶ p34

20MR11/NSP
▶ Architectural detailing
▶ Petite art/10' ceiling height or less
▶ Objet d'art
▶ Short throws
▶ Commercial, Retail

20MR11/WSP ▶ p35

20MR11/WSP
▶ Petite art/9' ceiling height or less
▶ Short throws
▶ Gallery, Hospitality, Residential

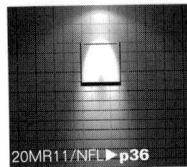  
20MR11/NFL ▶ p36

20MR11/NFL
▶ Petite art/9' ceiling height or less
▶ Short throws
▶ Commercial

 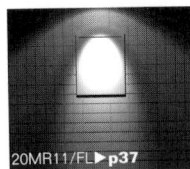 
20MR11/FL ▶ p37

20MR11/FL
▶ Small art/9' ceiling height or less
▶ Short throws
▶ Hospitality

35MR11/NSP ▶ p38

35MR11/NSP
▶ Architectural detailing
▶ Small art/13' ceiling height or less
▶ Sculpture
▶ Medium throws
▶ Hospitality

35MR11/VWSP ▶ p39

35MR11/VWSP
▶ Architectural accenting
▶ Small art/11' ceiling height or less
▶ Sculpture
▶ Medium throws
▶ Hospitality

35MR11/NFL ▶ p40

35MR11/NFL
▶ Soft accenting
▶ Feature downlighting
▶ Small art/10' ceiling height or less
▶ Short throws
▶ Hospitality

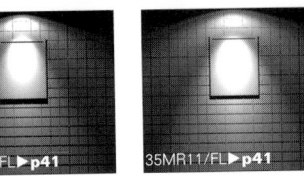 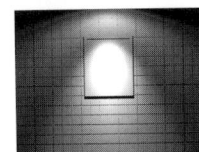
35MR11/FL ▶ p41

35MR11/FL
▶ General lighting
▶ Small art/9' ceiling height or less
▶ Short throws
▶ Hospitality

Halogen/MR11
Standard

▶Requires tempered cover glass.

Halogen
Low Voltage (12V)

4

Halogen Low Voltage MR Lamp Designation
20MR11/C/CG/NFL

Last set of letters identifies beam spread

FLOODS
- 24° to 32° for narrow flood (NFL)
- 33° to 44° for flood (FL)
- 45° to 54° for wide flood (WFL)
- 55° or greater for very wide flood (VWFL)

SPOTS
- 7° or less for very narrow spot (VNSP)
- 8° to 10° for narrow spot (NSP)
- 11° to 14° for spot (SP)
- 15° to 18° for wide spot (WSP)
- 19° to 23° for very wide spot (VWSP)

Ranges of beam spreads used for convenient/consistent reference in this text and do not necessarily correspond to each manufacturer's nor any ANSI or NEMA definitions.

"CG" for built–in cover glass or no designation for no cover glass

Reflector treatment code
- "C" for consistent color (hard dichroic coating) or no letter for inconsistent color
- "AL" for aluminized reflector eliminating back spill light through reflector

These two digits represent lamp diameter in eighths inches (51mm or 2 inches)

These letters indicate this is a multifaceted reflector (MR) lamp

First two digits represent lamp wattage

[This designation used for convenient reference throughout this *Time–Saver Standards for Architectural Lighting*—but not necessarily used by all lamp manufacturers (although it would be convenient if manufacturers would come to agreement on standard designations.]

Artwork Size Categories
⊃Petite = 2'–0" by 2'–0" or smaller
⊃Small = 2'–0" by 2'–0" to 2'–6" by 3'–0"
⊃Medium = 2'–6" by 3'–0" to 3'–0" by 4'–0"
⊃Large = 3'–0" by 4'–0" to 4'–0" by 5'–0"
⊃Extra Large = 4'–0" by 5'–0" or larger

Net Addresses/MR11 Track and Monopoints
http://www.ardeelighting.com/index.htm
http://www.artemide.com/
http://www.cooperlighting.com/
http://www.danalite.com/
http://www.lightolier.com/
http://www.lightproject.com/

CONNECT FOR MORE

▶*Requires tempered cover glass.*

Halogen 20MR11/NSP

Standard

Halogen Low Voltage (12V)

4

Uses
▶ Architectural detailing
▶ Petite art short throws
▶ Objet d'art
▶ Short throws
▶ Available in 12V

Specs
Beam spread/10°
Center beam candlepower/ 5500
Life/3000 hrs
▶**Osram Sylvania** (spot) (55109)

Design Tips
✔ One light best for petite art where mounting height of light is 10' or less.
✔ Intense light degrades some artwork rapidly and/or significantly—consider UV and IR filters or limit exposure time.
✔ Great for inert sculptural pieces and floral centerpieces.
✔ MR lamps are glary—consider shielding with hexcell louver or using deep luminaire trims or pinholes.
✔ Tempered cover glass required for safety.
✔ Required transformer also consumes 3 to 5 watts.

Application Key

Commercial
Gallery
Hospitality
Institutional
Manufacturing
Residential
Retail
Exterior

bold = primary application
partial fade = minimal application
fade = unlikely application

Performance Sketches

[All data outlined here is based on information from boldfaced manufacturer's published data for 12V version at time of manuscript preparation. Scale is ¼ inch equals 1 foot. **Note: At press time, actual photometry was unavailable for this lamp. Data were generated using Osram Sylvania 35MR11/NSP lamp and rerating candlepower—this is imprecise.**]

Art shown at 2' by 2' and centered 5'–6" AFF. Wall grid on 6" by 6" centers and scaled at ¼"=1'.

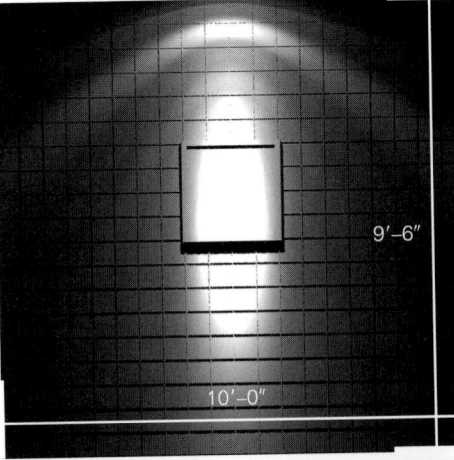

9'–6"

10'–0"

Lamp: 20MR11/NSP
Location: 1'–6" from wall
Center beam Aiming: 17° (at point 4'–6" AFF)
Luminaire: Monopoint or Recessed Adjustable
Illuminance: 27 fc avg, maintained (83 fc max)

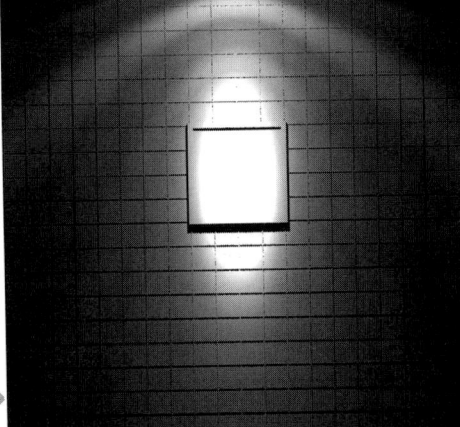

Lamp: 20MR11/NSP
Location: 2'–0" from wall
Center beam Aiming: 25° (at point 5'–3" AFF)
Luminaire: Monopoint or Recessed Adjustable
Illuminance: 37 fc avg, maintained (122 fc max)

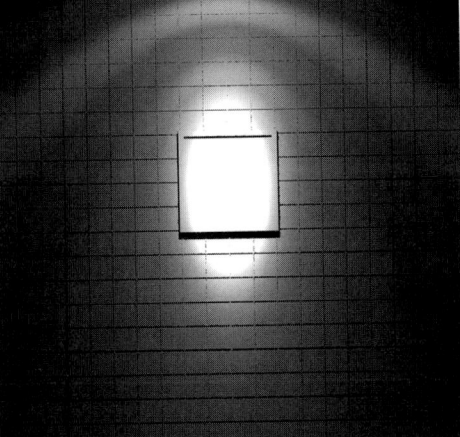

Lamp: 20MR11/NSP
Location: 2'–6" from wall
Center beam Aiming: 30° (at point 5'–3" AFF)
Luminaire: Monopoint or Recessed Adjustable
Illuminance: 39 fc avg, maintained (126 fc max)

Halogen 20MR11/WSP

Standard

► Requires tempered cover glass.

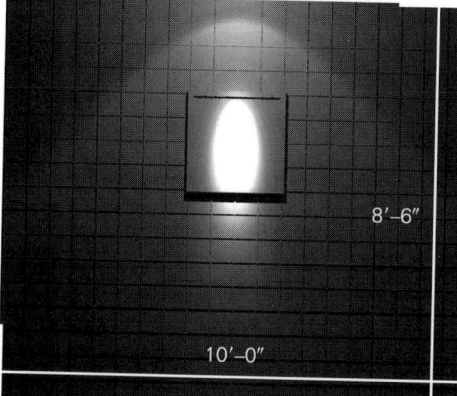

8'–6"

10'–0"

Performance Sketches

[All data outlined here is based on information from boldfaced manufacturer's published data for 12V version at time of manuscript preparation. Scale is ¼ inch equals 1 foot. **Note: At press time, actual photometry was unavailable for this lamp. Data were generated using GE 20MR16/C/WSP lamp and rerating candlepower—this is imprecise.**]

- -
Art shown at 2' by 2' and centered 5'–6" AFF. Wall grid on 6" by 6" centers and scaled at ¼"=1'.

Lamp: 20MR11/WSP
Location: 1'–6" from wall
Center beam Aiming: 25° (at point 5'–3" AFF)
Luminaire: Monopoint or Recessed Adjustable
Illuminance: 14 fc avg, maintained (80 fc max)

Lamp: 20MR11/WSP
Location: 2'–0" from wall
Center beam Aiming: 32° (at point 5'–3" AFF)
Luminaire: Monopoint or Recessed Adjustable
Illuminance: 14 fc avg, maintained (80 fc max)

Lamp: 20MR11/WSP
Location: 2'–6" from wall
Center beam Aiming: 38° (at point 5'–3" AFF)
Luminaire: Monopoint or Recessed Adjustable
Illuminance: 14 fc avg, maintained (76 fc max)

Uses
► Petite art with short throws
► Short throws
► Available in 12V

Specs
Beam spread/15°
Center beam
  candlepower/1760
Life/3500 hrs
► **GE** (30754)

Design Tips
✔ One light best for petite art where mounting height of light is 9' or less.
✔ Scallops/striations are most severe when light is closer to the wall (compare top image to bottom image).
✔ More grazing light creates prominent shadows.
✔ MR lamps are glary—consider shielding with hexcell louver or using deep luminaire trims or pinholes.
✔ Tempered cover glass required for safety.
✔ Required transformer also consumes 3 to 5 watts.

Application Key

Commercial
Gallery
Hospitality
Institutional
Manufacturing
Residential
Retail
Exterior

bold = primary application
partial fade = minimal application
fade = unlikely application

▶*Requires tempered cover glass.*

Halogen 20MR11/NFL

Standard

Halogen Low Voltage (12V)

4

Uses
▶Petite art with short throws
▶Short throws
▶Available in 12V

Specs
Beam spread/30°
Center beam
   candlepower/600
Life/3500 hrs
▶**GE** (30773)

Design Tips
✔One light best for petite art where mounting height of light is 9′ or less.
✔Scallops/striations are most severe when light is closer to the wall (compare top image to bottom image).
✔More grazing light creates prominent shadows.
✔MR lamps are glary—consider shielding with hexcell louver or using deep luminaire trims or pinholes.
✔Tempered cover glass required for safety.
✔Required transformer also consumes 3 to 5 watts.

Performance Sketches

[All data outlined here is based on information from boldfaced manufacturer's published data for 12V version at time of manuscript preparation. Scale is ¼ inch equals 1 foot.]

Art shown at 2′ by 2′ and centered 5′–6″ AFF. Wall grid on 6″ by 6″ centers and scaled at ¼″=1′.

8′–6″

10′–0″

Lamp: 20MR11/NFL
Location: 1′–6″ from wall
Center beam Aiming: 21° (at point 4′–6″ AFF)
Luminaire: Monopoint or Recessed Adjustable
Illuminance: 22 fc avg, maintained (62 fc max)

Lamp: 20MR11/NFL
Location: 2′–0″ from wall
Center beam Aiming: 30° (at point 5′–0″ AFF)
Luminaire: Monopoint or Recessed Adjustable
Illuminance: 26 fc avg, maintained (69 fc max)

Lamp: 20MR11/NFL
Location: 2′–6″ from wall
Center beam Aiming: 36° (at point 5′–0″ AFF)
Luminaire: Monopoint or Recessed Adjustable
Illuminance: 26 fc avg, maintained (60 fc max)

Application Key

Commercial
Gallery
Hospitality
Institutional
Manufacturing
Residential
Retail
Exterior

bold = primary application
partial fade = minimal application
fade = unlikely application

Halogen20MR11/FL

Standard

▶Requires tempered cover glass.

Performance Sketches

[All data outlined here is based on information from boldfaced manufacturer's published data for 12V version at time of manuscript preparation. Scale is ¼ inch equals 1 foot. **Note: At press time, actual photometry was unavailable for this lamp. Data were generated using Osram Sylvania 20MR16/AL/FL lamp and rerating candlepower—this is imprecise.**]

Art shown at 2'–6" by 3' and centered 5'–6" AFF. Wall grid on 6" by 6" centers and scaled at ¼"=1'.

Lamp: 20MR11/FL
Location: 1'–6" from wall
Center beam Aiming: 18° (at point 4'–0" AFF)
Luminaire: Monopoint or Recessed Adjustable
Illuminance: 14 fc avg, maintained (42 fc max)

Lamp: 20MR11/FL
Location: 2'–0" from wall
Center beam Aiming: 30° (at point 5'–0" AFF)
Luminaire: Monopoint or Recessed Adjustable
Illuminance: 18 fc avg, maintained (48 fc max)

Lamp: 20MR11/FL
Location: 2'–6" from wall
Center beam Aiming: 36° (at point 5'–0" AFF)
Luminaire: Monopoint or Recessed Adjustable
Illuminance: 18 fc avg, maintained (42 fc max)

Uses
▶Small art short throws
▶Short throws
▶Available in 12V

Specs
Beam spread/35°
**Center beam
candlepower/700**
Life/3000 hrs
▶**Osram Sylvania** (55107)

Design Tips
✔One light best for small art where mounting height of light is 9' or less.
✔Scallops/striations are most severe when light is closer to the wall (compare top image to bottom image).
✔More grazing light creates prominent shadows.
✔MR lamps are glary—consider shielding with hexcell louver or using deep luminaire trims or pinholes.
✔Tempered cover glass required for safety.
✔Required transformer also consumes 3 to 5 watts.

Halogen Low Voltage (12V)

4

Application Key

Commercial
Gallery
Hospitality
Institutional
Manufacturing
Residential
Retail
Exterior

bold = primary application
partial fade = minimal application
fade = unlikely application

▶ *Requires tempered cover glass.*

Halogen 35MR11/NSP

Standard

Performance Sketches

[All data outlined here is based on information from boldfaced manufacturer's published data for 12V version at time of manuscript preparation. Scale is ¼ inch equals 1 foot.]

Uses
▶ Architectural accenting
▶ Small art with medium throws
▶ Sculpture with medium throws
▶ Medium throws
▶ Available in 12V

Specs
Beam spread/10°
Center beam candlepower/ 5225
Life/3000 hrs
▶ **Osram Sylvania** (55113)

Design Tips
✔ One light best for small art where mounting height for light is 13' or less.
✔ Light may degrade artwork rapidly and/or significantly: consider UV and IR filters or limit exposure time.
✔ Great for glass pieces and floral centerpieces and where mounting height for light is 13' or less.
✔ More grazing light creates prominent shadows.
✔ MR lamps are glary—consider shielding with hexcell louver or using deep luminaire trims or pinholes.
✔ Tempered cover glass required for safety.
✔ Required transformer also consumes 3 to 5 watts.

Application Key

Commercial
Gallery
Hospitality
Institutional
Manufacturing
Residential
Retail
Exterior

bold = primary application
partial fade = minimal application
fade = unlikely application

Lamp: 35MR11/NSP
Location: 3'–0" from wall
Center beam Aiming: 21° (at point 4'–6" AFF)
Luminaire: Monopoint or Recessed Adjustable
Illuminance: 15 fc avg, maintained (30 fc max)

Lamp: 35MR11/NSP
Location: 4'–0" from wall
Center beam Aiming: 27° (at point 4'–6" AFF)
Luminaire: Monopoint or Recessed Adjustable
Illuminance: 16 fc avg, maintained (32 fc max)

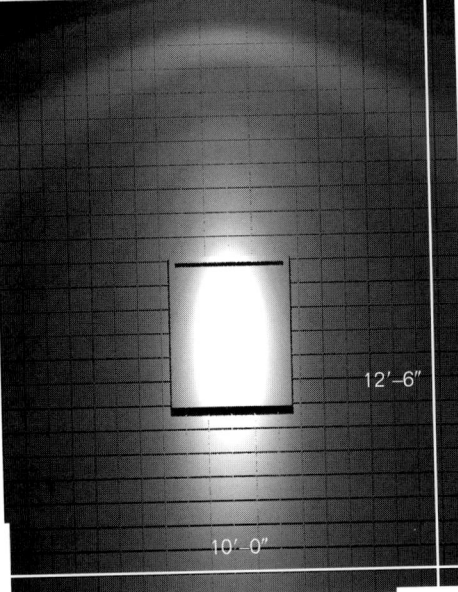

12'–6"

10'–0"

Art shown at 2'–6" by 3' and centered 5'–6" AFF. Wall grid on 6" by 6" centers and scaled at ¼"=1'.

Lamp: 35MR11/NSP
Location: 5'–0" from wall
Center beam Aiming: 32° (at point 4'–6" AFF)
Luminaire: Monopoint or Recessed Adjustable
Illuminance: 16 fc avg, maintained (33 fc max)

Halogen 35MR11/VWSP

Standard

Performance Sketches

[All data outlined here is based on information from boldfaced manufacturer's published data for 12V version at time of manuscript preparation. Scale is ¼ inch equals 1 foot.]

▶Requires tempered cover glass.

Halogen Low Voltage (12V)

4

Lamp: 35MR11/VWSP
Location: 2'–0" from wall
Center beam Aiming: 16° (at point 3'–6" AFF)
Luminaire: Monopoint or Recessed Adjustable
Illuminance: 16 fc avg, maintained (34 fc max)

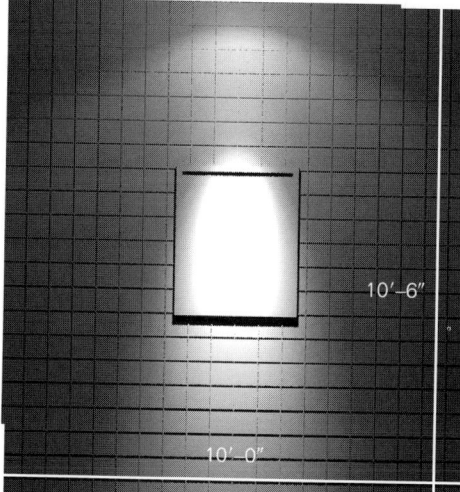

10'–6"

10'–0"

Lamp: 35MR11/VWSP
Location: 2'–6" from wall
Center beam Aiming: 24° (at point 5'–0" AFF)
Luminaire: Monopoint or Recessed Adjustable
Illuminance: 23 fc avg, maintained (50 fc max)

Uses

▶Architectural accenting
▶Small art with medium throws
▶Sculpture with medium throws
▶Medium throws
▶Available in 12V

Specs

Beam spread/20°
Center beam candlepower/ 3000
Life/3500 hrs
▶**GE** (spot) (30774)

Design Tips

✔One light best for small art and where mounting height for light is 11' or less.

✔Good for downlighting onto decorative floor materials— best with sconces and/or art accents and/or wallwashing to avoid cave effect and "I wish I had a visor" effect.

✔MR lamps are glary—consider shielding with hexcell louver or using deep luminaire trims or pinholes.

✔Tempered cover glass required for safety.

✔Required transformer also consumes 3 to 5 watts.

Art shown at 2'–6" by 3' and centered 5'–6" AFF. Wall grid on 6" by 6" centers and scaled at ¼"=1'.

Lamp: 35MR11/VWSP
Location: 3'–0" from wall
Center beam Aiming: 29° (at point 5'–0" AFF)
Luminaire: Monopoint or Recessed Adjustable
Illuminance: 24 fc avg, maintained (49 fc max)

Application Key

Commercial
Gallery
Hospitality
Institutional
Manufacturing
Residential
Retail
Exterior

bold = primary application
partial fade = minimal application
fade = unlikely application

Requires tempered cover glass.

Halogen 35MR11/NFL

Standard

Halogen Low Voltage (12V)

Uses
▶ Soft accenting
▶ Feature downlighting
▶ Small art with short throws
▶ Short throws
▶ Available in 12V

Specs
Beam spread/30°
Center beam candlepower/ 1300
Life/3500 hrs
▶ **GE** (30890)

4

Design Tips
✔ One light best for small art where mounting height for light is 10′ or less.
✔ Good for soft downlighting: best with sconces and/or art accents and/or wallwashing to avoid cave effect.
✔ More grazing light creates prominent shadows.
✔ MR lamps are glary—consider shielding with hexcell louver or using deep luminaire trims or pinholes.
✔ Tempered cover glass required for safety.
✔ Required transformer also consumes 3 to 5 watts.

Performance Sketches

[All data outlined here is based on information from boldfaced manufacturer's published data for 12V version at time of manuscript preparation. Scale is ¼ inch equals 1 foot.]

Art shown at 2′–6″ by 3′ and centered 5′–6″ AFF. Wall grid on 6″ by 6″ centers and scaled at ¼″=1′.

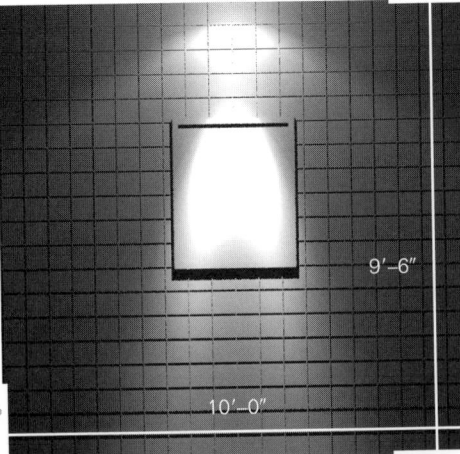
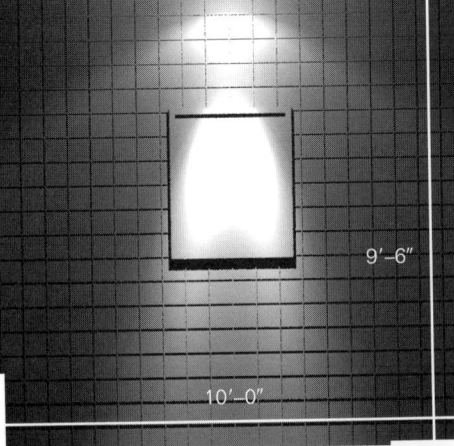

9′–6″

10′–0″

Lamp: 35MR11/NFL
Location: 1′–6″ from wall
Center beam Aiming: 14° (at point 3′–6″ AFF)
Luminaire: Monopoint or Recessed Adjustable
Illuminance: 17 fc avg, maintained (46 fc max)

Lamp: 35MR11/NFL
Location: 2′–0″ from wall
Center beam Aiming: 20° (at point 4′–0″ AFF)
Luminaire: Monopoint or Recessed Adjustable
Illuminance: 19 fc avg, maintained (45 fc max)

Lamp: 35MR11/NFL
Location: 2′–6″ from wall
Center beam Aiming: 27° (at point 4′–6″ AFF)
Luminaire: Monopoint or Recessed Adjustable
Illuminance: 21 fc avg, maintained (48 fc max)

Application Key

Commercial
Gallery
Hospitality
Institutional
Manufacturing
Residential
Retail
Exterior

bold = primary application
partial fade = minimal application
fade = unlikely application

Halogen 35MR11/FL
Standard

▶Requires tempered cover glass.

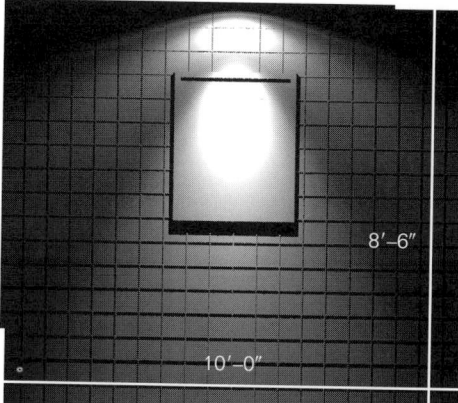

Performance Sketches

[All data outlined here is based on information from boldfaced manufacturer's published data for 12V version at time of manuscript preparation. Scale is ¼ inch equals 1 foot.]

Art shown at 2'–6"' by 3' and centered 5'–6" AFF. Wall grid on 6" by 6" centers and scaled at ¼"=1'.

8'–6"

10'–0"

Lamp: 35MR11/FL
Location: 1'–0" from wall
Center beam Aiming: 8° (at point 1'–0" AFF)
Luminaire: Monopoint or Recessed Adjustable
Illuminance: 13 fc avg, maintained (41 fc max)

Lamp: 35MR11/FL
Location: 1'–6" from wall
Center beam Aiming: 15° (at point 3'–0" AFF)
Luminaire: Monopoint or Recessed Adjustable
Illuminance: 15 fc avg, maintained (35 fc max)

Lamp: 35MR11/FL
Location: 2'–0" from wall
Center beam Aiming: 20° (at point 3'–0" AFF)
Luminaire: Monopoint or Recessed Adjustable
Illuminance: 14 fc avg, maintained (28 fc max)

Uses
▶General lighting
▶Small art with short throws
▶Short throws
▶Available in 12V

Specs
Beam spread/40°
Center beam candlepower/1035
Life/3000 hrs
▶**Osram Sylvania** (55111)

Design Tips
✔One light best for small art where mounting height of light is 9' or less.
✔Light may degrade artwork rapidly and/or significantly: consider UV and IR filters or limit exposure time.
✔Good for downlighting onto decorative floor materials—best with sconces and/or art accents and/or wallwashing to avoid cave effect and grazing glare (the "I wish I had a visor" effect).
✔More grazing light creates prominent shadows.
✔MR lamps are glary—consider shielding with hexcell louver or using deep luminaire trims or pinholes.
✔Tempered cover glass required for safety.
✔Required transformer also consumes 3 to 5 watts.

Application Key

Commercial
Gallery
Hospitality
Institutional
Manufacturing
Residential
Retail
Exterior

bold = primary application
partial fade = minimal application
fade = unlikely application

Halogen Low Voltage (12V)

4

▼ Has integral cover glass.

Halogen Low Voltage (12V)

4

Halogen/MR11/CG
Cover Glass

Guide
- ⟳ Longer life is better
- ⟳ Higher efficacy better for general lighting
- ⟳ Lower wattage is better
- ⟳ Lower color of light is warmer
- ⟳ Moderate cost range for halogen is about US$12.⁰⁰ to US$17.⁰⁰

Stats
Data varies manufacturer to manufacturer and changes from time to time

Life/3000 hours
A typical office environment might be occupied about 2600 hours each year.

Efficacy/Not reported
Excellent beam intensities and distributions available for the wattage.

Wattages/20 and 35

Color of Light/2925°K

Beam spreads/NSP, SP, NFL and FL
Narrow spot, spot, narrow flood and flood distributions are available.

Size/1⅜" L by 1⅜" Ø

Voltage/12V

Cost Magnitude/Low (2000)

Manufacturers
Philips

Net Addresses
http://www.lighting.philips.com/

CONNECT FOR MORE

Advantages
Crisp white light
Good intensities available
Very small size
Easily dimmed
Low initial cost
Alternative to less efficient R and halogen PAR lamps
Withstand extreme temperature range
Very little lamp lumen depreciation (light loss) over life

Disadvantages
Beam aberrations
♦ Striations
♦ Inconsistent beam shape
Color inconsistencies
♦ Lamp to lamp color variations
♦ Over time, color shifts—very evident with exposed–lamp monopoints or trackheads where back of lamp is visible
Integral cover glass
♦ Tempered safety glass may still be required (check with lamp manufacturer)
♦ May prevent use of auxiliary accessories (e.g., color filters, louvers, etc.)
High operating costs
♦ Low wattages may make these cost prohibitive in typical commercial applications
Hot to touch
Tempered glass lens may be required
Transformer required
♦ Additional cost and visual aspect of transformer must be accommodated
♦ Additional wattage requirement of transformer of 3 to 5 watts
♦ Potential for audible hum

For more efficient accent lighting, use HIR low voltage lamps (Section 6); **HIR PAR lamps** (Section 5) **or CMH PAR lamps** (Section 7).

Halogen/MR11/CG
Cover Glass

▶Has integral cover glass.

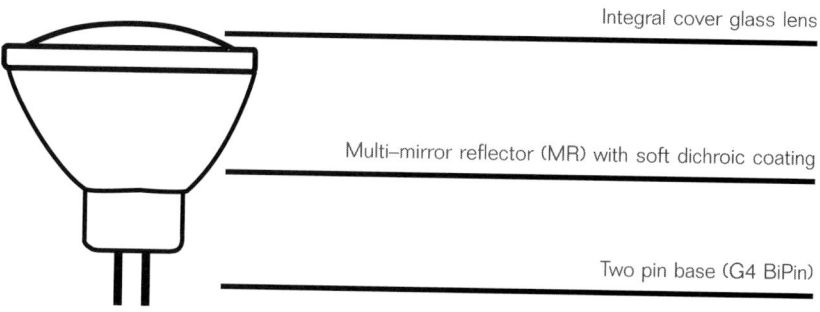

Integral cover glass lens

Multi–mirror reflector (MR) with soft dichroic coating

Two pin base (G4 BiPin)

Lamp shown actual size
Image copyright 2000/Philips

Uses

Architectural Accenting
Residential
Art Accenting
Residential
Downlighting
Residential

MR11 Lighting Design Tips

✔Best at lower mounting heights (less than 10′–0″).
✔Dimming increases life (typically doubling life), but shifts color warmer.
✔Very glary when used without much other general lighting or when aimed across long distances where ceilings are low.
✔Beam striations are most apparent on light–colored, nontextured flat surfaces and may be objectionable.
✔MR11 downlighting alone is harsh—combine several techniques for best results.
✔Use smaller scale (e.g., 4–inch diameter) downlights and adjustable accents for attractive, more human–scale approach—excellent in residential applications.
✔Integral cover glass may interfere with auxiliary add–on lenses, louvers or accessories.
✔Use in recessed luminaires or in details or in enclosed trackheads to avoid seeing color inconsistencies which are most visible from back of lamp.

Net Addresses/MR11 Track and Monopoints
http://www.ardeelighting.com/index.htm
http://www.artemide.com/
http://www.cooperlighting.com/
http://www.danalite.com/
http://www.lightolier.com/
http://www.lightproject.com/

CONNECT FOR MORE

▶**Has integral cover glass.**

Halogen/MR11/CG

MR11 Cover Glass Visual Index

Halogen
Low Voltage (12V)

4

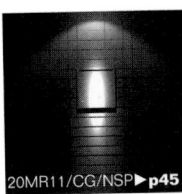
20MR11/CG/NSP▶**p45**

20MR11/CG/NSP▶**p45**

20MR11/CG/NSP▶**p45**

20MR11/CG/NSP
▶Architectural detailing
▶Petite art/10' ceiling height or less
▶Objet d'art
▶Gallery, Hospitality, Residential

20MR11/CG/NFL▶**p46**

20MR11/CG/NFL▶**p46**

20MR11/CG/NFL▶**p46**

20MR11/CG/NFL
▶Petite art/9' ceiling height or less
▶Short throws
▶Gallery, Residential

Throws/Ceilings
⊃Short = 7' to 10'
⊃Medium = 10' to 15'
⊃Long = 15' to 20'
⊃Very Long = 20' or more

Halogen Low Voltage MR Lamp Designation
20MR11/C/CG/NFL

Last set of letters identifies beam spread
FLOODS • 24° to 32° for narrow flood (NFL)
• 33° to 44° for flood (FL)
• 45° to 54° for wide flood (WFL)
• 55° or greater for very wide flood (VWFL)
SPOTS • 7° or less for very narrow spot (VNSP)
• 8° to 10° for narrow spot (NSP)
• 11° to 14° for spot (SP)
• 15° to 18° for wide spot (WSP)
• 19° to 23° for very wide spot (VWSP)

Ranges of beam spreads used for convenient/consistent reference in this text and do not necessarily correspond to each manufacturer's nor any ANSI or NEMA definitions.

"CG" for built–in cover glass or no designation for no cover glass

Reflector treatment code
• "C" for consistent color (hard dichroic coating) or no letter for inconsistent color
• "AL" for aluminized reflector eliminating back spill light through reflector

These two digits represent lamp diameter in eighths inches (51mm or 2 inches)

These letters indicate this is a multifaceted reflector (MR) lamp

First two digits represent lamp wattage

[This designation used for convenient reference throughout this *Time–Saver Standards for Architectural Lighting*—but not necessarily used by all lamp manufacturers (although it would be convenient if manufacturers would come to agreement on standard designations.]

Artwork Size Categories
⊃Petite = 2'–0" by 2'–0" or smaller
⊃Small = 2'–0" by 2'–0" to 2'–6" by 3'–0"
⊃Medium = 2'–6" by 3'–0" to 3'–0" by 4'–0"
⊃Large = 3'–0" by 4'–0" to 4'–0" by 5'–0"
⊃Extra Large = 4'–0" by 5'–0" or larger

Recommended intensities
for art accenting

Low (gallery, museum, residential)
5 to 10 fc average on art
Moderate (hospitality)
10 to 20 fc average on art
High (commercial—or higher for retail)
20 to 30 fc average on art

These are intended to be average, maintained values including effects of accent lighting and general room lighting. General room lighting contributes just a few footcandles in residential, gallery and museum settings, but contributes 5 to 10 fc in commercial settings. Maximums might range from 50 percent to 100 percent of the average value. Wherever artwork is highly valuable and/or sensitive to light, precautions should be taken to limit light exposure (maintaining low levels and/or switching lights off when art is not being viewed); and exposure to infrared and ultraviolet should be limited (using filters on lamps or placing art behind specially treated glass or acrylic).

Net Addresses/MR11 Recessed
http://www.ardeelighting.com/index.htm
http://www.artemide.com/
http://www.bartcolighting.com/index2.html

CONNECT FOR MORE

Halogen 20MR11/CG/NSP
Cover Glass

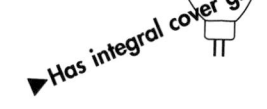
▶Has integral cover glass.

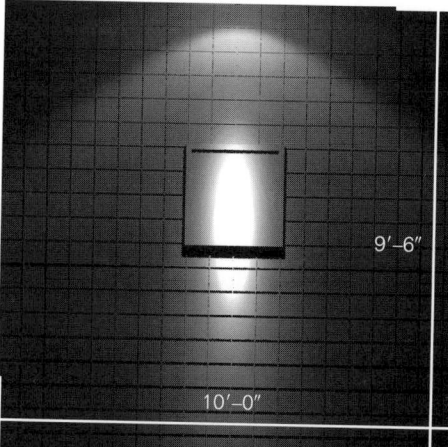

9'–6"

10'–0"

Performance Sketches

[All data outlined here is based on information from boldfaced manufacturer's published data for 12V version at time of manuscript preparation. Scale is ¼ inch equals 1 foot.]

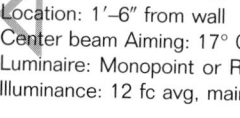
Art shown at 2' by 2' and centered 5'–6" AFF. Wall grid on 6" by 6" centers and scaled at ¼"=1'.

Lamp: 20MR11/CG/NSP
Location: 1'–6" from wall
Center beam Aiming: 17° (at point 4'–6" AFF)
Luminaire: Monopoint or Recessed Adjustable
Illuminance: 12 fc avg, maintained (57 fc max)

Lamp: 20MR11/CG/NSP
Location: 2'–0" from wall
Center beam Aiming: 24° (at point 5'–0" AFF)
Luminaire: Monopoint or Recessed Adjustable
Illuminance: 15 fc avg, maintained (85 fc max)

Lamp: 20MR11/CG/NSP
Location: 2'–6" from wall
Center beam Aiming: 29° (at point 5'–0" AFF)
Luminaire: Monopoint or Recessed Adjustable
Illuminance: 15 fc avg, maintained (88 fc max)

Halogen Low Voltage (12V)
4

Uses
▶Architectural detailing
▶Petite art with short throws
▶Objet d'art with short throws
▶Available in 12V

Specs
Beam spread/10°
Center beam candlepower/5500
Life/3000 hrs
▶**Philips** (30104–4)

Design Tips
✔One light best for small art where mounting height of light is 10' or less.
✔Intense light degrades some artwork rapidly and/or significantly—consider UV and IR filters or limit exposure time.
✔Great for inert sculptural pieces and floral centerpieces.
✔MR lamps are glary—consider shielding with hexcell louver or using deep luminaire trims or pinholes.
✔Required transformer also consumes 3 to 5 watts.

Application Key
Commercial
Gallery
Hospitality
Institutional
Manufacturing
Residential
Retail
Exterior

bold = primary application
partial fade = minimal application
fade = unlikely application

► Has integral cover glass.

Halogen 20MR11/CG/NFL

Cover Glass

Halogen Low Voltage (12V)

Uses
► Petite art
► Short throws
► Available in 12V

Specs
Beam spread/30°
Center beam candlepower/600
Life/3000 hrs
► **Philips** (flood) (30109–3)

Design Tips
✔ One light best for small art where mounting height of light is less than 9'.
✔ Scallops/striations are most severe when light is closer to the wall (compare top image to bottom image).
✔ More grazing light creates prominent shadows.
✔ MR lamps are glary—consider shielding with hexcell louver or using deep luminaire trims or pinholes.
✔ Required transformer also consumes 3 to 5 watts.

Performance Sketches

[All data outlined here is based on information from boldfaced manufacturer's published data for 12V version at time of manuscript preparation. Scale is ¼ inch equals 1 foot.]

Art shown at 2' by 2' and centered 5'–6" AFF. Wall grid on 6" by 6" centers and scaled at ¼"=1'.

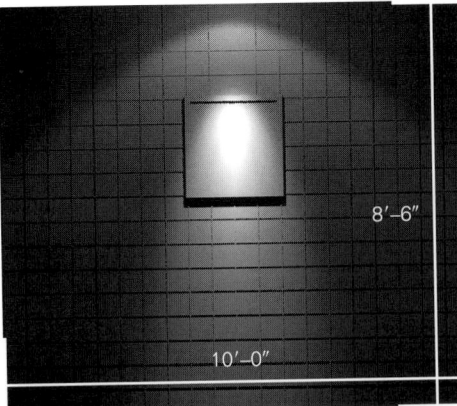

8'–6"

10'–0"

Lamp: 20MR11/CG/NFL
Location: 1'–6" from wall
Center beam Aiming: 21° (at point 4'–6" AFF)
Luminaire: Monopoint or Recessed Adjustable
Illuminance: 7 fc avg, maintained (18 fc max)

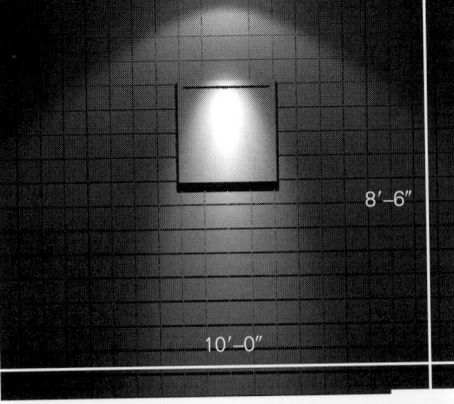

Lamp: 20MR11/CG/NFL
Location: 2'–0" from wall
Center beam Aiming: 30° (at point 5'–0" AFF)
Luminaire: Monopoint or Recessed Adjustable
Illuminance: 9 fc avg, maintained (21 fc max)

Application Key
Commercial
Gallery
Hospitality
Institutional
Manufacturing
Residential
Retail
Exterior

bold = primary application
partial fade = minimal application
fade = unlikely application

Lamp: 20MR11/CG/NFL
Location: 2'–6" from wall
Center beam Aiming: 36° (at point 5'–0" AFF)
Luminaire: Monopoint or Recessed Adjustable
Illuminance: 9 fc avg, maintained (20 fc max)

▶Requires tempered cover glass.

Halogen Low Voltage (12V)

4

Halogen/MR16

Standard

Stats<small>Data varies manufacturer to manufacturer and changes from time to time</small>

Life/2000 and 6000 hours

Varies manufacturer to manufacturer. A typical office environment might be occupied about 2600 hours each year.

Efficacy/Not reported

Excellent beam intensities and distributions available for the wattage.

Wattages/20 and 73

Color of Light/2900°K to 3200°K

Varies manufacturer to manufacturer and wattage to wattage. Typically the longer lived lamps have slightly lower color temperature and the higher wattage lamps have slightly higher color temperature.

Beam spreads/VNSP, NSP, SP, NFL, FL, WFL and VWFL

Narrow spot, spot, narrow flood and flood distributions are available.

Size/1⅞" L by 2" Ø

Voltage/12V

Cost Magnitude/Low (2000)

Manufacturers

General Electric
Osram Sylvania
Philips

Net Addresses

http://www.ge.com/lighting/business/index.htm
http://ecom.sylvania.com/osicatalog/
http://www.lighting.philips.com/

▸CONNECT FOR MORE

Advantages

Crisp white light
Good intensities available
Very small size
Easily dimmed
Low initial cost
Alternative to less efficient R and halogen PAR lamps
Withstand extreme temperature range
Very little lamp lumen depreciation (light loss) over life

Disadvantages

Beam aberrations
◆Striations
◆Inconsistent beam shape
Color inconsistencies
◆Lamp to lamp color variations
◆Over time, color shifts—very evident with exposed–lamp monopoints or trackheads where back of lamp is visible
Hot to touch
Tempered glass lens required
◆Prevents glass and filament shards from falling onto flammable surfaces/materials should a violent lamp failure occur (known as a non–passive failure)
Transformer required
◆Additional cost and visual aspect of transformer must be accommodated
◆Additional wattage requirement of transformer of 3 to 5 watts
◆Potential for audible hum

Guide

⊃Longer life is better
⊃Higher efficacy better for general lighting
⊃Lower wattage is better
⊃Lower color of light is warmer
⊃Very low cost range for halogen is less than US$7.⁵⁰

For more efficient accent lighting, use HIR low voltage lamps (Section 6); HIR PAR lamps (Section 5) or CMH PAR lamps (Section 7).

Halogen/MR16

Standard

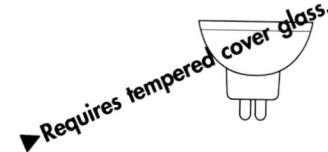

▶Requires tempered cover glass.

Halogen Low Voltage (12V)

4

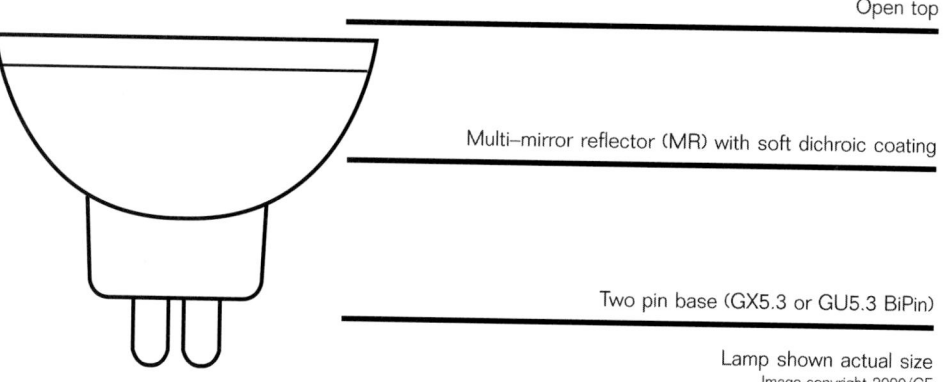

Open top

Multi–mirror reflector (MR) with soft dichroic coating

Two pin base (GX5.3 or GU5.3 BiPin)

Lamp shown actual size
Image copyright 2000/GE

Uses

Architectural Accenting

Art Accenting
Commercial
Gallery
Hospitality
Residential

Downlighting
Lobbies
Conference rooms
Hospitality
Residential
Retail

Merchandising

Surface Grazing
Specular/semispecular surfaces
◆Polished/honed stone
◆Polished/satin wood
◆Fritted/etched glass
◆Cut glass

■ ■ ■ ■ ■ ■ ■ ■ ■ ■ ■ ■ ■ ■ ■ ■

MR16 Lighting Design Tips

✔Dimming increases life (typically doubling life), but shifts color warmer.

✔Very glary when used without much other general lighting or when aimed across long distances where ceilings are low.

✔Beam striations are most apparent on light–colored, nontextured flat surfaces and may be objectionable.

✔MR16 downlighting alone is harsh—combine several techniques for best results.

✔Use smaller scale (e.g., 4–inch diameter) downlights and adjustable accents for attractive, more human–scale approach—excellent in residential applications.

✔Use tempered glass lens for safety.

✔Use in recessed luminaires or in details or in enclosed trackheads to avoid seeing color inconsistencies which are most visible from back of lamp.

Net Addresses/MR16 Track, Monopoints, Recessed and Exterior
See page 4–50

CONNECT FOR MORE

▶ Requires tempered cover glass.

Halogen/MR16

Standard

Halogen Low Voltage (12V)

4

Net Addresses/MR16 Track and Monopoints
http://www.ardeelighting.com/index.htm
http://www.artemide.com/
http://www.bartcolighting.com/index2.html
http://www.thomasltg.com/
http://www.con-techlighting.com/
http://www.cooperlighting.com/
http://www.flosusa.com/
http://www.hubbell-ltg.com/products.htm#Down&Track
http://www.danalite.com/
http://www.kramerlighting.com/
http://www.lightingservicesinc.com/
http://www.lightolier.com/
http://www.lightproject.com/
http://www.lithonia.com/
http://www.prescolite.com/
http://www.techlighting.com/pro1.html
http://www.tslight.com/

CONNECT FOR MORE

Net Addresses/MR16 Recessed
http://www.alkco.com/rtrak.htm
http://www.ardeelighting.com/index.htm
http://www.artemide.com/
http://www.bartcolighting.com/index2.html
http://www.thomasltg.com/
http://www.con-techlighting.com/
http://www.cooperlighting.com/
http://www.hubbell-ltg.com/products.htm#Down&Track
http://www.danalite.com/
http://www.kramerlighting.com/
http://www.lightolier.com/
http://www.lithonia.com/
http://www.prescolite.com/
http://www.progresslighting.com/

CONNECT FOR MORE

Net Addresses/Exterior Accent
http://www.bega-us.com/home/home.html
http://www.hadcolighting.com/html/bronzelite.htm
http://www.hydrel.com
http://www.kimlighting.com/ingrd1.html

CONNECT FOR MORE

Halogen/MR16

Standard

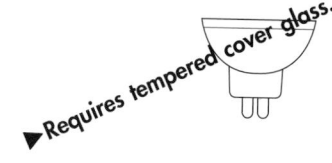

▶Requires tempered cover glass.

Halogen Low Voltage (12V)

4

Figure 4.4
Low voltage halogen 50MR16/NSP lamps are used in low voltage cable lights. These are essentially miniature trackheads spanning between two low voltage cables. The MR16 lamp is sufficiently small and lightweight that this concept is achievable. A tempered glass lens is held in an aluminum ring just in front of the lamp.Cable lights are Translite SF12V/ Byrdy.

Image copyright 2000 Robert Eovaldi

▶ Requires tempered cover glass.

Halogen Low Voltage (12V)

4

Halogen/MR16

Standard MR16 Visual Index

Throws/Ceilings
◗ Short = 7' to 10'
◗ Medium = 10' to 15'
◗ Long = 15' to 20'
◗ Very Long = 20' or more

20MR16/NSP8° ▶p55   20MR16/NSP8° ▶p55   20MR16/NSP8° ▶p55

20MR16/NSP8°
▶ Architectural detailing
▶ Petite art/10' ceiling height or less
▶ Objet d'art
▶ Commercial, Hospitality

20MR16/NSP10° ▶p56   20MR16/NSP10° ▶p56   20MR16/NSP10° ▶p56

20MR16/NSP10°
▶ Architectural detailing
▶ Small art/10' ceiling height or less
▶ Objet d'art
▶ Hospitality

20MR16/FL ▶p57   20MR16/FL ▶p57   20MR16/FL ▶p57

20MR16/FL
▶ General lighting
▶ Petite art/9' ceiling height or less
▶ Short throws
▶ Commercial, Hospitality

35MR16/NSP ▶p58   35MR16/NSP ▶p58   35MR16/NSP ▶p58

35MR16/NSP
▶ Architectural accenting
▶ Small art/13' ceiling height or less
▶ Sculpture
▶ Medium throws
▶ Commercial, Retail

35MR16/VWSP ▶p59   35MR16/VWSP ▶p59   35MR16/VWSP ▶p59

35MR16/VWSP
▶ Architectural accenting
▶ Small art/11' ceiling height or less
▶ Sculpture
▶ Medium throws
▶ Hospitality

Recommended intensities
for art accenting

Low (gallery, museum, residential)
5 to 10 fc average on art
Moderate (hospitality)
10 to 20 fc average on art
High (commercial—or higher for retail)
20 to 30 fc average on art

These are intended to be average, maintained values including effects of accent lighting and general room lighting. General room lighting contributes just a few footcandles in residential, gallery and museum settings, but contributes 5 to 10 fc in commercial settings. Maximums might range from 50 percent to 100 percent of the average value. Wherever artwork is highly valuable and/or sensitive to light, precautions should be taken to limit light exposure (maintaining low levels and/or switching lights off when art is not being viewed); and exposure to infrared and ultraviolet should be limited (using filters on lamps or placing art behind specially treated glass or acrylic).

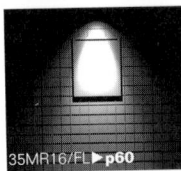  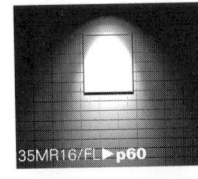

35MR16/FL ▶p60   35MR16/FL ▶p60   35MR16/FL ▶p60

35MR16/FL
▶ General lighting
▶ Small art/9' ceiling height or less
▶ Short throws
▶ Commercial, Retail

50MR16/NSP ▶p61   50MR16/NSP ▶p61   50MR16/NSP ▶p61

50MR16/NSP
▶ Architectural detailing
▶ Medium art/16' ceiling height or less
▶ Sculpture
▶ Long throws
▶ Hospitality

Halogen/MR16

Standard

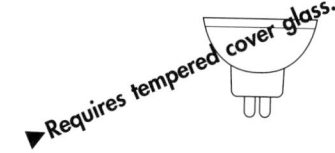
▶Requires tempered cover glass.

Halogen Low Voltage (12V)

4

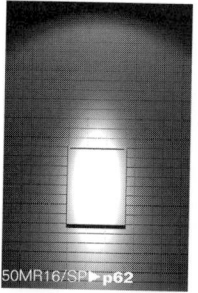  

50MR16/SP
- ▶Architectural detailing
- ▶Medium art/16' ceiling height or less
- ▶Sculpture
- ▶Long throws
- ▶Commercial, Hospitality, Retail

50MR16/NFL
- ▶Architectural accenting
- ▶Medium art/11' ceiling height or less
- ▶Sculpture
- ▶Medium throws
- ▶Commercial, Retail

  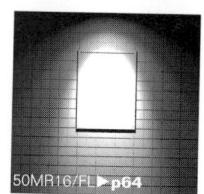

50MR16/FL
- ▶General lighting
- ▶Medium art/10' ceiling height or less
- ▶Short throws
- ▶Commercial, Retail

  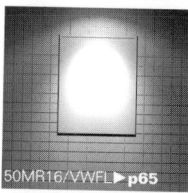

50MR16/VWFL
- ▶Soft general lighting
- ▶Wallwashing
- ▶Large art/10' ceiling height or less
- ▶Short throws
- ▶Hospitality

65MR16/NSP
- ▶Architectural detailing
- ▶Large art/19' ceiling height or less
- ▶Sculpture
- ▶Long throws
- ▶Hospitality

65MR16/NFL
- ▶Architectural accenting
- ▶Feature downlighting
- ▶Large art/13' ceiling height or less
- ▶Sculpture
- ▶Medium throws
- ▶Hospitality

Requires tempered cover glass.

Halogen/MR16

Standard

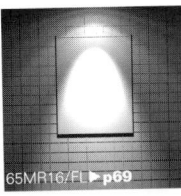 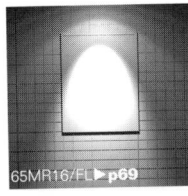 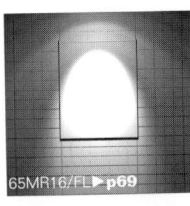

65MR16/FL▶p69   65MR16/FL▶p69   65MR16/FL▶p69

65MR16/FL
▶General lighting
▶Large art/10′ ceiling height or less
▶Short throws
▶Commercial, Hospitality

73MR16/SP▶p70   73MR16/SP▶p70   73MR16/SP▶p71

73MR16/SP
▶Architectural detailing
▶Large art/19′ ceiling height or less
▶Sculpture
▶Long throws
▶Hospitality

 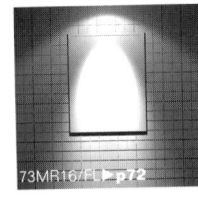 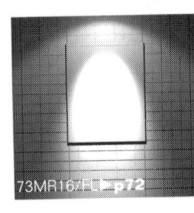

73MR16/FL▶p72   73MR16/FL▶p72   73MR16/FL▶p72

73MR16/FL
▶General lighting
▶Large art/10′ ceiling height or less
▶Short throws
▶Commercial, Retail

Halogen Low Voltage MR Lamp Designation

20MR16/C/CG/NFL

Last set of letters identifies beam spread

FLOODS
• 24° to 32° for narrow flood (NFL)
• 33° to 44° for flood (FL)
• 45° to 54° for wide flood (WFL)
• 55° or greater for very wide flood (VWFL)

SPOTS
• 7° or less for very narrow spot (VNSP)
• 8° to 10° for narrow spot (NSP)
• 11° to 14° for spot (SP)
• 15° to 18° for wide spot (WSP)
• 19° to 23° for very wide spot (VWSP)

Ranges of beam spreads used for convenient/consistent reference in this text and do not necessarily correspond to each manufacturer's nor any ANSI or NEMA definitions.

"CG" for built–in cover glass or no designation for no cover glass

Reflector treatment code
• "C" for consistent color (hard dichroic coating) or no letter for inconsistent color
• "AL" for aluminized reflector eliminating back spill light through reflector

These two digits represent lamp diameter in eighths inches (51mm or 2 inches)

These letters indicate this is a multifaceted reflector (MR) lamp

First two digits represent lamp wattage

[This designation used for convenient reference throughout this *Time–Saver Standards for Architectural Lighting*—but not necessarily used by all lamp manufacturers (although it would be convenient if manufacturers would come to agreement on standard designations.]

Artwork Size Categories
⟩Petite = 2′–0″ by 2′–0″ or smaller
⟩Small = 2′–0″ by 2′–0″ to 2′–6″ by 3′–0″
⟩Medium = 2′–6″ by 3′–0″ to 3′–0″ by 4′–0″
⟩Large = 3′–0″ by 4′–0″ to 4′–0″ by 5′–0″
⟩Extra Large = 4′–0″ by 5′–0″ or larger

Halogen Low Voltage (12V) 4

Halogen 20MR16/NSP8°
Standard

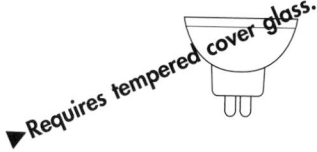

▶Requires tempered cover glass.

Halogen Low Voltage (12V)

4

Performance Sketches

[All data outlined here is based on information from boldfaced manufacturer's published data for 12V version at time of manuscript preparation. Scale is ¼ inch equals 1 foot.]

Art shown at 2' by 2' and centered 5'–6" AFF. Wall grid on 6" by 6" centers and scaled at ¼"=1'.

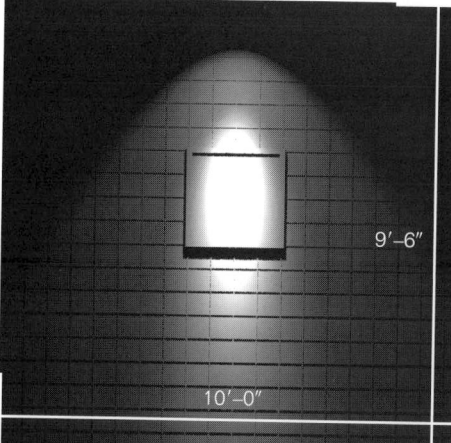

9'–6"

10'–0"

Lamp: 20MR16/NSP8°
Location: 1'–6" from wall
Center beam Aiming: 18° (at point 5'–0" AFF)
Luminaire: Monopoint or Recessed Adjustable
Illuminance: 20 fc avg, maintained (78 fc max)

Lamp: 20MR16/NSP8°
Location: 2'–0" from wall
Center beam Aiming: 24° (at point 5'–0" AFF)
Luminaire: Monopoint or Recessed Adjustable
Illuminance: 22 fc avg, maintained (85 fc max)

Lamp: 20MR16/NSP8°
Location: 2'–6" from wall
Center beam Aiming: 29° (at point 5'–0" AFF)
Luminaire: Monopoint or Recessed Adjustable
Illuminance: 23 fc avg, maintained (88 fc max)

Uses
▶Architectural detailing
▶Petite art with short throws
▶Objet d'art with short throws
▶Available in 12V

Specs
Beam spread/8°
Center beam
candlepower/5000
Life/4000 hrs
▶**Osram Sylvania** (58601)

Design Tips
✔One light best for petite art where mounting height of light is 10' or less.
✔Intense light degrades some artwork rapidly and/or significantly—consider UV and IR filters or limit exposure time.
✔Great for inert sculptural pieces and floral centerpieces.
✔MR lamps are glary—consider shielding with hexcell louver or using deep luminaire trims or pinholes.
✔Tempered cover glass required for safety.
✔Required transformer also consumes 3 to 5 watts.

Application Key

Commercial
Gallery
Hospitality
Institutional
Manufacturing
Residential
Retail
Exterior

bold = primary application
partial fade = minimal application
fade = unlikely application

▶ Requires tempered cover glass.

Halogen 20MR16/NSP10°

Standard

Halogen Low Voltage (12V)

4

Uses
▶ Architectural detailing
▶ Small art with short throws
▶ Objet d'art with short throws
▶ Available in 12V

Specs
Beam spread/10°
Center beam candlepower/4200
Life/4000 hrs
▶ **Philips** (28218–6)

Design Tips
✔ One light best for small art where mounting height of light is 10' or less.
✔ Intense light degrades some artwork rapidly and/or significantly—consider UV and IR filters or limit exposure time.
✔ Great for inert sculptural pieces and floral centerpieces.
✔ MR lamps are glary—consider shielding with hexcell louver or using deep luminaire trims or pinholes.
✔ Tempered cover glass required for safety.
✔ Required transformer also consumes 3 to 5 watts.

Performance Sketches

[All data outlined here is based on information from boldfaced manufacturer's published data for 12V version at time of manuscript preparation. Scale is ¼ inch equals 1 foot.]

Art shown at 2'–6" by 3' and centered 5'–6" AFF. Wall grid on 6" by 6" centers and scaled at ¼"=1'.

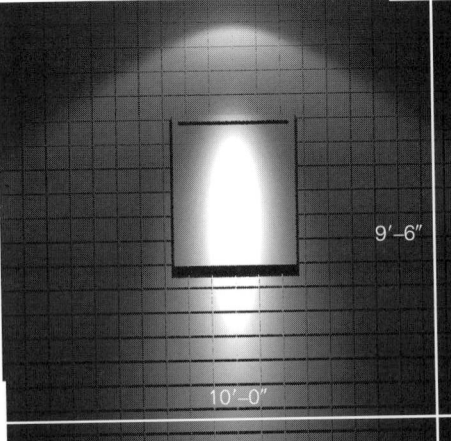

9'–6"

10'–0"

Lamp: 20MR16/NSP10°
Location: 1'–6" from wall
Center beam Aiming: 14° (at point 3'–6" AFF)
Luminaire: Monopoint or Recessed Adjustable
Illuminance: 11 fc avg, maintained (33 fc max)

Lamp: 20MR16/NSP10°
Location: 2'–0" from wall
Center beam Aiming: 24° (at point 5'–0" AFF)
Luminaire: Monopoint or Recessed Adjustable
Illuminance: 17 fc avg, maintained (70 fc max)

Lamp: 20MR16/NSP10°
Location: 2'–6" from wall
Center beam Aiming: 29° (at point 5'–0" AFF)
Luminaire: Monopoint or Recessed Adjustable
Illuminance: 18 fc avg, maintained (74 fc max)

Application Key

Commercial
Gallery
Hospitality
Institutional
Manufacturing
Residential
Retail
Exterior

bold = primary application
partial fade = minimal application
fade = unlikely application

Halogen 20MR16/FL
Standard

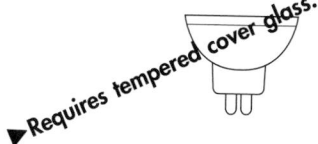

Halogen Low Voltage (12V)

4

Performance Sketches

[All data outlined here is based on information from boldfaced manufacturer's published data for 12V version at time of manuscript preparation. Scale is ¼ inch equals 1 foot.]

Art shown at 2' by 2' and centered 5'–6" AFF. Wall grid on 6" by 6" centers and scaled at ¼"=1'.

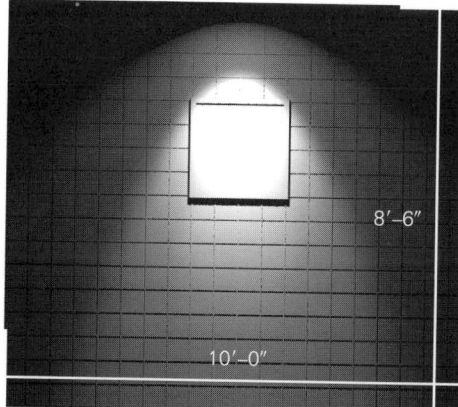

8'–6"

10'–0"

Lamp: 20MR16/FL
Location: 2'–0" from wall
Center beam Aiming: 27° (at point 4'–6" AFF)
Luminaire: Monopoint or Recessed Adjustable
Illuminance: 19 fc avg, maintained (34 fc max)

Lamp: 20MR16/FL
Location: 2'–6" from wall
Center beam Aiming: 32° (at point 4'–6" AFF)
Luminaire: Monopoint or Recessed Adjustable
Illuminance: 19 fc avg, maintained (29 fc max)

Lamp: 20MR16/FL
Location: 3'–0" from wall
Center beam Aiming: 41° (at point 5'–0" AFF)
Luminaire: Monopoint or Recessed Adjustable
Illuminance: 20 fc avg, maintained (30 fc max)

Uses
▶General lighting
▶Petite art with short throws
▶Available in 12V

Specs
Beam spread/38°
Center beam candlepower/700
Life/4000 hrs
▶**Osram Sylvania** (58600)
▶Philips

Design Tips
✔One light best for petite art where mounting height of light is 9' or less.
✔Multiple units can be used for wallwashing or lighting of larger artwork.
✔More grazing light creates prominent shadows.
✔MR lamps are glary—consider shielding with hexcell louver or using deep luminaire trims or pinholes.
✔Tempered cover glass required for safety.
✔Required transformer also consumes 3 to 5 watts.

Application Key

Commercial
Gallery
Hospitality
Institutional
Manufacturing
Residential
Retail
Exterior

bold = primary application
partial fade = minimal application
fade = unlikely application

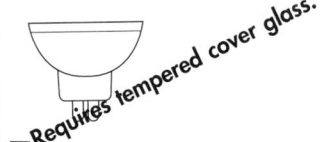

▶ *Requires tempered cover glass.*

Halogen Low Voltage (12V)

4

Halogen 35MR16/NSP

Standard

Uses
▶ Architectural accenting
▶ Small art with medium throws
▶ Sculpture with medium throws
▶ Medium throws
▶ Available in 12V

Specs
Beam spread/8°
Center beam candlepower/ 11000
Life/4000 hrs
▶ **Osram Sylvania** (58604)

Design Tips
✔ One light best for small art and where mounting height for light is 13' or less.
✔ Light may degrade artwork rapidly and/or significantly: consider UV and IR filters or limit exposure time.
✔ Scallops/striations are most severe when light is closer to the wall (compare top left image to bottom image).
✔ More grazing light creates prominent shadows.
✔ MR lamps are glary—consider shielding with hexcell louver or using deep luminaire trims or pinholes.
✔ Tempered cover glass required for safety.
✔ Required transformer also consumes 3 to 5 watts.

Application Key

Commercial
Gallery
Hospitality
Institutional
Manufacturing
Residential
Retail
Exterior

bold = primary application
partial fade = minimal application
fade = unlikely application

Performance Sketches

[All data outlined here is based on information from boldfaced manufacturer's published data for 12V version at time of manuscript preparation. Scale is ¼ inch equals 1 foot.]

Lamp: 35MR16/NSP
Location: 3'–0" from wall
Center beam Aiming: 21° (at point 4'–6" AFF)
Luminaire: Monopoint or Recessed Adjustable
Illuminance: 23 fc avg, maintained (57 fc max)

12'–6"

10'–0"

Lamp: 35MR16/NSP
Location: 4'–0" from wall
Center beam Aiming: 27° (at point 4'–6" AFF)
Luminaire: Monopoint or Recessed Adjustable
Illuminance: 25 fc avg, maintained (61 fc max)

Art shown at 2'–6" by 3' and centered 5'–6" AFF. Wall grid on 6" by 6" centers and scaled at ¼"=1'.

Lamp: 35MR16/NSP
Location: 5'–0" from wall
Center beam Aiming: 32° (at point 4'–6" AFF)
Luminaire: Monopoint or Recessed Adjustable
Illuminance: 25 fc avg, maintained (63 fc max)

Halogen 35MR16/VWSP
Standard

Performance Sketches

[All data outlined here is based on information from boldfaced manufacturer's published data for 12V version at time of manuscript preparation. Scale is ¼ inch equals 1 foot. **Note: At press time, actual photometry was unavailable for this lamp. Data were generated using GE 35MR16/C/VWSP lamp and rerating candlepower—this is imprecise.**]

Lamp: 35MR16/VWSP
Location: 1'–6" from wall
Center beam Aiming: 13° (at point 4'–0" AFF)
Luminaire: Monopoint or Recessed Adjustable
Illuminance: 13 fc avg, maintained (35 fc max)

Lamp: 35MR16/VWSP
Location: 2'–0" from wall
Center beam Aiming: 17° (at point 4'–0" AFF)
Luminaire: Monopoint or Recessed Adjustable
Illuminance: 14 fc avg, maintained (33 fc max)

10'–6"

10'–0"

Art shown at 2'–6" by 3' and centered 5'–6" AFF. Wall grid on 6" by 6" centers and scaled at ¼"=1'.

Lamp: 35MR16/VWSP
Location: 2'–6" from wall
Center beam Aiming: 23° (at point 4'–6" AFF)
Luminaire: Monopoint or Recessed Adjustable
Illuminance: 16 fc avg, maintained (38 fc max)

▶ *Requires tempered cover glass.*

Halogen Low Voltage (12V)

4

Uses
▶ Architectural accenting
▶ Small art with medium throws
▶ Sculpture with medium throws
▶ Medium throws
▶ Available in 12V

Specs
Beam spread/20°
Center beam candlepower/ 2800
Life/4000 hrs
▶ **Osram Sylvania** (58602)

Design Tips
✓ One light best for small art and where mounting height for light is 11' or less.
✓ Light may degrade artwork rapidly and/or significantly: consider UV and IR filters or limit exposure time.
✓ More grazing light creates prominent shadows.
✓ MR lamps are glary—consider shielding with hexcell louver or using deep luminaire trims or pinholes.
✓ Tempered cover glass required for safety.
✓ Required transformer also consumes 3 to 5 watts.

Application Key

Commercial
Gallery
Hospitality
Institutional
Manufacturing
Residential
Retail
Exterior

bold = primary application
partial fade = minimal application
fade = unlikely application

▶ Requires tempered cover glass.

Halogen 35MR16/FL

Standard

Halogen Low Voltage (12V) 4

Uses
▶ General lighting
▶ Small art with short throws
▶ Short throws
▶ Available in 12V

Specs
Beam spread/40°
Center beam
  candlepower/1400
Life/4000 hrs
▶ **Osram Sylvania** (58603)

Design Tips
✔ One light best for small art where mounting height of light is 9' or less.
✔ Light may degrade artwork rapidly and/or significantly: consider UV and IR filters or limit exposure time.
✔ Good for downlighting onto decorative floor materials—best with sconces and/or art accents and/or wallwashing to avoid cave effect and grazing glare (the "I wish I had a visor" effect).
✔ More grazing light creates prominent shadows.
✔ MR lamps are glary—consider shielding with hexcell louver or using deep luminaire trims or pinholes.
✔ Tempered cover glass required for safety.
✔ Required transformer also consumes 3 to 5 watts.

Application Key

Commercial
Gallery
Hospitality
Institutional
Manufacturing
Residential
Retail
Exterior

bold = primary application
partial fade = minimal application
fade = unlikely application

Performance Sketches

[All data outlined here is based on information from boldfaced manufacturer's published data for 12V version at time of manuscript preparation. Scale is ¼ inch equals 1 foot.]

Art shown at 2'–6" by 3' and centered 5'–6" AFF. Wall grid on 6" by 6" centers and scaled at ¼"=1'.

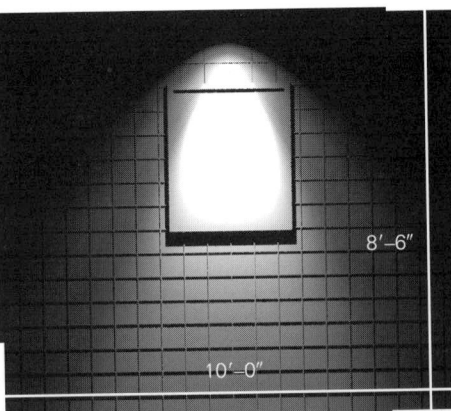

8'–6"

10'–0"

Lamp: 35MR16/FL
Location: 1'–0" from wall
Center beam Aiming: 11° (at point 3'–6" AFF)
Luminaire: Monopoint or Recessed Adjustable
Illuminance: 23 fc avg, maintained (92 fc max)

Lamp: 35MR16/FL
Location: 1'–6" from wall
Center beam Aiming: 18° (at point 4'–0" AFF)
Luminaire: Monopoint or Recessed Adjustable
Illuminance: 25 fc avg, maintained (75 fc max)

Lamp: 35MR16/FL
Location: 2'–0" from wall
Center beam Aiming: 27° (at point 4'–6" AFF)
Luminaire: Monopoint or Recessed Adjustable
Illuminance: 28 fc avg, maintained (72 fc max)

Halogen 50MR16/NSP
Standard

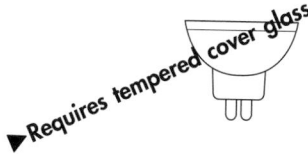

▶Requires tempered cover glass.

Halogen Low Voltage (12V)

4

Performance Sketches

[All data outlined here is based on information from boldfaced manufacturer's published data for 12V version at time of manuscript preparation. Scale is ¼ inch equals 1 foot.]

Lamp: 50MR16/NSP
Location: 4'–3" from wall
Center beam Aiming: 22° (at point 4'–6" AFF)
Luminaire: Monopoint or Recessed Adjustable
Illuminance: 16 fc avg, maintained (31 fc max)

Lamp: 50MR16/NSP
Location: 5'–3" from wall
Center beam Aiming: 27° (at point 4'–6" AFF)
Luminaire: Monopoint or Recessed Adjustable
Illuminance: 17 fc avg, maintained (33 fc max)

15'–0"

10'–0"

Art shown at 3' by 4' and centered 5'–6" AFF. Wall grid on 6" by 6" centers and scaled at ¼"=1'.

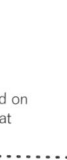

Lamp: 50MR16/NSP
Location: 6'–3" from wall
Center beam Aiming: 31° (at point 4'–6" AFF)
Luminaire: Monopoint or Recessed Adjustable
Illuminance: 17 fc avg, maintained (35 fc max)

Uses
▶Architectural detailing
▶Medium art with long throws
▶Sculpture with long throws
▶Long throws
▶Available in 12V

Specs
Beam spread/10°
Center beam candlepower/ 11000
Life/4000 hrs
▶**Philips** (28222–8)

Design Tips
✔One light best for medium art where mounting height for light is 16' or less.
✔Light may degrade artwork rapidly and/or significantly: consider UV and IR filters or limit exposure time.
✔Great for glass pieces and floral centerpieces and where mounting height for light is less than 16'.
✔These are strong accents and deserve judicious use.
✔Scallops, striations and frame shadows are most severe when light is closer to the wall (compare top left image to bottom image).
✔MR lamps are glary—consider shielding with hexcell louver or using deep luminaire trims or pinholes.
✔Tempered cover glass required for safety.
✔Required transformer consumes 3 to 5 watts.
✔Consider 37MR16/HIR/C/NSP.

Application Key

Commercial
Gallery
Hospitality
Institutional
Manufacturing
Residential
Retail
Exterior

bold = primary application
partial fade = minimal application
fade = unlikely application

▶ *Requires tempered cover glass.*

Halogen 50MR16/SP

Standard

Performance Sketches

[All data outlined here is based on information from boldfaced manufacturer's published data for 12V version at time of manuscript preparation. Scale is ¼ inch equals 1 foot.]

Halogen Low Voltage (12V)

4

Uses
▶ Architectural detailing
▶ Medium art with long throws
▶ Sculpture with long throws
▶ Long throws
▶ Available in 12V

Specs
Beam spread/12°
Center beam candlepower/ 11000
Life/4000 hrs
▶ **Osram Sylvania** (58608)

Design Tips
✔ One light best for medium art and where mounting height for light is 16' or less.
✔ These are strong accents and deserve judicious use. Light may degrade artwork rapidly and/or significantly: consider UV and IR filters or limit exposure time.
✔ Good for glass pieces and floral centerpieces and where mounting height for light is 16' or less.
✔ MR lamps are glary—consider shielding with hexcell louver or using deep luminaire trims or pinholes.
✔ Tempered cover glass required for safety.
✔ Required transformer consumes 3 to 5 watts.
✔ Consider 37MR16/HIR/C/NSP.

Application Key

Commercial
Gallery
Hospitality
Institutional
Manufacturing
Residential
Retail
Exterior

bold = primary application
partial fade = minimal application
fade = unlikely application

Lamp: 50MR16/SP
Location: 4'–3" from wall
Center beam Aiming: 22° (at point 4'–6" AFF)
Luminaire: Monopoint or Recessed Adjustable
Illuminance: 19 fc avg, maintained (37 fc max)

15'–0"

10'–0"

Lamp: 50MR16/SP
Location: 5'–3" from wall
Center beam Aiming: 27° (at point 4'–6" AFF)
Luminaire: Monopoint or Recessed Adjustable
Illuminance: 21 fc avg, maintained (39 fc max)

Art shown at 3' by 4' and centered 5'–6" AFF. Wall grid on 6" by 6" centers and scaled at ¼"=1'.

Lamp: 50MR16/SP
Location: 6'–3" from wall
Center beam Aiming: 31° (at point 4'–6" AFF)
Luminaire: Monopoint or Recessed Adjustable
Illuminance: 21 fc avg, maintained (40 fc max)

Halogen 50MR16/NFL
Standard

Halogen Low Voltage (12V)
4

Performance Sketches

[All data outlined here is based on information from boldfaced manufacturer's published data for 12V version at time of manuscript preparation. Scale is ¼ inch equals 1 foot.]

Lamp: 50MR16/NFL
Location: 3'–0" from wall
Center beam Aiming: 23° (at point 3'–6" AFF)
Luminaire: Monopoint or Recessed Adjustable
Illuminance: 20 fc avg, maintained (34 fc max)

Lamp: 50MR16/NFL
Location: 4'–0" from wall
Center beam Aiming: 34° (at point 4'–6" AFF)
Luminaire: Monopoint or Recessed Adjustable
Illuminance: 23 fc avg, maintained (37 fc max)

10'–6"

10'–0"

Art shown at 3' by 4' and centered 5'–6" AFF. Wall grid on 6" by 6" centers and scaled at ¼"=1'.

Lamp: 50MR16/NFL
Location: 5'–0" from wall
Center beam Aiming: 40° (at point 4'–6" AFF)
Luminaire: Monopoint or Recessed Adjustable
Illuminance: 22 fc avg, maintained (34 fc max)

Uses
▶Architectural accenting
▶Medium art with medium throws
▶Sculpture with medium throws
▶Medium throws
▶Available in 12V

Specs
Beam spread/25°
Center beam candlepower/ 3200
Life/4000 hrs
▶**Osram Sylvania** (58605)
▶**Philips**

Design Tips
✓One light best for medium art where mounting height for light is 11' or less.
✓Light may degrade artwork rapidly and/or significantly: consider UV and IR filters or limit exposure time.
✓Great for medium to large sculpture where mounting height for light is 11' or less.
✓Scallops, striations and frame shadows are most severe when light is closer to the wall (compare top left image to bottom image).
✓MR lamps are glary—consider shielding with hexcell louver or using deep luminaire trims or pinholes.
✓Tempered cover glass required for safety.
✓Required transformer consumes 3 to 5 watts.
✓Consider 37MR16/HIR/C/NFL.

Application Key
Commercial
Gallery
Hospitality
Institutional
Manufacturing
Residential
Retail
Exterior

bold = primary application
partial fade = minimal application
fade = unlikely application

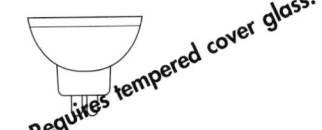

Requires tempered cover glass.

Halogen
Low Voltage (12V)

4

Uses
▶ General lighting
▶ Medium art with short throws
▶ Short throws
▶ Available in 12V

Specs
Beam spread/39°
Center beam candlepower/2000
Life/4000 hrs
▶ **Osram Sylvania** (58607)
▶ **Philips**

Design Tips
✔ One light best for medium art where mounting height for light is 10' or less.
✔ Light may degrade artwork rapidly and/or significantly: consider UV and IR filters or limit exposure time.
✔ Good for downlighting onto decorative floor materials—best with sconces and/or art accents and/or wallwashing to avoid cave effect and grazing glare (the "I wish I had a visor" effect).
✔ More grazing light creates prominent shadows.
✔ MR lamps are glary—consider shielding with hexcell louver or using deep luminaire trims or pinholes.
✔ Tempered cover glass required for safety.
✔ Required transformer consumes 3 to 5 watts.
✔ Consider 37MR16/HIR/C/FL.

Application Key

Commercial
Gallery
Hospitality
Institutional
Manufacturing
Residential
Retail
Exterior

bold = primary application
partial fade = minimal application
fade = unlikely application

Halogen 50MR16/FL
Standard

Performance Sketches
[All data outlined here is based on information from boldfaced manufacturer's published data for 12V version at time of manuscript preparation. Scale is ¼ inch equals 1 foot.]

Art shown at 3' by 4' and centered 5'–6" AFF. Wall grid on 6" by 6" centers and scaled at ¼"=1'.

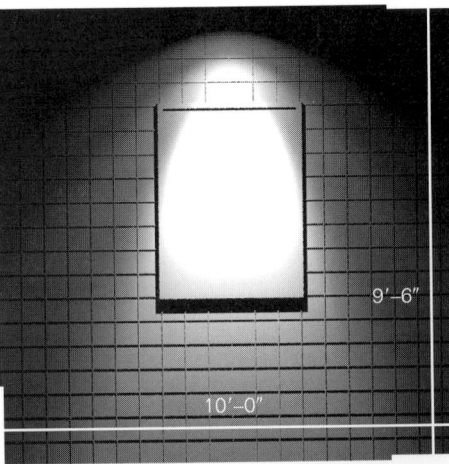

9'–6"

10'–0"

Lamp: 50MR16/FL
Location: 1'–6" from wall
Center beam Aiming: 14° (at point 3'–6" AFF)
Luminaire: Monopoint or Recessed Adjustable
Illuminance: 23 fc avg, maintained (84 fc max)

Lamp: 50MR16/FL
Location: 2'–0" from wall
Center beam Aiming: 20° (at point 4'–0" AFF)
Luminaire: Monopoint or Recessed Adjustable
Illuminance: 25 fc avg, maintained (72 fc max)

Lamp: 50MR16/FL
Location: 2'–6" from wall
Center beam Aiming: 27° (at point 4'–6" AFF)
Luminaire: Monopoint or Recessed Adjustable
Illuminance: 27 fc avg, maintained (71 fc max)

Halogen 50MR16/VWFL
Standard

▶"Requires tempered cover glass."

Halogen Low Voltage (12V)

4

Performance Sketches

[All data outlined here is based on information from boldfaced manufacturer's published data for 12V version at time of manuscript preparation. Scale is ¼ inch equals 1 foot. **Note: At press time, actual photometry was unavailable for this lamp. Data were generated using GE 50MR16/C/CG/VWFL lamp and rerating candlepower—this is imprecise.**]

. . . .

Art shown at 4' by 5' and centered 5'–6" AFF. Wall grid on 6" by 6" centers and scaled at ¼"=1'.

Lamp: 50MR16/VWFL
Location: 2'–0" from wall
Center beam Aiming: 8° (at point 0'–0" AFF)
Luminaire: Monopoint or Recessed Adjustable
Illuminance: 11 fc avg, maintained (26 fc max)

Lamp: 50MR16/VWFL
Location: 2'–6" from wall
Center beam Aiming: 16° (at point 1'–0" AFF)
Luminaire: Monopoint or Recessed Adjustable
Illuminance: 13 fc avg, maintained (30 fc max)

Lamp: 50MR16/VWFL
Location: 3'–0" from wall
Center beam Aiming: 22° (at point 2'–0" AFF)
Luminaire: Monopoint or Recessed Adjustable
Illuminance: 14 fc avg, maintained (29 fc max)

Uses
▶Soft general lighting
▶Wallwashing (see 11–117)
▶Large art with short throws
▶Short throws
▶Available in 12V

Specs
Beam spread/60°
Center beam candlepower/1200
Life/4000 hrs
▶**Osram Sylvania** (58606)

Design Tips
✔One light best for large art and where mounting height for light is 10' or less.
✔Good for soft downlighting—best with sconces and/or art accents and/or wallwashing to avoid cave effect.
✔MR lamps are glary—consider shielding with hexcell louver or using deep luminaire trims or pinholes.
✔Tempered cover glass required for safety.
✔Required transformer consumes 3 to 5 watts.

Application Key

Commercial
Gallery
Hospitality
Institutional
Manufacturing
Residential
Retail
Exterior

bold = primary application
partial fade = minimal application
fade = unlikely application

▼Requires tempered cover glass.

Halogen 65MR16/NSP

Standard

Performance Sketches

[All data outlined here is based on information from boldfaced manufacturer's published data for 12V version at time of manuscript preparation. Scale is ¼ inch equals 1 foot.]

Lamp: 65MR16/NSP
Location: 5'–0" from wall
Center beam Aiming: 20° (at point 4'–6" AFF)
Luminaire: Monopoint or Recessed Adjustable
Illuminance: 14 fc avg, maintained (29 fc max)

Lamp: 65MR16/NSP
Location: 6'–0" from wall
Center beam Aiming: 24° (at point 4'–6" AFF)
Luminaire: Monopoint or Recessed Adjustable
Illuminance: 15 fc avg, maintained (30 fc max)

18'–0"

10'–0"

Halogen Low Voltage (12V)

4

Uses
▶Architectural detailing
▶Large art with long throws
▶Sculpture with long throws
▶Long throws
▶Available in 12V

Specs
Beam spread/10°
Center beam candlepower/14000
Life/4000 hrs
▶**Osram Sylvania** (58563)

Design Tips
✔One light best for large art and where mounting height for light is 19' or less.
✔These are strong accents and deserve judicious use. Light may degrade artwork rapidly and/or significantly: consider UV and IR filters or limit exposure time.
✔Excellent for glass pieces and floral centerpieces and where mounting height for light is 19' or less.
✔Scallops, striations and frame shadows are most severe when light is closer to the wall (compare top left image to bottom image).
✔MR lamps are glary—consider shielding with hexcell louver or using deep luminaire trims or pinholes.
✔Tempered cover glass required for safety.
✔Required transformer consumes 3 to 5 watts.
✔Consider 50MR16/HIR/C/NSP.

Application Key

Commercial
Gallery
Hospitality
Institutional
Manufacturing
Residential
Retail
Exterior

bold = primary application
partial fade = minimal application
fade = unlikely application

Halogen65MR16/NSP

Standard

▶ Requires tempered cover glass.

Halogen
Low Voltage (12V)

4

Lamp: 65MR16/NSP
Location: 7'–0" from wall
Center beam Aiming: 27° (at point 4'–6" AFF)
Luminaire: Monopoint or Recessed Adjustable
Illuminance: 16 fc avg, maintained (30 fc max)

Art shown at 4' by 5' and
centered 5'–6" AFF. Wall grid on
6" by 6" centers and scaled at
¼"=1'.

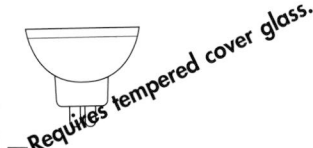

▶ Requires tempered cover glass.

Halogen 65MR16/NFL

Standard

Performance Sketches

[All data outlined here is based on information from boldfaced manufacturer's published data for 12V version at time of manuscript preparation. Scale is ¼ inch equals 1 foot.]

Uses
▶ Architectural accenting
▶ Feature downlighting
▶ Large art with medium throws
▶ Sculpture with medium throws
▶ Medium throws
▶ Available in 12V

Specs
Beam spread/25°
Center beam candlepower/ 4000
Life/4000 hrs
▶ **Osram Sylvania** (58565)

Design Tips
✔ One light best for large art where mounting height for light is 13' or less.
✔ Great for moderate to large sculpture where mounting height for light is 13' or less.
✔ Scallops, striations and frame shadows are most severe when light is closer to the wall (compare top left image to bottom image).
✔ MR lamps are glary—consider shielding with hexcell louver or using deep luminaire trims or pinholes.
✔ Tempered cover glass required for safety.
✔ Required transformer consumes 3 to 5 watts.

Application Key

Commercial
Gallery
Hospitality
Institutional
Manufacturing
Residential
Retail
Exterior

bold = primary application
partial fade = minimal application
fade = unlikely application

Lamp: 65MR16/NFL
Location: 3'–0" from wall
Center beam Aiming: 18° (at point 3'–6" AFF)
Luminaire: Monopoint or Recessed Adjustable
Illuminance: 13 fc avg, maintained (29 fc max)

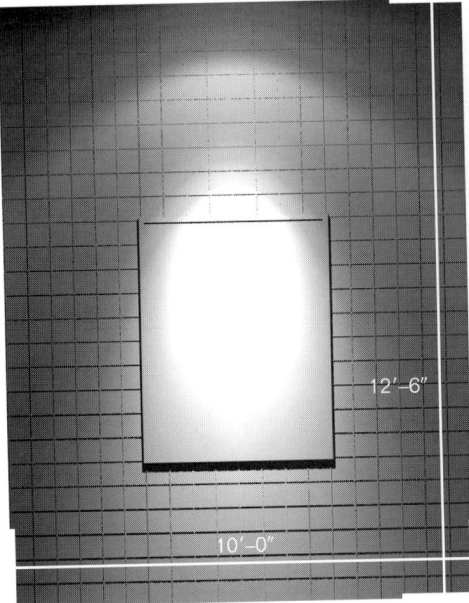

12'–6"

10'–0"

Lamp: 65MR16/NFL
Location: 4'–0" from wall
Center beam Aiming: 25° (at point 4'–0" AFF)
Luminaire: Monopoint or Recessed Adjustable
Illuminance: 15 fc avg, maintained (28 fc max)

Art shown at 4' by 5' and centered 5'–6" AFF. Wall grid on 6" by 6" centers and scaled at ¼"=1'.

Lamp: 65MR16/NFL
Location: 5'–0" from wall
Center beam Aiming: 30° (at point 4'–0" AFF)
Luminaire: Monopoint or Recessed Adjustable
Illuminance: 15 fc avg, maintained (26 fc max)

Halogen 65MR16/FL

Standard

▶*Requires tempered cover glass.*

Performance Sketches

[All data outlined here is based on information from boldfaced manufacturer's published data for 12V version at time of manuscript preparation. Scale is ¼ inch equals 1 foot.]

Art shown at 4' by 5' and centered 5'–6" AFF. Wall grid on 6" by 6" centers and scaled at ¼"=1'.

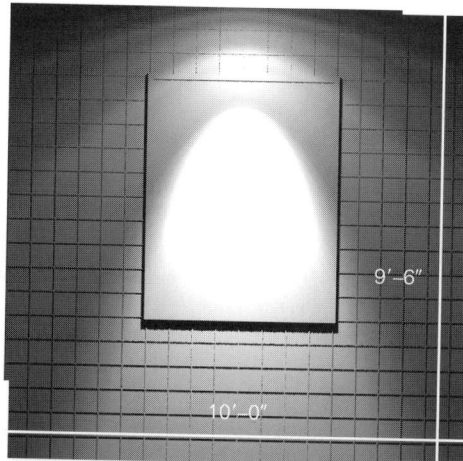

9'–6"

10'–0"

Lamp: 65MR16/FL
Location: 2'–0" from wall
Center beam Aiming: 17° (at point 3'–0" AFF)
Luminaire: Monopoint or Recessed Adjustable
Illuminance: 16 fc avg, maintained (60 fc max)

Lamp: 65MR16/FL
Location: 2'–6" from wall
Center beam Aiming: 24° (at point 4'–0" AFF)
Luminaire: Monopoint or Recessed Adjustable
Illuminance: 18 fc avg, maintained (62 fc max)

Lamp: 65MR16/FL
Location: 3'–0" from wall
Center beam Aiming: 31° (at point 4'–6" AFF)
Luminaire: Monopoint or Recessed Adjustable
Illuminance: 20 fc avg, maintained (60 fc max)

Uses
▶General lighting
▶Large art with short throws
▶Short throws
▶Available in 12V

Specs
Beam spread/40°
Center beam
 candlepower/2100
Life/4000 hrs
▶**Osram Sylvania** (58564)

Design Tips
✔One light best for large art where mounting height for light is 10' or less.
✔Good for downlighting onto decorative floor materials: best with sconces and/or art accents and/or wallwashing to avoid cave effect.
✔More grazing light creates prominent shadows.
✔MR lamps are glary—consider shielding with hexcell louver or using deep luminaire trims or pinholes.
✔Tempered cover glass required for safety.
✔Required transformer consumes 3 to 5 watts.
✔Consider 37MR16/HIR/C/FL.

Application Key

Commercial
Gallery
Hospitality
Institutional
Manufacturing
Residential
Retail
Exterior

bold = primary application
partial fade = minimal application
fade = unlikely application

Requires tempered cover glass.

Halogen 73MR16/SP

Standard

Performance Sketches

[All data outlined here is based on information from boldfaced manufacturer's published data for 12V version at time of manuscript preparation. Scale is ¼ inch equals 1 foot. **Note: At press time, actual photometry was unavailable for this lamp. Data were generated using Philips 50MR16/CG/NSP lamp and rerating candlepower—this is imprecise.**]

Lamp: 73MR16/SP
Location: 5'–0" from wall
Center beam Aiming: 20° (at point 4'–6" AFF)
Luminaire: Monopoint or Recessed Adjustable
Illuminance: 14 fc avg, maintained (25 fc max)

Lamp: 73MR16/SP
Location: 6'–0" from wall
Center beam Aiming: 24° (at point 4'–6" AFF)
Luminaire: Monopoint or Recessed Adjustable
Illuminance: 15 fc avg, maintained (26 fc max)

18'–0"

10'–0"

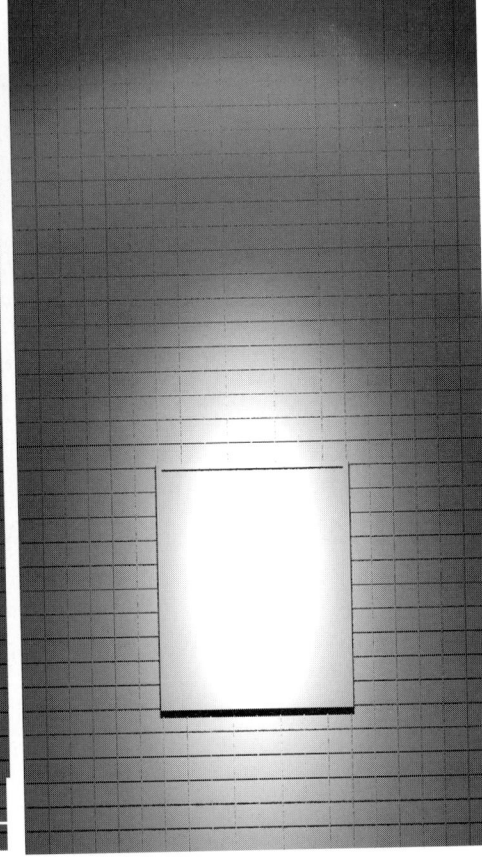

Halogen Low Voltage (12V)

Uses
▶ Architectural detailing
▶ Large art with long throws
▶ Sculpture with long throws
▶ Long throws
▶ Available in 12V

Specs
Beam spread/14°
Center beam
 candlepower/14000
Life/4000 hrs
▶ **Philips** (28232–7)

Design Tips
✔ One light best for large art and where mounting height for light is 19' or less.
✔ These are strong accents and deserve judicious use. Light may degrade artwork rapidly and/or significantly: consider UV and IR filters or limit exposure time.
✔ Excellent for glass pieces and floral centerpieces and where mounting height for light is 19' or less.
✔ Scallops, striations and frame shadows are most severe when light is closer to the wall (compare top left image to bottom image).
✔ MR lamps are glary—consider shielding with hexcell louver or using deep luminaire trims or pinholes.
✔ Tempered cover glass required for safety.
✔ Required transformer consumes 3 to 5 watts.
✔ Consider 50MR16/HIR/C/NSP.

Application Key

Commercial
Gallery
Hospitality
Institutional
Manufacturing
Residential
Retail
Exterior

bold = primary application
partial fade = minimal application
fade = unlikely application

Halogen 73MR16/SP

Standard

Halogen
Low Voltage (12V)

▶Requires tempered cover glass.

4

Lamp: 73MR16/SP
Location: 7'–0" from wall
Center beam Aiming: 27° (at point 4'–6" AFF)
Luminaire: Monopoint or Recessed Adjustable
Illuminance: 15 fc avg, maintained (27 fc max)

Art shown at 4' by 5' and
centered 5'–6" AFF. Wall grid on
6" by 6" centers and scaled at
¼"=1'.

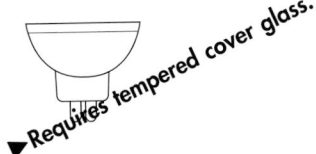

▶ Requires tempered cover glass.

Halogen 73MR16/FL

Standard

Uses
▶ General lighting
▶ Large art with short throws
▶ Short throws
▶ Available in 12V

Specs
Beam spread/36°
Center beam
  candlepower/2500
Life/4000 hrs
▶ **Philips** (28235–0)

Design Tips
✔ One light best for large art where mounting height for light is 10' or less.
✔ Good for downlighting onto decorative floor materials: best with sconces and/or art accents and/or wallwashing to avoid cave effect.
✔ More grazing light creates prominent shadows.
✔ MR lamps are glary—consider shielding with hexcell louver or using deep luminaire trims or pinholes.
✔ Tempered cover glass required for safety.
✔ Required transformer consumes 3 to 5 watts.

Application Key

Commercial
Gallery
Hospitality
Institutional
Manufacturing
Residential
Retail
Exterior

bold = primary application
partial fade = minimal application
fade = unlikely application

Performance Sketches

[All data outlined here is based on information from boldfaced manufacturer's published data for 12V version at time of manuscript preparation. Scale is ¼ inch equals 1 foot.]

Art shown at 4' by 5' and centered 5'–6" AFF. Wall grid on 6" by 6" centers and scaled at ¼"=1'.

9'–6"

10'–0"

Lamp: 73MR16/FL
Location: 1'–6" from wall
Center beam Aiming: 15° (at point 4'–0" AFF)
Luminaire: Monopoint or Recessed Adjustable
Illuminance: 22 fc avg, maintained (116 fc max)

Lamp: 73MR16/FL
Location: 2'–0" from wall
Center beam Aiming: 20° (at point 4'–0" AFF)
Luminaire: Monopoint or Recessed Adjustable
Illuminance: 23 fc avg, maintained (101 fc max)

Lamp: 73MR16/FL
Location: 2'–6" from wall
Center beam Aiming: 27° (at point 4'–6" AFF)
Luminaire: Monopoint or Recessed Adjustable
Illuminance: 25 fc avg, maintained (103 fc max)

Halogen/MR16

Standard

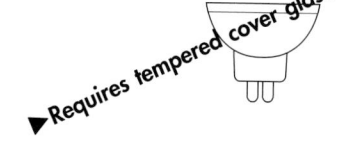

▶Requires tempered cover glass.

Halogen
Low Voltage (12V)

4

Net Addresses/MR16 Track and Monopoints
```
http://www.ardeelighting.com/index.htm
http://www.artemide.com/
http://www.bartcolighting.com/index2.html
http://www.thomasltg.com/
http://www.con-techlighting.com/
http://www.cooperlighting.com/
http://www.flosusa.com/
http://www.hubbell-ltg.com/products.htm#Down&Track
http://www.danalite.com/
http://www.kramerlighting.com/
http://www.lightingservicesinc.com/
http://www.lightolier.com/
http://www.lightproject.com/
http://www.lithonia.com/
http://www.prescolite.com/
http://www.techlighting.com/pro1.html
http://www.tslight.com/
```

CONNECT FOR MORE

Net Addresses/MR16 Recessed
```
http://www.alkco.com/rtrak.htm
http://www.ardeelighting.com/index.htm
http://www.artemide.com/
http://www.bartcolighting.com/index2.html
http://www.thomasltg.com/
http://www.con-techlighting.com/
http://www.cooperlighting.com/
http://www.hubbell-ltg.com/products.htm#Down&Track
http://www.danalite.com/
http://www.kramerlighting.com/
http://www.lightolier.com/
http://www.lithonia.com/
http://www.prescolite.com/
http://www.progresslighting.com/
```

CONNECT FOR MORE

Net Addresses/Exterior Accent
```
http://www.bega-us.com/home/home.html
http://www.hadcolighting.com/html/bronzelite.htm
http://www.hydrel.com
http://www.kimlighting.com/ingrd1.html
```

CONNECT FOR MORE

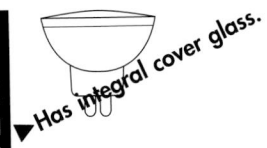

▶ Has integral cover glass.

Halogen
Low Voltage (12V)

4

Halogen/MR16/CG

Cover Glass

Guide
⊃ Longer life is better
⊃ Higher efficacy better for general lighting
⊃ Lower wattage is better
⊃ Lower color of light is warmer
⊃ Very low cost range for halogen is less than US$7.⁵⁰

Stats
Data varies manufacturer to manufacturer and changes from time to time

Life/4000 hours
Varies manufacturer to manufacturer. A typical office environment might be occupied about 2600 hours each year.

Efficacy/Not reported
Excellent beam intensities and distributions available for the wattage.

Wattages/20 and 50

Color of Light/2925°K to 3050°K
Varies manufacturer to manufacturer and wattage to wattage. Typically the longer lived lamps have slightly lower color temperature and the higher wattage lamps have slightly higher color temperature.

Beam spreads/NSP, NFL and FL
Narrow spot, spot, narrow flood and flood distributions are available.

Size/1¹⁵⁄₁₆″ L by 2″ Ø

Voltage/12V

Cost Magnitude/Low ⁽²⁰⁰⁰⁾

Manufacturers
Philips

Net Addresses
http://www.lighting.philips.com/

CONNECT FOR MORE

Advantages
Crisp white light
Very small size
Easily dimmed
Low initial cost
Alternative to less efficient R and halogen PAR lamps
Withstand extreme temperature range
Very little lamp lumen depreciation (light loss) over life

Disadvantages
Beam aberrations
◆ Striations
◆ Inconsistent beam shape
Color inconsistencies
◆ Lamp to lamp color variations
◆ Over time, color shifts—very evident with exposed–lamp monopoints or trackheads where back of lamp is visible
High operating costs
◆ Inefficiency may make this cost prohibitive in typical commercial applications
Hot to touch
Integral cover glass
◆ Tempered safety glass may still be required (check with lamp manufacturer)
◆ May prevent use of auxiliary accessories (e.g., color filters, louvers, etc.)
Transformer required
◆ Additional cost and visual aspect of transformer must be accommodated
◆ Additional wattage requirement of transformer of 3 to 5 watts
◆ Potential for audible hum

For more efficient accent lighting, use HIR low voltage lamps (Section 6)**; HIR PAR lamps** (Section 5) **or CMH PAR lamps** (Section 7).

Halogen/MR16/CG

Cover Glass

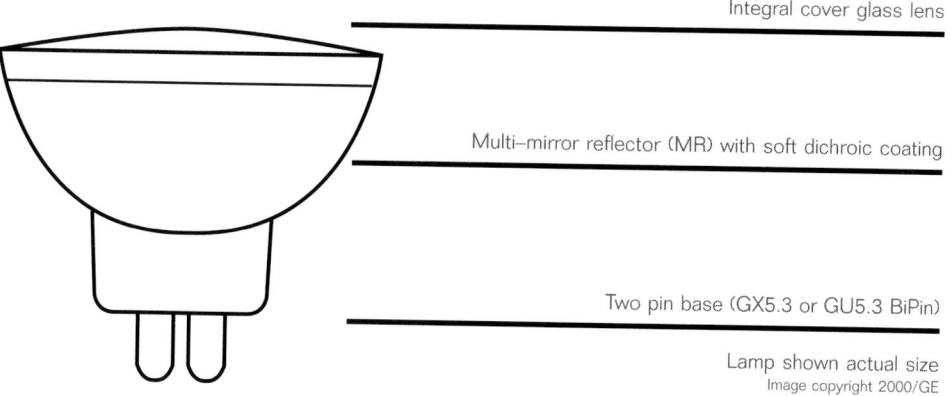

Integral cover glass lens

Multi–mirror reflector (MR) with soft dichroic coating

Two pin base (GX5.3 or GU5.3 BiPin)

Lamp shown actual size
Image copyright 2000/GE

Uses

Architectural Accenting
Art Accenting
  Commercial
  Gallery
  Hospitality
  Residential
Downlighting
  Lobbies
  Conference rooms
  Hospitality
  Residential
  Retail
Merchandising
Surface Grazing
  Specular/semispecular surfaces
  ◆Polished/honed stone
  ◆Polished/satin wood
  ◆Fritted/etched glass
  ◆Cut glass

MR16 Lighting Design Tips

✔Dimming increases life (typically doubling life), but shifts color warmer.
✔Very glary when used without much other general lighting or when aimed across long distances where ceilings are low.
✔Beam striations are most apparent on light–colored, nontextured flat surfaces and may be objectionable.
✔MR16 downlighting alone is harsh—combine several techniques for best results.
✔Use smaller scale (e.g., 4–inch diameter) downlights and adjustable accents for attractive, more human–scale approach.
✔Integral cover glass may interfere with auxiliary add–on lenses, louvers or accessories.
✔Use in recessed luminaires or in details or in enclosed trackheads to avoid seeing color inconsistencies which are most visible from back of lamp.

Net Addresses/MR16 Track, Monopoints, Recessed and Exterior
See page 4–73

CONNECT FOR MORE

▶Has integral cover glass.

Halogen Low Voltage (12V)

4

Halogen/MR16/CG

MR16 Cover Glass Visual Index

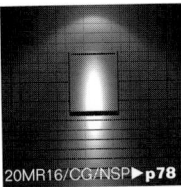

20MR16/CG/NSP▶p78   20MR16/CG/NSP▶p78   20MR16/CG/NSP▶p78

20MR16/CG/NSP
▶Architectural detailing
▶Small art/10' ceiling height or less
▶Objet d'art
▶Short throws
▶Gallery, Hospitality, Residential

Throws/Ceilings
◗Short = 7' to 10'
◗Medium = 10' to 15'
◗Long = 15' to 20'
◗Very Long = 20' or more

20MR16/CG/FL▶p79   20MR16/CG/FL▶p79   20MR16/CG/FL▶p79

20MR16/CG/FL
▶General lighting
▶Wallwashing
▶Small art/9' ceiling height or less
▶Short throws
▶Gallery, Residential

50MR16/CG/NSP▶p80   50MR16/CG/NSP▶p80   50MR16/CG/NSP▶p80

50MR16/CG/NSP
▶Architectural detailing
▶Medium art/16' ceiling height or less
▶Sculpture
▶Long throws
▶Commercial, Hospitality

50MR16/CG/NFL▶p81   50MR16/CG/NFL▶p81   50MR16/CG/NFL▶p81

50MR16/CG/NFL
▶Architectural accenting
▶Feature downlighting
▶Medium art/13' ceiling height or less
▶Sculpture
▶Medium throws
▶Hospitality

50MR16/CG/FL▶p82   50MR16/CG/FL▶p82   50MR16/CG/FL▶p82

50MR16/CG/FL
▶General lighting
▶Medium art/10' ceiling height or less
▶Short throws
▶Commercial, Retail

Recommended intensities
for art accenting

Low (gallery, museum, residential)
5 to 10 fc average on art
Moderate (hospitality)
10 to 20 fc average on art
High (commercial—or higher for retail)
20 to 30 fc average on art

These are intended to be average, maintained values including effects of accent lighting and general room lighting. General room lighting contributes just a few footcandles in residential, gallery and museum settings, but contributes 5 to 10 fc in commercial settings. Maximums might range from 50 percent to 100 percent of the average value. Wherever artwork is highly valuable and/or sensitive to light, precautions should be taken to limit light exposure (maintaining low levels and/or switching lights off when art is not being viewed); and exposure to infrared and ultraviolet should be limited (using filters on lamps or placing art behind specially treated glass or acrylic).

Artwork Size Categories
◗Petite = 2'–0" by 2'–0" or smaller
◗Small = 2'–0" by 2'–0" to 2'–6" by 3'–0"
◗Medium = 2'–6" by 3'–0" to 3'–0" by 4'–0"
◗Large = 3'–0" by 4'–0" to 4'–0" by 5'–0"
◗Extra Large = 4'–0" by 5'–0" or larger

Halogen/MR16/CG

Cover Glass

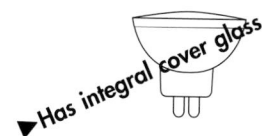

►Has integral cover glass.

Halogen Low Voltage (12V)

Halogen Low Voltage MR Lamp Designation

20MR16/C/CG/NFL

Last set of letters identifies beam spread

FLOODS
- 24° to 32° for narrow flood (NFL)
- 33° to 44° for flood (FL)
- 45° to 54° for wide flood (WFL)
- 55° or greater for very wide flood (VWFL)

SPOTS
- 7° or less for very narrow spot (VNSP)
- 8° to 10° for narrow spot (NSP)
- 11° to 14° for spot (SP)
- 15° to 18° for wide spot (WSP)
- 19° to 23° for very wide spot (VWSP)

Ranges of beam spreads used for convenient/consistent reference in this text and do not necessarily correspond to each manufacturer's nor any ANSI or NEMA definitions.

"CG" for built–in cover glass or no designation for no cover glass

Reflector treatment code
- "C" for consistent color (hard dichroic coating) or no letter for inconsistent color
- "AL" for aluminized reflector eliminating back spill light through reflector

These two digits represent lamp diameter in eighths inches (51mm or 2 inches)

These letters indicate this is a multifaceted reflector (MR) lamp

First two digits represent lamp wattage

[This designation used for convenient reference throughout this *Time–Saver Standards for Architectural Lighting*—but not necessarily used by all lamp manufacturers (although it would be convenient if manufacturers would come to agreement on standard designations.]

►Has integral cover glass.

Halogen 20MR16/CG/NSP

Cover Glass

Halogen Low Voltage (12V)

4

Uses
►Architectural detailing
►Small art with short throws
►Objet d'art
►Short throws
►Available in 12V

Specs
Beam spread/10°
Center beam
 candlepower/4200
Life/4000 hrs
►**Philips** (33303–9)

Design Tips
✔One light best for small art where mounting height for light is 10' or less.
✔More grazing light creates prominent shadows.
✔MR lamps are glary—consider shielding with hexcell louver or using deep luminaire trims or pinholes.
✔Required transformer consumes 3 to 5 watts.

Performance Sketches

[All data outlined here is based on information from boldfaced manufacturer's published data for 12V version at time of manuscript preparation. Scale is ¼ inch equals 1 foot.]

Art shown at 2'–6" by 3' and centered 5'–6" AFF. Wall grid on 6" by 6" centers and scaled at ¼"=1'.

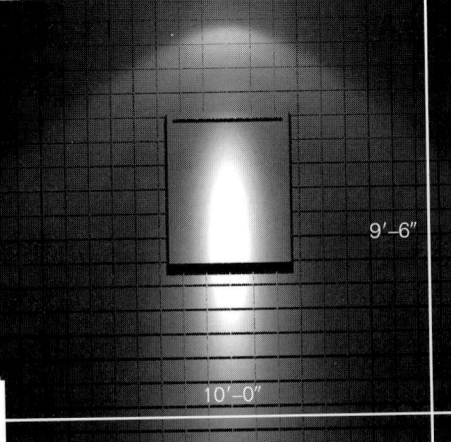

9'–6"

10'–0"

Lamp: 20MR16/CG/NSP
Location: 1'–6" from wall
Center beam Aiming: 14° (at point 3'–6" AFF)
Luminaire: Monopoint or Recessed Adjustable
Illuminance: 8 fc avg, maintained (29 fc max)

Lamp: 20MR16/CG/NSP
Location: 2'–0" from wall
Center beam Aiming: 24° (at point 5'–0" AFF)
Luminaire: Monopoint or Recessed Adjustable
Illuminance: 12 fc avg, maintained (65 fc max)

Lamp: 20MR16/CG/NSP
Location: 2'–6" from wall
Center beam Aiming: 29° (at point 5'–0" AFF)
Luminaire: Monopoint or Recessed Adjustable
Illuminance: 13 fc avg, maintained (68 fc max)

Application Key

Commercial
Gallery
Hospitality
Institutional
Manufacturing
Residential
Retail
Exterior

bold = primary application
partial fade = minimal application
fade = unlikely application

Halogen 20MR16/CG/FL
Cover Glass

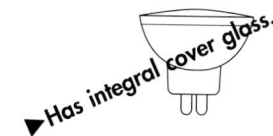

▶Has integral cover glass.

Halogen Low Voltage (12V)

4

Performance Sketches

[All data outlined here is based on information from boldfaced manufacturer's published data for 12V version at time of manuscript preparation. Scale is ¼ inch equals 1 foot.]

Art shown at 2'–6" by 3' and centered 5'–6" AFF. Wall grid on 6" by 6" centers and scaled at ¼"=1'.

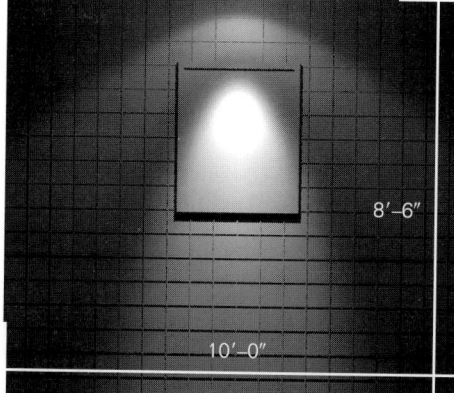

8'–6"

10'–0"

Lamp: 20MR16/CG/FL
Location: 1'–6" from wall
Center beam Aiming: 18° (at point 4'–0" AFF)
Luminaire: Monopoint or Recessed Adjustable
Illuminance: 7 fc avg, maintained (22 fc max)

Lamp: 20MR16/CG/FL
Location: 2'–0" from wall
Center beam Aiming: 27° (at point 4'–6" AFF)
Luminaire: Monopoint or Recessed Adjustable
Illuminance: 8 fc avg, maintained (23 fc max)

Lamp: 20MR16/CG/FL
Location: 2'–6" from wall
Center beam Aiming: 36° (at point 5'–0" AFF)
Luminaire: Monopoint or Recessed Adjustable
Illuminance: 10 fc avg, maintained (23 fc max)

Uses
▶General lighting
▶Wallwashing (see 11–109)
▶Small art with short throws
▶Available in 12V

Specs
Beam spread/36°
Center beam
 candlepower/560
Life/4000 hrs
▶**Philips** (33304–7)

Design Tips
✔One light best for small art where mounting height of light is 9' or less.
✔Multiple units can be used for wallwashing or lighting of larger artwork.
✔More grazing light creates prominent shadows.
✔MR lamps are glary—consider shielding with hexcell louver or using deep luminaire trims or pinholes.
✔Required transformer also consumes 3 to 5 watts.

Application Key

Commercial
Gallery
Hospitality
Institutional
Manufacturing
Residential
Retail
Exterior

bold = primary application
partial fade = minimal application
fade = unlikely application

▶Has integral cover glass.

Halogen 50MR16/CG/NSP

Cover Glass

Performance Sketches

[All data outlined here is based on information from boldfaced manufacturer's published data for 12V version at time of manuscript preparation. Scale is ¼ inch equals 1 foot.]

Halogen
Low Voltage (12V)

Uses
▶Architectural detailing
▶Medium art with long throws
▶Sculpture with long throws
▶Long throws
▶Available in 12V

Specs
Beam spread/10°
Center beam candlepower/ 11000
Life/4000 hrs
▶**Philips** (33305–4)

Design Tips
✔One light best for medium art where mounting height for light is 13' or less.
✔Light may degrade artwork rapidly and/or significantly: consider UV and IR filters or limit exposure time.
✔Great for glass pieces and floral centerpieces and where mounting height for light is 13' or less.
✔These are strong accents and deserve judicious use.
✔Scallops, striations and frame shadows are most severe when light is closer to the wall (compare top left image to bottom image).
✔MR lamps are glary—consider shielding with hexcell louver or using deep luminaire trims or pinholes.
✔Required transformer consumes 3 to 5 watts.
✔Consider 37MR16/HIR/C/NSP.

Lamp: 50MR16/CG/NSP
Location: 4'–3" from wall
Center beam Aiming: 20° (at point 3'–6" AFF)
Luminaire: Monopoint or Recessed Adjustable
Illuminance: 16 fc avg, maintained (27 fc max)

Lamp: 50MR16/CG/NSP
Location: 5'–3" from wall
Center beam Aiming: 27° (at point 4'–6" AFF)
Luminaire: Monopoint or Recessed Adjustable
Illuminance: 20 fc avg, maintained (35 fc max)

15'–0"

10'–0"

Application Key

Application Key

Commercial
Gallery
Hospitality
Institutional
Manufacturing
Residential
Retail
Exterior

bold = primary application
partial fade = minimal application
fade = unlikely application

Art shown at 3' by 4' and centered 5'–6" AFF. Wall grid on 6" by 6" centers and scaled at ¼"=1'.

Lamp: 50MR16/SP
Location: 6'–3" from wall
Center beam Aiming: 31° (at point 4'–6" AFF)
Luminaire: Monopoint or Recessed Adjustable
Illuminance: 20 fc avg, maintained (35 fc max)

Halogen 50MR16/CG/NFL
Cover Glass

▶Has integral cover glass.

Halogen Low Voltage (12V)

4

Performance Sketches

[All data outlined here is based on information from boldfaced manufacturer's published data for 12V version at time of manuscript preparation. Scale is ¼ inch equals 1 foot. **Note: At press time, actual photometry was unavailable for this lamp. Data were generated using Osram Sylvania 50MR16/NFL lamp and rerating candlepower—this is imprecise.**]

Lamp: 50MR16/CG/NFL
Location: 3'–0" from wall
Center beam Aiming: 16° (at point 2'–0" AFF)
Luminaire: Monopoint or Recessed Adjustable
Illuminance: 11 fc avg, maintained (19 fc max)

Lamp: 50MR16/CG/NFL
Location: 4'–0" from wall
Center beam Aiming: 23° (at point 3'–0" AFF)
Luminaire: Monopoint or Recessed Adjustable
Illuminance: 12 fc avg, maintained (18 fc max)

12'–6"

10'–0"

Art shown at 3' by 4' and centered 5'–6" AFF. Wall grid on 6" by 6" centers and scaled at ¼"=1'.

Lamp: 50MR16/CG/NFL
Location: 5'–0" from wall
Center beam Aiming: 29° (at point 3'–6" AFF)
Luminaire: Monopoint or Recessed Adjustable
Illuminance: 13 fc avg, maintained (17 fc max)

Uses
▶Architectural accenting
▶Feature downlighting
▶Medium art with medium throws
▶Sculpture with medium throws
▶Medium throws
▶Available in 12V

Specs
Beam spread/24°
Center beam candlepower/ 3100
Life/4000 hrs
▶**Philips** (33306-2)

Design Tips
✓One light best for medium art and where mounting height for light is 13' or less.
✓Good for downlighting onto decorative floor materials—best with sconces and/or art accents and/or wallwashing to avoid cave effect and "I wish I had a visor" effect.
✓MR lamps are glary—consider shielding with hexcell louver or using deep luminaire trims or pinholes.
✓Required transformer consumes 3 to 5 watts.
✓Consider 37MR16/HIR/C/NFL.

Application Key

Commercial
Gallery
Hospitality
Institutional
Manufacturing
Residential
Retail
Exterior

bold = primary application
partial fade = minimal application
fade = unlikely application

▶Has integral cover glass.

Halogen 50MR16/CG/FL
Cover Glass

Halogen Low Voltage (12V)

Uses
▶General lighting
▶Medium art with short throws
▶Short throws
▶Available in 12V

Specs
Beam spread/36°
Center beam candlepower/1800
Life/4000 hrs
▶Osram Sylvania
▶**Philips** (33307–0)

Design Tips
✔One light best for medium art where mounting height for light is 10' or less.
✔Light may degrade artwork rapidly and/or significantly: consider UV and IR filters or limit exposure time.
✔Good for downlighting onto decorative floor materials—best with sconces and/or art accents and/or wallwashing to avoid cave effect and grazing glare (the "I wish I had a visor" effect).
✔More grazing light creates prominent shadows.
✔MR lamps are glary—consider shielding with hexcell louver or using deep luminaire trims or pinholes.
✔Required transformer consumes 3 to 5 watts.
✔Consider 37MR16/HIR/C/FL.

Application Key

Commercial
Gallery
Hospitality
Institutional
Manufacturing
Residential
Retail
Exterior

bold = primary application
partial fade = minimal application
fade = unlikely application

Performance Sketches

[All data outlined here is based on information from boldfaced manufacturer's published data for 12V version at time of manuscript preparation. Scale is ¼ inch equals 1 foot.]

Art shown at 3' by 4' and centered 5'–6" AFF. Wall grid on 6" by 6" centers and scaled at ¼"=1'.

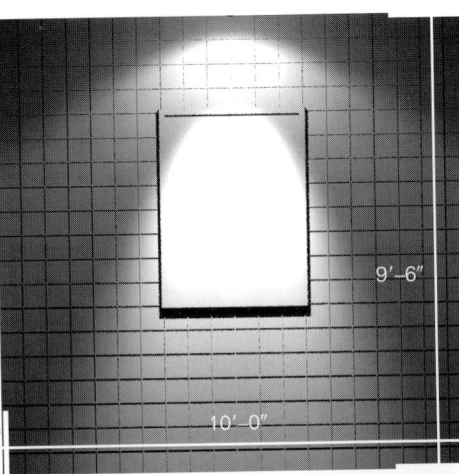

Lamp: 50MR16/CG/FL
Location: 2'–0" from wall
Center beam Aiming: 20° (at point 4'–0" AFF)
Luminaire: Monopoint or Recessed Adjustable
Illuminance: 22 fc avg, maintained (68 fc max)

Lamp: 50MR16/CG/FL
Location: 2'–6" from wall
Center beam Aiming: 27° (at point 4'–6" AFF)
Luminaire: Monopoint or Recessed Adjustable
Illuminance: 24 fc avg, maintained (66 fc max)

Lamp: 50MR16/CG/FL
Location: 3'–0" from wall
Center beam Aiming: 34° (at point 5'–0" AFF)
Luminaire: Monopoint or Recessed Adjustable
Illuminance: 27 fc avg, maintained (66 fc max)

▼*Requires tempered cover glass.*

Halogen Low Voltage (12V)

4

Halogen/MR16/AL

Aluminum Reflector

Stats
<small>Data varies manufacturer to manufacturer and changes from time to time</small>

Life/4000 hours
Varies manufacturer to manufacturer. A typical office environment might be occupied about 2600 hours each year.

Efficacy/Not reported
Excellent beam intensities and distributions available for the wattage.

Wattages/20, 35, 50 and 65

Color of Light/not reported (likely 2900°K to 3000°K)
Typically the higher wattage lamps have slightly higher color temperature.

Beam spreads/NSP, SP, NFL and FL
Narrow spot, spot, narrow flood and flood distributions are available.

Size/length not reported (likely 1⅞″ L) by 2″ Ø

Voltage/12V

Cost Magnitude/Low ⁽²⁰⁰⁰⁾

Manufacturers
Osram Sylvania

Advantages
Crisp white light
Very small size
Easily dimmed
Low initial cost
Alternative to less efficient R and halogen PAR lamps
Withstand extreme temperature range
No backlight spill
Less evident color variations and color shifts
Very little lamp lumen depreciation (light loss) over life

Disadvantages
Beam aberrations
◆Striations
◆Inconsistent beam shape
No backlight spill (sometimes this is part of the desired "look")
Hot to touch
Tempered glass lens required
◆Prevents glass and filament shards from falling onto flammable surfaces/ materials should a violent lamp failure occur (known as a non–passive failure)
Transformer required
◆Additional cost and visual aspect of transformer must be accommodated
◆Additional wattage requirement of transformer of 3 to 5 watts
◆Potential for audible hum

Net Addresses
http://ecom.sylvania.com/osicatalog/

CONNECT FOR MORE

For more efficient accent lighting, use HIR low voltage lamps (Section 6); **HIR PAR lamps** (Section 5) **or CMH PAR lamps** (Section 7).

Halogen/MR16/AL
Aluminum Reflector

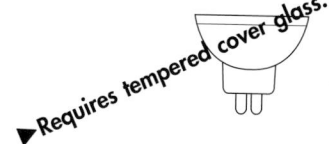

▶Requires tempered cover glass.

Halogen Low Voltage (12V)

4

Open top

Multi–mirror reflector (MR) with soft dichroic coating backed by aluminum reflector envelope

Two pin base (GU5.3 BiPin)

Lamp shown actual size
Image copyright 2000/GE

Uses

Architectural Accenting

Art Accenting
Commercial
Gallery
Hospitality
Residential

Downlighting
Lobbies
Conference rooms
Hospitality
Residential
Retail

Merchandising

Surface Grazing
Specular/semispecular surfaces
◆Polished/honed stone
◆Polished/satin wood
◆Fritted/etched glass
◆Cut glass

MR16 Lighting Design Tips
✔Dimming increases life (typically doubling life), but shifts color warmer.
✔Very glary when used without much other general lighting or when aimed across long distances where ceilings are low.
✔Beam striations are most apparent on light–colored, nontextured flat surfaces and may be objectionable.
✔MR16 downlighting alone is harsh—combine several techniques for best results.
✔Use smaller scale (e.g., 4–inch diameter) downlights and adjustable accents for attractive, more human–scale approach.
✔Use tempered glass lens for safety.

Net Addresses/MR16 Track, Monopoints, Recessed and Exterior
See page 4–73

CONNECT FOR MORE

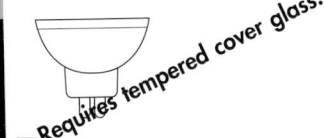

▶ *Requires tempered cover glass.*

Halogen/MR16/AL

MR16 Aluminum Reflector Visual Index

Halogen Low Voltage (12V)

4

20MR16/AL/NSP
▶ Architectural detailing
▶ Petite art/9′ ceiling height or less
▶ Objet d'art
▶ Short throws
▶ Hospitality

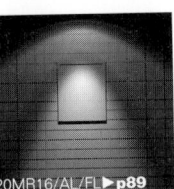
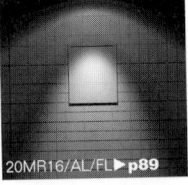

20MR16/AL/FL
▶ General lighting
▶ Wallwashing
▶ Small art/10′ ceiling height or less
▶ Short throws
▶ Gallery, Residential

Throws/Ceilings
◯ Short = 7′ to 10′
◯ Medium = 10′ to 15′
◯ Long = 15′ to 20′
◯ Very Long = 20′ or more

35MR16/AL/NSP
▶ Architectural detailing
▶ Medium art/13′ ceiling height or less
▶ Sculpture
▶ Medium throws
▶ Commercial, Hospitality, Retail

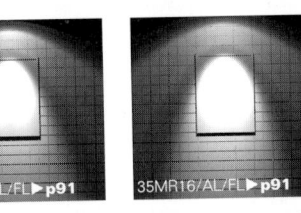

35MR16/AL/FL
▶ General lighting
▶ Medium art/10′ ceiling height or less
▶ Short throws
▶ Hospitality

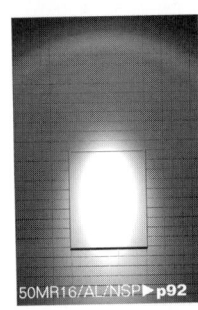

50MR16/AL/SP
▶ Architectural detailing
▶ Large art/16′ ceiling height or less
▶ Sculpture
▶ Long throws
▶ Hospitality

Recommended intensities
for art accenting

50MR16/AL/NFL
▶ Architectural accenting
▶ Feature downlighting
▶ Medium art/11′ ceiling height or less
▶ Sculpture
▶ Medium throws
▶ Commercial, Hospitality

Low (gallery, museum, residential)
5 to 10 fc average on art
Moderate (hospitality)
10 to 20 fc average on art
High (commercial—or higher for retail)
20 to 30 fc average on art

50MR16/AL/FL
▶ General lighting
▶ Medium art/10′ ceiling height or less
▶ Short throws
▶ Commercial, Hospitality

These are intended to be average, maintained values including effects of accent lighting and general room lighting. General room lighting contributes just a few footcandles in residential, gallery and museum settings, but contributes 5 to 10 fc in commercial settings. Maximums might range from 50 percent to 100 percent of the average value. Wherever artwork is highly valuable and/or sensitive to light, precautions should be taken to limit light exposure (maintaining low levels and/or switching lights off when art is not being viewed); and exposure to infrared and ultraviolet should be limited (using filters on lamps or placing art behind specially treated glass or acrylic).

Halogen/MR16/AL

Aluminum Reflector

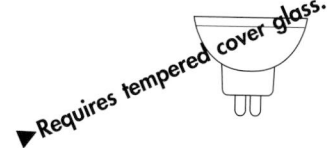

▶*Requires tempered cover glass.*

4

65MR16/AL/NSP
▶ Architectural accenting
▶ Medium art/16′ ceiling height or less
▶ Sculpture
▶ Long throws
▶ Commercial, Hospitality

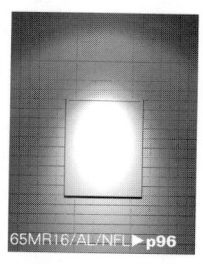

65MR16/AL/NFL
▶ Architectural accenting
▶ Feature downlighting
▶ Large art/13′ ceiling height or less
▶ Sculpture
▶ Medium throws
▶ Hospitality

65MR16/AL/FL
▶ General lighting
▶ Large art/10′ ceiling height or less
▶ Short throws
▶ Commercial, Hospitality

Halogen Low Voltage MR Lamp Designation

20MR16/C/CG/NFL

Last set of letters identifies beam spread

FLOODS
• 24° to 32° for narrow flood (NFL)
• 33° to 44° for flood (FL)
• 45° to 54° for wide flood (WFL)
• 55° or greater for very wide flood (VWFL)

SPOTS
• 7° or less for very narrow spot (VNSP)
• 8° to 10° for narrow spot (NSP)
• 11° to 14° for spot (SP)
• 15° to 18° for wide spot (WSP)
• 19° to 23° for very wide spot (VWSP)

Ranges of beam spreads used for convenient/consistent reference in this text and do not necessarily correspond to each manufacturer's nor any ANSI or NEMA definitions.

"CG" for built–in cover glass or no designation for no cover glass

Reflector treatment code
• "C" for consistent color (hard dichroic coating) or no letter for inconsistent color
• "AL" for aluminized reflector eliminating back spill light through reflector

These two digits represent lamp diameter in eighths inches (51mm or 2 inches)

These letters indicate this is a multifaceted reflector (MR) lamp

First two digits represent lamp wattage

[This designation used for convenient reference throughout this *Time–Saver Standards for Architectural Lighting*—but not necessarily used by all lamp manufacturers (although it would be convenient if manufacturers would come to agreement on standard designations.]

Artwork Size Categories
⊃Petite = 2′–0″ by 2′–0″ or smaller
⊃Small = 2′–0″ by 2′–0″ to 2′–6″ by 3′–0″
⊃Medium = 2′–6″ by 3′–0″ to 3′–0″ by 4′–0″
⊃Large = 3′–0″ by 4′–0″ to 4′–0″ by 5′–0″
⊃Extra Large = 4′–0″ by 5′–0″ or larger

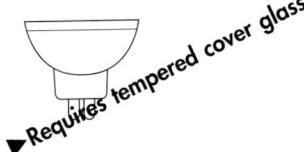

▶ *Requires tempered cover glass.*

Halogen 20MR16/AL/NSP

Aluminum Reflector

Performance Sketches

[All data outlined here is based on information from boldfaced manufacturer's published data for 12V version at time of manuscript preparation. Scale is ¼ inch equals 1 foot. **Note: At press time, actual photometry was unavailable for this lamp. Data were generated using Osram Sylvania 20MR16/NSP lamp and rerating candlepower—this is imprecise.**]

Art shown at 2' by 2' and centered 5'–6" AFF. Wall grid on 6" by 6" centers and scaled at ¼"=1'.

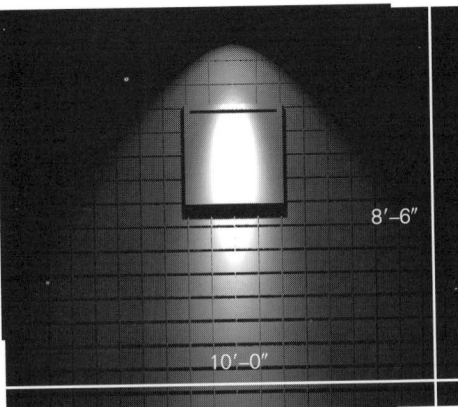

Lamp: 20MR16/AL/NSP
Location: 1'–0" from wall
Center beam Aiming: 14° (at point 4'–6" AFF)
Luminaire: Monopoint or Recessed Adjustable
Illuminance: 15 fc avg, maintained (71 fc max)

Lamp: 20MR16/AL/NSP
Location: 1'–6" from wall
Center beam Aiming: 21° (at point 4'–6" AFF)
Luminaire: Monopoint or Recessed Adjustable
Illuminance: 15 fc avg, maintained (80 fc max)

Lamp: 20MR16/AL/NSP
Location: 2'–0" from wall
Center beam Aiming: 30° (at point 5'–0" AFF)
Luminaire: Monopoint or Recessed Adjustable
Illuminance: 20 fc avg, maintained (115 fc max)

Halogen
Low Voltage (12V)

4

Uses
▶ Architectural detailing
▶ Petite art with short throws
▶ Objet d'art with short throws
▶ Short throws
▶ Available in 12V

Specs
Beam spread/8°
Center beam
   candlepower/3875
Life/4000 hrs
▶ **Osram Sylvania** (58589)

Design Tips
✔ One light best for petite art where mounting height of light is 9' or less.
✔ Intense light degrades artwork rapidly and/or significantly— consider UV and IR filters or limit exposure time.
✔ Great for glass pieces and floral centerpieces where mounting height of light is 9' or less.
✔ Scallops/striations are most severe when light is closer to the wall (compare top image to bottom image).
✔ More grazing light creates prominent shadows.
✔ MR lamps are glary—consider shielding with hexcell louver or using deep luminaire trims or pinholes.
✔ Tempered cover glass required for safety.
✔ Required transformer consumes 3 to 5 watts.

Application Key

Commercial
Gallery
Hospitality
Institutional
Manufacturing
Residential
Retail
Exterior

bold = primary application
partial fade = minimal application
fade = unlikely application

Halogen 20MR16/AL/FL
Aluminum Reflector

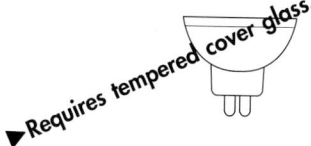

▶ Requires tempered cover glass.

Halogen Low Voltage (12V)

4

Performance Sketches

[All data outlined here is based on information from boldfaced manufacturer's published data for 12V version at time of manuscript preparation. Scale is ¼ inch equals 1 foot.]

Art shown at 2'–6" by 3' and centered 5'–6" AFF. Wall grid on 6" by 6" centers and scaled at ¼"=1'.

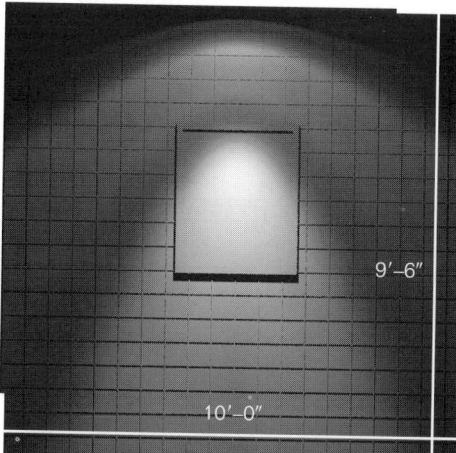

9'–6"

10'–0"

Lamp: 20MR16/AL/FL
Location: 2'–0" from wall
Center beam Aiming: 13° (at point 1'–0" AFF)
Luminaire: Monopoint or Recessed Adjustable
Illuminance: 7 fc avg, maintained (13 fc max)

Lamp: 20MR16/AL/FL
Location: 2'–6" from wall
Center beam Aiming: 18° (at point 2'–0" AFF)
Luminaire: Monopoint or Recessed Adjustable
Illuminance: 7 fc avg, maintained (13 fc max)

Lamp: 20MR16/AL/FL
Location: 3'–0" from wall
Center beam Aiming: 23° (at point 2'–6" AFF)
Luminaire: Monopoint or Recessed Adjustable
Illuminance: 7 fc avg, maintained (13 fc max)

Uses
▶ General lighting
▶ Wallwashing (see 11–115)
▶ Small art with short throws
▶ Available in 12V

Specs
Beam spread/35°
Center beam
  candlepower/650
Life/4000 hrs
▶ **Osram Sylvania** (58590)

Design Tips
✔ One light best for small art where mounting height of light is 10' or less.
✔ Multiple units can be used for wallwashing or lighting of larger artwork.
✔ More grazing light creates prominent shadows.
✔ MR lamps are glary—consider shielding with hexcell louver or using deep luminaire trims or pinholes.
✔ Tempered cover glass required for safety.
✔ Required transformer consumes 3 to 5 watts.

Application Key

Commercial
Gallery
Hospitality
Institutional
Manufacturing
Residential
Retail
Exterior

bold = primary application
partial fade = minimal application
fade = unlikely application

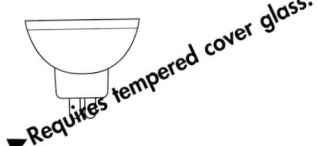

▶ Requires tempered cover glass.

Halogen Low Voltage (12V)

4

Halogen 35MR16/AL/NSP

Aluminum Reflector

Uses
▶ Architectural detailing
▶ Medium art with medium throws
▶ Sculpture with medium throws
▶ Medium throws
▶ Available in 12V

Specs
Beam spread/10°
Center beam candlepower/ 8700
Life/4000 hrs
▶ **Osram Sylvania** (58591)

Design Tips
✔ One light best for medium art where mounting height for light is 13' or less.
✔ Intense light degrades some artwork rapidly and/or significantly—consider UV and IR filters or limit exposure time.
✔ Great for glass pieces and floral centerpieces and where mounting height for light is 13' or less.
✔ Scallops, striations and frame shadows are most severe when light is closer to the wall (compare top left image to bottom image).
✔ MR lamps are glary—consider shielding with hexcell louver or using deep luminaire trims or pinholes.
✔ Tempered cover glass required for safety.
✔ Required transformer consumes 3 to 5 watts.

Application Key

Commercial
Gallery
Hospitality
Institutional
Manufacturing
Residential
Retail
Exterior

bold = primary application
partial fade = minimal application
fade = unlikely application

Performance Sketches

[All data outlined here is based on information from boldfaced manufacturer's published data for 12V version at time of manuscript preparation. Scale is ¼ inch equals 1 foot. **Note: At press time, actual photometry was unavailable for this lamp. Data were generated using Osram Sylvania 65MR16/NSP lamp and rerating candlepower—this is imprecise.**]

Lamp: 35MR16/AL/NSP
Location: 3'–0" from wall
Center beam Aiming: 18° (at point 3'–6" AFF)
Luminaire: Monopoint or Recessed Adjustable
Illuminance: 16 fc avg, maintained (40 fc max)

Lamp: 35MR16/AL/NSP
Location: 4'–0" from wall
Center beam Aiming: 27° (at point 4'–6" AFF)
Luminaire: Monopoint or Recessed Adjustable
Illuminance: 20 fc avg, maintained (54 fc max)

12'–6"

10'–0"

Art shown at 3' by 4' and centered 5'–6" AFF. Wall grid on 6" by 6" centers and scaled at ¼"=1'.

Lamp: 35MR16/AL/NSP
Location: 5'–0" from wall
Center beam Aiming: 32° (at point 4'–6" AFF)
Luminaire: Monopoint or Recessed Adjustable
Illuminance: 21 fc avg, maintained (54 fc max)

Halogen 35MR16/AL/FL
Aluminum Reflector

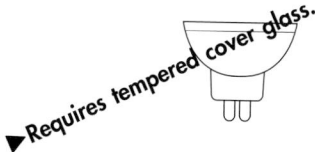

▶Requires tempered cover glass.

Performance Sketches

[All data outlined here is based on information from boldfaced manufacturer's published data for 12V version at time of manuscript preparation. Scale is ¼ inch equals 1 foot. **Note: At press time, actual photometry was unavailable for this lamp. Data were generated using Osram Sylvania 20MR16/AL/FL lamp and rerating candlepower—this is imprecise.**]

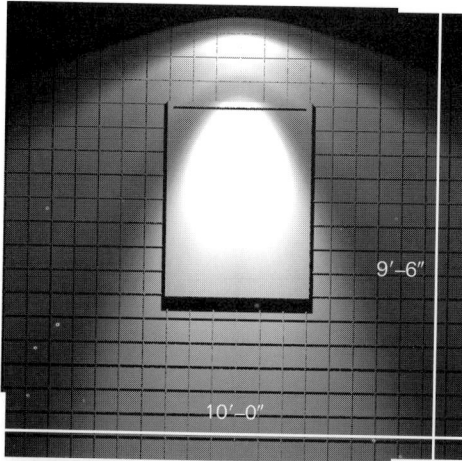

9'–6"

10'–0"

Art shown at 3' by 4' and centered 5'–6" AFF. Wall grid on 6" by 6" centers and scaled at ¼"=1'.

Lamp: 35MR16/AL/FL
Location: 1'–6" from wall
Center beam Aiming: 14° (at point 3'–6" AFF)
Luminaire: Monopoint or Recessed Adjustable
Illuminance: 15 fc avg, maintained (52 fc max)

Lamp: 35MR16/AL/FL
Location: 2'–0" from wall
Center beam Aiming: 20° (at point 4'–0" AFF)
Luminaire: Monopoint or Recessed Adjustable
Illuminance: 16 fc avg, maintained (48 fc max)

Lamp: 35MR16/AL/FL
Location: 2'–6" from wall
Center beam Aiming: 27° (at point 4'–6" AFF)
Luminaire: Monopoint or Recessed Adjustable
Illuminance: 18 fc avg, maintained (47 fc max)

Uses
▶General lighting
▶Medium art with short throws
▶Short throws
▶Available in 12V

Specs
Beam spread/35°
Center beam
 candlepower/1300
Life/4000 hrs
▶**Osram Sylvania** (58593)

Design Tips
✔One light best for medium art where mounting height for light is 10' or less.
✔Light may degrade artwork rapidly and/or significantly: consider UV and IR filters or limit exposure time.
✔Good for downlighting onto decorative floor materials—best with sconces and/or art accents and/or wallwashing to avoid cave effect and grazing glare (the "I wish I had a visor" effect).
✔More grazing light creates prominent shadows.
✔MR lamps are glary—consider shielding with hexcell louver or using deep luminaire trims or pinholes.
✔Tempered cover glass required for safety.
✔Required transformer consumes 3 to 5 watts.

Application Key

Commercial
Gallery
Hospitality
Institutional
Manufacturing
Residential
Retail
Exterior

bold = primary application
partial fade = minimal application
fade = unlikely application

▶ *Requires tempered cover glass.*

Halogen 50MR16/AL/SP

Aluminum Reflector

Halogen Low Voltage (12V)

4

Uses
▶ Architectural detailing
▶ Large art with long throws
▶ Sculpture with long throws
▶ Long throws
▶ Available in 12V

Specs
Beam spread/11°
Center beam candlepower/ 10500
Life/4000 hrs
▶ **Osram Sylvania** (58594)

Design Tips
✔ One light best for large art where mounting height for light is 16' or less.
✔ Light may degrade artwork rapidly and/or significantly: consider UV and IR filters or limit exposure time.
✔ Great for glass pieces and floral centerpieces where mounting height for light is 16' or less.
✔ These are strong accents and deserve judicious use.
✔ Scallops, striations and frame shadows are most severe when light is closer to the wall (compare top left image to bottom image).
✔ MR lamps are glary—consider shielding with hexcell louver or using deep luminaire trims or pinholes.
✔ Tempered cover glass required for safety.
✔ Required transformer consumes 3 to 5 watts.
✔ Consider 37MR16/HIR/C/NSP.

Application Key

Commercial
Gallery
Hospitality
Institutional
Manufacturing
Residential
Retail
Exterior

bold = primary application
partial fade = minimal application
fade = unlikely application

Performance Sketches

[All data outlined here is based on information from boldfaced manufacturer's published data for 12V version at time of manuscript preparation. Scale is ¼ inch equals 1 foot. **Note: At press time, actual photometry was unavailable for this lamp. Data were generated using Osram Sylvania 50MR16/AL/CG/SP lamp and rerating candlepower—this is imprecise.**]

Lamp: 50MR16/AL/SP
Location: 4'–3" from wall
Center beam Aiming: 20° (at point 3'–6" AFF)
Luminaire: Monopoint or Recessed Adjustable
Illuminance: 13 fc avg, maintained (27 fc max)

Lamp: 50MR16/AL/SP
Location: 5'–3" from wall
Center beam Aiming: 27° (at point 4'–6" AFF)
Luminaire: Monopoint or Recessed Adjustable
Illuminance: 15 fc avg, maintained (33 fc max)

Art shown at 4' by 5' and centered 5'–6" AFF. Wall grid on 6" by 6" centers and scaled at ¼"=1'.

Lamp: 50MR16/AL/SP
Location: 6'–3" from wall
Center beam Aiming: 31° (at point 4'–6" AFF)
Luminaire: Monopoint or Recessed Adjustable
Illuminance: 16 fc avg, maintained (33 fc max)

Halogen 50MR16/AL/NFL
Aluminum Reflector

▶ Requires tempered cover glass.

Performance Sketches

[All data outlined here is based on information from boldfaced manufacturer's published data for 12V version at time of manuscript preparation. Scale is ¼ inch equals 1 foot. **Note: At press time, actual photometry was unavailable for this lamp. Data were generated using Osram Sylvania 65MR16/NFL lamp and rerating candlepower—this is imprecise.**]

Lamp: 50MR16/AL/NFL
Location: 3'–0" from wall
Center beam Aiming: 25° (at point 4'–0" AFF)
Luminaire: Monopoint or Recessed Adjustable
Illuminance: 19 fc avg, maintained (37 fc max)

Lamp: 50MR16/AL/NFL
Location: 4'–0" from wall
Center beam Aiming: 34° (at point 4'–6" AFF)
Luminaire: Monopoint or Recessed Adjustable
Illuminance: 20 fc avg, maintained (38 fc max)

10'–6"

10'–0"

Art shown at 3' by 4' and centered 5'–6" AFF. Wall grid on 6" by 6" centers and scaled at ¼"=1'.

Lamp: 50MR16/AL/NFL
Location: 5'–0" from wall
Center beam Aiming: 40° (at point 4'–6" AFF)
Luminaire: Monopoint or Recessed Adjustable
Illuminance: 20 fc avg, maintained (33 fc max)

Uses
▶ Architectural accenting
▶ Feature downlighting
▶ Medium art with medium throws
▶ Sculpture with medium throws
▶ Medium throws
▶ Available in 12V

Specs
Beam spread/25°
Center beam candlepower/ 3000
Life/4000 hrs
▶ **Osram Sylvania** (58595)

Halogen
Low Voltage (12V)

4

Design Tips
✔ One light best for medium art where mounting height for light is 11' or less.
✔ Great for moderate to larger sculpture where mounting height for light is 11' or less.
✔ Scallops, striations and frame shadows are most severe when light is closer to the wall (compare top left image to bottom image).
✔ MR lamps are glary—consider shielding with hexcell louver or using deep luminaire trims or pinholes.
✔ Tempered cover glass required for safety.
✔ Required transformer consumes 3 to 5 watts.
✔ Consider 37MR16/HIR/C/NFL.

Application Key
Commercial
Gallery
Hospitality
Institutional
Manufacturing
Residential
Retail
Exterior

bold = primary application
partial fade = minimal application
fade = unlikely application

▶ Requires tempered cover glass.

Halogen 50MR16/AL/FL

Aluminum Reflector

Halogen
Low Voltage (12V)

4

Uses
▶ General lighting
▶ Medium art with short throws
▶ Short throws
▶ Available in 12V

Specs
Beam spread/35°
**Center beam
candlepower/1800**
Life/4000 hrs
▶ **Osram Sylvania** (58596)

Design Tips
✔ One light best for medium art where mounting height for light is 10′ or less.
✔ Light may degrade artwork rapidly and/or significantly: consider UV and IR filters or limit exposure time.
✔ Good for downlighting onto decorative floor materials—best with sconces and/or art accents and/or wallwashing to avoid cave effect and grazing glare (the "I wish I had a visor" effect).
✔ More grazing light creates prominent shadows.
✔ MR lamps are glary—consider shielding with hexcell louver or using deep luminaire trims or pinholes.
✔ Tempered cover glass required for safety.
✔ Required transformer consumes 3 to 5 watts.
✔ Consider 37MR16/HIR/C/FL.

Application Key

Commercial
Gallery
Hospitality
Institutional
Manufacturing
Residential
Retail
Exterior

bold = primary application
partial fade = minimal application
fade = unlikely application

Performance Sketches

[All data outlined here is based on information from boldfaced manufacturer's published data for 12V version at time of manuscript preparation. Scale is ¼ inch equals 1 foot. **Note: At press time, actual photometry was unavailable for this lamp. Data were generated using Osram Sylvania 20MR16/AL/FL lamp and rerating candlepower—this is imprecise.**]

Art shown at 3′ by 4′ and centered 5′–6″ AFF. Wall grid on 6″ by 6″ centers and scaled at ¼″=1′.

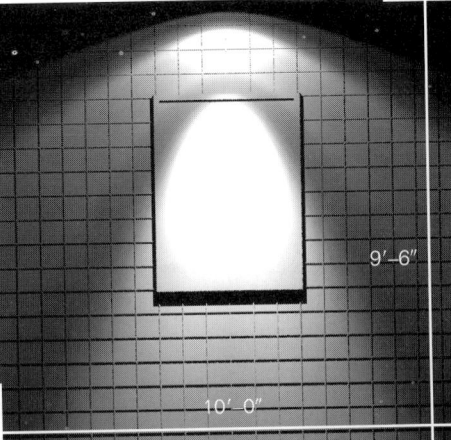

Lamp: 50MR16/AL/FL
Location: 1′–6″ from wall
Center beam Aiming: 11° (at point 2′–0″ AFF)
Luminaire: Monopoint or Recessed Adjustable
Illuminance: 18 fc avg, maintained (61 fc max)

Lamp: 50MR16/AL/FL
Location: 2′–0″ from wall
Center beam Aiming: 17° (at point 3′–0″ AFF)
Luminaire: Monopoint or Recessed Adjustable
Illuminance: 20 fc avg, maintained (55 fc max)

Lamp: 50MR16/AL/FL
Location: 2′–6″ from wall
Center beam Aiming: 24° (at point 4′–0″ AFF)
Luminaire: Monopoint or Recessed Adjustable
Illuminance: 23 fc avg, maintained (60 fc max)

Halogen 65MR16/AL/NSP
Aluminum Reflector

▶Requires tempered cover glass.

4

Performance Sketches

[All data outlined here is based on information from boldfaced manufacturer's published data for 12V version at time of manuscript preparation. Scale is ¼ inch equals 1 foot. **Note: At press time, actual photometry was unavailable for this lamp. Data were generated using Osram Sylvania 65MR16/NSP lamp and rerating candlepower—this is imprecise.**]

Lamp: 65MR16/AL/NSP
Location: 4'–3" from wall
Center beam Aiming: 20° (at point 3'–6" AFF)
Luminaire: Monopoint or Recessed Adjustable
Illuminance: 19 fc avg, maintained (36 fc max)

Lamp: 65MR16/AL/NSP
Location: 5'–3" from wall
Center beam Aiming: 27° (at point 4'–6" AFF)
Luminaire: Monopoint or Recessed Adjustable
Illuminance: 23 fc avg, maintained (45 fc max)

15'–0"

Art shown at 3' by 4' and centered 5'–6" AFF. Wall grid on 6" by 6" centers and scaled at ¼"=1'.

Lamp: 65MR16/AL/NSP
Location: 6'–3" from wall
Center beam Aiming: 31° (at point 4'–6" AFF)
Luminaire: Monopoint or Recessed Adjustable
Illuminance: 24 fc avg, maintained (45 fc max)

Uses
▶Architectural accenting
▶Medium art with long throws
▶Sculpture with long throws
▶Long throws
▶Available in 12V

Specs
Beam spread/10°
Center beam candlepower/ 12500
Life/4000 hrs
▶**Osram Sylvania** (58559)

Design Tips
✔One light best for medium art where mounting height for light is 16' or less.
✔Light may degrade artwork rapidly and/or significantly: consider UV and IR filters or limit exposure time.
✔Great for moderate to large scultpure where mounting height for light is 16' or less.
✔These are strong accents and deserve judicious use.
✔Scallops, striations and frame shadows are most severe when light is closer to the wall (compare top left image to bottom image).
✔MR lamps are glary—consider shielding with hexcell louver or using deep luminaire trims or pinholes.
✔Tempered cover glass required for safety.
✔Required transformer consumes 3 to 5 watts.
✔Consider 37MR16/HIR/C/NSP.

Application Key

Commercial
Gallery
Hospitality
Institutional
Manufacturing
Residential
Retail
Exterior

bold = primary application
partial fade = minimal application
fade = unlikely application

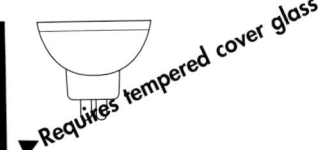

▶ Requires tempered cover glass.

Halogen 65MR16/AL/NFL

Aluminum Reflector

Halogen Low Voltage (12V)

4

Uses
▶ Architectural accenting
▶ Feature downlighting
▶ Large art with medium throws
▶ Sculpture with medium throws
▶ Medium throws
▶ Available in 12V

Specs
Beam spread/25°
Center beam candlepower/ 3600
Life/4000 hrs
▶ **Osram Sylvania** (58561)

Design Tips
✔ One light best for large art where mounting height for light is 13' or less.
✔ Great for moderate to large sculpture where mounting height for light is 13' or less.
✔ Scallops, striations and frame shadows are most severe when light is closer to the wall (compare top left image to bottom image).
✔ MR lamps are glary—consider shielding with hexcell louver or using deep luminaire trims or pinholes.
✔ Tempered cover glass required for safety.
✔ Required transformer consumes 3 to 5 watts.

Application Key

Commercial
Gallery
Hospitality
Institutional
Manufacturing
Residential
Retail
Exterior

bold = primary application
partial fade = minimal application
fade = unlikely application

Performance Sketches

[All data outlined here is based on information from boldfaced manufacturer's published data for 12V version at time of manuscript preparation. Scale is ¼ inch equals 1 foot. **Note: At press time, actual photometry was unavailable for this lamp. Data were generated using Osram Sylvania 65MR16/NFL lamp and rerating candlepower—this is imprecise.**]

Lamp: 65MR16/AL/NFL
Location: 3'–0" from wall
Center beam Aiming: 19° (at point 4'–0" AFF)
Luminaire: Monopoint or Recessed Adjustable
Illuminance: 13 fc avg, maintained (29 fc max)

Lamp: 65MR16/AL/NFL
Location: 4'–0" from wall
Center beam Aiming: 25° (at point 4'–0" AFF)
Luminaire: Monopoint or Recessed Adjustable
Illuminance: 13 fc avg, maintained (26 fc max)

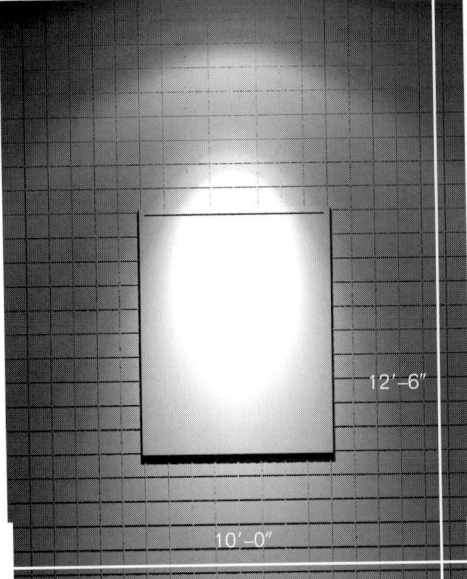

12'–6"

10'–0"

Art shown at 4' by 5' and centered 5'–6" AFF. Wall grid on 6" by 6" centers and scaled at ¼"=1'.

Lamp: 65MR16/AL/NFL
Location: 5'–0" from wall
Center beam Aiming: 30° (at point 4'–0" AFF)
Luminaire: Monopoint or Recessed Adjustable
Illuminance: 13 fc avg, maintained (24 fc max)

Halogen 65MR16/AL/FL
Aluminum Reflector

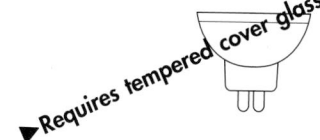

▶ Requires tempered cover glass.

Halogen Low Voltage (12V)

4

Performance Sketches

[All data outlined here is based on information from boldfaced manufacturer's published data for 12V version at time of manuscript preparation. Scale is ¼ inch equals 1 foot. **Note: At press time, actual photometry was unavailable for this lamp. Data were generated using Osram Sylvania 20MR16/AL/FL lamp and rerating candlepower—this is imprecise.**]

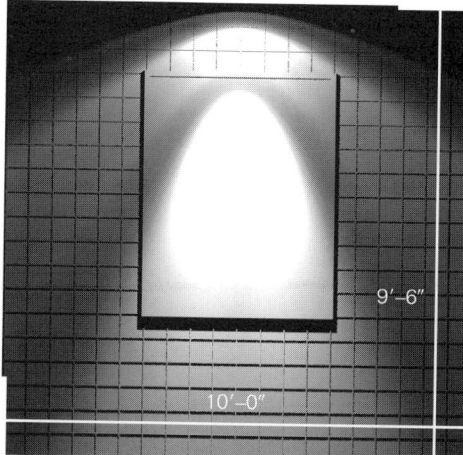

9'–6"

10'–0"

Art shown at 4' by 5' and centered 5'–6" AFF. Wall grid on 6" by 6" centers and scaled at ¼"=1'.

Lamp: 65MR16/AL/FL
Location: 2'–0" from wall
Center beam Aiming: 17° (at point 3'–0" AFF)
Luminaire: Monopoint or Recessed Adjustable
Illuminance: 16 fc avg, maintained (62 fc max)

Lamp: 65MR16/AL/FL
Location: 2'–6" from wall
Center beam Aiming: 24° (at point 4'–0" AFF)
Luminaire: Monopoint or Recessed Adjustable
Illuminance: 19 fc avg, maintained (67 fc max)

Lamp: 65MR16/AL/FL
Location: 3'–0" from wall
Center beam Aiming: 31° (at point 4'–6" AFF)
Luminaire: Monopoint or Recessed Adjustable
Illuminance: 21 fc avg, maintained (68 fc max)

Uses
▶ General lighting
▶ Large art with short throws
▶ Short throws
▶ Available in 12V

Specs
Beam spread/35°
Center beam candlepower/2100
Life/4000 hrs
▶ **Osram Sylvania** (58560)

Design Tips
✔ One light best for large art where mounting height for light is 10' or less.
✔ Light may degrade artwork rapidly and/or significantly: consider UV and IR filters or limit exposure time.
✔ Good for downlighting onto decorative floor materials—best with sconces and/or art accents and/or wallwashing to avoid cave effect and grazing glare (the "I wish I had a visor" effect).
✔ More grazing light creates prominent shadows.
✔ MR lamps are glary—consider shielding with hexcell louver or using deep luminaire trims or pinholes.
✔ Tempered cover glass required for safety.
✔ Required transformer consumes 3 to 5 watts.
✔ Consider 37MR16/HIR/C/FL

Application Key

Commercial
Gallery
Hospitality
Institutional
Manufacturing
Residential
Retail
Exterior

bold = primary application
partial fade = minimal application
fade = unlikely application

Has integral cover glass.

Halogen/MR16/AL/CG
Aluminum Reflector and Cover Glass

Halogen Low Voltage (12V)

4

Guide
➲Longer life is better
➲Higher efficacy better for general lighting
➲Lower wattage is better
➲Lower color of light is warmer
➲Low cost range for halogen is about US$7.⁵⁰ to US$12.⁰⁰

Stats
Data varies manufacturer to manufacturer and changes from time to time

Life/4000 hours
Varies manufacturer to manufacturer. A typical office environment might be occupied about 2600 hours each year.

Efficacy/Not reported
Excellent beam intensities and distributions available for the wattage.

Wattages/20 and 50
Color of Light/ 3000°K
Typically the higher wattage lamps have slightly higher color temperature.

Beam spreads/NSP, SP, NFL and FL
Narrow spot, spot, narrow flood and flood distributions are available.

Size/1⅞" L by 2" Ø
Voltage/12V
Cost Magnitude/Moderate (2000)
Manufacturers
  Osram Sylvania
  Philips

Advantages
  Crisp white light
  Very small size
  Easily dimmed
  Low initial cost
  Alternative to less efficient R and halogen PAR lamps
  Withstand extreme temperature range
  No backlight spill
  Less evident color variations and color shifts
  Very little lamp lumen depreciation (light loss) over life

Disadvantages
  Beam aberrations
  ◆Striations
  ◆Inconsistent beam shape
  No backlight spill (sometimes this is part of the desired "look")
  Hot to touch
  Integral cover glass
  ◆Tempered safety glass may still be required (check with lamp manufacturer)
  ◆May prevent use of auxiliary accessories (e.g., color filters, louvers, etc.)
  Transformer required
  ◆Additional cost and visual aspect of transformer must be accommodated
  ◆Additional wattage requirement of transformer of 3 to 5 watts
  ◆Potential for audible hum

Net Addresses
http://ecom.sylvania.com/osicatalog/
http://www.lighting.philips.com/

CONNECT FOR MORE

For more efficient accent lighting, use HIR low voltage lamps (Section 6); **HIR PAR lamps** (Section 5) **or CMH PAR lamps** (Section 7).

Halogen/MR16/AL/CG
Aluminum Reflector and Cover Glass

▶Has integral cover glass.

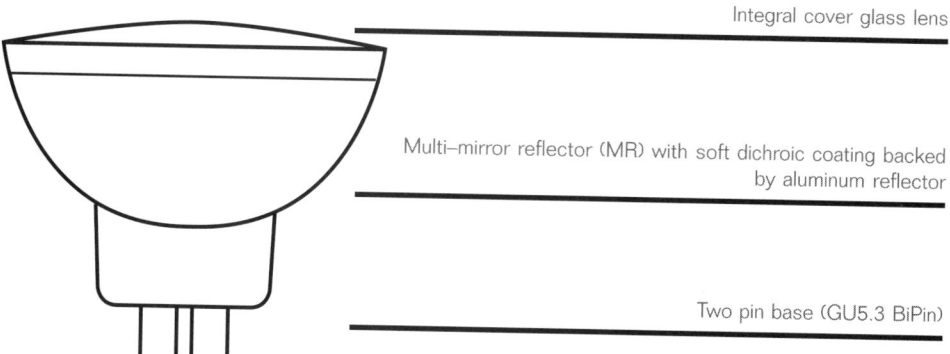

Integral cover glass lens

Multi–mirror reflector (MR) with soft dichroic coating backed by aluminum reflector

Two pin base (GU5.3 BiPin)

Lamp shown actual size
Image copyright 2000/GE

Uses

Architectural Accenting
Art Accenting
Commercial
Gallery
Hospitality
Residential
Downlighting
Lobbies
Conference rooms
Hospitality
Residential
Retail
Merchandising
Surface Grazing
Specular/semispecular surfaces
◆Polished/honed stone
◆Polished/satin wood
◆Fritted/etched glass
◆Cut glass

MR16 Lighting Design Tips
✔Dimming increases life (typically doubling life), but shifts color warmer.
✔Very glary when used without much other general lighting or when aimed across long distances where ceilings are low.
✔Beam striations are most apparent on light–colored, nontextured flat surfaces and may be objectionable.
✔MR16 downlighting alone is harsh—combine several techniques for best results.
✔Use smaller scale (e.g., 4–inch diameter) downlights and adjustable accents for attractive, more human–scale approach.
✔Integral cover glass may interfere with auxiliary add–on lenses, louvers or accessories.

Net Addresses/MR16 Track, Monopoints, Recessed and Exterior
See page 4–107

CONNECT FOR MORE

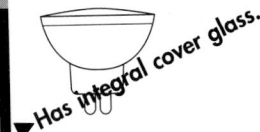
▶Has integral cover glass.

Halogen
Low Voltage (12V)

4

Halogen/MR16/AL/CG

MR16 Aluminum Reflector and Cover Glass Visual Index

20MR16/AL/CG/NSP
▶Architectural detailing
▶Petite art/9' ceiling height or less
▶Objet d'art
▶Short throws
▶Commercial, Hospitality

20MR16/AL/CG/FL
▶General lighting
▶Medium art/9' ceiling height or less
▶Short throws
▶Gallery, Residential

50MR16/AL/CG/SP
▶Architectural accenting
▶Medium art/16' ceiling height or less
▶Sculpture
▶Long throws
▶Hospitality

50MR16/AL/CG/NFL
▶Architectural accenting
▶Feature downlighting
▶Medium art/11' ceiling height or less
▶Sculpture
▶Medium throws
▶Commercial, Hospitality

50MR16/AL/CG/FL
▶General lighting
▶Wallwashing
▶Large art/11' ceiling height or less
▶Medium throws
▶Gallery, Hospitality, Residential

Throws/Ceilings
⊃Short = 7' to 10'
⊃Medium = 10' to 15'
⊃Long = 15' to 20'
⊃Very Long = 20' or more

Recommended intensities
for art accenting

Low (gallery, museum, residential)
5 to 10 fc average on art
Moderate (hospitality)
10 to 20 fc average on art
High (commercial—or higher for retail)
20 to 30 fc average on art

These are intended to be average, maintained values including effects of accent lighting and general room lighting. General room lighting contributes just a few footcandles in residential, gallery and museum settings, but contributes 5 to 10 fc in commercial settings. Maximums might range from 50 percent to 100 percent of the average value. Wherever artwork is highly valuable and/or sensitive to light, precautions should be taken to limit light exposure (maintaining low levels and/or switching lights off when art is not being viewed); and exposure to infrared and ultraviolet should be limited (using filters on lamps or placing art behind specially treated glass or acrylic).

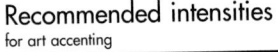

Artwork Size Categories
⊃Petite = 2'–0" by 2'–0" or smaller
⊃Small = 2'–0" by 2'–0" to 2'–6" by 3'–0"
⊃Medium = 2'–6" by 3'–0" to 3'–0" by 4'–0"
⊃Large = 3'–0" by 4'–0" to 4'–0" by 5'–0"
⊃Extra Large = 4'–0" by 5'–0" or larger

Halogen/MR16/AL/CG
Aluminum Reflector

 ▶Has integral cover glass.

Halogen Low Voltage (12V) · 4

Halogen Low Voltage MR Lamp Designation
20MR16/C/CG/NFL

Last set of letters identifies beam spread

FLOODS
• 24° to 32° for narrow flood (NFL)
• 33° to 44° for flood (FL)
• 45° to 54° for wide flood (WFL)
• 55° or greater for very wide flood (VWFL)

SPOTS
• 7° or less for very narrow spot (VNSP)
• 8° to 10° for narrow spot (NSP)
• 11° to 14° for spot (SP)
• 15° to 18° for wide spot (WSP)
• 19° to 23° for very wide spot (VWSP)

Ranges of beam spreads used for convenient/consistent reference in this text and do not necessarily correspond to each manufacturer's nor any ANSI or NEMA definitions.

"CG" for built–in cover glass or no designation for no cover glass

Reflector treatment code
• "C" for consistent color (hard dichroic coating) or no letter for inconsistent color
• "AL" for aluminized reflector eliminating back spill light through reflector

These two digits represent lamp diameter in eighths inches (51mm or 2 inches)

These letters indicate this is a multifaceted reflector (MR) lamp

First two digits represent lamp wattage

[This designation used for convenient reference throughout this *Time–Saver Standards for Architectural Lighting*—but not necessarily used by all lamp manufacturers (although it would be convenient if manufacturers would come to agreement on standard designations.]

▶ Has integral cover glass.

Halogen (Low Voltage (12V))

4

Halogen 20MR16/AL/CG/NSP

Aluminum Reflector and Cover Glass

Uses
▶ Architectural detailing
▶ Petite art with short throws
▶ Objet d'art with short throws
▶ Short throws
▶ Available in 12V

Specs
Beam spread/8°
Center beam
  candlepower/4400
Life/4000 hrs
▶ **Osram Sylvania** (58569)

Design Tips
✔ One light best for petite art where mounting height of light is 9' or less.
✔ Intense light degrades artwork rapidly and/or significantly—consider UV and IR filters or limit exposure time.
✔ Great for glass pieces and floral centerpieces where mounting height of light is 9' or less.
✔ Scallops/striations are most severe when light is closer to the wall (compare top image to bottom image).
✔ More grazing light creates prominent shadows.
✔ MR lamps are glary—consider shielding with hexcell louver or using deep luminaire trims or pinholes.
✔ Required transformer consumes 3 to 5 watts.

Application Key

Commercial
Gallery
Hospitality
Institutional
Manufacturing
Residential
Retail
Exterior

bold = primary application
partial fade = minimal application
fade = unlikely application

Performance Sketches

[All data outlined here is based on information from boldfaced manufacturer's published data for 12V version at time of manuscript preparation. Scale is ¼ inch equals 1 foot. **Note: At press time, actual photometry was unavailable for this lamp. Data were generated using Osram Sylvania 20MR16/VNSP lamp and rerating candlepower—this is imprecise.**]

Art shown at 2' by 2' and centered 5'–6" AFF. Wall grid on 6" by 6" centers and scaled at ¼"=1'.

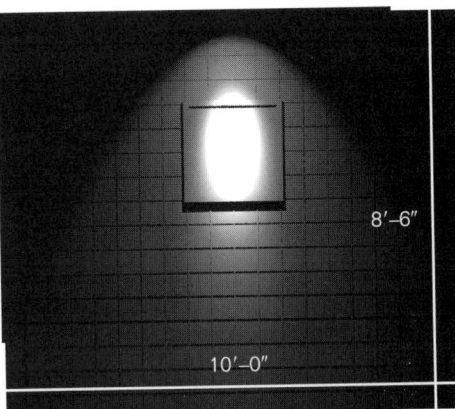

8'–6"

10'–0"

Lamp: 20MR16/AL/CG/NSP
Location: 1'–0" from wall
Center beam Aiming: 16° (at point 5'–0" AFF)
Luminaire: Monopoint or Recessed Adjustable
Illuminance: 17 fc avg, maintained (92 fc max)

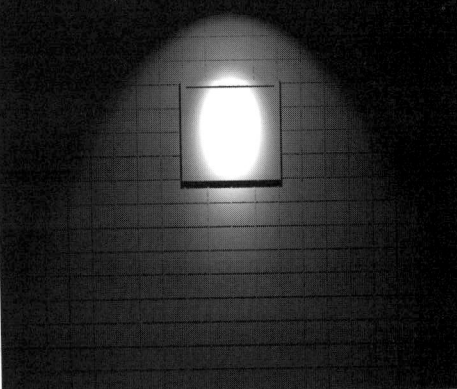

Lamp: 20MR16/AL/CG/NSP
Location: 1'–6" from wall
Center beam Aiming: 27° (at point 5'–6" AFF)
Luminaire: Monopoint or Recessed Adjustable
Illuminance: 20 fc avg, maintained (146 fc max)

Lamp: 20MR16/AL/CG/NSP
Location: 2'–0" from wall
Center beam Aiming: 34° (at point 5'–6" AFF)
Luminaire: Monopoint or Recessed Adjustable
Illuminance: 21 fc avg, maintained (150 fc max)

Halogen 20MR16/AL/CG/FL
Aluminum Reflector and Cover Glass

▶Has integral cover glass.

Halogen Low Voltage (12V)

4

Performance Sketches

[All data outlined here is based on information from boldfaced manufacturer's published data for 12V version at time of manuscript preparation. Scale is ¼ inch equals 1 foot. **Note: At press time, actual photometry was unavailable for this lamp. Data were generated using Osram Sylvania 20MR16/AL/FL lamp and rerating candlepower—this is imprecise.**]

Art shown at 3' by 4' and centered 5'–6" AFF. Wall grid on 6" by 6" centers and scaled at ¼"=1'.

Lamp: 20MR16/AL/CG/FL
Location: 1'–6" from wall
Center beam Aiming: 18° (at point 4'–0" AFF)
Luminaire: Monopoint or Recessed Adjustable
Illuminance: 9 fc avg, maintained (36 fc max)

Lamp: 20MR16/AL/CG/FL
Location: 2'–0" from wall
Center beam Aiming: 27° (at point 4'–6" AFF)
Luminaire: Monopoint or Recessed Adjustable
Illuminance: 10 fc avg, maintained (34 fc max)

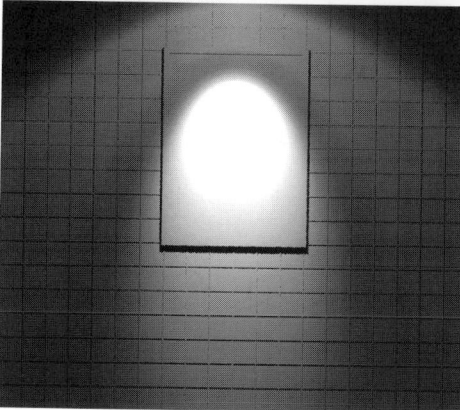

Lamp: 20MR16/AL/CG/FL
Location: 2'–6" from wall
Center beam Aiming: 36° (at point 5'–0" AFF)
Luminaire: Monopoint or Recessed Adjustable
Illuminance: 11 fc avg, maintained (36 fc max)

Uses
▶General lighting
▶Medium art with short throws
▶Available in 12V

Specs
Beam spread/35°
Center beam candlepower/600
Life/4000 hrs
▶**Osram Sylvania** (58570)

Design Tips
✔One light best for medium art where mounting height of light is 9' or less.
✔Multiple units can be used for wallwashing or lighting of larger artwork.
✔More grazing light creates prominent shadows.
✔MR lamps are glary—consider shielding with hexcell louver or using deep luminaire trims or pinholes.
✔Required transformer consumes 3 to 5 watts.

Application Key
Commercial
Gallery
Hospitality
Institutional
Manufacturing
Residential
Retail
Exterior

bold = primary application
partial fade = minimal application
fade = unlikely application

▶Has integral cover glass.

Halogen**50MR16/AL/CG/SP**

Aluminum Reflector and Cover Glass

Performance Sketches

[All data outlined here is based on information from boldfaced manufacturer's published data for 12V version at time of manuscript preparation. Scale is ¼ inch equals 1 foot.]

Halogen Low Voltage (12V)

Uses
▶Architectural accenting
▶Medium art with long throws
▶Sculpture with long throws
▶Long throws
▶Available in 12V

Specs
Beam spread/12°
Center beam candlepower/ 9500
Life/4000 hrs
▶**Osram Sylvania** (58574)
▶Philips

Design Tips
✔One light best for medium art where mounting height for light is 16' or less.
✔Light may degrade artwork rapidly and/or significantly: consider UV and IR filters or limit exposure time.
✔Great for glass pieces and floral centerpieces where mounting height for light is 16' or less.
✔These are strong accents and deserve judicious use.
✔Scallops, striations and frame shadows are most severe when light is closer to the wall (compare top left image to bottom image).
✔MR lamps are glary—consider shielding with hexcell louver or using deep luminaire trims or pinholes.
✔Required transformer consumes 3 to 5 watts.
✔Consider 37MR16/HIR/C/NSP.

Application Key

Commercial
Gallery
Hospitality
Institutional
Manufacturing
Residential
Retail
Exterior

bold = primary application
partial fade = minimal application
fade = unlikely application

Lamp: 50MR16/AL/CG/SP
Location: 4'–3" from wall
Center beam Aiming: 20° (at point 3'–6" AFF)
Luminaire: Monopoint or Recessed Adjustable
Illuminance: 15 fc avg, maintained (25 fc max)

15'–0"

10'–0"

Lamp: 50MR16/AL/CG/SP
Location: 5'–3" from wall
Center beam Aiming: 27° (at point 4'–6" AFF)
Luminaire: Monopoint or Recessed Adjustable
Illuminance: 19 fc avg, maintained (32 fc max)

Art shown at 3' by 4' and centered 5'–6" AFF. Wall grid on 6" by 6" centers and scaled at ¼"=1'.

Lamp: 50MR16/AL/CG/SP
Location: 6'–3" from wall
Center beam Aiming: 31° (at point 4'–6" AFF)
Luminaire: Monopoint or Recessed Adjustable
Illuminance: 19 fc avg, maintained (32 fc max)

Halogen50MR16/AL/CG/NFL
Aluminum Reflector and Cover Glass

►Has integral cover glass.

Halogen Low Voltage (12V)

Performance Sketches

[All data outlined here is based on information from boldfaced manufacturer's published data for 12V version at time of manuscript preparation. Scale is ¼ inch equals 1 foot.]

Lamp: 50MR16/AL/CG/NFL
Location: 3'–0" from wall
Center beam Aiming: 23° (at point 3'–6" AFF)
Luminaire: Monopoint or Recessed Adjustable
Illuminance: 17 fc avg, maintained (41 fc max)

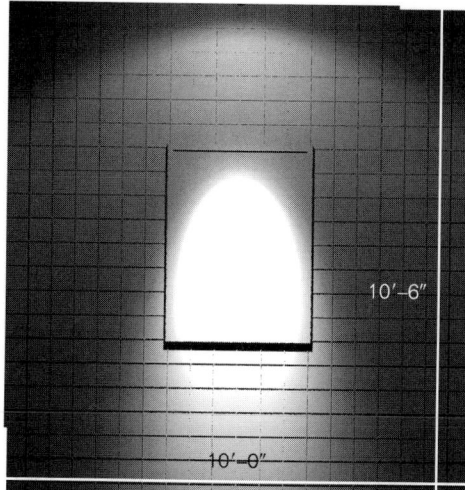

10'–6"

10'–0"

Lamp: 50MR16/AL/CG/NFL
Location: 4'–0" from wall
Center beam Aiming: 34° (at point 4'–6" AFF)
Luminaire: Monopoint or Recessed Adjustable
Illuminance: 21 fc avg, maintained (49 fc max)

Art shown at 3' by 4' and centered 5'–6" AFF. Wall grid on 6" by 6" centers and scaled at ¼"=1'.

Lamp: 50MR16/AL/CG/NFL
Location: 5'–0" from wall
Center beam Aiming: 40° (at point 4'–6" AFF)
Luminaire: Monopoint or Recessed Adjustable
Illuminance: 21 fc avg, maintained (43 fc max)

Uses
►Architectural accenting
►Feature downlighting
►Medium art with medium throws
►Sculpture with medium throws
►Medium throws
►Available in 12V

Specs
Beam spread/24°
Center beam candlepower/ 3200
Life/4000 hrs
►**Philips** (35640–2)

Design Tips
✔One light best for medium art where mounting height for light is 11' or less.
✔Great for moderate to larger sculpture where mounting height for light is 11' or less.
✔Scallops, striations and frame shadows are most severe when light is closer to the wall (compare top left image to bottom image).
✔MR lamps are glary—consider shielding with hexcell louver or using deep luminaire trims or pinholes.
✔Required transformer consumes 3 to 5 watts.
✔Consider 37MR16/HIR/C/NFL.

Application Key

Commercial
Gallery
Hospitality
Institutional
Manufacturing
Residential
Retail
Exterior

bold = primary application
partial fade = minimal application
fade = unlikely application

▶ Has integral cover glass.

Halogen Low Voltage (12V)

Halogen 50MR16/AL/CG/FL

Aluminum Reflector and Cover Glass

Uses
▶ General lighting
▶ Wallwashing (see 11–127)
▶ Large art with short throws
▶ Medium throws
▶ Available in 12V

Specs
Beam spread/38°
Center beam
  candlepower/1500
Life/4000 hrs
▶ Osram Sylvania
▶ **Philips** (35639–4)

Design Tips
✔ One light best for large art where mounting height for light is 11' or less.
✔ Light may degrade artwork rapidly and/or significantly: consider UV and IR filters or limit exposure time.
✔ Good for downlighting onto decorative floor materials—best with sconces and/or art accents and/or wallwashing to avoid cave effect and grazing glare (the "I wish I had a visor" effect).
✔ More grazing light creates prominent shadows.
✔ MR lamps are glary—consider shielding with hexcell louver or using deep luminaire trims or pinholes.
✔ Required transformer consumes 3 to 5 watts.
✔ Consider 37MR16/HIR/C/FL.

Application Key

Commercial
Gallery
Hospitality
Institutional
Manufacturing
Residential
Retail
Exterior

bold = primary application
partial fade = minimal application
fade = unlikely application

Performance Sketches

[All data outlined here is based on information from boldfaced manufacturer's published data for 12V version at time of manuscript preparation. Scale is ¼ inch equals 1 foot.]

Lamp: 50MR16/AL/CG/FL
Location: 4'–0" from wall
Center beam Aiming: 34° (at point 4'–6" AFF)
Luminaire: Monopoint or Recessed Adjustable
Illuminance: 14 fc avg, maintained (35 fc max)

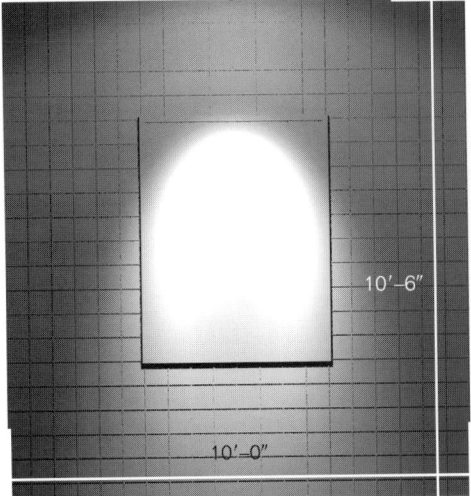

10'–6"

10'–0"

Lamp: 50MR16/AL/CG/FL
Location: 5'–0" from wall
Center beam Aiming: 42° (at point 5'–0" AFF)
Luminaire: Monopoint or Recessed Adjustable
Illuminance: 15 fc avg, maintained (32 fc max)

Art shown at 4' by 5' and centered 5'–6" AFF. Wall grid on 6" by 6" centers and scaled at ¼"=1'.

Lamp: 50MR16/AL/CG/FL
Location: 6'–0" from wall
Center beam Aiming: 50° (at point 5'–6" AFF)
Luminaire: Monopoint or Recessed Adjustable
Illuminance: 15 fc avg, maintained (30 fc max)
Application note: This aiming angle only achievable with track heads or monopoints.

Halogen/MR16
Aluminum Reflector and Cover Glass

▶Has integral cover glass.

Halogen Low Voltage (12V)

4

Net Addresses/MR16 Track and Monopoints
http://www.ardeelighting.com/index.htm
http://www.artemide.com/
http://www.bartcolighting.com/index2.html
http://www.thomasltg.com/
http://www.con-techlighting.com/
http://www.cooperlighting.com/
http://www.flosusa.com/
http://www.hubbell-ltg.com/products.htm#Down&Track
http://www.danalite.com/
http://www.kramerlighting.com/
http://www.lightingservicesinc.com/
http://www.lightolier.com/
http://www.lightproject.com/
http://www.lithonia.com/
http://www.prescolite.com/
http://www.techlighting.com/pro1.html
http://www.tslight.com/

CONNECT FOR MORE

Net Addresses/MR16 Recessed
http://www.alkco.com/rtrak.htm
http://www.ardeelighting.com/index.htm
http://www.artemide.com/
http://www.bartcolighting.com/index2.html
http://www.thomasltg.com/
http://www.con-techlighting.com/
http://www.cooperlighting.com/
http://www.hubbell-ltg.com/products.htm#Down&Track
http://www.danalite.com/
http://www.kramerlighting.com/
http://www.lightolier.com/
http://www.lithonia.com/
http://www.prescolite.com/
http://www.progresslighting.com/

CONNECT FOR MORE

Net Addresses/Exterior Accent
http://www.bega-us.com/home/home.html
http://www.hadcolighting.com/html/bronzelite.htm
http://www.hydrel.com
http://www.kimlighting.com/ingrd1.html

CONNECT FOR MORE

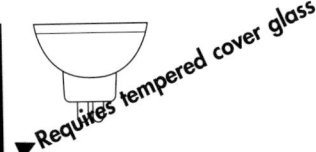

▶ *Requires tempered cover glass.*

Halogen Low Voltage (12V) **4**

Halogen/MR16/C

Hard Dichroic Coating

Guide
◗ Longer life is better
◗ Higher efficacy better for general lighting
◗ Lower wattage is better
◗ Lower color of light is warmer
◗ Low cost range for halogen is about US$7.⁵⁰ to US$12.⁰⁰

Stats
<small>Data varies manufacturer to manufacturer and changes from time to time</small>

Life/3000 to 6000 hours
Varies by wattage and manufacturer to manufacturer. A typical office environment might be occupied about 2600 hours each year.

Efficacy/Not reported
Excellent beam intensities and distributions available for the wattage.

Wattages/20, 35, 42, 50, 65 and 71

Color of Light/2900 to 3050°K
Typically the higher wattage lamps have slightly higher color temperature.

Beam spreads/VNSP, NSP, WSP, VWSP, NFL, FL and VWFL
Very narrow spot, narrow spot, wide spot, very wide spot, narrow flood, flood and very wide flood distributions are available.

Size/1⅞″ L by 2″ Ø

Voltage/12V

Cost Magnitude/Moderate ⁽²⁰⁰⁰⁾

Manufacturers
General Electric
Osram Sylvania

Net Addresses
http://www.ge.com/Lighting/business/index.htm
http://ecom.sylvania.com/osicatalog/

CONNECT FOR MORE

Advantages
Crisp white light
Very small size
Easily dimmed
Alternative to less efficient R and halogen PAR lamps
Withstand extreme temperature range
No backlight spill
No color variations and color shifts
Very little lamp lumen depreciation (light loss) over life

Disadvantages
Beam aberrations
◆ Striations
◆ Inconsistent beam shape
No backlight spill (sometimes this is part of the desired "look")
Hot to touch
Tempered glass lens required
◆ Prevents glass and filament chards from falling onto flammable surfaces/materials should a violent lamp failure occur (known as a non–passive failure)
Transformer required
◆ Additional cost and visual aspect of transformer must be accommodated
◆ Additional wattage requirement of transformer of 3 to 5 watts
◆ Potential for audible hum

For more efficient accent lighting, use HIR low voltage lamps (Section 6); HIR PAR lamps (Section 5) or CMH PAR lamps (Section 7).

Halogen/MR16/C
Hard Dichroic Coating

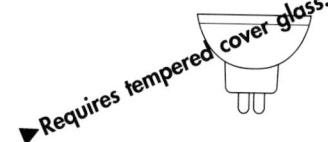
►Requires tempered cover glass.

Halogen Low Voltage (12V)

4

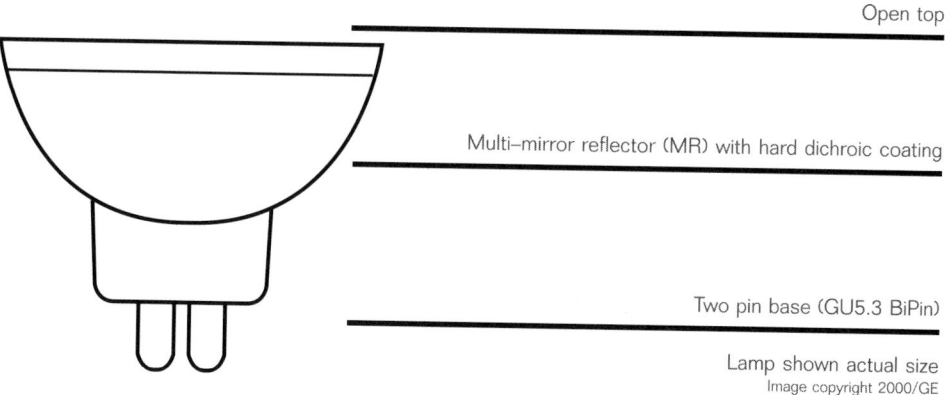

Open top

Multi–mirror reflector (MR) with hard dichroic coating

Two pin base (GU5.3 BiPin)

Lamp shown actual size
Image copyright 2000/GE

Uses

Architectural Accenting
Art Accenting
Commercial
Gallery
Hospitality
Residential
Downlighting
Lobbies
Conference rooms
Hospitality
Residential
Retail
Merchandising
Surface Grazing
Specular/semispecular surfaces
◆Polished/honed stone
◆Polished/satin wood
◆Fritted/etched glass
◆Cut glass

■ ■ ■ ■ ■ ■ ■ ■ ■ ■ ■ ■ ■ ■ ■ ■

MR16 Lighting Design Tips
✔Dimming increases life (typically doubling life), but shifts color warmer.
✔Very glary when used without much other general lighting or when aimed across long distances where ceilings are low.
✔Beam striations are most apparent on light–colored, nontextured flat surfaces and may be objectionable.
✔MR16 downlighting alone is harsh—combine several techniques for best results.
✔Use smaller scale (e.g., 4–inch diameter) downlights and adjustable accents for attractive, more human–scale approach.
✔Use tempered glass lens for safety.

Net Addresses/MR16 Track, Monopoints, Recessed and Exterior
See page 4–107

 CONNECT FOR MORE

▶ Requires tempered cover glass.

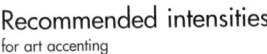

Halogen/MR16/C

MR16 Hard Dichroic Coating Visual Index

Halogen Low Voltage (12V)

4

Throws/Ceilings
⊃ Short = 7' to 10'
⊃ Medium = 10' to 15'
⊃ Long = 15' to 20'
⊃ Very Long = 20' or more

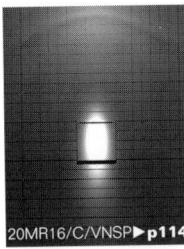  

20MR16/C/VNSP▶p114   20MR16/C/VNSP▶p114   20MR16/C/VNSP▶p114

20MR16/C/VNSP
▶ Architectural detailing
▶ Petite art/13' ceiling height or less
▶ Sculpture
▶ Medium throws
▶ Hospitality

20MR16/C/NSP▶p115   20MR16/C/NSP▶p115   20MR16/C/NSP▶p115

20MR16/C/NSP
▶ Architectural detailing
▶ Petite art/11' ceiling height or less
▶ Sculpture
▶ Medium throws
▶ Commercial, Hospitality, Retail

  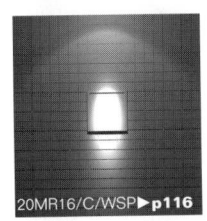

20MR16/C/WSP▶p116   20MR16/C/WSP▶p116   20MR16/C/WSP▶p116

20MR16/C/WSP
▶ Architectural accenting
▶ Petite art/11' ceiling height or less
▶ Sculpture
▶ Medium throws
▶ Commercial, Hospitality

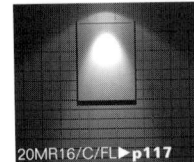 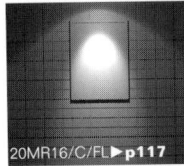 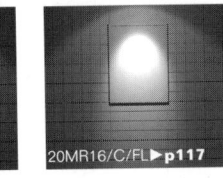

20MR16/C/FL▶p117   20MR16/C/FL▶p117   20MR16/C/FL▶p117

20MR16/C/FL
▶ General lighting
▶ Wallwashing
▶ Medium art/9' ceiling height or less
▶ Short throws
▶ Gallery, Residential

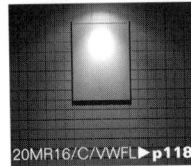   

20MR16/C/VWFL▶p118   20MR16/C/VWFL▶p118   20MR16/C/VWFL▶p118

20MR16/C/VWFL
▶ General lighting
▶ Wallwashing
▶ Medium art/9' ceiling height or less
▶ Short throws
▶ Gallery, Residential

Recommended intensities
for art accenting

Low (gallery, museum, residential)
5 to 10 fc average on art
Moderate (hospitality)
10 to 20 fc average on art
High (commercial—or higher for retail)
20 to 30 fc average on art

These are intended to be average, maintained values including effects of accent lighting and general room lighting. General room lighting contributes just a few footcandles in residential, gallery and museum settings, but contributes 5 to 10 fc in commercial settings. Maximums might range from 50 percent to 100 percent of the average value. Wherever artwork is highly valuable and/or sensitive to light, precautions should be taken to limit light exposure (maintaining low levels and/or switching lights off when art is not being viewed); and exposure to infrared and ultraviolet should be limited (using filters on lamps or placing art behind specially treated glass or acrylic).

35MR16/C/NSP▶p119   35MR16/C/NSP▶p119   35MR16/C/NSP▶p119

35MR16/C/NSP
▶ Architectural accenting
▶ Medium art/13' ceiling height or less
▶ Sculpture
▶ Medium throws
▶ Commercial, Retail

 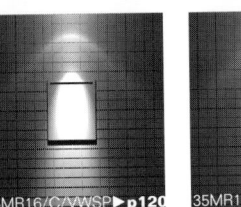  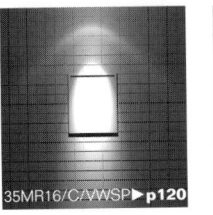 

35MR16/C/VWSP▶p120   35MR16/C/VWSP▶p120   35MR16/C/VWSP▶p120

35MR16/C/VWSP
▶ Architectural accenting
▶ Small art/11' ceiling height or less
▶ Sculpture
▶ Medium throws
▶ Commercial, Retail

Halogen/MR16/C

Hard Dichroic Coating

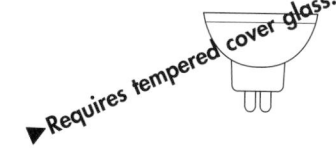

▶Requires tempered cover glass.

Halogen Low Voltage (12V)

4

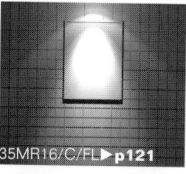 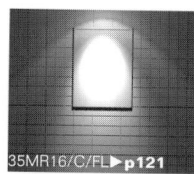 

35MR16/C/FL▶**p121**   35MR16/C/FL▶**p121**   35MR16/C/FL▶**p121**

35MR16/C/FL
▶General lighting
▶Medium art/9′ ceiling height or less
▶Short throws
▶Hospitality

35MR16/C/VWFL▶**p122**   35MR16/C/VWFL▶**p122**   35MR16/C/VWFL▶**p122**

35MR16/C/VWFL
▶General lighting
▶Wallwashing
▶Medium art/9′ ceiling height or less
▶Short throws
▶Hospitality

42MR16/C/NSP▶**p123**   42MR16/C/NSP▶**p123**   42MR16/C/NSP▶**p123**

42MR16/C/NSP
▶Architectural detailing
▶Small art/16′ ceiling height or less
▶Sculpture
▶Long throws
▶Commercial, Hospitality

50MR16/C/NSP▶**p124**   50MR16/C/NSP▶**p124**   50MR16/C/NSP▶**p124**

50MR16/C/NSP
▶Architectural detailing
▶Medium art/16′ ceiling height or less
▶Sculpture
▶Long throws
▶Commercial, Hospitality, Retail

50MR16/C/WSP▶**p125**   50MR16/C/WSP▶**p125**   50MR16/C/WSP▶**p125**

50MR16/C/WSP
▶Architectural accenting
▶Medium art/16′ ceiling height or less
▶Sculpture
▶Long throws
▶Commercial, Hospitality, Retail

50MR16/C/NFL25▶**p126**   50MR16/C/NFL25▶**p126**   50MR16/C/NFL25▶**p126**

50MR16/C/NFL25°
▶Architectural accenting
▶Feature downlighting
▶Medium art/11′ ceiling height or less
▶Sculpture
▶Medium throws
▶Commercial, Hospitality, Retail

▶Requires tempered cover glass.

4

Halogen/MR16/C

Hard Dichroic Coating

Throws/Ceilings

⊃Short = 7' to 10'
⊃Medium = 10' to 15'
⊃Long = 15' to 20'
⊃Very Long = 20' or more

50MR16/C/NFL30▶**p127**   50MR16/C/NFL30▶**p127**   50MR16/C/NFL30▶**p127**

50MR16/C/NFL30°
▶Architectural accenting
▶Feature downlighting
▶Medium/11' ceiling height or less
▶Sculpture
▶Medium throws
▶Commercial, Hospitality, Retail

 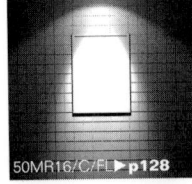 

50MR16/C/FL▶**p128**   50MR16/C/FL▶**p128**   50MR16/C/FL▶**p128**

50MR16/C/FL
▶General lighting
▶Medium art/10' ceiling height or less
▶Short throws
▶Commercial, Retail

50MR16/C/VWFL▶**p129**   50MR16/C/VWFL▶**p129**   50MR16/C/VWFL▶**p129**

50MR16/C/VWFL
▶Soft general lighting
▶Wallwashing
▶Large art/10' ceiling height or less
▶Short throws
▶Hospitality

65MR16/C/NSP▶**p130**   65MR16/C/NSP▶**p130**   65MR16/C/NSP▶**p131**

65MR16/C/NSP
▶Architectural detailing
▶Large art/19' ceiling height or less
▶Sculpture
▶Long throws
▶Hospitality

Recommended intensities
for art accenting

Low (gallery, museum, residential)
5 to 10 fc average on art
Moderate (hospitality)
10 to 20 fc average on art
High (commercial—or higher for retail)
20 to 30 fc average on art

These are intended to be average, maintained values including effects of accent lighting and general room lighting. General room lighting contributes just a few footcandles in residential, gallery and museum settings, but contributes 5 to 10 fc in commercial settings. Maximums might range from 50 percent to 100 percent of the average value. Wherever artwork is highly valuable and/or sensitive to light, precautions should be taken to limit light exposure (maintaining low levels and/or switching lights off when art is not being viewed); and exposure to infrared and ultraviolet should be limited (using filters on lamps or placing art behind specially treated glass or acrylic).

65MR16/C/NFL▶**p132**   65MR16/C/NFL▶**p132**   65MR16/C/NFL▶**p132**

65MR16/C/NFL
▶Architectural accenting
▶Feature downlighting
▶Large art/13' ceiling height or less
▶Sculpture
▶Medium throws
▶Hospitality

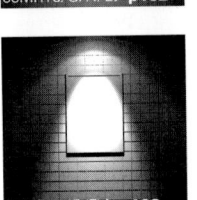  

65MR16/C/FL▶**p133**   65MR16/C/FL▶**p133**   65MR16/C/FL▶**p133**

65MR16/C/FL
▶General lighting
▶Medium art/10' ceiling height or less
▶Short throws
▶Commercial, Retail

65MR16/C/VWFL▶**p134**   65MR16/C/VWFL▶**p134**   65MR16/C/VWFL▶**p134**

65MR16/C/VWFL
▶Soft general lighting
▶Wallwashing
▶Large art/10' ceiling height or less
▶Short throws
▶Gallery, Hospitality, Residential

Halogen/MR16/C

Hard Dichroic Coating

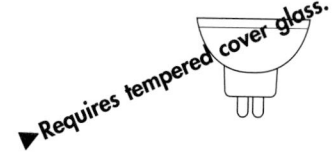

▶Requires tempered cover glass.

71MR16/C/WSP▶p135   71MR16/C/WSP▶p135   71MR16/C/WSP▶p135

71MR16/C/WSP
▶Architectural detailing
▶Large art/16′ ceiling height or less
▶Sculpture
▶Long throws
▶Commercial, Retail

71MR16/C/NFL▶p136   71MR16/C/NFL▶p136   71MR16/C/NFL▶p136

71MR16/C/NFL
▶Architectural accenting
▶Feature downlighting
▶Large art/13′ ceiling height or less
▶Sculpture
▶Medium throws
▶Hospitality

71MR16/C/FL▶p137   71MR16/C/FL▶p137   71MR16/C/FL▶p137

71MR16/C/FL
▶General lighting
▶Wallwashing
▶Large art/11′ ceiling height or less
▶Medium throws
▶Commercial

Halogen Low Voltage MR Lamp Designation

20MR16/C/CG/NFL

Last set of letters identifies beam spread

FLOODS
- 24° to 32° for narrow flood (NFL)
- 33° to 44° for flood (FL)
- 45° to 54° for wide flood (WFL)
- 55° or greater for very wide flood (VWFL)

SPOTS
- 7° or less for very narrow spot (VNSP)
- 8° to 10° for narrow spot (NSP)
- 11° to 14° for spot (SP)
- 15° to 18° for wide spot (WSP)
- 19° to 23° for very wide spot (VWSP)

Ranges of beam spreads used for convenient/consistent reference in this text and do not necessarily correspond to each manufacturer's nor any ANSI or NEMA definitions.

"CG" for built–in cover glass or no designation for no cover glass

Reflector treatment code
- "C" for consistent color (hard dichroic coating) or no letter for inconsistent color
- "AL" for aluminized reflector eliminating back spill light through reflector

These two digits represent lamp diameter in eighths inches (51mm or 2 inches)

These letters indicate this is a multifaceted reflector (MR) lamp

First two digits represent lamp wattage

[This designation used for convenient reference throughout this *Time–Saver Standards for Architectural Lighting*—but not necessarily used by all lamp manufacturers (although it would be convenient if manufacturers would come to agreement on standard designations.]

■ ■ ■ ■ ■ ■ ■ ■ ■ ■

Artwork Size Categories
⮑Petite = 2′–0″ by 2′–0″ or smaller
⮑Small = 2′–0″ by 2′–0″ to 2′–6″ by 3′–0″
⮑Medium = 2′–6″ by 3′–0″ to 3′–0″ by 4′–0″
⮑Large = 3′–0″ by 4′–0″ to 4′–0″ by 5′–0″
⮑Extra Large = 4′–0″ by 5′–0″ or larger

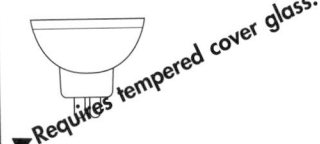

▶ *Requires tempered cover glass.*

Halogen 20MR16/C/VNSP

Hard Dichroic Coating

Halogen Low Voltage (12V)

4

Uses
▶ Architectural detailing
▶ Petite art with medium throws
▶ Sculpture with medium throws
▶ Medium throws
▶ Available in 12V

Specs
Beam spread/7°
Center beam candlepower/ 7400
Life/3000 hrs
▶ **GE** (20816)

Design Tips
✔ One light best for petite art and where mounting height for light is 13' or less.
✔ Light may degrade artwork rapidly and/or significantly: consider UV and IR filters or limit exposure time.
✔ Great for glass pieces and floral centerpieces where mounting height of light is 13' or less.
✔ Scallops/striations are most severe when light is closer to the wall (compare top left image to bottom image).
✔ More grazing light creates prominent shadows.
✔ MR lamps are glary—consider shielding with hexcell louver or using deep luminaire trims or pinholes.
✔ Tempered cover glass required for safety.
✔ Required transformer also consumes 3 to 5 watts.

Application Key

Commercial
Gallery
Hospitality
Institutional
Manufacturing
Residential
Retail
Exterior

bold = primary application
partial fade = minimal application
fade = unlikely application

Performance Sketches

[All data outlined here is based on information from boldfaced manufacturer's published data for 12V version at time of manuscript preparation. Scale is ¼ inch equals 1 foot.]

Lamp: 20MR16/C/VNSP
Location: 3'–0" from wall
Center beam Aiming: 22° (at point 5'–3" AFF)
Luminaire: Monopoint or Recessed Adjustable
Illuminance: 14 fc avg, maintained (48 fc max)

12'–6"

10'–0"

Lamp: 20MR16/C/VNSP
Location: 4'–0" from wall
Center beam Aiming: 29° (at point 5'–3" AFF)
Luminaire: Monopoint or Recessed Adjustable
Illuminance: 16 fc avg, maintained (53 fc max)

Art shown at 2' by 2' and centered 5'–6" AFF. Wall grid on 6" by 6" centers and scaled at ¼"=1'.

Lamp: 20MR16/C/VNSP
Location: 5'–0" from wall
Center beam Aiming: 35° (at point 5'–3" AFF)
Luminaire: Monopoint or Recessed Adjustable
Illuminance: 16 fc avg, maintained (53 fc max)

Halogen 20MR16/C/NSP
Hard Dichroic Coating

Performance Sketches

[All data outlined here is based on information from boldfaced manufacturer's published data for 12V version at time of manuscript preparation. Scale is ¼ inch equals 1 foot. **Note: At press time, actual photometry was unavailable for this lamp. Data were generated using GE 65MR16/NSP lamp and rerating candlepower—this is imprecise.**]

Lamp: 20MR16/C/NSP
Location: 1'–6" from wall
Center beam Aiming: 13° (at point 4'–0" AFF)
Luminaire: Monopoint or Recessed Adjustable
Illuminance: 18 fc avg, maintained (43 fc max)

Lamp: 20MR16/C/NSP
Location: 2'–0" from wall
Center beam Aiming: 20° (at point 5'–0" AFF)
Luminaire: Monopoint or Recessed Adjustable
Illuminance: 25 fc avg, maintained (64 fc max)

Art shown at 2' by 2' and centered 5'–6" AFF. Wall grid on 6" by 6" centers and scaled at ¼"=1'.

Lamp: 20MR16/C/NSP
Location: 2'–6" from wall
Center beam Aiming: 24° (at point 5'–0" AFF)
Luminaire: Monopoint or Recessed Adjustable
Illuminance: 27 fc avg, maintained (65 fc max)

▶ Requires tempered cover glass.

Halogen Low Voltage (12V)

4

Uses
- ▶ Architectural detailing
- ▶ Petite art with medium throws
- ▶ Sculpture with medium throws
- ▶ Medium throws
- ▶ Available in 12V

Specs
Beam spread/10°
Center beam candlepower/ 5000
Life/4000 hrs
▶ **Osram Sylvania** (58550)

Design Tips
✓ One light best for petite art and where mounting height for light is 11' or less.
✓ Light may degrade artwork rapidly and/or significantly: consider UV and IR filters or limit exposure time.
✓ Great for glass pieces and floral centerpieces where mounting height of light is 11' or less.
✓ More grazing light creates prominent shadows.
✓ MR lamps are glary—consider shielding with hexcell louver or using deep luminaire trims or pinholes.
✓ Tempered cover glass required for safety.
✓ Required transformer also consumes 3 to 5 watts.

Application Key

Commercial
Gallery
Hospitality
Institutional
Manufacturing
Residential
Retail
Exterior

bold = primary application
partial fade = minimal application
fade = unlikely application

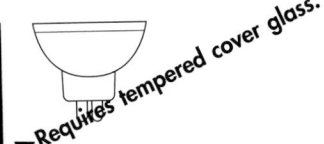

▶Requires tempered cover glass.

Halogen 20MR16/C/WSP

Hard Dichroic Coating

Performance Sketches

[All data outlined here is based on information from boldfaced manufacturer's published data for 12V version at time of manuscript preparation. Scale is ¼ inch equals 1 foot.]

Halogen Low Voltage (12V)

4

Uses
▶Architectural accenting
▶Petite art with medium throws
▶Sculpture with medium throws
▶Medium throws
▶Available in 12V

Specs
Beam spread/15°
Center beam candlepower/ 3600
Life/5000 hrs
▶**GE** (20815)

Design Tips
✔One light best for petite art and where mounting height for light is 11′ or less
✔Light may degrade artwork rapidly and/or significantly: consider UV and IR filters or limit exposure time.
✔More grazing light creates prominent shadows.
✔MR lamps are glary—consider shielding with hexcell louver or using deep luminaire trims or pinholes.
✔Tempered cover glass required for safety.
✔Required transformer also consumes 3 to 5 watts.

Lamp: 20MR16/C/WSP
Location: 1′–6″ from wall
Center beam Aiming: 13° (at point 4′–0″ AFF)
Luminaire: Monopoint or Recessed Adjustable
Illuminance: 13 fc avg, maintained (37 fc max)

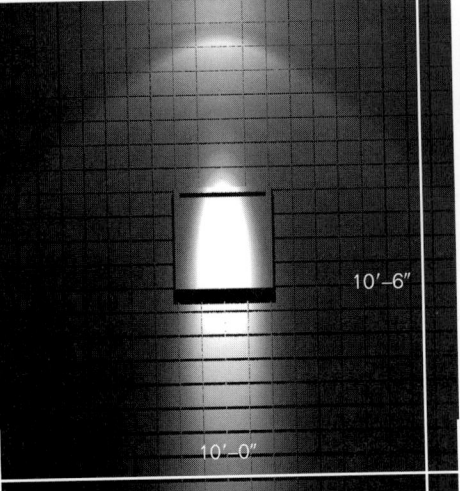

10′–6″

10′–0″

Lamp: 20MR16/C/WSP
Location: 2′–0″ from wall
Center beam Aiming: 20° (at point 5′–0″ AFF)
Luminaire: Monopoint or Recessed Adjustable
Illuminance: 19 fc avg, maintained (55 fc max)

Art shown at 2′ by 2′ and centered 5′–6″ AFF. Wall grid on 6″ by 6″ centers and scaled at ¼″=1′.

Application Key

Commercial
Gallery
Hospitality
Institutional
Manufacturing
Residential
Retail
Exterior

bold = primary application
partial fade = minimal application
fade = unlikely application

Lamp: 20MR16/C/WSP
Location: 2′–6″ from wall
Center beam Aiming: 24° (at point 5′–0″ AFF)
Luminaire: Monopoint or Recessed Adjustable
Illuminance: 20 fc avg, maintained (56 fc max)

Halogen 20MR16/C/FL
Hard Dichroic Coating

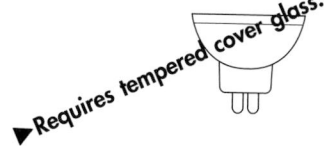

▶Requires tempered cover glass.

Halogen Low Voltage (12V)

4

Performance Sketches

[All data outlined here is based on information from boldfaced manufacturer's published data for 12V version at time of manuscript preparation. Scale is ¼ inch equals 1 foot.]

Art shown at 3' by 4' and centered 5'–6" AFF. Wall grid on 6" by 6" centers and scaled at ¼"=1'.

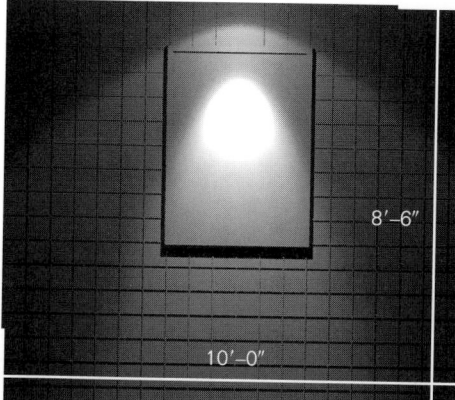

8'–6"

10'–0"

Lamp: 20MR16/C/FL
Location: 1'–6" from wall
Center beam Aiming: 18° (at point 4'–0" AFF)
Luminaire: Monopoint or Recessed Adjustable
Illuminance: 7 fc avg, maintained (29 fc max)

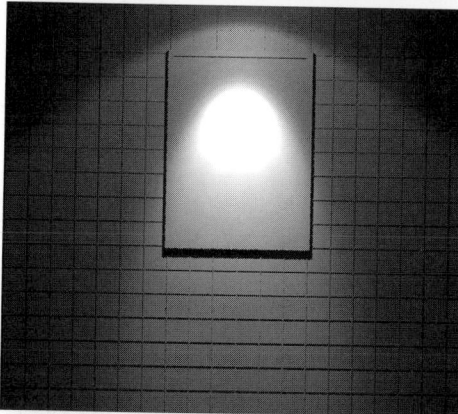

Lamp: 20MR16/C/FL
Location: 2'–0" from wall
Center beam Aiming: 27° (at point 4'–6" AFF)
Luminaire: Monopoint or Recessed Adjustable
Illuminance: 8 fc avg, maintained (28 fc max)

Lamp: 20MR16/C/FL
Location: 2'–6" from wall
Center beam Aiming: 36° (at point 5'–0" AFF)
Luminaire: Monopoint or Recessed Adjustable
Illuminance: 9 fc avg, maintained (26 fc max)

Uses
▶General lighting
▶Wallwashing (see 11–109)
▶Medium art with short throws
▶Available in 12V

Specs
Beam spread/40°
Center beam candlepower/525
Life/4500 hrs
▶**GE** (20814)
▶Osram Sylvania

Design Tips
✔One light best for medium art where mounting height of light is 9' or less.
✔Multiple units can be used for wallwashing or lighting of larger artwork.
✔More grazing light creates prominent shadows.
✔MR lamps are glary—consider shielding with hexcell louver or using deep luminaire trims or pinholes.
✔Tempered cover glass required for safety.
✔Required transformer also consumes 3 to 5 watts.

Application Key

Commercial
Gallery
Hospitality
Institutional
Manufacturing
Residential
Retail
Exterior

bold = primary application
partial fade = minimal application
fade = unlikely application

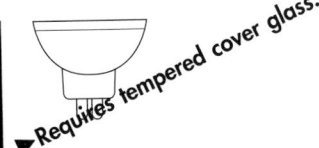

▶ *Requires tempered cover glass.*

Halogen (Low Voltage (12V))

4

Halogen 20MR16/C/VWFL

Hard Dichroic Coating

Performance Sketches

[All data outlined here is based on information from boldfaced manufacturer's published data for 12V version at time of manuscript preparation. Scale is ¼ inch equals 1 foot. **Note: At press time, actual photometry was unavailable for this lamp. Data were generated using GE 50MR16/C/CG/ VWFL lamp and rerating candlepower—this is imprecise.**]

Art shown at 3' by 4' and centered 5'–6" AFF. Wall grid on 6" by 6" centers and scaled at ¼"=1'.

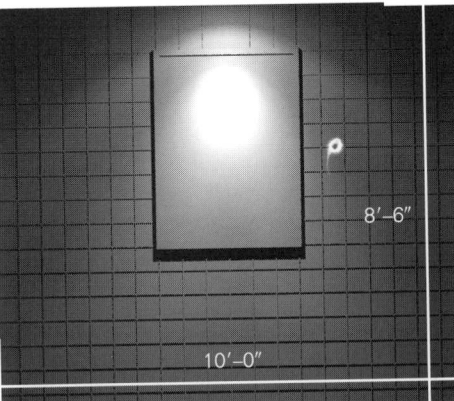

8'–6"

10'–0"

Lamp: 20MR16/C/VWFL
Location: 1'–6" from wall
Center beam Aiming: 14° (at point 2'–6" AFF)
Luminaire: Monopoint or Recessed Adjustable
Illuminance: 7 fc avg, maintained (21 fc max)

Uses
▶ General lighting
▶ Wallwashing (see 11–109)
▶ Medium art with short throws
▶ Available in 12V

Specs
Beam spread/60°
Center beam candlepower/350
Life/4000 hrs
▶ **Osram Sylvania** (58562)

Design Tips
✔ One light best for medium art where mounting height of light is 9' or less.
✔ Multiple units can be used for wallwashing or lighting of larger artwork.
✔ More grazing light creates prominent shadows.
✔ MR lamps are glary—consider shielding with hexcell louver or using deep luminaire trims or pinholes.
✔ Tempered cover glass required for safety.
✔ Required transformer also consumes 3 to 5 watts.

Lamp: 20MR16/C/VWFL
Location: 2'–0" from wall
Center beam Aiming: 22° (at point 3'–6" AFF)
Luminaire: Monopoint or Recessed Adjustable
Illuminance: 8 fc avg, maintained (19 fc max)

Application Key

Commercial
Gallery
Hospitality
Institutional
Manufacturing
Residential
Retail
Exterior

bold = primary application
partial fade = minimal application
fade = unlikely application

Lamp: 20MR16/C/VWFL
Location: 2'–6" from wall
Center beam Aiming: 29° (at point 4'–0" AFF)
Luminaire: Monopoint or Recessed Adjustable
Illuminance: 8 fc avg, maintained (18 fc max)

Halogen 35MR16/C/NSP
Hard Dichroic Coating

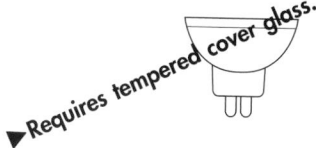

► Requires tempered cover glass.

Performance Sketches

[All data outlined here is based on information from boldfaced manufacturer's published data for 12V version at time of manuscript preparation. Scale is ¼ inch equals 1 foot. **Note: At press time, actual photometry was unavailable for this lamp. Data were generated using Osram Sylvania 20MR16/C/NSP lamp and rerating candlepower—this is imprecise.**]

Lamp: 35MR16/C/NSP
Location: 3'–0" from wall
Center beam Aiming: 21° (at point 4'–6" AFF)
Luminaire: Monopoint or Recessed Adjustable
Illuminance: 19 fc avg, maintained (52 fc max)

Lamp: 35MR16/C/NSP
Location: 4'–0" from wall
Center beam Aiming: 27° (at point 4'–6" AFF)
Luminaire: Monopoint or Recessed Adjustable
Illuminance: 20 fc avg, maintained (54 fc max)

12'–6"

10'–0"

Art shown at 3' by 4' and centered 5'–6" AFF. Wall grid on 6" by 6" centers and scaled at ¼"=1'.

Lamp: 35MR16/C/NSP
Location: 5'–0" from wall
Center beam Aiming: 32° (at point 4'–6" AFF)
Luminaire: Monopoint or Recessed Adjustable
Illuminance: 21 fc avg, maintained (54 fc max)

Uses

► Architectural accenting
► Medium art with medium throws
► Sculpture with medium throws
► Medium throws
► Available in 12V

Specs

Beam spread/10°
Center beam candlepower/ 8300
Life/4000 hrs
► **Osram Sylvania** (58558)

Design Tips

✔ One light best for medium art and where mounting height for light is 13' or less.
✔ Light may degrade artwork rapidly and/or significantly: consider UV and IR filters or limit exposure time.
✔ Scallops/striations are most severe when light is closer to the wall (compare top left image to bottom image).
✔ More grazing light creates prominent shadows.
✔ MR lamps are glary—consider shielding with hexcell louver or using deep luminaire trims or pinholes.
✔ Tempered cover glass required for safety.
✔ Required transformer also consumes 3 to 5 watts.

Application Key

Commercial
Gallery
Hospitality
Institutional
Manufacturing
Residential
Retail
Exterior

bold = primary application
partial fade = minimal application
fade = unlikely application

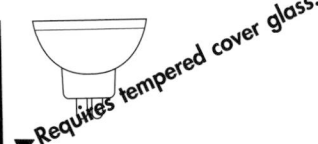

▶ *Requires tempered cover glass.*

Halogen 35MR16/C/VWSP

Hard Dichroic Coating

Performance Sketches

[All data outlined here is based on information from boldfaced manufacturer's published data for 12V version at time of manuscript preparation. Scale is ¼ inch equals 1 foot.]

Halogen Low Voltage (12V)

4

Uses
▶ Architectural accenting
▶ Small art with medium throws
▶ Sculpture with medium throws
▶ Medium throws
▶ Available in 12V

Specs
Beam spread/20°
Center beam candlepower/ 3900
Life/5000 hrs
▶ **GE** (20826)

Design Tips
✔ One light best for small art and where mounting height for light is 11' or less.
✔ Light may degrade artwork rapidly and/or significantly: consider UV and IR filters or limit exposure time.
✔ More grazing light creates prominent shadows.
✔ MR lamps are glary—consider shielding with hexcell louver or using deep luminaire trims or pinholes.
✔ Tempered cover glass required for safety.
✔ Required transformer also consumes 3 to 5 watts.

Lamp: 35MR16/C/VWSP
Location: 1'–6" from wall
Center beam Aiming: 12° (at point 3'–6" AFF)
Luminaire: Monopoint or Recessed Adjustable
Illuminance: 17 fc avg, maintained (42 fc max)

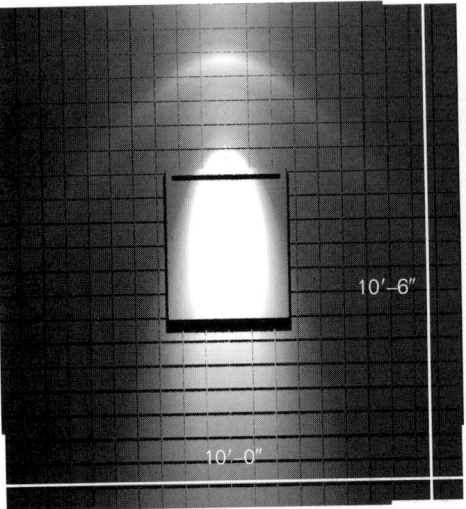

10'–6"

10'–0"

Lamp: 35MR16/C/VWSP
Location: 2'–0" from wall
Center beam Aiming: 20° (at point 5'–0" AFF)
Luminaire: Monopoint or Recessed Adjustable
Illuminance: 24 fc avg, maintained (63 fc max)

Art shown at 2'–6" by 3' and centered 5'–6" AFF. Wall grid on 6" by 6" centers and scaled at ¼"=1'.

Lamp: 35MR16/C/VWSP
Location: 2'–6" from wall
Center beam Aiming: 24° (at point 5'–0" AFF)
Luminaire: Monopoint or Recessed Adjustable
Illuminance: 25 fc avg, maintained (62 fc max)

Application Key

Commercial
Gallery
Hospitality
Institutional
Manufacturing
Residential
Retail
Exterior

bold = primary application
partial fade = minimal application
fade = unlikely application

Halogen 35MR16/C/FL
Hard Dichroic Coating

Requires tempered cover glass.

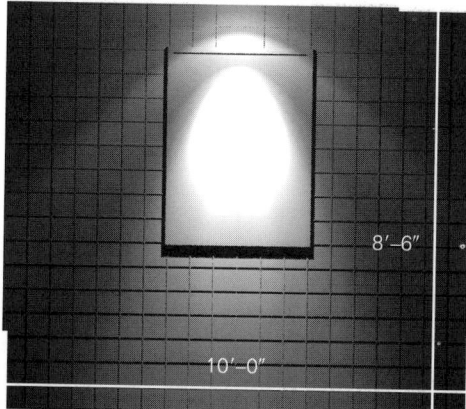

8'–6"

10'–0"

Performance Sketches

[All data outlined here is based on information from boldfaced manufacturer's published data for 12V version at time of manuscript preparation. Scale is ¼ inch equals 1 foot.]

Art shown at 3' by 4' and centered 5'–6" AFF. Wall grid on 6" by 6" centers and scaled at ¼"=1'.

Lamp: 35MR16/C/FL
Location: 1'–6" from wall
Center beam Aiming: 18° (at point 4'–0" AFF)
Luminaire: Monopoint or Recessed Adjustable
Illuminance: 14 fc avg, maintained (51 fc max)

Lamp: 35MR16/C/FL
Location: 2'–0" from wall
Center beam Aiming: 27° (at point 4'–6" AFF)
Luminaire: Monopoint or Recessed Adjustable
Illuminance: 15 fc avg, maintained (49 fc max)

Lamp: 35MR16/C/FL
Location: 2'–6" from wall
Center beam Aiming: 36° (at point 5'–0" AFF)
Luminaire: Monopoint or Recessed Adjustable
Illuminance: 17 fc avg, maintained (49 fc max)

Halogen Low Voltage (12V)

4

Uses
▶ General lighting
▶ Medium art with short throws
▶ Short throws
▶ Available in 12V

Specs
Beam spread/40°
Center beam candlepower/ 1000
Life/4500 hrs
▶ **GE** (20825)
▶ Osram Sylvania

Design Tips
✔ One light best for medium art where mounting height of light is 9' or less.
✔ Light may degrade artwork rapidly and/or significantly: consider UV and IR filters or limit exposure time.
✔ Good for downlighting onto decorative floor materials—best with sconces and/or art accents and/or wallwashing to avoid cave effect and grazing glare (the "I wish I had a visor" effect).
✔ More grazing light creates prominent shadows.
✔ MR lamps are glary—consider shielding with hexcell louver or using deep luminaire trims or pinholes.
✔ Tempered cover glass required for safety.
✔ Required transformer also consumes 3 to 5 watts.

Application Key

Commercial
Gallery
Hospitality
Institutional
Manufacturing
Residential
Retail
Exterior

bold = primary application
partial fade = minimal application
fade = unlikely application

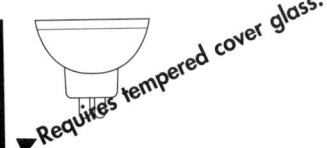

▶ *Requires tempered cover glass.*

Halogen 35MR16/C/VWFL

Hard Dichroic Coating

Halogen Low Voltage (12V)

4

Uses
▶ General lighting
▶ Wallwashing (see 11–111)
▶ Medium art with short throws
▶ Short throws
▶ Available in 12V

Specs
Beam spread/60°
Center beam candlepower/650
Life/4000 hrs
▶ **Osram Sylvania** (58552)

Design Tips
✔ One light best for medium art where mounting height of light is 9' or less.
✔ Light may degrade artwork rapidly and/or significantly: consider UV and IR filters or limit exposure time.
✔ Good for downlighting onto decorative floor materials—best with sconces and/or art accents and/or wallwashing to avoid cave effect and grazing glare (the "I wish I had a visor" effect).
✔ More grazing light creates prominent shadows.
✔ MR lamps are glary—consider shielding with hexcell louver or using deep luminaire trims or pinholes.
✔ Tempered cover glass required for safety.
✔ Required transformer also consumes 3 to 5 watts.

Application Key
Commercial
Gallery
Hospitality
Institutional
Manufacturing
Residential
Retail
Exterior

bold = primary application
partial fade = minimal application
fade = unlikely application

Performance Sketches

[All data outlined here is based on information from boldfaced manufacturer's published data for 12V version at time of manuscript preparation. Scale is ¼ inch equals 1 foot. **Note: At press time, actual photometry was unavailable for this lamp. Data were generated using GE 50MR16/C/CG/VWFL lamp and rerating candlepower—this is imprecise.**]

Art shown at 3' by 4' and centered 5'–6" AFF. Wall grid on 6" by 6" centers and scaled at ¼"=1'.

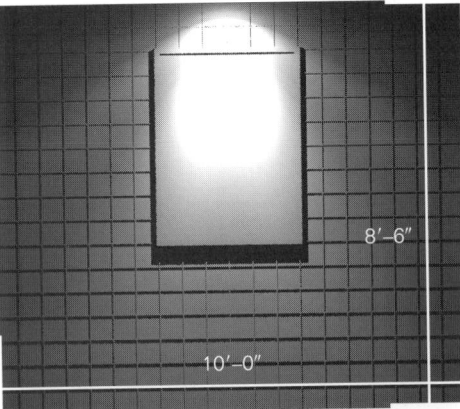

8'–6"

10'–0"

Lamp: 35MR16/C/VWFL
Location: 1'–0" from wall
Center beam Aiming: 11° (at point 3'–6" AFF)
Luminaire: Monopoint or Recessed Adjustable
Illuminance: 15 fc avg, maintained (78 fc max)

Lamp: 35MR16/C/VWFL
Location: 1'–6" from wall
Center beam Aiming: 18° (at point 4'–0" AFF)
Luminaire: Monopoint or Recessed Adjustable
Illuminance: 16 fc avg, maintained (53 fc max)

Lamp: 35MR16/C/VWFL
Location: 2'–0" from wall
Center beam Aiming: 27° (at point 4'–6" AFF)
Luminaire: Monopoint or Recessed Adjustable
Illuminance: 17 fc avg, maintained (46 fc max)

Halogen 42MR16/C/NSP
Hard Dichroic Coating

 ►Requires tempered cover glass.

Halogen Low Voltage (12V)

4

Performance Sketches

[All data outlined here is based on information from boldfaced manufacturer's published data for 12V version at time of manuscript preparation. Scale is ¼ inch equals 1 foot.]

Lamp: 42MR16/C/NSP
Location: 4'–3" from wall
Center beam Aiming: 23° (at point 5'–0" AFF)
Luminaire: Monopoint or Recessed Adjustable
Illuminance: 19 fc avg, maintained (43 fc max)

15'–0"

10'–0"

Lamp: 42MR16/C/NSP
Location: 5'–3" from wall
Center beam Aiming: 28° (at point 5'–0" AFF)
Luminaire: Monopoint or Recessed Adjustable
Illuminance: 20 fc avg, maintained (45 fc max)

Uses
► Architectural detailing
► Small art with long throws
► Sculpture with long throws
► Long throws
► Available in 12V

Specs
Beam spread/9°
Center beam candlepower/ 13100
Life/3500 hrs
► **GE** (20830)

Design Tips
✔ One light best for small art where mounting height for light is 16' or less.
✔ Light may degrade artwork rapidly and/or significantly: consider UV and IR filters or limit exposure time.
✔ Great for glass pieces and floral centerpieces and where mounting height for light is 16' or less.
✔ These are strong accents and deserve judicious use.
✔ Scallops, striations and frame shadows are most severe when light is closer to the wall (compare top left image to bottom image).
✔ MR lamps are glary—consider shielding with hexcell louver or using deep luminaire trims or pinholes.
✔ Tempered cover glass required for safety.
✔ Required transformer consumes 3 to 5 watts.
✔ Consider 37MR16/HIR/C/NSP. ⚡

Application Key

Commercial
Gallery
Hospitality
Institutional
Manufacturing
Residential
Retail
Exterior

bold = primary application
partial fade = minimal application
fade = unlikely application

Art shown at 2'–6" by 3' and centered 5'–6" AFF. Wall grid on 6" by 6" centers and scaled at ¼"=1'.

Lamp: 50MR16/C/NSP
Location: 6'–3" from wall
Center beam Aiming: 32° (at point 5'–0" AFF)
Luminaire: Monopoint or Recessed Adjustable
Illuminance: 20 fc avg, maintained (46 fc max)

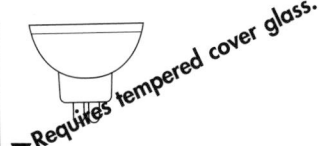

Requires tempered cover glass.

Halogen Low Voltage (12V)

4

Halogen 50MR16/C/NSP

Hard Dichroic Coating

Performance Sketches

[All data outlined here is based on information from boldfaced manufacturer's published data for 12V version at time of manuscript preparation. Scale is ¼ inch equals 1 foot. **Note: At press time, actual photometry was unavailable for this lamp. Data were generated using Osram Sylvania 65MR16/NSP lamp and rerating candlepower—this is imprecise.**]

Uses
▶ Architectural detailing
▶ Medium art with long throws
▶ Sculpture with long throws
▶ Long throws
▶ Available in 12V

Specs
Beam spread/10°
Center beam candlepower/ 11500
Life/4000 hrs
▶ **Osram Sylvania** (58556)

Design Tips
✔ One light best for medium art where mounting height for light is 16′ or less.
✔ Light may degrade artwork rapidly and/or significantly: consider UV and IR filters or limit exposure time.
✔ Great for glass pieces and floral centerpieces and where mounting height for light is 16′ or less.
✔ These are strong accents and deserve judicious use.
✔ Scallops, striations and frame shadows are most severe when light is closer to the wall (compare top left image to bottom image).
✔ MR lamps are glary—consider shielding with hexcell louver or using deep luminaire trims or pinholes.
✔ Tempered cover glass required for safety.
✔ Required transformer consumes 3 to 5 watts.
✔ Consider 37MR16/HIR/C/NSP.

Application Key

Commercial
Gallery
Hospitality
Institutional
Manufacturing
Residential
Retail
Exterior

bold = primary application
partial fade = minimal application
fade = unlikely application

Lamp: 50MR16/C/NSP
Location: 4′–3″ from wall
Center beam Aiming: 22° (at point 4′–6″ AFF)
Luminaire: Monopoint or Recessed Adjustable
Illuminance: 20 fc avg, maintained (40 fc max)

Lamp: 50MR16/C/NSP
Location: 5′–3″ from wall
Center beam Aiming: 28° (at point 5′–0″ AFF)
Luminaire: Monopoint or Recessed Adjustable
Illuminance: 23 fc avg, maintained (45 fc max)

15′–0″

10′–0″

Art shown at 3′ by 4′ and centered 5′–6″ AFF. Wall grid on 6″ by 6″ centers and scaled at ¼″=1′.

Lamp: 50MR16/C/NSP
Location: 6′–3″ from wall
Center beam Aiming: 32° (at point 5′–0″ AFF)
Luminaire: Monopoint or Recessed Adjustable
Illuminance: 24 fc avg, maintained (46 fc max)

Halogen 50MR16/C/WSP
Hard Dichroic Coating

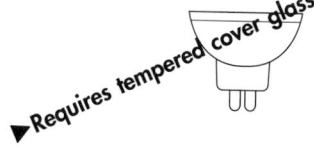
▶Requires tempered cover glass.

Performance Sketches

[All data outlined here is based on information from boldfaced manufacturer's published data for 12V version at time of manuscript preparation. Scale is ¼ inch equals 1 foot.]

Lamp: 50MR16/C/WSP
Location: 4'–3" from wall
Center beam Aiming: 22° (at point 4'–6" AFF)
Luminaire: Monopoint or Recessed Adjustable
Illuminance: 18 fc avg, maintained (36 fc max)

Lamp: 50MR16/C/WSP
Location: 5'–3" from wall
Center beam Aiming: 28° (at point 5'–0" AFF)
Luminaire: Monopoint or Recessed Adjustable
Illuminance: 21 fc avg, maintained (40 fc max)

Art shown at 3' by 4' and centered 5'–6" AFF. Wall grid on 6" by 6" centers and scaled at ¼"=1'.

Lamp: 50MR16/C/WSP
Location: 6'–3" from wall
Center beam Aiming: 32° (at point 5'–0" AFF)
Luminaire: Monopoint or Recessed Adjustable
Illuminance: 21 fc avg, maintained (40 fc max)

Uses
▶Architectural accenting
▶Medium art with long throws
▶Sculpture with long throws
▶Long throws
▶Available in 12V

Specs
Beam spread/15°
Center beam candlepower/ 10200
Life/6000 hrs
▶**GE** (20839)

Design Tips
✓One light best for medium art and where mounting height for light is 16' or less.
✓These are strong accents and deserve judicious use. Light may degrade artwork rapidly and/or significantly: consider UV and IR filters or limit exposure time.
✓Good for glass pieces and floral centerpieces and where mounting height for light is 16' or less.
✓MR lamps are glary—consider shielding with hexcell louver or using deep luminaire trims or pinholes.
✓Tempered cover glass required for safety.
✓Required transformer consumes 3 to 5 watts.

Application Key

Commercial
Gallery
Hospitality
Institutional
Manufacturing
Residential
Retail
Exterior

bold = primary application
partial fade = minimal application
fade = unlikely application

Halogen Low Voltage (12V)

4

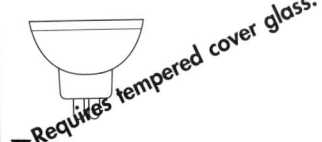

Requires tempered cover glass. ▶

Halogen (Low Voltage (12V))

4

Halogen 50MR16/C/NFL25°

Hard Dichroic Coating

Performance Sketches

[All data outlined here is based on information from boldfaced manufacturer's published data for 12V version at time of manuscript preparation. Scale is ¼ inch equals 1 foot.]

Uses
- ▶ Architectural accenting
- ▶ Feature downlighting
- ▶ Medium art with medium throws
- ▶ Sculpture with medium throws
- ▶ Medium throws
- ▶ Available in 12V

Specs
Beam spread/25°
Center beam candlepower/ 3400
Life/5000 hrs
- ▶ **GE** (20835)
- ▶ Osram Sylvania

Design Tips
- ✔ One light best for medium art where mounting height for light is 11' or less.
- ✔ Great for moderate to large sculpture where mounting height for light is 11' or less.
- ✔ Scallops, striations and frame shadows are most severe when light is closer to the wall (compare top left image to bottom image).
- ✔ More grazing light creates prominent shadows.
- ✔ MR lamps are glary—consider shielding with hexcell louver or using deep luminaire trims or pinholes.
- ✔ Tempered cover glass required for safety.
- ✔ Required transformer consumes 3 to 5 watts.
- ✔ Consider 37MR16/HIR/C/NFL.

Application Key

Commercial
Gallery
Hospitality
Institutional
Manufacturing
Residential
Retail
Exterior

bold = primary application
partial fade = minimal application
fade = unlikely application

Lamp: 50MR16/C/NFL25°
Location: 3'–0" from wall
Center beam Aiming: 23° (at point 3'–6" AFF)
Luminaire: Monopoint or Recessed Adjustable
Illuminance: 17 fc avg, maintained (32 fc max)

Lamp: 50MR16/C/NFL25°
Location: 4'–0" from wall
Center beam Aiming: 34° (at point 4'–6" AFF)
Luminaire: Monopoint or Recessed Adjustable
Illuminance: 20 fc avg, maintained (37 fc max)

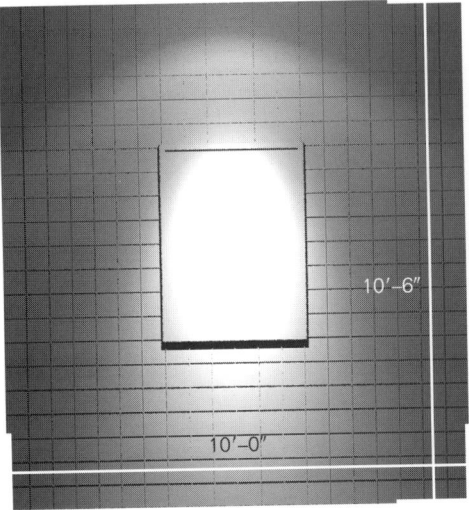

10'–6"

10'–0"

Art shown at 3' by 4' and centered 5'–6" AFF. Wall grid on 6" by 6" centers and scaled at ¼"=1'.

Lamp: 50MR16/C/NFL25°
Location: 5'–0" from wall
Center beam Aiming: 40° (at point 4'–6" AFF)
Luminaire: Monopoint or Recessed Adjustable
Illuminance: 20 fc avg, maintained (34 fc max)

Halogen 50MR16/C/NFL30°
Hard Dichroic Coating

▶Requires tempered cover glass.

Halogen Low Voltage (12V)

Performance Sketches

[All data outlined here is based on information from boldfaced manufacturer's published data for 12V version at time of manuscript preparation. Scale is ¼ inch equals 1 foot.]

Lamp: 50MR16/C/NFL30°
Location: 3'–0" from wall
Center beam Aiming: 23° (at point 3'–6" AFF)
Luminaire: Monopoint or Recessed Adjustable
Illuminance: 18 fc avg, maintained (33 fc max)

Lamp: 50MR16/C/NFL30°
Location: 4'–0" from wall
Center beam Aiming: 34° (at point 4'–6" AFF)
Luminaire: Monopoint or Recessed Adjustable
Illuminance: 21 fc avg, maintained (37 fc max)

Art shown at 3' by 4' and centered 5'–6" AFF. Wall grid on 6" by 6" centers and scaled at ¼"=1'.

Lamp: 50MR16/C/NFL30°
Location: 5'–0" from wall
Center beam Aiming: 40° (at point 4'–6" AFF)
Luminaire: Monopoint or Recessed Adjustable
Illuminance: 20 fc avg, maintained (32 fc max)

Uses
▶Architectural accenting
▶Feature downlighting
▶Medium art with medium throws
▶Sculpture with medium throws
▶Medium throws
▶Available in 12V

Specs
Beam spread/30°
Center beam candlepower/ 2450
Life/6000 hrs
▶**GE** (20834)

Design Tips
✔One light best for medium art where mounting height for light is 11' or less.
✔Great for moderate sculpture where mounting height for light is 11' or less.
✔Scallops, striations and frame shadows are most severe when light is closer to the wall (compare top left image to bottom image).
✔More grazing light creates prominent shadows.
✔MR lamps are glary—consider shielding with hexcell louver or using deep luminaire trims or pinholes.
✔Tempered cover glass required for safety.
✔Required transformer consumes 3 to 5 watts.
✔Consider 37MR16/HIR/C/NFL.

Application Key

Commercial
Gallery
Hospitality
Institutional
Manufacturing
Residential
Retail
Exterior

bold = primary application
partial fade = minimal application
fade = unlikely application

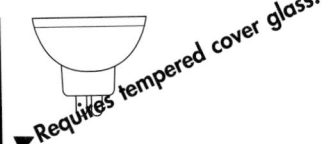

► *Requires tempered cover glass.*

Halogen — *Low Voltage (12V)*

4

Halogen 50MR16/C/FL

Hard Dichroic Coating

Performance Sketches

[All data outlined here is based on information from boldfaced manufacturer's published data for 12V version at time of manuscript preparation. Scale is ¼ inch equals 1 foot.]

Art shown at 3′ by 4′ and centered 5′–6″ AFF. Wall grid on 6″ by 6″ centers and scaled at ¼″=1′.

Uses
► General lighting
► Medium art with short throws
► Short throws
► Available in 12V

Specs
Beam spread/40°
Center beam candlepower/1850
Life/5000 hrs
► **GE** (20833)
► Osram Sylvania

Design Tips
✓ One light best for medium art where mounting height for light is 10′ or less.
✓ Light may degrade artwork rapidly and/or significantly: consider UV and IR filters or limit exposure time.
✓ Good for downlighting onto decorative floor materials—best with sconces and/or art accents and/or wallwashing to avoid cave effect and grazing glare (the "I wish I had a visor" effect).
✓ More grazing light creates prominent shadows.
✓ MR lamps are glary—consider shielding with hexcell louver or using deep luminaire trims or pinholes.
✓ Tempered cover glass required for safety.
✓ Required transformer consumes 3 to 5 watts.
✓ Consider 37MR16/HIR/C/FL.

Application Key

Commercial
Gallery
Hospitality
Institutional
Manufacturing
Residential
Retail
Exterior

bold = primary application
partial fade = minimal application
fade = unlikely application

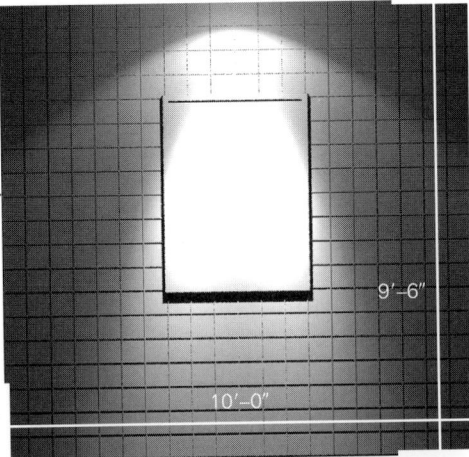

9′–6″

10′–0″

Lamp: 50MR16/C/FL
Location: 2′–0″ from wall
Center beam Aiming: 20° (at point 4′–0″ AFF)
Luminaire: Monopoint or Recessed Adjustable
Illuminance: 25 fc avg, maintained (75 fc max)

Lamp: 50MR16/C/FL
Location: 2′–6″ from wall
Center beam Aiming: 27° (at point 4′–6″ AFF)
Luminaire: Monopoint or Recessed Adjustable
Illuminance: 28 fc avg, maintained (74 fc max)

Lamp: 50MR16/C/FL
Location: 3′–0″ from wall
Center beam Aiming: 34° (at point 5′–0″ AFF)
Luminaire: Monopoint or Recessed Adjustable
Illuminance: 30 fc avg, maintained (74 fc max)

Halogen 50MR16/C/VWFL
Hard Dichroic Coating

▶*Requires tempered cover glass.*

Halogen Low Voltage (12V)

4

Performance Sketches
[All data outlined here is based on information from boldfaced manufacturer's published data for 12V version at time of manuscript preparation. Scale is ¼ inch equals 1 foot.]

Art shown at 4' by 5' and centered 5'–6" AFF. Wall grid on 6" by 6" centers and scaled at ¼"=1'.

Lamp: 50MR16/C/VWFL
Location: 2'–0" from wall
Center beam Aiming: 18° (at point 3'–6" AFF)
Luminaire: Monopoint or Recessed Adjustable
Illuminance: 12 fc avg, maintained (40 fc max)

Lamp: 50MR16/C/VWFL
Location: 2'–6" from wall
Center beam Aiming: 27° (at point 4'–6" AFF)
Luminaire: Monopoint or Recessed Adjustable
Illuminance: 13 fc avg, maintained (42 fc max)

Lamp: 50MR16/C/VWFL
Location: 3'–0" from wall
Center beam Aiming: 31° (at point 4'–6" AFF)
Luminaire: Monopoint or Recessed Adjustable
Illuminance: 13 fc avg, maintained (35 fc max)

Uses
▶Soft general lighting
▶Wallwashing (see 11–117)
▶Large art with short throws
▶Short throws
▶Available in 12V

Specs
Beam spread/58°
Center beam
   candlepower/900
Life/5000 hrs
▶**GE** (20832)
▶Osram Sylvania

Design Tips
✔One light best for large art and where mounting height for light is 10' or less.
✔Good for soft downlighting— best with sconces and/or art accents and/or wallwashing to avoid cave effect.
✔More grazing light creates prominent shadows.
✔MR lamps are glary—consider shielding with hexcell louver or using deep luminaire trims or pinholes.
✔Tempered cover glass required for safety.
✔Required transformer consumes 3 to 5 watts.

Application Key

Commercial
Gallery
Hospitality
Institutional
Manufacturing
Residential
Retail
Exterior

bold = primary application
partial fade = minimal application
fade = unlikely application

Requires tempered cover glass.

Halogen (Low Voltage · 12V)

4

Uses
▶ Architectural detailing
▶ Large art with long throws
▶ Sculpture with long throws
▶ Long throws
▶ Available in 12V

Specs
Beam spread/10°
Center beam
 candlepower/14000
Life/4000 hrs
▶ **Osram Sylvania** (58566)

Design Tips
✔ One light best for large art and where mounting height for light is 19' or less.
✔ These are strong accents and deserve judicious use. Light may degrade artwork rapidly and/or significantly: consider UV and IR filters or limit exposure time.
✔ Excellent for glass pieces and floral centerpieces and where mounting height for light is 19' or less.
✔ MR lamps are glary—consider shielding with hexcell louver or using deep luminaire trims or pinholes.
✔ Tempered cover glass required for safety.
✔ Required transformer consumes 3 to 5 watts.
✔ Consider 50MR16/HIR/C/NSP.

Application Key
Commercial
Gallery
Hospitality
Institutional
Manufacturing
Residential
Retail
Exterior

bold = primary application
partial fade = minimal application
fade = unlikely application

Halogen 65MR16/C/NSP
Hard Dichroic Coating

Performance Sketches

[All data outlined here is based on information from boldfaced manufacturer's published data for 12V version at time of manuscript preparation. Scale is ¼ inch equals 1 foot. **Note: At press time, actual photometry was unavailable for this lamp. Data were generated using Osram Sylvania 65MR16/NSP lamp and rerating candlepower—this is imprecise.**]

Lamp: 65MR16/C/NSP
Location: 5'–0" from wall
Center beam Aiming: 20° (at point 4'–6" AFF)
Luminaire: Monopoint or Recessed Adjustable
Illuminance: 14 fc avg, maintained (29 fc max)

Lamp: 65MR16/C/NSP
Location: 6'–0" from wall
Center beam Aiming: 24° (at point 4'–6" AFF)
Luminaire: Monopoint or Recessed Adjustable
Illuminance: 15 fc avg, maintained (30 fc max)

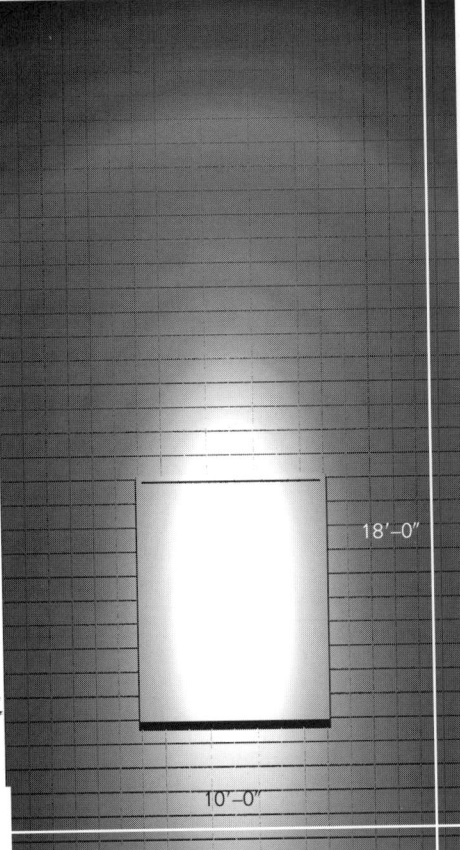

18'–0"

10'–0"

Halogen 65MR16/C/NSP

Hard Dichroic Coating

▶*Requires tempered cover glass.*

Lamp: 65MR16/C/NSP
Location: 7'–0" from wall
Center beam Aiming: 27° (at point 4'–6" AFF)
Luminaire: Monopoint or Recessed Adjustable
Illuminance: 16 fc avg, maintained (30 fc max)

Art shown at 4' by 5' and
centered 5'–6" AFF. Wall grid on
6" by 6" centers and scaled at
¼"=1'.

Halogen Low Voltage (12V)

4

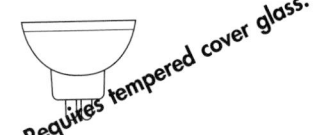

▶ Requires tempered cover glass.

Halogen 65MR16/C/NFL

Hard Dichroic Coating

Low Voltage (12V)

Halogen

4

Uses
▶ Architectural accenting
▶ Feature downlighting
▶ Large art with medium throws
▶ Sculpture with medium throws
▶ Medium throws
▶ Available in 12V

Specs
Beam spread/25°
Center beam candlepower/ 4000
Life/4000 hrs
▶ **Osram Sylvania** (58567)

Design Tips
✔ One light best for large art where mounting height for light is 13' or less.
✔ Great for moderate to large sculpture where mounting height for light is 13' or less.
✔ Scallops, striations and frame shadows are most severe when light is closer to the wall (compare top left image to bottom image).
✔ MR lamps are glary—consider shielding with hexcell louver or using deep luminaire trims or pinholes.
✔ Tempered cover glass required for safety.
✔ Required transformer consumes 3 to 5 watts.

Application Key

Commercial

Gallery

Hospitality

Institutional

Manufacturing

Residential

Retail

Exterior

bold = primary application
partial fade = minimal application
fade = unlikely application

Performance Sketches

[All data outlined here is based on information from boldfaced manufacturer's published data for 12V version at time of manuscript preparation. Scale is ¼ inch equals 1 foot. **Note: At press time, actual photometry was unavailable for this lamp. Data were generated using Osram Sylvania 65MR16/NFL lamp and rerating candlepower—this is imprecise.**]

Lamp: 65MR16/C/NFL
Location: 3'–0" from wall
Center beam Aiming: 18° (at point 3'–6" AFF)
Luminaire: Monopoint or Recessed Adjustable
Illuminance: 13 fc avg, maintained (29 fc max)

Lamp: 65MR16/C/NFL
Location: 4'–0" from wall
Center beam Aiming: 25° (at point 4'–0" AFF)
Luminaire: Monopoint or Recessed Adjustable
Illuminance: 15 fc avg, maintained (28 fc max)

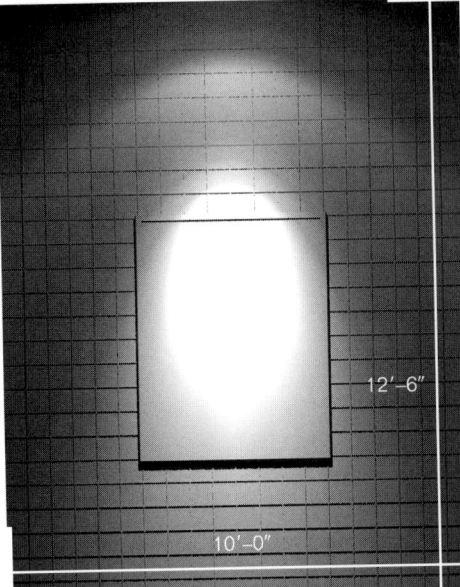

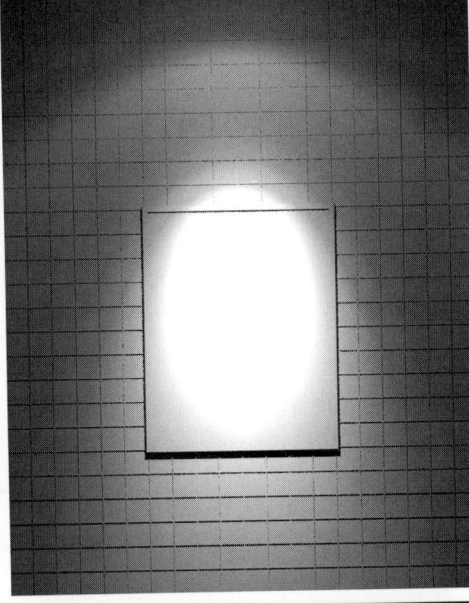

12'–6"

10'–0"

Art shown at 4' by 5' and centered 5'–6" AFF. Wall grid on 6" by 6" centers and scaled at ¼"=1'.

Lamp: 65MR16/C/NFL
Location: 5'–0" from wall
Center beam Aiming: 30° (at point 4'–0" AFF)
Luminaire: Monopoint or Recessed Adjustable
Illuminance: 15 fc avg, maintained (26 fc max)

Halogen65MR16/C/FL
Hard Dichroic Coating

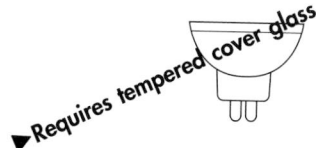

▶*Requires tempered cover glass.*

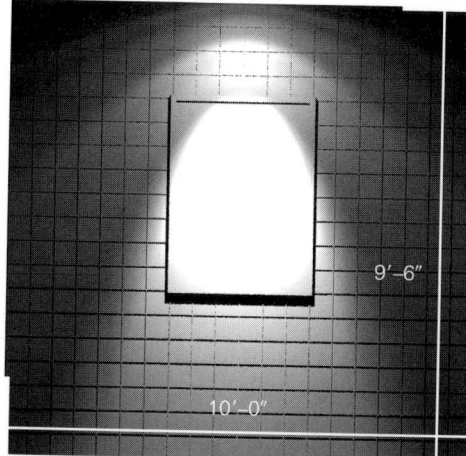

9'–6"

10'–0"

Performance Sketches

[All data outlined here is based on information from boldfaced manufacturer's published data for 12V version at time of manuscript preparation. Scale is ¼ inch equals 1 foot. **Note: At press time, actual photometry was unavailable for this lamp. Data were generated using Osram Sylvania 65MR16/FL lamp and rerating candlepower—this is imprecise.**]

. . . .

Art shown at 3' by 4' and centered 5'–6" AFF. Wall grid on 6" by 6" centers and scaled at ¼"=1'.

Lamp: 65MR16/C/FL
Location: 2'–0" from wall
Center beam Aiming: 20° (at point 4'–0" AFF)
Luminaire: Monopoint or Recessed Adjustable
Illuminance: 24 fc avg, maintained (72 fc max)

Lamp: 65MR16/C/FL
Location: 2'–6" from wall
Center beam Aiming: 27° (at point 4'–6" AFF)
Luminaire: Monopoint or Recessed Adjustable
Illuminance: 27 fc avg, maintained (70 fc max)

Lamp: 65MR16/C/FL
Location: 3'–0" from wall
Center beam Aiming: 34° (at point 5'–0" AFF)
Luminaire: Monopoint or Recessed Adjustable
Illuminance: 30 fc avg, maintained (69 fc max)

Halogen Low Voltage (12V)

4

Uses
▶General lighting
▶Medium art with short throws
▶Short throws
▶Available in 12V

Specs
Beam spread/40°
Center beam candlepower/ 2100
Life/4000 hrs
▶**Osram Sylvania** (58571)

Design Tips
✔One light best for medium art where mounting height for light is 10' or less.
✔Good for downlighting onto decorative floor materials: best with sconces and/or art accents and/or wallwashing to avoid cave effect.
✔More grazing light creates prominent shadows.
✔MR lamps are glary—consider shielding with hexcell louver or using deep luminaire trims or pinholes.
✔Tempered cover glass required for safety.
✔Required transformer consumes 3 to 5 watts.
✔Consider 37MR16/HIR/C/FL.

Application Key

Commercial
Gallery
Hospitality
Institutional
Manufacturing
Residential
Retail
Exterior

bold = primary application
partial fade = minimal application
fade = unlikely application

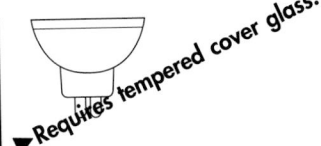

▶ Requires tempered cover glass.

Halogen 65MR16/C/VWFL

Hard Dichroic Coating

Halogen
Low Voltage (12V)

4

Uses
▶ Soft general lighting
▶ Wallwashing (see 11–119)
▶ Large art with short throws
▶ Short throws
▶ Available in 12V

Specs
Beam spread/60°
Center beam
  candlepower/1050
Life/4000 hrs
▶ **Osram Sylvania** (58572)

Design Tips
✔ One light best for large art and where mounting height for light is 10′ or less.
✔ Good for soft downlighting—best with sconces and/or art accents and/or wallwashing to avoid cave effect.
✔ More grazing light creates prominent shadows.
✔ MR lamps are glary—consider shielding with hexcell louver or using deep luminaire trims or pinholes.
✔ Tempered cover glass required for safety.
✔ Required transformer consumes 3 to 5 watts.

Performance Sketches

[All data outlined here is based on information from boldfaced manufacturer's published data for 12V version at time of manuscript preparation. Scale is ¼ inch equals 1 foot. **Note: At press time, actual photometry was unavailable for this lamp. Data were generated using GE 50MR16/C/CG/VWFL lamp and rerating candlepower—this is imprecise.**]

Art shown at 4′ by 5′ and centered 5′–6″ AFF. Wall grid on 6″ by 6″ centers and scaled at ¼″=1′.

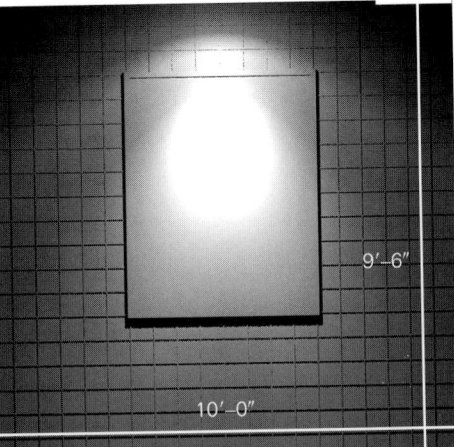

9′–6″

10′–0″

Lamp: 65MR16/C/VWFL
Location: 2′–0″ from wall
Center beam Aiming: 12° (at point 0′–0″ AFF)
Luminaire: Monopoint or Recessed Adjustable
Illuminance: 10 fc avg, maintained (24 fc max)

Lamp: 65MR16/C/VWFL
Location: 2′–6″ from wall
Center beam Aiming: 16° (at point 1′–0″ AFF)
Luminaire: Monopoint or Recessed Adjustable
Illuminance: 10 fc avg, maintained (21 fc max)

Application Key

Commercial
Gallery
Hospitality
Institutional
Manufacturing
Residential
Retail
Exterior

bold = primary application
partial fade = minimal application
fade = unlikely application

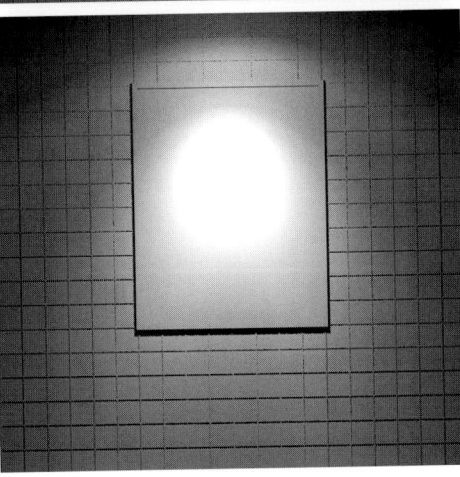

Lamp: 65MR16/C/VWFL
Location: 3′–0″ from wall
Center beam Aiming: 22° (at point 2′–0″ AFF)
Luminaire: Monopoint or Recessed Adjustable
Illuminance: 10 fc avg, maintained (20 fc max)

Halogen 71MR16/C/WSP
Hard Dichroic Coating

▶Requires tempered cover glass.

Halogen Low Voltage (12V)

4

Performance Sketches

[All data outlined here is based on information from boldfaced manufacturer's published data for 12V version at time of manuscript preparation. Scale is ¼ inch equals 1 foot.]

Lamp: 71MR16/C/WSP
Location: 5'–0" from wall
Center beam Aiming: 27° (at point 5'–0" AFF)
Luminaire: Monopoint or Recessed Adjustable
Illuminance: 19 fc avg, maintained (49 fc max)

Lamp: 71MR16/C/WSP
Location: 6'–0" from wall
Center beam Aiming: 31° (at point 5'–0" AFF)
Luminaire: Monopoint or Recessed Adjustable
Illuminance: 20 fc avg, maintained (49 fc max)

15'–0"

10'–0"

Art shown at 4' by 5' and centered 5'–6" AFF. Wall grid on 6" by 6" centers and scaled at ¼"=1'.

Lamp: 71MR16/C/WSP
Location: 7'–0" from wall
Center beam Aiming: 35° (at point 5'–0" AFF)
Luminaire: Monopoint or Recessed Adjustable
Illuminance: 20 fc avg, maintained (47 fc max)

Uses
▶Architectural detailing
▶Large art with long throws
▶Sculpture with long throws
▶Long throws
▶Available in 12V

Specs
Beam spread/15°
Center beam candlepower/ 12000
Life/4000 hrs
▶**GE** (20843)

Design Tips
✔One light best for large art and where mounting height for light is 16' or less.
✔These are strong accents and deserve judicious use. Light may degrade artwork rapidly and/or significantly: consider UV and IR filters or limit exposure time.
✔Great for moderate to large sculpture where mounting height for light is 16' or less.
✔MR lamps are glary—consider shielding with hexcell louver or using deep luminaire trims or pinholes.
✔Tempered cover glass required for safety.
✔Required transformer consumes 3 to 5 watts.

Application Key

Commercial
Gallery
Hospitality
Institutional
Manufacturing
Residential
Retail
Exterior

bold = primary application
partial fade = minimal application
fade = unlikely application

▶ *Requires tempered cover glass.*

Halogen 71MR16/C/NFL

Hard Dichroic Coating

Halogen Low Voltage (12V)

Uses
▶ Architectural accenting
▶ Feature downlighting
▶ Large art with medium throws
▶ Sculpture with medium throws
▶ Medium throws
▶ Available in 12V

Specs
Beam spread/25°
Center beam candlepower/ 4900
Life/4000 hrs
▶**GE** (20841)

Design Tips
✔ One light best for large art and where mounting height for light is 13' or less
✔ Good for moderate to larger sculpture where mounting height for light is 10' to 15'.
✔ MR lamps are glary—consider shielding with hexcell louver or using deep luminaire trims or pinholes.
✔ Tempered cover glass required for safety.
✔ Required transformer consumes 3 to 5 watts.
✔ Consider 50MR16/HIR/C/NFL.

Performance Sketches

[All data outlined here is based on information from boldfaced manufacturer's published data for 12V version at time of manuscript preparation. Scale is ¼ inch equals 1 foot.]

Lamp: 71MR16/C/NFL
Location: 3'–0" from wall
Center beam Aiming: 19° (at point 4'–0" AFF)
Luminaire: Monopoint or Recessed Adjustable
Illuminance: 17 fc avg, maintained (47 fc max)

12'–6"

10'–0"

Lamp: 71MR16/C/NFL
Location: 4'–0" from wall
Center beam Aiming: 25° (at point 4'–0" AFF)
Luminaire: Monopoint or Recessed Adjustable
Illuminance: 18 fc avg, maintained (43 fc max)

Art shown at 4' by 5' and centered 5'–6" AFF. Wall grid on 6" by 6" centers and scaled at ¼"=1'.

Lamp: 71MR16/C/NFL
Location: 5'–0" from wall
Center beam Aiming: 30° (at point 4'–0" AFF)
Luminaire: Monopoint or Recessed Adjustable
Illuminance: 18 fc avg, maintained (39 fc max)

Application Key

Commercial
Gallery
Hospitality
Institutional
Manufacturing
Residential
Retail
Exterior

bold = primary application
partial fade = minimal application
fade = unlikely application

Halogen 71MR16/C/FL
Hard Dichroic Coating

▶ Requires tempered cover glass.

Halogen Low Voltage (12V)

4

Performance Sketches

[All data outlined here is based on information from boldfaced manufacturer's published data for 12V version at time of manuscript preparation. Scale is ¼ inch equals 1 foot.]

Lamp: 71MR16/C/FL
Location: 3'–0" from wall
Center beam Aiming: 27° (at point 4'–6" AFF)
Luminaire: Monopoint or Recessed Adjustable
Illuminance: 20 fc avg, maintained (55 fc max)

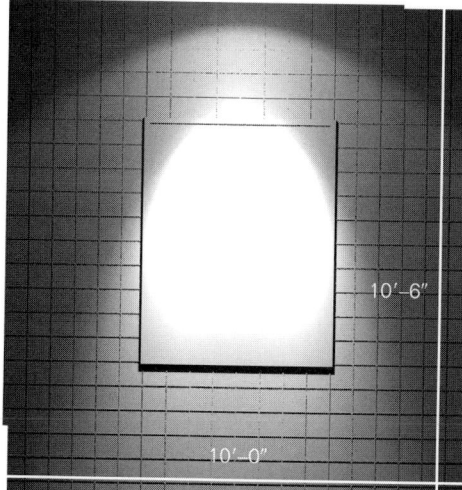

10'–6"

10'–0"

Lamp: 71MR16/C/FL
Location: 4'–0" from wall
Center beam Aiming: 36° (at point 5'–0" AFF)
Luminaire: Monopoint or Recessed Adjustable
Illuminance: 22 fc avg, maintained (51fc max)

Art shown at 4' by 5' and centered 5'–6" AFF. Wall grid on 6" by 6" centers and scaled at ¼"=1'.

Lamp: 71MR16/C/FL
Location: 5'–0" from wall
Center beam Aiming: 42° (at point 5'–0" AFF)
Luminaire: Monopoint or Recessed Adjustable
Illuminance: 21 fc avg, maintained (42 fc max)

Uses
▶ General lighting
▶ Wallwashing (see 11–127)
▶ Large art with medium throws
▶ Medium throws
▶ Available in 12V

Specs
Beam spread/40°
Center beam
   candlepower/2100
Life/4000 hrs
▶ **GE** (20840)

Design Tips
✔ One light best for large art where mounting height for light is 10' or less.
✔ Light may degrade artwork rapidly and/or significantly: consider UV and IR filters or limit exposure time.
✔ Good for downlighting onto decorative floor materials—best with sconces and/or art accents and/or wallwashing to avoid cave effect and grazing glare (the "I wish I had a visor" effect).
✔ More grazing light creates prominent shadows.
✔ MR lamps are glary—consider shielding with hexcell louver or using deep luminaire trims or pinholes.
✔ Tempered cover glass required for safety.
✔ Required transformer consumes 3 to 5 watts.
✔ Consider 37MR16/HIR/C/FL or 50MR16/HIR/C/FL.

Application Key

Commercial
Gallery
Hospitality
Institutional
Manufacturing
Residential
Retail
Exterior

bold = primary application
partial fade = minimal application
fade = unlikely application

▼Has integral cover glass.

Halogen Low Voltage (12V)

4

Halogen/MR16/C/CG

Hard Dichroic Coating and Cover Glass

Guide
↪Longer life is better
↪Higher efficacy better for general lighting
↪Lower wattage is better
↪Lower color of light is warmer
↪Moderate cost range for halogen is about US$12.⁰⁰ to US$17.⁰⁰

Stats Data varies manufacturer to manufacturer and changes from time to time

Life/4000 to 6000 hours
Varies by wattage and manufacturer to manufacturer. A typical office environment might be occupied about 2600 hours each year.

Efficacy/Not reported
Excellent beam intensities and distributions available for the wattage.

Wattages/20, 35, 50 and 71

Color of Light/2900 to 3200°K
Typically the higher wattage lamps have slightly higher color temperature.

Beam spreads/NSP, SP, WSP, VWSP, NFL, FL and VWFL
Very narrow spot, narrow spot, wide spot, very wide spot, narrow flood, flood and very wide flood distributions are available.

Size/1⅞″ L by 2″ Ø

Voltage/12V

Cost Magnitude/Moderate (2000)

Manufacturers
General Electric
Philips

Net Addresses
http://www.ge.com/lighting/business/index.htm
http://www.lighting.philips.com/

➤ CONNECT FOR MORE

Advantages
Crisp white light
Very small size
Easily dimmed
Alternative to less efficient R and halogen PAR lamps
Withstand extreme temperature range
No backlight spill
No color variations and color shifts
Very little lamp lumen depreciation (light loss) over life

Disadvantages
Beam aberrations
◆Striations
◆Inconsistent beam shape
No backlight spill (sometimes this is part of the desired "look")
Hot to touch
Integral cover glass
◆Tempered safety glass may still be required (check with lamp manufacturer)
◆May prevent use of auxiliary accessories (e.g., color filters, louvers, etc.)
Transformer required
◆Additional cost and visual aspect of transformer must be accommodated
◆Additional wattage requirement of transformer of 3 to 5 watts
◆Potential for audible hum

For more efficient accent lighting, use HIR low voltage lamps (Section 6); **HIR PAR lamps** (Section 5) **or CMH PAR lamps** (Section 7).

Halogen/MR16/C/CG
Hard Dichroic Coating and Cover Glass

►Has integral cover glass.

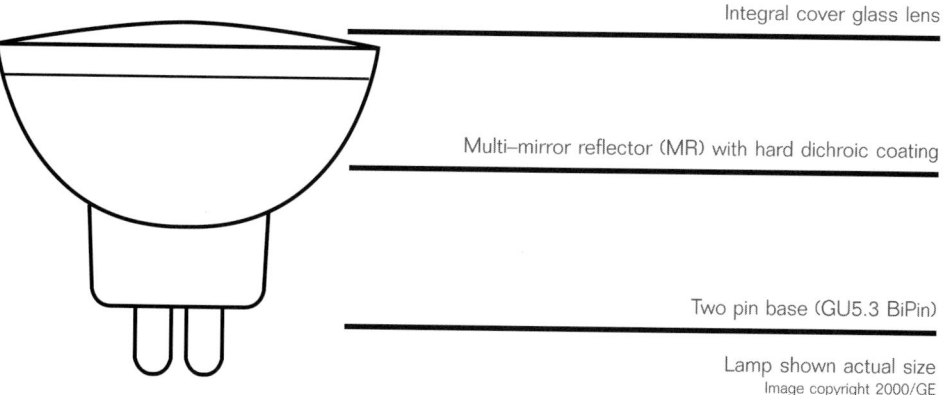

Integral cover glass lens

Multi–mirror reflector (MR) with hard dichroic coating

Two pin base (GU5.3 BiPin)

Lamp shown actual size
Image copyright 2000/GE

Uses

Architectural Accenting
Art Accenting
Commercial
Gallery
Hospitality
Residential
Downlighting
Lobbies
Conference rooms
Hospitality
Residential
Retail
Merchandising
Surface Grazing
Specular/semispecular surfaces
◆Polished/honed stone
◆Polished/satin wood
◆Fritted/etched glass
◆Cut glass

■ ■ ■ ■ ■ ■ ■ ■ ■ ■ ■ ■ ■ ■ ■

MR16 Lighting Design Tips
✔Dimming increases life (typically doubling life), but shifts color warmer.
✔Very glary when used without much other general lighting or when aimed across long distances where ceilings are low.
✔Beam striations are most apparent on light–colored, nontextured flat surfaces and may be objectionable.
✔MR16 downlighting alone is harsh—combine several techniques for best results.
✔Use smaller scale (e.g., 4–inch diameter) downlights and adjustable accents for attractive, more human–scale approach.
✔Integral cover glass may interfere with auxiliary add–on lenses, louvers or accessories.

Net Addresses/MR16 Track, Monopoints, Recessed and Exterior
See page 4–141

CONNECT FOR MORE

▶Has integral cover glass.

Halogen/MR16/C/CG

Hard Dichroic Coating and Cover Glass

Figure 4.5
Low voltage halogen 50MR16/C/CG/NFL lamps are used in shallow recessed luminaires under the drywall canopy to illuminate the display cases in the middle right of the photo. The small size of the MR16 lamp allows for design of luminaires with little volume and which can fit in smaller architectural elements. See Figure C.40 also. Recessed MR16 units are Reggiani.

Image copyright 2000 Stephen Graham

Halogen/MR16/C/CG
Hard Dichroic Coating and Cover Glass

▶Has integral cover glass.

Halogen Low Voltage (12V)

4

Net Addresses/MR16 Track and Monopoints
```
http://www.ardeelighting.com/index.htm
http://www.artemide.com/
http://www.bartcolighting.com/index2.html
http://www.thomasltg.com/
http://www.con-techlighting.com/
http://www.cooperlighting.com/
http://www.flosusa.com/
http://www.hubbell-ltg.com/products.htm#Down&Track
http://www.danalite.com/
http://www.kramerlighting.com/
http://www.lightingservicesinc.com/
http://www.lightolier.com/
http://www.lightproject.com/
http://www.lithonia.com/
http://www.prescolite.com/
http://www.techlighting.com/pro1.html
http://www.tslight.com/
```

CONNECT FOR MORE

Net Addresses/MR16 Recessed
```
http://www.alkco.com/rtrak.htm
http://www.ardeelighting.com/index.htm
http://www.artemide.com/
http://www.bartcolighting.com/index2.html
http://www.thomasltg.com/
http://www.con-techlighting.com/
http://www.cooperlighting.com/
http://www.hubbell-ltg.com/products.htm#Down&Track
http://www.danalite.com/
http://www.kramerlighting.com/
http://www.lightolier.com/
http://www.lithonia.com/
http://www.prescolite.com/
http://www.progresslighting.com/
```

CONNECT FOR MORE

Net Addresses/Exterior Accent
```
http://www.bega-us.com/home/home.html
http://www.hadcolighting.com/html/bronzelite.htm
http://www.hydrel.com
http://www.kimlighting.com/ingrd1.html
```

CONNECT FOR MORE

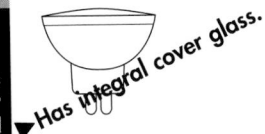

▶Has integral cover glass.

Halogen Low Voltage (12V)

MR16 Hard Dichroic Coating and Cover Glass Visual Index

Throws/Ceilings
◗Short = 7' to 10'
◗Medium = 10' to 15'
◗Long = 15' to 20'
◗Very Long = 20' or more

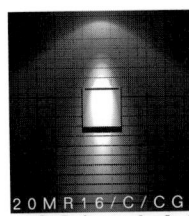

20MR16/C/CG/NSP
▶Architectural detailing
▶Petite art/11' ceiling height or less
▶Sculpture
▶Medium throws
▶Commercial, Hospitality

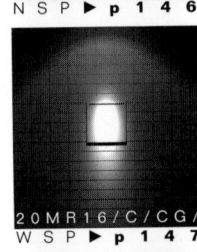

20MR16/C/CG/WSP
▶Architectural accenting
▶Petite art/11' ceiling height or less
▶Sculpture
▶Medium throws
▶Hospitality

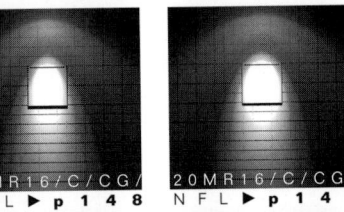

20MR16/C/CG/NFL
▶Feature downlighting
▶Petite art/10' ceiling height or less
▶Short throws
▶Commercial, Hospitality

20MR16/C/CG/FL (Philips)
▶General lighting
▶Petite art/9' ceiling height or less
▶Short throws
▶Hospitality

20MR16/C/CG/FL (GE)
▶General lighting
▶Wallwashing
▶Medium art/10' ceiling height or less
▶Short throws
▶Gallery, Residential

Recommended intensities
for art accenting

Low (gallery, museum, residential)
5 to 10 fc average on art
Moderate (hospitality)
10 to 20 fc average on art
High (commercial—or higher for retail)
20 to 30 fc average on art

These are intended to be average, maintained values including effects of accent lighting and general room lighting. General room lighting contributes just a few footcandles in residential, gallery and museum settings, but contributes 5 to 10 fc in commercial settings. Maximums might range from 50 percent to 100 percent of the average value. Wherever artwork is highly valuable and/or sensitive to light, precautions should be taken to limit light exposure (maintaining low levels and/or switching lights off when art is not being viewed); and exposure to infrared and ultraviolet should be limited (using filters on lamps or placing art behind specially treated glass or acrylic).

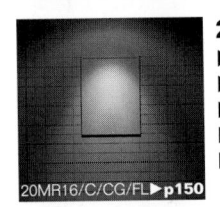

35MR16/C/CG/SP
▶Architectural accenting
▶Medium art/13' ceiling height or less
▶Sculpture
▶Medium throws
▶Hospitality

35MR16/C/CG/VWSP
▶Architectural accenting
▶Medium art/13' ceiling height or less
▶Sculpture
▶Medium throws
▶Gallery, Hospitality, Residential

Halogen/MR16/C/CG

Hard Dichroic Coating and Cover Glass

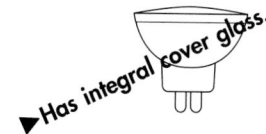
▶Has integral cover glass.

Halogen Low Voltage (12V)

4

 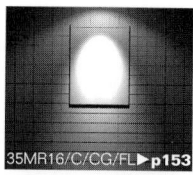 

35MR16/C/CG/FL
▶General lighting
▶Medium art/9' ceiling height or less
▶Short throws
▶Hospitality

50MR16/C/CG/NSP
▶Architectural detailing
▶Medium art/16' ceiling height or less
▶Sculpture
▶Long throws
▶Hospitality

50MR16/C/CG/WSP
▶Architectural accenting
▶Large art/16' ceiling height or less
▶Sculpture
▶Long throws
▶Hospitality

50MR16/C/CG/NFL (Philips)
▶Architectural accenting
▶Feature downlighting
▶Medium art/13' ceiling height or less
▶Medium throws
▶Commercial, Hospitality, Retail

50MR16/C/CG/NFL (GE)
▶Architectural accenting
▶Feature downlighting
▶Medium art/11' ceiling height or less
▶Sculpture
▶Medium throws
▶Commercial, Hospitality

50MR16/C/CG/FL (Philips)
▶General lighting
▶Medium art/10' ceiling height or less
▶Short throws
▶Commercial, Retail

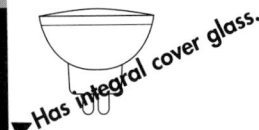

▶Has integral cover glass.

Halogen/MR16/C/CG

Hard Dichroic Coating and Cover Glass

Halogen Low Voltage (12V)

4

50MR16/C/CG/FL▶p159 50MR16/C/CG/FL▶p159 50MR16/C/CG/FL▶p159

50MR16/C/CG/FL (GE)
▶General lighting
▶Medium art/10′ ceiling height or less
▶Short throws
▶Commercial, Retail

Throws/Ceilings
◗Short = 7′ to 10′
◗Medium = 10′ to 15′
◗Long = 15′ to 20′
◗Very Long = 20′ or more

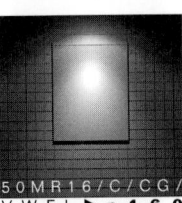  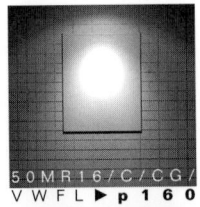

50MR16/C/CG/ VWFL▶p160 50MR16/C/CG/ VWFL▶p160 50MR16/C/CG/ VWFL▶p160

50MR16/C/CG/VWFL
▶Soft general lighting
▶Wallwashing
▶Large art/10′ ceiling height or less
▶Short throws
▶Gallery, Residential

71MR16/C/CG/ WSP▶p161 71MR16/C/CG/ WSP▶p161 71MR16/C/CG/ WSP▶p161

71MR16/C/CG/WSP
▶Architectural accenting
▶Large art/16′ ceiling height or less
▶Sculpture
▶Long throws
▶Hospitality

  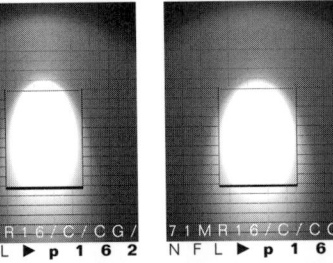 

71MR16/C/CG/ NFL▶p162 71MR16/C/CG/ NFL▶p162 71MR16/C/CG/ NFL▶p162

71MR16/C/CG/NFL
▶Architectural accenting
▶Feature downlighting
▶Large art/13′ ceiling height or less
▶Sculpture
▶Medium throws
▶Commercial, Retail

  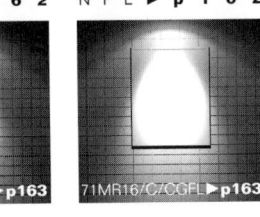

71MR16/C/CGFL▶p163 71MR16/C/CGFL▶p163 71MR16/C/CGFL▶p163

71MR16/C/CG/FL
▶General lighting
▶Wallwashing
▶Large art/10′ ceiling height or less
▶Short throws
▶Commercial, Hospitality

Recommended intensities
for art accenting

Low (gallery, museum, residential)
5 to 10 fc average on art
Moderate (hospitality)
10 to 20 fc average on art
High (commercial—or higher for retail)
20 to 30 fc average on art

These are intended to be average, maintained values including effects of accent lighting and general room lighting. General room lighting contributes just a few footcandles in residential, gallery and museum settings, but contributes 5 to 10 fc in commercial settings. Maximums might range from 50 percent to 100 percent of the average value. Wherever artwork is highly valuable and/or sensitive to light, precautions should be taken to limit light exposure (maintaining low levels and/or switching lights off when art is not being viewed); and exposure to infrared and ultraviolet should be limited (using filters on lamps or placing art behind specially treated glass or acrylic).

Artwork Size Categories
◗Petite = 2′–0″ by 2′–0″ or smaller
◗Small = 2′–0″ by 2′–0″ to 2′–6″ by 3′–0″
◗Medium = 2′–6″ by 3′–0″ to 3′–0″ by 4′–0″
◗Large = 3′–0″ by 4′–0″ to 4′–0″ by 5′–0″
◗Extra Large = 4′–0″ by 5′–0″ or larger

Halogen/MR16/C/CG
Hard Dichroic Coating and Cover Glass

►Has integral cover glass.

Halogen Low Voltage MR Lamp Designation

20MR16/C/CG/NFL

Last set of letters identifies beam spread

FLOODS
- 24° to 32° for narrow flood (NFL)
- 33° to 44° for flood (FL)
- 45° to 54° for wide flood (WFL)
- 55° or greater for very wide flood (VWFL)

SPOTS
- 7° or less for very narrow spot (VNSP)
- 8° to 10° for narrow spot (NSP)
- 11° to 14° for spot (SP)
- 15° to 18° for wide spot (WSP)
- 19° to 23° for very wide spot (VWSP)

Ranges of beam spreads used for convenient/consistent reference in this text and do not necessarily correspond to each manufacturer's nor any ANSI or NEMA definitions.

"CG" for built-in cover glass or no designation for no cover glass

Reflector treatment code
- "C" for consistent color (hard dichroic coating) or no letter for inconsistent color
- "AL" for aluminized reflector eliminating back spill light through reflector

These two digits represent lamp diameter in eighths inches (51mm or 2 inches)

These letters indicate this is a multifaceted reflector (MR) lamp

First two digits represent lamp wattage

[This designation used for convenient reference throughout this *Time–Saver Standards for Architectural Lighting*—but not necessarily used by all lamp manufacturers (although it would be convenient if manufacturers would come to agreement on standard designations.)]

▶Has integral cover glass.

Halogen **20MR16/C/CG/NSP**

Hard Dichroic Coating and Cover Glass

Performance Sketches

[All data outlined here is based on information from boldfaced manufacturer's published data for 12V version at time of manuscript preparation. Scale is ¼ inch equals 1 foot.]

Halogen Low Voltage (12V)

4

Uses
▶Architectural detailing
▶Petite art with medium throws
▶Sculpture with medium throws
▶Medium throws
▶Available in 12V

Specs
Beam spread/10°
Center beam candlepower/ 5350
Life/5000 hrs
▶**Philips** (26962–1)

Design Tips
✔One light best for petite art and where mounting height for light is 11′ or less.
✔Light may degrade artwork rapidly and/or significantly: consider UV and IR filters or limit exposure time.
✔Excellent for glass pieces and floral centerpieces and where mounting height for light is 11′ or less.
✔More grazing light creates prominent shadows.
✔MR lamps are glary—consider shielding with hexcell louver or using deep luminaire trims or pinholes.
✔Required transformer also consumes 3 to 5 watts.

Application Key

Commercial
Gallery
Hospitality
Institutional
Manufacturing
Residential
Retail
Exterior

bold = primary application
partial fade = minimal application
fade = unlikely application

Lamp: 20MR16/C/CG/NSP
Location: 1′–6″ from wall
Center beam Aiming: 14° (at point 4′–6″ AFF)
Luminaire: Monopoint or Recessed Adjustable
Illuminance: 15 fc avg, maintained (38 fc max)

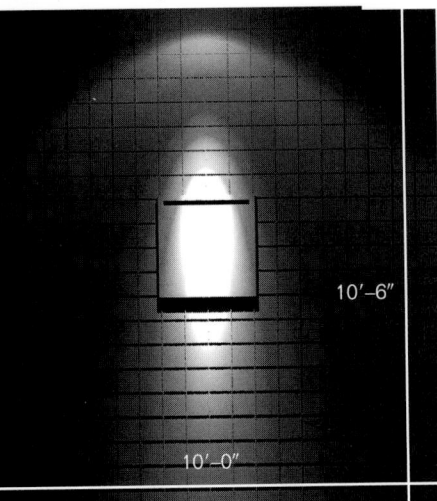

10′–6″

10′–0″

Lamp: 20MR16/C/CG/NSP
Location: 2′–0″ from wall
Center beam Aiming: 20° (at point 5′–0″ AFF)
Luminaire: Monopoint or Recessed Adjustable
Illuminance: 19 fc avg, maintained (52 fc max)

Art shown at 2′ by 2′ and centered 5′–6″ AFF. Wall grid on 6″ by 6″ centers and scaled at ¼″=1′.

Lamp: 20MR16/C/CG/NSP
Location: 2′–6″ from wall
Center beam Aiming: 24° (at point 5′–0″ AFF)
Luminaire: Monopoint or Recessed Adjustable
Illuminance: 20 fc avg, maintained (57 fc max)

Halogen 20MR16/C/CG/WSP

Hard Dichroic Coating and Cover Glass

▶Has integral cover glass.

Halogen Low Voltage (12V)

4

Performance Sketches

[All data outlined here is based on information from boldfaced manufacturer's published data for 12V version at time of manuscript preparation. Scale is ¼ inch equals 1 foot.]

Lamp: 20MR16/C/CG/WSP
Location: 1'–6" from wall
Center beam Aiming: 14° (at point 4'–6" AFF)
Luminaire: Monopoint or Recessed Adjustable
Illuminance: 11 fc avg, maintained (29 fc max)

Lamp: 20MR16/C/CG/WSP
Location: 2'–0" from wall
Center beam Aiming: 20° (at point 5'–0" AFF)
Luminaire: Monopoint or Recessed Adjustable
Illuminance: 14 fc avg, maintained (37 fc max)

Art shown at 2' by 2' and centered 5'–6" AFF. Wall grid on 6" by 6" centers and scaled at ¼"=1'.

Lamp: 20MR16/C/CG/WSP
Location: 2'–6" from wall
Center beam Aiming: 24° (at point 5'–0" AFF)
Luminaire: Monopoint or Recessed Adjustable
Illuminance: 15 fc avg, maintained (39 fc max)

Uses

▶Architectural accenting
▶Petite art with medium throws
▶Sculpture with medium throws
▶Medium throws
▶Available in 12V

Specs

Beam spread/15°
Center beam candlepower/ 3150
Life/5000 hrs
▶**GE** (20858)

Design Tips

✔One light best for petite art and where mounting height for light is 11' or less.
✔Light may degrade artwork rapidly and/or significantly: consider UV and IR filters or limit exposure time.
✔More grazing light creates prominent shadows.
✔MR lamps are glary—consider shielding with hexcell louver or using deep luminaire trims or pinholes.
✔Required transformer also consumes 3 to 5 watts.

Application Key

Commercial
Gallery
Hospitality
Institutional
Manufacturing
Residential
Retail
Exterior

bold = primary application
partial fade = minimal application
fade = unlikely application

▶Has integral cover glass.

Halogen Low Voltage (12V)

4

Halogen 20MR16/C/CG/NFL

Hard Dichroic Coating and Cover Glass

Uses
▶Feature downlighting
▶Petite art with short throws
▶Short throws
▶Available in 12V

Specs
Beam spread/24°
Center beam candlepower/2010
Life/5000 hrs
▶Philips (26966–2)

Design Tips
✔One light best for petite art where mounting height for light is 10' or less.
✔Light may degrade artwork rapidly and/or significantly: consider UV and IR filters or limit exposure time.
✔Good for downlighting onto decorative floor materials—best with sconces and/or art accents and/or wallwashing to avoid cave effect and grazing glare (the "I wish I had a visor" effect).
✔More grazing light creates prominent shadows.
✔MR lamps are glary—consider shielding with hexcell louver or using deep luminaire trims or pinholes.
✔Required transformer consumes 3 to 5 watts.

Application Key

Commercial
Gallery
Hospitality
Institutional
Manufacturing
Residential
Retail
Exterior

bold = primary application
partial fade = minimal application
fade = unlikely application

Performance Sketches

[All data outlined here is based on information from boldfaced manufacturer's published data for 12V version at time of manuscript preparation. Scale is ¼ inch equals 1 foot.]

Art shown at 2' by 2' and centered 5'–6" AFF. Wall grid on 6" by 6" centers and scaled at ¼"=1'.

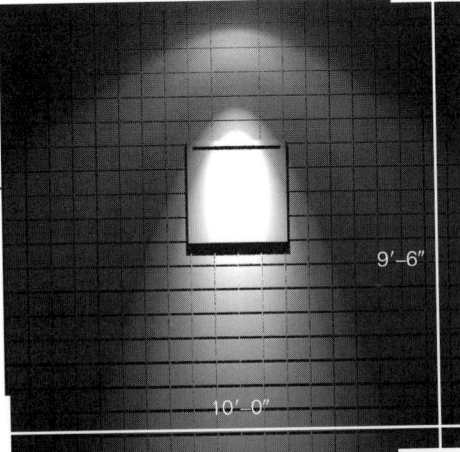

9'–6"

10'–0"

Lamp: 20MR16/C/CG/NFL
Location: 1'–6" from wall
Center beam Aiming: 14° (at point 3'–6" AFF)
Luminaire: Monopoint or Recessed Adjustable
Illuminance: 15 fc avg, maintained (30 fc max)

Lamp: 20MR16/C/CG/NFL
Location: 2'–0" from wall
Center beam Aiming: 20° (at point 4'–0" AFF)
Luminaire: Monopoint or Recessed Adjustable
Illuminance: 17 fc avg, maintained (33 fc max)

Lamp: 20MR16/C/CG/NFL
Location: 2'–6" from wall
Center beam Aiming: 27° (at point 4'–6" AFF)
Luminaire: Monopoint or Recessed Adjustable
Illuminance: 20 fc avg, maintained (39 fc max)

Halogen 20MR16/C/CG/FL (Philips)
Hard Dichroic Coating and Cover Glass

 ►Has integral cover glass.

Halogen Low Voltage (12V)

4

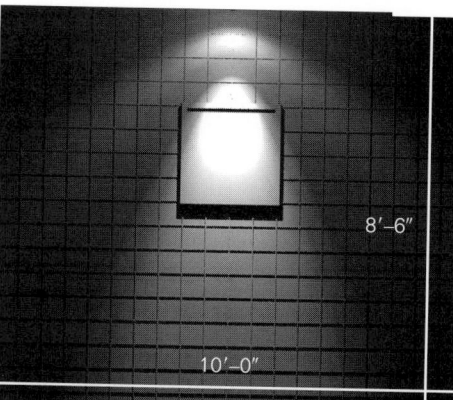

8'–6"

10'–0"

Performance Sketches

[All data outlined here is based on information from boldfaced manufacturer's published data for 12V version at time of manuscript preparation. Scale is ¼ inch equals 1 foot.]

Art shown at 2' by 2' and centered 5'–6" AFF. Wall grid on 6" by 6" centers and scaled at ¼"=1'.

Lamp: 20MR16/C/CG/FL (Philips)
Location: 1'–0" from wall
Center beam Aiming: 11° (at point 3'–6" AFF)
Luminaire: Monopoint or Recessed Adjustable
Illuminance: 12 fc avg, maintained (40 fc max)

Lamp: 20MR16/C/CG/FL (Philips)
Location: 1'–6" from wall
Center beam Aiming: 18° (at point 4'–0" AFF)
Luminaire: Monopoint or Recessed Adjustable
Illuminance: 14 fc avg, maintained (32 fc max)

Lamp: 20MR16/C/CG/FL (Philips)
Location: 2'–0" from wall
Center beam Aiming: 27° (at point 4'–6" AFF)
Luminaire: Monopoint or Recessed Adjustable
Illuminance: 16 fc avg, maintained (33 fc max)

Uses
►General lighting
►Petite art with short throws
►Available in 12V

Specs
Beam spread/38°
Center beam
 candlepower/700
Life/5000 hrs
►**Philips** (26970–4)

Design Tips
✔One light best for petite art where mounting height of light is 9' or less.
✔Multiple units can be used for wallwashing or lighting of larger artwork.
✔More grazing light creates prominent shadows.
✔MR lamps are glary—consider shielding with hexcell louver or using deep luminaire trims or pinholes.
✔Required transformer also consumes 3 to 5 watts.

Application Key

Commercial
Gallery
Hospitality
Institutional
Manufacturing
Residential
Retail
Exterior

bold = primary application
partial fade = minimal application
fade = unlikely application

▶Has integral cover glass.

Halogen **20MR16/C/CG/FL**(GE)

Hard Dichroic Coating and Cover Glass

Halogen Low Voltage (12V)

4

Uses
▶General lighting
▶Wallwashing (see 11–115)
▶Medium art with short throws
▶Available in 12V

Specs
Beam spread/40°
Center beam
   candlepower/490
Life/5000 hrs
▶**GE** (20857)

Design Tips
✔One light best for medium art where mounting height of light is 10′ or less.
✔Multiple units can be used for wallwashing or lighting of larger artwork.
✔More grazing light creates prominent shadows.
✔MR lamps are glary—consider shielding with hexcell louver or using deep luminaire trims or pinholes.
✔Required transformer also consumes 3 to 5 watts.

Performance Sketches
[All data outlined here is based on information from boldfaced manufacturer's published data for 12V version at time of manuscript preparation. Scale is ¼ inch equals 1 foot.]

Art shown at 3′ by 4′ and centered 5′–6″ AFF. Wall grid on 6″ by 6″ centers and scaled at ¼″=1′.

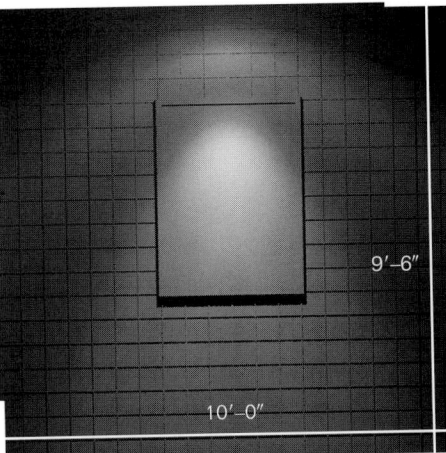

9′–6″

10′–0″

Lamp: 20MR16/C/CG/NFL
Location: 2′–0″ from wall
Center beam Aiming: 14° (at point 1′–6″ AFF)
Luminaire: Monopoint or Recessed Adjustable
Illuminance: 5 fc avg, maintained (10 fc max)

Lamp: 20MR16/C/CG/NFL
Location: 2′–6″ from wall
Center beam Aiming: 20° (at point 2′–6″ AFF)
Luminaire: Monopoint or Recessed Adjustable
Illuminance: 5 fc avg, maintained (10 fc max)

Lamp: 20MR16/C/CG/NFL
Location: 3′–0″ from wall
Center beam Aiming: 27° (at point 3′–6″ AFF)
Luminaire: Monopoint or Recessed Adjustable
Illuminance: 6 fc avg, maintained (11 fc max)

Application Key

Commercial
Gallery
Hospitality
Institutional
Manufacturing
Residential
Retail
Exterior

bold = primary application
partial fade = minimal application
fade = unlikely application

Halogen 35MR16/C/CG/SP
Hard Dichroic Coating and Cover Glass

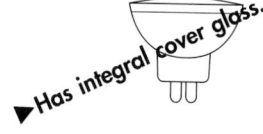
▶Has integral cover glass.

Performance Sketches

[All data outlined here is based on information from boldfaced manufacturer's published data for 12V version at time of manuscript preparation. Scale is ¼ inch equals 1 foot.]

Lamp: 35MR16/C/CG/SP
Location: 3'–0" from wall
Center beam Aiming: 22° (at point 5'–0" AFF)
Luminaire: Monopoint or Recessed Adjustable
Illuminance: 15 fc avg, maintained (48 fc max)

12'–6"

10'–0"

Lamp: 35MR16/C/CG/SP
Location: 4'–0" from wall
Center beam Aiming: 28° (at point 5'–0" AFF)
Luminaire: Monopoint or Recessed Adjustable
Illuminance: 16 fc avg, maintained (50 fc max)

Art shown at 3' by 4' and centered 5'–6" AFF. Wall grid on 6" by 6" centers and scaled at ¼"=1'.

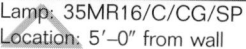

Lamp: 35MR16/C/CG/SP
Location: 5'–0" from wall
Center beam Aiming: 35° (at point 5'–3" AFF)
Luminaire: Monopoint or Recessed Adjustable
Illuminance: 17 fc avg, maintained (54 fc max)

Uses
▶Architectural accenting
▶Medium art with medium throws
▶Sculpture with medium throws
▶Medium throws
▶Available in 12V

Specs
Beam spread/12°
Center beam candlepower/ 7500
Life/5000 hrs
▶GE (20864)

Design Tips
✔One light best for medium art and where mounting height for light is 13' or less.
✔Light may degrade artwork rapidly and/or significantly: consider UV and IR filters or limit exposure time.
✔Scallops/striations are most severe when light is closer to the wall (compare top left image to bottom image).
✔More grazing light creates prominent shadows.
✔MR lamps are glary—consider shielding with hexcell louver or using deep luminaire trims or pinholes.
✔Required transformer also consumes 3 to 5 watts.

Application Key

Commercial
Gallery
Hospitality
Institutional
Manufacturing
Residential
Retail
Exterior

bold = primary application
partial fade = minimal application
fade = unlikely application

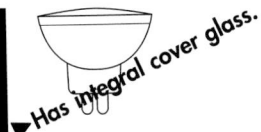

▶ Has integral cover glass.

Halogen 35MR16/C/CG/VWSP
Hard Dichroic Coating and Cover Glass

Halogen Low Voltage (12V)

4

Uses
▶ Architectural accenting
▶ Medium art with medium throws
▶ Sculpture with medium throws
▶ Medium throws
▶ Available in 12V

Specs
Beam spread/20°
Center beam candlepower/ 3300
Life/5000 hrs
▶ GE (20860)

Design Tips
✔ One light best for medium art and where mounting height for light is 13' or less.
✔ Light may degrade artwork rapidly and/or significantly: consider UV and IR filters or limit exposure time.
✔ More grazing light creates prominent shadows.
✔ MR lamps are glary—consider shielding with hexcell louver or using deep luminaire trims or pinholes.
✔ Required transformer also consumes 3 to 5 watts.

Performance Sketches
[All data outlined here is based on information from boldfaced manufacturer's published data for 12V version at time of manuscript preparation. Scale is ¼ inch equals 1 foot.]

Lamp: 35MR16/C/CG/VWSP
Location: 3'–0" from wall
Center beam Aiming: 18° (at point 3'–6" AFF)
Luminaire: Monopoint or Recessed Adjustable
Illuminance: 10 fc avg, maintained (19 fc max)

Lamp: 35MR16/C/CG/VWSP
Location: 4'–0" from wall
Center beam Aiming: 27° (at point 4'–6" AFF)
Luminaire: Monopoint or Recessed Adjustable
Illuminance: 13 fc avg, maintained (23 fc max)

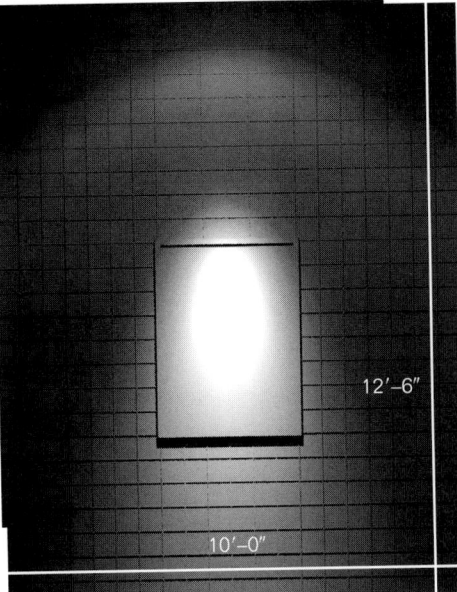

12'–6"

10'–0"

Art shown at 3' by 4' and centered 5'–6" AFF. Wall grid on 6" by 6" centers and scaled at ¼"=1'.

Lamp: 35MR16/C/CG/VWSP
Location: 5'–0" from wall
Center beam Aiming: 32° (at point 4'–6" AFF)
Luminaire: Monopoint or Recessed Adjustable
Illuminance: 13 fc avg, maintained (22 fc max)

Application Key

Commercial
Gallery
Hospitality
Institutional
Manufacturing
Residential
Retail
Exterior

bold = primary application
partial fade = minimal application
fade = unlikely application

Halogen 35MR16/C/CG/FL
Hard Dichroic Coating and Cover Glass

▶Has integral cover glass.

Halogen Low Voltage (12V)

4

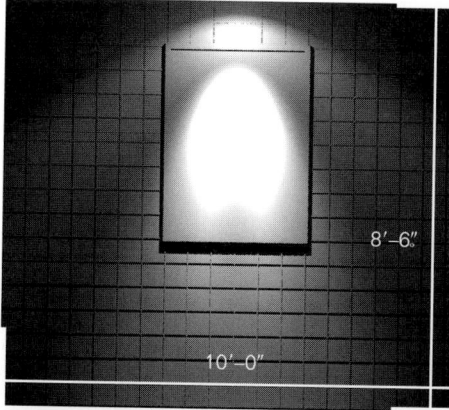

8'–6."

10'–0"

Performance Sketches

[All data outlined here is based on information from boldfaced manufacturer's published data for 12V version at time of manuscript preparation. Scale is ¼ inch equals 1 foot.]

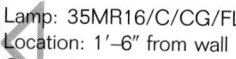

Art shown at 3' by 4' and centered 5'–6" AFF. Wall grid on 6" by 6" centers and scaled at ¼"=1'.

Lamp: 35MR16/C/CG/FL
Location: 1'–6" from wall
Center beam Aiming: 18° (at point 4'–0" AFF)
Luminaire: Monopoint or Recessed Adjustable
Illuminance: 14 fc avg, maintained (49 fc max)

Lamp: 35MR16/C/CG/FL
Location: 2'–0" from wall
Center beam Aiming: 27° (at point 4'–6" AFF)
Luminaire: Monopoint or Recessed Adjustable
Illuminance: 15 fc avg, maintained (49 fc max)

Lamp: 35MR16/C/CG/FL
Location: 2'–6" from wall
Center beam Aiming: 36° (at point 5'–0" AFF)
Luminaire: Monopoint or Recessed Adjustable
Illuminance: 17 fc avg, maintained (52 fc max)

Uses
▶General lighting
▶Medium art with short throws
▶Short throws
▶Available in 12V

Specs
Beam spread/40°
Center beam candlepower/ 900
Life/5000 hrs
▶**GE** (20859)

Design Tips
✔One light best for medium art where mounting height of light is 9' or less.
✔Light may degrade artwork rapidly and/or significantly: consider UV and IR filters or limit exposure time.
✔Good for downlighting onto decorative floor materials—best with sconces and/or art accents and/or wallwashing to avoid cave effect and grazing glare (the "I wish I had a visor" effect).
✔More grazing light creates prominent shadows.
✔MR lamps are glary—consider shielding with hexcell louver or using deep luminaire trims or pinholes.
✔Required transformer also consumes 3 to 5 watts.

Application Key

Commercial
Gallery
Hospitality
Institutional
Manufacturing
Residential
Retail
Exterior

bold = primary application
partial fade = minimal application
fade = unlikely application

▶Has integral cover glass.

Halogen 50MR16/C/CG/NSP

Hard Dichroic Coating and Cover Glass

Halogen Low Voltage (12V)

4

Uses
▶Architectural detailing
▶Medium art with long throws
▶Sculpture with long throws
▶Long throws
▶Available in 12V

Specs
Beam spread/10°
Center beam
candlepower/9650
Life/5000 hrs
▶**Philips** (26977–9)

Design Tips
✔One light best for medium art and where mounting height for light is 16′ or less.
✔These are strong accents and deserve judicious use. Light may degrade artwork rapidly and/or significantly: consider UV and IR filters or limit exposure time.
✔Excellent for glass pieces and floral centerpieces and where mounting height for light is 16′ or less.
✔Scallops, striations and frame shadows are most severe when light is closer to the wall (compare top left image to bottom image).
✔MR lamps are glary—consider shielding with hexcell louver or using deep luminaire trims or pinholes.
✔Required transformer consumes 3 to 5 watts.

Application Key

Commercial
Gallery
Hospitality
Institutional
Manufacturing
Residential
Retail
Exterior

bold = primary application
partial fade = minimal application
fade = unlikely application

Performance Sketches

[All data outlined here is based on information from boldfaced manufacturer's published data for 12V version at time of manuscript preparation. Scale is ¼ inch equals 1 foot.]

Lamp: 50MR16/C/CG/NSP
Location: 4′–3″ from wall
Center beam Aiming: 22° (at point 4′–6″ AFF)
Luminaire: Monopoint or Recessed Adjustable
Illuminance: 14 fc avg, maintained (27 fc max)

Lamp: 50MR16/C/CG/NSP
Location: 5′–3″ from wall
Center beam Aiming: 27° (at point 5′–0″ AFF)
Luminaire: Monopoint or Recessed Adjustable
Illuminance: 16 fc avg, maintained (32 fc max)

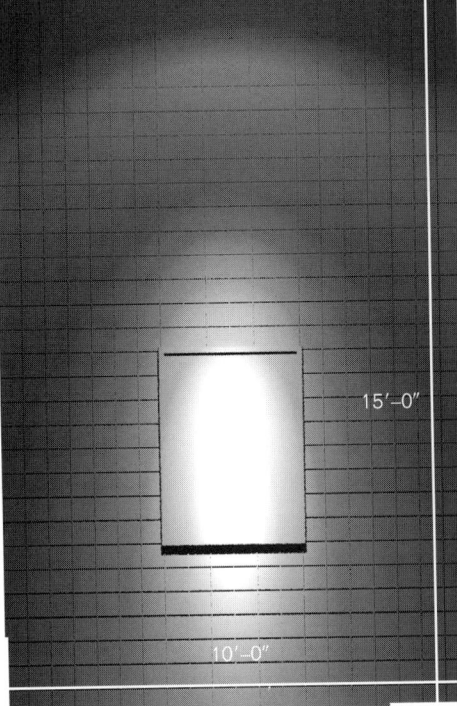

15′–0″

10′–0″

Art shown at 3′ by 4′ and centered 5′–6″ AFF. Wall grid on 6″ by 6″ centers and scaled at ¼″=1′.

Lamp: 50MR16/C/CG/NSP
Location: 6′–3″ from wall
Center beam Aiming: 32° (at point 5′–0″ AFF)
Luminaire: Monopoint or Recessed Adjustable
Illuminance: 16 fc avg, maintained (33 fc max)

Halogen 50MR16/C/CG/WSP

Hard Dichroic Coating and Cover Glass

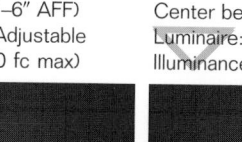

►Has integral cover glass.

Halogen Low Voltage (12V)

4

Performance Sketches

[All data outlined here is based on information from boldfaced manufacturer's published data for 12V version at time of manuscript preparation. Scale is ¼ inch equals 1 foot.]

Lamp: 50MR16/C/CG/WSP
Location: 4'–3" from wall
Center beam Aiming: 22° (at point 4'–6" AFF)
Luminaire: Monopoint or Recessed Adjustable
Illuminance: 11 fc avg, maintained (30 fc max)

Lamp: 50MR16/C/CG/WSP
Location: 5'–3" from wall
Center beam Aiming: 28° (at point 5'–0" AFF)
Luminaire: Monopoint or Recessed Adjustable
Illuminance: 13 fc avg, maintained (35 fc max)

Uses
►Architectural accenting
►Large art with long throws
►Sculpture with long throws
►Long throws
►Available in 12V

Specs
Beam spread/15°
Center beam candlepower/ 8400
Life/6000 hrs
►**GE** (20872)

Design Tips
✔One light best for large art and where mounting height for light is 16' or less.
✔These are strong accents and deserve judicious use. Light may degrade artwork rapidly and/or significantly: consider UV and IR filters or limit exposure time.
✔Good for glass pieces and floral centerpieces and where mounting height for light is 16' or less.
✔MR lamps are glary—consider shielding with hexcell louver or using deep luminaire trims or pinholes.
✔Required transformer consumes 3 to 5 watts.

Art shown at 4' by 5' and centered 5'–6" AFF. Wall grid on 6" by 6" centers and scaled at ¼"=1'.

Lamp: 50MR16/C/CG/WSP
Location: 6'–3" from wall
Center beam Aiming: 32° (at point 5'–0" AFF)
Luminaire: Monopoint or Recessed Adjustable
Illuminance: 13 fc avg, maintained (34 fc max)

Application Key

Commercial
Gallery
Hospitality
Institutional
Manufacturing
Residential
Retail
Exterior

bold = primary application
partial fade = minimal application
fade = unlikely application

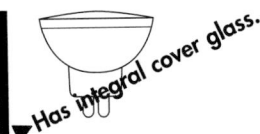

▶Has integral cover glass.

Halogen 50MR16/C/CG/NFL (Philips)

Hard Dichroic Coating and Cover Glass

Performance Sketches

[All data outlined here is based on information from boldfaced manufacturer's published data for 12V version at time of manuscript preparation. Scale is ¼ inch equals 1 foot.]

Halogen Low Voltage (12V)

4

Uses
▶Architectural accenting
▶Feature downlighting
▶Medium art with medium throws
▶Sculpture with medium throws
▶Medium throws
▶Available in 12V

Specs
Beam spread/24°
Center beam candlepower/ 5470
Life/5000 hrs
▶**Philips** (26981–1)

Design Tips
✔One light best for medium art where mounting height for light is 13' or less.
✔Great for moderate sculpture where mounting height for light is 13' or less.
✔Scallops, striations and frame shadows are most severe when light is closer to the wall (compare top left image to bottom image).
✔MR lamps are glary—consider shielding with hexcell louver or using deep luminaire trims or pinholes.
✔Required transformer consumes 3 to 5 watts.

Lamp: 50MR16/C/CG/NFL (Philips)
Location: 3'–0" from wall
Center beam Aiming: 18° (at point 3'–6" AFF)
Luminaire: Monopoint or Recessed Adjustable
Illuminance: 18 fc avg, maintained (32 fc max)

12'–6"

10'–0"

Lamp: 50MR16/C/CG/NFL (Philips)
Location: 4'–0" from wall
Center beam Aiming: 27° (at point 4'–6" AFF)
Luminaire: Monopoint or Recessed Adjustable
Illuminance: 22 fc avg, maintained (38 fc max)

Art shown at 3' by 4' and centered 5'–6" AFF. Wall grid on 6" by 6" centers and scaled at ¼"=1'.

Application Key

Commercial
Gallery
Hospitality
Institutional
Manufacturing
Residential
Retail
Exterior

bold = primary application
partial fade = minimal application
fade = unlikely application

Lamp: 50MR16/C/CG/NFL (Philips)
Location: 5'–0" from wall
Center beam Aiming: 32° (at point 4'–6" AFF)
Luminaire: Monopoint or Recessed Adjustable
Illuminance: 22 fc avg, maintained (36 fc max)

Halogen 50MR16/C/CG/NFL(GE)

Hard Dichroic Coating and Cover Glass

▶Has integral cover glass.

Halogen Low Voltage (12V)

4

Performance Sketches

[All data outlined here is based on information from boldfaced manufacturer's published data for 12V version at time of manuscript preparation. Scale is ¼ inch equals 1 foot.]

Lamp: 50MR16/C/CG/NFL (GE)
Location: 3'–0" from wall
Center beam Aiming: 23° (at point 3'–6" AFF)
Luminaire: Monopoint or Recessed Adjustable
Illuminance: 16 fc avg, maintained (31 fc max)

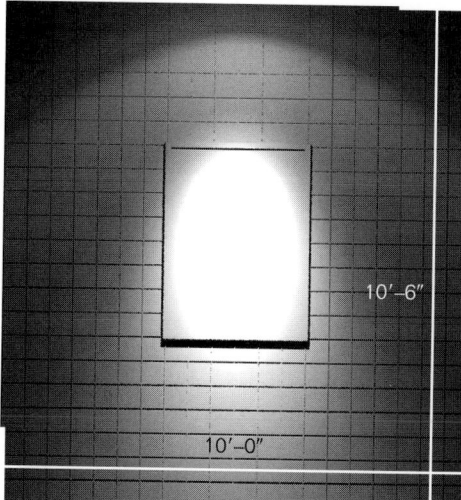

10'–6"

10'–0"

Lamp: 50MR16/C/CG/NFL (GE)
Location: 4'–0" from wall
Center beam Aiming: 34° (at point 4'–6" AFF)
Luminaire: Monopoint or Recessed Adjustable
Illuminance: 20 fc avg, maintained (36 fc max)

Art shown at 3' by 4' and centered 5'–6" AFF. Wall grid on 6" by 6" centers and scaled at ¼"=1'.

Lamp: 50MR16/C/CG/NFL (GE)
Location: 5'–0" from wall
Center beam Aiming: 40° (at point 4'–6" AFF)
Luminaire: Monopoint or Recessed Adjustable
Illuminance: 19 fc avg, maintained (33 fc max)

Uses

▶Architectural accenting
▶Feature downlighting
▶Medium art with medium throws
▶Sculpture with medium throws
▶Medium throws
▶Available in 12V

Specs

Beam spread/25°
Center beam candlepower/2900
Life/6000 hrs
▶**GE** (20871)

Design Tips

✔One light best for medium art where mounting height for light is 11' or less.
✔Great for moderate to large sculpture where mounting height for light is 11' or less.
✔Scallops, striations and frame shadows are most severe when light is closer to the wall (compare top left image to bottom image).
✔MR lamps are glary—consider shielding with hexcell louver or using deep luminaire trims or pinholes.
✔Required transformer consumes 3 to 5 watts.
✔Consider 37MR16/HIR/C/NFL.

Application Key

Commercial
Gallery
Hospitality
Institutional
Manufacturing
Residential
Retail
Exterior

bold = primary application
partial fade = minimal application
fade = unlikely application

▶Has integral cover glass.

Halogen **50MR16/C/CG/FL**(Philips)

Hard Dichroic Coating and Cover Glass

Halogen
Low Voltage (12V)

4

Uses
▶General lighting
▶Medium art with short throws
▶Short throws
▶Available in 12V

Specs
Beam spread/38°
Center beam candlepower/1890
Life/5000 hrs
▶**Philips** (27071–0)

Design Tips
✔One light best for medium art where mounting height for light is 10' or less.
✔Light may degrade artwork rapidly and/or significantly: consider UV and IR filters or limit exposure time.
✔Good for downlighting onto decorative floor materials—best with sconces and/or art accents and/or wallwashing to avoid cave effect and grazing glare (the "I wish I had a visor" effect).
✔More grazing light creates prominent shadows.
✔MR lamps are glary—consider shielding with hexcell louver or using deep luminaire trims or pinholes.
✔Required transformer consumes 3 to 5 watts.
✔Consider 37MR16/HIR/C/FL.

Application Key

Commercial
Gallery
Hospitality
Institutional
Manufacturing
Residential
Retail
Exterior

bold = primary application
partial fade = minimal application
fade = unlikely application

Performance Sketches

[All data outlined here is based on information from boldfaced manufacturer's published data for 12V version at time of manuscript preparation. Scale is ¼ inch equals 1 foot.]

Art shown at 3' by 4' and centered 5'–6" AFF. Wall grid on 6" by 6" centers and scaled at ¼"=1'.

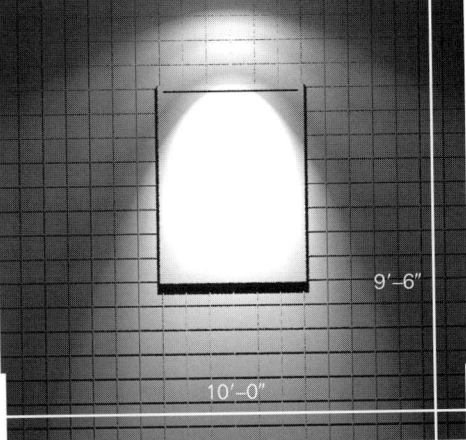

9'–6"

10'–0"

Lamp: 50MR16/C/CG/FL (Philips)
Location: 2'–0" from wall
Center beam Aiming: 20° (at point 4'–0" AFF)
Luminaire: Monopoint or Recessed Adjustable
Illuminance: 21 fc avg, maintained (62 fc max)

Lamp: 50MR16/C/CG/FL (Philips)
Location: 2'–6" from wall
Center beam Aiming: 27° (at point 4'–6" AFF)
Luminaire: Monopoint or Recessed Adjustable
Illuminance: 23 fc avg, maintained (61 fc max)

Lamp: 50MR16/C/CG/FL (Philips)
Location: 3'–0" from wall
Center beam Aiming: 34° (at point 5'–0" AFF)
Luminaire: Monopoint or Recessed Adjustable
Illuminance: 25 fc avg, maintained (61 fc max)

Halogen 50MR16/C/CG/FL(GE)
Hard Dichroic Coating and Cover Glass

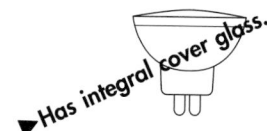
▶Has integral cover glass.

Performance Sketches

[All data outlined here is based on information from boldfaced manufacturer's published data for 12V version at time of manuscript preparation. Scale is ¼ inch equals 1 foot.]

Art shown at 3' by 4' and centered 5'–6" AFF. Wall grid on 6" by 6" centers and scaled at ¼"=1'.

Lamp: 50MR16/C/CG/FL (GE)
Location: 2'–0" from wall
Center beam Aiming: 20° (at point 4'–0" AFF)
Luminaire: Monopoint or Recessed Adjustable
Illuminance: 19 fc avg, maintained (52 fc max)

Lamp: 50MR16/C/CG/FL (GE)
Location: 2'–6" from wall
Center beam Aiming: 27° (at point 4'–6" AFF)
Luminaire: Monopoint or Recessed Adjustable
Illuminance: 21 fc avg, maintained (51 fc max)

Lamp: 50MR16/C/CG/FL (GE)
Location: 3'–0" from wall
Center beam Aiming: 34° (at point 5'–0" AFF)
Luminaire: Monopoint or Recessed Adjustable
Illuminance: 23 fc avg, maintained (52 fc max)

Halogen Low Voltage (12V)

4

Uses
- ▶General lighting
- ▶Medium art with short throws
- ▶Short throws
- ▶Available in 12V

Specs
Beam spread/40°
Center beam candlepower/1500
Life/6000 hrs
▶GE (20867)

Design Tips
✔One light best for medium art where mounting height for light is 10' or less.
✔Light may degrade artwork rapidly and/or significantly: consider UV and IR filters or limit exposure time.
✔Good for downlighting onto decorative floor materials—best with sconces and/or art accents and/or wallwashing to avoid cave effect and grazing glare (the "I wish I had a visor" effect).
✔More grazing light creates prominent shadows.
✔MR lamps are glary—consider shielding with hexcell louver or using deep luminaire trims or pinholes.
✔Required transformer consumes 3 to 5 watts.
✔Consider 37MR16/HIR/C/FL.

Application Key

Commercial
Gallery
Hospitality
Institutional
Manufacturing
Residential
Retail
Exterior

bold = primary application
partial fade = minimal application
fade = unlikely application

► Has integral cover glass.

Halogen 50MR16/C/CG/VWFL

Hard Dichroic Coating and Cover Glass

Halogen Low Voltage (12V)

4

Uses
► Soft general lighting
► Wallwashing (see 11–117)
► Large art with short throws
► Short throws
► Available in 12V

Specs
Beam spread/55°
Center beam candlepower/875
Life/6000 hrs
► **GE** (20865)

Design Tips
✔ One light best for large art and where mounting height for light is 10' or less.
✔ Good for soft downlighting— best with sconces and/or art accents and/or wallwashing to avoid cave effect.
✔ MR lamps are glary—consider shielding with hexcell louver or using deep luminaire trims or pinholes.
✔ Required transformer consumes 3 to 5 watts.

Performance Sketches
[All data outlined here is based on information from boldfaced manufacturer's published data for 12V version at time of manuscript preparation. Scale is ¼ inch equals 1 foot.]

Art shown at 4' by 5' and centered 5'–6" AFF. Wall grid on 6" by 6" centers and scaled at ¼"=1'.

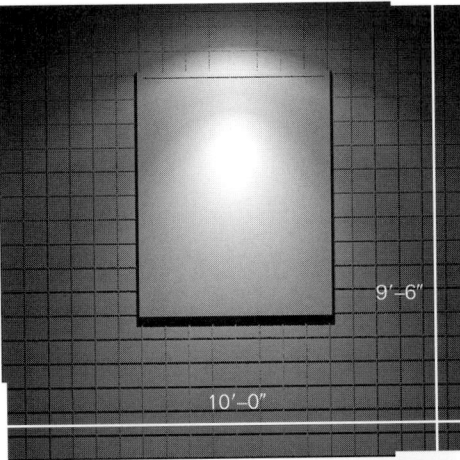

Lamp: 50MR16/C/CG/VWFL
Location: 2'–0" from wall
Center beam Aiming: 6° (at point 0'–0" AFF)
Luminaire: Monopoint or Recessed Adjustable
Illuminance: 7 fc avg, maintained (16 fc max)

Lamp: 50MR16/C/CG/VWFL
Location: 2'–6" from wall
Center beam Aiming: 16° (at point 1'–0" AFF)
Luminaire: Monopoint or Recessed Adjustable
Illuminance: 10 fc avg, maintained (21 fc max)

Application Key

Commercial
Gallery
Hospitality
Institutional
Manufacturing
Residential
Retail
Exterior

bold = primary application
partial fade = minimal application
fade = unlikely application

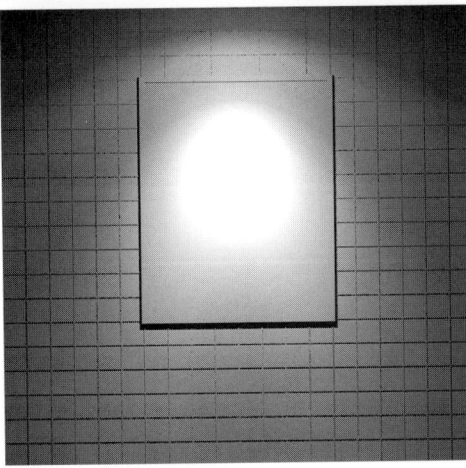

Lamp: 50MR16/C/CG/VWFL
Location: 3'–0" from wall
Center beam Aiming: 22° (at point 2'–0" AFF)
Luminaire: Monopoint or Recessed Adjustable
Illuminance: 10 fc avg, maintained (20 fc max)

Halogen 71MR16/C/CG/WSP
Hard Dichroic Coating and Cover Glass

▶Has integral cover glass.

Halogen Low Voltage (12V)

4

Performance Sketches

[All data outlined here is based on information from boldfaced manufacturer's published data for 12V version at time of manuscript preparation. Scale is ¼ inch equals 1 foot.]

Lamp: 71MR16/C/CG/WSP
Location: 5'–0" from wall
Center beam Aiming: 27° (at point 5'–0" AFF)
Luminaire: Monopoint or Recessed Adjustable
Illuminance: 18 fc avg, maintained (45 fc max)

15'–0"

10'–0"

Lamp: 71MR16/C/CG/WSP
Location: 6'–0" from wall
Center beam Aiming: 31° (at point 5'–0" AFF)
Luminaire: Monopoint or Recessed Adjustable
Illuminance: 19 fc avg, maintained (44 fc max)

Art shown at 4' by 5' and centered 5'–6" AFF. Wall grid on 6" by 6" centers and scaled at ¼"=1'.

Lamp: 71MR16/C/CG/WSP
Location: 7'–0" from wall
Center beam Aiming: 35° (at point 5'–0" AFF)
Luminaire: Monopoint or Recessed Adjustable
Illuminance: 19 fc avg, maintained (43 fc max)

Uses
▶Architectural accenting
▶Large art with long throws
▶Sculpture with long throws
▶Long throws
▶Available in 12V

Specs
Beam spread/15°
Center beam candlepower/10300
Life/4000 hrs
▶**GE** (20876)

Design Tips
✔One light best for large art and where mounting height for light is 16' or less.
✔These are strong accents and deserve judicious use. Light may degrade artwork rapidly and/or significantly: consider UV and IR filters or limit exposure time.
✔Good for glass pieces and floral centerpieces and where mounting height for light is 16' or less.
✔MR lamps are glary—consider shielding with hexcell louver or using deep luminaire trims or pinholes.
✔Required transformer consumes 3 to 5 watts.

Application Key

Commercial
Gallery
Hospitality
Institutional
Manufacturing
Residential
Retail
Exterior

bold = primary application
partial fade = minimal application
fade = unlikely application

▼ Has integral cover glass.

Halogen 71MR16/C/CG/NFL

Hard Dichroic Coating and Cover Glass

Halogen Low Voltage (12V)

Uses
▶ Architectural accenting
▶ Feature downlighting
▶ Large art with medium throws
▶ Sculpture with medium throws
▶ Medium throws
▶ Available in 12V

Specs
Beam spread/25°
Center beam candlepower/ 4550
Life/4000 hrs
▶ **GE** (20874)

Design Tips
✔ One light best for large art and where mounting height for light is 13′ or less.
✔ Good for moderate to larger sculpture where mounting height for light is 13′ or less.
✔ MR lamps are glary—consider shielding with hexcell louver or using deep luminaire trims or pinholes.
✔ Required transformer consumes 3 to 5 watts.

Performance Sketches

[All data outlined here is based on information from boldfaced manufacturer's published data for 12V version at time of manuscript preparation. Scale is ¼ inch equals 1 foot.]

Lamp: 71MR16/C/CG/NFL
Location: 3′–0″ from wall
Center beam Aiming: 21° (at point 4′–6″ AFF)
Luminaire: Monopoint or Recessed Adjustable
Illuminance: 18 fc avg, maintained (51 fc max)

12′–6″

10′–0″

Lamp: 71MR16/C/CG/NFL
Location: 4′–0″ from wall
Center beam Aiming: 28° (at point 5′–0″ AFF)
Luminaire: Monopoint or Recessed Adjustable
Illuminance: 21 fc avg, maintained (54 fc max)

Art shown at 4′ by 5′ and centered 5′–6″ AFF. Wall grid on 6″ by 6″ centers and scaled at ¼″=1′.

Lamp: 71MR16/C/CG/NFL
Location: 5′–0″ from wall
Center beam Aiming: 34° (at point 5′–0″ AFF)
Luminaire: Monopoint or Recessed Adjustable
Illuminance: 21 fc avg, maintained (49 fc max)

Application Key

Commercial
Gallery
Hospitality
Institutional
Manufacturing
Residential
Retail
Exterior

bold = primary application
partial fade = minimal application
fade = unlikely application

Halogen 71MR16/C/CG/FL
Hard Dichroic Coating and Cover Glass

▶Has integral cover glass.

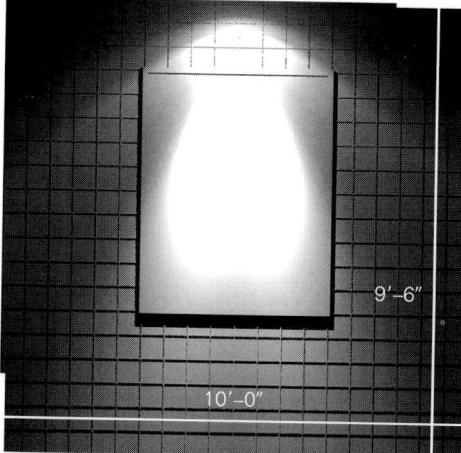

9'–6"

10'–0"

Performance Sketches

[All data outlined here is based on information from boldfaced manufacturer's published data for 12V version at time of manuscript preparation. Scale is ¼ inch equals 1 foot.]

Art shown at 4' by 5' and centered 5'–6" AFF. Wall grid on 6" by 6" centers and scaled at ¼"=1'.

Lamp: 71MR16/C/CG/FL
Location: 1'–6" from wall
Center beam Aiming: 11° (at point 2'–0" AFF)
Luminaire: Monopoint or Recessed Adjustable
Illuminance: 17 fc avg, maintained (67 fc max)

Lamp: 71MR16/C/CG/FL
Location: 2'–0" from wall
Center beam Aiming: 17° (at point 3'–0" AFF)
Luminaire: Monopoint or Recessed Adjustable
Illuminance: 18 fc avg, maintained (61 fc max)

Lamp: 71MR16/C/FL
Location: 2'–6" from wall
Center beam Aiming: 24° (at point 4'–0" AFF)
Luminaire: Monopoint or Recessed Adjustable
Illuminance: 21 fc avg, maintained (65 fc max)

Halogen Low Voltage (12V)

4

Uses
▶General lighting
▶Wallwashing (see 11–119)
▶Large art with short throws
▶Short throws
▶Available in 12V

Specs
Beam spread/40°
Center beam candlepower/2000
Life/4000 hrs
▶**GE** (20873)

Design Tips
✔One light best for large art where mounting height for light is 10' or less.
✔Light may degrade artwork rapidly and/or significantly: consider UV and IR filters or limit exposure time.
✔Good for downlighting onto decorative floor materials—best with sconces and/or art accents and/or wallwashing to avoid cave effect and grazing glare (the "I wish I had a visor" effect).
✔More grazing light creates prominent shadows.
✔MR lamps are glary—consider shielding with hexcell louver or using deep luminaire trims or pinholes.
✔Required transformer consumes 3 to 5 watts.
✔Consider 37MR16/HIR/C/FL.

Application Key

Commercial
Gallery
Hospitality
Institutional
Manufacturing
Residential
Retail
Exterior

bold = primary application
partial fade = minimal application
fade = unlikely application

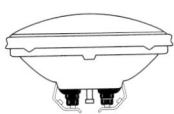

Halogen/PAR36

Screw Terminal Base

Halogen
Low Voltage (12V)

4

Guide
- Longer life is better
- Higher efficacy better for general lighting
- Lower wattage is better
- Lower color of light is warmer
- Moderate cost range for halogen is about US$12.⁰⁰ to US$17.⁰⁰

Stats
Data varies manufacturer to manufacturer and changes from time to time

Life/2000 to 4000 hours
Varies manufacturer to manufacturer. A typical office environment might be occupied about 2600 hours each year.

Efficacy/Not reported
Excellent beam intensities and distributions available for the wattage.

Wattages/35, 36 and 50

Color of Light/3050°K
Varies manufacturer to manufacturer and wattage to wattage. Typically the longer lived lamps have slightly lower color temperature and the higher wattage lamps have slightly higher color temperature.

Beam spreads/VNSP, NSP and NFL
Very narrow spot, narrow spot and narrow flood distributions are available.

Size/2¾″ L by 4½″ Ø

Voltage/12V

Cost Magnitude/Moderate (2000)

Manufacturers
GE
Osram Sylvania
Philips

Net Addresses
http://www.ge.com/lighting/business/index.htm
http://ecom.sylvania.com/osicatalog/
http://www.lighting.philips.com/

CONNECT FOR MORE

Advantages
Crisp white light
Very concentrated pinspot (on VNSP) unavailable in other lamps
Easily dimmed
No/Low Glare
- ◆Filament cap results in very little glare from most viewing angles
Relatively low initial cost
Withstand extreme temperature range
Very little lamp lumen depreciation (light loss) over life

Disadvantages
Beam aberrations
- ◆Striations
- ◆Inconsistent beam shape
Asymmetrical beam shape
- ◆Tends to be oblong beam spread
Hot to touch
Hum
- ◆Dimming especially causes audible filament hum or "singing"
Large size
Transformer required
- ◆Additional cost and visual aspect of transformer must be accommodated
- ◆Additional wattage requirement of transformer of 3 to 5 watts
- ◆Potential for audible hum

Halogen/PAR36

Screw Terminal Base

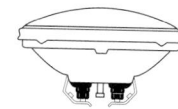

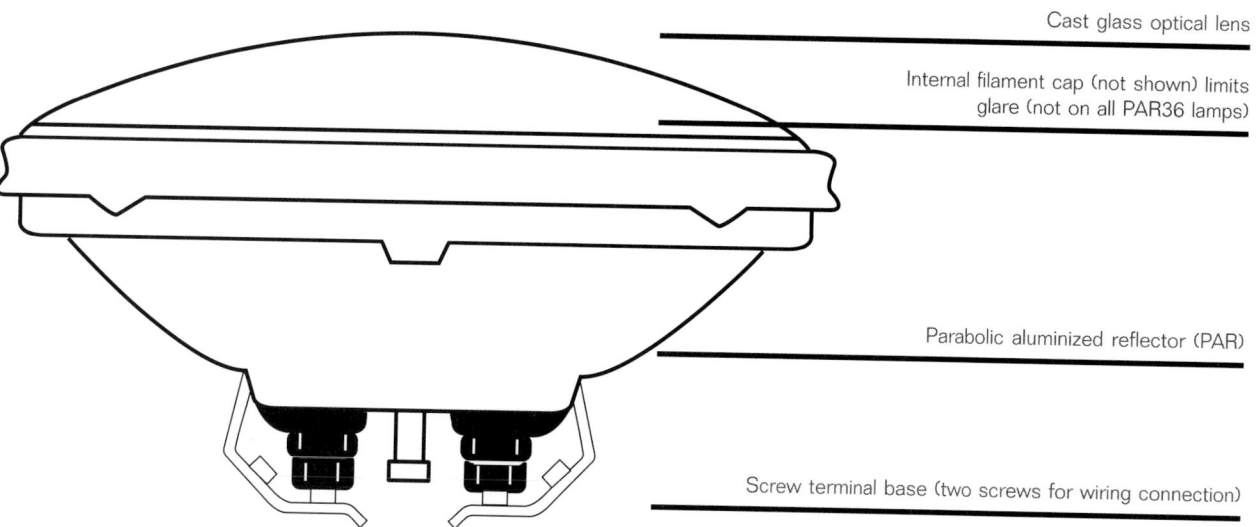

Cast glass optical lens

Internal filament cap (not shown) limits glare (not on all PAR36 lamps)

Parabolic aluminized reflector (PAR)

Screw terminal base (two screws for wiring connection)

Lamp shown actual size
Image copyright 2000/GE

Uses

Architectural Accenting
Art/Centerpiece Accenting

Commercial
Gallery
Hospitality
Residential

Merchandising

Feature display

PAR36 Lighting Design Tips

✔ Dimming increases life (typically doubling life), but shifts color warmer and causes filament hum on PAR36.

✔ Beam striations are most apparent on light–colored, nontextured flat surfaces and may be objectionable.

✔ Screw terminal base may create maintenance issues—look for luminaires with convenient "snap–in" socket.

✔ When using multiple lamps, aiming may be a challenge given oblong beam spreads.

Net Addresses/PAR36 Track and Monopoints

(see page 4–171 for Recessed)

http://www.cooperlighting.com/
http://www.hubbell-ltg.com/products.htm#Down&Track
http://www.danalite.com/
http://www.lightingservicesinc.com/
http://www.lightolier.com/
http://www.lithonia.com/
http://www.prescolite.com/
http://www.tslight.com/

CONNECT FOR MORE

Halogen/PAR36

Screw Terminal Base

Halogen Low Voltage (12V)

4

Halogen/PAR36
Screw Terminal Base

Figure 4.6 (left)

Low voltage halogen 50PAR36/H/VNSP5° lamps are used to highlight display cases and a centerpiece in one of the boutiques in a high ceiling retail space. The intense punch of light coupled with relatively long life and low wattage make this an excellent choice for accenting where ceiling heights are greater than 15 feet. See Figure C40 also. Adjustable accent lights are Kurt Versen.

Figure 4.7 (below)

A view of the main sales area shows the recessed round adjustable accents in this relatively high ceiling retail space. The VNSP5° lamps are aimed onto selected display vignettes on the display wall to the left. See also Figures C38 and C39 in the color section. Adjustable accent lights are Kurt Versen.

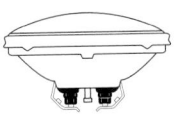

Halogen
Low Voltage (12V)

4

Halogen/PAR36
Screw Terminal Base

Figure 4.8
A low voltage halogen 35PAR36/H/NSP lamp is used in a single accent aimed onto the artwork. This lamp is permanently dimmed and thereby creates the aura that the table light in the Rockwell painting is itself electrified and providing light! Adjustable accent is Edison Price.

Image copyright 1999 Vance Roth/VRA PHotography

Halogen/PAR36

PAR36 Screw Terminal Base Visual Index

35PAR36/H/VNSP ▶p172 | 35PAR36/H/VNSP ▶p172 | 35PAR36/H/VNSP ▶p173

35PAR36/H/VNSP GE
▶Architectural detailing
▶Petite art/19′ ceiling height or less
▶Objet d'art
▶Long throws
▶Hospitality

36PAR36/H/VNSP ▶p174 | 36PAR36/H/VNSP ▶p174 | 36PAR36/H/VNSP ▶p174

36PAR36/H/VNSP Osram Sylvania
▶Architectural detailing
▶Petite art/16′ ceiling height or less
▶Objet d'art
▶Long throws
▶Hospitality

35PAR36/H/NSP ▶p176 | 35PAR36/H/NSP ▶p176 | 35PAR36/H/NSP ▶p177

35PAR36/H/NSP
▶Architectural detailing
▶Petite art/19′ ceiling height or less
▶Objet d'art
▶Long throws
▶Hospitality

35PAR36/H/SP ▶p178 | 35PAR36/H/SP ▶p178 | 35PAR36/H/SP ▶p178

36PAR36/H/SP
▶Architectural accenting
▶Petite art/11′ ceiling height or less
▶Objet d'art
▶Medium throws
▶Commercial, Hospitality, Retail

35PAR36/H/NFL ▶p179 | 35PAR36/H/NFL ▶p179 | 35PAR36/H/NFL ▶p179

35PAR36/H/NFL
▶Soft general lighting
▶Soft accenting
▶Medium art/9′ ceiling height or less
▶Gallery, Residential

Throws/Ceilings
⊃Short = 7′ to 10′
⊃Medium = 10′ to 15′
⊃Long = 15′ to 20′
⊃Very Long = 20′ or more

Halogen/PAR36

Screw Terminal Base

50PAR36/H/VNSP▶p180   50PAR36/H/VNSP▶p180   50PAR36/H/VNSP▶p181

50PAR36/H/VNSP5°
▶Architectural detailing
▶Petite art/21′ ceiling height or less
▶Objet d'art
▶Very long throws
▶Commercial, Hospitality, Retail

Throws/Ceilings
◗Short = 7′ to 10′
◗Medium = 10′ to 15′
◗Long = 15′ to 20′
◗Very Long = 20′ or more

50PAR36/H/VNSP▶p182   50PAR36/H/VNSP▶p182   50PAR36/H/VNSP▶p182

50PAR36/H/VNSP5.5°
▶Architectural detailing
▶Petite art/21′ ceiling height or less
▶Objet d'art
▶Very long throws
▶Commercial, Retail

50PAR36/H/VNSP6°
▶Architectural detailing
▶Petite art/21′ ceiling height or less
▶Objet d'art
▶Very long throws
▶Hospitality

50PAR36/H/VNSP▶p184   50PAR36/H/VNSP▶p184   50PAR36/H/VNSP▶p185

Recommended intensities
for art accenting

Low (gallery, museum, residential)
5 to 10 fc average on art
Moderate (hospitality)
10 to 20 fc average on art
High (commercial—or higher for retail)
20 to 30 fc average on art

These are intended to be average, maintained values including effects of accent lighting and general room lighting. General room lighting contributes just a few footcandles in residential, gallery and museum settings, but contributes 5 to 10 fc in commercial settings. Maximums might range from 50 percent to 100 percent of the average value. Wherever artwork is highly valuable and/or sensitive to light, precautions should be taken to limit light exposure (maintaining low levels and/or switching lights off when art is not being viewed); and exposure to infrared and ultraviolet should be limited (using filters on lamps or placing art behind specially treated glass or acrylic).

50PAR36/H/NSP▶p186   50PAR36/H/NSP▶p186   50PAR36/H/NSP▶p187

50PAR36/H/NSP
▶Architectural detailing
▶Petite art/21′ ceiling height or less
▶Objet d'art
▶Very long throws
▶Commercial, Hospitality, Retail

Halogen/PAR36

Screw Terminal Base

Halogen Low Voltage (12V)

4

  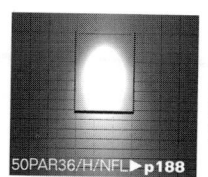

50PAR36/H/NFL▶**p188**   50PAR36/H/NFL▶**p188**   50PAR36/H/NFL▶**p188**

50PAR36/H/NFL
▶Soft general lighting
▶Medium art/9' ceiling height or less
▶Short throws
▶Residential

Net Addresses/PAR36 Recessed

http://www.cooperlighting.com/
http://www.danalite.com/
http://www.kramerlighting.com/
http://www.lightolier.com/
http://www.lithonia.com/
http://www.prescolite.com/

CONNECT FOR MORE

Halogen Low Voltage PAR Lamp Designation

35PAR36/H/NFL

Last set of letters identifies beam spread

FLOODS
- 24° to 32° for narrow flood (NFL)
- 33° to 44° for flood (FL)
- 45° to 54° for wide flood (WFL)
- 55° or greater for very wide flood (VWFL)

SPOTS
- 7° or less for very narrow spot (VNSP)
- 8° to 10° for narrow spot (NSP)
- 11° to 14° for spot (SP)
- 15° to 18° for wide spot (WSP)
- 19° to 23° for very wide spot (VWSP)

Ranges of beam spreads used for convenient/consistent reference in this text and do not necessarily correspond to each manufacturer's nor any ANSI or NEMA definitions.

This letter indicates "Halogen"

These two digits represent lamp diameter in eighths inches (51mm or 2 inches)

These letters indicate this is a multifaceted reflector (MR) lamp

First two digits represent lamp wattage

[This designation used for convenient reference throughout this *Time–Saver Standards for Architectural Lighting*—but not necessarily used by all lamp manufacturers (although it would be convenient if manufacturers would come to agreement on standard designations.]

▪ ▪ ▪ ▪ ▪ ▪ ▪ ▪ ▪ ▪

Artwork Size Categories
⊃ Petite = 2'–0" by 2'–0" or smaller
⊃ Small = 2'–0" by 2'–0" to 2'–6" by 3'–0"
⊃ Medium = 2'–6" by 3'–0" to 3'–0" by 4'–0"
⊃ Large = 3'–0" by 4'–0" to 4'–0" by 5'–0"
⊃ Extra Large = 4'–0" by 5'–0" or larger

Halogen 35PAR36/H/VNSP (GE)

Screw Terminal Base

Performance Sketches

[All data outlined here is based on information from boldfaced manufacturer's published data for 12V version at time of manuscript preparation. Scale is ¼ inch equals 1 foot.]

Halogen
Low Voltage (12V)

4

Uses
► Architectural detailing
► Petite art with long throws
► Objet d'art with long throws
► Long throws
► Available in 12V

Specs
Beam spread/5°
Center beam
  candlepower/25000
Life/4000 hrs
►**GE** (19873)

Design Tips
✓ One light best for petite art and where mounting height for light is 19′ or less.
✓ These are strong accents and deserve judicious use. Light may degrade artwork rapidly and/or significantly: consider UV and IR filters or limit exposure time.
✓ Excellent for glass pieces and floral centerpieces and where mounting height for light is 19′ or less.
✓ Scallops, striations and frame shadows are most severe when light is closer to the wall (compare top left image to bottom image).
✓ PAR36 lamps are typically low–glare because of the filament cap.
✓ Lamps may hum when dimmed.
✓ Required transformer consumes 3 to 5 watts.

Application Key

Commercial
Gallery
Hospitality
Institutional
Manufacturing
Residential
Retail
Exterior

bold = primary application
partial fade = minimal application
fade = unlikely application

Lamp: 35PAR36/H/VNSP (GE)
Location: 5′–0″ from wall
Center beam Aiming: 20° (at point 4′–0″ AFF)
Luminaire: Monopoint or Recessed Adjustable
Illuminance: 14 fc avg, maintained (35 fc max)

Lamp: 35PAR36/H/VNSP (GE)
Location: 6′–0″ from wall
Center beam Aiming: 24° (at point 4′–6″ AFF)
Luminaire: Monopoint or Recessed Adjustable
Illuminance: 17 fc avg, maintained (40 fc max)

18′–0″

10′–0″

Halogen35PAR36/H/VNSP(GE)
Screw Terminal Base

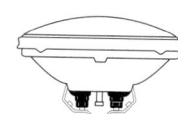

Lamp: 35PAR36/H/VNSP (GE)
Location: 7'–0" from wall
Center beam Aiming: 28° (at point 5'–0" AFF)
Luminaire: Monopoint or Recessed Adjustable
Illuminance: 21 fc avg, maintained (46 fc max)

Art shown at 2' by 2' and
centered 5'–6" AFF. Wall grid on
6" by 6" centers and scaled at
¼"=1'.

Halogen **36PAR36/H/VNSP** (Osram Sylvania)

Screw Terminal Base

Halogen Low Voltage (12V)

Uses
▶ Architectural detailing
▶ Petite art with long throws
▶ Objet d'art with long throws
▶ Long throws
▶ Available in 12V

Specs
Beam spread/5°
Center beam
  candlepower/17000
Life/4000 hrs
Wattage/36 (not 35)
▶ **Osram Sylvania** (55100)

Design Tips
✔ One light best for petite art and where mounting height for light is 16' or less.
✔ These are strong accents and deserve judicious use. Light may degrade artwork rapidly and/or significantly: consider UV and IR filters or limit exposure time.
✔ Excellent for glass pieces and floral centerpieces and where mounting height for light is 16' or less.
✔ Scallops, striations and frame shadows are most severe when light is closer to the wall (compare top left image to bottom image).
✔ PAR36 lamps are typically low–glare because of the filament cap.
✔ Lamps may hum when dimmed.
✔ Required transformer consumes 3 to 5 watts.

Application Key

Commercial
Gallery
Hospitality
Institutional
Manufacturing
Residential
Retail
Exterior

bold = primary application
partial fade = minimal application
fade = unlikely application

Performance Sketches

[All data outlined here is based on information from boldfaced manufacturer's published data for 12V version at time of manuscript preparation. Scale is ¼ inch equals 1 foot. **Note: At press time, actual photometry was unavailable for this lamp. Data were generated using GE 35PAR36/H/VNSP lamp and rerating candlepower—this is imprecise.**]

Lamp: 36PAR36/H/VNSP (Osram Sylvania/36w)
Location: 4'–0" from wall
Center beam Aiming: 21° (at point 4'–5" AFF)
Luminaire: Monopoint or Recessed Adjustable
Illuminance: 13 fc avg, maintained (42 fc max)

Lamp: 36PAR36/H/VNSP (Osram Sylvania/36w)
Location: 5'–0" from wall
Center beam Aiming: 27° (at point 5'–0" AFF)
Luminaire: Monopoint or Recessed Adjustable
Illuminance: 17 fc avg, maintained (52 fc max)

Art shown at 2' by 2' and centered 5'–6" AFF. Wall grid on 6" by 6" centers and scaled at ¼"=1'.

Lamp: 36PAR36/H/VNSP (Osram Sylvania/36w)
Location: 6'–0" from wall
Center beam Aiming: 32° (at point 5'–6" AFF)
Luminaire: Monopoint or Recessed Adjustable
Illuminance: 19 fc avg, maintained (61 fc max)

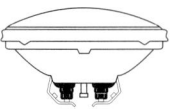

Halogen 35PAR36/H/NSP

Screw Terminal Base

Halogen Low Voltage (12V)

4

Uses
▶ Architectural detailing
▶ Petite art with long throws
▶ Objet d'art with long throws
▶ Long throws
▶ Available in 12V

Specs
Beam spread/8°
Center beam
  candlepower/19700
Life/4000 hrs
▶ **GE** (19876)

Design Tips
✔ One light best for petite art and where mounting height for light is 19' or less.
✔ These are strong accents and deserve judicious use. Light may degrade artwork rapidly and/or significantly: consider UV and IR filters or limit exposure time.
✔ Excellent for glass pieces and floral centerpieces and where mounting height for light is 19' or less.
✔ Scallops, striations and frame shadows are most severe when light is closer to the wall (compare top left image to bottom image).
✔ PAR36 lamps are typically low–glare because of the filament cap.
✔ Lamps may hum when dimmed.
✔ Required transformer consumes 3 to 5 watts.

Application Key

Commercial
Gallery
Hospitality
Institutional
Manufacturing
Residential
Retail
Exterior

bold = primary application
partial fade = minimal application
fade = unlikely application

Performance Sketches

[All data outlined here is based on information from boldfaced manufacturer's published data for 12V version at time of manuscript preparation. Scale is ¼ inch equals 1 foot.]

Lamp: 35PAR36/H/NSP
Location: 5'–0" from wall
Center beam Aiming: 21° (at point 5'–0" AFF)
Luminaire: Monopoint or Recessed Adjustable
Illuminance: 14 fc avg, maintained (32 fc max)

Lamp: 35PAR36/H/NSP
Location: 6'–0" from wall
Center beam Aiming: 27° (at point 5'–6" AFF)
Luminaire: Monopoint or Recessed Adjustable
Illuminance: 15 fc avg, maintained (39 fc max)

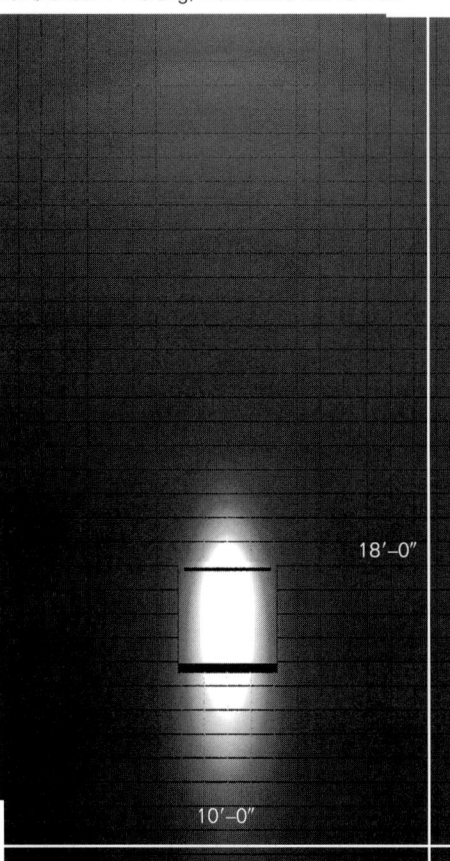

Halogen 35PAR36/H/NSP

Screw Terminal Base

Lamp: 35PAR36/H/NSP
Location: 7'–0" from wall
Center beam Aiming: 29° (at point 5'–6" AFF)
Luminaire: Monopoint or Recessed Adjustable
Illuminance: 16 fc avg, maintained (41 fc max)

Art shown at 2' by 2' and
centered 5'–6" AFF. Wall grid on
6" by 6" centers and scaled at
¼"=1'.

Halogen

Halogen 36PAR36/H/SP

Screw Terminal Base

Halogen Low Voltage (12V)

4

Uses
▶ Architectural accenting
▶ Petite art with medium throws
▶ Objet d'art
▶ Medium throws
▶ Available in 12V

Specs
Beam spread/13°
Center beam candlepower/ 3500
Life/4000 hrs
Wattage/36 (not 35)
▶ **Osram Sylvania** (55090)

Design Tips
✔ One light best for petite art and where mounting height for light is 11′ or less
✔ Light may degrade artwork rapidly and/or significantly: consider UV and IR filters or limit exposure time.
✔ More grazing light creates prominent shadows.
✔ PAR36 lamps are typically low–glare because of the filament cap.
✔ Lamps may hum when dimmed.
✔ Required transformer also consumes 3 to 5 watts.

Performance Sketches

[All data outlined here is based on information from boldfaced manufacturer's published data for 12V version at time of manuscript preparation. Scale is ¼ inch equals 1 foot. **Note: At press time, actual photometry was unavailable for this lamp. Data were generated using GE 80PAR38/HIR/SP lamp and rerating candlepower—this is imprecise.**]

Lamp: 36PAR36/H/SP
Location: 2′–0″ from wall
Center beam Aiming: 17° (at point 4′–0″ AFF)
Luminaire: Monopoint or Recessed Adjustable
Illuminance: 18 fc avg, maintained (44 fc max)

Lamp: 36PAR36/H/SP
Location: 2′–6″ from wall
Center beam Aiming: 23° (at point 4′–6″ AFF)
Luminaire: Monopoint or Recessed Adjustable
Illuminance: 22 fc avg, maintained (54 fc max)

10′–6″

10′–0″

Art shown at 2′ by 2′ and centered 5′–6″ AFF. Wall grid on 6″ by 6″ centers and scaled at ¼″=1′.

Lamp: 35PAR36/H/SP
Location: 3′–0″ from wall
Center beam Aiming: 29° (at point 5′–0″ AFF)
Luminaire: Monopoint or Recessed Adjustable
Illuminance: 27 fc avg, maintained (66 fc max)

Application Key

Commercial
Gallery
Hospitality
Institutional
Manufacturing
Residential
Retail
Exterior

bold = primary application
partial fade = minimal application
fade = unlikely application

Halogen 35PAR36/H/NFL

Screw Terminal Base

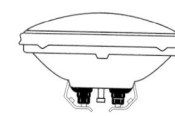

Performance Sketches

[All data outlined here is based on information from boldfaced manufacturer's published data for 12V version at time of manuscript preparation. Scale is ¼ inch equals 1 foot.]

Art shown at 3' by 4' and centered 5'–6" AFF. Wall grid on 6" by 6" centers and scaled at ¼"=1'.

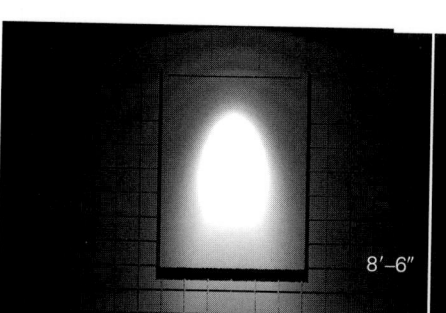

Lamp: 35PAR36/H/NFL
Location: 1'–6" from wall
Center beam Aiming: 15° (at point 3'–0" AFF)
Luminaire: Monopoint or Recessed Adjustable
Illuminance: 7 fc avg, maintained (41 fc max)

Lamp: 35PAR36/H/NFL
Location: 2'–0" from wall
Center beam Aiming: 22° (at point 3'–6" AFF)
Luminaire: Monopoint or Recessed Adjustable
Illuminance: 8 fc avg, maintained (39 fc max)

Lamp: 35PAR36/H/NFL
Location: 2'–6" from wall
Center beam Aiming: 29° (at point 4'–0" AFF)
Luminaire: Monopoint or Recessed Adjustable
Illuminance: 9 fc avg, maintained (39 fc max)

Halogen Low Voltage (12V)

4

Uses
► Soft general lighting
► Soft accenting
► Medium art with short throws
► Short throws
► Available in 12V

Specs
Beam spread/31°
Center beam candlepower/ 950
Life/4000 hours
► **GE** (19877)
► Osram Sylvania (36 watts)

Design Tips
✔ One light best for medium art where mounting height for light is 9' or less.
✔ Good for soft downlighting: best with sconces and/or art accents and/or wallwashing to avoid cave effect.
✔ More grazing light creates prominent shadows.
✔ PAR36 lamps are typically low–glare because of the filament cap.
✔ Lamps may hum when dimmed.
✔ Striations on light colored surfaces are likely.
✔ Required transformer also consumes 3 to 5 watts.

Application Key

Commercial
Gallery
Hospitality
Institutional
Manufacturing
Residential
Retail
Exterior

bold = primary application
partial fade = minimal application
fade = unlikely application

Halogen 50PAR36/H/VNSP5°

Screw Terminal Base

Performance Sketches

[All data outlined here is based on information from boldfaced manufacturer's published data for 12V version at time of manuscript preparation. Scale is ¼ inch equals 1 foot.]

Uses

▶ Architectural detailing
▶ Petite art with very long throws
▶ Objet d'art with very long throws
▶ Very long throws
▶ Available in 12V

Specs

Beam spread/5°
Center beam candlepower/ 40000
Life/4000 hrs
▶**GE** (19878)

Design Tips

✔ One light best for petite art and where mounting height for light is 21′ or less.
✔ These are strong accents and deserve judicious use. Light may degrade artwork rapidly and/or significantly: consider UV and IR filters or limit exposure time.
✔ Excellent for glass pieces and floral centerpieces and where mounting height for light is 21′ or less.
✔ Scallops, striations and frame shadows are most severe when light is closer to the wall (compare top left image to bottom image).
✔ PAR36 lamps are typically low–glare because of the filament cap.
✔ Lamps may hum when dimmed.
✔ Required transformer consumes 3 to 5 watts.

Application Key

Commercial
Gallery
Hospitality
Institutional
Manufacturing
Residential
Retail
Exterior

bold = primary application
partial fade = minimal application
fade = unlikely application

Lamp: 50PAR36/H/VNSP5°
Location: 6′–0″ from wall
Center beam Aiming: 21° (at point 4′–0″ AFF)
Luminaire: Monopoint or Recessed Adjustable
Illuminance: 16 fc avg, maintained (38 fc max)

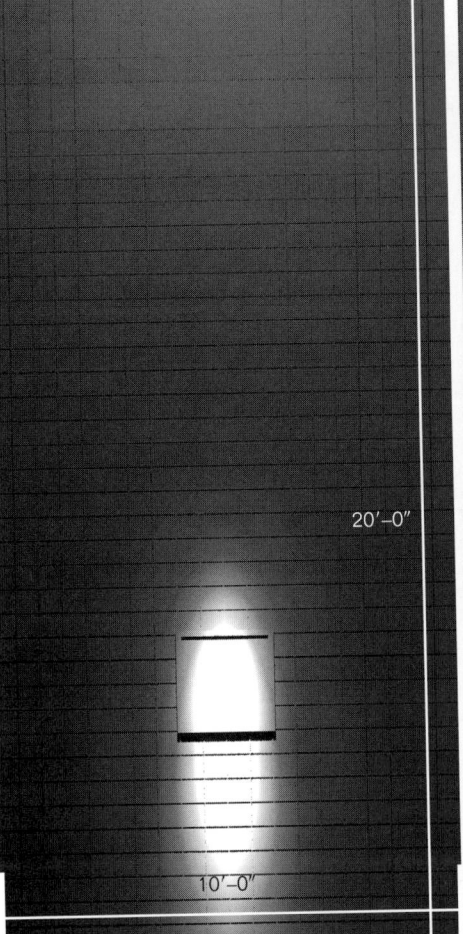

Lamp: 50PAR36/H/VNSP5°
Location: 7′–0″ from wall
Center beam Aiming: 24° (at point 4′–6″ AFF)
Luminaire: Monopoint or Recessed Adjustable
Illuminance: 20 fc avg, maintained (44 fc max)

Halogen 50PAR36/H/VNSP5°

Screw Terminal Base

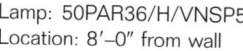

Lamp: 50PAR36/H/VNSP5°
Location: 8'–0" from wall
Center beam Aiming: 28° (at point 5'–0" AFF)
Luminaire: Monopoint or Recessed Adjustable
Illuminance: 23 fc avg, maintained (49 fc max)

Art shown at 2' by 2' and
centered 5'–6" AFF. Wall grid on
6" by 6" centers and scaled at
¼"=1'.

Halogen Low Voltage (12V)

4

Halogen (12V) Low Voltage

Halogen 50PAR36/H/VNSP5.5°

Screw Terminal Base

Performance Sketches

[All data outlined here is based on information from boldfaced manufacturer's published data for 12V version at time of manuscript preparation. Scale is ¼ inch equals 1 foot.]

Uses
▶ Architectural detailing
▶ Petite art with very long throws
▶ Objet d'art with very long throws
▶ Very long throws
▶ Available in 12V

Specs
Beam spread/5.5°
Center beam candlepower/ 50000
Life/2000 hrs
▶ **Philips** (31170–4)

Design Tips
✔ One light best for petite art and where mounting height for light is 21′ or less.
✔ These are strong accents and deserve judicious use. Light may degrade artwork rapidly and/or significantly: consider UV and IR filters or limit exposure time.
✔ Excellent for glass pieces and floral centerpieces and where mounting height for light is 21′ or less.
✔ Scallops, striations and frame shadows are most severe when light is closer to the wall (compare top left image to bottom image).
✔ PAR36 lamps are typically low–glare because of the filament cap.
✔ Lamps may hum when dimmed.
✔ Required transformer consumes 3 to 5 watts.

Application Key

Commercial
Gallery
Hospitality
Institutional
Manufacturing
Residential

Retail

Exterior

bold = primary application
partial fade = minimal application
fade = unlikely application

Lamp: 50PAR36/H/VNSP5.5°
Location: 6′–0″ from wall
Center beam Aiming: 21° (at point 4′–0″ AFF)
Luminaire: Monopoint or Recessed Adjustable
Illuminance: 28 fc avg, maintained (51 fc max)

Lamp: 50PAR36/H/VNSP5.5°
Location: 7′–0″ from wall
Center beam Aiming: 24° (at point 4′–6″ AFF)
Luminaire: Monopoint or Recessed Adjustable
Illuminance: 34 fc avg, maintained (61 fc max)

Halogen50PAR36/H/VNSP5.5°

Screw Terminal Base

Lamp: 50PAR36/H/VNSP5.5°
Location: 8'–0" from wall
Center beam Aiming: 28° (at point 5'–0" AFF)
Luminaire: Monopoint or Recessed Adjustable
Illuminance: 39 fc avg, maintained (68 fc max)

Art shown at 2' by 2' and
centered 5'–6" AFF. Wall grid on
6" by 6" centers and scaled at
¼"=1'.

Halogen
Low Voltage (12V)

4

Halogen 50PAR36/H/VNSP6°

Screw Terminal Base

Uses
▶ Architectural detailing
▶ Petite art with very long throws
▶ Objet d'art with very long throws
▶ Very long throws
▶ Available in 12V

Specs
Beam spread/6°
Center beam candlepower/ 25000
Life/4000 hrs
▶**Osram Sylvania** (55118)

Design Tips
✔ One light best for petite art and where mounting height for light is 21′ or less.
✔ These are strong accents and deserve judicious use. Light may degrade artwork rapidly and/or significantly: consider UV and IR filters or limit exposure time.
✔ Excellent for glass pieces and floral centerpieces and where mounting height for light is 21′ or less.
✔ Scallops, striations and frame shadows are most severe when light is closer to the wall (compare top left image to bottom image).
✔ PAR36 lamps are typically low-glare because of the filament cap.
✔ Lamps may hum when dimmed.
✔ Required transformer consumes 3 to 5 watts.

Application Key

Commercial
Gallery
Hospitality
Institutional
Manufacturing
Residential
Retail
Exterior

bold = primary application
partial fade = minimal application
fade = unlikely application

Performance Sketches

[All data outlined here is based on information from boldfaced manufacturer's published data for 12V version at time of manuscript preparation. Scale is ¼ inch equals 1 foot. **Note: At press time, actual photometry was unavailable for this lamp. Data were generated using GE 50PAR36/H/VNSP lamp and rerating candlepower—this is imprecise.**]

Lamp: 50PAR36/H/VNSP6°
Location: 6′–0″ from wall
Center beam Aiming: 21° (at point 4′–0″ AFF)
Luminaire: Monopoint or Recessed Adjustable
Illuminance: 13 fc avg, maintained (31 fc max)

Lamp: 50PAR36/H/VNSP6°
Location: 7′–0″ from wall
Center beam Aiming: 24° (at point 4′–6″ AFF)
Luminaire: Monopoint or Recessed Adjustable
Illuminance: 16 fc avg, maintained (35 fc max)

Halogen 50PAR36/H/VNSP6°
Screw Terminal Base

Lamp: 50PAR36/H/VNSP6°
Location: 8'–0" from wall
Center beam Aiming: 28° (at point 5'–0" AFF)
Luminaire: Monopoint or Recessed Adjustable
Illuminance: 19 fc avg, maintained (39fc max)

Art shown at 2' by 2' and
centered 5'–6" AFF. Wall grid on
6" by 6" centers and scaled at
¼"=1'.

Halogen
Low Voltage (12V)

4

Halogen 50PAR36/H/NSP

Screw Terminal Base

Halogen Low Voltage (12V)

4

Uses
▶ Architectural detailing
▶ Petite art with very long throws
▶ Objet d'art with very long throws
▶ Very long throws
▶ Available in 12V

Specs
Beam spread/8°
Center beam candlepower/ 30000
Life/4000 hrs
▶**GE** (19879)

Design Tips
✔ One light best for petite art and where mounting height for light is 21′ or less.
✔ These are strong accents and deserve judicious use. Light may degrade artwork rapidly and/or significantly: consider UV and IR filters or limit exposure time.
✔ Excellent for glass pieces and floral centerpieces and where mounting height for light is 21′ or less.
✔ Scallops, striations and frame shadows are most severe when light is closer to the wall (compare top left image to bottom image).
✔ PAR36 lamps are typically low–glare because of the filament cap.
✔ Lamps may hum when dimmed.
✔ Required transformer consumes 3 to 5 watts.

Application Key

Commercial
Gallery
Hospitality
Institutional
Manufacturing
Residential
Retail
Exterior

bold = primary application
partial fade = minimal application
fade = unlikely application

Performance Sketches

[All data outlined here is based on information from boldfaced manufacturer's published data for 12V version at time of manuscript preparation. Scale is ¼ inch equals 1 foot.]

Lamp: 50PAR36/H/NSP
Location: 6′–0″ from wall
Center beam Aiming: 21° (at point 4′–0″ AFF)
Luminaire: Monopoint or Recessed Adjustable
Illuminance: 17 fc avg, maintained (33 fc max)

Lamp: 50PAR36/H/NSP
Location: 7′–0″ from wall
Center beam Aiming: 24° (at point 4′–6″ AFF)
Luminaire: Monopoint or Recessed Adjustable
Illuminance: 21 fc avg, maintained (38 fc max)

20′–0″

10′–0″

Halogen 50PAR36/H/NSP
Screw Terminal Base

Lamp: 50PAR36/H/NSP
Location: 8'–0" from wall
Center beam Aiming: 28° (at point 5'–0" AFF)
Luminaire: Monopoint or Recessed Adjustable
Illuminance: 24 fc avg, maintained (43 fc max)

Art shown at 2' by 2' and
centered 5'–6" AFF. Wall grid on
6" by 6" centers and scaled at
¼"=1'.

Halogen 50PAR36/H/NFL

Screw Terminal Base

Halogen
Low Voltage (12V)

4

Uses
▶ Soft general lighting
▶ Medium art with short throws
▶ Short throws
▶ Available in 12V

Specs
Beam spread/30°
Center beam candlepower/
1300
Life/4000 hrs
▶ **GE** (19880)

Design Tips
✔ One light best for medium art where mounting height for light is 9' or less.
✔ Good for soft downlighting: best with sconces and/or art accents and/or wallwashing to avoid cave effect.
✔ More grazing light creates prominent shadows.
✔ PAR36 lamps are typically low–glare because of the filament cap.
✔ Lamps may hum when dimmed.
✔ Striations on light colored surfaces are likely.
✔ Required transformer also consumes 3 to 5 watts.

Application Key

Commercial
Gallery
Hospitality
Institutional
Manufacturing
Residential
Retail
Exterior

bold = primary application
partial fade = minimal application
fade = unlikely application

Performance Sketches

[All data outlined here is based on information from boldfaced manufacturer's published data for 12V version at time of manuscript preparation. Scale is ¼ inch equals 1 foot.]

Art shown at 3' by 4' and centered 5'–6" AFF. Wall grid on 6" by 6" centers and scaled at ¼"=1'.

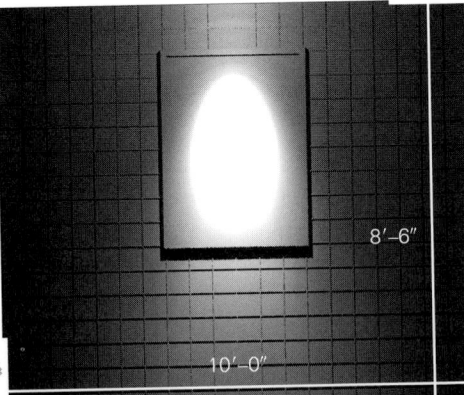

8'–6"

10'–0"

Lamp: 50PAR36/H/NFL
Location: 1'–6" from wall
Center beam Aiming: 15° (at point 3'–0" AFF)
Luminaire: Monopoint or Recessed Adjustable
Illuminance: 12 fc avg, maintained (50 fc max)

Lamp: 50PAR36/H/NFL
Location: 2'–0" from wall
Center beam Aiming: 20° (at point 3'–0" AFF)
Luminaire: Monopoint or Recessed Adjustable
Illuminance: 12 fc avg, maintained (40 fc max)

Lamp: 50PAR36/H/NFL
Location: 2'–6" from wall
Center beam Aiming: 29° (at point 4'–0" AFF)
Luminaire: Monopoint or Recessed Adjustable
Illuminance: 14 fc avg, maintained (48 fc max)

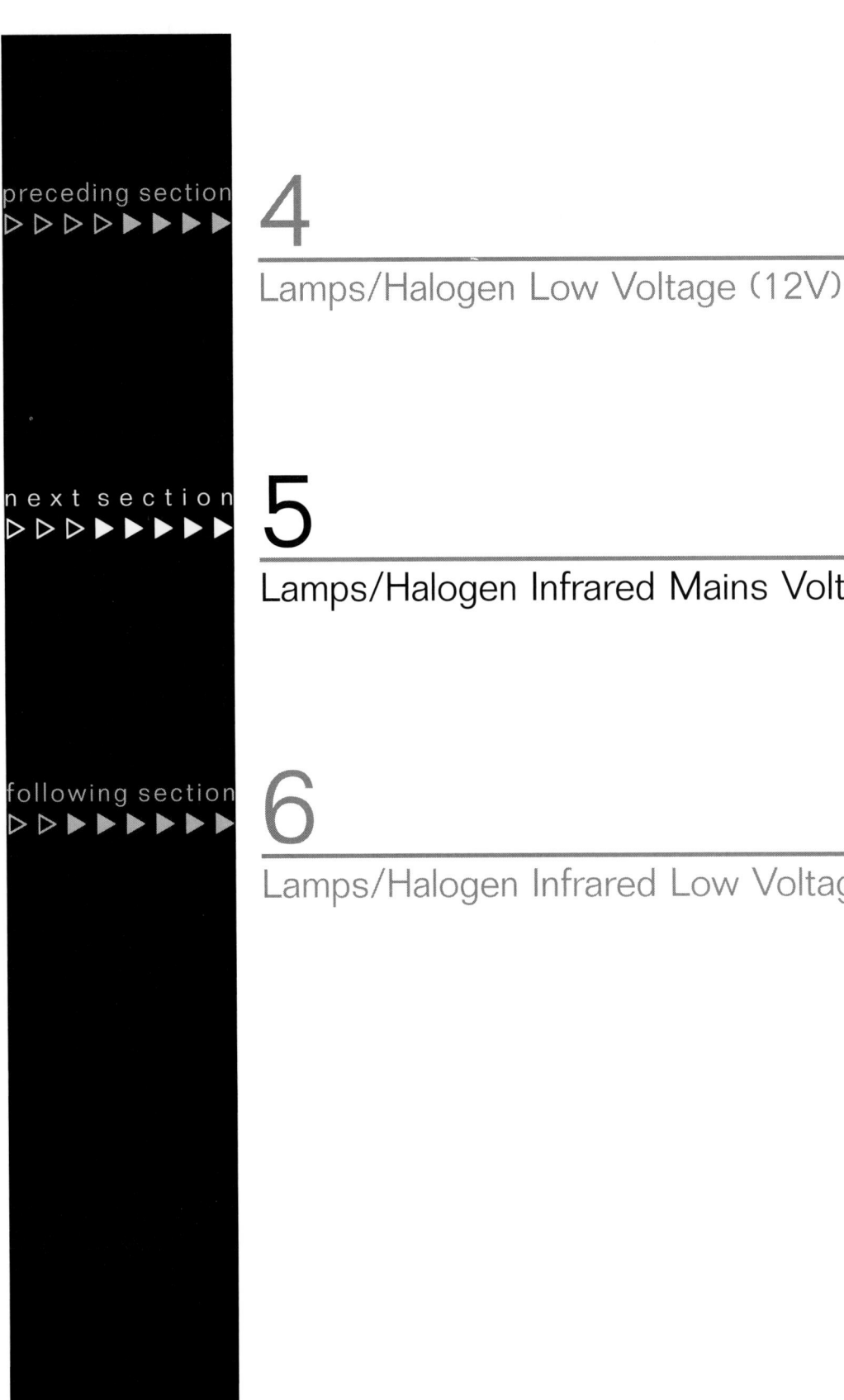

preceding section
▷ ▷ ▷ ▷ ▶ ▶ ▶ ▶ ▶

4

Lamps/Halogen Low Voltage (12V)

next section
▷ ▷ ▷ ▶ ▶ ▶ ▶ ▶ ▶

5

Lamps/Halogen Infrared Mains Voltage (120V)

following section
▷ ▷ ▶ ▶ ▶ ▶ ▶ ▶ ▶

6

Lamps/Halogen Infrared Low Voltage (12V)

HIR Lamps
Mains Voltage (120V)

5

This section addresses 120–volt (mains voltage) halogen infrared incandescent lamps. Halogen infrared (aka halogen IR or HIR) lamps are the most efficacious incandescent lamps available today for general architectural use. There are also low voltage (12–volt) HIR lamps (see *HIR Lamps—Low Voltage (12V)* in Section 6). In addition to being more efficacious, HIR mains voltage and HIR low voltage lamps are typically longer lived than halogen lamps.

Where incandescent lamps are a necessity, use HIR mains voltage or HIR low voltage lamps. This will result in the most efficient incandescent lighting application.

HIR Mains Voltage (120V)

5

HIR Lamps
Mains Voltage (120V)

HIR lamps are the most efficacious incandescent lamps available for general architectural applications.

HIR

Mains Voltage (120V)

5

Halogen infrared lamps are an extension of the incandescent lamp family. HIR lamps are more efficacious than standard halogen and in some instances are longer lived. An infrared coating on the filament capsule helps trap infrared radiation (heat) and reflect it back onto the filament. This additional heat causes the filament to get even hotter and thereby radiate visible radiation—all without using any additional electrical power. Therefore, if an incandescent solution has been deemed necessary for either focal accenting, for initial cost and/or for ease of dimming, then consideration should be given to using HIR lamps. HIR mains voltage lamps come in a variety of bulb envelope styles—linear and PAR. The PAR lamps are most appropriate for many interior applications and Table 5.1 outlines currently available HIR/PAR lamps. For even greater efficacy, compact fluorescent lamps offer significant energy and life cycle cost advantages over HIR. For even more efficient accenting, ceramic metal halide PAR lamps are recommended over HIR/PAR lamps, although these are not dimmable (or at least look awful when dimmed).

HIR lamp color characteristics

Color of light (color temperature), color rendering (how colors look under the light in question) and easy, low–cost dimming are advantages of HIR lamps. Color temperature of most HIR lamps is "crisp

Table 5.1 HIR Mains Voltage (120V) Lamps

HIR Lamp	Use/Replaces	Lamp Profile
PAR30S	New • Halogen PAR30S • Standard Incandescent R30 • Standard Incandescent R40 Retrofit • Halogen PAR30S	
PAR38	New or Retrofit • Halogen PAR38 • Standard Incandescent PAR38 • Standard Incandescent R40	

Lamp images courtesy of and copyright by GE 2000.

HIR Lamps
Mains Voltage (120V)

white" which is good for contemporary interiors. Where HIR lamps are used for accent lighting, this whiteness adds to the sense of brightness and attraction.

HIR lamps have excellent color rendering—perceived by most folks as rendering all colors "true." As such, HIR/PAR lamps are best for highlighting merchandise, art and most natural–material surfaces.

HIR dimming

Dimming HIR lamps can increase lamp life and also will warm the color of light somewhat. This helps to "take the edge" off the light and makes HIR an efficient alternative in more traditional residential and hospitality settings. Since HIR/PAR lamps are surrounded by cast glass enclosures and use small filaments which themselves are enclosed in a small capsule, there is little hum from dimming that is more common with standard tungsten filament incandescent lamps.

Just as with halogen lamps, the halogen cycle in HIR lamps results in improved efficacy and long life. This same cycle, however, can be "short circuited" with continual dimming. Hence, where dimming is used, occasionally the lamps must be operated at full bright to allow the halogen cycle to operate.

■ ■ ■ ■ ■ ■ ■ ■ ■ ■ ■ ■

HIR dimming

❶Increases lamp life—doubling it, but only if lamps are brightened to full output occasionally.
❶Yellows color of light—more cozy, intimate appearance.
❶Results in little or no hum (which is common in standard tungsten filament incandescent).

HIR hazards

The hazards of halogen lamps have been highly publicized in recent years. The very reason a halogen lamp operates so efficiently is the same reason it can be a hazard—the high temperature operation of the filament. Early halogen lamps were actually comprised of quartz (literally clear stone) envelopes—this stone was the only substance at the time that could withstand the high temperatures of the halogen gas and filament. These first lamps were linear, typically several inches in length and of relatively high wattage. This linear version of the halogen lamp is still available today and is available in high wattage HIR versions. These linear lamps may be found in some varieties of

Cost magnitude²⁰⁰⁰
for halogen and HIR lamps

Low
US$7.⁵⁰ to US$12.⁰⁰
Moderate
US$12.⁰⁰ to US$17.⁰⁰
High
US$17.⁰⁰–plus

Costs vary based on quantities, distributor and contractor markups, and market conditions, manufacturing situations, and annual inflation. These values are for preliminary, magnitude budgeting and do not represent quotes nor actual final pricing.

HIR Lamps
Mains Voltage (120V)

HIR
Mains Voltage (120V)

5

floor torchieres. If the hot quartz or hot surrounding metal components come in contact with flammable materials (e.g., drapes or
sheers), fire can certainly result. There have been cases where the quartz envelope fails violently—in this situation shards of hot quartz and filament bits explode onto surrounding surfaces. Again, flammability is an issue with room/furniture finishes. If the torchiere falls while the lamps are energized, the breaking

■ ■ ■ ■ ■ ■ ■ ■ ■ ■ ■ ■

HIR hazards
⊘Linear HIR (aka quartz) lamps should only be used in
  luminaires that are UL–listed and labeled and fitted with
  tempered glass lenses.
⊘Portable HIR luminaires should be well balanced and heavily
  weighted to prevent tipping.
⊘Off–brand HIR lamps may not offer the same protective
  quality of cast glass enclosure as the brand–name lamps.
⊘Luminaires with HIR lamps typically are hot to touch.
⊘Ultraviolet light from HIR lamps can be damaging to humans
  and artwork unless UV filters are used.

halogen lamp is likely to cause fire. Because of their very portability,
torchieres can be unstable and/or are easily placed too close to
ceilings, walls, drapes and the like.

This flammability hazard can be reduced if linear HIR lamps are
used in well–constructed, well–balanced luminaires fitted with tempered glass lenses; the luminaire is located away from circulation
paths; and clear, safe distances (at least 1 foot from any part of the
torchiere) are maintained between the torchiere and any materials of
any kind. If HIR/PAR lamps are used from any of the name–brand
manufacturers, then the lamp failure hazard has been addressed in
manufacturing with thick, cast or pressed glass enclosures surrounding a halogen capsule. Nevertheless, lamps should be used
with care. Maintain sufficient distance from lamp to temperature–
sensitive surfaces or materials.

Ultraviolet light output is higher in HIR lamps than in standard
tungsten filament incandescent lamps. Artwork requires protection
from this UV, otherwise degradation of the pigments, colors and/or
base media (e.g., paper) may be significant and irreparable over a
fairly short period of time. Similarly, people require protection from
this UV. Tempered glass lenses help reduce UV light output from
halogen lamps. For sensitive art, however, UV reduction lenses or
filters should be used with halogen lamps.

HIR Lamps
Mains Voltage (120V)

When to use HIR

HIR lamps are appropriate on projects where construction budgets are relatively low, life cycle budgets are less important or intended use is not prolonged. Typically, residential projects fit this profile. Further, where fine, low–level dimming is a critical criterion or where the budget cannot support fluorescent dimming, HIR/PAR lamps are justified. Where focused accenting is a requirement along with dimming or instant–on characteristics, HIR/PAR lamps are appropriate. Religious facilities, hospitality facilities (e.g., banquet facilities, meeting facilities) and some portions of commercial facilities fit this profile. Finally, where historic restoration is required and where lamping is behind frosted or otherwise nonimage–preserving glass or faux alabaster type media, HIR lamps may be appropriate— offering better efficacy than halogen lamps and offering improved maintenance cycles.

Environments where temperature extremes exist or where the temperature is likely to remain at or below 45°F are candidates for HIR lamps. HIR lamps are particularly useful in cold settings if dimming and/or instant "bright" or "full–on" conditions are a requirement of the lighting operation.

Where flashing of lights is a requirement, either for effect or for signaling purposes, 120V HIR lamps deserve consideration. Rapid flashing, however, is likely to mitigate the action of the infrared coating.

When not to use HIR

Many areas of commercial, institutional and retail projects deserve compact fluorescent, linear fluorescent and/or ceramic metal halide lamp solutions for general lighting. HIR lamps do not offer the low energy use and long life that the majority of lighting in these facilities requires to be cost–competitive and energy efficient. If accent lighting is required in these kinds of facilities, then HIR lamps should be considered, with ceramic metal halide lamps

■ ■ ■ ■ ■ ■ ■ ■ ■ ■ ■ ■ ■ ■ ■ ■

Halogen cycle

❶ Heated filament metal boils away into vapor (as happens in standard tungsten filament incandescent lamps).

❷ Vaporized filament metal is redirected by halogen gas back onto filament—the filament is regenerated!

❸ Yields longer life lamp or more efficacious lamp or a lamp with some increase in life and some increase in light output.

HIR Mains Voltage (120V)

HIR Lamps
Mains Voltage (120V)

offering much more accent capability in smaller lamp sizes and wattages. HIR lamps should not be used where very close proximity to people for prolonged periods is expected (e.g., task lighting)— given their high operating temperatures (resulting in higher ambient temperatures of nearby air and yielding potential for skin burns) and ultraviolet radiation. Table 5.2 outlines more efficacious alternatives to HIR lamps.

Table 5.2 Efficacious Alternatives to HIR Lamps

HIR Lamp	Efficacious Alternative	Hardware Application
PAR30S	• CMH PAR20 (ceramic metal halide)❶	• Accent Luminaires • Downlight Luminaires
PAR38	• CMH PAR20 (ceramic metal halide)❶ • CMH PAR30 (ceramic metal halide)❶ • CMH PAR38 (ceramic metal halide)❶	• Accent Luminaires • Downlight Luminaires • Wallwash Luminaires

❶This alternative lamp is not dimmable inexpensively, if at all.

Triple–tube compact fluorescent lamps (Section 8) **are the efficient alternative to HIR lamps for general lighting. For efficient accent lighting, consider ceramic metal halide (CMH) PAR lamps** (Section 7).

HIR Mains Voltage (120V)

5

HIR/PAR

AR lamps—PAR is an acronym for parabolic aluminized reflector—have been effective accent lamps since the 1950's. With the HIR technology, the PAR lamp provides the most light at the least wattage with a longer life compared to any other incandescent lamp for architectural applications. A small halogen capsule with an infrared coating is enclosed by a relatively thick, cast/pressed glass envelope. The reflector portion of this envelope is coated with specular (polished) aluminum. The front of the lamp has a specially designed glass lens which refracts the light concentrated by the reflector into a pattern—from narrow spot (NSP) to wide flood (WFL). Some manufacturers offer watertight versions for direct exposure to weather. Tables 5.3 through 5.10 are lamp guides—outlining HIR/PAR lamps that might be considered for various tasks and effects in several applications and ceiling heights. More detailed information about these specific lamps and information about their intensities and coverages can be found on the pages cited.

■ ■ ■ ■ ■ ■ ■ ■ ■ ■ ■ ■

Most Efficient Incandescent Lighting
❶ Don't use R lamps.
❶ Don't use BR or Krypton–filled lamps.
❶ Consider Halogen Infrared/PAR (HIR/PAR) lamps.

Less efficient reflector or R lamps were inexpensive and therefore popular prior to the 1992 Energy Policy Act (EPACT), at which point the lamps were essentially legislated out of existence. Standard incandescent R lamps cannot meet the legislated efficacy requirements of the EPACT. Today, some variations on R lamps remain available—typically known as BR lamps or Krypton–filled R lamps. For best efficacy in accenting, however, HIR/PAR lamps should be used. HIR/PAR lamps should be the common replacement for R lamps, with PAR20s replacing R20s; PAR30Ss and PAR30Ls replacing R30s; and PAR38s replacing R40s. The figures on the facing page illustrate various R and BR lamps with notes regarding the better–advised PAR lamp alternatives. PAR lamp replacements are ghosted background images to illustrate relative size. The PAR lamps offer smaller bulb envelopes and therefore can be fit into smaller housings, or in retrofit situations will eliminate the glary and unsightly exposed bulb envelope of the R lamp bulging from the luminaire. HIR/PAR lamps are excellent retrofit lamps for standard incandescent PAR lamps. Power reductions of at least 50 percent are achievable while maintaining or actually increasing light output!

HIR Mains Voltage (120V)

5

HIR/PAR lamps are excellent retrofits for standard tungsten incandescent PAR lamps—reducing power by as much as 50 percent!

HIR/PAR

R40 Lamp/Replaced by HIR/PAR38

R30 Lamp/Replaced by HIR/PAR30S

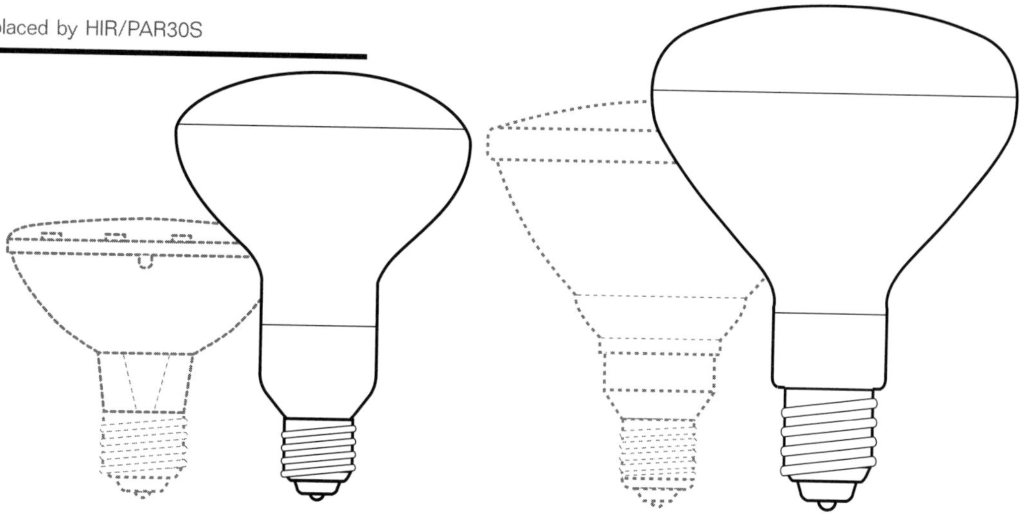

BR40 Lamp/Consider HIR/PAR38

BR30 Lamp/Consider HIR/PAR30

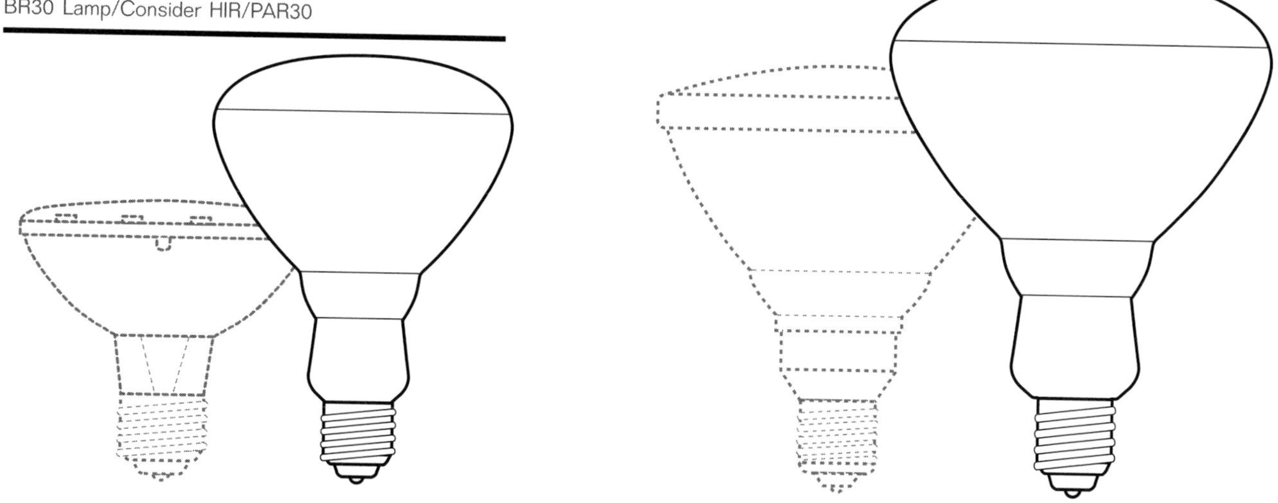

HIR Mains Voltage (120V)

5

▶The HIR/PAR lamps improved optics along with the smaller "point source" created by the halogen capsule allow for shallower lamps.
▶Old inefficient R and slightly improved BR lamps shown in solid outline for convenient reference when looking to retrofit existing luminaires.
▶HIR/PAR lamp replacements shown as dashed background.
▶All lamps shown to rough relative scale.
Images copyright 2000/GE

HIR/PAR

HIR Mains Voltage (120V) 5

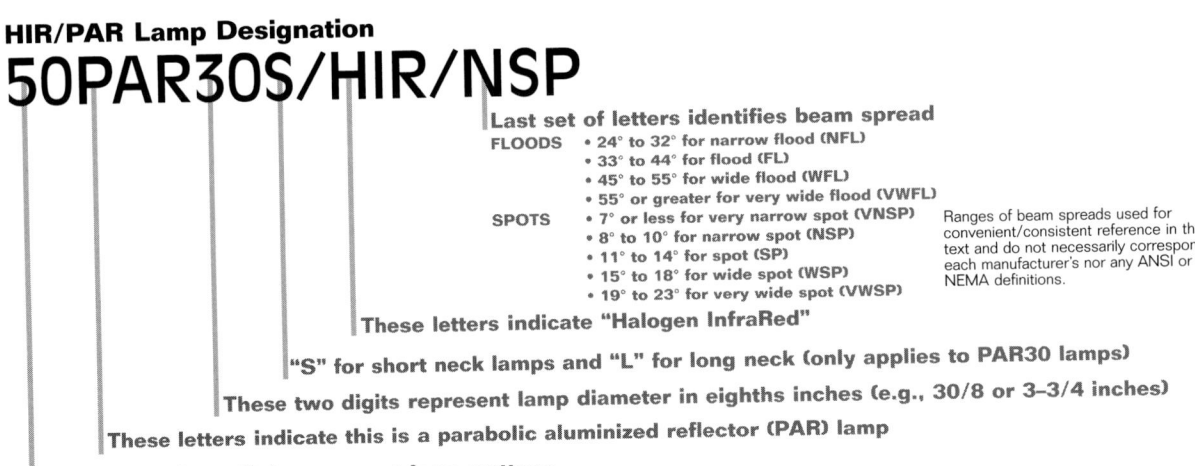

HIR/PAR Lamp Designation

50PAR30S/HIR/NSP

Last set of letters identifies beam spread
FLOODS
- 24° to 32° for narrow flood (NFL)
- 33° to 44° for flood (FL)
- 45° to 55° for wide flood (WFL)
- 55° or greater for very wide flood (VWFL)

SPOTS
- 7° or less for very narrow spot (VNSP)
- 8° to 10° for narrow spot (NSP)
- 11° to 14° for spot (SP)
- 15° to 18° for wide spot (WSP)
- 19° to 23° for very wide spot (VWSP)

Ranges of beam spreads used for convenient/consistent reference in this text and do not necessarily correspond to each manufacturer's nor any ANSI or NEMA definitions.

These letters indicate "Halogen InfraRed"

"S" for short neck lamps and "L" for long neck (only applies to PAR30 lamps)

These two digits represent lamp diameter in eighths inches (e.g., 30/8 or 3–3/4 inches)

These letters indicate this is a parabolic aluminized reflector (PAR) lamp

First two or three digits represent lamp wattage

[This designation used for convenient reference throughout this *Time–Saver Standards for Architectural Lighting*—but not necessarily used by all lamp manufacturers (although it would be convenient if manufacturers would come to agreement on standard designations.]

HIR/PAR lamp designation

HIR/PAR lamps are typically identified with an alphanumeric designation outlined above. Since each manufacturer has slightly different means for lamp designation, the designation method presented above is solely for convenient reference throughout the text. Actual lamp manufacturers' designations will need to be used in final specifications. Perhaps the lamp manufacturers will standardize lamp designations in the near future in an effort to ease lamp cross–referencing and specifications.

HIR/PAR lamps for accenting

HIR/PAR lamps are good for art accenting—remembering that their improved efficacy over halogen/PAR lamps will result in either more light or, better yet, the need for a lower wattage lamp. With HIR/PAR lamps, however, there is not as large a selection of wattage and beam spreads as with halogen/PAR lamps. For the HIR/PAR30S and HIR/PAR38 lamps, a series of Performance Sketches™ are presented which offers quick visual assessment of what lamp wattage and beam spread might be appropriate for given ceiling heights and artwork sizes.

HIR/PAR

■ ■ ■ ■ ■ ■ ■ ■ ■ ■

Manufacturers' Update Data:

ⓘ Periodically due to materials' technology
ⓘ Periodically due to manufacturing techniques
ⓘ Periodically due to governmental regulations
ⓘ Performance changes may result

Performance Sketch™

A quarter inch scale rendering illustrates lighting effect for a typical situation. Note how the area of coverage, intensity and frame shadow change with beam spread, ceiling height and distance from wall. Higher or lower ceilings and longer or shorter distances from walls will impact intensities, light patterns and application(s).

Performance Details

Details such as lamp distance from wall, aiming angle, point above floor at which lamp is aimed, along with maintained average and maximum illuminance are reported. The aiming angle should be used to select a luminaire capable of achieving such an angle. Maximum illuminances are calculated maintained over time and are reported to help assess likelihood of artwork degradation. Average illuminance is calculated maintained over time just for the size of the artwork shown.

Suggested Uses/Lamp Specs

A bullet list outlines uses for which the lamp seems best suited at cited ceiling height. Lamp specifications are outlined for quick reference and to assist in specification writing. Where two or more offer nearly identical products, only one manufacturer's data (boldfaced) has been used for development of the sketches. An SKU–number or product code is included parenthetically. Data cited in the sidebar "Specs" is averaged from all available manufacturers' data.

Energy Tip

The energy bolt ⚡ is used to denote an energy saving option.

Design Tips

Key considerations are offered for the specific lamp.

Application Key

Those applications for the cited ceiling height where the specific lamp may be most useful are highlighted in bold. Applications where the lamp may be of lesser use are faded—the softer the fade, the less appropriate is the lamp for that application in cited ceiling height. Application appropriateness will change with ceiling heights.

Artwork Outline

An outline of two dimensional artwork of size that might be acceptably lighted by lamp cited.

Inset sample page:

Time-Saver Standards for Architectural Lighting **L a m p s**

HIR 60PAR38/HIR/WFL

Performance Sketches

(All data outlined here is based on information from boldfaced manufacturer's published data for 120V version at time of manuscript preparation. Scale is ¼ inch equals 1 foot.)

Art shown at 3' by 4' and centered 5'–6" AFF. Wall grid on 6" by 6" centers and scaled at ¼".

Lamp: 60PAR38/HIR/WFL
Location: 2'–0" from wall
Center beam Aiming: 12' (at point 0' AFF)
Luminaire: Monopoint or Recessed Adjustable
Illuminance: 15 fc avg. maintained (36 fc max)

Lamp: 60PAR38/HIR/WFL
Location: 2'–6" from wall
Center beam Aiming: 15° (at point 0' AFF)
Luminaire: Monopoint or Recessed Adjustable
Illuminance: 14 fc avg. maintained (26 fc max)

Lamp: 60PAR38/HIR/WFL
Location: 3'–0" from wall
Center beam Aiming: 19° (at point 1'–0" AFF)
Luminaire: Monopoint or Recessed Adjustable
Illuminance: 14 fc avg. maintained (23 fc max)

Uses
► Soft general lighting
► Wallwashing (see 11–129)
► Medium art with short throws
► Short throws
► Available in 120V and 130V

Specs
Beam spread/50°
Center beam
  candlepower/1250
Life/3000 hrs
► GE (20947)

Design Tips
✔ One light best for medium art and where mounting height for light is 10' or less.
✔ Multiple units can be used for wallwashing.
✔ Good for soft downlighting—best with sconces and/or art accents and/or wallwashing to avoid cave effect.

Application Key
Commercial
Gallery
Hospitality
Institutional
Manufacturing
Residential
Retail
Exterior

bold = primary application
partial fade = minimal application
fade = unlikely application

HIR Mains Voltage (120V)
5

5–45 ▲

HIR Mains Voltage (120V)
5

HIR/PAR

HIR Mains Voltage (120V)

5

Performance Sketches™ are shown at a scale of ¼ inch equals 1 foot. Tracing for evaluation of various art sizes and elevations is encouraged. Copying is permitted (according to the conventions outlined in "Licensing Agreement" in the Introduction of this book). Each Sketch includes an outline of artwork of a size category which might be lighted by the cited lamp.

Radiance (lighting rendering software)
http://www.radsite.lbl.gov/radiance/refer/long.html

CONNECT FOR MORE

This is to assist in judgement of area of coverage. Location or distance from wall is listed with each Sketch as are center beam aiming angle and average maintained light level over the area of the artwork outline. Maximum maintained light intensity is also listed. For museum quality artwork where conservation of art is a priority, intensities should be 5 to 10 footcandles average, maintained, with maximums of perhaps 10 to 15 fc. In other applications, average maintained intensities on art will depend on the general or ambient lighting within the space(s). In typical hospitality applications, artwork accenting might exhibit average maintained intensities of 10 to 20 fc, with maximum intensities of 20 to 30 fc. In commercial applications, average maintained intensities on artwork might range from 20 to 30 fc, with maximums of 30 to 40 fc. These Sketches are based on the cited manufacturer's data at time of manuscript preparation. Where two or more manufacturers offer nearly identical products, only one manufacturer's data has been used for development of the sketches. Data cited in the sidebar "Specs" is averaged from all available manufacturers' data. Where any manufacturer's data did not fall within 15 percent of this averaged data, the respective manufacturer's lamp is cited separately along with separate sketches. Manufacturers' catalog data may be of one date and for one set of lamps while photometric data available from manufacturers may be based on another set of lamps of another date (this may have occurred with some data in this reference). Sketches were generated with this data in Radiance (a Unix–based lighting simulation and rendering program available free of charge at www.radsite.lbl.gov/radiance/refer/long.html).

These Performance Sketches™ illustrate a specific condition. As ceiling heights change and focal objects' sizes change, the cited lamp may be quite appropriate for other applications. Generally, in

Recommended intensities
for art accenting

Low (gallery, museum, residential)
5 to 10 fc average on art
Moderate (hospitality)
10 to 20 fc average on art
High (commercial—or higher for retail)
20 to 30 fc average on art

These are intended to be average, maintained values including effects of accent lighting and general room lighting. General room lighting contributes just a few footcandles in residential, gallery and museum settings, but contributes 5 to 10 fc in commercial settings. Maximums might range from 50 percent to 100 percent of the average value. Wherever artwork is highly valuable and/or sensitive to light, precautions should be taken to limit light exposure (maintaining low levels and/or switching lights off when art is not being viewed); and exposure to infrared and ultraviolet should be limited (using filters on lamps or placing art behind specially treated glass or acrylic).

HIR/PAR

lower ceiling settings, light intensities will be greater and area of coverage will be smaller. In higher ceiling settings, light intensities will be less and area of coverage will be larger.

Lamp data changes from time to time. As manufacturing techniques change, as materials' technologies change, and as the market changes, lamp manufacturers will revise wattages, beam spreads and intensities. Manufacturers' data should be checked periodically for any possible updates.

Performance Sketch™ thumbnails

At the beginning of each HIR/PAR lamp section (HIR/PAR30S and HIR/PAR38) there are thumbnails of the Performance Sketches™ to further help provide quick visual assessment of various lamp beam spreads and intensities.

■ ■ ■ ■ ■ ■ ■ ■ ■ ■ ■ ■

Performance Sketches™ and Information
❶ Typical application is shown
❶ Other applications may also be acceptable
❶ Other applications may depend on ceiling heights
❶ Maintained values based on 0.9 maintenance factor
❶ Cited data is for a single lamp—two or more lamps may be necessary for other applications

Uses

For convenient reference, a shaded sidebar box entitled "Uses" appears in the upper corner of each page detailing the more likely uses for which the lamp seems suited. Artwork sizes are based on the rough dimensions as outlined in the tip box on this page (to the left). Lamp "Specs" are also included in the shaded box. These specs are averages of the listed manufacturers' data at time of manuscript preparation. Since products are constantly upgraded or deleted, a check of the current status of the listed lamp is recommended before including it in specifications.

Lighting characteristics

In addition to beam spread and intensity, these images also illustrate beam striations and cutoff patterns. Beam striations and aberrations may appear more pronounced in reality since Performance Sketches™ herein are based on finite candlepower reports (typically at 2.5° increments—a bright spot or dead spot might occur at an increment not reported on the photometry but which could be seen visually on mockup). Color of light cannot be illustrated in this black and white format. These Performance Sketches™ will help narrow selections so that appropriate lamp mockups can be made to assess actual beam striations and color of light.

HIR Mains Voltage (120V)

5

HIR/PAR

Luminaires

PAR lamps are used in recessed downlights, recessed adjustable accent lights and track and monopoint adjustable accents. Since the PAR lamp is a self–contained optical package, track and monopoint luminaires do not need to offer much more than a lamp socket and perhaps a decorative shroud to hide the bare PAR lamp. With downlights and adjustable accents, the reflector cone in the luminaire can help to control some light spill and/or minimize direct glare. Nevertheless, the lamp's optics provide the bulk of the lighting performance. Hence, the Performance Sketches™ shown here illustrate expected results from PAR lamps in track and monopoint adjustable accents as well as recessed adjustable accents.

Selection guides

Selection guides are offered on the next several pages for various uses in various applications. These guides are intended to help limit the designer's, engineer's or facility engineer's search and offer a good starting point from which design or alternative analyses can progress. Lamps which have similar beam spreads and intensities may be suggested for different ceiling heights and art sizes simply to show the variety of situations in which lamps might be used. Uses for HIR/PAR lamps include:

▶pinspot accent
   An intense, well–controlled, focused lighting effect.
▶accent/medium art
   Relatively intense lighting effect on medium piece of art.
▶accent/large art
   Relatively intense lighting effect on large piece of art.
▶wallwashing/matte materials
   Relatively uniform wash of light on wall materials exhibiting no sheen, polish or specularity.
▶wallwashing/polished materials
   Relatively uniform wash of light on polished or specular wall materials.
▶feature downlighting
   Somewhat intense lighting effect on flooring material details.
▶general downlighting
   Relatively uniform wash of light at lap height, table height or on floor.

While there are exceptions, Halogen PAR lamps are most appropriate in the following applications:

▶commercial/retail
▶gallery/residential
▶hospitality

HIR/PAR

For low ceiling conditions—less than 10 feet—see Tables 5.3 (Commercial or Retail), 5.4 (Gallery or Residential) and 5.5 (Hospitality). For high ceiling conditions—10 feet or greater in height—see Tables 5.6 (Commercial or Retail), 5.7 (Gallery or Residential) and 5.8 (Hospitality).

Extrapolating information

There will be situations where perhaps twice as much light is desired (to provide a significant focal point) or where half as much light is desired (very sensitive artworks—paper–based pieces or media which fades easily). Find a ceiling height condition and artwork size which matches the planned situation. Then, to almost double light intensities, consider using two lights instead of a single light. Lights should be spaced on center at about the same distance they are spaced from the wall. So, if a single light spaced 2 feet/6 inches from the wall provides 17 fc average on a piece of art that is 3 feet wide by 4 feet long, then two such lights spaced 2 feet/6 inches from the wall and each spaced 2 feet/6 inches on center from the other centered on the art will provide about 30 fc average on the piece of art.

Another way to double the light is to find a lamp with similar beam spread (within 5 degrees) but twice the center beam candlepower. A search over several lamp families is suggested (e.g., consider halogen low voltage MR lamps). This is generally less costly than adding a second light, both initially and operationally. Most fine art installations, however, will have two lights aimed onto each piece for maximum viewing quality from most any angle—eliminating the harsh veiling reflections (particularly problematic with oils and acrylics) which are evident with single light accents.

To halve the light level, look for lamps of similar beam spreads (within 5 degrees) but with half the center beam candlepower. Light intensities can be reduced with neutral density filters, although this does waste energy and thus should be a last resort.

■ ■ ■ ■ ■ ■ ■ ■ ■ ■ ■

Artwork Size Categories
⊃Petite = 2′–0″ by 2′–0″ or smaller
⊃Small = 2′–0″ by 2′–0″ to 2′–6″ by 3′–0″
⊃Medium = 2′–6″ by 3′–0″ to 3′–0″ by 4′–0″
⊃Large = 3′–0″ by 4′–0″ to 4′–0″ by 5′–0″
⊃Extra Large = 4′–0″ by 5′–0″ or larger

HIR Mains Voltage (120V)

5

HIR PAR Lamp Selection Guide

Table 5.3 Commercial or Retail Low Ceiling (less than 10 feet)❶❷

	◀Smaller lamp/smaller luminaire	Larger lamp/larger luminaire▶
Lighting Task	**PAR30S**	**PAR38**
Pinspot Accent	Consider (1)l or more depending on desired impact	
• Petite to small art • Feature merchandise • Small merchandise	• 50/NSP p5–26	• 50/NSP p5–38
Key Area Accent	Consider (2) or more depending on desired impact	
• Small to medium art • Feature display • Moderately sized merchandisee	• 50/NFL p5–28	• 50/NFL p5–39
Feature Lighting	Consider (2) or more depending on size of featured area	
• Medium to large art • Floor feature • Destination focus (e.g., cashier)	• 50/NFL p5–28	• 50/NFL p5–39 • 60/FL p5–44
General Downlighting	Consider symmetric arrangement❸	
• Overall lighting of lap, table or floor	• 50/FL p5–29	• 60/WFL p5–45

beam spread
wattage

❶These lamp guides are intended to direct your attention to lamps which might meet your needs. If daylighting is prevalent, more lamps may be necessary to provide a visual focus to the artwork or merchandise. Design analysis is required to finalize lamp selections for projects.

❷For retail applications it is reasonable to also use lamps cited in Table 5.6—intensities will be significantly greater in low ceiling spaces.

❸Spacing between lights should not exceed 0.75 to 1.0 of the distance between the surface being downlighted and the ceiling.

Artwork Size Categories
❍Petite = 2′–0″ by 2′–0″ or smaller
❍Small = 2′–0″ by 2′–0″ to 2′–6″ by 3′–0″
❍Medium = 2′–6″ by 3′–0″ to 3′–0″ by 4′–0″
❍Large = 3′–0″ by 4′–0″ to 4′–0″ by 5′–0″
❍Extra Large = 4′–0″ by 5′–0″ or larger

HIR Mains Voltage (120V)

5

HIR PAR Lamp Selection Guide

Table 5.4 Gallery or Residential Low Ceiling (less than 10 feet)❶

◀Smaller lamp/smaller luminaire   Larger lamp/larger luminaire▶

Lighting Task	PAR30S	PAR38
Pinspot Accent	Consider Halogen/PAR (Section 3) or Halogen/MR (Section 4)	
• Petite art	HIR/PAR lamps are too intense for petite gallery art at these mounting heights	
Pinspot Accent	Consider (1)	
• Objet d'art❷ • Centerpieces❷	• 50/NSP p5–26	• 50/NSP p5–38
Accent/Small Artwork	Consider Halogen/PAR (Section 3) or Halogen/MR (Section 4)	
• Small art	HIR/PAR lamps are too intense for small gallery art at these mounting heights	
Accent/Medium Artwork	Consider (1)	
• Medium art	• 50/FL p5–29❸	
Accent/Large Artwork	Consider Halogen/PAR (Section 3) or Halogen/MR (Section 4)	
• Large art • Large sculpture • Larger architectural detail or feature		
General Downlighting	Consider symmetric arrangement❹	
• Overall lighting of lap, table or floor		

wattage / beam spread

❶These lamp guides are intended to direct your attention to lamps which might meet your needs. Design analysis is required to finalize lamp selections for projects.
❷HIR/PAR NSP lamps are very intense at these mounting heights, yet provide a crisp, small area of highlight appropriate for inert objet d'art pieces and centerpieces. Fading will occur on centerpieces and/or highlighted furnishings.
❸Neutral density filter will be required in front of halogen lamp to limit intensity to between 5 and 10 fc if art preservation is paramount.
❹Spacing between lights should not exceed 0.75 to 1.0 of the distance between the surface being downlighted and the ceiling.

HIR Mains Voltage (120V)

5

HIR PAR Lamp Selection Guide

Table 5.5 Hospitality Low Ceiling (less than 10 feet)❶❷

◀Smaller lamp/smaller luminaire   Larger lamp/larger luminaire▶

Lighting Task	PAR30S	PAR38
Pinspot Accent	Consider Halogen/PAR (Section 3) or Halogen/MR (Section 4)	
• Petite art	HIR/PAR lamps are too intense for petite commercial and hospitality art at these mounting heights	
Pinspot Accent	Consider (1)	
• Objet d'art❸ • Centerpieces❸ • Petite or fine architectural detail	• 50/NSP p5–26	• 50/NSP p5–40
Accent/Small Artwork	Consider Halogen/PAR (Section 3) or Halogen/MR (Section 4)	
• Small art	HIR/PAR lamps are too intense for small commercial and hospitality art at these mounting heights	
Accent/Medium Artwork	Consider (1)	
• Medium art • Medium sculpture • Architectural detail or feature	• 50/FL p5–29	• 60/WFL p5–45
Accent/Large Artwork	Consider Halogen/PAR (Section 3) or Halogen/MR (Section 4)	
• Large art • Large sculpture • Larger architectural detail or feature		
Wallwashing/Matte Materials	Continuous row of units 2 to 3 feet from wall and 2 to 3 feet on center	
• Flat, frontal lighting❹❺		• 60/WFL p5–45
Wallwashing/Polished Materials	Architectural slot with sockets 6 inches from wall and 6 to 12 inches on center	
• Grazing wash light❻❼	See PAR lamp wallslots, Section 11, pp11–214 to 11–221	
Feature Lighting	Consider (1) or more depending on size of featured area	
• Floor feature • Pattern highlight • Destination focus (e.g., reception desk)	• 50/FL p5–29	
General Downlighting	Consider symmetric arrangement❽	
• Overall lighting of lap, table or floor		• 60/WFL p5–45

Annotations: "wattage" and "beam spread" labels point to the entry "60/WFL p5–45" in the Accent/Medium Artwork row.

❶These lamp guides are intended to direct your attention to lamps which might meet your needs. Design analysis is required to finalize lamp selections for projects.

❷Higher light levels on artwork and features are typically expected in hospitality spaces compared to residential and gallery spaces. Where artwork preservation is a priority, neutral density filters will be required in front of these lamps or lamps suggested for Gallery or Residential Applications should be used.

❸HIR/PAR NSP lamps are very intense at these mounting heights, yet provide a crisp, small area of highlight appropriate for inert objet d'art pieces and centerpieces. Fading will occur on centerpieces and/or highlighted furnishings.

❹Energy intensive—only consider for exclusive zones of small size in highly visible public areas. Also see spreadlens wallwashers, pp11–136 to 11–143.

❺High–wattage lamps not listed—for high intensities, more efficacious sources should be considered.

❻Extraordinarily energy intensive and should be reserved for very exclusive zones of small size in highly visible public areas.

❼This approach accentuates any wall imperfections.

❽Spacing between lights should not exceed 0.75 to 1.0 of the distance between the surface being downlighted and the ceiling.

HIR PAR Lamp Selection Guide

Table 5.6 Commercial or Retail High Ceiling (10 feet or greater)❶❷

◄Smaller lamp/smaller luminaire   Larger lamp/larger luminaire►

Lighting Task	PAR30S	PAR38
Pinspot Accent	Consider (1) or more depending on ceiling height and desired impact	
• Petite to small art • Feature merchandise • Small merchandise	• 50/NSP p5–26	• 50/NSP p5–38 • 60/NSP p5–40 • 80/NSP p5–46 • 100/NSP p5–52
Key Area Accent	Consider (2) or more depending on ceiling height and desired impact	
• Small to medium art • Feature display • Moderately sized merchandise	• 50/NFL p5–28	• 50/NSP p5–38 • 60/NSP p5–40 • 60/FL p5–44 • 80/NSP p5–46 • 80/SP p5–48 • 100/NSP p5–52
Feature Lighting	Consider (2) or more depending on size of featured area and ceiling height	
• Medium to large art • Floor feature • Destination focus (e.g., cashier)	• 50/NFL p5–28 beam spread wattage	• 60/NFL25° p5–42 • 80/NFL p5–50 • 100/NFL p5–54 • 100/FL p5–55
General Downlighting	Consider symmetric arrangement❸	
• Overall lighting of table or floor		• 60/FL p5–44 • 100/FL p5–55

Typically dimmer▲ / ▼Typically brighter

❶These lamp guides are intended to direct your attention to lamps which might meet your needs. If daylighting is prevalent, more lamps may be necessary to provide a visual focus to the artwork or merchandise. Design analysis is required to finalize lamp selections for projects.
❷Some lamps' performance sketches which are referenced here are not optimized for retail applications. Lower ceiling heights will result in higher intensities on merchandise.
❸Spacing between lights should not exceed 0.75 to 1.0 of the distance between the surface being downlighted and the ceiling.

■ ■ ■ ■ ■ ■ ■ ■ ■ ■

Artwork Size Categories

⊃Petite = 2'–0" by 2'–0" or smaller
⊃Small = 2'–0" by 2'–0" to 2'–6" by 3'–0"
⊃Medium = 2'–6" by 3'–0" to 3'–0" by 4'–0"
⊃Large = 3'–0" by 4'–0" to 4'–0" by 5'–0"
⊃Extra Large = 4'–0" by 5'–0" or larger

HIR Mains Voltage (120V)

5

HIR PAR Lamp Selection Guide

Table 5.7 Gallery or Residential High Ceiling (10 feet or greater)[1][2]

◀Smaller lamp/smaller luminaire   Larger lamp/larger luminaire▶

Lighting Task	PAR30S	PAR38
Pinspot Accent	Consider lower wattage lamps[2]	
• Petite art • Objet d'art • Centerpieces	• Halogen low voltage, low wattage MR16 VNSP • Halogen low voltage, low wattage PAR36 VNSP	
Accent/Small Artwork	Consider lower wattage lamps[2]	
• Small art	• Halogen low voltage, low wattage MR16 VNSP • Halogen low voltage, low wattage PAR36 VNSP	
Accent/Medium Artwork	Consider (1)	
• Medium art • Medium sculpture • Architectural detail or feature		• 60/FL p5–44
Accent/Large Artwork	Consider (1)	
• Large art • Large sculpture • Larger architectural detail or feature	• 50/NSP p5–26[3] • 50/NFL p5–28[3]	• 50/NSP p5–38[3] • 50/NFL p5–39[3] • 60/NSP p5–40
General Downlighting	Consider symmetric arrangement[5]	
• Overall lighting of lap, table or floor	• 50/FL p5–29	• 60/WFL p5–45

wattage

beam spread

Typically dimmer▲
Typically brighter▼

[1] These lamp guides are intended to direct your attention to lamps which might meet your needs. Design analysis is required to finalize lamp selections for projects.

[2] To achieve small, well–controlled beam spreads, VNSP lamps in the halogen low voltage MR16 and PAR36 are recommended. The intensities, however, are significant and therefore should be considered for inert artwork or use of dimmers and/or neutral density filters will be necessary to limit intensities.

[3] Neutral density filter required in front of lamp to limit intensity to between 5 and 10 fc for art preservation.

[4] Spacing between lights should not exceed 0.75 to 1.0 of the distance between the surface being downlighted and the ceiling.

Footnotes to Table 5.8 (opposite page)▶

[1] These lamp guides are intended to direct your attention to lamps which might meet your needs. Design analysis is required to finalize lamp selections for projects.

[2] Higher light levels on artwork and features are expected in hospitality spaces compared to residential and gallery spaces. If art preservation is a priority, neutral density filters will be required in front of these lamps or lamps suggested for Gallery or Residential Applications should be used.

[3] To achieve small, well–controlled beam spreads, VNSP lamps in the halogen low voltage MR16 and PAR36 are recommended. The intensities, however, are significant and therefore should be considered for inert artwork or use of dimmers and/or neutral density filters will be necessary to limit intensities.

[4] Energy intensive—only consider for exclusive zones of small size in highly visible public areas. Also see spreadlens wallwashers, pp11–136 to 11–143.

[5] High–wattage lamps not listed—for high intensities, more efficacious sources should be considered.

[6] Extraordinarily energy intensive and should be reserved for very exclusive zones of small size in highly visible public areas.

[7] This approach accentuates any wall imperfections.

[8] Spacing between lights should not exceed 0.75 to 1.0 of the distance between the surface being downlighted and the ceiling.

HIR PAR Lamp Selection Guide

Table 5.8 Hospitality High Ceiling (10 feet or greater)❶❷

◀Smaller lamp/smaller luminaire   Larger lamp/larger luminaire▶

Lighting Task	PAR30S	PAR38
Pinspot Accent	Consider more efficient spot lamps❸	
• Petite art • Objet d'art	• Halogen low voltage, low wattage MR16 VNSP • Halogen low voltage, low wattage PAR36 VNSP	
Accent/Small Artwork	Consider more efficient spot lamps❸	
• Small art • Small sculpture	• Halogen low voltage, low wattage MR16 VNSP • Halogen low voltage, low wattage PAR36 VNSP	
Accent/Medium Artwork	Consider (1)	
• Medium art • Medium sculpture • Architectural detail or feature	• 50/NFL p5–28	• 50/NSP p5–38 • 60/FL p5–44
Accent/Large Artwork	Consider (1)	

• Large art
• Large sculpture
• Larger architectural detail or feature
	• 50/NSP p5–26
	• 50/NFL p5–28
	    beam spread
	    wattage
		• 50/NSP p5–38
		• 50/NFL p5–39
		• 60/NSP p5–40
		• 60/NFL25° p5–42
		• 60/NFL30° p5–43
		• 80/NSP p5–46
		• 80/SP p5–48
		• 100/NSP p5–52
		• 100/NFL p5–54
		• 100/FL p5–55

Wallwashing/Matte Materials	Continuous row of units 2 to 3 feet from wall and 2 to 3 feet on center	
• Flat, frontal lighting❹❺		• 60/WFL p5–45
Wallwashing/Polished Materials	Architectural slot with sockets 6 inches from wall and 6 to 12 inches on center	
• Grazing wash light❻❼	See PAR lamp wallslots, Section 11, pp11–214 to 11–221	
Feature Downlighting	Consider (1 or more depending on size of featured area and ceiling height)	
• Floor feature • Pattern highlight • Destination focus (e.g., reception desk)	• 50/NFL p5–28	• 50/NFL p5–39 • 60/NFL30° p5–43 • 60/FL p5–44 • 100/FL p5–55
General Downlighting❺	Consider symmetric arrangement❽	
• Overall lighting of lap, table or floor	• 50/FL p5–29	• 60/NFL p5–45

Typically dimmer ▲
▼ Typically brighter

HIR Mains Voltage (120V)

5

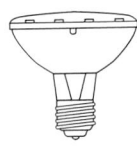

HIR/PAR30S

Short Neck

HIR Mains Voltage (120V)

5

Stats
Data varies manufacturer to manufacturer and changes from time to time

Life/3000
A typical office environment might be occupied about 2600 hours each year.

Efficacy/15 LPW
Good beam intensities and distributions available.

Wattages/ 50

Color of Light/2810°K

Beam spreads/NSP, NFL and FL
Narrow spot, narrow flood and wide flood distributions are available.

Size/3⅝" L by 3¾" Ø

Voltage/120V and 130V

Cost Magnitude/Moderate (2000)

Manufacturers
General Electric

Net Addresses
http://www.ge.com/lighting/business/index.htm

CONNECT FOR MORE

Advantages
Crisp white light
Excellent intensities available for wattage
Relatively small size
Short neck versions fit into shallower/shorter lights
Standard voltage (no transformers)
Easily dimmed
Relatively low initial cost compared to fluorescent and metal halide
Alternative to less efficient R lamps
Alternative to less efficient halogen/PAR lamps
Withstand extreme temperature range
Very little lamp lumen depreciation (light loss) over life

Disadvantages
Beam aberrations
◆Striations
◆Inconsistent beam shape
High operating costs
◆Inefficiency and short life may make this cost prohibitive in typical commercial applications, depending on ease of changing lamps and maintenance staffing.
Hot to touch
Very hot-shock sensitive
◆Lamp may fail if sudden/mild vibration occurs while lamp is energized (e.g., aiming can be problematic).

For more distinct accent light or more focused light, use CMH PAR lamps (Section 7).

HIR/PAR30S

Short Neck

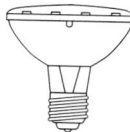

Pressed/cast glass optical lens

Parabolic aluminized reflector (PAR)

Medium screw base

Lamp shown actual size
Image copyright 2000/GE

Uses

Architectural Accenting

Art Accenting

Commercial
Gallery
Hospitality
Residential

Downlighting

Circulation
Conference rooms
Hospitality
Residential
Retail

Merchandise Accenting

Surface Grazing

Specular/semispecular surfaces
◆Polished/honed stone
◆Polished/satin wood
◆Fritted/etched glass
◆Cut glass

■ ■ ■ ■ ■ ■ ■ ■ ■ ■ ■ ■ ■ ■ ■ ■ ■ ■ ■

HIR/PAR30S Lighting Design Tips

✔Work well in tall ceilings (up to 18′–0″).
✔Dimming increases life (typically doubling life), but shifts color warmer.
✔Glary when used without much other general lighting or when aimed across long distances where ceilings are low.
✔Beam striations are most apparent on light–colored, nontextured, flat surfaces and may be objectionable.
✔HIR/PAR30S Downlighting alone is harsh—combine several techniques for best results.
✔Use smaller scale (e.g., 5–inch diameter) downlights and adjustable accents for attractive, more human scale approach.

Net Addresses/PAR20 Track, Monopoints, Recessed and Exterior
See page 5–37

CONNECT FOR MORE

HIR Mains Voltage (120V)

5

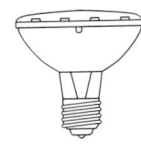

HIR/PAR30S

Short Neck

Throws/Ceilings

⊃Short = 7' to 10'
⊃Medium = 10' to 15'
⊃Long = 15' to 20'
⊃Very Long = 20' or more

HIR Mains Voltage (120V)

5

Recommended intensities
for art accenting

Low (gallery, museum, residential)
5 to 10 fc average on art
Moderate (hospitality)
10 to 20 fc average on art
High (commercial—or higher for retail)
20 to 30 fc average on art

These are intended to be average, maintained values including effects of accent lighting and general room lighting. General room lighting contributes just a few footcandles in residential, gallery and museum settings, but contributes 5 to 10 fc in commercial settings. Maximums might range from 50 percent to 100 percent of the average value. Wherever artwork is highly valuable and/or sensitive to light, precautions should be taken to limit light exposure (maintaining low levels and/or switching lights off when art is not being viewed); and exposure to infrared and ultraviolet should be limited (using filters on lamps or placing art behind specially treated glass or acrylic).

Images copyright 2000 Hedrich-Blessing/Quesada

Figure 5.1 (above) and 5.2 (right)

The sanctuary space at St. Philip the Apostle Church (shown above) features a tapestry wall. This wall or niche is accented with an array of monopoints each with a 50/PAR30S/HIR/NFL lamp. With the nearly 12 foot mounting height and skewed aiming, these lamps illuminate the tapestry in undulating fashion. In Figure 5.2 (right), the icon is lighted with a single 50/PAR30S/HIR/NFL lamp on dimmer—hence the soft focal. Monopoints are Litelab. Downlights and adjustable accents are Prescolite.

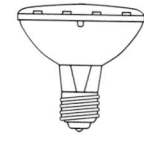

HIR/PAR30S

HIR PAR30S Visual Index

50PAR30S/HIR/NSP ▶**p26** 50PAR30S/HIR/NSP ▶**p26** 50PAR30S/HIR/NSP ▶**p27**

50PAR30S/HIR/NSP
▶ Architectural detailing
▶ Large art/18'–0" ceiling height
▶ Sculpture
▶ Long throws
▶ Hospitality

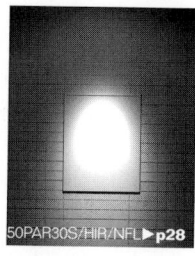

50PAR30S/HIR/NFL ▶**p28** 50PAR30S/HIR/NFL ▶**p28** 50PAR30S/HIR/NFL ▶**p28**

50PAR30S/HIR/NFL
▶ Feature downlighting
▶ Large art/12'–6" ceiling height
▶ Sculpture
▶ Medium throws
▶ Hospitality

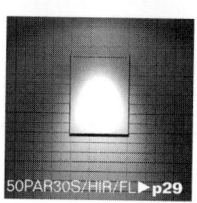

50PAR30S/HIR/FL ▶**p29** 50PAR30S/HIR/FL ▶**p29** 50PAR30S/HIR/FL ▶**p29**

50PAR30S/HIR/FL
▶ General lighting
▶ Medium art/9'–6" ceiling height
▶ Short throws
▶ Hospitality

HIR/PAR Lamp Designation

50PAR30S/HIR/NSP

Last set of letters identifies beam spread

FLOODS
- 24° to 32° for narrow flood (NFL)
- 33° to 44° for flood (FL)
- 45° to 55° for wide flood (WFL)
- 55° or greater for very wide flood (VWFL)

SPOTS
- 7° or less for very narrow spot (VNSP)
- 8° to 10° for narrow spot (NSP)
- 11° to 14° for spot (SP)
- 15° to 18° for wide spot (WSP)
- 19° to 23° for very wide spot (VWSP)

Ranges of beam spreads used for convenient/consistent reference in this text and do not necessarily correspond to each manufacturer's nor any ANSI or NEMA definitions.

These letters indicate "Halogen InfraRed"

"S" for short neck lamps and "L" for long neck (only applies to PAR30 lamps)

These two digits represent lamp diameter in eighths inches (e.g., 30/8 or 3–3/4 inches)

These letters indicate this is a parabolic aluminized reflector (PAR) lamp

First two or three digits represent lamp wattage

[This designation used for convenient reference throughout this *Time–Saver Standards for Architectural Lighting*—but not necessarily used by all lamp manufacturers (although it would be convenient if manufacturers would come to agreement on standard designations.]

Artwork Size Categories
⤵ Petite = 2'–0" by 2'–0" or smaller
⤵ Small = 2'–0" by 2'–0" to 2'–6" by 3'–0"
⤵ Medium = 2'–6" by 3'–0" to 3'–0" by 4'–0"
⤵ Large = 3'–0" by 4'–0" to 4'–0" by 5'–0"
⤵ Extra Large = 4'–0" by 5'–0" or larger

HIR Mains Voltage (120V)

5

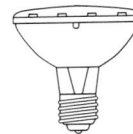

HIR50PAR30S/HIR/NSP

Short Neck

Performance Sketches

[All data outlined here is based on information from boldfaced manufacturer's published data for 120V version at time of manuscript preparation. Scale is ¼ inch equals 1 foot.]

Uses

► Architectural detailing
► Large art with long throws
► Sculpture with long throws
► Long throws
► Available in 120V and 130V

Specs

Beam spread/8°
**Center beam
  candlepower/17000**
Life/3000 hrs
► **GE** (19902)

Design Tips

✓ One light best for large art where mounting height for light is 19' or less.
✓ Intense light degrades some artwork rapidly and/or significantly—consider UV and IR filters or limit exposure time.
✓ Great for glass pieces and floral centerpieces and where mounting height for light is 19' or less.
✓ Shadows are less severe with flatter light (compare bottom image to top image).
✓ Consider halogen low voltage 35PAR36/H/NSP.

Lamp: 50PAR30S/HIR/NSP
Location: 5'–0" from wall
Center beam Aiming: 21° (at point 4'–9" AFF)
Luminaire: Monopoint or Recessed Adjustable
Illuminance: 12 fc, avg, maintained (30 fc max)

Lamp: 50PAR30S/HIR/NSP
Location: 6'–0" from wall
Center beam Aiming: 24° (at point 4'–9" AFF)
Luminaire: Monopoint or Recessed Adjustable
Illuminance: 13 fc avg, maintained (33 fc max)

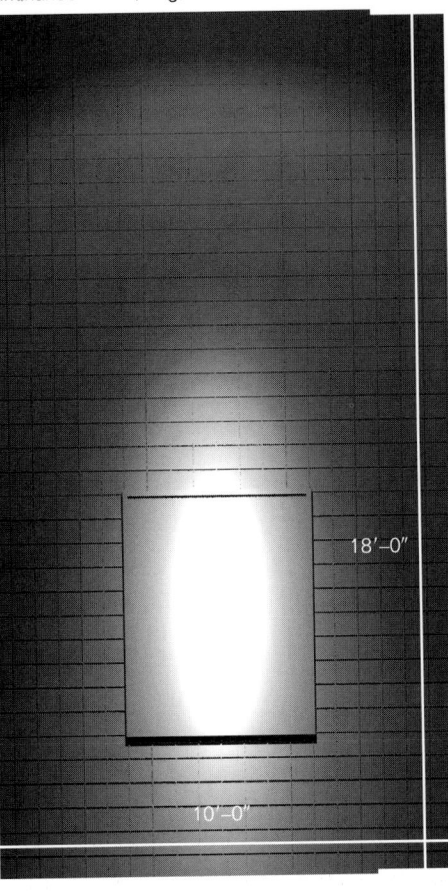

18'–0"

10'–0"

HIR Mains Voltage (120V)

5

Application Key

Commercial
Gallery
Hospitality
Institutional
Manufacturing
Residential
Retail
Exterior

bold = primary application
partial fade = minimal application
fade = unlikely application

HIR50PAR30S/HIR/NSP

Short Neck

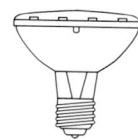

Lamp: 50PAR30S/HIR/NSP
Location: 7'–0" from wall
Center beam Aiming: 28° (at point 4'–9" AFF)
Luminaire: Monopoint or Recessed Adjustable
Illuminance: 13 fc avg, maintained (34 fc max)

Art shown at 4' by 5' and
centered 5'–6" AFF. Wall grid on
6" by 6" centers and scaled at
¼"=1'.

HIR Mains Voltage (120V)

5

HIR50PAR30S/HIR/NFL

Short Neck

Performance Sketches

[All data outlined here is based on information from boldfaced manufacturer's published data for 120V version at time of manuscript preparation. Scale is ¼ inch equals 1 foot.]

Uses
▶ Feature downlighting
▶ Large art with medium throws
▶ Sculpture with long throws
▶ Medium throws
▶ Available in 120V and 130V

Specs
Beam spread/26°
Center beam candlepower/3000
Life/3000 hrs
▶ **GE** (19901)

Design Tips
✔ One light best for large art where mounting height for light is 13′ or less.
✔ Great for moderate to larger sculpture where mounting height for light is 13′or less.
✔ Intense light degrades some artwork rapidly and/or significantly—consider UV and IR filters or limit exposure time.
✔ Good for downlighting onto decorative floor materials—best with sconces and/or art accents and/or wallwashing to avoid cave effect and grazing glare (the "I wish I had a visor" effect).
✔ Shadows are less severe with flatter light (compare bottom image to top image).

HIR — Mains Voltage (120V)
5

Application Key

Commercial
Gallery
Hospitality
Institutional
Manufacturing
Residential
Retail
Exterior

bold = primary application
partial fade = minimal application
fade = unlikely application

Lamp: 50PAR30S/HIR/NFL
Location: 3′–0″ from wall
Center beam Aiming: 18° (at point 3′–6″ AFF)
Luminaire: Monopoint or Recessed Adjustable
Illuminance: 11 fc avg, maintained (26 fc max)

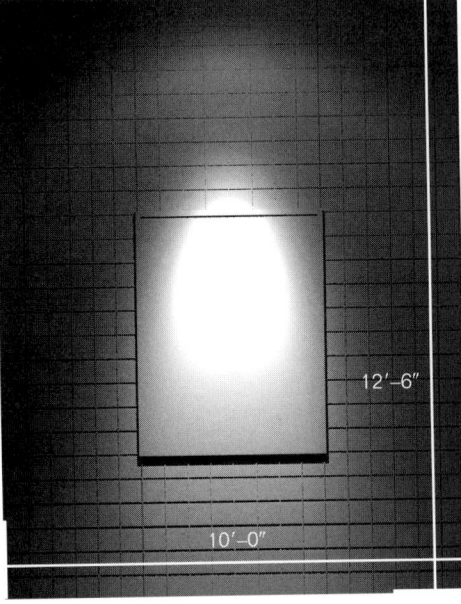

12′–6″

10′–0″

Lamp: 50PAR30S/HIR/NFL
Location: 4′–0″ from wall
Center beam Aiming: 25° (at point 4′–0″ AFF)
Luminaire: Monopoint or Recessed Adjustable
Illuminance: 12 fc avg, maintained (25 fc max)

Art shown at 4′ by 5′ and centered 5′–6″ AFF. Wall grid on 6″ by 6″ centers and scaled at ¼″=1′.

Lamp: 50PAR30S/HIR/NFL
Location: 5′–0″ from wall
Center beam Aiming: 30° (at point 4′–0″ AFF)
Luminaire: Monopoint or Recessed Adjustable
Illuminance: 12 fc avg, maintained (24 fc max)

HIR 50PAR30S/HIR/FL

Short Neck

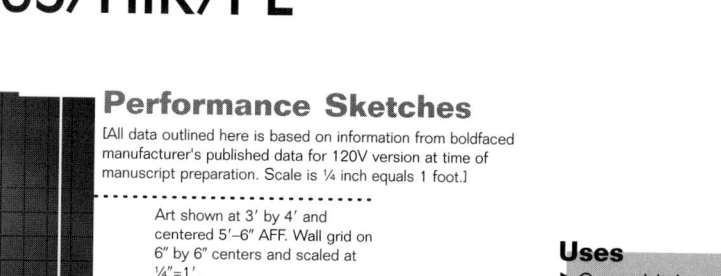

Performance Sketches

[All data outlined here is based on information from boldfaced manufacturer's published data for 120V version at time of manuscript preparation. Scale is ¼ inch equals 1 foot.]

Art shown at 3' by 4' and centered 5'–6" AFF. Wall grid on 6" by 6" centers and scaled at ¼"=1'.

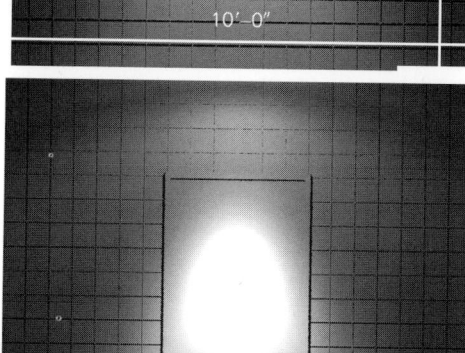

9'–6"

10'–0"

Lamp: 50PAR30S/HIR/FL
Location: 2'–0" from wall
Center beam Aiming: 13° (at point 1'–0" AFF)
Luminaire: Monopoint or Recessed Adjustable
Illuminance: 12 fc avg, maintained (29 fc max)

Lamp: 50PAR30S/HIR/FL
Location: 2'–6" from wall
Center beam Aiming: 16° (at point 1'–0" AFF)
Luminaire: Monopoint or Recessed Adjustable
Illuminance: 11 fc avg, maintained (24 fc max)

Lamp: 50PAR30S/HIR/FL
Location: 3'–0" from wall
Center beam Aiming: 22° (at point 2'–0" AFF)
Luminaire: Monopoint or Recessed Adjustable
Illuminance: 12 fc avg, maintained (25 fc max)

Uses

► General lighting
► Medium art with short throws
► Short throws
► Available in 120V and 130V

Specs

Beam spread/35°
Center beam
candlepower/1600
Life/3000 hrs
► **GE** (19900)

Design Tips

✔ One light best for medium art where mounting height for light is 10' or less.
✔ Light may degrade artwork rapidly and/or significantly: consider UV and IR filters or limit exposure time.
✔ Good for downlighting onto decorative floor materials— best with sconces and/or art accents and/or wallwashing to avoid cave effect and grazing glare (the "I wish I had a visor" effect).
✔ Larger protruding frames create more prominent shadows.

HIR Mains Voltage (120V)

5

Application Key

Commercial
Gallery
Hospitality
Institutional
Manufacturing
Residential
Retail
Exterior

bold = primary application
partial fade = minimal application
fade = unlikely application

HIR/PAR38

Stats
<small>Data varies manufacturer to manufacturer and changes from time to time</small>

Life/3000 hours
Different versions from different manufacturers. A typical office environment might be occupied about 2600 hours each year.

Efficacy/17 to 21 LPW
Varies by wattage. Good to excellent beam intensities and distributions available.

Wattages/ 50, 60, 80 and 100
Color of Light/2810°K to 2900°K
Beam spreads/NSP, SP, NFL, FL and WFL
Narrow spot, spot, narrow flood, flood and wide flood distributions are available.

Size/5⁵⁄₁₆″ L by 4³⁄₄″ Ø
Voltage/120V and 130V
Cost Magnitude/Moderate ⁽²⁰⁰⁰⁾

Manufacturers
General Electric
Osram Sylvania

Net Addresses
http://www.ge.com/lighting/business/index.htm
http://ecom.sylvania.com/osicatalog/

CONNECT FOR MORE

Advantages
Crisp white light
Good to excellent intensities available
Standard voltage (no transformers)
Easily dimmed
Low initial cost
Alternative to less efficient R lamps
Alternative to less efficient halogen/PAR lamps
Withstand extreme temperature range
Withstand moisture contact
◆When used in wet–rated sockets or wet–rated open luminaires.
Very little lamp lumen depreciation (light loss) over life

Disadvantages
Beam aberrations
◆Striations
◆Inconsistent beam shape
High operating costs
◆Inefficiency and short life may make this cost prohibitive in typical commercial applications, depending on ease of changing lamps and maintenance staffing.
Hot to touch
Hot-shock sensitive
◆Lamp may fail if rough aiming–adjustment or sharp vibration occurs while lamp is energized.
Large/bulky
◆HIRPAR30S lamps are smaller.

For more distinct accent light, use 12–volt Halogen PAR36 lamps or CMH PAR lamps.

HIR/PAR38

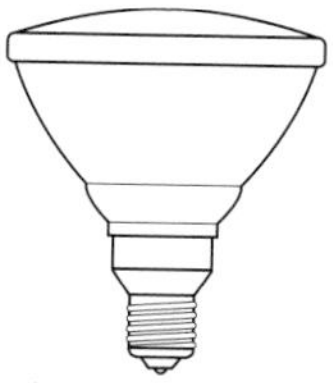

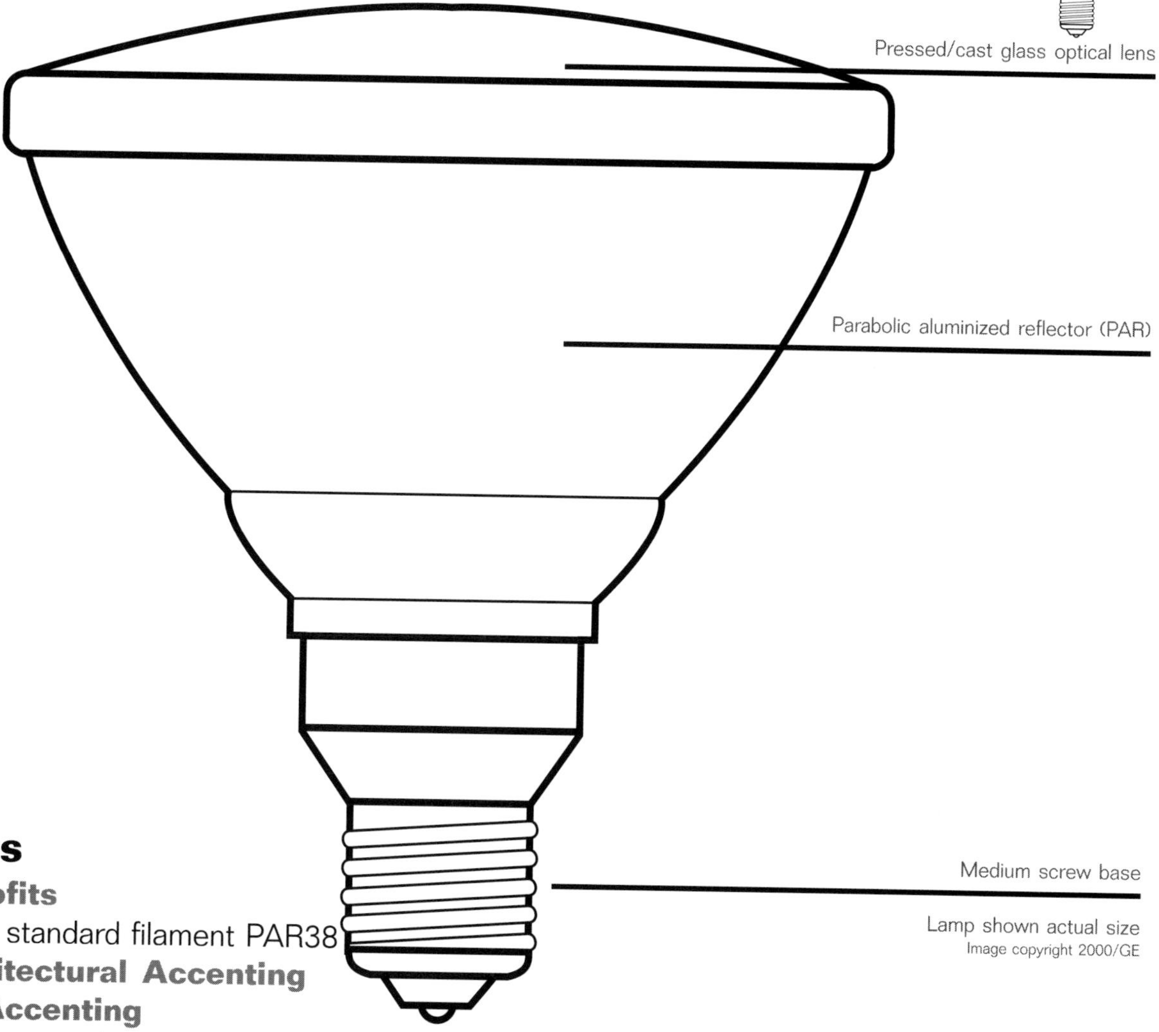

Pressed/cast glass optical lens

Parabolic aluminized reflector (PAR)

Medium screw base

Lamp shown actual size
Image copyright 2000/GE

HIR Mains Voltage (120V)

5

Uses

Retrofits

Any standard filament PAR38

Architectural Accenting

Art Accenting

Commercial

Gallery

Hospitality

Residential

Downlighting

Circulation

Conference rooms

Hospitality

Residential

Retail

Merchandise Accenting

Surface Grazing

Specular/semispecular surfaces

◆Polished/honed stone

◆Polished/satin wood

◆Fritted/etched glass

◆Cut glass

HIR/PAR38 Lighting Design Tips

✔HIR/PAR38 spot versions best at high mounting heights (15'–0" to 20'–0").
✔HIR/PAR38 narrow flood versions best at moderate mounting heights (12'–0" to 15'–0").
✔HIR/PAR38 flood versions best at low mounting heights (8'–6" to 10'–0").
✔Dimming increases life (typically doubling life), but shifts color warmer.
✔Glary when used without much other general lighting or when aimed across long distances where ceilings are low.
✔Beam striations are most apparent on light–colored, nontextured, flat surfaces and may be objectionable.
✔HIR/PAR38 downlighting alone is harsh—combine several techniques for best results.
✔HIR/PAR38 typically rated for exposure to moisture when used in wet–rated sockets or wet–rated open luminaires—confirm with manufacturers' latest datasheets—so can be used in many open exterior utility lights, sign lights and above–grade uplights.

Net Addresses/PAR20 Track, Monopoints, Recessed and Exterior
See page 5–37

HIR/PAR38

Figure 5.3
This hotel lobby and circulation area has many natural materials and key focal elements. To maximize the visual experience which can be experienced by all visitors, halogen infrared PAR lamps were used in a variety of applications. Downlights and the continuous wall slot (above the water wall) use 100PAR38/HIR/FL lamps while adjustable accents and water wall uplights use 100PAR38/HIR/NSP lamps. Downlights and adjustable accents are Lightolier. Wall slot socket strips and water wall wet–rated socket strip are Hubbell/Sterner. Water wall uplights are Lithonia/Hydrel.

Figure 5.4 (opposite page, bottom)
Water wall uplights are on close spacings (about 1 foot on center) and are hidden in a trough at the bottom of the wall. Each uplight is lamped with a 100PAR38/HIR/NSP lamp. Every other uplight has a clear lens, with the intermediate uplights fitted with blue lenses. The top of each uplight is mounted to within about 3 inches of the waterline to minimize the loss of light through water. The slot at the juncture of the wall/ceiling houses wet–rated sockets on about 6–inch centers. A third of the lamps are amber (dipped in an amber gel); a third are red (dipped in red); and third are blue (dipped in blue). Water wall uplights are Lithonia/Hydrel. Wall slot wet–rated sockets are Hubbell/Sterner. Colored lamps were dipped by FX.

HIR Mains Voltage (120V)

5

HIR/PAR38

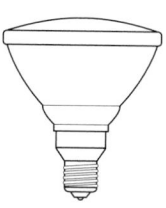

Figure 5.5
The sculptural panels at the registration desk are lighted by 100PAR38/HIR/NSP lamps in trackheads hidden in a continuous slot several feet in front of the wall. These highlight specific features of each panel without washing out the entire piece. Track and trackheads are Litelab.

Image copyright 2000 C.M. Korab

Image copyright 2000 C.M. Korab

HIR Mains Voltage (120V)

5

HIR/PAR38

HIR PAR30SVisual Index

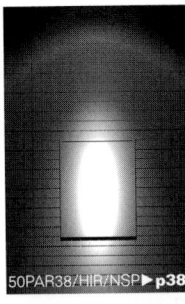  

50PAR38/HIR/NSP
► Architectural detailing
► Large art/15'–0" ceiling height
► Sculpture
► Long throws
► Hospitality

50PAR38/HIR/NFL
► Feature downlighting
► Large art/12'–6" ceiling height
► Sculpture
► Medium throws
► Hospitality

Throws/Ceilings
- ⊃ Short = 7' to 10'
- ⊃ Medium = 10' to 15'
- ⊃ Long = 15' to 20'
- ⊃ Very Long = 20' or more

60PAR38/HIR/NSP
► Architectural detailing
► Large art/18'–0" ceiling height
► Sculpture
► Long throws
► Hospitality

60PAR38/HIR/NFL25°
► Architectural accenting
► Feature downlighting
► Large art/12'–6" ceiling height
► Sculpture
► Medium throws
► Commercial, Hospitality, Retail

Recommended intensities
for art accenting

Low (gallery, museum, residential)
5 to 10 fc average on art
Moderate (hospitality)
10 to 20 fc average on art
High (commercial—or higher for retail)
20 to 30 fc average on art

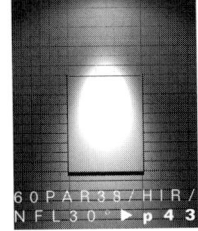 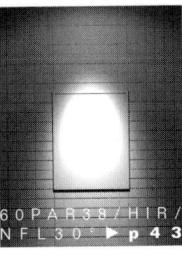 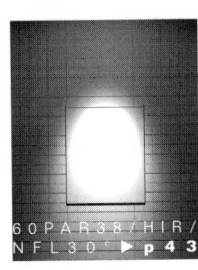

60PAR38/HIR/NFL30°
► Feature downlighting
► Large art/12'–6" ceiling height
► Sculpture
► Medium throws
► Hospitality

These are intended to be average, maintained values including effects of accent lighting and general room lighting. General room lighting contributes just a few footcandles in residential, gallery and museum settings, but contributes 5 to 10 fc in commercial settings. Maximums might range from 50 percent to 100 percent of the average value. Wherever artwork is highly valuable and/or sensitive to light, precautions should be taken to limit light exposure (maintaining low levels and/or switching lights off when art is not being viewed); and exposure to infrared and ultraviolet should be limited (using filters on lamps or placing art behind specially treated glass or acrylic).

HIR/PAR38

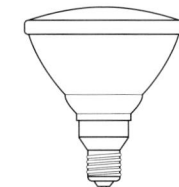

 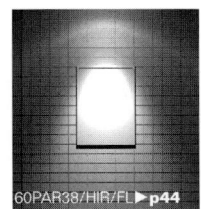 

60PAR38/HIR/FL
► General lighting
► Medium art/10'–6" ceiling height
► Medium throws
► Hospitality

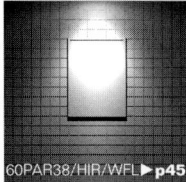 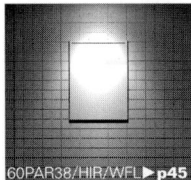 

60PAR38/HIR/WFL
► Soft general lighting
► Wallwashing
► Medium art/9'–6" ceiling height
► Short throws
► Hospitality

  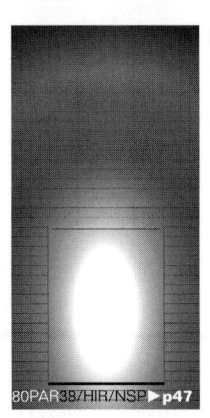

80PAR38/HIR/NSP
► Strong architectural detailing
► Extra large art/20'–0" ceiling height
► Sculpture
► Very long throws
► Hospitality

 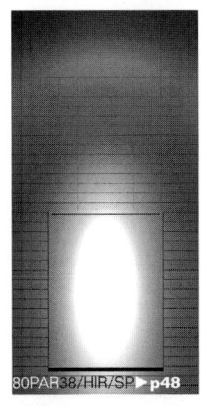 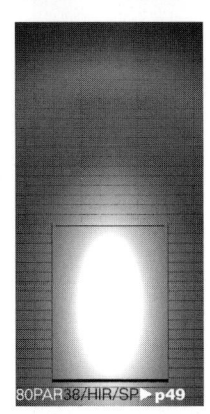

80PAR38/HIR/SP
► Strong architectural detailing
► Extra large art/20'–0" ceiling height
► Sculpture
► Very long throws
► Hospitality

80PAR38/HIR/NFL
► Feature downlighting
► Large art/12'–6" ceiling height
► Sculpture
► Medium throws
► Commercial, Retail

HIR Mains Voltage (120V)

5

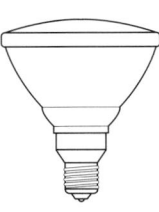

HIR/PAR38

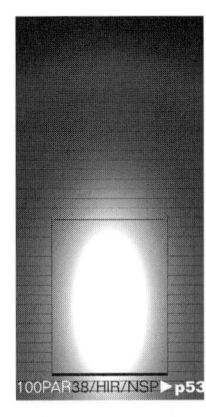

100PAR38/HIR/NSP
▶Strong architectural detailing
▶Extra large art/20'–0" ceiling height
▶Sculpture
▶Very long throws
▶Hospitality

Throws/Ceilings
⊃Short = 7' to 10'
⊃Medium = 10' to 15'
⊃Long = 15' to 20'
⊃Very Long = 20' or more

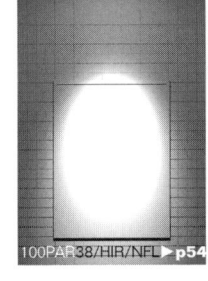

100PAR38/HIR/NFL
▶Feature downlighting
▶Extra large art/15'–0" ceiling height
▶Sculpture
▶Long throws
▶Hospitality

100PAR38/HIR/FL
▶Feature downlighting
▶Large art/12'–6" ceiling height
▶Sculpture
▶Medium throws
▶Hospitality

HIR Mains Voltage (120V)

5

HIR/PAR Lamp Designation
50PAR38/HIR/NSP

Last set of letters identifies beam spread

FLOODS
• 24° to 32° for narrow flood (NFL)
• 33° to 44° for flood (FL)
• 45° to 55° for wide flood (WFL)
• 55° or greater for very wide flood (VWFL)

SPOTS
• 7° or less for very narrow spot (VNSP)
• 8° to 10° for narrow spot (NSP)
• 11° to 14° for spot (SP)
• 15° to 18° for wide spot (WSP)
• 19° to 23° for very wide spot (VWSP)

Ranges of beam spreads used for convenient/consistent reference in this text and do not necessarily correspond to each manufacturer's nor any ANSI or NEMA definitions.

These letters indicate "Halogen InfraRed"

These two digits represent lamp diameter in eighths inches (e.g., 30/8 or 3–3/4 inches)

These letters indicate this is a parabolic aluminized reflector (PAR) lamp

First two or three digits represent lamp wattage

[This designation used for convenient reference throughout this *Time–Saver Standards for Architectural Lighting*—but not necessarily used by all lamp manufacturers (although it would be convenient if manufacturers would come to agreement on standard designations.]

HIR/PAR38

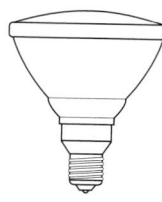

Net Addresses/PAR30S and PAR38 Track and Monopoints
http://www.bartcolighting.com/index2.html
http://www.thomasltg.com/
http://www.con-techlighting.com/
http://www.cooperlighting.com/
http://www.hubbell-ltg.com/products.htm#Down&Track
http://www.danalite.com/
http://www.kramerlighting.com/
http://www.lightingservicesinc.com/
http://www.lightolier.com/
http://www.lithonia.com/
http://www.prescolite.com/
http://www.tslight.com/

CONNECT FOR MORE

Net Addresses/PAR30S and PAR38 Recessed
http://www.thomasltg.com/
http://www.con-techlighting.com/
http://www.cooperlighting.com/
http://www.hubbell-ltg.com/products.htm#Down&Track
http://www.danalite.com/
http://www.kramerlighting.com/
http://www.lightolier.com/
http://www.lithonia.com/
http://www.prescolite.com/

CONNECT FOR MORE

Net Addresses/Exterior Accent
http://www.hadcolighting.com/html/bronzelite.htm
http://www.hydrel.com
http://www.kimlighting.com/ingrd1.html

CONNECT FOR MORE

HIR Mains Voltage (120V)

5

Recommended intensities
for art accenting

Low (gallery, museum, residential)
5 to 10 fc average on art
Moderate (hospitality)
10 to 20 fc average on art
High (commercial—or higher for retail)
20 to 30 fc average on art

These are intended to be average, maintained values including effects of accent lighting and general room lighting. General room lighting contributes just a few footcandles in residential, gallery and museum settings, but contributes 5 to 10 fc in commercial settings. Maximums might range from 50 percent to 100 percent of the average value. Wherever artwork is highly valuable and/or sensitive to light, precautions should be taken to limit light exposure (maintaining low levels and/or switching lights off when art is not being viewed); and exposure to infrared and ultraviolet should be limited (using filters on lamps or placing art behind specially treated glass or acrylic).

HIR 50PAR38/HIR/NSP

Performance Sketches

[All data outlined here is based on information from boldfaced manufacturer's published data for 120V version at time of manuscript preparation. Scale is ¼ inch equals 1 foot.]

Uses
▶ Architectural detailing
▶ Large art with long throws
▶ Sculpture with long throws
▶ Long throws
▶ Available in 120V and 130V

Specs
Beam spread/9°
Center beam
  candlepower/14000
Life/3000 hrs
▶**GE** (12396)

Design Tips
✓ One light best for large art and where mounting height for light is 16' or less.
✓ These are strong accents at lower mounting heights and deserve judicious use. Light may degrade artwork rapidly and/or significantly: consider UV and IR filters or limit exposure time.
✓ Great for glass pieces and floral centerpieces and where mounting height for light is 16' or less.
✓ Consider halogen low voltage 35PAR36/H/NSP.

HIR Mains Voltage (120V)

5

Application Key

Commercial
Gallery
Hospitality
Institutional
Manufacturing
Residential
Retail
Exterior

bold = primary application
partial fade = minimal application
fade = unlikely application

Lamp: 50PAR38/HIR/NSP
Location: 4'–3" from wall
Center beam Aiming: 22° (at point 4'–6" AFF)
Luminaire: Monopoint or Recessed Adjustable
Illuminance: 13 fc avg, maintained (42 fc max)

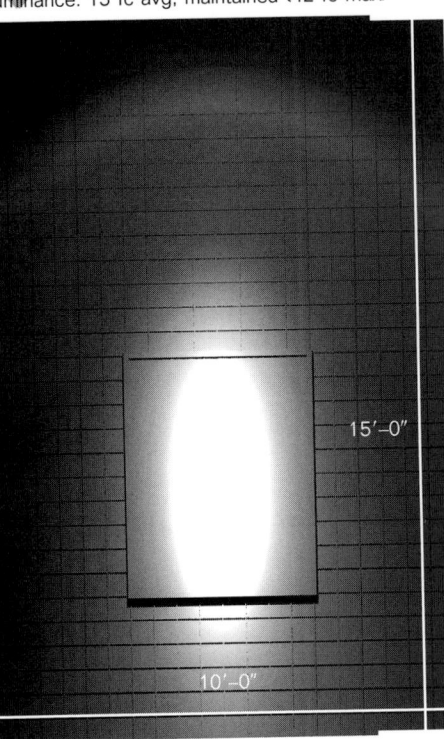

15'–0"

10'–0"

Lamp: 50PAR38/HIR/NSP
Location: 5'–3" from wall
Center beam Aiming: 27° (at point 4'–6" AFF)
Luminaire: Monopoint or Recessed Adjustable
Illuminance: 14 fc avg, maintained (44 fc max)

Art shown at 4' by 5' and centered 5'–6" AFF. Wall grid on 6" by 6" centers and scaled at ¼"=1'.

Lamp: 50PAR38/HIR/NSP
Location: 6'–3" from wall
Center beam Aiming: 31° (at point 4'–6" AFF)
Luminaire: Monopoint or Recessed Adjustable
Illuminance: 14 fc avg, maintained (46 fc max)

HIR50PAR38/HIR/NFL

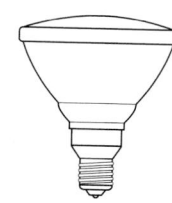

Performance Sketches

[All data outlined here is based on information from boldfaced manufacturer's published data for 120V version at time of manuscript preparation. Scale is ¼ inch equals 1 foot.]

Lamp: 50PAR38/HIR/NFL
Location: 3'–0" from wall
Center beam Aiming: 19° (at point 4'–0" AFF)
Luminaire: Monopoint or Recessed Adjustable
Illuminance: 12 fc avg, maintained (28 fc max)

12'–6"

10'–0"

Lamp: 50PAR38/HIR/NFL
Location: 4'–0" from wall
Center beam Aiming: 25° (at point 4'–0" AFF)
Luminaire: Monopoint or Recessed Adjustable
Illuminance: 12 fc avg, maintained (25 fc max)

Uses
▶ Feature downlighting
▶ Large art with medium throws
▶ Sculpture with medium to long throws
▶ Medium throws
▶ Available in 120V

Specs
Beam spread/27°
Center beam candlepower/ 3000
Life/3000 hrs
▶ **GE** (12397)

Design Tips
✔ One light best for large art and where mounting height for light is 13' or less.
✔ Light may degrade artwork rapidly and/or significantly: consider UV and IR filters or limit exposure time.
✔ Scallops/striations are most severe when light is closer to the wall (compare top left image to bottom image).
✔ Larger frames create more prominent shadows.

Art shown at 4' by 5' and centered 5'–6" AFF. Wall grid on 6" by 6" centers and scaled at ¼"=1'.

Lamp: 50PAR38/HIR/NFL
Location: 5'–0" from wall
Center beam Aiming: 30° (at point 4'–0" AFF)
Luminaire: Monopoint or Recessed Adjustable
Illuminance: 12 fc avg, maintained (24 fc max)

Application Key

Commercial
Gallery
Hospitality
Institutional
Manufacturing
Residential
Retail
Exterior

bold = primary application
partial fade = minimal application
fade = unlikely application

HIR Mains Voltage (120V)

5

HIR 60PAR38/HIR/NSP

Uses
▶ Architectural detailing
▶ Large art with long throws
▶ Sculpture with long throws
▶ Long throws
▶ Available in 120V and 130V

Specs
Beam spread/9°
Center beam
   candlepower/18250
Life/3000 hrs
▶ **GE** (spot) (18627)
▶ Osram Sylvania (spot)

Design Tips
✔ One light best for large art and where mounting height for light is 19' or less.
✔ These are strong accents at lower mounting heights and deserve judicious use. Light may degrade artwork rapidly and/or significantly: consider UV and IR filters or limit exposure time.
✔ Excellent for glass pieces and floral centerpieces and where mounting height for light is 19' or less.
✔ Consider halogen low voltage 35PAR36/H/NSP or ceramic metal halide 35PAR20/CMH/NSP.

Application Key

Commercial
Gallery
Hospitality
Institutional
Manufacturing
Residential
Retail
Exterior

bold = primary application
partial fade = minimal application
fade = unlikely application

HIR Mains Voltage (120V)

5

Performance Sketches

[All data outlined here is based on information from boldfaced manufacturer's published data for 120V version at time of manuscript preparation. Scale is ¼ inch equals 1 foot.]

Lamp: 60PAR38/HIR/NSP
Location: 5'–0" from wall
Center beam Aiming: 20° (at point 4'–6" AFF)
Luminaire: Monopoint or Recessed Adjustable
Illuminance: 11 fc avg, maintained (26 fc max)

Lamp: 60PAR38/HIR/NSP
Location: 6'–0" from wall
Center beam Aiming: 24° (at point 4'–6" AFF)
Luminaire: Monopoint or Recessed Adjustable
Illuminance: 11 fc avg, maintained (28 fc max)

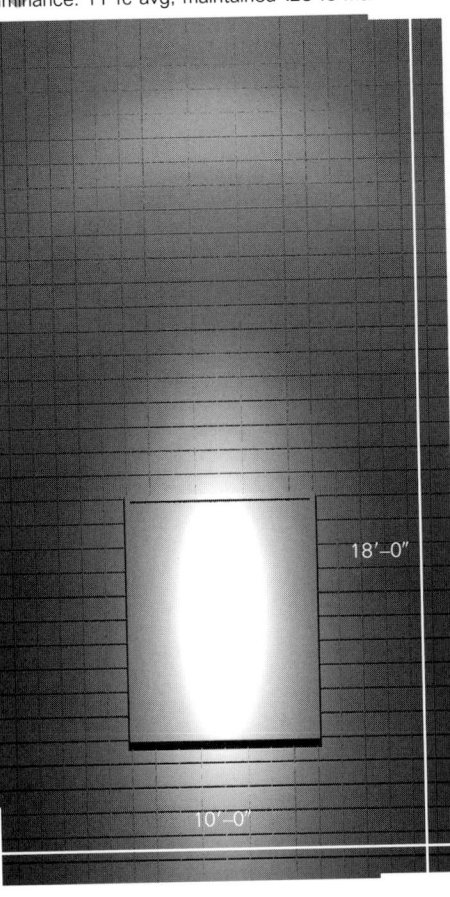

18'–0"

10'–0"

HIR60PAR38/HIR/NSP

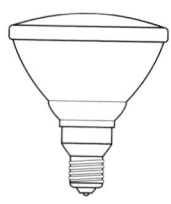

Lamp: 60PAR38/HIR/NSP
Location: 7'–0" from wall
Center beam Aiming: 27° (at point 4'–6" AFF)
Luminaire: Monopoint or Recessed Adjustable
Illuminance: 12 fc avg, maintained (29 fc max)

Art shown at 4' by 5' and
centered 5'–6" AFF. Wall grid on
6" by 6" centers and scaled at
¼"=1'.

HIR Mains Voltage (120V)

5

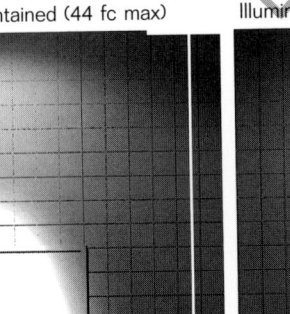

HIR60PAR38/HIR/NFL25°

Uses
► Architectural accenting
► Feature downlighting
► Large art with medium throws
► Sculpture with medium throws
► Medium throws
► Available in 120V

Specs
Beam spread/25°
Center beam candlepower/ 5100
Life/2500 hrs
► **Osram Sylvania** (14442)

Design Tips
✓ One light best for large art and where mounting height for light is 13' or less.
✓ These are strong accents and deserve judicious use. Light may degrade artwork rapidly and/or significantly: consider UV and IR filters or limit exposure time.
✓ Good for glass pieces and floral centerpieces and where mounting height for light is 13' or less.
✓ Good for downlighting onto decorative floor materials— best with sconces and/or art accents and/or wallwashing to avoid cave effect and grazing glare (the "I wish I had a visor" effect)

(vertical text, left margin:) **HIR** Mains Voltage (120V) 5

Application Key

Commercial
Gallery
Hospitality
Institutional
Manufacturing
Residential
Retail
Exterior

bold = primary application
partial fade = minimal application
fade = unlikely application

Performance Sketches

[All data outlined here is based on information from boldfaced manufacturer's published data for 120V version at time of manuscript preparation. Scale is ¼ inch equals 1 foot. **Note: At press time, actual photometry was unavailable for this lamp. Data were generated using GE's 60PAR38/HIR/NFL30° lamp and rerating candlepower—this is imprecise.**]

Lamp: 60PAR38/HIR/NFL25°
Location: 3'–0" from wall
Center beam Aiming: 18° (at point 3'–6" AFF)
Luminaire: Monopoint or Recessed Adjustable
Illuminance: 19 fc avg, maintained (44 fc max)

Lamp: 60PAR38/HIR/NFL25°
Location: 4'–0" from wall
Center beam Aiming: 25° (at point 4'–0" AFF)
Luminaire: Monopoint or Recessed Adjustable
Illuminance: 21 fc avg, maintained (43 fc max)

12'–6"

10'–0"

Art shown at 4' by 5' and centered 5'–6" AFF. Wall grid on 6" by 6" centers and scaled at ¼"=1'.

Lamp: 60PAR38/HIR/NFL25°
Location: 5'–0" from wall
Center beam Aiming: 32° (at point 4'–6" AFF)
Luminaire: Monopoint or Recessed Adjustable
Illuminance: 22 fc avg, maintained (41 fc max)

HIR60PAR38/HIR/NFL30°

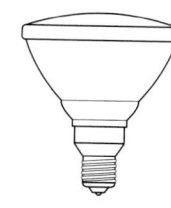

Performance Sketches

[All data outlined here is based on information from boldfaced manufacturer's published data for 120V version at time of manuscript preparation. Scale is ¼ inch equals 1 foot.]

Lamp: 60PAR38/HIR/NFL30°
Location: 3'–0" from wall
Center beam Aiming: 18° (at point 3'–6" AFF)
Luminaire: Monopoint or Recessed Adjustable
Illuminance: 13 fc avg, maintained (31 fc max)

Lamp: 60PAR38/HIR/NFL30°
Location: 4'–0" from wall
Center beam Aiming: 25° (at point 4'–0" AFF)
Luminaire: Monopoint or Recessed Adjustable
Illuminance: 15 fc avg, maintained (31 fc max)

Art shown at 4' by 5' and centered 5'–6" AFF. Wall grid on 6" by 6" centers and scaled at ¼"=1'.

Lamp: 60PAR38/HIR/NFL30°
Location: 5'–0" from wall
Center beam Aiming: 32° (at point 4'–6" AFF)
Luminaire: Monopoint or Recessed Adjustable
Illuminance: 16 fc avg, maintained (30 fc max)

Uses

▶ Feature downlighting
▶ Large art with medium throws
▶ Sculpture with medium throws
▶ Medium throws
▶ Available in 120V and 130V

Specs

Beam spread/30°
Center beam candlepower/ 3600
Life/2750 hrs
▶ **GE** (flood) (18626)
▶ Osram Sylvania (flood)

Design Tips

✓ One light best for large art where mounting height for light is 13' or less.
✓ Great for moderate to large sculpture where mounting height for light is 13' or less.
✓ Scallops, striations and frame shadows are most severe when light is closer to the wall (compare top left image to bottom image).
✓ Good for downlighting onto decorative floor materials— best with sconces and/or art accents and/or wallwashing to avoid cave effect and grazing glare (the "I wish I had a visor" effect)

Application Key

Commercial
Gallery
Hospitality
Institutional
Manufacturing
Residential
Retail
Exterior

bold = primary application
partial fade = minimal application
fade = unlikely application

HIR Mains Voltage (120V)

5

HIR60PAR38/HIR/FL

Performance Sketches

[All data outlined here is based on information from boldfaced manufacturer's published data for 120V version at time of manuscript preparation. Scale is ¼ inch equals 1 foot.]

Uses
▶ General lighting
▶ Medium art with medium throws
▶ Medium throws
▶ Available in 120V

Specs
Beam spread/40°
Center beam candlepower/2000
Life/3000 hrs
▶**GE** (10467)

Design Tips
✔ One light best for medium art and where mounting height for light is 11' or less.
✔ Good for downlighting—best with sconces and/or art accents and/or wallwashing to avoid cave effect.

Lamp: 60PAR38/HIR/FL
Location: 2'–0" from wall
Center beam Aiming: 12° (at point 1'–0" AFF)
Luminaire: Monopoint or Recessed Adjustable
Illuminance: 16 fc avg, maintained (37 fc max)

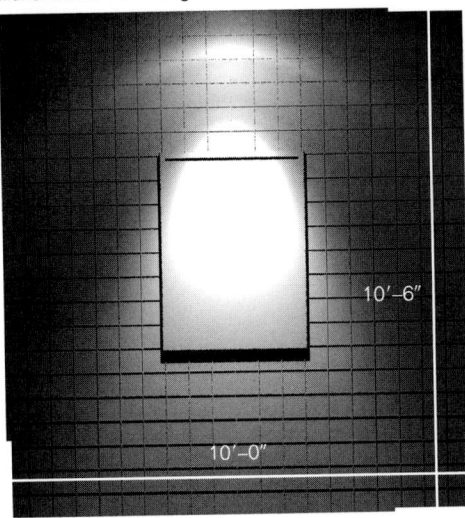

Lamp: 60PAR38/HIR/FL
Location: 2'–6" from wall
Center beam Aiming: 15° (at point 1'–0" AFF)
Luminaire: Monopoint or Recessed Adjustable
Illuminance: 15 fc avg, maintained (30 fc max)

HIR Mains Voltage (120V)

5

Art shown at 3' by 4' and centered 5'–6" AFF. Wall grid on 6" by 6" centers and scaled at ¼"=1'.

Lamp: 60PAR38/HIR/FL
Location: 3'–0" from wall
Center beam Aiming: 19° (at point 2'–0" AFF)
Luminaire: Monopoint or Recessed Adjustable
Illuminance: 16 fc avg, maintained (29 fc max)

Application Key

Commercial
Gallery
Hospitality
Institutional
Manufacturing
Residential
Retail
Exterior

bold = primary application
partial fade = minimal application
fade = unlikely application

HIR60PAR38/HIR/WFL

Performance Sketches

[All data outlined here is based on information from boldfaced manufacturer's published data for 120V version at time of manuscript preparation. Scale is ¼ inch equals 1 foot.]

Art shown at 3' by 4' and centered 5'-6" AFF. Wall grid on 6" by 6" centers and scaled at ¼"=1'.

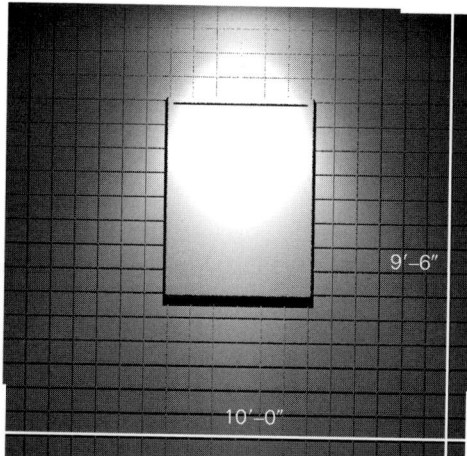
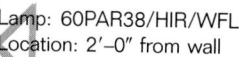

Lamp: 60PAR38/HIR/WFL
Location: 2'-0" from wall
Center beam Aiming: 12° (at point 0' AFF)
Luminaire: Monopoint or Recessed Adjustable
Illuminance: 15 fc avg, maintained (36 fc max)

Lamp: 60PAR38/HIR/WFL
Location: 2'-6" from wall
Center beam Aiming: 15° (at point 0' AFF)
Luminaire: Monopoint or Recessed Adjustable
Illuminance: 14 fc avg, maintained (26 fc max)

Lamp: 60PAR38/HIR/WFL
Location: 3'-0" from wall
Center beam Aiming: 19° (at point 1'-0" AFF)
Luminaire: Monopoint or Recessed Adjustable
Illuminance: 14 fc avg, maintained (23 fc max)

Uses

▶ Soft general lighting
▶ Wallwashing (see 11–129)
▶ Medium art with short throws
▶ Short throws
▶ Available in 120V and 130V

Specs

Beam spread/50°
Center beam candlepower/1250
Life/3000 hrs
▶**GE** (20947)

Design Tips

✔ One light best for medium art and where mounting height for light is 10' or less.
✔ Multiple units can be used for wallwashing.
✔ Good for soft downlighting— best with sconces and/or art accents and/or wallwashing to avoid cave effect.

HIR Mains Voltage (120V)

5

Application Key

Commercial
Gallery
Hospitality
Institutional
Manufacturing
Residential
Retail
Exterior

bold = primary application
partial fade = minimal application
fade = unlikely application

HIR80PAR38/HIR/NSP

Performance Sketches

[All data outlined here is based on information from boldfaced manufacturer's published data for 120V version at time of manuscript preparation. Scale is ¼ inch equals 1 foot.]

Uses
▶Strong architectural detailing
▶Extra large art with long throws
▶Sculpture with long throws
▶Very long throws
▶Available in 120V

Specs
Beam spread/10°
Center beam
   candlepower/25000
Life/3000 hrs
▶**GE** (spot) (27216)

Design Tips
✓One light best for extra large art and where mounting height for light is 21' or less.
✓These are strong accents at lower mounting heights and deserve judicious use. Light may degrade artwork rapidly and/or significantly: consider UV and IR filters or limit exposure time.
✓Excellent for glass pieces and floral centerpieces and where mounting height for light is 9' to 12'.
✓Good for dimmable downlighting in ceiling heights of 25' to 35'. [Recognize relamping issues at such heights—lifts are a necessity or top relamping is desired]
✓Consider ceramic metal halide 35PAR20/CMH/NSP.

HIR Mains Voltage (120V)

5

Application Key

Commercial
Gallery
Hospitality
Institutional
Manufacturing
Residential
Retail
Exterior

bold = primary application
partial fade = minimal application
fade = unlikely application

Lamp: 80PAR38/HIR/NSP
Location: 7'–0" from wall
Center beam Aiming: 24° (at point 4'–6" AFF)
Luminaire: Monopoint or Recessed Adjustable
Illuminance: 11 fc avg, maintained (35 fc max)

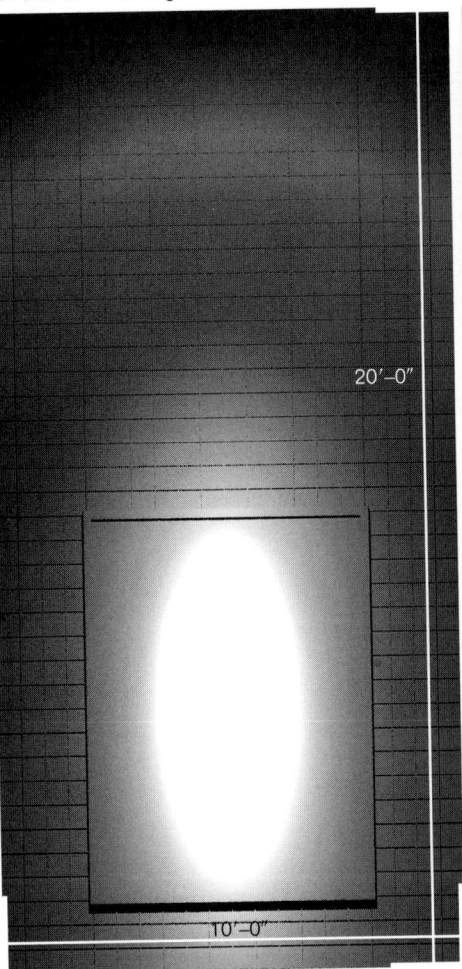

20'–0"

10'–0"

Lamp: 80PAR38/HIR/NSP
Location: 8'–0" from wall
Center beam Aiming: 27° (at point 4'–6" AFF)
Luminaire: Monopoint or Recessed Adjustable
Illuminance: 12 fc avg, maintained (36 fc max)

HIR80PAR38/HIR/NSP

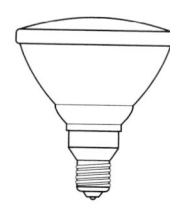

Lamp: 80PAR38/HIR/NSP
Location: 9'–0" from wall
Center beam Aiming: 30° (at point 4'–6" AFF)
Luminaire: Monopoint or Recessed Adjustable
Illuminance: 12 fc avg, maintained (36 fc max)

Art shown at 6' by 8' and centered 5'–6" AFF. Wall grid on 6" by 6" centers and scaled at ¼"=1'.

HIR Mains Voltage (120V)

5

HIR80PAR38/HIR/SP

Performance Sketches

[All data outlined here is based on information from boldfaced manufacturer's published data for 120V version at time of manuscript preparation. Scale is ¼ inch equals 1 foot.]

Uses
► Strong architectural detailing
► Extra large art with long throws
► Sculpture with long throws
► Very long throws
► Available in 120V

Specs
Beam spread/12°
Center beam candlepower/19000
Life/3000 hrs
► **GE** (27217)

Design Tips
✔ One light best for extra large art and where mounting height for light is 21′ or less.
✔ These are strong accents at lower mounting heights and deserve judicious use. Light may degrade artwork rapidly and/or significantly: consider UV and IR filters or limit exposure time.
✔ Excellent for glass pieces and floral centerpieces and where mounting height for light is 21′ or less.
✔ Good for dimmable downlighting in ceiling heights of 25′ to 35′. [Recognize relamping issues at such heights—lifts are a necessity or top relamping is desired]
✔ Consider ceramic metal halide 35PAR20/CMH/NSP.

HIR Mains Voltage (120V)

5

Application Key

Commercial
Gallery
Hospitality
Institutional
Manufacturing
Residential
Retail
Exterior

bold = primary application
partial fade = minimal application
fade = unlikely application

Lamp: 80PAR38/HIR/SP
Location: 7′–0″ from wall
Center beam Aiming: 24° (at point 4′–6″ AFF)
Luminaire: Monopoint or Recessed Adjustable
Illuminance: 12 fc avg, maintained (30 fc max)

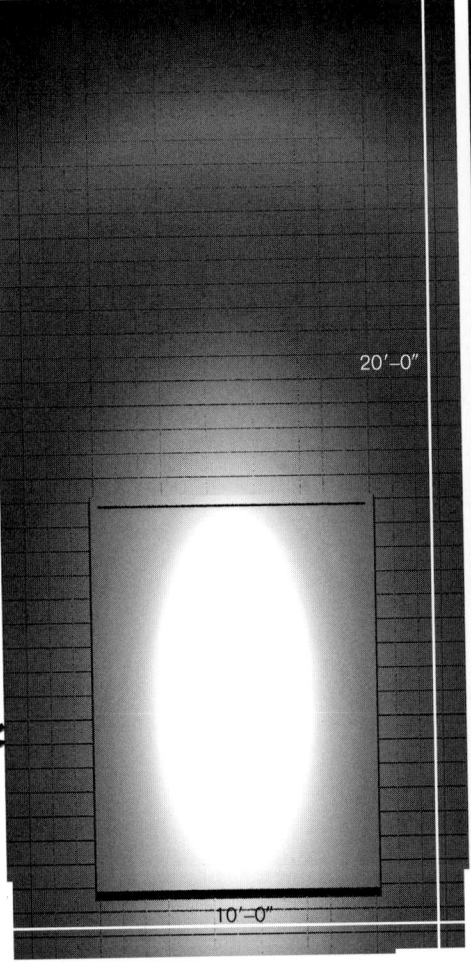

Lamp: 80PAR38/HIR/NSP
Location: 8′–0″ from wall
Center beam Aiming: 27° (at point 4′–6″ AFF)
Luminaire: Monopoint or Recessed Adjustable
Illuminance: 12 fc avg, maintained (31 fc max)

HIR80PAR38/HIR/SP

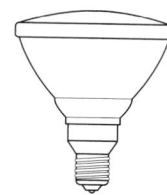

Lamp: 80PAR38/HIR/NSP
Location: 9′–0″ from wall
Center beam Aiming: 30° (at point 4′–6″ AFF)
Luminaire: Monopoint or Recessed Adjustable
Illuminance: 12 fc avg, maintained (31 fc max)

Art shown at 6′ by 8′ and
centered 5′–6″ AFF. Wall grid on
6″ by 6″ centers and scaled at
¼″=1′.

HIR Mains Voltage (120V)

5

HIR80PAR38/HIR/NFL

Performance Sketches

[All data outlined here is based on information from boldfaced manufacturer's published data for 120V version at time of manuscript preparation. Scale is ¼ inch equals 1 foot.]

Uses
▶ Feature downlighting
▶ Large art with medium throws
▶ Sculpture with medium throws
▶ Medium throws
▶ Available in 120V

Specs
Beam spread/25°
Center beam candlepower/ 5500
Life/3000 hrs
▶ **GE** (flood) (27218)

Design Tips
✔ One light best for large art where mounting height for light is 13′ or less.
✔ Great for moderate to large sculpture where mounting height for light is 13′ or less.
✔ Scallops, striations and frame shadows are most severe when light is closer to the wall (compare top left image to bottom image).

HIR Mains Voltage (120V)

5

Application Key

Commercial
Gallery
Hospitality
Institutional
Manufacturing
Residential
Retail
Exterior

bold = primary application
partial fade = minimal application
fade = unlikely application

Lamp: 80PAR38/HIR/NFL
Location: 4′–0″ from wall
Center beam Aiming: 25° (at point 4′–0″ AFF)
Luminaire: Monopoint or Recessed Adjustable
Illuminance: 21 fc avg, maintained (42 fc)

12′–6″

10′–0″

Lamp: 80PAR38/HIR/NFL
Location: 5′–0″ from wall
Center beam Aiming: 30° (at point 4′–0″ AFF)
Luminaire: Monopoint or Recessed Adjustable
Illuminance: 21 fc avg, maintained (38 fc max)

Art shown at 4′ by 5′ and centered 5′–6″ AFF. Wall grid on 6″ by 6″ centers and scaled at ¼″=1′.

Lamp: 80PAR38/HIR/NFL
Location: 6′–0″ from wall
Center beam Aiming: 35° (at point 4′–0″ AFF)
Luminaire: Monopoint or Recessed Adjustable
Illuminance: 21 fc avg, maintained (35 fc max)

HIR100PAR38/HIR/NSP

Uses
▶Strong architectural detailing
▶Extra large art with very long throws
▶Sculpture with very long throws
▶Very long throws
▶Available in 120V and 130V

Specs
Beam spread/10°
Center beam candlepower/29000
Life/3000 hrs
▶**GE** (spot) (18635)

Design Tips
✔One light best for extra large art and where mounting height for light is 21′ or less.
✔These are very strong accents at lower mounting heights and deserve judicious use. Light may degrade artwork rapidly and/or significantly—consider UV and IR filters or limit exposure time.
✔Excellent for glass pieces and floral centerpieces and where mounting height for light is 21′ or less.
✔Good for dimmable downlighting in ceiling heights of 30′ to 40′. [Recognize relamping issues at such heights—lifts are a necessity or top relamping is desired]

HIR Mains Voltage (120V)

5

Application Key

Commercial
Gallery
Hospitality
Institutional
Manufacturing
Residential
Retail
Exterior

bold = primary application
partial fade = minimal application
fade = unlikely application

Performance Sketches

[All data outlined here is based on information from boldfaced manufacturer's published data for 120V version at time of manuscript preparation. Scale is ¼ inch equals 1 foot.]

Lamp: 100PAR38/HIR/NSP
Location: 7′–0″ from wall
Center beam Aiming: 24° (at point 4′–6″ AFF)
Luminaire: Monopoint or Recessed Adjustable
Illuminance: 14 fc avg, maintained (42 fc max)

Lamp: 100PAR38/HIR/NSP
Location: 8′–0″ from wall
Center beam Aiming: 27° (at point 4′–6″ AFF)
Luminaire: Monopoint or Recessed Adjustable
Illuminance: 15 fc avg, maintained (44 fc max)

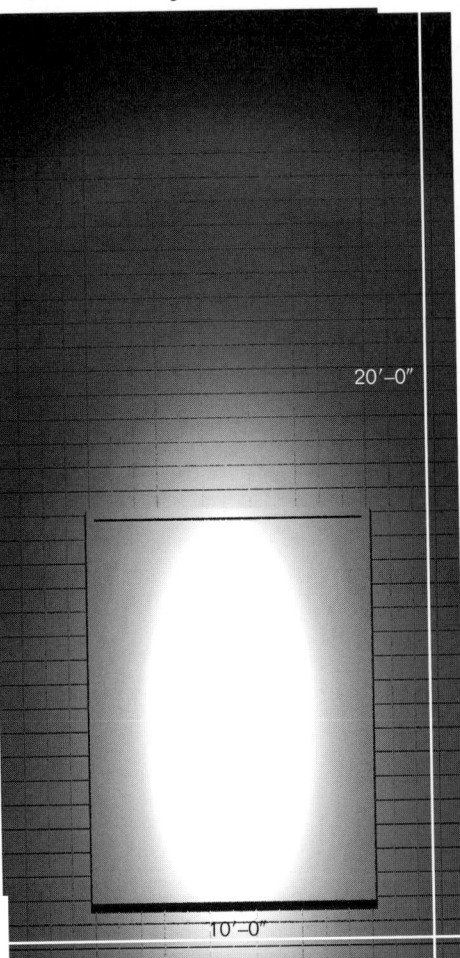

HIR100PAR38/HIR/NSP

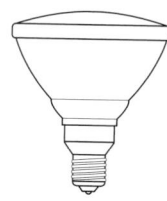

Lamp: 100PAR38/HIR/NSP
Location: 9'–0" from wall
Center beam Aiming: 30° (at point 4'–6" AFF)
Luminaire: Monopoint or Recessed Adjustable
Illuminance: 15 fc avg, maintained (45 fc max)

Art shown at 6' by 8' and
centered 5'–6" AFF. Wall grid on
6" by 6" centers and scaled at
¼"=1'.

HIR Mains Voltage (120V)

5

HIR 100PAR38/HIR/NFL

Uses

► Feature downlighting
► Extra large art with long throws
► Sculpture with long throws
► Long throws
► Available in 120V and 130V

Specs

Beam spread/27°
Center beam candlepower/
7500
Life/3000 hrs
► **GE** (flood) (18631)

Design Tips

✓ One light best for large art and where mounting height for light is 16' or less.
✓ Good for moderate to larger sculpture where mounting height for light is 16' or less.
✓ Good for downlighting onto decorative floor materials— best with sconces and/or art accents and/or wallwashing to avoid cave effect and grazing glare (the "I wish I had a visor" effect)
✓ Consider ceramic metal halide 35PAR30/CMH/NFL.

Application Key

Commercial
Gallery
Hospitality
Institutional
Manufacturing
Residential
Retail
Exterior

bold = primary application
partial fade = minimal application
fade = unlikely application

Performance Sketches

[All data outlined here is based on information from boldfaced manufacturer's published data for 120V version at time of manuscript preparation. Scale is ¼ inch equals 1 foot.]

Lamp: 100PAR38/HIR/NFL
Location: 4'–0" from wall
Center beam Aiming: 21° (at point 4'–6" AFF)
Luminaire: Monopoint or Recessed Adjustable
Illuminance: 15 fc avg, maintained (42 fc max)

15'–0"

10'–0"

Lamp: 100PAR38/HIR/NFL
Location: 5'–0" from wall
Center beam Aiming: 27° (at point 5'–0" AFF)
Luminaire: Monopoint or Recessed Adjustable
Illuminance: 17 fc avg, maintained (43 fc max)

Art shown at 6' by 8' and centered 5'–6" AFF. Wall grid on 6" by 6" centers and scaled at ¼"=1'.

Lamp: 100PAR38/HIR/NFL
Location: 6'–0" from wall
Center beam Aiming: 31° (at point 5'–0" AFF)
Luminaire: Monopoint or Recessed Adjustable
Illuminance: 17 fc avg, maintained (40 fc max)

HIR Mains Voltage (120V)

5

HIR100PAR38/HIR/FL

Performance Sketches

[All data outlined here is based on information from boldfaced manufacturer's published data for 120V version at time of manuscript preparation. Scale is ¼ inch equals 1 foot.]

Lamp: 100PAR38/HIR/FL
Location: 3'–0" from wall
Center beam Aiming: 14° (at point 0' AFF)
Luminaire: Monopoint or Recessed Adjustable
Illuminance: 16 fc avg, maintained (33 fc max)

Lamp: 100PAR38/HIR/FL
Location: 3'–6" from wall
Center beam Aiming: 16° (at point 0' AFF)
Luminaire: Monopoint or Recessed Adjustable
Illuminance: 16 fc avg, maintained (29 fc max)

12'–6"

10'–0"

Art shown at 4' by 5' and centered 5'–6" AFF. Wall grid on 6" by 6" centers and scaled at ¼"=1'.

Lamp: 100PAR38/HIR/FL
Location: 4'–0" from wall
Center beam Aiming: 19° (at point 1'–0" AFF)
Luminaire: Monopoint or Recessed Adjustable
Illuminance: 17 fc avg, maintained (29 fc max)

Uses
▶Feature downlighting
▶Large art with medium throws
▶Sculpture with medium throws
▶Medium throws
▶Available in 120V

Specs
Beam spread/40°
Center beam candlepower/3400
Life/3000 hrs
▶**GE** (10473)

Design Tips
✔One light best for large art and where mounting height for light is 13' or less.
✔Light may degrade artwork rapidly and/or significantly: consider UV and IR filters or limit exposure time.
✔Larger protruding frames create more prominent shadows.

HIR Mains Voltage (120V)

5

Application Key

Commercial
Gallery
Hospitality
Institutional
Manufacturing
Residential
Retail
Exterior

bold = primary application
partial fade = minimal application
fade = unlikely application

HIR Lamps
Low Voltage (12V)

6

This section covers 12–volt (low voltage) halogen infrared incandescent lamps. Halogen infrared (aka halogen IR or HIR) low voltage lamps are among the most efficacious incandescent lamps available for general architectural use. There are also mains voltage (120–volt) HIR lamps (see *HIR Lamps—Mains Voltage (120V)* in Section 5). HIR low voltage lamps are typically longer lived than halogen and HIR mains voltage lamps. Although this lamp category is just emerging, look for HIR low voltage lamps to become widely used as available wattage and beamspread options grow and as energy efficiency and sustainability pressures increase.

Where incandescent lamps are a necessity, use mains voltage or low voltage HIR lamps. This will result in the most efficient incandescent lighting application.

HIR Lamps
Low Voltage (12V)

Low voltage halogen infrared lamps are a result of marrying the relatively new halogen infrared technology with very low voltages (presently 12V). The halogen technology, more completely discussed in Section 3, offers excellent efficacy and life advantages over standard tungsten incandescent lamps. The halogen infrared technology, discussed in Section 5, improves this efficacy by nearly 40 percent. Operating lamps at low voltages allow for miniaturization of the filament and halogen capsule, which allows for much better optical control of the light, thereby leading to excellent beam control and very efficient focusing—hence more punch, candlepower or intensity per watt.

Low voltage HIR lamps are used in architectural lighting in commercial, gallery, hospitality residential and retail projects as accents for artwork, architectural detailing or for merchandising.

HIR low voltage lamp color characteristics

Color of light (color temperature) of low voltage HIR lamps is "crisp white" like their mains voltage counterparts. These lamps are very appropriate for contemporary interiors. When used with straw color filters or when dimmed, however, they are also appropriate in more traditional interiors. Indeed, given their typically small–aperture housings, these lamps may be more appropriate for traditional interiors than their mains voltage counterparts.

Color rendering of the low voltage HIR lamps is excellent—rated a perfect 100 by industry standards. As such, low voltage MR16 HIR lamps are excellent for highlighting art.

HIR low voltage dimming

Dimming low voltage HIR lamps can increase lamp life and also will warm the color of light somewhat, although type of dimmer selection is important (see *Low voltage transformers* below). Dimming helps to take the edge off the light and makes HIR low voltage an efficient alternative in more traditional residential and hospitality settings. Since low voltage HIR lamps have small filaments which are surrounded by a capsule, there is little hum from dimming that is more common with standard tungsten filament incandescent lamps.

The halogen cycle, for which the lamp is named, results in improved efficacy and long life in HIR lamps over standard tungsten incandescent lamps. This same cycle, however, can be "short circuited" with continual dimming. Hence, where dimming is em-

HIR Lamps
Low Voltage (12V)

Table 6.1 Low Voltage (12V) HIR Lamps

Low Voltage HIR Lamp	Use/Replaces	Lamp Profile
MR16	New • Halogen MR16 • Standard Incandescent R20 • Standard Incandescent R30 • Standard Incandescent 12v PAR36 • Standard Incandescent R40 Retrofit • Halogen MR16	

Lamp image courtesy of and copyright by GE 2000.

ployed, occasionally the lamps must be operated at full bright to allow the halogen cycle to operate.

Low voltage transformers

For most applications, low voltage lamps are powered by transformers—devices which transform the mains voltage (e.g., 120 volts) to the lamp's operating voltage (e.g., 12 volts). There are two types of transformers—electromagnetic and electronic. Both are essentially "black box" devices which contain wire coils and/or electronic components sealed away from view. Transformers require additional power—typically 3 to 5 watts per lamp.

Electronic transformers are typically smaller than electromagnetic. Where dimming is involved, however, care must be taken in specifying a dimmer which is compatible with the transformer—and electromagnetic transformers are consistently more reliable than electronic versions when dimming is involved. This aspect cannot be overemphasized. For electromagnetic transformers, use dimmers which are designed specifically to operate electromagnetic transformers. Similarly, for electronic transformers, use only dimmers rated for control of electronic transformers. Otherwise, expect significant system dysfunction—from poor dimming range and non–smooth dimming to humming, short lamp life and/or short transformer life.

Low voltage transformers can be designed to operate a single lamp and therefore are small enough to be integral to the luminaire. For track fittings, however, the transformers are usually larger than

Ceramic metal halide (CMH) PAR lamps (Section 7) are more efficacious than HIR low voltage lamps.

HIR Low Voltage (12V)

HIR Lamps
Low Voltage (12V)

the lamp and socket assembly. For these applications, it may be desirable to specify a larger transformer to operate several lamps on a track which is rated for low voltage operation. These larger transformers need to be remotely located, but within proximity of the lamps—otherwise voltage drop (the loss of voltage over length of wire run) is sufficient to operate the lamps in a permanently dimmed (and lower output) state. Further, these remote transformers need to be located in accessible, appropriately ventilated and sound isolated areas. Access is necessary for maintenance replacement. Ventilation is required to limit the amount of ambient heat in the area of the transformer (the higher the ambient temperature, typically the less efficient the transformer operation and the more likely the transformer will fail prematurely) and to prevent overheating of the transformer. Sound isolation is a necessity, as the larger the transformer the more likely humming or buzzing will occur. In any event, installation must conform to manufacturer's instructions and code requirements.

■ ■ ■ ■ ■ ■ ■ ■ ■ ■ ■ ■ ■

Low voltage halogen dimming

❶ Increases lamp life—doubling it, but only if lamps are brightened to full output occasionally.
❶ Yellows color of light—more cozy, intimate appearance.
❶ Results in little or no filament hum (which is common in standard tungsten filament incandescent).
❶ Electromagnetic transformers are historically more reliable and typically quieter than electronic transformers counterparts.
❶ Only successful when dimmer switch is matched to the transformer type—electromagnetic or electronic.

Low voltage HIR lamp hazards

The hazards of halogen lamps have been highly publicized in recent years. The very reason a halogen lamp operates so efficiently is the same reason it can be a hazard—the high–temperature, pressurized operation of the filament. While HIR low voltage lamps are primarily available in wattages up to 50 watts—much less than the 300– and 500–watt mains voltage linear halogen lamps used in many torchieres—their small size makes them a serious hazard. Low voltage HIR lamps' flammability hazard due to very infrequent "non–passive failure"—exploding—can be reduced with tempered cover glass filters. Additional precaution should be taken by maintaining appropriate distances between lamps and any materials and surfaces.

Tempered cover glass filters help contain any hot filament bits and capsule shards should the lamp fail violently. Common sense should be exercised when placing lamps/luminaires—this follows for any lamp type. Spot lamps of higher wattage act similarly to focusing the sun's rays through a magnifying glass—surface materials within a few inches or even a foot or so of a high wattage spot lamp could

HIR Low Voltage (12V)

6

HIR Lamps
Low Voltage (12V)

ignite depending on the type of material and the ambient conditions. This suggests that millwork–integrated lighting must be limited in wattage and beam spread and that installation must conform to lamp and luminaire manufacturers' instructions and code requirements.

Ultraviolet light output is higher in low voltage halogen lamps than in standard tungsten filament incandescent lamps. Artwork requires protection from this UV, otherwise degradation of the pigments, colors and/or base media (e.g., paper) may be significant and irreparable over a fairly short period of time. Similarly, people require protection from this UV. Tempered glass lenses help reduce UV light output from halogen lamps. For sensitive art, however, UV reduction lenses or filters should be used with halogen lamps.

When to use low voltage HIR

Low voltage HIR lamps are appropriate on a variety of projects. Commercial, hospitality, residential and retail spaces can all be appropriate applications for low voltage HIR lamps. In commercial spaces, art accenting helps to minimize the "haze" or "fog" of the uniform fluorescent lighting. Here, HIR/MR16 lamps are appropriate. Projects where small, dimmable accent lights are required and/or where construction budgets are moderate, life cycle budgets are of less importance or intended use is not prolonged are candidates for the HIR/MR16 lamps. Further, where fine, low–level dimming is a critical criterion or where the budget cannot support fluorescent dimming, low voltage HIR/MR lamps are justified. Some hospitality facilities (e.g., restaurants, banquet facilities, meeting facilities) and commercial facilities fit this profile. Executive office suites fitted with motion sensors (to limit operational time) and retail spaces are also candidates for low voltage HIR lighting. Finally, where historic restoration is required and where architectural or fixture detailing accommodates the small low voltage lamp, it can be used as supplemental accent and/or functional lighting without betraying its presence.

Low voltage HIR hazards

- Lamps may fail violently—exploding. Only use luminaires that are UL–listed and labeled and fitted with tempered glass lenses.
- Portable low voltage halogen luminaires should be well balanced and heavily weighted to prevent tipping.
- Portable luminaires should be used infrequently and users should be warned of fire and burn hazards.
- Lamps are extremely hot—sufficient distance from people, furnishings and architectural elements is a must.
- The spottier (more focused) the beam spread, the greater the distance necessary to avoid rapid surface material degradation and/or fire.
- Ultraviolet light from low voltage halogen lamps can be damaging to humans and artwork—use UV filters.

HIR Low Voltage (12V)

6

6–5 ▲

HIR Lamps
Low Voltage (12V)

When not to use low voltage HIR

Most commercial and institutional projects deserve compact fluorescent and/or ceramic metal halide lamp solutions for the majority of the lighting. Low voltage HIR lamps do not offer the long life that these facilities require for the majority of their lighting needs in order to be cost–competitive. Where accent lighting is required in these kinds of facilities, then low voltage halogen infrared lamps or, depending on mounting heights and coverages, ceramic metal halide lamps should be considered. Table 6.2 outlines more efficacious alternatives to low voltage HIR lamps. Low voltage HIR lamps should not be used where very close proximity to people for prolonged periods is expected (e.g., task lighting)—given their high operating temperatures (resulting in higher ambient temperatures of nearby air and yielding potential for skin burns) and ultraviolet radiation. Portable task lights and floor lights using HIR lamps should be avoided in children–intensive settings, particularly where adult supervision is unlikely or only periodic.

Halogen cycle
✔ Heated filament metal boils away into vapor (as happens in standard tungsten filament incandescent lamps).
✔ Vaporized filament metal is redirected by halogen gas back onto the filament—the filament is regenerated!
✔ Results in either construction of longer life lamp or more efficient lamp or a lamp with some increase in life and some increase in light output.

Triple–tube compact fluorescent lamps (Section 8) are the efficient alternative to HIR low voltage lamps for general lighting. For efficient accent lighting in high–ceiling spaces, consider ceramic metal halide (CMH) PAR lamps (Section 7).

Table 6.2 Efficacious Alternatives to HIR Low Voltage Lamps

Low Voltage HIR Lamp	Efficacious Alternative	Hardware Application
MR16	• CMH PAR20 (ceramic metal halide)❶❷ • CMH PAR30 (ceramic metal halide)❶❷	• Accent Luminaires • Decorative Luminaires❸ • Downlight Luminaires

Note: Due to lamp's very small size, no other lamp is a reasonable alternative —larger luminaires will be necessary in order to use alternative.

❶ This alternative lamp is not dimmable inexpensively, if at all.
❷ CMH lamps will be more efficacious where very long throws occur or where very intense accenting is required.
❸ Decorative Luminaires include pendents.

HIR Low Voltage (12V)

6

HIR/MR
Low Voltage (12V)

This is a new lamp technology—combining low voltage and HIR technologies. At publication, this technology is only available in MR lamp envelopes. MR is an acronym for multifaceted reflector—are essentially slide projector lamps modified for architectural lighting applications. A very tiny filament possible because of the low voltage (12 volts) operation in a halogen capsule surrounded by a multifaceted dichroic reflector makes for a tiny, intense light source. The glass reflector has a dichroic coating which selectively transmits infrared radiation (through the back of the lamp) while reflecting visible radiation through the front of the lamp. As such, multifaceted (or multi–mirror) reflectors (MR) built around these small halogen capsules make for potent light sources at relatively low wattages— 20 to 50 watts. For architectural lighting and accenting, the MR16 low voltage HIR lamps are appropriate. Given their wattage, candle-power (punch or intensity) and beamspread, HIR/MR lamps are very efficient for accenting.

Low voltage HIR/MR lamps have life ratings of 4000 hours—this is even longer life than that of halogen and HIR mains voltage lamps. The size and shape of the facets establishes the intensity and spread of the light—a narrow spot (NSP), a narrow flood (NFL) or a flood (FL). HIR/MR16 lamps are 2 inches in diameter. Presently, these HIR/MR lamps are only available in uncovered (or open) versions—the halogen capsule is exposed and open. Hence tempered glass lenses are required over the lamp.

MR lamps, while capable of providing good intensity, are notorious for creating odd striations and scallops on walls, light colored

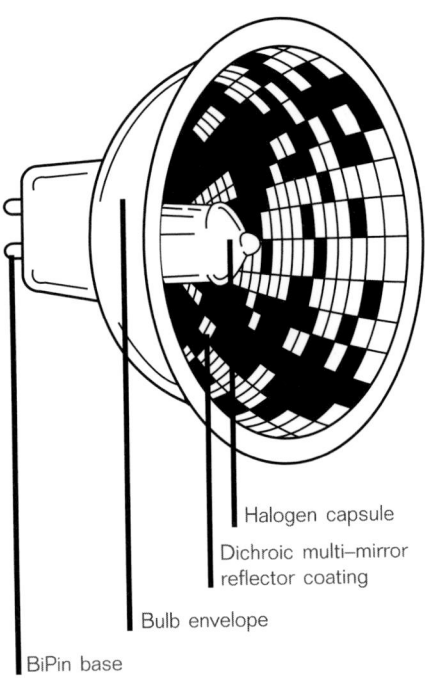

Halogen capsule

Dichroic multi–mirror reflector coating

Bulb envelope

BiPin base

MR16 lamp shown more or less actual size
Image copyright 2000/GE

Figure 6.4
MR16 lamp detail.

HIR Low Voltage (12V)

HIR/MR
Low Voltage (12V)

Table 6.3 HIR/MR Low Voltage (12V) Lamps

MR Lamp Type	Advantages	Disadvantages	Brand Names❶
HIR Hard Dichroic MR16 (MR16/HIR/C)	• 2 inch diameter • Better beam control • Excellent efficacy • 4000 hour life • Consistent color of light • Consistent color backlight	• BiPin base may be unreliable • Requires tempered glass cover	• Osram Sylvania Tru–Aim™

❶Brand names where so noted (® or ™) are registered by and/or trademarks of respective manufacturers.

artwork or merchandise. This can be mitigated with the use of tempered soft diffusion lenses.

HIR/MR low voltage lamp designation

HIR/MR low voltage lamps are typically identified with an alphanumeric designation outlined on the following page. To maintain consistency with the designations previously used in Section 4, the designation method presented on the following page is solely for convenient and consistent reference throughout the text. Actual lamp manufacturers' designations will need to be used in final specifications. Perhaps the lamp manufacturers will standardize lamp designations in the near future in an effort to ease lamp cross–referencing and specifications.

HIR/MR low voltage lamps for accenting

HIR/MR lamps are great for accenting artwork, architectural details or features and objets d'art. Since this lamp technology is quite new as of the date of this publication, there is significant potential for greater selection of wattages and beam spreads in the near future. A series of Performance Sketches™ are presented which offers quick visual assessment of what lamp wattage and beam spread might be appropriate for given ceiling heights and artwork sizes.

Performance Sketches™ are shown at a scale of ¼ inch equals 1 foot. Tracing for evaluation of various elevations is encouraged. Copying is permitted (according to the conventions outlined in "Licensing Agreement" in the Introduction of this book). Each Sketch includes an outline of artwork of the size category which

Cost magnitude[2000]
for halogen and HIR lamps

Low
US$7.⁵⁰ to US$12.⁰⁰
Moderate
US$12.⁰⁰ to US$17.⁰⁰
High
US$17.⁰⁰–plus

Costs vary based on quantities, distributor and contractor markups, and market conditions, manufacturing situations, and annual inflation. These values are for preliminary, magnitude budgeting and do not represent quotes nor actual final pricing.

HIR Low Voltage (12V)

HIR/MR

Low Voltage (12V)

■ ■ ■ ■ ■ ■ ■ ■ ■ ■

See Manufacturers' Updates:
❶Periodically due to materials' technology
❶Periodically due to manufacturing techniques
❶Periodically due to governmental regulations
❶May result in performance changes

may be lighted by the cited lamp. This is to assist in judgement of area of coverage. Location or distance from wall is listed with each Sketch as are center beam aiming angle and average light level over the area of the artwork outline. Maximum maintained intensity is also listed. For museum quality artwork where conservation of art is a priority, intensities should be 5 to 10 footcandles average, maintained, with maximums of perhaps 10 to 15 fc. In other applications, average maintained intensities on art will depend on the general or ambient lighting within the space(s). In typical hospitality applications, artwork accenting might exhibit average maintained intensities of 10 to 20 fc, with maximum intensities of 20 to 30 fc. In commercial applications, average maintained intensities on artwork might range from 20 to 30 fc, with maximums of 30 to 40 fc. These Sketches are based on the cited manufacturer's data at time of manuscript preparation. Where two or more manufacturers offer nearly identical products, only one manufacturer's data has been used for development of the sketches. Data cited in the sidebar "Specs" is averaged from all available manufacturers' data. Where any manufacturer's data did not fall within 15 percent of this aver-

HIR/MR Low Voltage Lamp Designation

37MR16/HIR/C/NFL

Last set of letters identifies beam spread

FLOODS
• 24° to 32° for narrow flood (NFL)
• 33° to 44° for flood (FL)
• 45° to 55° for wide flood (WFL)
• 55° or greater for very wide flood (VWFL)

SPOTS
• 7° or less for very narrow spot (VNSP)
• 8° to 10° for narrow spot (NSP)
• 11° to 14° for spot (SP)
• 15° to 18° for wide spot (WSP)
• 19° to 23° for very wide spot (VWSP)

Ranges of beam spreads used for convenient/consistent reference in this text and do not necessarily correspond to each manufacturer's nor any ANSI or NEMA definitions.

Reflector treatment code
• "C" for consistent color (hard dichroic coating) or no letter for inconsistent color

These letters indicate "Halogen InfraRed"

These two digits represent lamp diameter in eighths inches (51mm or 2 inches)

These letters indicate this is a multifaceted reflector (MR) lamp

First two digits represent lamp wattage

[This designation used for convenient reference throughout this *Time–Saver Standards for Architectural Lighting*—but not necessarily used by all lamp manufacturers (although it would be convenient if manufacturers would come to agreement on standard designations.]

HIR

Low Voltage (12V)

6

HIR/MR
Low Voltage (12V)

aged data, the respective manufacturer's lamp is cited separately along with separate sketches. Sketches were generated with this data in Radiance (a Unix–based lighting simulation and rendering program available free of charge at www.radsite.lbl.gov/radiance/refer/long.html).

Radiance (lighting rendering software)
http://www.radsite.lbl.gov/radiance/refer/long.html

CONNECT FOR MORE

Performance Sketch
A quarter inch scale rendering illustrates lighting effect for a typical situation. Note how the area of coverage, intensity and frame shadow change with beam spread, ceiling height and distance from wall. Higher or lower ceilings and longer or shorter distances from walls will impact intensities and light patterns.

Performance Details
Details such as lamp distance from wall, aiming angle, point above floor at which lamp is aimed, and maintained average and maximum illuminance are reported. The aiming angle should be used to select a luminaire capable of achieving such an angle. Maximum illuminances are calculated maintained over time and are reported to help assess likelihood of artwork degradation. Average illuminance is calculated maintained over time just for the size of the artwork shown.

Suggested Uses/Lamp Specs
A bullet list outlines uses for which the lamp seems best suited. Lamp specifications are outlined for quick reference and to assist in specification writing. Where two or more offer nearly identical products, only one manufacturer's data (boldfaced) has been used for development of the sketches. An SKU–number or product code is included parenthetically—no such product codes were available at time of publishing on 50MR16/HIR/C lamps. Data cited in the sidebar "Specs" is averaged from all available manufacturers' data.

Design Tips
Key considerations are offered for the specific lamp.

Application Key
Those applications where the specific lamp may be most useful are highlighted in bold, while applications where the lamp may be of lesser use are faded—the softer the fade, the less appropriate is the lamp for that application.

Artwork Outline
An outline of two dimensional artwork of size that might be acceptably lighted by lamp cited.

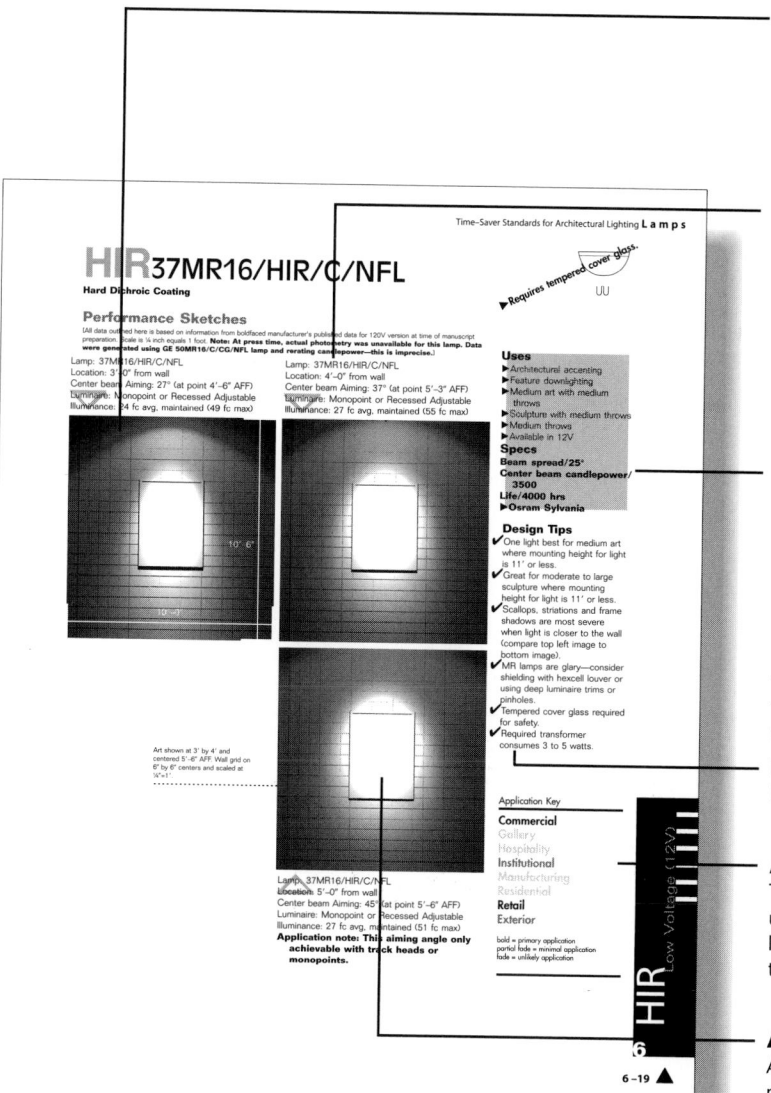

Time–Saver Standards for Architectural Lighting **L a m p s**

HIR 37MR16/HIR/C/NFL
Hard Dichroic Coating

Performance Sketches
[All data outlined here is based on information from boldfaced manufacturer's published data for 120V version at time of manuscript preparation. Scale is ¼ inch equals 1 foot. **Note: At press time, actual photometry was unavailable for this lamp. Data were generated using GE 50MR16/C/CG/NFL lamp and rerating candlepower—this is imprecise.**]

Lamp: 37MR16/HIR/C/NFL
Location: 3'-0" from wall
Center beam Aiming: 27° (at point 4'-6" AFF)
Luminaire: Monopoint or Recessed Adjustable
Illuminance: 24 fc avg, maintained (49 fc max)

Lamp: 37MR16/HIR/C/NFL
Location: 4'-0" from wall
Center beam Aiming: 37° (at point 5'-3" AFF)
Luminaire: Monopoint or Recessed Adjustable
Illuminance: 27 fc avg, maintained (55 fc max)

Art shown at 3' by 4' and centered 5'-6" AFF. Wall grid on 6' by 6' centers and scaled at ¼"=1'

Lamp: 37MR16/HIR/C/NFL
Location: 5'-0" from wall
Center beam Aiming: 45° (at point 5'-6" AFF)
Luminaire: Monopoint or Recessed Adjustable
Illuminance: 27 fc avg, maintained (51 fc max)
Application note: This aiming angle only achievable with track heads or monopoints.

▶ Requires tempered cover glass. UU

Uses
▶Architectural accenting
▶Feature downlighting
▶Medium art with medium throws
▶Sculpture with medium throws
▶Medium throws
▶Available in 12V

Specs
Beam spread/25°
Center beam candlepower/3500
Life/4000 hrs
▶Osram Sylvania

Design Tips
✔One light best for medium art where mounting height for light is 11' or less.
✔Great for moderate to large sculpture where mounting height for light is 11' or less.
✔Scallops, striations and frame shadows are most severe when light is closer to the wall (compare top left image to bottom image).
✔MR lamps are glary—consider shielding with hexcell louver or using deep luminaire trims or pinholes.
✔Tempered cover glass required for safety.
✔Required transformer consumes 3 to 5 watts.

Application Key

Commercial
Gallery
Hospitality
Institutional
Manufacturing
Residential
Retail
Exterior

bold = primary application
partial fade = minimal application
fade = unlikely application

HIR Low Voltage (12V)

6

6–19 ▲

HIR Low Voltage (12V)

6

HIR/MR
Low Voltage (12V)

Performance Sketch™ thumbnails

At the beginning of the HIR/MR16 lamp section there are thumbnails of the lighting images to further help provide quick visual assessment of various lamp beam spreads and intensities.

Lighting characteristics

In addition to beam spread and intensity, these images also illustrate beam striations and cutoff patterns. Beam striations and aberrations may appear more pronounced in reality since performance sketches herein are based on finite candlepower reports (typically at 2.5° increments—a bright spot or dead spot might occur at an increment not reported on the photometry but which could be seen visually on mockup). Color of light cannot be illustrated in this black and white format. These performance sketches will help narrow selections so that appropriate lamp mockups can be made to asess actual beam striations and color of light.

HIR/MR Lamp Applications
- ✔ Commercial
- ✔ Gallery
- ✔ Hospitality
- ✔ Residential
- ✔ Retail

Uses

For convenient reference, a shaded box entitled "Uses" appears in the upper corner of each page detailing the more likely uses for which the lamp seems suited. Artwork sizes are based on the rough dimensions as outlined in the tip box on page 6–13. Lamp "Specs" are also included in the shaded box. These specs are averages of the listed manufacturers' data at time of manuscript preparation. Since products are constantly upgraded or deleted, a check of the current status of the listed lamp is recommended before including it in specifications. Uses for HIR/MR16 lamps include:

▶ dramatic accent/medium art
 Relatively intense lighting effect on medium piece of art.

▶ dramatic accent/large art
 Relatively intense lighting effect on large piece of art.

▶ wallwashing/matte materials
 Relatively uniform wash of light on wall materials exhibiting little or no sheen, polish or specularity.

▶ wallwashing/polished materials
 Relatively uniform vertical wash of light on large areas of polished or specular wall materials.

▶ feature downlighting
 Somewhat intense lighting effect on flooring material details.

HIR/MR
Low Voltage (12V)

▶general downlighting
Relatively uniform horizontal wash of light at lap height, table height or on floor.

While there are exceptions, HIR/MR16 low voltage lamps are most appropriate for accent lighting. Small size, low wattage, and long life make these candidates for:

▶commercial

▶hospitality

▶retail

HIR/MR16 lamps are, at time of publication, of sufficient wattage and intensity that they are most appropriate for medium and long throws. Because of the limited number of lamps available at this time which use the HIR technology, there is no need for selection guides. Refer to page 6–16 for thumbnail Performance Sketches™ and application citations.

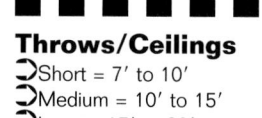

Throws/Ceilings
⊃Short = 7′ to 10′
⊃Medium = 10′ to 15′
⊃Long = 15′ to 20′
⊃Very Long = 20′ or more

Recommended intensities
for art accenting

Low (gallery, museum, residential)
5 to 10 fc average on art
Moderate (hospitality)
10 to 20 fc average on art
High (commercial—or higher for retail)
20 to 30 fc average on art

These are intended to be average, maintained values including effects of accent lighting and general room lighting. General room lighting contributes just a few footcandles in residential, gallery and museum settings, but contributes 5 to 10 fc in commercial settings. Maximums might range from 50 percent to 100 percent of the average value. Wherever artwork is highly valuable and/or sensitive to light, precautions should be taken to limit light exposure (maintaining low levels and/or switching lights off when art is not being viewed); and exposure to infrared and ultraviolet should be limited (using filters on lamps or placing art behind specially treated glass or acrylic).

HIR Low Voltage (12V)

6

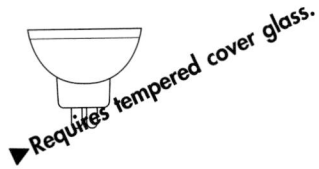

▶ Requires tempered cover glass.

HIR/MR16/C
Hard Dichroic Coating

Stats_{Data varies manufacturer to manufacturer and changes from time to time}

Life/4000

A typical office environment might be occupied about 2600 hours each year.

Efficacy/Not reported

Best beam intensities and distributions available for the wattage in the incandescent family.

Wattages/20, 37 and 50

Color of Light/Not Reported

Typically the higher wattage lamps have slightly higher color temperature.

Beam spreads/NSP, NFL and FL

Narrow spot, narrow flood and flood are available.

Size/1–7/8″ L by 2″ Ø

Voltage/12V

Cost Magnitude/High (2000)

Manufacturers

Osram Sylvania

Advantages

Excellent efficacy

Crisp white light

Very small size

Easily dimmed

Alternative to less efficient R, halogen PAR, halogen MR lamps

Withstand extreme temperature range

No color variations and color shifts

Very little lamp lumen depreciation (light loss) over life

Disadvantages

Beam aberrations
◆Striations
◆Inconsistent beam shape

Hot to touch

Odd pattern/odd coloration backlight spill
◆May be undesirable on open back track heads or monopoints

Tempered glass lens required
◆Prevents glass and filament shards from falling onto flammable surfaces/materials should a violent lamp failure occur (known as a non–passive failure)

Transformer required
◆Additional cost and visual aspect of transformer must be accommodated
◆Additional wattage requirement of transformer of 3 to 5 watts

Net Addresses
http://ecom.sylvania.com/osicatalog/

CONNECT FOR MORE

HIR Low Voltage (12V)

6

HIR/MR16/C

Hard Dichroic Coating

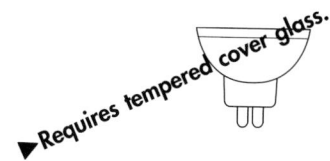
▶Requires tempered cover glass.

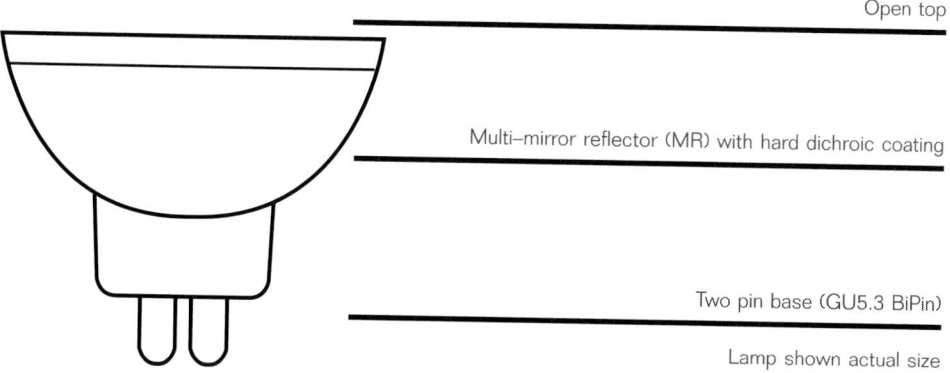

Open top

Multi–mirror reflector (MR) with hard dichroic coating

Two pin base (GU5.3 BiPin)

Lamp shown actual size
Image copyright 2000/GE

Uses

Architectural Accenting
Art Accenting
  Commercial
  Gallery
  Hospitality
  Residential
Downlighting
  Lobbies
  Conference rooms
  Hospitality
  Residential
  Retail
Merchandising
Surface Grazing
  Specular/semispecular surfaces
  ◆Polished/honed stone
  ◆Polished/satin wood
  ◆Fritted/etched glass
  ◆Cut glass

HIR/MR16 Lighting Design Tips
✔Dimming increases life (typically doubling life), but shifts color warmer.
✔Very glary when used without much other general lighting or when aimed across long distances where ceilings are low.
✔Beam striations are most apparent on light–colored, nontextured flat surfaces and may be objectionable.
✔MR16 downlighting alone is harsh—combine several techniques for best results.
✔Use smaller scale (e.g., 4–inch diameter) downlights and adjustable accents for attractive, more human scale approach.
✔Use tempered glass lens for safety.

Net Addresses/MR16 Track, Monopoints and Recessed
See page 6–21

CONNECT FOR MORE

HIR Low Voltage (12V)

6

► Requires tempered cover glass.

HIR/MR16/C

MR16 Hard Dichroic Coating Visual Index

37MR16/HIR/C/NSP
► Architectural detailing
► Medium art/16' ceiling height or less
► Sculpture
► Long throws
► Commercial, Retail

■ ■ ■ ■ ■ ■

Throws/Ceilings
⊃Short = 7' to 10'
⊃Medium = 10' to 15'
⊃Long = 15' to 20'
⊃Very Long = 20' or more

37MR16/HIR/C/NFL
► Architectural accenting
► Feature downlighting
► Medium art/11' ceiling height or less
► Sculpture
► Medium throws
► Commercial, Retail

  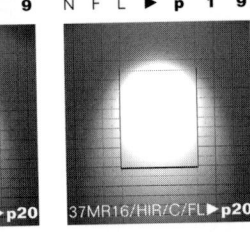

37MR16/HIR/C/FL
► General lighting
► Large art/11' ceiling height or less
► Medium throws
► Hospitality

50MR16/HIR/C/NSP
► Architectural detailing
► Large art/19' ceiling height or less
► Sculpture
► Long throws
► Hospitality

Recommended intensities
for art accenting

Low (gallery, museum, residential)
5 to 10 fc average on art
Moderate (hospitality)
10 to 20 fc average on art
High (commercial—or higher for retail)
20 to 30 fc average on art

These are intended to be average, maintained values including effects of accent lighting and general room lighting. General room lighting contributes just a few footcandles in residential, gallery and museum settings, but contributes 5 to 10 fc in commercial settings. Maximums might range from 50 percent to 100 percent of the average value. Wherever artwork is highly valuable and/or sensitive to light, precautions should be taken to limit light exposure (maintaining low levels and/or switching lights off when art is not being viewed); and exposure to infrared and ultraviolet should be limited (using filters on lamps or placing art behind specially treated glass or acrylic).

 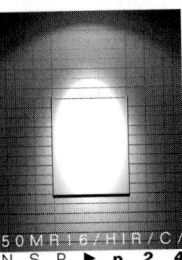 

50MR16/HIR/C/NFL
► Architectural accenting
► Feature downlighting
► Large art/13' ceiling height or less
► Sculpture
► Medium throws
► Commercial, Hospitality, Retail

50MR16/HIR/C/FL
► General lighting
► Large art/11' ceiling height or less
► Medium throws
► Commercial, Retail

HIR Low Voltage (12V)

6

HIR/MR16/C
Hard Dichroic Coating

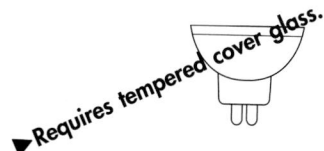
▶ Requires tempered cover glass.

HIR/MR Low Voltage Lamp Designation
37MR16/HIR/C/NFL

Last set of letters identifies beam spread

FLOODS
- 24° to 32° for narrow flood (NFL)
- 33° to 44° for flood (FL)
- 45° to 55° for wide flood (WFL)
- 55° or greater for very wide flood (VWFL)

SPOTS
- 7° or less for very narrow spot (VNSP)
- 8° to 10° for narrow spot (NSP)
- 11° to 14° for spot (SP)
- 15° to 18° for wide spot (WSP)
- 19° to 23° for very wide spot (VWSP)

Ranges of beam spreads used for convenient/consistent reference in this text and do not necessarily correspond to each manufacturer's nor any ANSI or NEMA definitions.

Reflector treatment code
- "C" for consistent color (hard dichroic coating) or no letter for inconsistent color

These two digits represent lamp diameter in eighths inches (51mm or 2 inches)

These letters indicate this is a multifaceted reflector (MR) lamp

First two digits represent lamp wattage

[This designation used for convenient reference throughout this *Time–Saver Standards for Architectural Lighting*—but not necessarily used by all lamp manufacturers (although it would be convenient if manufacturers would come to agreement on standard designations.]

Artwork Size Categories
⊃ Petite = 2'–0" by 2'–0" or smaller
⊃ Small = 2'–0" by 2'–0" to 2'–6" by 3'–0"
⊃ Medium = 2'–6" by 3'–0" to 3'–0" by 4'–0"
⊃ Large = 3'–0" by 4'–0" to 4'–0" by 5'–0"
⊃ Extra Large = 4'–0" by 5'–0" or larger

Low voltage HIR lamps will yield the most efficient incandescent lighting application.

HIR Low Voltage (12V)

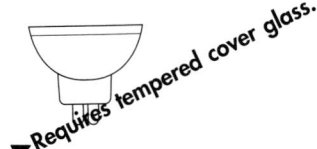

▶Requires tempered cover glass.

HIR37MR16/HIR/C/NSP

Hard Dichroic Coating

Performance Sketches

[All data outlined here is based on information from boldfaced manufacturer's published data for 120V version at time of manuscript preparation. Scale is ¼ inch equals 1 foot. **Note: At press time, actual photometry was unavailable for this lamp. Data were generated using Osram Sylvania65MR16/NSP lamp and rerating candlepower—this is imprecise.**]

Uses
▶Architectural detailing
▶Medium art with long throws
▶Sculpture with long throws
▶Long throws
▶Available in 12V

Specs
Beam spread/10°
Center beam candlepower/11500
Life/4000 hrs
▶**Osram Sylvania** (58641)

Design Tips
✔One light best for medium art where mounting height for light is 16' or less.
✔Light may degrade artwork rapidly and/or significantly: consider UV and IR filters or limit exposure time.
✔Great for glass pieces and floral centerpieces and where mounting height for light is 16' or less.
✔These are strong accents and deserve judicious use.
✔Scallops, striations and frame shadows are most severe when light is closer to the wall (compare top left image to bottom image).
✔MR lamps are glary—consider shielding with hexcell louver or using deep luminaire trims or pinholes.
✔Tempered cover glass required for safety.
✔Required transformer consumes 3 to 5 watts.

Application Key

Commercial
Gallery
Hospitality
Institutional
Manufacturing
Residential
Retail
Exterior

bold = primary application
partial fade = minimal application
fade = unlikely application

Lamp: 37MR16/HIR/C/NSP
Location: 4'–3" from wall
Center beam Aiming: 23° (at point 5'–0" AFF)
Luminaire: Monopoint or Recessed Adjustable
Illuminance: 22 fc avg, maintained (45 fc max)

Lamp: 37MR16/HIR/C/NSP
Location: 5'–3" from wall
Center beam Aiming: 28° (at point 5'–3" AFF)
Luminaire: Monopoint or Recessed Adjustable
Illuminance: 24 fc avg, maintained (46 fc max)

15'–0"

10'–0"

Art shown at 3' by 4' and centered 5'–6" AFF. Wall grid on 6" by 6" centers and scaled at ¼"=1'.

Lamp: 37MR16/HIR/C/NSP
Location: 6'–3" from wall
Center beam Aiming: 32° (at point 5'–6" AFF)
Luminaire: Monopoint or Recessed Adjustable
Illuminance: 25 fc avg, maintained (51 fc max)

HIR Low Voltage (12V)

HIR37MR16/HIR/C/NFL

Hard Dichroic Coating

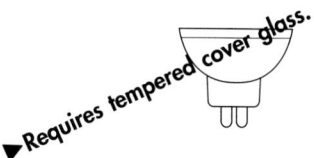

▶Requires tempered cover glass.

Performance Sketches

[All data outlined here is based on information from boldfaced manufacturer's published data for 120V version at time of manuscript preparation. Scale is ¼ inch equals 1 foot. **Note: At press time, actual photometry was unavailable for this lamp. Data were generated using GE 50MR16/C/CG/NFL lamp and rerating candlepower—this is imprecise.**]

Lamp: 37MR16/HIR/C/NFL
Location: 3'–0" from wall
Center beam Aiming: 27° (at point 4'–6" AFF)
Luminaire: Monopoint or Recessed Adjustable
Illuminance: 24 fc avg, maintained (49 fc max)

Lamp: 37MR16/HIR/C/NFL
Location: 4'–0" from wall
Center beam Aiming: 37° (at point 5'–3" AFF)
Luminaire: Monopoint or Recessed Adjustable
Illuminance: 27 fc avg, maintained (55 fc max)

10'–6"

10'–0"

Art shown at 3' by 4' and centered 5'–6" AFF. Wall grid on 6" by 6" centers and scaled at ¼"=1'.

Lamp: 37MR16/HIR/C/NFL
Location: 5'–0" from wall
Center beam Aiming: 45° (at point 5'–6" AFF)
Luminaire: Monopoint or Recessed Adjustable
Illuminance: 27 fc avg, maintained (51 fc max)
Application note: This aiming angle only achievable with track heads or monopoints.

Uses

▶Architectural accenting
▶Feature downlighting
▶Medium art with medium throws
▶Sculpture with medium throws
▶Medium throws
▶Available in 12V

Specs

Beam spread/25°
Center beam candlepower/ 3500
Life/4000 hrs
▶**Osram Sylvania** (58634)

Design Tips

✔One light best for medium art where mounting height for light is 11' or less.
✔Great for moderate to large sculpture where mounting height for light is 11' or less.
✔Scallops, striations and frame shadows are most severe when light is closer to the wall (compare top left image to bottom image).
✔MR lamps are glary—consider shielding with hexcell louver or using deep luminaire trims or pinholes.
✔Tempered cover glass required for safety.
✔Required transformer consumes 3 to 5 watts.

Application Key

Commercial
Gallery
Hospitality
Institutional
Manufacturing
Residential
Retail
Exterior

bold = primary application
partial fade = minimal application
fade = unlikely application

HIR Low Voltage (12V)

6

▶ Requires tempered cover glass.

HIR 37MR16/HIR/C/FL

Hard Dichroic Coating

Performance Sketches

[All data outlined here is based on information from boldfaced manufacturer's published data for 12V version at time of manuscript preparation. Scale is ¼ inch equals 1 foot. **Note: At press time, actual photometry was unavailable for this lamp. Data were generated using GE 50MR16/C/CG/FL lamp and rerating candlepower—this is imprecise.**]

Uses
▶ General lighting
▶ Large art with medium throws
▶ Medium throws
▶ Available in 12V

Specs
Beam spread/40°
Center beam candlepower/2050
Life/4000 hrs
▶ **Osram Sylvania** (58633)

Design Tips
✔ One light best for medium art where mounting height for light is 11' or less
✔ Light may degrade artwork rapidly and/or significantly: consider UV and IR filters or limit exposure time.
✔ Good for downlighting onto decorative floor materials—best with sconces and/or art accents and/or wallwash lighting to avoid cave effect and grazing glare (the "I wish I had a visor" effect).
✔ Larger protruding frames create more prominent shadows.
✔ MR lamps are glary—consider shielding with hexcell louver or using deep luminaire trims or pinholes.
✔ Tempered cover glass required for safety.
✔ Required transformer consumes 3 to 5 watts.

Application Key

Commercial
Gallery
Hospitality
Institutional
Manufacturing
Residential
Retail
Exterior

bold = primary application
partial fade = minimal application
fade = unlikely application

Lamp: 37MR16/HIR/C/FL
Location: 4'–0" from wall
Center beam Aiming: 34° (at point 4'–6" AFF)
Luminaire: Monopoint or Recessed Adjustable
Illuminance: 18 fc avg, maintained (39 fc max)

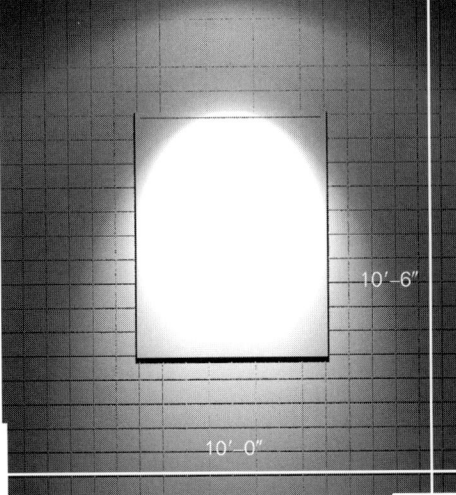

10'–6"

10'–0"

Lamp: 37MR16/HIR/C/FL
Location: 5'–0" from wall
Center beam Aiming: 42° (at point 5'–0" AFF)
Luminaire: Monopoint or Recessed Adjustable
Illuminance: 19 fc avg, maintained (37 fc max)

Art shown at 4' by 5' and centered 5'–6" AFF. Wall grid on 6" by 6" centers and scaled at ¼"=1'.

Lamp: 37MR16/HIR/C/FL
Location: 6'–0" from wall
Center beam Aiming: 47° (at point 5'–0" AFF)
Luminaire: Monopoint or Recessed Adjustable
Illuminance: 18 fc avg, maintained (31 fc max)
Application note: This aiming angle only achievable with track heads or monopoints.

HIR Low Voltage (12V)

HIR/MR16/C
Hard Dichroic Coating

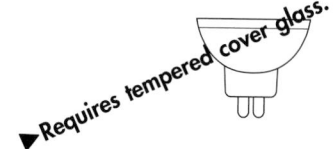

▶Requires tempered cover glass.

Net Addresses/**MR16 Track and Monopoints**
```
http://www.ardeelighting.com/index.htm
http://www.artemide.com/
http://www.bartcolighting.com/index2.html
http://www.thomasltg.com/
http://www.con-techlighting.com/
http://www.cooperlighting.com/
http://www.flosusa.com/
http://www.hubbell-ltg.com/products.htm#Down&Track
http://www.danalite.com/
http://www.kramerlighting.com/
http://www.lightingservicesinc.com/
http://www.lightolier.com/
http://www.lightproject.com/
http://www.lithonia.com/
http://www.prescolite.com/
http://www.techlighting.com/pro1.html
http://www.tslight.com/
```

CONNECT FOR MORE

Net Addresses/**MR16 Recessed**
```
http://www.alkco.com/rtrak.htm
http://www.ardeelighting.com/index.htm
http://www.artemide.com/
http://www.bartcolighting.com/index2.html
http://www.thomasltg.com/
http://www.con-techlighting.com/
http://www.cooperlighting.com/
http://www.hubbell-ltg.com/products.htm#Down&Track
http://www.danalite.com/
http://www.kramerlighting.com/
http://www.lightolier.com/
http://www.lithonia.com/
http://www.prescolite.com/
http://www.progresslighting.com/
```

CONNECT FOR MORE

Net Addresses/**Exterior Accent**
```
http://www.bega-us.com/home/home.html
http://www.hadcolighting.com/html/bronzelite.htm
http://www.hydrel.com
http://www.kimlighting.com/ingrd1.html
```

CONNECT FOR MORE

HIR Low Voltage (12V)

6

▶ Requires tempered cover glass.

HIR50MR16/HIR/C/NSP

Hard Dichroic Coating

Performance Sketches

[All data outlined here is based on information averaged/rounded from listed manufacturers' published data for 12V version at time of manuscript preparation. Scale is ¼ inch equals 1 foot. **Note: At press time, actual photometry was unavailable for this lamp. Data were generated using Osram Sylvania 65MR16/NSP lamp and rerating candlepower—this is imprecise.**]

Lamp: 50MR16/HIR/C/NSP
Location: 5'–0" from wall
Center beam Aiming: 20° (at point 4'–6" AFF)
Luminaire: Monopoint or Recessed Adjustable
Illuminance: 15 fc avg, maintained (31 fc max)

Lamp: 50MR16/HIR/C/NSP
Location: 6'–0" from wall
Center beam Aiming: 24° (at point 4'–6" AFF)
Luminaire: Monopoint or Recessed Adjustable
Illuminance: 16 fc avg, maintained (32 fc max)

18'–0"

10'–0"

Uses

▶ Architectural detailing
▶ Large art with long throws
▶ Sculpture with long throws
▶ Long throws
▶ Available in 12V

Specs

Beam spread/10°
Center beam
  candlepower/17100
Life/4000 hrs
▶ **Osram Sylvania** (code na)

Design Tips

✔ One light best for large art where mounting height for light is 19' or less.
✔ These are strong accents and deserve judicious use. Light may degrade artwork rapidly and/or significantly: consider UV and IR filters or limit exposure time.
✔ Excellent for glass pieces and floral centerpieces and where mounting height for light is 19' or less.
✔ MR lamps are glary—consider shielding with hexcell louver or using deep luminaire trims or pinholes.
✔ Tempered cover glass required for safety.
✔ Required transformer consumes 3 to 5 watts.

Application Key

Commercial
Gallery
Hospitality
Institutional
Manufacturing
Residential
Retail
Exterior

bold = primary application
partial fade = minimal application
fade = unlikely application

HIR Low Voltage (12V)

6

HIR50MR16/HIR/C/NSP

Hard Dichroic Coating

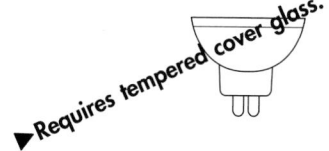

▶Requires tempered cover glass.

Lamp: 50MR16/HIR/C/NSP
Location: 7'–0" from wall
Center beam Aiming: 27° (at point 4'–6" AFF)
Luminaire: Monopoint or Recessed Adjustable
Illuminance: 17 fc avg, maintained (32 fc max)

Art shown at 4' by 5' and centered 5'–6" AFF. Wall grid on 6" by 6" centers and scaled at ¼"=1'.

HIR Low Voltage (12V)

6

► Requires tempered cover glass.

HIR50MR16/HIR/C/NFL

Hard Dichroic Coating

Uses

► Architectural accenting
► Feature downlighting
► Large art with medium throws
► Sculpture with medium throws
► Medium throws
► Available in 12V

Specs

Beam spread/25°
Center beam candlepower/ 5400
Life/4000 hrs
► **Osram Sylvania** (code na)

Design Tips

✔ One light best for large art and where mounting height for light is 13' or less
✔ Good for moderate to larger sculpture where mounting height for light is 13' or less.
✔ MR lamps are glary—consider shielding with hexcell louver or using deep luminaire trims or pinholes.
✔ Tempered cover glass required for safety.
✔ Required transformer consumes 3 to 5 watts.

Performance Sketches

[All data outlined here is based on information from boldfaced manufacturer's published data for 12V version at time of manuscript preparation. Scale is ¼ inch equals 1 foot. **Note: At press time, actual photometry was unavailable for this lamp. Data were generated using GE 50MR16/C/CG/NFL lamp and rerating candlepower—this is imprecise.**]

Lamp: 50MR16/HIR/C/NFL
Location: 3'–0" from wall
Center beam Aiming: 21° (at point 4'–6" AFF)
Luminaire: Monopoint or Recessed Adjustable
Illuminance: 19 fc avg, maintained (44 fc max)

12'–6"

10'–0"

Lamp: 50MR16/HIR/C/NFL
Location: 4'–0" from wall
Center beam Aiming: 27° (at point 4'–6" AFF)
Luminaire: Monopoint or Recessed Adjustable
Illuminance: 20 fc avg, maintained (39 fc max)

Art shown at 4' by 5' and centered 5'–6" AFF. Wall grid on 6" by 6" centers and scaled at ¼"=1'.

Lamp: 50MR16/HIR/C/NFL
Location: 5'–0" from wall
Center beam Aiming: 32° (at point 4'–6" AFF)
Luminaire: Monopoint or Recessed Adjustable
Illuminance: 20 fc avg, maintained (37 fc max)

HIR Low Voltage (12V)

Requires tempered cover glass.

HIR 50MR16/HIR/C/FL
Hard Dichroic Coating

Performance Sketches

[All data outlined here is based on information from boldfaced manufacturer's published data for 120V version at time of manuscript preparation. Scale is ¼ inch equals 1 foot. **Note: At press time, actual photometry was unavailable for this lamp. Data were generated using GE 50MR16/C/CG/FL lamp and rerating candlepower—this is imprecise.**]

Lamp: 50MR16/HIR/C/FL
Location: 4'–0" from wall
Center beam Aiming: 34° (at point 4'–6" AFF)
Luminaire: Monopoint or Recessed Adjustable
Illuminance: 22 fc avg, maintained (48 fc max)

Lamp: 50MR16/HIR/C/FL
Location: 5'–0" from wall
Center beam Aiming: 42° (at point 5'–0" AFF)
Luminaire: Monopoint or Recessed Adjustable
Illuminance: 23 fc avg, maintained (45 fc max)

10'–6"

10'–0"

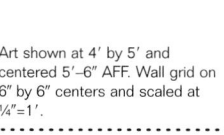

Art shown at 4' by 5' and centered 5'–6" AFF. Wall grid on 6" by 6" centers and scaled at ¼"=1'.

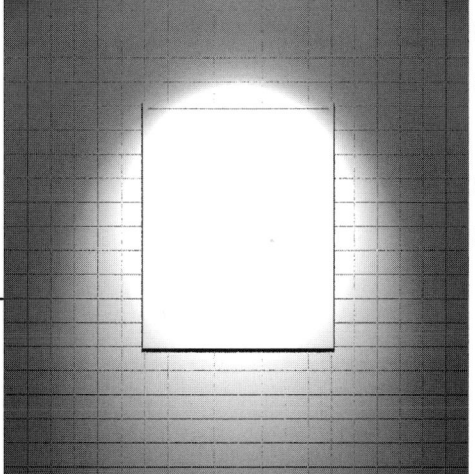

Lamp: 50MR16/HIR/C/FL
Location: 6'–0" from wall
Center beam Aiming: 47° (at point 5'–0" AFF)
Luminaire: Monopoint or Recessed Adjustable
Illuminance: 22 fc avg, maintained (38 fc max)
Application note: This aiming angle only achievable with track heads or monopoints.

Uses
▶ General lighting
▶ Large art with medium throws
▶ Medium throws
▶ Available in 12V

Specs
Beam spread/40°
Center beam candlepower/ 3000
Life/4000 hrs
▶ **Osram Sylvania** (code na)

Design Tips
✔ One light best for large art where mounting height for light is 11' or less.
✔ Light may degrade artwork rapidly and/or significantly: consider UV and IR filters or limit exposure time.
✔ Good for downlighting onto decorative floor materials—best with sconces and/or art accents and/or wallwash lighting to avoid cave effect and grazing glare (the "I wish I had a visor" effect).
✔ More grazing light creates prominent shadows.
✔ MR lamps are glary—consider shielding with hexcell louver or using deep luminaire trims or pinholes.
✔ Tempered cover glass required for safety.
✔ Required transformer consumes 3 to 5 watts.

Application Key

Commercial
Gallery
Hospitality
Institutional
Manufacturing
Residential
Retail
Exterior

bold = primary application
partial fade = minimal application
fade = unlikely application

HIR Low Voltage (12V)

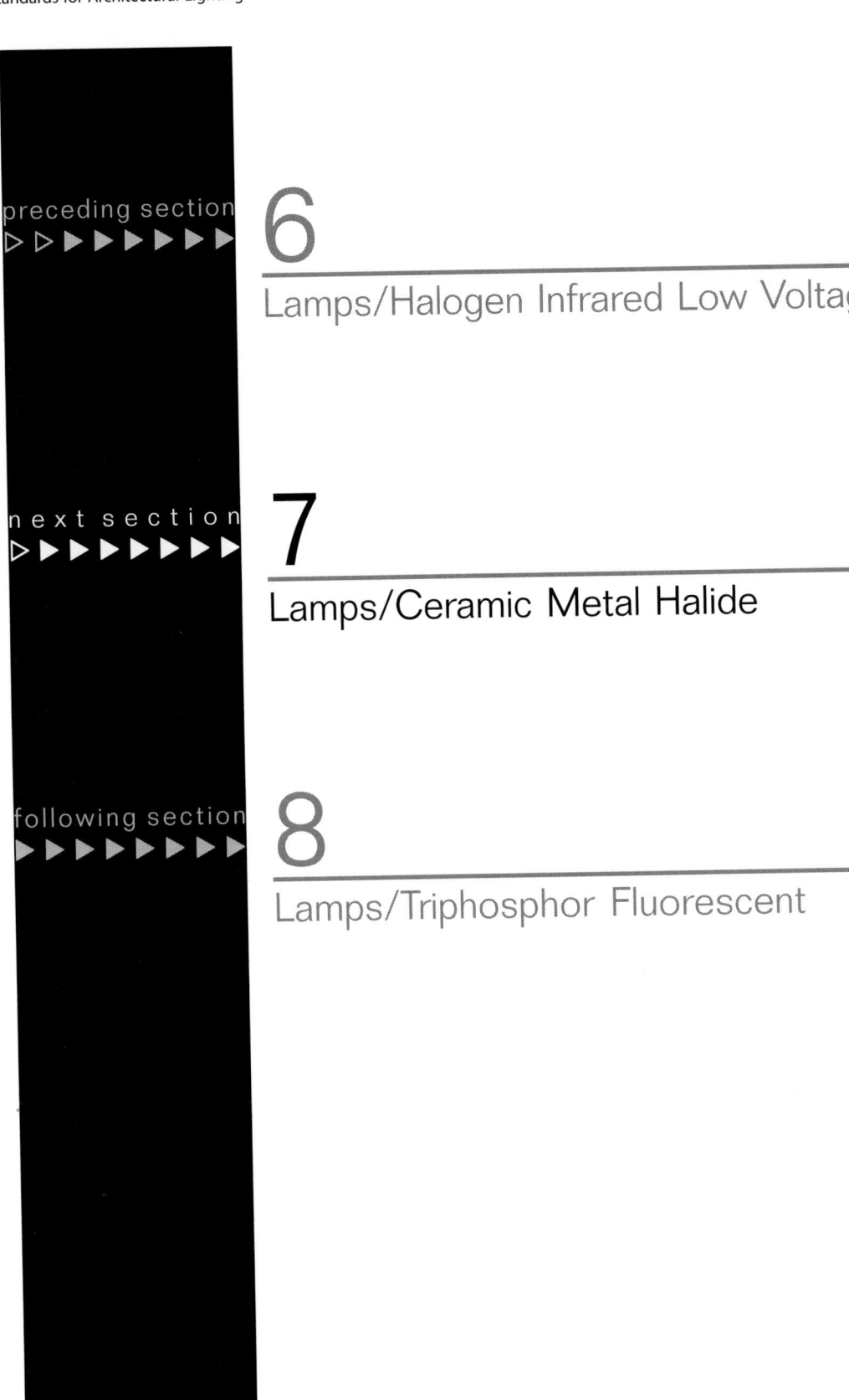

Ceramic Metal Halide Lamps

7

This section addresses ceramic metal halide lamps. For purposes of this reference book, these ceramic metal halide lamps are considered the definitive high intensity discharge (HID) lamp of the day. Standard (quartz arc tube) metal halide lamps while offering "white" light, generally provide less consistent color quality and poorer color rendering. High pressure sodium lamps, because of their awful yellow (some would say "gold" or "warm white"—both of which are too kind) color, are not addressed in this reference (HPS lamps are of little use except in storage yards, perhaps). If the world is to progress to efficient lighting which is visually effective for people, then that lighting must exhibit, at the very least, quality color characteristics and must approximate incandescent light in appearance. Only ceramic metal halide offers such quality in the HID family. At the moment, then, these ceramic metal halide lamps are the most efficient, widely applicable (across a range of interior and exterior applications) high intensity discharge lamps for accenting and downlighting.

Ceramic Metal Halide Lamps

Ceramic metal halide lamps are the efficient alternative to halogen mains voltage, halogen low voltage, halogen infrared mains voltage and halogen infrared low voltage lamps.

Ceramic metal halide lamps are an extension of the metal halide lamp family. Because of new manufacturing techniques and the advent of ceramic arc tubes, color consistency, color rendering, efficacy and life advantages have quickly propelled ceramic metal halide lamps to popularity. Ceramic metal halide (CMH) lamps are now specified in new applications instead of halogen or halogen infrared lamps. With energy costs, cooling load costs and the pollution concerns associated with electricity production, it is difficult to justify the use of halogen and halogen infrared lamps in commercial applications. Ceramic metal halide lamps should be considered the common accent lamp—the norm where dimming is not required. These lamps come in a variety of shapes for different hardware and design applications (see Table 7.1).

Table 7.1 Ceramic Metal Halide Lamps

CMH Lamp	Use/Replaces	Lamp Profile
ED17	New or Retrofit • Standard metal halide ED17	
ED17P	New or Retrofit • Standard metal halide ED17 • Standard metal halide ED17P New • High wattage incandescent A–lamps	
T6	New • High–wattage quartz halogen T3 • High–wattage quartz halogen T4	
PAR20	New • High–wattage halogen PAR38 • High–wattage halogen infrared PAR38	
PAR30L	New • High–wattage halogen PAR38 • High–wattage halogen infrared PAR38	
PAR38	New • Several high–wattage halogen PAR38 • Several high–wattage halogen infrared PAR38	

Lamp images courtesy of and copyright by Philips 2000.

Metal Halide Lamps

Ceramic metal halide lamp color characteristics

Color of light (color temperature), color rendering (how colors look under the light in question) along with excellent efficacy and relatively long life are the advantages of CMH lamps. Ceramic metal halide lamps are available in a variety of lamp shapes and wattages. Some wattages are available in warm tone and cool tone varieties. Color temperature of the warm tone CMH lamps is "crisp white" (approximating halogen and halogen infrared) at 3000°K, which is good for contemporary interiors. Where CMH lamps are used for accent lighting, this whiteness adds to the sense of brightness and attraction. There are times, particularly where videotaping and/or television broadcasting is a possibility, when a cooler toned light is desired. For such applications, consider the CMH cool tone lamp (4000°K).

■ ■ ■ ■ ■ ■ ■ ■ ■ ■ ■ ■ ■

CMH dimming

❶ Very costly

❶ Yields poor color.

❶ May yield lamp life reduction.

❶ Only available over a short "range" (typically 40 percent to 100 percent).

❶ Not suggested.

CMH lamps have excellent color rendering—perceived by most folks as rendering colors "true." As such, CMH PAR lamps are best for highlighting merchandise, art and architectural facades.

CMH dimming

Dimming CMH lamps, while a technical possibility, is not recommended. Very high system costs, operational stability of the system, color characteristics and lamp life are all likely to suffer..

CMH ballasts

CMH lamps are powered by ballasts—devices which provide sufficient electrical energy to start the lamp and maintain a constant flow of current to limit the lamp's wattage and light output to rated quantity. There are two types of ballasts—electromagnetic (also known as coil and core) and electronic. Both are essentially "black box" devices which contain wire coils and/or electronic components sealed away from view. Ballasts require additional power, which can be significant for CMH lamps. Electronic ballasts are preferred due to low energy consumption and quiet operation. Electronic ballasts

Cost magnitude^{2000}
for ceramic metal halide lamps

Low
US$25.^{00} to US$50.^{00}
Moderate
US$50.^{00} to US$75.^{00}
High
US$75.^{00} to US$100.^{00}

Costs vary based on quantities, distributor and contractor markups, and market conditions, manufacturing situations and annual inflation. These values are for preliminary, magnitude budgeting and do not represent quotes nor actual final pricing.

Ceramic Metal Halide Lamps

typically constitute an additional load of about 10 percent of the lamp watts. Hence, a 39–watt CMH lamp will, with ballast, use a total connected load of 44 watts. A 70–watt CMH lamp uses a total connected load of 78 watts and a 100–watt CMH lamps uses a total connected load of 110 watts.

There are several disadvantages introduced by ballasts. First—how to hide them! These are relatively large compared to the size of the lamp and typically must be located within 15 feet of the lamp. Each lamp requires a separate ballast. Second—ballasts are very specific to lamp wattage. So, it is not possible to "downsize" if too much light is achieved with, say, 100 watt lamps without changing ballasts and sockets. To dim CMH, neutral density filters must be used (glass or screen cloth which cuts light intensity). Conversely, it is not possible to "upsize" if too little light is achieved.

Net Addresses/Ballasts
http://www.aromat.com/metalhal.htm
http://www.wpigroup.com/

CONNECT FOR MORE

CMH hazards

While quite infrequent, metal halide lamps have been known to experience non–passive failure (in other words, a violent rupture or exploding of the lamp). Some CMH lamps are available with special glass envelopes which are intended to contain any internal arc tube rupture. These lamps are typically referred to as "Protected" and are limited to versions of PAR lamps and elliptical (ED) lamps. Any ceramic metal halide lamp not identified as "Protected" requires an enclosed luminaire around it for safe use. Hence, a non–protected lamp cannot be used in an open (unlensed) downlight.

Ultraviolet light output is higher in ceramic metal halide lamps than in standard tungsten filament incandescent lamps. Artwork requires protection from this UV, otherwise degradation of the pigments, colors and/or base media (e.g., paper) may be significant and irreparable over a fairly short period of time. Similarly, people require protection from this UV. Tempered glass lenses help reduce UV light output from halogen lamps. For sensitive art, however, UV reduction lenses or filters should be used with CMH lamps.

Ceramic
Metal Halide Lamps

When to use ceramic metal halide

CMH lamps are appropriate on projects where system longevity, energy effectiveness and color quality are important and dimming is not required and egress lighting is handled by some lighting hardware other than the CMH lighting. Typically, commercial, hospitality, institutional, retail and exterior hardscape and landscape projects fit this profile. Environments where temperature extremes exist or where the temperature is likely to remain at or below 45°F are candidates for CMH lamps.

When not to use ceramic metal halide

Most galleries (museums), residential and audiovisual intensive projects where dimming is paramount are typically best served by halogen and/or halogen infrared lamps on dimmers. Any project which requires careful dimming, continuous dimming and/or flashing of lights, or repetitive on/off switching of lights throughout the day or night are better served by halogen, halogen infrared and/or fluorescent lamps.

■ ■ ■ ■ ■ ■ ■ ■ ■ ■ ■ ■ ■ ■

CMH hazards

⊘Nonprotected CMH lamps should only be used in luminaires that are fitted with tempered glass lensing and UL–listed and labeled as appropriate for nonprotected lamps.

⊘Ultraviolet light from metal halide lamps can be damaging to humans and artwork unless UV filters are used.

⊘Where lamps operate continuously, switch off lamps for at least 15 minutes once each week (or as directed by lamp manufacturer) to minimize the potential for violent failure of lamp (this is the case with all metal halide lamps).

⊘Relamp luminaires at or prior to end of rated life (or as directed my lamp manufacturer) to reduce risk of violent failure of lamp (this is the case with all metal halide lamps).

⊘Visually inspect lamps on a periodic basis and replace any with cracked outer envelopes to prevent severe UV radiation exposure (this is the case with all metal halide lamps).

Metal Halide

Ceramic

Ceramic Metal Halide/ED17

for Enclosed Luminaires ONLY

Stats
Data varies manufacturer to manufacturer and changes from time to time

Life/10000 and 15000 hours
Dependent on color temperature and wattage. A typical office environment might be occupied about 2600 hours each year.

Efficacy/75 to 85 LPW
Wattage dependent (higher wattage/higher efficacy) (values include electronic ballast loss)

Wattages/70 and 100

Color of Light/3000°K (warm tone) and 4000°K (cool tone)

Beam spreads/Not Applicable
This is a general service lamp intended for downlights where an all–around glow of light is desired. Also appropriate for in–ground uplights and in floodlights, pedestrian postlights and parking lot lights. Do not use where distinct accent light or focused light is desired—for accent use halogen or HIR PAR lamps (Sections 3 and 5), halogen or HIR MR lamps (Sections 4 and 6) or CMH PAR lamps (this section).

Size/5⁷⁄₁₆″ L by 2⅛″ Ø

Voltage/120V and 277V
Ballast dependent (check ballast vendors for other voltages)

Cost Magnitude/Moderate ⁽²⁰⁰⁰⁾

Manufacturers
Philips

Advantages
Crisp white light (3000°K)
Much longer life than halogen TB lamps
Better alternative to less efficient standard filament A lamps
Withstand extreme temperature range

Disadvantages
Ballast required
Must be used in enclosed luminaires
◆Tempered glass lens required
Noticeable lamp lumen depreciation (light loss) over life
◆15 to 20 percent
Not dimmable
◆Without great expense
◆Without sacrificing color temperature and color rendering
Not easily integrated into egress system
Not instant–on
◆Cold start requires a few minutes for full light output
◆Power interruption initiates restrike cycle of fifteen to twenty minutes

Net Addresses/ED17
See page 7–11
CONNECT FOR MORE

For distinct accent light or focused light, use CMH PAR lamps.

Metal Halide/ED17
Ceramic
for Enclosed Luminaires ONLY

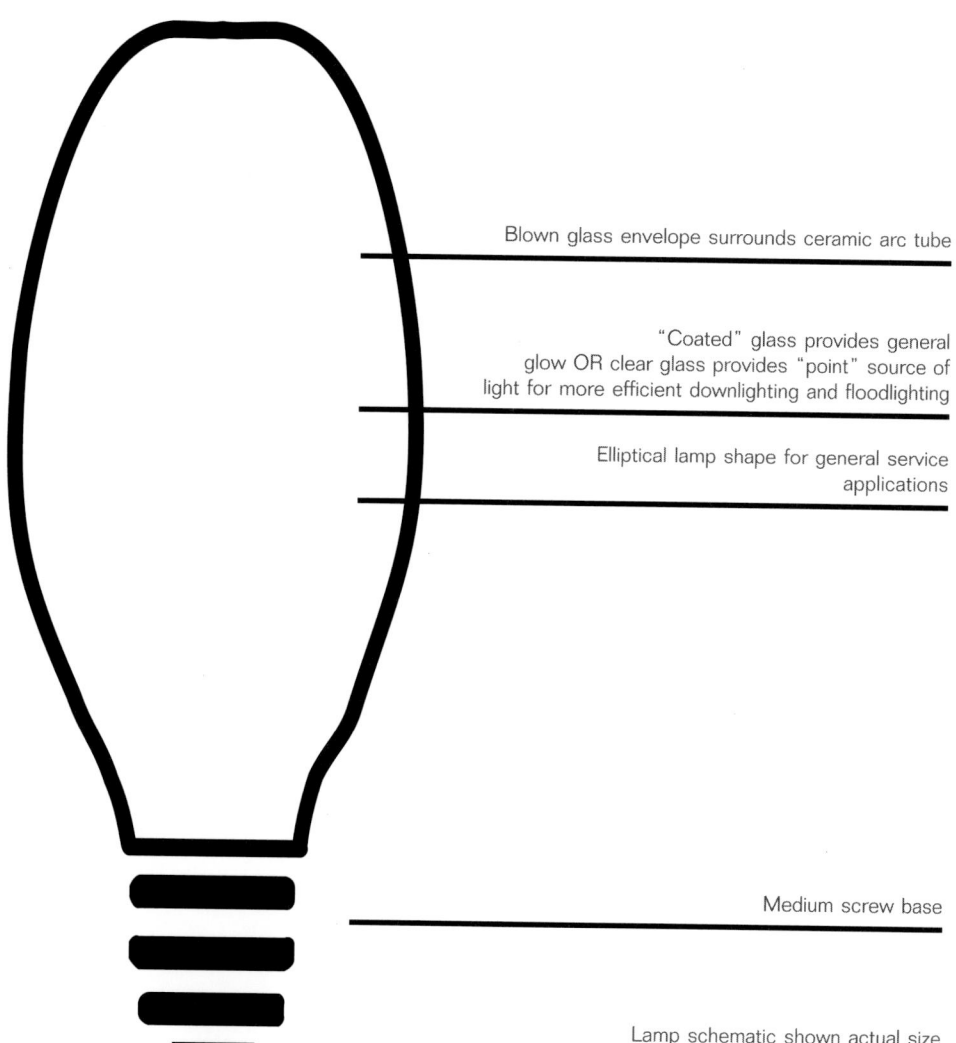

Blown glass envelope surrounds ceramic arc tube

"Coated" glass provides general glow OR clear glass provides "point" source of light for more efficient downlighting and floodlighting

Elliptical lamp shape for general service applications

Medium screw base

Lamp schematic shown actual size
Image copyright 2000/Philips

Metal Halide
c e r a m i c

7

Uses

Downlighting
Commercial
Institutional
Manufacturing
Retail

Facade Accenting

Tree Uplighting

■ ■ ■ ■ ■ ■ ■ ■ ■ ■ ■ ■ ■

CMH ED17 Lighting Design Tips
✔ Best in moderate to higher ceiling spaces (15'–0" or greater).
✔ Use only in enclosed luminaires (tempered glass lenses necessary).
✔ Coated lamps are less glary than clear lamps.
✔ Clear lamps operate most efficiently in most luminaires.

For general downlighting in low ceilings, consider compact fluorescent triple tube lamps (Section 8).

Metal Halide/ED17

Ceramic

for Enclosed Luminaires ONLY

Lamp must be used in an enclosed luminaire

Ceramic Metal Halide ED (Unprotected) Lamp Designation

70ED17/CMH/3K/C/U

Last letter identifies allowable base orientation
- **base down (BD)** [lamp base must be pointed straight down]
- **base up (BU)** [lamp base must be pointed straight up]
- **horizontal (HOR)** [lamp base must be pointed horizontally]
- **universal (U)** [lamp base may be pointed in any orientation]

This letter (if present) indicates lamp is coated or "frosted"
[no letter indicates lamp is clear]

This number/letter combination indicates color temperature
- **3K is shorthand for 3000°K (warm tone)**

This series of letters indicates "Ceramic Metal Halide"

These two digits represent lamp diameter in eighths inches (e.g., 17/8 or 2–1/8 inches)

These letters indicate this is an ellipsoidal envelope (ED) lamp

First two or three digits represent lamp wattage

[This designation used for convenient reference throughout this *Time–Saver Standards for Architectural Lighting*—but not necessarily used by all lamp manufacturers (although it would be convenient if manufacturers would come to agreement on standard designations.]

Ceramic Metal Halide/ED17

for Enclosed Luminaires ONLY

Performance Data

[All data outlined here is based on information from boldfaced manufacturer's published data at time of manuscript preparation.]

▷ **70 watt/3000°K/Clear (70ED17/CMH/3K/U)** ◀ *Best efficacy*
Lumens/6200
Life/10000 hrs
Application notes
  • Protective tempered glass lens required (use in enclosed luminaire)
  • Clear lamp is glary without diffuser and/or well–designed reflector
  • Clear lamp most optically precise
  • Electronic ballast is most quiet and most efficient
Suggested uses/downlighting, pedestrian postlight, parking lot postlight, facade floodlighting, tree uplight (consider pale blue lens for this app)
▶**Philips** (MasterColor®)

▷ **70 watt/3000°K/Coated (70ED17/CMH/3K/C/U)** ◀ *Softer light quality*
Lumens/6000
Life/10000 hrs
Application notes
  • Protective tempered glass lens required
  • Coated lamp is less glary
  • Coated lamp provides softer, but less–precise light distribution
  • Electronic ballast is most quiet and most efficient
Suggested uses/downlighting, pedestrian postlight, parking lot postlight, facade floodlighting, tree uplight (consider pale blue lens for this app)
▶**Philips** (MasterColor®)

▷ **70 watt/4000°K/Clear (70ED17/CMH/4K/U)** ◀ *Longest life; Cool tone*
Lumens/6000
Life/15000 hrs
Application notes
  • Protective tempered glass lens required
  • Clear lamp is glary without diffuser and/or well–designed reflector
  • Clear lamp most optically precise
  • Electronic ballast is most quiet and most efficient
Suggested uses/downlighting for so–called daylight compatability, facade floodlighting, tree uplight (consider pale blue lens for this app)
▶**Philips** (MasterColor®)

▷ **70 watt/4000°K/Coated (70ED17/CMH/4K/C/U)** ◀ *Longest life; Cool tone; Softer light quality*
Lumens/5800
Life/10000 hrs
Application notes
  • Protective tempered glass lens required
  • Coated lamp is less glary
  • Coated lamp provides softer, but less–precise light distribution
  • Electronic ballast is most quiet and most efficient
Suggested uses/downlighting for so–called daylight compatability, facade floodlighting, tree uplight (consider pale blue lens for this app)
▶**Philips** (MasterColor®)

Uses
▶Interior lensed downlighting
▶Pedestrian postlights
▶Parking garage lights
▶Parking lot postlights
▶Exterior facade lighting
▶Tree uplighting

Design Tips

✔In interiors, best when used in relatively high ceilings and in conjunction with other lighting techniques to avoid cave effect.

✔Use in enclosed luminaires to avoid hazards of violent lamp failure.

✔Coated lamps offer softest look and least glare.

Application Key

Commercial
Gallery
Hospitality
Institutional
Manufacturing
Residential
Retail
Exterior

bold = primary application
partial fade = minimal application
fade = unlikely application

MasterColor® is a registered trademark of Philips Lighting Company

Ceramic Metal Halide/ED17

for Enclosed Luminaires ONLY

Performance Data

[All data outlined here is based on information from boldfaced
manufacturer's published data at time of manuscript preparation.]

Uses
► Interior lensed downlighting
► Pedestrian postlights
► Parking garage lights
► Parking lot postlights
► Exterior facade lighting
► Tree uplighting

Design Tips
✔ In interiors, best when used in relatively high ceilings and in conjunction with other lighting techniques to avoid cave effect.
✔ Use in enclosed luminaires to avoid hazards of violent lamp failure.
✔ Coated lamps offer softest look and least glare.

Application Key

Commercial
Gallery
Hospitality
Institutional
Manufacturing
Residential
Retail
Exterior

bold = primary application
partial fade = minimal application
fade = unlikely application

▷ **100 watt/3000°K/Clear (100ED17/CMH/3K/U)** ◀ *Best efficacy*
Lumens/9300
Life/12500 hrs
Application notes
- Protective tempered glass lens required (use in enclosed luminaire)
- Clear lamp is glary without diffuser and/or well–designed reflector
- Clear lamp most optically precise
- Electronic ballast is most quiet and most efficient

Suggested uses/downlighting, pedestrian postlight, parking lot postlight, facade floodlighting, tree uplight (consider pale blue lens for this app)
►**Philips** (MasterColor®)

▷ **100 watt/3000°K/Coated (100ED17/CMH/3K/C/U)** ◀ *Softer light quality*
Lumens/9000
Life/12500 hrs
Application notes
- Protective tempered glass lens required
- Coated lamp is less glary
- Coated lamp provides softer, but less–precise light distribution
- Electronic ballast is most quiet and most efficient

Suggested uses/downlighting, pedestrian postlight, parking lot postlight, facade floodlighting, tree uplight (consider pale blue lens for this app)
►**Philips** (MasterColor®)

▷ **100 watt/4000°K/Clear (100ED17/CMH/4K/U)** ◀ *Longest life; Cool tone*
Lumens/9000
Life/15000 hrs
Application notes
- Protective tempered glass lens required
- Clear lamp is glary without diffuser and/or well–designed reflector
- Clear lamp most optically precise
- Electronic ballast is most quiet and most efficient

Suggested uses/downlighting for so–called daylight compatability, facade floodlighting, tree uplight (consider pale blue lens for this app)
►**Philips** (MasterColor®)

▷ **100 watt/4000°K/Coated (100ED17/CMH/4K/C/U)** ◀ *Longest life; Cool tone; Softer light quality*
Lumens/8700
Life/15000 hrs
Application notes
- Protective tempered glass lens required
- Coated lamp is less glary
- Coated lamp provides softer, but less–precise light distribution
- Electronic ballast is most quiet and most efficient

Suggested uses/downlighting for so–called daylight compatability, facade floodlighting, tree uplight (consider pale blue lens for this app)
►**Philips** (MasterColor®)

Metal Halide/ED17

Ceramic

for Enclosed Luminaires ONLY

Metal Halide
Ceramic

7

Net Addresses/ED17 Recessed

```
http://www.thomasltg.com/
http://www.cooperlighting.com/
http://www.hubbell-ltg.com/products.htm#Down&Track
http://www.danalite.com/
http://www.kramerlighting.com/
http://www.lightolier.com/
http://www.lithonia.com/
http://www.prescolite.com/
```

CONNECT FOR MORE

Net Addresses/ED17 Exterior

```
http://www.bega-us.com/home/home.html
http://www.cooperlighting.com/
http://www.emcolighting.com/emco/
http://www.sitelighting.com/gardco/
http://www.guth.com/outdoor.htm
http://www.hadcolighting.com/html/bronzelite.htm
http://www.holophane.com/Product/OutArch/Welcome.htm
http://www.hydrel.com
http://www.kimlighting.com/ingrd1.html
http://www.lithonia.com/
http://www.mcphilbenoutdoor.com/mcphilben/
```

CONNECT FOR MORE

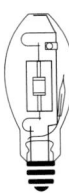

Metal Halide
Ceramic

7

Ceramic Metal Halide/ED17P

Protected—for Enclosed OR Open Luminaires

Stats <small>Data varies manufacturer to manufacturer and changes from time to time</small>

Life/10000 and 12500 hours
Dependent on wattage. A typical office environment might be occupied about 2600 hours each year.

Efficacy/75 to 80 LPW<small>Wattage dependent (higher wattage/higher efficacy) (values include electronic ballast loss)</small>

Wattages/70 and 100

Color of Light/3000°K (warm tone)

Beam spreads/Not Applicable
This is a general service lamp intended for downlights where an all–around glow of light is desired. Also appropriate for in–ground uplights and in floodlights, pedestrian postlights and parking lot lights. Do not use where distinct accent light or focused light is desired—for accent use halogen or HIR PAR lamps (Sections 3 and 5), halogen or HIR MR lamps (Sections 4 and 6) or CMH PAR lamps (this section).

Size/5⁷⁄₁₆″ L by 2⅛″ Ø

Voltage/120V and 277V<small>Ballast dependent (check ballast vendors for other voltages)</small>

Cost Magnitude/Moderate <small>(2000)</small>

Manufacturers
Philips

Advantages
Can be used in enclosed OR open luminaires<small>(must be enclosed for damp/wet apps)</small>
Crisp white light
Much longer life than halogen TB lamps
Better alternative to less efficient standard filament A lamps
Withstand extreme temperature range

Disadvantages
Ballast required
Noticeable lamp lumen depreciation (light loss) over life
 ◆15 to 20 percent
Not dimmable
 ◆Without great expense
 ◆Without sacrificing color temperature and color rendering
Not easily integrated into egress system
Not instant–on
 ◆Cold start requires a few minutes for full light output
 ◆Power interruption initiates restrike cycle of fifteen to twenty minutes

Stats Guide
⊃Longer life is better
⊃Higher efficacy is better
⊃Lower wattage is better
⊃Lower color of light is warmer
⊃Moderate cost range for CMH is about US$50.⁰⁰ to US$75.⁰⁰

For distinct accent light or focused light, use CMH PAR lamps.

Ceramic Metal Halide/ED17P
Protected—for Enclosed OR Open Luminaires

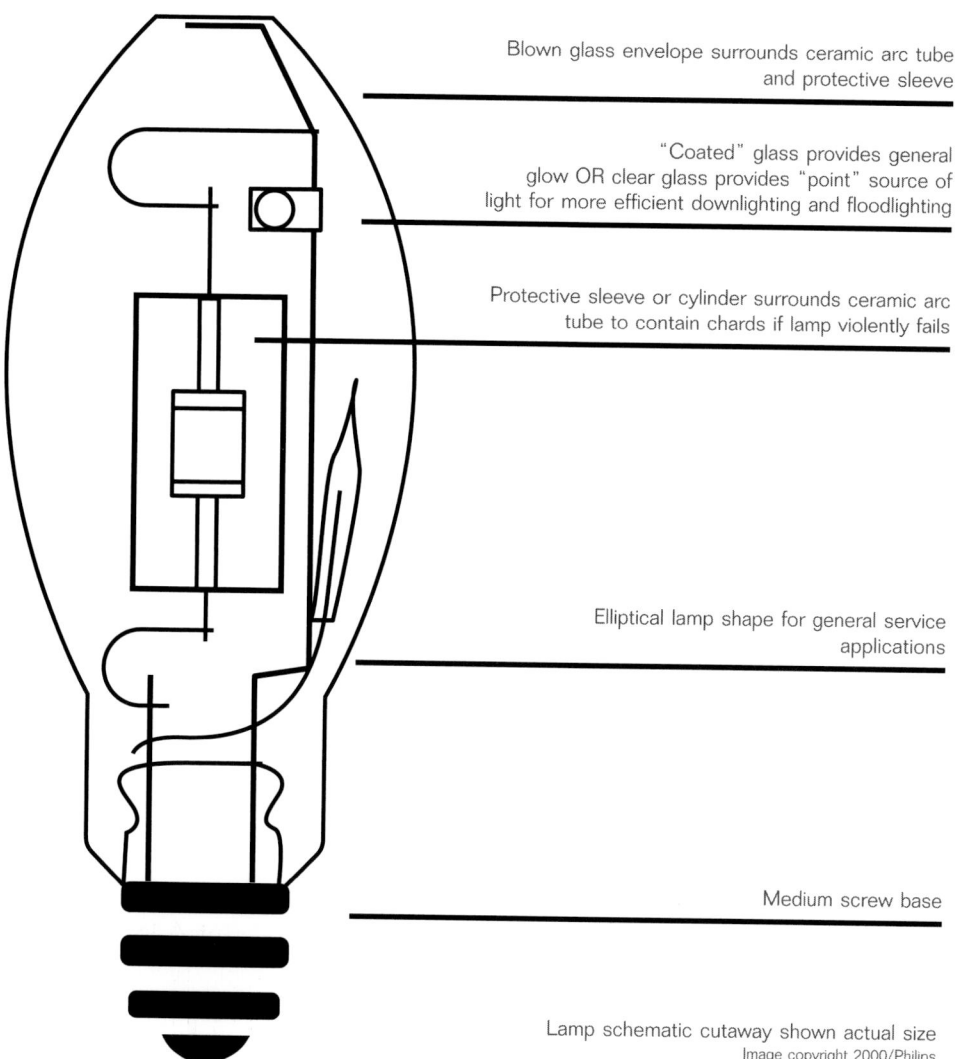

Blown glass envelope surrounds ceramic arc tube and protective sleeve

"Coated" glass provides general glow OR clear glass provides "point" source of light for more efficient downlighting and floodlighting

Protective sleeve or cylinder surrounds ceramic arc tube to contain chards if lamp violently fails

Elliptical lamp shape for general service applications

Medium screw base

Lamp schematic cutaway shown actual size
Image copyright 2000/Philips

Uses
Downlighting
Commercial
Institutional
Manufacturing
Retail
Facade Accenting
Tree Uplighting

CMH ED17P Lighting Design Tips
✔ Best in moderate to higher ceiling spaces (15'–0" or greater).
✔ Coated lamps are less glary than clear lamps.
✔ Clear lamps operate most efficiently in most luminaires.

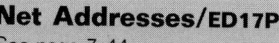

Net Addresses/ED17P
See page 7–11

CONNECT FOR MORE

For general downlighting in low ceilings, consider compact fluorescent triple tube lamps (Section 8).

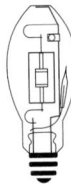

Metal Halide/ED17P

Protected—for Enclosed OR Open Luminaires

Lamp may be used in enclosed OR open luminaires

Ceramic Metal Halide ED Protected Lamp Designation

70ED17P/CMH/3K/C/U

Last letter identifies allowable base orientation
- **base down (BD)** [lamp base must be pointed straight down]
- **base up (BU)** [lamp base must be pointed straight up]
- **horizontal (HOR)** [lamp base must be pointed horizontally]
- **universal (U)** [lamp base may be pointed in any orientation]

This letter (if present) indicates lamp is coated or "frosted"
[no letter indicates lamp is clear]

This number/letter combination indicates color temperature
- 3K is shorthand for 3000°K (warm tone)

This series of letters indicates "Ceramic Metal Halide"

This letter "P" indicates lamp has integral protection from violent failures

These two digits represent lamp diameter in eighths inches (e.g., 17/8 or 2–1/8 inches)

These letters indicate this is an ellipsoidal envelope (ED) lamp

First two or three digits represent lamp wattage

[This designation used for convenient reference throughout this *Time–Saver Standards for Architectural Lighting*—but not necessarily used by all lamp manufacturers (although it would be convenient if manufacturers would come to agreement on standard designations.]

Ceramic Halide/ED17P

Protected—for Enclosed OR Open Luminaires

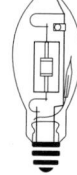

 Metal Halide
ceramic

7

Performance Data

[All data outlined here is based on information from boldfaced
manufacturer's published data at time of manuscript preparation.]

▷ ### 70 watt/3000°K/Clear (70ED17P/CMH/3K/U)
Lumens/5900
Life/10000 hrs
Application notes

 *Best efficacy*

- Use in enclosed OR open luminaire in any orientation
- Clear lamp is glary without diffuser and/or well–designed reflector
- Clear lamp most optically precise
- Electronic ballast is most quiet and most efficient

Suggested uses/downlighting, pedestrian postlight, parking lot postlight, facade floodlighting, tree uplight (consider pale blue lens for this app)
▶**Philips** (MasterColor®)

▷ ### 70 watt/3000°K/Coated (70ED17P/CMH/3K/C/U)
Lumens/5700
Life/10000 hrs
Application notes

 *Softer light quality*

- Use in enclosed OR open luminaire in any orientation
- Coated lamp is less glary
- Coated lamp provides softer, but less–precise light distribution
- Electronic ballast is most quiet and most efficient

Suggested uses/downlighting, pedestrian postlight, parking lot postlight, facade floodlighting, tree uplight (consider pale blue lens for this app)
▶**Philips** (MasterColor®)

▷ ### 100 watt/3000°K/Clear (100ED17P/CMH/3K/U)
Lumens/8800
Life/12500 hrs
Application notes

 *Longest life; Cool tone*

- Use in enclosed OR open luminaire in any orientation
- Clear lamp is glary without diffuser and/or well–designed reflector
- Clear lamp most optically precise
- Electronic ballast is most quiet and most efficient

Suggested uses/downlighting for so–called daylight compatability, facade floodlighting, tree uplight (consider pale blue lens for this app)
▶**Philips** (MasterColor®)

▷ ### 100 watt/3000°K/Coated (100ED17P/CMH/3K/C/U)
Lumens/8700
Life/12500 hrs
Application notes

Longest life; Cool tone; Softer light quality

- Use in enclosed OR open luminaire in any orientation
- Coated lamp is less glary
- Coated lamp provides softer, but less–precise light distribution
- Electronic ballast is most quiet and most efficient

Suggested uses/downlighting for so–called daylight compatability, facade floodlighting, tree uplight (consider pale blue lens for this app)
▶**Philips** (MasterColor®)

MasterColor® is a registered trademark of Philips Lighting Company

Uses
▶Interior lensed downlighting
▶Pedestrian postlights
▶Parking garage lights
▶Parking lot postlights
▶Exterior facade lighting
▶Tree uplighting

Design Tips
✔ In interiors, best when used in relatively high ceilings and in conjunction with other lighting techniques to avoid cave effect.
✔ Use in enclosed luminaires to avoid hazards of violent lamp failure.
✔ Coated lamps offer softest look and least glare.

Application Key

Commercial
Gallery
Hospitality
Institutional
Manufacturing
Residential
Retail
Exterior

bold = primary application
partial fade = minimal application
fade = unlikely application

Ceramic Metal Halide/T6

Single Ended for Enclosed Luminaires ONLY

Stats Data varies manufacturer to manufacturer and changes from time to time

Life/10000
A typical office environment might be occupied about 2600 hours each year.

Efficacy/77 to 80 LPW Wattage dependent (higher wattage/higher efficacy) (values include electronic ballast loss)

Wattages/39, 70 and 150

Color of Light/3000°K (warm tone)

Beam spreads/Not Applicable
This is a general service lamp intended for precise, optically designed luminaires—downlights, wallwashers, accents, theatrical/studio spots/floods and facade floodlights.

Size/Varies by Wattage
- 39–watt is $3^{15}/_{16}$" L by $\frac{3}{4}$" Ø
- 70–watt is $3^{15}/_{16}$" L by $\frac{3}{4}$" Ø
- 150–watt is $4^{11}/_{32}$" L by $\frac{3}{4}$" Ø

Voltage/120V and 277V Ballast dependent (check ballast vendors for other voltages)

Cost Magnitude/Moderate–to–High (2000)

Manufacturers
Philips

Net Addresses
http://www.lighting.philips.com/
CONNECT FOR MORE

Advantages
Crisp white light
Very small size (yields small luminaires and/or excellent optics)
Much longer life than halogen TB lamps
Better alternative to less efficient standard filament A lamps
Better alternative to less efficient single–ended and double–ended quartz halogen lamps
Withstand extreme temperature range

Disadvantages
Ballast required
Must be used in enclosed luminaires
- Tempered glass lens required
Noticeable lamp lumen depreciation (light loss) over life
- 15 to 20 percent
Not dimmable
- Without great expense
- Without sacrificing color temperature and color rendering
Not easily integrated into egress system
Not instant–on
- Cold start requires a few minutes for full light output
- Power interruption initiates restrike cycle of fifteen to twenty minutes

For distinct accent light or focused light, use CMH PAR lamps.

Metal Halide/T6
Ceramic
Single Ended for Enclosed Luminaires ONLY

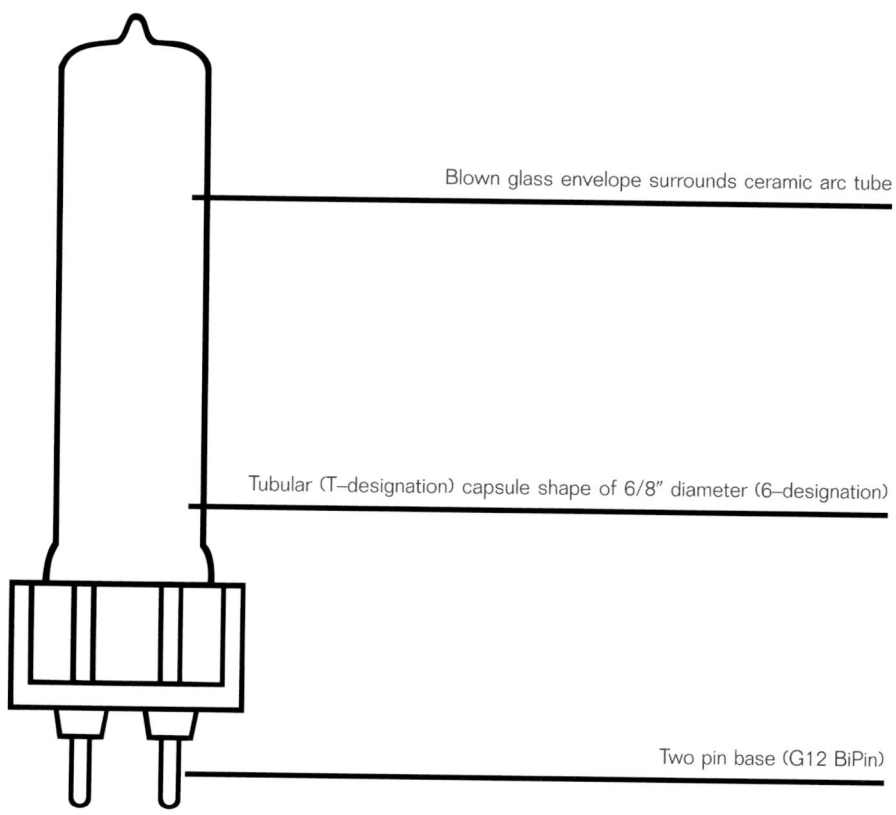

Blown glass envelope surrounds ceramic arc tube

Tubular (T–designation) capsule shape of 6/8" diameter (6–designation)

Two pin base (G12 BiPin)

Lamp schematic shown actual size (for 39– and 70–watt versions)
Image copyright 2000/Philips

Uses
Architectural Accenting
Art Accenting
Downlighting
   Commercial
   Institutional
   Manufacturing
   Retail
Facade Accenting
Merchandise Accenting
Studio Lighting
   Broadcast
   Special effects projectors
   Theatrical accenting
   Videotaping
Tree Uplighting

■ ■ ■ ■ ■ ■ ■ ■ ■ ■ ■ ■

CMH T6 Lighting Design Tips
✔ Wattage variety allows for application in variety of ceiling heights.
✔ Glare potential is significant—carefully shield light source.

For general downlighting in low ceilings, consider compact fluorescent triple tube lamps (Section 8).

Net Addresses/T6 Track, Monopoints and Recessed
See page 7–18

Ceramic Metal Halide/T6
Single Ended for Enclosed Luminaires ONLY

Ceramic Metal Halide T6 Lamp Designation
70T6/CMH/3K

This number/letter combination indicates color temperature
• 3K is shorthand for 3000°K (warm tone)

This series of letters indicates "Ceramic Metal Halide"

This digit represents lamp diameter in eighths inches (e.g., 6/8 or 3/4 inches)

This letter indicates this is a tubular envelope (T) lamp

First two or three digits represent lamp wattage

[This designation used for convenient reference throughout this *Time–Saver Standards for Architectural Lighting*—but not necessarily used by all lamp manufacturers (although it would be convenient if manufacturers would come to agreement on standard designations.]

Net Addresses/T6 Track and Monopoints
http://www.cooperlighting.com/
http://www.katiegroup.com [Note: Altman Architectural Master Ellipse and Star PAR]
http://www.tslight.com/

CONNECT FOR MORE

Net Addresses/T6 Recessed, Semi–recessed and Surface Mount
http://www.elliptipar.com/home/Contents.htm
http://www.cooperlighting.com/

CONNECT FOR MORE

Net Addresses/T6 Exterior Accent
http://www.elliptipar.com/home/Contents.htm
http://www.hydrel.com

CONNECT FOR MORE

Metal Halide/T6
Ceramic

Single Ended for Enclosed Luminaires ONLY

Performance Data

[All data outlined here is based on information from boldfaced
manufacturer's published data at time of manuscript preparation.]

39 watt/3000°K/Clear (39T6/CMH/3K)
Lumens/3400
Life/10000 hrs
Application notes
  • Protective tempered glass lens required (use in enclosed luminaire)
  • Clear lamp is glary without diffuser and/or well–designed reflector
  • Electronic ballast is most quiet and most efficient
**Suggested uses/downlighting in relatively low ceilings, wallwashing,
accenting, facade feature accenting, facade floodlighting, tree uplight
(consider pale blue lens for this app)**
▶**Philips** (MasterColor®)

70 watt/3000°K/Clear (70T6/CMH/3K)
Lumens/6600
Life/10000 hrs
Application notes
  • Protective tempered glass lens required (use in enclosed luminaire)
  • Clear lamp is glary without diffuser and/or well–designed reflector
  • Electronic ballast is most quiet and most efficient
**Suggested uses/downlighting in moderate to high ceilings, wallwashing,
accenting, facade feature accenting, facade floodlighting, tree uplight
(consider pale blue lens for this app)**
▶**Philips** (MasterColor®)

150 watt/3000°K/Clear (150T6/CMH/3K)
Lumens/14000
Life/10000 hrs
Application notes
  • Protective tempered glass lens required (use in enclosed luminaire)
  • Clear lamp is glary without diffuser and/or well–designed reflector
  • Electronic ballast is most quiet and most efficient
**Suggested uses/downlighting in high to very high ceilings, wallwashing,
accenting, facade feature accenting, facade floodlighting**
▶**Philips** (MasterColor®)

Uses
▶Interior lensed downlighting
▶Interior lensed wallwashing
▶Interior lensed accent lighting
▶Exterior facade lighting
▶Tree uplighting

Design Tips
✔Use in enclosed luminaires to
avoid hazards of violent lamp
failure.

Application Key

Commercial
Gallery
Hospitality
Institutional
Manufacturing
Residential
Retail
Exterior

bold = primary application
partial fade = minimal application
fade = unlikely application

MasterColor® is a registered trademark of Philips Lighting Company

Ceramic Metal Halide/PAR

PAR lamps—PAR is an acronym for parabolic aluminized reflector—have been effective accent lamps since the 1950's. Until very recently, however, PAR lamps appropriate for many commercial, hospitality, institutional, retail and exterior applications were halogen or halogen infrared. Philips Lighting developed a technique for metal halide lamps (ceramic arc tubes as opposed to quartz arc tubes) that offer a very consistent color quality as well as a very incandescent–like color. Couple this with the usual advantages of metal halide lamps—long life and excellent efficacy—and a new family of lamps appropriate to many applications is now available. These lamps—known as ceramic metal halide or CMH lamps—are much more efficacious than halogen or halogen infrared lamps. Two distinct disadvantages will limit their use somewhat. First, these metal halide lamps are not instant–on.

■ ■ ■ ■ ■ ■ ■ ■ ■ ■ ■ ■

Ceramic metal halide PAR lamps

❶ Not instant–on
❶ Not easily, conveniently nor inexpensively dimmed
❶ Excellent efficiencies for accenting
❶ Excellent color quality

From a cold start (i.e., the lamps have been off for at least half an hour), it takes ceramic metal halide lamps a few minutes to provide full light output. From a hot start (i.e., lamps are in operation when a power interruption occurs), these lamps must cool down and then restrike. This process takes about fifteen minutes. Second, metal halide lamps cannot be dimmed. At least not effectively, yet. Dimming can be done at great expense, with limited range (dimming from about 40 percent output to 100 percent output) and with serious color shifting (once the lamp is dimmed, color tends toward a ghostly blue–white).

A further problem, albeit one which should not arise if design programming, assessment and client review are addressed, is dedicated lamping with CMH. Because ballasts which operate the lamp can only accommodate a single size/wattage lamp, it is not possible to "downsize" after the fact. In other words, if too much light is achieved with, say, 100 watt lamps, then to downsize to a lower wattage lamp requires changing ballasts and sockets—a labor intensive and thus costly operation. To dim CMH, neutral density filters must be used (glass or screen cloth which cuts light intensity). Conversely, it is not possible to "upsize" if too little light is achieved.

Metal Halide/PAR

CMH/PAR lamp designation

CMH/PAR lamps are typically identified with an alphanumeric designation outlined below. Since each manufacturer has slightly different means for lamp designation, the designation method presented here is solely for convenient reference throughout the text. Actual lamp manufacturers' designations will need to be used in final specifications. Perhaps the lamp manufacturers will standardize lamp designations in the near future in an effort to ease lamp cross referencing and specifications.

CMH/PAR lamps for accenting

PAR lamps are great for art accenting. With so many choices on the spread of the light pattern (beam spread), on the intensity of the pattern (candlepower) and on wattage, accenting can be achieved from a wide range of ceiling heights and lateral distances from the objects to be highlighted. For the PAR20, PAR30L and PAR38

Ceramic Metal Halide PAR Lamp Designation

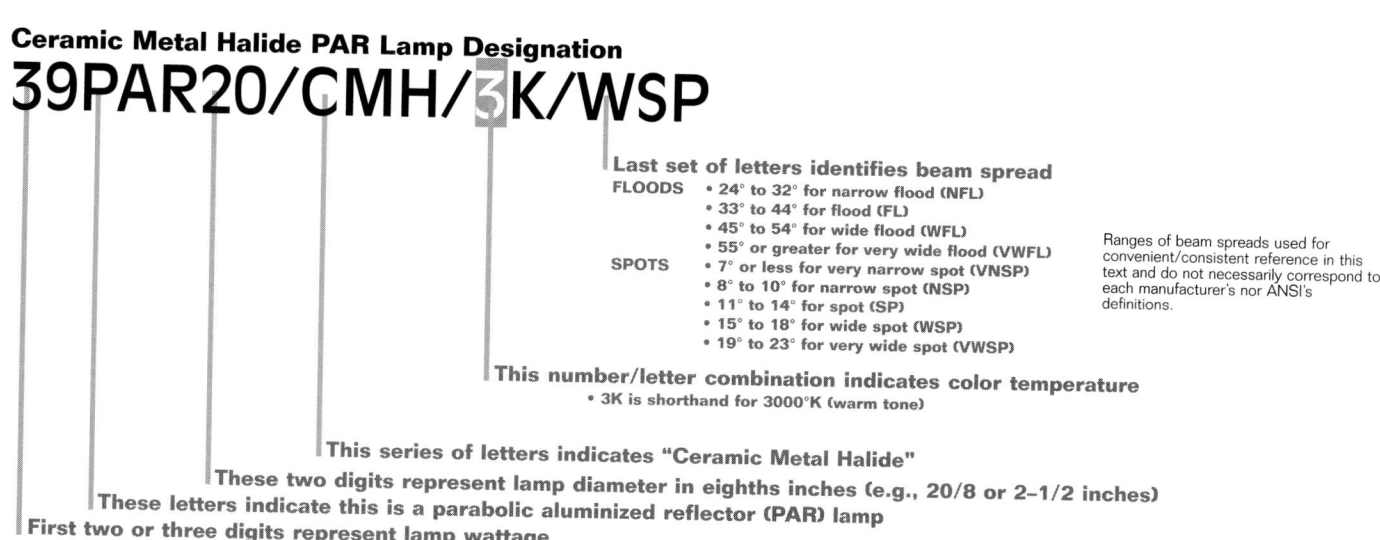

39PAR20/CMH/3K/WSP

Last set of letters identifies beam spread

FLOODS
- 24° to 32° for narrow flood (NFL)
- 33° to 44° for flood (FL)
- 45° to 54° for wide flood (WFL)
- 55° or greater for very wide flood (VWFL)

SPOTS
- 7° or less for very narrow spot (VNSP)
- 8° to 10° for narrow spot (NSP)
- 11° to 14° for spot (SP)
- 15° to 18° for wide spot (WSP)
- 19° to 23° for very wide spot (VWSP)

Ranges of beam spreads used for convenient/consistent reference in this text and do not necessarily correspond to each manufacturer's nor ANSI's definitions.

This number/letter combination indicates color temperature
- 3K is shorthand for 3000°K (warm tone)

This series of letters indicates "Ceramic Metal Halide"

These two digits represent lamp diameter in eighths inches (e.g., 20/8 or 2-1/2 inches)

These letters indicate this is a parabolic aluminized reflector (PAR) lamp

First two or three digits represent lamp wattage

[This designation used for convenient reference throughout this *Time–Saver Standards for Architectural Lighting*—but not necessarily used by all lamp manufacturers (although it would be convenient if manufacturers would come to agreement on standard designations.]

Metal Halide/PAR
Ceramic

lamps, a series of Performance Sketches™ are presented which offers quick visual assessment of what lamp wattage and beam spread might be appropriate for given ceiling heights and artwork sizes.

Performance Sketches™ are shown at a scale of ¼ inch equals 1 foot. Tracing for evaluation of various elevations is encouraged. Copying is permitted (according to the conventions outlined in "Licensing Agreement" in the Introduction of this book). Each Sketch includes an outline of artwork of the size category which may be lighted by the cited lamp. This is to assist in judgement of area of coverage. Location or distance from wall is listed with each Sketch as are center beam aiming angle and average light level over the area of the artwork outline. Maximum maintained intensity on artwork outline is also listed. For museum quality artwork where conservation of art is a priority, intensities should be 5 to 10 footcandles average, maintained, with maximums of perhaps 10 to 15 fc. In other applications, average maintained intensities on art will depend on the general or ambient lighting within the space(s). In typical hospitality applications, artwork accenting might exhibit average maintained intensities of 10 to 20 fc, with maximum intensities of 20 to 40 fc. In commercial applications, average maintained intensities on artwork might range from 20 to 30 fc, with maximums of 40 fc and greater. These Sketches are based on Philips' data at time of manuscript preparation. Data cited in the sidebar "Specs" is Philips' data. Manufacturer's catalog data may be of one date and for one set of lamps while photometric data available from manufacturers may be based on another set of lamps of another date (this may have occurred with some data in this reference). Sketches were generated with this data in Radiance (a Unix–based lighting simulation and rendering program available free of charge at www.radsite.lbl.gov/radiance/refer/long.html).

These Performance Sketches™ illustrate a specific condition. As ceiling heights change and focal objects' sizes change, the cited lamp may be quite appropriate for other applications. Generally, in lower ceiling settings, light intensities will be greater and area of coverage will be smaller. In higher ceiling settings, light intensities will be less and area of coverage will be larger. Lamp data changes

Radiance (lighting rendering software)
http://www.radsite.lbl.gov/radiance/refer/
   long.html

CONNECT FOR MORE

Recommended intensities
for art accenting

Low (gallery, museum, residential)
5 to 10 fc average on art
Moderate (hospitality)
10 to 20 fc average on art
High (commercial—or higher for retail)
20 to 30 fc average on art

These are intended to be average, maintained values including effects of accent lighting and general room lighting. General room lighting contributes just a few footcandles in residential, gallery and museum settings, but contributes 5 to 10 fc in commercial settings. Maximums might range from 50 percent to 100 percent of the average value. Wherever artwork is highly valuable and/or sensitive to light, precautions should be taken to limit light exposure (maintaining low levels and/or switching lights off when art is not being viewed); and exposure to infrared and ultraviolet should be limited (using filters on lamps or placing art behind specially treated glass or acrylic).

Metal Halide/PAR

from time to time. As manufacturing techniques change, as materials' technologies change, and as the market changes, lamp manufacturers will revise wattages, beam spreads and intensities. Manufacturers' data should be checked periodically for any possible updates.

Performance Sketch™ thumbnails

At the beginning of each PAR lamp section (PAR20, PAR30L and PAR38) there are thumbnails of the lighting images to further help provide quick visual assessment of various lamp beam spreads and intensities. Thumbnails serve as an index to the lamps in that category. The pages containing thumbnails are highlighted with gray margin bleeds.

Uses

For convenient reference, a shaded box entitled "Uses" appears in the upper corner of each page detailing the more likely uses for which the lamp seems suited. Artwork sizes are based on the rough dimensions as outlined in the tip box on this page (to the left). Lamp "Specs" are also included in the shaded box. These specs are averages of the listed manufacturers' data at time of manuscript preparation. Since products are constantly upgraded or deleted, a check of the current status of the listed lamp is recommended before including it in specifications.

■ ■ ■ ■ ■ ■ ■ ■ ■ ■ ■ ■ ■ ■

Performance Sketches™ and Information
❶ Typical application is shown
❷ Other applications may also be acceptable
❸ Other applications may depend on ceiling heights
❹ Maintained values based on 0.8 maintenance factor
❺ Cited data is for a single lamp—two or more lamps may

Lighting characteristics

In addition to beam spread and intensity, these images illustrate beam striations and cutoff patterns to some extent. When lamps are being considered for a specific project, actual samples should be obtained and mocked up for visual inspection of quality of light patterning. Beam striations and aberrations may appear more pronounced in reality since Performance Sketches™ herein are based on finite candlepower reports (typically at 2.5° increments—a bright spot or dead spot might occur at an increment not reported on the photometry but which could be seen visually on mockup). Color of

Metal Halide/PAR

Performance Sketch

A quarter inch scale rendering illustrates lighting effect for a typical situation. Note how the area of coverage, intensity and frame shadow change with beam spread, ceiling height and distance from wall. Higher or lower ceilings and longer or shorter distances from walls will impact intensities, light patterns and application(s).

Performance Details

Details such as lamp distance from wall, aiming angle, point above floor at which lamp is aimed, along with maintained average and maximum illuminance are reported. The aiming angle should be used to select a luminaire capable of achieving such an angle. Maximum illuminances are calculated maintained (0.8 maintenance factor) over time and are reported to help assess likelihood of artwork degradation. Average illuminance is calculated maintained over time just for the size of the artwork shown.

Suggested Uses/Lamp Specs

A bullet list outlines uses for which the lamp seems best suited at cited ceiling height. Lamp specifications are outlined for quick reference and to assist in specification writing. Where two or more offer nearly identical products, only one manufacturer's data (boldfaced) has been used for development of the sketches. An SKU–number or product code is included parenthetically. Data cited in the sidebar "Specs" is averaged from all available manufacturers' data.

Design Tips

Key considerations are offered for the specific lamp.

Application Key

Those applications for the cited ceiling height where the specific lamp may be most useful are highlighted in bold. Applications where the lamp may be of lesser use are faded—the softer the fade, the less appropriate is the lamp for that application in cited ceiling height. Application appropriateness will change with ceiling heights.

Artwork Outline

An outline of two dimensional artwork of size that might be acceptably lighted by lamp cited.

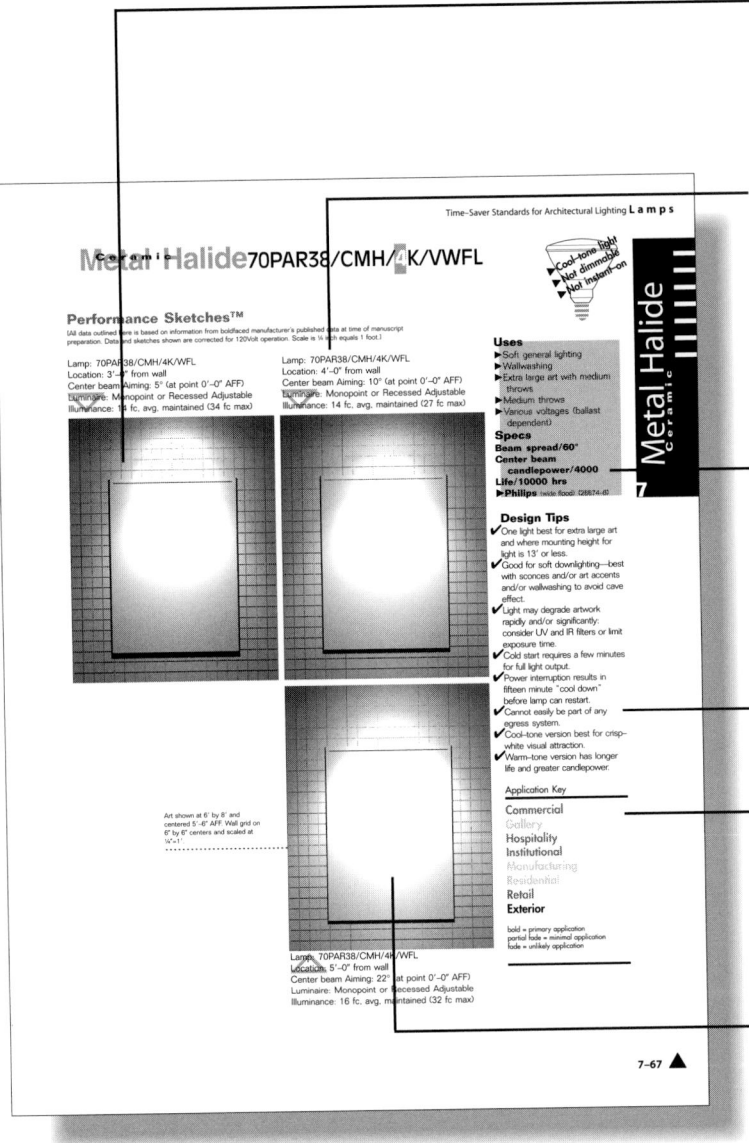

Time–Saver Standards for Architectural Lighting **L a m p s**

Metal Halide 70PAR38/CMH/4K/VWFL

Performance Sketches™

[All data outlined here is based on information from boldfaced manufacturer's published data at time of manuscript preparation. Data and sketches shown are corrected for 120Volt operation. Scale is ¼ inch equals 1 foot.]

Lamp: 70PAR38/CMH/4K/WFL
Location: 3'–0" from wall
Center beam Aiming: 5' (at point 0'–0" AFF)
Luminaire: Monopoint or Recessed Adjustable
Illuminance: 14 fc. avg. maintained (34 fc max)

Lamp: 70PAR38/CMH/4K/WFL
Location: 4'–0" from wall
Center beam Aiming: 10' (at point 0'–0" AFF)
Luminaire: Monopoint or Recessed Adjustable
Illuminance: 14 fc. avg. maintained (27 fc max)

Art shown at 6' by 8' and centered 5'–6" AFF Wall grid on 6' by 8" centers and scaled at ¼"=1'.

Lamp: 70PAR38/CMH/4K/WFL
Location: 5'–0" from wall
Center beam Aiming: 22' (at point 0'–0" AFF)
Luminaire: Monopoint or Recessed Adjustable
Illuminance: 16 fc. avg. maintained (32 fc max)

Cool-tone light
► Not dimmable
► Not instant-on

Uses
► Soft general lighting
► Wallwashing
► Extra large art with medium throws
► Medium throws
► Various voltages (ballast dependent)

Specs
Beam spread/60°
Center beam candlepower/4000
Life/10000 hrs
► **Philips** (wide flood) (28874-8)

Design Tips
✓ One light best for extra large art and where mounting height for light is 13' or less.
✓ Good for soft downlighting—best with sconces and/or art accents and/or wallwashing to avoid cave effect.
✓ Light may degrade artwork rapidly and/or significantly: consider UV and IR filters or limit exposure time.
✓ Cold start requires a few minutes for full light output.
✓ Power interruption results in fifteen minute "cool down" before lamp can restart.
✓ Cannot easily be part of any egress system.
✓ Cool-tone version best for crisp-white visual attraction.
✓ Warm-tone version has longer life and greater candlepower.

Application Key
Commercial
Gallery
Hospitality
Institutional
Manufacturing
Residential
Retail
Exterior

bold = primary application
partial fade = minimal application
fade = unlikely application

7–67 ▲

Artwork Size Categories

⊃Petite = 2'–0" by 2'–0" or smaller
⊃Small = 2'–0" by 2'–0" to 2'–6" by 3'–0"
⊃Medium = 2'–6" by 3'–0" to 3'–0" by 4'–0"
⊃Large = 3'–0" by 4'–0" to 4'–0" by 5'–0"
⊃Extra Large = 4'–0" by 5'–0" or larger

Metal Halide/PAR
Ceramic

light cannot be illustrated in this black and white format. These Performance Sketches™ will help narrow selections so that appropriate lamp mockups can be made to assess actual beam striations and color of light.

Luminaires

PAR lamps are used in recessed downlights, recessed adjustable accent lights and track and monopoint adjustable accents. Since the PAR lamp is a self–contained optical package, track and monopoint luminaires do not need to offer much more than a lamp socket and perhaps a decorative shroud to hide the bare PAR lamp. With downlights and adjustable accents, the reflector cone in the luminaire can help to control some light spill and/or minimize direct glare. Nevertheless, the lamp's optics provide the bulk of the lighting performance. Hence, the Performance Sketches™ shown here illustrate expected results from PAR lamps in track and monopoint adjustable accents as well as recessed adjustable accents.

Manufacturers' Update Data:
❶Periodically due to materials' technology
❶Periodically due to manufacturing techniques
❶Periodically due to governmental regulations
❶Performance changes may result

Selection guides

Selection guides are offered on the next several pages for various uses in various applications. These guides are intended to help limit the designer's, engineer's or facility engineer's search and offer a good starting point from which design or alternative analyses can progress. Lamps which have similar beam spreads and intensities may be suggested for different ceiling heights and art sizes simply to show the variety of situations in which lamps might be used. Uses for CMH/PAR lamps include:

▶pinspot accent

An intense, well–controlled, focused lighting effect. Use the 39PAR20/ CMH/3K/NSP in lower ceilings (10– to 15 feet).

▶accent/medium art

Very intense lighting effect on medium piece of art.

▶accent/large art

Relatively intense lighting effect on large piece of art.

▶wallwashing/matte materials

Relatively uniform wash of light on wall materials exhibiting little or no sheen, polish or specularity.

▶wallwashing/polished materials

Relatively uniform wash of light on polished or specular wall materials.

▶feature downlighting

Somewhat intense lighting effect on flooring material details.

Metal Halide/PAR
Ceramic

▶general downlighting
Relatively uniform wash of light at lap height, table height or on floor.
While there are exceptions, CMH/PAR lamps are appropriate in the following interior applications:

▶commercial /retail

▶hospitality

CMH/PAR lamps are best suited for ceilings of 10 feet in height or greater—see Tables 7.2 (Commercial/Retail) and 7.3 (Hospitality). For lower ceilings, consider halogen infrared lamps.

Exterior applications are also quite appropriate for CMH/PAR lamps. In more urban settings, somewhat higher light levels on hardscape and softscape materials can be appropriate. Similarly, on facades and structural features, somewhat higher light levels are acceptable. For urban settings, see Table 7.4 (Exterior/Urban Facades and Landscapes). In more suburban and country settings, lower light intensities are appropriate. See Table 7.5 (Exterior/ Suburban/Country Facades and Landscapes).

Extrapolating information
There will be situations where perhaps twice as much light is desired (to provide a significant focal point) or where half as much light is desired (very sensitive artworks—paper–based pieces or media which fades easily). Find a ceiling height condition and artwork size which matches the planned situation. Then, to almost double light intensities, consider using two lights instead of a single light. Lights should be spaced on center at about the same distance they are spaced from the wall. So, if a single light spaced 2 feet/6 inches from the wall provides 17 fc average on a piece of art that is 3 feet wide by 4 feet long, then two such lights spaced 2 feet/6 inches from the wall and each spaced 2 feet/6 inches on center from the other centered on the art will provide about 30 fc average on the piece of art.

Another way to double the light is to find a lamp with similar beam spread (within 5 degrees) but twice the center beam candlepower. A search over several lamp families is suggested (e.g., perhaps a halogen low voltage MR lamp exists with nearly identical beam spread and twice the center beam candlepower as a halogen mains voltage PAR lamp). This is generally less costly than adding a second light, both initially and operationally. Most fine art installations,

Metal Halide
Ceramic
7

Metal Halide/PAR
Ceramic

however, will have two lights aimed onto each piece for maximum viewing quality from most any angle—eliminating the harsh veiling reflections (particularly problematic with oils and acrylics) which are evident with single light accents.

To halve the light level, look for lamps of similar beam spreads (within 5 degrees) but with half the center beam candlepower. Light intensities can be reduced with neutral density filters, although this does waste energy and thus should be a last resort.

Net Addresses/PAR/CMH Track and Monopoints
```
http://www.bartcolighting.com/index2.html
http://www.thomasltg.com/
http://www.cooperlighting.com/
http://www.hubbell-ltg.com/products.htm#Down&Track
http://www.danalite.com/
http://www.kramerlighting.com/
http://www.lightingservicesinc.com/
http://www.lightolier.com/
http://www.lithonia.com/
http://www.prescolite.com/
http://www.tslight.com/
```
CONNECT FOR MORE

Net Addresses/PAR/CMH Recessed
```
http://www.thomasltg.com/
http://www.cooperlighting.com/
http://www.hubbell-ltg.com/products.htm#Down&Track
http://www.danalite.com/
http://www.kramerlighting.com/
http://www.lightolier.com/
http://www.lithonia.com/
http://www.prescolite.com/
```
CONNECT FOR MORE

Net Addresses/PAR/CMH Exterior Accent
```
http://www.bega-us.com/home/home.html
http://www.hadcolighting.com/html/bronzelite.htm
http://www.hydrel.com
http://www.kimlighting.com/ingrd1.html
```
CONNECT FOR MORE

Metal Halide Ceramic

7

Ceramic **Metal Halide** PAR Lamp Selection Guide

Table 7.2 Commercial or Retail High Ceiling (10 feet or greater)[1][2]

◄Smaller lamp/smaller luminaire   Larger lamp/larger luminaire►

Lighting Task	PAR20	PAR30L	PAR38
Pinspot Accent	Consider (1) or more depending on ceiling height and desired impact		
• Petite to small art[3] • Feature merchandise • Small merchandise	• 39/NSP p7–36	• 39/NSP p7–44 • 70/NSP p7–48	• 70/3K/WSP p7–58 • 70/4K/WSP p7–60 • 100/3K/WSP p7–68[4] • 100/4K/WSP p7–70[4]
Key Area Accent	Consider (2) or more depending on ceiling height and desired impact		
• Small to medium art[3] • Feature display • Moderately sized merchandise	• 39/NSP p7–36	• 39/NSP p7–44 • 70/NSP p7–48	• 70/3K/WSP p7–58 • 70/4K/WSP p7–60 • 100/3K/WSP p7–68[4] • 100/4K/WSP p7–70[4]
Feature Lighting	Consider (2) or more depending on size of featured area and ceiling height		
• Medium to larger art[3] • Floor feature • Destination focus (e.g., cashier)	• 39/NFL p7–38	• 39/NFL p7–46	• 70/3K/NFL p7–62 • 70/4K/NFL p7–64 • 100/3K/NFL p7–72 • 100/4K/NFL p7–74
General Downlighting	Consider symmetric arrangement[5]		
• Overall lighting of table or floor		• 70/FL p7–50	• 70/3K/VWFL p7–66 • 70/4K/VWFL p7–67 • 100/3K/VWFL p7–76 • 100/4K/VWFL p7–77

beam spread

wattage

Typically dimmer▲
▼Typically brighter

[1]These lamp guides are intended to direct your attention to lamps which might meet your needs. If daylighting is prevalent, more lamps may be necessary to provide a visual focus to the artwork or merchandise. Design analysis is required to finalize lamp selections for projects.

[2]Some lamps' performance sketches which are referenced here are not optimized for retail applications. Lower ceiling heights will result in higher intensities on merchandise.

[3]Area of light coverage will be larger than the art. To achieve small, well –controlled beam spreads, VNSP lamps in the halogen low voltage MR16 and PAR36 families are recommended. The intensities, however, are significant and therefore should be considered for inert artwork or use of dimmers and/or neutral density filters will be necessary to limit intensities.

[4]These are very powerful lamps typically intended for high ceilings and/or where very strong intensities are required.

[5]Spacing between lights should not exceed 0.75 to 1.0 of the distance between the surface being downlighted and the ceiling.

Recommended intensities
for art accenting

Low (gallery, museum, residential)
5 to 10 fc average on art
Moderate (hospitality)
10 to 20 fc average on art
High (commercial—or higher for retail)
20 to 30 fc average on art

These are intended to be average, maintained values including effects of accent lighting and general room lighting. General room lighting contributes just a few footcandles in residential, gallery and museum settings, but contributes 5 to 10 fc in commercial settings. Maximums might range from 50 percent to 100 percent of the average value. Wherever artwork is highly valuable and/or sensitive to light, precautions should be taken to limit light exposure (maintaining low levels and/or switching lights off when art is not being viewed); and exposure to infrared and ultraviolet should be limited (using filters on lamps or placing art behind specially treated glass or acrylic).

Artwork Size Categories
➲Petite = 2'–0" by 2'–0" or smaller
➲Small = 2'–0" by 2'–0" to 2'–6" by 3'–0"
➲Medium = 2'–6" by 3'–0" to 3'–0" by 4'–0"
➲Large = 3'–0" by 4'–0" to 4'–0" by 5'–0"
➲Extra Large = 4'–0" by 5'–0" or larger

Ceramic Metal Halide PAR Lamp Selection Guide

Table 7.3 Hospitality High Ceiling (10 feet or greater)❶❷

◀Smaller lamp/smaller luminaire Larger lamp/larger luminaire▶

Lighting Task	PAR20	PAR30L	PAR38
Pinspot Accent	Consider tighter spot lamps❸		
• Petite art • Objet d'art	• Halogen low voltage MR16 VNSP • Halogen low voltage PAR36 VNSP		
Accent/Small Artwork	Consider tighter spot lamps❸		
• Small art • Small sculpture	• Halogen low voltage MR16 VNSP • Halogen low voltage PAR36 VNSP		
Accent/Medium Artwork	Consider less efficacious lamps		
• Medium art • Medium sculpture • Architectural detail or feature	• Halogen PAR • HIR PAR		
Accent/Large Artwork	Consider (1)		
• Large art • Large sculpture • Larger architectural detail or feature		• 70/FL p7–50	• 70/3K/VWFL p7–66 • 100/3K/VWFL p7–76
Wallwashing/Matte Materials	Continuous row of adjustables 2– to 3–feet from wall and 2– to 3–feet on center		
• Flat, frontal lighting❹			• 70/3K/VWFL p7–66 • 100/3K/VWFL p7–76
Wallwashing/Polished Materials	Architectural slot with sockets 6–inches from wall and 6– to 12–inches on center		
• Grazing wash light❺❻	See Section 11		
Feature Downlighting	Consider (1) or more depending on size of featured area		
• Floor feature • Pattern highlight • Destination focus (e.g., reception desk)	• 39/NFL p7–38　beam spread　wattage		
General Downlighting	Consider symmetric arrangement❼		
• Overall lighting of lap, table or floor		• 70/FL p7–50	• 70/3K/VWFL p7–66 • 100/3K/VWFL p7–76

Typically dimmer▲ / ▼Typically brighter

❶These lamp guides are intended to direct your attention to lamps which might meet your needs. Design analysis is required to finalize lamp selections for projects.
❷Higher light levels on artwork and features are expected in hospitality spaces compared to residential and gallery spaces. If art preservation is a priority, neutral density filters will be required in front of these lamps.
❸To achieve small, well–controlled beam spreads, VNSP lamps in halogen low voltage MR and PAR36 families are recommended. The intensities, however, are significant and therefore should be considered for inert artwork or use of neutral density filters will be necessary to limit intensities.
❹Energy intensive and should only be considered for exclusive zones of small size in highly visible public areas.
❺Extraordinarily energy intensive and should be reserved for very exclusive zones of small size in highly visible public areas.
❻This approach accentuates any wall imperfections.
❼Spacing between lights should not exceed 0.75 to 1.0 of the distance between the surface being downlighted and the ceiling.

Metal Halide ceramic / 7

Metal Halide Ceramic PAR Lamp Selection Guide

Table 7.4 Exterior/Urban Facades and Landscapes[1]

◀Smaller lamp/smaller luminaire   Larger lamp/larger luminaire▶

Lighting Task	PAR20	PAR30L	PAR38
Accent	Consider (1) or more depending on ceiling height and desired impact		
• Hardscape details • Sculpture	• 39/NSP p7–36 • 39/NFL p7–38	• 39/NSP p7–44 • 70/NSP p7–48	• 100/3K/WSP p7–68[2] • 100/4K/WSP p7–70[2]
Feature	Consider (2) or more depending on ceiling height and desired impact		
• Hardscape area or zone • Facade elements		• 70/FL p7–50	• 100/3K/NFL p7–72[2] • 100/4K/NFL p7–74[2]
Wash	Consider (2) or more depending on size of featured area and ceiling height		
• General area lighting • Whole or partial facade			• 100/3K/VWFL p7–76 • 100/4K/VWFL p7–77
Tree/Coniferous	Consider symmetric arrangement of (3) or (4)[3]		
• Frontal wash (consider pale blue filter)			• 70/3K/VWFL p7–66 • 70/4K/VWFL p7–67 • 100/3K/VWFL p7–76 • 100/4K/VWFL p7–77
Tree/Coniferous and Deciduous	Consider symmetric arrangement of (3) or (4)[3]		
• Uplight (consider pale blue filter)			• 70/3K/VWFL p7–66 • 70/4K/VWFL p7–67 • 100/3K/VWFL p7–76 • 100/4K/VWFL p7–77
Tree/Flowering Ornamental	Consider symmetric arrangement of (2) or (3)[3]		
• Uplight			• 70/3K/VWFL p7–66 • 100/3K/VWFL p7–76

Typically dimmer▲
▼Typically brighter

[1] These lamp guides are intended to direct your attention to lamps which might meet your needs. Design analysis is required to finalize lamp selections for projects.

[2] These are very powerful lamps typically intended for features where very strong intensities are required.

[3] Smaller trees require fewer lights. For washing effect, use in–ground wallwash uplights spaced a few feet from tree canopy edge and spaced 3 to 4 feet on center around the tree. For uplighting effect, use in–ground uplights spaced roughly half the distance to two–thirds the distance from the trunk to the canopy edge.

Metal Halide PAR Lamp Selection Guide

Table 7.5 Exterior/Suburban and Country Facades and Landscapes[1][2]

	◄Smaller lamp/smaller luminaire		Larger lamp/larger luminaire►
Lighting Task	**PAR20**	**PAR30L**	**PAR38**
Accent	Consider (1) or more depending on feature height and desired impact		
• Hardscape details • Sculpture	• 39/NSP p7–36	• 39/NSP p7–44	• 70/3K/WSP p7–58[2] • 70/4K/WSP p7–60[2]
Feature	Consider (1) or more depending on feature height and desired impact		
• Hardscape area or zone • Facade elements	• 39/NFL p7–38	• 39/NFL p7–46 • 70/FL p7–50	• 70/3K/NFL p7–62[2] • 70/4K/NFL p7–64[2]
Wash	Consider (1) or more depending on size of featured area and desired impact		
• General area lighting • Whole or partial facade			• 70/3K/VWFL p7–66 • 70/4K/VWFL p7–67
Tree/Coniferous	Consider symmetric arrangement of (2) or (3)[3]		
• Frontal wash (consider pale blue filter)			• 70/3K/VWFL p7–66 • 70/4K/VWFL p7–67
Tree/Coniferous and Deciduous	Consider symmetric arrangement of (2) or (3)[3]		
• Uplight (consider pale blue filter)			• 70/3K/VWFL p7–66 • 70/4K/VWFL p7–67
Tree/Flowering Ornamental	Consider symmetric arrangement of (2) or (3)[3]		
• Uplight			• 70/3K/VWFL p7–66

[1] These lamp guides are intended to direct your attention to lamps which might meet your needs. Design analysis is required to finalize lamp selections for projects.

[2] These are very powerful lamps typically intended for features where very strong intensities are required.

[3] Smaller trees require fewer lights. For washing effect, use in–ground wallwash uplights spaced a few feet from tree canopy edge and spaced 3 to 4 feet on center around the tree. For uplighting effect, use in–ground uplights spaced roughly half the distance to two–thirds the distance from the trunk to the canopy edge.

Metal Halide
Ceramic

▶ Not dimmable
▶ Not instant–on

Metal Halide
Ceramic

7

Ceramic Halide/PAR20

Guide
⊃ Longer life is better
⊃ Higher efficacy is better and/or higher intensity from beam spread is better
⊃ Lower wattage is better
⊃ Lower color of light is yellower
⊃ Moderate cost range for CMH is about US$50.⁰⁰ to US$75.⁰⁰

Stats
Data varies manufacturer to manufacturer and changes from time to time

Life/9000 hours
A typical office environment might be occupied about 2600 hours each year.

Efficacy/44 LPW Based on electronic ballast (includes ballast loss)
Excellent beam intensities and distributions available given lamp size.

Wattages/39

Color of Light/3000°K

Beam spreads/NSP, and NFL
Narrow spot and narrow flood distributions are available.

Size/3¾" L by 2½" Ø

Voltage/120V and 277V Ballast dependent (check ballast vendors for other voltages)

Cost Magnitude/Moderate–to–High (2000)

Manufacturers
Philips

Net Addresses
http://www.lighting.philips.com/
CONNECT FOR MORE

Advantages
Crisp white light
Excellent intensities available
Excellent alternative to less efficient halogen and HIR lamps
Small size
Withstand extreme temperature range

Disadvantages
Ballast required
Beam aberrations
◆Striations
◆Inconsistent beam shape
Hot to touch
Not dimmable
◆Without great expense
◆Without sacrificing color temperature and color rendering
Not easily integrated into egress system
Not instant–on
◆Cold start requires a few minutes for full light output
◆Power interruption initiates restrike cycle of fifteen to twenty minutes

Net Addresses/PAR20/CMH Track, Monopoints and Recessed
See page 7–27
CONNECT FOR MORE

CMH PAR lamps offer the most efficient accent lighting.

Ceramic Halide/PAR20

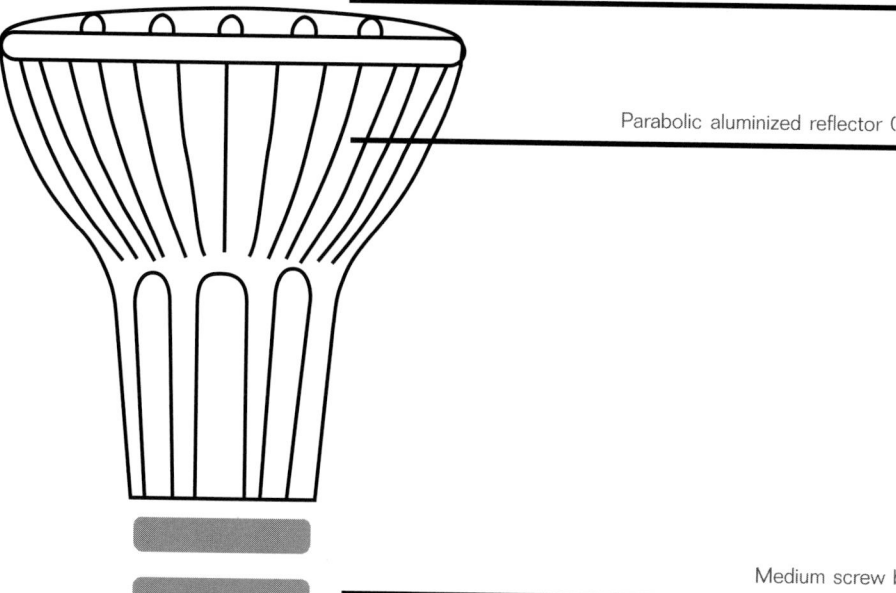

Pressed/cast glass optical lens

Parabolic aluminized reflector (PAR)

Medium screw base

Lamp schematic shown actual size
Image copyright 2000/Philips

Metal Halide
Ceramic

7

Uses

Architectural Accenting
Art Accenting
Commercial
Hospitality
Residential
Feature downlighting
Commercial
Hospitality
Retail
Merchandise Accenting
Surface Grazing
Specular/semispecular surfaces
◆Polished/honed stone
◆Polished/satin wood
◆Fritted/etched glass
◆Cut glass

PAR20 Lighting Design Tips
✔Use smaller scale (e.g., 4–inch diameter) downlights and adjustable accents for attractive, more human scale approach—excellent in residential applications.
✔PAR20 narrow spot versions best at moderate mounting heights (15′–0″ to 20′–0″).
✔PAR20 narrow flood versions best at low mounting heights (12′–6″ to 15′–0″).
✔Glary when used without much other general lighting or when aimed across long distances where ceilings are low.
✔Beam striations are most apparent on light–colored, nontextured, flat surfaces and may be objectionable.
✔PAR20 downlighting alone is harsh—combine several techniques for best results.
✔Ballasts are relatively large and must be accommodated in luminaire or within a few feet of luminaire in separate ballast box.
✔Not suggested for areas where switching lights on/off frequently is common.
✔If lamps are intended to operate continuously (24x7), then at least once every seven days all metal halide lamps should be extinguished (shut off) for at least fifteen minutes (or as instructed by lamp manufacturer) to limit non–passive failures (failures where lamps fail violently—exploding hot glass chards or at least cracking lamp envelope and creating UV exposure hazard).

▶Not dimmable
▶Not instant–on

Not dimmable
Not instant–on

Metal Halide/PAR20

Ceramic Metal Halide PAR20 Visual Index

39PAR20/CMH/3K/NSP
▶Strong architectural detailing
▶Large art/20'–0" ceiling height
▶Sculpture
▶Very long throws
▶Commercial, Hospitality, Retail, Exterior

39PAR20/CMH/3K/NFL
▶Architectural accenting
▶Feature downlighting
▶Large art/12'–6" ceiling height
▶Medium throws
▶Commercial, Retail, Exterior

Throws/Ceilings
⊃Short = 7' to 10'
⊃Medium = 10' to 15'
⊃Long = 15' to 20'
⊃Very Long = 20' or more

Recommended intensities
for art accenting

Low (gallery, museum, residential)
5 to 10 fc average on art
Moderate (hospitality)
10 to 20 fc average on art
High (commercial—or higher for retail)
20 to 30 fc average on art

These are intended to be average, maintained values including effects of accent lighting and general room lighting. General room lighting contributes just a few footcandles in residential, gallery and museum settings, but contributes 5 to 10 fc in commercial settings. Maximums might range from 50 percent to 100 percent of the average value. Wherever artwork is highly valuable and/or sensitive to light, precautions should be taken to limit light exposure (maintaining low levels and/or switching lights off when art is not being viewed); and exposure to infrared and ultraviolet should be limited (using filters on lamps or placing art behind specially treated glass or acrylic).

Ceramic Metal Halide/PAR20

▶ Not dimmable
▶ Not instant-on

Metal Halide
C e r a m i c

7

Ceramic Metal Halide PAR Lamp Designation

39PAR20/CMH/3K/WSP

Last set of letters identifies beam spread

FLOODS
- 24° to 32° for narrow flood (NFL)
- 33° to 44° for flood (FL)
- 45° to 54° for wide flood (WFL)
- 55° or greater for very wide flood (VWFL)

SPOTS
- 7° or less for very narrow spot (VNSP)
- 8° to 10° for narrow spot (NSP)
- 11° to 14° for spot (SP)
- 15° to 18° for wide spot (WSP)
- 19° to 23° for very wide spot (VWSP)

Ranges of beam spreads used for convenient/consistent reference in this text and do not necessarily correspond to each manufacturer's nor ANSI's definitions.

This number/letter combination indicates color temperature
- 3K is shorthand for 3000°K (warm tone)

This series of letters indicates "Ceramic Metal Halide"

These two digits represent lamp diameter in eighths inches (e.g., 20/8 or 2–1/2 inches)

These letters indicate this is a parabolic aluminized reflector (PAR) lamp

First two or three digits represent lamp wattage

[This designation used for convenient reference throughout this *Time–Saver Standards for Architectural Lighting*—but not necessarily used by all lamp manufacturers (although it would be convenient if manufacturers would come to agreement on standard designations.]

Artwork Size Categories

- Petite = 2'–0" by 2'–0" or smaller
- Small = 2'–0" by 2'–0" to 2'–6" by 3'–0"
- Medium = 2'–6" by 3'–0" to 3'–0" by 4'–0"
- Large = 3'–0" by 4'–0" to 4'–0" by 5'–0"
- Extra Large = 4'–0" by 5'–0" or larger

Not dimmable
Not instant-on

Metal Halide 39PAR20/CMH/3K/NSP

Metal Halide
Ceramic

Uses
▶ Strong architectural detailing
▶ Large art with very long throws
▶ Sculpture with very long throws
▶ Very long throws
▶ Various voltages (ballast dependent)

Specs
Beam spread/10°
Center beam candlepower/23000
Life/9000 hrs
▶ **Philips** (spot) (23365–0)

Design Tips
✔ One light best for large art and where mounting height for light is 21' or less.
✔ These are very strong accents and deserve judicious use. Light may degrade artwork rapidly and/or significantly— consider UV and IR filters or limit exposure time.
✔ Excellent for glass pieces and floral centerpieces and where mounting height for light is 21' or less.
✔ Cold start requires a few minutes for full light output.
✔ Power interruption results in fifteen minute "cool down" before lamp can restart.
✔ Cannot easily be part of any egress system.
✔ Warm–tone lamp very close match to incandescent.

Application Key

Commercial
Gallery
Hospitality
Institutional
Manufacturing
Residential
Retail
Exterior

bold = primary application
partial fade = minimal application
fade = unlikely application

Performance Sketches™

[All data outlined here is based on information from boldfaced manufacturer's published data at time of manuscript preparation. Scale is ¼ inch equals 1 foot.]

Lamp: 39PAR20/CMH/3K/NSP
Location: 7'–0" from wall
Center beam Aiming: 24° (at point 4'–6" AFF)
Luminaire: Monopoint or Recessed Adjustable
Illuminance: 17 fc, avg, maintained (28 fc max)

Lamp: 39PAR20/CMH/3K/NSP
Location: 8'–0" from wall
Center beam Aiming: 27° (at point 4'–6" AFF)
Luminaire: Monopoint or Recessed Adjustable
Illuminance: 17 fc, avg, maintained (29 fc max)

20'–0"

10'–0"

Metal Halide 39PAR20/CMH/3K/NSP
Ceramic

►Not dimmable
►Not instant-on

Metal Halide
ceramic

7

Lamp: 39PAR20/CMH/3K/NSP
Location: 9'–0" from wall
Center beam Aiming: 30° (at point 4'–6" AFF)
Luminaire: Monopoint or Recessed Adjustable
Illuminance: 18 fc, avg, maintained (30 fc max)

Art shown at 4' by 5' and
centered 5'–6" AFF. Wall grid on
6" by 6" centers and scaled at
¼"=1'.

▶ Not dimmable
▶ Not instant-on

Metal Halide 39PAR20/CMH/3K/NFL
Ceramic

Uses
▶ Architectural accenting
▶ Feature downlighting
▶ Large art with medium throws
▶ Sculpture with medium throws
▶ Medium throws
▶ Various voltages (ballast dependent)

Specs
Beam spread/30°
Center beam candlepower/ 5000
Life/9000 hrs
▶ **Philips** (flood) (23364–6)

Design Tips
✔ One light best for large art and where mounting height for light is 13' or less.
✔ Good for moderate to larger sculpture where mounting height for light is 13' or less.
✔ Light may degrade artwork rapidly and/or significantly: consider UV and IR filters or limit exposure time.
✔ Cold start requires a few minutes for full light output.
✔ Power interruption results in fifteen minute "cool down" before lamp can restart.
✔ Cannot easily be part of any egress system.
✔ Warm–tone lamp very close match to incandescent.

Application Key

Commercial
Gallery
Hospitality
Institutional
Manufacturing
Residential
Retail
Exterior

bold = primary application
partial fade = minimal application
fade = unlikely application

Performance Sketches™

[All data outlined here is based on information from boldfaced manufacturer's published data at time of manuscript preparation. Scale is ¼ inch equals 1 foot.

Lamp: 39PAR20/CMH/3K/NFL
Location: 3'–0" from wall
Center beam Aiming: 17° (at point 2'–6" AFF)
Luminaire: Monopoint or Recessed Adjustable
Illuminance: 19 fc, avg, maintained (40 fc max)

12'–6"

10'–0"

Lamp: 39PAR20/CMH/3K/NFL
Location: 4'–0" from wall
Center beam Aiming: 25° (at point 4'–0" AFF)
Luminaire: Monopoint or Recessed Adjustable
Illuminance: 22 fc, avg, maintained (44 fc max)

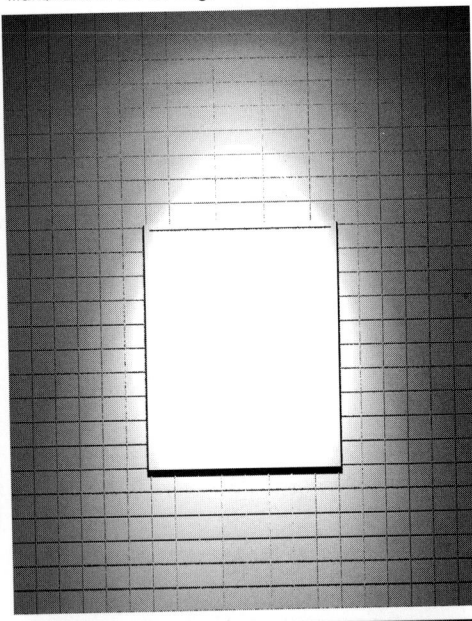

Art shown at 4' by 5' and centered 5'–6" AFF. Wall grid on 6" by 6" centers and scaled at ¼"=1'.

Lamp: 39PAR20/CMH/3K/NFL
Location: 5'–0" from wall
Center beam Aiming: 32° (at point 4'–6" AFF)
Luminaire: Monopoint or Recessed Adjustable
Illuminance: 24 fc, avg, maintained (44 fc max)

Metal Halide Lamps

Figure 7.1
Ceramic metal halide lamps are used in two applications here. The series of slabs appear to pierce the ceiling plane. In the slot created by this piercing, 39PAR30L/CMH/3K/NFL lamps are used in monopoints to graze the short face of the slabs. In the background, in–floor uplights with wallwash optics, asymmetric louvers and 70ED17P/CMH/3K/C/U lamps accent the wall (which is two stories in height). Monopoints are Indy Lighting and uplights are Lithonia/Hydrel.

Not dimmable
▼ Not instant-on

Ceramic Metal Halide/PAR30L
Long Neck

Metal Halide
ceramic

7

Guide
⟩Longer life is better
⟩Higher efficacy is better and/or higher intensity from beam spread is better
⟩Lower wattage is better
⟩Lower color of light is yellower
⟩Moderate cost range for CMH is about US$50.⁰⁰ to US$75.⁰⁰

Stats Data varies manufacturer to manufacturer and changes from time to time

Life/9000 hours
A typical office environment might be occupied about 2600 hours each year.

Efficacy/45 to 60 LPW Wattage dependent (higher wattage/higher efficacy) (values include electronic ballast loss)
Excellent beam intensities and distributions available.

Wattages/39 and 70

Color of Light/3000°K

Beam spreads/NSP, FL
Wide spot, narrow flood, and very wide flood distributions are available.

Size/4¾" L by 3¾" Ø

Voltage/120V and 277V Ballast dependent (check ballast vendors for other voltages)

Cost Magnitude/Moderate–to–High (2000)
Costs are easily recovered over the life of the lamp based on savings from maintenance replacement and energy reductions.

Manufacturers
Philips

Advantages
Crisp white light
Excellent intensities available
Excellent alternative to less efficient halogen and HIR lamps
Withstand extreme temperature range

Net Addresses
http://www.lighting.philips.com/

CONNECT FOR MORE

Disadvantages
Ballast required
Beam aberrations
◆Striations
◆Inconsistent beam shape
Hot to touch
Somewhat bulky
◆PAR20 lamps are smaller
Noticeable lamp lumen depreciation (light loss) over life
◆15 to 20 percent
Not dimmable
◆Without great expense
◆Without sacrificing color temperature and color rendering
Not easily integrated into egress system
Not instant–on
◆Cold start requires a few minutes for full light output
◆Power interruption initiates restrike cycle of fifteen to twenty minutes

CMH PAR lamps offer the most efficient accent lighting.

Net Addresses/PAR30L/CMH Track, Monopoints and Recessed
See page 7–27

CONNECT FOR MORE

Metal Halide/PAR30L
Long Neck

▶ Not dimmable
▶ Not instant-on

Pressed/cast glass optical lens

Parabolic aluminized reflector (PAR)

Medium screw base

Lamp schematic shown actual size
Image copyright 2000/Philips

Uses
Architectural Accenting
Art Accenting
Commercial
Hospitality
Retail
Downlighting
Commercial
Institutional
Retail
Facade Accenting
Merchandise Accenting
Surface Grazing
Specular/semispecular surfaces
◆Polished/honed stone
◆Polished/satin wood
◆Fritted/etched glass
◆Cut glass

Ceramic Metal Halide PAR30L Lighting Design Tips
✔ PAR30L narrow spot versions best at high mounting heights (20'–0" to 25'–0").
✔ PAR30L flood versions best at moderate mounting heights (15'–0" to 20'–0").
✔ Glary when used without much other general lighting or when aimed across long distances where ceilings are low.
✔ Beam striations are most apparent on light–colored, nontextured, flat surfaces and may be objectionable.
✔ PAR30L downlighting alone is harsh—combine several techniques for best results.
✔ Ballasts are relatively large and must be accommodated in luminaire or within a few feet of luminaire in separate ballast box.
✔ Not suggested for areas where switching lights on/off frequently is common.
✔ If lamps are intended to operate continuously (24x7), then at least once every seven days all metal halide lamps should be extinguished (shut off) for at least fifteen minutes (or as instructed by lamp manufacturer) to limit non–passive failures (failures where lamps fail violently—exploding hot glass chards or at least cracking lamp envelope and creating UV exposure hazard).

▶ Not dimmable
▶ Not instant-on

Ceramic Metal Halide/PAR30L

Ceramic Metal Halide PAR30L Visual Index

250PAR38/H/SP ▶ **p44**
250PAR38/H/SP ▶ **p44**
250PAR38/H/SP ▶ **p45**

39PAR30L/CMH/3K/NSP
▶ Very strong architectural detailing
▶ Large art/25′–0″ ceiling height
▶ Sculpture
▶ Very long throws
▶ Commercial, Retail, Exterior

Throws/Ceilings
◗ Short = 7′ to 10′
◗ Medium = 10′ to 15′
◗ Long = 15′ to 20′
◗ Very Long = 20′ or more

250PAR38/H/NFL ▶ **p46**
250PAR38/H/NFL ▶ **p46**
250PAR38/H/NFL ▶ **p47**

39PAR30L/CMH/3K/NFL
▶ Architectural accenting
▶ Extra large art/18′–0″ ceiling height
▶ Sculpture
▶ Long throws
▶ Hospitality, Exterior

70PAR30L/CMH/3K/NSP
▶ Strong architectural detailing
▶ Extra large art/25′–0″ ceiling height
▶ Sculpture
▶ Very long throws
▶ Commercial, Retail, Exterior

Recommended intensities
for art accenting

Low (gallery, museum, residential)
5 to 10 fc average on art
Moderate (hospitality)
10 to 20 fc average on art
High (commercial—or higher for retail)
20 to 30 fc average on art

These are intended to be average, maintained values including effects of accent lighting and general room lighting. General room lighting contributes just a few footcandles in residential, gallery and museum settings, but contributes 5 to 10 fc in commercial settings. Maximums might range from 50 percent to 100 percent of the average value. Wherever artwork is highly valuable and/or sensitive to light, precautions should be taken to limit light exposure (maintaining low levels and/or switching lights off when art is not being viewed); and exposure to infrared and ultraviolet should be limited (using filters on lamps or placing art behind specially treated glass or acrylic).

70PAR38/CMH/4K/ WSP ▶ **p48**
70PAR38/CMH/4K/ WSP ▶ **p48**
70PAR38/CMH/4K/ WSP ▶ **p49**

Metal Halide/PAR30L

Long Neck

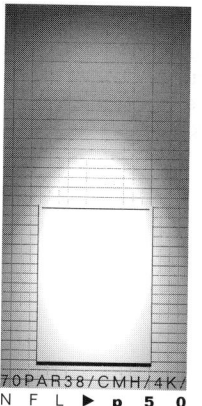 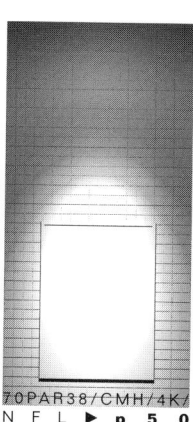 

70PAR38/CMH/4K/
NFL ▶ **p 50**

70PAR38/CMH/4K/
NFL ▶ **p 50**

70PAR38/CMH/4K/
NFL ▶ **p 51**

70PAR30L/CMH/3K/FL

▶General lighting
▶Extra large art/20'–0" ceiling height
▶Sculpture
▶Very long throws
▶Hospitality, Exterior

▶Not dimmable
▶Not instant-on

Ceramic Metal Halide PAR Lamp Designation

39PAR30L/CMH/3K/WSP

Last set of letters identifies beam spread

FLOODS
• 24° to 32° for narrow flood (NFL)
• 33° to 44° for flood (FL)
• 45° to 54° for wide flood (WFL)
• 55° or greater for very wide flood (VWFL)

SPOTS
• 7° or less for very narrow spot (VNSP)
• 8° to 10° for narrow spot (NSP)
• 11° to 14° for spot (SP)
• 15° to 18° for wide spot (WSP)
• 19° to 23° for very wide spot (VWSP)

Ranges of beam spreads used for convenient/consistent reference in this text and do not necessarily correspond to each manufacturer's nor ANSI's definitions.

This number/letter combination indicates color temperature
• 3K is shorthand for 3000°K (warm tone)

This series of letters indicates "Ceramic Metal Halide"

"L" for long neck (only applies to PAR30 lamps)

These two digits represent lamp diameter in eighths inches (e.g., 30/8 or 3–3/4 inches)

These letters indicate this is a parabolic aluminized reflector (PAR) lamp

First two or three digits represent lamp wattage

[This designation used for convenient reference throughout this *Time–Saver Standards for Architectural Lighting*—but not necessarily used by all lamp manufacturers (although it would be convenient if manufacturers would come to agreement on standard designations.]

Metal Halide
ceramic

7

▶ Not dimmable
▶ Not instant-on

Metal Halide 39PAR30L/CMH/3K/NSP
Long Neck

Ceramic

Uses
▶ Very strong architectural detailing
▶ Large art with very long throws
▶ Sculpture with very long throws
▶ Very long throws
▶ Various voltages (ballast dependent)

Specs
Beam spread/10°
Center beam candlepower/ 44000
Life/9000 hrs
▶ **Philips** (spot) (22329–7)

Design Tips
✔ One light best for large art and where mounting height for light is less than 26'.
✔ These are very strong accents and deserve judicious use. Light may degrade artwork rapidly and/or significantly: consider UV and IR filters or limit exposure time.
✔ Cold start requires a few minutes for full light output.
✔ Power interruption results in fifteen minute "cool down" before lamp can restart.
✔ Cannot easily be part of any egress system.
✔ Warm–tone lamp very close match to incandescent.
✔ Excellent for public art where maintenance and energy are priorities and where mounting height for light is 26' or less.

Application Key

Commercial
Gallery
Hospitality
Institutional
Manufacturing
Residential
Retail
Exterior

bold = primary application
partial fade = minimal application
fade = unlikely application

Performance Sketches™

[All data outlined here is based on information from boldfaced manufacturer's published data at time of manuscript preparation. Scale is ¼ inch equals 1 foot.]

Lamp: 39PAR30L/CMH/3K/NSP
Location: 8'–0" from wall
Center beam Aiming: 21° (at point 4'–6" AFF)
Luminaire: Monopoint or Recessed Adjustable
Illuminance: 18 fc, avg, maintained (28 fc max)

Lamp: 39PAR30L/CMH/3K/NSP
Location: 9'–0" from wall
Center beam Aiming: 24° (at point 5'–0" AFF)
Luminaire: Monopoint or Recessed Adjustable
Illuminance: 19 fc, avg, maintained (31 fc max)

25'–0"

10'–0"

Metal Halide 39PAR30L/CMH/3K/NSP
Ceramic
Long Neck

▶Not dimmable
▶Not instant-on

Metal Halide
Ceramic

7

Lamp: 39PAR30L/CMH/3K/NSP
Location: 10'–0" from wall
Center beam Aiming: 27° (at point 5'–0" AFF)
Luminaire: Monopoint or Recessed Adjustable
Illuminance: 20 fc, avg, maintained (32 fc max)

Art shown at 4' by 5' and centered 5'–6" AFF. Wall grid on 6" by 6" centers and scaled at ¼"=1'.

▶ Not dimmable
▶ Not instant-on

Metal Halide 39PAR30L/CMH/3K/NFL
Long Neck

Ceramic

Metal Halide
Ceramic

7

Uses
▶Architectural accenting
▶Extra large art with long throws
▶Sculpture with long throws
▶Long throws
▶Various voltages (ballast dependent)

Specs
Beam spread/30°
Center beam candlepower/ 7400
Life/9000 hrs
▶**Philips** (22320–5)

Design Tips
✔One light best for large art and where mounting height for light is 19′ or less.
✔Good for moderate to larger sculpture where mounting height for light is 19′ or less.
✔Cold start requires a few minutes for full light output.
✔Power interruption results in fifteen minute "cool down" before lamp can restart.
✔Cannot easily be part of any egress system.
✔Warm–tone lamp very close match to incandescent.

Performance Sketches™
[All data outlined here is based on information from boldfaced manufacturer's published data at time of manuscript preparation. Scale is ¼ inch equals 1 foot.]

Lamp: 39PAR30L/CMH/3K/NFL
Location: 7′–0″ from wall
Center beam Aiming: 27° (at point 4′–6″ AFF)
Luminaire: Monopoint or Recessed Adjustable
Illuminance: 12 fc, avg, maintained (23 fc max)

Lamp: 39PAR30L/CMH/3K/NFL
Location: 8′–0″ from wall
Center beam Aiming: 31° (at point 4′–6″ AFF)
Luminaire: Monopoint or Recessed Adjustable
Illuminance: 12 fc, avg, maintained (22 fc max)

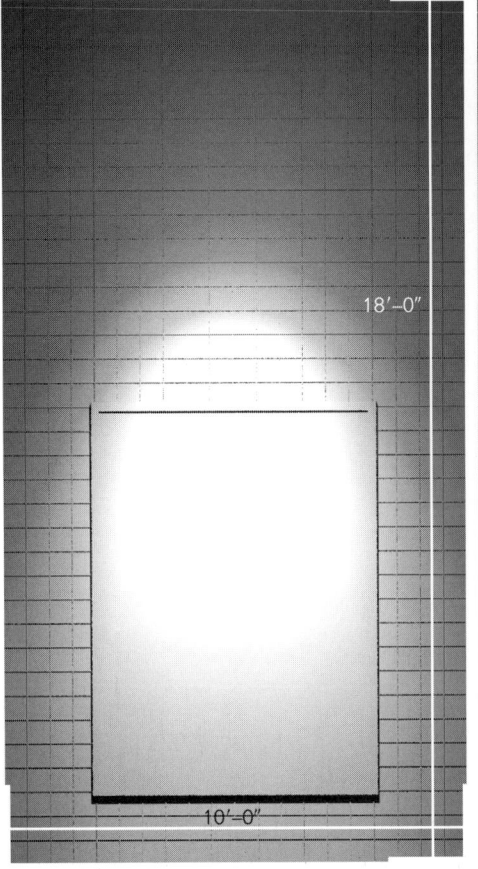

18′–0″

10′–0″

Application Key

Commercial
Gallery
Hospitality
Institutional
Manufacturing
Residential
Retail
Exterior

bold = primary application
partial fade = minimal application
fade = unlikely application

Metal Halide 39PAR30L/CMH/3K/NFL
Ceramic
Long Neck

▶ Not dimmable
▶ Not instant-on

Metal Halide
Ceramic

7

Lamp: 39PAR30L/CMH/3K/NFL
Location: 9'–0" from wall
Center beam Aiming: 34° (at point 4'–9" AFF)
Luminaire: Monopoint or Recessed Adjustable
Illuminance: 12 fc, avg, maintained (21 fc max)

Art shown at 6' by 8' and
centered 5'–6" AFF. Wall grid on
6" by 6" centers and scaled at
¼"=1'.

Metal Halide 70PAR30L/CMH/3K/NSP
Ceramic
Long Neck

Uses
- ►Strong architectural detailing
- ►Extra large art with very long throws
- ►Sculpture with very long throws
- ►Very long throws
- ►Various voltages (ballast dependent)

Specs
Beam spread/10°
Center beam candlepower/ 68000
Life/9000 hrs
►**Philips** (23224–9)

Design Tips
✔One light best for large art and where mounting height for light is less than 26'.

✔These are strong accents and deserve judicious use. Light may degrade artwork rapidly and/or significantly: consider UV and IR filters or limit exposure time.

✔Cold start requires a few minutes for full light output.

✔Power interruption results in fifteen minute "cool down" before lamp can restart.

✔Cannot easily be part of any egress system.

✔Warm–tone lamp very close match to incandescent.

✔Excellent for public art where maintenance and energy are priorities and where mounting height for light is 26' or less.

Application Key

Commercial
Gallery
Hospitality
Institutional
Manufacturing
Residential
Retail
Exterior

bold = primary application
partial fade = minimal application
fade = unlikely application

Performance Sketches™

[All data outlined here is based on information from boldfaced manufacturer's published data at time of manuscript preparation. Scale is ¼ inch equals 1 foot.]

Lamp: 70PAR30L/CMH/3K/NSP
Location: 8'–0" from wall
Center beam Aiming: 21° (at point 4'–0" AFF)
Luminaire: Monopoint or Recessed Adjustable
Illuminance: 22 fc, avg, maintained (40 fc max)

Lamp: 70PAR30L/CMH/3K/NSP
Location: 9'–0" from wall
Center beam Aiming: 23° (at point 4'–0" AFF)
Luminaire: Monopoint or Recessed Adjustable
Illuminance: 23 fc, avg, maintained (43 fc max)

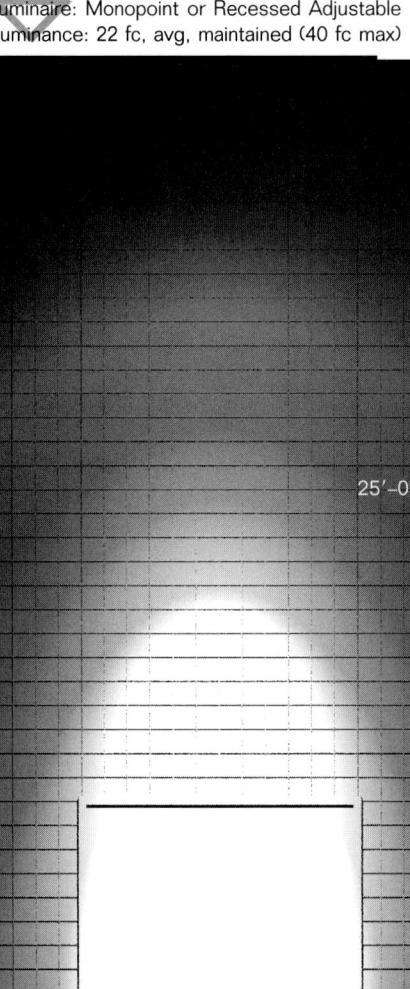

Metal Halide 70PAR30L/CMH/3K/NSP

Ceramic
Long Neck

▶ Not dimmable
▶ Not instant-on

Lamp: 70PAR30L/CMH/3K/NSP
Location: 10'–0" from wall
Center beam Aiming: 26° (at point 4'–0" AFF)
Luminaire: Monopoint or Recessed Adjustable
Illuminance: 24 fc, avg, maintained (45 fc max)

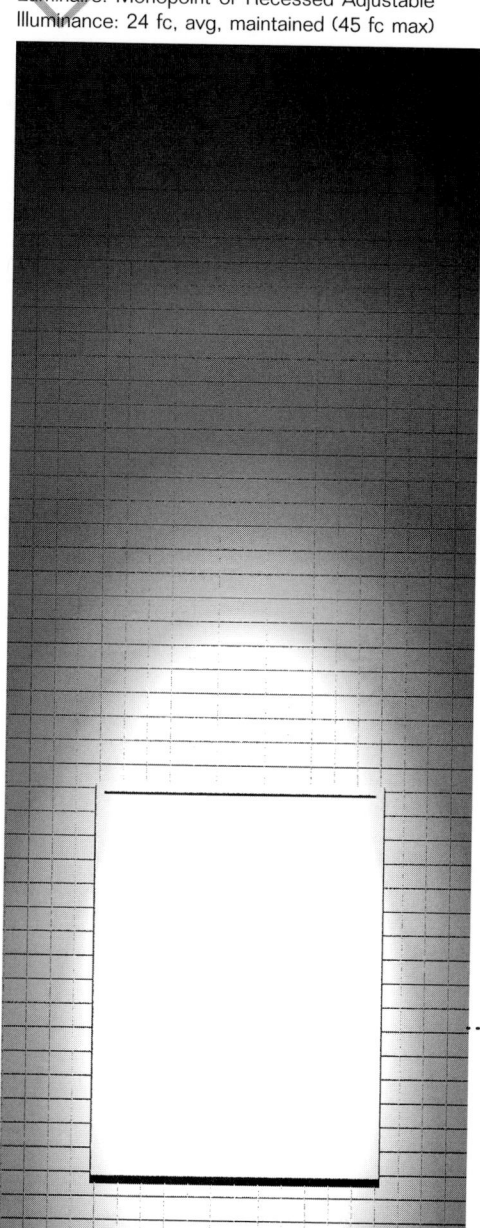

Art shown at 6' by 8' and centered 5'–6" AFF. Wall grid on 6" by 6" centers and scaled at ¼"=1'.

Metal Halide
ceramic

7

▶ Not dimmable
▶ Not instant–on

Metal Halide
Ceramic

7

Uses
▶ General lighting
▶ Extra large art with very long throws
▶ Large sculpture with very long throws
▶ Very long throws
▶ Various voltages (ballast dependent)

Specs
Beam spread/40°
Center beam candlepower/10000
Life/9000 hrs
▶ **Philips** (23221–5)

Design Tips
✔ One light best for extra large art and where mounting height for light is 21′ or less.
✔ Light may degrade artwork rapidly and/or significantly—consider UV and IR filters or limit exposure time.
✔ Cold start requires a few minutes for full light output.
✔ Power interruption results in fifteen minute "cool down" before lamp can restart.
✔ Cannot easily be part of any egress system.
✔ Warm–tone lamp very close match to incandescent.

Application Key

Commercial
Gallery
Hospitality
Institutional
Manufacturing
Residential
Retail
Exterior

bold = primary application
partial fade = minimal application
fade = unlikely application

Ceramic Halide 70PAR30L/CMH/3K/FL
Long Neck

Performance Sketches™

[All data outlined here is based on information from boldfaced manufacturer's published data at time of manuscript preparation. Scale is ¼ inch equals 1 foot.]

Lamp: 70PAR30L/CMH/3K/FL
Location: 7′–0″ from wall
Center beam Aiming: 13° (at point 0′–0″ AFF)
Luminaire: Monopoint or Recessed Adjustable
Illuminance: 10 fc, avg, maintained (14 fc max)

Lamp: 70PAR30L/CMH/3K/FL
Location: 8′–0″ from wall
Center beam Aiming: 18° (at point 0′–0″ AFF)
Luminaire: Monopoint or Recessed Adjustable
Illuminance: 12 fc, avg, maintained (15 fc max)

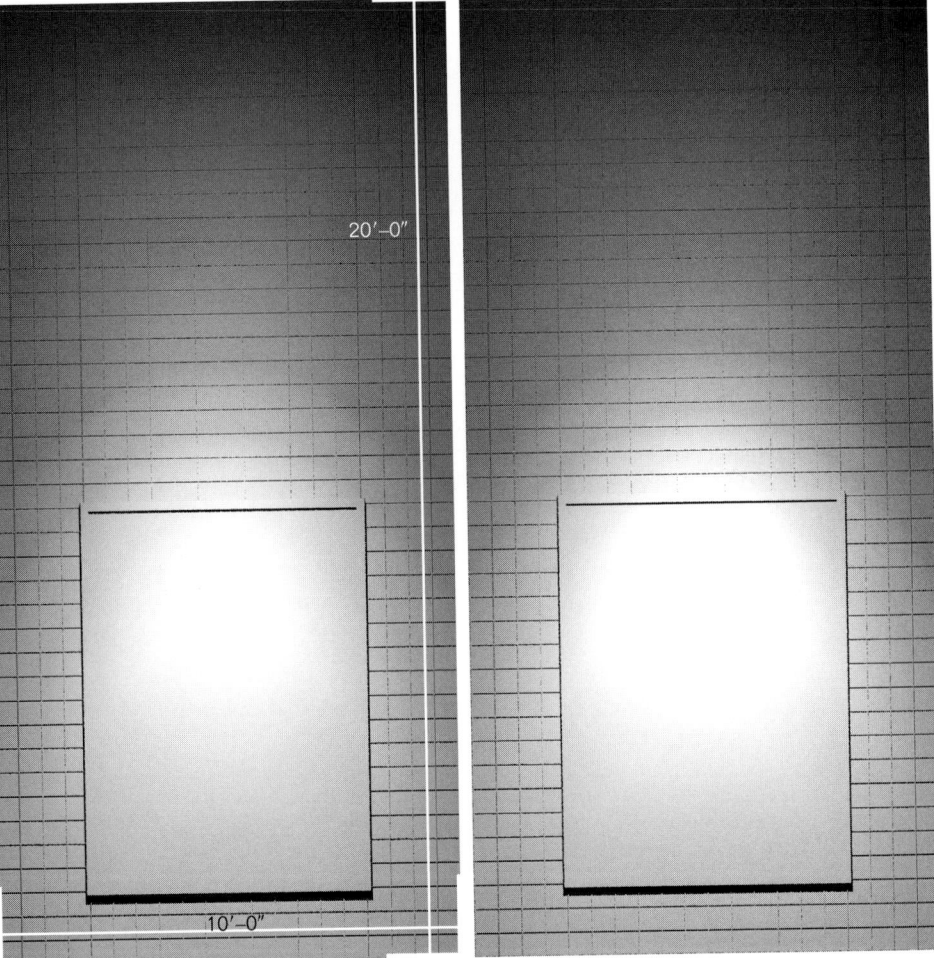

20′–0″

10′–0″

Metal Halide 70PAR30L/CMH/3K/FL

Ceramic

Long Neck

Lamp: 70 PAR30L/CMH/3K/FL
Location: 9'–0" from wall
Center beam Aiming: 24° (at point 0'–0" AFF)
Luminaire: Monopoint or Recessed Adjustable
Illuminance: 14 fc, avg, maintained (20 fc max)

Art shown at 6' by 8' and
centered 5'–6" AFF. Wall grid on
6" by 6" centers and scaled at
¼"=1'.

▶Not dimmable
▶Not instant-on

Metal Halide
c e r a m i c

7

Not dimmable
Not instant-on

Ceramic Halide/PAR38

Metal Halide
Ceramic

7

Guide
⊃ Longer life is better
⊃ Higher efficacy is better and/or higher intensity from beam spread is better
⊃ Lower wattage is better
⊃ Lower color of light is yellower
⊃ Moderate cost range for CMH is about US$50.⁰⁰ to US$75.⁰⁰

Stats
Data varies manufacturer to manufacturer and changes from time to time

Life/10000 and 12500 hours
Dependent on color temperature and wattage. A typical office environment might be occupied about 2600 hours each year.

Efficacy/50 to 60 LPW
Wattage dependent (higher wattage/higher efficacy) (values include electronic ballast loss)
Excellent beam intensities and distributions available.

Wattages/ 70 and 100

Color of Light/3000°K (warm tone) and 4000°K (cool tone)

Beam spreads/WSP, NFL, VWFL
Wide spot, narrow flood, and very wide flood distributions are available.

Size/5⁷⁄₁₆″ L by 4³⁄₄″ Ø

Voltage/120V and 277V
Ballast dependent (check ballast vendors for other voltages)

Cost Magnitude/Moderate (2000)
Costs are easily recovered over the life of the lamp based on savings from maintenance replacement and energy reductions.

Manufacturers
Philips

Net Addresses
http://www.lighting.philips.com/

CONNECT FOR MORE

Advantages
Crisp white light
Excellent intensities available
Excellent alternative to less efficient halogen and HIR lamps
Withstand extreme temperature range
PAR38 versions are TCLP–compliant (Toxic Characteristic Leaching Procedure)
◆ Can be disposed in traditional landfill

Disadvantages
Ballast required
Beam aberrations
◆ Striations
◆ Inconsistent beam shape
Hot to touch
Large/bulky
◆ PAR30L lamps are smaller
Noticeable lamp lumen depreciation (light loss) over life
◆ 15 to 20 percent
Not dimmable
◆ Without great expense
◆ Without sacrificing color temperature and color rendering
Not easily integrated into egress system
Not instant–on
◆ Cold start requires a few minutes for full light output
◆ Power interruption initiates restrike cycle of fifteen to twenty minutes

CMH PAR lamps offer the most efficient accent lighting.

Net Addresses/PAR38/CMH Track, Monopoints and Recessed
See page 7–27

CONNECT FOR MORE

Metal Halide/PAR38
Ceramic

Not dimmable
Not instant-on

Metal Halide
Ceramic

7

Pressed/cast glass optical lens

Parabolic aluminized reflector (PAR)

Medium screw base

Lamp schematic shown actual size
Image copyright 2000/Philips

Uses
Architectural Accenting
Art Accenting
Commercial
Hospitality
Retail
Downlighting
Commercial
Institutional
Retail
Facade Accenting
Merchandise Accenting
Surface Grazing
Specular/semispecular surfaces
◆Polished/honed stone
◆Polished/satin wood
◆Fritted/etched glass
◆Cut glass

Ceramic Metal Halide PAR38 Lighting Design Tips
✔PAR38 wide spot versions best at high mounting heights (20'–0" to 25'–0").
✔PAR38 narrow flood versions best at moderate mounting heights (15'–0" to 20'–0").
✔PAR38 wide flood versions best at low mounting heights (12'–6" to 15'–0").
✔Glary when used without much other general lighting or when aimed across long distances where ceilings are low.
✔Beam striations are most apparent on light–colored, nontextured, flat surfaces and may be objectionable.
✔PAR38 downlighting alone is harsh—combine several techniques for best results.
✔Ballasts are relatively large and must be accommodated in luminaire or within a few feet of luminaire in separate ballast box.
✔Not suggested for areas where switching lights on/off frequently is common.
✔If lamps are intended to operate continuously (24x7), then at least once every seven days all metal halide lamps should be extinguished (shut off) for at least fifteen minutes (or as instructed by lamp manufacturer) to limit non–passive failures (failures where lamps fail violently—exploding hot glass chards or at least cracking lamp envelope and creating UV exposure hazard).

▶ Not dimmable
▶ Not instant-on

Metal Halide
Ceramic

7

Ceramic
Metal Halide/PAR38

Ceramic Metal Halide PAR38 Visual Index

70PAR38/CMH/3K/WSP

▶ Strong architectural detailing
▶ Extra large art/25'–0" ceiling height
▶ Sculpture
▶ Very long throws
▶ Commercial, Retail, Exterior

70PAR38/CMH/3K/   70PAR38/CMH/3K/   70PAR38/CMH/3K/
W S P ▶ **p 5 8**   W S P ▶ **p 5 8**   W S P ▶ **p 5 9**

■ ■ ■ ■ ■

Throws/Ceilings
⟩ Short = 7' to 10'
⟩ Medium = 10' to 15'
⟩ Long = 15' to 20'
⟩ Very Long = 20' or more

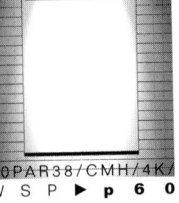

70PAR38/CMH/4K/WSP

▶ Strong architectural detailing
▶ Extra large art/25'–0" ceiling height
▶ Sculpture
▶ Very long throws
▶ Commercial, Retail, Exterior

70PAR38/CMH/4K/   70PAR38/CMH/4K/   70PAR38/CMH/4K/
W S P ▶ **p 6 0**   W S P ▶ **p 6 0**   W S P ▶ **p 6 1**

Recommended intensities
for art accenting

Low (gallery, museum, residential)
5 to 10 fc average on art
Moderate (hospitality)
10 to 20 fc average on art
High (commercial—or higher for retail)
20 to 30 fc average on art

These are intended to be average, maintained values including effects of accent lighting and general room lighting. General room lighting contributes just a few footcandles in residential, gallery and museum settings, but contributes 5 to 10 fc in commercial settings. Maximums might range from 50 percent to 100 percent of the average value. Wherever artwork is highly valuable and/or sensitive to light, precautions should be taken to limit light exposure (maintaining low levels and/or switching lights off when art is not being viewed); and exposure to infrared and ultraviolet should be limited (using filters on lamps or placing art behind specially treated glass or acrylic).

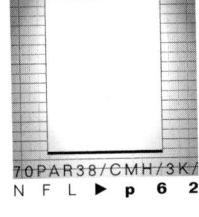

70PAR38/CMH/3K/NFL

▶ Architectural accenting
▶ Extra large art/20'–0" ceiling height
▶ Sculpture
▶ Very long throws
▶ Commercial, Retail, Exterior

70PAR38/CMH/3K/   70PAR38/CMH/3K/   70PAR38/CMH/3K/
N F L ▶ **p 6 2**   N F L ▶ **p 6 2**   N F L ▶ **p 6 3**

Metal Halide/PAR38

▶ Not dimmable
▶ Not instant-on

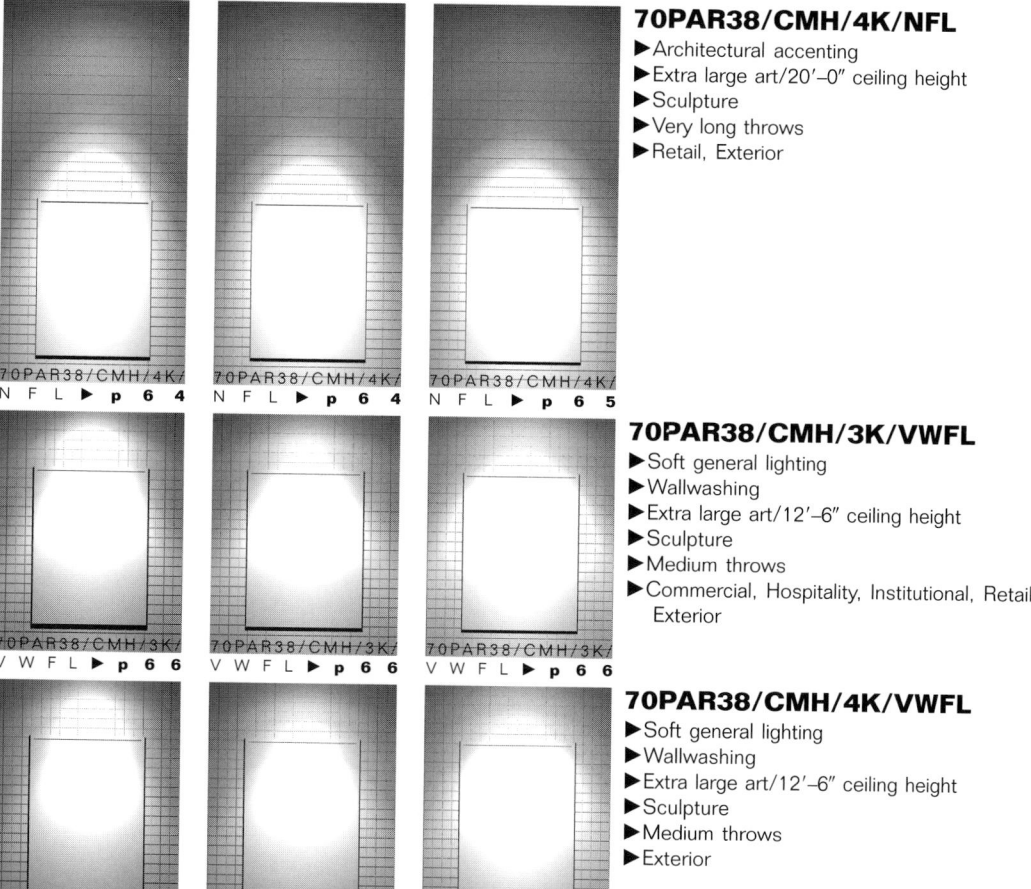

70PAR38/CMH/4K/NFL
▶ Architectural accenting
▶ Extra large art/20'–0" ceiling height
▶ Sculpture
▶ Very long throws
▶ Retail, Exterior

70PAR38/CMH/4K/NFL ▶ **p 6 4**
70PAR38/CMH/4K/NFL ▶ **p 6 4**
70PAR38/CMH/4K/NFL ▶ **p 6 5**

70PAR38/CMH/3K/VWFL
▶ Soft general lighting
▶ Wallwashing
▶ Extra large art/12'–6" ceiling height
▶ Sculpture
▶ Medium throws
▶ Commercial, Hospitality, Institutional, Retail, Exterior

70PAR38/CMH/3K/VWFL ▶ **p 6 6**
70PAR38/CMH/3K/VWFL ▶ **p 6 6**
70PAR38/CMH/3K/VWFL ▶ **p 6 6**

70PAR38/CMH/4K/VWFL
▶ Soft general lighting
▶ Wallwashing
▶ Extra large art/12'–6" ceiling height
▶ Sculpture
▶ Medium throws
▶ Exterior

70PAR38/CMH/4K/VWFL ▶ **p 6 7**
70PAR38/CMH/4K/VWFL ▶ **p 6 7**
70PAR38/CMH/4K/VWFL ▶ **p 6 7**

Ceramic Metal Halide PAR Lamp Designation

70PAR38/CMH/3K/WSP

Last set of letters identifies beam spread

FLOODS
- 24° to 32° for narrow flood (NFL)
- 33° to 44° for flood (FL)
- 45° to 54° for wide flood (WFL)
- 55° or greater for very wide flood (VWFL)

SPOTS
- 7° or less for very narrow spot (VNSP)
- 8° to 10° for narrow spot (NSP)
- 11° to 14° for spot (SP)
- 15° to 18° for wide spot (WSP)
- 19° to 23° for very wide spot (VWSP)

Ranges of beam spreads used for convenient/consistent reference in this text and do not necessarily correspond to each manufacturer's nor ANSI's definitions.

This number/letter combination indicates color temperature
- 3K is shorthand for 3000°K (warm tone)
- 4K is shorthand for 4000°K (cool tone)

This series of letters indicates "Ceramic Metal Halide"

These two digits represent lamp diameter in eighths inches (e.g., 38/8 or 4–3/4 inches)

These letters indicate this is a parabolic aluminized reflector (PAR) lamp

First two or three digits represent lamp wattage

[This designation used for convenient reference throughout this *Time–Saver Standards for Architectural Lighting*—but not necessarily used by all lamp manufacturers (although it would be convenient if manufacturers would come to agreement on standard designations.]

▶ Not dimmable
▶ Not instant-on

Ceramic Metal Halide/PAR38

100PAR38/CMH/3K/WSP

▶ Very strong architectural detailing
▶ Extra large art/25'–0" ceiling height
▶ Sculpture
▶ Very long throws
▶ Commercial, Retail, Exterior

100PAR38/CMH/3K/ W S P ▶ p 6 8
100PAR38/CMH/3K/ W S P ▶ p 6 8
100PAR38/CMH/3K/ W S P ▶ p 6 9

100PAR38/CMH/4K/WSP

▶ Very strong architectural detailing
▶ Extra large art/25'–0" ceiling height
▶ Sculpture
▶ Very long throws
▶ Commercial, Retail, Exterior

100PAR38/CMH/4K/ W S P ▶ p 7 0
100PAR38/CMH/4K/ W S P ▶ p 7 0
100PAR38/CMH/4K/ W S P ▶ p 7 1

100PAR38/CMH/3K/NFL

▶ Strong architectural accenting
▶ Extra large art/20'–0" ceiling height
▶ Sculpture
▶ Very long throws
▶ Commercial, Retail, Exterior

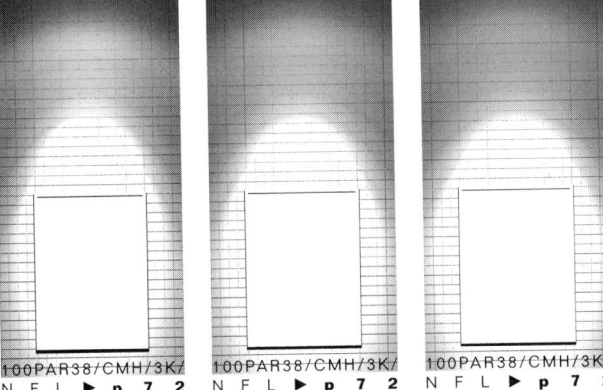

100PAR38/CMH/3K/ N F L ▶ p 7 2
100PAR38/CMH/3K/ N F L ▶ p 7 2
100PAR38/CMH/3K/ N F L ▶ p 7 2

■ ■ ■ ■ ■ ■

Throws/Ceilings

⊃ Short = 7' to 10'
⊃ Medium = 10' to 15'
⊃ Long = 15' to 20'
⊃ Very Long = 20' or more

Recommended intensities
for art accenting

Low (gallery, museum, residential)
5 to 10 fc average on art
Moderate (hospitality)
10 to 20 fc average on art
High (commercial—or higher for retail)
20 to 30 fc average on art

These are intended to be average, maintained values including effects of accent lighting and general room lighting. General room lighting contributes just a few footcandles in residential, gallery and museum settings, but contributes 5 to 10 fc in commercial settings. Maximums might range from 50 percent to 100 percent of the average value. Wherever artwork is highly valuable and/or sensitive to light, precautions should be taken to limit light exposure (maintaining low levels and/or switching lights off when art is not being viewed); and exposure to infrared and ultraviolet should be limited (using filters on lamps or placing art behind specially treated glass or acrylic).

Metal Halide
Ceramic

7

Ceramic Halide/PAR38

▶Not dimmable
▶Not instant-on

Metal Halide
C e r a m i c

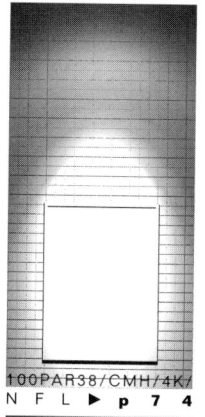

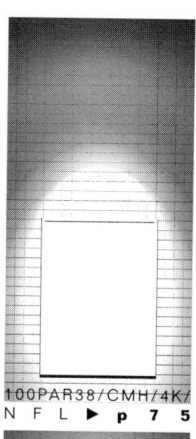

100PAR38/CMH/4K/
N F L ▶ **p 7 4**
100PAR38/CMH/4K/
N F L ▶ **p 7 4**
100PAR38/CMH/4K/
N F L ▶ **p 7 5**

100PAR38/CMH/4K/NFL
▶Strong architectural accenting
▶Extra large art/20'–0" ceiling height
▶Sculpture
▶Very long throws
▶Retail, Exterior

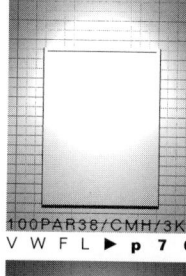

100PAR38/CMH/3K/
V W F L ▶ **p 7 6**
100PAR38/CMH/3K/
V W F L ▶ **p 7 6**
100PAR38/CMH/3K/
V W F L ▶ **p 7 6**

100PAR38/CMH/3K/VWFL
▶General lighting
▶Wallwashing
▶Extra large art/15'–0" ceiling height
▶Sculpture
▶Long throws
▶Commercial, Hospitality, Institutional, Retail, Exterior

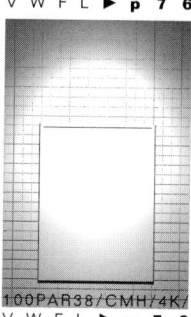

100PAR38/CMH/4K/
V W F L ▶ **p 7 6**
100PAR38/CMH/4K/
V W F L ▶ **p 7 6**
100PAR38/CMH/4K/
V W F L ▶ **p 7 6**

100PAR38/CMH/4K/VWFL
▶General lighting
▶Wallwashing
▶Extra large art/15'–0" ceiling height
▶Sculpture
▶Long throws
▶Retail, Exterior

Artwork Size Categories
⊃Petite = 2'–0" by 2'–0" or smaller
⊃Small = 2'–0" by 2'–0" to 2'–6" by 3'–0"
⊃Medium = 2'–6" by 3'–0" to 3'–0" by 4'–0"
⊃Large = 3'–0" by 4'–0" to 4'–0" by 5'–0"
⊃Extra Large = 4'–0" by 5'–0" or larger

Not dimmable
Not instant-on

Metal Halide 70PAR38/CMH/3K/WSP

Ceramic

Metal Halide Ceramic

Uses
▶Strong architectural detailing
▶Extra large art with very long throws
▶Sculpture with very long throws
▶Very long throws
▶Various voltages (ballast dependent)

Specs
Beam spread/15°
Center beam candlepower/ 50000
Life/10000 hrs
▶**Philips** (22250–5)

Design Tips
✔One light best for large art and where mounting height for light is less than 26′.
✔These are strong accents and deserve judicious use. Light may degrade artwork rapidly and/or significantly: consider UV and IR filters or limit exposure time.
✔Cold start requires a few minutes for full light output.
✔Power interruption results in fifteen minute "cool down" before lamp can restart.
✔Cannot easily be part of any egress system.
✔Warm–tone lamp very close match to incandescent.
✔Excellent for public art where maintenance and energy are priorities and where mounting height for light is 26′ or less.

Application Key

Commercial
Gallery
Hospitality
Institutional
Manufacturing
Residential
Retail
Exterior

bold = primary application
partial fade = minimal application
fade = unlikely application

Performance Sketches™

[All data outlined here is based on information from boldfaced manufacturer's published data at time of manuscript preparation. Scale is ¼ inch equals 1 foot.]

Lamp: 70PAR38/CMH/3K/WSP
Location: 8′–0″ from wall
Center beam Aiming: 21° (at point 4′–0″ AFF)
Luminaire: Monopoint or Recessed Adjustable
Illuminance: 20 fc, avg, maintained (34 fc max)

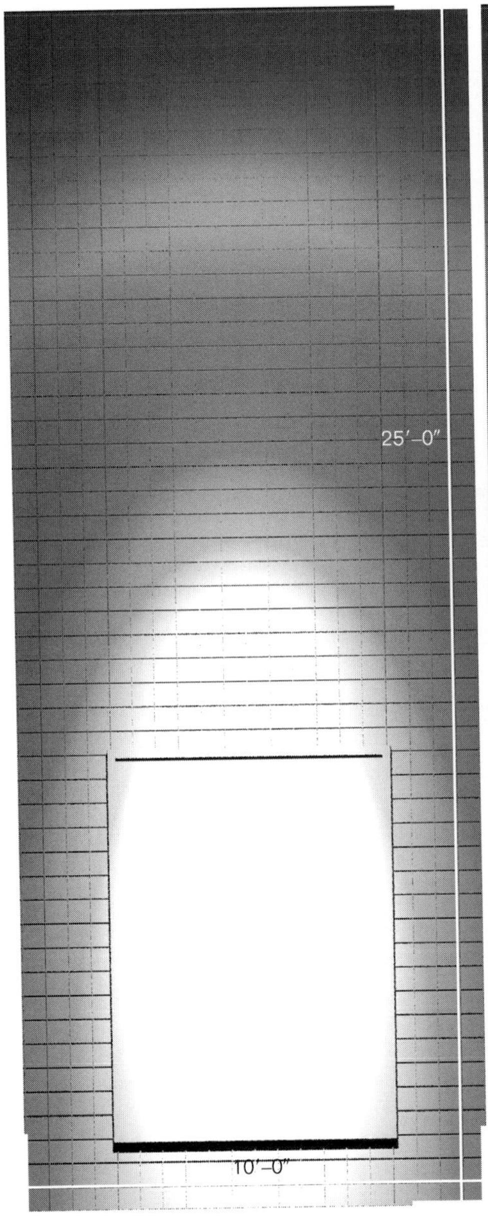

25′–0″

10′–0″

Lamp: 70PAR38/CMH/3K/WSP
Location: 9′–0″ from wall
Center beam Aiming: 23° (at point 4′–0″ AFF)
Luminaire: Monopoint or Recessed Adjustable
Illuminance: 21 fc, avg, maintained (34 fc max)

Metal Halide 70PAR38/CMH/3K/WSP

Metal Halide
ceramic

7

Not dimmable
Not instant-on

Lamp: 70PAR38/CMH/3K/WSP
Location: 10'–0" from wall
Center beam Aiming: 26° (at point 4'–0" AFF)
Luminaire: Monopoint or Recessed Adjustable
Illuminance: 22 fc, avg, maintained (34 fc max)

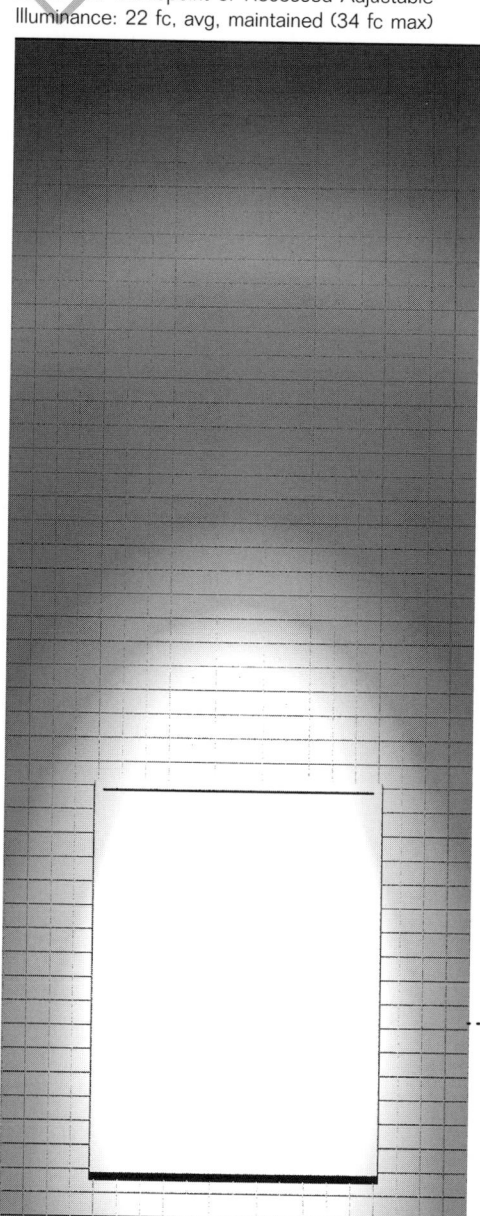

Art shown at 6' by 8' and
centered 5'–6" AFF. Wall grid on
6" by 6" centers and scaled at
¼"=1'.

Cool–tone light
▶ Not dimmable
▶ Not instant–on

Metal Halide 70PAR38/CMH/4K/WSP
Ceramic

Uses
▶ Strong architectural detailing
▶ Extra large art with very long throws
▶ Sculpture with very long throws
▶ Very long throws
▶ Various voltages (ballast dependent)

Specs
Beam spread/15°
Center beam candlepower/ 42000
Life/10000 hrs
▶ **Philips** (28872–0)

Design Tips
✔ One light best for large art and where mounting height for light is less than 26'.
✔ These are strong accents and deserve judicious use. Light may degrade artwork rapidly and/or significantly: consider UV and IR filters or limit exposure time.
✔ Cold start requires a few minutes for full light output.
✔ Power interruption results in fifteen minute "cool down" before lamp can restart.
✔ Cannot easily be part of any egress system.
✔ Cool–tone lamp best for crisp–white visual attraction.
✔ Consider for special–focus in stores, particularly where most lighting is warm–tone.
✔ Warm–tone version has greater candlepower.

Application Key

Commercial
Gallery
Hospitality
Institutional
Manufacturing
Residential
Retail
Exterior

bold = primary application
partial fade = minimal application
fade = unlikely application

Performance Sketches™

[All data outlined here is based on information from boldfaced manufacturer's published data at time of manuscript preparation. Scale is ¼ inch equals 1 foot.]

Lamp: 70PAR38/CMH/4K/WSP
Location: 8'–0" from wall
Center beam Aiming: 21° (at point 4'–0" AFF)
Luminaire: Monopoint or Recessed Adjustable
Illuminance: 19 fc, avg, maintained (32 fc max)

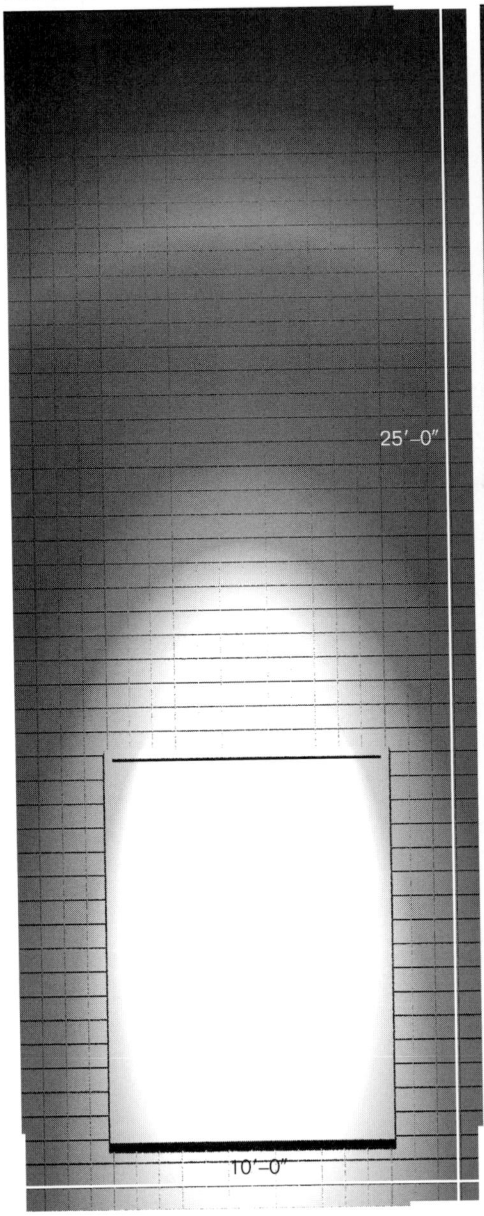

Lamp: 70PAR38/CMH/4K/WSP
Location: 9'–0" from wall
Center beam Aiming: 23° (at point 4'–0" AFF)
Luminaire: Monopoint or Recessed Adjustable
Illuminance: 20 fc, avg, maintained (32 fc max)

25'–0"

10'–0"

Metal Halide 70PAR38/CMH/4K/WSP

Ceramic

▶ Cool-tone light
▶ Not dimmable
▶ Not instant-on

Lamp: 70PAR38/CMH/4K/WSP
Location: 10'–0" from wall
Center beam Aiming: 26° (at point 4'–0" AFF)
Luminaire: Monopoint or Recessed Adjustable
Illuminance: 21 fc, avg, maintained (32 fc max)

Art shown at 6' by 8' and
centered 5'–6" AFF. Wall grid on
6" by 6" centers and scaled at
¼"=1'.

▶ Not dimmable
▶ Not instant-on

Metal Halide 70PAR38/CMH/3K/NFL

Ceramic • **Metal Halide** 70PAR38/CMH/3K/NFL

Uses
▶ Architectural accenting
▶ Extra large art with very long throws
▶ Large sculpture with very long throws
▶ Very long throws
▶ Various voltages (ballast dependent)

Specs
Beam spread/25°
Center beam candlepower/18000
Life/10000 hrs
▶ **Philips** (flood) (22249–7)

Design Tips
✔ One light best for extra large art and where mounting height for light is 21′ or less.
✔ Light may degrade artwork rapidly and/or significantly—consider UV and IR filters or limit exposure time.
✔ Cold start requires a few minutes for full light output.
✔ Power interruption results in fifteen minute "cool down" before lamp can restart.
✔ Cannot easily be part of any egress system.
✔ Warm–tone lamp very close match to incandescent.
✔ Excellent for public art where maintenance and energy are priorities and where mounting height for light is 21′ or less.

Application Key

Commercial
Gallery
Hospitality
Institutional
Manufacturing
Residential
Retail
Exterior

bold = primary application
partial fade = minimal application
fade = unlikely application

Performance Sketches™

[All data outlined here is based on information from boldfaced manufacturer's published data at time of manuscript preparation. Scale is ¼ inch equals 1 foot.]

Lamp: 70PAR38/CMH/3K/NFL
Location: 7′–0″ from wall
Center beam Aiming: 22° (at point 3′–0″ AFF)
Luminaire: Monopoint or Recessed Adjustable
Illuminance: 19 fc, avg, maintained (31 fc max)

Lamp: 70PAR38/CMH/3K/NFL
Location: 8′–0″ from wall
Center beam Aiming: 25° (at point 3′–0″ AFF)
Luminaire: Monopoint or Recessed Adjustable
Illuminance: 19 fc, avg, maintained (30 fc max)

20′–0″

10′–0″

Metal Halide70PAR38/CMH/3K/NFL

C e r a m i c

Not dimmable
Not instant-on

Metal Halide
c e r a m i c

7

Lamp: 70 PAR38/CMH/3K/NFL
Location: 9'–0" from wall
Center beam Aiming: 29° (at point 3'–6" AFF)
Luminaire: Monopoint or Recessed Adjustable
Illuminance: 20 fc, avg, maintained (30 fc max)

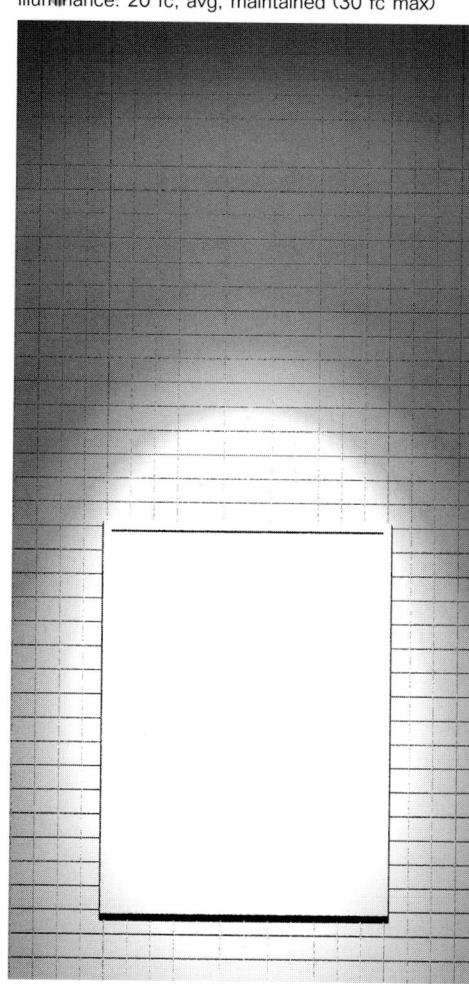

Art shown at 6' by 8' and
centered 5'–6" AFF. Wall grid on
6" by 6" centers and scaled at
¼"=1'.

▶ Cool–tone light
▶ Not dimmable
▶ Not instant–on

Metal Halide 70PAR38/CMH/4K/NFL
Ceramic

Uses
▶ Architectural accenting
▶ Extra large art with very long throws
▶ Large sculpture with very long throws
▶ Very long throws
▶ Various voltages (ballast dependent)

Specs
Beam spread/25°
Center beam candlepower/16000
Life/10000 hrs
▶**Philips** (flood) (28873–8)

Design Tips
✔ One light best for extra large art and where mounting height for light is 21' or less.
✔ Light may degrade artwork rapidly and/or significantly— consider UV and IR filters or limit exposure time.
✔ Cold start requires a few minutes for full light output.
✔ Power interruption results in fifteen minute "cool down" before lamp can restart.
✔ Cannot easily be part of any egress system.
✔ Cool–tone lamp best for crisp–white visual attraction.
✔ Consider for cooler tone stone surface lighting.
✔ Warm–tone version has greater candlepower.

Application Key

Commercial
Gallery
Hospitality
Institutional
Manufacturing
Residential
Retail
Exterior

bold = primary application
partial fade = minimal application
fade = unlikely application

Performance Sketches™

[All data outlined here is based on information from boldfaced manufacturer's published data at time of manuscript preparation. Scale is ¼ inch equals 1 foot.]

Lamp: 70PAR38/CMH/4K/NFL
Location: 7'–0" from wall
Center beam Aiming: 22° (at point 3'–0" AFF)
Luminaire: Monopoint or Recessed Adjustable
Illuminance: 16 fc, avg, maintained (27 fc max)

Lamp: 70PAR38/CMH/4K/NFL
Location: 8'–0" from wall
Center beam Aiming: 25° (at point 3'–0" AFF)
Luminaire: Monopoint or Recessed Adjustable
Illuminance: 17 fc, avg, maintained (26 fc max)

20'–0"

10'–0"

Metal·Halide 70PAR38/CMH/4K/NFL

Ceramic

▶ Cool-tone light
▶ Not dimmable
▶ Not instant-on

Lamp: 70 PAR38/CMH/4K/NFL
Location: 9'–0" from wall
Center beam Aiming: 29° (at point 3'–6" AFF)
Luminaire: Monopoint or Recessed Adjustable
Illuminance: 17 fc, avg, maintained (26 fc max)

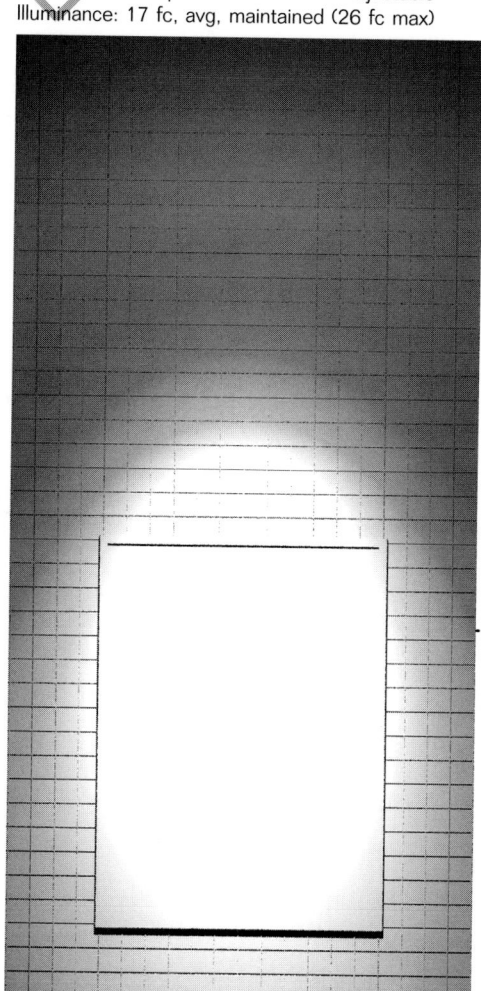

Art shown at 6' by 8' and
centered 5'–6" AFF. Wall grid on
6" by 6" centers and scaled at
¼"=1'.

Metal Halide
ceramic

7

Not dimmable
Not instant–on

Metal Halide 70PAR38/CMH/3K/VWFL

Ceramic

Uses
▶ Soft general lighting
▶ Wallwashing (see 11–131)
▶ Extra large art with medium throws
▶ Medium throws
▶ Various voltages (ballast dependent)

Specs
Beam spread/60°
Center beam candlepower/5000
Life/10000 hrs
▶ **Philips** (wide flood) (23216–5)

Design Tips
✔ One light best for extra large art and where mounting height for light is 13' or less.
✔ Good for soft downlighting—best with sconces and/or art accents and/or wallwashing to avoid cave effect.
✔ Light may degrade artwork rapidly and/or significantly: consider UV and IR filters or limit exposure time.
✔ Cold start requires a few minutes for full light output.
✔ Power interruption results in fifteen minute "cool down" before lamp can restart.
✔ Cannot easily be part of any egress system.
✔ Warm–tone lamp very close match to incandescent.

Application Key

Commercial
Gallery
Hospitality
Institutional
Manufacturing
Residential
Retail
Exterior

bold = primary application
partial fade = minimal application
fade = unlikely application

Performance Sketches™

[All data outlined here is based on information from boldfaced manufacturer's published data at time of manuscript preparation. Scale is ¼ inch equals 1 foot.]

Lamp: 70PAR38/CMH/3K/WFL
Location: 3'–0" from wall
Center beam Aiming: 5° (at point 0'–0" AFF)
Luminaire: Monopoint or Recessed Adjustable
Illuminance: 16 fc, avg, maintained (43 fc max)

Lamp: 70PAR38/CMH/3K/WFL
Location: 4'–0" from wall
Center beam Aiming: 10° (at point 0'–0" AFF)
Luminaire: Monopoint or Recessed Adjustable
Illuminance: 15 fc, avg, maintained (34 fc max)

Art shown at 6' by 8' and centered 5'–6" AFF. Wall grid on 6" by 6" centers and scaled at ¼"=1'.

Lamp: 70PAR38/CMH/3K/WFL
Location: 5'–0" from wall
Center beam Aiming: 22° (at point 0'–0" AFF)
Luminaire: Monopoint or Recessed Adjustable
Illuminance: 19 fc, avg, maintained (39 fc max)

Ceramic Metal·Halide 70PAR38/CMH/4K/VWFL

▶Cool–tone light
▶Not dimmable
▶Not instant–on

Metal Halide ceramic

Performance Sketches™

[All data outlined here is based on information from boldfaced manufacturer's published data at time of manuscript preparation. Data and sketches shown are corrected for 120Volt operation. Scale is ¼ inch equals 1 foot.]

Lamp: 70PAR38/CMH/4K/WFL
Location: 3'–0" from wall
Center beam Aiming: 5° (at point 0'–0" AFF)
Luminaire: Monopoint or Recessed Adjustable
Illuminance: 14 fc, avg, maintained (34 fc max)

Lamp: 70PAR38/CMH/4K/WFL
Location: 4'–0" from wall
Center beam Aiming: 10° (at point 0'–0" AFF)
Luminaire: Monopoint or Recessed Adjustable
Illuminance: 14 fc, avg, maintained (27 fc max)

Art shown at 6' by 8' and centered 5'–6" AFF. Wall grid on 6" by 6" centers and scaled at ¼"=1'.

Lamp: 70PAR38/CMH/4K/WFL
Location: 5'–0" from wall
Center beam Aiming: 22° (at point 0'–0" AFF)
Luminaire: Monopoint or Recessed Adjustable
Illuminance: 16 fc, avg, maintained (32 fc max)

Uses

▶Soft general lighting
▶Wallwashing
▶Extra large art with medium throws
▶Medium throws
▶Various voltages (ballast dependent)

Specs

Beam spread/60°
Center beam candlepower/4000
Life/10000 hrs
▶**Philips** (wide flood) (28874–6)

Design Tips

✔One light best for extra large art and where mounting height for light is 13' or less.
✔Good for soft downlighting—best with sconces and/or art accents and/or wallwashing to avoid cave effect.
✔Light may degrade artwork rapidly and/or significantly: consider UV and IR filters or limit exposure time.
✔Cold start requires a few minutes for full light output.
✔Power interruption results in fifteen minute "cool down" before lamp can restart.
✔Cannot easily be part of any egress system.
✔Cool–tone version best for crisp–white visual attraction.
✔Warm–tone version has longer life and greater candlepower.

Application Key

Commercial
Gallery
Hospitality
Institutional
Manufacturing
Residential
Retail
Exterior

bold = primary application
partial fade = minimal application
fade = unlikely application

▶ Not dimmable
▶ Not instant-on

Metal Halide 100PAR38/CMH/3K/WSP

Ceramic

Metal Halide
Ceramic

7

Uses
▶Very strong architectural detailing
▶Extra large art with very long throws
▶Sculpture with very long throws
▶Very long throws
▶Various voltages (ballast dependent)

Specs
Beam spread/15°
Center beam candlepower/ 70000
Life/12500 hrs
▶**Philips** (24477–2)

Design Tips
✔One light best for large art and where mounting height for light is less than 26'.
✔These are very strong accents and deserve judicious use. Light may degrade artwork rapidly and/or significantly: consider UV and IR filters or limit exposure time.
✔Cold start requires a few minutes for full light output.
✔Power interruption results in fifteen minute "cool down" before lamp can restart.
✔Cannot easily be part of any egress system.
✔Warm–tone lamp very close match to incandescent.
✔Excellent for public art where maintenance and energy are priorities and where mounting height for light is 26' or less.

Application Key

Commercial
Gallery
Hospitality
Institutional
Manufacturing
Residential
Retail
Exterior

bold = primary application
partial fade = minimal application
fade = unlikely application

Performance Sketches™

[All data outlined here is based on information from boldfaced manufacturer's published data at time of manuscript preparation. Scale is ¼ inch equals 1 foot.]

Lamp: 100PAR38/CMH/3K/WSP
Location: 8'–0" from wall
Center beam Aiming: 21° (at point 4'–0" AFF)
Luminaire: Monopoint or Recessed Adjustable
Illuminance: 28 fc, avg, maintained (48 fc max)

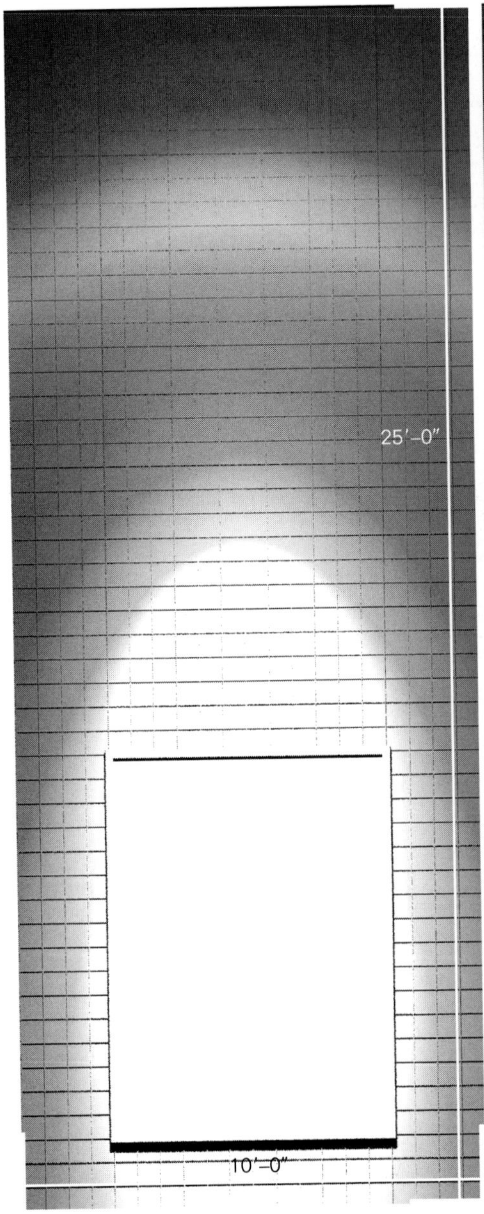

Lamp: 100PAR38/CMH/3K/WSP
Location: 9'–0" from wall
Center beam Aiming: 23° (at point 4'–0" AFF)
Luminaire: Monopoint or Recessed Adjustable
Illuminance: 30 fc, avg, maintained (48 fc max)

25'–0"

10'–0"

Metal Halide **Ceramic** 100PAR38/CMH/3K/WSP

Not dimmable
Not instant-on

Lamp: 100PAR38/CMH/3K/WSP
Location: 10'–0" from wall
Center beam Aiming: 26° (at point 4'–0" AFF)
Luminaire: Monopoint or Recessed Adjustable
Illuminance: 31 fc, avg, maintained (48 fc max)

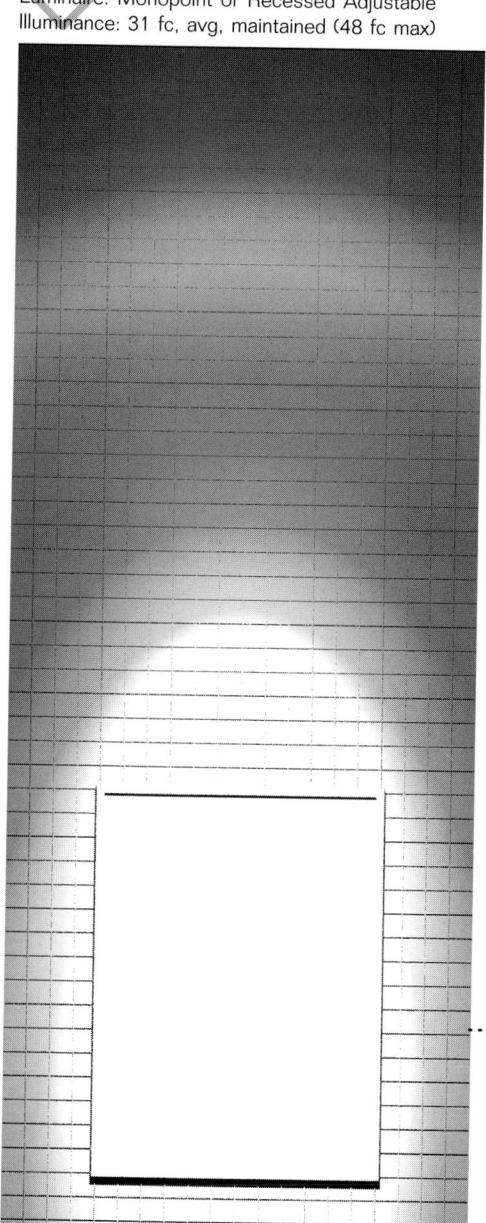

Art shown at 6' by 8' and
centered 5'–6" AFF. Wall grid on
6" by 6" centers and scaled at
¼"=1'.

Metal Halide
ceramic

7

▼ Cool–tone light
▼ Not dimmable
▼ Not instant–on

Uses
► Very strong architectural detailing
► Extra large art with very long throws
► Sculpture with very long throws
► Very long throws
► Various voltages (ballast dependent)

Specs
Beam spread/15°
Center beam candlepower/ 54000
Life/10000 hrs
► **Philips** (28876–1)

Design Tips
✔ One light best for large art and where mounting height for light is less than 26'.
✔ These are very strong accents and deserve judicious use. Light may degrade artwork rapidly and/or significantly: consider UV and IR filters or limit exposure time.
✔ Cold start requires a few minutes for full light output.
✔ Power interruption results in fifteen minute "cool down" before lamp can restart.
✔ Cannot easily be part of any egress system.
✔ Cool–tone lamp best for crisp–white visual attraction.
✔ Consider for special–focus in stores, particularly where most lighting is warm–tone.
✔ Warm–tone version has longer life and greater candlepower.

Application Key

Commercial
Gallery
Hospitality
Institutional
Manufacturing
Residential
Retail
Exterior

bold = primary application
partial fade = minimal application
fade = unlikely application

Metal Halide 100PAR38/CMH/4K/WSP

Performance Sketches™

[All data outlined here is based on information from boldfaced manufacturer's published data at time of manuscript preparation. Scale is ¼ inch equals 1 foot.]

Lamp: 100PAR38/CMH/4K/WSP
Location: 8'–0" from wall
Center beam Aiming: 21° (at point 4'–0" AFF)
Luminaire: Monopoint or Recessed Adjustable
Illuminance: 24 fc, avg, maintained (40 fc max)

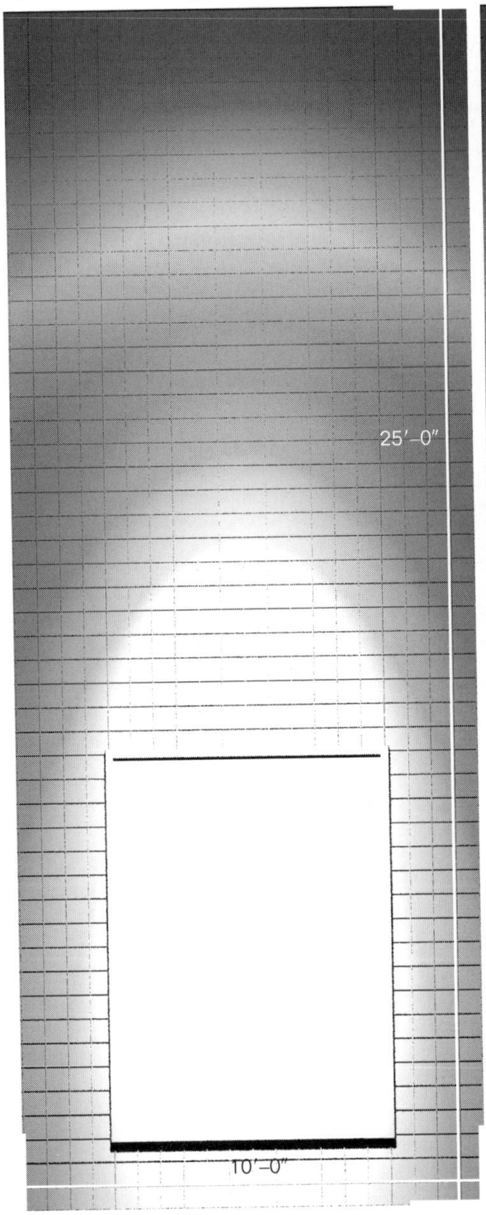

Lamp: 100PAR38/CMH/4K/WSP
Location: 9'–0" from wall
Center beam Aiming: 23° (at point 4'–0" AFF)
Luminaire: Monopoint or Recessed Adjustable
Illuminance: 26 fc, avg, maintained (40 fc max)

Metal Halide 100PAR38/CMH/4K/WSP

Ceramic

▼ Cool-tone light
▼ Not dimmable
▼ Not instant-on

Lamp: 100PAR38/CMH/4K/WSP
Location: 10'–0" from wall
Center beam Aiming: 26° (at point 4'–0" AFF)
Luminaire: Monopoint or Recessed Adjustable
Illuminance: 27 fc, avg, maintained (41 fc max)

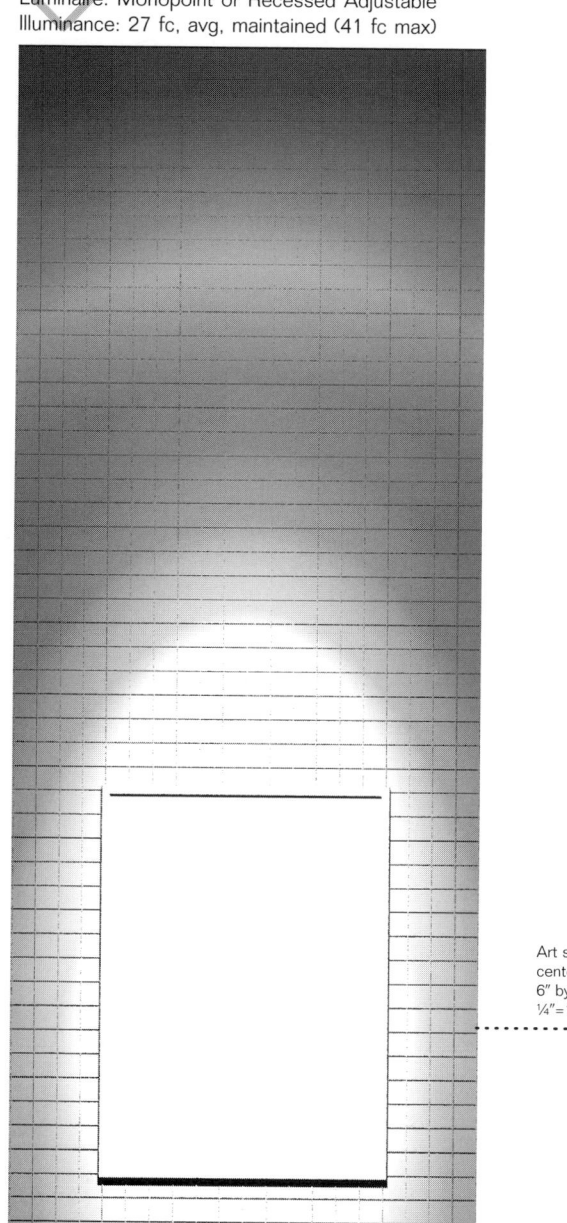

Art shown at 6' by 8' and
centered 5'–6" AFF. Wall grid on
6" by 6" centers and scaled at
¼"=1'.

Metal Halide ceramic

7

▶ Not dimmable
▶ Not instant-on

Metal Halide Ceramic

7

Ceramic Halide 100PAR38/CMH/3K/NFL

Uses
▶ Strong architectural accenting
▶ Extra large art with very long throws
▶ Large sculpture with very long throws
▶ Very long throws
▶ Various voltages (ballast dependent)

Specs
Beam spread/25°
Center beam candlepower/25000
Life/12500 hrs
▶ **Philips** (flood) (24476–4)

Design Tips
✔ One light best for extra large art and where mounting height for light is 21' or less.
✔ These are very strong accents and deserve judicious use. Light may degrade artwork rapidly and/or significantly— consider UV and IR filters or limit exposure time.
✔ Cold start requires a few minutes for full light output.
✔ Power interruption results in fifteen minute "cool down" before lamp can restart.
✔ Cannot easily be part of any egress system.
✔ Warm–tone lamp very close match to incandescent.
✔ Excellent for public art where maintenance and energy are priorities and where mounting height for light is 21' or less.

Application Key

Commercial
Gallery
Hospitality
Institutional
Manufacturing
Residential
Retail
Exterior

bold = primary application
partial fade = minimal application
fade = unlikely application

Performance Sketches™

[All data outlined here is based on information from boldfaced manufacturer's published data at time of manuscript preparation. Scale is ¼ inch equals 1 foot.]

Lamp: 100PAR38/CMH/3K/NFL
Location: 7'–0" from wall
Center beam Aiming: 23° (at point 3'–6" AFF)
Luminaire: Monopoint or Recessed Adjustable
Illuminance: 24 fc, avg, maintained (43 fc max)

Lamp: 100PAR38/CMH/3K/NFL
Location: 8'–0" from wall
Center beam Aiming: 26° (at point 3'–6" AFF)
Luminaire: Monopoint or Recessed Adjustable
Illuminance: 25 fc, avg, maintained (40 fc max)

20'–0"

10'–0"

Metal Halide 100PAR38/CMH/3K/NFL

Metal Halide
Ceramic

7

Lamp: 100PAR38/CMH/3K/NFL
Location: 9'–0" from wall
Center beam Aiming: 29° (at point 4'–0" AFF)
Luminaire: Monopoint or Recessed Adjustable
Illuminance: 26 fc, avg, maintained (40 fc max)

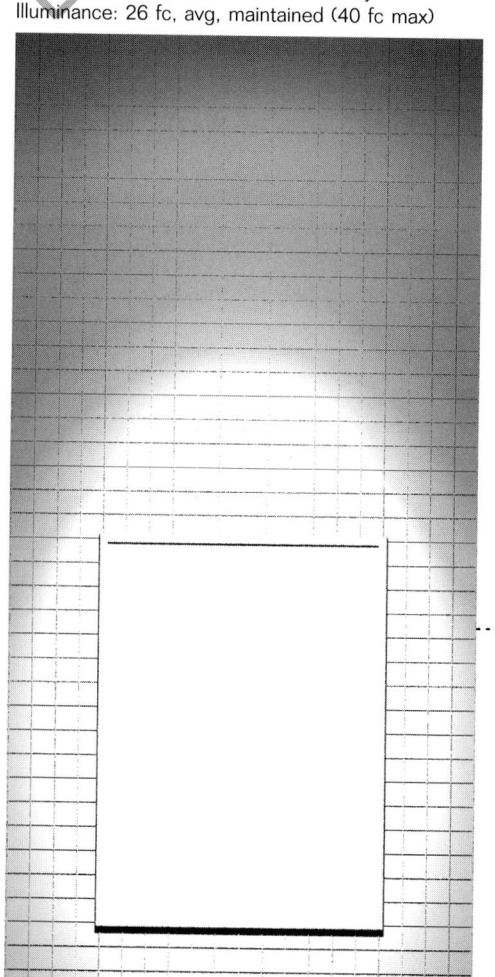

Art shown at 6' by 8' and
centered 5'–6" AFF. Wall grid on
6" by 6" centers and scaled at
¼"=1'.

Cool–tone light
▶ Not dimmable
▶ Not instant–on

Metal Halide 100PAR38/CMH/4K/NFL

Uses
▶ Strong architectural accenting
▶ Extra large art with very long throws
▶ Large sculpture with very long throws
▶ Very long throws
▶ Various voltages (ballast dependent)

Specs
Beam spread/25°
Center beam candlepower/20000
Life/10000 hrs
▶**Philips** (flood) (28878–7)

Design Tips
✔ One light best for extra large art and where mounting height for light is 21′ or less.
✔ These are very strong accents and deserve judicious use. Light may degrade artwork rapidly and/or significantly— consider UV and IR filters or limit exposure time.
✔ Cold start requires a few minutes for full light output.
✔ Power interruption results in fifteen minute "cool down" before lamp can restart.
✔ Cannot easily be part of any egress system.
✔ Cool–tone lamp best for crisp–white visual attraction.
✔ Consider for cooler toned stone surface lighting.
✔ Warm–tone version has longer life and greater candlepower.

Application Key

Commercial
Gallery
Hospitality
Institutional
Manufacturing
Residential
Retail
Exterior

bold = primary application
partial fade = minimal application
fade = unlikely application

Performance Sketches™

[All data outlined here is based on information from boldfaced manufacturer's published data at time of manuscript preparation. Scale is ¼ inch equals 1 foot.]

Lamp: 100PAR38/CMH/4K/NFL
Location: 7′–0″ from wall
Center beam Aiming: 23° (at point 3′–6″ AFF)
Luminaire: Monopoint or Recessed Adjustable
Illuminance: 23 fc, avg, maintained (40 fc max)

Lamp: 100PAR38/CMH/4K/NFL
Location: 8′–0″ from wall
Center beam Aiming: 26° (at point 3′–6″ AFF)
Luminaire: Monopoint or Recessed Adjustable
Illuminance: 24 fc, avg, maintained (38 fc max)

20′–0″

10′–0″

Metal Halide 100PAR38/CMH/4K/NFL

▶ Cool-tone light
▶ Not dimmable
▶ Not instant-on

Lamp: 100PAR38/CMH/4K/NFL
Location: 9'–0" from wall
Center beam Aiming: 29° (at point 4'–0" AFF)
Luminaire: Monopoint or Recessed Adjustable
Illuminance: 25 fc, avg, maintained (39 fc max)

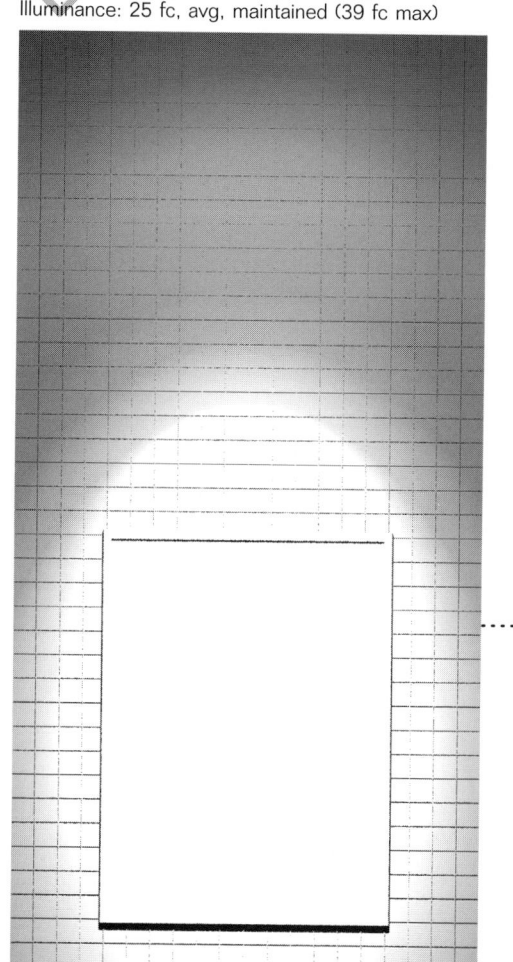

Art shown at 6' by 8' and
centered 5'–6" AFF. Wall grid on
6" by 6" centers and scaled at
¼"=1'.

Not dimmable
Not instant–on

Metal Halide 100PAR38/CMH/3K/VWFL
Ceramic

Uses
► General lighting
► Wallwashing (see 11–133)
► Extra large art with long throws
► Sculpture with long throws
► Long throws
► Various voltages (ballast dependent)

Specs
Beam spread/60°
Center beam candlepower/ 7000
Life/12500 hrs
► **Philips** (wide flood) (24478–0)

Design Tips
✔ One light best for extra large art where mounting height for light is 16' or less.
✔ Light may degrade artwork rapidly and/or significantly: consider UV and IR filters or limit exposure time.
✔ Cold start requires a few minutes for full light output.
✔ Power interruption results in fifteen minute "cool down" before lamp can restart.
✔ Cannot easily be part of any egress system.
✔ Warm–tone lamp very close match to incandescent.
✔ Consider for general lighting in high–ceiling stores, atria, auditoria, large gathering spaces or transition spaces where dimming is not required.

Application Key

Commercial
Gallery
Hospitality
Institutional
Manufacturing
Residential
Retail
Exterior

bold = primary application
partial fade = minimal application
fade = unlikely application

Performance Sketches™

[All data outlined here is based on information from boldfaced manufacturer's published data at time of manuscript preparation. Scale is ¼ inch equals 1 foot.]

Lamp: 100PAR38/CMH/3K/VWFL
Location: 5'–0" from wall
Center beam Aiming: 7° (at point 0'–0" AFF)
Luminaire: Monopoint or Recessed Adjustable
Illuminance: 14 fc, avg, maintained (24 fc max)

Lamp: 100PAR38/CMH/3K/VWFL
Location: 6'–0" from wall
Center beam Aiming: 14° (at point 0'–0" AFF)
Luminaire: Monopoint or Recessed Adjustable
Illuminance: 14 fc, avg, maintained (24 fc max)

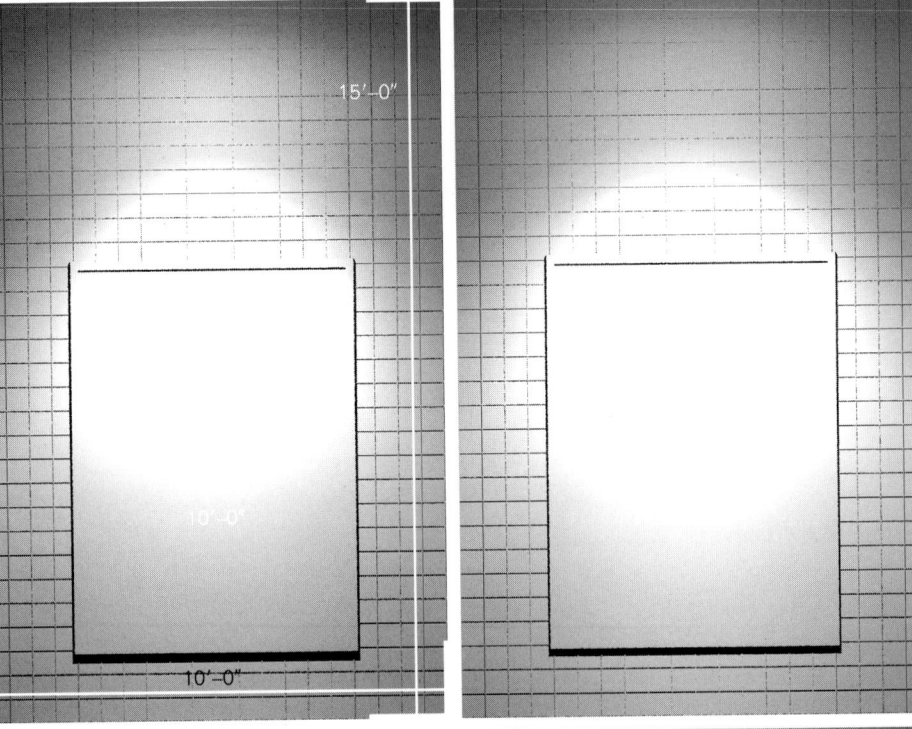

15'–0"
10'–0"
10'–0"

Art shown at 6' by 8' and centered 5'–6" AFF. Wall grid on 6" by 6" centers and scaled at ¼"=1'.

Lamp: 100PAR38/CMH/3K/VWFL
Location: 7'–0" from wall
Center beam Aiming: 25° (at point 0'–0" AFF)
Luminaire: Monopoint or Recessed Adjustable
Illuminance: 18 fc, avg, maintained (30 fc max)

Ceramic Metal Halide 100PAR38/CMH/4K/VWFL

▶ Cool–tone light
▶ Not dimmable
▶ Not instant–on

Performance Sketches™

[All data outlined here is based on information from boldfaced manufacturer's published data at time of manuscript preparation. Scale is ¼ inch equals 1 foot.]

Lamp: 100PAR38/CMH/4K/VWFL
Location: 5'–0" from wall
Center beam Aiming: 7° (at point 0'–0" AFF)
Luminaire: Monopoint or Recessed Adjustable
Illuminance: 12 fc, avg, maintained (21 fc max)

Lamp: 100PAR38/CMH/4K/VWFL
Location: 6'–0" from wall
Center beam Aiming: 14° (at point 0'–0" AFF)
Luminaire: Monopoint or Recessed Adjustable
Illuminance: 13 fc, avg, maintained (22 fc max)

Art shown at 6' by 8' and centered 5'–6" AFF. Wall grid on 6" by 6" centers and scaled at ¼"=1'.

Lamp: 100PAR38/CMH/4K/VWFL
Location: 7'–0" from wall
Center beam Aiming: 25° (at point 0'–0" AFF)
Luminaire: Monopoint or Recessed Adjustable
Illuminance: 15 fc, avg, maintained (26 fc max)

Uses
▶ General lighting
▶ Wallwashing
▶ Extra large art with long throws
▶ Sculpture with long throws
▶ Long throws
▶ Various voltages (ballast dependent)

Specs
Beam spread/60°
Center beam candlepower/ 5000
Life/10000 hrs
▶ **Philips** (wide flood) (28880–3)

Design Tips
✔ One light best for extra large art where mounting height for light is 16' or less.
✔ Light may degrade artwork rapidly and/or significantly: consider UV and IR filters or limit exposure time.
✔ Cold start requires a few minutes for full light output.
✔ Power interruption results in fifteen minute "cool down" before lamp can restart.
✔ Cannot easily be part of any egress system.
✔ Cool–tone lamp best for crisp–white visual attraction.
✔ Consider for special–focus in stores, particularly where most lighting is warm–tone.
✔ Warm–tone version has longer life and greater candlepower.

Application Key

Commercial
Gallery
Hospitality
Institutional
Manufacturing
Residential
Retail
Exterior

bold = primary application
partial fade = minimal application
fade = unlikely application

Metal Halide ceramic

7

Fluorescent
Triphosphor

8

Triphosphor fluorescent lamps are covered in this section. These lamps are considered the common fluorescent lamp of the day. Triphosphor fluorescent lamps are available in suffcient variety of sizes and wattages that they can be used for general lighting, for gentle lighting details and for efficient, long–life table lights and wall sconces. Indeed, the great story here is that the breadth of lamps enables the designer to develop a consistent lamp color palette on many projects. The old halophosphate, standard fluorescent lamps were, by comparison, energy intensive—and exhibited lousy color. Further, these old standard lamps operated on electromagnetic ballasts which hummed and buzzed, all the while flickering the lamps on/off rapidly, but to some folks noticeably. This is the experience to which most folks relate when fluorescent lighting is mentioned. Today, triphosphor fluorescent lamps combined with electronic ballasts offer a new standard for common lighting practices.

Fluorescent Triphosphor

Only deluxe triphosphor fluorescent lamps offer high color rendering, warm color tone, highest efficacy of any white light source, and longest life.

Fluorescent lamps have evolved significantly over the past years. New phosphor coatings, manufacturing techniques and the advent of electronics have enabled lamp makers to offer fluorescent lighting with significant color consistency, color rendering, efficacy and life advantages. Linear lamps and compact lamps (in small and large sizes) give the designer options of using fluorescent lamps in the usual work–a–day luminaires to the Italian decorative lights so popular as eye candy. With energy costs, cooling load costs and the pollution concerns associated with electricity production, it is difficult to justify the use of halogen and halogen infrared lamps in many applications. Triphosphor fluorescent lamps should be considered the common lamp for most every application—including many of those where dimming is considered a necessity. Table 8.1 outlines those triphosphor fluorescent lamps which are emerging as standards.

Table 8.1 Triphosphor Fluorescent Lamp Categories

Category	Characteristics	Lamp Profile	Brand Names❶
Triple–Tube (CFTriple)	• Small size • Dimmable (with serious drawbacks) • Cold temperature operation • Optimized for downlights and sconces		• GE Triple Biax® • Osram Sylvania Dulux® T/E/IN • Philips PL–T
Compact Single–Tube (CFLong)	• Moderate size • Dimmable (rapid start versions only—with some drawbacks) • Lots of light from relatively small package		• GE Biax®/RS • Osram Sylvania Dulux® L/RS • Philips PL–L/RS
Linear/Miniature (T2)	• Miniature diameter (¼″) • Moderate lengths	(no art available)	• Osram Sylvania Subminiature
Linear/Small (T5)	• Small diameter (just over ½″) • Moderate to long lengths • Standard or high output versions • Dimmable (with some drawbacks)		• Osram Sylvania Pentron • Philips Silhouette™
Linear/Medium (T8)	• Moderate diameter (1″) • Moderate to long lengths • Dimmable (with some drawbacks)		• GE Starcoat® and Ecolux® • Osram Sylvania Octron® and ECOLOGIC • Philips Standard T8, Hi–Vision® and ALTO®

❶Brand names, whether so noted (® or ™) or not, are registered by and/or trademarks of respective manufacturers.
Lamp images courtesy of and copyright by GE 2000.

Fluorescent Triphosphor

8

Fluorescent Triphosphor

Triphosphor fluorescent lamp color characteristics

Color of light (color temperature), color rendering (how colors look under the light in question) along with highest efficacy of any white light source and long life are characteristics of deluxe triphosphor fluorescent lamps. Deluxe triphosphor (XT) fluorescent lamps are available in a wide variety of lamp shapes and wattages. All wattages are available in warm tone (3000°K) and neutral (3500°K) varieties. The warm tone lamps approximate halogen and warm tone ceramic metal halide. Good for contemporary interiors. Most suitable applications are general lighting, wallwashing and cove lighting. The low wattage lamps are excellent for task lighting—very little heat and a diffusion which minimizes glare. All of the deluxe triphosphor lamps addressed in this section are dimmable (with the use of dimming ballasts; and some with serious drawbacks as outlined for each lamp). This section only addresses those deluxe triphosphor lamps considered a reasonable and current cross section of the technology which will address most all application needs. Over the past few years, the manufacturers have come to terms with base configuration standards and efficiency standards that has resulted in a recent category of compact fluorescent lamps—the Triple–Tube—which appear to have broad application in many luminaire types and which work well in many ambient temperatures. There exist many more standard and deluxe triphosphor fluorescent lamps, however, their wattages, sizes, base configurations and operational limitations now make them obsolete except for niche applications. This section attempts to address the more efficacious, readily standardized fluorescent lamps. To minimize maintenance issues, purchasing issues and cost issues, it is suggested that the lamps outlined herein be considered the lamps of choice for fluorescent lighting.

Linear deluxe triphosphor fluorescent lamps have revolutionized the fluorescent lamp over the past fifteen years. Today, T8 lamps are standard in new and retrofit construction. T12 lamps are obsolete (although millions of lamp sockets remain in place and must be relamped for another generation). T5 lamps (the rapid start/programmed start versions—and not preheat or switch start versions),

■ ■ ■ ■ ■ ■ ■ ■ ■ ■ ■

Deluxe triphosphor (XT) fluorescent
- Best efficacy white light source
- Excellent color rendering
- Dimmable
- Very long life (wattage dependent)
- Excellent breadth of wattages and sizes
- Can work in all applications

Cost magnitude[2000]
for deluxe triphoshpor fluorescent lamps

Low
US$3.⁰⁰ to US$8.⁰⁰
Moderate
US$8.⁰⁰ to US$16.⁰⁰
High
US$16.⁰⁰ to US$30.⁰⁰

Costs vary based on quantities, distributor and contractor markups, and market conditions, manufacturing situations and annual inflation. These values are for preliminary, magnitude budgeting and do not represent quotes nor actual final pricing.

Fluorescent Triphosphor

Fluorescent
Triphosphor
8

while a staple in Europe, are significantly improving efficiency opportunities in the States. A side benefit is the fact that these lamps use less glass and less rare–earth phosphor coating (about 37 percent less) given the smaller diameter of the lamp.

XT fluorescent dimming

Dimming deluxe triphosphor fluorescent lamps is technically possible and is a practical reality. Electronic dimming ballasts enable these lamps to dim to as low as 1 percent of full light output. Versions of electronic dimming ballasts are now available in small case sizes for use in relatively small luminaires. Some brands of these ballasts (ESI SuperDim™) are also addressable.

A significant flaw, however, results in short lamp life and/or short ballast life. This flaw is most prominent on the smaller, CFTriple fluorescent XT lamps, somewhat prominent on the CFLong XT fluorescent lamps, and least prominent on the linear T5 and T8 XT lamps. If lamps are not seasoned (burned–in) at 100 percent output for at least 100 hours prior to dimming, then the cathodes apparently fail prematurely. This seasoning poses significant expense and logistic issues for clients, particularly for spot relamping and group relamping. Keeping track of which lamps were installed when, and establishing a method for the 100 percent burn–in without disrupting normal operations are increasingly serious issues as energy codes and sustainability become more restrictive.

Fluorescent ballasts

XT fluorescent lamps are powered by ballasts—devices which provide sufficient electrical energy to start the lamp and maintain a constant flow of current to limit the lamp's wattage and light output to rated quantity. There are two types of ballasts—electromagnetic (also known as coil and core) and electronic. Both are essentially "black box" devices which contain wire coils and/or electronic components sealed away from view. Ballasts require additional power, which can be relatively minimal for fluorescent lamps. Electronic ballasts are preferred due to the very low energy consumption, quiet operation and lack of flicker.

Net Addresses/Ballasts
http://www.advancetransformer.com/
http://www.magnetek.com/transLighting.html
http://www.mot.com/ies/MLI/
http://www.sylvania.com/ballast/phops.htm

CONNECT FOR MORE

SuperDim is a registered trademark of ESI, Inc.

Fluorescent Triphosphor

An arbitrary industry standard on fluorescent ballasts has set the common ballast to power the intended lamp to 88 percent of the full light output of the lamp. That's right. Specify any standard fluorescent luminarie with any standard electronic ballast, and the end user will get 12 percent less light than you/he/she expected (unless, of course, this factor is considered in calculations). This also means that the lamp is consuming full power—so that the total electrical power necessary to operate the lamp and ballast together ends up being equal to the rated wattage of the lamp. So, a 32–watt fluorescent lamp operating on a common electronic ballast will consume about 32 watts for the lamp and ballast together and will produce about 88 percent of the rated light of the lamp. This 88 percent factor is known as ballast factor or BF. Fortunately, there is a cadre of electronic ballasts exhibiting differing ballast factors.

For XT/T8 fluorescent lamps, available BFs are 0.78, 0.88, 0.98 and 1.18. This enables the specifier to tune the light levels to meet criteria while not overlighting a space and thereby not wasting energy.

The single disadvantage of ballasts is how to hide them! Fortunately, for fluorescent lamps, electronic ballasts are relatively small and can typically be fit inside the luminaire housing without causing excessive volume or odd "growths" to house the ballast. For XT/T2, T5 and T8 lamps, ballasts for one–, two– or three–lamp operation are available for many wattages. For the other XT fluorescent lamps, ballasts exist for one– and two–lamp operation. Ballasts operating more than one lamp help limit initial costs, maintenance costs and reduce solid waste.

XT/T8 Fluorescent Ballast Factors

- 0.78/produces about 78 percent of the light and consumes about 82 percent of the energy of lamp ratings
- 0.88 (industry standard)/produces 88 percent of the light and consumes 94 percent of the energy of lamp ratings
- 0.98/produces 98 percent of the light and consumes about 105 percent of the energy of lamp ratings
- 1.18/produces 118 percent of the light and consumes about 123 percent of the energy of lamp ratings

Net Addresses/Fluorescent Hazards

http://www.epa.gov/epaoswer/hazwaste/id/merc-emi/merc-emi.htm
http://www.facilitiesnet.com/NS/NS3b6lb.html

CONNECT FOR MORE

Fluorescent hazards

There are small (very small—in the order of 5 to 20mg) amounts of mercury used in many of today's XT fluorescent lamps. Manufacturers are progressing toward the 5 mg value. Nevertheless, since

Fluorescent
Triphosphor

more an half a billion 4–foot lamps alone in the US are disposed every year, the mercury content is a problem. As such, consideration should be given to reclamation recycling. Depending on the type and age of lamp(s) being disposed, regulations may impact the disposal process. Additional information is available at the cited websites and at lamp manufacturers' websites.

When to use XT fluorescent lamps
Deluxe triphosphor lamps are appropriate on projects where system longevity, energy effectiveness and color quality are important. Typically, commercial, hospitality, institutional, retail and exterior hardscape and landscape projects fit this profile.

Even environments where temperature extremes exist or where the temperature is likely to remain at or below 45°F are candidates for fluorescent lamps—providing that the ballasts are rated for 0°F (or lower) starting temperature and that the lamps are enclosed (e.g., glass or acrylic lenses downlights, jelly–jars, bollards, and so on).

When not to use fluorescent lamps
Most galleries (museums) and residential settings—where the richness of color and sparkle of light is important to the experience are best served by halogen or halogen infrared lamps. Audio–visual intensive projects demanding very low level, no–flicker dimming, are typically best served by halogen and/or halogen infrared lamps on dimmers. Any project which requires flashing of lights, or repetitive on/off switching of lights on frequent timing (once every thirty minutes or less) throughout the day or night are better served by halogen or halogen infrared lamps. However, as fluorescent lamp wattages and physical sizes are reduced and dimming capabilities for fluorescent lighting evolve, look for fluorescent lamps to migrate into the galleries, residences and AV intensive settings.

Other fluorescent lamps
There are an overwhelming variety of fluorescent lamps currently available. Many of these are legacy items—from previous eras and yet the luminaires which use these lamps remain operational. Unlike automobiles or home appliances, the generational cycles on lamps

Fluorescent Triphosphor

are much longer. *Time–Saver Standards for Architectural Lighting* is addressing the most current, most efficient of the fluorescent lamps now available. Continued specification of other fluorescent lamps will only serve to exacerbate market shifts to more efficient, better performing lamps. Indeed, the more efficient lamps will cost less the more frequently they are specified!

Degrees of whiteness

TSSAL addresses the 3000°K and 3500°K deluxe triphosphor fluorescent lamps (warm tone and neutral white respectively). Here's a brief, however, on other varieties of whiteness of fluorescent lamps.

In an attempt to appease every potential user of triphosphor fluorescent lamps—and with some unfound desire to offer some resemblance to the lousy cool white legacy color of whiteness—manufacturers typically offer four color temperatures of whiteness. Two of these are 3000°K and 3500°K. The other two are odd colors of whiteness for most interiors. A very warm white, pinkish color of light is produced by the 2700°K variety (also tagged as 827). This is an unfortunate color of white. Too pink to match standard incandescent (which apparently it was to replace) and much too pink against halogen sources, this lamp color should simply be deprecated. Suppliers and maintenance crews accidentally supply these lamps from time to time when 3000°K lamps are specified—hence mixing these with the 3000°K lamps and ultimately leading to a project which looks more like a carnival and less like a professional application.

At the other extreme is 4100°K. A perfect match to the old cool white fluorescent—which nearly every user loves to hate. Again, a lamp worthy of deprecation. If specifiers wish to provide a visually cool interior or a "relief" accent of very cool toned light, then consideration should be given to the super deluxe triphosphor 5000°K lamps (also tagged as 950)—and these should be used quite sparingly for some "surprise." In more garish settings, these lamps might be used in signage for "attraction." Where subtle signage is the desire, however, consideration should be given to lighting the signs with lamps of the same color of whiteness as the overall environmental lighting.

Fluorescent/Triple
Triphosphor

for electronic ballasts ONLY

Triple–tube compact fluorescent lamps have emerged in the past three years as the workhorse lamp for most downlighting and wallwashing applications in most any kind of facility with ceiling heights less than 15 feet. These lamps are available in a range of wattages and physical sizes that meet a number of design tenets—relatively small luminaires, relatively low glare and, with the proper ballast, dimmable. Additionally, their shape is consisten across wattages and is similar to that of a halogen TB lamp—making an excellent candidate for vertical–lamp downlights and thereby providing the look of a typical incandescent downlight application. To

Table 8.2 Deluxe Triphosphor Triple Lamps

Fluorescent Lamp	Uses	Lamp Profile
F13Triple	New • Small downlights • Sconces • Table lights • Task lights	
F18Triple	New • Small downlights • Sconces • Table lights	
F26Triple	New • Medium downlights • Sconces • Table lights	
F32Triple	New/Exterior • Bollards • Sconces • Uplights New/Interior • Medium downlights • Sconces • Table lights	
F42Triple	New/Exterior • Bollards • Sconces • Uplights New/Interior • Large downlights • Sconces • Table lights	

Lamp images courtesy of and copyright by GE 2000.

Fluorescent/Triple
Triphosphor
for electronic ballasts ONLY

maintain high efficacies, Triple lamps are only available in the deluxe triphosphor versions (or abbreviated here as XT (for delu**X**e **T**riphosphor)) . Table 8.2 outlines the variety of XT/Triple lamps presently available.

While these lamps may seem expensive, the overall cost of light over the life of the lamp/ lighting system must be considered. Many of these Triple lamps have rated lives of 10,000 hours. This is three to five times that of many mains voltage halogen and halogen infrared lamps and is two to three times that of many low voltage halogen and halogen infrared lamps. Very simply, the Triple lamps could/should cost two to five times that of halogen or halogen infrared lamps. And this doesn't account for the savings in energy costs from lower wattages, nor more significantly, in personnel time necessary to change a lamp.

Presumably, decorative luminaire manufacturers will soon get with the program—introducing myriad styles and sizes of table lights, floor lights, task lights, sconces, pendents and the like which take advantage of the Triple lamp. Recognize that many decorative lights accept the screw–base compact fluorescent lamps. For energy efficiency and lamp replacement/maintenance consistency, however, designers should specify dedicated socket systems—luminaires with sockets which accept just one type/wattage lamp. This is a necessity if energy efficiency and maintenance consistency are priorities.

Since ballasts are now available to dim many of these lamps, their use in more "moody" and less worklike settings should increase— providing lamp and ballast manufacturers address the drawbacks of lamp seasoning or reduced lamp and ballast life. The color rendering is excellent with XT/Triple lamps and warm tone lamps approximate halogen in whiteness.

One distinct disadvantage will require a learning curve on their use. To provide all of the appropriate operational characteristics such as optimal performance under various temperatures, use in any orientation and use in exterior/cold weather settings, XT/Triple lamps use a technology that results in a warm up period. While light is instantaneous with the flick of a switch, the light will continue to brighten during a warm up period which can take from five to fifteen minutes, depending on the ambient temperature.

XT/Triple Lamps
ⓘInstant–on, but require warm–up to full brightness
ⓘDimmable
ⓘElectronic operation results in no hum and no flicker
ⓘVarious sizes/wattages for all applications

Fluorescent
Triphosphor

8

Fluorescent/Triple

Triphosphor

for electronic ballasts ONLY

XT/Triple lamp designation

XT/Triple lamps are typically identified with an alphanumeric designation outlined to the right. Since each manufacturer has slightly different means for lamp designation, the designation method presented here is solely for convenient reference throughout the text. Actual lamp manufacturers' designations will need to be used in final specifications. Perhaps the lamp manufacturers will standardize lamp designations in the near future in an effort to ease lamp cross referencing and specifications—especially problematic with the compact fluorescent lamps.

XT/Triple lamps for general lighting

Compact fluorescent lamps work very well in general lighting applications where light levels are not too intense. The shape and physical size of these lamps make the comparable to standard A–lamp and halogen/TB19 lamps. Downlight luminaires are available from a wide range of manufacturers in a wide range of aperture sizes and trim/reflector finishes.

XT/Triple lamps for wallwashing

Spreadlens wallwashers are reasonably well suited for compact fluorescent lamps. Within the past few years, manufacturers have developed spreadlens luminaires in several aperture sizes and to accommodate several XT/Triple lamp wattages. Again, a range of trim and reflector finishes are available.

XT/Triple lamps for decorative lighting

Essentially unexplored territory by luminaire manufacturers, XT/Triple lamps make excellent candidates in sconces, pendents, table lights and floor lights. Hospitality facilties, especially, could take advantage of the wonderful color rendering, long life and energy efficiency aspects of XT/Triple lamps and limit lamp inventory for maintenance. Recent experience shows that use of traditional shade materials (e.g., parchment–like Nomex®, amber frosted glass and the like) and XT/Triple lamps yield finished lights which are virtually indistinguishable from incandescent or halogen counterparts. Compact electronic dimmable ballasts further blur the differences between XT/Triple lamps and halogen counterparts in these applications.

Nomex is a registered trademark of E. I. du Pont de Nemours and Company

Fluorescent/Triple
Triphosphor
for electronic ballasts ONLY

Deluxe Triphosphor Triple–Tube Compact Fluorescent Lamp Designation

F32Triple/830

These last two digits indicate color temperature
- 30 is shorthand for 3000°K (warm tone)
- 35 is shorthand for 3500°K (neutral)

This number represents the "deluxe" nature of the lamp indicating a color rendering index of greater than 80 (where 100 is perfect)

This signifies that this is a triple–tube compact fluorescent lamp

These two digits represent lamp wattage

Signifies that the lamp is a fluorescent (F) lamp

[This designation used for convenient reference throughout this *Time–Saver Standards for Architectural Lighting*—but not necessarily used by all lamp manufacturers (although it would be convenient if manufacturers would come to agreement on standard designations.]

Net Addresses/Triple Recessed, Semi–recessed and Surface
```
http://www.elliptipar.com/home/Contents.htm
http://www.elplighting.com/product.html
http://www.thomasltg.com/
http://www.cooperlighting.com/
http://www.hubbell-ltg.com/products.htm#Down&Track
http://www.danalite.com/
http://www.kramerlighting.com/
http://www.lightolier.com/
http://www.lithonia.com/
http://www.prescolite.com/
http://www.visalighting.com/1a_main.html
http://www.zumtobelstaff.co.at/   [Note: German language]
```

CONNECT FOR MORE

Net Addresses/Triple Exterior
```
http://www.elliptipar.com/home/Contents.htm
http://www.guth.com/outdoor.htm
http://www.hydrel.com
http://www.kimlighting.com/boll1.html
http://www.lithonia.com/
http://www.mcphilbenoutdoor.com/mcphilben/
```

CONNECT FOR MORE

Net Addresses/Ballasts
```
http://www.advancetransformer.com/
http://www.magnetek.com/transLighting.html
http://www.mot.com/ies/MLI/
http://www.lutron.com/hilume/   [for dimming ballasts only]
```

CONNECT FOR MORE

Fluorescent
Triphosphor
8

Cost magnitude[2000]
for deluxe triphoshor fluorescent lamps

Low
US$3.⁰⁰ to US$8.⁰⁰
Moderate
US$8.⁰⁰ to US$16.⁰⁰
High
US$16.⁰⁰ to US$30.⁰⁰

Costs vary based on quantities, distributor and contractor markups, and market conditions, manufacturing situations and annual inflation. These values are for preliminary, magnitude budgeting and do not represent quotes nor actual final pricing.

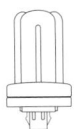

Fluorescent F13Triple/830
Triphosphor
for electronic ballasts ONLY

Stats Guide
⊃Longer life is better
⊃Higher efficacy is better
⊃Lower wattage is better
⊃Lower color of light is warmer
⊃Moderate cost range for Triple is about US$8.⁰⁰ to US$16.⁰⁰

Stats
Data varies manufacturer to manufacturer and changes from time to time

Life/10000 hours
A typical office environment might be occupied about 2600 hours each year.

Efficacy/55 LPW (values include electronic ballast loss)

Wattage and Output/13 watts and 900 lumens

Color of Light/3000°K (warm tone) and 3500°K (neutral)

Beam spreads/Not Applicable
This is a general service lamp intended for downlights, sconces and task lights where an all around glow of light is desired or where a luminaire reflector will provide optical control. Do not use where distinct accent light or focused light is desired—for accent use halogen or HIR PAR lamps (Section 3 and 5), halogen or HIR MR lamps (Sections 4 and 6) or CMH PAR lamps (Section 7).

Size/4³⁄₁₆″ L by 2³⁄₈″ Ø (diameter is approximate)

Voltage/120V and 277V Ballast dependent (check ballast vendors for other voltages)

Cost Magnitude/Moderate (2000)

Manufacturers
GE

Net Addresses
http://www.ge.com/lighting/business/index.htm

CONNECT FOR MORE

Advantages
Crisp white light (3000°K)
Dimmable (see disadvantages, however)
Electronic operation yields no flicker and no hum
Instant–on
Much longer life than halogen/TB19 lamps
Better alternative to less efficient standard filament A lamps
Withstand reasonable temperature range

Disadvantages
Ballast required
◆Additional wattage requirement of ballast of 3 watts
Dimming reduces lamp life and ballast life significantly
◆Unless all lamps are seasoned at full output for 100 hours prior to installation
◆Very tedious to season lamps for spot and group relamping
◆Lamp life and ballast life can be reduced to just several weeks of operation
Some lamp lumen depreciation (light loss) over life
◆5 to 10 percent
Some warm–up time to full brightness
◆Instant–on operation, but 5 to 15 minutes to full output

For efficient distinct accent light or focused light, use HIR/MR or CMH PAR lamps.

Net Addresses/Triple Recessed and Exterior
See page 8–11

CONNECT FOR MORE

Fluorescent Triphosphor **8**

Fluorescent Triphosphor F13Triple/830

for electronic ballasts ONLY

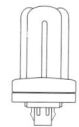

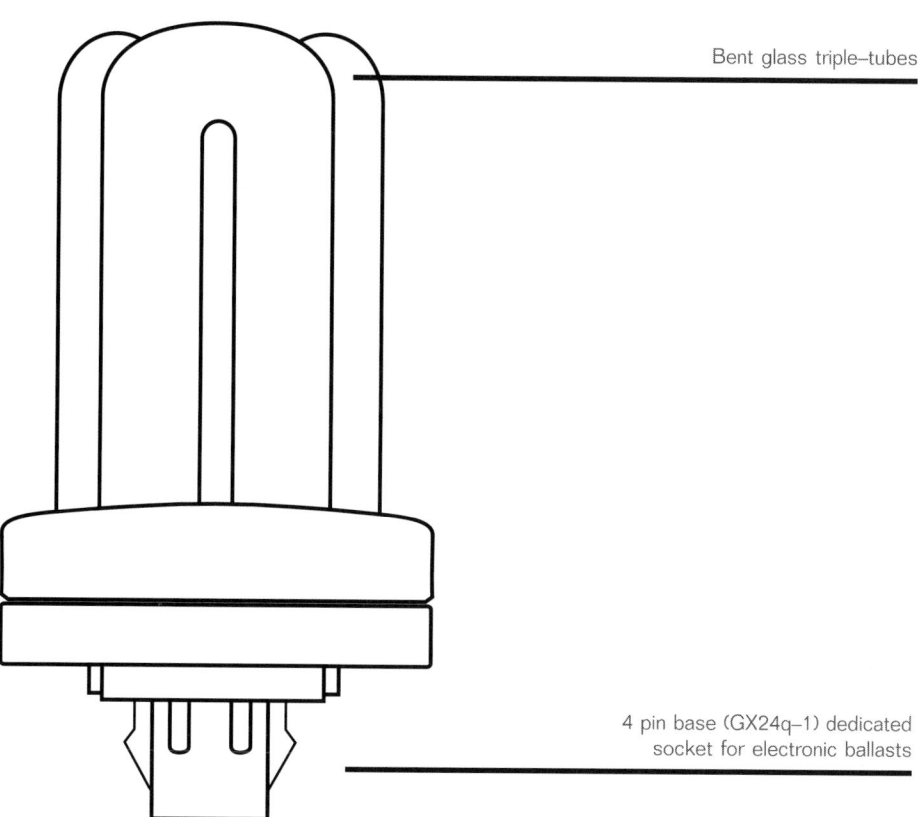

Bent glass triple–tubes

4 pin base (GX24q–1) dedicated socket for electronic ballasts

Lamp schematic shown actual size
Image copyright 2000/GE

Power budgeting
for 1–lamp luminaires

BF 0.98 yields 16 watts

Ballast vendors include Energy Savings, Inc. (ESI) and Magnetek. Above factor based on ESI.

Uses

Soft Decorative
♦ Floor lights ❶
♦ Table lights ❶
♦ Sconces ❶

Soft Downlighting
Break areas
Circulation
Secondary lobbies
Waiting areas

Task Lighting

■ ■ ■ ■ ■ ■ ■ ■ ■ ■ ■ ■ ■

F13Triple Lighting Design Tips
✔ Best in low ceiling spaces (9'–0" or less).
✔ Best for low light situations
✔ Best for soft glow approach
✔ Excellent for "eye candy" lights in corporate settings
✔ Excellent in single–lamp, vertical–orientation downlights (mimic incandescent downlight appearance)
✔ Use in 4– and 5–inch diameter downlights for less institutional look

Application Key

Commercial low level
Gallery
Hospitality low level
Institutional low level
Manufacturing
Residential
Retail
Exterior

bold = primary application
partial fade = minimal application
fade = unlikely application

Fluorescent Triphosphor

8

❶ At time of publication, many of these luminaires require customization to accept 13Triple lamping.

Fluorescent F18Triple/830

Triphosphor

for electronic ballasts ONLY

Stats
Data varies manufacturer to manufacturer and changes from time to time

Life/10000 hours
A typical office environment might be occupied about 2600 hours each year.

Efficacy/63 LPW
(values include electronic ballast loss)

Wattage and Output/18 watts and 1200 lumens

Color of Light/3000°K (warm tone) and 3500°K (neutral)

Beam spreads/Not Applicable
This is a general service lamp intended for downlights and sconces where an all around glow of light is desired or where a luminaire reflector will provide optical control. Do not use where distinct accent light or focused light is desired—for accent use halogen or HIR PAR lamps (Section 3 and 5), halogen or HIR MR lamps (Sections 4 and 6) or CMH PAR lamps (Section 7).

Size/4⅝" L by 2⅜" Ø
(length varies somewhat by manufacturer; diameter is approximate)

Voltage/120V and 277V
Ballast dependent (check ballast vendors for other voltages)

Cost Magnitude/Moderate (2000)

Manufacturers
GE
Osram Sylvania
Philips

Net Addresses
http://www.ge.com/lighting/business/index.htm
http://ecom.sylvania.com/osicatalog/
http://www.lighting.philips.com/

CONNECT FOR MORE

Advantages
Crisp white light (3000°K)
Dimmable (see disadvantages, however)
Electronic operation yields no flicker and no hum
Instant–on
Much longer life than halogen/TB19 lamps
Better alternative to less efficient standard filament A lamps
Withstand reasonable temperature range

Disadvantages
Ballast required
◆Additional wattage requirement of ballast of 1 watt (BF of 0.98)
Dimming reduces lamp life and ballast life significantly
◆Unless all lamps are seasoned at full output for 100 hours prior to installation
◆Very tedious to season lamps for spot and group relamping
◆Lamp life and ballast life can be reduced to just several weeks of operation
Some lamp lumen depreciation (light loss) over life
◆5 to 10 percent
Some warm–up time to full brightness
◆Instant–on operation, but 5 to 15 minutes to full output

Net Addresses/Triple Recessed and Exterior
See page 8–11

CONNECT FOR MORE

For efficient distinct accent light or focused light, use HIR/MR or CMH PAR lamps.

Fluorescent Triphosphor **F18Triple/830**

for electronic ballasts ONLY

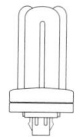

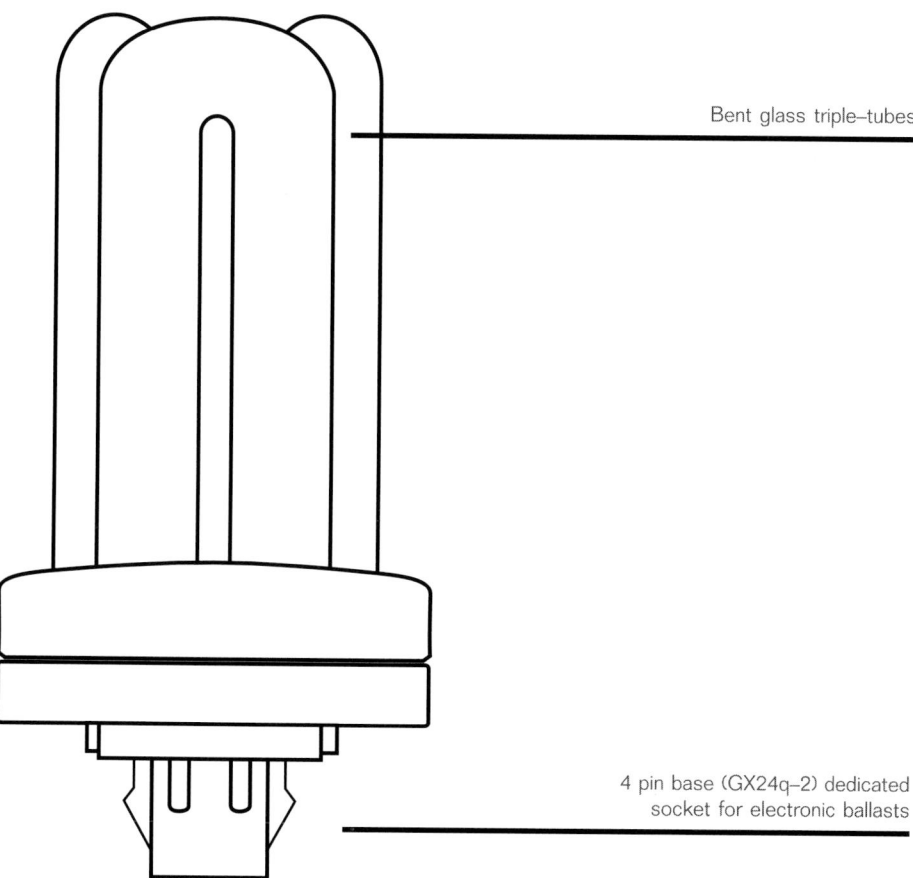

Bent glass triple–tubes

4 pin base (GX24q–2) dedicated
socket for electronic ballasts

Lamp schematic shown actual size
Image copyright 2000/GE

Power budgeting

for 1–lamp luminaires

BF 0.98 yields 19 watts

Ballast vendors include Energy Savings,
Inc. (ESI) and Magnetek. Above factor
based on ESI.

Uses

Decorative

◆Sconces❶

Soft Downlighting

Break areas
Circulation
Secondary lobbies
Waiting areas

■ ■ ■ ■ ■ ■ ■ ■ ■ ■ ■ ■ ■

F18Triple Lighting Design Tips

✔Best in low ceiling spaces (9′–0″ or less).
✔Best for low light situations
✔Best for soft glow approach
✔Excellent in single–lamp, vertical–orientation downlights
(mimic incandescent downlight appearance)
✔Use in 4– and 5–inch diameter downlights for less
institutional look

Application Key

Commerciallow level
Gallery
Hospitalitylow level
Institutionallow level
Manufacturing
Residential
Retail
Exterior

bold = primary application
partial fade = minimal application
fade = unlikely application

Fluorescent Triphosphor

8

❶ At time of publication, many of these luminaires require customization to accept 18Triple lamping.

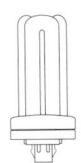

Fluorescent **Triphosphor** F26Triple/830

for electronic ballasts ONLY

Stats Guide
➲Longer life is better
➲Higher efficacy is better
➲Lower wattage is better
➲Lower color of light is warmer
➲High cost range for Triple is about US$16.⁰⁰ to US$30.⁰⁰

Stats
Data varies manufacturer to manufacturer and changes from time to time

Life/10000 hours
A typical office environment might be occupied about 2600 hours each year.

Efficacy/67 LPW (values include electronic ballast loss)

Wattage and Output/26 watts and 1800 lumens

Color of Light/3000°K (warm tone) and 3500°K (neutral)

Beam spreads/Not Applicable
This is a general service lamp intended for downlights and sconces where an all around glow of light is desired or where a luminaire reflector will provide optical control. Do not use where distinct accent light or focused light is desired—for accent use halogen or HIR PAR lamps (Section 3 and 5), halogen or HIR MR lamps (Sections 4 and 6) or CMH PAR lamps (Section 7).

Size/5″ L by 2⅜″ Ø (length varies somewhat by manufacturer; diameter is approximate)

Voltage/120V and 277V Ballast dependent (check ballast vendors for other voltages)

Cost Magnitude/High⁽²⁰⁰⁰⁾

Manufacturers
GE
Osram Sylvania
Philips

Net Addresses
http://www.ge.com/lighting/business/index.htm
http://ecom.sylvania.com/osicatalog/
http://www.lighting.philips.com/

CONNECT FOR MORE

Advantages
Crisp white light (3000°K)
Dimmable (see disadvantages, however)
Electronic operation yields no flicker and no hum
Instant–on
Much longer life than halogen/TB19 lamps
Better alternative to less efficient standard filament A lamps
Withstand reasonable temperature range
Interchangeable with F32Triple (same base configuration)
◆Enable quick change between lamps for more/less light (recognize circuiting and power budget implications)

Disadvantages
Ballast required
◆Additional wattage requirement of ballast of 1 to 4 watts (BF of 0.98)
Dimming reduces lamp life and ballast life significantly
◆Unless all lamps are seasoned at full output for 100 hours prior to installation
◆Very tedious to season lamps for spot and group relamping
◆Lamp life and ballast life can be reduced to just several weeks of operation
Some lamp lumen depreciation (light loss) over life
◆5 to 10 percent
Some warm–up time to full brightness
◆Instant–on operation, but 5 to 15 minutes to full output

For efficient distinct accent light or focused light, use HIR/MR or CMH PAR lamps.

Net Addresses/Triple Recessed and Exterior
See page 8–11

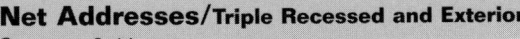

CONNECT FOR MORE

Fluorescent F26Triple/830
Triphosphor
for electronic ballasts ONLY

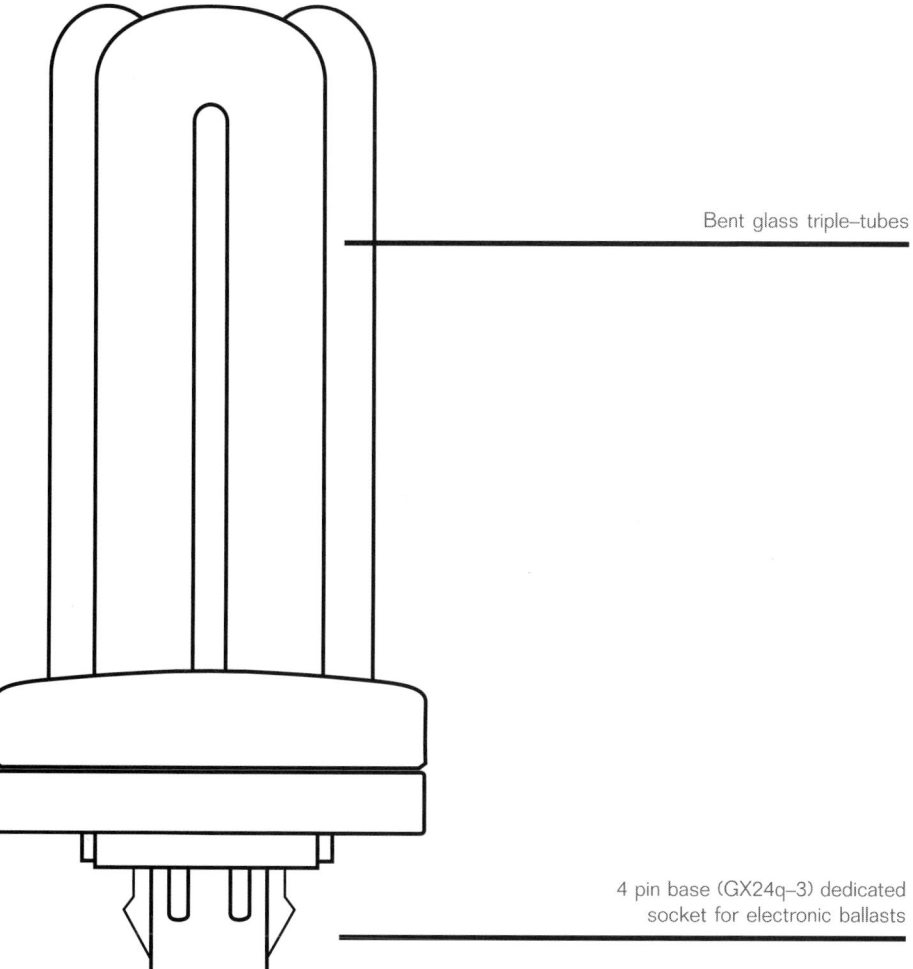

Bent glass triple–tubes

4 pin base (GX24q–3) dedicated socket for electronic ballasts

Lamp schematic shown actual size
Image copyright 2000/GE

Power budgeting
for 1–lamp luminaires

BF 0.98 yields 27 watts

Ballast vendors include Advance, Energy Savings, Inc. (ESI), Magnetek and Motorola. Above factor based on ESI.

Uses
Decorative
◆Sconces
Downlighting
Circulation
Secondary lobbies
Waiting areas
Exterior
Bollards
Downlights
Sconces
Uplights

■ ■ ■ ■ ■ ■ ■ ■ ■ ■ ■ ■ ■ ■ ■

F26Triple Lighting Design Tips
✔Best in low to moderate ceiling spaces (10'–6" or less).
✔Best for general moderate light situations
✔Best for more functional, less decorative lighting
✔Excellent in single–lamp, vertical–orientation downlights (mimic incandescent downlight appearance)
✔Use in 5– and 6–inch diameter downlights for less institutional look

Application Key

Commercialmoderate level
Gallery
Hospitalitymoderate level
Institutionalmoderate level
Manufacturing
Residential
Retail
Exterior

bold = primary application
partial fade = minimal application
fade = unlikely application

Fluorescent
Triphosphor

8

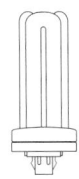

Fluorescent Triphosphor F32Triple/830

for electronic ballasts ONLY

Stats Data varies manufacturer to manufacturer and changes from time to time

Life/10000 hours
A typical office environment might be occupied about 2600 hours each year.

Efficacy/64 LPW (values include electronic ballast loss)

Wattage and Output/32 watts and 2350 lumens

Color of Light/3000°K (warm tone) and 3500°K (neutral)

Beam spreads/Not Applicable
This is a general service lamp intended for downlights, sconces and spreadlens wallwashers where an all around glow of light is desired or where a luminaire reflector will provide optical control. Do not use where distinct accent light or focused light is desired—for accent use halogen or HIR PAR lamps (Section 3 and 5), halogen or HIR MR lamps (Sections 4 and 6) or CMH PAR lamps (Section 7).

Size/5⅝" L by 2⅜" Ø (length varies somewhat by manufacturer; diameter is approximate)

Voltage/120V and 277V Ballast dependent (check ballast vendors for other voltages)

Cost Magnitude/High⁽²⁰⁰⁰⁾

Manufacturers
GE
Osram Sylvania
Philips

Net Addresses
http://www.ge.com/lighting/business/index.htm
http://ecom.sylvania.com/osicatalog/
http://www.lighting.philips.com/

CONNECT FOR MORE

Advantages
Crisp white light (3000°K)
Dimmable (see disadvantages, however)
Electronic operation yields no flicker and no hum
Instant–on
Much longer life than halogen TB lamps
Better alternative to less efficient standard filament A lamps
Withstand reasonable temperature range
Interchangeable with F26Triple (same base configuration)
◆Enable quick change between lamps for more/less light (recognize circuiting and power budget implications)

Disadvantages
Ballast required
◆Additional wattage requirement of ballast of 1 to 4 watts (BF of 0.98)
Dimming reduces lamp life and ballast life significantly
◆Unless all lamps are seasoned at full output for 100 hours prior to installation
◆Very tedious to season lamps for spot and group relamping
◆Lamp life and ballast life can be reduced to just several weeks of operation
Some lamp lumen depreciation (light loss) over life
◆5 to 10 percent
Some warm–up time to full brightness
◆Instant–on operation, but 5 to 15 minutes to full output

For efficient distinct accent light or focused light, use HIR/MR or CMH PAR lamps.

Net Addresses/Triple Recessed and Exterior
See page 8–11

CONNECT FOR MORE

Fluorescent F32Triple/830
Triphosphor
for electronic ballasts ONLY

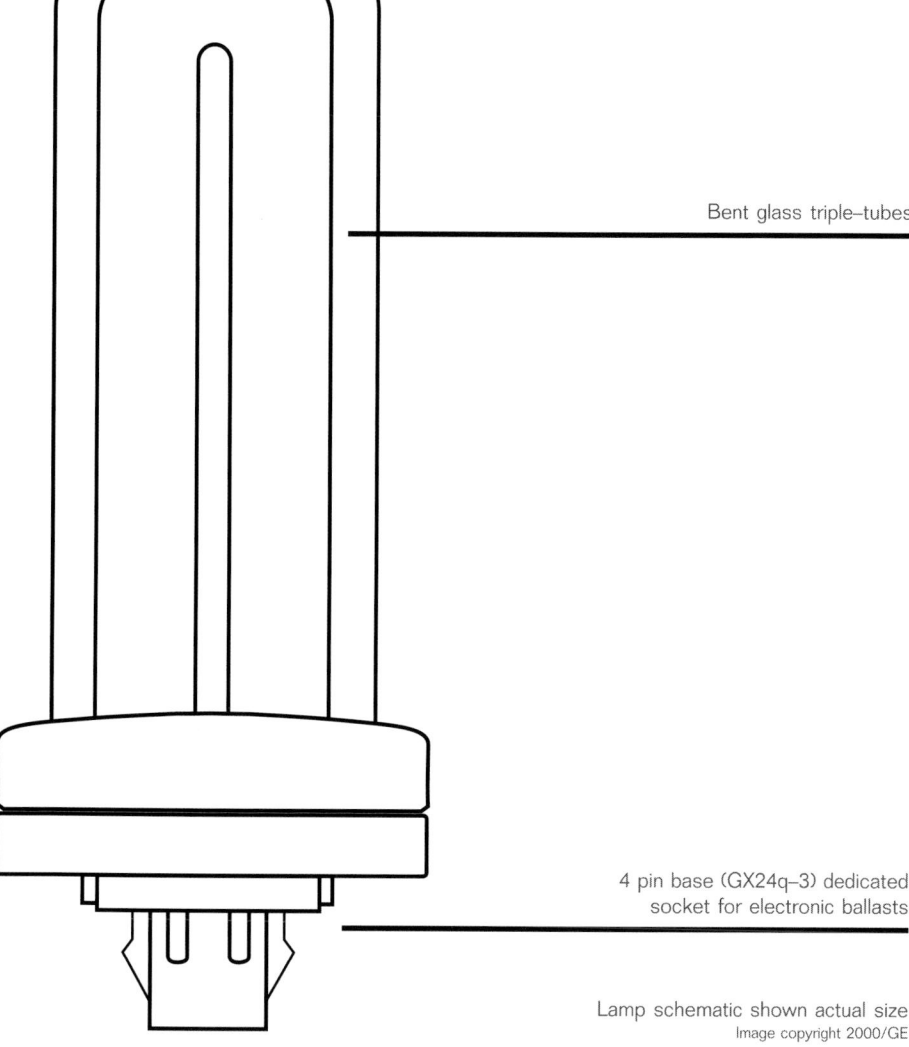

Bent glass triple–tubes

4 pin base (GX24q–3) dedicated socket for electronic ballasts

Lamp schematic shown actual size
Image copyright 2000/GE

Power budgeting
for 1–lamp luminaires

BF 0.98 yields 37 watts

Ballast vendors include Advance, Energy Savings, Inc. (ESI), Magnetek and Motorola. Above factor based on ESI.

Fluorescent Triphosphor

8

Uses
Decorative
◆Sconces
Downlighting
Task Areas
Exterior
Bollards
Downlights
Sconces
Uplights
Wallwashing
See Section 11

■ ■ ■ ■ ■ ■ ■ ■ ■ ■ ■

F32Triple Lighting Design Tips
✔ Best in moderate ceiling spaces (10'–6" or less).
✔ Best for task/area lighting situations
✔ Best for functional, less decorative lighting
✔ Excellent in single–lamp, vertical–orientation downlights (mimic incandescent downlight appearance)
✔ Excellent for exterior, fully enclosed lights
✔ Use in 6– and 7–inch diameter downlights for less institutional look
✔ Available in matching aperture downlights and spreadlens wallwashers

Application Key

Commercialmoderate to high level
Gallery
Hospitalitymoderate to high level
Institutionalmoderate to high level
Manufacturing
Residential
Retail
Exterior

bold = primary application
partial fade = minimal application
fade = unlikely application

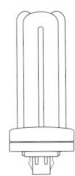

Fluorescent **Triphosphor** **F32Triple/830**

for electronic ballasts ONLY

Image copyright 2000 and courtesy of Steelcase Inc.

Fluorescent F32Triple/830
Triphosphor
for electronic ballasts ONLY

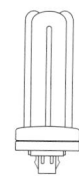

Figure 8.1

Indirect lighting was introduced into this waiting area with inverted steplights. Each steplight is lamped with an F32Triple/830 lamp. This lamp offered the most lumens for the smallest size while still allowing instant–on capabilities (some lights can then be used on the egress and/or nightlight system). The luminaires are centered 6 feet AFF to provide a relatively uniform wash across the ceiling and providing sufficient lighting for facial recognition and circulation. With the exception of some accent lights aimed onto the wall (unseen to the right side of photo), there is no other light in the area. Inverted steplights are Bega.

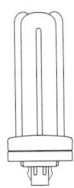

Fluorescent _{Triphosphor} F42Triple/830

for electronic ballasts ONLY

Stats _{Data varies manufacturer to manufacturer and changes from time to time}

Life/10000 hours

A typical office environment might be occupied about 2600 hours each year.

Efficacy/68 LPW _(values include electronic ballast loss)

Wattage and Output/42 watts and 3200 lumens

Color of Light/3000°K (warm tone) and 3500°K (neutral)

Beam spreads/Not Applicable

This is a general service lamp intended for downlights, sconces and spreadlens wallwashers where an all around glow of light is desired or where a luminaire reflector will provide optical control. Do not use where distinct accent light or focused light is desired—for accent use halogen or HIR PAR lamps (Section 3 and 5), halogen or HIR MR lamps (Sections 4 and 6) or CMH PAR lamps (Section 7).

Size/6" L by 2⅜" Ø _(length varies somewhat by manufacturer; diameter is approximate)

Voltage/120V and 277V _{Ballast dependent (check ballast vendors for other voltages)}

Cost Magnitude/High⁽²⁰⁰⁰⁾

Manufacturers

GE
Osram Sylvania
Philips

Net Addresses
http://www.ge.com/lighting/business/index.htm
http://ecom.sylvania.com/osicatalog/
http://www.lighting.philips.com/

CONNECT FOR MORE

Advantages

Crisp white light (3000°K)
Dimmable (see disadvantages, however)
Electronic operation yields no flicker and no hum
Instant–on
Much longer life than halogen TB lamps
Better alternative to less efficient standard filament A lamps
Withstand reasonable temperature range

Disadvantages

Ballast required
♦ Additional wattage requirement of ballast of 1 to 4 watts (BF of 0.98)
Dimming reduces lamp life and ballast life significantly
♦ Unless all lamps are seasoned at full output for 100 hours prior to installation
♦ Very tedious to season lamps for spot and group relamping
♦ Lamp life and ballast life can be reduced to just several weeks of operation
Some lamp lumen depreciation (light loss) over life
♦ 5 to 10 percent
Some warm–up time to full brightness
♦ Instant–on operation, but 5 to 15 minutes to full output

Net Addresses/Triple Recessed and Exterior
See page 8–11

CONNECT FOR MORE

For efficient distinct accent light or focused light, use HIR/MR or CMH PAR lamps.

Fluorescent F42Triple/830
Triphosphor
for electronic ballasts ONLY

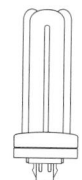

Bent glass triple–tubes

Power budgeting
for 1–lamp luminaires

BF 0.98 yields 47 watts

Ballast vendors include Advance, Energy Savings, Inc. (ESI), Magnetek and Motorola. Above factor based on ESI.

4 pin base (GX24q–4) dedicated socket for electronic ballasts

Lamp schematic shown actual size
Image copyright 2000/GE

Uses

Decorative
♦ Sconces

Downlighting
Task Areas

Exterior
Bollards
Downlights
Sconces
Uplights

Wallwashing
See Section 11

■ ■ ■ ■ ■ ■ ■ ■ ■ ■ ■ ■ ■ ■

F42Triple Lighting Design Tips
✔ Best in moderate to high ceiling spaces (12'–6" or less).
✔ Best for task/area lighting situations
✔ Best for functional, less decorative lighting
✔ Excellent in single–lamp, vertical–orientation downlights (mimic incandescent downlight appearance)
✔ Excellent for exterior, fully enclosed lights
✔ Potentially glary (use in well–shielded units or deep reflectors)
✔ Use in 7–inch diameter downlights for less institutional look
✔ Available in matching aperture downlights and spreadlens wallwashers

Application Key

Commercialhigh level
Gallery
Hospitalityhigh level
Institutionalhigh level
Manufacturing
Residential
Retail
Exterior

bold = primary application
partial fade = minimal application
fade = unlikely application

Fluorescent
Triphosphor

8

Fluorescent/LongCompact

for electronic ballasts ONLY

LongCompact fluorescent lamps—so named for their relatively long length single bent tube—have been a staple in the "high lumen package" fluorescent lamp market for more than ten years. These lamps offer significant light output over in a relatively compact lamp, but these lamps are also longer than the original compact fluorescent, low wattage lamps—hence the reference to "long." These lamps are available in a range of wattages and physical sizes that meet a number of design tenets—relatively small luminaires, relatively low glare and, with the proper ballast, dimmable. Because of their relatively small, linear shape, these lamps are best suited for

Table 8.3 Deluxe Triphosphor LongCompact Lamps

Fluorescent Lamp	Use/Replaces	Lamp Profile
F18Long/RS	New • Cove lights • Sconces • Wallwashers	
F39Long/RS	New • Cove lights • Sconces • Wallwashers	
F40Long/RS	New • Cove lights • Sconces • Wallwashers	
F50Long/RS	New • Cove lights • Sconces • Wallwashers	

Lamp images courtesy of and copyright by GE 2000.

cove uplighting applications, some sconce applications and merchandise accenting and wallwashing applications. To maintain high efficacies, LongCompact lamps are only available in the deluxe triphosphor (XT) versions. Table 8.3 outlines several key XT/LongCompact lamps presently available.

While these lamps may seem expensive, the overall cost of light over the life of the lamp/lighting system must be considered. Many of these LongCompact lamps have rated lives of 12,000 to 20,000

Fluorescent/LongCompact
Triphosphor
for electronic ballasts ONLY

hours. This is six to ten times that of many mains voltage halogen and halogen infrared lamps and is three to four times that of many low voltage halogen and halogen infrared lamps. Very simply, the LongCompact lamps rightfully could/should cost two to five times that of halogen or halogen infrared lamps. And this doesn't account for the savings in energy costs from lower wattages, nor more significantly, in personnel time necessary to change a lamp.

Several decorative luminaire manufacturers have available various styles and sizes of sconces and pendents which take advantage of the LongCompact lamp. Recognize that many decorative lights accept the screw–base compact fluorescent lamps. For energy efficiency and lamp replacement/maintenance consistency, however, designers should specify dedicated socket systems—luminaires with sockets which accept just one type/wattage lamp. This is a necessity if energy efficiency and maintenance consistency are priorities.

Since ballasts are now available to dim these LongCompact lamps, their use in more "moody" and less worklike settings should increase—providing lamp and ballast manufacturers address the drawbacks of lamp seasoning or reduced lamp and ballast life. The color rendering is excellent with XT/LongCompact 3000°K lamps approximating halogen in whiteness.

■ ■ ■ ■ ■ ■ ■ ■ ■ ■ ■

XT/LongCompact Lamps
❶ Instant–on, but require warm–up to full brightness
❶ Dimmable
❶ Electronic operation results in no hum and no flicker
❶ Various sizes/wattages for all applications

XT/LongCompact lamp designation

XT/LongCompact lamps are typically identified with an alphanumeric designation outlined on page 8–27. Since each manufacturer has slightly different means for lamp designation, the designation method presented here is solely for convenient reference throughout the text. Actual lamp manufacturers' designations will need to be used in final specifications. Perhaps the lamp manufacturers will standardize lamp designations in the near future in an effort to ease lamp cross referencing and specifications.

XT/LongCompact lamps for general lighting

XT/LongCompact lamps, except for the 18–watt unit, are high lumen lamps and therefore are glary. These lamps are not recom-

Fluorescent/LongCompact

Triphosphor

for electronic ballasts ONLY

mended for use in recessed parabolic luminaires for general lighting situations. XT/LongCompact lamps do work well in pendent uplights where higher–than–normal light intensities are required. In private offices, corridors and reception areas and conference rooms, XT/LongCompact lamps work well in the recently popular recessed/indirect perforated basket luminaires.

XT/LongCompact lamps for cove lighting

LongCompact fluorescent lamps work well in architectural cove details (see Figures 8.2 and 8.3). The high lumen output of the lamp enables designers to develop architectural cove details which provide sufficient light intensities for circulation, casual work and conversation activity areas. The shallow depth of these lamps, combined with their relatively short lengths allow them to be used in simple, low profile striplights and in more optically efficient low profile reflector systems. With the exception of the 18LongCompact/830/RS lamp, these lamps are typically too glary for use in direct, recessed luminaires.

XT/LongCompact lamps for decorative lighting

LongCompact lamps make excellent candidates in sconces and pendents where more functional light intensities are required. Used in striplights, which are relatively small but yield lots of light, XT/LongCompact lamps can help create dramatic details (see Figures 8.2, 8.3 and 8.4).

XT/LongCompact lamps for wallwashing

Some of the most efficient wallwash lighting is achieved with XT/LongCompact lamps in recessed, semi–recessed and surface mounted wallwash luminaires (see Figures 8.2 and 8.3). These lights are rectilinear and generally have asymmetric reflectors and/or baffles with open apertures to direct light onto walls.

Fluorescent/LongCompact
Triphosphor
for electronic ballasts ONLY

Deluxe Triphosphor LongCompact Fluorescent Lamp Designation

F50Long/830/RS

These last two letters indicate if the lamp is Rapid Start (typically offers longer life and necessary for dimming)

These two digits indicate color temperature
- **30 is shorthand for 3000°K (warm tone)**
- **35 is shorthand for 3500°K (neutral)**

This number represents the "deluxe" nature of the lamp indicating a color rendering index of greater than 80 (where 100 is perfect)

This signifies that this is a LongCompact fluorescent lamp

These two digits represent lamp wattage

Signifies that the lamp is a fluorescent (F) lamp

[This designation used for convenient reference throughout this *Time–Saver Standards for Architectural Lighting*—but not necessarily used by all lamp manufacturers (although it would be convenient if manufacturers would come to agreement on standard designations.]

XT/LongCompact lamps for merchandising

Using the kind of luminaires described for wallwashing, lighting of large merchandise areas or vertical merchandise displays can be quite effective with XT/LongCompact lamps. Intensities with the higher wattage lamps (39, 40 and 50 watt versions) are sufficient for merchandise accenting. The excellent color rendering, energy efficiency and long life exhibited by these lamps are significant benefits for merchandising.

Net Addresses/Ballasts
http://www.advancetransformer.com/
http://www.magnetek.com/transLighting.html
http://www.mot.com/ies/MLI/
http://www.sylvania.com/ballast/phops.htm
http://www.lutron.com/hilume/ [for dimming ballasts only]

CONNECT FOR MORE

Cost magnitude[2000]
for deluxe triphoshpor fluorescent lamps

Low
US$3.⁰⁰ to US$8.⁰⁰
Moderate
US$8.⁰⁰ to US$16.⁰⁰
High
US$16.⁰⁰ to US$30.⁰⁰

Costs vary based on quantities, distributor and contractor markups, and market conditions, manufacturing situations and annual inflation. These values are for preliminary, magnitude budgeting and do not represent quotes nor actual final pricing.

Fluorescent
Triphosphor
8

Fluorescent/LongCompact
Triphosphor
for electronic ballasts ONLY

Figure 8.2
The "floating wall" is infused with light—most of it a result of using F50Long lamps. Striplights at the top and bottom thirds of the wall (and arranged for access through the top slot and wall cutouts) create a back glow. Also see Figures C38 and C39. Linear elements in ceiling are related to HVAC.

Image copyright 2000 Stephen Graham

Fluorescent
Triphosphor
8

Fluorescent/LongCompact
for electronic ballasts ONLY

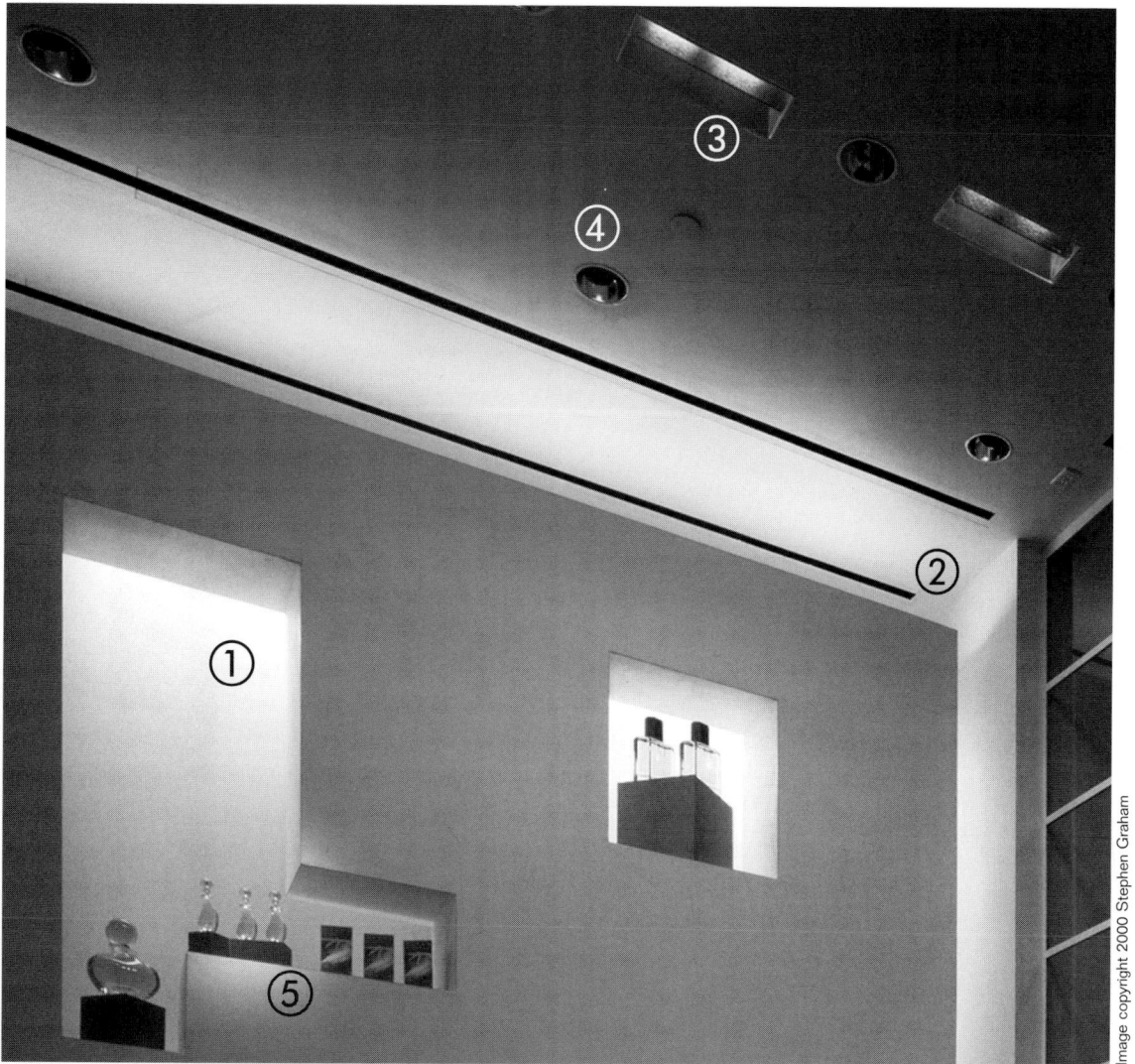

Fluorescent
Triphosphor

8

Figure 8.3

In addition to the back glow ① created by the striplights discussed in Figure 8.2, an asymmetric uplight on top of the wall ② distributes more function light across the ceiling and into the room—contributing to the ambient lighting in the store. The asymmetric uplight is lamped with F40Long lamps. The front of the wall is softly lighted with fluorescent wallwashers ③ recessed in the ceiling. For key focals, recessed adjustable accents ④ with 50PAR36/H/VNSP5° lamps (see Section 4 and Figures 4.6 and 4.7) are used (effect shown at ⑤). See Section 11 for asymmetric uplight details and for wallwashing. Asymmetric uplight is Lithonia/ Peerless. Recessed adjustable accents are Kurt Versen. Wallwashers are Columbia Parawash.

Triphosphor
Fluorescent/LongCompact
for electronic ballasts ONLY

Triphosphor

Fluorescent
Triphosphor

8

Image copyright 2000 Stephen Graham

Figure 8.4
The vaulted ceiling elements are uplighted from an architectural detail fitted with striplights using F40Long lamps. The intensity of the lamps along with the tall, sweeping arch of the vault help to distribute light in a relatively uniformly fashion up/across the ceiling.

Fluorescent/LongCompact
Triphosphor

for electronic ballasts ONLY

Net Addresses/LongCompact Recessed–Indirect
http://www.columbia-ltg.com/products/recessed-surface.html
http://www.thomasltg.com/
http://www.focalpointlights.com/
http://www.cooperlighting.com/
http://www.hubbell-ltg.com/products.htm#Down&Track
http://www.lightolier.com/
http://www.litecontrol2.com/
http://www.lithonia.com/
http://www.neoray.com/CATALOG/catalog.html
http://www.zumtobelstaff.co.at/   [Note: German language]

CONNECT FOR MORE

Net Addresses/LongCompact Cove
http://www.bartcolighting.com/index2.html
http://www.belfer.com/prodsi.html
http://www.columbia-ltg.com/products/recessed-surface.html
http://www.thomasltg.com/
http://www.elliptipar.com/home/Contents.htm
http://www.elplighting.com
http://www.focalpointlights.com/
http://www.cooperlighting.com/
http://www.hubbell-ltg.com/products.htm#Down&Track
http://www.danalite.com/
http://www.ledalite.com/
http://www.lightolier.com/
http://www.litecontrol2.com/
http://www.lithonia.com/
http://www.neoray.com/CATALOG/catalog.html
http://www.peerless-lighting.com/

CONNECT FOR MORE

Net Addresses/LongCompact Decorative
(Sconces and/or Pendents)
http://www.boydlighting.com/
http://www.lightolier.com/
http://www.litecontrol2.com/
http://www.visalighting.com/1a_main.html
http://www.peerless-lighting.com/

CONNECT FOR MORE

Net Addresses/LongCompact Wallwash/Merchandising
http://www.alkco.com/flare.htm
http://www.columbia-ltg.com/products/recessed-surface.html
http://www.thomasltg.com/
http://www.elliptipar.com/home/Contents.htm
http://www.elplighting.com
http://www.focalpointlights.com/
http://www.cooperlighting.com/
http://www.hubbell-ltg.com/products.htm#Down&Track
http://www.danalite.com/
http://www.ledalite.com/
http://www.lightolier.com/
http://www.litecontrol2.com/
http://www.lithonia.com/
http://www.neoray.com/CATALOG/catalog.html
http://www.peerless-lighting.com/

CONNECT FOR MORE

Fluorescent
Triphosphor

8

Fluorescent Triphosphor F18Long/830/RS

for electronic ballasts ONLY

Stats Guide
- ⊃Longer life is better
- ⊃Higher efficacy is better
- ⊃Lower wattage is better
- ⊃Lower color of light is warmer
- ⊃Moderate cost range for LongCompact is about US$8.⁰⁰ to US$16.⁰⁰

Stats Data varies manufacturer to manufacturer and changes from time to time

Life/20000 hours
A typical office environment might be occupied about 2600 hours each year.

Efficacy/66 LPW (values include electronic ballast loss)

Wattage and Output/18 watts and 1250 lumens

Color of Light/3000°K (warm tone) and 3500°K (neutral)

Beam spreads/Not Applicable
This is a general service lamp intended for striplights or for wallwashers, sconces and task lights where a luminaire reflector will provide optical control. Do not use where distinct accent light or focused light is desired—for accent use halogen or HIR PAR lamps (Section 3 and 5), halogen or HIR MR lamps (Sections 4 and 6) or CMH PAR lamps (Section 7).

Size/10½" L by 1¾" W (width is approximate)

Voltage/120V and 277V Ballast dependent (check ballast vendors for other voltages)

Cost Magnitude/Moderate (2000)

Manufacturers
GE

Osram Sylvania

Net Addresses
http://www.ge.com/lighting/business/index.htm
http://ecom.sylvania.com/osicatalog/

CONNECT FOR MORE

Advantages
Crisp white light (3000°K)

Dimmable (see disadvantages, however)

Electronic operation yields no flicker and no hum

Instant–on

Much longer life than halogen TB lamps

Better alternative to less efficient standard filament A lamps

Withstand reasonable temperature range

Disadvantages
Ballast required

◆Additional wattage requirement of ballast of 1 to 4 watts (BF of 0.98)

Dimming reduces lamp life and ballast life

◆Unless all lamps are seasoned at full output for 100 hours prior to installation

◆Very tedious to season lamps for spot and group relamping

◆Lamp life and ballast life can be reduced to just several months of operation

Some lamp lumen depreciation (light loss) over life

◆5 to 10 percent

Some warm–up time to full brightness

◆Instant–on operation, but several minutes to full output

For efficient distinct accent light or focused light, use HIR/MR or CMH PAR lamps.

Net Addresses/LongCompact
See page 8–31

CONNECT FOR MORE

Fluorescent
Triphosphor F18Long/830/RS
for electronic ballasts ONLY

Bent glass single long tube

4 pin base (2G11) dedicated socket
for electronic ballasts

Lamp schematic shown about one–third actual size
Image copyright 2000/GE

Power budgeting
for 1–lamp luminaires

BF 0.98 yields 19 watts

Ballast vendors include Advance, Energy
Savings, Inc. (ESI), Magnetek and
Motorola. Above factor based on ESI.

Fluorescent
Triphosphor

8

Uses

Decorative
- ◆Architectural details
- ◆Coves
- ◆Sconces

Wallwashing
See Section 11

■ ■ ■ ■ ■ ■ ■ ■ ■ ■ ■ ■

F18Long Lighting Design Tips
✔Best in low ceiling spaces (9′–0″ or less).
✔Best for soft glow approach
✔Excellent in single–lamp, horizontal–orientation
wallwashers
✔Great in segmented striplights for circular or curvilinear
cove shapes

Application Key

Commercial
Gallery
Hospitality
Institutional
Manufacturing
Residential
Retail
Exterior

bold = primary application
partial fade = minimal application
fade = unlikely application

Fluorescent F39Long/830
Triphosphor
for electronic ballasts ONLY

Fluorescent
Triphosphor
8

Stats Guide
◗Longer life is better
◗Higher efficacy is better
◗Lower wattage is better
◗Lower color of light is warmer
◗Moderate cost range for LongCompact is about US$8.⁰⁰ to US$16.⁰⁰

Stats Data varies manufacturer to manufacturer and changes from time to time

Life/12000 hours
A typical office environment might be occupied about 2600 hours each year.

Efficacy/74 LPW (values include electronic ballast loss)

Wattage and Output/36 and 39 (GE) watts and 2885 lumens

Color of Light/3000°K (warm tone) and 3500°K (neutral)

Beam spreads/Not Applicable
This is a high output general service lamp intended for striplights or for wallwashers, relatively large sconces and pendents where an all around glow of light is desired or where a luminaire reflector will provide optical control. Do not use where distinct accent light or focused light is desired—for accent use halogen or HIR PAR lamps (Section 3 and 5), halogen or HIR MR lamps (Sections 4 and 6) or CMH PAR lamps (Section 7).

Size/16½" L (varies by manufacturer) by 1¾" W (width is approximate)

Voltage/120V and 277V Ballast dependent (check ballast vendors for other voltages)

Cost Magnitude/Moderate to High (2000)

Manufacturers
GE
Osram Sylvania
Philips

Net Addresses
http://www.ge.com/lighting/business/index.htm
http://ecom.sylvania.com/osicatalog/
http://www.lighting.philips.com/

CONNECT FOR MORE

Advantages
Crisp white light (3000°K)
Dimmable (see disadvantages, however)
Electronic operation yields no flicker and no hum
Instant–on
Much longer life than halogen TB lamps
Better alternative to less efficient standard filament A lamps
Withstand reasonable temperature range

Disadvantages
Ballast required
◆Additional wattage requirement of ballast of 1 to 4 watts (BF of 0.98)
Dimming reduces lamp life and ballast life
◆Unless all lamps are seasoned at full output for 100 hours prior to installation
◆Very tedious to season lamps for spot and group relamping
◆Lamp life and ballast life can be reduced to just several months of operation
Some lamp lumen depreciation (light loss) over life
◆5 to 10 percent
Some warm–up time to full brightness
◆Instant–on operation, but several minutes to full output

Net Addresses/LongCompact
See page 8–31

CONNECT FOR MORE

For efficient distinct accent light or focused light, use HIR/MR or CMH PAR lamps.

Fluorescent F39Long/830

Triphosphor

for electronic ballasts ONLY

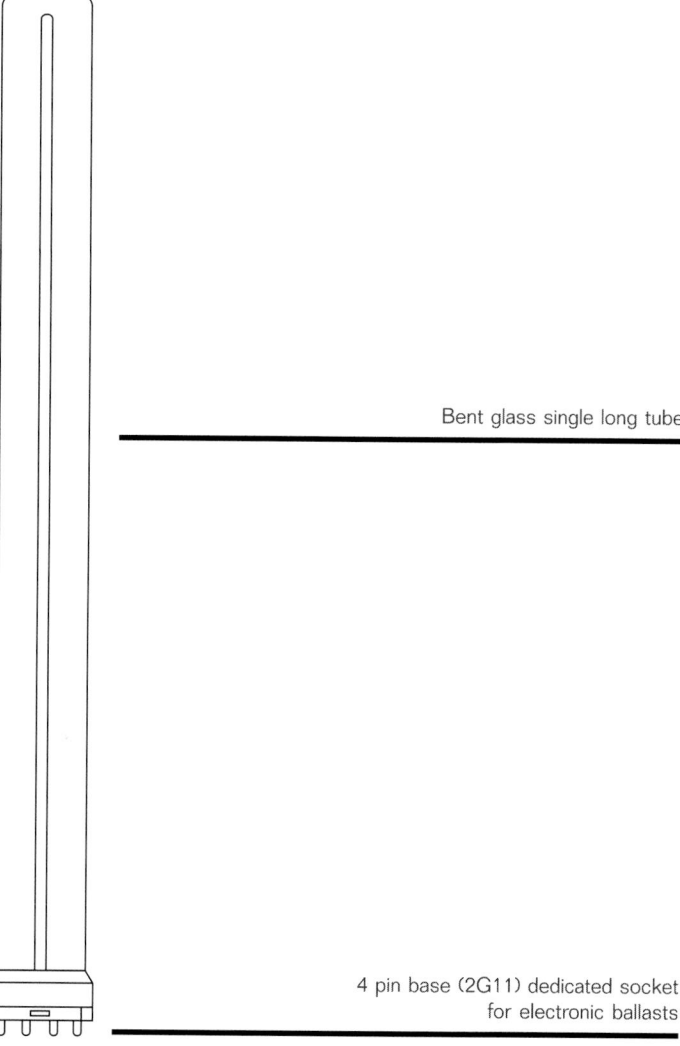

Bent glass single long tube

Power budgeting
for 1–lamp luminaires

BF 0.98 yields 39 watts

Ballast vendors include Advance, Energy Savings, Inc. (ESI), Magnetek and Motorola. Above factor based on ESI.

4 pin base (2G11) dedicated socket
for electronic ballasts

Lamp schematic shown about one–third actual size
Image copyright 2000/GE

Uses

Decorative/Functional
- ◆Coves
- ◆Pendents
- ◆Sconces

Merchandise Accenting

Wallwashing
See Section 11

■ ■ ■ ■ ■ ■ ■ ■ ■ ■ ■ ■ ■

F39Long Lighting Design Tips
✔ Best in moderate ceiling spaces (10′–6″ or less).
✔ Excellent in single–lamp, horizontal–orientation wallwashers or trackheads or monopoints
✔ Great in segmented striplights for circular or curvilinear cove shapes

Application Key

Commercial
Gallery
Hospitality
Institutional
Manufacturing
Residential
Retail
Exterior

bold = primary application
partial fade = minimal application
fade = unlikely application

Fluorescent
Triphosphor

8

Fluorescent Triphosphor F40Long/830/RS

for electronic ballasts ONLY

Stats Guide
⊃ Longer life is better
⊃ Higher efficacy is better
⊃ Lower wattage is better
⊃ Lower color of light is warmer
⊃ High cost range for LongCompact is about US$16.00 to US$30.00

Stats
Data varies manufacturer to manufacturer and changes from time to time

Life/20000 hours
A typical office environment might be occupied about 2600 hours each year.

Efficacy/78 LPW (values include electronic ballast loss)

Wattage and Output/38 (Philips) and 40 watts and 3200 lumens

Color of Light/3000°K (warm tone) and 3500°K (neutral)

Beam spreads/Not Applicable
This is a high output general service lamp intended for striplights or for wallwashers, relatively large sconces and pendents where an all around glow of light is desired or where a luminaire reflector will provide optical control. Do not use where distinct accent light or focused light is desired—for accent use halogen or HIR PAR lamps (Section 3 and 5), halogen or HIR MR lamps (Sections 4 and 6) or CMH PAR lamps (Section 7).

Size/22½" L by 1¾" W (width is approximate)

Voltage/120V and 277V Ballast dependent (check ballast vendors for other voltages)

Cost Magnitude/High (2000)

Manufacturers
GE
Osram Sylvania
Philips

Net Addresses
http://www.ge.com/lighting/business/index.htm
http://ecom.sylvania.com/osicatalog/
http://www.lighting.philips.com/

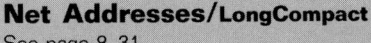
CONNECT FOR MORE

Advantages
Crisp white light (3000°K)
Dimmable (see disadvantages, however)
Electronic operation yields no flicker and no hum
Instant–on
Much longer life than halogen TB lamps
Better alternative to less efficient standard filament A lamps
Withstand reasonable temperature range

Disadvantages
Ballast required
♦ Additional wattage requirement of ballast of 1 to 4 watts (BF of 0.98)
Dimming reduces lamp life and ballast life
♦ Unless all lamps are seasoned at full output for 100 hours prior to installation
♦ Very tedious to season lamps for spot and group relamping
♦ Lamp life and ballast life can be reduced to just several months of operation
Some lamp lumen depreciation (light loss) over life
♦ 5 to 10 percent
Some warm–up time to full brightness
♦ Instant–on operation, but several minutes to full output

For efficient distinct accent light or focused light, use HIR/MR or CMH PAR lamps.

Net Addresses/LongCompact
See page 8–31

CONNECT FOR MORE

Fluorescent Triphosphor **F40Long/830/RS**
for electronic ballasts ONLY

Bent glass single long tube

Power budgeting
for 1–lamp luminaires

BF 0.98 yields 41 watts

Ballast vendors include Advance, Energy Savings, Inc. (ESI), Magnetek and Motorola. Above factor based on ESI.

■ ■ ■ ■ ■ ■ ■ ■ ■ ■ ■ ■ ■

F40Long Lighting Design Tips
✔ Best in moderate ceiling spaces (10'–6" or less).
✔ Excellent in single–lamp, horizontal–orientation wallwashers or trackheads or monopoints
✔ Great in segmented striplights for circular or curvilinear cove shapes
✔ Great where more decorative lighting techniques are desired but lighting intensities must be significant
✔ Very glary lamp—not recommended for general, direct lighting applications (even in parabolic luminaires)

Fluorescent Triphosphor

8

Application Key

Commercial
Gallery
Hospitality
Institutional
Manufacturing
Residential
Retail
Exterior

bold = primary application
partial fade = minimal application
fade = unlikely application

4 pin base (2G11) dedicated socket for electronic ballasts

Lamp schematic shown about one–third actual size
Image copyright 2000/GE

Uses
Decorative/Functional
 ◆ Coves
 ◆ Pendents
 ◆ Sconces
Merchandise Accenting
Wallwashing
 See Section 11

Fluorescent Triphosphor F50Long/830/RS

for electronic ballasts ONLY

Stats Guide
⊃Longer life is better
⊃Higher efficacy is better
⊃Lower wattage is better
⊃Lower color of light is warmer
⊃High cost range for LongCompact is about US$16.⁰⁰ to US$30.⁰⁰

Stats Data varies manufacturer to manufacturer and changes from time to time

Life/20000 hours
A typical office environment might be occupied about 2600 hours each year.

Efficacy/75 LPW (values include electronic ballast loss)

Wattage and Output/50 watts and 4150 lumens

Color of Light/3000°K (warm tone) and 3500°K (neutral)

Beam spreads/Not Applicable
This is a very high output general service lamp intended for striplights or for wallwashers, relatively large sconces and pendents where an all around glow of light is desired or where a luminaire reflector will provide optical control. Do not use where distinct accent light or focused light is desired—for accent use halogen or HIR PAR lamps (Section 3 and 5), halogen or HIR MR lamps (Sections 4 and 6) or CMH PAR lamps (Section 7).

Size/22½" L by 1¾" W (width is approximate)

Voltage/120V and 277V Ballast dependent (check ballast vendors for other voltages)

Cost Magnitude/High (2000)

Manufacturers
GE
Philips

Net Addresses
http://www.ge.com/Lighting/business/index.htm
http://www.lighting.philips.com/

CONNECT FOR MORE

Advantages
Crisp white light (3000°K)
Dimmable (see disadvantages, however)
Electronic operation yields no flicker and no hum
Instant–on
Much longer life than halogen TB lamps
Better alternative to less efficient standard filament A lamps
Withstand reasonable temperature range

Disadvantages
Ballast required
◆Additional wattage requirement of ballast of 1 to 4 watts (BF of 0.98)
Dimming reduces lamp life and ballast life
◆Unless all lamps are seasoned at full output for 100 hours prior to installation
◆Very tedious to season lamps for spot and group relamping
◆Lamp life and ballast life can be reduced to just several months of operation
Some lamp lumen depreciation (light loss) over life
◆5 to 10 percent
Some warm–up time to full brightness
◆Instant–on operation, but several minutes to full output

Net Addresses/LongCompact
See page 8–31

For efficient distinct accent light or focused light, use HIR/MR or CMH PAR lamps.

Fluorescent
Triphosphor
F50Long/830/RS
for electronic ballasts ONLY

Bent glass single long tube

Power budgeting
for 1–lamp luminaires

BF 0.98 yields 55 watts

Ballast vendors include Advance, Energy Savings, Inc. (ESI), Magnetek and Motorola. Above factor based on ESI.

■ ■ ■ ■ ■ ■ ■ ■ ■ ■ ■ ■

F50Long Lighting Design Tips
✔ Best in moderate to high ceiling spaces (18'–0" or less).
✔ Excellent in single–lamp, horizontal–orientation wallwashers or trackheads or monopoints
✔ Great in segmented striplights for circular or curvilinear cove shapes
✔ Great where more decorative lighting techniques are desired but lighting intensities must be significant
✔ Very glary lamp—not recommended for general, direct lighting applications (even in parabolic luminaires)

4 pin base (2G11) dedicated socket for electronic ballasts

Lamp schematic shown about one–third actual size
Image copyright 2000/GE

Uses
Decorative/Functional
♦ Coves
♦ Pendents
♦ Sconces
Merchandise Accenting
Wallwashing
See Section 11

Application Key

Commercial
Gallery
Hospitality
Institutional
Manufacturing
Residential
Retail
Exterior

bold = primary application
partial fade = minimal application
fade = unlikely application

Fluorescent
Triphosphor

8

Fluorescent/Tubular
Triphosphor
for electronic ballasts ONLY

Fluorescent tubular lamps are the common fluorescent lamp—a long, linear, tubular lamp. In the last fifteen years, tubular (T) lamps have gotten smaller in diameter. This allows for more efficient operation of the lamp (hence increasing light output from the lamp). Further, this results in better optical performance when a reflector is placed around the lamp. An additional benefit is the use of less material resources in making a smaller diameter tube—less glass and less phosphor along with less mercury. Very recently, another step in miniaturization of tubular lamps has occurred—to the point where fluorescent lamps are now available in diameters the

Table 8.4 Deluxe Triphosphor Tubular Lamps

Fluorescent Lamp	Uses	Lamp Profile
T2	New • Architectural details • Cove lights • Display cases	(no art available)
T5/SE	New • Ambient/Recessed ❶ • Ambient/Pendent • Architectural details • Cove lights • Display cases • Wallwashers	
T5/HO	New • Ambient/Pendent • Architectural details • Cove lights • Display cases • Wallwashers	
T8	New • Ambient/Recessed • Ambient/Pendent • Architectural details • Cove lights • Display cases • Wallwashers	

Lamp images courtesy of and copyright by GE 2000.

Fluorescent/Tubular
Triphosphor
for electronic ballasts ONLY

equivalent of a pencil. The smaller diameter lamps are only made possible by using deluxe triphosphor (XT) coatings. Although the T8 versions are available in standard triphosphor, the XT varieties exhibit the best efficacies. Also, since the Tubular lamps are typically used in settings in which Triple and LongCompact fluorescent lamps are used, then to match color characteristics requires the use of XT/Tubular versions. Hence only XT/Tubular lamps are considered here.

These lamps are available in a range of wattages and physical sizes. Because of their small diameters, these lamps can be fit into relatively shallow luminaire designs and details. These lamps are best suited for cove uplighting applications, merchandise accenting and wallwashing applications, display case lighting, task lighting and ambient recessed and pendent lighting.

While these lamps may seem expensive, the overall cost of light over the life of the lamp/lighting system must be considered. These tubular lamps have rated lives of 10,000 to 20,000 hours. This is five to ten times that of many mains voltage halogen and halogen infrared lamps and is two to four times that of many low voltage halogen and halogen infrared lamps.

Ballasts are now readily available to dim all but the T2 versions of these tubular lamps—although lamp and ballast manufacturers must address the drawbacks of lamp seasoning or reduced lamp and ballast life. The color rendering is excellent with XT/Tubular 3000°K lamps approximating halogen in whiteness.

■ ■ ■ ■ ■ ■ ■ ■ ■ ■ ■ ■

XT/Tubular Lamps
❶ Instant–on, but require warm–up to full brightness
❶ Dimmable (except T2)
❶ Electronic operation results in no hum and no flicker
❶ Various sizes/wattages for all applications

XT/Tubular lamp designation

XT/Tubular lamps are typically identified with an alphanumeric designation outlined on page 8–43. Since each manufacturer has slightly different means for lamp designation, the designation method presented here is solely for convenient reference throughout the text. Actual lamp manufacturers' designations will need to be used in final specifications. Perhaps the lamp manufacturers will standardize lamp designations in the near future in an effort to ease lamp cross referencing and specifications.

Fluorescent
Triphosphor

8

Fluorescent/Tubular
Triphosphor

for electronic ballasts ONLY

XT/T2 lamps (metric)

T2 lamps are just ¼ inch in diameter! Metric length means they fit into US Customary–length details quite readily. Since the lamp bulb wall does get hot, however, sufficient airspace around the lamp is necessary. These lamps are well suited for undershelf lighting, display lighting and discrete signage lighting. The higher wattage versions could work well in wallwash applications—but no luminaires are known to exist at press time which take advantage of this lamp's size.

■ ■ ■ ■ ■ ■ ■ ■ ■ ■ ■

XT/T5
❶Standard metric Programmed Start versions recommended
❶Not to be confused with Preheat T5
❶Electronic operation results in no hum and no flicker
❶Various sizes/wattages for all applications

XT/T5 lamps (metric)

XT/T5 lamps are just ⅝ inch in diameter! Metric length means they fit into US Customary–length details quite readily. Since the lamp bulb wall does get hot, however, sufficient airspace around the lamp is necessary. These lamps are sometimes labeled HE for High Efficacy, but this is extraordinarly confusing with the HO version (see below). This XT/T5 lamp is a more efficient light source than the HO version—particularly important when meeting energy codes. These lamps are well suited for ambient pendent indirect lighting, cove lighting details, display lighting, signage lighting and, if properly fitted with lenses/reflectors and low–output or dimmable ballasts, task lighting. Advisory: these XT/T5 metric lamps may also be referred to as Programmed Start, Premium or Rapid Start but are not to be confused with Preheat T5 lamps—which are inefficient and typically flicker and hum.

XT/T5/HO lamps (metric)

T5/HO lamps are just ⅝ inch in diameter! Metric length means they fit into US Customary–length details quite readily. Since the lamp bulb wall does get hot, however, sufficient airspace around the lamp is necessary. HO represents High Output. The standard XT/T5 cited in the previous paragraph is a more efficient light source than the HO version—particularly important when meeting energy codes. The HO lamps provide more light, but consuming significantly more

Fluorescent/Tubular
Triphosphor

for electronic ballasts ONLY

Deluxe Triphosphor Tubular Fluorescent Lamp Designation

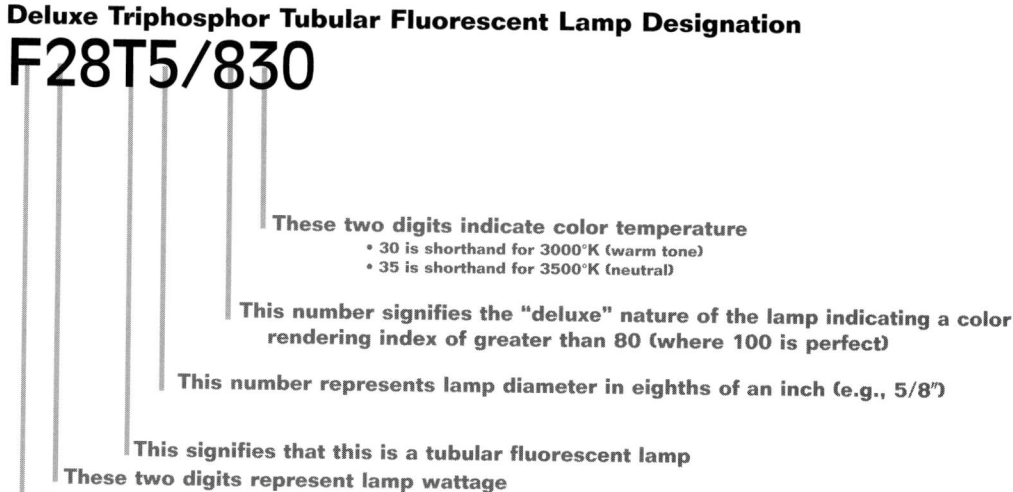

F28T5/830

These two digits indicate color temperature
• 30 is shorthand for 3000°K (warm tone)
• 35 is shorthand for 3500°K (neutral)

This number signifies the "deluxe" nature of the lamp indicating a color rendering index of greater than 80 (where 100 is perfect)

This number represents lamp diameter in eighths of an inch (e.g., 5/8")

This signifies that this is a tubular fluorescent lamp

These two digits represent lamp wattage

Signifies that the lamp is a fluorescent (F) lamp

[This designation used for convenient reference throughout this *Time–Saver Standards for Architectural Lighting*—but not necessarily used by all lamp manufacturers (although it would be convenient if manufacturers would come to agreement on standard designations.]

energy. As such, HO lamps are well suited for high light level applications—manufacturing, clean rooms, laboratory spaces and the like. Unfortunately, while highly touted by lamp, ballast and luminaire manufacturers, the T5HO lamp seems to have little application in spaces with relatively low ceilings, where only moderate ambient light levels are necessary and where energy use is a concern. Certainly applications exist, but they are not mainstream situations.

XT/Tubular lamps for general lighting

XT/Tubular lamps are high lumen lamps. Combined with their small tube diameter, and many of these lamps are very glary when viewed directly. T2 and T5 lamps are not recommended for use in recessed parabolic luminaires for general lighting situations. T5/HO (high output) lamps do work well in pendent uplights where higher–than–normal light intensities are required. In most typical open plan offices, private offices, corridors, reception areas and conference rooms, standard XT/T5 and T8 lamps are better choices. XT/T5 and T8 lamps work well in pendent luminaires and the recently popular recessed/indirect perforated basket luminaires. T8 lamps also work

Cost magnitude[2000]
for deluxe triphoshpor fluorescent lamps

Low
US$3.⁰⁰ to US$8.⁰⁰
Moderate
US$8.⁰⁰ to US$16.⁰⁰
High
US$16.⁰⁰ to US$30.⁰⁰

Costs vary based on quantities, distributor and contractor markups, and market conditions, manufacturing situations and annual inflation. These values are for preliminary, magnitude budgeting and do not represent quotes nor actual final pricing.

Fluorescent
Triphosphor

8

Fluorescent/Tubular
Triphosphor

for electronic ballasts ONLY

well in parabolic luminaires. See Section 9 for application information on XT/T8 lamps in ambient recessed luminaires. See Section 10 for application information on XT/T5, XT/T5/HO and XT/T8 lamps in ambient pendent luminaires.

XT/Tubular lamps for cove lighting

XT/Tubular fluorescent lamps work very well in architectural cove details (see Section 11). The high lumen output of the lamp enables designers to develop architectural cove details which provide sufficient light intensities for circulation, casual work and conversation activity areas. The small diameter of these lamps allows them to be used in simple, low profile striplights and in more optically efficient low profile reflector systems.

XT/Tubular lamps for decorative lighting

XT/Tubular lamps are not good candidates for decorativej lighting. The high lumen output of these lamps combined with their relatively long lengths make them poor candidates for lamping behind decorative/colored acrylic and/or glass materials.

XT/Tubular lamps for wallwashing

The most efficient wallwash lighting is achieved with XT/Tubular lamps in recessed, semi–recessed and surface mounted wallwash luminaires or in wallslots. These lights are rectilinear and generally have asymmetric reflectors and/or baffles with open apertures to direct light onto walls. Wallslot details are shown in Section 11.

XT/Tubular lamps for merchandising

Using the kind of luminaires described for wallwashing, lighting of large merchandise areas or vertical merchandise displays can be quite effective with XT/Tubular lamps. Intensities with the higher wattage T5 and T8 lamps are sufficient for merchandise accenting. The excellent color rendering, energy efficiency and long life exhibited by these lamps are significant benefits for merchandising.

▶**Metric lamps**

Fluorescent/T2
Triphosphor
for electronic ballasts ONLY

Stats Guide
⊃Longer life is better
⊃Higher efficacy is better
⊃Lower wattage is better
⊃Lower color of light is warmer
⊃High cost range for Tubular is about US$16.⁰⁰ to US$30.⁰⁰

Stats
Data varies manufacturer to manufacturer and changes from time to time

Life/10000 hours
A typical office environment might be occupied about 2600 hours each year.

Efficacy/45 LPW
(value includes electronic ballast loss and is average across all wattages)

Wattages/6, 8, 11 and 13
(see Light Output Table opposite page)

Color of Light/3000°K (warm tone)

Beam spreads/Not Applicable
This is a moderate output general service lamp intended for striplights or task lights or where a luminaire reflector will provide optical control. Do not use where distinct accent light or focused light is desired—for accent use halogen or HIR PAR lamps (Section 3 and 5), halogen or HIR MR lamps (Sections 4 and 6) or CMH PAR lamps (Section 7).

Size/Lengths of 8½″, 12½″, 16½″ and 20½″ by ¼″Ø
Lengths approximate. Shortest length corresponds to lowest wattage and so on.

Voltage/120V and 277V
Ballast dependent (check ballast vendors for other voltages)

Cost Magnitude/High⁽²⁰⁰⁰⁾

Manufacturers
Osram Sylvania

Net Addresses
http://ecom.sylvania.com/osicatalog/
CONNECT FOR MORE

Advantages
Crisp white light (3000°K)
Electronic operation yields no flicker and no hum
Instant–on
Much longer life than halogen TB lamps
Better alternative to less efficient standard filament A lamps
Withstand reasonable temperature range

Disadvantages
Ballast required
♦Additional wattage requirement of ballast of 3 to 4 watts (BF of 0.98)
♦Relative to lamp size, ballast is very large
Some lamp lumen depreciation (light loss) over life
♦5 to 10 percent
Some warm–up time to full brightness
♦Instant–on operation, but several minutes to full output
Not dimmable
Bulb wall at lamp ends is hot during operation

For efficient distinct accent light or focused light, use HIR/MR or CMH PAR lamps.

⚡

Fluorescent/T2
Triphosphor

for electronic ballasts ONLY

▶Metric lamps

No art available

Net Addresses/T2 Striplights and Display Case Lights
http://www.bartcolighting.com/index2.html
http://www.danalite.com/spec/danalite/danaindx.htm

CONNECT FOR MORE

Net Addresses/T2 Undershelf Lights
http://www.alkco.com/
http://www.bartcolighting.com/index2.html
http://www.danalite.com/spec/danalite/danaindx.htm

CONNECT FOR MORE

Net Addresses/Ballasts
http://www.sylvania.com/ballast/phops.htm

CONNECT FOR MORE

Power budgeting
for 1–lamp luminaires

BF 0.98 for 6W yields 9 watts
BF 0.98 for 8W yields 11 watts
BF 0.98 for 11W yields 14 watts
BF 0.98 for 13W yields 17 watts

Ballast vendors include Energy Savings, Inc. (ESI) and Osram Sylvania. Above factors based on ESI.

Light output

6W yields 310 lumens
8W yields 500 lumens
11W yields 680 lumens
13W yields 860 lumens

Fluorescent
Triphosphor

8

T2 Lighting Design Tips
✔Best in small luminaires/limited space applications
✔Excellent in single–lamp, horizontal–orientation striplights or mini–reflectors with remote ballasts
✔Metric length allows excellent fit into US Customary architectural openings, millwork and details

Uses

Decorative/Functional
◆Coves
◆Undershelf

Merchandise Accenting
◆Display cases

Task Lighting

Wallwashing

Application Key

Commercial
Gallery
Hospitality
Institutional
Manufacturing
Residential
Retail
Exterior

bold = primary application
partial fade = minimal application
fade = unlikely application

▶**Metric lamps**

Fluorescent/T5
Triphosphor

for electronic ballasts ONLY

(vertical sidebar) **Fluorescent** Triphosphor **8**

Stats<small>Data varies manufacturer to manufacturer and changes from time to time</small>

Stats Guide
⊃Longer life is better
⊃Higher efficacy is better
⊃Lower wattage is better
⊃Lower color of light is warmer
⊃Moderate cost range for Tubular is about US$8.⁰⁰ to US$16.⁰⁰

Life/20000 hours
A typical office environment might be occupied about 2600 hours each year.

Efficacy/86 LPW<small>(value includes electronic ballast loss and is average across all wattages)</small>

Wattages/14, 21, 28 and 35<small>(see Light Output Table opposite page)</small>

Color of Light/3000°K (warm tone) and 3500°K (neutral)

Beam spreads/Not Applicable
This is a high output general service lamp intended for striplights or for wallwashers, recessed/indirect (perforated basket) luminaires, pendents or other lights where a luminaire reflector will provide optical control. Do not use where distinct accent light or focused light is desired—for accent use halogen or HIR PAR lamps (Section 3 and 5), halogen or HIR MR lamps (Sections 4 and 6) or CMH PAR lamps (Section 7).

Size/Lengths of 22″, 34″, 46″ and 58″ by ⅝″⌀
Lengths approximate. Shortest length corresponds to lowest wattage and so on.

Voltage/120V and 277V<small>Ballast dependent (check ballast vendors for other voltages)</small>

Cost Magnitude/High⁽²⁰⁰⁰⁾

Manufacturers
Osram Sylvania
Philips

Net Addresses
http://ecom.sylvania.com/osicatalog/
http://www.lighting.philips.com/

CONNECT FOR MORE

Advantages
Crisp white light (3000°K)
Dimmable (see disadvantages, however)
Electronic operation yields no flicker and no hum
Instant–on
Much longer life than halogen TB lamps
Better alternative to less efficient standard filament A lamps
Withstand reasonable temperature range

Disadvantages
Ballast required
◆Additional wattage requirement of ballast of 3 to 6 watts (BF of 0.98)
◆Relative to lamp size, ballast is very large
Dimming may reduce lamp life and ballast life
◆Unless all lamps are seasoned at full output for 100 hours prior to installation
◆Very tedious to season lamps for spot and group relamping
◆Lamp life and ballast life can be reduced to just several years of operation
Some lamp lumen depreciation (light loss) over life
◆5 percent
Some warm–up time to full brightness
◆Instant–on operation, but several minutes to full output
Bulb wall at lamp ends is hot during operation

For efficient distinct accent light or focused light, use HIR/MR or CMH PAR lamps.

Net Addresses/Ballasts and Luminaires
See Section 10 throughout

CONNECT FOR MORE

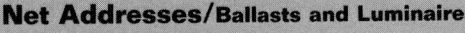

Fluorescent/T5
Triphosphor
for electronic ballasts ONLY

▼Metric lamps

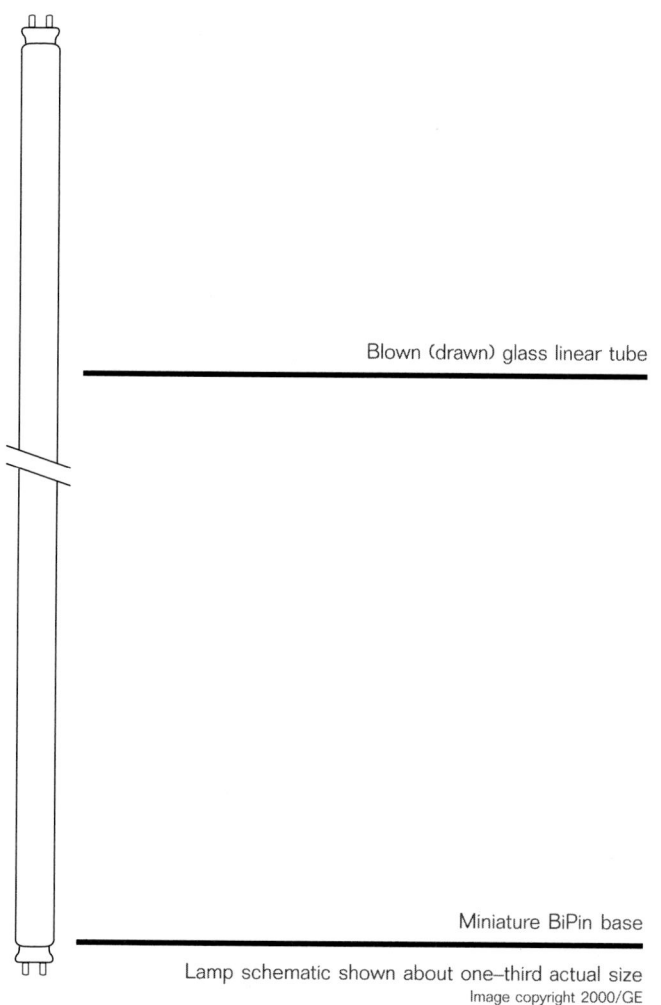

Blown (drawn) glass linear tube

Miniature BiPin base

Lamp schematic shown about one–third actual size
Image copyright 2000/GE

Power budgeting
for 1–lamp luminaires

BF 0.98 for 14W yields 17 watts
BF 0.98 for 21W yields 24 watts
BF 0.98 for 28W yields 33 watts
BF 0.98 for 35W yields 41 watts

Ballast vendors include Energy Savings, Inc. (ESI) and Osram Sylvania. Above factors based on ESI. 2–lamp ballasts available for all but the 35 watt lamp.

Light output

14W yields 1350 lumens
21W yields 2100 lumens
28W yields 2900 lumens
35W yields 3650 lumens

Fluorescent
Triphosphor

8

Uses

Decorative/Functional
◆Coves
◆Undershelf

General Lighting
◆Pendent (see Section 10)

Merchandise Accenting
◆Display cases

Task lighting
Wallwashing

■ ■ ■ ■ ■ ■ ■ ■ ■ ■ ■ ■ ■

T5 Lighting Design Tips
✔Best in small luminaires/limited space applications
✔Excellent in single–lamp, horizontal–orientation striplights or mini–reflectors with remote ballasts
✔Metric length allows excellent fit into US Customary architectural openings, millwork and details

Application Key

Commercial
Gallery
Hospitality
Institutional
Manufacturing
Residential
Retail
Exterior

bold = primary application
partial fade = minimal application
fade = unlikely application

▶ Metric lamps
▶ Very high output

Fluorescent/T5/HO
Triphosphor
for electronic ballasts ONLY

Fluorescent
Triphosphor

8

For efficient distinct accent light or focused light, use HIR/MR or CMH PAR lamps.

Stats Guide
⊃ Longer life is better
⊃ Higher efficacy is better
⊃ Lower wattage is better
⊃ Lower color of light is warmer
⊃ High cost range for Tubular is about US$16.⁰⁰ to US$30.⁰⁰

Stats
Data varies manufacturer to manufacturer and changes from time to time

Life/20000 hours
A typical office environment might be occupied about 2600 hours each year.

Efficacy/81 LPW
(value includes electronic ballast loss and is average across all wattages)

Wattages/24, 39 and 54
(see Light Output Table opposite page)

Color of Light/3000°K (warm tone) and 3500°K (neutral)

Beam spreads/Not Applicable
This is a VERY high output general service lamp intended for striplights or for wallwashers and pendents or other lights where a luminaire reflector will provide optical control. Standard XT/T5 lamp on previous double page spread is more efficacious and more appropriate for most commercial, institutional and retail applications.

Size/Lengths of 22″, 34″ and 46″ by ⅝″Ø
Lengths approximate. Shortest length corresponds to lowest wattage and so on.

Voltage/120V and 277V
Ballast dependent (check ballast vendors for other voltages)

Cost Magnitude/High⁽²⁰⁰⁰⁾

Manufacturers
Osram Sylvania
Philips

Net Addresses
http://ecom.sylvania.com/osicatalog/
http://www.lighting.philips.com/

CONNECT FOR MORE

Advantages
Crisp white light (3000°K)
Dimmable (steep premium, but lamp too bright for most applications)
Electronic operation yields no flicker and no hum
Instant–on
Much longer life than halogen TB lamps
Better alternative to less efficient standard filament A lamps
Withstand reasonable temperature range

Disadvantages
Ballast required
◆Additional wattage requirement of ballast of 3 to 6 watts (BF of 0.98)
◆Relative to lamp size, ballast is very large
Dimming may reduce lamp life and ballast life
◆Unless all lamps are seasoned at full output for 100 hours prior to installation
◆Very tedious to season lamps for spot and group relamping
◆Lamp life and ballast life can be reduced to just several years of operation
Too much light for most applications/expending too much energy
Some lamp lumen depreciation (light loss) over life
◆5 percent
Some warm–up time to full brightness
◆Instant–on operation, but several minutes to full output
Bulb wall at lamp ends is quite hot during operation

Net Addresses/Ballasts and Luminaires
See Section 10 throughout

CONNECT FOR MORE

Fluorescent/T5/HO
Triphosphor
for electronic ballasts ONLY

▼Metric lamps
▼Very high output

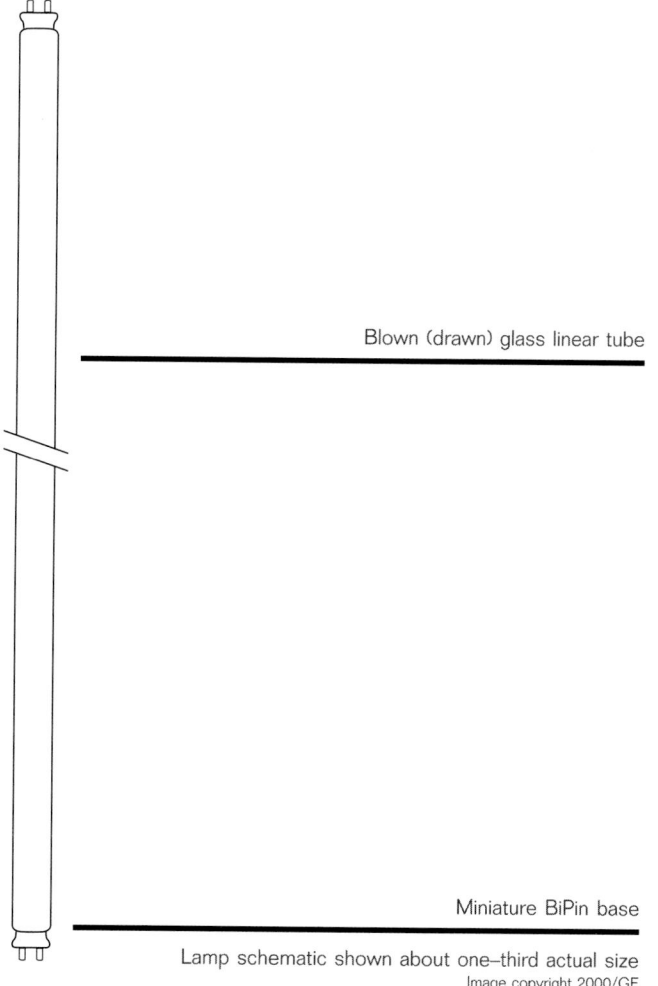

Blown (drawn) glass linear tube

Miniature BiPin base

Lamp schematic shown about one–third actual size
Image copyright 2000/GE

Power budgeting
for 1–lamp luminaires

BF 0.98 for 24W yields 28 watts
BF 0.98 for 39W yields 42 watts
BF 0.98 for 54W yields 64 watts

Ballast vendors include Energy Savings, Inc. (ESI) and Osram Sylvania. Above factors based on ESI. 2–lamp ballasts also available.

Light output

24W yields 1950 lumens
39W yields 3500 lumens
54W yields 5000 lumens

Fluorescent
Triphosphor

8

Uses

(Limited) Decorative/Functional
♦Coves

(Limited) General Lighting
♦Pendent (see Section 10)

Merchandise Accenting

Wallwashing

■■■■■■■■■■■■■

T5 /HOLighting Design Tips
✔Best in small luminaires/limited space applications
✔Excellent in single–lamp, horizontal–orientation striplights or mini–reflectors with remote ballasts
✔Metric length allows excellent fit into US Customary architectural openings, millwork and details
✔Requires glare control and may require dimming for most applications

Application Key

Commercial
Gallery
Hospitality
Institutional
Manufacturing
Residential
Retail
Exterior

bold = primary application
partial fade = minimal application
fade = unlikely application

Triphosphor
Fluorescent/T8
for electronic ballasts ONLY

Fluorescent
Triphosphor

8

Stats Data varies manufacturer to manufacturer and changes from time to time

Life/20000 hours
A typical office environment might be occupied about 2600 hours each year.

Efficacy/79 LPW (value includes electronic ballast loss and is average across all wattages)

Wattages/17, 25 and 32 (see Light Output Table opposite page)

Color of Light/3000°K (warm tone) and 3500°K (neutral)

Beam spreads/Not Applicable
This is a high output general service lamp intended for striplights or for wallwashers, recessed/indirect (perforated basket) luminaires, pendents or other lights where a luminaire reflector will provide optical control. Do not use where distinct accent light or focused light is desired—for accent use halogen or HIR PAR lamps (Section 3 and 5), halogen or HIR MR lamps (Sections 4 and 6) or CMH PAR lamps (Section 7).

Size/Lengths of 24″, 36″ and 48″ by 1″Ø
Lengths approximate. Shortest length corresponds to lowest wattage and so on.

Voltage/120V and 277V Ballast dependent (check ballast vendors for other voltages)

Cost Magnitude/Low⁽²⁰⁰⁰⁾

Manufacturers
GE
Osram Sylvania
Philips

Net Addresses
http://www.ge.com/lighting/business/index.htm
http://ecom.sylvania.com/osicatalog/
http://www.lighting.philips.com/

CONNECT FOR MORE

Advantages
Crisp white light (3000°K)
Dimmable (see disadvantages, however)
Electronic operation yields no flicker and no hum
Instant–on
Much longer life than halogen TB lamps
Better alternative to less efficient standard filament A lamps
Withstand reasonable temperature range

Disadvantages
Ballast required
◆Additional wattage requirement of ballast of 3 to 4 watts (BF of 0.98)
Dimming may reduce lamp life and ballast life
◆Unless all lamps are seasoned at full output for 100 hours prior to installation
◆Very tedious to season lamps for spot and group relamping
◆Lamp life and ballast life can be reduced to just several years of operation
Some lamp lumen depreciation (light loss) over life
◆5 to 10 percent
Some warm–up time to full brightness
◆Instant–on operation, but several minutes to full output

For efficient distinct accent light or focused light, use HIR/MR or CMH PAR lamps.

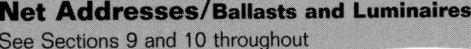

Net Addresses/Ballasts and Luminaires
See Sections 9 and 10 throughout

Fluorescent/T8
Triphosphor
for electronic ballasts ONLY

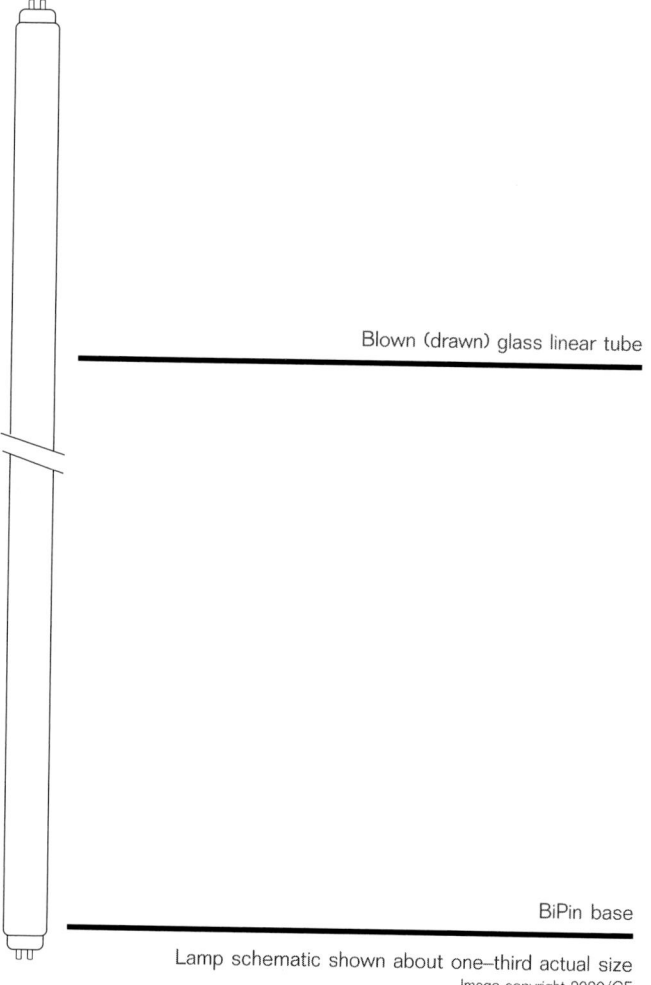

Blown (drawn) glass linear tube

BiPin base

Lamp schematic shown about one–third actual size

Image copyright 2000/GE

Power budgeting
for 1–lamp luminaires

BF 0.92 for F17T8 yields 18 watts
BF 0.98 for F25T8 yields 28 watts
BF 0.98 for F32T8 yields 35 watts

Ballast vendors include Advance, Energy Savings, Inc. (ESI), Magnetek, Motorola. Not all ballast factors available from all ballast vendors. 2–lamp ballasts typically use fewer watts and are therefore more efficient. Above factors based on Magnetek. Other ballast factors cited in Sections 9 and 10 as necessary to achieve lighting criteria.

Light output

17W yields 1385 lumens
25W yields 2200 lumens
32W yields 2950 lumens

Fluorescent
Triphosphor

8

Uses

Decorative/Functional
- ◆Coves
- ◆Undershelf

General Lighting
- ◆Pendent (see Section 10)
- ◆Recessed (see Section 9)

Merchandise Accenting
- ◆Display cases

Task lighting

Wallwashing

Application Key

Commercial
Gallery
Hospitality
Institutional
Manufacturing
Residential
Retail
Exterior

bold = primary application
partial fade = minimal application
fade = unlikely application

■ ■ ■ ■ ■ ■ ■ ■ ■ ■ ■ ■
T8 Lighting Design Tips
✔ Good general lighting light source—low cost and high availability
✔ Excellent in dimming situations—dimming ballasts readily available and reasonably priced

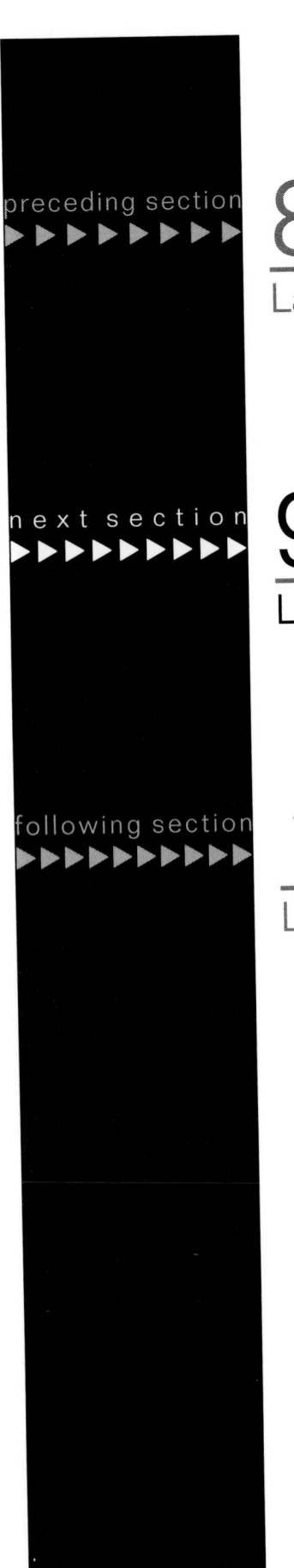

Ambient

Recessed Fluorescent Lighting for Commercial Facilities

9

This section outlines recessed ambient lighting using fluorescent lamps. Various lamp/luminaire combinations are reviewed for typical commercial open office lighting applications and schematic layouts are offered for design development consideration.

Ambient
Recessed Fluorescent Lighting for Commercial Facilities

The most energy efficient lighting installations are based on a balance of three lighting layers—ambient lighting, task lighting and accent lighting.

Ambient (or general) lighting can be achieved through a variety of lighting techniques. Ambient lighting is a strategic method for meeting code restrictions on energy use in commercial facilities. When used with well balanced and equally energy efficient task lighting and accent lighting layers, ambient lighting will be successful for users/workers as well as for facility owners and managers.

This section addresses ambient lighting for commercial facilities from recessed fluorescent lighting equipment. Since ambient lighting works only in combination with task and accent lighting, some discussion is offered here for choosing successful task and accent lighting.

The latest energy standards (e.g., California Title 24/1999 and ASHRAE/IESNA 90.1/1999) limit connected loads for lighting in most office spaces to between 1.1 and 1.4 watts per square foot (w/sf). As such, ambient lighting might typically be designed to 0.7 to 1.0 w/sf. Task lighting might be designed to 0.25 w/sf and accent lighting might be designed to 0.25 w/sf. While the designer has flexibility to trade-off between ambient, task and accent lighting, the range is relatively narrow. Clearly, energy efficient sources must be used in energy efficient luminaires and with little tolerance in calculated quantities (so–called design safety factors should be limited to the barest).

Lighting layers

For office environments of the day, electronic tasks are primary with paperwork as secondary or equal. So, ambient lighting is typically developed to address the lowest common denominator—lighting intensities and brightnesses which allow best viewing conditions for computer tasks. Task lighting is then used to supplement the ambient lighting so that the aggregate lighting intensities—lighting is additive, hence aggregate lighting is the combined intensities of ambient and task—provide good viewing conditions for paperwork. Finally, to maintain some brightness balance between paper tasks and four–color, bright background computer screens and distant architectural surfaces, accent lighting is layered into the overall lighting plan. Task lighting is further discussed in the following paragraphs. Accent lighting is well documented in Sections 3, 4, 5 and 6.

Local task lighting

Local task lighting is that lighting which is generated by a localized luminaire (in close proximity to the task or work area) and which

Ambient
Recessed Fluorescent Lighting for Commercial Facilities

supplements the ambient lighting to a level sufficient for performing the tasks at hand. When carefully planned, task and ambient lighting can combine to provide both effective, comfortable lighting for work and good energy efficiencies. Connected loads alone can be reduced with these techniques. Localized task lights on workstation motion sensors can further reduce energy consumption.

Task lighting is achieved one of three ways (or a combination of these). Binder bin mounted, partition or rail mounted, and freestanding. Where binder bins exist over work or reference surfaces, task lights should be integrated into the bottom of the binder bins. This eliminates harsh shadows, dingy work zones and transient adaptation effects folks can encounter when viewing from brighter areas to darker areas. Table 9.1 outlines key suggestions for planning and implementing task lights under binder bins. Partition or rail mounted lights are appropriate where work and reference surfaces are bounded by 54–inch or taller partitions and where no binder bins

Table 9.1 Binder Bin Mounted Local Task Lighting

Task Lighting	Suggestions
Binder bin mount • T5 or T8 Linear fluorescent • Reduced output or dimmable❶ • Optically designed❷	• Use wherever binder bin occurs and work is likely to take place • Additional average illuminance from task light over task zone should be 25 to 35 fc • Additional maximum illuminance from task light should not exceed 50 fc • Task light intensity gradient should increase from front to back of worksurface • Use integral switch on task light located convenient to user • Use local motion sensor(s) to control task lights on a workstation level • Total connected load for all task lighting is typically 0.25 w/sf over entire office space • Review cord management and power plug–in locations for task lights • Use lights which integrate into binder bin bottom to minimize clutter and optical issues • Use electronic ballasts (50 to 65 percent output; two– or three–level switching or dimmable) • Use lamps of same color temperature and color rendering as ambient lighting • Use lamps of same wattage as ambient, if practical • Use only newer technology triphosphor rapid start T5 or T8 lamps • Do not use old–style preheat or switch start T5 or T8 lamps • Do not use sealed task lights with no obvious means of changing ballasts • Do not use striplights • Do not use unlensed white boxes • Avoid anything with aperture width of less than 2½ inches • Check warranty on coverage and limitations • Use only UL or ETL labeled products tested/listed to UL requirements • Pricing through electrical distributor/contractor may be most competitive

❶On higher wattage units (exceeding 18 watts), consider reduced output versions (2– or 3–level switching) or dimmable versions. Otherwise light output is too great, resulting in reflected glare, veiling reflections and transient adaptation effects (which occur when changing view from brighter areas to darker areas).
❷Specify units which have been optically designed with appropriate reflector and/or lens media to best direct the light onto task areas in a soft, uniform manner. Bare lamps in simple white or glossy aluminum metal boxes should be avoided as these create glare and do not cover a broad area softly. Standard stippled lenses should be avoided as these do not appropriately distribute light to the back tack surface nor across the worksurface.

Ambient
Recessed Fluorescent Lighting for Commercial Facilities

exist. Table 9.2 offers suggestions for planning task lights on partitions or rails. Freestanding task lights are intended for work or reference surfaces which are not bounded by tall partitions or which are themselves freestanding (e.g., tables, desks, peninsula tops). See Table 9.3 for suggestions on planning freestanding task lights.

With any task lighting technique, the key is not to overlight or highly focus (pinspot) the task area. This kind of brightness exacerbates adaptation (adapting from one brightness to another) as workers scan the work setting. Further, no consideration should be given to incandescent, halogen or xenon lamp task lights. These task lights are inefficient (and will not permit compliance with energy codes), hot to touch, generate excessive heat and are typically quite glary. If such a light were to tip over, it could result in fire. While these task lights may offer a "fashion statement" or "eye candy," they do not offer users convenient, functional, efficient lighting. Further, lamp life is short with incandescent and halogen lamp task lights.

Table 9.2 Partition or Rail Mount Task Lighting

Task Lighting	Suggestions
Partition or rail mount • T5 or T8 Linear fluorescent • Reduced output or dimmable❶ • Optically designed❷	• Use wherever no binder bin occurs and work is likely to take place • Additional average illuminance from task light over task zone should be 20 to 30 fc • Additional maximum illuminance from task light should not exceed 45 fc • Task light intensity gradient should increase from front to back of worksurface • Use integral switch on task light located convenient to user • Use local motion sensor(s) to control task lights on a workstation level • Total connected load for all task lighting is typically 0.25 w/sf over entire office space • Review cord management and power plug–in locations for task lights • Use lights which integrate into partition or auxiliary accessories rail to minimize clutter • Use electronic ballasts (50 to 65 percent output; two– or three–level switching or dimmable) • Use lamps of same color temperature and color rendering as ambient lighting • Use lamps of same wattage as ambient, if practical • Use only newer technology triphosphor rapid start T5 or T8 lamps • Do not use old–style preheat or switch start T5 or T8 lamps • Do not use sealed task lights with no obvious means of changing ballasts • Do not use striplights • Do not use unlensed white boxes • Avoid anything with aperture width of less than 2½ inches • Check warranty on coverage and limitations • Use only UL or ETL labeled products tested/listed to UL requirements • Pricing through electrical distributor/contractor may be most competitive

❶On higher wattage units (exceeding 18 watts), consider reduced output versions (2– or 3–level switching) or dimmable versions. Otherwise light output is too great, resulting in reflected glare, veiling reflections and transient adaptation effects (which occur when changing view from brighter areas to darker areas.
❷Specify units which have been optically designed with appropriate reflector and/or lens media to best direct the light onto task areas in a soft, uniform manner. Bare lamps in simple white or glossy aluminum metal boxes should be avoided as these create glare and do not cover a broad area softly.

Ambient
Recessed Fluorescent Lighting for Commercial Facilities

Table 9.3 Freestanding Task Lighting

Task Lighting	Suggestions
Freestanding • Small compact fluorescent • T2 or T5 Linear fluorescent • Reduced output or dimmable❶ • Optically designed❷	• Use wherever binder bin and/or partition– or rail–mount lights cannot be accommodated • Additional average illuminance from task light over task zone should be 20 to 30 fc • Additional maximum illuminance from task light should not exceed 45 fc • Use integral switch on task light located convenient to user • Use local motion sensor(s) to control task lights on a workstation level • Total connected load for all task lighting is typically 0.25 w/sf over entire office space • Review cord management and power plug–in locations for task lights • Review ballast location (sometimes cheap ballasts are mounted on end of cord causing wiring problems through grommets or furniture wire management systems) • Use electronic ballasts (50 to 65 percent output; two– or three–level switching or dimmable • Use lamps of same color temperature and color rendering as ambient lighting • Use only newer technology triphosphor lamps • Do not use old–style preheat or switch start T5 or T8 lamps • Do not use sealed task lights with no obvious means of changing ballasts • Do not use unlensed white boxes • Check warranty on coverage and limitations • Use only UL, ETL, or other recognized lab–labeled products tested/listed to UL requirements • Pricing through electrical distributor/contractor may be most competitive

❶On higher wattage units (exceeding 18 watts), consider reduced output versions (2– or 3–level switching) or dimmable versions. Otherwise light output is too great, resulting in reflected glare, veiling reflections and transient adaptation effects (which occur when changing view from brighter areas to darker areas.
❷Specify units which have been optically designed with appropriate reflector and/or lens media to best direct the light onto task areas in a soft, uniform manner. Bare lamps in simple white or glossy aluminum metal boxes should be avoided as these create glare and do not cover a broad area softly.

Task lighting purchasing chain

Purchasing task lights can be problematic and this issue deserves the designer's and facility manager's attention. In general, furniture manufacturers have available binder bin mounted task lights and/or partition mount task lights and/or freestanding task lights for their respective furniture systems. Unfortunately, many of these lights are inferior. Either one, several or all of the following parameters are inappropriately addressed or not addressed by furniture manufacturers—photometric distribution and intensity; lamping; ballasting; range of lamp and ballast options to meet task requirements for specific situations and to meet the needs of the facility's lamp/ballast stock; variety of lengths of binder bin lights necessary to provide relatively continuous coverage over worksurfaces. When good quality task lights are available, these may be priced quite high as "specials" or "high–end features." Unfortunately, manufacturers' list pricing, net pricing (net to whom—distributor, dealer, contractor, retailer, etc.) and actual costs vary quite dramatically. As much as 100 percent or more in markups are not uncommon. As such, consideration might be given to seeking alternative quotes for task lights from both the electrical and the furniture distribution chains.

Ambient
Recessed Fluorescent Lighting for Commercial Facilities

Figure 9.1
Single–lamp, F32T8/830, VDT–sensitive, 1x4 parabolic luminaires with 9 cell baffle are used to provide ambient light in this open office where paper and computer tasks are equally important. Task lighting under the binder bins is necessary to avoid high–contrast shadowing and dingy work areas. Total connected load is 1.17 w/sf.

Ambient
Recessed Fluorescent Lighting for Commercial Facilities

Figure 9.2
This illustrates the pattern of 1x4 parabolic luminaires—continuous rows on 8– to 10–foot centers (depending on the location of the workstations, which are essentially fixed in place). Parabolic luminaires are Lithonia Optimax® (VDT sensitive). Wall lighting is used to balance task brightness and enhance impressions of spaciousness and brightness. Wallwash luminaires are Columbia Parawash (each using (1) F18Long/830/RS lamp).

Image copyright 2000 Robert Eovaldi

Ambient
Recessed Fluorescent Lighting for Commercial Facilities

Vendors not affiliated with the furniture manufacturers and which have available some quality versions of task lights include Alkco, Garcy/SLP, Lithonia/Peerless, Luxo and Waldmann. Steelcase Inc. is a furniture manufacturer which offers some quality task lights (the Advanced Model Shelf Light with 50 percent output ballast or with dimming ballast) along with readily available data "open" to anyone interested in the product, not just to company sales-people or to select architects/interior designers.

Ambient recessed fluorescent lighting

Fluorescent lamps are most appropriate for energy efficient ambient lighting. T8 fluorescent lamps are, at this time, considered the most appropriate lamp for direct ambient lighting from recessed luminaires. These lamps (linear and U–shaped T8—see Section 8), exhibit direct brightnesses which are at the limit of comfort for most users working on most tasks. Recessed fluorescent luminaires include lensed units, parabolic louvered units and the newer recessed "indirect" or perfo-rated "basket" luminaires. At this time, only parabolic louvered lumi-naires meet criteria considered appropriate to large, open workspaces in which people use computers extensively. The recessed "indirect" units only fair well in private offices and are not covered here. Lensed units are extraordinarily glary and inefficient and are not covered here.

Ambient lighting is most efficient in relatively large, open areas where partitions are low and where room finish reflectances are high. Large open areas (greater than 900 square feet (sf)) take better advantage of the overlapping light effects of many luminaires. Smaller rooms exhibit relatively more wall area compared to ceiling area—and the walls ab-sorb light. Nevertheless, the concept of lower ambient light with local-ized task light works in all sizes of office spaces.

Luminaires using one or two lamps of typically available sizes and wattages have a greater chance of meeting most or all of the lighting criteria outlined in Table 9.4. Many 3– and 4–lamp luminaires are simply too intense—creating glare and actually consuming more wattage than necessary to light typical open office areas. This is most evident where partitions and/or overhead storage create shadows, thereby requiring the addition of task lighting and the consumption of energy for that task lighting. Hence, large, 3– and 4–lamp, high–wattage luminaires are not included here—analyses show that these luminaires will not meet the criteria outlined in Table 9.4.

Lower partitions result in less light–absorbing surface area. This is an especially significant issue with lower ceilings. In spaces with 10–foot

Ambient
Recessed Fluorescent Lighting for Commercial Facilities

ceiling heights, 66–inch partitions are not as problematic as they would be in spaces with 8– or 9–foot ceiling heights. To some extent, this problem can be reduced with light colored partitions. If 66–inch partitions are necessary in a space with 9–foot ceilings, then consider selecting very light colored partitions.

Room surface finishes greatly influence lighting efficiencies. Horizontal surfaces tend to have the greatest impact on efficiency in larger open spaces. In small spaces, the vertical surfaces have a significant impact on lighting efficiencies. New ceiling manufacturing techniques and coatings now allow for ceilings of 89 percent reflectance. Providing all other room surfaces are kept light in color, this is a preferred reflectance for ceilings. Wall reflectances should range from 30 percent to 60 percent. Furniture partitions should range from 30 percent to 50 percent. Floors should never fall below 10 percent reflectance. All of these surfaces' reflectances will greatly improve the efficiency and effectiveness of ambient electric lighting and also of daylighting.

Ambient lighting criteria

It is undoubtedly difficult to remain focused on why a building is being newly constructed, renovated or retrofitted. Many times cost issues or maintenance issues drive a design. While these are important, they should not overshadow that the building will be used to further commerce and the life experience. As such, attention should be given to meeting the users' needs. For ambient lighting in today's commercial facilities where people work extensively on computers, lighting criteria should reduce or eliminate glare reflections from computer monitors; reduce or eliminate the veil of light on computer monitors; and reduce or eliminate harsh contrasts.

Several criteria for recessed ambient lighting for typical open office areas in commercial facilities, outlined in Table 9.4, are horizontal illuminance and uniformity; vertical illuminance and luminaire luminance. As noted previously, connected load or power budget is also a requisite criterion. Figure 9.1 illustrates an installation where connected load is 1.17 w/sf while meeting criteria outlined in Table 9.4. Success depends on luminaire selection and layout. Lamping also influences success— typically lower lumen packages (relatively low output lamps or few lamps per luminaire to minimize glare while using more luminaires for uniformity) are most successful.

Based on the criteria in Table 9.4 and on the recessed luminaires typically available on the market, Concept Starters™ for ambient lighting in commercial facilities are offered on the following pages. With each Concept Starter™, a product specification is outlined along with

Room surface finishes significantly influence lighting efficiencies. Lighter finishes can effectively reduce lighting energy by as much as 20 percent.

Ambient

Recessed Fluorescent Lighting for Commercial Facilities

Table 9.4 Ambient Lighting Criteria for Concept Starters™

Lighting	Horizontal Illuminance❶❷	Uniformity❸	Vertical Illuminance❹	Luminaire Luminance❺	Power Budget
Open Plan Office					
• Paper tasks	35 fc	• avg:min of 2:1	≤20 fc	250 fL at 55°	≤1.0 w/sf❼
• VDT tasks		• avg:max of 1:1.3			
• Ballast factor as required					
• Lamp lumen depreciation 0.95					
• Luminaire dirt depreciation 0.95					
• Room surface dirt depreciation 0.95					
• Including partition factor❻					
• Task lighting anticipated					

❶This represents just the amount of light from the ambient lighting. Total light on task area should average 50 fc, with maximum of 75 fc.
❷Criteria to be calculated average, maintained over the worksurface, including effects of partitions All calculations performed for Concept Starters included effects of workstation partitions of heights described for respective Concept Starter™ layout. Workstation density was set at rows of 6x8 workstations back–to–back with 4–foot aisles. Room size was set at 50 feet by 50 feet by ceiling height described in respective Concept Starter™ layout. Compliance was achieved if calculated values are within 15 percent of criteria.
❸Criteria to be uniformity of the horizontal illuminance calculated over the whole room (to accommodate flexibility).
❹Criteria to be calculated average, maintained at 4 feet above finished floor in four cardinal viewing directions—related to computer screen viewing.
❺Criteria to be calculated/photometered in the lab average luminance of exposed lamp and/or reflector at specified viewing angle—related to direct glare (much higher than this value, and reflector will become uncomfortably glary).
❻Partition heights, color and density will affect light losses. Typically, partition factors range between 0.85 and 0.60 (light losses of 15% to 40%).
❼Based on latest proposed ASHRAE/IES 90.1/1999. This is a rough average of what power budget likely needs to be to meet the code, but this can vary based on overall building use, size and lighting layouts. Solely for ambient lighting. Task lighting typically adds 0.25 w/sf.

Criteria references
1. M.S. Rea, ed., Lighting Handbook Reference and Application (New York: Illuminating Engineering Society of North America, 2000).
2. American National Standards Institute/Illuminating Engineering Society of North America. American National Standard Practice for Office Lighting (RP–1) (New York: Illuminating Engineering Society of North America, 1993.
3. Illuminating Engineering Society of North America. RP–24: IESNA Recommended Practice for Lighting Offices Containing Computer Visual Display Terminals (New York: Illuminating Engineering Society of North America, 1990).
4. Gary R. Steffy, Lighting the Electronic Office (New York: Van Nostrand Reinhold, 1995).
5. Gary R. Steffy, Architectural Lighting Design (New York: Van Nostrand Reinhold, 1990).

design tips, power budgeting and cost magnitude. Tables 9.5 through 9.17 are a quick reference to the variety of luminaires reviewed as Concept Starters™ and their suggested application and layout options. Concept Starters™ are just that. These are not recommended final layouts. Actual project conditions must be considered and final calculations with intended–to–be–specified luminaires under these actual conditions need to be made prior to contract document layout and specification. Lumen Micro™ 6.0 and 7.5 software was used for the calculations. Typical large–room setups consisted of a footprint of 50 feet by 50 feet with small partitioned workstations (see Figure 9.3). Ceiling heights varied as outlined in each table shown for each luminaire layout, as did room surface reflectances. Lamp lumen depreciation (LLD) was set at 0.95. Luminaire dirt depreciation

Ambient
Recessed Fluorescent Lighting for Commercial Facilities

(LDD) was set at 0.95. Room surface dirt depreciation (RSDD) was set at 0.95. Ballast factors were set as needed to meet criteria and these are cited in the appropriate tabular summaries so that the most efficient system can be specified.

An excerpted example of a Concept Starter™ follows on the next two pages with annotations about the information presented therein. Concept Starters™ can be used to:

▶ establish a preliminary lighting plan
▶ establish a preliminary cost budget for lighting hardware
▶ establish a preliminary power budget for code compliance
▶ establish a preliminary ceiling height
▶ establish a direction on partition heights
▶ establish a direction on surface finishes
▶ compare various direct and pendent systems quickly

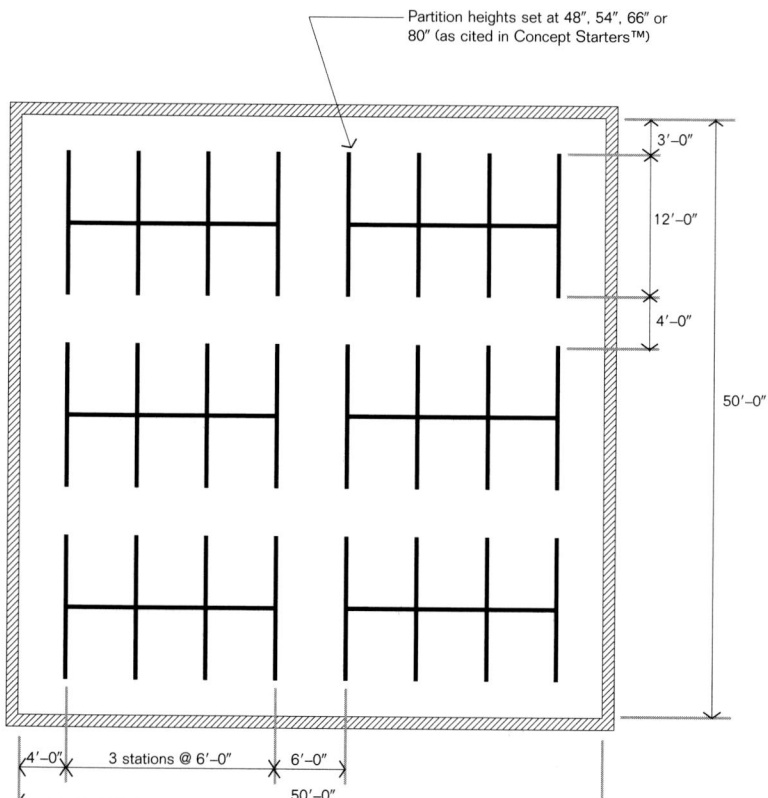

Partition heights set at 48″, 54″, 66″ or 80″ (as cited in Concept Starters™)

Figure 9.3
This illustrates the open office room plan that was used as a basis for all lighting analyses. Partition heights, ceiling heights and surface reflectances were varied to provide a large cross section of situations with various luminaire options. See Tables 9.5 through 9.17 for a summary of luminaire options for which Concept Starters™ are offered.

Lumen Micro™ is a registered trademark of Lighting Technologies, Inc.

ConceptStarters™

Ambient
Recessed Fluorescent Lighting for Commercial Facilities

Using the Concept Starters™

Luminaire Type

Each double page spread is devoted to a single luminaire type. The header outlines the luminaire family (1x4, 2x2, 2x4) and cites lamping (number and type of lamps); intended use (VDT sensitive settings for areas where VDT tasks are predominant or Standard setting for areas where paperwork is most intense); and louver finish (specular (or shiny) and diffuse (or matte)). **Advisory**: read the header carefully—sometimes only the lamping changes; other times only the louver finish changes; etc.

Luminaire Sketch

For quick visual reference, a plan view, cross section and side view of the luminaire of interest are shown. Since some luminaire profiles vary from manufacturer to manufacturer, a note on some indicates which manufacturer's profile data were used for the sketch. **Advisory**: check manufacturers actual datasheets or cutsheets for exact dimensions and profiles.

Design Tips

Tips are offered about the luminaire. Salient advantages and/or disadvantages are cited. Application notes are offered.

Concept Starters™/Reading the Charts

① Ceiling height options are 8', 9' and 10'. Select the column identifying the ceiling height for your project.

② For each ceiling height, surface reflectance categories are light, medium and dark. Specific reflectance values are cited in each category. Select the column of reflectances most closely matching your project's conditions.

③ Four workstation partition height options are evaluated—48", 54", 66" and 80". Select the partition height most predominant in your project. Now, for your ceiling height, surface reflectances and partition heights, look for a shaded cell(s).

④ Shaded cells indicate that the luminaire layout under the circumstances cited (ceiling height, reflectances and partition heights) will likely meet the lighting criteria outlined in Table 9.4. Each cell is inscribed with a ballast factor—very important specification information to provide an energy efficient layout.

⑤ Depending on the luminaire version (VDT Sensitive or Standard), a note will indicate whether all or most of the criteria outlined in Table 9.4 will be met when cells are shaded. Since VDT Sensitive environments are more problematic, all lighting criteria in Table 9.4 is considered applicable. For the lesser demanding Standard environments, most lighting criteria in Table 9.4 is considered applicable—excepting luminaire luminance.

⑥ Layout icons illustrate four different layouts. Use those layouts which offer shaded cells corresponding to your specific project's ceiling height, surface reflectances and partition heights. Amount of area covered by each luminaire cited for convenient preliminary power budgeting and preliminary cost magnitude.

Ambient Recessed Fluorescent

9

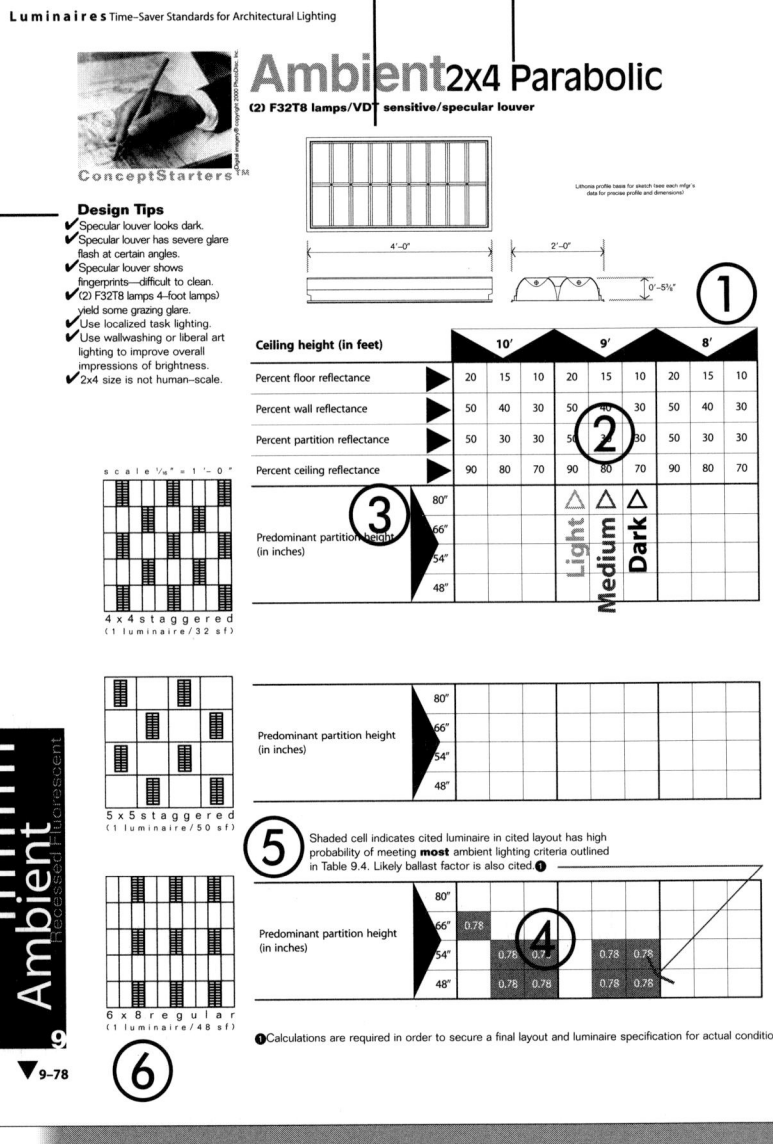

Ambient

Recessed Fluorescent Lighting for Commercial Facilities

ConceptStarters™

Using the Concept Starters™

Luminaire Specification

An outline of the salient specifics on the luminaire featured on the double page spread. Use this information when contacting potential vendors for more specific technical and cost data or when asking manufacturers to provide project–specific calculations.

Luminaire Vendors

A collection of vendors that have or had luminaires fitting the specification and performance characteristics used in developing the Concept Starters™ on the double page spread. Boldfaced manufacturers' photometric data were averaged and used for all calculation analyses on the double page spread. Only luminaires having similar profiles and photometry were used. **Advisory**: in the interest of time, the vendors' search was limited. Other vendors likely can be found with similar performing products.

Ballast Factors

If lamps are the engines for luminaires, then ballasts are the gas peddles. The higher the ballast factor, the more light output from the lamp and the higher the wattage for operation. Ballast factors represent the percentage of light that will be produced by a lamp attached to that ballast (e.g., BF of 0.78 means a lamp on this ballast will produce 78 percent of the lamp manufacturer's rated light output). Ballast factors allow for the fine tuning of a lighting system in order to get just the right quantity and quality of light without using more energy than necessary. **Advisory**: each shaded cell cites the ballast factor used to achieve criteria compliance—so specifying the correct ballast factor is critical to meeting criteria and limiting energy consumption.

[Reproduced sample spread at left:]

Time–Saver Standards for Architectural Lighting **L u m i n a i r e s**

Ambient 2x4 Parabolic

(2) F32T8 lamps/VDT sensitive/specular louver

Specification

⊃18–cell parabolic louver contoured up to and/or around lamp
⊃VDT–sensitive design (exhibits least glare)
⊃Specular, low iridescent clear aluminum
⊃2–F32T8/830 rapid start lamps (2950 lumens each)
⊃Electronic instant start ballast (for best efficiency)
⊃See power budgeting sidebar to right
⊃Cost magnitude: moderate
⊃Possible vendors: Columbia, DayBrite, **Lightolier**, **Lithonia**, Metalux, Zumtobel/Staff

Boldfaced manufacturers' photometric data were averaged and input to Lumen Micro 7.5 in order to assess probability of meeting ambient lighting criteria cited in Table 9.4. Use the above specification list when contacting manufacturers about potential luminaire options for a specific project. Dimensions, specifications and performance are somewhat generic. Check with specific manufacturers' data for nominal dimensions and performance. Data is subject to change.

10′			9′			8′			Ceiling height (in feet)
20	15	10	20	15	10	20	15	10	Percent floor reflectance
50	40	30	50	40	30	50	40	30	Percent wall reflectance
50	30	30	50	30	30	50	30	30	Percent partition reflectance
90	80	70	90	80	70	90	80	70	Percent ceiling reflectance
									80″
									66″ — Predominant partition height (in inches)
									54″
0.78	0.88								48″

Open cell indicates cited luminaire in cited layout has low probability of meeting **most** ambient lighting criteria outlined in Table 9.4. ❶

Power budgeting
for these 2–lamp luminaires

BF 0.78 yields 52 watts
BF 0.88 yields 58 watts
BF 0.98 yields 68 watts
BF 1.18 yields 76 watts

Power budget for ambient lighting should be less than 1.0 w/sf. Ballast vendors include Advance, Energy Savings, Inc. (ESI), Magnetek, Motorola. Not all ballast factors available from all ballast vendors. Above factors based on Magnetek.

8 x 8 r e g u l a r
(1 luminaire/64 sf)

Cost magnitude²⁰⁰⁰
for recessed fluorescent luminaires

Low
US$70. to US$90.
Moderate
US$90. to US$120.
High
US$120. and greater

Quality ambient lighting for today's electronic office environment typically runs US$3.00 to US$5.00/sf, hardware only (luminaires, lamps and ballasts), excluding installation costs. Costs vary based on distributor and contractor markups, luminaire variations such as louver style/finish, lamping, ballasting and options and accessories, and market conditions, manufacturing situations and annual inflation. These values are for preliminary, magnitude budgeting and do not represent quotes nor actual final pricing to client.

9

Ambient
Recessed Fluorescent

9–79 ▲

Net Addresses/Luminaires
http://www.columbia-ltg.com/products/parabolics.html
http://www.thomaslighting.com/daybrite/default.htm
http://www.lightolier.com/html/v4home.htm
http://www.lithonia.com/
http://www.cooperlighting.com/
http://www.zumtobelstaff.co.at/ [Note: German language]

Net Addresses/Ballasts
http://www.advancetransformer.com/
http://www.magnetek.com/transLighting.html
http://www.mot.com/ies/MLI/

CONNECT FOR MORE

Cost Magnitude

For each luminaire on a double page spread, a cost magnitude is reported under **Specification**. The cost magnitude table in the lower right corner can then be referenced to establish preliminary budget information.

Internet Addresses

For more information, and in some cases downloadable cutsheets and installation examples, website addresses are noted. The list is not exhaustive and includes those checked for depth of material available. New sites are added daily—so a global search is advisable from time to time.

Ambient
Recessed Fluorescent

9

ConceptStarters™

Ambient
Recessed Fluorescent Lighting Selection Guide

Using the Selection Guide

Descriptor/Icons
A brief description of louvers and lamping along with reference icons are provided for quick assessment.

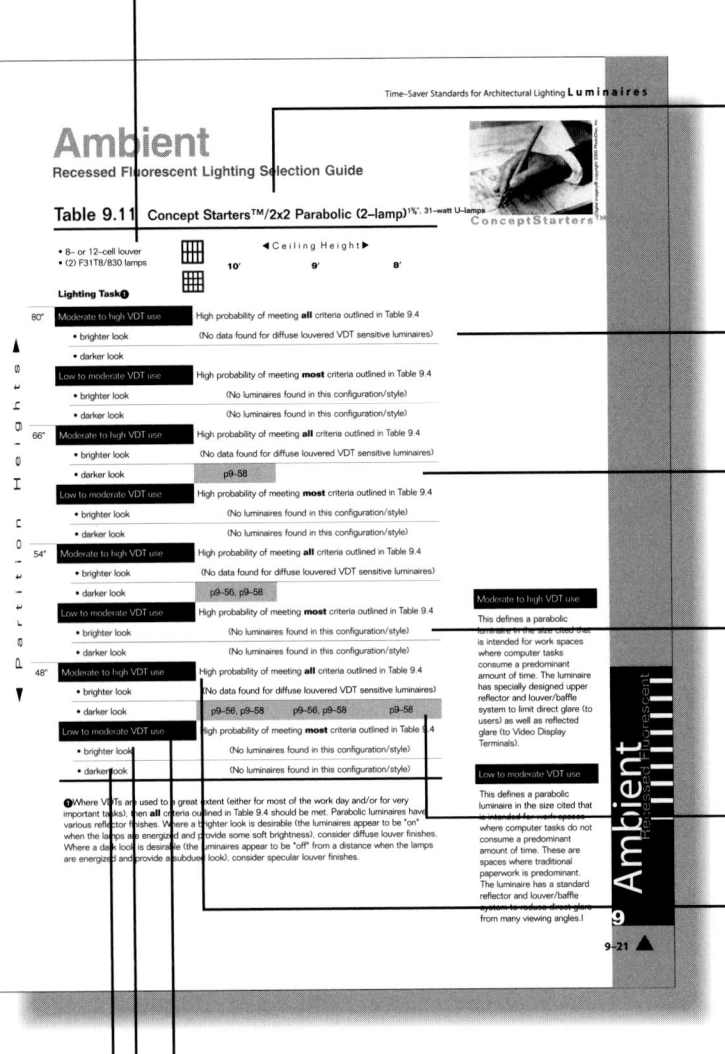

Luminaire
Three luminaire types are assessed here in Section 9—1x4, 2x2 and 2x4 parabolic. For each type, there are variations in lamping and louvering. Within these variations there are "VDT Sensitive" luminaires (intended for moderate to high VDT use) and "Standard" luminaires (intended for low to moderate VDT use).

No Data Found
The luminaire version exists, but no photometric data were found and therefore no analyses could be made. If no photometric data are available for a given luminaire, then it is wise to not specify the product.

Blank Cells
Blank cells indicate that, while these luminaires exist and photometry was available, analyses show that none met criteria outlined in Table 9.4 for the respective partition height, ceiling height and luminaire version.

No Luminaires Found
For this particular version, no luminaires could be found exhibiting these characteristics. Sometimes product lines are specifically intended for one type of setting or environment—VDT Sensitive, for example. Hence, no similar versions are offered in the Standard (for low to moderate VDT use).

Shaded Cell
Check the cited page for Concept Starters™.

VDT Sensitive
For environments where folks will be using VDTs regularly for many tasks, then VDT Sensitive luminaires should be used—look for "Moderate to high VDT use" category under the appropriate partition height category. Shaded cells indicate page number on which Concept Starters™ can be found which meet **all** criteria outlined in Table 9.4.

Standard
For environments where VDTs are used only occasionally, then Standard luminaires might be acceptable—look for "Low to moderate VDT use" category under the appropriate partition height category. Shaded cells indicate page number on which Concept Starters™ can be found which meet **most** criteria outlined in Table 9.4, excepting luminaire luminance (hence luminaire will be more glary).

Brighter Look
These luminaires use diffuse louvers and offer a somewhat brighter look to the ceiling.

Darker Look
These luminaires use specular (shiny) louvers and offer a dark look to the ceiling.

Ambient
Recessed Fluorescent Lighting Selection Guide

ConceptStarters™

Table 9.5 Concept Starters™/1x4 Parabolic (1–lamp)

- 8– or 9–cell baffle
- (1) F32T8/830 lamp

◄ C e i l i n g H e i g h t ►

Partition Heights	Lighting Task❶	**10′**	**9′**	**8′**
80″	Moderate to high VDT use — High probability of meeting **all** criteria outlined in Table 9.4			
	• brighter look	p9–28		
	• darker look	p9–30		
	Low to moderate VDT use — High probability of meeting **most** criteria outlined in Table 9.4			
	• brighter look			
	• darker look			
66″	Moderate to high VDT use — High probability of meeting **all** criteria outlined in Table 9.4			
	• brighter look	p9–28	p9–28	
	• darker look	p9–30	p9–30	
	Low to moderate VDT use — High probability of meeting **most** criteria outlined in Table 9.4			
	• brighter look	p9–32	p9–32	
	• darker look	p9–34	p9–34	
54″	Moderate to high VDT use — High probability of meeting **all** criteria outlined in Table 9.4			
	• brighter look	p9–28	p9–28	p9–28
	• darker look	p9–30	p9–30	p9–30
	Low to moderate VDT use — High probability of meeting **most** criteria outlined in Table 9.4			
	• brighter look	p9–32	p9–32	p9–32
	• darker look	p9–34	p9–34	
48″	Moderate to high VDT use — High probability of meeting **all** criteria outlined in Table 9.4			
	• brighter look	p9–28	p9–28	p9–28
	• darker look		p9–30	p9–30
	Low to moderate VDT use — High probability of meeting **most** criteria outlined in Table 9.4			
	• brighter look	p9–32	p9–32	p9–32
	• darker look	p9–34	p9–34	p9–34

❶Where VDTs are used to a great extent (either for most of the work day and/or for very important tasks), then **all** criteria outlined in Table 9.4 should be met. Parabolic luminaires have various reflector finishes. Where a brighter look is desirable (the luminaires appear to be "on" when the lamps are energized and provide some soft brightness), consider diffuse louver finishes. Where a dark look is desirable (the luminaires appear to be "off" from a distance when the lamps are energized and provide a subdued look), consider specular louver finishes.

Using Table 9.5: select a ceiling height (along top) and select predominant partition heights (along left), establish task and luminaire finish. Shaded boxes indicate reference pages of luminaire layout concept starters that are likely to meet **some** or **all** (as noted) criteria cited in Table 9.4. Blank boxes indicate that no concept starters were found in this luminaire configuration that would likely meet criteria outlined in Table 9.4. Where no luminaire options are available for certain luminaire configurations, this is so noted.

Moderate to high VDT use

This defines a parabolic luminaire in the size cited that is intended for work spaces where computer tasks consume a predominant amount of time. The luminaire has specially designed upper reflector and louver/baffle system to limit direct glare (to users) as well as reflected glare (to Video Display Terminals).

Low to moderate VDT use

This defines a parabolic luminaire in the size cited that is intended for work spaces where computer tasks do not consume a predominant amount of time. These are spaces where traditional paperwork is predominant. The luminaire has a standard reflector and louver/baffle system to reduce direct glare from many viewing angles.l

ConceptStarters™

Ambient
Recessed Fluorescent Lighting Selection Guide

Table 9.6 Concept Starters™/1x4 Parabolic (2–lamp)

- 8– or 9–cell baffle
- (2) F32T8/830 lamps

◀ C e i l i n g   H e i g h t ▶

	10′	9′	8′

Lighting Task❶

▲ Partition Heights

80″

Moderate to high VDT use	High probability of meeting **all** criteria outlined in Table 9.4
• brighter look	(No data found for diffuse louvered VDT sensitive luminaires)
• darker look	

Low to moderate VDT use	High probability of meeting **most** criteria outlined in Table 9.4
• brighter look	
• darker look	

66″

Moderate to high VDT use	High probability of meeting **all** criteria outlined in Table 9.4
• brighter look	(No data found for diffuse louvered VDT sensitive luminaires)
• darker look	p9–36

Low to moderate VDT use	High probability of meeting **most** criteria outlined in Table 9.4
• brighter look	p9–38
• darker look	

54″

Moderate to high VDT use	High probability of meeting **all** criteria outlined in Table 9.4
• brighter look	(No data found for diffuse louvered VDT sensitive luminaires)
• darker look	p9–36

Low to moderate VDT use	High probability of meeting **most** criteria outlined in Table 9.4
• brighter look	p9–38	p9–38
• darker look	

48″

Moderate to high VDT use	High probability of meeting **all** criteria outlined in Table 9.4
• brighter look	(No data found for diffuse louvered VDT sensitive luminaires)
• darker look	p9–36	p9–36

Low to moderate VDT use	High probability of meeting **most** criteria outlined in Table 9.4
• brighter look	p9–38, p9–39	p9–38	p9–38
• darker look	p9–40, p9–41

Using Tables 9.6 and 9.7: select a ceiling height (along top) and select predominant partition heights (along left), establish task and luminaire finish. Shaded boxes indicate reference pages of luminaire layout concept starters that are likely to meet **some** or **all** (as noted) criteria cited in Table 9.4. Blank boxes indicate that no luminaires of this type could be configured in a layout to meet some or all of the criteria outlined in Table 9.4. Where no luminaire options are available for certain luminaire configurations, this is so noted.

❶Where VDTs are used to a great extent (either for most of the work day and/or for very important tasks), then **all** criteria outlined in Table 9.4 should be met. Parabolic luminaires have various reflector finishes. Where a brighter look is desirable (the luminaires appear to be "on" when the lamps are energized and provide some soft brightness), consider diffuse louver finishes. Where a dark look is desirable (the luminaires appear to be "off" from a distance when the lamps are energized and provide a subdued look), consider specular louver finishes.

Ambient Recessed Fluorescent

9

Ambient
Recessed Fluorescent Lighting Selection Guide

ConceptStarters™

—Digital imagery® copyright 2000 PhotoDisc, Inc.

Table 9.7 Concept Starters™/2x2 Parabolic (2–lamp)²⁻foot, 17–watt lamps

- 8– or 12–cell louver
- (2) F17T8/830 lamps

◀ C e i l i n g   H e i g h t ▶

	10′	**9′**	**8′**

Lighting Task❶

P a r t i t i o n   H e i g h t s

80″

Moderate to high VDT use	High probability of meeting **all** criteria outlined in Table 9.4		
• brighter look	(No data found for diffuse louvered VDT sensitive luminaires)		
• darker look	p9–42, p9–44	p9–42, p9–44	
Low to moderate VDT use	High probability of meeting **most** criteria outlined in Table 9.4		
• brighter look	(No luminaires found in this configuration/style)		
• darker look	(No luminaires found in this configuration/style)		

66″

Moderate to high VDT use	High probability of meeting **all** criteria outlined in Table 9.4		
• brighter look	(No data found for diffuse louvered VDT sensitive luminaires)		
• darker look	p9–42, p9–44	p9–42, p9–44	p9–44
Low to moderate VDT use	High probability of meeting **most** criteria outlined in Table 9.4		
• brighter look	(No luminaires found in this configuration/style)		
• darker look	(No luminaires found in this configuration/style)		

54″

Moderate to high VDT use	High probability of meeting **all** criteria outlined in Table 9.4		
• brighter look	(No data found for diffuse louvered VDT sensitive luminaires)		
• darker look	p9–42, p9–44	p9–42, p9–44	p9–42, p9–44
Low to moderate VDT use	High probability of meeting **most** criteria outlined in Table 9.4		
• brighter look	(No luminaires found in this configuration/style)		
• darker look	(No luminaires found in this configuration/style)		

48″

Moderate to high VDT use	High probability of meeting **all** criteria outlined in Table 9.4		
• brighter look	(No data found for diffuse louvered VDT sensitive luminaires)		
• darker look	p9–42, p9–44	p9–42, p9–44	p9–42, p9–44
Low to moderate VDT use	High probability of meeting **most** criteria outlined in Table 9.4		
• brighter look	(No luminaires found in this configuration/style)		
• darker look	(No luminaires found in this configuration/style)		

❶Where VDTs are used to a great extent (either for most of the work day and/or for very important tasks), then **all** criteria outlined in Table 9.4 should be met. Parabolic luminaires have various reflector finishes. Where a brighter look is desirable (the luminaires appear to be "on" when the lamps are energized and provide some soft brightness), consider diffuse louver finishes. Where a dark look is desirable (the luminaires appear to be "off" from a distance when the lamps are energized and provide a subdued look), consider specular louver finishes.

Moderate to high VDT use

This defines a parabolic luminaire in the size cited that is intended for work spaces where computer tasks consume a predominant amount of time. The luminaire has specially designed upper reflector and louver/baffle system to limit direct glare (to users) as well as reflected glare (to Video Display Terminals).

Low to moderate VDT use

This defines a parabolic luminaire in the size cited that is intended for work spaces where computer tasks do not consume a predominant amount of time. These are spaces where traditional paperwork is predominant. The luminaire has a standard reflector and louver/baffle system to reduce direct glare from many viewing angles.l

Ambient
Recessed Fluorescent

9

Digital imagery® copyright 2000 PhotoDisc, Inc.

ConceptStarters™

Ambient
Recessed Fluorescent Lighting Selection Guide

Table 9.8 Concept Starters™/2x2 Parabolic (3–lamp)²⁻ᶠᵒᵒᵗ, ¹⁷⁻ʷᵃᵗᵗ ˡᵃᵐᵖˢ

- 9–cell louver
- (3) F17T8/830 lamps

	◀ Ceiling Height ▶		
	10′	**9′**	**8′**

Lighting Task❶

Partition Heights

80″	**Moderate to high VDT use**	High probability of meeting **all** criteria outlined in Table 9.4		
	• brighter look			
	• darker look	(No data found for specular louvered VDT sensitive luminaires)		
	Low to moderate VDT use	High probability of meeting **most** criteria outlined in Table 9.4		
	• brighter look	p9–48		
	• darker look			
66″	**Moderate to high VDT use**	High probability of meeting **all** criteria outlined in Table 9.4		
	• brighter look			
	• darker look	(No data found for specular louvered VDT sensitive luminaires)		
	Low to moderate VDT use	High probability of meeting **most** criteria outlined in Table 9.4		
	• brighter look	p9–48		
	• darker look	p9–50		
54″	**Moderate to high VDT use**	High probability of meeting **all** criteria outlined in Table 9.4		
	• brighter look	p9–46		
	• darker look	(No data found for specular louvered VDT sensitive luminaires)		
	Low to moderate VDT use	High probability of meeting **most** criteria outlined in Table 9.4		
	• brighter look	p9–48		
	• darker look	p9–50	p9–50	
48″	**Moderate to high VDT use**	High probability of meeting **all** criteria outlined in Table 9.4		
	• brighter look	p9–46	p9–46	
	• darker look	(No data found for specular louvered VDT sensitive luminaires)		
	Low to moderate VDT use	High probability of meeting **most** criteria outlined in Table 9.4		
	• brighter look	p9–48	p9–48	p9–48
	• darker look	p9–50		

Using Tables 9.8 and 9.9: select a ceiling height (along top) and select predominant partition heights (along left), establish task and luminaire finish. Shaded boxes indicate reference pages of luminaire layout concept starters that are likely to meet **some** or **all** (as noted) criteria cited in Table 9.4. Blank boxes indicate that no luminaires of this type could be configured in a layout to meet some or all of the criteria outlined in Table 9.4. Where no luminaire options are available for certain luminaire configurations, this is so noted.

❶Where VDTs are used to a great extent (either for most of the work day and/or for very important tasks), then **all** criteria outlined in Table 9.4 should be met. Parabolic luminaires have various reflector finishes. Where a brighter look is desirable (the luminaires appear to be "on" when the lamps are energized and provide some soft brightness), consider diffuse louver finishes. Where a dark look is desirable (the luminaires appear to be "off" from a distance when the lamps are energized and provide a subdued look), consider specular louver finishes.

Ambient Recessed Fluorescent

Ambient
Recessed Fluorescent Lighting Selection Guide

ConceptStarters™

Table 9.9 Concept Starters™/2x2 Parabolic (3–lamp)²⁻foot, 17–watt lamps

- 12–cell louver
- (3) F17T8/830 lamps

◀Ceiling Height▶

	10′	**9′**	**8′**

Lighting Task❶

Partition Heights

80″	Moderate to high VDT use	High probability of meeting **all** criteria outlined in Table 9.4
	• brighter look	(No data found for diffuse louvered VDT sensitive luminaires)
	• darker look	
	Low to moderate VDT use	High probability of meeting **most** criteria outlined in Table 9.4
	• brighter look	(No luminaires found in this configuration/style)
	• darker look	(No luminaires found in this configuration/style)
66″	Moderate to high VDT use	High probability of meeting **all** criteria outlined in Table 9.4
	• brighter look	(No data found for diffuse louvered VDT sensitive luminaires)
	• darker look	
	Low to moderate VDT use	High probability of meeting **most** criteria outlined in Table 9.4
	• brighter look	(No luminaires found in this configuration/style)
	• darker look	(No luminaires found in this configuration/style)
54″	Moderate to high VDT use	High probability of meeting **all** criteria outlined in Table 9.4
	• brighter look	(No data found for diffuse louvered VDT sensitive luminaires)
	• darker look	p9–52
	Low to moderate VDT use	High probability of meeting **most** criteria outlined in Table 9.4
	• brighter look	(No luminaires found in this configuration/style)
	• darker look	(No luminaires found in this configuration/style)
48″	Moderate to high VDT use	High probability of meeting **all** criteria outlined in Table 9.4
	• brighter look	(No data found for diffuse louvered VDT sensitive luminaires)
	• darker look	p9–52 p9–52
	Low to moderate VDT use	High probability of meeting **most** criteria outlined in Table 9.4
	• brighter look	(No luminaires found in this configuration/style)
	• darker look	(No luminaires found in this configuration/style)

❶Where VDTs are used to a great extent (either for most of the work day and/or for very important tasks), then **all** criteria outlined in Table 9.4 should be met. Parabolic luminaires have various reflector finishes. Where a brighter look is desirable (the luminaires appear to be "on" when the lamps are energized and provide some soft brightness), consider diffuse louver finishes. Where a dark look is desirable (the luminaires appear to be "off" from a distance when the lamps are energized and provide a subdued look), consider specular louver finishes.

Moderate to high VDT use

This defines a parabolic luminaire in the size cited that is intended for work spaces where computer tasks consume a predominant amount of time. The luminaire has specially designed upper reflector and louver/baffle system to limit direct glare (to users) as well as reflected glare (to Video Display Terminals).

Low to moderate VDT use

This defines a parabolic luminaire in the size cited that is intended for work spaces where computer tasks do not consume a predominant amount of time. These are spaces where traditional paperwork is predominant. The luminaire has a standard reflector and louver/baffle system to reduce direct glare from many viewing angles.l

Ambient
Recessed Fluorescent

9

ConceptStarters™

Ambient
Recessed Fluorescent Lighting Selection Guide

Table 9.10 Concept Starters™/2x2 Parabolic (4–lamp) 2–foot, 17–watt lamps

- 16–cell louver
- (4) F17T8/830 lamps

◀ C e i l i n g   H e i g h t ▶

| | 10′ | 9′ | 8′ |

Lighting Task❶

Partition Heights		10′	9′	8′
80″	**Moderate to high VDT use**	High probability of meeting **all** criteria outlined in Table 9.4		
	• brighter look	(No luminaires found in this configuration/style)		
	• darker look	(No luminaires found in this configuration/style)		
	Low to moderate VDT use	High probability of meeting **most** criteria outlined in Table 9.4		
	• brighter look			
	• darker look	(No data found for specular louvered standard luminaires)		
66″	**Moderate to high VDT use**	High probability of meeting **all** criteria outlined in Table 9.4		
	• brighter look	(No luminaires found in this configuration/style)		
	• darker look	(No luminaires found in this configuration/style)		
	Low to moderate VDT use	High probability of meeting **most** criteria outlined in Table 9.4		
	• brighter look			
	• darker look	(No data found for specular louvered standard luminaires)		
54″	**Moderate to high VDT use**	High probability of meeting **all** criteria outlined in Table 9.4		
	• brighter look	(No luminaires found in this configuration/style)		
	• darker look	(No luminaires found in this configuration/style)		
	Low to moderate VDT use	High probability of meeting **most** criteria outlined in Table 9.4		
	• brighter look		p9–54	
	• darker look	(No data found for specular louvered standard luminaires)		
48″	**Moderate to high VDT use**	High probability of meeting **all** criteria outlined in Table 9.4		
	• brighter look	(No luminaires found in this configuration/style)		
	• darker look	(No luminaires found in this configuration/style)		
	Low to moderate VDT use	High probability of meeting **most** criteria outlined in Table 9.4		
	• brighter look		p9–54	p9–54
	• darker look	(No data found for specular louvered standard luminaires)		

Using Tables 9.10 and 9.11: select a ceiling height (along top) and select predominant partition heights (along left), establish task and luminaire finish. Shaded boxes indicate reference pages of luminaire layout concept starters that are likely to meet **some** or **all** (as noted) criteria cited in Table 9.4. Blank boxes indicate that no luminaires of this type could be configured in a layout to meet some or all of the criteria outlined in Table 9.4. Where no luminaire options are available for certain luminaire configurations, this is so noted.

❶Where VDTs are used to a great extent (either for most of the work day and/or for very important tasks), then **all** criteria outlined in Table 9.4 should be met. Parabolic luminaires have various reflector finishes. Where a brighter look is desirable (the luminaires appear to be "on" when the lamps are energized and provide some soft brightness), consider diffuse louver finishes. Where a dark look is desirable (the luminaires appear to be "off" from a distance when the lamps are energized and provide a subdued look), consider specular louver finishes.

Ambient
Recessed Fluorescent

Ambient
Recessed Fluorescent Lighting Selection Guide

ConceptStarters™

Table 9.11 Concept Starters™/2x2 Parabolic (2–lamp)¹⅝", 31–watt U–lamps

- 8– or 12–cell louver
- (2) F31T8/830 lamps

◀ C e i l i n g H e i g h t ▶

| | | 10' | 9' | 8' |

Lighting Task❶

P a r t i t i o n H e i g h t s

80"	Moderate to high VDT use	High probability of meeting **all** criteria outlined in Table 9.4		
	• brighter look	(No data found for diffuse louvered VDT sensitive luminaires)		
	• darker look			
	Low to moderate VDT use	High probability of meeting **most** criteria outlined in Table 9.4		
	• brighter look	(No luminaires found in this configuration/style)		
	• darker look	(No luminaires found in this configuration/style)		
66"	Moderate to high VDT use	High probability of meeting **all** criteria outlined in Table 9.4		
	• brighter look	(No data found for diffuse louvered VDT sensitive luminaires)		
	• darker look	p9–58		
	Low to moderate VDT use	High probability of meeting **most** criteria outlined in Table 9.4		
	• brighter look	(No luminaires found in this configuration/style)		
	• darker look	(No luminaires found in this configuration/style)		
54"	Moderate to high VDT use	High probability of meeting **all** criteria outlined in Table 9.4		
	• brighter look	(No data found for diffuse louvered VDT sensitive luminaires)		
	• darker look	p9–56, p9–58		
	Low to moderate VDT use	High probability of meeting **most** criteria outlined in Table 9.4		
	• brighter look	(No luminaires found in this configuration/style)		
	• darker look	(No luminaires found in this configuration/style)		
48"	Moderate to high VDT use	High probability of meeting **all** criteria outlined in Table 9.4		
	• brighter look	(No data found for diffuse louvered VDT sensitive luminaires)		
	• darker look	p9–56, p9–58	p9–56, p9–58	p9–58
	Low to moderate VDT use	High probability of meeting **most** criteria outlined in Table 9.4		
	• brighter look	(No luminaires found in this configuration/style)		
	• darker look	(No luminaires found in this configuration/style)		

❶Where VDTs are used to a great extent (either for most of the work day and/or for very important tasks), then **all** criteria outlined in Table 9.4 should be met. Parabolic luminaires have various reflector finishes. Where a brighter look is desirable (the luminaires appear to be "on" when the lamps are energized and provide some soft brightness), consider diffuse louver finishes. Where a dark look is desirable (the luminaires appear to be "off" from a distance when the lamps are energized and provide a subdued look), consider specular louver finishes.

Moderate to high VDT use

This defines a parabolic luminaire in the size cited that is intended for work spaces where computer tasks consume a predominant amount of time. The luminaire has specially designed upper reflector and louver/baffle system to limit direct glare (to users) as well as reflected glare (to Video Display Terminals).

Low to moderate VDT use

This defines a parabolic luminaire in the size cited that is intended for work spaces where computer tasks do not consume a predominant amount of time. These are spaces where traditional paperwork is predominant. The luminaire has a standard reflector and louver/baffle system to reduce direct glare from many viewing angles.l

Ambient
Recessed Fluorescent

9

Ambient

Recessed Fluorescent Lighting Selection Guide

Table 9.12 Concept Starters™/2x2 Parabolic (2–lamp)[60, 32–watt U–lamps]

- 9–cell louver
- (2) F32T8/830 lamps

◄ C e i l i n g   H e i g h t ►

			10'	**9'**	**8'**

Lighting Task❶

Partition Heights			10'	9'	8'
80"	**Moderate to high VDT use**	High probability of meeting **all** criteria outlined in Table 9.4			
	• brighter look	(No data found for diffuse louvered VDT sensitive luminaires)			
	• darker look				
	Low to moderate VDT use	High probability of meeting **most** criteria outlined in Table 9.4			
	• brighter look				
	• darker look				
66"	**Moderate to high VDT use**	High probability of meeting **all** criteria outlined in Table 9.4			
	• brighter look	(No data found for diffuse louvered VDT sensitive luminaires)			
	• darker look				
	Low to moderate VDT use	High probability of meeting **most** criteria outlined in Table 9.4			
	• brighter look		p9–62		
	• darker look				
54"	**Moderate to high VDT use**	High probability of meeting **all** criteria outlined in Table 9.4			
	• brighter look	(No data found for diffuse louvered VDT sensitive luminaires)			
	• darker look		p9–60		
	Low to moderate VDT use	High probability of meeting **most** criteria outlined in Table 9.4			
	• brighter look		p9–62	p9–62	
	• darker look		p9–64		
48"	**Moderate to high VDT use**	High probability of meeting **all** criteria outlined in Table 9.4			
	• brighter look	(No data found for diffuse louvered VDT sensitive luminaires)			
	• darker look		p9–60	p9–60	
	Low to moderate VDT use	High probability of meeting **most** criteria outlined in Table 9.4			
	• brighter look		p9–62, p9–63	p9–62	
	• darker look		p9–64	p9–64	

Using Tables 9.12 and 9.13: select a ceiling height (along top) and select predominant partition heights (along left), establish task and luminaire finish. Shaded boxes indicate reference pages of luminaire layout concept starters that are likely to meet **some** or **all** (as noted) criteria cited in Table 9.4. Blank boxes indicate that no luminaires of this type could be configured in a layout to meet some or all of the criteria outlined in Table 9.4. Where no luminaire options are available for certain luminaire configurations, this is so noted.

❶Where VDTs are used to a great extent (either for most of the work day and/or for very important tasks), then **all** criteria outlined in Table 9.4 should be met. Parabolic luminaires have various reflector finishes. Where a brighter look is desirable (the luminaires appear to be "on" when the lamps are energized and provide some soft brightness), consider diffuse louver finishes. Where a dark look is desirable (the luminaires appear to be "off" from a distance when the lamps are energized and provide a subdued look), consider specular louver finishes.

Ambient
Recessed Fluorescent Lighting Selection Guide

ConceptStarters™

Table 9.13 Concept Starters™/2x2 Parabolic (2–lamp)6[0, 32–watt U–lamps]

- 16–cell louver
- (2) F32T8/830 lamps

◀ C e i l i n g   H e i g h t ▶

10′　　　　　　　**9′**　　　　　　　**8′**

Lighting Task❶

Partition Heights		10′	9′	8′
80″	Moderate to high VDT use	High probability of meeting **all** criteria outlined in Table 9.4		
	• brighter look	(No data found for diffuse louvered VDT sensitive luminaires)		
	• darker look			
	Low to moderate VDT use	High probability of meeting **most** criteria outlined in Table 9.4		
	• brighter look			
	• darker look			
66″	Moderate to high VDT use	High probability of meeting **all** criteria outlined in Table 9.4		
	• brighter look	(No data found for diffuse louvered VDT sensitive luminaires)		
	• darker look			
	Low to moderate VDT use	High probability of meeting **most** criteria outlined in Table 9.4		
	• brighter look			
	• darker look			
54″	Moderate to high VDT use	High probability of meeting **all** criteria outlined in Table 9.4		
	• brighter look	(No data found for diffuse louvered VDT sensitive luminaires)		
	• darker look			
	Low to moderate VDT use	High probability of meeting **most** criteria outlined in Table 9.4		
	• brighter look	p9–68		
	• darker look	p9–70		
48″	Moderate to high VDT use	High probability of meeting **all** criteria outlined in Table 9.4		
	• brighter look	(No data found for diffuse louvered VDT sensitive luminaires)		
	• darker look	p9–66		
	Low to moderate VDT use	High probability of meeting **most** criteria outlined in Table 9.4		
	• brighter look	p9–68		
	• darker look	p9–70	p9–70	

❶Where VDTs are used to a great extent (either for most of the work day and/or for very important tasks), then **all** criteria outlined in Table 9.4 should be met. Parabolic luminaires have various reflector finishes. Where a brighter look is desirable (the luminaires appear to be "on" when the lamps are energized and provide some soft brightness), consider diffuse louver finishes. Where a dark look is desirable (the luminaires appear to be "off" from a distance when the lamps are energized and provide a subdued look), consider specular louver finishes.

Moderate to high VDT use

This defines a parabolic luminaire in the size cited that is intended for work spaces where computer tasks consume a predominant amount of time. The luminaire has specially designed upper reflector and louver/baffle system to limit direct glare (to users) as well as reflected glare (to Video Display Terminals).

Low to moderate VDT use

This defines a parabolic luminaire in the size cited that is intended for work spaces where computer tasks do not consume a predominant amount of time. These are spaces where traditional paperwork is predominant. The luminaire has a standard reflector and louver/baffle system to reduce direct glare from many viewing angles.l

Ambient
Recessed Fluorescent

9

Digital imagery® copyright 2000 PhotoDisc, Inc.

ConceptStarters™

Ambient
Recessed Fluorescent Lighting Selection Guide

Table 9.14 Concept Starters™/2x4 Parabolic (2–lamp)

- 12–cell louver
- (2) F32T8/830 lamps

◀ Ceiling Height ▶

				10'	9'	8'

Lighting Task❶

Partition Heights

80"	**Moderate to high VDT use**		High probability of meeting **all** criteria outlined in Table 9.4		
	• brighter look	(No luminaires found in this configuration/style)			
	• darker look	(No luminaires found in this configuration/style)			
	Low to moderate VDT use	High probability of meeting **most** criteria outlined in Table 9.4			
	• brighter look				
	• darker look				
66"	**Moderate to high VDT use**	High probability of meeting **all** criteria outlined in Table 9.4			
	• brighter look	(No luminaires found in this configuration/style)			
	• darker look	(No luminaires found in this configuration/style)			
	Low to moderate VDT use	High probability of meeting **most** criteria outlined in Table 9.4			
	• brighter look				
	• darker look				
54"	**Moderate to high VDT use**	High probability of meeting **all** criteria outlined in Table 9.4			
	• brighter look	(No luminaires found in this configuration/style)			
	• darker look	(No luminaires found in this configuration/style)			
	Low to moderate VDT use	High probability of meeting **most** criteria outlined in Table 9.4			
	• brighter look		p9–72	p9–72	
	• darker look		p9–74, p9–75		
48"	**Moderate to high VDT use**	High probability of meeting **all** criteria outlined in Table 9.4			
	• brighter look	(No luminaires found in this configuration/style)			
	• darker look	(No luminaires found in this configuration/style)			
	Low to moderate VDT use	High probability of meeting **most** criteria outlined in Table 9.4			
	• brighter look			p9–72	
	• darker look		p9–74, p9–75	p9–74, p9–75	p9–74, p9–75

Using Tables 9.14 and 9.15: select a ceiling height (along top) and select predominant partition heights (along left), establish task and luminaire finish. Shaded boxes indicate reference pages of luminaire layout concept starters that are likely to meet **some** or **all** (as noted) criteria cited in Table 9.4. Blank boxes indicate that no luminaires of this type could be configured in a layout to meet some or all of the criteria outlined in Table 9.4. Where no luminaire options are available for certain luminaire configurations, this is so noted.

❶Where VDTs are used to a great extent (either for most of the work day and/or for very important tasks), then **all** criteria outlined in Table 9.4 should be met. Parabolic luminaires have various reflector finishes. Where a brighter look is desirable (the luminaires appear to be "on" when the lamps are energized and provide some soft brightness), consider diffuse louver finishes. Where a dark look is desirable (the luminaires appear to be "off" from a distance when the lamps are energized and provide a subdued look), consider specular louver finishes.

Ambient
Recessed Fluorescent

9

Ambient
Recessed Fluorescent Lighting Selection Guide

C o n c e p t S t a r t e r s ™

Table 9.15 Concept Starters™/2x4 Parabolic (2–lamp)

- 18–cell louver
- (2) F32T8/830 lamps

◀ C e i l i n g H e i g h t ▶

	10′	9′	8′

Lighting Task❶

P a r t i t i o n H e i g h t s ▲▼

	Lighting Task	10′	9′	8′
80″	**Moderate to high VDT use** — High probability of meeting **all** criteria outlined in Table 9.4			
	• brighter look			
	• darker look			
	Low to moderate VDT use — High probability of meeting **most** criteria outlined in Table 9.4			
	• brighter look			
	• darker look			
66″	**Moderate to high VDT use** — High probability of meeting **all** criteria outlined in Table 9.4			
	• brighter look		p9–76	
	• darker look		p9–78	
	Low to moderate VDT use — High probability of meeting **most** criteria outlined in Table 9.4			
	• brighter look			
	• darker look			
54″	**Moderate to high VDT use** — High probability of meeting **all** criteria outlined in Table 9.4			
	• brighter look	p9–76	p9–76	
	• darker look	p9–78	p9–78	
	Low to moderate VDT use — High probability of meeting **most** criteria outlined in Table 9.4			
	• brighter look	p9–80	p9–80	
	• darker look	p9–82	p9–82	
48″	**Moderate to high VDT use** — High probability of meeting **all** criteria outlined in Table 9.4			
	• brighter look	p9–76	p9–76	
	• darker look	p9–78, p9–79	p9–78	
	Low to moderate VDT use — High probability of meeting **most** criteria outlined in Table 9.4			
	• brighter look	p9–80, p9–81	p9–80	
	• darker look	p9–82, p9–83	p9–82	

❶Where VDTs are used to a great extent (either for most of the work day and/or for very important tasks), then **all** criteria outlined in Table 9.4 should be met. Parabolic luminaires have various reflector finishes. Where a brighter look is desirable (the luminaires appear to be "on" when the lamps are energized and provide some soft brightness), consider diffuse louver finishes. Where a dark look is desirable (the luminaires appear to be "off" from a distance when the lamps are energized and provide a subdued look), consider specular louver finishes.

Moderate to high VDT use

This defines a parabolic luminaire in the size cited that is intended for work spaces where computer tasks consume a predominant amount of time. The luminaire has specially designed upper reflector and louver/baffle system to limit direct glare (to users) as well as reflected glare (to Video Display Terminals).

Low to moderate VDT use

This defines a parabolic luminaire in the size cited that is intended for work spaces where computer tasks do not consume a predominant amount of time. These are spaces where traditional paperwork is predominant. The luminaire has a standard reflector and louver/baffle system to reduce direct glare from many viewing angles.l

Ambient
Recessed Fluorescent

9

Digital imagery® copyright 2000 PhotoDisc, Inc.

ConceptStarters™

Ambient

Recessed Fluorescent Lighting Selection Guide

Table 9.16 Concept Starters™/2x4 Parabolic (2–lamp)

- 24– or 27–cell louver
- (2) F32T8/830 lamps

◀ C e i l i n g   H e i g h t ▶

	10'	**9'**	**8'**

Lighting Task❶

P a r t i t i o n   H e i g h t s ▲ ▼

80"

Moderate to high VDT use	High probability of meeting **all** criteria outlined in Table 9.4
• brighter look	(No data found for diffuse louvered VDT sensitive luminaires)
• darker look	

Low to moderate VDT use	High probability of meeting **most** criteria outlined in Table 9.4
• brighter look	(No luminaires found in this configuration/style)
• darker look	(No luminaires found in this configuration/style)

66"

Moderate to high VDT use	High probability of meeting **all** criteria outlined in Table 9.4
• brighter look	(No data found for diffuse louvered VDT sensitive luminaires)
• darker look	p9–84

Low to moderate VDT use	High probability of meeting **most** criteria outlined in Table 9.4
• brighter look	(No luminaires found in this configuration/style)
• darker look	(No luminaires found in this configuration/style)

54"

Moderate to high VDT use	High probability of meeting **all** criteria outlined in Table 9.4
• brighter look	(No data found for diffuse louvered VDT sensitive luminaires)
• darker look	p9–84

Low to moderate VDT use	High probability of meeting **most** criteria outlined in Table 9.4
• brighter look	(No luminaires found in this configuration/style)
• darker look	(No luminaires found in this configuration/style)

48"

Moderate to high VDT use	High probability of meeting **all** criteria outlined in Table 9.4
• brighter look	(No data found for diffuse louvered VDT sensitive luminaires)
• darker look	p9–84 (at 9') p9–84 (at 8')

Low to moderate VDT use	High probability of meeting **most** criteria outlined in Table 9.4
• brighter look	(No luminaires found in this configuration/style)
• darker look	(No luminaires found in this configuration/style)

Using Tables 9.16 and 9.17: select a ceiling height (along top) and select predominant partition heights (along left), establish task and luminaire finish. Shaded boxes indicate reference pages of luminaire layout concept starters that are likely to meet **some** or **all** (as noted) criteria cited in Table 9.4. Blank boxes indicate that no luminaires of this type could be configured in a layout to meet some or all of the criteria outlined in Table 9.4. Where no luminaire options are available for certain luminaire configurations, this is so noted.

❶ Where VDTs are used to a great extent (either for most of the work day and/or for very important tasks), then **all** criteria outlined in Table 9.4 should be met. Parabolic luminaires have various reflector finishes. Where a brighter look is desirable (the luminaires appear to be "on" when the lamps are energized and provide some soft brightness), consider diffuse louver finishes. Where a dark look is desirable (the luminaires appear to be "off" from a distance when the lamps are energized and provide a subdued look), consider specular louver finishes.

Ambient Recessed Fluorescent

Ambient
Recessed Fluorescent Lighting Selection Guide

ConceptStarters™

Table 9.17 Concept Starters™/2x4 Parabolic (2–lamp)

- 32–cell louver
- (2) F32T8/830 lamps

◄ C e i l i n g   H e i g h t ►

	10′	**9′**	**8′**

Lighting Task❶

Partition Heights ▲ ▼

80″	Moderate to high VDT use	High probability of meeting **all** criteria outlined in Table 9.4		
	• brighter look	(No luminaires found in this configuration/style)		
	• darker look	(No luminaires found in this configuration/style)		
	Low to moderate VDT use	High probability of meeting **most** criteria outlined in Table 9.4		
	• brighter look			
	• darker look			
66″	Moderate to high VDT use	High probability of meeting **all** criteria outlined in Table 9.4		
	• brighter look	(No luminaires found in this configuration/style)		
	• darker look	(No luminaires found in this configuration/style)		
	Low to moderate VDT use	High probability of meeting **most** criteria outlined in Table 9.4		
	• brighter look			
	• darker look			
54″	Moderate to high VDT use	High probability of meeting **all** criteria outlined in Table 9.4		
	• brighter look	(No luminaires found in this configuration/style)		
	• darker look	(No luminaires found in this configuration/style)		
	Low to moderate VDT use	High probability of meeting **most** criteria outlined in Table 9.4		
	• brighter look	p9–86	p9–86	
	• darker look	p9–88	p9–88	
48″	Moderate to high VDT use	High probability of meeting **all** criteria outlined in Table 9.4		
	• brighter look	(No luminaires found in this configuration/style)		
	• darker look	(No luminaires found in this configuration/style)		
	Low to moderate VDT use	High probability of meeting **most** criteria outlined in Table 9.4		
	• brighter look	p9–86		p9–86
	• darker look	p9–88, p9–89	p9–88	p9–88

❶Where VDTs are used to a great extent (either for most of the work day and/or for very important tasks), then **all** criteria outlined in Table 9.4 should be met. Parabolic luminaires have various reflector finishes. Where a brighter look is desirable (the luminaires appear to be "on" when the lamps are energized and provide some soft brightness), consider diffuse louver finishes. Where a dark look is desirable (the luminaires appear to be "off" from a distance when the lamps are energized and provide a subdued look), consider specular louver finishes.

Moderate to high VDT use

This defines a parabolic luminaire in the size cited that is intended for work spaces where computer tasks consume a predominant amount of time. The luminaire has specially designed upper reflector and louver/baffle system to limit direct glare (to users) as well as reflected glare (to Video Display Terminals).

Low to moderate VDT use

This defines a parabolic luminaire in the size cited that is intended for work spaces where computer tasks do not consume a predominant amount of time. These are spaces where traditional paperwork is predominant. The luminaire has a standard reflector and louver/baffle system to reduce direct glare from many viewing angles.l

Ambient
Recessed Fluorescent

9

ConceptStarters™

Ambient 1x4 Parabolic

(1) F32T8 lamp/VDT sensitive/diffuse baffle

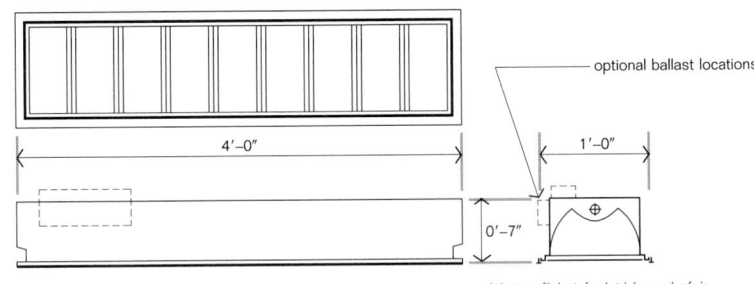

optional ballast locations

4'–0"   1'–0"   0'–7"

Lithonia profile basis for sketch (see each mfgr's
data for precise profile and dimensions)

Design Tips

✔ Diffuse baffle improves overall impression of brightness.

✔ Diffuse baffle masks fingerprints.

✔ Use localized task lighting.

✔ Use wallwashing or liberal art lighting to improve overall impressions of brightness.

✔ Ceiling integration takes greater effort—1x4 size is not typical tile module.

scale ¹⁄₁₆″ = 1'–0"

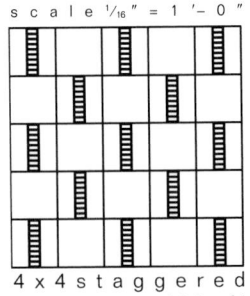

4 x 4 staggered
(1 luminaire / 32 sf)

Ceiling height (in feet)		10'			9'			8'		
Percent floor reflectance		20	15	10	20	15	10	20	15	10
Percent wall reflectance		50	40	30	50	40	30	50	40	30
Percent partition reflectance		50	30	30	50	30	30	50	30	30
Percent ceiling reflectance		90	80	70	90	80	70	90	80	70
Predominant partition height (in inches)	80″	0.78	0.78	0.78						
	66″	0.78	0.78	0.78	0.78	0.78	0.78			
	54″	0.78	0.78	0.78	0.78	0.78	0.78	0.78	0.78	0.78
	48″	0.78	0.78	0.78	0.78	0.78	0.78	0.78	0.78	0.78

Shaded cell indicates cited luminaire in cited layout has high probability of meeting **all** ambient lighting criteria outlined in Table 9.4. Likely ballast factor is also cited.❶

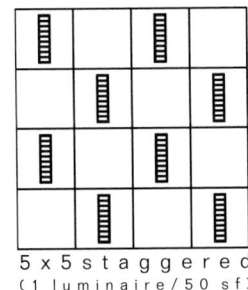

5 x 5 staggered
(1 luminaire / 50 sf)

Predominant partition height (in inches)	80″									
	66″	1.18								
	54″	0.98	1.18	1.18						
	48″	0.98	1.18	1.18	0.98	1.18	1.18	0.98		

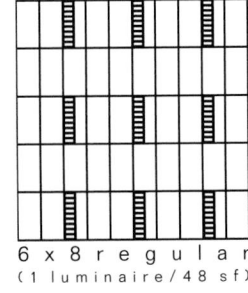

6 x 8 regular
(1 luminaire / 48 sf)

Predominant partition height (in inches)	80″									
	66″									
	54″									
	48″									

❶ Calculations are required in order to secure a final layout and luminaire specification for actual conditions.

Ambient Recessed Fluorescent

Ambient 1x4 Parabolic

(1) F32T8 lamp/VDT sensitive/diffuse baffle

Specification

- 8– or 9–cell parabolic baffle contoured up to and/or around lamp
- VDT–sensitive design (exhibits least glare)
- Diffuse, low iridescent clear aluminum
- 1–F32T8/830 rapid start lamp (2900 lumens)
- Electronic instant start ballast (for best efficiency)
- See power budgeting sidebar to right
- Cost magnitude: moderate
- Possible vendors: **Columbia**, DayBrite, Lightolier, Lithonia, Metalux, Zumtobel/Staff

Boldfaced manufacturers' photometric data were averaged and input to Lumen Micro 7.5 in order to assess probability of meeting ambient lighting criteria cited in Table 9.4. Use the above specification list when contacting manufacturers about potential luminaire options for a specific project. Dimensions, specifications and performance are somewhat generic. Check with specific manufacturers' data for nominal dimensions and performance. Data is subject to change.

C o n c e p t S t a r t e r s ™

Power budgeting
for these 1–lamp luminaires

BF 0.78 yields 26 watts ①
BF 0.88 yields 30 watts ①②
BF 0.98 yields 35 watts ①②
BF 1.18 yields 38 watts ①

Power budget for ambient lighting should be less than 1.0 w/sf. Ballast vendors include Advance, Energy Savings, Inc. (ESI), Magnetek, Motorola. Not all ballast factors available from all ballast vendors. Above factors based on Magnetek. ①Requires tandem lamp operation—using one ballast to operate two single–lamp luminaires. Wattage reported is only for one of the tandem–wired lamps. ②Single lamp ballasts are available for these ballast factors.

10'			9'			8'			Ceiling height (in feet)
20	15	10	20	15	10	20	15	10	◀ Percent floor reflectance
50	40	30	50	40	30	50	40	30	◀ Percent wall reflectance
50	30	30	50	30	30	50	30	30	◀ Percent partition reflectance
90	80	70	90	80	70	90	80	70	◀ Percent ceiling reflectance
									80"
									66" — Predominant partition height (in inches)
									54"
									48"

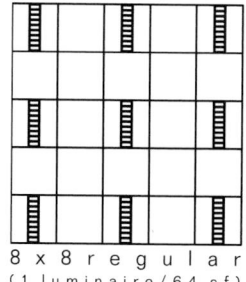

8 x 8   r e g u l a r
(1 l u m i n a i r e / 6 4 s f)

Open cell indicates cited luminaire in cited layout has low probability of meeting **all** ambient lighting criteria outlined in Table 9.4. ❶

Cost magnitude²⁰⁰⁰
for recessed fluorescent luminaires

Low
US$70. to US$90.
Moderate
US$90. to US$120.
High
US$120. and greater

Quality ambient lighting for today's electronic office environment typically runs US$3.00 to US$5.00/sf, hardware only (luminaires, lamps and ballasts), excluding installation costs. Costs vary based on distributor and contractor markups, luminaire variations such as louver style/finish, lamping, ballasting and options and accessories, and market conditions, manufacturing situations and annual inflation. These values are for preliminary, magnitude budgeting and do not represent quotes nor actual final pricing to client.

Ambient Recessed Fluorescent

ConceptStarters™

Ambient 1x4 Parabolic
(1) F32T8 lamp/VDT sensitive/specular baffle

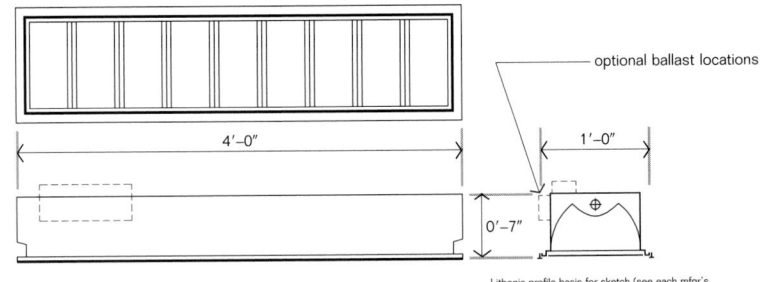

4'–0"   1'–0"   optional ballast locations   0'–7"

Lithonia profile basis for sketch (see each mfgr's
data for precise profile and dimensions)

Design Tips
✔ Specular baffle looks dark.
✔ Specular baffle has severe glare flash at certain angles.
✔ Specular baffle shows fingerprints—difficult to clean.
✔ Use localized task lighting.
✔ Use wallwashing or liberal art lighting to improve overall impressions of brightness.
✔ Ceiling integration takes greater effort—1x4 size is not typical tile module.

Ceiling height (in feet)			10'			9'			8'		
Percent floor reflectance			20	15	10	20	15	10	20	15	10
Percent wall reflectance			50	40	30	50	40	30	50	40	30
Percent partition reflectance			50	30	30	50	30	30	50	30	30
Percent ceiling reflectance			90	80	70	90	80	70	90	80	70
Predominant partition height (in inches)	80"		0.78		0.88						
	66"			0.78	0.78		0.78	0.78			
	54"				0.78						0.78
	48"							0.78			0.78

Shaded cell indicates cited luminaire in cited layout has high probability of meeting **all** ambient lighting criteria outlined in Table 9.4. Likely ballast factor is also cited.❶

scale ¹⁄₁₆" = 1'–0"

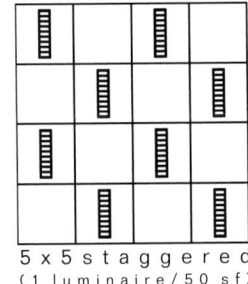

4 x 4 staggered
(1 luminaire / 32 sf)

Predominant partition height (in inches)	80"										
	66"										
	54"		0.98	0.98	1.18						
	48"		0.98	0.98	1.18	0.98	0.98	1.18			

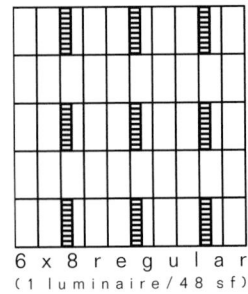

5 x 5 staggered
(1 luminaire / 50 sf)

Predominant partition height (in inches)	80"										
	66"										
	54"										
	48"										

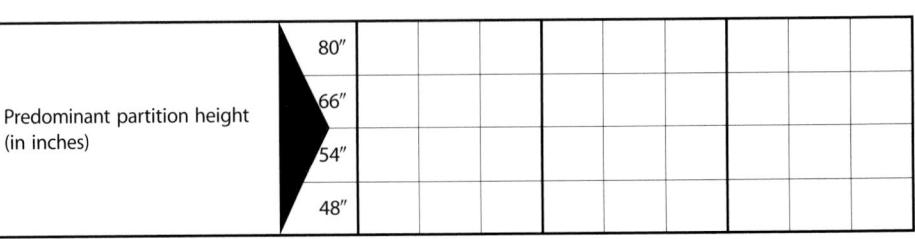

6 x 8 regular
(1 luminaire / 48 sf)

❶ Calculations are required in order to secure a final layout and luminaire specification for actual conditions.

Ambient Recessed Fluorescent

Ambient 1x4 Parabolic

(1) F32T8 lamp/VDT sensitive/specular baffle

Specification

- ➲8– or 9–cell parabolic baffle contoured up to and/or around lamp
- ➲VDT–sensitive design (exhibits least glare)
- ➲Specular, low iridescent clear aluminum
- ➲1–F32T8/830 rapid start lamp (2900 lumens)
- ➲Electronic instant start ballast (for best efficiency)
- ➲See power budgeting sidebar to right
- ➲Cost magnitude: moderate
- ➲Possible vendors: **Columbia**, DayBrite, Lightolier, **Lithonia**, Metalux, Zumtobel/Staff

Boldfaced manufacturers' photometric data were averaged and input to Lumen Micro 7.5 in order to assess probability of meeting ambient lighting criteria cited in Table 9.4. Use the above specification list when contacting manufacturers about potential luminaire options for a specific project. Dimensions, specifications and performance are somewhat generic. Check with specific manufacturers' data for nominal dimensions and performance. Data is subject to change.

ConceptStarters™

Power budgeting
for these 1–lamp luminaires

BF 0.78 yields 26 watts①
BF 0.88 yields 30 watts①②
BF 0.98 yields 35 watts①②
BF 1.18 yields 38 watts①

Power budget for ambient lighting should be less than 1.0 w/sf. Ballast vendors include Advance, Energy Savings, Inc. (ESI), Magnetek, Motorola. Not all ballast factors available from all ballast vendors. Above factors based on Magnetek. ①Requires tandem lamp operation—using one ballast to operate two single–lamp luminaires. Wattage reported is only for one of the tandem–wired lamps. ②Single lamp ballasts are available for these ballast factors.

10'			9'			8'			Ceiling height (in feet)
20	15	10	20	15	10	20	15	10	Percent floor reflectance
50	40	30	50	40	30	50	40	30	Percent wall reflectance
50	30	30	50	30	30	50	30	30	Percent partition reflectance
90	80	70	90	80	70	90	80	70	Percent ceiling reflectance
									80″
									66″
									Predominant partition height (in inches) 54″
									48″

Open cell indicates cited luminaire in cited layout has low probability of meeting **all** ambient lighting criteria outlined in Table 9.4.❶

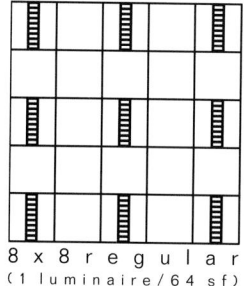

8 x 8 r e g u l a r
(1 l u m i n a i r e / 6 4 s f)

Cost magnitude²⁰⁰⁰
for recessed fluorescent luminaires

Low
US$70. to US$90.
Moderate
US$90. to US$120.
High
US$120. and greater

Quality ambient lighting for today's electronic office environment typically runs US$3.00 to US$5.00/sf, hardware only (luminaires, lamps and ballasts), excluding installation costs. Costs vary based on distributor and contractor markups, luminaire variations such as louver style/finish, lamping, ballasting and options and accessories, and market conditions, manufacturing situations and annual inflation. These values are for preliminary, magnitude budgeting and do not represent quotes nor actual final pricing to client.

Ambient Recessed Fluorescent

9

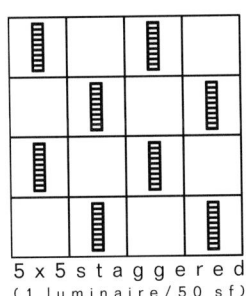

Ambient 1x4 Parabolic

(1) F32T8 lamp/standard/diffuse baffle

Lithonia profile basis for sketch (see each mfgr's data for precise profile and dimensions)

4'–0"   1'–0"   0'–7"   0'–3"

ConceptStarters™

Design Tips

✔ Diffuse baffle improves overall impression of brightness.
✔ Diffuse baffle masks fingerprints.
✔ Use localized task lighting.
✔ Use wallwashing or liberal art lighting to improve overall impressions of brightness.
✔ Ceiling integration takes greater effort—1x4 size is not typical tile module.
✔ Only consider where computer tasks are secondary, intermittent and not critical.

scale ¹/₁₆" = 1'–0"

4 x 4 staggered
(1 luminaire / 32 sf)

5 x 5 staggered
(1 luminaire / 50 sf)

6 x 8 regular
(1 luminaire / 48 sf)

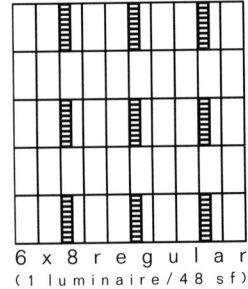

Ceiling height (in feet)	10'			9'			8'		
Percent floor reflectance	20	15	10	20	15	10	20	15	10
Percent wall reflectance	50	40	30	50	40	30	50	40	30
Percent partition reflectance	50	30	30	50	30	30	50	30	30
Percent ceiling reflectance	90	80	70	90	80	70	90	80	70

Predominant partition height (in inches)

	10' (20)	10' (15)	10' (10)	9' (20)	9' (15)	9' (10)	8' (20)	8' (15)	8' (10)
80"									
66"		0.78	0.78		0.78	0.88			
54"			0.78			0.78			
48"									

Shaded cell indicates cited luminaire in cited layout has high probability of meeting **most** ambient lighting criteria outlined in Table 9.4. Likely ballast factor is also cited. ❶

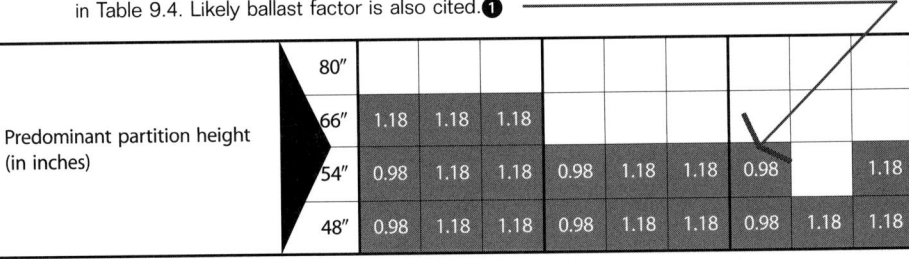

Predominant partition height (in inches)

	10' (20)	10' (15)	10' (10)	9' (20)	9' (15)	9' (10)	8' (20)	8' (15)	8' (10)
80"									
66"	1.18	1.18	1.18						
54"	0.98	1.18	1.18	0.98	1.18	1.18	0.98		1.18
48"	0.98	1.18	1.18	0.98	1.18	1.18	0.98	1.18	1.18

Predominant partition height (in inches)

80"									
66"									
54"									
48"									

Ambient Recessed Fluorescent

9

❶ Calculations are required in order to secure a final layout and luminaire specification for actual conditions.

Ambient 1x4 Parabolic

(1) F32T8 lamp/standard/diffuse baffle

Specification

↪ 8– or 9–cell parabolic baffle
↪ Standard design (glare is more ubiquitous)
↪ Diffuse, low iridescent clear aluminum
↪ 1–F32T8/830 rapid start lamp (2900 lumens)
↪ Electronic instant start ballast (for best efficiency)
↪ See power budgeting sidebar to right
↪ Cost magnitude: low
↪ Possible vendors: **Columbia**, DayBrite, Lightolier, **Lithonia**, Metalux, Zumtobel/Staff

Boldfaced manufacturers' photometric data were averaged and input to Lumen Micro 7.5 in order to assess probability of meeting ambient lighting criteria cited in Table 9.4. Use the above specification list when contacting manufacturers about potential luminaire options for a specific project. Dimensions, specifications and performance are somewhat generic. Check with specific manufacturers' data for nominal dimensions and performance. Data is subject to change.

ConceptStarters™

Power budgeting
for these 1–lamp luminaires

BF 0.78 yields 26 watts①
BF 0.88 yields 30 watts①②
BF 0.98 yields 35 watts①②
BF 1.18 yields 38 watts①

Power budget for ambient lighting should be less than 1.0 w/sf. Ballast vendors include Advance, Energy Savings, Inc. (ESI), Magnetek, Motorola. Not all ballast factors available from all ballast vendors. Above factors based on Magnetek. ①Requires tandem lamp operation—using one ballast to operate two single–lamp luminaires. Wattage reported is only for one of the tandem–wired lamps. ②Single lamp ballasts are available for these ballast factors.

10'			9'			8'			**Ceiling height (in feet)**
20	15	10	20	15	10	20	15	10	Percent floor reflectance
50	40	30	50	40	30	50	40	30	Percent wall reflectance
50	30	30	50	30	30	50	30	30	Percent partition reflectance
90	80	70	90	80	70	90	80	70	Percent ceiling reflectance
								80"	
								66"	Predominant partition height (in inches)
								54"	
								48"	

Open cell indicates cited luminaire in cited layout has low probability of meeting **most** ambient lighting criteria outlined in Table 9.4. ❶

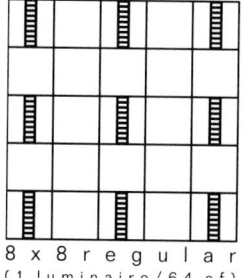

8 x 8 regular
(1 luminaire / 64 sf)

Cost magnitude²⁰⁰⁰
for recessed fluorescent luminaires

Low
US$70. to US$90.
Moderate
US$90. to US$120.
High
US$120. and greater

Quality ambient lighting for today's electronic office environment typically runs US$3.00 to US$5.00/sf, hardware only (luminaires, lamps and ballasts), excluding installation costs. Costs vary based on distributor and contractor markups, luminaire variations such as louver style/finish, lamping, ballasting and options and accessories, and market conditions, manufacturing situations and annual inflation. These values are for preliminary, magnitude budgeting and do not represent quotes nor actual final pricing to client.

Ambient
Recessed Fluorescent

9

Ambient 1x4 Parabolic

(1) F32T8 lamp/standard/specular baffle

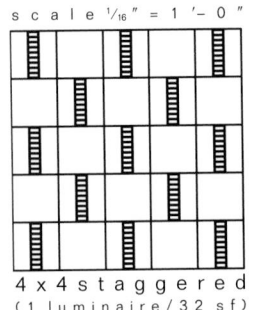

ConceptStarters™

Design Tips

✔ Specular baffle looks dark.

✔ Low performance specular baffle has severe glare flash at many angles.

✔ Specular baffle shows fingerprints—difficult to clean.

✔ Use localized task lighting.

✔ Use wallwashing or liberal art lighting to improve overall impressions of brightness.

✔ Ceiling integration takes greater effort—1x4 size is not typical tile module.

✔ Only consider where computer tasks are secondary, intermittent and not critical.

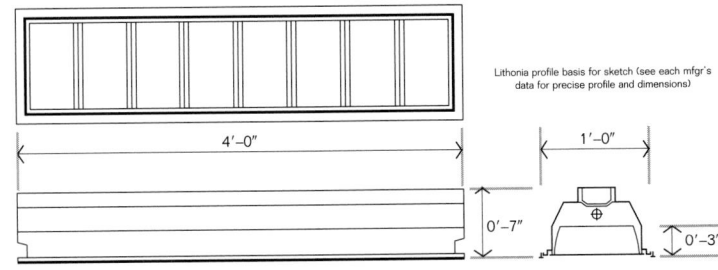

Lithonia profile basis for sketch (see each mfgr's data for precise profile and dimensions)

4'-0"   1'-0"   0'-7"   0'-3"

scale ¹⁄₁₆" = 1'–0"

4 x 4 staggered
(1 luminaire / 32 sf)

Ceiling height (in feet)			10'			9'			8'		
Percent floor reflectance			20	15	10	20	15	10	20	15	10
Percent wall reflectance			50	40	30	50	40	30	50	40	30
Percent partition reflectance			50	30	30	50	30	30	50	30	30
Percent ceiling reflectance			90	80	70	90	80	70	90	80	70
Predominant partition height (in inches)	80"										
	66"			0.78	0.78		0.78	0.78			
	54"				0.78						
	48"										

Shaded cell indicates cited luminaire in cited layout has high probability of meeting **most** ambient lighting criteria outlined in Table 9.4. Likely ballast factor is also cited.❶

5 x 5 staggered
(1 luminaire / 50 sf)

Predominant partition height (in inches)	80"										
	66"		0.98	1.18	1.18	0.98					
	54"		0.98	0.98	0.98	0.98	0.98	0.98			
	48"		0.98	0.98	0.98	0.98	0.98	0.98	0.98		

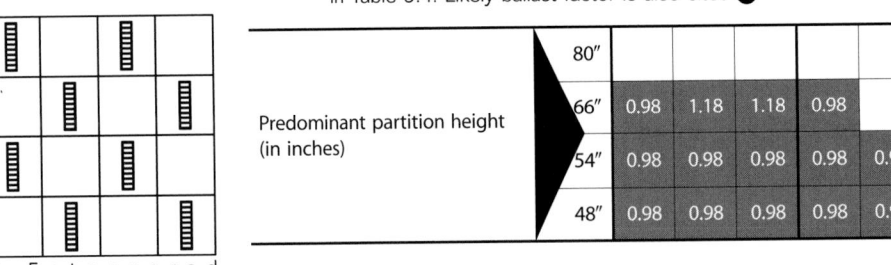

6 x 8 regular
(1 luminaire / 48 sf)

Predominant partition height (in inches)	80"				
	66"				
	54"				
	48"				

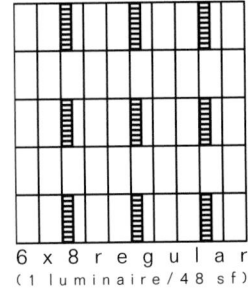

❶ Calculations are required in order to secure a final layout and luminaire specification for actual conditions.

Ambient 1x4 Parabolic

(1) F32T8 lamp/standard/specular baffle

Specification

- ⟳ 8– or 9–cell parabolic baffle
- ⟳ Standard design (glare is more ubiquitous)
- ⟳ Specular, low iridescent clear aluminum
- ⟳ 1–F32T8/830 rapid start lamp (2900 lumens)
- ⟳ Electronic instant start ballast (for best efficiency)
- ⟳ See power budgeting sidebar to right
- ⟳ Cost magnitude: low
- ⟳ Possible vendors: Columbia, DayBrite, **Lightolier**, **Lithonia**, Metalux, Zumtobel/Staff

Boldfaced manufacturers' photometric data were averaged and input to Lumen Micro 7.5 in order to assess probability of meeting ambient lighting criteria cited in Table 9.4. Use the above specification list when contacting manufacturers about potential luminaire options for a specific project. Dimensions, specifications and performance are somewhat generic. Check with specific manufacturers' data for nominal dimensions and performance. Data is subject to change.

ConceptStarters™

Power budgeting
for these 1–lamp luminaires

BF 0.78 yields 26 watts ①
BF 0.88 yields 30 watts ①②
BF 0.98 yields 35 watts ①②
BF 1.18 yields 38 watts ①

Power budget for ambient lighting should be less than 1.0 w/sf. Ballast vendors include Advance, Energy Savings, Inc. (ESI), Magnetek, Motorola. Not all ballast factors available from all ballast vendors. Above factors based on Magnetek.
① Requires tandem lamp operation—using one ballast to operate two single–lamp luminaires. Wattage reported is only for one of the tandem–wired lamps.
② Single lamp ballasts are available for these ballast factors.

10'			9'			8'			Ceiling height (in feet)
20	15	10	20	15	10	20	15	10	Percent floor reflectance
50	40	30	50	40	30	50	40	30	Percent wall reflectance
50	30	30	50	30	30	50	30	30	Percent partition reflectance
90	80	70	90	80	70	90	80	70	Percent ceiling reflectance
								80"	Predominant partition height (in inches)
								66"	
								54"	
								48"	

Open cell indicates cited luminaire in cited layout has low probability of meeting **most** ambient lighting criteria outlined in Table 9.4. ①

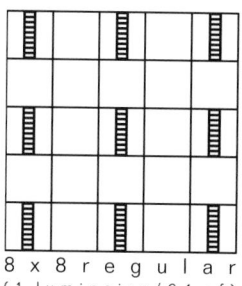

8 x 8 r e g u l a r
(1 luminaire/64 sf)

Cost magnitude²⁰⁰⁰
for recessed fluorescent luminaires

Low
US$70. to US$90.
Moderate
US$90. to US$120.
High
US$120. and greater

Quality ambient lighting for today's electronic office environment typically runs US$3.00 to US$5.00/sf, hardware only (luminaires, lamps and ballasts), excluding installation costs. Costs vary based on distributor and contractor markups, luminaire variations such as louver style/finish, lamping, ballasting and options and accessories, and market conditions, manufacturing situations and annual inflation. These values are for preliminary, magnitude budgeting and do not represent quotes nor actual final pricing to client.

Ambient Recessed Fluorescent

9

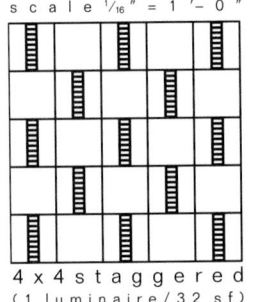

Ambient 1x4 Parabolic

(2) F32T8 lamps/VDT sensitive/specular baffle❶

ConceptStarters™

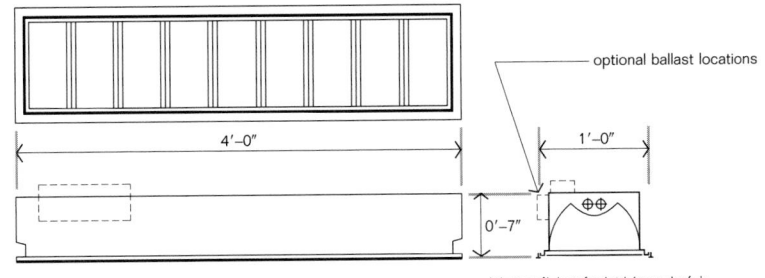

optional ballast locations

4'–0"   1'–0"

0'–7"

Lithonia profile basis for sketch (see each mfgr's data for precise profile and dimensions)

Design Tips

✔ Specular baffle looks dark.
✔ (2) F32T8 lamps in such a small luminaire yield grazing glare.
✔ Specular baffle has severe glare flash at certain angles.
✔ Specular baffle shows fingerprints—difficult to clean.
✔ Use localized task lighting.
✔ Use wallwashing or liberal art lighting to improve overall impressions of brightness.
✔ Ceiling integration takes greater effort—1x4 size is not typical tile module.

Ceiling height (in feet)		10'			9'			8'		
Percent floor reflectance		20	15	10	20	15	10	20	15	10
Percent wall reflectance		50	40	30	50	40	30	50	40	30
Percent partition reflectance		50	30	30	50	30	30	50	30	30
Percent ceiling reflectance		90	80	70	90	80	70	90	80	70
Predominant partition height (in inches)	80"									
	66"									
	54"									
	48"									

scale ¹⁄₁₆" = 1'–0"

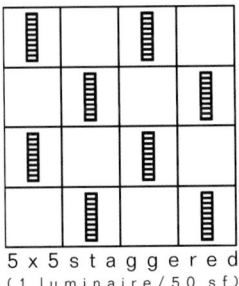

4 x 4 staggered
(1 luminaire / 32 sf)

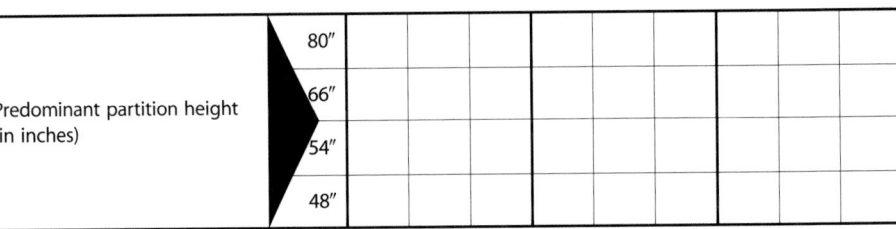

Predominant partition height (in inches)	80"									
	66"									
	54"									
	48"									

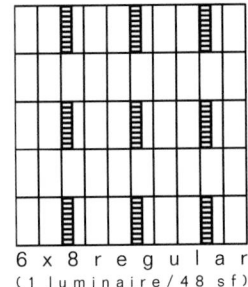

5 x 5 staggered
(1 luminaire / 50 sf)

Shaded cell indicates cited luminaire in cited layout has high probability of meeting **all** ambient lighting criteria outlined in Table 9.4. Likely ballast factor is also cited.❷

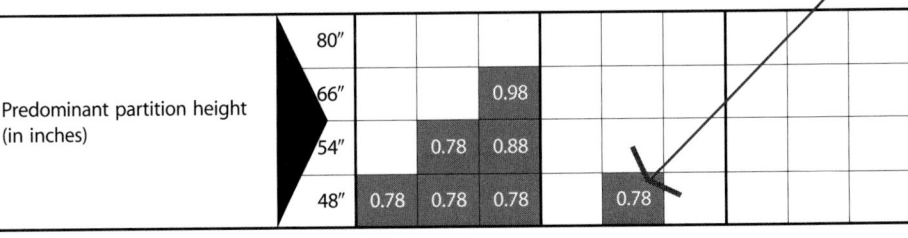

Predominant partition height (in inches)	80"									
	66"			0.98						
	54"		0.78	0.88						
	48"	0.78	0.78	0.78		0.78				

6 x 8 regular
(1 luminaire / 48 sf)

❶ No photometry was available at publication time for diffuse louver option, hence this option is not covered.
❷ Calculations are required in order to secure a final layout and luminaire specification for actual conditions.

Ambient Recessed Fluorescent

Ambient 1x4 Parabolic

(2) F32T8 lamps/VDT sensitive/specular baffle❶

Specification

⟳8– or 9–cell parabolic baffle contoured up to and/or around lamp
⟳VDT–sensitive design (exhibits least glare)
⟳Specular, low iridescent clear aluminum
⟳2–F32T8/830 rapid start lamps (2900 lumens each)
⟳Electronic instant start ballast (for best efficiency)
⟳See power budgeting sidebar to right
⟳Cost magnitude: moderate
⟳Possible vendors: **Columbia**, DayBrite, Lightolier, **Lithonia**, Metalux, Zumtobel/Staff

Boldfaced manufacturers' photometric data were averaged and input to Lumen Micro 7.5 in order to assess probability of meeting ambient lighting criteria cited in Table 9.4. Use the above specification list when contacting manufacturers about potential luminaire options for a specific project. Dimensions, specifications and performance are somewhat generic. Check with specific manufacturers' data for nominal dimensions and performance. Data is subject to change.

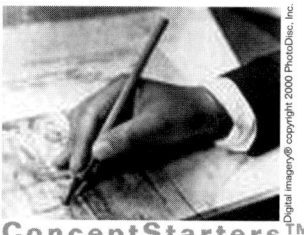
ConceptStarters™

Power budgeting
for these 2–lamp luminaires

BF 0.78 yields 52 watts
BF 0.88 yields 58 watts
BF 0.98 yields 68 watts
BF 1.18 yields 76 watts

Power budget for ambient lighting should be less than 1.0 w/sf. Ballast vendors include Advance, Energy Savings, Inc. (ESI), Magnetek, Motorola. Not all ballast factors available from all ballast vendors. Above factors based on Magnetek.

10'			9'			8'			Ceiling height (in feet)
20	15	10	20	15	10	20	15	10	◀ Percent floor reflectance
50	40	30	50	40	30	50	40	30	◀ Percent wall reflectance
50	30	30	50	30	30	50	30	30	◀ Percent partition reflectance
90	80	70	90	80	70	90	80	70	◀ Percent ceiling reflectance
									80"
									66" Predominant partition height (in inches)
									54"
									48"

Open cell indicates cited luminaire in cited layout has low probability of meeting **all** ambient lighting criteria outlined in Table 9.4.❷

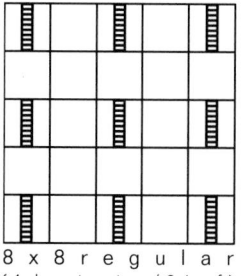

8 x 8 r e g u l a r
(1 l u m i n a i r e / 6 4 s f)

Cost magnitude²⁰⁰⁰
for recessed fluorescent luminaires

Low
US$70. to US$90.
Moderate
US$90. to US$120.
High
US$120. and greater

Quality ambient lighting for today's electronic office environment typically runs US$3.00 to US$5.00/sf, hardware only (luminaires, lamps and ballasts), excluding installation costs. Costs vary based on distributor and contractor markups, luminaire variations such as louver style/ finish, lamping, ballasting and options and accessories, and market conditions, manufacturing situations and annual inflation. These values are for preliminary, magnitude budgeting and do not represent quotes nor actual final pricing to client.

Ambient Recessed Fluorescent

ConceptStarters™

Digital Imagery® copyright 2000 PhotoDisc, Inc.

Ambient 1x4 Parabolic

(2) F32T8 lamps/standard/diffuse baffle

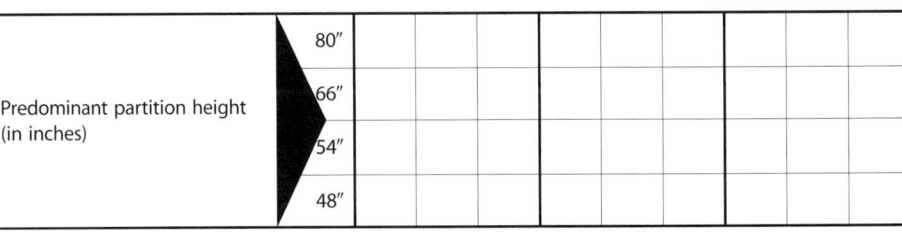

Lithonia profile basis for sketch (see each mfgr's data for precise profile and dimensions)

4'–0"   1'–0"   0'–7"   0'–3"

Design Tips

✔ Diffuse baffle improves overall impression of brightness.

✔ Diffuse baffle masks fingerprints.

✔ (2) F32T8 lamps in such a small luminaire yield grazing glare.

✔ Use localized task lighting.

✔ Use wallwashing or liberal art lighting to improve overall impressions of brightness.

✔ Ceiling integration takes greater effort—1x4 size is not typical tile module.

✔ Only consider where computer tasks are infrequent and/or unimportant.

scale ¹/₁₆" = 1'–0"

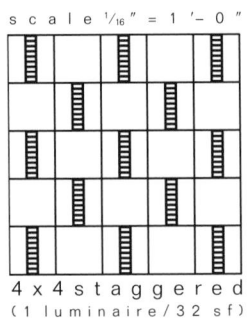

4 x 4 staggered
(1 luminaire / 32 sf)

Ceiling height (in feet)	10'			9'			8'		
Percent floor reflectance	20	15	10	20	15	10	20	15	10
Percent wall reflectance	50	40	30	50	40	30	50	40	30
Percent partition reflectance	50	30	30	50	30	30	50	30	30
Percent ceiling reflectance	90	80	70	90	80	70	90	80	70
Predominant partition height (in inches) 80" 66" 54" 48"									

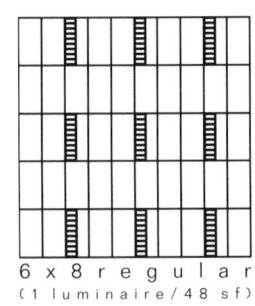

5 x 5 staggered
(1 luminaire / 50 sf)

Predominant partition height (in inches) 80" 66" 54" 48"									

Shaded cell indicates cited luminaire in cited layout has high probability of meeting **most** ambient lighting criteria outlined in Table 9.4. Likely ballast factor is also cited. ❶

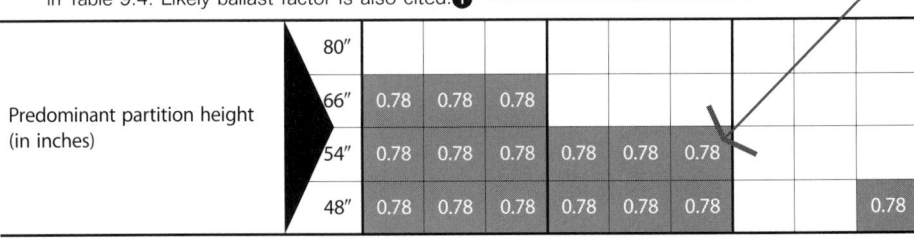

Predominant partition height (in inches)	80"								
66"	0.78	0.78	0.78						
54"	0.78	0.78	0.78	0.78	0.78	0.78			
48"	0.78	0.78	0.78	0.78	0.78	0.78			0.78

6 x 8 regular
(1 luminaire / 48 sf)

❶ Calculations are required in order to secure a final layout and luminaire specification for actual conditions.

Ambient Recessed Fluorescent

9

Ambient 1x4 Parabolic

(2) F32T8 lamps/standard/diffuse baffle

Specification

➲8– or 9–cell parabolic baffle
➲Standard design (glare is more ubiquitous)
➲Diffuse, low iridescent clear aluminum
➲2–F32T8/830 rapid start lamps (2900 lumens each)
➲Electronic instant start ballast (for best efficiency)
➲See power budgeting sidebar to right
➲Cost magnitude: low
➲Possible vendors: **Columbia**, DayBrite, Lightolier, **Lithonia**, Metalux, Zumtobel/Staff

Boldfaced manufacturers' photometric data were averaged and input to Lumen Micro 7.5 in order to assess probability of meeting ambient lighting criteria cited in Table 9.4. Use the above specification list when contacting manufacturers about potential luminaire options for a specific project. Dimensions, specifications and performance are somewhat generic. Check with specific manufacturers' data for nominal dimensions and performance. Data is subject to change.

ConceptStarters™

Power budgeting
for these 2–lamp luminaires

BF 0.78 yields 52 watts
BF 0.88 yields 59 watts
BF 0.98 yields 68 watts
BF 1.18 yields 76 watts

Power budget for ambient lighting should be less than 1.0 w/sf. Ballast vendors include Advance, Energy Savings, Inc. (ESI), Magnetek, Motorola. Not all ballast factors available from all ballast vendors. Above factors based on Magnetek.

10′			9′			8′			Ceiling height (in feet)
20	15	10	20	15	10	20	15	10	◄ Percent floor reflectance
50	40	30	50	40	30	50	40	30	◄ Percent wall reflectance
50	30	30	50	30	30	50	30	30	◄ Percent partition reflectance
90	80	70	90	80	70	90	80	70	◄ Percent ceiling reflectance
									80″
									66″ Predominant partition height (in inches)
									54″
	0.88	0.98							48″

Open cell indicates cited luminaire in cited layout has low probability of meeting **most** ambient lighting criteria outlined in Table 9.4.❶

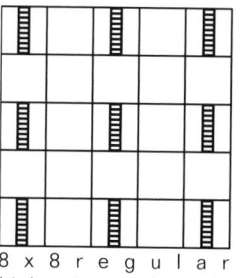

8 x 8 r e g u l a r
(1 l u m i n a i r e / 6 4 s f)

Cost magnitude²⁰⁰⁰
for recessed fluorescent luminaires

Low
US$70. to US$90.
Moderate
US$90. to US$120.
High
US$120. and greater

Quality ambient lighting for today's electronic office environment typically runs US$3.00 to US$5.00/sf, hardware only (luminaires, lamps and ballasts), excluding installation costs. Costs vary based on distributor and contractor markups, luminaire variations such as louver style/finish, lamping, ballasting and options and accessories, and market conditions, manufacturing situations and annual inflation. These values are for preliminary, magnitude budgeting and do not represent quotes nor actual final pricing to client.

Ambient Recessed Fluorescent

9

ConceptStarters™

Ambient 1x4 Parabolic

(2) F32T8 lamps/standard/specular baffle

Lithonia profile basis for sketch (see each mfgr's data for precise profile and dimensions)

4'–0" 1'–0" 0'–7" 0'–3"

Design Tips

✔ Specular baffle looks dark.
✔ Low performance specular baffle has severe glare flash at many angles.
✔ Specular baffle shows fingerprints—difficult to clean.
✔ (2) F32T8 lamps in such a small luminaire yield grazing glare.
✔ Use localized task lighting.
✔ Use wallwashing or liberal art lighting to improve overall impressions of brightness.
✔ Ceiling integration takes greater effort—1x4 size is not typical tile module.
✔ Only consider where computer tasks are infrequent and/or unimportant.

s c a l e 1/16" = 1'–0"

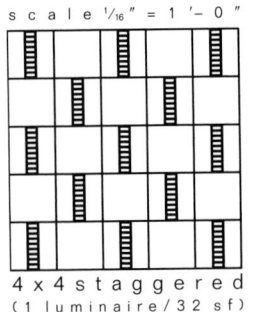

4 x 4 s t a g g e r e d
(1 l u m i n a i r e / 3 2 s f)

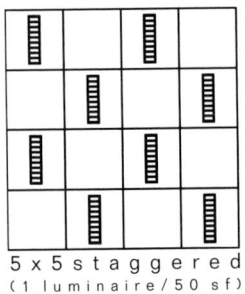

5 x 5 s t a g g e r e d
(1 l u m i n a i r e / 5 0 s f)

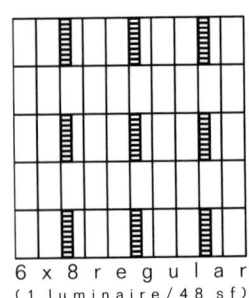

6 x 8 r e g u l a r
(1 l u m i n a i r e / 4 8 s f)

Ceiling height (in feet)		10'			9'			8'		
Percent floor reflectance		20	15	10	20	15	10	20	15	10
Percent wall reflectance		50	40	30	50	40	30	50	40	30
Percent partition reflectance		50	30	30	50	30	30	50	30	30
Percent ceiling reflectance		90	80	70	90	80	70	90	80	70

Predominant partition height (in inches): 80" 66" 54" 48"

Predominant partition height (in inches): 80" 66" 54" 48"

Shaded cell indicates cited luminaire in cited layout has high probability of meeting **most** ambient lighting criteria outlined in Table 9.4. Likely ballast factor is also cited. ❶

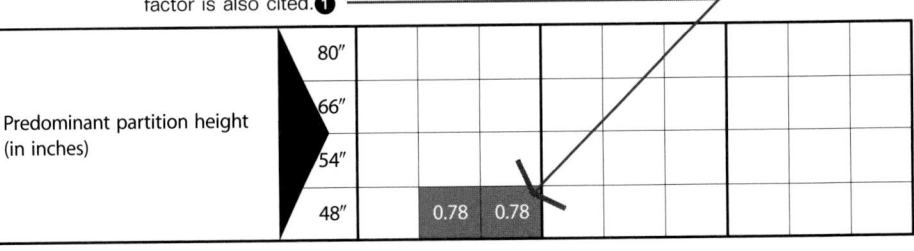

Predominant partition height (in inches): 80" 66" 54" 48"   0.78 0.78

❶ Calculations are required in order to secure a final layout and luminaire specification for actual conditions.

Ambient Recessed Fluorescent

Ambient 1x4 Parabolic

(2) F32T8 lamps/standard/specular baffle

Specification

⊃8– or 9–cell parabolic baffle
⊃Standard design (glare is more ubiquitous)
⊃Specular, low iridescent clear aluminum
⊃2–F32T8/830 rapid start lamps (2900 lumens each)
⊃Electronic instant start ballast (for best efficiency)
⊃See power budgeting sidebar to right
⊃Cost magnitude: low
⊃Possible vendors: Columbia, DayBrite, **Lightolier**, Lithonia, Metalux, Zumtobel/Staff

Boldfaced manufacturers' photometric data were averaged and input to Lumen Micro 7.5 in order to assess probability of meeting ambient lighting criteria cited in Table 9.4. Use the above specification list when contacting manufacturers about potential luminaire options for a specific project. Dimensions, specifications and performance are somewhat generic. Check with specific manufacturers' data for nominal dimensions and performance. Data is subject to change.

ConceptStarters™

Power budgeting
for these 2–lamp luminaires

BF 0.78 yields 52 watts
BF 0.88 yields 59 watts
BF 0.98 yields 68 watts
BF 1.18 yields 76 watts

Power budget for ambient lighting should be less than 1.0 w/sf. Ballast vendors include Advance, Energy Savings, Inc. (ESI), Magnetek, Motorola. Not all ballast factors available from all ballast vendors. Above factors based on Magnetek.

10'			9'			8'			Ceiling height (in feet)
20	15	10	20	15	10	20	15	10	Percent floor reflectance
50	40	30	50	40	30	50	40	30	Percent wall reflectance
50	30	30	50	30	30	50	30	30	Percent partition reflectance
90	80	70	90	80	70	90	80	70	Percent ceiling reflectance
									80"
									66" Predominant partition height (in inches)
									54"
	0.88	0.88							48"

Open cell indicates cited luminaire in cited layout has low probability of meeting **most** ambient lighting criteria outlined in Table 9.4. ❶

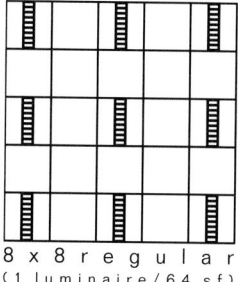

8 x 8 r e g u l a r
(1 luminaire/64 sf)

Cost magnitude^{2000}
for recessed fluorescent luminaires

Low
US$70. to US$90.
Moderate
US$90. to US$120.
High
US$120. and greater

Quality ambient lighting for today's electronic office environment typically runs US$3.00 to US$5.00/sf, hardware only (luminaires, lamps and ballasts), excluding installation costs. Costs vary based on distributor and contractor markups, luminaire variations such as louver style/finish, lamping, ballasting and options and accessories, and market conditions, manufacturing situations and annual inflation. These values are for preliminary, magnitude budgeting and do not represent quotes nor actual final pricing to client.

Ambient Recessed Fluorescent

ConceptStarters™

Digital imagery® copyright 2000 PhotoDisc, Inc.

Ambient 2x2 Parabolic

(2) F17T8 lamps/VDT sensitive/specular louver❶

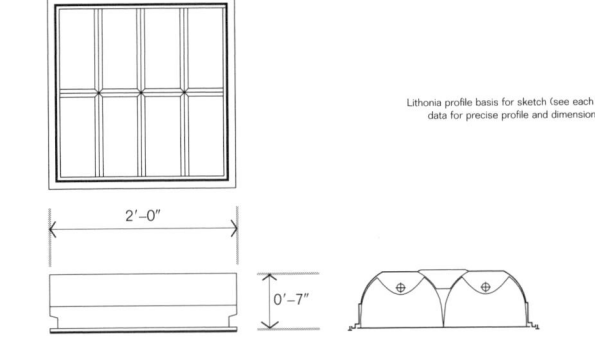

2'–0"

0'–7"

Lithonia profile basis for sketch (see each mfgr's data for precise profile and dimensions)

Design Tips

✔ Specular louver looks dark.
✔ (2) F17T8 lamps (2–foot lamps) exhibit little glare.
✔ Specular louver has glare flash at certain angles.
✔ Specular louver shows fingerprints— difficult to clean.
✔ Use localized task lighting.
✔ Use wallwashing or liberal art lighting to improve overall impressions of brightness.
✔ 8–cell, 2x2 luminaire has directionality.

Ceiling height (in feet)		10'			9'			8'		
Percent floor reflectance		20	15	10	20	15	10	20	15	10
Percent wall reflectance		50	40	30	50	40	30	50	40	30
Percent partition reflectance		50	30	30	50	30	30	50	30	30
Percent ceiling reflectance		90	80	70	90	80	70	90	80	70
Predominant partition height (in inches)	80"	0.82	0.82	0.82	0.82	0.82	0.82			
	66"	0.82	0.82	0.82	0.82	0.82	0.82			
	54"		0.82	0.82		0.82	0.82		0.82	0.82
	48"		0.82	0.82		0.82	0.82		0.82	0.82

scale ¹⁄₁₆" = 1'–0"

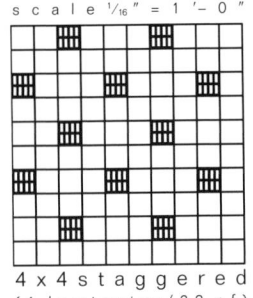

4 x 4 staggered
(1 luminaire / 32 sf)

Shaded cell indicates cited luminaire in cited layout has high probability of meeting **all** ambient lighting criteria outlined in Table 9.4. Likely ballast factor is also cited.❷

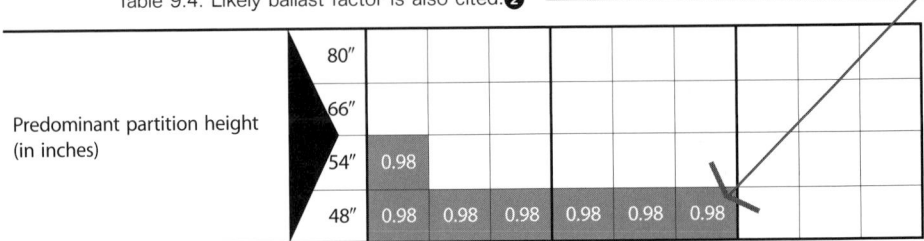

Predominant partition height (in inches)	80"									
	66"									
	54"	0.98								
	48"	0.98	0.98	0.98	0.98	0.98	0.98			

5 x 5 staggered
(1 luminaire / 50 sf)

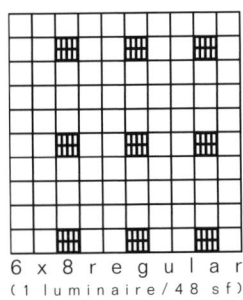

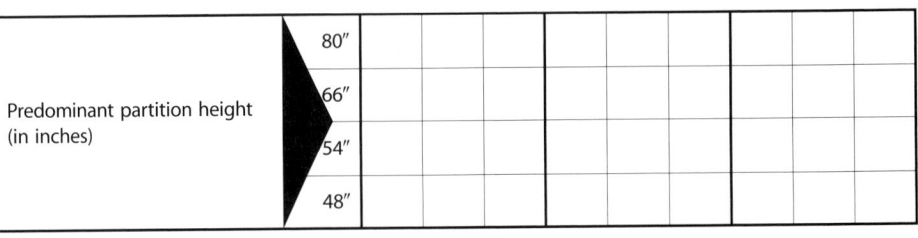

Predominant partition height (in inches)	80"									
	66"									
	54"									
	48"									

6 x 8 regular
(1 luminaire / 48 sf)

❶ No photometry was available at publication time for diffuse louver option, hence this option is not covered.
❷ Calculations are required in order to secure a final layout and luminaire specification for actual conditions.

Ambient Recessed Fluorescent

9

Ambient 2x2 Parabolic
(2) F17T8 lamps/VDT sensitive/specular louver❶

Specification

⟳8–cell parabolic louver contoured up to and/or around lamp
⟳VDT–sensitive design (exhibits least glare)
⟳Specular, low iridescent clear aluminum
⟳2–F17T8/830 rapid start lamps (1350 lumens each)
⟳Electronic instant start ballast (for best efficiency)
⟳See power budgeting sidebar to right
⟳Cost magnitude: high
⟳Possible vendors: Columbia, DayBrite, Lightolier, **Lithonia**, Metalux, Zumtobel/Staff

Boldfaced manufacturers' photometric data were averaged and input to Lumen Micro 7.5 in order to assess probability of meeting ambient lighting criteria cited in Table 9.4. Use the above specification list when contacting manufacturers about potential luminaire options for a specific project. Dimensions, specifications and performance are somewhat generic. Check with specific manufacturers' data for nominal dimensions and performance. Data is subject to change.

ConceptStarters™

Power budgeting
for these 2–lamp luminaires

BF 0.82 yields 29 watts
BF 0.88 yields 34 watts
BF 0.98 yields 35 watts
BF 1.18 not available for F17

Power budget for ambient lighting should be less than 1.0 w/sf. Ballast vendors include Advance, Energy Savings, Inc. (ESI), Magnetek, Motorola. Not all ballast factors available from all ballast vendors. Above factors based on Magnetek.

10'			9'			8'			Ceiling height (in feet)
20	15	10	20	15	10	20	15	10	Percent floor reflectance
50	40	30	50	40	30	50	40	30	Percent wall reflectance
50	30	30	50	30	30	50	30	30	Percent partition reflectance
90	80	70	90	80	70	90	80	70	Percent ceiling reflectance
									80″
									66″ — Predominant partition height (in inches)
									54″
									48″

Open cell indicates cited luminaire in cited layout has low probability of meeting **all** ambient lighting criteria outlined in Table 9.4.❷

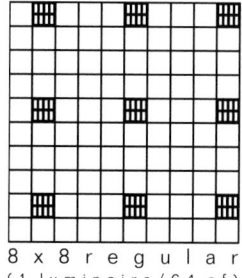

8 x 8 r e g u l a r
(1 l u m i n a i r e / 6 4 s f)

Cost magnitude²⁰⁰⁰
for recessed fluorescent luminaires

Low
US$70. to US$90.
Moderate
US$90. to US$120.
High
US$120. and greater

Quality ambient lighting for today's electronic office environment typically runs US$3.00 to US$5.00/sf, hardware only (luminaires, lamps and ballasts), excluding installation costs. Costs vary based on distributor and contractor markups, luminaire variations such as louver style/finish, lamping, ballasting and options and accessories, and market conditions, manufacturing situations and annual inflation. These values are for preliminary, magnitude budgeting and do not represent quotes nor actual final pricing to client.

Ambient Recessed Fluorescent

9

ConceptStarters™

Ambient 2x2 Parabolic

(2) F17T8 lamps/VDT sensitive/specular louver❶

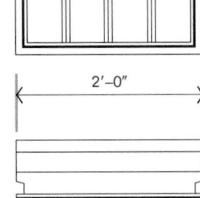

2'–0"

0'–6¾"

Lithonia profile basis for sketch (see each mfgr's data for precise profile and dimensions)

Design Tips

✔ Specular louver looks dark.
✔ (2) F17T8 lamps (2–foot lamps) exhibit little glare.
✔ Specular louver has glare flash at certain angles.
✔ Specular louver shows fingerprints— difficult to clean.
✔ Use localized task lighting.
✔ Use wallwashing or liberal art lighting to improve overall impressions of brightness.
✔ 12–cell, 2x2 luminaire has directionality.

scale ¹⁄₁₆" = 1'–0"

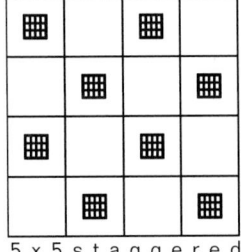

4 x 4 staggered
(1 luminaire / 32 sf)

Ceiling height (in feet)		10'			9'			8'		
Percent floor reflectance		20	15	10	20	15	10	20	15	10
Percent wall reflectance		50	40	30	50	40	30	50	40	30
Percent partition reflectance		50	30	30	50	30	30	50	30	30
Percent ceiling reflectance		90	80	70	90	80	70	90	80	70
Predominant partition height (in inches)	80"	0.88		0.98	0.88		0.98			
	66"	0.82	0.88	0.88	0.82	0.88	0.88	0.82	0.88	0.88
	54"	0.82	0.82	0.82	0.82	0.82	0.82	0.82	0.82	0.82
	48"	0.82	0.82	0.82	0.82	0.82	0.82	0.82	0.82	0.82

Shaded cell indicates cited luminaire in cited layout has high probability of meeting **all** ambient lighting criteria outlined in Table 9.4. Likely ballast factor is also cited.❷

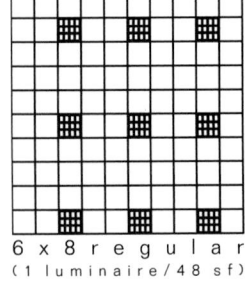

5 x 5 staggered
(1 luminaire / 50 sf)

Predominant partition height (in inches)	80"			
	66"			
	54"			
	48"	0.98	0.98	0.98

6 x 8 regular
(1 luminaire / 48 sf)

Predominant partition height (in inches)	80"			
	66"			
	54"			
	48"			

Ambient Recessed Fluorescent

9

❶ No photometry was available at publication time for diffuse louver option, hence this option is not covered.
❷ Calculations are required in order to secure a final layout and luminaire specification for actual conditions.

Ambient 2x2 Parabolic

(2) F17T8 lamps/VDT sensitive/specular louver❶

Specification

⊃ 12–cell parabolic louver contoured up to and/or around lamp

⊃ VDT–sensitive design (exhibits least glare)

⊃ Specular, low iridescent clear aluminum

⊃ 2–F17T8/830 rapid start lamps (1350 lumens each)

⊃ Electronic instant start ballast (for best efficiency)

⊃ See power budgeting sidebar to right

⊃ Cost magnitude: high

⊃ Possible vendors: Columbia, DayBrite, Lightolier, **Lithonia**, Metalux, Zumtobel/Staff

Boldfaced manufacturers' photometric data were averaged and input to Lumen Micro 7.5 in order to assess probability of meeting ambient lighting criteria cited in Table 9.4. Use the above specification list when contacting manufacturers about potential luminaire options for a specific project. Dimensions, specifications and performance are somewhat generic. Check with specific manufacturers' data for nominal dimensions and performance. Data is subject to change.

10'			9'			8'			Ceiling height (in feet)
20	15	10	20	15	10	20	15	10	◀ Percent floor reflectance
50	40	30	50	40	30	50	40	30	◀ Percent wall reflectance
50	30	30	50	30	30	50	30	30	◀ Percent partition reflectance
90	80	70	90	80	70	90	80	70	◀ Percent ceiling reflectance
									80"
									66" — Predominant partition height (in inches)
									54"
									48"

Open cell indicates cited luminaire in cited layout has low probability of meeting **all** ambient lighting criteria outlined in Table 9.4.❷

Net Addresses/Luminaires

http://www.columbia-ltg.com/products/parabolics.html
http://www.thomaslighting.com/daybrite/default.htm
http://www.lightolier.com/html/v4home.htm
http://www.lithonia.com/
http://www.cooperlighting.com/
http://www.zumtobelstaff.co.at/ [Note: German language]

Net Addresses/Ballasts

http://www.advancetransformer.com/
http://www.magnetek.com/transLighting.html
http://www.mot.com/ies/MLI/

CONNECT FOR MORE

ConceptStarters™

Power budgeting
for these 2–lamp luminaires

BF 0.82 yields 29 watts
BF 0.88 yields 34 watts
BF 0.98 yields 35 watts
BF 1.18 not available for F17

Power budget for ambient lighting should be less than 1.0 w/sf. Ballast vendors include Advance, Energy Savings, Inc. (ESI), Magnetek, Motorola. Not all ballast factors available from all ballast vendors. Above factors based on Magnetek.

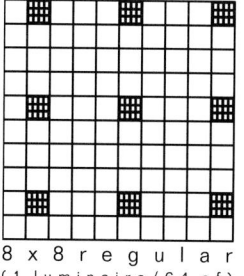

8 x 8 r e g u l a r
(1 luminaire/64 sf)

Cost magnitude²⁰⁰⁰
for recessed fluorescent luminaires

Low
US$70. to US$90.
Moderate
US$90. to US$120.
High
US$120. and greater

Quality ambient lighting for today's electronic office environment typically runs US$3.00 to US$5.00/sf, hardware only (luminaires, lamps and ballasts), excluding installation costs. Costs vary based on distributor and contractor markups, luminaire variations such as louver style/ finish, lamping, ballasting and options and accessories, and market conditions, manufacturing situations and annual inflation. These values are for preliminary, magnitude budgeting and do not represent quotes nor actual final pricing to client.

Ambient
Recessed Fluorescent

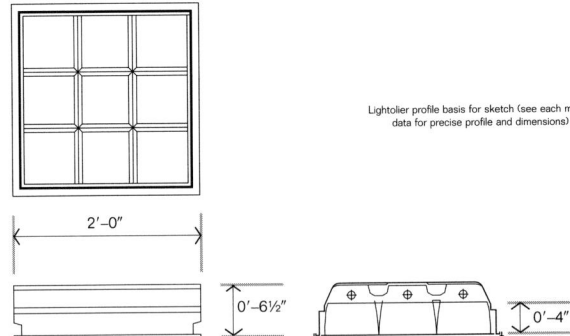

Ambient 2x2 Parabolic

3) F17T8 lamps/VDT sensitive/diffuse louver❶

Lightolier profile basis for sketch (see each mfgr's
data for precise profile and dimensions)

2'–0"

0'–6½"   0'–4"

ConceptStarters™

Design Tips

✔ Diffuse louver improves overall impression of brightness.
✔ (3) F17T8 lamps (2–foot lamps) exhibit little glare.
✔ Diffuse louver masks fingerprints.
✔ Use localized task lighting.
✔ Use wallwashing or liberal art lighting to improve overall impressions of brightness.

Ceiling height (in feet)		10'			9'			8'		
Percent floor reflectance		20	15	10	20	15	10	20	15	10
Percent wall reflectance		50	40	30	50	40	30	50	40	30
Percent partition reflectance		50	30	30	50	30	30	50	30	30
Percent ceiling reflectance		90	80	70	90	80	70	90	80	70

Predominant partition height (in inches)	80"									
	66"									
	54"									
	48"									

scale ¹⁄₁₆" = 1'–0"

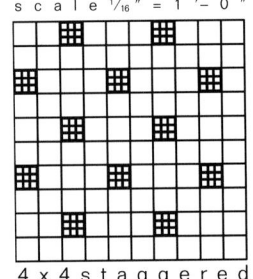

4 x 4 staggered
(1 luminaire / 32 sf)

Shaded cell indicates cited luminaire in cited layout has high probability of meeting **all** ambient lighting criteria outlined in Table 9.4. Likely ballast factor is also cited.❷

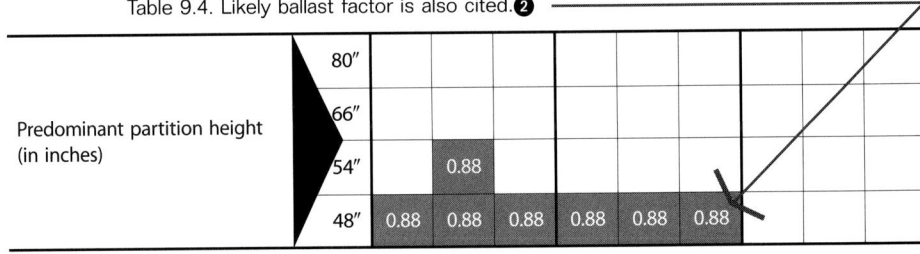

Predominant partition height (in inches)	80"									
	66"									
	54"		0.88							
	48"	0.88	0.88	0.88	0.88	0.88	0.88			

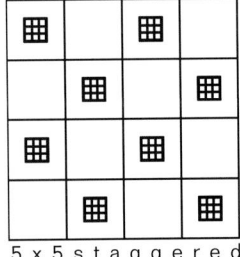

5 x 5 staggered
(1 luminaire / 50 sf)

Predominant partition height (in inches)	80"									
	66"									
	54"									
	48"									

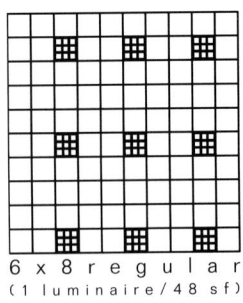

6 x 8 regular
(1 luminaire / 48 sf)

❶No photometry was available at publication time for specular louver option, hence this option is not covered.
❷Calculations are required in order to secure a final layout and luminaire specification for actual conditions.

Ambient Recessed Fluorescent

Ambient 2x2 Parabolic

3) F17T8 lamps/VDT sensitive/diffuse louver❶

Specification

⊃9–cell parabolic louver contoured up to and/or around lamp
⊃VDT–sensitive design (exhibits least glare)
⊃Diffuse, low iridescent clear aluminum
⊃3–F17T8/830 rapid start lamps (1350 lumens each)
⊃Electronic instant start ballast (for best efficiency)
⊃See power budgeting sidebar to right
⊃Cost magnitude: high
⊃Possible vendors: Columbia, DayBrite, **Lightolier**, Lithonia, Metalux, Zumtobel/Staff

Boldfaced manufacturers' photometric data were averaged and input to Lumen Micro 7.5 in order to assess probability of meeting ambient lighting criteria cited in Table 9.4. Use the above specification list when contacting manufacturers about potential luminaire options for a specific project. Dimensions, specifications and performance are somewhat generic. Check with specific manufacturers' data for nominal dimensions and performance. Data is subject to change.

ConceptStarters™

Power budgeting
for these 3–lamp luminaires

BF 0.80 yields 43 watts
BF 0.88 yields 46 watts
BF 0.99 yields 52 watts
BF 1.18 not available for F17

Power budget for ambient lighting should be less than 1.0 w/sf. Ballast vendors include Advance, Energy Savings, Inc. (ESI), Magnetek, Motorola. Not all ballast factors available from all ballast vendors. Above factors based on Magnetek.

10'			9'			8'			Ceiling height (in feet)
20	15	10	20	15	10	20	15	10	Percent floor reflectance
50	40	30	50	40	30	50	40	30	Percent wall reflectance
50	30	30	50	30	30	50	30	30	Percent partition reflectance
90	80	70	90	80	70	90	80	70	Percent ceiling reflectance
									80"
									66" Predominant partition height (in inches)
									54"
									48"

Open cell indicates cited luminaire in cited layout has low probability of meeting **all** ambient lighting criteria outlined in Table 9.4.❷

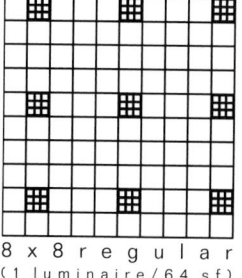

8 x 8 r e g u l a r
(1 l u m i n a i r e / 6 4 s f)

Cost magnitude²⁰⁰⁰
for recessed fluorescent luminaires

Low
US$70. to US$90.
Moderate
US$90. to US$120.
High
US$120. and greater

Quality ambient lighting for today's electronic office environment typically runs US$3.00 to US$5.00/sf, hardware only (luminaires, lamps and ballasts), excluding installation costs. Costs vary based on distributor and contractor markups, luminaire variations such as louver style/finish, lamping, ballasting and options and accessories, and market conditions, manufacturing situations and annual inflation. These values are for preliminary, magnitude budgeting and do not represent quotes nor actual final pricing to client.

Ambient Recessed Fluorescent

ConceptStarters™

Ambient 2x2 Parabolic

(3) F17T8 lamps/standard/diffuse louver

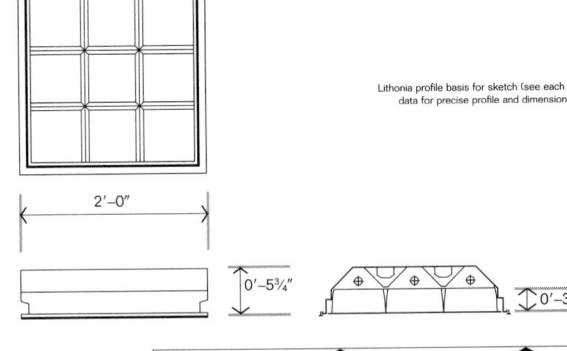

2'–0"

0'–5¾"   0'–3"

Lithonia profile basis for sketch (see each mfgr's
data for precise profile and dimensions)

Design Tips
✔ Diffuse louver improves overall impression of brightness.
✔ Diffuse louver masks fingerprints.
✔ (3) F17T8 lamps in such a standard louver design will yield some glare.
✔ Use localized task lighting.
✔ Use wallwashing or liberal art lighting to improve overall impressions of brightness.
✔ Only consider where computer tasks are infrequent and/or unimportant.

Ceiling height (in feet)	10'			9'			8'		
Percent floor reflectance	20	15	10	20	15	10	20	15	10
Percent wall reflectance	50	40	30	50	40	30	50	40	30
Percent partition reflectance	50	30	30	50	30	30	50	30	30
Percent ceiling reflectance	90	80	70	90	80	70	90	80	70

scale ¹⁄₁₆" = 1'–0"

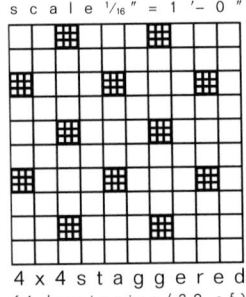

4 x 4 staggered
(1 luminaire/32 sf)

Predominant partition height (in inches)									
80"									
66"									
54"									
48"									

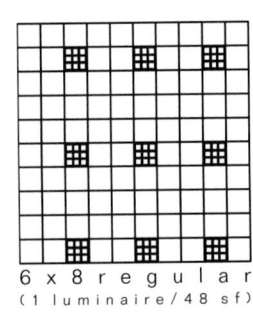

5 x 5 staggered
(1 luminaire/50 sf)

Predominant partition height (in inches)									
80"	0.80								
66"	0.80	0.80							
54"		0.80							
48"						0.80		0.80	0.80

Shaded cell indicates cited luminaire in cited layout has high probability of meeting **most** ambient lighting criteria outlined in Table 9.4. Likely ballast factor is also cited. ❶

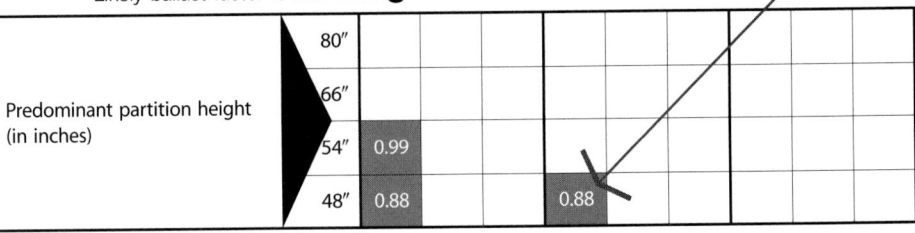

6 x 8 regular
(1 luminaire/48 sf)

Predominant partition height (in inches)									
80"									
66"									
54"	0.99								
48"	0.88			0.88					

❶ Calculations are required in order to secure a final layout and luminaire specification for actual conditions.

Ambient Recessed Fluorescent

Ambient 2x2 Parabolic

(3) F17T8 lamps/standard/diffuse louver

Specification

⊃9–cell parabolic louver
⊃Standard design (glare is more ubiquitous)
⊃Diffuse, low iridescent clear aluminum
⊃3–F17T8/830 rapid start lamps (1350 lumens each)
⊃Electronic instant start ballast (for best efficiency)
⊃See power budgeting sidebar to right
⊃Cost magnitude: moderate
⊃Possible vendors: Columbia, DayBrite, Lightolier, **Lithonia**, Metalux, Zumtobel/Staff

Boldfaced manufacturers' photometric data were averaged and input to Lumen Micro 7.5 in order to assess probability of meeting ambient lighting criteria cited in Table 9.4. Use the above specification list when contacting manufacturers about potential luminaire options for a specific project. Dimensions, specifications and performance are somewhat generic. Check with specific manufacturers' data for nominal dimensions and performance. Data is subject to change.

ConceptStarters™

Power budgeting
for these 3–lamp luminaires

BF 0.80 yields 43 watts
BF 0.88 yields 46 watts
BF 0.99 yields 52 watts
BF 1.18 not available for F17

Power budget for ambient lighting should be less than 1.0 w/sf. Ballast vendors include Advance, Energy Savings, Inc. (ESI), Magnetek, Motorola. Not all ballast factors available from all ballast vendors. Above factors based on Magnetek.

10'			9'			8'			Ceiling height (in feet)
20	15	10	20	15	10	20	15	10	◀ Percent floor reflectance
50	40	30	50	40	30	50	40	30	◀ Percent wall reflectance
50	30	30	50	30	30	50	30	30	◀ Percent partition reflectance
90	80	70	90	80	70	90	80	70	◀ Percent ceiling reflectance
									80″ ◀
									66″ ◀ Predominant partition height (in inches)
									54″ ◀
									48″ ◀

Open cell indicates cited luminaire in cited layout has low probability of meeting **most** ambient lighting criteria outlined in Table 9.4. ❶

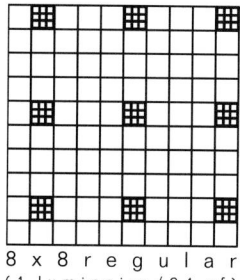

8 x 8 r e g u l a r
(1 l u m i n a i r e / 6 4 s f)

Cost magnitude²⁰⁰⁰
for recessed fluorescent luminaires

Low
US$70. to US$90.
Moderate
US$90. to US$120.
High
US$120. and greater

Quality ambient lighting for today's electronic office environment typically runs US$3.00 to US$5.00/sf, hardware only (luminaires, lamps and ballasts), excluding installation costs. Costs vary based on distributor and contractor markups, luminaire variations such as louver style/ finish, lamping, ballasting and options and accessories, and market conditions, manufacturing situations and annual inflation. These values are for preliminary, magnitude budgeting and do not represent quotes nor actual final pricing to client.

Ambient Recessed Fluorescent

9

ConceptStarters™

Ambient 2x2 Parabolic

(3) F17T8 lamps/standard/specular louver

Lithonia profile basis for sketch (see each mfgr's data for precise profile and dimensions)

2'–0"

0'–5¾"   0'–3"

Design Tips

✔ Specular louver looks dark.
✔ Low performance specular louver has glare flash at many angles.
✔ Specular louver shows fingerprints—difficult to clean.
✔ (3) F17T8 lamps in such a standard louver design will yield some glare.
✔ Use localized task lighting.
✔ Use wallwashing or liberal art lighting to improve overall impressions of brightness.
✔ Only consider where computer tasks are infrequent and/or unimportant.

scale ¹⁄₁₆" = 1'–0"

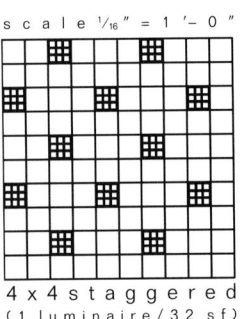

4 x 4 staggered
(1 luminaire / 32 sf)

Ceiling height (in feet)		10'			9'			8'		
Percent floor reflectance		20	15	10	20	15	10	20	15	10
Percent wall reflectance		50	40	30	50	40	30	50	40	30
Percent partition reflectance		50	30	30	50	30	30	50	30	30
Percent ceiling reflectance		90	80	70	90	80	70	90	80	70
Predominant partition height (in inches)	80"									
	66"									
	54"									
	48"									

Shaded cell indicates cited luminaire in cited layout has high probability of meeting **most** ambient lighting criteria outlined in Table 9.4. Likely ballast factor is also cited. ❶

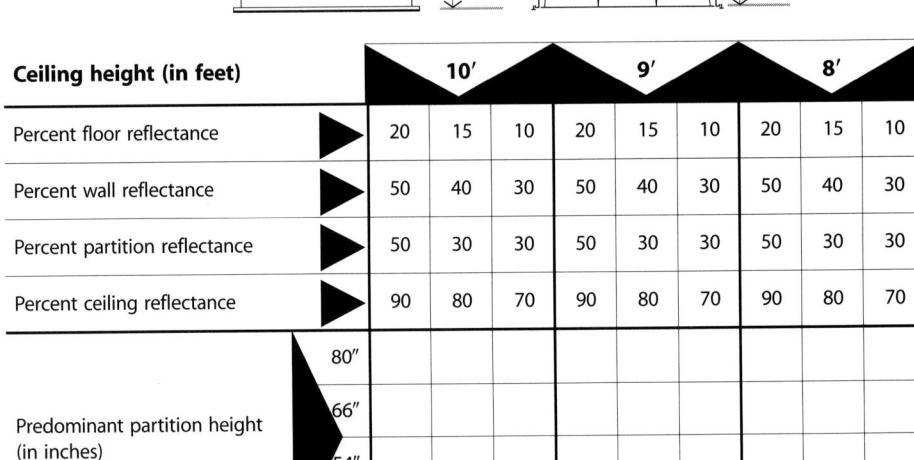

Predominant partition height (in inches)	80"									
	66"	0.80								
	54"		0.80	0.80		0.80				
	48"		0.80	0.80						

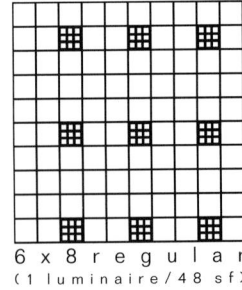

5 x 5 staggered
(1 luminaire / 50 sf)

Predominant partition height (in inches)	80"									
	66"									
	54"	0.99								
	48"	0.88								

6 x 8 regular
(1 luminaire / 48 sf)

❶ Calculations are required in order to secure a final layout and luminaire specification for actual conditions.

Ambient Recessed Fluorescent

9

Ambient 2x2 Parabolic

(3) F17T8 lamps/standard/specular louver

Specification
- 9–cell parabolic louver
- Standard design (glare is more ubiquitous)
- Specular, low iridescent clear aluminum
- 3–F17T8/830 rapid start lamps (1350 lumens each)
- Electronic instant start ballast (for best efficiency)
- See power budgeting sidebar to right
- Cost magnitude: moderate
- Possible vendors: **Columbia**, DayBrite, Lightolier, Lithonia, Metalux, Zumtobel/Staff

Boldfaced manufacturers' photometric data were averaged and input to Lumen Micro 7.5 in order to assess probability of meeting ambient lighting criteria cited in Table 9.4. Use the above specification list when contacting manufacturers about potential luminaire options for a specific project. Dimensions, specifications and performance are somewhat generic. Check with specific manufacturers' data for nominal dimensions and performance. Data is subject to change.

ConceptStarters™

Power budgeting
for these 3–lamp luminaires

BF 0.80 yields 43 watts
BF 0.88 yields 46 watts
BF 0.99 yields 52 watts
BF 1.18 not available for F17

Power budget for ambient lighting should be less than 1.0 w/sf. Ballast vendors include Advance, Energy Savings, Inc. (ESI), Magnetek, Motorola. Not all ballast factors available from all ballast vendors. Above factors based on Magnetek.

10'			9'			8'			Ceiling height (in feet)
20	15	10	20	15	10	20	15	10	Percent floor reflectance
50	40	30	50	40	30	50	40	30	Percent wall reflectance
50	30	30	50	30	30	50	30	30	Percent partition reflectance
90	80	70	90	80	70	90	80	70	Percent ceiling reflectance
								80"	
								66"	Predominant partition height (in inches)
								54"	
								48"	

Open cell indicates cited luminaire in cited layout has low probability of meeting **most** ambient lighting criteria outlined in Table 9.4. ❶

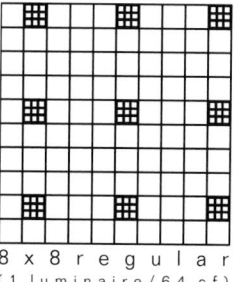

8 x 8 regular
(1 luminaire/64 sf)

Cost magnitude[2000]
for recessed fluorescent luminaires

Low
US$70. to US$90.
Moderate
US$90. to US$120.
High
US$120. and greater

Quality ambient lighting for today's electronic office environment typically runs US$3.00 to US$5.00/sf, hardware only (luminaires, lamps and ballasts), excluding installation costs. Costs vary based on distributor and contractor markups, luminaire variations such as louver style/finish, lamping, ballasting and options and accessories, and market conditions, manufacturing situations and annual inflation. These values are for preliminary, magnitude budgeting and do not represent quotes nor actual final pricing to client.

Ambient Recessed Fluorescent

9

ConceptStarters™

Ambient 2x2 Parabolic

(3) F17T8 lamps/VDT sensitive/specular louver❶

Lithonia profile basis for sketch (see each mfgr's data for precise profile and dimensions)

2'–0"

0'–6"

Design Tips

✔ Specular louver looks dark.
✔ (3) F17T8 lamps (2–foot lamps) exhibit little glare.
✔ Specular louver has glare flash at certain angles.
✔ Specular louver shows fingerprints— difficult to clean.
✔ Use localized task lighting.
✔ Use wallwashing or liberal art lighting to improve overall impressions of brightness.
✔ 12–cell, 2x2 luminaire has directionality.

Ceiling height (in feet)		10'			9'			8'		
Percent floor reflectance		20	15	10	20	15	10	20	15	10
Percent wall reflectance		50	40	30	50	40	30	50	40	30
Percent partition reflectance		50	30	30	50	30	30	50	30	30
Percent ceiling reflectance		90	80	70	90	80	70	90	80	70

scale ¹⁄₁₆" = 1'–0"

Predominant partition height (in inches)	80"									
	66"									
	54"									
	48"									

4 x 4 staggered
(1 luminaire / 32 sf)

Shaded cell indicates cited luminaire in cited layout has high probability of meeting **all** ambient lighting criteria outlined in Table 9.4. Likely ballast factor is also cited.❷

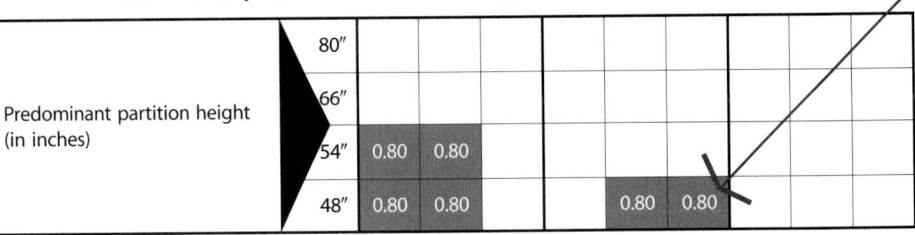

Predominant partition height (in inches)	80"									
	66"									
	54"	0.80	0.80							
	48"	0.80	0.80			0.80	0.80			

5 x 5 staggered
(1 luminaire / 50 sf)

Predominant partition height (in inches)	80"									
	66"									
	54"	0.88								
	48"	0.88	0.99							

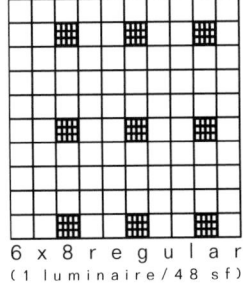

6 x 8 regular
(1 luminaire / 48 sf)

❶ No photometry was available at publication time for diffuse louver option, hence this option is not covered.
❷ Calculations are required in order to secure a final layout and luminaire specification for actual conditions.

Ambient Recessed Fluorescent

9

Ambient 2x2 Parabolic
(3) F17T8 lamps/VDT sensitive/specular louver❶

Specification
⊃ 12–cell parabolic louver contoured up to and/or around lamp
⊃ VDT–sensitive design (exhibits least glare)
⊃ Specular, low iridescent clear aluminum
⊃ 3–F17T8/830 rapid start lamps (1350 lumens each)
⊃ Electronic instant start ballast (for best efficiency)
⊃ See power budgeting sidebar to right
⊃ Cost magnitude: high
⊃ Possible vendors: **Columbia**, DayBrite, **Lightolier**, **Lithonia**, Metalux, Zumtobel/Staff

Boldfaced manufacturers' photometric data were averaged and input to Lumen Micro 7.5 in order to assess probability of meeting ambient lighting criteria cited in Table 9.4. Use the above specification list when contacting manufacturers about potential luminaire options for a specific project. Dimensions, specifications and performance are somewhat generic. Check with specific manufacturers' data for nominal dimensions and performance. Data is subject to change.

ConceptStarters™

Power budgeting
for these 3–lamp luminaires

BF 0.80 yields 43 watts
BF 0.88 yields 46 watts
BF 0.99 yields 52 watts
BF 1.18 not available for F17

Power budget for ambient lighting should be less than 1.0 w/sf. Ballast vendors include Advance, Energy Savings, Inc. (ESI), Magnetek, Motorola. Not all ballast factors available from all ballast vendors. Above factors based on Magnetek.

10'			9'			8'			**Ceiling height (in feet)**
20	15	10	20	15	10	20	15	10	◀ Percent floor reflectance
50	40	30	50	40	30	50	40	30	◀ Percent wall reflectance
50	30	30	50	30	30	50	30	30	◀ Percent partition reflectance
90	80	70	90	80	70	90	80	70	◀ Percent ceiling reflectance
									80″
									66″ ◀ Predominant partition height (in inches)
									54″
									48″

Open cell indicates cited luminaire in cited layout has low probability of meeting **all** ambient lighting criteria outlined in Table 9.4.❷

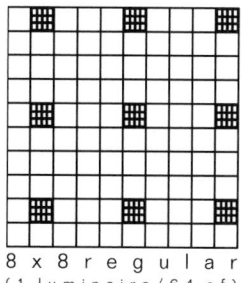

8 x 8 r e g u l a r
(1 luminaire/64 sf)

Cost magnitude[2000]
for recessed fluorescent luminaires

Low
US$70. to US$90.
Moderate
US$90. to US$120.
High
US$120. and greater

Quality ambient lighting for today's electronic office environment typically runs US$3.00 to US$5.00/sf, hardware only (luminaires, lamps and ballasts), excluding installation costs. Costs vary based on distributor and contractor markups, luminaire variations such as louver style/finish, lamping, ballasting and options and accessories, and market conditions, manufacturing situations and annual inflation. These values are for preliminary, magnitude budgeting and do not represent quotes nor actual final pricing to client.

Ambient Recessed Fluorescent

ConceptStarters™

Ambient2x2 Parabolic

(4) F17T8 lamps/standard/diffuse louver❶

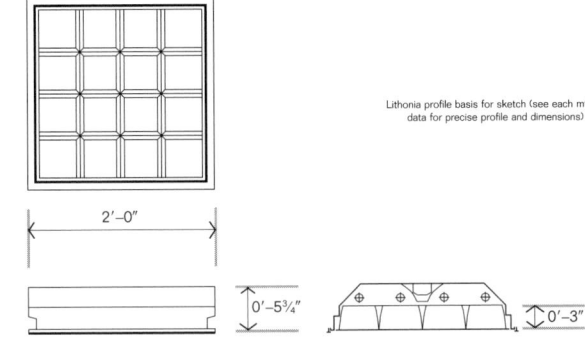

Lithonia profile basis for sketch (see each mfgr's data for precise profile and dimensions)

2'–0"

0'–5¾"   0'–3"

Design Tips

✔ Diffuse louver improves overall impression of brightness.
✔ Diffuse louver masks fingerprints.
✔ (4) F17T8 lamps in such a standard louver design will yield glare.
✔ Use localized task lighting.
✔ Use wallwashing or liberal art lighting to improve overall impressions of brightness.
✔ Only consider where computer tasks are infrequent and/or unimportant.

Ceiling height (in feet)		10'			9'			8'		
Percent floor reflectance		20	15	10	20	15	10	20	15	10
Percent wall reflectance		50	40	30	50	40	30	50	40	30
Percent partition reflectance		50	30	30	50	30	30	50	30	30
Percent ceiling reflectance		90	80	70	90	80	70	90	80	70

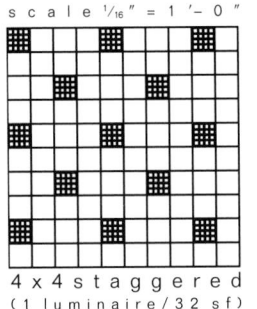

s c a l e ¹⁄₁₆" = 1'– 0"

4 x 4 s t a g g e r e d
(1 luminaire / 32 sf)

Predominant partition height (in inches)	80"									
	66"									
	54"									
	48"									

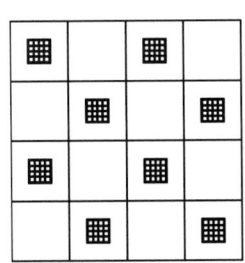

5 x 5 s t a g g e r e d
(1 luminaire / 50 sf)

Predominant partition height (in inches)	80"					
	66"					
	54"					
	48"					

Shaded cell indicates cited luminaire in cited layout has high probability of meeting **most** ambient lighting criteria outlined in Table 9.4. Likely ballast factor is also cited.❷

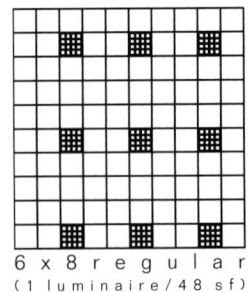

6 x 8 r e g u l a r
(1 luminaire / 48 sf)

Predominant partition height (in inches)	80"					
	66"					
	54"	0.82	0.88	0.98		
	48"	0.82	0.88	0.88	0.88	0.88

❶ No photometry was available at publication time for specular louver option, hence this option is not covered.
❷ Calculations are required in order to secure a final layout and luminaire specification for actual conditions.

Ambient Recessed Fluorescent

Ambient 2x2 Parabolic

(4) F17T8 lamps/standard/diffuse louver ❶

Specification

⊃16–cell parabolic louver
⊃Standard design (glare is more ubiquitous)
⊃Diffuse, low iridescent clear aluminum
⊃4–F17T8/830 rapid start lamps (1350 lumens each)
⊃Electronic instant start ballast (for best efficiency)
⊃See power budgeting sidebar to right
⊃Cost magnitude: high
⊃Possible vendors: Columbia, DayBrite, Lightolier, **Lithonia**, Metalux, Zumtobel/Staff

Boldfaced manufacturers' photometric data were averaged and input to Lumen Micro 7.5 in order to assess probability of meeting ambient lighting criteria cited in Table 9.4. Use the above specification list when contacting manufacturers about potential luminaire options for a specific project. Dimensions, specifications and performance are somewhat generic. Check with specific manufacturers' data for nominal dimensions and performance. Data is subject to change.

ConceptStarters™

Power budgeting
for these 4–lamp luminaires

BF 0.82 yields 56 watts
BF 0.88 yields 63 watts
BF 0.95 yields 68 watts①
BF 1.18 not available for F17

Power budget for ambient lighting should be less than 1.0 w/sf. Ballast vendors include Advance, Energy Savings, Inc. (ESI), Magnetek, Motorola. Not all ballast factors available from all ballast vendors. Above factors based on Magnetek.
①Requires two 2–lamp ballasts.

10'			9'			8'			Ceiling height (in feet)
20	15	10	20	15	10	20	15	10	Percent floor reflectance
50	40	30	50	40	30	50	40	30	Percent wall reflectance
50	30	30	50	30	30	50	30	30	Percent partition reflectance
90	80	70	90	80	70	90	80	70	Percent ceiling reflectance
						80"			
						66"			Predominant partition height (in inches)
						54"			
						48"			

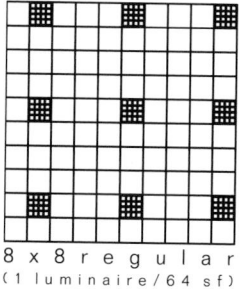

8 x 8 r e g u l a r
(1 luminaire/64 sf)

Open cell indicates cited luminaire in cited layout has low probability of meeting **most** ambient lighting criteria outlined in Table 9.4. ❷

Net Addresses/Luminaires
http://www.columbia-ltg.com/products/parabolics.html
http://www.thomaslighting.com/daybrite/default.htm
http://www.lightolier.com/html/v4home.htm
http://www.lithonia.com/
http://www.cooperlighting.com/
http://www.zumtobelstaff.co.at/   [Note: German language]

Net Addresses/Ballasts
http://www.advancetransformer.com/
http://www.magnetek.com/transLighting.html
http://www.mot.com/ies/MLI/

CONNECT FOR MORE

Cost magnitude²⁰⁰⁰
for recessed fluorescent luminaires

Low
US$70. to US$90.
Moderate
US$90. to US$120.
High
US$120. and greater

Quality ambient lighting for today's electronic office environment typically runs US$3.00 to US$5.00/sf, hardware only (luminaires, lamps and ballasts), excluding installation costs. Costs vary based on distributor and contractor markups, luminaire variations such as louver style/finish, lamping, ballasting and options and accessories, and market conditions, manufacturing situations and annual inflation. These values are for preliminary, magnitude budgeting and do not represent quotes nor actual final pricing to client.

Ambient Recessed Fluorescent

9

ConceptStarters™

Ambient 2x2 Parabolic

(2) F31T8/U/1⅝ lamps/VDT sensitive/specular louver❶

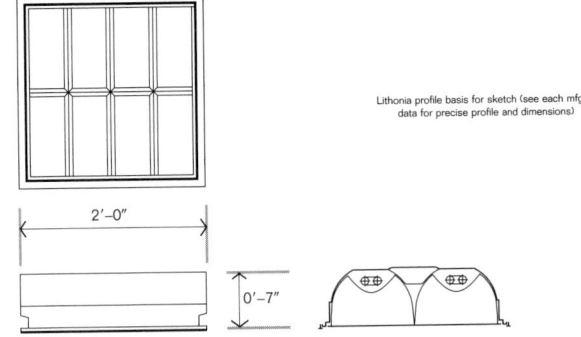

Lithonia profile basis for sketch (see each mfgr's
data for precise profile and dimensions)

2'–0"

0'–7"

Design Tips

✔Specular louver looks dark.
✔(2) F31T8/U/1⅝ lamps (2–foot U–bent lamps) yield grazing glare.
✔Specular louver has severe glare flash at certain angles.
✔Specular louver shows fingerprints—difficult to clean.
✔Use localized task lighting.
✔Use wallwashing or liberal art lighting to improve overall impressions of brightness.
✔8–cell, 2x2 luminaire has directionality.

scale ¹/₁₆" = 1'–0"

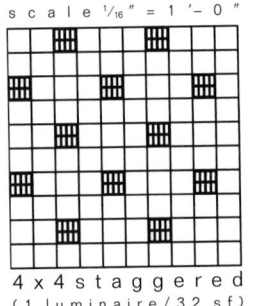

4 x 4 staggered
(1 luminaire / 32 sf)

Ceiling height (in feet)		10'			9'			8'		
Percent floor reflectance		20	15	10	20	15	10	20	15	10
Percent wall reflectance		50	40	30	50	40	30	50	40	30
Percent partition reflectance		50	30	30	50	30	30	50	30	30
Percent ceiling reflectance		90	80	70	90	80	70	90	80	70
Predominant partition height (in inches)	80"									
	66"									
	54"									
	48"									

5 x 5 staggered
(1 luminaire / 50 sf)

Predominant partition height (in inches)	80"									
	66"									
	54"									
	48"									

Shaded cell indicates cited luminaire in cited layout has high probability of meeting **all** ambient lighting criteria outlined in Table 9.4. Likely ballast factor is also cited.❷

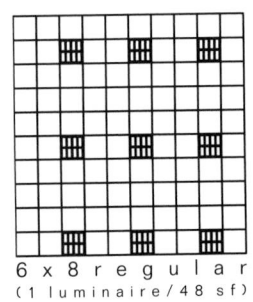

6 x 8 regular
(1 luminaire / 48 sf)

Predominant partition height (in inches)	80"									
	66"									
	54"			0.88	0.98					
	48"		0.78	0.88	0.98		0.88			

❶No photometry was available at publication time for diffuse louver option, hence this option is not covered.
❷Calculations are required in order to secure a final layout and luminaire specification for actual conditions.

Ambient Recessed Fluorescent

9

Ambient 2x2 Parabolic
(2) F31T8/U/1⅝ lamps/VDT sensitive/specular louver❶

Specification

⭮ 8–cell parabolic louver contoured up to and/or around lamp
⭮ VDT–sensitive design (exhibits least glare)
⭮ Specular, low iridescent clear aluminum
⭮ 2–F31T8/U/1⅝/830 rapid start lamps (2725 lumens each)
⭮ Electronic instant start ballast (for best efficiency)
⭮ See power budgeting sidebar to right
⭮ Cost magnitude: high
⭮ Possible vendors: Columbia, DayBrite, Lightolier, **Lithonia**, Metalux, Zumtobel/Staff

Boldfaced manufacturers' photometric data were averaged and input to Lumen Micro 7.5 in order to assess probability of meeting ambient lighting criteria cited in Table 9.4. Use the above specification list when contacting manufacturers about potential luminaire options for a specific project. Dimensions, specifications and performance are somewhat generic. Check with specific manufacturers' data for nominal dimensions and performance. Data is subject to change.

ConceptStarters™

Power budgeting
for these 2–lamp luminaires

BF 0.78 yields 52 watts
BF 0.88 yields 58 watts
BF 0.98 yields 68 watts
BF 1.18 yields 76 watts

Power budget for ambient lighting should be less than 1.0 w/sf. Ballast vendors include Advance, Energy Savings, Inc. (ESI), Magnetek, Motorola. Not all ballast factors available from all ballast vendors. Above factors based on Magnetek.

10'			9'			8'			Ceiling height (in feet)
20	15	10	20	15	10	20	15	10	Percent floor reflectance
50	40	30	50	40	30	50	40	30	Percent wall reflectance
50	30	30	50	30	30	50	30	30	Percent partition reflectance
90	80	70	90	80	70	90	80	70	Percent ceiling reflectance
									80"
									66" Predominant partition height (in inches)
									54"
									48"

Open cell indicates cited luminaire in cited layout has low probability of meeting **all** ambient lighting criteria outlined in Table 9.4.❷

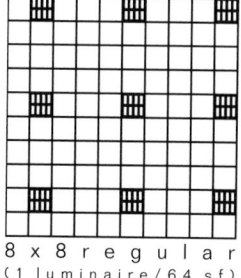
8 x 8 r e g u l a r
(1 l u m i n a i r e / 6 4 s f)

Cost magnitude²⁰⁰⁰
for recessed fluorescent luminaires

Low
US$70. to US$90.
Moderate
US$90. to US$120.
High
US$120. and greater

Quality ambient lighting for today's electronic office environment typically runs US$3.00 to US$5.00/sf, hardware only (luminaires, lamps and ballasts), excluding installation costs. Costs vary based on distributor and contractor markups, luminaire variations such as louver style/finish, lamping, ballasting and options and accessories, and market conditions, manufacturing situations and annual inflation. These values are for preliminary, magnitude budgeting and do not represent quotes nor actual final pricing to client.

Ambient Recessed Fluorescent

ConceptStarters™

Ambient 2x2 Parabolic

(2) F31T8/U/1⅝ lamps/VDT sensitive/specular louver❶

Lithonia profile basis for sketch (see each mfgr's data for precise profile and dimensions)

2'–0"

0'–6¾"

Design Tips

✔Specular louver looks dark.
✔(2) F31T8/U/1⅝ lamps (2–foot U–bent lamps) yield grazing glare.
✔Specular louver has severe glare flash at certain angles.
✔Specular louver shows fingerprints—difficult to clean.
✔Use localized task lighting.
✔Use wallwashing or liberal art lighting to improve overall impressions of brightness.
✔12–cell, 2x2 luminaire has directionality.

Ceiling height (in feet)		10'			9'			8'		
Percent floor reflectance		20	15	10	20	15	10	20	15	10
Percent wall reflectance		50	40	30	50	40	30	50	40	30
Percent partition reflectance		50	30	30	50	30	30	50	30	30
Percent ceiling reflectance		90	80	70	90	80	70	90	80	70

scale ¹⁄₁₆ " = 1'–0"

Predominant partition height (in inches)	80"									
	66"									
	54"									
	48"									

4 x 4 staggered
(1 luminaire/32 sf)

Predominant partition height (in inches)	80"									
	66"	0.78								
	54"		0.78	0.78						
	48"		0.78	0.78		0.78	0.78		0.78	0.78

5 x 5 staggered
(1 luminaire/50 sf)

Shaded cell indicates cited luminaire in cited layout has high probability of meeting **all** ambient lighting criteria outlined in Table 9.4. Likely ballast factor is also cited.❷ ——————

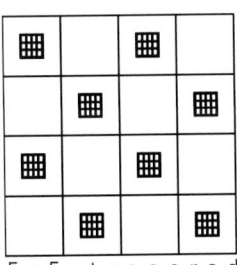

Predominant partition height (in inches)	80"									
	66"									
	54"			0.98	1.18					
	48"	0.98	0.98	0.98				0.98		

6 x 8 regular
(1 luminaire/48 sf)

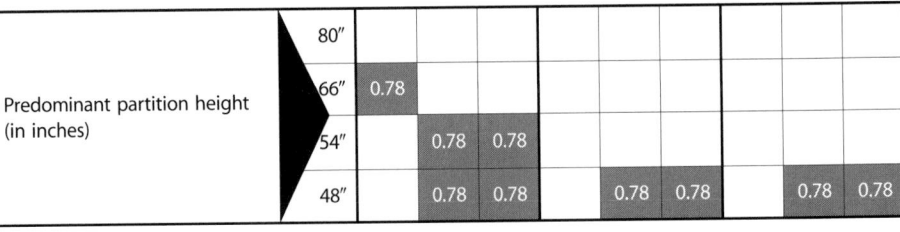

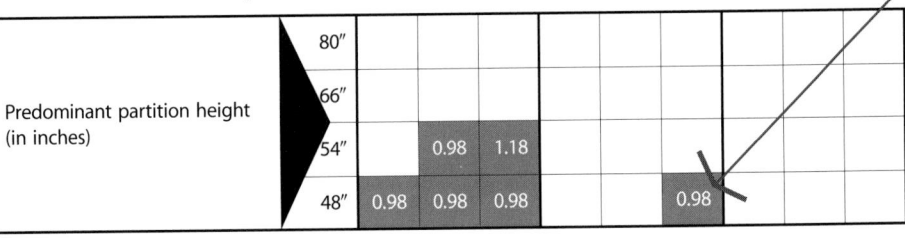

❶No photometry was available at publication time for diffuse louver option, hence this option is not covered.
❷Calculations are required in order to secure a final layout and luminaire specification for actual conditions.

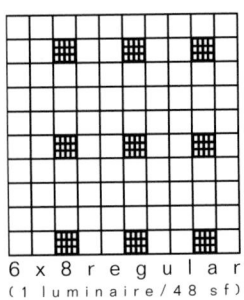

Ambient
Recessed Fluorescent

9

Ambient 2x2 Parabolic

(2) F31T8/U/1⅝ lamps/VDT sensitive/specular louver❶

Specification

⮫12–cell parabolic louver contoured up to and/or around lamp
⮫VDT–sensitive design (exhibits least glare)
⮫Specular, low iridescent clear aluminum
⮫2–F31T8/U/1⅝/830 rapid start lamps (2725 lumens each)
⮫Electronic instant start ballast (for best efficiency)
⮫See power budgeting sidebar to right
⮫Cost magnitude: high
⮫Possible vendors: Columbia, DayBrite, Lightolier, **Lithonia**, Metalux, Zumtobel/Staff

Boldfaced manufacturers' photometric data were averaged and input to Lumen Micro 7.5 in order to assess probability of meeting ambient lighting criteria cited in Table 9.4. Use the above specification list when contacting manufacturers about potential luminaire options for a specific project. Dimensions, specifications and performance are somewhat generic. Check with specific manufacturers' data for nominal dimensions and performance. Data is subject to change.

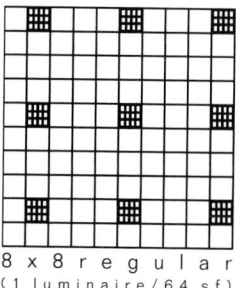

ConceptStarters™

Power budgeting
for these 2–lamp luminaires

BF 0.78 yields 52 watts
BF 0.88 yields 58 watts
BF 0.98 yields 68 watts
BF 1.18 yields 76 watts

Power budget for ambient lighting should be less than 1.0 w/sf. Ballast vendors include Advance, Energy Savings, Inc. (ESI), Magnetek, Motorola. Not all ballast factors available from all ballast vendors. Above factors based on Magnetek.

10'			9'			8'			**Ceiling height (in feet)**
20	15	10	20	15	10	20	15	10	◀ Percent floor reflectance
50	40	30	50	40	30	50	40	30	◀ Percent wall reflectance
50	30	30	50	30	30	50	30	30	◀ Percent partition reflectance
90	80	70	90	80	70	90	80	70	◀ Percent ceiling reflectance

									80"	
									66"	Predominant partition height (in inches)
									54"	
									48"	

Open cell indicates cited luminaire in cited layout has low probability of meeting **all** ambient lighting criteria outlined in Table 9.4.❷

8 x 8 regular
(1 luminaire/64 sf)

Cost magnitude²⁰⁰⁰
for recessed fluorescent luminaires

Low
US$70. to US$90.
Moderate
US$90. to US$120.
High
US$120. and greater

Quality ambient lighting for today's electronic office environment typically runs US$3.00 to US$5.00/sf, hardware only (luminaires, lamps and ballasts), excluding installation costs. Costs vary based on distributor and contractor markups, luminaire variations such as louver style/ finish, lamping, ballasting and options and accessories, and market conditions, manufacturing situations and annual inflation. These values are for preliminary, magnitude budgeting and do not represent quotes nor actual final pricing to client.

Ambient Recessed Fluorescent

9

ConceptStarters™

Ambient 2x2 Parabolic

(2) F32T8/U/6 lamps/VDT sensitive/specular louver❶

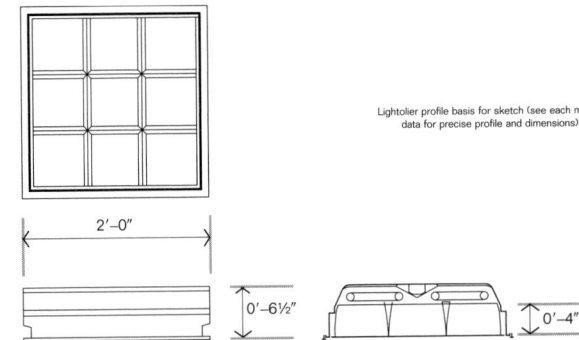

2'–0"

0'–6½"

0'–4"

Lightolier profile basis for sketch (see each mfgr's data for precise profile and dimensions)

Design Tips

✔ Specular louver looks dark.
✔ (2) F32T8/U/6 lamps (2–foot U–bent lamps) yield grazing glare.
✔ Specular louver has severe glare flash at certain angles.
✔ Specular louver shows fingerprints—difficult to clean.
✔ Use localized task lighting.
✔ Use wallwashing or liberal art lighting to improve overall impressions of brightness.

Ceiling height (in feet)		10'			9'			8'		
Percent floor reflectance	▶	20	15	10	20	15	10	20	15	10
Percent wall reflectance	▶	50	40	30	50	40	30	50	40	30
Percent partition reflectance	▶	50	30	30	50	30	30	50	30	30
Percent ceiling reflectance	▶	90	80	70	90	80	70	90	80	70

scale ¹⁄₁₆" = 1'–0"

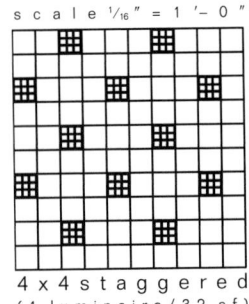

4 x 4 staggered
(1 luminaire / 32 sf)

Predominant partition height (in inches)	80"									
	66"									
	54"									
	48"									

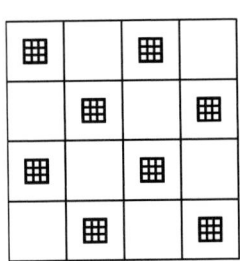

5 x 5 staggered
(1 luminaire / 50 sf)

Predominant partition height (in inches)	80"									
	66"									
	54"									
	48"									

Shaded cell indicates cited luminaire in cited layout has high probability of meeting **all** ambient lighting criteria outlined in Table 9.4. Likely ballast factor is also cited.❷

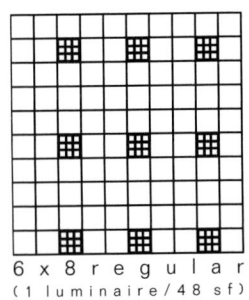

6 x 8 regular
(1 luminaire / 48 sf)

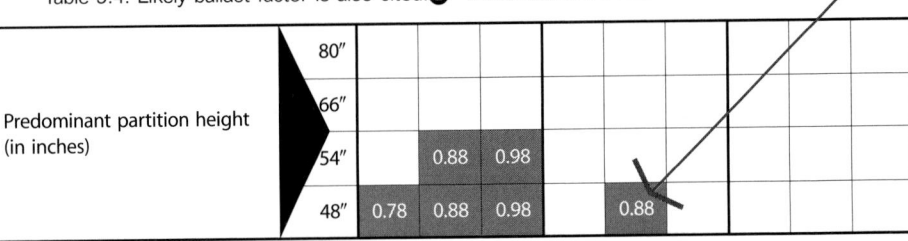

Predominant partition height (in inches)	80"									
	66"									
	54"			0.88	0.98					
	48"	0.78	0.88	0.98				0.88		

❶ No photometry was available at publication time for diffuse louver option, hence this option is not covered.
❷ Calculations are required in order to secure a final layout and luminaire specification for actual conditions.

Ambient Recessed Fluorescent

9

Ambient 2x2 Parabolic

(2) F32T8/U/6 lamps/VDT sensitive/specular louver❶

Specification

➲ 9–cell parabolic louver contoured up to and/or around lamp
➲ VDT–sensitive design (exhibits least glare)
➲ Specular, low iridescent clear aluminum
➲ 2–F32T8/U/6/830 rapid start lamps (2800 lumens each)
➲ Electronic instant start ballast (for best efficiency)
➲ See power budgeting sidebar to right
➲ Cost magnitude: high
➲ Possible vendors: Columbia, DayBrite, **Lightolier**, Lithonia, Metalux, Zumtobel/Staff

Boldfaced manufacturers' photometric data were averaged and input to Lumen Micro 7.5 in order to assess probability of meeting ambient lighting criteria cited in Table 9.4. Use the above specification list when contacting manufacturers about potential luminaire options for a specific project. Dimensions, specifications and performance are somewhat generic. Check with specific manufacturers' data for nominal dimensions and performance. Data is subject to change.

ConceptStarters™

Power budgeting
for these 2–lamp luminaires

BF 0.78 yields 52 watts
BF 0.88 yields 58 watts
BF 0.98 yields 68 watts
BF 1.18 yields 76 watts

Power budget for ambient lighting should be less than 1.0 w/sf. Ballast vendors include Advance, Energy Savings, Inc. (ESI), Magnetek, Motorola. Not all ballast factors available from all ballast vendors. Above factors based on Magnetek.

10'			9'			8'			Ceiling height (in feet)
20	15	10	20	15	10	20	15	10	◀ Percent floor reflectance
50	40	30	50	40	30	50	40	30	◀ Percent wall reflectance
50	30	30	50	30	30	50	30	30	◀ Percent partition reflectance
90	80	70	90	80	70	90	80	70	◀ Percent ceiling reflectance
									80″
									66″ ◀ Predominant partition height (in inches)
									54″
									48″

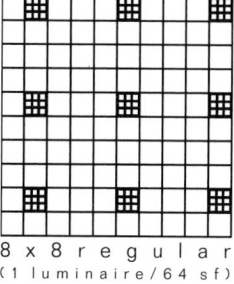

8 x 8 r e g u l a r
(1 luminaire/64 sf)

Open cell indicates cited luminaire in cited layout has low probability of meeting **all** ambient lighting criteria outlined in Table 9.4.❷

Cost magnitude[2000]
for recessed fluorescent luminaires

Low
US$70. to US$90.
Moderate
US$90. to US$120.
High
US$120. and greater

Quality ambient lighting for today's electronic office environment typically runs US$3.00 to US$5.00/sf, hardware only (luminaires, lamps and ballasts), excluding installation costs. Costs vary based on distributor and contractor markups, luminaire variations such as louver style/finish, lamping, ballasting and options and accessories, and market conditions, manufacturing situations and annual inflation. These values are for preliminary, magnitude budgeting and do not represent quotes nor actual final pricing to client.

Ambient Recessed Fluorescent

9

ConceptStarters™

Ambient 2x2 Parabolic

(2) F32T8/U/6 lamps/standard/diffuse louver

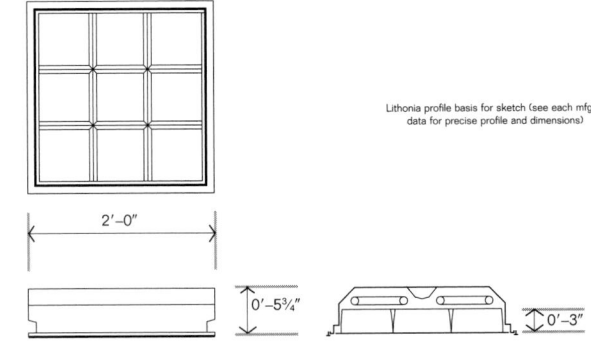

Lithonia profile basis for sketch (see each mfg'r's data for precise profile and dimensions)

2'–0"

0'–5¾"

0'–3"

Design Tips
✔ Diffuse louver improves overall impression of brightness.
✔ Diffuse louver masks fingerprints.
✔ (2) F32T8/U/6 lamps (2–foot U–bent lamps) yield grazing glare.
✔ Use localized task lighting.
✔ Use wallwashing or liberal art lighting to improve overall impressions of brightness.
✔ Only consider where computer tasks are infrequent and/or unimportant.

Ceiling height (in feet)		10'			9'			8'		
Percent floor reflectance		20	15	10	20	15	10	20	15	10
Percent wall reflectance		50	40	30	50	40	30	50	40	30
Percent partition reflectance		50	30	30	50	30	30	50	30	30
Percent ceiling reflectance		90	80	70	90	80	70	90	80	70
Predominant partition height (in inches)	80"									
	66"									
	54"									
	48"									

scale ¹⁄₁₆" = 1'–0"

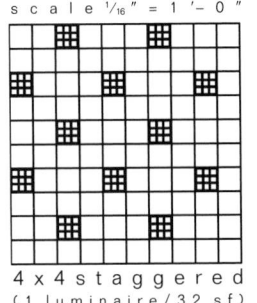

4 x 4 staggered
(1 luminaire / 32 sf)

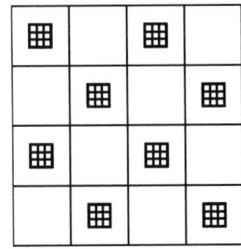

5 x 5 staggered
(1 luminaire / 50 sf)

Predominant partition height (in inches)	80"									
	66"									
	54"									
	48"									

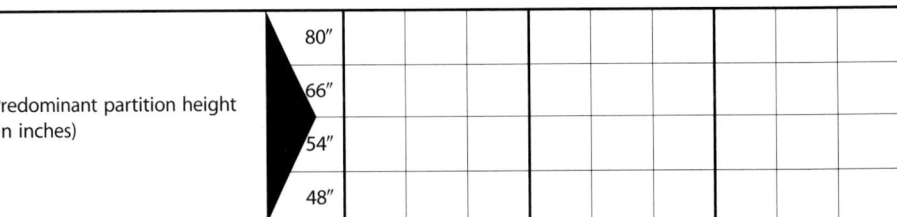

Shaded cell indicates cited luminaire in cited layout has high probability of meeting **all** ambient lighting criteria outlined in Table 9.4. Likely ballast factor is also cited.❶

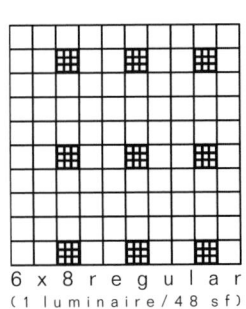

6 x 8 regular
(1 luminaire / 48 sf)

Predominant partition height (in inches)	80"									
	66"			0.88	0.98					
	54"		0.78	0.78	0.88		0.78	0.88		
	48"		0.78	0.78	0.88	0.78	0.78	0.98		

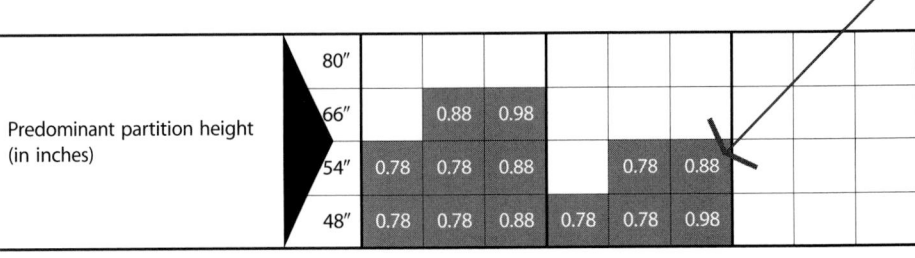

❶ Calculations are required in order to secure a final layout and luminaire specification for actual conditions.

Ambient Recessed Fluorescent

9

Ambient 2x2 Parabolic

(2) F32T8/U/6 lamps/standard/diffuse louver

Specification
- 9–cell parabolic louver
- Standard design (glare is more ubiquitous)
- Diffuse, low iridescent clear aluminum
- 2–F32T8/U/6/830 rapid start lamps (2800 lumens each)
- Electronic instant start ballast (for best efficiency)
- See power budgeting sidebar to right
- Cost magnitude: moderate
- Possible vendors: **Columbia**, DayBrite, Lightolier, **Lithonia**, Metalux, Zumtobel/Staff

Boldfaced manufacturers' photometric data were averaged and input to Lumen Micro 7.5 in order to assess probability of meeting ambient lighting criteria cited in Table 9.4. Use the above specification list when contacting manufacturers about potential luminaire options for a specific project. Dimensions, specifications and performance are somewhat generic. Check with specific manufacturers' data for nominal dimensions and performance. Data is subject to change.

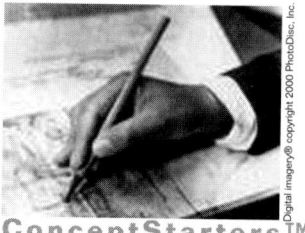

ConceptStarters™

Power budgeting
for these 2–lamp luminaires

BF 0.78 yields 52 watts
BF 0.88 yields 58 watts
BF 0.98 yields 68 watts
BF 1.18 yields 76 watts

Power budget for ambient lighting should be less than 1.0 w/sf. Ballast vendors include Advance, Energy Savings, Inc. (ESI), Magnetek, Motorola. Not all ballast factors available from all ballast vendors. Above factors based on Magnetek.

10′			9′			8′			Ceiling height (in feet)
20	15	10	20	15	10	20	15	10	Percent floor reflectance
50	40	30	50	40	30	50	40	30	Percent wall reflectance
50	30	30	50	30	30	50	30	30	Percent partition reflectance
90	80	70	90	80	70	90	80	70	Percent ceiling reflectance
									80″
									66″ Predominant partition height (in inches)
									54″
	0.98	1.18							48″

Open cell indicates cited luminaire in cited layout has low probability of meeting **all** ambient lighting criteria outlined in Table 9.4. ❶

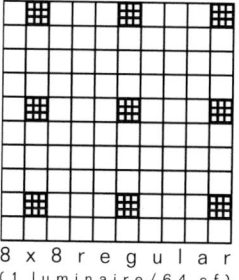

8 x 8 r e g u l a r
(1 l u m i n a i r e / 6 4 s f)

Cost magnitude[2000]
for recessed fluorescent luminaires

Low
US$70. to US$90.
Moderate
US$90. to US$120.
High
US$120. and greater

Quality ambient lighting for today's electronic office environment typically runs US$3.00 to US$5.00/sf, hardware only (luminaires, lamps and ballasts), excluding installation costs. Costs vary based on distributor and contractor markups, luminaire variations such as louver style/ finish, lamping, ballasting and options and accessories, and market conditions, manufacturing situations and annual inflation. These values are for preliminary, magnitude budgeting and do not represent quotes nor actual final pricing to client.

Ambient Recessed Fluorescent

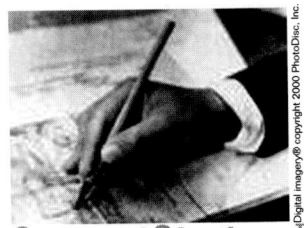

ConceptStarters™

Ambient 2x2 Parabolic

(2) F32T8/U/6 lamps/standard/specular louver

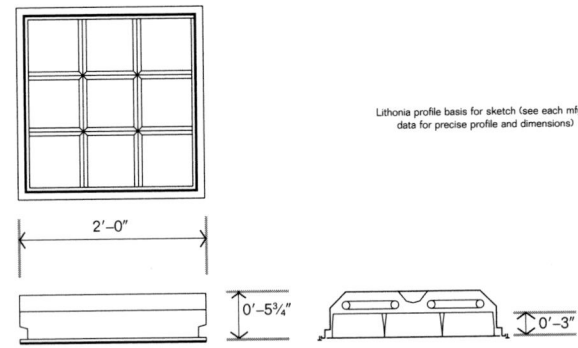

Lithonia profile basis for sketch (see each mfgr's
data for precise profile and dimensions)

2'–0"

0'–5¾"   0'–3"

Design Tips

✔ Specular louver looks dark.
✔ Low performance specular louver has severe glare flash at many angles.
✔ Specular louver shows fingerprints—difficult to clean.
✔ (2) F32T8/U/6 lamps (2-foot U–bent lamps) yield grazing glare.
✔ Use localized task lighting.
✔ Use wallwashing or liberal art lighting to improve overall impressions of brightness.
✔ Only consider where computer tasks are infrequent and/or unimportant.

Ceiling height (in feet)		10'			9'			8'		
Percent floor reflectance		20	15	10	20	15	10	20	15	10
Percent wall reflectance		50	40	30	50	40	30	50	40	30
Percent partition reflectance		50	30	30	50	30	30	50	30	30
Percent ceiling reflectance		90	80	70	90	80	70	90	80	70
Predominant partition height (in inches)	80"									
	66"									
	54"									
	48"									

scale ¹⁄₁₆" = 1'–0"

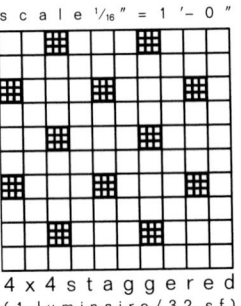

4 x 4 staggered
(1 luminaire / 32 sf)

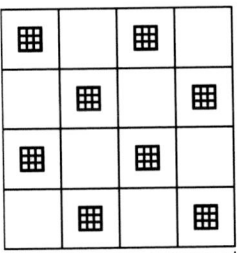

5 x 5 staggered
(1 luminaire / 50 sf)

Predominant partition height (in inches)	80"									
	66"									
	54"									
	48"									

Shaded cell indicates cited luminaire in cited layout has high probability of meeting **most** ambient lighting criteria outlined in Table 9.4. Likely ballast factor is also cited. ❶

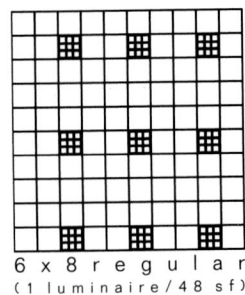

6 x 8 regular
(1 luminaire / 48 sf)

Predominant partition height (in inches)	80"									
	66"									
	54"			0.78	0.88					
	48"			0.78	0.88		0.78	0.88		

❶ Calculations are required in order to secure a final layout and luminaire specification for actual conditions.

Ambient 2x2 Parabolic

(2) F32T8/U/6 lamps/standard/specular louver

Specification

⊃ 9–cell parabolic louver
⊃ Standard design (glare is more ubiquitous)
⊃ Specular, low iridescent clear aluminum
⊃ 2–F32T8/U/6/830 rapid start lamps (2800 lumens each)
⊃ Electronic instant start ballast (for best efficiency)
⊃ See power budgeting sidebar to right
⊃ Cost magnitude: moderate
⊃ Possible vendors: **Columbia**, DayBrite, Lightolier, Lithonia, Metalux, Zumtobel/Staff

Boldfaced manufacturers' photometric data were averaged and input to Lumen Micro 7.5 in order to assess probability of meeting ambient lighting criteria cited in Table 9.4. Use the above specification list when contacting manufacturers about potential luminaire options for a specific project. Dimensions, specifications and performance are somewhat generic. Check with specific manufacturers' data for nominal dimensions and performance. Data is subject to change.

ConceptStarters™

Power budgeting
for these 2–lamp luminaires

BF 0.78 yields 52 watts
BF 0.88 yields 58 watts
BF 0.98 yields 68 watts
BF 1.18 yields 76 watts

Power budget for ambient lighting should be less than 1.0 w/sf. Ballast vendors include Advance, Energy Savings, Inc. (ESI), Magnetek, Motorola. Not all ballast factors available from all ballast vendors. Above factors based on Magnetek.

10'			9'			8'			**Ceiling height (in feet)**
20	15	10	20	15	10	20	15	10	◀ Percent floor reflectance
50	40	30	50	40	30	50	40	30	◀ Percent wall reflectance
50	30	30	50	30	30	50	30	30	◀ Percent partition reflectance
90	80	70	90	80	70	90	80	70	◀ Percent ceiling reflectance
									80"
									66" — Predominant partition height (in inches)
									54"
									48"

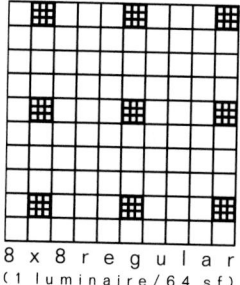

8 x 8 r e g u l a r
(1 l u m i n a i r e / 6 4 s f)

Open cell indicates cited luminaire in cited layout has low probability of meeting **most** ambient lighting criteria outlined in Table 9.4. ❶

Cost magnitude²⁰⁰⁰
for recessed fluorescent luminaires

Low
US$70. to US$90.
Moderate
US$90. to US$120.
High
US$120. and greater

Quality ambient lighting for today's electronic office environment typically runs US$3.00 to US$5.00/sf, hardware only (luminaires, lamps and ballasts), excluding installation costs. Costs vary based on distributor and contractor markups, luminaire variations such as louver style/finish, lamping, ballasting and options and accessories, and market conditions, manufacturing situations and annual inflation. These values are for preliminary, magnitude budgeting and do not represent quotes nor actual final pricing to client.

Ambient Recessed Fluorescent

ConceptStarters™

Ambient 2x2 Parabolic

(2) F32T8/U/6 lamps/VDT sensitive/specular louver❶

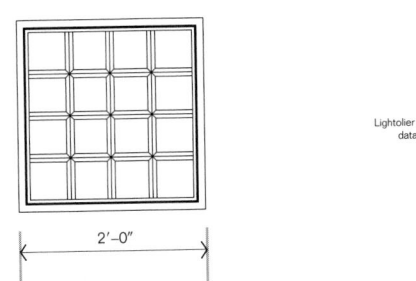

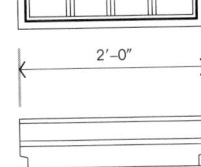

2'-0" 0'-6½" 0'-4"

Lightolier profile basis for sketch (see each mfgr's data for precise profile and dimensions)

Design Tips
✔ Specular louver looks dark.
✔ (2) F32T8/U/6 lamps (2–foot U–bent lamps) yield grazing glare.
✔ Specular louver has severe glare flash at certain angles.
✔ Specular louver shows fingerprints—difficult to clean.
✔ Use localized task lighting.
✔ Use wallwashing or liberal art lighting to improve overall impressions of brightness.

Ceiling height (in feet)		10'			9'			8'		
Percent floor reflectance		20	15	10	20	15	10	20	15	10
Percent wall reflectance		50	40	30	50	40	30	50	40	30
Percent partition reflectance		50	30	30	50	30	30	50	30	30
Percent ceiling reflectance		90	80	70	90	80	70	90	80	70

s c a l e ¹⁄₁₆ " = 1 ' - 0 "

Predominant partition height (in inches)	80"									
	66"									
	54"									
	48"									

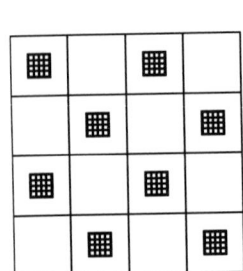

4 x 4 s t a g g e r e d
(1 luminaire / 32 sf)

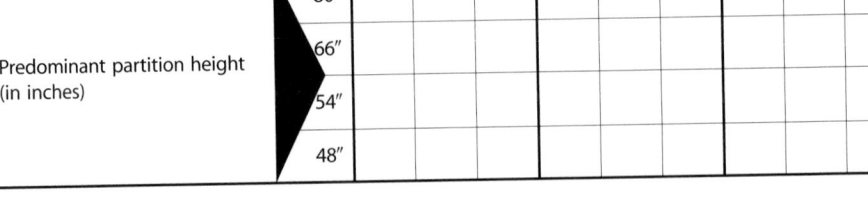

Predominant partition height (in inches)	80"									
	66"									
	54"									
	48"									

5 x 5 s t a g g e r e d
(1 luminaire / 50 sf)

Shaded cell indicates cited luminaire in cited layout has high probability of meeting **all** ambient lighting criteria outlined in Table 9.4. Likely ballast factor is also cited.❷

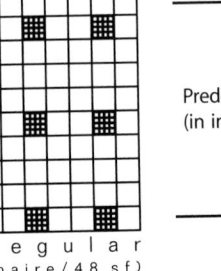

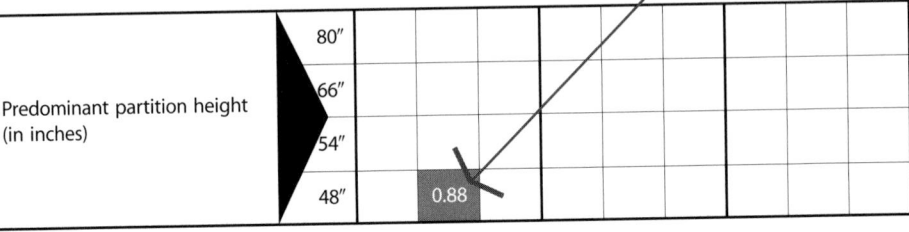

Predominant partition height (in inches)	80"									
	66"									
	54"									
	48"		0.88								

6 x 8 r e g u l a r
(1 luminaire / 48 sf)

❶ No photometry was available at publication time for diffuse louver option, hence this option is not covered.
❷ Calculations are required in order to secure a final layout and luminaire specification for actual conditions.

Ambient 2x2 Parabolic

(2) F32T8/U/6 lamps/VDT sensitive/specular louver❶

Specification

➲16–cell parabolic louver contoured up to and/or around lamp
➲VDT–sensitive design (exhibits least glare)
➲Specular, low iridescent clear aluminum
➲2–F32T8/U/6/830 rapid start lamps (2800 lumens each)
➲Electronic instant start ballast (for best efficiency)
➲See power budgeting sidebar to right
➲Cost magnitude: high
➲Possible vendors: Columbia, DayBrite, **Lightolier**, Lithonia, Metalux, Zumtobel/Staff

Boldfaced manufacturers' photometric data were averaged and input to Lumen Micro 7.5 in order to assess probability of meeting ambient lighting criteria cited in Table 9.4. Use the above specification list when contacting manufacturers about potential luminaire options for a specific project. Dimensions, specifications and performance are somewhat generic. Check with specific manufacturers' data for nominal dimensions and performance. Data is subject to change.

ConceptStarters™

Power budgeting
for these 2–lamp luminaires

BF 0.78 yields 52 watts
BF 0.88 yields 58 watts
BF 0.98 yields 68 watts
BF 1.18 yields 76 watts

Power budget for ambient lighting should be less than 1.0 w/sf. Ballast vendors include Advance, Energy Savings, Inc. (ESI), Magnetek, Motorola. Not all ballast factors available from all ballast vendors. Above factors based on Magnetek.

10'			9'			8'			Ceiling height (in feet)
20	15	10	20	15	10	20	15	10	Percent floor reflectance
50	40	30	50	40	30	50	40	30	Percent wall reflectance
50	30	30	50	30	30	50	30	30	Percent partition reflectance
90	80	70	90	80	70	90	80	70	Percent ceiling reflectance
									80"
									66" — Predominant partition height (in inches)
									54"
									48"

Open cell indicates cited luminaire in cited layout has low probability of meeting **all** ambient lighting criteria outlined in Table 9.4.❷

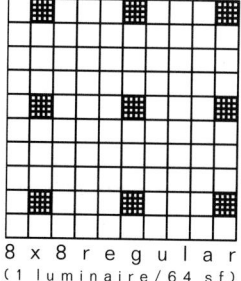

8 x 8 r e g u l a r
(1 luminaire/64 sf)

Cost magnitude[2000]
for recessed fluorescent luminaires

Low
US$70. to US$90.
Moderate
US$90. to US$120.
High
US$120. and greater

Quality ambient lighting for today's electronic office environment typically runs US$3.00 to US$5.00/sf, hardware only (luminaires, lamps and ballasts), excluding installation costs. Costs vary based on distributor and contractor markups, luminaire variations such as louver style/finish, lamping, ballasting and options and accessories, and market conditions, manufacturing situations and annual inflation. These values are for preliminary, magnitude budgeting and do not represent quotes nor actual final pricing to client.

Ambient Recessed Fluorescent

9

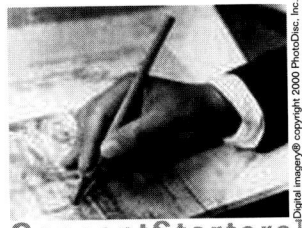

ConceptStarters™

Ambient 2x2 Parabolic

(2) F32T8/U/6 lamps/standard/diffuse louver❶

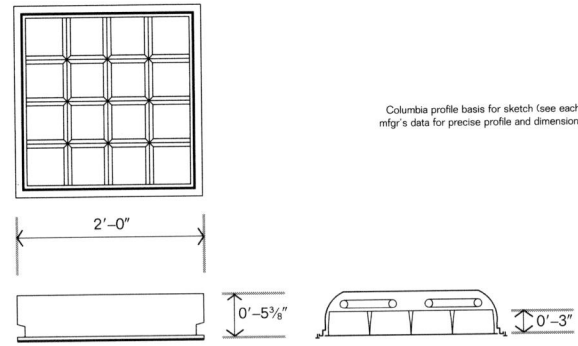

Columbia profile basis for sketch (see each mfgr's data for precise profile and dimensions)

2'–0"

0'–5³/₈"

0'–3"

Design Tips
✔ Diffuse louver improves overall impression of brightness.
✔ Diffuse louver masks fingerprints.
✔ (2) F32T8/U/6 lamps (2–foot U–bent lamps) yield grazing glare.
✔ Use localized task lighting.
✔ Use wallwashing or liberal art lighting to improve overall impressions of brightness.
✔ Only consider where computer tasks are infrequent and/or unimportant.

Ceiling height (in feet)		10'			9'			8'		
Percent floor reflectance		20	15	10	20	15	10	20	15	10
Percent wall reflectance		50	40	30	50	40	30	50	40	30
Percent partition reflectance		50	30	30	50	30	30	50	30	30
Percent ceiling reflectance		90	80	70	90	80	70	90	80	70
Predominant partition height (in inches)	80"									
	66"									
	54"									
	48"									

scale ¹/₁₆" = 1'–0"

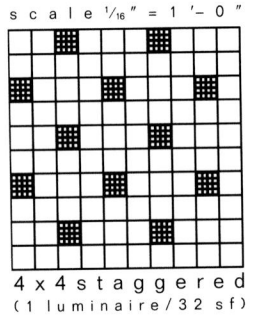

4 x 4 staggered
(1 luminaire/32 sf)

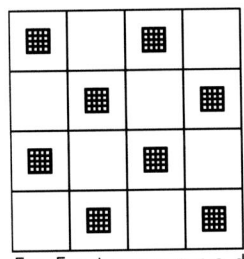

5 x 5 staggered
(1 luminaire/50 sf)

Predominant partition height (in inches)	80"									
	66"									
	54"									
	48"									

Shaded cell indicates cited luminaire in cited layout has high probability of meeting **most** ambient lighting criteria outlined in Table 9.4. Likely ballast factor is also cited.❷

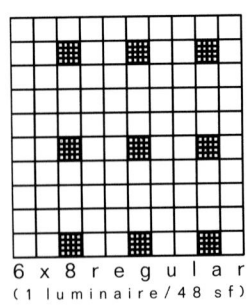

6 x 8 regular
(1 luminaire/48 sf)

Predominant partition height (in inches)	80"									
	66"									
	54"			0.88	0.98					
	48"	0.78	0.88	0.88						

Ambient Recessed Fluorescent

❶ No photometry was available at publication time for specular louver option, hence this option is not covered.
❷ Calculations are required in order to secure a final layout and luminaire specification for actual conditions.

Ambient 2x2 Parabolic
(2) F32T8/U/6 lamps/standard/diffuse louver❶

Specification
➲ 16–cell parabolic louver
➲ Standard design (glare is more ubiquitous)
➲ Diffuse, low iridescent clear aluminum
➲ 2–F32T8/U/6/830 rapid start lamps (2800 lumens each)
➲ Electronic instant start ballast (for best efficiency)
➲ See power budgeting sidebar to right
➲ Cost magnitude: moderate
➲ Possible vendors: **Columbia**, DayBrite, Lightolier, **Lithonia**, Metalux, Zumtobel/Staff

Boldfaced manufacturers' photometric data were averaged and input to Lumen Micro 7.5 in order to assess probability of meeting ambient lighting criteria cited in Table 9.4. Use the above specification list when contacting manufacturers about potential luminaire options for a specific project. Dimensions, specifications and performance are somewhat generic. Check with specific manufacturers' data for nominal dimensions and performance. Data is subject to change.

ConceptStarters™

Power budgeting
for these 2–lamp luminaires

BF 0.78 yields 52 watts
BF 0.88 yields 58 watts
BF 0.98 yields 68 watts
BF 1.18 yields 76 watts

Power budget for ambient lighting should be less than 1.0 w/sf. Ballast vendors include Advance, Energy Savings, Inc. (ESI), Magnetek, Motorola. Not all ballast factors available from all ballast vendors. Above factors based on Magnetek.

10'			9'			8'			**Ceiling height (in feet)**
20	15	10	20	15	10	20	15	10	◀ Percent floor reflectance
50	40	30	50	40	30	50	40	30	◀ Percent wall reflectance
50	30	30	50	30	30	50	30	30	◀ Percent partition reflectance
90	80	70	90	80	70	90	80	70	◀ Percent ceiling reflectance
									80"
									66" Predominant partition height (in inches)
									54"
									48"

Open cell indicates cited luminaire in cited layout has low probability of meeting **most** ambient lighting criteria outlined in Table 9.4.❷

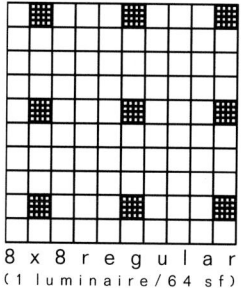
8 x 8 r e g u l a r
(1   l u m i n a i r e / 6 4   s f)

Cost magnitude²⁰⁰⁰
for recessed fluorescent luminaires

Low
US$70. to US$90.
Moderate
US$90. to US$120.
High
US$120. and greater

Quality ambient lighting for today's electronic office environment typically runs US$3.00 to US$5.00/sf, hardware only (luminaires, lamps and ballasts), excluding installation costs. Costs vary based on distributor and contractor markups, luminaire variations such as louver style/finish, lamping, ballasting and options and accessories, and market conditions, manufacturing situations and annual inflation. These values are for preliminary, magnitude budgeting and do not represent quotes nor actual final pricing to client.

Ambient Recessed Fluorescent

ConceptStarters™

Ambient 2x2 Parabolic

(2) F32T8/U/6 lamps/standard/specular louver

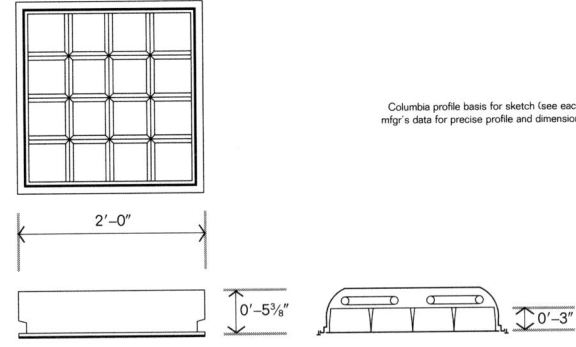

Columbia profile basis for sketch (see each
mfgr's data for precise profile and dimensions)

2'–0"

0'–5³⁄₈" 0'–3"

Design Tips
✔ Specular louver looks dark.
✔ Low performance specular louver has severe glare flash at many angles.
✔ Specular louver shows fingerprints—difficult to clean.
✔ (2) F32T8/U/6 lamps (2–foot U–bent lamps) yield grazing glare.
✔ Use localized task lighting.
✔ Use wallwashing or liberal art lighting to improve overall impressions of brightness.
✔ Only consider where computer tasks are infrequent and/or unimportant.

Ceiling height (in feet)		10'			9'			8'		
Percent floor reflectance		20	15	10	20	15	10	20	15	10
Percent wall reflectance		50	40	30	50	40	30	50	40	30
Percent partition reflectance		50	30	30	50	30	30	50	30	30
Percent ceiling reflectance		90	80	70	90	80	70	90	80	70

scale ¹⁄₁₆" = 1'–0"

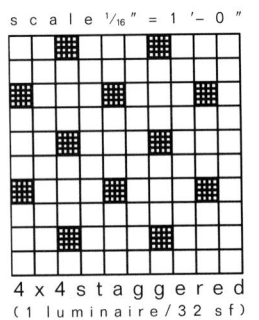

4 x 4 staggered
(1 luminaire / 32 sf)

Predominant partition height (in inches)	80"									
	66"									
	54"									
	48"									

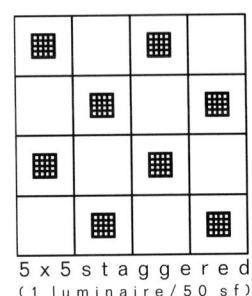

5 x 5 staggered
(1 luminaire / 50 sf)

Predominant partition height (in inches)	80"									
	66"									
	54"									
	48"									

Shaded cell indicates cited luminaire in cited layout has high probability of meeting **most** ambient lighting criteria outlined in Table 9.4. Likely ballast factor is also cited.❶

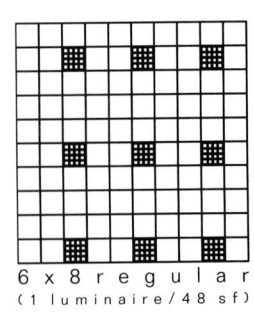

6 x 8 regular
(1 luminaire / 48 sf)

Predominant partition height (in inches)	80"									
	66"									
	54"			0.88	0.98					
	48"	0.78	0.88	0.88		0.88	0.88			

❶Calculations are required in order to secure a final layout and luminaire specification for actual conditions.

Ambient Recessed Fluorescent

Ambient 2x2 Parabolic

(2) F32T8/U/6 lamps/standard/specular louver

Specification

➲16–cell parabolic louver
➲Standard design (glare is more ubiquitous)
➲Specular, low iridescent clear aluminum
➲2–F32T8/U/6/830 rapid start lamps (2800 lumens each)
➲Electronic instant start ballast (for best efficiency)
➲See power budgeting sidebar to right
➲Cost magnitude: moderate
➲Possible vendors: **Columbia**, DayBrite, **Lightolier**, Lithonia, Metalux, Zumtobel/Staff

Boldfaced manufacturers' photometric data were averaged and input to Lumen Micro 7.5 in order to assess probability of meeting ambient lighting criteria cited in Table 9.4. Use the above specification list when contacting manufacturers about potential luminaire options for a specific project. Dimensions, specifications and performance are somewhat generic. Check with specific manufacturers' data for nominal dimensions and performance. Data is subject to change.

ConceptStarters™

Power budgeting
for these 2–lamp luminaires

BF 0.78 yields 52 watts
BF 0.88 yields 58 watts
BF 0.98 yields 68 watts
BF 1.18 yields 76 watts

Power budget for ambient lighting should be less than 1.0 w/sf. Ballast vendors include Advance, Energy Savings, Inc. (ESI), Magnetek, Motorola. Not all ballast factors available from all ballast vendors. Above factors based on Magnetek.

10'			9'			8'			Ceiling height (in feet)
20	15	10	20	15	10	20	15	10	Percent floor reflectance
50	40	30	50	40	30	50	40	30	Percent wall reflectance
50	30	30	50	30	30	50	30	30	Percent partition reflectance
90	80	70	90	80	70	90	80	70	Percent ceiling reflectance
									80"
									66" Predominant partition height (in inches)
									54"
									48"

Open cell indicates cited luminaire in cited layout has low probability of meeting **most** ambient lighting criteria outlined in Table 9.4.❶

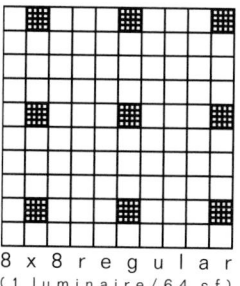

8 x 8 regular
(1 luminaire/64 sf)

Cost magnitude[2000]
for recessed fluorescent luminaires

Low
US$70. to US$90.
Moderate
US$90. to US$120.
High
US$120. and greater

Quality ambient lighting for today's electronic office environment typically runs US$3.00 to US$5.00/sf, hardware only (luminaires, lamps and ballasts), excluding installation costs. Costs vary based on distributor and contractor markups, luminaire variations such as louver style/finish, lamping, ballasting and options and accessories, and market conditions, manufacturing situations and annual inflation. These values are for preliminary, magnitude budgeting and do not represent quotes nor actual final pricing to client.

Ambient Recessed Fluorescent

ConceptStarters™

Ambient 2x4 Parabolic

(2) F32T8 lamps/standard/diffuse louver

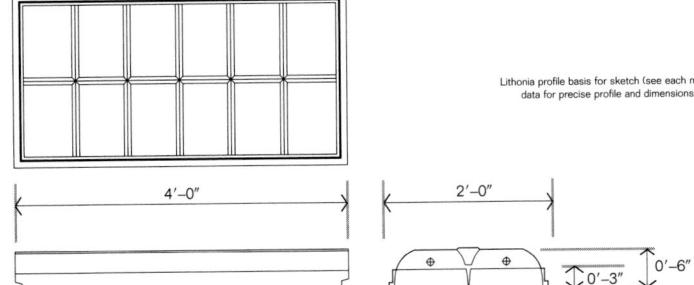

Lithonia profile basis for sketch (see each mfgr's data for precise profile and dimensions)

4'–0" 2'–0" 0'–3" 0'–6"

Design Tips

✔ Diffuse louver improves overall impression of brightness.
✔ Diffuse louver masks fingerprints.
✔ (2) F32T8 lamps 4–foot lamps) yield some grazing glare.
✔ Use localized task lighting.
✔ Use wallwashing or liberal art lighting to improve overall impressions of brightness.
✔ Only consider where computer tasks are infrequent and/or unimportant.
✔ 2x4 size is not human scale.

scale ¹⁄₁₆" = 1'–0"

4 x 4 staggered
(1 luminaire / 32 sf)

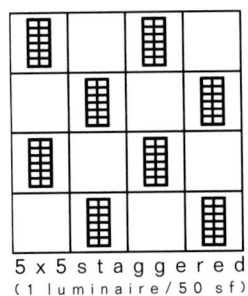

5 x 5 staggered
(1 luminaire / 50 sf)

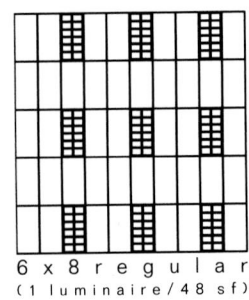

6 x 8 regular
(1 luminaire / 48 sf)

Ceiling height (in feet)		10'			9'			8'		
Percent floor reflectance		20	15	10	20	15	10	20	15	10
Percent wall reflectance		50	40	30	50	40	30	50	40	30
Percent partition reflectance		50	30	30	50	30	30	50	30	30
Percent ceiling reflectance		90	80	70	90	80	70	90	80	70

Predominant partition height (in inches)										
80"										
66"										
54"										
48"										

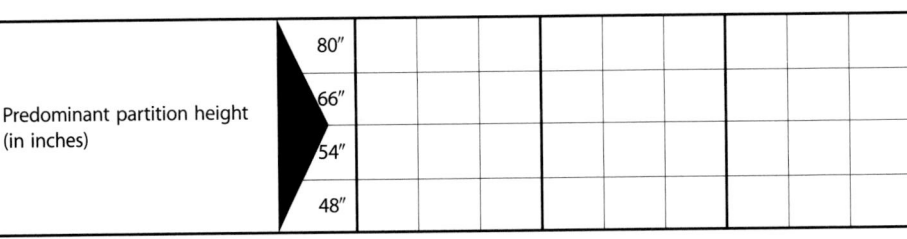

Predominant partition height (in inches)										
80"										
66"										
54"										
48"										

Shaded cell indicates cited luminaire in cited layout has high probability of meeting **most** ambient lighting criteria outlined in Table 9.4. Likely ballast factor is also cited.❶

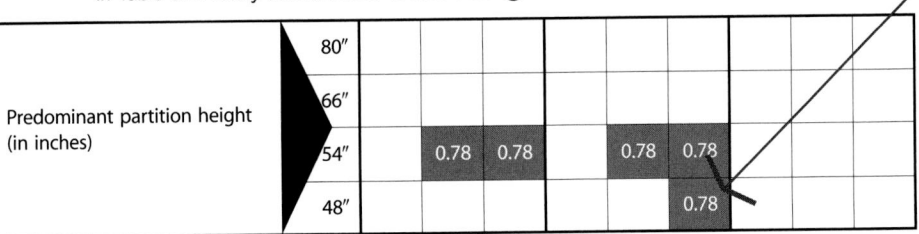

Predominant partition height (in inches)										
80"										
66"										
54"				0.78	0.78		0.78	0.78		
48"								0.78		

❶ Calculations are required in order to secure a final layout and luminaire specification for actual conditions.

Ambient Recessed Fluorescent

Ambient 2x4 Parabolic

(2) F32T8 lamps/standard/diffuse louver

Specification

⊃12–cell parabolic louver
⊃Standard design (glare is more ubiquitous)
⊃Diffuse, low iridescent clear aluminum
⊃2–F32T8/830 rapid start lamps (2950 lumens each)
⊃Electronic instant start ballast (for best efficiency)
⊃See power budgeting sidebar to right
⊃Cost magnitude: low
⊃Possible vendors: **Columbia**, DayBrite, Lightolier, **Lithonia**, Metalux, Zumtobel/Staff

Boldfaced manufacturers' photometric data were averaged and input to Lumen Micro 7.5 in order to assess probability of meeting ambient lighting criteria cited in Table 9.4. Use the above specification list when contacting manufacturers about potential luminaire options for a specific project. Dimensions, specifications and performance are somewhat generic. Check with specific manufacturers' data for nominal dimensions and performance. Data is subject to change.

ConceptStarters™

Power budgeting
for these 2–lamp luminaires

BF 0.78 yields 52 watts
BF 0.88 yields 58 watts
BF 0.98 yields 68 watts
BF 1.18 yields 76 watts

Power budget for ambient lighting should be less than 1.0 w/sf. Ballast vendors include Advance, Energy Savings, Inc. (ESI), Magnetek, Motorola. Not all ballast factors available from all ballast vendors. Above factors based on Magnetek.

10'			9'			8'			Ceiling height (in feet)
20	15	10	20	15	10	20	15	10	Percent floor reflectance
50	40	30	50	40	30	50	40	30	Percent wall reflectance
50	30	30	50	30	30	50	30	30	Percent partition reflectance
90	80	70	90	80	70	90	80	70	Percent ceiling reflectance
									80"
									66"
									54" Predominant partition height (in inches)
									48"

Open cell indicates cited luminaire in cited layout has low probability of meeting **most** ambient lighting criteria outlined in Table 9.4. ❶

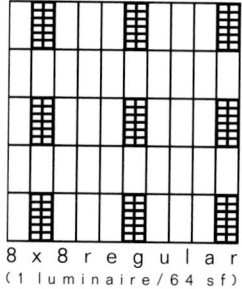

8 x 8 r e g u l a r
(1 l u m i n a i r e / 6 4 s f)

Cost magnitude²⁰⁰⁰
for recessed fluorescent luminaires

Low
US$70. to US$90.
Moderate
US$90. to US$120.
High
US$120. and greater

Quality ambient lighting for today's electronic office environment typically runs US$3.00 to US$5.00/sf, hardware only (luminaires, lamps and ballasts), excluding installation costs. Costs vary based on distributor and contractor markups, luminaire variations such as louver style/ finish, lamping, ballasting and options and accessories, and market conditions, manufacturing situations and annual inflation. These values are for preliminary, magnitude budgeting and do not represent quotes nor actual final pricing to client.

Ambient Recessed Fluorescent

9

ConceptStarters™

Ambient 2x4 Parabolic

(2) F32T8 lamps/standard/specular louver

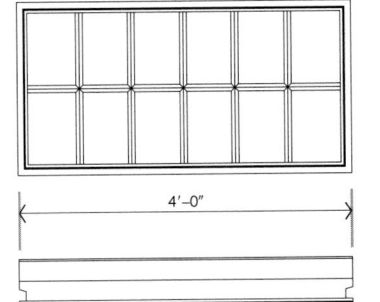

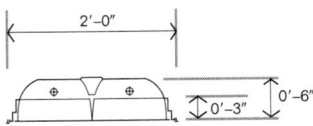

Lithonia profile basis for sketch (see each mfgr's data for precise profile and dimensions)

4'–0"

2'–0"

0'–3" 0'–6"

Design Tips

✔ Specular louver looks dark.
✔ Low performance specular louver has severe glare flash at many angles.
✔ Specular louver shows fingerprints—difficult to clean.
✔ (2) F32T8 lamps 4–foot lamps) yield some grazing glare.
✔ Use localized task lighting.
✔ Use wallwashing or liberal art lighting to improve overall impressions of brightness.
✔ Only consider where computer tasks are infrequent and/or unimportant.
✔ 2x4 size is not human–scale.

scale ¹/₁₆" = 1'–0"

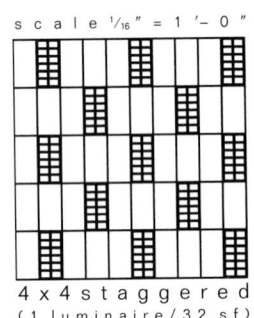

4 x 4 staggered
(1 luminaire / 32 sf)

Ceiling height (in feet)		10'			9'			8'		
Percent floor reflectance		20	15	10	20	15	10	20	15	10
Percent wall reflectance		50	40	30	50	40	30	50	40	30
Percent partition reflectance		50	30	30	50	30	30	50	30	30
Percent ceiling reflectance		90	80	70	90	80	70	90	80	70
Predominant partition height (in inches)	80"									
	66"									
	54"									
	48"									

5 x 5 staggered
(1 luminaire / 50 sf)

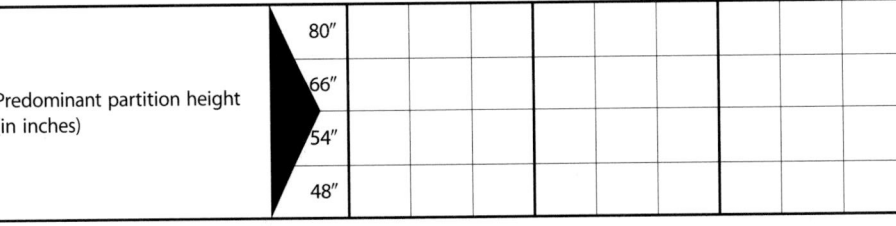

Predominant partition height (in inches)	80"									
	66"									
	54"									
	48"									

Shaded cell indicates cited luminaire in cited layout has high probability of meeting **most** ambient lighting criteria outlined in Table 9.4. Likely ballast factor is also cited.❶

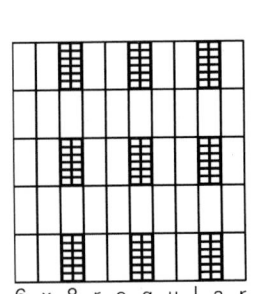

6 x 8 regular
(1 luminaire / 48 sf)

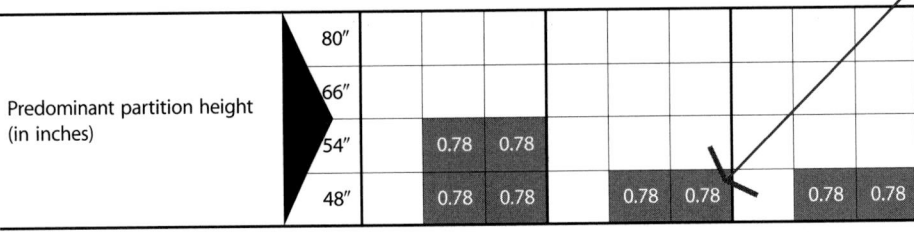

Predominant partition height (in inches)	80"									
	66"									
	54"			0.78	0.78						
	48"			0.78	0.78		0.78	0.78		0.78	0.78

❶ Calculations are required in order to secure a final layout and luminaire specification for actual conditions.

Ambient Recessed Fluorescent

9

Ambient 2x4 Parabolic

(2) F32T8 lamps/standard/specular louver

Specification

➲ 12–cell parabolic louver
➲ Standard design (glare is more ubiquitous)
➲ Specular, low iridescent clear aluminum
➲ 2–F32T8/830 rapid start lamps (2950 lumens each)
➲ Electronic instant start ballast (for best efficiency)
➲ See power budgeting sidebar to right
➲ Cost magnitude: low
➲ Possible vendors: **Columbia**, DayBrite, **Lightolier**, **Lithonia**, Metalux, Zumtobel/Staff

Boldfaced manufacturers' photometric data were averaged and input to Lumen Micro 7.5 in order to assess probability of meeting ambient lighting criteria cited in Table 9.4. Use the above specification list when contacting manufacturers about potential luminaire options for a specific project. Dimensions, specifications and performance are somewhat generic. Check with specific manufacturers' data for nominal dimensions and performance. Data is subject to change.

ConceptStarters™

Digital imagery® copyright 2000 PhotoDisc, Inc.

Power budgeting
for these 2–lamp luminaires

BF 0.78 yields 52 watts
BF 0.88 yields 58 watts
BF 0.98 yields 68 watts
BF 1.18 yields 76 watts

Power budget for ambient lighting should be less than 1.0 w/sf. Ballast vendors include Advance, Energy Savings, Inc. (ESI), Magnetek, Motorola. Not all ballast factors available from all ballast vendors. Above factors based on Magnetek.

10'			9'			8'			Ceiling height (in feet)
20	15	10	20	15	10	20	15	10	Percent floor reflectance
50	40	30	50	40	30	50	40	30	Percent wall reflectance
50	30	30	50	30	30	50	30	30	Percent partition reflectance
90	80	70	90	80	70	90	80	70	Percent ceiling reflectance
									80"
									66"
0.78	0.98	0.98							54" — Predominant partition height (in inches)
0.78	0.88	0.88	0.78	0.88			0.88		48"

Open cell indicates cited luminaire in cited layout has low probability of meeting **most** ambient lighting criteria outlined in Table 9.4. ➊

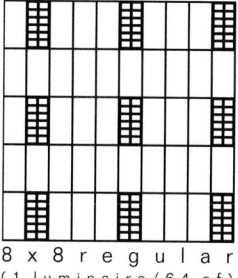

8 x 8 r e g u l a r
(1 l u m i n a i r e / 6 4 s f)

Net Addresses/Luminaires
http://www.columbia-ltg.com/products/parabolics.html
http://www.thomaslighting.com/daybrite/default.htm
http://www.lightolier.com/html/v4home.htm
http://www.lithonia.com/
http://www.cooperlighting.com/
http://www.zumtobelstaff.co.at/ [Note: German language]

Net Addresses/Ballasts
http://www.advancetransformer.com/
http://www.magnetek.com/transLighting.html
http://www.mot.com/ies/MLI/

CONNECT FOR MORE

Cost magnitude²⁰⁰⁰
for recessed fluorescent luminaires

Low
US$70. to US$90.
Moderate
US$90. to US$120.
High
US$120. and greater

Quality ambient lighting for today's electronic office environment typically runs US$3.00 to US$5.00/sf, hardware only (luminaires, lamps and ballasts), excluding installation costs. Costs vary based on distributor and contractor markups, luminaire variations such as louver style/finish, lamping, ballasting and options and accessories, and market conditions, manufacturing situations and annual inflation. These values are for preliminary, magnitude budgeting and do not represent quotes nor actual final pricing to client.

Ambient Recessed Fluorescent

9

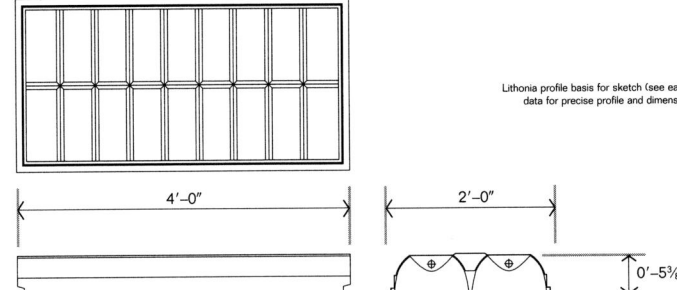

Ambient 2x4 Parabolic

(2) F32T8 lamps/VDT sensitive/diffuse louver

ⒸDigital imagery® copyright 2000 PhotoDisc, Inc.

ConceptStarters™

Lithonia profile basis for sketch (see each mfgr's data for precise profile and dimensions)

4'–0"   2'–0"   0'–5⅜"

Design Tips

✔ Diffuse louver improves overall impression of brightness.
✔ Diffuse louver masks fingerprints.
✔ (2) F32T8 lamps 4–foot lamps) yield some grazing glare.
✔ Use localized task lighting.
✔ Use wallwashing or liberal art lighting to improve overall impressions of brightness.
✔ 2x4 size is not human–scale.

Ceiling height (in feet)		10'			9'			8'		
Percent floor reflectance	▶	20	15	10	20	15	10	20	15	10
Percent wall reflectance	▶	50	40	30	50	40	30	50	40	30
Percent partition reflectance	▶	50	30	30	50	30	30	50	30	30
Percent ceiling reflectance	▶	90	80	70	90	80	70	90	80	70

scale 1/16" = 1'–0"

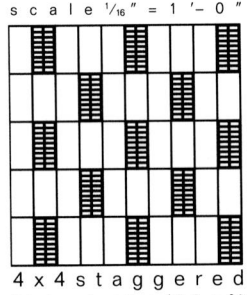

4 x 4 staggered
(1 luminaire / 32 sf)

Predominant partition height (in inches)										
80"										
66"										
54"										
48"										

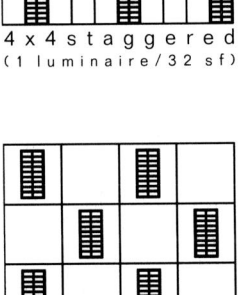

5 x 5 staggered
(1 luminaire / 50 sf)

Predominant partition height (in inches)										
80"										
66"										
54"										
48"										

Shaded cell indicates cited luminaire in cited layout has high probability of meeting **most** ambient lighting criteria outlined in Table 9.4. Likely ballast factor is also cited. ❶

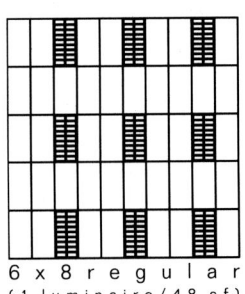

6 x 8 regular
(1 luminaire / 48 sf)

Predominant partition height (in inches)		20	15	10	20	15	10	20	15	10
80"										
66"		0.78	0.88	0.88						
54"		0.78	0.78	0.78		0.78	0.88			
48"		0.78	0.78	0.78	0.78	0.78	0.78			

❶ Calculations are required in order to secure a final layout and luminaire specification for actual conditions.

Ambient 2x4 Parabolic

(2) F32T8 lamps/VDT sensitive/diffuse louver

Specification

⊃18–cell parabolic louver contoured up to and/or around lamp
⊃VDT–sensitive design (exhibits least glare)
⊃Diffuse, low iridescent clear aluminum
⊃2–F32T8/830 rapid start lamps (2950 lumens each)
⊃Electronic instant start ballast (for best efficiency)
⊃See power budgeting sidebar to right
⊃Cost magnitude: moderate
⊃Possible vendors: Columbia, DayBrite, Lightolier, **Lithonia**, Metalux, Zumtobel/Staff

Boldfaced manufacturers' photometric data were averaged and input to Lumen Micro 7.5 in order to assess probability of meeting ambient lighting criteria cited in Table 9.4. Use the above specification list when contacting manufacturers about potential luminaire options for a specific project. Dimensions, specifications and performance are somewhat generic. Check with specific manufacturers' data for nominal dimensions and performance. Data is subject to change.

ConceptStarters™

Power budgeting
for these 2–lamp luminaires

BF 0.78 yields 52 watts
BF 0.88 yields 58 watts
BF 0.98 yields 68 watts
BF 1.18 yields 76 watts

Power budget for ambient lighting should be less than 1.0 w/sf. Ballast vendors include Advance, Energy Savings, Inc. (ESI), Magnetek, Motorola. Not all ballast factors available from all ballast vendors. Above factors based on Magnetek.

10'			9'			8'			Ceiling height (in feet)
20	15	10	20	15	10	20	15	10	Percent floor reflectance
50	40	30	50	40	30	50	40	30	Percent wall reflectance
50	30	30	50	30	30	50	30	30	Percent partition reflectance
90	80	70	90	80	70	90	80	70	Percent ceiling reflectance
								80"	Predominant partition height (in inches)
								66"	
								54"	
								48"	

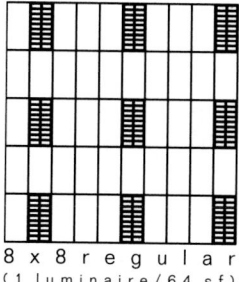

8 x 8 r e g u l a r
(1 luminaire / 64 sf)

Open cell indicates cited luminaire in cited layout has low probability of meeting **most** ambient lighting criteria outlined in Table 9.4. ❶

Cost magnitude²⁰⁰⁰
for recessed fluorescent luminaires

Low
US$70. to US$90.
Moderate
US$90. to US$120.
High
US$120. and greater

Quality ambient lighting for today's electronic office environment typically runs US$3.00 to US$5.00/sf, hardware only (luminaires, lamps and ballasts), excluding installation costs. Costs vary based on distributor and contractor markups, luminaire variations such as louver style/ finish, lamping, ballasting and options and accessories, and market conditions, manufacturing situations and annual inflation. These values are for preliminary, magnitude budgeting and do not represent quotes nor actual final pricing to client.

Ambient Recessed Fluorescent

Ambient 2x4 Parabolic
(2) F32T8 lamps/VDT sensitive/specular louver

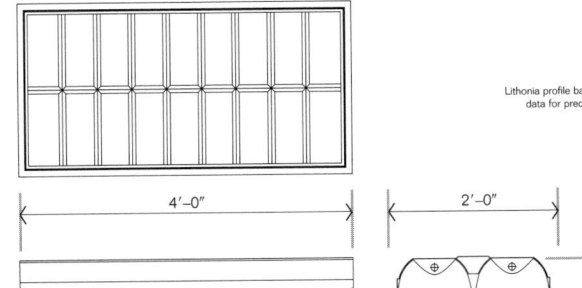

ConceptStarters™

Lithonia profile basis for sketch (see each mfgr's data for precise profile and dimensions)

4'–0"   2'–0"   0'–5⅜"

Design Tips
✔ Specular louver looks dark.
✔ Specular louver has severe glare flash at certain angles.
✔ Specular louver shows fingerprints—difficult to clean.
✔ (2) F32T8 lamps 4–foot lamps) yield some grazing glare.
✔ Use localized task lighting.
✔ Use wallwashing or liberal art lighting to improve overall impressions of brightness.
✔ 2x4 size is not human–scale.

Ceiling height (in feet)		10'			9'			8'		
Percent floor reflectance		20	15	10	20	15	10	20	15	10
Percent wall reflectance		50	40	30	50	40	30	50	40	30
Percent partition reflectance		50	30	30	50	30	30	50	30	30
Percent ceiling reflectance		90	80	70	90	80	70	90	80	70
Predominant partition height (in inches)	80"									
	66"									
	54"									
	48"									

s c a l e ¹/₁₆" = 1'–0"

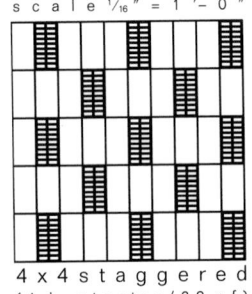

4 x 4 s t a g g e r e d
(1 luminaire / 32 sf)

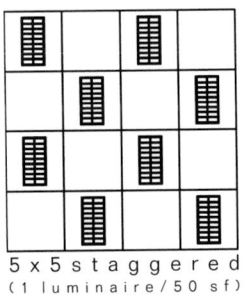

5 x 5 s t a g g e r e d
(1 luminaire / 50 sf)

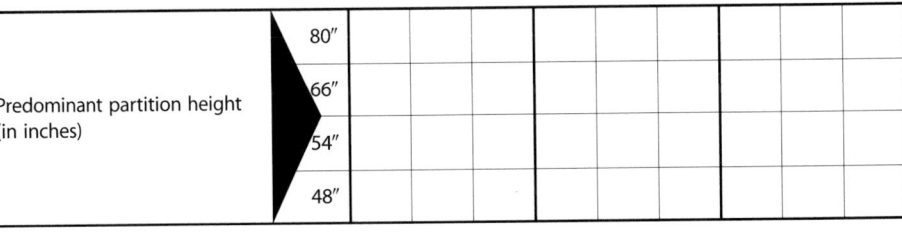

Predominant partition height (in inches)	80"									
	66"									
	54"									
	48"									

Shaded cell indicates cited luminaire in cited layout has high probability of meeting **most** ambient lighting criteria outlined in Table 9.4. Likely ballast factor is also cited. ❶

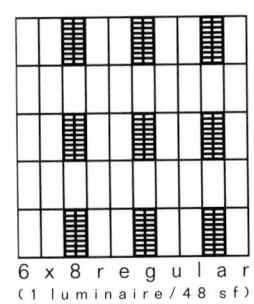

6 x 8 r e g u l a r
(1 luminaire / 48 sf)

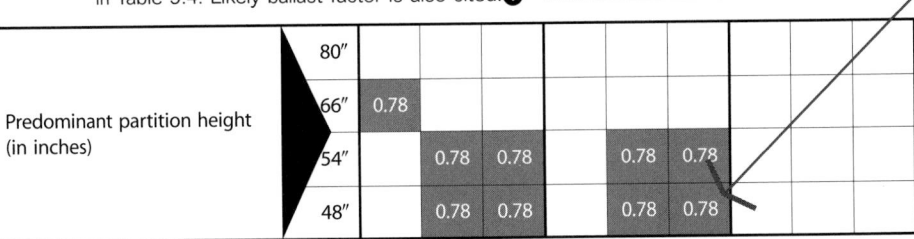

Predominant partition height (in inches)	80"									
	66"	0.78								
	54"		0.78	0.78		0.78	0.78			
	48"		0.78	0.78		0.78	0.78			

❶ Calculations are required in order to secure a final layout and luminaire specification for actual conditions.

Ambient Recessed Fluorescent

Ambient 2x4 Parabolic

(2) F32T8 lamps/VDT sensitive/specular louver

Specification

⟳ 18–cell parabolic louver contoured up to and/or around lamp
⟳ VDT–sensitive design (exhibits least glare)
⟳ Specular, low iridescent clear aluminum
⟳ 2–F32T8/830 rapid start lamps (2950 lumens each)
⟳ Electronic instant start ballast (for best efficiency)
⟳ See power budgeting sidebar to right
⟳ Cost magnitude: moderate
⟳ Possible vendors: Columbia, DayBrite, **Lightolier**, **Lithonia**, Metalux, Zumtobel/Staff

Boldfaced manufacturers' photometric data were averaged and input to Lumen Micro 7.5 in order to assess probability of meeting ambient lighting criteria cited in Table 9.4. Use the above specification list when contacting manufacturers about potential luminaire options for a specific project. Dimensions, specifications and performance are somewhat generic. Check with specific manufacturers' data for nominal dimensions and performance. Data is subject to change.

ConceptStarters™

Power budgeting
for these 2–lamp luminaires

BF 0.78 yields 52 watts
BF 0.88 yields 58 watts
BF 0.98 yields 68 watts
BF 1.18 yields 76 watts

Power budget for ambient lighting should be less than 1.0 w/sf. Ballast vendors include Advance, Energy Savings, Inc. (ESI), Magnetek, Motorola. Not all ballast factors available from all ballast vendors. Above factors based on Magnetek.

10'			9'			8'			Ceiling height (in feet)
20	15	10	20	15	10	20	15	10	◀ Percent floor reflectance
50	40	30	50	40	30	50	40	30	◀ Percent wall reflectance
50	30	30	50	30	30	50	30	30	◀ Percent partition reflectance
90	80	70	90	80	70	90	80	70	◀ Percent ceiling reflectance
									80″
									66″ — Predominant partition height (in inches)
									54″
	0.78	0.88							48″

Open cell indicates cited luminaire in cited layout has low probability of meeting **most** ambient lighting criteria outlined in Table 9.4.❶

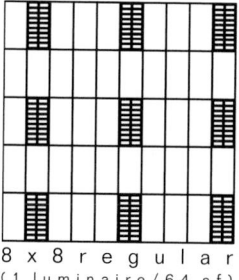

8 x 8 r e g u l a r
(1 luminaire/64 sf)

Cost magnitude²⁰⁰⁰
for recessed fluorescent luminaires

Low
US$70. to US$90.
Moderate
US$90. to US$120.
High
US$120. and greater

Quality ambient lighting for today's electronic office environment typically runs US$3.00 to US$5.00/sf, hardware only (luminaires, lamps and ballasts), excluding installation costs. Costs vary based on distributor and contractor markups, luminaire variations such as louver style/finish, lamping, ballasting and options and accessories, and market conditions, manufacturing situations and annual inflation. These values are for preliminary, magnitude budgeting and do not represent quotes nor actual final pricing to client.

Ambient Recessed Fluorescent

9

ConceptStarters™

Ambient 2x4 Parabolic

(2) F32T8 lamps/standard/diffuse louver

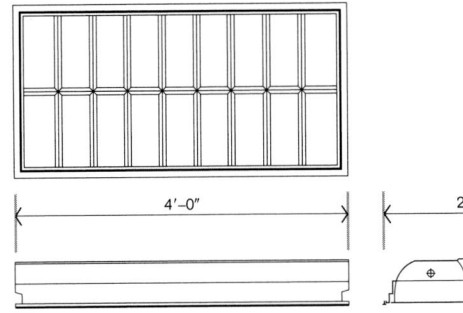

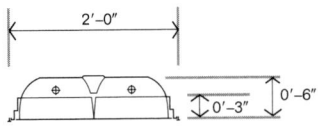

Lithonia profile basis for sketch (see each mfgr's data for precise profile and dimensions)

4'-0"   2'-0"   0'-6"   0'-3"

Design Tips

✔Diffuse louver improves overall impression of brightness.
✔Diffuse louver masks fingerprints.
✔(2) F32T8 lamps 4–foot lamps) yield grazing glare.
✔Use localized task lighting.
✔Use wallwashing or liberal art lighting to improve overall impressions of brightness.
✔Only consider where computer tasks are infrequent and/or unimportant.
✔2x4 size is not human scale.

scale ¹⁄₁₆" = 1'-0"

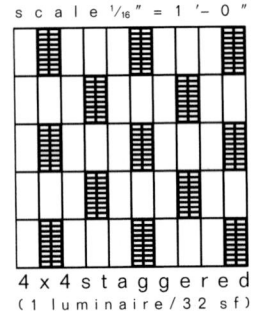

4 x 4 staggered
(1 luminaire / 32 sf)

Ceiling height (in feet)		10'			9'			8'		
Percent floor reflectance		20	15	10	20	15	10	20	15	10
Percent wall reflectance		50	40	30	50	40	30	50	40	30
Percent partition reflectance		50	30	30	50	30	30	50	30	30
Percent ceiling reflectance		90	80	70	90	80	70	90	80	70
Predominant partition height (in inches)	80"									
	66"									
	54"									
	48"									

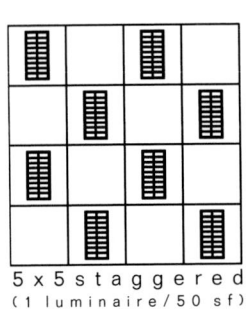

5 x 5 staggered
(1 luminaire / 50 sf)

Predominant partition height (in inches)	80"									
	66"									
	54"									
	48"									

Shaded cell indicates cited luminaire in cited layout has high probability of meeting **most** ambient lighting criteria outlined in Table 9.4. Likely ballast factor is also cited.❶

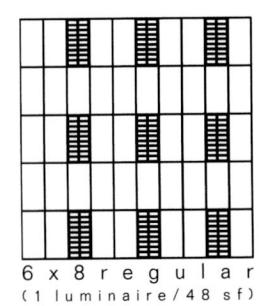

6 x 8 regular
(1 luminaire / 48 sf)

Predominant partition height (in inches)	80"									
	66"									
	54"		0.78	0.78	0.88		0.78	0.88		
	48"			0.78	0.78		0.78	0.78		

❶Calculations are required in order to secure a final layout and luminaire specification for actual conditions.

Ambient Recessed Fluorescent

9

Ambient 2x4 Parabolic

(2) F32T8 lamps/standard/diffuse louver

Specification

➲ 18–cell parabolic louver
➲ Standard design (glare is more ubiquitous)
➲ Diffuse, low iridescent clear aluminum
➲ 2–F32T8/830 rapid start lamps (2950 lumens each)
➲ Electronic instant start ballast (for best efficiency)
➲ See power budgeting sidebar to right
➲ Cost magnitude: low
➲ Possible vendors: **Columbia**, DayBrite, Lightolier, **Lithonia**, Metalux, Zumtobel/Staff

Boldfaced manufacturers' photometric data were averaged and input to Lumen Micro 7.5 in order to assess probability of meeting ambient lighting criteria cited in Table 9.4. Use the above specification list when contacting manufacturers about potential luminaire options for a specific project. Dimensions, specifications and performance are somewhat generic. Check with specific manufacturers' data for nominal dimensions and performance. Data is subject to change.

ConceptStarters™

Power budgeting
for these 2–lamp luminaires

BF 0.78 yields 52 watts
BF 0.88 yields 58 watts
BF 0.98 yields 68 watts
BF 1.18 yields 76 watts

Power budget for ambient lighting should be less than 1.0 w/sf. Ballast vendors include Advance, Energy Savings, Inc. (ESI), Magnetek, Motorola. Not all ballast factors available from all ballast vendors. Above factors based on Magnetek.

10'			9'			8'			Ceiling height (in feet)
20	15	10	20	15	10	20	15	10	Percent floor reflectance
50	40	30	50	40	30	50	40	30	Percent wall reflectance
50	30	30	50	30	30	50	30	30	Percent partition reflectance
90	80	70	90	80	70	90	80	70	Percent ceiling reflectance
									80"
									66" — Predominant partition height (in inches)
									54"
0.88	0.88								48"

Open cell indicates cited luminaire in cited layout has low probability of meeting **most** ambient lighting criteria outlined in Table 9.4.❶

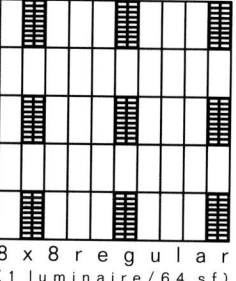

8 x 8 regular
(1 luminaire/64 sf)

Cost magnitude²⁰⁰⁰
for recessed fluorescent luminaires

Low
US$70. to US$90.
Moderate
US$90. to US$120.
High
US$120. and greater

Quality ambient lighting for today's electronic office environment typically runs US$3.00 to US$5.00/sf, hardware only (luminaires, lamps and ballasts), excluding installation costs. Costs vary based on distributor and contractor markups, luminaire variations such as louver style/finish, lamping, ballasting and options and accessories, and market conditions, manufacturing situations and annual inflation. These values are for preliminary, magnitude budgeting and do not represent quotes nor actual final pricing to client.

Ambient Recessed Fluorescent

ConceptStarters™

iDigital imagery® copyright 2000 PhotoDisc, Inc.

Ambient 2x4 Parabolic

(2) F32T8 lamps/standard/specular louver

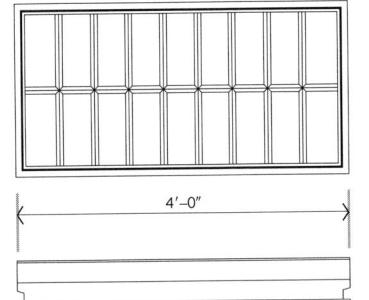

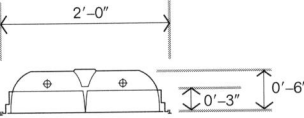

Lithonia profile basis for sketch (see each mfgr's data for precise profile and dimensions)

4'–0"   2'–0"   0'–6"   0'–3"

Design Tips

✔ Specular louver looks dark.
✔ Low performance specular louver has severe glare flash at many angles.
✔ Specular louver shows fingerprints—difficult to clean.
✔ (2) F32T8 lamps 4–foot lamps) yield some grazing glare.
✔ Use localized task lighting.
✔ Use wallwashing or liberal art lighting to improve overall impressions of brightness.
✔ Only consider where computer tasks are infrequent and/or unimportant.
✔ 2x4 size is not human–scale.

Ceiling height (in feet)	10'			9'			8'		
Percent floor reflectance	20	15	10	20	15	10	20	15	10
Percent wall reflectance	50	40	30	50	40	30	50	40	30
Percent partition reflectance	50	30	30	50	30	30	50	30	30
Percent ceiling reflectance	90	80	70	90	80	70	90	80	70

scale 1/16" = 1'–0"

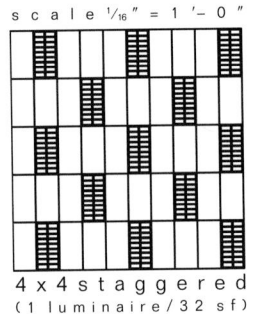

4 x 4 staggered
(1 luminaire / 32 sf)

Predominant partition height (in inches) — 80", 66", 54", 48"

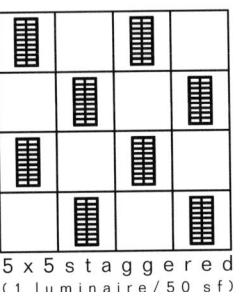

5 x 5 staggered
(1 luminaire / 50 sf)

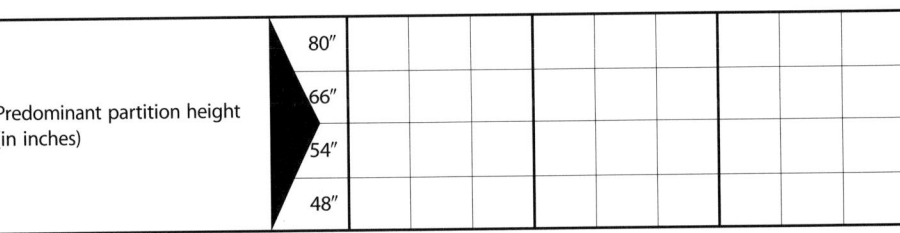

Predominant partition height (in inches) — 80", 66", 54", 48"

Shaded cell indicates cited luminaire in cited layout has high probability of meeting **most** ambient lighting criteria outlined in Table 9.4. Likely ballast factor is also cited. ❶

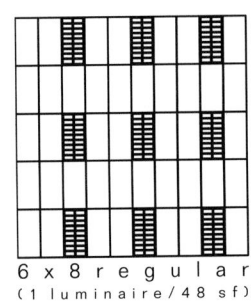

6 x 8 regular
(1 luminaire / 48 sf)

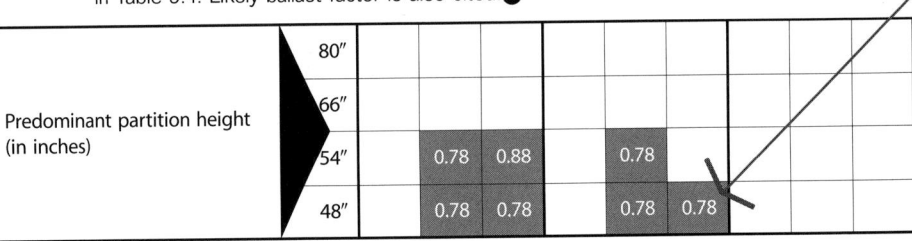

Predominant partition height (in inches)

	80"								
66"									
54"		0.78	0.88		0.78				
48"		0.78	0.78		0.78	0.78			

❶ Calculations are required in order to secure a final layout and luminaire specification for actual conditions.

Ambient Recessed Fluorescent

9

Ambient 2x4 Parabolic

(2) F32T8 lamps/standard/specular louver

Specification

➲18–cell parabolic louver
➲Standard design (glare is more ubiquitous)
➲Specular, low iridescent clear aluminum
➲2–F32T8/830 rapid start lamps (2950 lumens each)
➲Electronic instant start ballast (for best efficiency)
➲See power budgeting sidebar to right
➲Cost magnitude: low
➲Possible vendors: Columbia, DayBrite, Lightolier, **Lithonia**, Metalux, Zumtobel/Staff

Boldfaced manufacturers' photometric data were averaged and input to Lumen Micro 7.5 in order to assess probability of meeting ambient lighting criteria cited in Table 9.4. Use the above specification list when contacting manufacturers about potential luminaire options for a specific project. Dimensions, specifications and performance are somewhat generic. Check with specific manufacturers' data for nominal dimensions and performance. Data is subject to change.

C o n c e p t S t a r t e r s™

Power budgeting
for these 2–lamp luminaires

BF 0.78 yields 52 watts
BF 0.88 yields 58 watts
BF 0.98 yields 68 watts
BF 1.18 yields 76 watts

Power budget for ambient lighting should be less than 1.0 w/sf. Ballast vendors include Advance, Energy Savings, Inc. (ESI), Magnetek, Motorola. Not all ballast factors available from all ballast vendors. Above factors based on Magnetek.

10'			9'			8'			Ceiling height (in feet)
20	15	10	20	15	10	20	15	10	Percent floor reflectance
50	40	30	50	40	30	50	40	30	Percent wall reflectance
50	30	30	50	30	30	50	30	30	Percent partition reflectance
90	80	70	90	80	70	90	80	70	Percent ceiling reflectance
									80″
									66″ Predominant partition height (in inches)
									54″
	0.78	0.88							48″

Open cell indicates cited luminaire in cited layout has low probability of meeting **most** ambient lighting criteria outlined in Table 9.4.❶

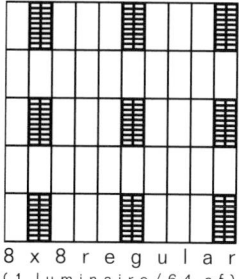
8 x 8 r e g u l a r
(1 luminaire / 64 sf)

Cost magnitude²⁰⁰⁰
for recessed fluorescent luminaires

Low
US$70. to US$90.
Moderate
US$90. to US$120.
High
US$120. and greater

Quality ambient lighting for today's electronic office environment typically runs US$3.00 to US$5.00/sf, hardware only (luminaires, lamps and ballasts), excluding installation costs. Costs vary based on distributor and contractor markups, luminaire variations such as louver style/finish, lamping, ballasting and options and accessories, and market conditions, manufacturing situations and annual inflation. These values are for preliminary, magnitude budgeting and do not represent quotes nor actual final pricing to client.

Ambient Recessed Fluorescent

ConceptStarters™

Ambient 2x4 Parabolic

(2) F32T8 lamps/VDT sensitive/specular louver❶

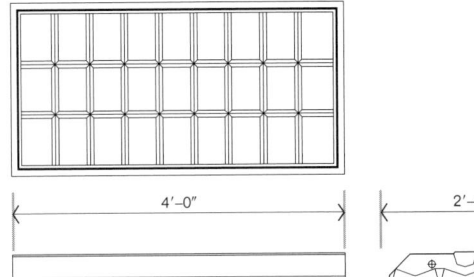

Lithonia profile basis for sketch (see each mfgr's data for precise profile and dimensions)

4'–0" 2'–0" 0'–6¾"

Design Tips

✔ Specular louver looks dark.

✔ Specular louver has severe glare flash at certain angles.

✔ Specular louver shows fingerprints—difficult to clean.

✔ (2) F32T8 lamps 4–foot lamps) yield some grazing glare.

✔ Use localized task lighting.

✔ Use wallwashing or liberal art lighting to improve overall impressions of brightness.

✔ 2x4 size is not human–scale.

scale ¹⁄₁₆" = 1'–0"

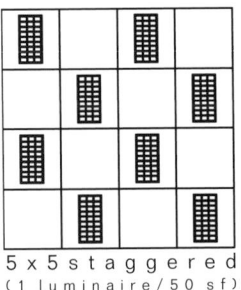

4 x 4 staggered
(1 luminaire / 32 sf)

Ceiling height (in feet)		10'			9'			8'		
Percent floor reflectance		20	15	10	20	15	10	20	15	10
Percent wall reflectance		50	40	30	50	40	30	50	40	30
Percent partition reflectance		50	30	30	50	30	30	50	30	30
Percent ceiling reflectance		90	80	70	90	80	70	90	80	70
Predominant partition height (in inches)	80"									
	66"									
	54"									
	48"									

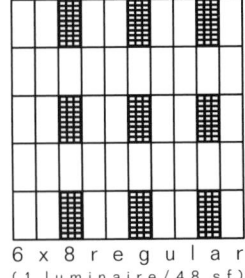

5 x 5 staggered
(1 luminaire / 50 sf)

Predominant partition height (in inches)	80"									
	66"									
	54"									
	48"									

Shaded cell indicates cited luminaire in cited layout has high probability of meeting **most** ambient lighting criteria outlined in Table 9.4. Likely ballast factor is also cited.❷

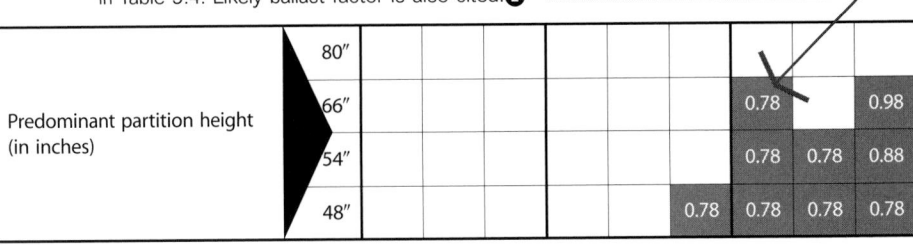

Predominant partition height (in inches)	80"									
	66"								0.78		0.98
	54"								0.78	0.78	0.88
	48"							0.78	0.78	0.78	0.78

6 x 8 regular
(1 luminaire / 48 sf)

❶ No photometry was available at publication time for diffuse louver option, hence this option is not covered.

❷ Calculations are required in order to secure a final layout and luminaire specification for actual conditions.

Ambient Recessed Fluorescent

9

Ambient 2x4 Parabolic

(2) F32T8 lamps/VDT sensitive/specular louver❶

Specification

⊃24– or 27–cell parabolic louver contoured up to and/or around lamp
⊃VDT–sensitive design (exhibits least glare)
⊃Specular, low iridescent clear aluminum
⊃2–F32T8/830 rapid start lamps (2950 lumens each)
⊃Electronic instant start ballast (for best efficiency)
⊃See power budgeting sidebar to right
⊃Cost magnitude: moderate
⊃Possible vendors: **Columbia**, DayBrite, Lightolier, Lithonia, Metalux, Zumtobel/Staff

Boldfaced manufacturers' photometric data were averaged and input to Lumen Micro 7.5 in order to assess probability of meeting ambient lighting criteria cited in Table 9.4. Use the above specification list when contacting manufacturers about potential luminaire options for a specific project. Dimensions, specifications and performance are somewhat generic. Check with specific manufacturers' data for nominal dimensions and performance. Data is subject to change.

ConceptStarters™

Power budgeting
for these 2–lamp luminaires

BF 0.78 yields 52 watts
BF 0.88 yields 58 watts
BF 0.98 yields 68 watts
BF 1.18 yields 76 watts

Power budget for ambient lighting should be less than 1.0 w/sf. Ballast vendors include Advance, Energy Savings, Inc. (ESI), Magnetek, Motorola. Not all ballast factors available from all ballast vendors. Above factors based on Magnetek.

10'			9'			8'			**Ceiling height (in feet)**
20	15	10	20	15	10	20	15	10	◀ Percent floor reflectance
50	40	30	50	40	30	50	40	30	◀ Percent wall reflectance
50	30	30	50	30	30	50	30	30	◀ Percent partition reflectance
90	80	70	90	80	70	90	80	70	◀ Percent ceiling reflectance
									80″
									66″ — Predominant partition height (in inches)
									54″
									48″

Open cell indicates cited luminaire in cited layout has low probability of meeting **most** ambient lighting criteria outlined in Table 9.4.❷

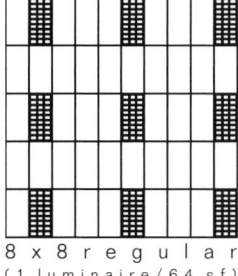

8 x 8 r e g u l a r
(1 l u m i n a i r e / 6 4 s f)

Cost magnitude²⁰⁰⁰
for recessed fluorescent luminaires

Low
US$70. to US$90.
Moderate
US$90. to US$120.
High
US$120. and greater

Quality ambient lighting for today's electronic office environment typically runs US$3.00 to US$5.00/sf, hardware only (luminaires, lamps and ballasts), excluding installation costs. Costs vary based on distributor and contractor markups, luminaire variations such as louver style/ finish, lamping, ballasting and options and accessories, and market conditions, manufacturing situations and annual inflation. These values are for preliminary, magnitude budgeting and do not represent quotes nor actual final pricing to client.

Ambient
Recessed Fluorescent

ConceptStarters™

Ambient 2x4 Parabolic

(2) F32T8 lamps/standard/diffuse louver

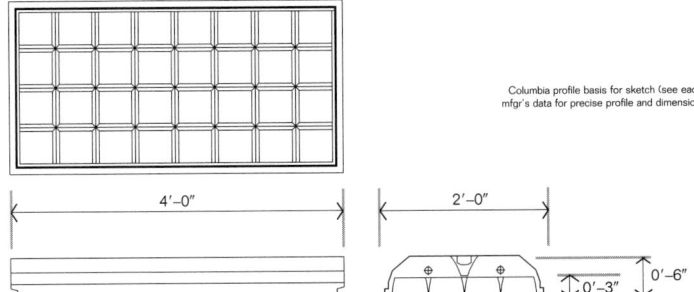

Columbia profile basis for sketch (see each mfgr's data for precise profile and dimensions)

4'–0"   2'–0"   0'–3"   0'–6"

Design Tips

✔ Diffuse louver improves overall impression of brightness.
✔ Diffuse louver masks fingerprints.
✔ (2) F32T8 lamps 4–foot lamps) yield some grazing glare.
✔ Use localized task lighting.
✔ Use wallwashing or liberal art lighting to improve overall impressions of brightness.
✔ Only consider where computer tasks are infrequent and/or unimportant.
✔ 2x4 size is not human scale.

Ceiling height (in feet)	10'			9'			8'		
Percent floor reflectance	20	15	10	20	15	10	20	15	10
Percent wall reflectance	50	40	30	50	40	30	50	40	30
Percent partition reflectance	50	30	30	50	30	30	50	30	30
Percent ceiling reflectance	90	80	70	90	80	70	90	80	70

scale ¹/₁₆" = 1'–0"

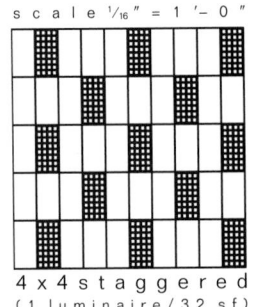

4 x 4 staggered
(1 luminaire / 32 sf)

Predominant partition height (in inches)	80"								
	66"								
	54"								
	48"								

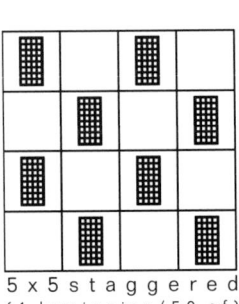

5 x 5 staggered
(1 luminaire / 50 sf)

Predominant partition height (in inches)	80"								
	66"								
	54"								
	48"								

Shaded cell indicates cited luminaire in cited layout has high probability of meeting **most** ambient lighting criteria outlined in Table 9.4. Likely ballast factor is also cited.❶

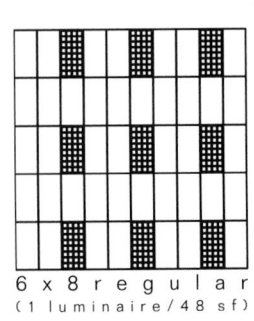

6 x 8 regular
(1 luminaire / 48 sf)

Predominant partition height (in inches)	80"								
	66"								
	54"	0.78	0.88	0.88	0.78	0.88			
	48"		0.78	0.78				0.78	0.78	0.78

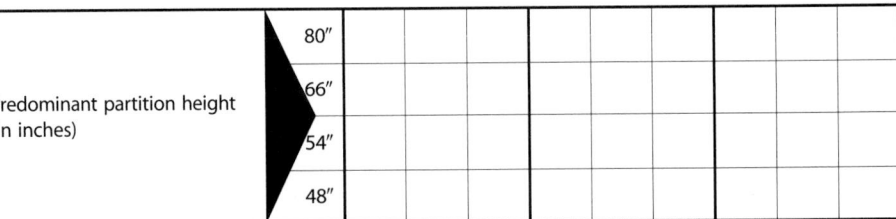
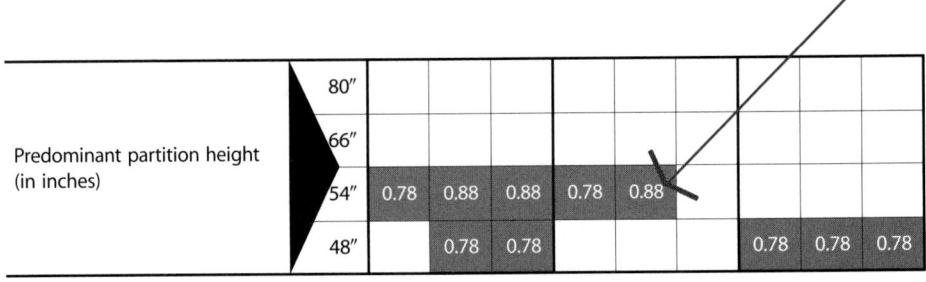

❶ Calculations are required in order to secure a final layout and luminaire specification for actual conditions.

Ambient Recessed Fluorescent

9

Ambient 2x4 Parabolic

(2) F32T8 lamps/standard/diffuse louver

Specification

⊃32–cell parabolic louver
⊃Standard design (glare is more ubiquitous)
⊃Diffuse, low iridescent clear aluminum
⊃2–F32T8/830 rapid start lamps (2950 lumens each)
⊃Electronic instant start ballast (for best efficiency)
⊃See power budgeting sidebar to right
⊃Cost magnitude: low
⊃Possible vendors: **Columbia**, DayBrite, Lightolier, Lithonia, Metalux, Zumtobel/Staff

Boldfaced manufacturers' photometric data were averaged and input to Lumen Micro 7.5 in order to assess probability of meeting ambient lighting criteria cited in Table 9.4. Use the above specification list when contacting manufacturers about potential luminaire options for a specific project. Dimensions, specifications and performance are somewhat generic. Check with specific manufacturers' data for nominal dimensions and performance. Data is subject to change.

ConceptStarters™

Digital Imagery® copyright 2000 PhotoDisc, Inc.

Power budgeting
for these 2–lamp luminaires

BF 0.78 yields 52 watts
BF 0.88 yields 58 watts
BF 0.98 yields 68 watts
BF 1.18 yields 76 watts

Power budget for ambient lighting should be less than 1.0 w/sf. Ballast vendors include Advance, Energy Savings, Inc. (ESI), Magnetek, Motorola. Not all ballast factors available from all ballast vendors. Above factors based on Magnetek.

	10'			9'			8'		Ceiling height (in feet)
20	15	10	20	15	10	20	15	10		Percent floor reflectance
50	40	30	50	40	30	50	40	30		Percent wall reflectance
50	30	30	50	30	30	50	30	30		Percent partition reflectance
90	80	70	90	80	70	90	80	70		Percent ceiling reflectance
									80"	Predominant partition height (in inches)
									66"	
									54"	
									48"	

Open cell indicates cited luminaire in cited layout has low probability of meeting **most** ambient lighting criteria outlined in Table 9.4.❶

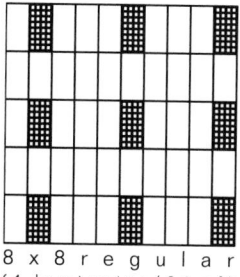

8 x 8 r e g u l a r
(1 luminaire / 64 sf)

Cost magnitude[2000]
for recessed fluorescent luminaires

Low
US$70. to US$90.
Moderate
US$90. to US$120.
High
US$120. and greater

Quality ambient lighting for today's electronic office environment typically runs US$3.00 to US$5.00/sf, hardware only (luminaires, lamps and ballasts), excluding installation costs. Costs vary based on distributor and contractor markups, luminaire variations such as louver style/finish, lamping, ballasting and options and accessories, and market conditions, manufacturing situations and annual inflation. These values are for preliminary, magnitude budgeting and do not represent quotes nor actual final pricing.

Ambient Recessed Fluorescent

Ambient 2x4 Parabolic

(2) F32T8 lamps/standard/specular louver

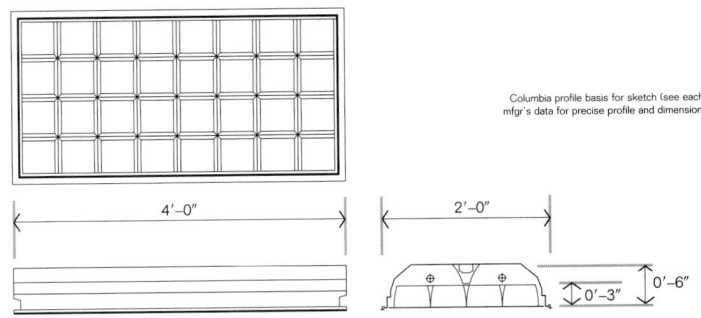

Columbia profile basis for sketch (see each mfgr's data for precise profile and dimensions)

4'–0" 2'–0" 0'–6" 0'–3"

Design Tips

✔ Specular louver looks dark.
✔ Low performance specular louver has severe glare flash at many angles.
✔ Specular louver shows fingerprints—difficult to clean.
✔ (2) F32T8 lamps 4–foot lamps) yield some grazing glare.
✔ Use localized task lighting.
✔ Use wallwashing or liberal art lighting to improve overall impressions of brightness.
✔ Only consider where computer tasks are infrequent and/or unimportant.
✔ 2x4 size is not human–scale.

s c a l e ¹⁄₁₆ " = 1 ' – 0 "

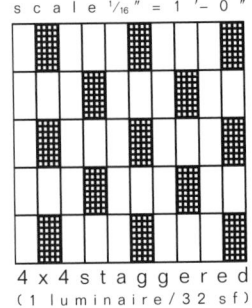

4 x 4 staggered
(1 luminaire / 32 sf)

Ceiling height (in feet)		10'			9'			8'		
Percent floor reflectance		20	15	10	20	15	10	20	15	10
Percent wall reflectance		50	40	30	50	40	30	50	40	30
Percent partition reflectance		50	30	30	50	30	30	50	30	30
Percent ceiling reflectance		90	80	70	90	80	70	90	80	70
Predominant partition height (in inches)	80"									
	66"									
	54"									
	48"									

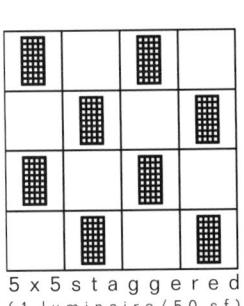

5 x 5 staggered
(1 luminaire / 50 sf)

Predominant partition height (in inches)	80"									
	66"									
	54"									
	48"									

Shaded cell indicates cited luminaire in cited layout has high probability of meeting **most** ambient lighting criteria outlined in Table 9.4. Likely ballast factor is also cited. ❶

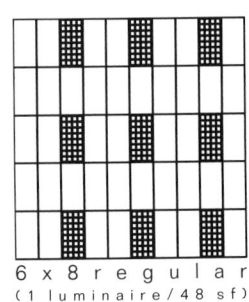

6 x 8 regular
(1 luminaire / 48 sf)

Predominant partition height (in inches)	80"									
	66"									
	54"		0.88	0.88	0.78	0.88	0.88				
	48"		0.78	0.78		0.78	0.78	0.78	0.78	0.78

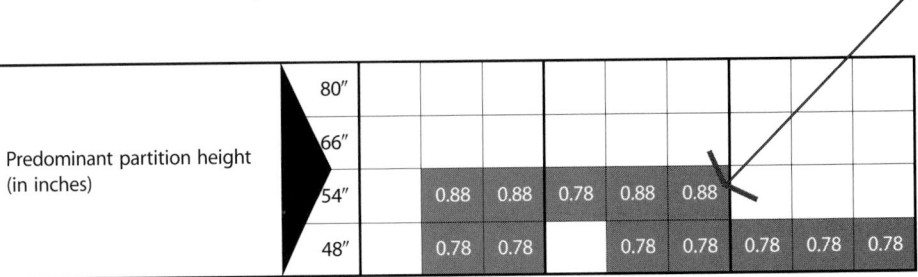

❶ Calculations are required in order to secure a final layout and luminaire specification for actual conditions.

Ambient Recessed Fluorescent

Ambient 2x4 Parabolic

(2) F32T8 lamps/standard/specular louver

Specification

◗32–cell parabolic louver
◗Standard design (glare is more ubiquitous)
◗Specular, low iridescent clear aluminum
◗2–F32T8/830 rapid start lamps (2950 lumens each)
◗Electronic instant start ballast (for best efficiency)
◗See power budgeting sidebar to right
◗Cost magnitude: low
◗Possible vendors: **Columbia**, DayBrite, Lightolier, Lithonia, Metalux, Zumtobel/Staff

Boldfaced manufacturers' photometric data were averaged and input to Lumen Micro 7.5 in order to assess probability of meeting ambient lighting criteria cited in Table 9.4. Use the above specification list when contacting manufacturers about potential luminaire options for a specific project. Dimensions, specifications and performance are somewhat generic. Check with specific manufacturers' data for nominal dimensions and performance. Data is subject to change.

ConceptStarters™

Power budgeting
for these 2–lamp luminaires

BF 0.78 yields 52 watts
BF 0.88 yields 58 watts
BF 0.98 yields 68 watts
BF 1.18 yields 76 watts

Power budget for ambient lighting should be less than 1.0 w/sf. Ballast vendors include Advance, Energy Savings, Inc. (ESI), Magnetek, Motorola. Not all ballast factors available from all ballast vendors. Above factors based on Magnetek.

10'			9'			8'			Ceiling height (in feet)
20	15	10	20	15	10	20	15	10	◀ Percent floor reflectance
50	40	30	50	40	30	50	40	30	◀ Percent wall reflectance
50	30	30	50	30	30	50	30	30	◀ Percent partition reflectance
90	80	70	90	80	70	90	80	70	◀ Percent ceiling reflectance
									80″
									66″ Predominant partition height (in inches)
									54″
	0.88								48″

Open cell indicates cited luminaire in cited layout has low probability of meeting **most** ambient lighting criteria outlined in Table 9.4.❶

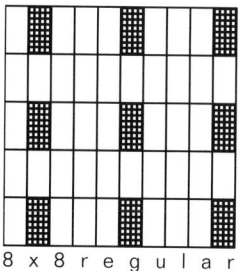

8 x 8 r e g u l a r
(1 luminaire/64 sf)

Cost magnitude²⁰⁰⁰
for recessed fluorescent luminaires

Low
US$70. to US$90.
Moderate
US$90. to US$120.
High
US$120. and greater

Quality ambient lighting for today's electronic office environment typically runs US$3.00 to US$5.00/sf, hardware only (luminaires, lamps and ballasts), excluding installation costs. Costs vary based on distributor and contractor markups, luminaire variations such as louver style/finish, lamping, ballasting and options and accessories, and market conditions, manufacturing situations and annual inflation. These values are for preliminary, magnitude budgeting and do not represent quotes nor actual final pricing to client.

Ambient Recessed Fluorescent

9

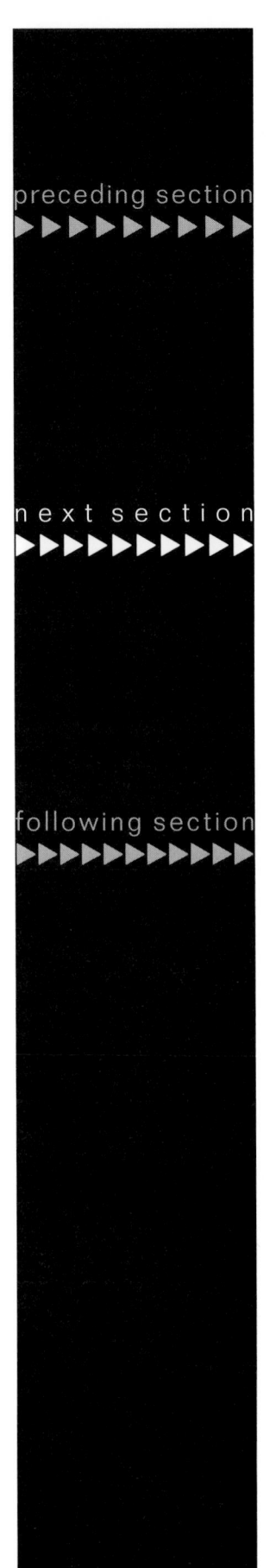

Ambient

Pendent Fluorescent Lighting for Commercial Facilities

10

Pendent ambient fluorescent lighting is addressed in this section. Various lamp/luminaire combinations are reviewed for typical commercial open office lighting applications and schematic layouts are offered for design development consideration.

Ambient
Pendent Fluorescent Lighting for Commercial Facilities

The most energy efficient lighting installations are based on a balance of three lighting layers— ambient lighting, task lighting and accent lighting.

In Section 9, ambient lighting from recessed fluorescent luminaires is addressed. Another option typically considered for ambient lighting is pendent fluorescent lighting, which is addressed here in Section 10. Ambient lighting—whether recessed or pendent—is a strategic method for meeting code restrictions on energy use in commercial facilities. When used with well balanced and equally energy efficient task lighting and accent lighting layers, ambient lighting will be successful for users/workers as well as for facility owners and managers.

The latest energy standards (e.g., California Title 24/1999 and ASHRAE/IESNA 90.1/1999) limit connected loads for lighting in most office spaces to between 1.1 and 1.4 watts per square foot (w/sf). As such, ambient lighting might typically be designed to 0.7 to 1.0 w/sf. Task lighting might be designed to 0.25 w/sf and accent lighting might be designed to 0.25 w/sf. While the designer has flexibility to trade-off between ambient, task and accent lighting, the range is relatively narrow. Clearly, energy efficient sources must be used in energy efficient luminaires and with little tolerance in calculated quantities (so–called design safety factors should be limited to the barest).

Considering ambient pendent indirect or semi–indirect fluorescent lighting as inefficient compared to ambient recessed lighting is a fallacy. It is possible to find situations where ambient recessed lighting offers more illuminance than pendent indirect or semi–indirect lighting. If systems are attempting to meet all of the criteria outlined in Table 9.4 and Table 10.4, then there may be a few percentage points difference in system efficiencies. However, the subjective aspects associated with each approach are important considerations. Recessed ambient lighting uses parabolic louvers to address glare and screen reflection issues in open spaces populated by users with VDT–intensive tasks. Parabolic louvers offer a darker, more "hushed" look than indirect lighting. For some users this is desirable, but for many, the cave–like look is considered a disadvantage. Indirect systems offer a more "open, bright" appearance. Few, if any, users can "see" the difference between 32 footcandles (say, from an indirect system) and 35 footcandles (say, from a direct, recessed system), however, the brightness typically introduced by indirect systems results in perceptions of more light.

Ambient
Pendent Fluorescent Lighting for Commercial Facilities

More on pendent lights

This chapter will address pendent lights (exemplified in Figures 10.1 and 10.2) which direct all light upward. These lights are commonly referenced as indirect. Another pendent light type addressed here is one which directs most light upward and some light downward. These are identified as semi–indirect. Specifically here, the semi–indirect light directs between 75 and 80 percent of the light upward and the remainder (20 to 25 percent) of the light downward.

Pendent lights are often criticized as creating secondary ceiling planes. That is, the bottoms of the pendents, when viewed across their length from most any point in a large room, create the effect of a lower plane. The phenomena seems most problematic for designers and architects and much less so for users. Ways to minimize this effect include using smaller profile luminaires, using rounded

Table 10.1 Binder Bin Mounted Local Task Lighting

Task Lighting	Suggestions
Binder bin mount • T5 or T8 Linear fluorescent • Reduced output or dimmable❶ • Optically designed❷	• Use wherever binder bin occurs and work is likely to take place • Additional average illuminance from task light over task zone should be 25 to 35 fc • Additional maximum illuminance from task light should not exceed 50 fc • Task light intensity gradient should increase from front to back of worksurface • Use integral switch on task light located convenient to user • Use local motion sensor(s) to control task lights on a workstation level • Total connected load for all task lighting is typically 0.25 w/sf over entire office space • Review cord management and power plug–in locations for task lights • Use lights which integrate into binder bin bottom to minimize clutter and optical issues • Use electronic ballasts (50 to 65 percent output; two– or three–level switching or dimmable) • Use lamps of same color temperature and color rendering as ambient lighting • Use lamps of same wattage as ambient, if practical • Use only newer technology triphosphor rapid start T5 or T8 lamps • Do not use old–style preheat or switch start T5 or T8 lamps • Do not use sealed task lights with no obvious means of changing ballasts • Do not use striplights • Do not use unlensed white boxes • Avoid anything with aperture width of less than 2½ inches • Check warranty on coverage and limitations • Use only UL or ETL labeled products tested/listed to UL requirements • Pricing through electrical distributor/contractor may be most competitive

❶On higher wattage units (exceeding 18 watts), consider reduced output versions (2– or 3–level switching) or dimmable versions. Otherwise light output is too great, resulting in reflected glare, veiling reflections and transient adaptation effects (which occur when changing view from brighter areas to darker areas).

❷Specify units which have been optically designed with appropriate reflector and/or lens media to best direct the light onto task areas in a soft, uniform manner. Bare lamps in simple white or glossy aluminum metal boxes should be avoided as these create glare and do not cover a broad area softly. Standard stippled lenses should be avoided as these do not appropriately distribute light to the back tack surface nor across the worksurface.

Ambient
Pendent Fluorescent Lighting for Commercial Facilities

luminaires, using light colored luminaires (preferably matching the ceiling), using relatively short suspension lengths (although this will likely mean using closer spacing of pendent rows) or spacing luminaires quite far apart (but also recognizing that suspension will need to be quite low to allow for proper spread of light across the ceiling).

Lighting layers
For office environments of the day, electronic tasks are primary with paperwork as secondary or equal. So, ambient lighting is typically developed to address the lowest common denominator—lighting intensities and brightnesses which allow best viewing conditions for computer tasks. Task lighting is then used to supplement the ambient lighting so that the aggregate lighting intensities—lighting is additive, hence aggregate lighting is the combined intensities of ambient and task—provide good viewing conditions for paperwork.

Table 10.2 Partition or Rail Mount Task Lighting

Task Lighting	Suggestions
Partition or rail mount • T5 or T8 Linear fluorescent • Reduced output or dimmable❶ • Optically designed❷	• Use wherever no binder bin occurs and work is likely to take place • Additional average illuminance from task light over task zone should be 20 to 30 fc • Additional maximum illuminance from task light should not exceed 45 fc • Task light intensity gradient should increase from front to back of worksurface • Use integral switch on task light located convenient to user • Use local motion sensor(s) to control task lights on a workstation level • Total connected load for all task lighting is typically 0.25 w/sf over entire office space • Review cord management and power plug–in locations for task lights • Use lights which integrate into partition or auxiliary accessories rail to minimize clutter • Use electronic ballasts (50 to 65 percent output; two– or three–level switching or dimmable) • Use lamps of same color temperature and color rendering as ambient lighting • Use lamps of same wattage as ambient, if practical • Use only newer technology triphosphor rapid start T5 or T8 lamps • Do not use old–style preheat or switch start T5 or T8 lamps • Do not use sealed task lights with no obvious means of changing ballasts • Do not use striplights • Do not use unlensed white boxes • Avoid anything with aperture width of less than 2½ inches • Check warranty on coverage and limitations • Use only UL or ETL labeled products tested/listed to UL requirements • Pricing through electrical distributor/contractor may be most competitive

❶On higher wattage units (exceeding 18 watts), consider reduced output versions (2– or 3–level switching) or dimmable versions. Otherwise light output is too great, resulting in reflected glare, veiling reflections and transient adaptation effects (which occur when changing view from brighter areas to darker areas.

❷Specify units which have been optically designed with appropriate reflector and/or lens media to best direct the light onto task areas in a soft, uniform manner. Bare lamps in simple white or glossy aluminum metal boxes should be avoided as these create glare and do not cover a broad area softly.

Ambient

Pendent Fluorescent Lighting for Commercial Facilities

Table 10.3 Freestanding Task Lighting

Task Lighting	Suggestions
Freestanding • Small compact fluorescent • T2 or T5 Linear fluorescent • Reduced output or dimmable❶ • Optically designed❷	• Use wherever binder bin and/or partition– or rail–mount lights cannot be accommodated • Additional average illuminance from task light over task zone should be 20 to 30 fc • Additional maximum illuminance from task light should not exceed 45 fc • Use integral switch on task light located convenient to user • Use local motion sensor(s) to control task lights on a workstation level • Total connected load for all task lighting is typically 0.25 w/sf over entire office space • Review cord management and power plug–in locations for task lights • Review ballast location (sometimes cheap ballasts are mounted on end of cord causing wiring problems through grommets or furniture wire management systems) • Use electronic ballasts (50 to 65 percent output; two– or three–level switching or dimmable) • Use lamps of same color temperature and color rendering as ambient lighting • Use only newer technology triphosphor lamps • Do not use old–style preheat or switch start T5 or T8 lamps • Do not use sealed task lights with no obvious means of changing ballasts • Do not use unlensed white boxes • Check warranty on coverage and limitations • Use only UL, ETL, or other recognized lab–labeled products tested/listed to UL requirements • Pricing through electrical distributor/contractor may be most competitive

❶On higher wattage units (exceeding 18 watts), consider reduced output versions (2– or 3–level switching) or dimmable versions. Otherwise light output is too great, resulting in reflected glare, veiling reflections and transient adaptation effects (which occur when changing view from brighter areas to darker areas.

❷Specify units which have been optically designed with appropriate reflector and/or lens media to best direct the light onto task areas in a soft, uniform manner. Bare lamps in simple white or glossy aluminum metal boxes should be avoided as these create glare and do not cover a broad area softly.

Finally, to maintain some brightness balance between paper tasks and four–color, bright background computer screens and distant architectural surfaces, accent lighting is layered into the overall lighting plan. Task lighting is further discussed in the following paragraphs. Accent lighting is well documented in Sections 3, 4, 5 and 6.

Local task lighting

Local task lighting is that lighting which is generated by a localized luminaire (in close proximity to the task or work area) and which supplements the ambient lighting to a level sufficient for performing the tasks at hand. When carefully planned, task and ambient lighting can combine to provide both effective, comfortable lighting for work and good energy efficiencies. Connected loads alone can be reduced with these techniques. Localized task lights on workstation motion sensors can further reduce energy consumption.

Task lighting is achieved one of three ways (or a combination of these). Binder bin mounted, partition or rail mounted, and freestanding. Where binder bins exist over work or reference surfaces, task

Ambient
Pendent Fluorescent Lighting for Commercial Facilities

Figure 10.1
In this space, ceiling height is 9 feet/6 inches and luminaires have a clear suspension of 1 foot/9 inches. The ambient/indirect, rounded fluorescent luminaires use a single F32T8 lamp in cross–section. Luminaires are 8 feet on center in continuous rows (a core area is to the left, hence rows stop short). Lighting meets the criteria outlined in Table 10.4. Luminaires are Lithonia/Peerless.

Image copyright 2000 Robert Eovaldi

Ambient
Pendent Fluorescent Lighting for Commercial Facilities

Figure 10.2
Task lighting is used to provide additional light (between 50 and 75 fc) on the task area. Note the uniformity of ceiling luminances and also the balance of luminance from the background windows to the ceiling. The rounded shape helps minimize the contrast between the bottom of the luminaire and the ceiling plane. This is most evident on distant views of the luminaires—which have a cross–section of 7¾ inches in width by 2¾ inches in height. Luminaires are Lithonia/Peerless.

Ambient
Pendent Fluorescent Lighting for Commercial Facilities

lights should be integrated into the bottom of the binder bins. This eliminates harsh shadows, dingy work zones and transient adaptation effects folks can encounter when viewing from brighter areas to darker areas. Table 10.1 outlines key suggestions for planning and implementing task lights under binder bins. Partition or rail mounted lights are appropriate where work and reference surfaces are bounded by 54 inch or taller partitions and where no binder bins exist. Table 10.2 offers suggestions for planning task lights on partitions or rails. Freestanding task lights are intended for work or reference surfaces which are not bounded by tall partitions or which are themselves freestanding (e.g., tables, desks, peninsula tops). See Table 10.3 for suggestions on planning freestanding task lights.

With any task lighting technique, the key is not to overlight or highly focus (pinspot) the task area. This kind of brightness exacerbates adaptation (adapting from one brightness to another) as workers scan the work setting. Further, no consideration should be given to incandescent, halogen or xenon lamp task lights. These task lights are inefficient (and will not permit compliance with energy codes), hot to touch, generate excessive heat and are typically quite glary. If such a light were to tip over, it could result in fire. While these task lights may offer a "fashion statement" or "eye candy," they do not offer users convenient, functional, efficient lighting. Further, lamp life is short with incandescent and halogen lamp task lights.

In work areas where users are highly mobile and/or transient, then consider using motion sensors for task lighting control. This is most effective in small, well–defined workstations where 66–inch and 80–inch partitions are common. If, however, users are in and out of the workstation frequently and not gone for more than ten or fifteen minutes at a time, motion sensors may result in shortened lamp life.

Task lighting purchasing chain

Purchasing task lights can be problematic and this issue deserves the designer's and facility manager's attention. In general, furniture manufacturers have available binder bin mounted task lights and/or partition mount task lights and/or freestanding task lights for their

Net Addresses/Local Task Lights
http://www.alkco.com/cabinet.htm
http://www.luxo.com/
http://www.peerless-lighting.com/
http://www.steelcase.com/knowledgebase/
http://www.waldmannlighting.com/page1/

CONNECT FOR MORE

Ambient

Pendent Fluorescent Lighting for Commercial Facilities

respective furniture systems. Unfortunately, most of these lights are inferior. Either one, several or all of the following parameters are inappropriately addressed or not addressed by furniture manufacturers—photometric distribution and intensity; lamping; ballasting; range of lamp and ballast options to meet task requirements for specific situations and to meet the needs of the facility's lamp/ballast stock; variety of lengths of binder bin lights necessary to provide relatively continuous coverage over worksurfaces. When good quality task lights are available, these may be priced quite high as "specials" or "high–end features." Unfortunately, manufacturers' list pricing, net pricing (net to whom—distributor, dealer, contractor, retailer, etc.) and actual costs vary quite dramatically. As much as 100 percent or more in markups are not uncommon. As such, consideration might be given to seeking alternative quotes for task lights from both the electrical and the furniture distribution chains. Vendors not affiliated with the furniture manufacturers and which have available some quality versions of task lights include Alkco, Garcy/SLP, Lithonia/Peerless, Luxo and Waldmann. Steelcase Inc. is a furniture manufacturer which offers some quality task lights (the Advanced Model Shelf Light with 50 percent output ballast or with dimming ballast) along with readily available data "open" to anyone interested in the product, not just to company salespeople or to select architects/interior designers.

Room surface finishes significantly influence lighting efficiencies— higher is better.

Ambient pendent fluorescent lighting

Fluorescent lamps are most appropriate for energy efficient ambient lighting. Large compact fluorescent, T5 and T8 fluorescent lamps are, at time of publication, considered the most appropriate lamps for ambient lighting from pendent luminaires. Large compact fluorescent lamps and T5 lamps exhibit sufficient brightness that they are glary in direct lighting applications, but can be acceptable in indirect applications.

Pendent fluorescent lighting is typically not suitable for applications with ceiling heights less than 9 feet/0 inches. Better efficiencies are achieved with relatively higher ceilings (up to 11 feet/0 inches) and relatively long suspensions to allow a good spread of light across the ceiling.

Ambient lighting is most efficient in relatively large, open areas where partitions are low and where room finish reflectances are

Ambient
Pendent Fluorescent Lighting for Commercial Facilities

Table 10.4 Ambient Lighting Criteria for Concept Starters™

Lighting	Horizontal Illuminance❶❷	Uniformity❸	Vertical Illuminance❹	Ceiling Luminance❺	Luminance Ratios❼	Power Budget
Open Plan Office						
• Paper tasks	35 fc	• avg:min of 2:1	≤20 fc	250 fL	5:1	≤1.0 w/sf❽
• VDT tasks		• avg:max of 1:1.3		(250 fL @ 55°)		
• Ballast factor as required						
• Lamp lumen depreciation 0.95						
• Luminaire dirt depreciation 0.95						
• Room surface dirt depreciation 0.95						
• Including partition factor❻						
• Task lighting anticipated						

❶This represents just the amount of light from the ambient lighting. Total light on task area should average 50 fc, with maximum of 75 fc.
❷Criteria to be calculated average, maintained over the worksurface, including effects of partitions All calculations performed for Concept Starters™ included effects of workstation partitions of heights described for respective Concept Starter™ layout. Workstation density was set at rows of 6x8 workstations back–to–back with 4–foot aisles. Room size was set at 50–feet by 50–feet by ceiling height described in respective Concept Starter layout. Compliance was achieved if calculated values are within 15 percent of criteria.
❸Criteria to be uniformity of the horizontal illuminance calculated over the whole room (to accommodate flexibility).
❹Criteria to be calculated average, maintained at 4 feet above finished floor in four cardinal viewing directions—related to computer screen viewing.
❺Criteria to be calculated maximum on ceiling. For indirect/direct luminaires, then direct portion of luminaire shall exhibit max luminance of 250 fL at 55° AND maximum ceiling luminance shall be 250 fL or less.
❻Partition heights, color and density will affect light losses. Typically, partition factors range between 0.85 and 0.60 (light losses of 15% to 40%).
❼Luminance ratios on ceiling are calculated maximum–to–minimum over a relatively small area.
❽Based on latest proposed ASHRAE/IES 90.1/1999. This is a rough average of what power budget likely needs to be to meet code, but this can vary based on overall building use, size and lighting layouts. Solely for ambient lighting. Task lighting typically adds 0.25 w/sf.

Criteria references
1. M.S. Rea, ed., Lighting Handbook Reference and Application (New York: Illuminating Engineering Society of North America, 2000).
2. American National Standards Institute/Illuminating Engineering Society of North America. American National Standard Practice for Office Lighting (RP–1) (New York: Illuminating Engineering Society of North America, 1993.
3. Illuminating Engineering Society of North America. RP–24: IESNA Recommended Practice for Lighting Offices Containing Computer Visual Display Terminals (New York: Illuminating Engineering Society of North America, 1990).
4. Gary R. Steffy, Lighting the Electronic Office (New York: Van Nostrand Reinhold, 1995).
5. Gary R. Steffy, Architectural Lighting Design (New York: Van Nostrand Reinhold, 1990).

high. Large open areas (greater than 900 square feet (sf)) take better advantage of the overlapping of light effects of many luminaires. Smaller rooms exhibit relatively more wall area compared to ceiling area—and the walls absorb light. Nevertheless, the concept of ambient light works in all sizes of office spaces.

Lower partitions result in less light–absorbing surface area. This is an especially significant issue with lower ceilings. In spaces with 10–foot ceiling heights, 66–inch partitions are not as problematic as they would be in spaces with 9–foot ceiling heights. To some extent,

Ambient
Pendent Fluorescent Lighting for Commercial Facilities

this problem can be reduced with light colored partitions. If 66–inch partitions are necessary in a space with 9–foot ceilings, then consider selecting very light colored partitions.

Room surface finishes greatly influence lighting efficiencies. Horizontal surfaces tend to have the greatest impact on efficiency in larger open spaces. In small spaces, the vertical surfaces have a significant impact on lighting efficiencies. In any event, all surfaces should have relatively matte finishes. Contrary to popular belief, glossy or specular surfaces do not inherently reflect more light than matte surfaces. Further, matte surfaces reflect light in a diffuse manner, which results in more uniform light distribution and less glare potential. New ceiling manufacturing techniques and coatings now allow for ceilings of 89 percent reflectance. Providing all other room surfaces are kept light in color, this is a preferred reflectance for ceilings. Wall reflectances should range from 30 percent to 60 percent. Furniture partitions should range from 30 percent to 50 percent. Floors should never fall below 10 percent reflectance. All of these surfaces' reflectances will greatly improve the efficiency and effectiveness of ambient electric lighting and also of daylighting.

If efficiency is an issue (it should be in every situation), and if user productivity and comfort are issues (as they should be in every situation—since productivity is so singly important to return on investment in facilities), then trying to use large lumen packages (2– and 3–lamp luminaires) isn't an option in typical situations. Such large lumen packages cannot meet reasonable lighting criteria while maintaining an efficient operation unless very odd ballast and/or dimming configurations are also involved. Hence, these large lumen solutions are not covered here—they aren't appropriate.

Ambient lighting criteria
It is undoubtedly difficult to remain focused on why a building is being newly constructed, renovated or retrofitted. Many times cost issues or maintenance issues drive a design. While these are important, they should not overshadow that the building will be used to further commerce and the life experience. As such, attention should be given to meeting users' needs. For ambient lighting in today's commercial facilities where people work extensively on computers,

T5 fluorescent lamps are sufficiently new that ballast variations are limited.

Ambient
Pendent Fluorescent Lighting for Commercial Facilities

Figure 10.3
These ambient/indirect, rounded fluorescent luminaires use a single F32T8 lamp in cross–section; are 8 feet on center in continuous rows to meet the criteria in Table 10.4. These luminaires have a more subtle, oval cross–section compared to those shown in Figures 10.1 and 10.2. A gullwing–style lens helps hide the lamp while giving the appearance of a shallower luminaire. Compact fluorescent, semi–recessed downlights provide an efficient method of highlighting the core/circulation area. The downlights use compact fluorescent lamps. Pendent luminaires are Lithonia/Peerless. Downlights by Louis Poulsen.

Ambient

Pendent Fluorescent Lighting for Commercial Facilities

lighting criteria should reduce or eliminate glare reflections from computer monitors; reduce or eliminate the veil of light on computer monitors; and reduce or eliminate harsh contrasts.

Several criteria for pendent fluorescent ambient lighting for typical open office areas in commercial facilities, outlined in Table 10.4, are horizontal illuminance and uniformity; vertical illuminance; ceiling luminance and ceiling luminance uniformity. As noted previously, connected load or power budget is also a requisite criterion. Figure 10.3 illustrates an installation where connected load is 1.14 w/sf while meeting criteria outlined in Table 10.4. Success depends on luminaire selection and layout. Lamping also influences success— typically lower lumen packages (relatively low output lamps or few lamps per luminaire) are most successful in providing uniform, relatively soft lighting.

Based on the criteria in Table 10.4 and on the pendent luminaires typically available on the market, Concept Starters™ for ambient pendent fluorescent lighting in commercial facilities are offered on the following pages. For each Concept Starter™ series, a product specification is outlined along with design tips, power budgeting and cost magnitude. Tables 10.5 through 10.19 are a quick reference to the variety of luminaires reviewed as Concept Starters™ and their suggested application and layout options.

Concept Starters™ are just that. These are not recommended final layouts. Actual project conditions must be considered and final calculations with intended–to–be–specified luminaires under these actual conditions need to be made prior to contract document layout and specification. Lumen Micro™ 7.5 software was used for the calculations. Typical large–room setups consisted of a footprint of 50 feet by 50 feet. Ceiling heights varied as outlined in each table shown for each luminaire layout, as did room surface reflectances. Lamp lumen depreciation (LLD) was set at 0.95. Luminaire dirt depreciation (LDD) was set at 0.95. Room surface dirt depreciation (RSDD) was set at 0.95. Ballast factors were set as needed to meet criteria and these are recorded in the appropriate tabular summaries.

Make no mistake, luminaires from different manufacturers are not identical and should not be considered equals. The intent of the Concept Starters™ is to establish preliminary or concept lighting

Lumen Micro™ is a registered trademark of Lighting Technologies, Inc.

Ambient
Pendent Fluorescent Lighting for Commercial Facilities

layouts for various classes of pendent lights. Cited manufacturers for each Concept Starter™ series offer products with some similarities. Final product dimensions, exact performance, appearance, leadtime and/or cost require professional review before establishing a final specification of one or more manufacturers' hardware for a given situation.

Interpolation throughout the tables is an acceptable method of assessing situations not explicitly cited. For example, a Concept Starter™ which is likely to meet criteria for a 9–foot/6–inch ceiling height and a 10–foot ceiling height will likely meet criteria for a 9–foot/9 inch ceiling height.

Since T5 triphosphor, high efficiency and high output fluorescent lamps are relatively new, there are few ballasts available for them. Hence, there are few Concept Starters™ for T5–lamped luminaires. And where T5 Concept Starters™ are cited, there may be few situations where criteria is likely to be met. When/if ballast manufacturers offer a more complete line of ballasts with varying outputs and power requirements, then more T5 Concept Starters™ will likely meet criteria.

Ambient

Pendent Fluorescent Lighting for Commercial Facilities

ConceptStarters™

Using the Concept Starters™

Luminaire Type

Each set of four pages is devoted to a single luminaire type. The header outlines the luminaire family (indirect or semi–indirect), cross section (beveled, oval, V or square) and size and also cites lamping (number and type of lamps). Pendent lighting success is dependent on the relationship between ceiling height and suspension length. Three ceiling height variations are cited (9′, 9′–6″ and 10′) as are three suspension length options (1′–6″, 2′–0″ and 2′–6″). A larger view of one double page spread along with annotations follows on the next two pages (10–16 and 10–17) to help in understanding the Concept Starters™.

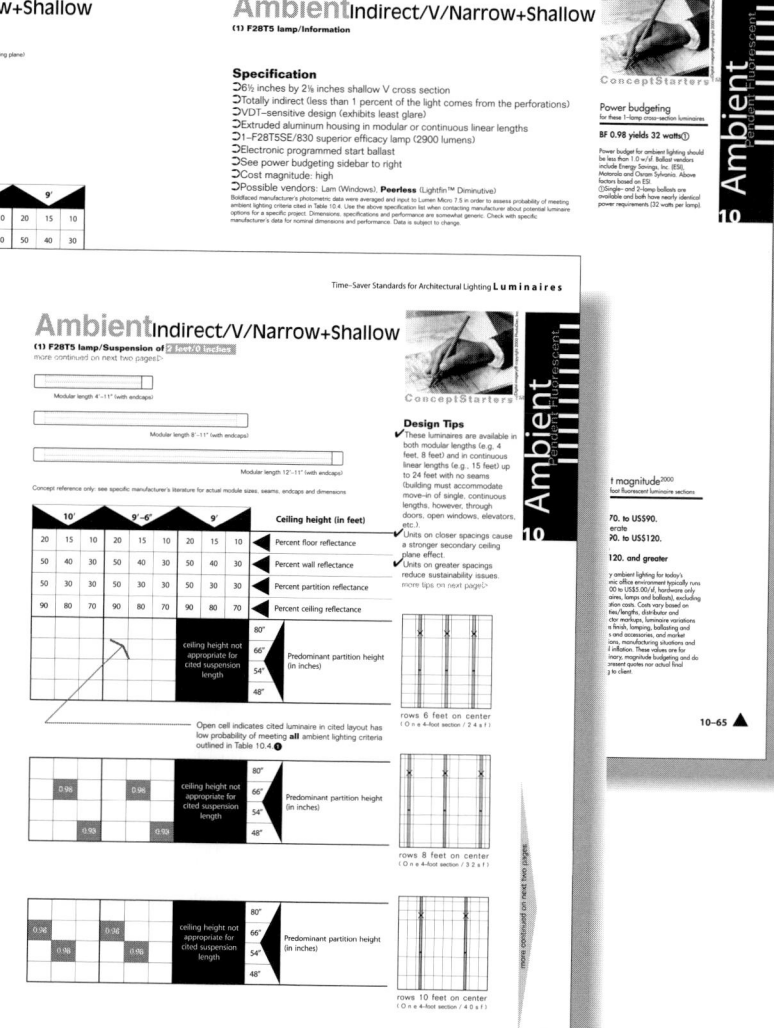

ConceptStarters™

Ambient
Pendent Fluorescent Lighting for Commercial Facilities

Using the Concept Starters™

Four Pages
Four pages are devoted to each luminaire review. This illustrates the second double page spread for each luminaire.

Luminaire Type
Each set of four pages is devoted to a single luminaire type. The header outlines the luminaire family (indirect or semi–indirect), cross section (beveled, oval, V or square) and size and also cites lamping (number and type of lamps).

Luminaire Sketch
For quick visual reference, a plan view, cross section and side view of the luminaire of interest are shown. Since some luminaire profiles vary from manufacturer to manufacturer, a note on some indicates which manufacturer's profile data were used for the sketch.
Advisory: check manufacturers actual datasheets or cutsheets for exact dimensions and profiles.

Design Tips
Tips are offered about the luminaire. Salient advantages and/or disadvantages are cited. Application notes are offered.

Concept Starters™/Reading the Charts
① Ceiling height options are 9', 9'–6" and 10'. Select the column identifying the ceiling height for your project.

② For each ceiling height, surface reflectance categories are light, medium and dark. Specific reflectance values are cited in each category. Select the column of reflectances most closely matching your project's conditions.

③ Four workstation partition height options are evaluated—48", 54", 66" and 80". Select the partition height most predominant in your project. Now, for your ceiling height, surface reflectances and partition heights, look for a shaded cell(s).

④ Shaded cells indicate that the luminaire layout under the circumstances cited (ceiling height, reflectances and partition heights) will likely meet the lighting criteria outlined in Table 10.4. Each cell is inscribed with a ballast factor—very important specification information to provide an energy efficient layout.

⑤ Layout icons illustrate three different layouts. Use those layouts which offer shaded cells corresponding to your specific project's ceiling height, surface reflectances and partition heights. Amount of area covered by each luminaire is cited for convenient preliminary power budgeting and preliminary cost magnitude.

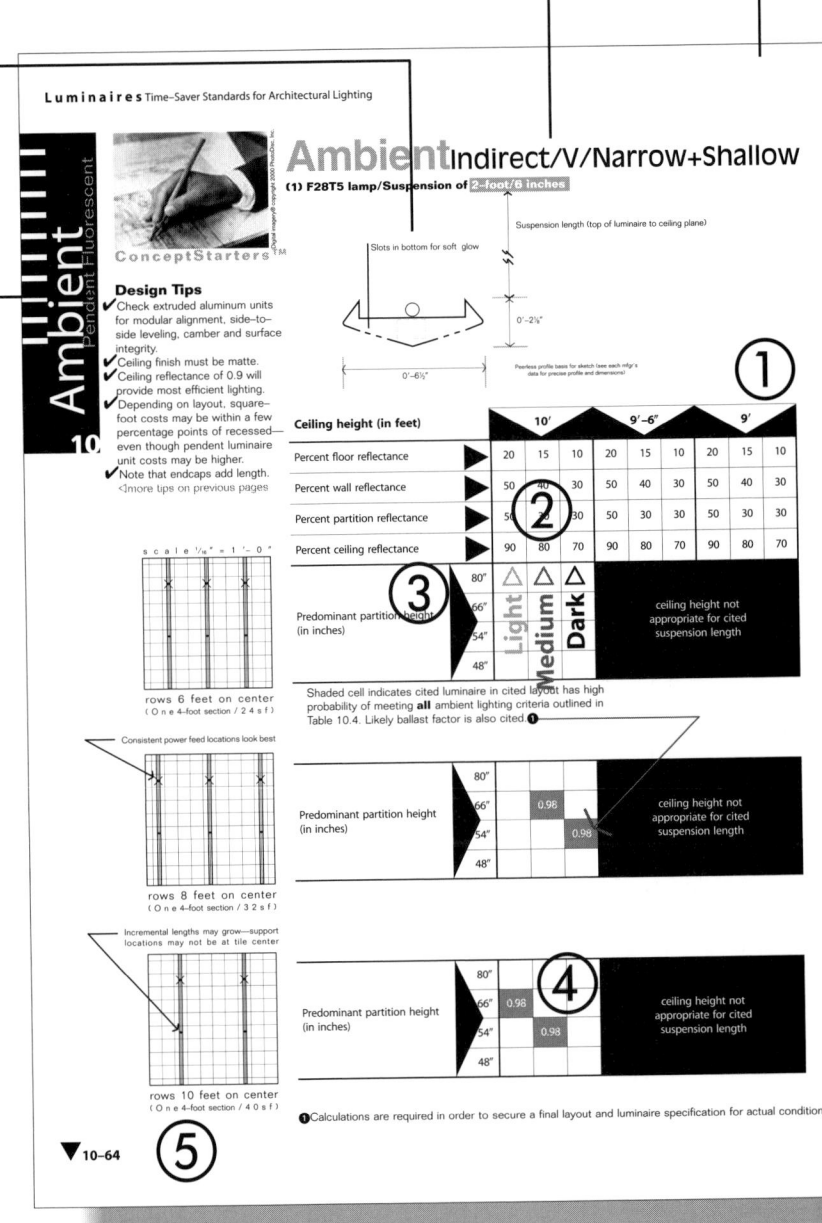

Ambient

Pendent Fluorescent Lighting for Commercial Facilities

Using the Concept Starters™

Luminaire Specification

An outline of the salient specifics on the luminaire featured on the four pages. Use this information when contacting potential vendors for more specific technical and cost data or when asking manufacturers to provide project–specific calculations.

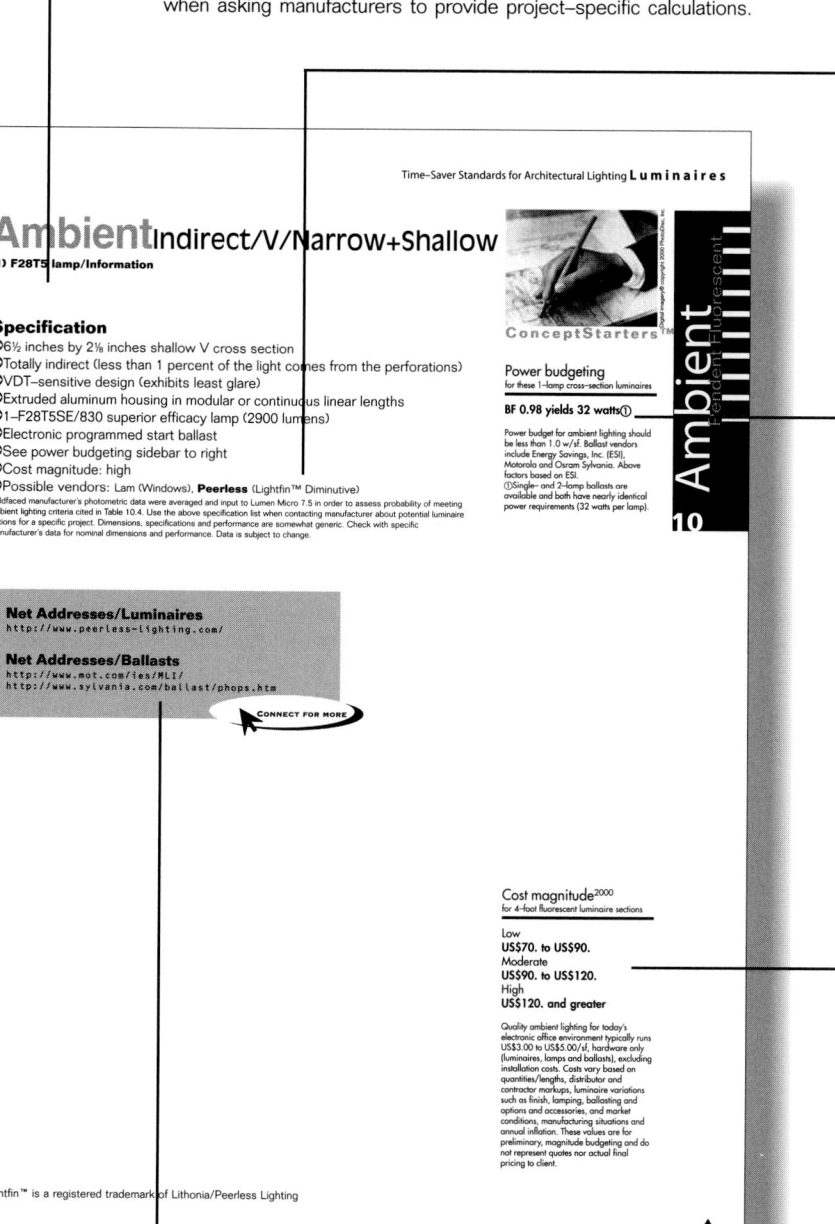

Ambient Indirect/V/Narrow+Shallow
(1) F28T5 lamp/Information

Specification
- 6½ inches by 2⅛ inches shallow V cross section
- Totally indirect (less than 1 percent of the light comes from the perforations)
- VDT–sensitive design (exhibits least glare)
- Extruded aluminum housing in modular or continuous linear lengths
- 1–F28T5SE/830 superior efficacy lamp (2900 lumens)
- Electronic programmed start ballast
- See power budgeting sidebar to right
- Cost magnitude: high
- Possible vendors: Lam (Windows), **Peerless** (Lightfin™ Diminutive)

Boldfaced manufacturer's photometric data were averaged and input to Lumen Micro 7.5 in order to assess probability of meeting ambient lighting criteria cited in Table 10.4. Use the above specification list when contacting manufacturer about potential luminaire options for a specific project. Dimensions, specifications and performance are somewhat generic. Check with specific manufacturer's data for nominal dimensions and performance. Data is subject to change.

Net Addresses/Luminaires
http://www.peerless-lighting.com/

Net Addresses/Ballasts
http://www.mot.com/ies/MLI/
http://www.sylvania.com/ballast/phops.htm

CONNECT FOR MORE

Power budgeting
for these 1–lamp cross–section luminaires

BF 0.98 yields 32 watts①

Power budget for ambient lighting should be less than 1.0 w/sf. Ballast vendors include Energy Savings, Inc. (ESI), Motorola and Osram Sylvania. Above factors based on ESI.
① Single– and 2–lamp ballasts are available and both have nearly identical power requirements (32 watts per lamp).

Cost magnitude²⁰⁰⁰
for 4–foot fluorescent luminaire sections

Low
US$70. to US$90.
Moderate
US$90. to US$120.
High
US$120. and greater

Quality ambient lighting for today's electronic office environment typically runs US$3.00 to US$5.00/sf, hardware only (luminaires, lamps and ballasts), excluding installation costs. Costs vary based on quantities/lengths, distributor and contractor markups, luminaire variations such as finish, lamping, ballasting and options and accessories, and market conditions, manufacturing situations and annual inflation. These values are for preliminary, magnitude budgeting and do not represent quotes nor actual final pricing to client.

Lightfin™ is a registered trademark of Lithonia/Peerless Lighting.

10–65 ▲

Luminaire Vendors

A collection of vendors that have or had luminaires fitting the specification and performance characteristics used in developing the Concept Starters™ on the four pages. Boldfaced manufacturers' photometric data were averaged and used for all calculation analyses on the four pages. Only luminaires having similar profiles and photometry were used. **Advisory**: in the interest of time, the vendors' search was limited. Other vendors may be found with similar performing products.

Ballast Factors

If lamps are the engines for luminaires, then ballasts are the gas peddles. The higher the ballast factor, the more light output from the lamp and the higher the wattage for operation. Ballast factors represent the percentage of light that will be produced by a lamp attached to that ballast (e.g., BF of 0.78 means a lamp on this ballast will produce 78 percent of the lamp manufacturer's rated light output). Ballast factors allow for the fine tuning of a lighting system in order to get just the right quantity and quality of light without using more energy than necessary. **Advisory**: each shaded cell cites the ballast factor used to achieve criteria compliance—so specifying the correct ballast factor is critical to meeting criteria and limiting energy consumption. For T5 lamps only 0.98 BFs are presently available (and therefore T5 lamps have limited application success).

Cost Magnitude

For each luminaire on a double page spread, a cost magnitude is reported under **Specification**. The cost magnitude table in the lower right corner can then be referenced to establish preliminary budget information.

Internet Addresses

For more information, and in some cases downloadable cutsheets and installation examples, website addresses are noted. The list is not exhaustive and includes those checked for depth of material available. New sites are added daily—so a global search is advisable from time to time.

Ambient

Pendent Fluorescent Lighting Selection Guide

ConceptStarters™

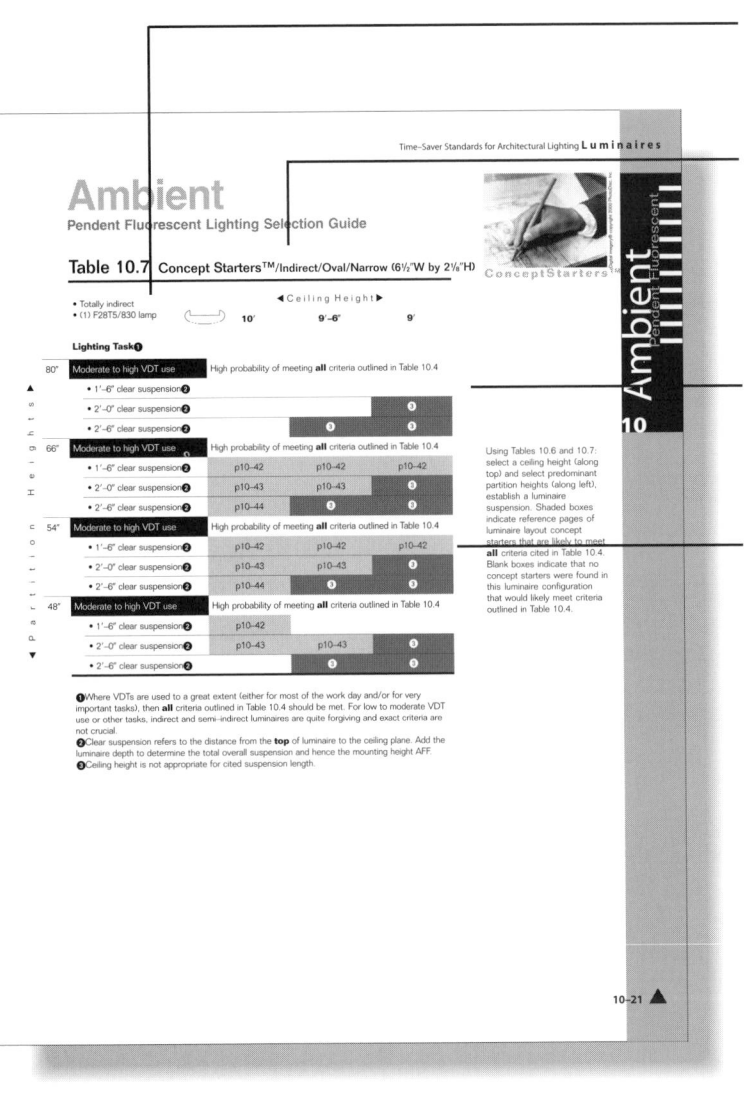

Using the Selection Guide

Descriptor/Icons
A brief description of light distribution (indirect or semi–indirect) and lamping along with reference icons are provided for quick assessment.

Luminaire
Two distribution types (indirect and semi–indirect) and four cross section shapes (beveled, oval, V and square) are assessed here in Section 10. For each type, there are variations in lamping and cross section size.

Blank Cells
Blank cells indicate that analyses show that none met criteria outlined in Table 10.4 for the respective partition height, ceiling height and luminaire version.

Shaded Cell
Check the cited page for Concept Starters™.

Ambient
Pendent Fluorescent Lighting Selection Guide

ConceptStarters™

Table 10.5 Concept Starters™/Indirect/Beveled/Narrow (7″Wx2¹/₂″H)

- Totally indirect
- (1) F28T5/830 lamp

◀ C e i l i n g H e i g h t ▶

Lighting Task❶	10′	9′–6″	9′
80″ Moderate to high VDT use	High probability of meeting **all** criteria outlined in Table 10.4		
• 1′–6″ clear suspension❷			
• 2′–0″ clear suspension❷			❸
• 2′–6″ clear suspension❷		❸	❸
66″ Moderate to high VDT use	High probability of meeting **all** criteria outlined in Table 10.4		
• 1′–6″ clear suspension❷	p10–34	p10–34	p10–34
• 2′–0″ clear suspension❷	p10–35	p10–35	❸
• 2′–6″ clear suspension❷		❸	❸
54″ Moderate to high VDT use	High probability of meeting **all** criteria outlined in Table 10.4		
• 1′–6″ clear suspension❷	p10–34	p10–34	p10–34
• 2′–0″ clear suspension❷	p10–35	p10–35	❸
• 2′–6″ clear suspension❷	p10–36	❸	❸
48″ Moderate to high VDT use	High probability of meeting **all** criteria outlined in Table 10.4		
• 1′–6″ clear suspension❷	p10–34	p10–34	p10–34
• 2′–0″ clear suspension❷	p10–35	p10–35	❸
• 2′–6″ clear suspension❷	p10–36	❸	❸

Partition Heights ▲ ▼

Using Table 10.5: select a ceiling height (along top) and select predominant partition heights (along left), establish a luminaire suspension. Shaded boxes indicate reference pages of luminaire layout concept starters that are likely to meet **all** criteria cited in Table 10.4. Blank boxes indicate that no concept starters were found in this luminaire configuration that would likely meet criteria outlined in Table 10.4.

❶Where VDTs are used to a great extent (either for most of the work day and/or for very important tasks), then **all** criteria outlined in Table 10.4 should be met. For low to moderate VDT use or other tasks, indirect and semi–indirect luminaires are quite forgiving and exact criteria are not crucial.
❷Clear suspension refers to the distance from the **top** of luminaire to the ceiling plane. Add the luminaire depth to determine the total overall suspension and hence the mounting height AFF.
❸Ceiling height is not appropriate for cited suspension length.

ConceptStarters™

Digital imagery® copyright 2000 PhotoDisc, Inc.

Ambient
Pendent Fluorescent Lighting Selection Guide

Table 10.6 Concept Starters™/Indirect/Beveled/ExtraWide (10"Wx3"H)

- Totally indirect
- (1) F32T8/830 lamp

◀ Ceiling Height ▶

Lighting Task❶	10'	9'–6"	9'
80" Moderate to high VDT use	High probability of meeting **all** criteria outlined in Table 10.4		
• 1'–6" clear suspension❷			
• 2'–0" clear suspension❷			③
• 2'–6" clear suspension❷		③	③
66" Moderate to high VDT use	High probability of meeting **all** criteria outlined in Table 10.4		
• 1'–6" clear suspension❷	p10–38	p10–38	p10–38
• 2'–0" clear suspension❷	p10–39	p10–39	③
• 2'–6" clear suspension❷	p10–40	③	③
54" Moderate to high VDT use	High probability of meeting **all** criteria outlined in Table 10.4		
• 1'–6" clear suspension❷	p10–38	p10–38	p10–38
• 2'–0" clear suspension❷	p10–39	p10–39	③
• 2'–6" clear suspension❷	p10–40	③	③
48" Moderate to high VDT use	High probability of meeting **all** criteria outlined in Table 10.4		
• 1'–6" clear suspension❷	p10–38	p10–38	p10–38
• 2'–0" clear suspension❷	p10–39	p10–39	③
• 2'–6" clear suspension❷	p10–40	③	③

◀ Partition Heights ▶

❶Where VDTs are used to a great extent (either for most of the work day and/or for very important tasks), then **all** criteria outlined in Table 10.4 should be met. For low to moderate VDT use or other tasks, indirect and semi–indirect luminaires are quite forgiving and exact criteria are not crucial.

❷Clear suspension refers to the distance from the **top** of luminaire to the ceiling plane. Add the luminaire depth to determine the total overall suspension and hence the mounting height AFF.

❸Ceiling height is not appropriate for cited suspension length.

Ambient

Pendent Fluorescent Lighting Selection Guide

Table 10.7 Concept Starters™/Indirect/Oval/Narrow (6½"W by 2⅛"H)

- Totally indirect
- (1) F28T5/830 lamp

◀ C e i l i n g H e i g h t ▶

10′ **9′–6″** **9′**

Lighting Task❶

Partition Heights	Lighting Task		10′	9′–6″	9′
80″	Moderate to high VDT use	High probability of meeting **all** criteria outlined in Table 10.4			
	• 1′–6″ clear suspension❷				
	• 2′–0″ clear suspension❷				❸
	• 2′–6″ clear suspension❷			❸	❸
66″	Moderate to high VDT use	High probability of meeting **all** criteria outlined in Table 10.4			
	• 1′–6″ clear suspension❷		p10–42	p10–42	p10–42
	• 2′–0″ clear suspension❷		p10–43	p10–43	❸
	• 2′–6″ clear suspension❷		p10–44	❸	❸
54″	Moderate to high VDT use	High probability of meeting **all** criteria outlined in Table 10.4			
	• 1′–6″ clear suspension❷		p10–42	p10–42	p10–42
	• 2′–0″ clear suspension❷		p10–43	p10–43	❸
	• 2′–6″ clear suspension❷		p10–44	❸	❸
48″	Moderate to high VDT use	High probability of meeting **all** criteria outlined in Table 10.4			
	• 1′–6″ clear suspension❷		p10–42		
	• 2′–0″ clear suspension❷		p10–43	p10–43	❸
	• 2′–6″ clear suspension❷			❸	❸

Using Tables 10.6 and 10.7: select a ceiling height (along top) and select predominant partition heights (along left), establish a luminaire suspension. Shaded boxes indicate reference pages of luminaire layout concept starters that are likely to meet **all** criteria cited in Table 10.4. Blank boxes indicate that no concept starters were found in this luminaire configuration that would likely meet criteria outlined in Table 10.4.

❶Where VDTs are used to a great extent (either for most of the work day and/or for very important tasks), then **all** criteria outlined in Table 10.4 should be met. For low to moderate VDT use or other tasks, indirect and semi–indirect luminaires are quite forgiving and exact criteria are not crucial.

❷Clear suspension refers to the distance from the **top** of luminaire to the ceiling plane. Add the luminaire depth to determine the total overall suspension and hence the mounting height AFF.

❸Ceiling height is not appropriate for cited suspension length.

ConceptStarters™

Ambient
Pendent Fluorescent Lighting Selection Guide

Table 10.8 Concept Starters™/Indirect/Oval/Wide(8″W by 3⅛″H)

- Totally indirect
- (1) F32T8/830 lamp

◀ Ceiling Height ▶

			10′	9′–6″	9′

Lighting Task❶

			10′	9′–6″	9′
80″	Moderate to high VDT use	High probability of meeting **all** criteria outlined in Table 10.4			
	• 1′–6″ clear suspension❷		p10–46		
	• 2′–0″ clear suspension❷			p10–47	❸
	• 2′–6″ clear suspension❷			❸	❸
66″	Moderate to high VDT use	High probability of meeting **all** criteria outlined in Table 10.4			
	• 1′–6″ clear suspension❷		p10–46	p10–46	p10–46
	• 2′–0″ clear suspension❷		p10–47	p10–47	❸
	• 2′–6″ clear suspension❷		p10–48	❸	❸
54″	Moderate to high VDT use	High probability of meeting **all** criteria outlined in Table 10.4			
	• 1′–6″ clear suspension❷		p10–46	p10–46	p10–46
	• 2′–0″ clear suspension❷		p10–47	p10–47	❸
	• 2′–6″ clear suspension❷		p10–48	❸	❸
48″	Moderate to high VDT use	High probability of meeting **all** criteria outlined in Table 10.4			
	• 1′–6″ clear suspension❷		p10–46	p10–46	p10–46
	• 2′–0″ clear suspension❷		p10–47	p10–47	❸
	• 2′–6″ clear suspension❷		p10–48	❸	❸

❶ Where VDTs are used to a great extent (either for most of the work day and/or for very important tasks), then **all** criteria outlined in Table 10.4 should be met. For low to moderate VDT use or other tasks, indirect and semi–indirect luminaires are quite forgiving and exact criteria are not crucial.

❷ Clear suspension refers to the distance from the **top** of luminaire to the ceiling plane. Add the luminaire depth to determine the total overall suspension and hence the mounting height AFF.

❸ Ceiling height is not appropriate for cited suspension length.

Ambient
Pendent Fluorescent Lighting Selection Guide

ConceptStarters™

Table 10.9 Concept Starters™/Indirect/Oval/ExtraWide (9½″W by 2⅜″H)

- Totally indirect
- (1) F28T5/830 lamp

◀ C e i l i n g H e i g h t ▶

10′ **9′–6″** **9′**

Lighting Task❶

P a r t i t i o n H e i g h t s ▲ ▼

Partition Height		10′	9′–6″	9′
80″	**Moderate to high VDT use**	High probability of meeting **all** criteria outlined in Table 10.4		
	• 1′–6″ clear suspension❷			
	• 2′–0″ clear suspension❷			③
	• 2′–6″ clear suspension❷		③	③
66″	**Moderate to high VDT use**	High probability of meeting **all** criteria outlined in Table 10.4		
	• 1′–6″ clear suspension❷	p10–50	p10–50	p10–50
	• 2′–0″ clear suspension❷	p10–51	p10–51	③
	• 2′–6″ clear suspension❷		③	③
54″	**Moderate to high VDT use**	High probability of meeting **all** criteria outlined in Table 10.4		
	• 1′–6″ clear suspension❷	p10–50	p10–50	p10–50
	• 2′–0″ clear suspension❷		p10–51	③
	• 2′–6″ clear suspension❷	p10–52	③	③
48″	**Moderate to high VDT use**	High probability of meeting **all** criteria outlined in Table 10.4		
	• 1′–6″ clear suspension❷	p10–50	p10–50	p10–50
	• 2′–0″ clear suspension❷	p10–51	p10–51	③
	• 2′–6″ clear suspension❷	p10–52	③	③

Using Tables 10.8 and 10.9: select a ceiling height (along top) and select predominant partition heights (along left), establish a luminaire suspension. Shaded boxes indicate reference pages of luminaire layout concept starters that are likely to meet **all** criteria cited in Table 10.4. Blank boxes indicate that no concept starters were found in this luminaire configuration that would likely meet criteria outlined in Table 10.4.

❶Where VDTs are used to a great extent (either for most of the work day and/or for very important tasks), then **all** criteria outlined in Table 10.4 should be met. For low to moderate VDT use or other tasks, indirect and semi–indirect luminaires are quite forgiving and exact criteria are not crucial.

❷Clear suspension refers to the distance from the **top** of luminaire to the ceiling plane. Add the luminaire depth to determine the total overall suspension and hence the mounting height AFF.

❸Ceiling height is not appropriate for cited suspension length.

ConceptStarters™

Ambient
Pendent Fluorescent Lighting Selection Guide

Table 10.10 Concept Starters™/Indirect/Oval/ExtraWide (9½"W by 2⅜"H)

- Totally indirect
- (1) F54T5/HO/830 lamp

◀ Ceiling Height ▶

Lighting Task❶

Partition Heights

	Lighting Task		10'	9'–6"	9'
80"	Moderate to high VDT use	High probability of meeting **all** criteria outlined in Table 10.4			
	• 1'–6" clear suspension❷				
	• 2'–0" clear suspension❷				③
	• 2'–6" clear suspension❷			③	③
66"	Moderate to high VDT use	High probability of meeting **all** criteria outlined in Table 10.4			
	• 1'–6" clear suspension❷		p10–54	p10–54	p10–54
	• 2'–0" clear suspension❷		p10–55		③
	• 2'–6" clear suspension❷			③	③
54"	Moderate to high VDT use	High probability of meeting **all** criteria outlined in Table 10.4			
	• 1'–6" clear suspension❷		p10–54	p10–54	p10–54
	• 2'–0" clear suspension❷		p10–55	p10–55	③
	• 2'–6" clear suspension❷		p10–56	③	③
48"	Moderate to high VDT use	High probability of meeting **all** criteria outlined in Table 10.4			
	• 1'–6" clear suspension❷				
	• 2'–0" clear suspension❷			p10–55	③
	• 2'–6" clear suspension❷			③	③

❶Where VDTs are used to a great extent (either for most of the work day and/or for very important tasks), then **all** criteria outlined in Table 10.4 should be met. For low to moderate VDT use or other tasks, indirect and semi–indirect luminaires are quite forgiving and exact criteria are not crucial.
❷Clear suspension refers to the distance from the **top** of luminaire to the ceiling plane. Add the luminaire depth to determine the total overall suspension and hence the mounting height AFF.
❸Ceiling height is not appropriate for cited suspension length.

Ambient

Pendent Fluorescent Lighting Selection Guide

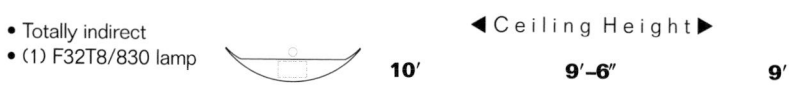

ConceptStarters™

Table 10.11 Concept Starters™/Indirect/Oval/ExtraWide (9½"W by 2³/₈"H)

- Totally indirect
- (1) F32T8/830 lamp

◀ Ceiling Height ▶

Lighting Task❶	10'	9'–6"	9'
80" Moderate to high VDT use	High probability of meeting **all** criteria outlined in Table 10.4		
• 1'–6" clear suspension❷	p10–58		
• 2'–0" clear suspension❷		p10–59	❸
• 2'–6" clear suspension❷		❸	❸
66" Moderate to high VDT use	High probability of meeting **all** criteria outlined in Table 10.4		
• 1'–6" clear suspension❷	p10–58	p10–58	p10–58
• 2'–0" clear suspension❷	p10–59	p10–59	❸
• 2'–6" clear suspension❷	p10–60	❸	❸
54" Moderate to high VDT use	High probability of meeting **all** criteria outlined in Table 10.4		
• 1'–6" clear suspension❷	p10–58	p10–58	p10–58
• 2'–0" clear suspension❷	p10–59	p10–59	❸
• 2'–6" clear suspension❷	p10–60	❸	❸
48" Moderate to high VDT use	High probability of meeting **all** criteria outlined in Table 10.4		
• 1'–6" clear suspension❷	p10–58	p10–58	p10–58
• 2'–0" clear suspension❷	p10–59	p10–59	❸
• 2'–6" clear suspension❷	p10–60	❸	❸

◀▲ Partition Heights ▼▶

Using Tables 10.10 and 10.11: select a ceiling height (along top) and select predominant partition heights (along left), establish a luminaire suspension. Shaded boxes indicate reference pages of luminaire layout concept starters that are likely to meet **all** criteria cited in Table 10.4. Blank boxes indicate that no concept starters were found in this luminaire configuration that would likely meet criteria outlined in Table 10.4.

❶ Where VDTs are used to a great extent (either for most of the work day and/or for very important tasks), then **all** criteria outlined in Table 10.4 should be met. For low to moderate VDT use or other tasks, indirect and semi–indirect luminaires are quite forgiving and exact criteria are not crucial.

❷ Clear suspension refers to the distance from the **top** of luminaire to the ceiling plane. Add the luminaire depth to determine the total overall suspension and hence the mounting height AFF.

❸ Ceiling height is not appropriate for cited suspension length.

ConceptStarters™

Ambient

Pendent Fluorescent Lighting Selection Guide

Table 10.12 Concept Starters™/ Indirect/V/Narrow+Shallow (6½″W by 2⅛″H)

- Totally indirect
- (1) F28T5/830 lamp

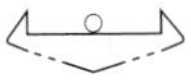

Partition Heights	Lighting Task❶	Ceiling Height 10′	9′–6″	9′
80″	**Moderate to high VDT use**	High probability of meeting **all** criteria outlined in Table 10.4		
	• 1′–6″ clear suspension❷			
	• 2′–0″ clear suspension❷			❸
	• 2′–6″ clear suspension❷		❸	❸
66″	**Moderate to high VDT use**	High probability of meeting **all** criteria outlined in Table 10.4		
	• 1′–6″ clear suspension❷	p10–62	p10–62	p10–62
	• 2′–0″ clear suspension❷	p10–63	p10–63	❸
	• 2′–6″ clear suspension❷	p10–64	❸	❸
54″	**Moderate to high VDT use**	High probability of meeting **all** criteria outlined in Table 10.4		
	• 1′–6″ clear suspension❷	p10–62	p10–62	p10–62
	• 2′–0″ clear suspension❷	p10–63	p10–63	❸
	• 2′–6″ clear suspension❷	p10–64	❸	❸
48″	**Moderate to high VDT use**	High probability of meeting **all** criteria outlined in Table 10.4		
	• 1′–6″ clear suspension❷	p10–62		
	• 2′–0″ clear suspension❷	p10–63	p10–63	❸
	• 2′–6″ clear suspension❷		❸	❸

❶Where VDTs are used to a great extent (either for most of the work day and/or for very important tasks), then **all** criteria outlined in Table 10.4 should be met. For low to moderate VDT use or other tasks, indirect and semi–indirect luminaires are quite forgiving and exact criteria are not crucial.
❷Clear suspension refers to the distance from the **top** of luminaire to the ceiling plane. Add the luminaire depth to determine the total overall suspension and hence the mounting height AFF.
❸Ceiling height is not appropriate for cited suspension length.

Ambient

Pendent Fluorescent Lighting Selection Guide

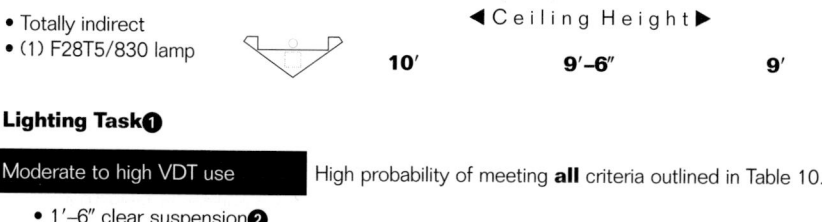

ConceptStarters™

Table 10.13 Concept Starters™/Indirect/V/Narrow (7"W by 2⅞"H)

- Totally indirect
- (1) F28T5/830 lamp

◀ Ceiling Height ▶

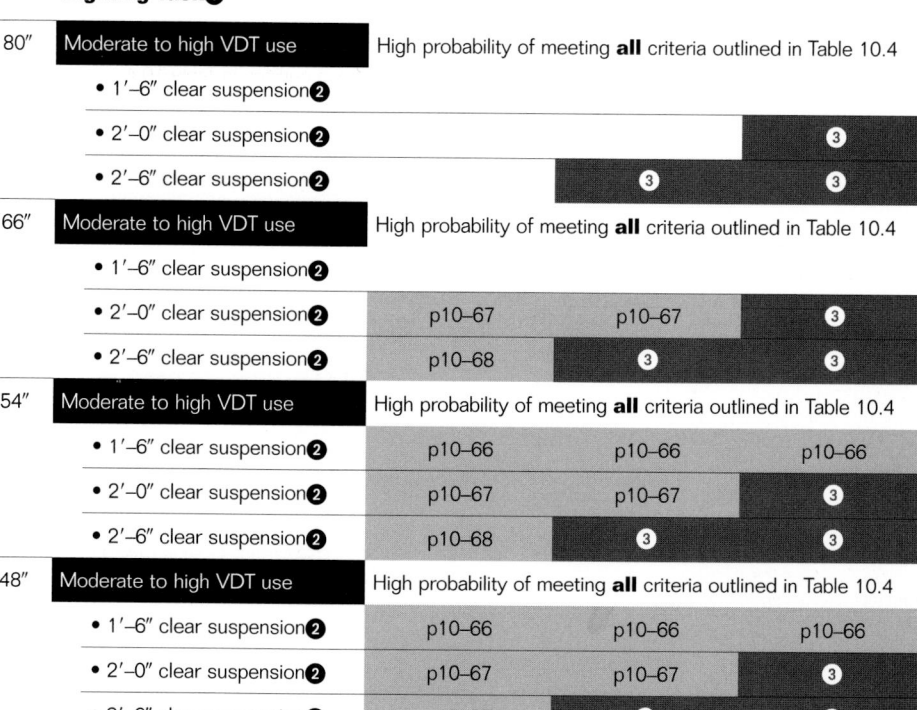

Partition Heights	Lighting Task ❶	10'	9'–6"	9'
80"	**Moderate to high VDT use**	High probability of meeting **all** criteria outlined in Table 10.4		
	• 1'–6" clear suspension ❷			
	• 2'–0" clear suspension ❷			③
	• 2'–6" clear suspension ❷		③	③
66"	**Moderate to high VDT use**	High probability of meeting **all** criteria outlined in Table 10.4		
	• 1'–6" clear suspension ❷			
	• 2'–0" clear suspension ❷	p10–67	p10–67	③
	• 2'–6" clear suspension ❷	p10–68	③	③
54"	**Moderate to high VDT use**	High probability of meeting **all** criteria outlined in Table 10.4		
	• 1'–6" clear suspension ❷	p10–66	p10–66	p10–66
	• 2'–0" clear suspension ❷	p10–67	p10–67	③
	• 2'–6" clear suspension ❷	p10–68	③	③
48"	**Moderate to high VDT use**	High probability of meeting **all** criteria outlined in Table 10.4		
	• 1'–6" clear suspension ❷	p10–66	p10–66	p10–66
	• 2'–0" clear suspension ❷	p10–67	p10–67	③
	• 2'–6" clear suspension ❷	p10–68	③	③

❶Where VDTs are used to a great extent (either for most of the work day and/or for very important tasks), then **all** criteria outlined in Table 10.4 should be met. For low to moderate VDT use or other tasks, indirect and semi–indirect luminaires are quite forgiving and exact criteria are not crucial.

❷Clear suspension refers to the distance from the **top** of luminaire to the ceiling plane. Add the luminaire depth to determine the total overall suspension and hence the mounting height AFF.

❸Ceiling height is not appropriate for cited suspension length.

Using Tables 10.12 and 10.13: select a ceiling height (along top) and select predominant partition heights (along left), establish a luminaire suspension. Shaded boxes indicate reference pages of luminaire layout concept starters that are likely to meet **all** criteria cited in Table 10.4. Blank boxes indicate that no concept starters were found in this luminaire configuration that would likely meet criteria outlined in Table 10.4.

Digital imagery® copyright 2000 PhotoDisc, Inc.

ConceptStarters™

Ambient
Pendent Fluorescent Lighting Selection Guide

Table 10.14 Concept Starters™/Indirect/V/Wide+Deep (8½"W by 4"H)

- Totally indirect
- (1) F32T8/830 lamp

◀ Ceiling Height ▶

		10'	9'–6"	9'
Lighting Task ❶				
80"	Moderate to high VDT use	High probability of meeting **all** criteria outlined in Table 10.4		
	• 1'–6" clear suspension ❷			p10–70
	• 2'–0" clear suspension ❷		p10–71	❸
	• 2'–6" clear suspension ❷	p10–72	❸	❸
66"	Moderate to high VDT use	High probability of meeting **all** criteria outlined in Table 10.4		
	• 1'–6" clear suspension ❷	p10–70	p10–70	p10–70
	• 2'–0" clear suspension ❷	p10–71	p10–71	❸
	• 2'–6" clear suspension ❷	p10–72	❸	❸
54"	Moderate to high VDT use	High probability of meeting **all** criteria outlined in Table 10.4		
	• 1'–6" clear suspension ❷	p10–70	p10–70	p10–70
	• 2'–0" clear suspension ❷	p10–71	p10–71	❸
	• 2'–6" clear suspension ❷	p10–72	❸	❸
48"	Moderate to high VDT use	High probability of meeting **all** criteria outlined in Table 10.4		
	• 1'–6" clear suspension ❷	p10–70	p10–70	p10–70
	• 2'–0" clear suspension ❷	p10–71	p10–71	❸
	• 2'–6" clear suspension ❷	p10–72	❸	❸

Partition Heights (vertical label at left)

❶ Where VDTs are used to a great extent (either for most of the work day and/or for very important tasks), then **all** criteria outlined in Table 10.4 should be met. For low to moderate VDT use or other tasks, indirect and semi–indirect luminaires are quite forgiving and exact criteria are not crucial.

❷ Clear suspension refers to the distance from the **top** of luminaire to the ceiling plane. Add the luminaire depth to determine the total overall suspension and hence the mounting height AFF.

❸ Ceiling height is not appropriate for cited suspension length.

Ambient
Pendent Fluorescent Lighting Selection Guide

ConceptStarters™

Table 10.15 Concept Starters™/
Indirect/Square/Miniature (2³/₄"W by 2¹/₂"H)

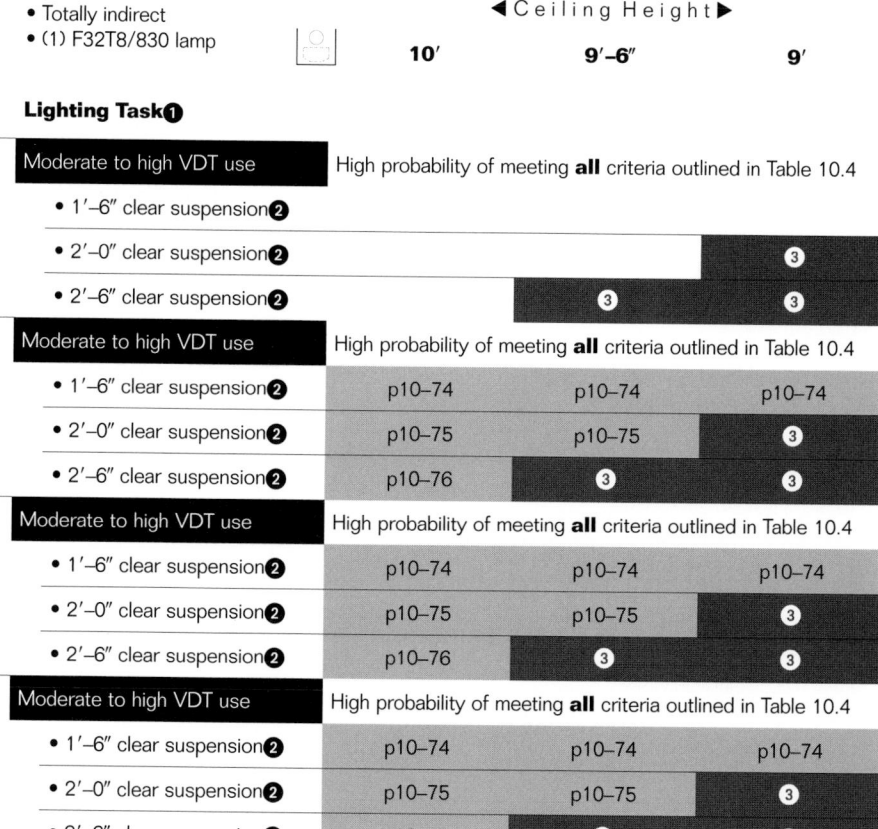

- Totally indirect
- (1) F32T8/830 lamp

◄ C e i l i n g   H e i g h t ►

10' **9'–6"** **9'**

Lighting Task ❶

Partition Heights	Lighting Task		10'	9'–6"	9'
80"	Moderate to high VDT use	High probability of meeting **all** criteria outlined in Table 10.4			
	• 1'–6" clear suspension ❷				
	• 2'–0" clear suspension ❷				❸
	• 2'–6" clear suspension ❷			❸	❸
66"	Moderate to high VDT use	High probability of meeting **all** criteria outlined in Table 10.4			
	• 1'–6" clear suspension ❷		p10–74	p10–74	p10–74
	• 2'–0" clear suspension ❷		p10–75	p10–75	❸
	• 2'–6" clear suspension ❷		p10–76	❸	❸
54"	Moderate to high VDT use	High probability of meeting **all** criteria outlined in Table 10.4			
	• 1'–6" clear suspension ❷		p10–74	p10–74	p10–74
	• 2'–0" clear suspension ❷		p10–75	p10–75	❸
	• 2'–6" clear suspension ❷		p10–76	❸	❸
48"	Moderate to high VDT use	High probability of meeting **all** criteria outlined in Table 10.4			
	• 1'–6" clear suspension ❷		p10–74	p10–74	p10–74
	• 2'–0" clear suspension ❷		p10–75	p10–75	❸
	• 2'–6" clear suspension ❷		p10–76	❸	❸

Using Tables 10.14 and 10.15: select a ceiling height (along top) and select predominant partition heights (along left), establish a luminaire suspension. Shaded boxes indicate reference pages of luminaire layout concept starters that are likely to meet **all** criteria cited in Table 10.4. Blank boxes indicate that no concept starters were found in this luminaire configuration that would likely meet criteria outlined in Table 10.4.

❶Where VDTs are used to a great extent (either for most of the work day and/or for very important tasks), then **all** criteria outlined in Table 10.4 should be met. For low to moderate VDT use or other tasks, indirect and semi–indirect luminaires are quite forgiving and exact criteria are not crucial.

❷Clear suspension refers to the distance from the **top** of luminaire to the ceiling plane. Add the luminaire depth to determine the total overall suspension and hence the mounting height AFF.

❸Ceiling height is not appropriate for cited suspension length.

Digital imagery® copyright 2000 PhotoDisc, Inc.

ConceptStarters™

Ambient
Pendent Fluorescent Lighting Selection Guide

Table 10.16 Concept Starters™/Indirect/Square/Large (7″W by 6¼″H)

- Totally indirect
- (1) F32T8/830 lamp

◄ Ceiling Height ►

Lighting Task❶

Partition Heights	Lighting Task	10′	9′–6″	9′
80″	Moderate to high VDT use	High probability of meeting **all** criteria outlined in Table 10.4		
	• 1′–6″ clear suspension❷			
	• 2′–0″ clear suspension❷			❸
	• 2′–6″ clear suspension❷		❸	❸
66″	Moderate to high VDT use	High probability of meeting **all** criteria outlined in Table 10.4		
	• 1′–6″ clear suspension❷		p10–78	p10–78
	• 2′–0″ clear suspension❷	p10–79	p10–79	❸
	• 2′–6″ clear suspension❷	p10–80	❸	❸
54″	Moderate to high VDT use	High probability of meeting **all** criteria outlined in Table 10.4		
	• 1′–6″ clear suspension❷	p10–78	p10–78	p10–78
	• 2′–0″ clear suspension❷	p10–79	p10–79	❸
	• 2′–6″ clear suspension❷	p10–80	❸	❸
48″	Moderate to high VDT use	High probability of meeting **all** criteria outlined in Table 10.4		
	• 1′–6″ clear suspension❷	p10–78	p10–78	p10–78
	• 2′–0″ clear suspension❷	p10–79	p10–79	❸
	• 2′–6″ clear suspension❷	p10–80	❸	❸

❶Where VDTs are used to a great extent (either for most of the work day and/or for very important tasks), then **all** criteria outlined in Table 10.4 should be met. For low to moderate VDT use or other tasks, indirect and semi–indirect luminaires are quite forgiving and exact criteria are not crucial.

❷Clear suspension refers to the distance from the **top** of luminaire to the ceiling plane. Add the luminaire depth to determine the total overall suspension and hence the mounting height AFF.

❸Ceiling height is not appropriate for cited suspension length.

Ambient

Pendent Fluorescent Lighting Selection Guide

Digital imagery® copyright 2000 PhotoDisc, Inc.

ConceptStarters™

Table 10.17 Concept Starters™/Semi–indirect/Beveled/Narrow ($7^1/_4$"W by $2^1/_2$"H)

- Semi–indirect (25%▽ and 75%△)
- (1) F28T5/830 lamp

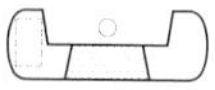

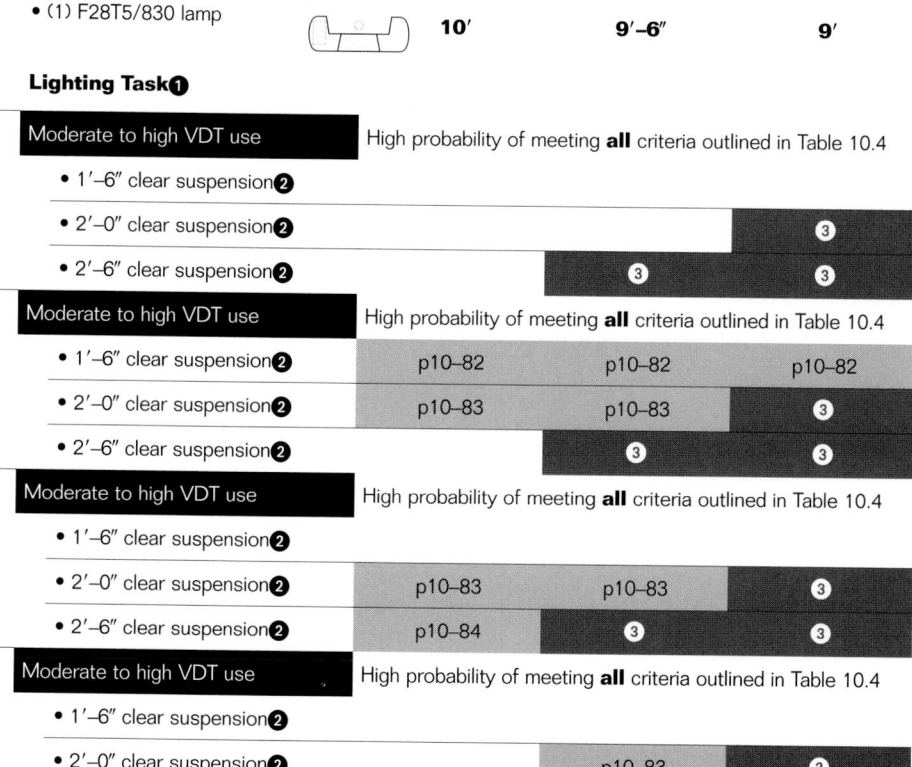

◀ Ceiling Height ▶

Partition Heights	Lighting Task❶	10′	9′–6″	9′
80″	**Moderate to high VDT use**	High probability of meeting **all** criteria outlined in Table 10.4		
	• 1′–6″ clear suspension❷			
	• 2′–0″ clear suspension❷			③
	• 2′–6″ clear suspension❷		③	③
66″	**Moderate to high VDT use**	High probability of meeting **all** criteria outlined in Table 10.4		
	• 1′–6″ clear suspension❷	p10–82	p10–82	p10–82
	• 2′–0″ clear suspension❷	p10–83	p10–83	③
	• 2′–6″ clear suspension❷		③	③
54″	**Moderate to high VDT use**	High probability of meeting **all** criteria outlined in Table 10.4		
	• 1′–6″ clear suspension❷			
	• 2′–0″ clear suspension❷	p10–83	p10–83	③
	• 2′–6″ clear suspension❷	p10–84	③	③
48″	**Moderate to high VDT use**	High probability of meeting **all** criteria outlined in Table 10.4		
	• 1′–6″ clear suspension❷			
	• 2′–0″ clear suspension❷		p10–83	③
	• 2′–6″ clear suspension❷	p10–84	③	③

❶Where VDTs are used to a great extent (either for most of the work day and/or for very important tasks), then **all** criteria outlined in Table 10.4 should be met. For low to moderate VDT use or other tasks, indirect and semi–indirect luminaires are quite forgiving and exact criteria are not crucial.
❷Clear suspension refers to the distance from the **top** of luminaire to the ceiling plane. Add the luminaire depth to determine the total overall suspension and hence the mounting height AFF.
❸Ceiling height is not appropriate for cited suspension length.

Using Tables 10.16 and 10.17: select a ceiling height (along top) and select predominant partition heights (along left), establish a luminaire suspension. Shaded boxes indicate reference pages of luminaire layout concept starters that are likely to meet **all** criteria cited in Table 10.4. Blank boxes indicate that no concept starters were found in this luminaire configuration that would likely meet criteria outlined in Table 10.4.

ConceptStarters™

Ambient
Pendent Fluorescent Lighting Selection Guide

Table 10.18 Concept Starters™/Semi–indirect/Beveled/Wide (8¼"W by 3½"H)

- Semi–indirect (30%⌄ and 70%⌃)
- (1) F32T8/830 lamp

	Lighting Task❶		◄ Ceiling Height ►		
			10′	9′–6″	9′
80″	Moderate to high VDT use	High probability of meeting **all** criteria outlined in Table 10.4			
	• 1′–6″ clear suspension❷				
	• 2′–0″ clear suspension❷				❸
	• 2′–6″ clear suspension❷		p10–88	❸	❸
66″	Moderate to high VDT use	High probability of meeting **all** criteria outlined in Table 10.4			
	• 1′–6″ clear suspension❷		p10–86	p10–86	p10–86
	• 2′–0″ clear suspension❷		p10–87	p10–87	❸
	• 2′–6″ clear suspension❷			❸	❸
54″	Moderate to high VDT use	High probability of meeting **all** criteria outlined in Table 10.4			
	• 1′–6″ clear suspension❷		p10–86	p10–86	p10–86
	• 2′–0″ clear suspension❷		p10–87	p10–87	❸
	• 2′–6″ clear suspension❷		p10–88	❸	❸
48″	Moderate to high VDT use	High probability of meeting **all** criteria outlined in Table 10.4			
	• 1′–6″ clear suspension❷		p10–86	p10–86	p10–86
	• 2′–0″ clear suspension❷		p10–87	p10–87	❸
	• 2′–6″ clear suspension❷		p10–88	❸	❸

◄ Partition Heights ►

❶Where VDTs are used to a great extent (either for most of the work day and/or for very important tasks), then **all** criteria outlined in Table 10.4 should be met. For low to moderate VDT use or other tasks, indirect and semi–indirect luminaires are quite forgiving and exact criteria are not crucial.
❷Clear suspension refers to the distance from the **top** of luminaire to the ceiling plane. Add the luminaire depth to determine the total overall suspension and hence the mounting height AFF.
❸Ceiling height is not appropriate for cited suspension length.

Ambient

Pendent Fluorescent Lighting Selection Guide

ConceptStarters™

Table 10.19 Concept Starters™/Semi–indirect/Square/Large (7″W by 6¼″H)

- Semi–indirect (25%⌄ and 75%⌃)
- (1) F32T8/830 lamp

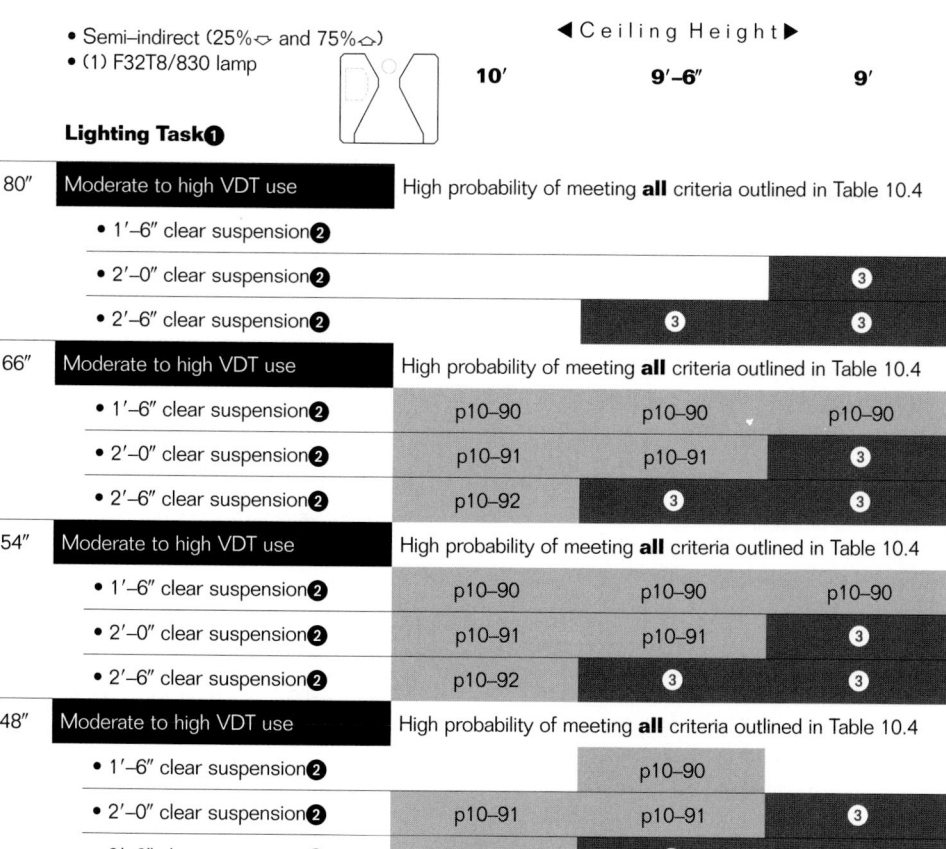

◀ Ceiling Height ▶

Lighting Task ❶

Partition Heights	Lighting Task	10′	9′–6″	9′
80″	Moderate to high VDT use	High probability of meeting **all** criteria outlined in Table 10.4		
	• 1′–6″ clear suspension ❷			
	• 2′–0″ clear suspension ❷			❸
	• 2′–6″ clear suspension ❷		❸	❸
66″	Moderate to high VDT use	High probability of meeting **all** criteria outlined in Table 10.4		
	• 1′–6″ clear suspension ❷	p10–90	p10–90	p10–90
	• 2′–0″ clear suspension ❷	p10–91	p10–91	❸
	• 2′–6″ clear suspension ❷	p10–92	❸	❸
54″	Moderate to high VDT use	High probability of meeting **all** criteria outlined in Table 10.4		
	• 1′–6″ clear suspension ❷	p10–90	p10–90	p10–90
	• 2′–0″ clear suspension ❷	p10–91	p10–91	❸
	• 2′–6″ clear suspension ❷	p10–92	❸	❸
48″	Moderate to high VDT use	High probability of meeting **all** criteria outlined in Table 10.4		
	• 1′–6″ clear suspension ❷		p10–90	
	• 2′–0″ clear suspension ❷	p10–91	p10–91	❸
	• 2′–6″ clear suspension ❷	p10–92	❸	❸

Partition Heights ▲ ▼

Using Tables 10.18 and 10.19: select a ceiling height (along top) and select predominant partition heights (along left), establish a luminaire suspension. Shaded boxes indicate reference pages of luminaire layout concept starters that are likely to meet **all** criteria cited in Table 10.4. Blank boxes indicate that no concept starters were found in this luminaire configuration that would likely meet criteria outlined in Table 10.4.

❶Where VDTs are used to a great extent (either for most of the work day and/or for very important tasks), then **all** criteria outlined in Table 10.4 should be met. For low to moderate VDT use or other tasks, indirect and semi–indirect luminaires are quite forgiving and exact criteria are not crucial.
❷Clear suspension refers to the distance from the **top** of luminaire to the ceiling plane. Add the luminaire depth to determine the total overall suspension and hence the mounting height AFF.
❸Ceiling height is not appropriate for cited suspension length.

ConceptStarters™

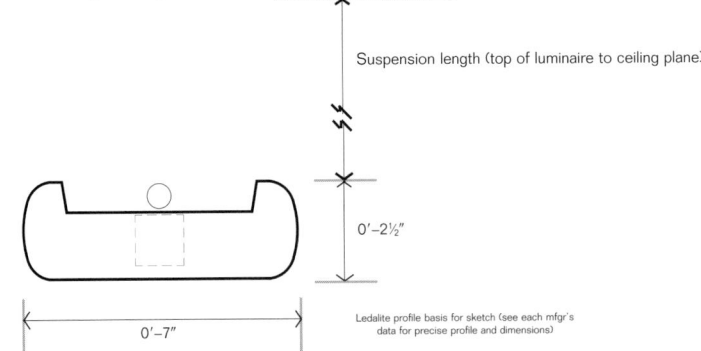

Ambient Indirect/Beveled/Narrow

(1) F28T5 lamp/Suspension of 1 foot/6 inches

Suspension length (top of luminaire to ceiling plane)

0'–2½"

0'–7"

Ledalite profile basis for sketch (see each mfg's data for precise profile and dimensions)

Design Tips

✔ Flat bottom results in contrast with ceiling—harsh look and more likely "streaks" in VDTs.
✔ Flat, dark bottom creates secondary ceiling plane.
✔ Use localized task lighting.
✔ Use art accenting for visual interest.
more tips on next page▷

10

s c a l e ¹⁄₁₆ ″ = 1 ′ – 0 ″

rows 6 feet on center
(O n e 4–foot section / 2 4 s f)

Consistent power feed locations look best

rows 8 feet on center
(O n e 4–foot section / 3 2 s f)

Incremental lengths may grow—support locations may not be at tile center

rows 10 feet on center
(O n e 4–foot section / 4 0 s f)

Ceiling height (in feet)	10'			9'–6"			9'		
Percent floor reflectance	20	15	10	20	15	10	20	15	10
Percent wall reflectance	50	40	30	50	40	30	50	40	30
Percent partition reflectance	50	30	30	50	30	30	50	30	30
Percent ceiling reflectance	90	80	70	90	80	70	90	80	70

Predominant partition height (in inches)	10'			9'–6"			9'		
80"									
66"									
54"									
48"									

Shaded cell indicates cited luminaire in cited layout has high probability of meeting **all** ambient lighting criteria outlined in Table 10.4. Likely ballast factor is also cited.❶

Predominant partition height (in inches)	10'			9'–6"			9'		
80"									
66"									
54"		0.98			0.98			0.98	
48"			0.98			0.98			0.98

Predominant partition height (in inches)	10'			9'–6"			9'		
80"									
66"	0.98			0.98	0.98	0.98	0.98		
54"									
48"									0.98

❶Calculations are required in order to secure a final layout and luminaire specification for actual conditions.

Ambient Indirect/Beveled/Narrow

(1) F28T5 lamp/Suspension of 2-foot/0 inches

more continued on next two pages▷

ConceptStarters™

Modular length 4'–2"

Modular length 8'–2"

Modular length 12'–2"

Concept reference only: see specific manufacturer's literature for actual module sizes, seams, endcaps and dimensions

Ambient Pendent Fluorescent

10

Design Tips

✔ Many of these luminaires exhibit incremental growth in length of several inches from suspension point to suspension point—so every 8 to 12 feet it may be necessary to add 2 to 6 inches. Critical for final layout/ceiling integration.

✔ Units on closer spacings cause a stronger secondary ceiling plane effect.

✔ Units on greater spacings reduce sustainability issues.

more tips on next page▷

10'			9'–6"			9'			Ceiling height (in feet)
20	15	10	20	15	10	20	15	10	Percent floor reflectance
50	40	30	50	40	30	50	40	30	Percent wall reflectance
50	30	30	50	30	30	50	30	30	Percent partition reflectance
90	80	70	90	80	70	90	80	70	Percent ceiling reflectance

						80"	
			ceiling height not appropriate for cited suspension length	66"	Predominant partition height (in inches)
				54"	
				48"	

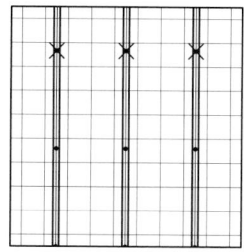

rows 6 feet on center
(O n e 4–foot section / 2 4 s f)

Open cell indicates cited luminaire in cited layout has low probability of meeting **all** ambient lighting criteria outlined in Table 10.4.❶

						80"	
	0.98			0.98		ceiling height not appropriate for cited suspension length	66"	Predominant partition height (in inches)
		0.98			0.98	54"	
						48"	

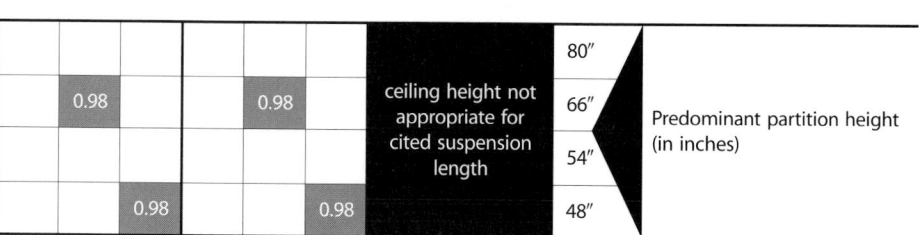

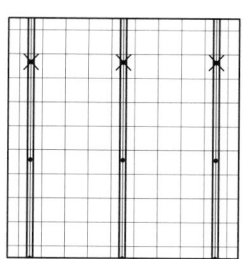

rows 8 feet on center
(O n e 4–foot section / 3 2 s f)

						80"	
						ceiling height not appropriate for cited suspension length	66"	Predominant partition height (in inches)
0.98			0.98			54"	
						48"	

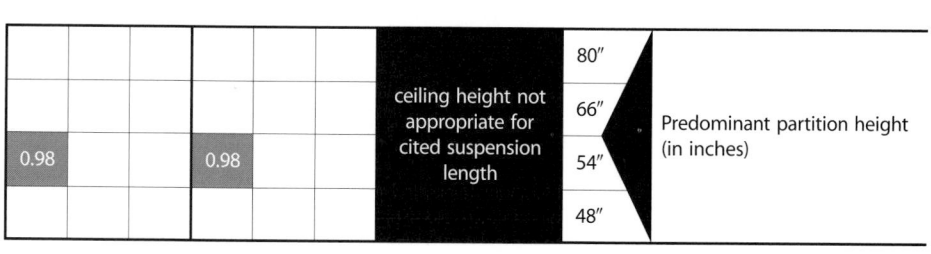

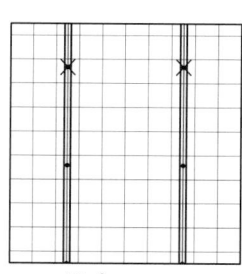

rows 10 feet on center
(O n e 4–foot section / 4 0 s f)

more continued on next two pages

ConceptStarters™

Ambient Indirect/Beveled/Narrow

(1) F28T5 lamp/Suspension of 2–foot/6 inches

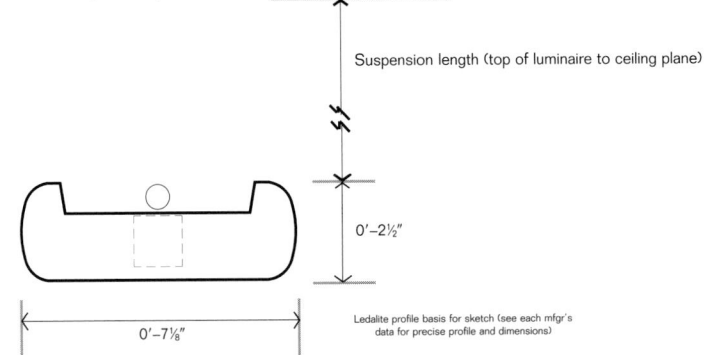

Suspension length (top of luminaire to ceiling plane)

0'–2½"

0'–7⅛"

Ledalite profile basis for sketch (see each mfgr's data for precise profile and dimensions)

Design Tips

✓ Check extruded aluminum units for modular alignment, side–to–side leveling, camber and surface integrity.

✓ Ceiling finish must be matte.

✓ Ceiling reflectance of 0.9 will provide most efficient lighting.

✓ Depending on layout, square–foot costs may be within a few percentage points of recessed—even though pendent luminaire unit costs may be higher.

✓ Note that endcaps add length.
◁more tips on previous pages

Ceiling height (in feet)			10'			9'–6"			9'		
Percent floor reflectance			20	15	10	20	15	10	20	15	10
Percent wall reflectance			50	40	30	50	40	30	50	40	30
Percent partition reflectance			50	30	30	50	30	30	50	30	30
Percent ceiling reflectance			90	80	70	90	80	70	90	80	70

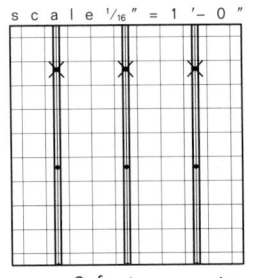

scale ¹⁄₁₆" = 1'–0"

rows 6 feet on center
(One 4–foot section / 24 sf)

Predominant partition height (in inches)							
80"							
66"				ceiling height not appropriate for cited suspension length			
54"		0.98					
48"							

Shaded cell indicates cited luminaire in cited layout has high probability of meeting **all** ambient lighting criteria outlined in Table 10.4. Likely ballast factor is also cited. ❶

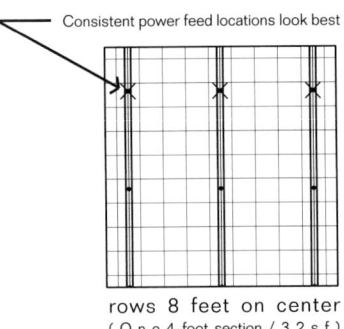

Consistent power feed locations look best

rows 8 feet on center
(One 4–foot section / 32 sf)

Predominant partition height (in inches)							
80"							
66"				ceiling height not appropriate for cited suspension length			
54"		0.98					
48"			0.98				

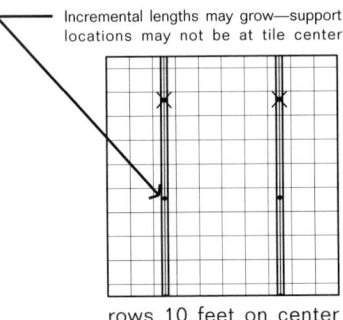

Incremental lengths may grow—support locations may not be at tile center

rows 10 feet on center
(One 4–foot section / 40 sf)

Predominant partition height (in inches)							
80"							
66"				ceiling height not appropriate for cited suspension length			
54"	0.98						
48"							

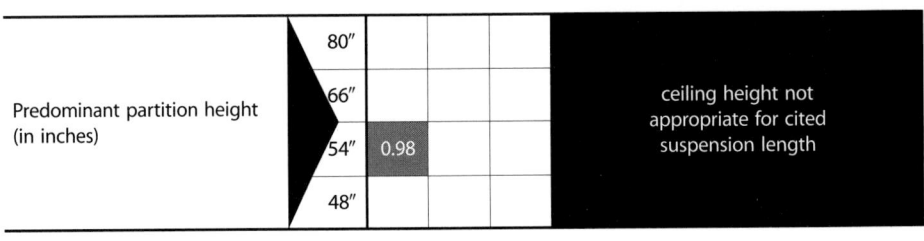

❶Calculations are required in order to secure a final layout and luminaire specification for actual conditions.

Ambient**Indirect/Beveled/Narrow**

(1) F28T5 lamp/Information

ConceptStarters™

Specification
⊃7 inches by 2½ inches rounded or beveled sides/flat–bottom cross section
⊃Totally indirect (uplight only)
⊃VDT–sensitive design (exhibits least glare)
⊃Extruded aluminum housing in modular lengths
⊃1–F28T5SE/830 superior efficacy lamp (2900 lumens)
⊃Electronic programmed start ballast
⊃See power budgeting sidebar to right
⊃Cost magnitude: high
⊃Possible vendors: **Ledalite** (Minuet)

Boldfaced manufacturer's photometric data were averaged and input to Lumen Micro 7.5 in order to assess probability of meeting ambient lighting criteria cited in Table 10.4. Use the above specification list when contacting manufacturer about potential luminaire options for a specific project. Dimensions, specifications and performance are somewhat generic. Check with specific manufacturer's data for nominal dimensions and performance. Data is subject to change.

Power budgeting
for these 1–lamp cross–section luminaires

BF 0.98 yields 32 watts①

Power budget for ambient lighting should be less than 1.0 w/sf. Ballast vendors include Energy Savings, Inc. (ESI), Motorola and Osram Sylvania. Above factors based on ESI.
①Single– and 2–lamp ballasts are available and both have nearly identical power requirements (32 watts per lamp).

Net Addresses/Luminaires
http://www.ledalite.com/

Net Addresses/Ballasts
http://www.mot.com/ies/MLI/
http://www.sylvania.com/ballast/phops.htm

CONNECT FOR MORE

Cost magnitude²⁰⁰⁰
for 4–foot fluorescent luminaire sections

Low
US$70. to US$90.
Moderate
US$90. to US$120.
High
US$120. and greater

Quality ambient lighting for today's electronic office environment typically runs US$3.00 to US$5.00/sf, hardware only (luminaires, lamps and ballasts), excluding installation costs. Costs vary based on quantities/lengths, distributor and contractor markups, luminaire variations such as finish, lamping, ballasting and options and accessories, and market conditions, manufacturing situations and annual inflation. These values are for preliminary, magnitude budgeting and do not represent quotes nor actual final pricing to client.

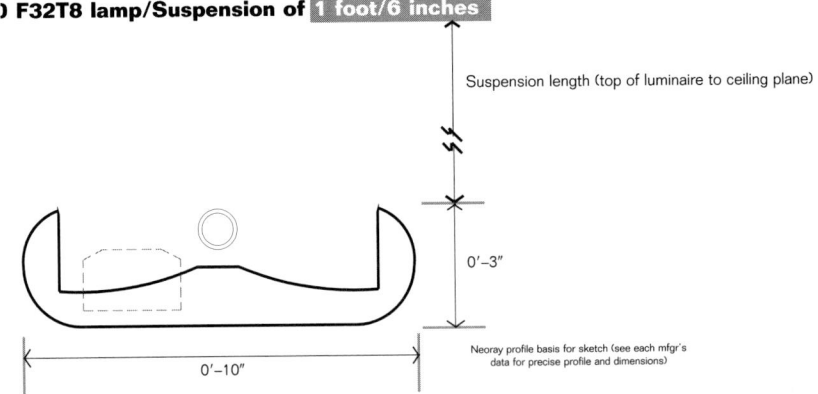

AmbientIndirect/Beveled/ExtraWide

(1) F32T8 lamp/Suspension of `1 foot/6 inches`

ConceptStarters™

Design Tips

✔Check steel linear units for modular alignment, straightness, sag, side–to–side leveling and body integrity.
✔Flat bottom creates contrast with ceiling—harsh look and more likely "streaks" in VDTs.
✔Wide, flat, dark bottom creates strong secondary ceiling plane.
✔Use localized task lighting.
✔Use art accenting for visual interest.
more tips on next page▷

Suspension length (top of luminaire to ceiling plane)

0'–3"

0'–10"

Neoray profile basis for sketch (see each mfgr's data for precise profile and dimensions)

Ceiling height (in feet)		10'			9'–6"			9'		
Percent floor reflectance		20	15	10	20	15	10	20	15	10
Percent wall reflectance		50	40	30	50	40	30	50	40	30
Percent partition reflectance		50	30	30	50	30	30	50	30	30
Percent ceiling reflectance		90	80	70	90	80	70	90	80	70
Predominant partition height (in inches)	80"									
	66"			0.88		0.78	0.88		0.78	0.88
	54"			0.78			0.78			0.78
	48"			0.78			0.78			0.78

Shaded cell indicates cited luminaire in cited layout has high probability of meeting **all** ambient lighting criteria outlined in Table 10.4. Likely ballast factor is also cited.❶

		10'			9'–6"			9'		
Predominant partition height (in inches)	80"									
	66"	0.78	0.88	0.98	0.78	0.88	0.98	0.78	0.88	0.98
	54"		0.78	0.98		0.78	0.98		0.78	0.98
	48"		0.78	0.88		0.78	0.88		0.78	0.88

		10'			9'–6"			9'		
Predominant partition height (in inches)	80"									
	66"	0.88			0.88	0.98	1.18	0.88	1.18	1.18
	54"	0.88	0.98	1.18	0.88	0.98	1.18	0.88	0.98	1.18
	48"	0.78	0.88	1.18	0.78	0.88	1.18	0.78	0.88	1.18

s c a l e ¹/₁₆" = 1'– 0"

rows 6 feet on center
(O n e 4–foot section / 2 4 s f)

Consistent power feed locations look best

rows 8 feet on center
(O n e 4–foot section / 3 2 s f)

Incremental lengths may grow—support locations may not be at tile center

rows 10 feet on center
(O n e 4–foot section / 4 0 s f)

❶Calculations are required in order to secure a final layout and luminaire specification for actual conditions.

Ambient Indirect/Beveled/ExtraWide

(1) F32T8 lamp/Suspension of `2 feet/0 inches`

more continued on next two pages▷

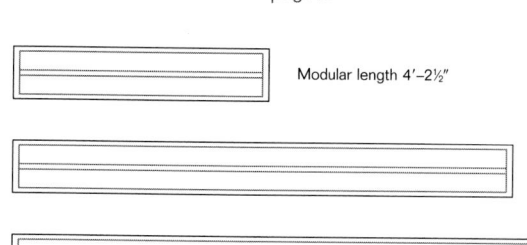

Modular length 4'–2½"

Modular length 8'–2½"

Modular length 12'–2½"

Concept reference only: see specific manufacturer's literature for actual module sizes, seams, endcaps and dimensions

ConceptStarters™

Design Tips

✔ Many of these luminaires exhibit incremental growth in length of several inches from suspension point to suspension point—so every 8 to 12 feet it may be necessary to add 2 to 6 inches. Critical for final layout/ceiling integration.

✔ Units on closer spacings cause a stronger secondary ceiling plane effect.

✔ Units on greater spacings reduce sustainability issues.

more tips on next page▷

Ambient
Pendent Fluorescent

10

10'			9'–6"			9'			Ceiling height (in feet)
20	15	10	20	15	10	20	15	10	Percent floor reflectance
50	40	30	50	40	30	50	40	30	Percent wall reflectance
50	30	30	50	30	30	50	30	30	Percent partition reflectance
90	80	70	90	80	70	90	80	70	Percent ceiling reflectance

rows 6 feet on center

10'			9'–6"				Predominant partition height (in inches)
				0.78	0.88	ceiling height not appropriate for cited suspension length	80" / 66"
		0.88			0.78		54"
		0.78			0.78		48"

Open cell indicates cited luminaire in cited layout has low probability of meeting **all** ambient lighting criteria outlined in Table 10.4. ❶

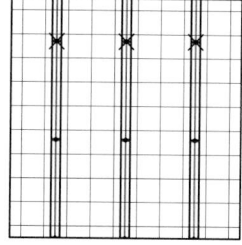

rows 6 feet on center
(O n e 4–foot section / 2 4 s f)

rows 8 feet on center

10'			9'–6"				Predominant partition height (in inches)
						ceiling height not appropriate for cited suspension length	80"
0.78	0.88	0.98	0.78	0.88	0.98		66"
	0.88	0.98		0.88	0.98		54"
	0.78	0.88		0.78	0.88		48"

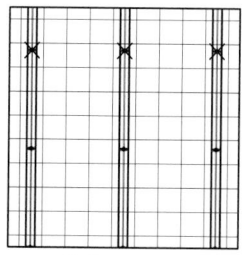

rows 8 feet on center
(O n e 4–foot section / 3 2 s f)

rows 10 feet on center

10'			9'–6"				Predominant partition height (in inches)
						ceiling height not appropriate for cited suspension length	80"
0.88	1.18		0.88	1.18	1.18		66"
0.78	0.98	1.18	0.78	0.98	1.18		54"
0.78	0.88	0.98	0.78	0.88	0.98		48"

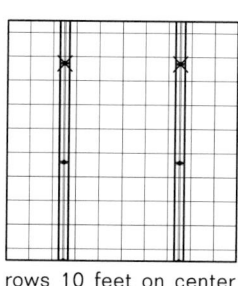

rows 10 feet on center
(O n e 4–foot section / 4 0 s f)

more continued on next two pages

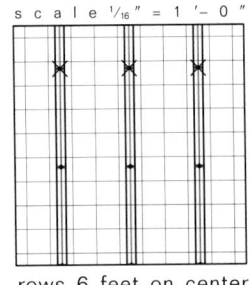

ConceptStarters™

Digital imagery® copyright 2000 PhotoDisc, Inc.

Ambient Pendent Fluorescent

10

Ambient Indirect/Beveled/ExtraWide

(1) F32T8 lamp/Suspension of 2 feet/6 inches

Suspension length (top of luminaire to ceiling plane)

0'–3"

0'–10"

Neoray profile basis for sketch (see each mfgr's data for precise profile and dimensions)

Design Tips

✔ Check extruded aluminum units for modular alignment, side–to–side leveling, camber and surface integrity.
✔ Ceiling finish must be matte.
✔ Ceiling reflectance of 0.9 will provide most efficient lighting.
✔ Note that endcaps add length.
✔ Depending on layout, square–foot costs may be within a few percentage points of recessed—even though pendent luminaire unit costs may be higher.
◁ more tips on previous pages

Ceiling height (in feet)	10'			9'–6"			9'		
Percent floor reflectance	20	15	10	20	15	10	20	15	10
Percent wall reflectance	50	40	30	50	40	30	50	40	30
Percent partition reflectance	50	30	30	50	30	30	50	30	30
Percent ceiling reflectance	90	80	70	90	80	70	90	80	70

scale ¹⁄₁₆" = 1'–0"

rows 6 feet on center
(One 4–foot section / 2 4 s f)

Predominant partition height (in inches)	10'			9'–6" / 9'
80"				ceiling height not appropriate for cited suspension length
66"		0.78		
54"			0.88	
48"			0.78	

Consistent power feed locations look best

rows 8 feet on center
(One 4–foot section / 3 2 s f)

Shaded cell indicates cited luminaire in cited layout has high probability of meeting **all** ambient lighting criteria outlined in Table 10.4. Likely ballast factor is also cited. ❶

Predominant partition height (in inches)	10'			9'–6" / 9'
80"				ceiling height not appropriate for cited suspension length
66"	0.78	0.98	1.18	
54"		0.98	0.?8	
48"		0.88	0.88	

Incremental lengths may grow—support locations may not be at tile center

rows 10 feet on center
(One 4–foot section / 4 0 s f)

Predominant partition height (in inches)	10'			9'–6" / 9'
80"				ceiling height not appropriate for cited suspension length
66"	1.18	1.18	1.18	
54"	0.88	1.18	1.18	
48"	0.88	0.98	1.18	

❶ Calculations are required in order to secure a final layout and luminaire specification for actual conditions.

Ambient Indirect/Beveled/ExtraWide

(1) F32T8 lamp/Information

ConceptStarters™

Specification

- ↪ 10 inches by 3 inches rounded or beveled sides/flat–bottom cross section
- ↪ Totally indirect (uplight only)
- ↪ VDT–sensitive design (exhibits least glare)
- ↪ Extruded aluminum or 20 gauge steel housing in modular lengths
- ↪ 1–F32T8/830 rapid start lamp (2950 lumens)
- ↪ Electronic instant start ballast
- ↪ See power budgeting sidebar to right
- ↪ Cost magnitude: moderate to high
- ↪ Possible vendors: **LITECONTROL** (PI300)**❷**, **Neoray** (67IP)**❸**, **Peerless** (LD3)**❸**

Boldfaced manufacturers' photometric data were averaged and input to Lumen Micro 7.5 in order to assess probability of meeting ambient lighting criteria cited in Table 10.4. Use the above specification list when contacting manufacturers about potential luminaire options for a specific project. Dimensions, specifications and performance are somewhat generic. Check with specific manufacturers' data for nominal dimensions and performance. Data is subject to change.

Power budgeting
for these 1–lamp cross–section luminaires

BF 0.78 yields 26 watts①
BF 0.88 yields 30 watts①②
BF 0.98 yields 35 watts①②
BF 1.18 yields 38 watts①

Power budget for ambient lighting should be less than 1.0 w/sf. Ballast vendors include Advance, Energy Savings, Inc. (ESI), Magnetek, Motorola. Not all ballast factors available from all ballast vendors. Above factors based on Magnetek. ①Based on tandem lamp operation— using one ballast to operate two lamps. Wattage reported is only for one of the tandem–wired lamps. ②Single lamp ballasts are available for these ballast factors.

Ambient
Pendent Fluorescent

10

Net Addresses/Luminaires
http://www.litecontrol.com/products/index.html
http://www.peerless-lighting.com/

Net Addresses/Ballasts
http://www.magnetek.com/transLighting.html
http://www.mot.com/ies/MLI/
http://www.sylvania.com/ballast/phops.htm

CONNECT FOR MORE

Cost magnitude²⁰⁰⁰
for 4–foot fluorescent luminaire sections

Low
US$70. to US$90.
Moderate
US$90. to US$120.
High
US$120. and greater

Quality ambient lighting for today's electronic office environment typically runs US$3.00 to US$5.00/sf, hardware only (luminaires, lamps and ballasts), excluding installation costs. Costs vary based on quantities/lengths, distributor and contractor markups, luminaire variations such as finish, lamping, ballasting and options and accessories, and market conditions, manufacturing situations and annual inflation. These values are for preliminary, magnitude budgeting and do not represent quotes nor actual final pricing to client.

❷ 20 gauge steel housing (typically available in shorter modular lengths than extruded aluminum and require more frequent suspension points).
❸ Extruded aluminum housing (typically available in greater lengths than steel housings).

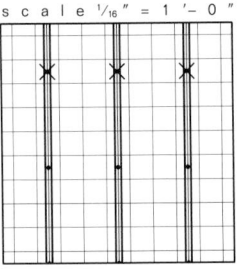

ConceptStarters™

Ambient Indirect/Oval/Narrow

(1) F28T5 lamp/Suspension of 1 foot/6 inches

Suspension length (top of luminaire to ceiling plane)

Slots in bottom for soft glow

0'–2⅛"

0'–6½"

Peerless profile basis for sketch (see each mfgr's data for precise profile and dimensions)

Design Tips

✔ Oval shape helps limit high contrast between luminaire bottom and ceiling—only if luminaire finish is white or very light.

✔ Available in solid bottom and perforated bottom versions.

✔ Slots in bottom further reduces contrast between luminaire bottom and ceiling. Also minimizes haziness of entirely indirect lighting.

✔ Use localized task lighting.

✔ Use art accenting for visual interest.

more tips on next page▷

Ceiling height (in feet)	10'			9'–6"			9'		
Percent floor reflectance	20	15	10	20	15	10	20	15	10
Percent wall reflectance	50	40	30	50	40	30	50	40	30
Percent partition reflectance	50	30	30	50	30	30	50	30	30
Percent ceiling reflectance	90	80	70	90	80	70	90	80	70

scale ¹⁄₁₆" = 1'–0"

Predominant partition height (in inches)									
80"									
66"									
54"									
48"									

rows 6 feet on center
(One 4–foot section / 2 4 s f)

Shaded cell indicates cited luminaire in cited layout has high probability of meeting **all** ambient lighting criteria outlined in Table 10.4. Likely ballast factor is also cited.❶

Consistent power feed locations look best

Predominant partition height (in inches)									
80"									
66"		0.98			0.98	0.98		0.98	0.98
54"			0.98						
48"									

rows 8 feet on center
(One 4–foot section / 3 2 s f)

Incremental lengths may grow—support locations may not be at tile center

Predominant partition height (in inches)									
80"									
66"	0.98								
54"		0.98			0.98			0.98	
48"		0.98							

rows 10 feet on center
(One 4–foot section / 4 0 s f)

❶ Calculations are required in order to secure a final layout and luminaire specification for actual conditions.

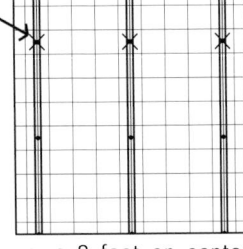

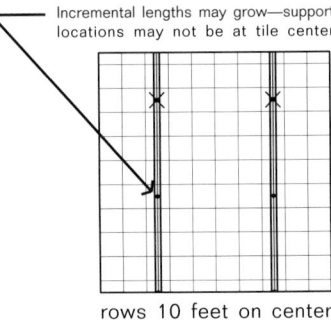

Ambient Pendent Fluorescent

10

Ambient Indirect/Oval/Narrow

(1) F28T5 lamp/Suspension of `2 feet/0 inches`

more continued on next two pages▷

Modular length 4'–11" (with endcaps)

Modular length 8'–11" (with endcaps)

Modular length 12'–11" (with endcaps)

Concept reference only: see specific manufacturer's literature for actual module sizes, seams, endcaps and dimensions

ConceptStarters™

Design Tips

✔ These luminaires are available in both modular lengths (e.g, 4 feet, 8 feet) and in continuous linear lengths (e.g., 15 feet) up to 24 feet with no seams (building must accommodate move–in of single, continuous lengths, however, through doors, open windows, elevators, etc.).

✔ Units on closer spacings cause a stronger secondary ceiling plane effect.

✔ Units on greater spacings reduce sustainability issues.

more tips on next page▷

10'			9'–6"			9'			Ceiling height (in feet)
20	15	10	20	15	10	20	15	10	Percent floor reflectance
50	40	30	50	40	30	50	40	30	Percent wall reflectance
50	30	30	50	30	30	50	30	30	Percent partition reflectance
90	80	70	90	80	70	90	80	70	Percent ceiling reflectance

						80"	Predominant partition height (in inches)
					ceiling height not appropriate for cited suspension length	66"	
						54"	
						48"	

Open cell indicates cited luminaire in cited layout has low probability of meeting **all** ambient lighting criteria outlined in Table 10.4.❶

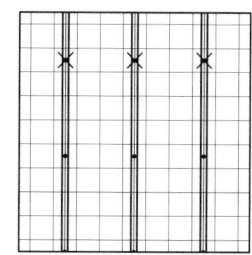

rows 6 feet on center
(O n e 4–foot section / 2 4 s f)

						80"	Predominant partition height (in inches)
	0.98			0.98	ceiling height not appropriate for cited suspension length	66"	
						54"	
		0.98			0.98	48"	

rows 8 feet on center
(O n e 4–foot section / 3 2 s f)

						80"	Predominant partition height (in inches)
0.98			0.98		ceiling height not appropriate for cited suspension length	66"	
	0.98			0.98		54"	
						48"	

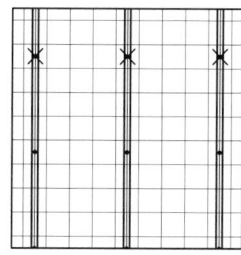

rows 10 feet on center
(O n e 4–foot section / 4 0 s f)

more continued on next two pages

ConceptStarters™

Design Tips

✔ Check extruded aluminum units for modular alignment, side–to–side leveling, camber and surface integrity.

✔ Ceiling finish must be matte.

✔ Ceiling reflectance of 0.9 will provide most efficient lighting.

✔ Depending on layout, square–foot costs may be within a few percentage points of recessed—even though pendent luminaire unit costs may be higher.

✔ Note that endcaps add length.

◁ more tips on previous pages

Ambient Indirect/Oval/Narrow

(1) F28T5 lamp/Suspension of 2–foot/6 inches

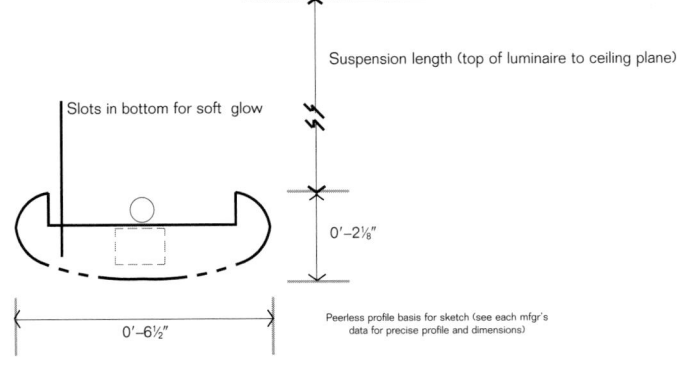

Suspension length (top of luminaire to ceiling plane)

Slots in bottom for soft glow

0'–2⅛"

0'–6½"

Peerless profile basis for sketch (see each mfgr's data for precise profile and dimensions)

Ceiling height (in feet)		10'			9'–6"			9'		
Percent floor reflectance		20	15	10	20	15	10	20	15	10
Percent wall reflectance		50	40	30	50	40	30	50	40	30
Percent partition reflectance		50	30	30	50	30	30	50	30	30
Percent ceiling reflectance		90	80	70	90	80	70	90	80	70

scale ¹/₁₆" = 1'–0"

rows 6 feet on center
(O n e 4–foot section / 2 4 s f)

Predominant partition height (in inches)	80"			ceiling height not appropriate for cited suspension length
	66"			
	54"			
	48"			

Shaded cell indicates cited luminaire in cited layout has high probability of meeting **all** ambient lighting criteria outlined in Table 10.4. Likely ballast factor is also cited.❶

Consistent power feed locations look best

rows 8 feet on center
(O n e 4–foot section / 3 2 s f)

Predominant partition height (in inches)	80"			ceiling height not appropriate for cited suspension length
	66"		0.98	
	54"			0.98
	48"			

Incremental lengths may grow—support locations may not be at tile center

rows 10 feet on center
(O n e 4–foot section / 4 0 s f)

Predominant partition height (in inches)	80"			ceiling height not appropriate for cited suspension length
	66"	0.98		
	54"		0.98	
	48"			

❶ Calculations are required in order to secure a final layout and luminaire specification for actual conditions.

Ambient Indirect/Oval/Narrow
(1) F28T5 lamp/Information

ConceptStarters™

Specification
⊃ 6½ inches by 2⅛ inches oval cross section
⊃ Totally indirect (less than 1 percent of the light comes from the slots in bottom)
⊃ VDT–sensitive design (exhibits least glare)
⊃ Extruded aluminum housing in modular or continuous linear lengths
⊃ 1–F28T5SE/830 superior efficacy lamp (2900 lumens)
⊃ Electronic programmed start ballast
⊃ See power budgeting sidebar to right
⊃ Cost magnitude: high
⊃ Possible vendors: **Peerless** (Lightduct™ Diminutive)

Boldfaced manufacturer's photometric data were averaged and input to Lumen Micro 7.5 in order to assess probability of meeting ambient lighting criteria cited in Table 10.4. Use the above specification list when contacting manufacturer about potential luminaire options for a specific project. Dimensions, specifications and performance are somewhat generic. Check with specific manufacturer's data for nominal dimensions and performance. Data is subject to change.

Power budgeting
for these 1–lamp cross–section luminaires

BF 0.98 yields 32 watts①

Power budget for ambient lighting should be less than 1.0 w/sf. Ballast vendors include Energy Savings, Inc. (ESI), Motorola and Osram Sylvania. Above factors based on ESI.
① Single– and 2–lamp ballasts are available and both have nearly identical power requirements (32 watts per lamp).

Ambient Pendent Fluorescent

10

Net Addresses/Luminaires
http://www.peerless-lighting.com/

Net Addresses/Ballasts
http://www.mot.com/ies/MLI/
http://www.sylvania.com/ballast/phops.htm

CONNECT FOR MORE

Cost magnitude²⁰⁰⁰
for 4–foot fluorescent luminaire sections

Low
US$70. to US$90.
Moderate
US$90. to US$120.
High
US$120. and greater

Quality ambient lighting for today's electronic office environment typically runs US$3.00 to US$5.00/sf, hardware only (luminaires, lamps and ballasts), excluding installation costs. Costs vary based on quantities/lengths, distributor and contractor markups, luminaire variations such as finish, lamping, ballasting and options and accessories, and market conditions, manufacturing situations and annual inflation. These values are for preliminary, magnitude budgeting and do not represent quotes nor actual final pricing to client.

Lightduct™ is a registered trademark of Lithonia/Peerless Lighting

ConceptStarters™

©Digital Imagery® copyright 2000 PhotoDisc, Inc.

AmbientIndirect/Oval/Wide

(1) F32T8 lamp/Suspension of `1 foot/6 inches`

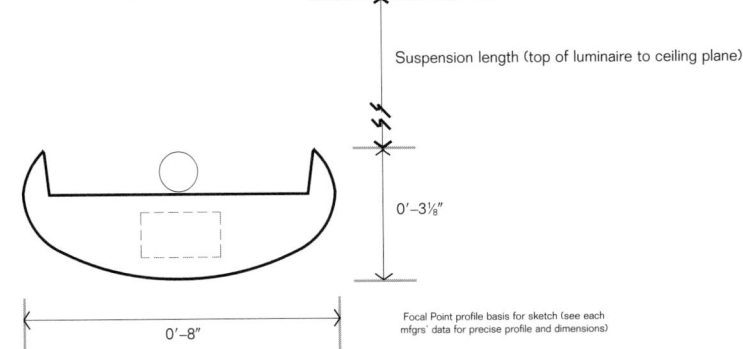

Suspension length (top of luminaire to ceiling plane)

0'–3⅛"

0'–8"

Focal Point profile basis for sketch (see each mfgrs' data for precise profile and dimensions)

Design Tips

✔ Oval shape helps limit high contrast between luminaire bottom and ceiling—only if luminaire finish is white or very light. Narrow oval is better.
✔ Use localized task lighting.
✔ Use art accenting for visual interest.

more tips on next page▷

Ceiling height (in feet)		10'			9'–6"			9'		
Percent floor reflectance	▶	20	15	10	20	15	10	20	15	10
Percent wall reflectance	▶	50	40	30	50	40	30	50	40	30
Percent partition reflectance	▶	50	30	30	50	30	30	50	30	30
Percent ceiling reflectance	▶	90	80	70	90	80	70	90	80	70

scale ¹⁄₁₆" = 1'–0"

Predominant partition height (in inches)	10' (20)	10' (15)	10' (10)	9'–6" (20)	9'–6" (15)	9'–6" (10)	9' (20)	9' (15)	9' (10)
80"	0.78								
66"		0.78	0.78		0.78	0.78		0.78	0.78
54"		0.78	0.78			0.78			0.78
48"						0.78			0.78

rows 6 feet on center
(O n e 4–foot section / 2 4 s f)

Shaded cell indicates cited luminaire in cited layout has high probability of meeting **all** ambient lighting criteria outlined in Table 10.4. Likely ballast factor is also cited.❶

Consistent power feed locations look best

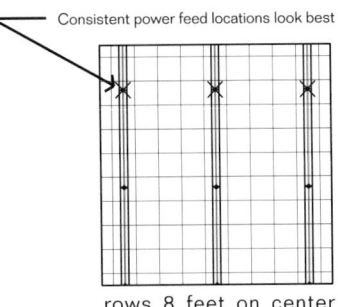

Predominant partition height (in inches)	10' (20)	10' (15)	10' (10)	9'–6" (20)	9'–6" (15)	9'–6" (10)	9' (20)	9' (15)	9' (10)
80"									
66"	0.78	0.88	1.18	0.78	0.88	0.98	0.78	0.88	0.98
54"	0.78	0.78	0.98	0.78	0.78	0.88			0.88
48"		0.78	0.88		0.78	0.88		0.78	0.88

rows 8 feet on center
(O n e 4–foot section / 3 2 s f)

Incremental lengths may grow—support locations may not be at tile center

Predominant partition height (in inches)	10' (20)	10' (15)	10' (10)	9'–6" (20)	9'–6" (15)	9'–6" (10)	9' (20)	9' (15)	9' (10)
80"									
66"	0.88	1.18		0.88	1.18	1.18	0.88	1.18	
54"	0.78	0.98	1.18	0.78	0.98	1.18	0.78	0.98	1.18
48"	0.78	0.98	1.18	0.78	0.98	0.98	0.78	0.98	0.98

rows 10 feet on center
(O n e 4–foot section / 4 0 s f)

❶Calculations are required in order to secure a final layout and luminaire specification for actual conditions.

Ambient Indirect/Oval/Wide

(1) F32T8 lamp/Suspension of `2 feet/0 inches`

more continued on next two pages▷

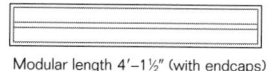

Modular length 4'–1½" (with endcaps)

Modular length 8'–1½" (with endcaps)

Modular length 12'–3" (with endcaps)

Concept reference only: see specific manufacturer's literature for actual module sizes, seams, endcaps and dimensions

ConceptStarters™

Design Tips

✔ These luminaires are available in both modular lengths (e.g, 4 feet, 8 feet) and in continuous linear lengths (e.g., 15 feet) up to 24 feet with no seams depending on manufacturer (building must accommodate move–in of single, continuous lengths, however, through doors, open windows, elevators, etc.).

✔ Units on closer spacings cause a stronger secondary ceiling plane effect.

✔ Units on greater spacings reduce sustainability issues.

more tips on next page▷

Ambient Pendent Fluorescent

10

10'			9'–6"			9'			Ceiling height (in feet)
20	15	10	20	15	10	20	15	10	◀ Percent floor reflectance
50	40	30	50	40	30	50	40	30	◀ Percent wall reflectance
50	30	30	50	30	30	50	30	30	◀ Percent partition reflectance
90	80	70	90	80	70	90	80	70	◀ Percent ceiling reflectance

10'			9'–6"			9'	Predominant partition height (in inches)
				0.78	0.98	ceiling height not appropriate for cited suspension length	80"
							66"
	0.78	0.88			0.88		54"
		0.78			0.78		48"

Open cell indicates cited luminaire in cited layout has low probability of meeting **all** ambient lighting criteria outlined in Table 10.4.❶

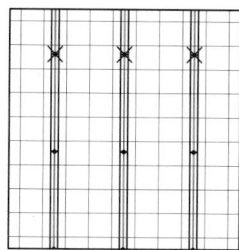

rows 6 feet on center
(O n e 4–foot section / 2 4 s f)

10'			9'–6"			9'	Predominant partition height (in inches)
		0.88				ceiling height not appropriate for cited suspension length	80"
	0.98	1.18	0.78	0.98	1.18		66"
	0.88	0.98		0.88	0.98		54"
	0.78	0.88		0.78	0.88		48"

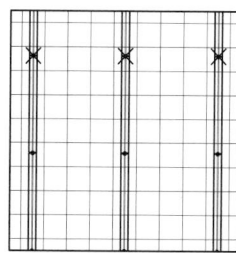

rows 8 feet on center
(O n e 4–foot section / 3 2 s f)

10'			9'–6"			9'	Predominant partition height (in inches)
		0.98				ceiling height not appropriate for cited suspension length	80"
0.88	1.18		0.88	1.18	1.18		66"
0.78	0.98	1.18	0.78	0.98	1.18		54"
0.78	0.88	1.18	0.78	0.88	0.98		48"

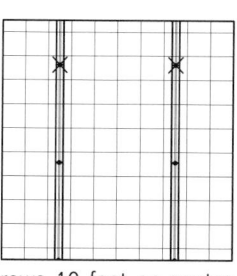

rows 10 feet on center
(O n e 4–foot section / 4 0 s f)

more continued on next two pages

ConceptStarters™

Ambient Indirect/Oval/Wide

(1) F32T8 lamp/Suspension of 2 feet/6 inches

Suspension length (top of luminaire to ceiling plane)

0'–3⅛"

0'–8"

Focal Point profile basis for sketch (see each mfgrs' data for precise profile and dimensions)

Design Tips

✔ Check extruded aluminum units for modular alignment, side–to–side leveling, camber and surface integrity.

✔ Ceiling finish must be matte.

✔ Ceiling reflectance of 0.9 will provide most efficient lighting.

✔ Depending on layout, square–foot costs may be within a few percentage points of recessed— even though pendent luminaire unit costs may be higher.

✔ Note that endcaps add length.
◁more tips on previous pages

Ceiling height (in feet)		10'			9'–6"			9'		
Percent floor reflectance		20	15	10	20	15	10	20	15	10
Percent wall reflectance		50	40	30	50	40	30	50	40	30
Percent partition reflectance		50	30	30	50	30	30	50	30	30
Percent ceiling reflectance		90	80	70	90	80	70	90	80	70

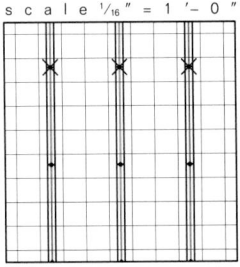

scale 1/16" = 1'–0"

rows 6 feet on center
(O n e 4–foot section / 2 4 s f)

Predominant partition height (in inches)	20	15	10	9'–6" / 9'
80"				ceiling height not appropriate for cited suspension length
66"				
54"		0.78	0.88	
48"			0.78	

Shaded cell indicates cited luminaire in cited layout has high probability of meeting **all** ambient lighting criteria outlined in Table 10.4. Likely ballast factor is also cited. ❶

Consistent power feed locations look best

rows 8 feet on center
(O n e 4–foot section / 3 2 s f)

Predominant partition height (in inches)	20	15	10	9'–6" / 9'
80"				ceiling height not appropriate for cited suspension length
66"	0.78	1.18	1.18	
54"	0.78	0.98	0.98	
48"		0.88	0.88	

Incremental lengths may grow—support locations may not be at tile center

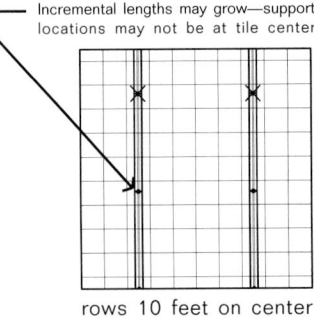

rows 10 feet on center
(O n e 4–foot section / 4 0 s f)

Predominant partition height (in inches)	20	15	10	9'–6" / 9'
80"				ceiling height not appropriate for cited suspension length
66"		1.18		
54"		0.98	1.18	
48"	0.88	0.88	1.18	

❶Calculations are required in order to secure a final layout and luminaire specification for actual conditions.

Ambient Indirect/Oval/Wide

(1) F32T8 lamp/Information

Specification

- ➲ 8 inches by 3⅛ inches oval cross section
- ➲ Totally indirect
- ➲ VDT–sensitive design (exhibits least glare)
- ➲ Extruded aluminum housing in modular or continuous linear lengths
- ➲ 1–F32T8/830 rapid start lamp (2950 lumens)
- ➲ Electronic instant start ballast (for best efficiency)
- ➲ See power budgeting sidebar to right
- ➲ Cost magnitude: high
- ➲ Possible vendors: **Focal Point** (Optec), Lam (Widelume), **Ledalite** (Flexxa), Linear (SlimLite 38), **Peerless** (Envision™), **Precision Architectural Lighting** (Ellipse)

Boldfaced manufacturers' photometric data were averaged and input to Lumen Micro 7.5 in order to assess probability of meeting ambient lighting criteria cited in Table 10.4. Use the above specification list when contacting manufacturers about potential luminaire options for a specific project. Dimensions, specifications and performance are somewhat generic. Check with specific manufacturer's data for nominal dimensions and performance. Data is subject to change.

Net Addresses/Luminaires
http://www.focalpointlights.com/
http://www.ledalite.com/
http://www.peerless-lighting.com/

Net Addresses/Ballasts
http://www.magnetek.com/transLighting.html
http://www.mot.com/ies/MLI/
http://www.sylvania.com/ballast/phops.htm

CONNECT FOR MORE

ConceptStarters™

Power budgeting
for these 1–lamp cross–section luminaires

BF 0.78 yields 26 watts①
BF 0.88 yields 30 watts①②
BF 0.98 yields 35 watts①②
BF 1.18 yields 38 watts①

Power budget for ambient lighting should be less than 1.0 w/sf. Ballast vendors include Advance, Energy Savings, Inc. (ESI), Magnetek, Motorola. Not all ballast factors available from all ballast vendors. Above factors based on Magnetek.
① Based on tandem lamp operation— using one ballast to operate two lamps. Wattage reported is only for one of the tandem–wired lamps.
② Single lamp ballasts are available for these ballast factors.

Ambient
Pendent Fluorescent

10

Cost magnitude²⁰⁰⁰
for 4–foot fluorescent luminaire sections

Low
US$70. to US$90.
Moderate
US$90. to US$120.
High
US$120. and greater

Quality ambient lighting for today's electronic office environment typically runs US$3.00 to US$5.00/sf, hardware only (luminaires, lamps and ballasts), excluding installation costs. Costs vary based on quantities/lengths, distributor and contractor markups, luminaire variations such as finish, lamping, ballasting and options and accessories, and market conditions, manufacturing situations and annual inflation. These values are for preliminary, magnitude budgeting and do not represent quotes nor actual final pricing to client.

Envision™ is a registered trademark of Lithonia/Peerless Lighting

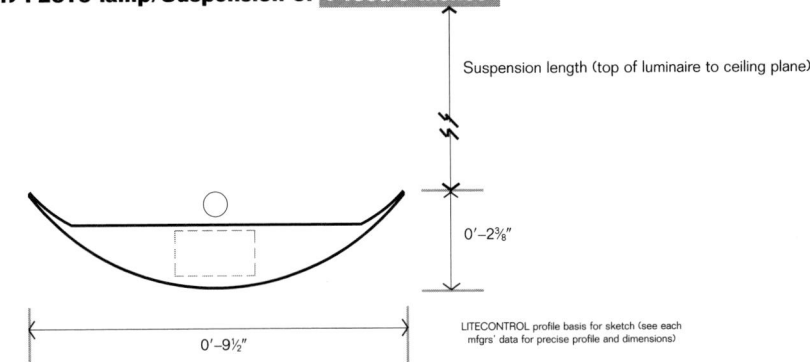

Ambient Indirect/Oval/Extra Wide

(1) F28T5 lamp/Suspension of `1 foot/6 inches`

Suspension length (top of luminaire to ceiling plane)

0'–2⅜"

0'–9½"

LITECONTROL profile basis for sketch (see each mfgrs' data for precise profile and dimensions)

Design Tips
✔ Oval shape helps limit high contrast between luminaire bottom and ceiling—only if luminaire finish is white or very light. Narrow oval is better.
✔ Use localized task lighting.
✔ Use art accenting for visual interest.

more tips on next page▷

10

Ceiling height (in feet)		10'			9'–6"			9'		
Percent floor reflectance	▶	20	15	10	20	15	10	20	15	10
Percent wall reflectance	▶	50	40	30	50	40	30	50	40	30
Percent partition reflectance	▶	50	30	30	50	30	30	50	30	30
Percent ceiling reflectance	▶	90	80	70	90	80	70	90	80	70
Predominant partition height (in inches)	80"									
	66"	0.98			0.98			0.98		
	54"		0.98			0.98			0.98	
	48"		0.98							

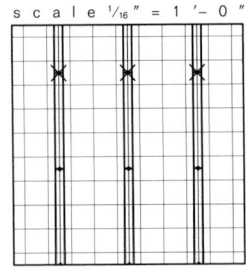

scale ¹⁄₁₆" = 1'–0"

rows 6 feet on center
(One 4–foot section / 2 4 s f)

Shaded cell indicates cited luminaire in cited layout has high probability of meeting **all** ambient lighting criteria outlined in Table 10.4. Likely ballast factor is also cited.❶

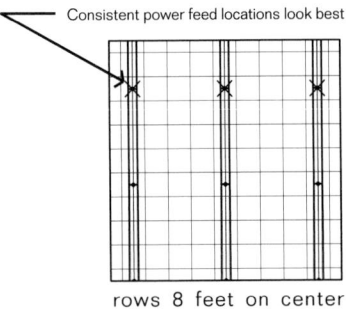

Consistent power feed locations look best

Predominant partition height (in inches)	80"									
	66"									
	54"									
	48"						0.98		0.98	

rows 8 feet on center
(One 4–foot section / 3 2 s f)

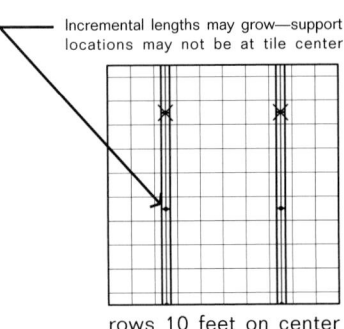

Incremental lengths may grow—support locations may not be at tile center

Predominant partition height (in inches)	80"									
	66"									
	54"									
	48"									

rows 10 feet on center
(One 4–foot section / 4 0 s f)

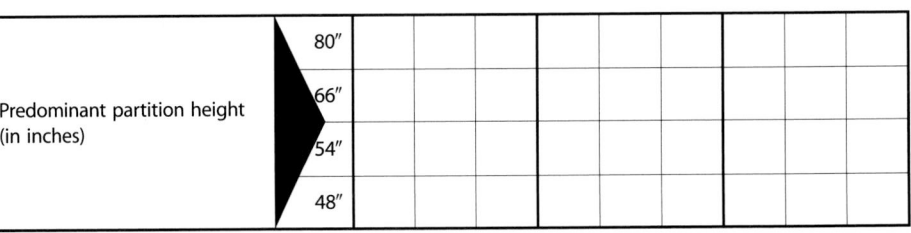

❶Calculations are required in order to secure a final layout and luminaire specification for actual conditions.

Ambient Indirect/Oval/Extra Wide

(1) F28T5 lamp/Suspension of `2 feet/0 inches`

more continued on next two pages▷

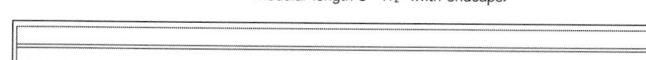

Modular length 4'–1½" (with endcaps)

Modular length 8'–1½" (with endcaps)

Modular length 12'–3" (with endcaps)

Concept reference only: see specific manufacturer's literature for actual module sizes, seams, endcaps and dimensions

ConceptStarters™

Design Tips

✔ These luminaires are available in modular lengths (e.g. 4 feet, 8 feet and 12 feet).

✔ Units on closer spacings cause a stronger secondary ceiling plane effect.

✔ Units on greater spacings reduce sustainability issues.

more tips on next page▷

Ambient Pendent Fluorescent

10

10'			9'–6"			9'			Ceiling height (in feet)
20	15	10	20	15	10	20	15	10	Percent floor reflectance
50	40	30	50	40	30	50	40	30	Percent wall reflectance
50	30	30	50	30	30	50	30	30	Percent partition reflectance
90	80	70	90	80	70	90	80	70	Percent ceiling reflectance

						Predominant partition height (in inches)	
						80"
0.98			0.98			66"
				0.98		54"
	0.98			0.98		48"

(ceiling height not appropriate for cited suspension length)

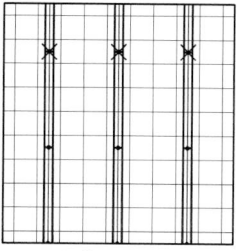

rows 6 feet on center
(O n e 4–foot section / 2 4 s f)

Open cell indicates cited luminaire in cited layout has low probability of meeting **all** ambient lighting criteria outlined in Table 10.4.❶

					Predominant partition height (in inches)	
					80"
					66"
					54"
					48"

(ceiling height not appropriate for cited suspension length)

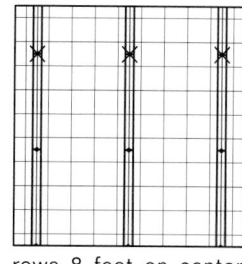

rows 8 feet on center
(O n e 4–foot section / 3 2 s f)

					Predominant partition height (in inches)	
					80"
					66"
					54"
					48"

(ceiling height not appropriate for cited suspension length)

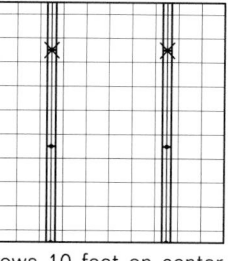

rows 10 feet on center
(O n e 4–foot section / 4 0 s f)

more continued on next two pages

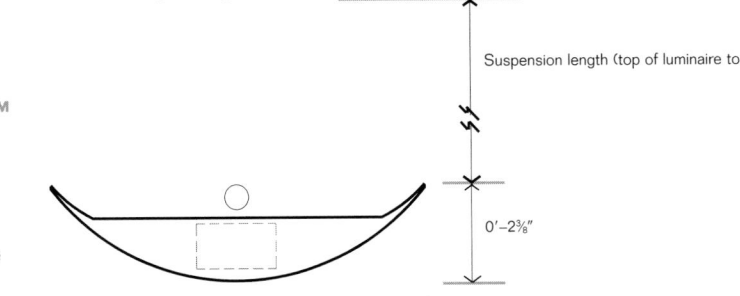

Ambient Indirect/Oval/Extra Wide

ConceptStarters™

(1) F28T5 lamp/Suspension of 2 feet/6 inches

Design Tips

✔Check extruded aluminum units for modular alignment, side–to–side leveling, camber and surface integrity.

✔Ceiling finish must be matte.

✔Ceiling reflectance of 0.9 will provide most efficient lighting.

✔Depending on layout, square–foot costs may be within a few percentage points of recessed—even though pendent luminaire unit costs may be higher.

✔Note that endcaps add length.

◁more tips on previous pages

Suspension length (top of luminaire to ceiling plane)

0'–2⅜"

0'–9½"

LITECONTROL profile basis for sketch (see each mfgrs' data for precise profile and dimensions)

Ceiling height (in feet)		10'			9'–6"			9'		
Percent floor reflectance	▶	20	15	10	20	15	10	20	15	10
Percent wall reflectance	▶	50	40	30	50	40	30	50	40	30
Percent partition reflectance	▶	50	30	30	50	30	30	50	30	30
Percent ceiling reflectance	▶	90	80	70	90	80	70	90	80	70

scale ¹⁄₁₆" = 1'–0"

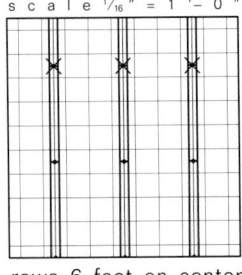

rows 6 feet on center
(O n e 4–foot section / 2 4 s f)

Predominant partition height (in inches)	80"			ceiling height not appropriate for cited suspension length					
	66"								
	54"	0.98							
	48"		0.98						

Shaded cell indicates cited luminaire in cited layout has high probability of meeting **all** ambient lighting criteria outlined in Table 10.4. Likely ballast factor is also cited.❶

Consistent power feed locations look best

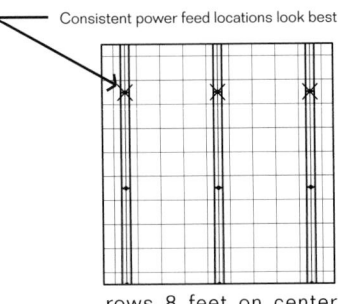

rows 8 feet on center
(O n e 4–foot section / 3 2 s f)

Predominant partition height (in inches)	80"			ceiling height not appropriate for cited suspension length		
	66"					
	54"					
	48"					

Incremental lengths may grow—support locations may not be at tile center

rows 10 feet on center
(O n e 4–foot section / 4 0 s f)

Predominant partition height (in inches)	80"			ceiling height not appropriate for cited suspension length		
	66"					
	54"					
	48"					

❶Calculations are required in order to secure a final layout and luminaire specification for actual conditions.

10 Ambient Pendent Fluorescent

Ambient Indirect/Oval/Extra Wide

(1) F28T5 lamp/Information

Specification

⊃9½ inches by 2⅜ inches oval cross section
⊃Totally indirect
⊃VDT–sensitive design (exhibits least glare)
⊃Extruded aluminum housing in modular lengths
⊃1–F28T5SE/830 superior efficacy lamp (2900 lumens)
⊃Electronic programmed start ballast (for best efficiency)
⊃See power budgeting sidebar to right
⊃Cost magnitude: high
⊃Possible vendors: **LITECONTROL** (Arcos™)

Boldfaced manufacturer's photometric data were averaged and input to Lumen Micro 7.5 in order to assess probability of meeting ambient lighting criteria cited in Table 10.4. Use the above specification list when contacting manufacturer about potential luminaire options for a specific project. Dimensions, specifications and performance are somewhat generic. Check with specific manufacturer's data for nominal dimensions and performance. Data is subject to change.

ConceptStarters™

Power budgeting
for these 1–lamp cross–section luminaires

BF 0.98 yields 32 watts①

Power budget for ambient lighting should be less than 1.0 w/sf. Ballast vendors include Energy Savings, Inc. (ESI), Motorola and Osram Sylvania. Above factors based on ESI.
①Single– and 2–lamp ballasts are available and both have nearly identical power requirements (32 watts per lamp).

Ambient Pendent Fluorescent
10

Net Addresses/Luminaires
http://www.litecontrol.com/products/index.html

Net Addresses/Ballasts
http://www.mot.com/ies/MLI/
http://www.sylvania.com/ballast/phops.htm

CONNECT FOR MORE

Cost magnitude²⁰⁰⁰
for 4–foot fluorescent luminaire sections

Low
US$70. to US$90.
Moderate
US$90. to US$120.
High
US$120. and greater

Quality ambient lighting for today's electronic office environment typically runs US$3.00 to US$5.00/sf, hardware only (luminaires, lamps and ballasts), excluding installation costs. Costs vary based on quantities/lengths, distributor and contractor markups, luminaire variations such as finish, lamping, ballasting and options and accessories, and market conditions, manufacturing situations and annual inflation. These values are for preliminary, magnitude budgeting and do not represent quotes nor actual final pricing to client.

Arcos™ is a registered trademark of LITECONTROL

ConceptStarters™

Digital imagery© copyright 2000 PhotoDisc, Inc.

Ambient
Pendent Fluorescent
10

Design Tips

✓ Oval shape helps limit high contrast between luminaire bottom and ceiling—only if luminaire finish is white or very light. Narrow oval is better.

✓ Use localized task lighting.

✓ Use art accenting for visual interest.

more tips on next page▷

Ambient Indirect/Oval/Extra Wide

(1) F54T5/HO lamp/Suspension of 1 foot/6 inches

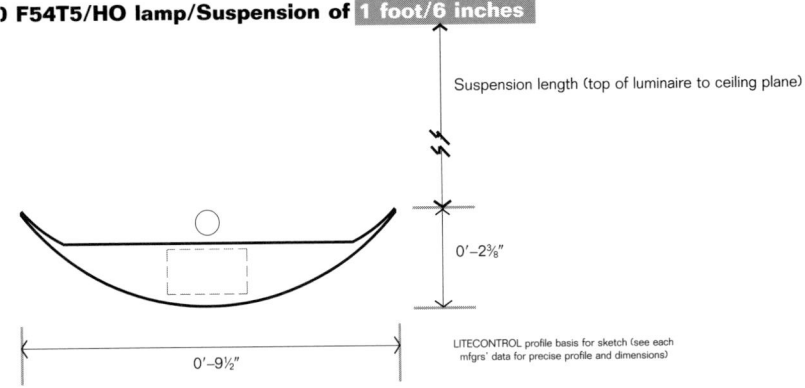

Suspension length (top of luminaire to ceiling plane)

0'–2⅜"

0'–9½"

LITECONTROL profile basis for sketch (see each mfgrs' data for precise profile and dimensions)

scale ¹⁄₁₆″ = 1'–0"

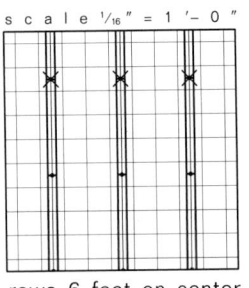

rows 6 feet on center
(O n e 4–foot section / 2 4 s f)

Consistent power feed locations look best

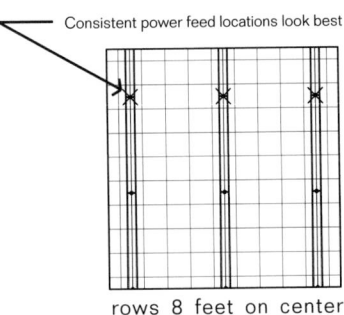

rows 8 feet on center
(O n e 4–foot section / 3 2 s f)

Incremental lengths may grow—support locations may not be at tile center

rows 10 feet on center
(O n e 4–foot section / 4 0 s f)

Ceiling height (in feet)		10'			9'–6"			9'		
Percent floor reflectance		20	15	10	20	15	10	20	15	10
Percent wall reflectance		50	40	30	50	40	30	50	40	30
Percent partition reflectance		50	30	30	50	30	30	50	30	30
Percent ceiling reflectance		90	80	70	90	80	70	90	80	70
Predominant partition height (in inches)	80"									
	66"									
	54"									
	48"									

Shaded cell indicates cited luminaire in cited layout has high probability of meeting **all** ambient lighting criteria outlined in Table 10.4. Likely ballast factor is also cited.❶

Predominant partition height (in inches)	80"									
	66"	0.98			0.98			0.98		
	54"		0.98			0.98				0.98
	48"									

Predominant partition height (in inches)	80"									
	66"									
	54"									
	48"									

❶Calculations are required in order to secure a final layout and luminaire specification for actual conditions.

Ambient Indirect/Oval/Extra Wide

(1) F54T5/HO lamp/Suspension of 2 feet/0 inches

more continued on next two pages▷

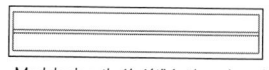

Modular length 4'–1½" (with endcaps)

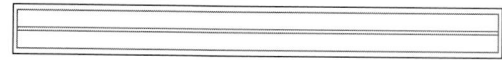

Modular length 8'–1½" (with endcaps)

Modular length 12'–3" (with endcaps)

Concept reference only: see specific manufacturer's literature for actual module sizes, seams, endcaps and dimensions

ConceptStarters™

Design Tips

✔These luminaires are available in modular lengths (e.g. 4 feet, 8 feet and 12 feet).

✔Units on closer spacings cause a stronger secondary ceiling plane effect.

✔Units on greater spacings reduce sustainability issues.

more tips on next page▷

Ambient
Pendent Fluorescent

10

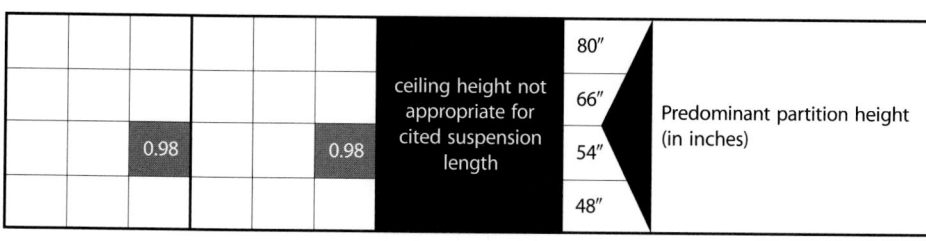

10'			9'–6"			9'			Ceiling height (in feet)
20	15	10	20	15	10	20	15	10	Percent floor reflectance
50	40	30	50	40	30	50	40	30	Percent wall reflectance
50	30	30	50	30	30	50	30	30	Percent partition reflectance
90	80	70	90	80	70	90	80	70	Percent ceiling reflectance

						80"	Predominant partition height (in inches)
						66"	
					ceiling height not appropriate for cited suspension length	54"	
						48"	

Open cell indicates cited luminaire in cited layout has low probability of meeting **all** ambient lighting criteria outlined in Table 10.4.❶

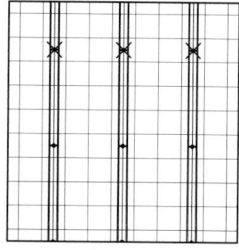

rows 6 feet on center
(O n e 4–foot section / 2 4 s f)

						80"	Predominant partition height (in inches)
						66"	
	0.98			0.98	ceiling height not appropriate for cited suspension length	54"	
						48"	

rows 8 feet on center
(O n e 4–foot section / 3 2 s f)

						80"	Predominant partition height (in inches)
	0.98					66"	
			0.98	0.98	ceiling height not appropriate for cited suspension length	54"	
			0.98	0.98		48"	

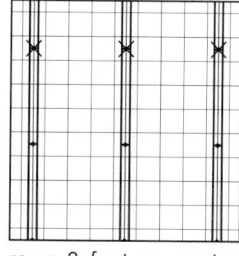

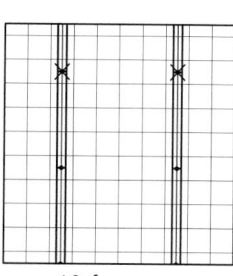

rows 10 feet on center
(O n e 4–foot section / 4 0 s f)

more continued on next two pages

ConceptStarters™

Ambient Indirect/Oval/Extra Wide
(1) F54T5/HO lamp/Suspension of `2 feet/6 inches`

Design Tips

✔ Check extruded aluminum units for modular alignment, side–to–side leveling, camber and surface integrity.

✔ Ceiling finish must be matte.

✔ Ceiling reflectance of 0.9 will provide most efficient lighting.

✔ Depending on layout, square–foot costs may be within a few percentage points of recessed—even though pendent luminaire unit costs may be higher.

✔ Note that endcaps add length.

◁ more tips on previous pages

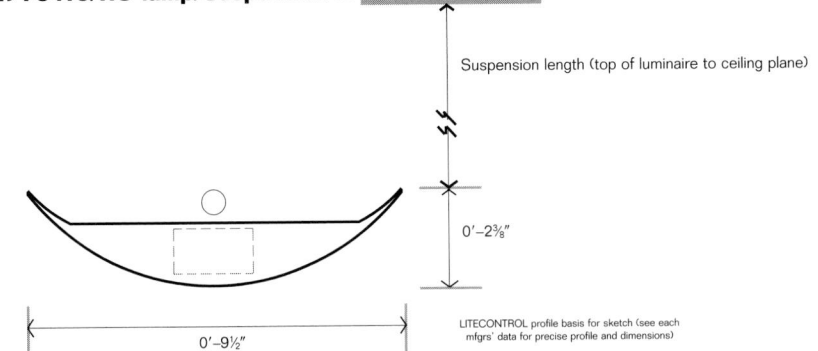

Suspension length (top of luminaire to ceiling plane)

0'–2⅜"

0'–9½"

LITECONTROL profile basis for sketch (see each mfgrs' data for precise profile and dimensions)

Ceiling height (in feet)		10'			9'–6"			9'		
Percent floor reflectance		20	15	10	20	15	10	20	15	10
Percent wall reflectance		50	40	30	50	40	30	50	40	30
Percent partition reflectance		50	30	30	50	30	30	50	30	30
Percent ceiling reflectance		90	80	70	90	80	70	90	80	70

scale ¹⁄₁₆" = 1'–0"

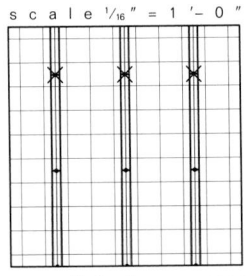

rows 6 feet on center
(One 4–foot section / 24 s f)

Predominant partition height (in inches)	80"				ceiling height not appropriate for cited suspension length		
	66"						
	54"						
	48"						

Shaded cell indicates cited luminaire in cited layout has high probability of meeting **all** ambient lighting criteria outlined in Table 10.4. Likely ballast factor is also cited. ❶

Consistent power feed locations look best

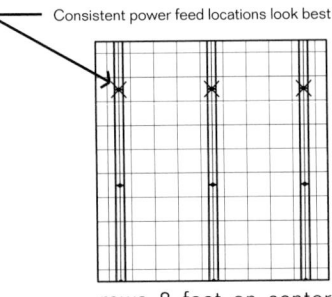

rows 8 feet on center
(One 4–foot section / 32 s f)

Predominant partition height (in inches)	80"				ceiling height not appropriate for cited suspension length		
	66"						
	54"		0.98	0.98			
	48"						

Incremental lengths may grow—support locations may not be at tile center

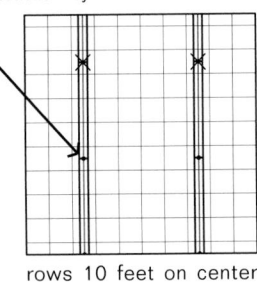

rows 10 feet on center
(One 4–foot section / 40 s f)

Predominant partition height (in inches)	80"				ceiling height not appropriate for cited suspension length		
	66"						
	54"		0.98				
	48"						

❶ Calculations are required in order to secure a final layout and luminaire specification for actual conditions.

Ambient Indirect/Oval/Extra Wide

(1) F54T5/HO lamp/Information

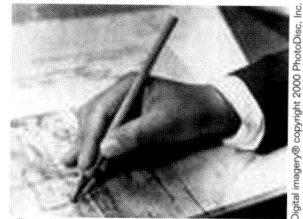

Specification

- 9½ inches by 2⅜ inches oval cross section
- Totally indirect
- VDT–sensitive design (exhibits least glare)
- Extruded aluminum housing in modular lengths
- 1–F54T5/830 high output lamp (5000 lumens)
- Electronic programmed start ballast (for best efficiency)
- See power budgeting sidebar to right
- Cost magnitude: high
- Possible vendors: **LITECONTROL** (Arcos™)

Boldfaced manufacturer's photometric data were averaged and input to Lumen Micro 7.5 in order to assess probability of meeting ambient lighting criteria cited in Table 10.4. Use the above specification list when contacting manufacturer about potential luminaire options for a specific project. Dimensions, specifications and performance are somewhat generic. Check with specific manufacturer's data for nominal dimensions and performance. Data is subject to change.

Power budgeting
for these 1–lamp cross–section luminaires

BF 0.98 yields 58 watts①

Power budget for ambient lighting should be less than 1.0 w/sf. Ballast vendors include Energy Savings, Inc. (ESI) and Osram Sylvania. Above factors based on ESI.

① Single– and 2–lamp ballasts are available. Above data based on 2–lamp ballast. Single–lamp ballast requires about 64 watts per lamp, depending on project voltage.

Net Addresses/Luminaires
http://www.litecontrol.com/products/index.html

Net Addresses/Ballasts
http://www.sylvania.com/ballast/phops.htm

CONNECT FOR MORE

Cost magnitude²⁰⁰⁰
for 4–foot fluorescent luminaire sections

Low
US$70. to US$90.
Moderate
US$90. to US$120.
High
US$120. and greater

Quality ambient lighting for today's electronic office environment typically runs US$3.00 to US$5.00/sf, hardware only (luminaires, lamps and ballasts), excluding installation costs. Costs vary based on quantities/lengths, distributor and contractor markups, luminaire variations such as finish, lamping, ballasting and options and accessories, and market conditions, manufacturing situations and annual inflation. These values are for preliminary, magnitude budgeting and do not represent quotes nor actual final pricing to client.

Arcos™ is a registered trademark of LITECONTROL

Ambient Pendent Fluorescent

10

Ambient Indirect/Oval/Extra Wide

(1) F32T8 lamp/Suspension of `1 foot/6 inches`

ConceptStarters™

Digital imagery® copyright 2000 PhotoDisc, Inc.

Design Tips
✔ Oval shape helps limit high contrast between luminaire bottom and ceiling—only if luminaire finish is white or very light. Narrow oval is better.
✔ Use localized task lighting.
✔ Use art accenting for visual interest.
more tips on next page▷

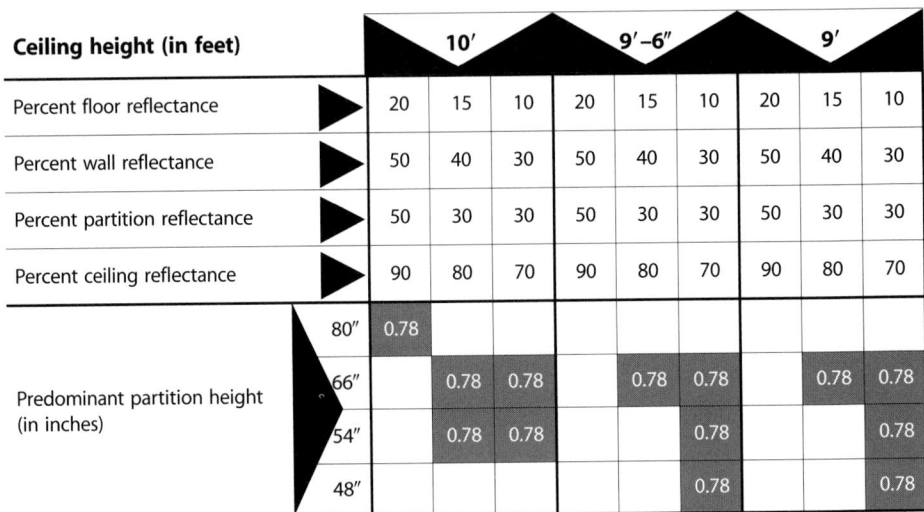

Suspension length (top of luminaire to ceiling plane)

0'–2⅜"

0'–9½"

LITECONTROL profile basis for sketch (see each mfgrs' data for precise profile and dimensions)

Ceiling height (in feet)		10'			9'–6"			9'		
Percent floor reflectance		20	15	10	20	15	10	20	15	10
Percent wall reflectance		50	40	30	50	40	30	50	40	30
Percent partition reflectance		50	30	30	50	30	30	50	30	30
Percent ceiling reflectance		90	80	70	90	80	70	90	80	70
Predominant partition height (in inches)	80"	0.78								
	66"		0.78	0.78		0.78	0.78		0.78	0.78
	54"		0.78	0.78			0.78			0.78
	48"						0.78			0.78

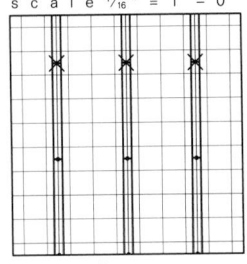

scale 1/16" = 1'–0"

rows 6 feet on center
(O n e 4–foot section / 2 4 s f)

Shaded cell indicates cited luminaire in cited layout has high probability of meeting **all** ambient lighting criteria outlined in Table 10.4. Likely ballast factor is also cited.❶

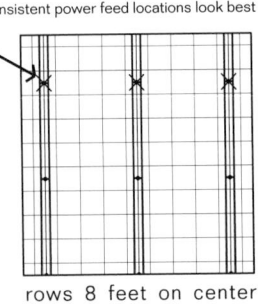

Consistent power feed locations look best

Predominant partition height (in inches)	80"									
	66"	0.78	0.88	1.18	0.78	0.88	0.98	0.78	0.88	0.98
	54"	0.78	0.78	0.98	0.78	0.78	0.88			0.88
	48"		0.78	0.88		0.78	0.88		0.78	0.88

rows 8 feet on center
(O n e 4–foot section / 3 2 s f)

Incremental lengths may grow—support locations may not be at tile center

Predominant partition height (in inches)	80"									
	66"	0.88	1.18		0.88	1.18	1.18	0.88	1.18	
	54"	0.78	0.98	1.18	0.78	0.98	1.18	0.78	0.98	1.18
	48"	0.78	0.98	1.18	0.78	0.98	0.98	0.78	0.98	0.98

rows 10 feet on center
(O n e 4–foot section / 4 0 s f)

❶ Calculations are required in order to secure a final layout and luminaire specification for actual conditions.

Ambient Indirect/Oval/Extra Wide

(1) F32T8 lamp/Suspension of `2 feet/0 inches`

more continued on next two pages▷

Modular length 4′–1½″ (with endcaps)

Modular length 8′–1½″ (with endcaps)

Modular length 12′–3″ (with endcaps)

Concept reference only: see specific manufacturer's literature for actual module sizes, seams, endcaps and dimensions

ConceptStarters™

Design Tips

✔ These luminaires are available in modular lengths (e.g, 4 feet, 8 feet and 12 feet, etc.) (building must accommodate move–in of single, continuous lengths, however, through doors, open windows, elevators, etc.).

✔ Units on closer spacings cause a stronger secondary ceiling plane effect.

✔ Units on greater spacings reduce sustainability issues.
more tips on next page▷

Ambient Pendent Fluorescent

10

10′			9′–6″			9′			Ceiling height (in feet)
20	15	10	20	15	10	20	15	10	Percent floor reflectance
50	40	30	50	40	30	50	40	30	Percent wall reflectance
50	30	30	50	30	30	50	30	30	Percent partition reflectance
90	80	70	90	80	70	90	80	70	Percent ceiling reflectance

rows 6 feet on center
(O n e 4–foot section / 2 4 s f)

10′			9′–6″			9′			Predominant partition height (in inches)
				0.78	0.98	ceiling height not appropriate for cited suspension length			80″
									66″
	0.78	0.88			0.88				54″
		0.78			0.78				48″

Open cell indicates cited luminaire in cited layout has low probability of meeting **all** ambient lighting criteria outlined in Table 10.4.❶

rows 8 feet on center
(O n e 4–foot section / 3 2 s f)

10′			9′–6″			9′			Predominant partition height (in inches)
			0.88			ceiling height not appropriate for cited suspension length			80″
	0.98	1.18	0.78	0.98	1.18				66″
	0.88	0.98		0.88	0.98				54″
	0.78	0.88		0.78	0.88				48″

rows 10 feet on center
(O n e 4–foot section / 4 0 s f)

10′			9′–6″			9′			Predominant partition height (in inches)
			0.98			ceiling height not appropriate for cited suspension length			80″
0.88	1.18		0.88	1.18	1.18				66″
0.78	0.98	1.18	0.78	0.98	1.18				54″
0.78	0.88	1.18	0.78	0.88	0.98				48″

more continued on next two pages

Ambient Indirect/Oval/Extra Wide

(1) F32T8 lamp/Suspension of `2 feet/6 inches`

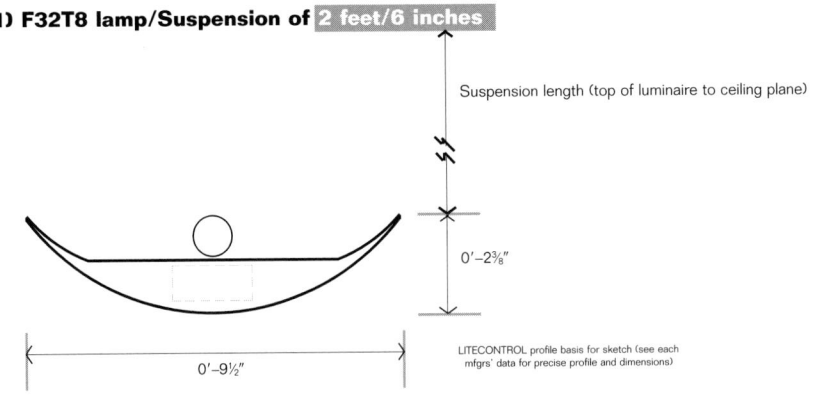

Suspension length (top of luminaire to ceiling plane)

0'–2⅜"

0'–9½"

LITECONTROL profile basis for sketch (see each mfgrs' data for precise profile and dimensions)

Digital imagery® copyright 2000 PhotoDisc, Inc.

ConceptStarters™

Design Tips

✔ Check extruded aluminum units for modular alignment, side–to–side leveling, camber and surface integrity.
✔ Ceiling finish must be matte.
✔ Ceiling reflectance of 0.9 will provide most efficient lighting.
✔ Depending on layout, square–foot costs may be within a few percentage points of recessed—even though pendent luminaire unit costs may be higher.
✔ Note that endcaps add length.
◁more tips on previous pages

Ambient Pendent Fluorescent

10

Ceiling height (in feet)	10'			9'–6"			9'		
Percent floor reflectance	20	15	10	20	15	10	20	15	10
Percent wall reflectance	50	40	30	50	40	30	50	40	30
Percent partition reflectance	50	30	30	50	30	30	50	30	30
Percent ceiling reflectance	90	80	70	90	80	70	90	80	70

scale 1/16" = 1'–0"

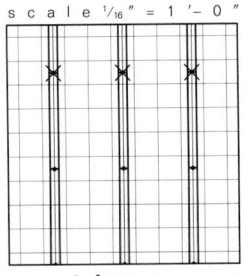

rows 6 feet on center
(O n e 4–foot section / 2 4 s f)

Predominant partition height (in inches)	10'			9'–6" / 9'
80"				ceiling height not appropriate for cited suspension length
66"				
54"		0.78	0.88	
48"			0.78	

Shaded cell indicates cited luminaire in cited layout has high probability of meeting **all** ambient lighting criteria outlined in Table 10.4. Likely ballast factor is also cited.❶

Consistent power feed locations look best

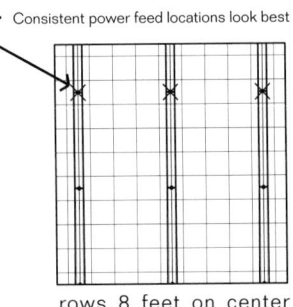

rows 8 feet on center
(O n e 4–foot section / 3 2 s f)

Predominant partition height (in inches)	10'			9'–6" / 9'
80"				ceiling height not appropriate for cited suspension length
66"	0.78	1.18	1.18	
54"	0.78	0.98	0.98	
48"		0.88	0.88	

Incremental lengths may grow—support locations may not be at tile center

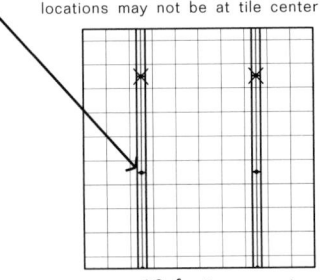

rows 10 feet on center
(O n e 4–foot section / 4 0 s f)

Predominant partition height (in inches)	10'			9'–6" / 9'
80"				ceiling height not appropriate for cited suspension length
66"		1.18		
54"		0.98	1.18	
48"	0.88	0.88	1.18	

❶Calculations are required in order to secure a final layout and luminaire specification for actual conditions.

Ambient Indirect/Oval/Extra Wide

(1) F32T8 lamp/Information

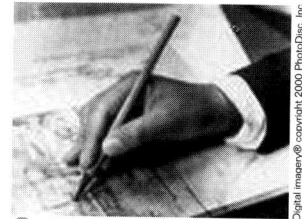

ConceptStarters™

Specification

⊃ 9½ inches by 2⅜ inches oval cross section
⊃ Totally indirect
⊃ VDT–sensitive design (exhibits least glare)
⊃ Extruded aluminum housing in modular lengths
⊃ 1–F32T8/830 rapid start lamp (2950 lumens)
⊃ Electronic instant start ballast (for best efficiency)
⊃ See power budgeting sidebar to right
⊃ Cost magnitude: high
⊃ Possible vendors: **LITECONTROL** (Arcos™)

Boldfaced manufacturers' photometric data were averaged and input to Lumen Micro 7.5 in order to assess probability of meeting ambient lighting criteria cited in Table 10.4. Use the above specification list when contacting manufacturer about potential luminaire options for a specific project. Dimensions, specifications and performance are somewhat generic. Check with specific manufacturer's data for nominal dimensions and performance. Data is subject to change.

Net Addresses/Luminaires

http://www.litecontrol.com/products/index.html

Net Addresses/Ballasts

http://www.magnetek.com/transLighting.html
http://www.mot.com/ies/MLI/
http://www.sylvania.com/ballast/phops.htm

CONNECT FOR MORE

Power budgeting

for these 1–lamp cross–section luminaires

BF 0.78 yields 26 watts①
BF 0.88 yields 30 watts①②
BF 0.98 yields 35 watts①②
BF 1.18 yields 38 watts①

Power budget for ambient lighting should be less than 1.0 w/sf. Ballast vendors include Advance, Energy Savings, Inc. (ESI), Magnetek, Motorola. Not all ballast factors available from all ballast vendors. Above factors based on Magnetek.
① Based on tandem lamp operation— using one ballast to operate two lamps. Wattage reported is only for one of the tandem–wired lamps.
② Single lamp ballasts are available for these ballast factors.

Cost magnitude[2000]

for 4–foot fluorescent luminaire sections

Low
US$70. to US$90.
Moderate
US$90. to US$120.
High
US$120. and greater

Quality ambient lighting for today's electronic office environment typically runs US$3.00 to US$5.00/sf, hardware only (luminaires, lamps and ballasts), excluding installation costs. Costs vary based on quantities/lengths, distributor and contractor markups, luminaire variations such as finish, lamping, ballasting and options and accessories, and market conditions, manufacturing situations and annual inflation. These values are for preliminary, magnitude budgeting and do not represent quotes nor actual final pricing to client.

Arcos™ is a registered trademark of LITECONTROL

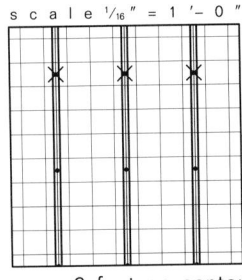

ConceptStarters™

Design Tips

✔ Shallow V shape helps limit high contrast between luminaire bottom and ceiling—only if luminaire finish is white or very light.

✔ Available in solid bottom and perforated bottom versions.

✔ Slots in bottom further reduces contrast between luminaire bottom and ceiling. Also minimizes haziness of entirely indirect lighting.

✔ Use localized task lighting.

✔ Use art accenting for visual interest.

more tips on next page▷

Ambient Indirect/V/Narrow+Shallow

(1) F28T5 lamp/Suspension of `1 foot/6 inches`

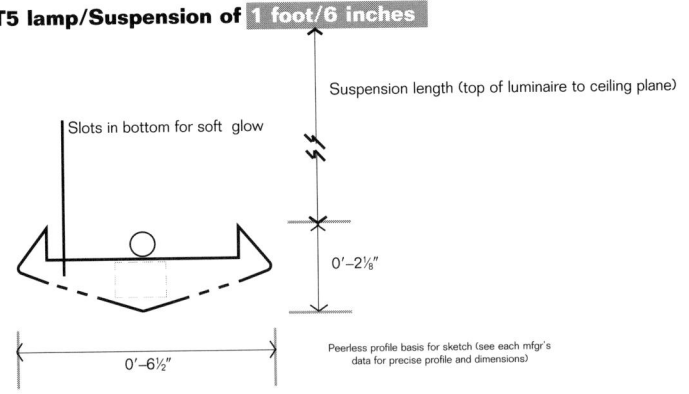

Suspension length (top of luminaire to ceiling plane)

Slots in bottom for soft glow

0'–2⅛"

0'–6½"

Peerless profile basis for sketch (see each mfgr's data for precise profile and dimensions)

scale ¹⁄₁₆" = 1'–0"

rows 6 feet on center
(O n e 4–foot section / 2 4 s f)

Ceiling height (in feet)		10'			9'–6"			9'		
Percent floor reflectance		20	15	10	20	15	10	20	15	10
Percent wall reflectance		50	40	30	50	40	30	50	40	30
Percent partition reflectance		50	30	30	50	30	30	50	30	30
Percent ceiling reflectance		90	80	70	90	80	70	90	80	70
Predominant partition height (in inches)	80"									
	66"									
	54"									
	48"									

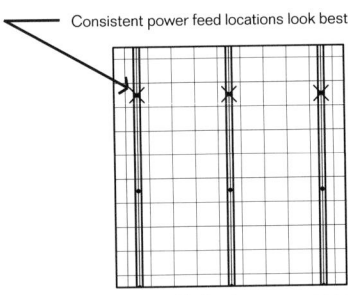

Consistent power feed locations look best

rows 8 feet on center
(O n e 4–foot section / 3 2 s f)

Shaded cell indicates cited luminaire in cited layout has high probability of meeting **all** ambient lighting criteria outlined in Table 10.4. Likely ballast factor is also cited. ❶

Predominant partition height (in inches)	80"									
	66"		0.98			0.98	0.98		0.98	0.98
	54"			0.98						
	48"									

Incremental lengths may grow—support locations may not be at tile center

rows 10 feet on center
(O n e 4–foot section / 4 0 s f)

Predominant partition height (in inches)	80"									
	66"	0.98								
	54"		0.98				0.98			0.98
	48"		0.98							

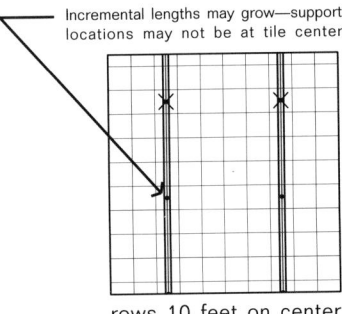

❶Calculations are required in order to secure a final layout and luminaire specification for actual conditions.

Ambient Indirect/V/Narrow+Shallow

(1) F28T5 lamp/Suspension of `2 feet/0 inches`

more continued on next two pages▷

Modular length 4'–11" (with endcaps)

Modular length 8'–11" (with endcaps)

Modular length 12'–11" (with endcaps)

Concept reference only: see specific manufacturer's literature for actual module sizes, seams, endcaps and dimensions

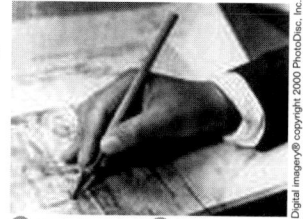

ConceptStarters™

Ambient Pendent Fluorescent

10

Design Tips

✔These luminaires are available in both modular lengths (e.g, 4 feet, 8 feet) and in continuous linear lengths (e.g., 15 feet) up to 24 feet with no seams (building must accommodate move–in of single, continuous lengths, however, through doors, open windows, elevators, etc.).

✔Units on closer spacings cause a stronger secondary ceiling plane effect.

✔Units on greater spacings reduce sustainability issues.

more tips on next page▷

10'			9'–6"			9'			Ceiling height (in feet)
20	15	10	20	15	10	20	15	10	Percent floor reflectance
50	40	30	50	40	30	50	40	30	Percent wall reflectance
50	30	30	50	30	30	50	30	30	Percent partition reflectance
90	80	70	90	80	70	90	80	70	Percent ceiling reflectance

						80"	
			ceiling height not appropriate for cited suspension length			66"	Predominant partition height (in inches)
						54"	
						48"	

Open cell indicates cited luminaire in cited layout has low probability of meeting **all** ambient lighting criteria outlined in Table 10.4.❶

						80"	
	0.98			0.98		66"	Predominant partition height (in inches)
			ceiling height not appropriate for cited suspension length			54"	
		0.98			0.98	48"	

						80"	
						80"	
0.98			0.98			66"	Predominant partition height (in inches)
	0.98			0.98	ceiling height not appropriate for cited suspension length	54"	
						48"	

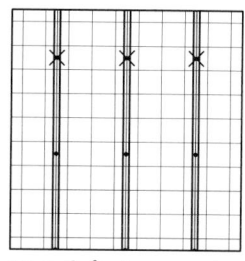

rows 6 feet on center
(O n e 4–foot section / 2 4 s f)

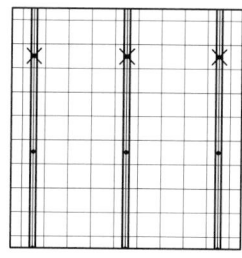

rows 8 feet on center
(O n e 4–foot section / 3 2 s f)

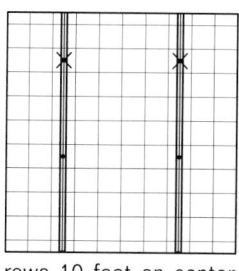

rows 10 feet on center
(O n e 4–foot section / 4 0 s f)

more continued on next two pages

ConceptStarters™

Design Tips

✔ Check extruded aluminum units for modular alignment, side–to–side leveling, camber and surface integrity.

✔ Ceiling finish must be matte.

✔ Ceiling reflectance of 0.9 will provide most efficient lighting.

✔ Depending on layout, square–foot costs may be within a few percentage points of recessed—even though pendent luminaire unit costs may be higher.

✔ Note that endcaps add length.

◁more tips on previous pages

Ambient Indirect/V/Narrow+Shallow

(1) F28T5 lamp/Suspension of `2-foot/6 inches`

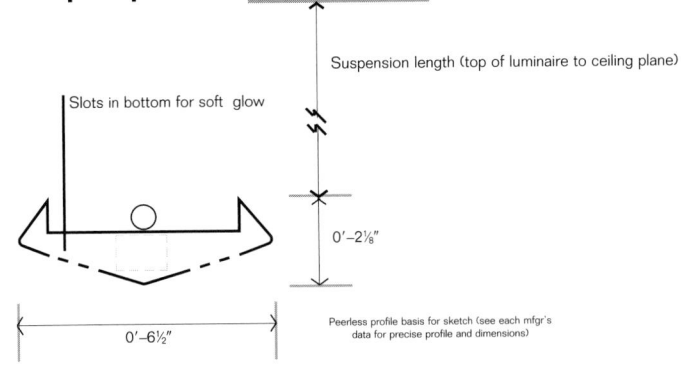

Suspension length (top of luminaire to ceiling plane)

Slots in bottom for soft glow

0'–2⅛"

0'–6½"

Peerless profile basis for sketch (see each mfgr's data for precise profile and dimensions)

Ceiling height (in feet)		10'			9'–6"			9'		
Percent floor reflectance		20	15	10	20	15	10	20	15	10
Percent wall reflectance		50	40	30	50	40	30	50	40	30
Percent partition reflectance		50	30	30	50	30	30	50	30	30
Percent ceiling reflectance		90	80	70	90	80	70	90	80	70

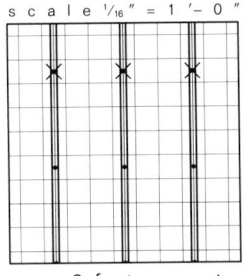

scale ¹⁄₁₆" = 1'–0"

rows 6 feet on center
(O n e 4–foot section / 2 4 s f)

Predominant partition height (in inches)	80"				ceiling height not appropriate for cited suspension length		
	66"						
	54"						
	48"						

Shaded cell indicates cited luminaire in cited layout has high probability of meeting **all** ambient lighting criteria outlined in Table 10.4. Likely ballast factor is also cited.❶

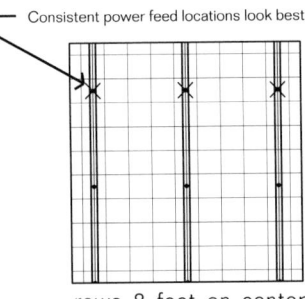

Consistent power feed locations look best

rows 8 feet on center
(O n e 4–foot section / 3 2 s f)

Predominant partition height (in inches)	80"				ceiling height not appropriate for cited suspension length		
	66"		0.98				
	54"			0.98			
	48"						

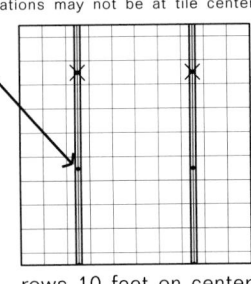

Incremental lengths may grow—support locations may not be at tile center

rows 10 feet on center
(O n e 4–foot section / 4 0 s f)

Predominant partition height (in inches)	80"				ceiling height not appropriate for cited suspension length		
	66"	0.98					
	54"		0.98				
	48"						

❶Calculations are required in order to secure a final layout and luminaire specification for actual conditions.

Ambient Indirect/V/Narrow+Shallow

(1) F28T5 lamp/Information

Specification

⊃6½ inches by 2⅛ inches shallow V cross section

⊃Totally indirect (less than 1 percent of the light comes from the perforations)

⊃VDT–sensitive design (exhibits least glare)

⊃Extruded aluminum housing in modular or continuous linear lengths

⊃1–F28T5SE/830 superior efficacy lamp (2900 lumens)

⊃Electronic programmed start ballast

⊃See power budgeting sidebar to right

⊃Cost magnitude: high

⊃Possible vendors: Lam (Windows), **Peerless** (Lightfin™ Diminutive)

Boldfaced manufacturer's photometric data were averaged and input to Lumen Micro 7.5 in order to assess probability of meeting ambient lighting criteria cited in Table 10.4. Use the above specification list when contacting manufacturer about potential luminaire options for a specific project. Dimensions, specifications and performance are somewhat generic. Check with specific manufacturer's data for nominal dimensions and performance. Data is subject to change.

ConceptStarters™

Power budgeting
for these 1–lamp cross–section luminaires

BF 0.98 yields 32 watts①

Power budget for ambient lighting should be less than 1.0 w/sf. Ballast vendors include Energy Savings, Inc. (ESI), Motorola and Osram Sylvania. Above factors based on ESI.
①Single– and 2–lamp ballasts are available and both have nearly identical power requirements (32 watts per lamp).

Ambient *Pendent Fluorescent*

10

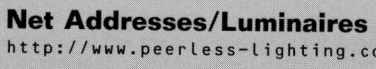

Net Addresses/Luminaires
http://www.peerless-lighting.com/

Net Addresses/Ballasts
http://www.mot.com/ies/MLI/
http://www.sylvania.com/ballast/phops.htm

CONNECT FOR MORE

Cost magnitude²⁰⁰⁰
for 4–foot fluorescent luminaire sections

Low
US$70. to US$90.
Moderate
US$90. to US$120.
High
US$120. and greater

Quality ambient lighting for today's electronic office environment typically runs US$3.00 to US$5.00/sf, hardware only (luminaires, lamps and ballasts), excluding installation costs. Costs vary based on quantities/lengths, distributor and contractor markups, luminaire variations such as finish, lamping, ballasting and options and accessories, and market conditions, manufacturing situations and annual inflation. These values are for preliminary, magnitude budgeting and do not represent quotes nor actual final pricing to client.

Lightfin™ is a registered trademark of Lithonia/Peerless Lighting

Ambient Pendent Fluorescent

10

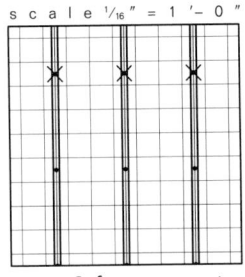

ConceptStarters™

Ambient Indirect/V/Narrow

(1) F28T5 lamp/Suspension of 1 foot/6 inches

Slots along side for soft glow

Suspension length (top of luminaire to ceiling plane)

0'–2⅞"

0'–7"

LITECONTROL profile basis for sketch (see each mfgr's data for precise profile and dimensions)

Design Tips

✔ Shallow V shape helps limit high contrast between luminaire bottom and ceiling—only if luminaire finish is white or very light.

✔ Available in solid side and slotted/perforated side versions.

✔ Slots in bottom further reduces contrast between luminaire bottom and ceiling. Also minimizes haziness of entirely indirect lighting.

✔ Use localized task lighting.

✔ Use art accenting for visual interest.

more tips on next page▷

Ceiling height (in feet)		10'			9'–6"			9'		
Percent floor reflectance		20	15	10	20	15	10	20	15	10
Percent wall reflectance		50	40	30	50	40	30	50	40	30
Percent partition reflectance		50	30	30	50	30	30	50	30	30
Percent ceiling reflectance		90	80	70	90	80	70	90	80	70
Predominant partition height (in inches)	80"									
	66"									
	54"									
	48"									

scale ¹⁄₁₆" = 1'-0"

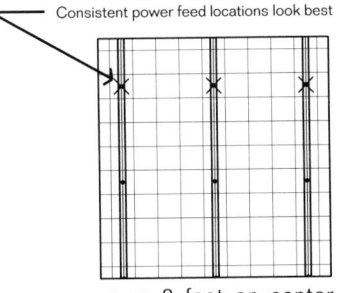

rows 6 feet on center
(O n e 4–foot section / 2 4 s f)

Shaded cell indicates cited luminaire in cited layout has high probability of meeting **all** ambient lighting criteria outlined in Table 10.4. Likely ballast factor is also cited.❶

Consistent power feed locations look best

Predominant partition height (in inches)	80"									
	66"									
	54"		0.98			0.98			0.98	
	48"			0.98			0.98			0.98

rows 8 feet on center
(O n e 4–foot section / 3 2 s f)

Incremental lengths may grow—support locations may not be at tile center

Predominant partition height (in inches)	80"									
	66"									
	54"	0.98			0.98			0.98		
	48"		0.98			0.98			0.98	

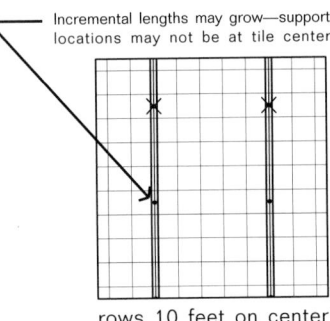

rows 10 feet on center
(O n e 4–foot section / 4 0 s f)

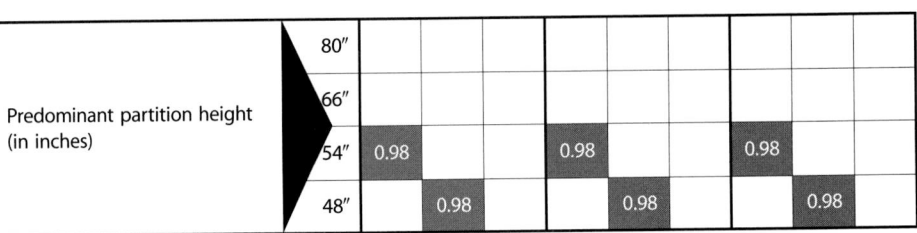

❶ Calculations are required in order to secure a final layout and luminaire specification for actual conditions.

Ambient Indirect/V/Narrow

(1) F28T5 lamp/Suspension of `2 feet/0 inches`

more continued on next two pages▷

Modular length 4'–2" (with endcaps)

Modular length 8'–3" (with endcaps)

Modular length 12'–3" (with endcaps)

Concept reference only: see specific manufacturer's literature for actual module sizes, seams, endcaps and dimensions

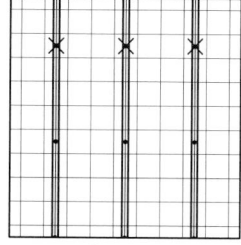

ConceptStarters™

Digital imagery® copyright 2000 PhotoDisc, Inc.

Design Tips

✓ These luminaires are available in both modular lengths (e.g, 4 feet, 8 feet) (building must accommodate move–in of single, continuous lengths, however, through doors, open windows, elevators, etc.).

✓ Units on closer spacings cause a stronger secondary ceiling plane effect.

✓ Units on greater spacings reduce sustainability issues.

more tips on next page▷

Ambient Pendent Fluorescent

10

10'			9'–6"			9'			Ceiling height (in feet)
20	15	10	20	15	10	20	15	10	Percent floor reflectance
50	40	30	50	40	30	50	40	30	Percent wall reflectance
50	30	30	50	30	30	50	30	30	Percent partition reflectance
90	80	70	90	80	70	90	80	70	Percent ceiling reflectance

									Predominant partition height (in inches)
						ceiling height not appropriate for cited suspension length		80"	
								66"	
								54"	
								48"	

Open cell indicates cited luminaire in cited layout has low probability of meeting **all** ambient lighting criteria outlined in Table 10.4.❶

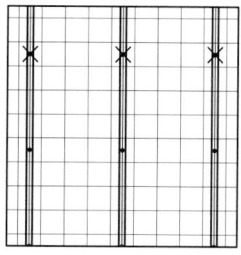

rows 6 feet on center
(O n e 4–foot section / 2 4 s f)

									Predominant partition height (in inches)
	0.98			0.98		ceiling height not appropriate for cited suspension length		80"	
		0.98			0.98			66"	
								54"	
								48"	

rows 8 feet on center
(O n e 4–foot section / 3 2 s f)

									Predominant partition height (in inches)
								80"	
						ceiling height not appropriate for cited suspension length		66"	
0.98			0.98					54"	
	0.98			0.98				48"	

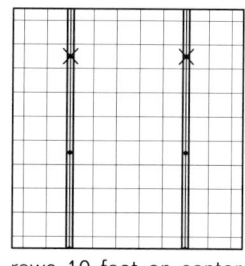

rows 10 feet on center
(O n e 4–foot section / 4 0 s f)

more continued on next two pages

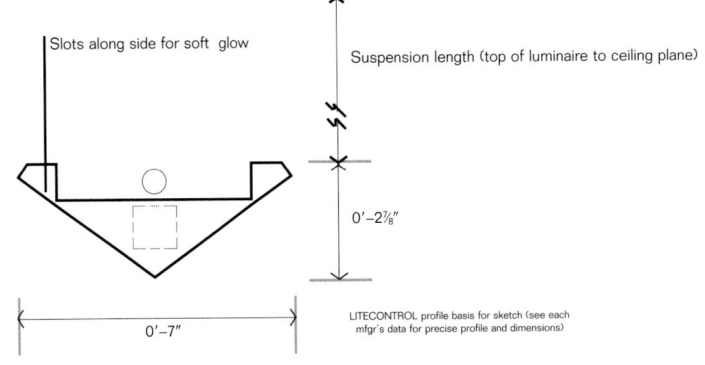

ConceptStarters™

Ambient Indirect/V/Narrow

(1) F28T5 lamp/Suspension of `2-foot/6 inches`

Slots along side for soft glow

Suspension length (top of luminaire to ceiling plane)

0'–2⅞"

0'–7"

LITECONTROL profile basis for sketch (see each mfgr's data for precise profile and dimensions)

Design Tips

✓ Check extruded aluminum units for modular alignment, side–to–side leveling, camber and surface integrity.
✓ Ceiling finish must be matte.
✓ Ceiling reflectance of 0.9 will provide most efficient lighting.
✓ Depending on layout, square-foot costs may be within a few percentage points of recessed—even though pendent luminaire unit costs may be higher.
✓ Note that endcaps add length.
◁ more tips on previous pages

Ceiling height (in feet)	10'			9'–6"			9'		
Percent floor reflectance	20	15	10	20	15	10	20	15	10
Percent wall reflectance	50	40	30	50	40	30	50	40	30
Percent partition reflectance	50	30	30	50	30	30	50	30	30
Percent ceiling reflectance	90	80	70	90	80	70	90	80	70

scale 1/16" = 1'–0"

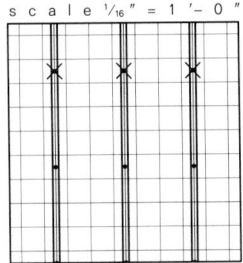

rows 6 feet on center
(One 4–foot section / 24 s f)

Predominant partition height (in inches)					
80"			ceiling height not appropriate for cited suspension length		
66"					
54"					
48"					

Shaded cell indicates cited luminaire in cited layout has high probability of meeting **all** ambient lighting criteria outlined in Table 10.4. Likely ballast factor is also cited. ❶

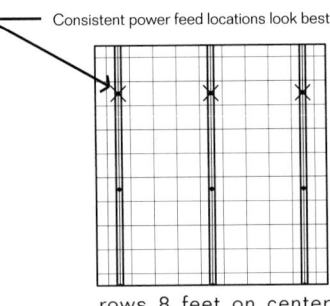

Consistent power feed locations look best

rows 8 feet on center
(One 4–foot section / 32 s f)

Predominant partition height (in inches)					
80"				ceiling height not appropriate for cited suspension length	
66"	0.98				
54"		0.98			
48"		0.98			

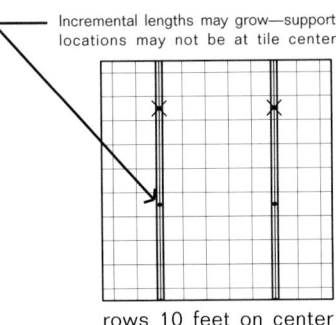

Incremental lengths may grow—support locations may not be at tile center

rows 10 feet on center
(One 4–foot section / 40 s f)

Predominant partition height (in inches)					
80"				ceiling height not appropriate for cited suspension length	
66"					
54"	0.98				
48"		0.98			

❶ Calculations are required in order to secure a final layout and luminaire specification for actual conditions.

Ambient Indirect/V/Narrow

(1) F28T5 lamp/Information

Specification

⊃7 inches by 2⅞ inches shallow V cross section

⊃Totally indirect (less than 1 percent of the light comes from the perforations)

⊃VDT–sensitive design (exhibits least glare)

⊃Extruded aluminum housing in modular or continuous linear lengths

⊃1–F28T5SE/830 superior efficacy lamp (2900 lumens)

⊃Electronic programmed start ballast

⊃See power budgeting sidebar to right

⊃Cost magnitude: high

⊃Possible vendors: **LITECONTROL** (Aviva)

Boldfaced manufacturer's photometric data were averaged and input to Lumen Micro 7.5 in order to assess probability of meeting ambient lighting criteria cited in Table 10.4. Use the above specification list when contacting manufacturer about potential luminaire options for a specific project. Dimensions, specifications and performance are somewhat generic. Check with specific manufacturer's data for nominal dimensions and performance. Data is subject to change.

ConceptStarters™

Power budgeting
for these 1–lamp cross–section luminaires

BF 0.98 yields 32 watts①

Power budget for ambient lighting should be less than 1.0 w/sf. Ballast vendors include Energy Savings, Inc. (ESI), Motorola and Osram Sylvania. Above factors based on ESI.
①Single– and 2–lamp ballasts are available and both have nearly identical power requirements (32 watts per lamp).

Ambient Pendent Fluorescent

10

Net Addresses/Luminaires
http://www.litecontrol.com/products/index.html

Net Addresses/Ballasts
http://www.mot.com/ies/MLI/
http://www.sylvania.com/ballast/phops.htm

CONNECT FOR MORE

Cost magnitude²⁰⁰⁰
for 4–foot fluorescent luminaire sections

Low
US$70. to US$90.
Moderate
US$90. to US$120.
High
US$120. and greater

Quality ambient lighting for today's electronic office environment typically runs US$3.00 to US$5.00/sf, hardware only (luminaires, lamps and ballasts), excluding installation costs. Costs vary based on quantities/lengths, distributor and contractor markups, luminaire variations such as finish, lamping, ballasting and options and accessories, and market conditions, manufacturing situations and annual inflation. These values are for preliminary, magnitude budgeting and do not represent quotes nor actual final pricing to client.

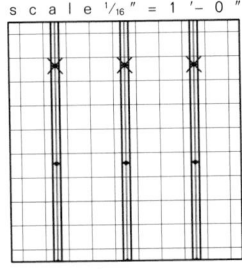

ConceptStarters™

AmbientIndirect/V/Wide+Deep

(1) F32T8 lamp/Suspension of `1 foot/6 inches`

Suspension length (top of luminaire to ceiling plane)

0'–4"

0'–8½"

Peerless profile basis for sketch (see each mfg's data for precise profile and dimensions)

Digital imagery® copyright 2000 PhotoDisc, Inc.

Design Tips

✔ V shape not as bulky as rectilinear cross section, but is more obtrusive than narrow oval cross section. Narrower and shallower V is better.

✔ V shape exhibits less contrast between luminaire and ceiling compared to rounded flat bottom versions and oval versions.

✔ Use localized task lighting.

✔ Use art accenting for visual interest.

more tips on next page▷

10.

Ceiling height (in feet)	10'			9'–6"			9'		
Percent floor reflectance	20	15	10	20	15	10	20	15	10
Percent wall reflectance	50	40	30	50	40	30	50	40	30
Percent partition reflectance	50	30	30	50	30	30	50	30	30
Percent ceiling reflectance	90	80	70	90	80	70	90	80	70

scale ¹⁄₁₆" = 1'–0"

rows 6 feet on center
(O n e 4–foot section / 2 4 s f)

Predominant partition height (in inches)

height	20	15	10	20	15	10	20	15	10
80"							0.78		
66"		0.78	0.88		0.78	0.88		0.78	0.78
54"		0.78	0.78			0.78			0.78
48"			0.78			0.78			0.78

Shaded cell indicates cited luminaire in cited layout has high probability of meeting **all** ambient lighting criteria outlined in Table 10.4. Likely ballast factor is also cited.❶

Consistent power feed locations look best

rows 8 feet on center
(O n e 4–foot section / 3 2 s f)

Predominant partition height (in inches)

height	20	15	10	20	15	10	20	15	10
80"									
66"	0.78	0.88	1.18	0.78	0.88	0.98	0.78	0.88	0.98
54"	0.78	0.78	0.98		0.88	0.88		0.78	0.88
48"		0.78	0.88	0.78	0.78	0.88		0.78	0.88

Incremental lengths may grow—support locations may not be at tile center

rows 10 feet on center
(O n e 4–foot section / 4 0 s f)

Predominant partition height (in inches)

height	20	15	10	20	15	10	20	15	10
80"									
66"	0.88			0.88	1.18	1.18	0.88	1.18	
54"	0.78	0.98	1.18	0.78	0.98	1.18	0.78	0.98	1.18
48"	0.78	0.98	1.18	0.78	0.98	0.98	0.78	0.98	0.98

❶Calculations are required in order to secure a final layout and luminaire specification for actual conditions.

Ambient Indirect/V/Wide+Deep

(1) F32T8 lamp/Suspension of `2 feet/0 inches`

more continued on next two pages▷

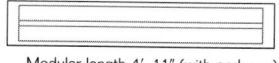

Modular length 4'–11" (with endcaps)

Modular length 8'–11" (with endcaps)

Modular length 12'–11" (with endcaps)

Concept reference only: see specific manufacturer's literature for actual module sizes, seams, endcaps and dimensions

ConceptStarters™

Design Tips

✔ These luminaires are available in both modular lengths (e.g, 4 feet, 8 feet) and in continuous linear lengths (e.g., 15 feet) up to 24 feet with no seams depending on manufacturer (building must accommodate move–in of single, continuous lengths, however, through doors, open windows, elevators, etc.).

✔ Units on closer spacings cause a stronger secondary ceiling plane effect.

✔ Units on greater spacings reduce sustainability issues.

more tips on next page▷

Reflectance / Ceiling height

	10'			9'-6"			9'			Ceiling height (in feet)
Percent floor reflectance	20	15	10	20	15	10	20	15	10	
Percent wall reflectance	50	40	30	50	40	30	50	40	30	
Percent partition reflectance	50	30	30	50	30	30	50	30	30	
Percent ceiling reflectance	90	80	70	90	80	70	90	80	70	

First layout (rows 6 feet on center): Predominant partition height (in inches) — 9' section: ceiling height not appropriate for cited suspension length

Partition height	10' (20)	10' (15)	10' (10)	9'-6" (20)	9'-6" (15)	9'-6" (10)
80"				0.78		
66"					0.78	0.88
54"		0.78	0.78		0.78	0.78
48"			0.78			0.78

Open cell indicates cited luminaire in cited layout has low probability of meeting **all** ambient lighting criteria outlined in Table 10.4. ❶

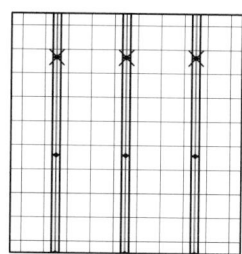

rows 6 feet on center
(O n e 4–foot section / 2 4 s f)

Second layout (rows 8 feet on center): Predominant partition height (in inches) — 9' section: ceiling height not appropriate for cited suspension length

Partition height	10' (20)	10' (15)	10' (10)	9'-6" (20)	9'-6" (15)	9'-6" (10)
80"						
66"	0.78	0.98	1.18	0.78	0.98	1.18
54"	0.78	0.88	0.98	0.78	0.88	0.98
48"		0.78	0.88		0.78	0.88

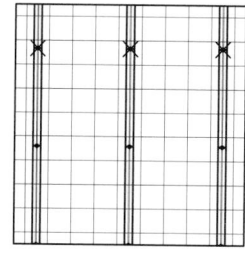

rows 8 feet on center
(O n e 4–foot section / 3 2 s f)

Third layout (rows 10 feet on center): Predominant partition height (in inches) — 9' section: ceiling height not appropriate for cited suspension length

Partition height	10' (20)	10' (15)	10' (10)	9'-6" (20)	9'-6" (15)	9'-6" (10)
80"						
66"	0.88	1.18	1.18	0.88	1.18	1.18
54"	0.88	0.98	1.18	0.78	0.98	1.18
48"	0.78	0.88	1.18	0.78	0.88	1.18

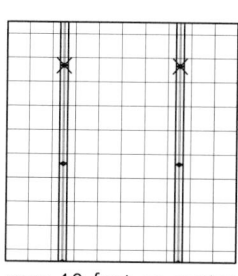

rows 10 feet on center
(O n e 4–foot section / 4 0 s f)

more continued on next two pages

Ambient Pendent Fluorescent

10

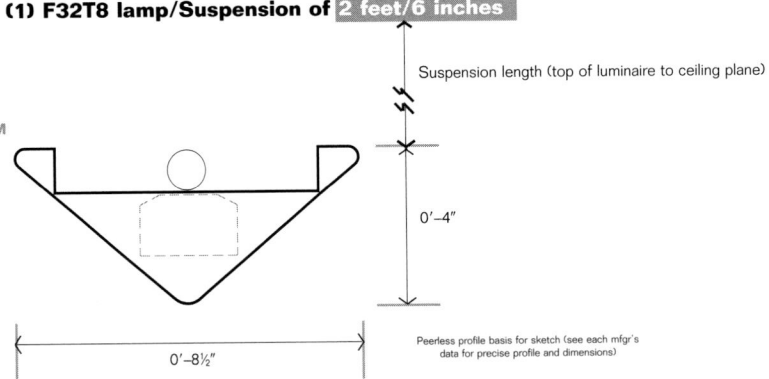

Ambient Indirect/V/Wide+Deep

(1) F32T8 lamp/Suspension of 2 feet/6 inches

ConceptStarters™

Design Tips

✔ Check extruded aluminum units for modular alignment, side–to–side leveling, camber and surface integrity.

✔ Ceiling finish must be matte.

✔ Ceiling reflectance of 0.9 will provide most efficient lighting.

✔ Depending on layout, square–foot costs may be within a few percentage points of recessed—even though pendent luminaire unit costs may be higher.

✔ Note that endcaps add length.

◁ more tips on previous pages

Suspension length (top of luminaire to ceiling plane)

0'–4"

0'–8½"

Peerless profile basis for sketch (see each mfgr's data for precise profile and dimensions)

Ceiling height (in feet)	10'			9'–6"			9'		
Percent floor reflectance	20	15	10	20	15	10	20	15	10
Percent wall reflectance	50	40	30	50	40	30	50	40	30
Percent partition reflectance	50	30	30	50	30	30	50	30	30
Percent ceiling reflectance	90	80	70	90	80	70	90	80	70

s c a l e ¹⁄₁₆ " = 1 ' – 0 "

Predominant partition height (in inches)	10'			9'–6" etc.					
80"									
66"									
54"		0.78	0.88						
48"			0.78						

ceiling height not appropriate for cited suspension length

rows 6 feet on center
(O n e 4–foot section / 2 4 s f)

Shaded cell indicates cited luminaire in cited layout has high probability of meeting **all** ambient lighting criteria outlined in Table 10.4. Likely ballast factor is also cited. ❶

Consistent power feed locations look best

Predominant partition height (in inches)									
80"									
66"	0.78	1.18	1.18						
54"	0.78	0.98	0.98						
48"		0.88	0.88						

ceiling height not appropriate for cited suspension length

rows 8 feet on center
(O n e 4–foot section / 3 2 s f)

Incremental lengths may grow—support locations may not be at tile center

Predominant partition height (in inches)									
80"	1.18								
66"	0.98	1.18							
54"	0.88	0.98							
48"	0.88	0.88							

ceiling height not appropriate for cited suspension length

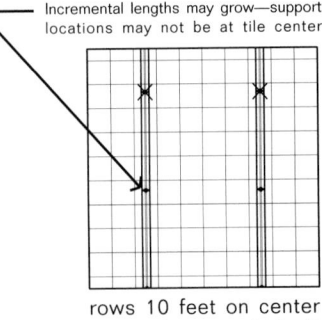

rows 10 feet on center
(O n e 4–foot section / 4 0 s f)

❶ Calculations are required in order to secure a final layout and luminaire specification for actual conditions.

Ambient Indirect/V/Wide+Deep
(1) F32T8 lamp/Information

ConceptStarters™

Specification
⊃8½ inches by 4 inches soft–edge V cross section
⊃Totally indirect
⊃VDT–sensitive design (exhibits least glare)
⊃Extruded aluminum housing in modular or continuous linear lengths
⊃1–F32T8/830 rapid start lamp (2950 lumens)
⊃Electronic instant start ballast (for best efficiency)
⊃See power budgeting sidebar to right
⊃Cost magnitude: high
⊃Possible vendors: **Focal Point** (Pinnacle), **Ledalite** (Vector II), **Peerless** (Envision™), Linear (TriLite 491), **Precision Architectural Lighting** (AV)

Boldfaced manufacturers' photometric data were averaged and input to Lumen Micro 7.5 in order to assess probability of meeting ambient lighting criteria cited in Table 10.4. Use the above specification list when contacting manufacturers about potential luminaire options for a specific project. Dimensions, specifications and performance are somewhat generic. Check with specific manufacturer's data for nominal dimensions and performance. Data is subject to change.

Net Addresses/Luminaires
http://www.focalpointlights.com/
http://www.ledalite.com/
http://www.peerless-lighting.com/

Net Addresses/Ballasts
http://www.magnetek.com/transLighting.html
http://www.mot.com/ies/MLI/
http://www.sylvania.com/ballast/phops.htm

CONNECT FOR MORE

Power budgeting
for these 1–lamp cross–section luminaires

BF 0.78 yields 26 watts①
BF 0.88 yields 30 watts①②
BF 0.98 yields 35 watts①②
BF 1.18 yields 38 watts①

Power budget for ambient lighting should be less than 1.0 w/sf. Ballast vendors include Advance, Energy Savings, Inc. (ESI), Magnetek, Motorola. Not all ballast factors available from all ballast vendors. Above factors based on Magnetek.
①Based on tandem lamp operation—using one ballast to operate two lamps. Wattage reported is only for one of the tandem–wired lamps.
②Single lamp ballasts are available for these ballast factors.

Ambient Pendent Fluorescent

10

Cost magnitude²⁰⁰⁰
for 4–foot fluorescent luminaire sections

Low
US$70. to US$90.
Moderate
US$90. to US$120.
High
US$120. and greater

Quality ambient lighting for today's electronic office environment typically runs US$3.00 to US$5.00/sf, hardware only (luminaires, lamps and ballasts), excluding installation costs. Costs vary based on quantities/lengths, distributor and contractor markups, luminaire variations such as finish, lamping, ballasting and options and accessories, and market conditions, manufacturing situations and annual inflation. These values are for preliminary, magnitude budgeting and do not represent quotes nor actual final pricing to client.

Envision™ is a registered trademark of Lithonia/Peerless Lighting

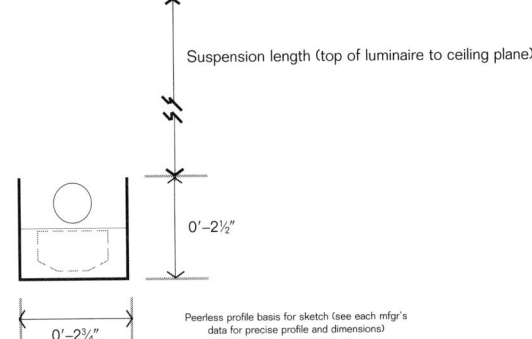

ConceptStarters™

Digital imagery® copyright 2000 PhotoDisc, Inc.

Ambient Indirect/Square/Miniature

(1) F32T8 lamp/Suspension of `1 foot/6 inches`

Suspension length (top of luminaire to ceiling plane)

0'–2½"

0'–2¾"

Peerless profile basis for sketch (see each mfgr's data for precise profile and dimensions)

Design Tips

✔ Miniature square shape creates fine line look.

✔ Miniature size exhibits less contrast between luminaire and ceiling compared to rounded flat bottom versions and larger square versions.

✔ Use localized task lighting.

✔ Use art accenting for visual interest.

more tips on next page▷

Ceiling height (in feet)		10'			9'–6"			9'		
Percent floor reflectance		20	15	10	20	15	10	20	15	10
Percent wall reflectance		50	40	30	50	40	30	50	40	30
Percent partition reflectance		50	30	30	50	30	30	50	30	30
Percent ceiling reflectance		90	80	70	90	80	70	90	80	70
Predominant partition height (in inches)	80"									
	66"	0.78			0.78	0.88	0.98	0.78	0.88	0.98
	54"	0.78	0.88	0.88		0.88	0.88		0.88	0.88
	48"		0.78	0.88		0.78	0.78		0.78	0.88

Shaded cell indicates cited luminaire in cited layout has high probability of meeting **all** ambient lighting criteria outlined in Table 10.4. Likely ballast factor is also cited.❶

scale ¹⁄₁₆" = 1'–0"

rows 6 feet on center
(O n e 4–foot section / 2 4 s f)

Consistent power feed locations look best

Predominant partition height (in inches)	80"									
	66"									
	54"									
	48"									

rows 8 feet on center
(O n e 4–foot section / 3 2 s f)

Incremental lengths may grow—support locations may not be at tile center

Predominant partition height (in inches)	80"									
	66"									
	54"									
	48"									

rows 10 feet on center
(O n e 4–foot section / 4 0 s f)

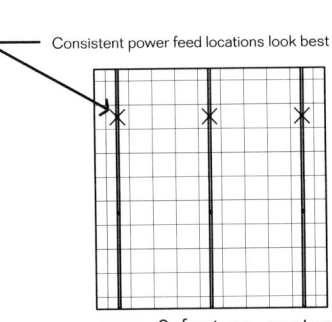

❶Calculations are required in order to secure a final layout and luminaire specification for actual conditions.

Ambient Indirect/Square/Miniature

(1) F32T8 lamp/Suspension of 2 feet/0 inches

more continued on next two pages▷

Modular length 4'–2¼" (with endcaps)

Modular length 8'–2¼" (with endcaps)

Modular length 12'–2¼" (with endcaps)

Concept reference only: see specific manufacturer's literature for actual module sizes, seams, endcaps and dimensions

ConceptStarters™

Design Tips

✔ These luminaires are available in continuous linear lengths (e.g., 15 feet) up to 24 feet with no seams (building must accommodate move–in of single, continuous lengths, however, through doors, open windows, elevators, etc.).

✔ Units on closer spacings cause a stronger secondary ceiling plane effect.

✔ Units on greater spacings reduce sustainability issues.

more tips on next page▷

	10'			9'–6"			9'			Ceiling height (in feet)
Percent floor reflectance	20	15	10	20	15	10	20	15	10	
Percent wall reflectance	50	40	30	50	40	30	50	40	30	
Percent partition reflectance	50	30	30	50	30	30	50	30	30	
Percent ceiling reflectance	90	80	70	90	80	70	90	80	70	

Data table 1 — Predominant partition height (in inches):

Partition height	10'			9'–6"			9'
80"							ceiling height not appropriate for cited suspension length
66"	0.78	0.88		0.78	0.88	0.98	
54"	0.78	0.88	0.88		0.88	0.88	
48"		0.78	0.78		0.78	0.78	

Open cell indicates cited luminaire in cited layout has low probability of meeting **all** ambient lighting criteria outlined in Table 10.4.❶

Data table 2 — Predominant partition height (in inches):

Partition height	10'			9'–6"			9'
80"							ceiling height not appropriate for cited suspension length
66"	0.98			0.98	1.18		
54"	0.88	0.98	1.18	0.88	0.98	1.18	
48"	0.78	0.98	1.18	0.78	0.88	1.18	

Data table 3 — Predominant partition height (in inches):

Partition height	10'			9'–6"			9'
80"							ceiling height not appropriate for cited suspension length
66"							
54"							
48"							

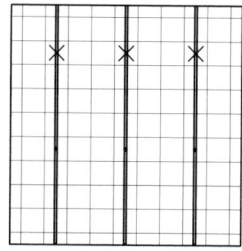

rows 6 feet on center
(One 4–foot section / 2 4 s f)

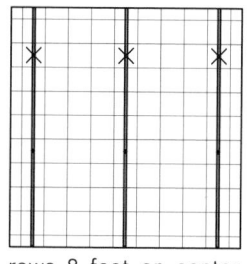

rows 8 feet on center
(One 4–foot section / 3 2 s f)

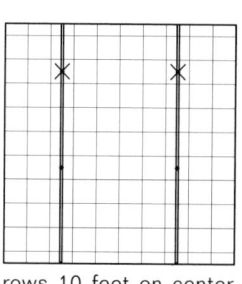

rows 10 feet on center
(One 4–foot section / 4 0 s f)

Ambient Pendent Fluorescent

10

more continued on next two pages

ConceptStarters™

©Digital imagery/® copyright 2000 PhotoDisc, Inc.

Ambient Indirect/Square/Miniature

(1) F32T8 lamp/Suspension of `2 feet/6 inches`

Suspension length (top of luminaire to ceiling plane)

0'–2½"

0'–2¾"

Peerless profile basis for sketch (see each mfgr's data for precise profile and dimensions)

Design Tips

✔ Check extruded aluminum units for modular alignment, side–to–side leveling, camber and surface integrity.

✔ Ceiling finish must be matte.

✔ Ceiling reflectance of 0.9 will provide most efficient lighting.

✔ Depending on layout, square–foot costs may be within a few percentage points of recessed—even though pendent luminaire unit costs may be higher.

✔ Note that endcaps add length.

◁ more tips on previous pages

Ambient Pendent Fluorescent

10

Ceiling height (in feet)		10'			9'–6"			9'		
Percent floor reflectance		20	15	10	20	15	10	20	15	10
Percent wall reflectance		50	40	30	50	40	30	50	40	30
Percent partition reflectance		50	30	30	50	30	30	50	30	30
Percent ceiling reflectance		90	80	70	90	80	70	90	80	70

scale 1/16" = 1'–0"

Predominant partition height (in inches)	80"				
	66"	0.78	0.88		ceiling height not appropriate for cited suspension length
	54"	0.78	0.88	0.88	
	48"		0.78	0.88	

rows 6 feet on center
(O n e 4–foot section / 2 4 s f)

Shaded cell indicates cited luminaire in cited layout has high probability of meeting **all** ambient lighting criteria outlined in Table 10.4. Likely ballast factor is also cited. ❶

Consistent power feed locations look best

Predominant partition height (in inches)	80"				
	66"	0.98	1.18		ceiling height not appropriate for cited suspension length
	54"	0.88	0.98	1.18	
	48"	0.78	0.98	1.18	

rows 8 feet on center
(O n e 4–foot section / 3 2 s f)

Incremental lengths may grow—support locations may not be at tile center

Predominant partition height (in inches)	80"			
	66"	1.18		ceiling height not appropriate for cited suspension length
	54"	1.18	1.18	
	48"	0.98	1.18	

rows 10 feet on center
(O n e 4–foot section / 4 0 s f)

❶ Calculations are required in order to secure a final layout and luminaire specification for actual conditions.

Ambient Indirect/Square/Miniature

(1) F32T8 lamp/Information

ConceptStarters™

Specification

⊃ 2½ inches by 2¾ inches square cross section
⊃ Totally indirect
⊃ VDT–sensitive design (exhibits least glare)
⊃ Extruded aluminum housing in continuous linear lengths
⊃ 1–F32T8/830 rapid start lamp (2950 lumens)
⊃ Electronic instant start ballast (for best efficiency)
⊃ See power budgeting sidebar to right
⊃ Cost magnitude: moderate to high (quantity and length dependent)
⊃ Possible vendors: **Peerless** (LL1)

Boldfaced manufacturer's photometric data were averaged and input to Lumen Micro 7.5 in order to assess probability of meeting ambient lighting criteria cited in Table 10.4. Use the above specification list when contacting manufacturer about potential luminaire options for a specific project. Dimensions, specifications and performance are somewhat generic. Check with specific manufacturer's data for nominal dimensions and performance. Data is subject to change.

Net Addresses/Luminaires
http://www.peerless-lighting.com/

Net Addresses/Ballasts
http://www.magnetek.com/transLighting.html
http://www.mot.com/ies/MLI/
http://www.sylvania.com/ballast/phops.htm

CONNECT FOR MORE

Ambient
Pendent Fluorescent

10

Power budgeting
for these 1–lamp cross–section luminaires

BF 0.78 yields 26 watts①
BF 0.88 yields 30 watts①②
BF 0.98 yields 35 watts①②
BF 1.18 yields 38 watts①

Power budget for ambient lighting should be less than 1.0 w/sf. Ballast vendors include Advance, Energy Savings, Inc. (ESI), Magnetek, Motorola. Not all ballast factors available from all ballast vendors. Above factors based on Magnetek.
①Based on tandem lamp operation— using one ballast to operate two lamps. Wattage reported is only for one of the tandem–wired lamps.
②Single lamp ballasts are available for these ballast factors.

Cost magnitude[2000]
for 4–foot fluorescent luminaire sections

Low
US$70. to US$90.
Moderate
US$90. to US$120.
High
US$120. and greater

Quality ambient lighting for today's electronic office environment typically runs US$3.00 to US$5.00/sf, hardware only (luminaires, lamps and ballasts), excluding installation costs. Costs vary based on quantities/lengths, distributor and contractor markups, luminaire variations such as finish, lamping, ballasting and options and accessories, and market conditions, manufacturing situations and annual inflation. These values are for preliminary, magnitude budgeting and do not represent quotes nor actual final pricing to client.

ConceptStarters™

Digital Imagery® copyright 2000 PhotoDisc, Inc.

Ledalite profile basis for sketch (see each mfgr's data for precise profile and dimensions)

Design Tips

✔ Flat bottom results in contrast with ceiling—harsh look and more likely "streaks" in VDTs.

✔ Flat, dark bottom creates secondary ceiling plane.

✔ Use localized task lighting.

✔ Use art accenting for visual interest.

more tips on next page▷

Ambient Indirect/Square/Large

(1) F32T8 lamp/Suspension of 1 foot/6 inches

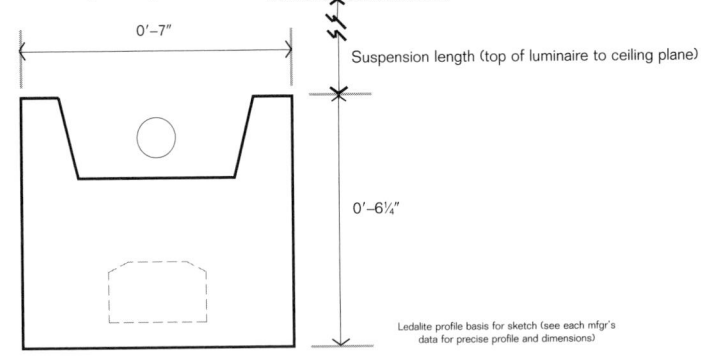

0'–7"

Suspension length (top of luminaire to ceiling plane)

0'–6¼"

scale ¹⁄₁₆" = 1'–0"

Ceiling height (in feet)	10'			9'–6"			9'		
Percent floor reflectance	20	15	10	20	15	10	20	15	10
Percent wall reflectance	50	40	30	50	40	30	50	40	30
Percent partition reflectance	50	30	30	50	30	30	50	30	30
Percent ceiling reflectance	90	80	70	90	80	70	90	80	70
Predominant partition height (in inches) 80"									
66"					0.78	0.88		0.88	0.98
54"		0.78	0.88		0.78	0.88		0.78	0.88
48"		0.78	0.88		0.78	0.88			0.88

rows 6 feet on center
(O n e 4–foot section / 2 4 s f)

Shaded cell indicates cited luminaire in cited layout has high probability of meeting **all** ambient lighting criteria outlined in Table 10.4. Likely ballast factor is also cited.❶

Consistent power feed locations look best

Predominant partition height (in inches) 80"									
66"									
54"									
48"									

rows 8 feet on center
(O n e 4–foot section / 3 2 s f)

Incremental lengths may grow—support locations may not be at tile center

Predominant partition height (in inches) 80"									
66"									
54"									
48"									

rows 10 feet on center
(O n e 4–foot section / 4 0 s f)

❶Calculations are required in order to secure a final layout and luminaire specification for actual conditions.

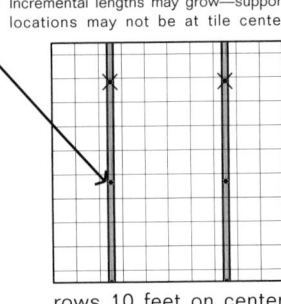

Ambient Indirect/Square/Large

(1) F32T8 lamp/Suspension of `2 feet/0 inches`

more continued on next two pages▷

Modular length 4'–4¼" (with endcaps)

Modular length 8'–4¼" (with endcaps)

Modular length 12'–4¼" (with endcaps)

Concept reference only: see specific manufacturer's literature for actual module sizes, seams, endcaps and dimensions

ConceptStarters™

Design Tips

✔ These luminaires are available in both modular lengths (e.g, 4 feet, 8 feet) and in continuous linear lengths (e.g., 15 feet) up to 24 feet with no seams depending on manufacturer (building must accommodate move–in of single, continuous lengths, however, through doors, open windows, elevators, etc.).

✔ Units on closer spacings cause a stronger secondary ceiling plane effect.

✔ Units on greater spacings reduce sustainability issues.
more tips on next page▷

10'			9'–6"			9'			Ceiling height (in feet)
20	15	10	20	15	10	20	15	10	Percent floor reflectance
50	40	30	50	40	30	50	40	30	Percent wall reflectance
50	30	30	50	30	30	50	30	30	Percent partition reflectance
90	80	70	90	80	70	90	80	70	Percent ceiling reflectance

							Predominant partition height (in inches)
						80"	
0.78	0.88	0.88	0.78	0.88	0.88	66"	
	0.78	0.78		0.78	0.78	54"	
	0.78	0.78		0.78		48"	

(center column: ceiling height not appropriate for cited suspension length)

Open cell indicates cited luminaire in cited layout has low probability of meeting **all** ambient lighting criteria outlined in Table 10.4. ❶

							Predominant partition height (in inches)
						80"	
0.98			0.98	1.18	1.18	66"	
0.78	0.98	1.18	0.78	0.98	1.18	54"	
0.78	0.88	1.18	0.78	0.88	1.18	48"	

(center column: ceiling height not appropriate for cited suspension length)

					Predominant partition height (in inches)
				80"	
				66"	
				54"	
				48"	

(center column: ceiling height not appropriate for cited suspension length)

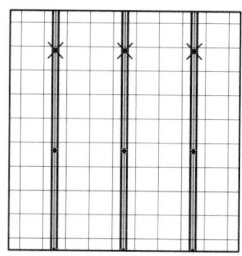

rows 6 feet on center
(O n e 4–foot section / 2 4 s f)

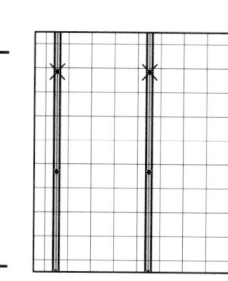

rows 8 feet on center
(O n e 4–foot section / 3 2 s f)

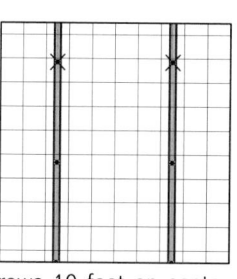

rows 10 feet on center
(O n e 4–foot section / 4 0 s f)

Ambient Pendent Fluorescent

10

more continued on next two pages

ConceptStarters™

Ambient Indirect/Square/Large

(1) F32T8 lamp/Suspension of 2 feet/6 inches

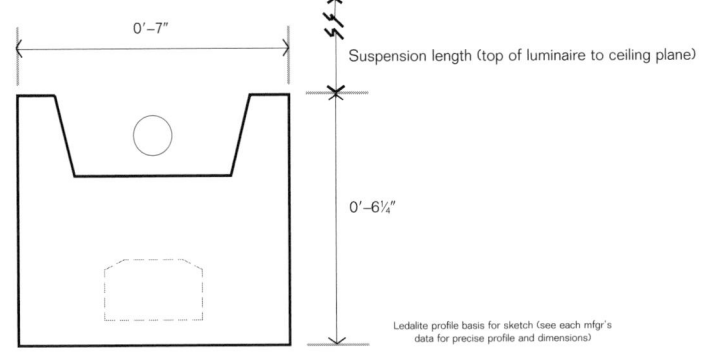

0'–7"

Suspension length (top of luminaire to ceiling plane)

0'–6¼"

Ledalite profile basis for sketch (see each mfgr's data for precise profile and dimensions)

Design Tips

✔ Check extruded aluminum units for modular alignment, side–to–side leveling, camber and surface integrity.

✔ Ceiling finish must be matte.

✔ Ceiling reflectance of 0.9 will provide most efficient lighting.

✔ Depending on layout, square–foot costs may be within a few percentage points of recessed—even though pendent luminaire unit costs may be higher.

✔ Note that endcaps add length.

◁more tips on previous pages

Ceiling height (in feet)		10'			9'–6"			9'		
Percent floor reflectance		20	15	10	20	15	10	20	15	10
Percent wall reflectance		50	40	30	50	40	30	50	40	30
Percent partition reflectance		50	30	30	50	30	30	50	30	30
Percent ceiling reflectance		90	80	70	90	80	70	90	80	70

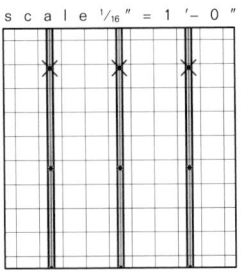

scale ¹⁄₁₆" = 1'–0"

rows 6 feet on center
(O n e 4–foot section / 2 4 s f)

Predominant partition height (in inches)	80"				ceiling height not appropriate for cited suspension length
	66"	0.78	0.88	0.98	
	54"		0.88	0.88	
	48"		0.78	0.88	

Shaded cell indicates cited luminaire in cited layout has high probability of meeting **all** ambient lighting criteria outlined in Table 10.4. Likely ballast factor is also cited. ❶

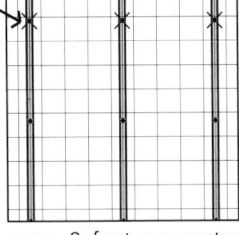

Consistent power feed locations look best

rows 8 feet on center
(O n e 4–foot section / 3 2 s f)

Predominant partition height (in inches)	80"				ceiling height not appropriate for cited suspension length
	66"	0.98	1.18	1.18	
	54"	0.88	0.98	1.18	
	48"	0.78	0.88	1.18	

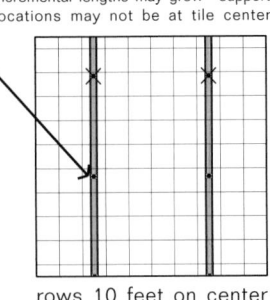

Incremental lengths may grow—support locations may not be at tile center

rows 10 feet on center
(O n e 4–foot section / 4 0 s f)

Predominant partition height (in inches)	80"				ceiling height not appropriate for cited suspension length
	66"	1.18			
	54"	0.98	1.18		
	48"	0.88	1.18	1.18	

❶Calculations are required in order to secure a final layout and luminaire specification for actual conditions.

Ambient Indirect/Square/Large

(1) F32T8 lamp/Information

Specification

➲7 inches by 6¼ inches (nearly) square cross section
➲Totally indirect
➲VDT–sensitive design (exhibits least glare)
➲Extruded aluminum housing in continuous linear lengths
➲1–F32T8/830 rapid start lamp (2950 lumens)
➲Electronic instant start ballast (for best efficiency)
➲See power budgeting sidebar to right
➲Cost magnitude: moderate to high (quantity and length dependent)
➲Possible vendors: **Ledalite** (RLS), Linear (SquareLite 661), **Precision Architectural Lighting** (AS)

Boldfaced manufacturers' photometric data were averaged and input to Lumen Micro 7.5 in order to assess probability of meeting ambient lighting criteria cited in Table 10.4. Use the above specification list when contacting manufacturers about potential luminaire options for a specific project. Dimensions, specifications and performance are somewhat generic. Check with specific manufacturer's data for nominal dimensions and performance. Data is subject to change.

ConceptStarters™

ⓘDigital imagery® copyright 2000 PhotoDisc, Inc.

Power budgeting
for these 1–lamp cross–section luminaires

BF 0.78 yields 26 watts①
BF 0.88 yields 30 watts①②
BF 0.98 yields 35 watts①②
BF 1.18 yields 38 watts①

Power budget for ambient lighting should be less than 1.0 w/sf. Ballast vendors include Advance, Energy Savings, Inc. (ESI), Magnetek, Motorola. Not all ballast factors available from all ballast vendors. Above factors based on Magnetek.
①Based on tandem lamp operation— using one ballast to operate two lamps. Wattage reported is only for one of the tandem–wired lamps.
②Single lamp ballasts are available for these ballast factors.

Ambient Pendent Fluorescent

10

Cost magnitude[2000]
for 4–foot fluorescent luminaire sections

Low
US$70. to US$90.
Moderate
US$90. to US$120.
High
US$120. and greater

Quality ambient lighting for today's electronic office environment typically runs US$3.00 to US$5.00/sf, hardware only (luminaires, lamps and ballasts), excluding installation costs. Costs vary based on quantities/lengths, distributor and contractor markups, luminaire variations such as finish, lamping, ballasting and options and accessories, and market conditions, manufacturing situations and annual inflation. These values are for preliminary, magnitude budgeting and do not represent quotes nor actual final pricing to client.

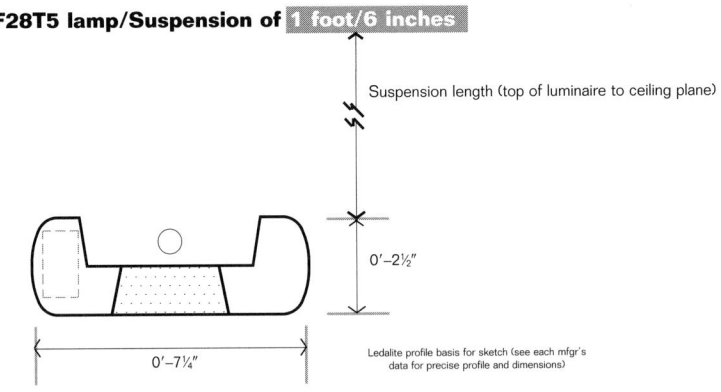

Ambient Semi–indirect/Beveled/Narrow

(1) F28T5 lamp/Suspension of `1 foot/6 inches`

Suspension length (top of luminaire to ceiling plane)

0'–2½"

0'–7¼"

Ledalite profile basis for sketch (see each mfgr's data for precise profile and dimensions)

Digital Imagery® copyright 2000 PhotoDisc, Inc.

ConceptStarters™

Design Tips

✔ Open flat bottom results in less contrast with ceiling when compared to totally indirect flat bottom luminaires

✔ Downlight helps to minimize the "haze" often attributed to totally indirect lighting.

✔ Downlight may introduce glare and an undesirable "veil" when seated directly under unit.

✔ Use localized task lighting.

✔ Use art accenting for visual interest.

more tips on next page▷

Ceiling height (in feet)		10'			9'–6"			9'		
Percent floor reflectance		20	15	10	20	15	10	20	15	10
Percent wall reflectance		50	40	30	50	40	30	50	40	30
Percent partition reflectance		50	30	30	50	30	30	50	30	30
Percent ceiling reflectance		90	80	70	90	80	70	90	80	70
Predominant partition height (in inches)	80"									
	66"									
	54"									
	48"									

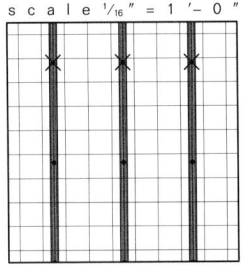

scale ¹⁄₁₆" = 1'–0"

rows 6 feet on center
(O n e 4–foot section / 2 4 s f)

Consistent power feed locations look best

rows 8 feet on center
(O n e 4–foot section / 3 2 s f)

Shaded cell indicates cited luminaire in cited layout has high probability of meeting **all** ambient lighting criteria outlined in Table 10.4. Likely ballast factor is also cited.❶

Predominant partition height (in inches)	80"									
	66"			0.98			0.98			0.98
	54"									
	48"									

Incremental lengths may grow—support locations may not be at tile center

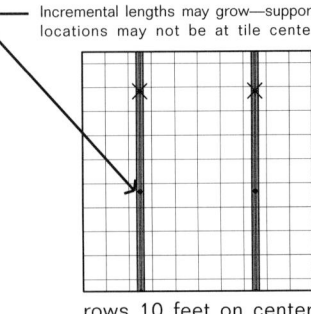

rows 10 feet on center
(O n e 4–foot section / 4 0 s f)

Predominant partition height (in inches)	80"									
	66"									
	54"									
	48"									

❶Calculations are required in order to secure a final layout and luminaire specification for actual conditions.

Ambient Pendent Fluorescent

10

Ambient Semi–indirect/Beveled/Narrow

(1) F28T5 lamp/Suspension of `2 feet/0 inches`

more continued on next two pages▷

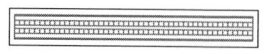

Modular length 4′–4¼″ (with endcaps)

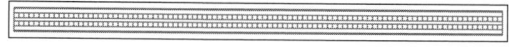

Modular length 8′–4¼″ (with endcaps)

Modular length 12′–4¼″ (with endcaps)

Concept reference only: see specific manufacturer's literature for actual module sizes, seams, endcaps and dimensions

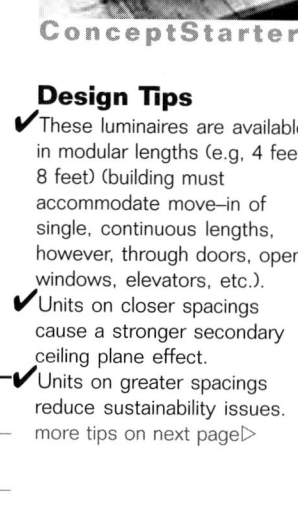

ConceptStarters™

Ambient Pendent Fluorescent

10

Design Tips

✔ These luminaires are available in modular lengths (e.g, 4 feet, 8 feet) (building must accommodate move–in of single, continuous lengths, however, through doors, open windows, elevators, etc.).

✔ Units on closer spacings cause a stronger secondary ceiling plane effect.

✔ Units on greater spacings reduce sustainability issues.

more tips on next page▷

10′			9′–6″			9′			Ceiling height (in feet)
20	15	10	20	15	10	20	15	10	Percent floor reflectance
50	40	30	50	40	30	50	40	30	Percent wall reflectance
50	30	30	50	30	30	50	30	30	Percent partition reflectance
90	80	70	90	80	70	90	80	70	Percent ceiling reflectance

							Predominant partition height (in inches)
			ceiling height not appropriate for cited suspension length			80″	
						66″	
						54″	
						48″	

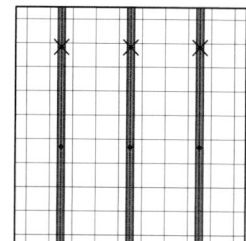

rows 6 feet on center
(O n e 4–foot section / 2 4 s f)

Open cell indicates cited luminaire in cited layout has low probability of meeting **all** ambient lighting criteria outlined in Table 10.4.❶

							Predominant partition height (in inches)
	0.98	0.98		0.98		ceiling height not appropriate for cited suspension length → 80″	
						66″	
						54″	
						48″	

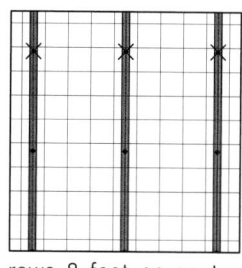

rows 8 feet on center
(O n e 4–foot section / 3 2 s f)

							Predominant partition height (in inches)
						80″	
			ceiling height not appropriate for cited suspension length			66″	
	0.98	0.98		0.98	0.98	54″	
					0.98	48″	

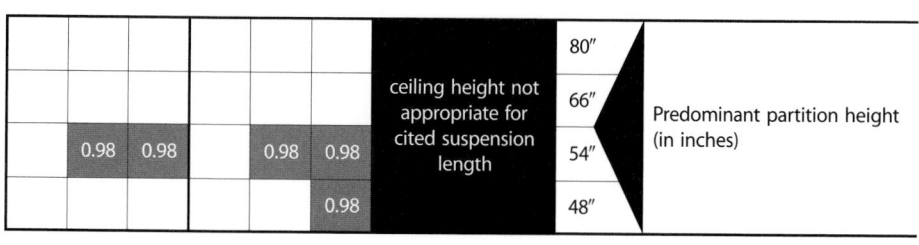

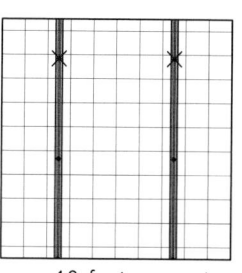

rows 10 feet on center
(O n e 4–foot section / 4 0 s f)

more continued on next two pages

ConceptStarters™

Ambient Semi–indirect/Beveled/Narrow

(1) F28T5 lamp/Suspension of `2 feet/6 inches`

Suspension length (top of luminaire to ceiling plane)

0'–2½"

0'–7¼"

Ledalite profile basis for sketch (see each mfgr's data for precise profile and dimensions)

Design Tips

✔ Check extruded aluminum units for modular alignment, side–to–side leveling, camber and surface integrity.

✔ Ceiling finish must be matte.

✔ Ceiling reflectance of 0.9 will provide most efficient lighting.

✔ Depending on layout, square–foot costs may be within a few percentage points of recessed—even though pendent luminaire unit costs may be higher.

✔ Note that endcaps add length.

◁more tips on previous pages

Ceiling height (in feet)			10'			9'–6"			9'		
Percent floor reflectance			20	15	10	20	15	10	20	15	10
Percent wall reflectance			50	40	30	50	40	30	50	40	30
Percent partition reflectance			50	30	30	50	30	30	50	30	30
Percent ceiling reflectance			90	80	70	90	80	70	90	80	70

scale ¹⁄₁₆" = 1'–0"

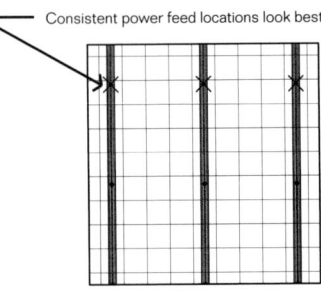

rows 6 feet on center
(O n e 4–foot section / 2 4 s f)

Predominant partition height (in inches)											
		80"									
		66"				ceiling height not appropriate for cited suspension length					
		54"									
		48"									

Shaded cell indicates cited luminaire in cited layout has high probability of meeting **all** ambient lighting criteria outlined in Table 10.4. Likely ballast factor is also cited. ❶

Consistent power feed locations look best

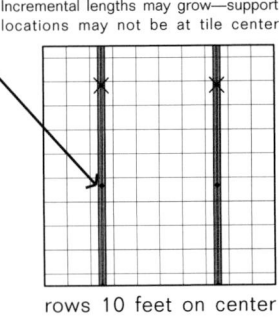

rows 8 feet on center
(O n e 4–foot section / 3 2 s f)

Predominant partition height (in inches)											
		80"									
		66"				ceiling height not appropriate for cited suspension length					
		54"			0.98						
		48"									

Incremental lengths may grow—support locations may not be at tile center

rows 10 feet on center
(O n e 4–foot section / 4 0 s f)

Predominant partition height (in inches)											
		80"									
		66"				ceiling height not appropriate for cited suspension length					
		54"		0.98	0.98						
		48"		0.98							

❶ Calculations are required in order to secure a final layout and luminaire specification for actual conditions.

Ambient Semi–indirect/Beveled/Narrow

(1) F28T5 lamp/Information

Specification

⊃7¹/₄ inches by 2¹/₂ inches rounded or beveled sides/flat–bottom cross section

⊃25 percent direct (downlight)/75 percent indirect

⊃Open bottom fitted with parabolic baffle

⊃Extruded aluminum housing in continuous linear lengths

⊃1–F28T5SE/830 superior efficacy lamp (2900 lumens)

⊃Electronic programmed start ballast

⊃See power budgeting sidebar to right

⊃Cost magnitude: moderate to high (quantity and length dependent)

⊃Possible vendors: **Ledalite** (Minuet), **Metalumen** (Helix)

Boldfaced manufacturers' photometric data were averaged and input to Lumen Micro 7.5 in order to assess probability of meeting ambient lighting criteria cited in Table 10.4. Use the above specification list when contacting manufacturers about potential luminaire options for a specific project. Dimensions, specifications and performance are somewhat generic. Check with specific manufacturer's data for nominal dimensions and performance. Data is subject to change.

ConceptStarters™

Power budgeting
for these 1–lamp cross–section luminaires

BF 0.98 yields 32 watts①

Power budget for ambient lighting should be less than 1.0 w/sf. Ballast vendors include Energy Savings, Inc. (ESI), Motorola and Osram Sylvania. Above factors based on ESI.
①Single– and 2–lamp ballasts are available and both have nearly identical power requirements (32 watts per lamp).

Ambient Pendent Fluorescent

10

Net Addresses/Luminaires
http://www.ledalite.com/
http://www.metalumen.com/

Net Addresses/Ballasts
http://www.mot.com/ies/MLI/
http://www.sylvania.com/ballast/phops.htm

CONNECT FOR MORE

Cost magnitude[2000]
for 4–foot fluorescent luminaire sections

Low
US$70. to US$90.
Moderate
US$90. to US$120.
High
US$120. and greater

Quality ambient lighting for today's electronic office environment typically runs US$3.00 to US$5.00/sf, hardware only (luminaires, lamps and ballasts), excluding installation costs. Costs vary based on quantities/lengths, distributor and contractor markups, luminaire variations such as finish, lamping, ballasting and options and accessories, and market conditions, manufacturing situations and annual inflation. These values are for preliminary, magnitude budgeting and do not represent quotes nor actual final pricing to client.

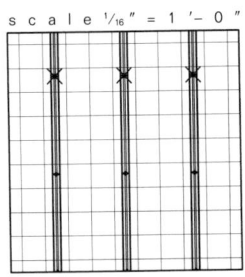

Ambient Pendent Fluorescent
10

Digital Imagery® copyright 2000 PhotoDisc, Inc.

ConceptStarters™

Ambient Semi–indirect/Beveled/Wide

(1) F32T8 lamp/Suspension of `1 foot/6 inches`

Design Tips

✓ Check steel linear units for modular alignment, straightness, sag, side–to–side leveling and body integrity.

✓ Open flat bottom results in less contrast with ceiling when compared to totally indirect flat bottom luminaires

✓ Downlight helps to minimize the "haze" often attributed to totally indirect lighting.

✓ Downlight may introduce glare and an undesirable "veil" when seated directly under unit.

more tips on next page▷

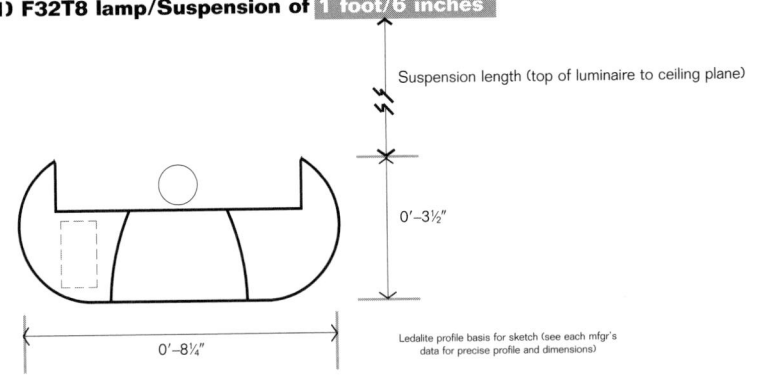

Suspension length (top of luminaire to ceiling plane)

0'–3½"

0'–8¼"

Ledalite profile basis for sketch (see each mfgr's data for precise profile and dimensions)

Ceiling height (in feet)	10'			9'–6"			9'		
Percent floor reflectance	20	15	10	20	15	10	20	15	10
Percent wall reflectance	50	40	30	50	40	30	50	40	30
Percent partition reflectance	50	30	30	50	30	30	50	30	30
Percent ceiling reflectance	90	80	70	90	80	70	90	80	70

Predominant partition height (in inches)

	10'-20	10'-15	10'-10	9'6"-20	9'6"-15	9'6"-10	9'-20	9'-15	9'-10
80"									
66"									
54"		0.78			0.78				0.78
48"									

scale 1/16" = 1'–0"

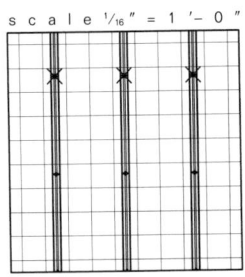

rows 6 feet on center
(One 4–foot section / 24 sf)

Shaded cell indicates cited luminaire in cited layout has high probability of meeting **all** ambient lighting criteria outlined in Table 10.4. Likely ballast factor is also cited.❶

Consistent power feed locations look best

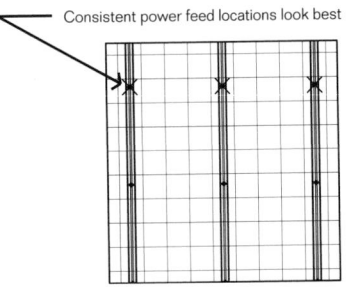

Predominant partition height (in inches)

	10'-20	10'-15	10'-10	9'6"-20	9'6"-15	9'6"-10	9'-20	9'-15	9'-10
80"									
66"	0.78			0.78	0.88	0.98	0.78	0.88	0.98
54"	0.78	0.78	0.88	0.78	0.78	0.88	0.78	0.78	0.88
48"		0.78	0.88	0.78	0.78	0.88		0.78	0.88

rows 8 feet on center
(One 4–foot section / 32 sf)

Incremental lengths may grow—support locations may not be at tile center

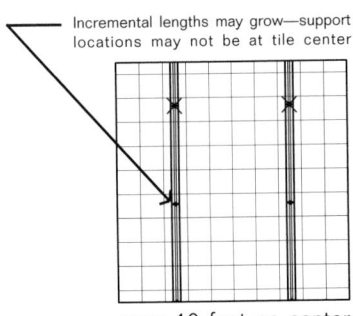

Predominant partition height (in inches)

	10'-20	10'-15	10'-10	9'6"-20	9'6"-15	9'6"-10	9'-20	9'-15	9'-10
80"									
66"					0.98				
54"	0.78				0.88		0.78		
48"	0.78				0.78		0.78		

rows 10 feet on center
(One 4–foot section / 40 sf)

❶ Calculations are required in order to secure a final layout and luminaire specification for actual conditions.

Ambient Semi–indirect/Beveled/Wide

(1) F32T8 lamp/Suspension of `2 feet/0 inches`

more continued on next two pages▷

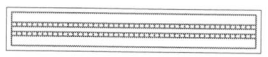

Modular length 4'–4" (with endcaps)

Modular length 8'–4" (with endcaps)

Modular length 12'–4" (with endcaps)

Concept reference only: see specific manufacturer's literature for actual module sizes, seams, endcaps and dimensions

ConceptStarters™

Ambient Pendent Fluorescent

10

Design Tips

✔ Use localized task lighting.
✔ Use art accenting for visual interest.
✔ These luminaires are available in modular lengths (e.g, 4 feet, 8 feet) (building must accommodate move–in of single, continuous lengths, however, through doors, open windows, elevators, etc.).
✔ Units on closer spacings cause a stronger secondary ceiling plane effect.
✔ Units on greater spacings reduce sustainability issues.
more tips on next page▷

Table — Ceiling heights

10'			9'–6"			9'			Ceiling height (in feet)
20	15	10	20	15	10	20	15	10	Percent floor reflectance
50	40	30	50	40	30	50	40	30	Percent wall reflectance
50	30	30	50	30	30	50	30	30	Percent partition reflectance
90	80	70	90	80	70	90	80	70	Percent ceiling reflectance

rows 6 feet on center

									Predominant partition height (in inches)
									80"
		0.88			0.78	*ceiling height not appropriate for cited suspension length*			66"
									54"
									48"

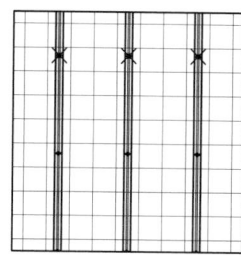

rows 6 feet on center
(O n e 4–foot section / 2 4 s f)

Open cell indicates cited luminaire in cited layout has low probability of meeting **all** ambient lighting criteria outlined in Table 10.4.❶

rows 8 feet on center

									Predominant partition height (in inches)
									80"
0.78	0.98	0.98	0.78	0.98	0.98	*ceiling height not appropriate for cited suspension length*			66"
0.78	0.88	0.98	0.78	0.88	0.98				54"
		0.88			0.88				48"

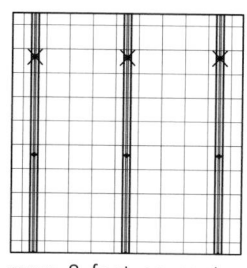

rows 8 feet on center
(O n e 4–foot section / 3 2 s f)

rows 10 feet on center

									Predominant partition height (in inches)
									80"
						ceiling height not appropriate for cited suspension length			66"
0.78	0.88		0.78	0.88					54"
0.78		1.18	0.78		0.98				48"

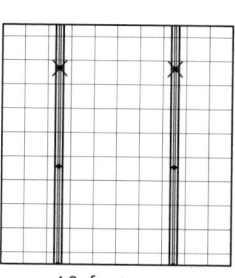

rows 10 feet on center
(O n e 4–foot section / 4 0 s f)

more continued on next two pages

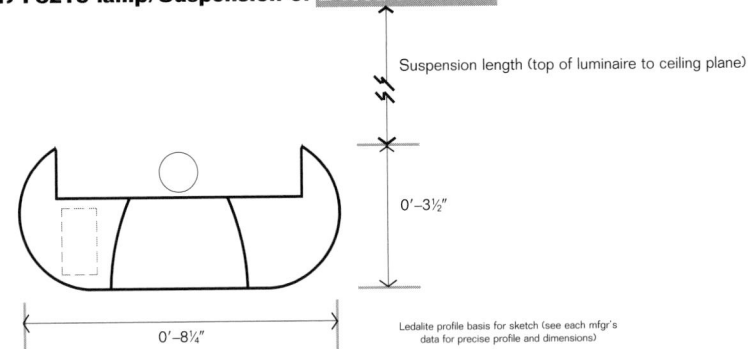

Ambient Semi–indirect/Beveled/Narrow

(1) F32T8 lamp/Suspension of 2 feet/6 inches

Suspension length (top of luminaire to ceiling plane)

0'–3½"

0'–8¼"

Ledalite profile basis for sketch (see each mfgr's data for precise profile and dimensions)

Ambient Pendent Fluorescent

10

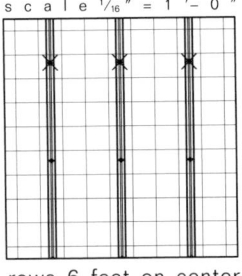

ConceptStarters™

Design Tips

✔ Check extruded aluminum units for modular alignment, side–to–side leveling, camber and surface integrity.

✔ Ceiling finish must be matte.

✔ Ceiling reflectance of 0.9 will provide most efficient lighting.

✔ Depending on layout, square–foot costs may be within a few percentage points of recessed—even though pendent luminaire unit costs may be higher.

✔ Note that endcaps add length.

◁more tips on previous pages

Ceiling height (in feet)		10'			9'–6"			9'		
Percent floor reflectance		20	15	10	20	15	10	20	15	10
Percent wall reflectance		50	40	30	50	40	30	50	40	30
Percent partition reflectance		50	30	30	50	30	30	50	30	30
Percent ceiling reflectance		90	80	70	90	80	70	90	80	70

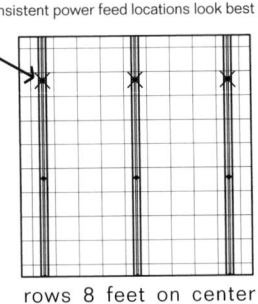

scale 1/16" = 1'–0"

rows 6 feet on center
(O n e 4–foot section / 2 4 s f)

Predominant partition height (in inches)						
80"				ceiling height not appropriate for cited suspension length		
66"						
54"			0.78			
48"			0.78			

Shaded cell indicates cited luminaire in cited layout has high probability of meeting **all** ambient lighting criteria outlined in Table 10.4. Likely ballast factor is also cited.❶

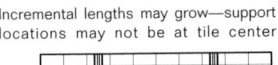

Consistent power feed locations look best

rows 8 feet on center
(O n e 4–foot section / 3 2 s f)

Predominant partition height (in inches)						
80"					ceiling height not appropriate for cited suspension length	
66"	0.78	0.88	0.98			
54"	0.78	0.88	0.98			
48"		0.88	0.88			

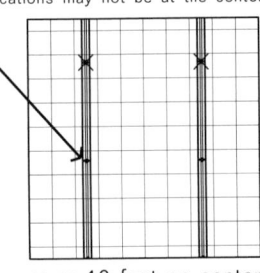

Incremental lengths may grow—support locations may not be at tile center

rows 10 feet on center
(O n e 4–foot section / 4 0 s f)

Predominant partition height (in inches)						
80"	1.18				ceiling height not appropriate for cited suspension length	
66"						
54"	0.98					
48"	0.88		1.18			

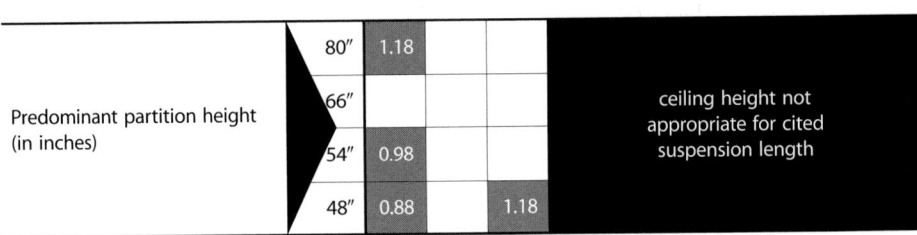

❶Calculations are required in order to secure a final layout and luminaire specification for actual conditions.

Ambient Semi–indirect/Beveled/Narrow

(1) F32T8 lamp/Information

ConceptStarters™

Specification

◗8¼ to 9½ inches by 3½ inches rounded or beveled sides/flat–bottom cross section

◗About 30 percent direct (downlight)/70 percent indirect

◗Open bottom fitted with parabolic baffle

◗Extruded aluminum or 20 gauge steel housing in modular lengths

◗1–F32T8/830 rapid start lamp (2950 lumens)

◗Electronic instant start ballast

◗See power budgeting sidebar to right

◗Cost magnitude: moderate to high (steel vs. extruded aluminum and quantity and length dependent)

◗Possible vendors: **Focal Point** (Formula I)❷, **Ledalite** (RLS)❸

Boldfaced manufacturers' photometric data were averaged and input to Lumen Micro 7.5 in order to assess probability of meeting ambient lighting criteria cited in Table 10.4. Use the above specification list when contacting manufacturers about potential luminaire options for a specific project. Dimensions, specifications and performance are somewhat generic. Check with specific manufacturer's data for nominal dimensions and performance. Data is subject to change.

Power budgeting
for these 1–lamp cross–section luminaires

BF 0.78 yields 26 watts①
BF 0.88 yields 30 watts①②
BF 0.98 yields 35 watts①②
BF 1.18 yields 38 watts①

Power budget for ambient lighting should be less than 1.0 w/sf. Ballast vendors include Advance, Energy Savings, Inc. (ESI), Magnetek, Motorola. Not all ballast factors available from all ballast vendors. Above factors based on Magnetek.
①Based on tandem lamp operation— using one ballast to operate two lamps. Wattage reported is only for one of the tandem–wired lamps.
②Single lamp ballasts are available for these ballast factors.

Ambient Pendent Fluorescent

10

Net Addresses/Luminaires
http://www.focalpointlights.com/
http://www.ledalite.com/

Net Addresses/Ballasts
http://www.magnetek.com/transLighting.html
http://www.mot.com/ies/MLI/
http://www.sylvania.com/ballast/phops.htm

CONNECT FOR MORE

Cost magnitude²⁰⁰⁰
for 4–foot fluorescent luminaire sections

Low
US$70. to US$90.
Moderate
US$90. to US$120.
High
US$120. and greater

Quality ambient lighting for today's electronic office environment typically runs US$3.00 to US$5.00/sf, hardware only (luminaires, lamps and ballasts), excluding installation costs. Costs vary based on quantities/lengths, distributor and contractor markups, luminaire variations such as finish, lamping, ballasting and options and accessories, and market conditions, manufacturing situations and annual inflation. These values are for preliminary, magnitude budgeting and do not represent quotes nor actual final pricing to client.

---

❷20 gauge steel housing exhibiting angular/beveled sides (typically available in shorter modular lengths than extruded aluminum and require more frequent suspension points).

❸Extruded aluminum housing exhibiting round/beveled sides as shown (typically available in greater lengths than steel housings).

ConceptStarters™

Design Tips
✔ Open flat bottom results in less contrast with ceiling when compared to totally indirect flat bottom luminaires
✔ Downlight helps to minimize the "haze" often attributed to totally indirect lighting.
✔ Downlight may introduce glare and an undesirable "veil" when seated directly under unit.
more tips on next page▷

Ambient Semi–indirect/Square/Large
(1) F32T8 lamp/Suspension of 1 foot/6 inches

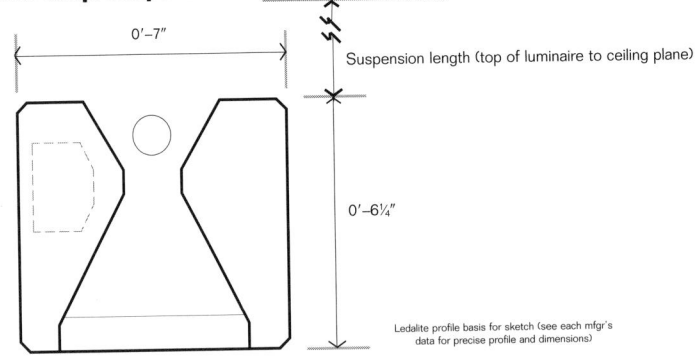

Suspension length (top of luminaire to ceiling plane)

0'–7"

0'–6¼"

Ledalite profile basis for sketch (see each mfgr's data for precise profile and dimensions)

Ceiling height (in feet)		10'			9'–6"			9'		
Percent floor reflectance		20	15	10	20	15	10	20	15	10
Percent wall reflectance		50	40	30	50	40	30	50	40	30
Percent partition reflectance		50	30	30	50	30	30	50	30	30
Percent ceiling reflectance		90	80	70	90	80	70	90	80	70
Predominant partition height (in inches)	80"									
	66"		0.78	0.78			0.78		0.78	0.78
	54"			0.78			0.78			0.78
	48"						0.78			

Shaded cell indicates cited luminaire in cited layout has high probability of meeting **all** ambient lighting criteria outlined in Table 10.4. Likely ballast factor is also cited.❶

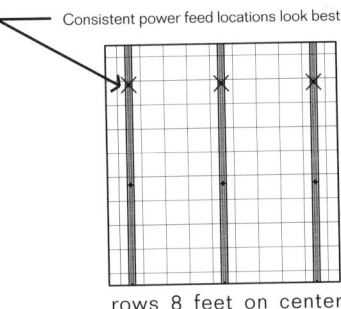

scale ¹⁄₁₆" = 1'–0"

rows 6 feet on center
(O n e 4–foot section / 2 4 s f)

Consistent power feed locations look best

rows 8 feet on center
(O n e 4–foot section / 3 2 s f)

Predominant partition height (in inches)	80"			
	66"			
	54"			
	48"			

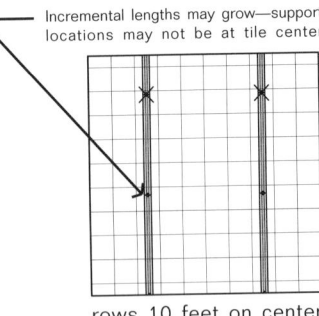

Incremental lengths may grow—support locations may not be at tile center

rows 10 feet on center
(O n e 4–foot section / 4 0 s f)

Predominant partition height (in inches)	80"			
	66"			
	54"			
	48"			

❶Calculations are required in order to secure a final layout and luminaire specification for actual conditions.

Ambient Semi–indirect/Square/Large

(1) F32T8 lamp/Suspension of `2 feet/0 inches`

more continued on next two pages▷

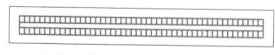

Modular length 4'–4" (with endcaps)

Modular length 8'–4" (with endcaps)

Modular length 12'–4" (with endcaps)

Concept reference only: see specific manufacturer's literature for actual module sizes, seams, endcaps and dimensions

ConceptStarters™

Design Tips
✔ Use localized task lighting.
✔ Use art accenting for visual interest.
✔ These luminaires are available in modular lengths (e.g. 4 feet, 8 feet) (building must accommodate move–in of single, continuous lengths, however, through doors, open windows, elevators, etc.).
✔ Units on closer spacings cause a stronger secondary ceiling plane effect.
✔ Units on greater spacings reduce sustainability issues.
more tips on next page▷

Ambient Pendent Fluorescent

10

10'			9'–6"			9'			Ceiling height (in feet)
20	15	10	20	15	10	20	15	10	Percent floor reflectance
50	40	30	50	40	30	50	40	30	Percent wall reflectance
50	30	30	50	30	30	50	30	30	Percent partition reflectance
90	80	70	90	80	70	90	80	70	Percent ceiling reflectance

10'			9'–6"			9'	Predominant partition height (in inches)
							80"
	0.78	0.78		0.78	0.78	ceiling height not appropriate for cited suspension length	66"
		0.78			0.78		54"
							48"

Open cell indicates cited luminaire in cited layout has low probability of meeting **all** ambient lighting criteria outlined in Table 10.4.❶

10'			9'–6"			9'	Predominant partition height (in inches)
							80"
0.78						ceiling height not appropriate for cited suspension length	66"
0.78	0.78	0.88		0.78	0.88		54"
	0.78	0.78		0.78	0.78		48"

10'			9'–6"			9'	Predominant partition height (in inches)
							80"
						ceiling height not appropriate for cited suspension length	66"
							54"
							48"

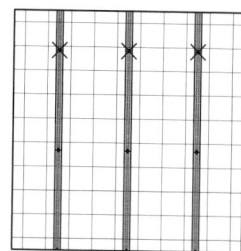

rows 6 feet on center
(O n e 4–foot section / 2 4 s f)

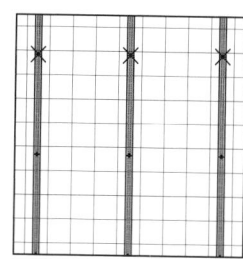

rows 8 feet on center
(O n e 4–foot section / 3 2 s f)

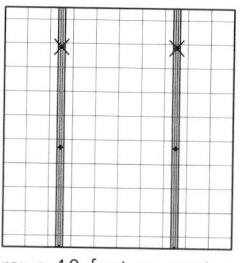

rows 10 feet on center
(O n e 4–foot section / 4 0 s f)

more continued on next two pages

ConceptStarters™

Design Tips

✔ Check extruded aluminum units for modular alignment, side–to–side leveling, camber and surface integrity.

✔ Ceiling finish must be matte.

✔ Ceiling reflectance of 0.9 will provide most efficient lighting.

✔ Depending on layout, square–foot costs may be within a few percentage points of recessed—even though pendent luminaire unit costs may be higher.

✔ Note that endcaps add length.

◁ more tips on previous pages

Ambient Semi–indirect/Square/Large

(1) F32T8 lamp/Suspension of `2 feet/6 inches`

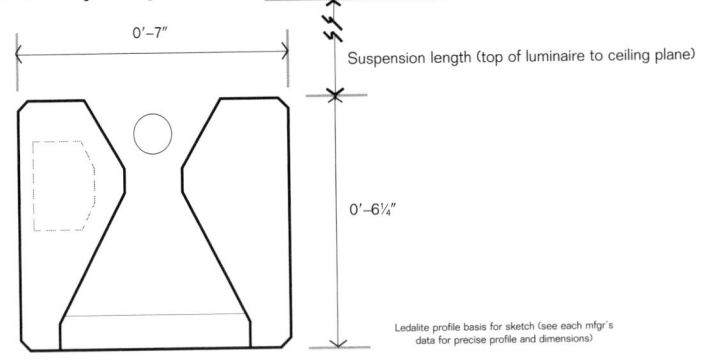

Suspension length (top of luminaire to ceiling plane)

0'–7"

0'–6¼"

Ledalite profile basis for sketch (see each mfgr´s data for precise profile and dimensions)

Ceiling height (in feet)		10'			9'–6"			9'		
Percent floor reflectance		20	15	10	20	15	10	20	15	10
Percent wall reflectance		50	40	30	50	40	30	50	40	30
Percent partition reflectance		50	30	30	50	30	30	50	30	30
Percent ceiling reflectance		90	80	70	90	80	70	90	80	70

scale ¹⁄₁₆″ = 1'–0″

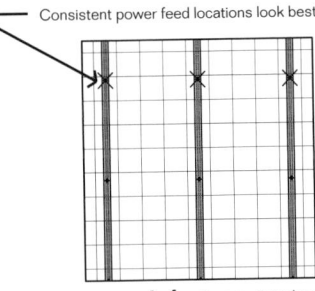

rows 6 feet on center
(O n e 4–foot section / 2 4 s f)

Consistent power feed locations look best

Predominant partition height (in inches)	80"							
	66"		0.78	0.78	ceiling height not appropriate for cited suspension length			
	54"			0.78				
	48"			0.78				

Shaded cell indicates cited luminaire in cited layout has high probability of meeting **all** ambient lighting criteria outlined in Table 10.4. Likely ballast factor is also cited.❶

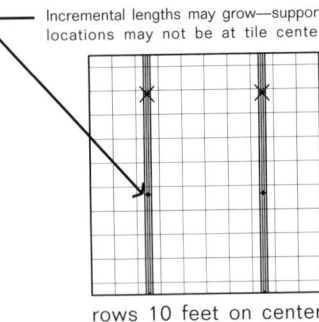

rows 8 feet on center
(O n e 4–foot section / 3 2 s f)

Incremental lengths may grow—support locations may not be at tile center

Predominant partition height (in inches)	80"					
	66"				ceiling height not appropriate for cited suspension length	
	54"	0.78	0.78	0.78		
	48"		0.78	0.78		

rows 10 feet on center
(O n e 4–foot section / 4 0 s f)

Predominant partition height (in inches)	80"			ceiling height not appropriate for cited suspension length
	66"			
	54"			
	48"			

❶ Calculations are required in order to secure a final layout and luminaire specification for actual conditions.

Ambient Semi–indirect/Square/Large
(1) F32T8 lamp/Information

ConceptStarters™

Specification

- 7 inches by 6¼ inches rounded or beveled sides/flat–bottom cross section
- 25 percent direct (downlight)/75 percent indirect
- Open bottom fitted with parabolic baffle
- Extruded aluminum housing in modular lengths
- 1–F32T8/830 rapid start lamp (2950 lumens)
- Electronic instant start ballast
- See power budgeting sidebar to right
- Cost magnitude: moderate to high (quantity and length dependent)
- Possible vendors: **Ledalite** (RLS)

Boldfaced manufacturer's photometric data were averaged and input to Lumen Micro 7.5 in order to assess probability of meeting ambient lighting criteria cited in Table 10.4. Use the above specification list when contacting manufacturer about potential luminaire options for a specific project. Dimensions, specifications and performance are somewhat generic. Check with specific manufacturer's data for nominal dimensions and performance. Data is subject to change.

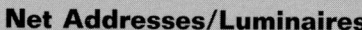

Net Addresses/Luminaires
http://www.ledalite.com/

Net Addresses/Ballasts
http://www.magnetek.com/transLighting.html
http://www.mot.com/ies/MLI/
http://www.sylvania.com/ballast/phops.htm

CONNECT FOR MORE

Power budgeting
for these 1–lamp cross–section luminaires

BF 0.78 yields 26 watts①
BF 0.88 yields 30 watts①②
BF 0.98 yields 35 watts①②
BF 1.18 yields 38 watts①

Power budget for ambient lighting should be less than 1.0 w/sf. Ballast vendors include Advance, Energy Savings, Inc. (ESI), Magnetek, Motorola. Not all ballast factors available from all ballast vendors. Above factors based on Magnetek.
①Based on tandem lamp operation—using one ballast to operate two lamps. Wattage reported is only for one of the tandem–wired lamps.
②Single lamp ballasts are available for these ballast factors.

Cost magnitude[2000]
for 4–foot fluorescent luminaire sections

Low
US$70. to US$90.
Moderate
US$90. to US$120.
High
US$120. and greater

Quality ambient lighting for today's electronic office environment typically runs US$3.00 to US$5.00/sf, hardware only (luminaires, lamps and ballasts), excluding installation costs. Costs vary based on quantities/lengths, distributor and contractor markups, luminaire variations such as finish, lamping, ballasting and options and accessories, and market conditions, manufacturing situations and annual inflation. These values are for preliminary, magnitude budgeting and do not represent quotes nor actual final pricing to client.

10 Ambient Pendent Fluorescent

Details Concepts and Effects

11

Many lighting effects—for mood, function or both—are a result of developing architectural details. Still other effects are from the selection and layout of the lighting hardware—lamps, ballasts and luminaires. This section illustrates detail concepts for cove lighting, wallwashers and wall slots by identifying the kind of lamps and luminaires that might be considered and the resulting anticipated effects.

The first segment which follows addresses cove lighting. The second segment, beginning on page 11–96, addresses wall lighting.

Details Concepts and Effects

Cove Lighting

Cove lighting is traditionally a decorative effect used to "lift" the ceiling visually by creating an oculus–like raised ceiling and then uplighting that raised ceiling. Used more for effect in residential and hospitality environments, it is now a mainstay in many interior environments. The advent of deluxe triphosphor compact fluorescent lamps and the small profile T5 and T8 deluxe triphosphor linear fluorescent lamps now enable the designer to use cove lighting which is functional, energy efficient and maintenance–reasonable.

Coves for accenting

Coves can be used to help identify special areas, zones or "destinations." In retail applications, coves can be used to draw attention to certain departments or areas within departments. In commercial applications, coves can be used to identify receptionists, waiting areas, conference centers and the like. Even if the cove is intended to be primarily an accent or decorative element, the light it produces should be considered as part of the overall ambient lighting in the area.

■ ■ ■ ■ ■ ■ ■ ■ ■ ■ ■ ■ ■ ■ ■ ■

Coves
- Pleasant aesthetic
- Functional—use light intensities from coves to conserve energy
- Helps soften harshness of downlighting or art accenting
- Provides improved sense of overall brightness in environment

This will minimize energy consumption. If coves are intended solely for aesthetic appearance, then dimmers or low output ballasts (ballasts with low ballast factors) might be used to reduce light output.

Coves for function

Coves can and should be considered one tool in the kit of parts to develop functional lighting. Even if coves are intended to be a decorative feature, not taking advantage of the illuminance they produce is essentially wasting energy and resources. The coves outlined herein have been assessed for their appearance and their ability to produce light. Generally, the greater the cove height and the more optically–active is the luminaire (e.g., a special reflector is used), the more efficient the cove lighting technique. With so many choices on

Details Concepts and Effects

Cove Lighting

cove opening sizes (the plan view of the cove), on the cove height, on lamping, on ballasts and ballast factors, on luminaire style—optically inactive (such as a striplight) or optically active (such as a specially designed reflector)—and on wattage, cove lighting results vary substantially. For the T5/T8, T5/HO and LongCompact lamps, a series of Performance Perspec-tives™ are presented which offers quick visual assessment of the results of various cove param-eters, luminaire style options and lamp type and wattage.

Radiance (lighting rendering software)
http://www.radsite.lbl.gov/radiance/refer/
long.html

CONNECT FOR MORE

Performance Perspectives™ are shown (no scale) to help in assessment of function and aesthetic. Nuances in lighting are not well represented in the cove images. The software was incapable of representing socket shadows which are prominent in striplight solutions. Further, the software did not render the smooth unifor-mity typical of asymmetric cove luminaires. Please read the Design Tips. Copying is permitted (according to the conventions outlined in "Licensing Agreement" in the Introduction of this book). Average light levels on the floor plane directly below the cove are cited for each cove along with maximum illuminance. This helps in the as-sessment of functionality and uniformity (a maximum to average ratio of 2:1 will appear uniform). For museum (gallery) and residen-tial settings, intensites on the floor should be less than 10 fc aver-age maintained (otherwise the space will appear too bright and commercial–like). This will also minimize the quantity of ambient light ultimately falling onto sensitive artwork. In typical hospitality applica-tions (including more casual commercial and institutional spaces), light intensity on the floor from a cove should be in the 10 to 20 fc range. In commercial, retail and institutional applications, average, maintained intensities on the floor could be as much as 40 fc. A special application note, however: if coves are being used as deco-rative elements where light intensities are not of significance, then intensities from cove lighting may be much less than cited.

These Performance Perspectives™ are based on luminaire/lamp data available at time of manuscript preparation and is cited with each cove. Manufacturers' catalog data may be of one date and for

Details Concepts and Effects

Cove Lighting

one set of lamps while photometric data available from manufacturers may be based on another set of lamps of another date (this may have occurred with some data in this reference). Performance Perspectives™ were generated with this data in Radiance (a Unix–based lighting simulation and rendering program available free of charge at www.radsite.lbl.gov/radiance/refer/long.html).

These Performance Perspectives™ illustrate a specific condition. As ceiling heights change, as cove parameters change, the cited lamp may be quite appropriate for other applications.

■ ■ ■ ■ ■ ■ ■ ■ ■ ■ ■ ■ ■ ■ ■

Performance Perspectives™ and Information
❶ Typical application is shown
❶ Other applications may also be acceptable
❶ Other applications may depend on ceiling heights
❶ Maintained values based on 0.86 maintenance factor
❶ Cited data is for a single lamp cross section luminaire
❶ Performance data cited based on ceiling reflectance of 0.9; walls of 0.5 and floor of 0.2

Recommended intensities
for ambient lighting from coves

Low (gallery, museum, residential)
5 to 10 fc average on floor
Moderate (hospitality)
10 to 20 fc average on floor
High (commercial—or higher for retail)
20 to 30 fc average on floor

These are intended to be average, maintained values and will vary based on actual task requirements. These are suggested intensities from cove lighting alone. Other lighting techniques (task lighting, accent lighting) will contribute additional light to an area.

Selection guides
Selection guides are offered on the next several pages for various coves in various applications. These guides are intended to help limit the designer's, engineer's or facility engineer's search and offer a good starting point from which design or alternative analyses can progress.

While there are exceptions, coves are appropriate in the following interior applications:
▶commercial /institutional/retail
▶gallery and residential
▶hospitality

Extrapolating information
There will be situations where perhaps twice as much light is desired (to provide a significant focal point or task lighting intensities) or where less light is desired. For cove situations where more light is desired, consider using 2–lamp cross sections. This typically increases light intensities by 75 percent (there are losses resulting from the tightness of the lamps' fit with one another). Alternatively, review other cove/luminaire parameters herein, as some situations will result in nearly twice as much light over some situations. Where less light is desired, consider using low output ballasts (0.78 ballast factors will reduce output by 22 percent; 0.88 ballast factors will reduce output by 12 percent).

Details Concepts and Effects
Cove Lighting

Net Addresses/Striplights
```
http://www.bartcolighting.com/index2.html
http://www.belfer.com/prodsi.html
http://www.columbia-ltg.com/products/recessed-surface.html
http://www.thomasltg.com/
http://www.lightolier.com/
http://www.lithonia.com/
http://www.cooperlighting.com/
```

CONNECT FOR MORE

Net Addresses/Asymmetric Cove Luminaires
```
http://www.elliptipar.com/home/Contents.htm
http://www.elplighting.com
http://www.focalpointlights.com/
http://www.ledalite.com/
http://www.lightolier.com/
http://www.litecontrol.com/products/index.html
http://www.neoray.com/CATALOG/catalog.html
http://www.peerless-lighting.com/
```

CONNECT FOR MORE

Details Concepts and Effects

11

Details Cove Lighting

DetailConcepts™

Using the Detail Concepts™

Two Pages
Two pages are devoted to each cove "style" review. This illustrates the double page spread for a cove style.

Luminaire Type
Each double page spread is devoted to a single cove style. The header outlines the luminaire family (striplights or asymmetric cove luminaires), ceiling height, cove opening and clear cove height. Advisory: ceiling height is from the floor to the central ceiling ("inside" the cove opening).

Performance Perspectives™
For each lamping configuration (T8/T5; T5/HO; and LongCompact), a Performance Perspective™ is offered for general guidance in considering size and overall visual impact of the cove (these are not well–detailed images). You won't see socket shadows nor the extent of uniformity—see Design Tips for commentary on socket shadows and uniformity.

Concept Details™/Information Presented
Cove information is reiterated next to each Performance Perspective™. Perimeter ceiling height is also noted—and this will vary based on final structural detailing of the cove. Anticipated average, maintained (0.90 maintenance factor) intensity on the floor is cited along with the maximum. Advisory: values are based on high reflectance ceiling material 90 percent reflectance); light–toned walls (50 percent reflectance); and neutral flooring (20 percent reflectance). No scale.

(Side tab:) **Details** Concepts and Effects **11**

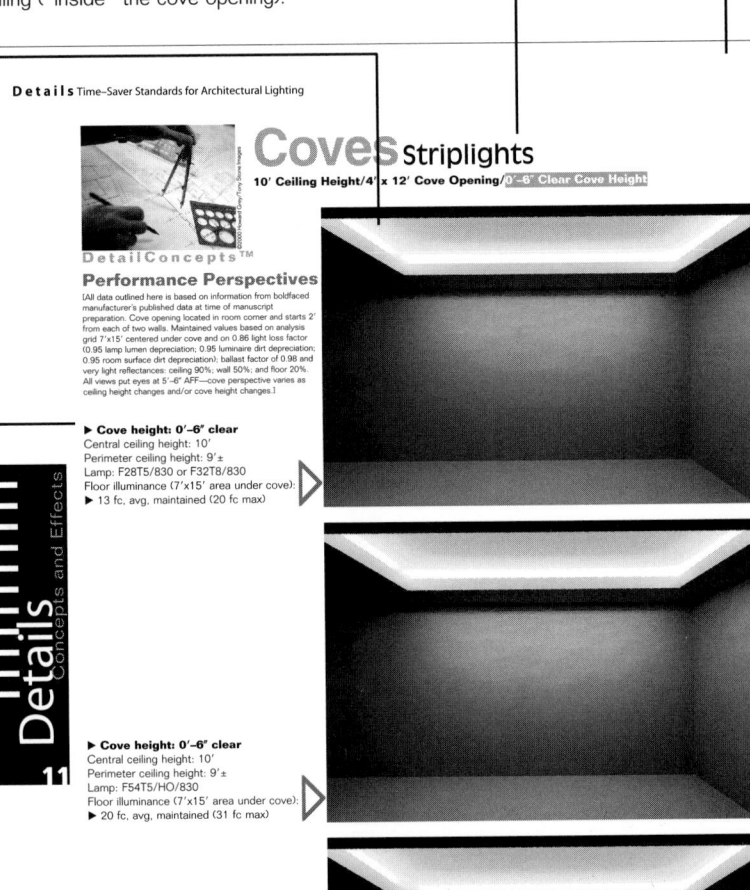

Details Time–Saver Standards for Architectural Lighting

DetailConcepts™

Coves Striplights

10' Ceiling Height/4' x 12' Cove Opening/0'–6" Clear Cove Height

Performance Perspectives

[All data outlined here is based on information from boldfaced manufacturer's published data at time of manuscript preparation. Cove opening located in room corner and starts 2' from each of two walls. Maintained values based on analysis grid 7'x15' centered under cove and on 0.86 light loss factor (0.95 lamp lumen depreciation; 0.95 luminaire dirt depreciation; 0.95 room surface dirt depreciation); ballast factor of 0.98 and very light reflectances: ceiling 90%; wall 50%; and floor 20%. All views put eyes at 5'–6" AFF—cove perspective varies as ceiling height changes and/or cove height changes.]

▶ **Cove height: 0'–6" clear**
Central ceiling height: 10'
Perimeter ceiling height: 9'±
Lamp: F28T5/830 or F32T8/830
Floor illuminance (7'x15' area under cove):
▶ 13 fc, avg, maintained (20 fc max)

▶ **Cove height: 0'–6" clear**
Central ceiling height: 10'
Perimeter ceiling height: 9'±
Lamp: F54T5/HO/830
Floor illuminance (7'x15' area under cove):
▶ 20 fc, avg, maintained (31 fc max)

▶ **Cove height: 0'–6" clear**
Central ceiling height: 10'
Perimeter ceiling height: 9'±
Lamp: F40Long/830
Floor illuminance (7'x15' area under cove):
▶ 25 fc, avg, maintained (41 fc max)

(Side tab:) **Details** Concepts and Effects **11**

Details Cove Lighting

Using the Detail Concepts™

DetailConcepts™

Detail Concept™
The concept outlines the salient specifics related to lighting—showing luminaire location/ orientation and dimensional criteria.

Notes
Key points related to lighting are noted, including the surface reflectance parameters; intended finishes; and application issues.

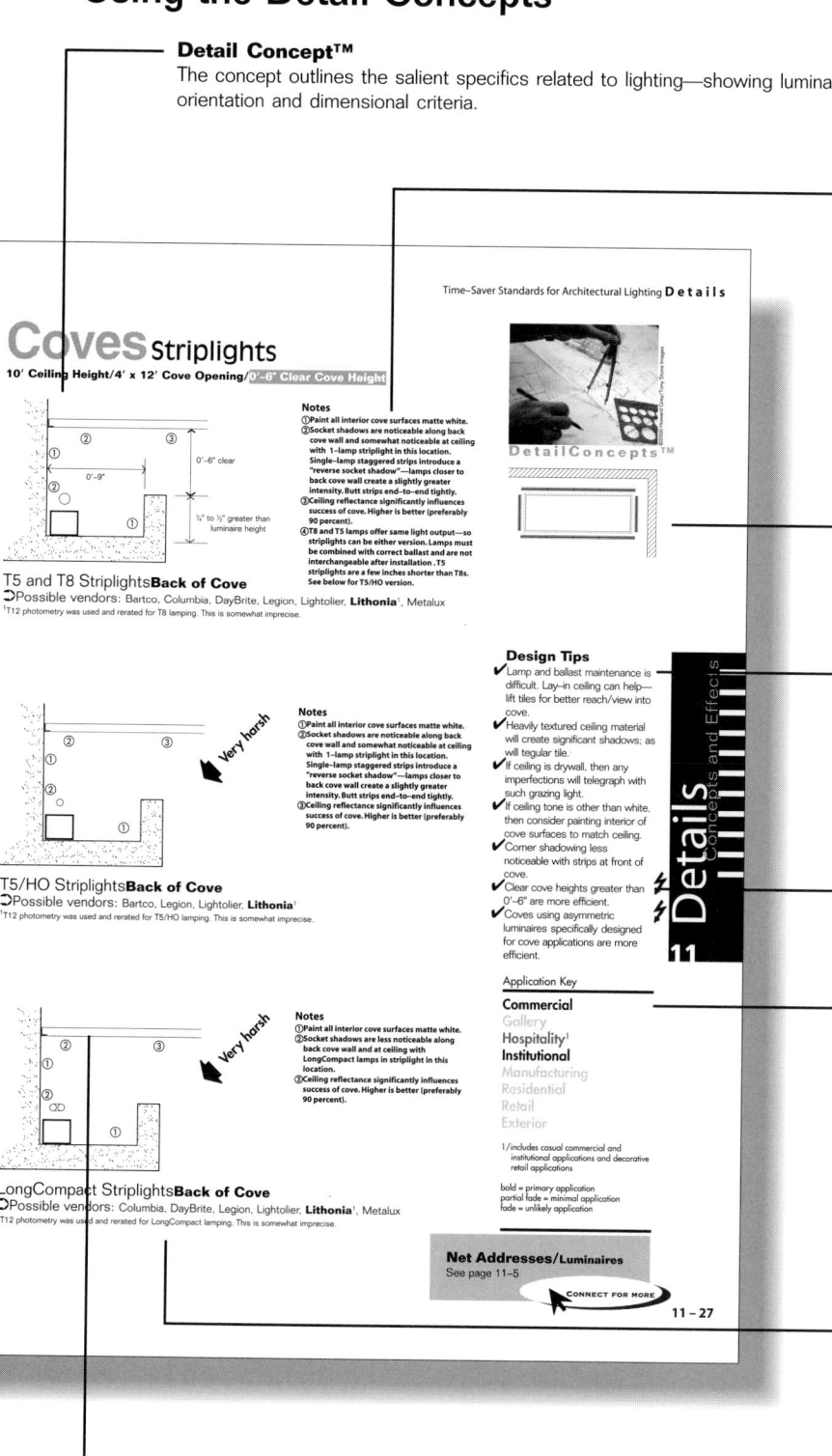

Layout Icons
Icons represent cove and luminaire layouts used in generating the respective images. No scale.

Design Tips
Key considerations are offered for the specific cove style.

Energy Tip
The energy bolt ⚡ is used to denote an energy saving option.

Application Key
Those applications for the cited cove style where the specific lamping configuration(s) may be most useful are highlighted in bold. Applications where only one of the lamping configurations may be most useful is partially faded. Applications where cited cove style may not be appropriate are faded out. Application appropriateness will change with ceiling heights, cove heights, room surface finishes, ballast factors and the like.

Luminaire Vendors
A collection of vendors that have or had luminaires fitting the performance characteristics used in developing the Detail Concepts™ on the double page spread. Boldfaced manufacturers' photometric data were used for the calculation analysis for the cited cove style/lamping configuration. Advisory: in the interest of time, the vendors' search was limited. Other vendors may be found with similar performing products.

Ceiling Height
The ceiling height cited is from the floor to the central ceiling ("inside" the cove opening). The leader line here shows the ceiling plane for which the ceiling height is cited.

Details Cove Lighting Selection Guide

Cove Lighting

DetailConcepts™

Using the Selection Guide

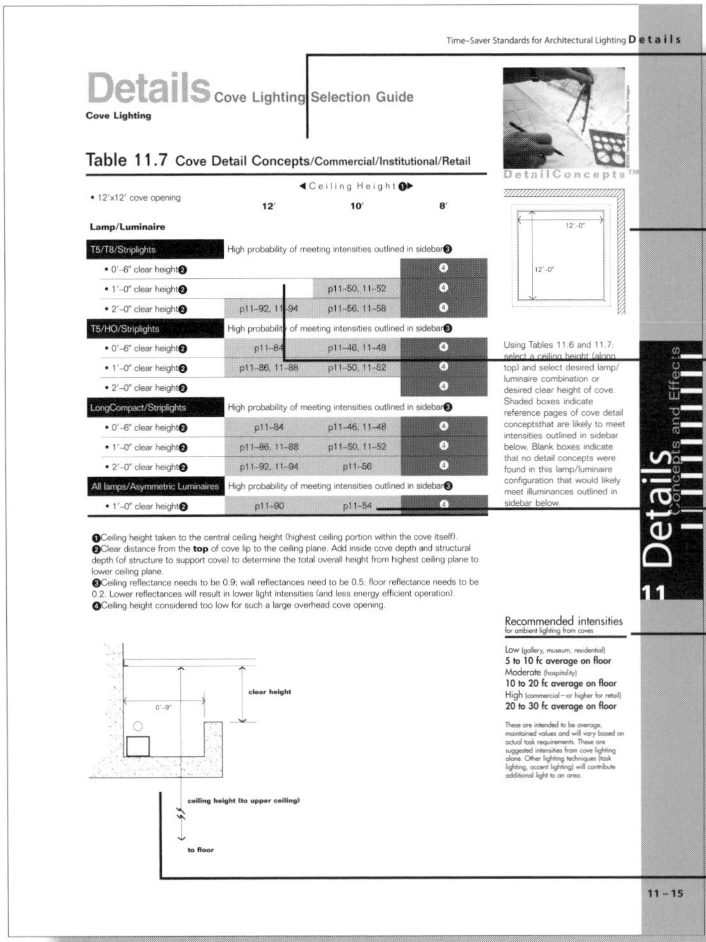

Time–Saver Standards for Architectural Lighting **Details**

Details Cove Lighting Selection Guide

Cove Lighting

Table 11.7 Cove Detail Concepts/Commercial/Institutional/Retail

• 12'x12' cove opening

◄ Ceiling Height ❶►

Lamp/Luminaire	12'	10'	8'
T5/T8/Striplights	High probability of meeting intensities outlined in sidebar ❸		
• 0'–6" clear height ❷			❹
• 1'–0" clear height ❷		p11–50, 11–52	❹
• 2'–0" clear height ❷	p11–92, 11–94	p11–56, 11–58	❹
T5/HO/Striplights	High probability of meeting intensities outlined in sidebar ❸		
• 0'–6" clear height ❷	p11–84	p11–46, 11–48	❹
• 1'–0" clear height ❷	p11–86, 11–88	p11–50, 11–52	❹
• 2'–0" clear height ❷			❹
LongCompact/Striplights	High probability of meeting intensities outlined in sidebar ❸		
• 0'–6" clear height ❷	p11–84	p11–46, 11–48	❹
• 1'–0" clear height ❷	p11–86, 11–88	p11–50, 11–52	❹
• 2'–0" clear height ❷	p11–92, 11–94	p11–56	❹
All lamps/Asymmetric Luminaires	High probability of meeting intensities outlined in sidebar ❸		
• 1'–0" clear height ❷	p11–90	p11–54	❹

❶Ceiling height taken to the central ceiling height (highest ceiling portion within the cove itself).
❷Clear distance from the **top** of cove lip to the ceiling plane. Add inside cove depth and structural depth (of structure to support cove) to determine the total overall height from highest ceiling plane to lower ceiling plane.
❸Ceiling reflectance needs to be 0.9; wall reflectances need to be 0.5; floor reflectance needs to be 0.2. Lower reflectances will result in lower light intensities (and less energy efficient operation).
❹Ceiling height considered too low for such a large overhead cove opening.

Using Tables 11.6 and 11.7: select a ceiling height (along top) and select desired lamp/ luminaire combination or desired clear height of cove. Shaded boxes indicate reference pages of cove detail conceptsthat are likely to meet intensities outlined in sidebar below. Blank boxes indicate that no detail concepts were found in this lamp/luminaire configuration that would likely meet illuminances outlined in sidebar below.

Recommended intensities
for ambient lighting from cove

Low (gallery, museum, residential)
5 to 10 fc average on floor
Moderate (hospitality)
10 to 20 fc average on floor
High (commercial—or higher for retail)
20 to 30 fc average on floor

These are intended to be average, maintained values and will vary based on actual task requirements. These are suggested intensities from cove lighting alone. Other lighting techniques (task lighting, accent lighting) will contribute additional light to an area.

clear height

0'–9"

ceiling height (to upper ceiling)

to floor

11 – 15

Applications
Selection guides are based on likely appropriate applications. Find the application of interest here in the table title.

Reference Icon
An icon provides quick visual reference of the cove size cited in the selection guide.

Blank Cells
Blank cells indicate that analyses show that this particular cove style/lamping/ceiling configuration did not meet the suggested criteria intensities (cited in the lower right corner).

Shaded Cell
Check the cited page for Concept Starters™.

Recommended Intensities
Lighting intensities which may be reasonable and desirable from cove lighting are suggested for various applications.

Cove Profile
A profile of one of the cove types is shown for quick visual reference.

Details
Cove Lighting Selection Guide

Cove Lighting

DetailConcepts™

Table 11.1 Cove Detail Concepts/Commercial/Institutional/Retail

• 4'x12' cove opening

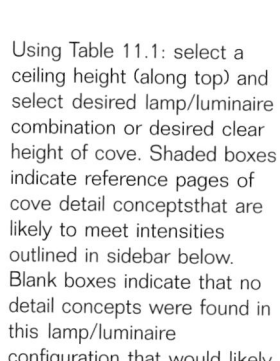

◄ Ceiling Height ❶ ►

Lamp/Luminaire	12'	10'	8'
T5/T8/Striplights	High probability of meeting intensities outlined in sidebar❸		
• 0'–6" clear height❷			p11–20
• 1'–0" clear height❷		p11–30, 11–32	❹
• 2'–0" clear height❷			❹
T5/HO/Striplights	High probability of meeting intensities outlined in sidebar❸		
• 0'–6" clear height❷	p11–62	p11–26, 11–28	p11–18, 11–20
• 1'–0" clear height❷	p11–64, 11–66	p11–30, 11–32	❹
• 2'–0" clear height❷			❹
LongCompact/Striplights	High probability of meeting intensities outlined in sidebar❸		
• 0'–6" clear height❷	p11–62	p11–26, 11–28	p11–18, 11–20
• 1'–0" clear height❷	p11–64, 11–66	p11–30, 11–32	❹
• 2'–0" clear height❷			❹
All lamps/Asymmetric Luminaires	High probability of meeting intensities outlined in sidebar❸		
• 1'–0" clear height❷	p11–68	p11–34	❹

❶Ceiling height taken to the central ceiling height (highest ceiling portion within the cove itself).
❷Clear distance from the **top** of cove lip to the ceiling plane. Add inside cove depth and structural depth (of structure to support cove) to determine the total overall height from highest ceiling plane to lower ceiling plane.
❸Ceiling reflectance needs to be 0.9; wall reflectances need to be 0.5; floor reflectance needs to be 0.2. Lower reflectances will result in lower light intensities (and less energy efficient operation).
❹Ceiling height cannot accommodate cove height (lower ceiling plane elevation will be below 7'.)

Using Table 11.1: select a ceiling height (along top) and select desired lamp/luminaire combination or desired clear height of cove. Shaded boxes indicate reference pages of cove detail conceptsthat are likely to meet intensities outlined in sidebar below. Blank boxes indicate that no detail concepts were found in this lamp/luminaire configuration that would likely meet illuminances outlined in sidebar below.

Details
Concepts and Effects

11

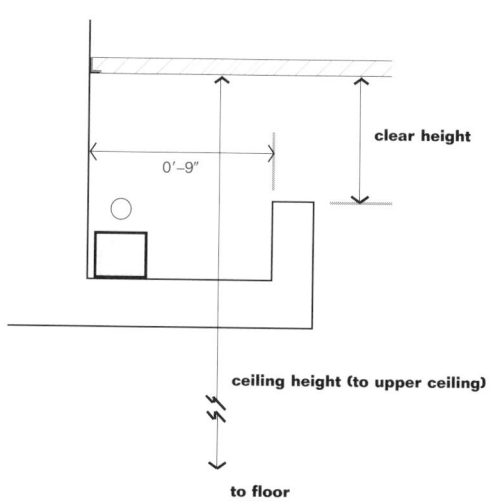

clear height

0'–9"

ceiling height (to upper ceiling)

to floor

Recommended intensities
for ambient lighting from coves

Low (gallery, museum, residential)
5 to 10 fc average on floor
Moderate (hospitality)
10 to 20 fc average on floor
High (commercial—or higher for retail)
20 to 30 fc average on floor

These are intended to be average, maintained values and will vary based on actual task requirements. These are suggested intensities from cove lighting alone. Other lighting techniques (task lighting, accent lighting) will contribute additional light to an area.

Details Cove Lighting Selection Guide

Cove Lighting

DetailConcepts™

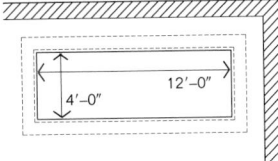

Table 11.2 Cove Detail Concepts/Gallery/Residential

• 4'x12' cove opening

Lamp/Luminaire	◄ Ceiling Height ❶ ►		
	12'	10'	8'
T5/T8/Striplights	High probability of meeting intensities outlined in sidebar❸		
• 0'–6" clear height❷	p11–60		
• 1'–0" clear height❷			❹
• 2'–0" clear height❷			❹
T5/HO/Striplights	High probability of meeting intensities outlined in sidebar❸		
• 0'–6" clear height❷			
• 1'–0" clear height❷			❹
• 2'–0" clear height❷			❹
LongCompact/Striplights	High probability of meeting intensities outlined in sidebar❸		
• 0'–6" clear height❷			
• 1'–0" clear height❷			❹
• 2'–0" clear height❷			❹
All lamps/Asymmetric Luminaires	High probability of meeting intensities outlined in sidebar❸		
• 1'–0" clear height❷			❹

❶ Ceiling height taken to the central ceiling height (highest ceiling portion within the cove itself).

❷ Clear distance from the **top** of cove lip to the ceiling plane. Add inside cove depth and structural depth (of structure to support cove) to determine the total overall height from highest ceiling plane to lower ceiling plane.

❸ Ceiling reflectance needs to be 0.9; wall reflectances need to be 0.5; floor reflectance needs to be 0.2. Lower reflectances will result in lower light intensities (and less energy efficient operation).

❹ Ceiling height cannot accommodate cove height (lower ceiling plane elevation will be below 7'.)

Design Tips

✔ Fluorescent coves in gallery and residential applications are most successful when using dimmable ballasts (e.g., ESI or Lutron) or using low output ballasts. All cove data shown here based on full output ballasts (ballast factor of 0.98).

✔ Ballast factor of 0.78 will reduce reported intensities herein by 22 percent.

✔ Dimmable ballasts will permit output as low as 5 percent or so of reported intensities herein.

Cost magnitude²⁰⁰⁰

per linear foot of fluorescent luminaire

T8 Striplights
US$10./ft to US$12./ft
T5 and T5/HO Striplights
US$15./ft to US$20./ft
LongCompact Striplights
US$20./ft to US$25./ft
Asymmetric Luminaires
US$25./ft to US$35./ft

Cost magnitude for hardware only (luminaires, lamps and ballasts—with full output ballasts), excluding drywall/structure/millwork work and installation costs. For dimming ballasts, add US$15./ft to US$50./ft depending on brand and/or performance. Costs vary based on quantities/lengths, distributor and contractor markups, market conditions, manufacturing situations and annual inflation. These values are for preliminary, magnitude budgeting and do not represent quotes nor actual final pricing to client.

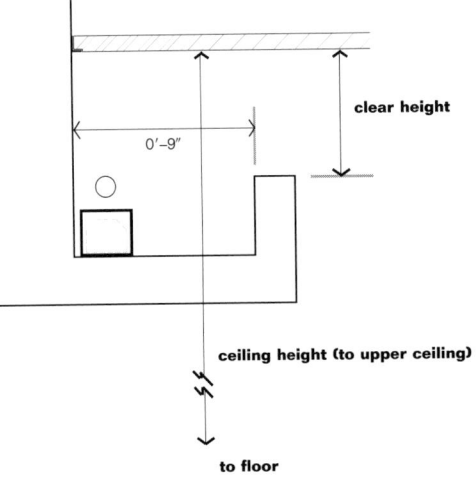

clear height

0'–9"

ceiling height (to upper ceiling)

to floor

Details Cove Lighting Selection Guide

Cove Lighting

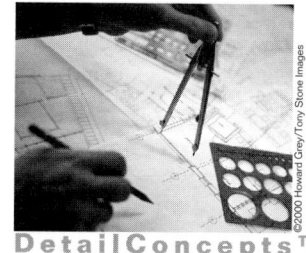

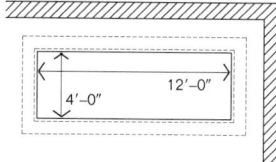

D e t a i l C o n c e p t s ™

Table 11.3 Cove Detail Concepts/Hospitality

- 4'x12' cove opening

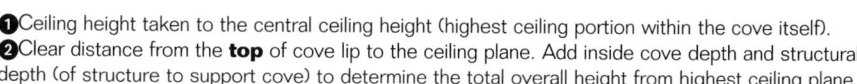

Lamp/Luminaire

	◄ C e i l i n g H e i g h t ❶▶		
	12'	10'	8'
T5/T8/Striplights	High probability of meeting intensities outlined in sidebar❸		
• 0'–6" clear height❷	p11–62	p11–26, 11–28	p11–18
• 1'–0" clear height❷	p11–64, p11–66		❹
• 2'–0" clear height❷			❹
T5/HO/Striplights	High probability of meeting intensities outlined in sidebar❸		
• 0'–6" clear height❷	p11–60		
• 1'–0" clear height❷			❹
• 2'–0" clear height❷			❹
LongCompact/Striplights	High probability of meeting intensities outlined in sidebar❸		
• 0'–6" clear height❷	p11–60		
• 1'–0" clear height❷			❹
• 2'–0" clear height❷			❹
All lamps/Asymmetric Luminaires	High probability of meeting intensities outlined in sidebar❸		
• 1'–0" clear height❷	p11–68		❹

❶Ceiling height taken to the central ceiling height (highest ceiling portion within the cove itself).
❷Clear distance from the **top** of cove lip to the ceiling plane. Add inside cove depth and structural depth (of structure to support cove) to determine the total overall height from highest ceiling plane to lower ceiling plane.
❸Ceiling reflectance needs to be 0.9; wall reflectances need to be 0.5; floor reflectance needs to be 0.2. Lower reflectances will result in lower light intensities (and less energy efficient operation).
❹Ceiling height cannot accommodate cove height (lower ceiling plane elevation will be below 7'.)

Using Tables 11.2 and 11.3: select a ceiling height (along top) and select desired lamp/ luminaire combination or desired clear height of cove. Shaded boxes indicate reference pages of cove detail conceptsthat are likely to meet intensities outlined in sidebar below. Blank boxes indicate that no detail concepts were found in this lamp/luminaire configuration that would likely meet illuminances outlined in sidebar below.

Details
Concepts and Effects

11

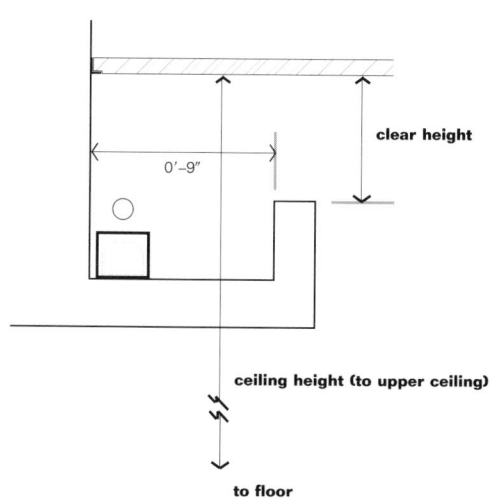

clear height

0'–9"

ceiling height (to upper ceiling)

to floor

Recommended intensities
for ambient lighting from coves

Low (gallery, museum, residential)
5 to 10 fc average on floor
Moderate (hospitality)
10 to 20 fc average on floor
High (commercial—or higher for retail)
20 to 30 fc average on floor

These are intended to be average, maintained values and will vary based on actual task requirements. These are suggested intensities from cove lighting alone. Other lighting techniques (task lighting, accent lighting) will contribute additional light to an area.

Details Cove Lighting Selection Guide

Cove Lighting

DetailConcepts™

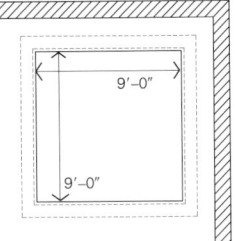

• 9'x9' cove opening

Table 11.4 Cove Detail Concepts/Commercial/Institutional/Retail

Lamp/Luminaire	◀ Ceiling Height ❶ ▶		
	12'	10'	8'
T5/T8/Striplights	High probability of meeting intensities outlined in sidebar❸		
• 0'–6" clear height❷			p11–24
• 1'–0" clear height❷		p11–40, 11–42	❹
• 2'–0" clear height❷			❹
T5/HO/Striplights	High probability of meeting intensities outlined in sidebar❸		
• 0'–6" clear height❷	p11–74	p11–36, 11–38	p11–22, 11–24
• 1'–0" clear height❷	p11–76, 11–78	p11–40, 11–42	❹
• 2'–0" clear height❷			❹
LongCompact/Striplights	High probability of meeting intensities outlined in sidebar❸		
• 0'–6" clear height❷	p11–74	p11–36, 11–38	p11–22, 11–24
• 1'–0" clear height❷	p11–76, 11–78	p11–40, 11–42	❹
• 2'–0" clear height❷			❹
All lamps/Asymmetric Luminaires	High probability of meeting intensities outlined in sidebar❸		
• 1'–0" clear height❷	p11–80	p11–44	❹

Using Tables 11.4 and 11.5: select a ceiling height (along top) and select desired lamp/luminaire combination or desired clear height of cove. Shaded boxes indicate reference pages of cove detail conceptsthat are likely to meet intensities outlined in sidebar below. Blank boxes indicate that no detail concepts were found in this lamp/luminaire configuration that would likely meet illuminances outlined in sidebar below.

❶Ceiling height taken to the central ceiling height (highest ceiling portion within the cove itself).
❷Clear distance from the **top** of cove lip to the ceiling plane. Add inside cove depth and structural depth (of structure to support cove) to determine the total overall height from highest ceiling plane to lower ceiling plane.
❸Ceiling reflectance needs to be 0.9; wall reflectances need to be 0.5; floor reflectance needs to be 0.2. Lower reflectances will result in lower light intensities (and less energy efficient operation).
❹Ceiling height cannot accommodate cove height (lower ceiling plane elevation will be below 7'.)

Cost magnitude²⁰⁰⁰
per linear foot of fluorescent luminaire

T8 Striplights
US$10./ft to US$12./ft
T5 and T5/HO Striplights
US$15./ft to US$20./ft
LongCompact Striplights
US$20./ft to US$25./ft
Asymmetric Luminaires
US$25./ft to US$35./ft

Cost magnitude for hardware only (luminaires, lamps and ballasts—with full output ballasts), excluding drywall/structure/millwork work and installation costs. For dimming ballasts, add US$15./ft to US$50./ft depending on brand and/or performance. Costs vary based on quantities/lengths, distributor and contractor markups, market conditions, manufacturing situations and annual inflation. These values are for preliminary, magnitude budgeting and do not represent quotes nor actual final pricing to client.

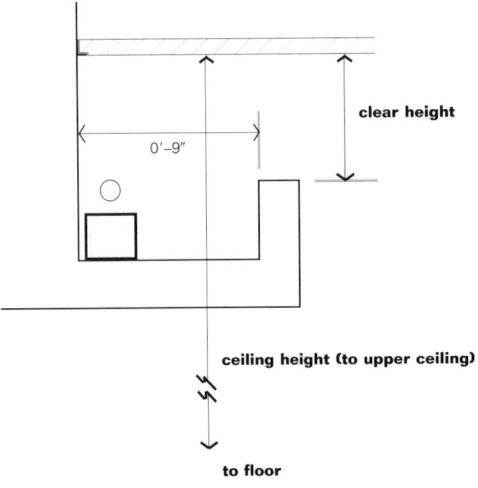

Details Cove Lighting Selection Guide

Cove Lighting

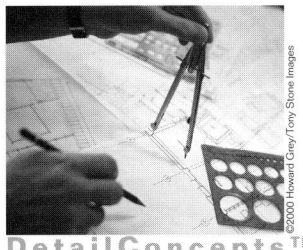
©2000 Howard Grey/Tony Stone Images

DetailConcepts™

Table 11.5 Cove Detail Concepts/Gallery/Residential

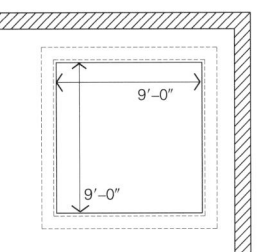
9'–0"
9'–0"

• 9'x9' cove opening

Lamp/Luminaire	◀ Ceiling Height ❶ ▶		
	12'	10'	8'
T5/T8/Striplights	High probability of meeting intensities outlined in sidebar❸		
• 0'–6" clear height❷	p11–72		
• 1'–0" clear height❷			❹
• 2'–0" clear height❷			❹
T5/HO/Striplights	High probability of meeting intensities outlined in sidebar❸		
• 0'–6" clear height❷			
• 1'–0" clear height❷			❹
• 2'–0" clear height❷			❹
LongCompact/Striplights	High probability of meeting intensities outlined in sidebar❸		
• 0'–6" clear height❷			
• 1'–0" clear height❷			❹
• 2'–0" clear height❷			❹
All lamps/Asymmetric Luminaires	High probability of meeting intensities outlined in sidebar❸		
• 1'–0" clear height❷			❹

❶ Ceiling height taken to the central ceiling height (highest ceiling portion within the cove itself).
❷ Clear distance from the **top** of cove lip to the ceiling plane. Add inside cove depth and structural depth (of structure to support cove) to determine the total overall height from highest ceiling plane to lower ceiling plane.
❸ Ceiling reflectance needs to be 0.9; wall reflectances need to be 0.5; floor reflectance needs to be 0.2. Lower reflectances will result in lower light intensities (and less energy efficient operation).
❹ Ceiling height cannot accommodate cove height (lower ceiling plane elevation will be below 7'.)

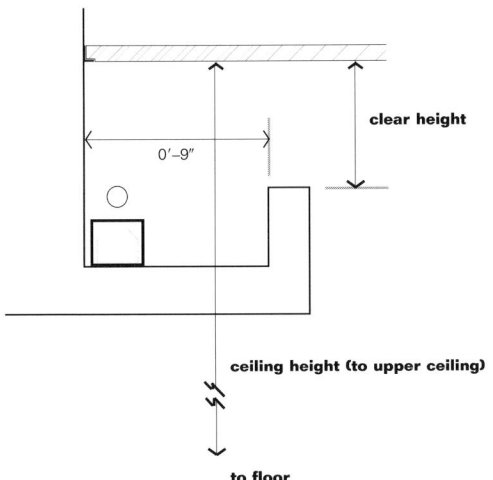

clear height

0'–9"

ceiling height (to upper ceiling)

to floor

Design Tips

✔ Fluorescent coves in gallery and residential applications are most successful when using dimmable ballasts (e.g., ESI or Lutron) or using low output ballasts. All cove data shown here based on full output ballasts (ballast factor of 0.98).
✔ Ballast factor of 0.78 will reduce reported intensities herein by 22 percent.
✔ Dimmable ballasts will permit output as low as 5 percent or so of reported intensities herein.

Recommended intensities
for ambient lighting from coves

Low (gallery, museum, residential)
5 to 10 fc average on floor
Moderate (hospitality)
10 to 20 fc average on floor
High (commercial—or higher for retail)
20 to 30 fc average on floor

These are intended to be average, maintained values and will vary based on actual task requirements. These are suggested intensities from cove lighting alone. Other lighting techniques (task lighting, accent lighting) will contribute additional light to an area.

Details
Concepts and Effects

11

Details Cove Lighting Selection Guide

Cove Lighting

DetailConcepts™

©2000 Howard Grey/Tony Stone Images

Table 11.6 Cove Detail Concepts/Hospitality

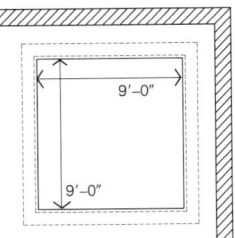

9'–0"

9'–0"

• 9'x9' cove opening

Concepts and Effects

Details

11

◄ Ceiling Height ❶ ►

Lamp/Luminaire	12'	10'	8'
T5/T8/Striplights	High probability of meeting intensities outlined in sidebar ❸		
• 0'–6" clear height ❷	p11–72, 11–74	p11–36, 11–38	p11–22
• 1'–0" clear height ❷	p11–76, 11–78		❹
• 2'–0" clear height ❷			❹
T5/HO/Striplights	High probability of meeting intensities outlined in sidebar ❸		
• 0'–6" clear height ❷	p11–72		
• 1'–0" clear height ❷			❹
• 2'–0" clear height ❷			❹
LongCompact/Striplights	High probability of meeting intensities outlined in sidebar ❸		
• 0'–6" clear height ❷	p11–72		
• 1'–0" clear height ❷			❹
• 2'–0" clear height ❷			❹
All lamps/Asymmetric Luminaires	High probability of meeting intensities outlined in sidebar ❸		
• 1'–0" clear height ❷			❹

❶Ceiling height taken to the central ceiling height (highest ceiling portion within the cove itself).
❷Clear distance from the **top** of cove lip to the ceiling plane. Add inside cove depth and structural depth (of structure to support cove) to determine the total overall height from highest ceiling plane to lower ceiling plane.
❸Ceiling reflectance needs to be 0.9; wall reflectances need to be 0.5; floor reflectance needs to be 0.2. Lower reflectances will result in lower light intensities (and less energy efficient operation).
❹Ceiling height cannot accommodate cove height (lower ceiling plane elevation will be below 7'.)

Cost magnitude²⁰⁰⁰
per linear foot of fluorescent luminaire

T8 Striplights
US$10./ft to US$12./ft
T5 and T5/HO Striplights
US$15./ft to US$20./ft
LongCompact Striplights
US$20./ft to US$25./ft
Asymmetric Luminaires
US$25./ft to US$35./ft

Cost magnitude for hardware only (luminaires, lamps and ballasts—with full output ballasts), excluding drywall/ structure/millwork work and installation costs. For dimming ballasts, add US$15./ ft to US$50./ft depending on brand and/ or performance. Costs vary based on quantities/lengths, distributor and contractor markups, market conditions, manufacturing situations and annual inflation. These values are for preliminary, magnitude budgeting and do not represent quotes nor actual final pricing to client.

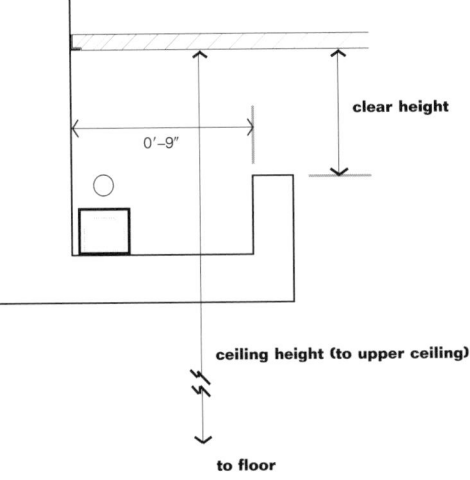

clear height

0'–9"

ceiling height (to upper ceiling)

to floor

Details Cove Lighting Selection Guide

Cove Lighting

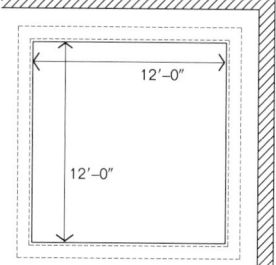

DetailConcepts™

Table 11.7 Cove Detail Concepts/Commercial/Institutional/Retail

- 12'x12' cove opening

Lamp/Luminaire	◀ Ceiling Height ❶ ▶		
	12'	10'	8'
T5/T8/Striplights	High probability of meeting intensities outlined in sidebar❸		
• 0'–6" clear height❷			❹
• 1'–0" clear height❷		p11–50, 11–52	❹
• 2'–0" clear height❷	p11–92, 11–94	p11–56, 11–58	❹
T5/HO/Striplights	High probability of meeting intensities outlined in sidebar❸		
• 0'–6" clear height❷	p11–84	p11–46, 11–48	❹
• 1'–0" clear height❷	p11–86, 11–88	p11–50, 11–52	❹
• 2'–0" clear height❷			❹
LongCompact/Striplights	High probability of meeting intensities outlined in sidebar❸		
• 0'–6" clear height❷	p11–84	p11–46, 11–48	❹
• 1'–0" clear height❷	p11–86, 11–88	p11–50, 11–52	❹
• 2'–0" clear height❷	p11–92, 11–94	p11–56	❹
All lamps/Asymmetric Luminaires	High probability of meeting intensities outlined in sidebar❸		
• 1'–0" clear height❷	p11–90	p11–54	❹

❶Ceiling height taken to the central ceiling height (highest ceiling portion within the cove itself).
❷Clear distance from the **top** of cove lip to the ceiling plane. Add inside cove depth and structural depth (of structure to support cove) to determine the total overall height from highest ceiling plane to lower ceiling plane.
❸Ceiling reflectance needs to be 0.9; wall reflectances need to be 0.5; floor reflectance needs to be 0.2. Lower reflectances will result in lower light intensities (and less energy efficient operation).
❹Ceiling height considered too low for such a large overhead cove opening.

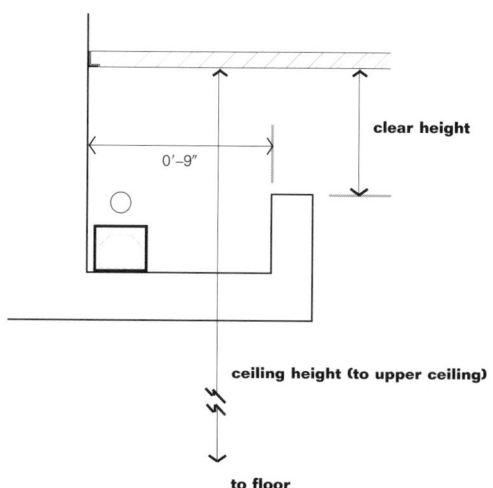

Using Tables 11.6 and 11.7: select a ceiling height (along top) and select desired lamp/luminaire combination or desired clear height of cove. Shaded boxes indicate reference pages of cove detail conceptsthat are likely to meet intensities outlined in sidebar below. Blank boxes indicate that no detail concepts were found in this lamp/luminaire configuration that would likely meet illuminances outlined in sidebar below.

Details Concepts and Effects

11

Recommended intensities
for ambient lighting from coves

Low (gallery, museum, residential)
5 to 10 fc average on floor
Moderate (hospitality)
10 to 20 fc average on floor
High (commercial—or higher for retail)
20 to 30 fc average on floor

These are intended to be average, maintained values and will vary based on actual task requirements. These are suggested intensities from cove lighting alone. Other lighting techniques (task lighting, accent lighting) will contribute additional light to an area.

©2000 Howard Grey/Tony Stone Images

Details Cove Lighting Selection Guide

Cove Lighting

DetailConcepts™

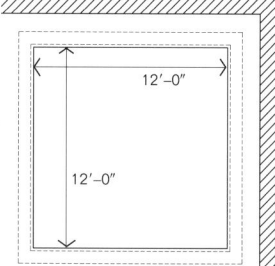

Table 11.8 Cove Detail Concepts/Gallery/Residential

- 12'x12' cove opening

◀ C e i l i n g H e i g h t ❶ ▶

Lamp/Luminaire	12'	10'	8'
T5/T8/Striplights	High probability of meeting intensities outlined in sidebar ❸		
• 0'–6" clear height ❷			❹
• 1'–0" clear height ❷		p11–50, 11–52	❹
• 2'–0" clear height ❷	p11–92, 11–94	p11–56, 11–58	❹
T5/HO/Striplights	High probability of meeting intensities outlined in sidebar ❸		
• 0'–6" clear height ❷	p11–84	p11–46, 11–48	❹
• 1'–0" clear height ❷	p11–86, 11–88	p11–50, 11–52	❹
• 2'–0" clear height ❷	p11–92, 11–94	p11–56, 11–58	❹
LongCompact/Striplights	High probability of meeting intensities outlined in sidebar ❸		
• 0'–6" clear height ❷	p11–84	p11–46, 11–48	❹
• 1'–0" clear height ❷	p11–86, 11–88	p11–50, 11–52	❹
• 2'–0" clear height ❷	p11–92, 11–94	p11–56	❹
All lamps/Asymmetric Luminaires	High probability of meeting intensities outlined in sidebar ❸		
• 1'–0" clear height ❷	p11–90	p11–54	❹

❶Ceiling height taken to the central ceiling height (highest ceiling portion within the cove itself).
❷Clear distance from the **top** of cove lip to the ceiling plane. Add inside cove depth and structural depth (of structure to support cove) to determine the total overall height from highest ceiling plane to lower ceiling plane.
❸Ceiling reflectance needs to be 0.9; wall reflectances need to be 0.5; floor reflectance needs to be 0.2. Lower reflectances will result in lower light intensities (and less energy efficient operation).
❹Ceiling height considered too low for such a large overhead cove opening.

Design Tips

✓Fluorescent coves in gallery and residential applications are most successful when using dimmable ballasts (e.g., ESI or Lutron) or using low output ballasts. All cove data shown here based on full output ballasts (ballast factor of 0.98).
✓Ballast factor of 0.78 will reduce reported intensities herein by 22 percent.
✓Dimmable ballasts will permit output as low as 5 percent or so of reported intensities herein.

Details Concepts and Effects

11

Cost magnitude²⁰⁰⁰
per linear foot of fluorescent luminaire

T8 Striplights
US$10./ft to US$12./ft
T5 and T5/HO Striplights
US$15./ft to US$20./ft
LongCompact Striplights
US$20./ft to US$25./ft
Asymmetric Luminaires
US$25./ft to US$35./ft

Cost magnitude for hardware only (luminaires, lamps and ballasts—with full output ballasts), excluding drywall/ structure/millwork work and installation costs. For dimming ballasts, add US$15./ ft to US$50./ft depending on brand and/ or performance. Costs vary based on quantities/lengths, distributor and contractor markups, market conditions, manufacturing situations and annual inflation. These values are for preliminary, magnitude budgeting and do not represent quotes nor actual final pricing to client.

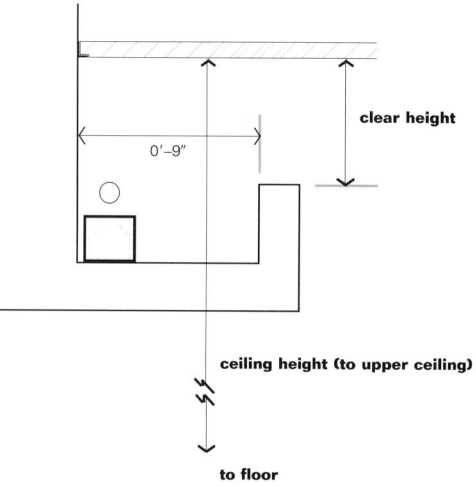

Details Cove Lighting Selection Guide

Cove Lighting

D e t a i l C o n c e p t s ™

Table 11.9 Cove Detail Concepts/Hospitality

• 12'x12' cove opening

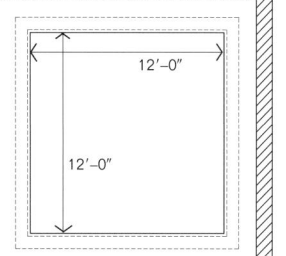

Lamp/Luminaire	◀ Ceiling Height ❶ ▶		
	12'	**10'**	**8'**
T5/T8/Striplights	High probability of meeting intensities outlined in sidebar ❸		
• 0'–6" clear height ❷	p11–82, 11–84	p11–46, 11–48	❹
• 1'–0" clear height ❷	p11–86, 11–88		❹
• 2'–0" clear height ❷			❹
T5/HO/Striplights	High probability of meeting intensities outlined in sidebar ❸		
• 0'–6" clear height ❷	p11–82		❹
• 1'–0" clear height ❷			❹
• 2'–0" clear height ❷			❹
LongCompact/Striplights	High probability of meeting intensities outlined in sidebar ❸		
• 0'–6" clear height ❷	p11–82		❹
• 1'–0" clear height ❷			❹
• 2'–0" clear height ❷			❹
All lamps/Asymmetric Luminaires	High probability of meeting intensities outlined in sidebar ❸		
• 1'–0" clear height ❷			❹

❶ Ceiling height taken to the central ceiling height (highest ceiling portion within the cove itself).
❷ Clear distance from the **top** of cove lip to the ceiling plane. Add inside cove depth and structural depth (of structure to support cove) to determine the total overall height from highest ceiling plane to lower ceiling plane.
❸ Ceiling reflectance needs to be 0.9; wall reflectances need to be 0.5; floor reflectance needs to be 0.2. Lower reflectances will result in lower light intensities (and less energy efficient operation).
❹ Ceiling height considered too low for such a large overhead cove opening.

Using Tables 11.8 and 11.9: select a ceiling height (along top) and select desired lamp/ luminaire combination or desired clear height of cove. Shaded boxes indicate reference pages of cove detail conceptsthat are likely to meet intensities outlined in sidebar below. Blank boxes indicate that no detail concepts were found in this lamp/luminaire configuration that would likely meet illuminances outlined in sidebar below.

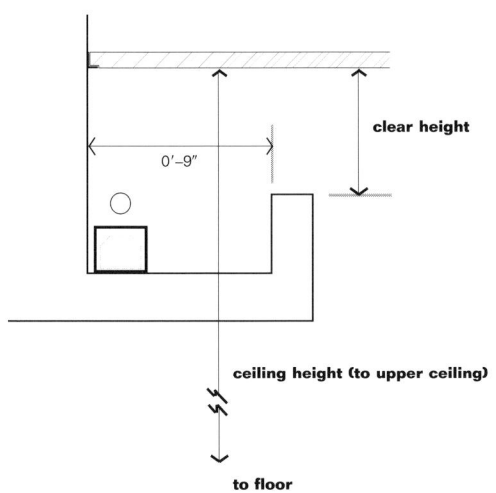

clear height

0'–9"

ceiling height (to upper ceiling)

to floor

Recommended intensities
for ambient lighting from coves

Low (gallery, museum, residential)
5 to 10 fc average on floor
Moderate (hospitality)
10 to 20 fc average on floor
High (commercial—or higher for retail)
20 to 30 fc average on floor

These are intended to be average, maintained values and will vary based on actual task requirements. These are suggested intensities from cove lighting alone. Other lighting techniques (task lighting, accent lighting) will contribute additional light to an area.

Coves Striplights

8′ Ceiling Height/4′ x 12′ Cove Opening/0′–6″ Clear Cove Height

DetailConcepts™

Performance Perspectives

[All data outlined here is based on information from boldfaced manufacturer's published data at time of manuscript preparation. Cove opening located in room corner and starts 2′ from each of two walls. Maintained values based on analysis grid 7′x15′ centered under cove and on 0.86 light loss factor (0.95 lamp lumen depreciation; 0.95 luminaire dirt depreciation; 0.95 room surface dirt depreciation); ballast factor of 0.98 and very light reflectances: ceiling 90%; wall 50%; and floor 20%. All views put eyes at 5′–6″ AFF—cove perspective varies as ceiling height changes and/or cove height changes.]

▶ **Cove height: 0′–6″ clear**
Central ceiling height: 8′
Perimeter ceiling height: 7′±
Lamp: F28T5/830 or F32T8/830
Floor illuminance (7′x15′ area under cove):
▶ 16 fc, avg, maintained (29 fc max)

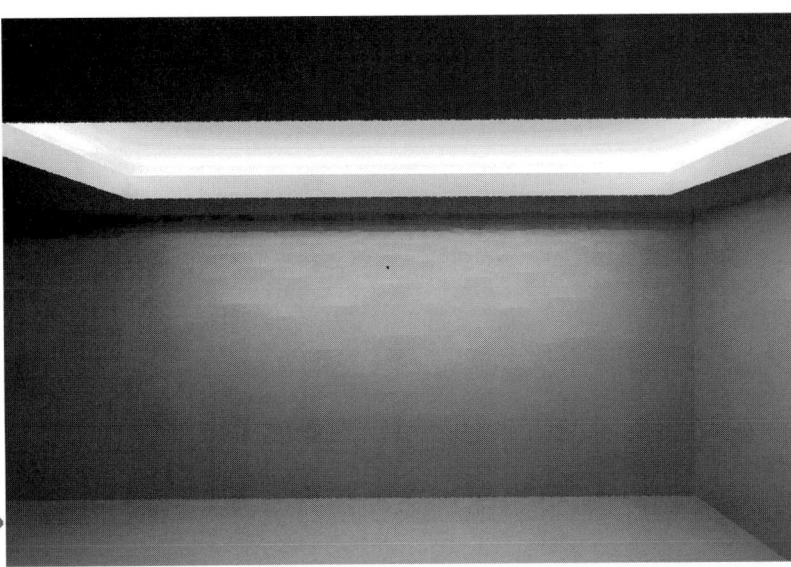

▶ **Cove height: 0′–6″ clear**
Central ceiling height: 8′
Perimeter ceiling height: 7′±
Lamp: F54T5/HO/830
Floor illuminance (7′x15′ area under cove):
▶ 27 fc, avg, maintained (42 fc max)

▶ **Cove height: 0′–6″ clear**
Central ceiling height: 8′
Perimeter ceiling height: 7′±
Lamp: F40Long/830
Floor illuminance (7′x15′ area under cove):
▶ 31 fc, avg, maintained (49 fc max)

Details
Concepts and Effects

11

Coves Striplights

8' Ceiling Height/4' x 12' Cove Opening/0'–6" Clear Cove Height

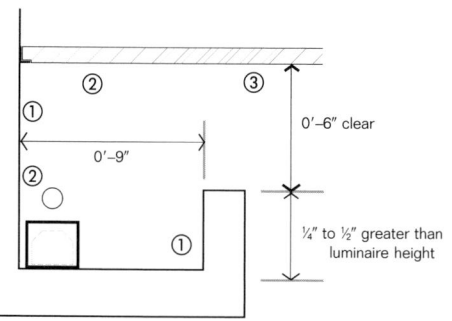

0'–9"

0'–6" clear

¼" to ½" greater than luminaire height

T5 and T8 Striplights**Back of Cove**

⊃Possible vendors: Bartco, Columbia, DayBrite, Legion, Lightolier, **Lithonia**[1], Metalux

[1]T12 photometry was used and rerated for T8 lamping. This is somewhat imprecise.

Notes
①Paint all interior cove surfaces matte white.
②Socket shadows are noticeable along back cove wall and somewhat noticeable at ceiling with 1–lamp striplight in this location. Single–lamp staggered strips introduce a "reverse socket shadow"—lamps closer to back cove wall create a slightly greater intensity. Butt strips end–to–end tightly.
③Ceiling reflectance significantly influences success of cove. Higher is better (preferably 90 percent).
④T8 and T5 lamps offer same light output—so striplights can be either version. Lamps must be combined with correct ballast and are not interchangeable after installation .T5 striplights are a few inches shorter than T8s. See below for T5/HO version.

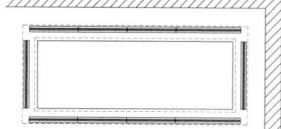
DetailConcepts™

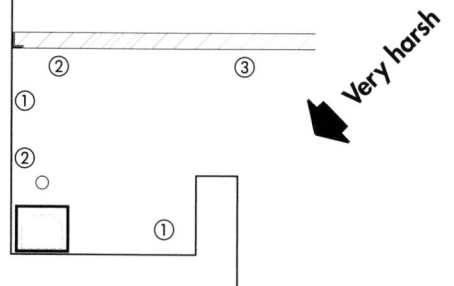

Very harsh

T5/HO Striplights**Back of Cove**

⊃Possible vendors: Bartco, Legion, Lightolier, **Lithonia**[1]

[1]T12 photometry was used and rerated for T5/HO lamping. This is somewhat imprecise.

Notes
①Paint all interior cove surfaces matte white.
②Socket shadows are noticeable along back cove wall and somewhat noticeable at ceiling with 1–lamp striplight in this location. Single–lamp staggered strips introduce a "reverse socket shadow"—lamps closer to back cove wall create a slightly greater intensity. Butt strips end–to–end tightly.
③Ceiling reflectance significantly influences success of cove. Higher is better (preferably 90 percent).

Very harsh

LongCompact Striplights**Back of Cove**

⊃Possible vendors: Columbia, DayBrite, Legion, Lightolier, **Lithonia**[1], Metalux

[1]T12 photometry was used and rerated for LongCompact lamping. This is somewhat imprecise.

Notes
①Paint all interior cove surfaces matte white.
②Socket shadows are less noticeable along back cove wall and at ceiling with LongCompact lamps in striplight in this location.
③Ceiling reflectance significantly influences success of cove. Higher is better (preferably 90 percent).

Design Tips
✔Lamp and ballast maintenance is difficult. Lay–in ceiling can help—lift tiles for better reach/view into cove.
✔Heavily textured ceiling material will create significant shadows; as will tegular tile.
✔If ceiling is drywall, then any imperfections will telegraph with such grazing light.
✔If ceiling tone is other than white, then consider painting interior of cove surfaces to match ceiling.
✔Corner shadowing less noticeable with strips at front of cove.
✔Clear cove heights greater than 0'–6" are more efficient.
✔Coves using asymmetric luminaires specifically designed for cove applications are more efficient.

Details Concepts and Effects 11

Application Key

Commercial
Gallery
Hospitality[1]
Institutional
Manufacturing
Residential
Retail
Exterior

1/includes casual commercial and institutional applications and decorative retail applications

bold = primary application
partial fade = minimal application
fade = unlikely application

Net Addresses/Luminaires
See page 11–5

CONNECT FOR MORE

©2000 Howard Grey/Tony Stone Images

Coves Striplights

8' Ceiling Height/4' x 12' Cove Opening/0'–6" Clear Cove Height

DetailConcepts™

Performance Perspectives

[All data outlined here is based on information from boldfaced manufacturer's published data at time of manuscript preparation. Cove opening located in room corner and starts 2' from each of two walls. Maintained values based on analysis grid 7'x15' centered under cove and on 0.86 light loss factor (0.95 lamp lumen depreciation; 0.95 luminaire dirt depreciation; 0.95 room surface dirt depreciation); ballast factor of 0.98 and very light reflectances: ceiling 90%; wall 50%; and floor 20%. All views put eyes at 5'–6" AFF—cove perspective varies as ceiling height changes and/or cove height changes.]

▶ **Cove height: 0'–6" clear**
Central ceiling height: 8'
Perimeter ceiling height: 7'±
Lamp: F28T5/830 or F32T8/830
Floor illuminance (7'x15' area under cove):
▶ 22 fc, avg, maintained (37 fc max)

▶ **Cove height: 0'–6" clear**
Central ceiling height: 8'
Perimeter ceiling height: 7'±
Lamp: F54T5/HO/830
Floor illuminance (7'x15' area under cove):
▶ 37 fc, avg, maintained (57 fc max)

▶ **Cove height: 0'–6" clear**
Central ceiling height: 8'
Perimeter ceiling height: 7'±
Lamp: F40Long/830
Floor illuminance (7'x15' area under cove):
▶ 44 fc, avg, maintained (72 fc max)

Details Concepts and Effects

11

Coves Striplights

8' Ceiling Height/4' x 12' Cove Opening/`0'–6" Clear Cove Height`

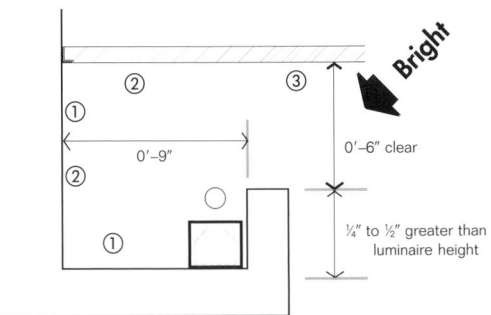

DetailConcepts™

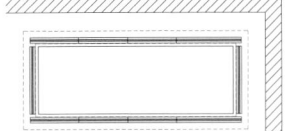

Notes
①Paint all interior cove surfaces matte white.
②Socket shadows are less noticeable with 1–lamp striplight in this location. Shadow line on ceiling may be visible a foot or so out from front edge of cove lip, however.
③Ceiling reflectance significantly influences success of cove. Higher is better (preferably 90 percent).
④T8 and T5 lamps offer same light output—so striplights can be either version. Lamps must be combined with correct ballast and are not interchangeable after installation. See below for T5/HO version.

Bright

0'–9"

0'–6" clear

¼" to ½" greater than luminaire height

T5 and T8 Striplights **Front of Cove**
⊃Possible vendors: Bartco, Columbia, DayBrite, Legion, Lightolier, **Lithonia**[1], Metalux
[1]T12 photometry was used and rerated for T8 lamping. This is somewhat imprecise.

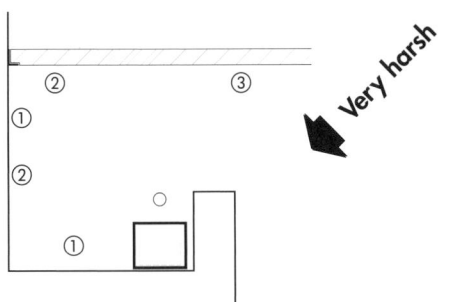

Notes
①Paint all interior cove surfaces matte white.
②Socket shadows are less noticeable with 1–lamp striplight in this location. Shadow line on ceiling may be visible a foot or so out from front edge of cove lip, however.
③Ceiling reflectance significantly influences success of cove. Higher is better (preferably 90 percent).

Very harsh

T5/HO Striplights **Front of Cove**
⊃Possible vendors: Bartco, Legion, Lightolier, **Lithonia**[1]
[1]T12 photometry was used and rerated for T5/HO lamping. This is somewhat imprecise.

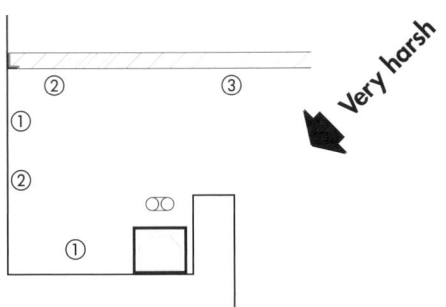

Notes
①Paint all interior cove surfaces matte white.
②Socket shadows are not noticeable with LongCompact lamps in striplight in this location.
③Ceiling reflectance significantly influences success of cove. Higher is better (preferably 90 percent).

Very harsh

LongCompact Striplights **Front of Cove**
⊃Possible vendors: Columbia, DayBrite, Legion, Lightolier, **Lithonia**[1], Metalux
[1]T12 photometry was used and rerated for LongCompact lamping. This is somewhat imprecise.

Design Tips
✔Lamp and ballast maintenance is difficult. Lay–in ceiling can help—lift tiles for better reach/view into cove.
✔Heavily textured ceiling material will create significant shadows; as will tegular tile.
✔If ceiling is drywall, then any imperfections will telegraph with such grazing light.
✔If ceiling tone is other than white, then consider painting interior of cove surfaces to match ceiling.
✔Clear cove heights greater than 0'–6" are more efficient.
✔Coves using asymmetric luminaires specifically designed for cove applications are more efficient.

Details
Concepts and Effects

11

Application Key

Commercial
Gallery
Hospitality[1]
Institutional
Manufacturing
Residential
Retail
Exterior

1/includes casual commercial and institutional applications and decorative retail applications

bold = primary application
partial fade = minimal application
fade = unlikely application

Net Addresses/Luminaires
See page 11–5

CONNECT FOR MORE

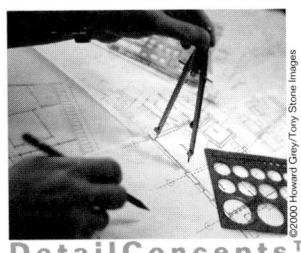

Coves Striplights

8′ Ceiling Height/9′ x 9′ Cove Opening/0′–6″ Clear Cove Height

DetailConcepts™

Performance Perspectives

[All data outlined here is based on information from boldfaced manufacturer's published data at time of manuscript preparation. Cove opening located in room corner and starts 2′ from each of two walls. Maintained values based on analysis grid 12′x12′ centered under cove and on 0.86 light loss factor (0.95 lamp lumen depreciation; 0.95 luminaire dirt depreciation; 0.95 room surface dirt depreciation); ballast factor of 0.98 and very light reflectances: ceiling 90%; wall 50%; and floor 20%. All views put eyes at 5′–6″ AFF—cove perspective varies as ceiling height changes and/or cove height changes.]

▶ **Cove height: 0′–6″ clear**
Central ceiling height: 8′
Perimeter ceiling height: 7′±
Lamp: F28T5/830 or F32T8/830
Floor illuminance (12′x12′ area under cove)
▶ 15 fc, avg, maintained (27 fc max)

▶ **Cove height: 0′–6″ clear**
Central ceiling height: 8′
Perimeter ceiling height: 7′±
Lamp: F54T5/HO/830
Floor illuminance (12′x12′ area under cove)
▶ 27 fc, avg, maintained (50 fc max)

▶ **Cove height: 0′–6″ clear**
Central ceiling height: 8′
Perimeter ceiling height: 7′±
Lamp: F40Long/830
Floor illuminance (12′x12′ area under cove)
▶ 32 fc, avg, maintained (54 fc max)

Details
Concepts and Effects

11

Coves Striplights

8' Ceiling Height/9' x 9' Cove Opening/0'–6" Clear Cove Height

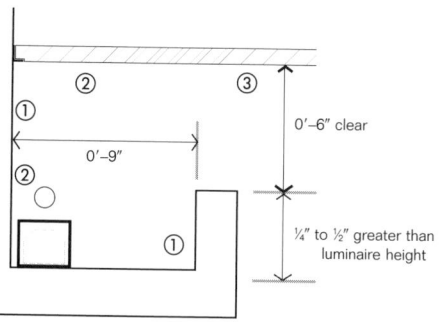

0'–6" clear

0'–9"

¼" to ½" greater than
luminaire height

DetailConcepts™

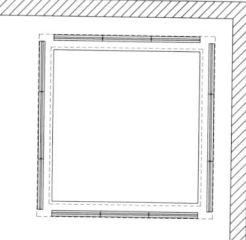

Notes

①Paint all interior cove surfaces matte white.
②Socket shadows are noticeable along back cove wall and somewhat noticeable at ceiling with 1–lamp striplight in this location. Single–lamp staggered strips introduce a "reverse socket shadow"—lamps closer to back cove wall create a slightly greater intensity. Butt strips end–to–end tightly.
③Ceiling reflectance significantly influences success of cove. Higher is better (preferably 90 percent).
④T8 and T5 lamps offer same light output—so striplights can be either version. Lamps must be combined with correct ballast and are not interchangeable after installation . T5 striplights are a few inches shorter than T8s. See below for T5/HO version.

T5 and T8 Striplights**Back of Cove**

⊃Possible vendors: Bartco, Columbia, DayBrite, Legion, Lightolier, **Lithonia**[1], Metalux

[1]T12 photometry was used and rerated for T8 lamping. This is somewhat imprecise.

Very harsh

Notes

①Paint all interior cove surfaces matte white.
②Socket shadows are noticeable along back cove wall and somewhat noticeable at ceiling with 1–lamp striplight in this location. Single–lamp staggered strips introduce a "reverse socket shadow"—lamps closer to back cove wall create a slightly greater intensity. Butt strips end–to–end tightly.
③Ceiling reflectance significantly influences success of cove. Higher is better (preferably 90 percent).

T5/HO Striplights**Back of Cove**

⊃Possible vendors: Bartco, Legion, Lightolier, **Lithonia**[1]

[1]T12 photometry was used and rerated for T5/HO lamping. This is somewhat imprecise.

Very harsh

Notes

①Paint all interior cove surfaces matte white.
②Socket shadows are less noticeable along back cove wall and at ceiling with LongCompact lamps in striplight in this location.
③Ceiling reflectance significantly influences success of cove. Higher is better (preferably 90 percent).

LongCompact Striplights**Back of Cove**

⊃Possible vendors: Columbia, DayBrite, Legion, Lightolier, **Lithonia**[1], Metalux

[1]T12 photometry was used and rerated for LongCompact lamping. This is somewhat imprecise.

Design Tips

✔Lamp and ballast maintenance is difficult. Lay–in ceiling can help—lift tiles for better reach/view into cove.

✔Heavily textured ceiling material will create significant shadows; as will tegular tile.

✔If ceiling is drywall, then any imperfections will telegraph with such grazing light.

✔If ceiling tone is other than white, then consider painting interior of cove surfaces to match ceiling.

✔Corner shadowing less noticeable with strips at front of cove.

✔Clear cove heights greater than 0'–6" are more efficient.

✔Coves using asymmetric luminaires specifically designed for cove applications are more efficient.

Details Concepts and Effects

11

Application Key

Commercial
Gallery
Hospitality[1]
Institutional
Manufacturing
Residential
Retail
Exterior

1/includes casual commercial and institutional applications and decorative retail applications

bold = primary application
partial fade = minimal application
fade = unlikely application

Net Addresses/Luminaires
See page 11–5

CONNECT FOR MORE

DetailConcepts™

Coves striplights

8′ Ceiling Height/9′ x 9′ Cove Opening/0′–6″ Clear Cove Height

Performance Perspectives

[All data outlined here is based on information from boldfaced manufacturer's published data at time of manuscript preparation. Cove opening located in room corner and starts 2′ from each of two walls. Maintained values based on analysis grid 12′x12′ centered under cove and on 0.86 light loss factor (0.95 lamp lumen depreciation; 0.95 luminaire dirt depreciation; 0.95 room surface dirt depreciation); ballast factor of 0.98 and very light reflectances: ceiling 90%; wall 50%; and floor 20%. All views put eyes at 5′–6″ AFF—cove perspective varies as ceiling height changes and/or cove height changes.]

▶ **Cove height: 0′–6″ clear**
Central ceiling height: 8′
Perimeter ceiling height: 7′±
Lamp: F28T5/830 or F32T8/830
Floor illuminance (12′x12′ area under cove):
▶ 21 fc, avg, maintained (40 fc max)

▶ **Cove height: 0′–6″ clear**
Central ceiling height: 8′
Perimeter ceiling height: 7′±
Lamp: F54T5/HO/830
Floor illuminance (12′x12′ area under cove):
▶ 36 fc, avg, maintained (67 fc max)

▶ **Cove height: 0′–6″ clear**
Central ceiling height: 8′
Perimeter ceiling height: 7′±
Lamp: F40Long/830
Floor illuminance (12′x12′ area under cove):
▶ 45 fc, avg, maintained (83 fc max)

Details Concepts and Effects

11

Coves Striplights

8′ Ceiling Height/9′ x 9′ Cove Opening/0′–6″ Clear Cove Height

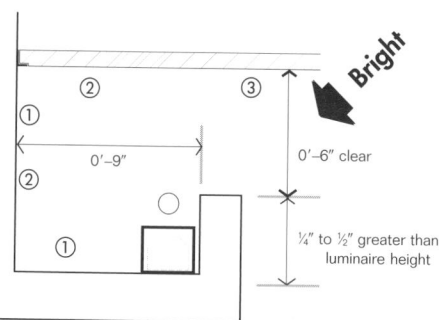

Bright

Notes
① Paint all interior cove surfaces matte white.
② Socket shadows are less noticeable with 1-lamp striplight in this location. Shadow line on ceiling may be visible a foot or so out from front edge of cove lip, however.
③ Ceiling reflectance significantly influences success of cove. Higher is better (preferably 90 percent).
④ T8 and T5 lamps offer same light output—so striplights can be either version. Lamps must be combined with correct ballast and are not interchangeable after installation. See below for T5/HO version.

DetailConcepts™

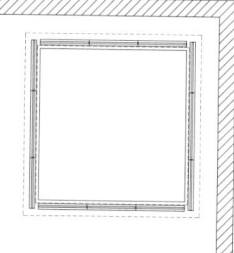

T5 and T8 Striplights**Front of Cove**
➲ Possible vendors: Bartco, Columbia, DayBrite, Legion, Lightolier, **Lithonia**[1], Metalux
[1] T12 photometry was used and rerated for T8 lamping. This is somewhat imprecise.

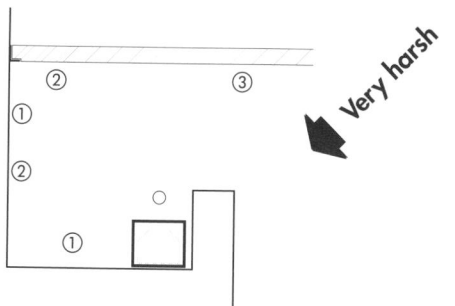

Very harsh

Notes
① Paint all interior cove surfaces matte white.
② Socket shadows are less noticeable with 1-lamp striplight in this location. Shadow line on ceiling may be visible a foot or so out from front edge of cove lip, however.
③ Ceiling reflectance significantly influences success of cove. Higher is better (preferably 90 percent).

Design Tips
✔ Lamp and ballast maintenance is difficult. Lay–in ceiling can help—lift tiles for better reach/ view into cove.
✔ Heavily textured ceiling material will create significant shadows; as will tegular tile.
✔ If ceiling is drywall, then any imperfections will telegraph with such grazing light.
✔ If ceiling tone is other than white, then consider painting interior of cove surfaces to match ceiling.
✔ Clear cove heights greater than 0′–6″ are more efficient.
✔ Coves using asymmetric luminaires specifically designed for cove applications are more efficient.

T5/HO Striplights**Front of Cove**
➲ Possible vendors: Bartco, Legion, Lightolier, **Lithonia**[1]
[1] T12 photometry was used and rerated for T5/HO lamping. This is somewhat imprecise.

Details Concepts and Effects

11

Application Key

Commercial
Gallery
Hospitality[1]
Institutional
Manufacturing
Residential
Retail
Exterior

1/includes casual commercial and institutional applications and decorative retail applications

bold = primary application
partial fade = minimal application
fade = unlikely application

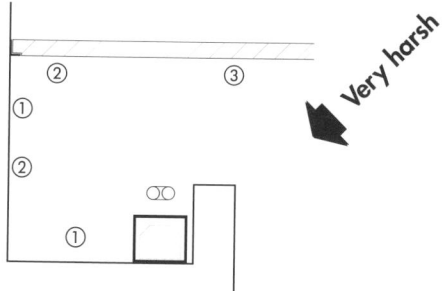

Very harsh

Notes
① Paint all interior cove surfaces matte white.
② Socket shadows are not noticeable with LongCompact lamps in striplight in this location.
③ Ceiling reflectance significantly influences success of cove. Higher is better (preferably 90 percent).

LongCompact Striplights**Front of Cove**
➲ Possible vendors: Columbia, DayBrite, Legion, Lightolier, **Lithonia**[1], Metalux
[1] T12 photometry was used and rerated for LongCompact lamping. This is somewhat imprecise.

Net Addresses/Luminaires
See page 11–5

CONNECT FOR MORE

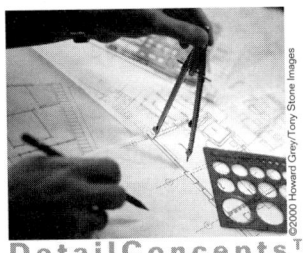

DetailConcepts™

Coves Striplights

10′ Ceiling Height/4′ x 12′ Cove Opening/0′–6″ Clear Cove Height

Performance Perspectives

[All data outlined here is based on information from boldfaced manufacturer's published data at time of manuscript preparation. Cove opening located in room corner and starts 2′ from each of two walls. Maintained values based on analysis grid 7′x15′ centered under cove and on 0.86 light loss factor (0.95 lamp lumen depreciation; 0.95 luminaire dirt depreciation; 0.95 room surface dirt depreciation); ballast factor of 0.98 and very light reflectances: ceiling 90%; wall 50%; and floor 20%. All views put eyes at 5′–6″ AFF—cove perspective varies as ceiling height changes and/or cove height changes.]

▶ **Cove height: 0′–6″ clear**
Central ceiling height: 10′
Perimeter ceiling height: 9′±
Lamp: F28T5/830 or F32T8/830
Floor illuminance (7′x15′ area under cove):
  ▶ 13 fc, avg, maintained (20 fc max)

▶ **Cove height: 0′–6″ clear**
Central ceiling height: 10′
Perimeter ceiling height: 9′±
Lamp: F54T5/HO/830
Floor illuminance (7′x15′ area under cove):
  ▶ 20 fc, avg, maintained (31 fc max)

▶ **Cove height: 0′–6″ clear**
Central ceiling height: 10′
Perimeter ceiling height: 9′±
Lamp: F40Long/830
Floor illuminance (7′x15′ area under cove):
  ▶ 25 fc, avg, maintained (41 fc max)

Coves Striplights

10' Ceiling Height/4' x 12' Cove Opening/0'–6" Clear Cove Height

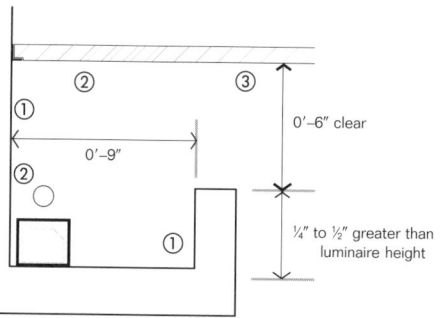

0'–6" clear

0'–9"

¼" to ½" greater than luminaire height

T5 and T8 Striplights**Back of Cove**

↪Possible vendors: Bartco, Columbia, DayBrite, Legion, Lightolier, **Lithonia**[1], Metalux

[1]T12 photometry was used and rerated for T8 lamping. This is somewhat imprecise.

Notes
①Paint all interior cove surfaces matte white.
②Socket shadows are noticeable along back cove wall and somewhat noticeable at ceiling with 1–lamp striplight in this location. Single–lamp staggered strips introduce a "reverse socket shadow"—lamps closer to back cove wall create a slightly greater intensity. Butt strips end–to–end tightly.
③Ceiling reflectance significantly influences success of cove. Higher is better (preferably 90 percent).
④T8 and T5 lamps offer same light output—so striplights can be either version. Lamps must be combined with correct ballast and are not interchangeable after installation .T5 striplights are a few inches shorter than T8s. See below for T5/HO version.

D e t a i l C o n c e p t s™

©2000 Howard Grey/Tony Stone Images

Very harsh

Notes
①Paint all interior cove surfaces matte white.
②Socket shadows are noticeable along back cove wall and somewhat noticeable at ceiling with 1–lamp striplight in this location. Single–lamp staggered strips introduce a "reverse socket shadow"—lamps closer to back cove wall create a slightly greater intensity. Butt strips end–to–end tightly.
③Ceiling reflectance significantly influences success of cove. Higher is better (preferably 90 percent).

T5/HO Striplights**Back of Cove**

↪Possible vendors: Bartco, Legion, Lightolier, **Lithonia**[1]

[1]T12 photometry was used and rerated for T5/HO lamping. This is somewhat imprecise.

Design Tips
✔Lamp and ballast maintenance is difficult. Lay–in ceiling can help—lift tiles for better reach/view into cove.
✔Heavily textured ceiling material will create significant shadows; as will tegular tile.
✔If ceiling is drywall, then any imperfections will telegraph with such grazing light.
✔If ceiling tone is other than white, then consider painting interior of cove surfaces to match ceiling.
✔Corner shadowing less noticeable with strips at front of cove.
✔Clear cove heights greater than 0'–6" are more efficient.
✔Coves using asymmetric luminaires specifically designed for cove applications are more efficient.

Details Concepts and Effects **11**

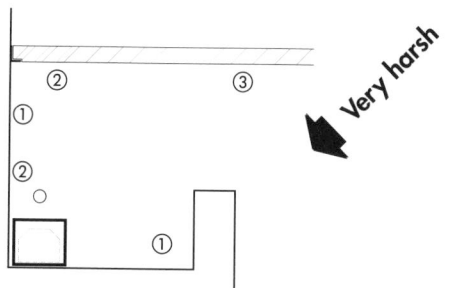

Very harsh

Notes
①Paint all interior cove surfaces matte white.
②Socket shadows are less noticeable along back cove wall and at ceiling with LongCompact lamps in striplight in this location.
③Ceiling reflectance significantly influences success of cove. Higher is better (preferably 90 percent).

LongCompact Striplights**Back of Cove**

↪Possible vendors: Columbia, DayBrite, Legion, Lightolier, **Lithonia**[1], Metalux

[1]T12 photometry was used and rerated for LongCompact lamping. This is somewhat imprecise.

Application Key

Commercial
Gallery
Hospitality[1]
Institutional
Manufacturing
Residential
Retail
Exterior

1/includes casual commercial and institutional applications and decorative retail applications

bold = primary application
partial fade = minimal application
fade = unlikely application

Net Addresses/Luminaires
See page 11–5

CONNECT FOR MORE

©2000 Howard Grey/Tony Stone Images

DetailConcepts™

Coves Striplights

10′ Ceiling Height/4′ x 12′ Cove Opening/0′–6″ Clear Cove Height

Performance Perspectives

[All data outlined here is based on information from boldfaced manufacturer's published data at time of manuscript preparation. Cove opening located in room corner and starts 2′ from each of two walls. Maintained values based on analysis grid 7′x15′ centered under cove and on 0.86 light loss factor (0.95 lamp lumen depreciation; 0.95 luminaire dirt depreciation; 0.95 room surface dirt depreciation); ballast factor of 0.98 and very light reflectances: ceiling 90%; wall 50%; and floor 20%. All views put eyes at 5′–6″ AFF—cove perspective varies as ceiling height changes and/or cove height changes.]

▶ **Cove height: 0′–6″ clear**
Central ceiling height: 10′
Perimeter ceiling height: 9′±
Lamp: F28T5/830 or F32T8/830
Floor illuminance (7′x15′ area under cove):
▶ 18 fc, avg, maintained (31 fc max)

▶ **Cove height: 0′–6″ clear**
Central ceiling height: 10′
Perimeter ceiling height: 9′±
Lamp: F54T5/HO/830
Floor illuminance (7′x15′ area under cove):
▶ 28 fc, avg, maintained (42 fc max)

▶ **Cove height: 0′–6″ clear**
Central ceiling height: 10′
Perimeter ceiling height: 9′±
Lamp: F40Long/830
Floor illuminance (7′x15′ area under cove):
▶ 35 fc, avg, maintained (66 fc max)

Details Concepts and Effects

11

Coves Striplights
10' Ceiling Height/4' x 12' Cove Opening/0'–6" Clear Cove Height

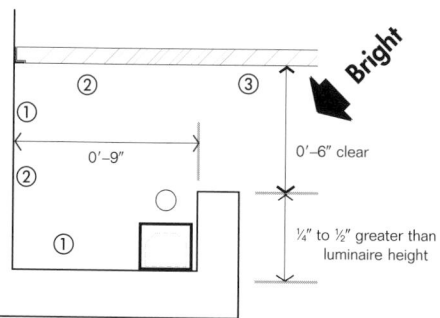

DetailConcepts™

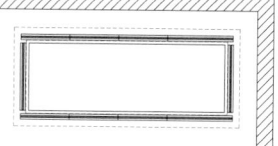

Notes
①Paint all interior cove surfaces matte white.
②Socket shadows are less noticeable with 1–lamp striplight in this location. Shadow line on ceiling may be visible a foot or so out from front edge of cove lip, however.
③Ceiling reflectance significantly influences success of cove. Higher is better (preferably 90 percent).
④T8 and T5 lamps offer same light output—so striplights can be either version. Lamps must be combined with correct ballast and are not interchangeable after installation. See below for T5/HO version.

T5 and T8 Striplights **Front of Cove**
⟳Possible vendors: Bartco, Columbia, DayBrite, Legion, Lightolier, **Lithonia**[1], Metalux
[1]T12 photometry was used and rerated for T8 lamping. This is somewhat imprecise.

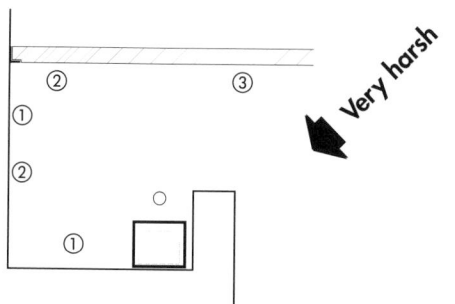

Notes
①Paint all interior cove surfaces matte white.
②Socket shadows are less noticeable with 1–lamp striplight in this location. Shadow line on ceiling may be visible a foot or so out from front edge of cove lip, however.
③Ceiling reflectance significantly influences success of cove. Higher is better (preferably 90 percent).

T5/HO Striplights **Front of Cove**
⟳Possible vendors: Bartco, Legion, Lightolier, **Lithonia**[1]
[1]T12 photometry was used and rerated for T5/HO lamping. This is somewhat imprecise.

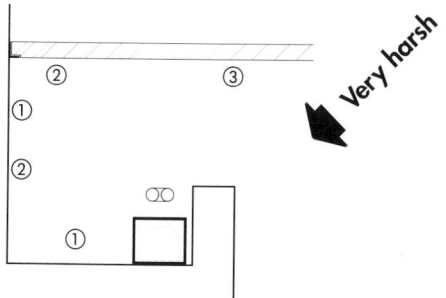

Notes
①Paint all interior cove surfaces matte white.
②Socket shadows are not noticeable with LongCompact lamps in striplight in this location.
③Ceiling reflectance significantly influences success of cove. Higher is better (preferably 90 percent).

LongCompact Striplights **Front of Cove**
⟳Possible vendors: Columbia, DayBrite, Legion, Lightolier, **Lithonia**[1], Metalux
[1]T12 photometry was used and rerated for LongCompact lamping. This is somewhat imprecise.

Design Tips
✔Lamp and ballast maintenance is difficult. Lay–in ceiling can help—lift tiles for better reach/view into cove.
✔Heavily textured ceiling material will create significant shadows; as will tegular tile.
✔If ceiling is drywall, then any imperfections will telegraph with such grazing light.
✔If ceiling tone is other than white, then consider painting interior of cove surfaces to match ceiling.
✔Clear cove heights greater than 0'–6" are more efficient.
✔Coves using asymmetric luminaires specifically designed for cove applications are more efficient.

Details Concepts and Effects **11**

Application Key
Commercial
Gallery
Hospitality[1]
Institutional
Manufacturing
Residential
Retail
Exterior

1/includes casual commercial and institutional applications and decorative retail applications

bold = primary application
partial fade = minimal application
fade = unlikely application

Net Addresses/Luminaires
See page 11–5

CONNECT FOR MORE

Coves Striplights

10' Ceiling Height/4' x 12' Cove Opening/1'–0" Clear Cove Height

DetailConcepts™

Performance Perspectives

[All data outlined here is based on information from boldfaced manufacturer's published data at time of manuscript preparation. Cove opening located in room corner and starts 2' from each of two walls. Maintained values based on analysis grid 7'x15' centered under cove and on 0.86 light loss factor (0.95 lamp lumen depreciation; 0.95 luminaire dirt depreciation; 0.95 room surface dirt depreciation); ballast factor of 0.98 and very light reflectances: ceiling 90%; wall 50%; and floor 20%. All views put eyes at 5'–6" AFF—cove perspective varies as ceiling height changes and/or cove height changes.]

▶ **Cove height: 1'–0" clear**
Central ceiling height: 10'
Perimeter ceiling height: 8'–6"±
Lamp: F28T5/830 or F32T8/830
Floor illuminance (7'x15' area under cove):
▶ 20 fc, avg, maintained (33 fc max)

▶ **Cove height: 1'–0" clear**
Central ceiling height: 10'
Perimeter ceiling height: 8'–6"±
Lamp: F54T5/HO/830
Floor illuminance (7'x15' area under cove):
▶ 32 fc, avg, maintained (51 fc max)

▶ **Cove height: 1'–0" clear**
Central ceiling height: 10'
Perimeter ceiling height: 8'–6"±
Lamp: F40Long/830
Floor illuminance (7'x15' area under cove):
▶ 39 fc, avg, maintained (54 fc max)

Details Concepts and Effects

11

Coves Striplights

10′ Ceiling Height/4′ x 12′ Cove Opening/ 1′–0″ Clear Cove Height

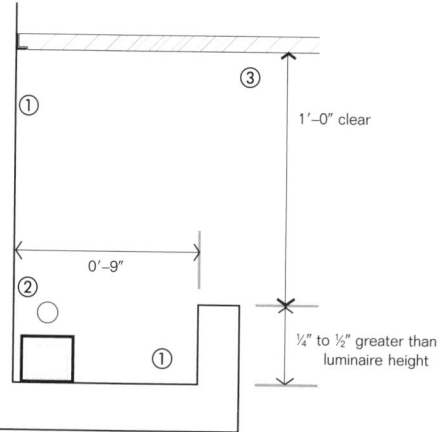

DetailConcepts™

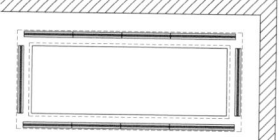

Notes
① Paint all interior cove surfaces matte white.
② Socket shadows are noticeable along back cove wall with 1-lamp striplight in this location. Single–lamp staggered strips introduce a "reverse socket shadow"—lamps closer to back cove wall create a slightly greater intensity. Butt strips end–to–end tightly.
③ Ceiling reflectance significantly influences success of cove. Higher is better (preferably 90 percent).
④ T8 and T5 lamps offer same light output—so striplights can be either version. Lamps must be combined with correct ballast and are not interchangeable after installation. See below for T5/HO version.

T5 and T8 Striplights **Back of Cove**
⊃Possible vendors: Bartco, Columbia, DayBrite, Legion, Lightolier, **Lithonia**¹, Metalux
¹T12 photometry was used and rerated for T8 lamping. This is somewhat imprecise.

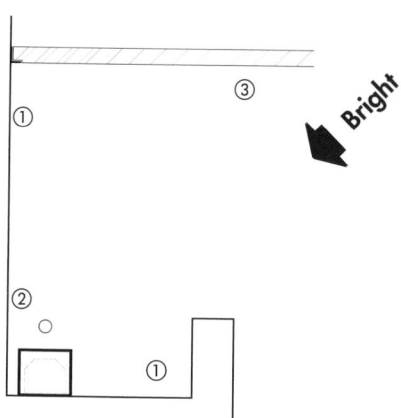

Notes
① Paint all interior cove surfaces matte white.
② Socket shadows are noticeable along back cove wall with 1-lamp striplight in this location. Single–lamp staggered strips introduce a "reverse socket shadow"—lamps closer to back cove wall create a slightly greater intensity. Butt strips end–to–end tightly.
③ Ceiling reflectance significantly influences success of cove. Higher is better (preferably 90 percent).

T5/HO Striplights **Back of Cove**
⊃Possible vendors: Bartco, Legion, Lightolier, **Lithonia**¹
¹T12 photometry was used and rerated for T5/HO lamping. This is somewhat imprecise.

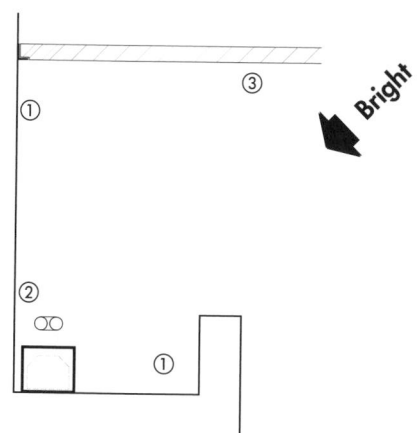

Notes
① Paint all interior cove surfaces matte white.
② Socket shadows are less noticeable along back cove wall and at ceiling with LongCompact lamps in striplight in this location.
③ Ceiling reflectance significantly influences success of cove. Higher is better (preferably 90 percent).

LongCompact Striplights **Back of Cove**
⊃Possible vendors: Columbia, DayBrite, Legion, Lightolier, **Lithonia**¹, Metalux
¹T12 photometry was used and rerated for LongCompact lamping. This is somewhat imprecise.

Design Tips
✔ Heavily textured ceiling material will create some shadows; as will tegular tile.
✔ If ceiling tone is other than white, then consider painting interior of cove surfaces to match ceiling.
✔ Corner shadowing less noticeable with strips at front of cove.
✔ Clear cove heights greater than 1′–0″ are more efficient.
✔ Coves using asymmetric luminaires specifically designed for cove applications are more efficient. See 11–35.

Details Concepts and Effects **11**

Application Key

Commercial
Gallery
Hospitality¹
Institutional
Manufacturing
Residential
Retail
Exterior

1/includes casual commercial and institutional applications and decorative retail applications

bold = primary application
partial fade = minimal application
fade = unlikely application

Net Addresses/Luminaires
See page 11–5

CONNECT FOR MORE

Coves Striplights

10′ Ceiling Height/4′ x 12′ Cove Opening/1′–0″ Clear Cove Height

DetailConcepts™

Performance Perspectives

[All data outlined here is based on information from boldfaced manufacturer's published data at time of manuscript preparation. Cove opening located in room corner and starts 2′ from each of two walls. Maintained values based on analysis grid 7′x15′ centered under cove and on 0.86 light loss factor (0.95 lamp lumen depreciation; 0.95 luminaire dirt depreciation; 0.95 room surface dirt depreciation); ballast factor of 0.98 and very light reflectances: ceiling 90%; wall 50%; and floor 20%. All views put eyes at 5′–6″ AFF—cove perspective varies as ceiling height changes and/or cove height changes.]

▶ **Cove height: 1′–0″ clear**
Central ceiling height: 10′
Perimeter ceiling height: 8′–6″±
Lamp: F28T5/830 or F32T8/830
Floor illuminance (7′x15′ area under cove):
▶ 23 fc, avg, maintained (35 fc max)

▶ **Cove height: 1′–0″ clear**
Central ceiling height: 10′
Perimeter ceiling height: 8′–6″±
Lamp: F54T5/HO/830
Floor illuminance (7′x15′ area under cove):
▶ 40 fc, avg, maintained (63 fc max)

▶ **Cove height: 1′–0″ clear**
Central ceiling height: 10′
Perimeter ceiling height: 8′–6″±
Lamp: F40Long/830
Floor illuminance (7′x15′ area under cove):
▶ 48 fc, avg, maintained (73 fc max)

Coves Striplights

10' Ceiling Height/4' x 12' Cove Opening/ 1'–0" Clear Cove Height

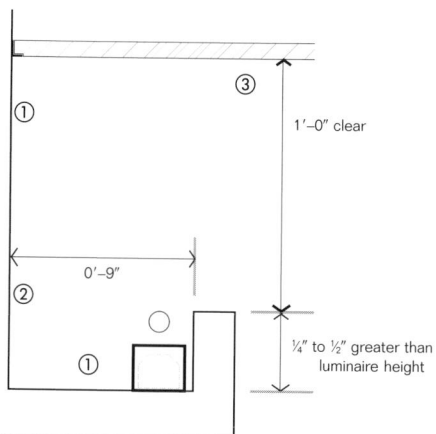

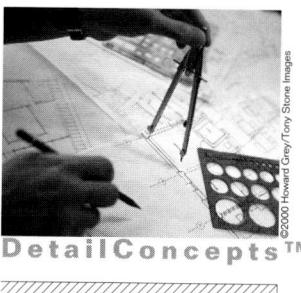

DetailConcepts™

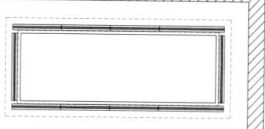

Notes
① Paint all interior cove surfaces matte white.
② Socket shadows are less noticeable with 1–lamp striplight in this location. Shadow line on ceiling may be visible a few feet or so out from front edge of cove lip, however.
③ Ceiling reflectance significantly influences success of cove. Higher is better (preferably 90 percent).
④ T8 and T5 lamps offer same light output—so striplights can be either version. Lamps must be combined with correct ballast and are not interchangeable after installation. See below for T5/HO version.

1'–0" clear
0'–9"
¼" to ½" greater than luminaire height

T5 and T8 Striplights **Front of Cove**
⊃ Possible vendors: Bartco, Columbia, DayBrite, Legion, Lightolier, **Lithonia**[1], Metalux
[1]T12 photometry was used and rerated for T8 lamping. This is somewhat imprecise.

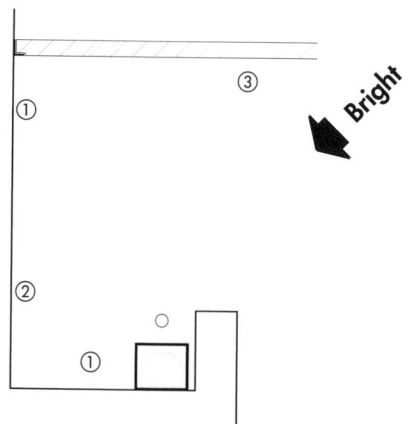

Bright

Notes
① Paint all interior cove surfaces matte white.
② Socket shadows are less noticeable with 1–lamp striplight in this location. Shadow line on ceiling may be visible a few feet or so out from front edge of cove lip, however.
③ Ceiling reflectance significantly influences success of cove. Higher is better (preferably 90 percent).

Design Tips
✔ Heavily textured ceiling material will create some shadows; as will tegular tile.
✔ If ceiling tone is other than white, then consider painting interior of cove surfaces to match ceiling.
✔ Clear cove heights greater than 0'–6" are more efficient.
✔ Coves using asymmetric luminaires specifically designed for cove applications are more efficient. See 11–35.

Details
Concepts and Effects
11

T5/HO Striplights **Front of Cove**
⊃ Possible vendors: Bartco, Legion, Lightolier, **Lithonia**[1]
[1]T12 photometry was used and rerated for T5/HO lamping. This is somewhat imprecise.

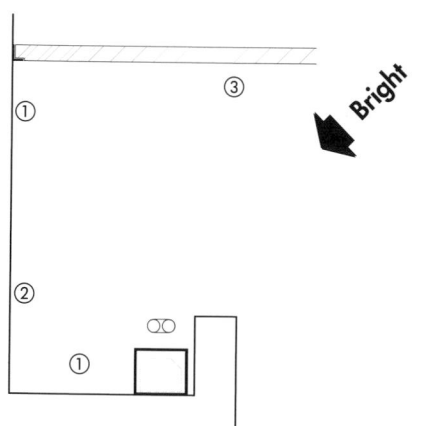

Bright

Notes
① Paint all interior cove surfaces matte white.
② Socket shadows are not noticeable with LongCompact lamps in striplight in this location. Shadow line on ceiling may be visible a few feet or so out from front edge of cove lip, however.
③ Ceiling reflectance significantly influences success of cove. Higher is better (preferably 90 percent).

Application Key

Commercial
Gallery
Hospitality[1]
Institutional
Manufacturing
Residential
Retail
Exterior

1/includes casual commercial and institutional applications and decorative retail applications

bold = primary application
partial fade = minimal application
fade = unlikely application

LongCompact Striplights **Front of Cove**
⊃ Possible vendors: Columbia, DayBrite, Legion, Lightolier, **Lithonia**[1], Metalux
[1]T12 photometry was used and rerated for LongCompact lamping. This is somewhat imprecise.

Net Addresses/Luminaires
See page 11–5

CONNECT FOR MORE

DetailConcepts™

Coves Asymmetric Luminaires

10′ Ceiling Height/4′ x 12′ Cove Opening/1′–0″ Clear Cove Height

Performance Perspectives

[All data outlined here is based on information from boldfaced manufacturer's published data at time of manuscript preparation. Cove opening located in room corner and starts 2′ from each of two walls. Maintained values based on analysis grid 7′x15′ centered under cove and on 0.86 light loss factor (0.95 lamp lumen depreciation; 0.95 luminaire dirt depreciation; 0.95 room surface dirt depreciation); ballast factor of 0.98 and very light reflectances: ceiling 90%; wall 50%; and floor 20%. All views put eyes at 5′–6″ AFF—cove perspective varies as ceiling height changes and/or cove height changes.]

▶ **Cove height: 1′–0″ clear**
Central ceiling height: 10′
Perimeter ceiling height: 8′–6″±
Lamp: F32T8/830
Floor illuminance (7′x15′ area under cove):
▶ 26 fc, avg, maintained (40 fc max)

▶ **Cove height: 1′–0″ clear**
Central ceiling height: 10′
Perimeter ceiling height: 8′–6″±
Lamp: F54T5/HO/830
Floor illuminance (7′x15′ area under cove):
▶ 46 fc, avg, maintained (68 fc max)

▶ **Cove height: 1′–0″ clear**
Central ceiling height: 10′
Perimeter ceiling height: 8′–6″±
Lamp: F40Long/830
Floor illuminance (7′x15′ area under cove):
▶ 57 fc, avg, maintained (87 fc max)

Details Concepts and Effects

11

Coves Asymmetric Luminaires

10' Ceiling Height/4' x 12' Cove Opening/ 1'–0" Clear Cove Height

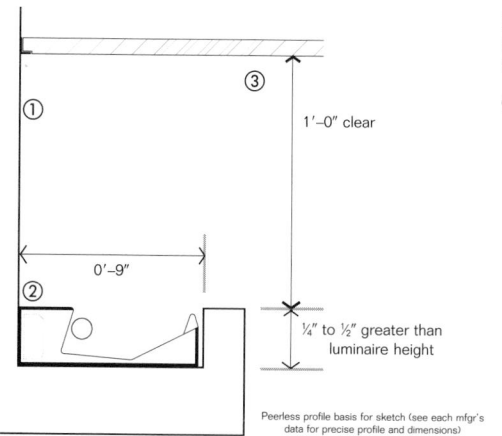

Notes
① Paint all interior cove surfaces matte white.
② Socket shadows are not noticeable with optically designed asymmetric luminaires.
③ Ceiling reflectance significantly influences success of cove. Higher is better (preferably 90 percent).

1'–0" clear
0'–9"
¼" to ½" greater than luminaire height

Peerless profile basis for sketch (see each mfgr's data for precise profile and dimensions)

D e t a i l C o n c e p t s ™

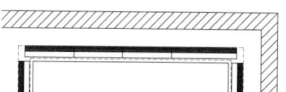

T8 Asymmetric Luminaires

⊃Possible vendors: Columbia (PIC), Elliptipar (301, 306/T8), ELP (CL series), Ledalite (In–Cove Asymmetry™), Lightolier (Covelite™), LITECONTROL (Cove–45), Neoray (74IC), **Peerless** (ECX®)

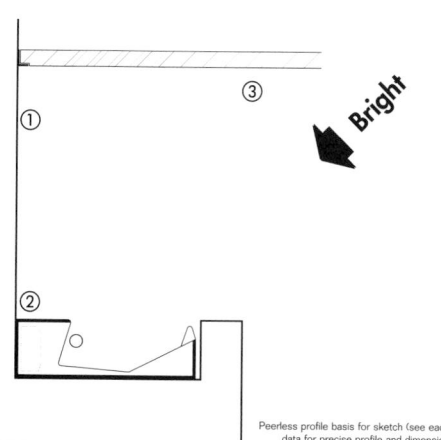

Bright

Notes
① Paint all interior cove surfaces matte white.
② Socket shadows are not noticeable with optically designed asymmetric luminaires.
③ Ceiling reflectance significantly influences success of cove. Higher is better (preferably 90 percent).
④ Standard T5 lamp/ballast will produce 42% less light and use 45% less energy.

Peerless profile basis for sketch (see each mfgr's data for precise profile and dimensions)

Design Tips
✔ Asymmetric luminaires will perform better than striplights—better uniformity across the ceiling and no socket shadows.
✔ Asymmetric luminaires are more efficient in coves than striplights.

T5/HO Asymmetric Luminaires

⊃Possible vendors: Elliptipar (303, 305), Lightolier (Covelite™), **Peerless** (ECX®)

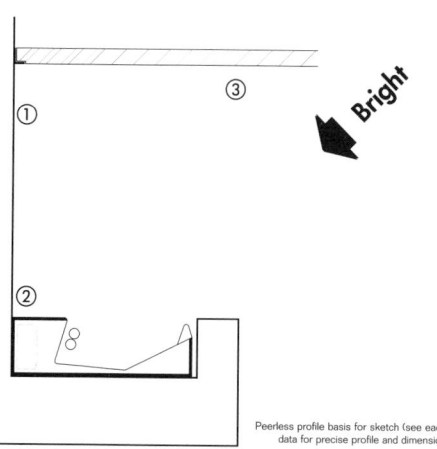

Bright

Notes
① Paint all interior cove surfaces matte white.
② Socket shadows are not noticeable with optically designed asymmetric luminaires.
③ Ceiling reflectance significantly influences success of cove. Higher is better (preferably 90 percent).

Peerless profile basis for sketch (see each mfgr's data for precise profile and dimensions)

Application Key

Commercial
Gallery
Hospitality[1]
Institutional
Manufacturing
Residential
Retail
Exterior

1/includes casual commercial and institutional applications and decorative retail applications

bold = primary application
partial fade = minimal application
fade = unlikely application

LongCompact Asymmetric Luminaires

⊃Possible vendors: Columbia (PIC), Elliptipar (302), ELP (CL series), Ledalite (In–Cove Asymmetry™), Lightolier (Covelite™), LITECONTROL (Cove–45), Neoray (74IC), **Peerless** (ECX®)

Trademarks, service marks and product names are owned and registered by respective manufacturers.

Net Addresses/Luminaires
See page 11–5

CONNECT FOR MORE

Details
Concepts and Effects

11

DetailConcepts™

Coves Striplights

10′ Ceiling Height/9′ x 9′ Cove Opening/0′–6″ Clear Cove Height

Performance Perspectives

[All data outlined here is based on information from boldfaced manufacturer's published data at time of manuscript preparation. Cove opening located in room corner and starts 2′ from each of two walls. Maintained values based on analysis grid 12′x12′ centered under cove and on 0.86 light loss factor (0.95 lamp lumen depreciation; 0.95 luminaire dirt depreciation; 0.95 room surface dirt depreciation); ballast factor of 0.98 and very light reflectances: ceiling 90%; wall 50%; and floor 20%. All views put eyes at 5′–6″ AFF—cove perspective varies as ceiling height changes and/or cove height changes.]

▶ **Cove height: 0′–6″ clear**
Central ceiling height: 10′
Perimeter ceiling height: 9′±
Lamp: F28T5/830 or F32T8/830
Floor illuminance (12′x12′ area under cove):
▶ 12 fc, avg, maintained (26 fc max)

▶ **Cove height: 0′–6″ clear**
Central ceiling height: 10′
Perimeter ceiling height: 9′±
Lamp: F54T5/HO/830
Floor illuminance (12′x12′ area under cove):
▶ 22 fc, avg, maintained (47 fc max)

▶ **Cove height: 0′–6″ clear**
Central ceiling height: 10′
Perimeter ceiling height: 9′±
Lamp: F40Long/830
Floor illuminance (12′x12′ area under cove):
▶ 27 fc, avg, maintained (44 fc max)

Details Concepts and Effects

11

Coves Striplights

10′ Ceiling Height/9′ x 9′ Cove Opening/0′–6″ Clear Cove Height

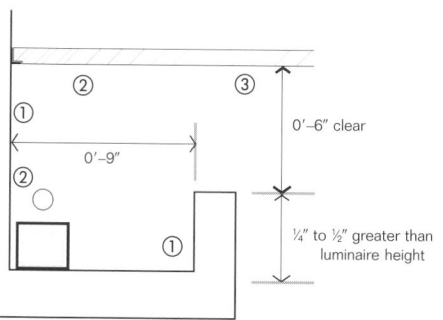

0′–6″ clear

0′–9″

¼″ to ½″ greater than luminaire height

T5 and T8 Striplights**Back of Cove**

⊃Possible vendors: Bartco, Columbia, DayBrite, Legion, Lightolier, **Lithonia**[1], Metalux

[1]T12 photometry was used and rerated for T8 lamping. This is somewhat imprecise.

Notes

①Paint all interior cove surfaces matte white.
②Socket shadows are noticeable along back cove wall and somewhat noticeable at ceiling with 1–lamp striplight in this location. Single–lamp staggered strips introduce a "reverse socket shadow"—lamps closer to back cove wall create a slightly greater intensity. Butt strips end–to–end tightly.
③Ceiling reflectance significantly influences success of cove. Higher is better (preferably 90 percent).
④T8 and T5 lamps offer same light output—so striplights can be either version. Lamps must be combined with correct ballast and are not interchangeable after installation .T5 striplights are a few inches shorter than T8s. See below for T5/HO version.

DetailConcepts™

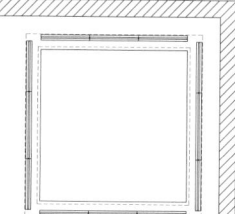

Very harsh

Notes

①Paint all interior cove surfaces matte white.
②Socket shadows are noticeable along back cove wall and somewhat noticeable at ceiling with 1–lamp striplight in this location. Single–lamp staggered strips introduce a "reverse socket shadow"—lamps closer to back cove wall create a slightly greater intensity. Butt strips end–to–end tightly.
③Ceiling reflectance significantly influences success of cove. Higher is better (preferably 90 percent).

T5/HO Striplights**Back of Cove**

⊃Possible vendors: Bartco, Legion, Lightolier, **Lithonia**[1]

[1]T12 photometry was used and rerated for T5/HO lamping. This is somewhat imprecise.

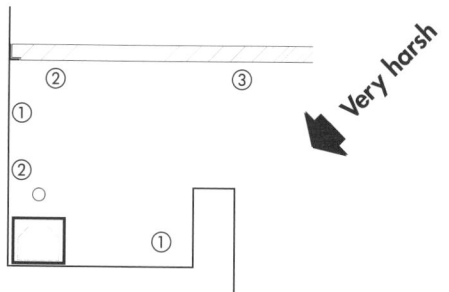

Very harsh

Notes

①Paint all interior cove surfaces matte white.
②Socket shadows are less noticeable along back cove wall and at ceiling with LongCompact lamps in striplight in this location.
③Ceiling reflectance significantly influences success of cove. Higher is better (preferably 90 percent).

LongCompact Striplights**Back of Cove**

⊃Possible vendors: Columbia, DayBrite, Legion, Lightolier, **Lithonia**[1], Metalux

[1]T12 photometry was used and rerated for LongCompact lamping. This is somewhat imprecise.

Design Tips

✔Lamp and ballast maintenance is difficult. Lay–in ceiling can help—lift tiles for better reach/view into cove.

✔Heavily textured ceiling material will create significant shadows; as will tegular tile.

✔If ceiling is drywall, then any imperfections will telegraph with such grazing light.

✔If ceiling tone is other than white, then consider painting interior of cove surfaces to match ceiling.

✔Corner shadowing less noticeable with strips at front of cove.

✔Clear cove heights greater than 0′–6″ are more efficient.

✔Coves using asymmetric luminaires specifically designed for cove applications are more efficient.

Application Key

Commercial
Gallery
Hospitality[1]
Institutional
Manufacturing
Residential
Retail
Exterior

1/includes casual commercial and institutional applications and decorative retail applications

bold = primary application
partial fade = minimal application
fade = unlikely application

Details Concepts and Effects

11

Net Addresses/Luminaires
See page 11–5

CONNECT FOR MORE

©2000 Howard Grey/Tony Stone Images

DetailConcepts™

Coves Striplights

10′ Ceiling Height/9′ x 9′ Cove Opening/0′–6″ Clear Cove Height

Performance Perspectives

[All data outlined here is based on information from boldfaced manufacturer's published data at time of manuscript preparation. Cove opening located in room corner and starts 2′ from each of two walls. Maintained values based on analysis grid 12′x12′ centered under cove and on 0.86 light loss factor (0.95 lamp lumen depreciation; 0.95 luminaire dirt depreciation; 0.95 room surface dirt depreciation); ballast factor of 0.98 and very light reflectances: ceiling 90%; wall 50%; and floor 20%. All views put eyes at 5′–6″ AFF—cove perspective varies as ceiling height changes and/or cove height changes.]

▶ **Cove height: 0′–6″ clear**
Central ceiling height: 10′
Perimeter ceiling height: 9′±
Lamp: F28T5/830 or F32T8/830
Floor illuminance (12′x12′ area under cove)
  ▶ 18 fc, avg, maintained (40 fc max)

▶ **Cove height: 0′–6″ clear**
Central ceiling height: 10′
Perimeter ceiling height: 9′±
Lamp: F54T5/HO/830
Floor illuminance (12′x12′ area under cove)
  ▶ 30 fc, avg, maintained (54 fc max)

▶ **Cove height: 0′–6″ clear**
Central ceiling height: 10′
Perimeter ceiling height: 9′±
Lamp: F40Long/830
Floor illuminance (12′x12′ area under cove)
  ▶ 37 fc, avg, maintained (77 fc max)

Details Concepts and Effects

11

Coves Striplights

10′ Ceiling Height/9′ x 9′ Cove Opening/ 0′–6″ Clear Cove Height

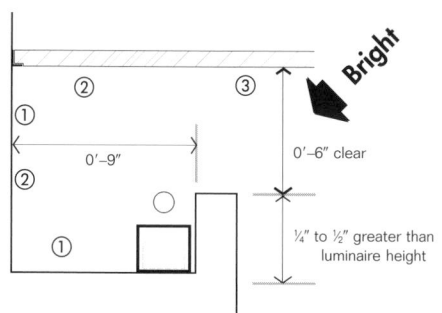

DetailConcepts™

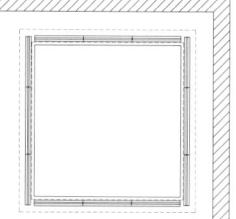

Notes
①Paint all interior cove surfaces matte white.
②Socket shadows are less noticeable with 1–lamp striplight in this location. Shadow line on ceiling may be visible a foot or so out from front edge of cove lip, however.
③Ceiling reflectance significantly influences success of cove. Higher is better (preferably 90 percent).
④T8 and T5 lamps offer same light output—so striplights can be either version. Lamps must be combined with correct ballast and are not interchangeable after installation. See below for T5/HO version.

0′–9″

0′–6″ clear

¼″ to ½″ greater than luminaire height

Bright

T5 and T8 Striplights**Front of Cove**
⊃Possible vendors: Bartco, Columbia, DayBrite, Legion, Lightolier, **Lithonia**[1], Metalux
[1]T12 photometry was used and rerated for T8 lamping. This is somewhat imprecise.

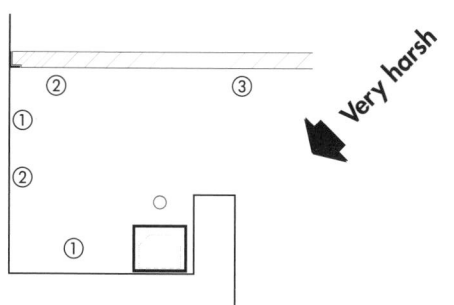

Notes
①Paint all interior cove surfaces matte white.
②Socket shadows are less noticeable with 1–lamp striplight in this location. Shadow line on ceiling may be visible a foot or so out from front edge of cove lip, however.
③Ceiling reflectance significantly influences success of cove. Higher is better (preferably 90 percent).

Very harsh

T5/HO Striplights**Front of Cove**
⊃Possible vendors: Bartco, Legion, Lightolier, **Lithonia**[1]
[1]T12 photometry was used and rerated for T5/HO lamping. This is somewhat imprecise.

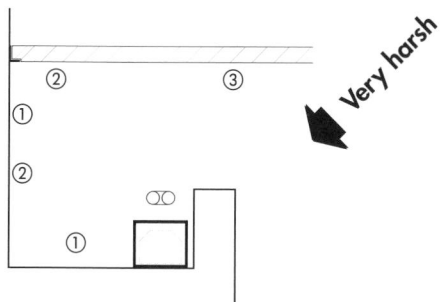

Notes
①Paint all interior cove surfaces matte white.
②Socket shadows are not noticeable with LongCompact lamps in striplight in this location.
③Ceiling reflectance significantly influences success of cove. Higher is better (preferably 90 percent).

Very harsh

LongCompact Striplights**Front of Cove**
⊃Possible vendors: Columbia, DayBrite, Legion, Lightolier, **Lithonia**[1], Metalux
[1]T12 photometry was used and rerated for LongCompact lamping. This is somewhat imprecise.

Details Concepts and Effects

11

Design Tips
✔Lamp and ballast maintenance is difficult. Lay–in ceiling can help—lift tiles for better reach/view into cove.
✔Heavily textured ceiling material will create significant shadows; as will tegular tile.
✔If ceiling is drywall, then any imperfections will telegraph with such grazing light.
✔If ceiling tone is other than white, then consider painting interior of cove surfaces to match ceiling.
✔Clear cove heights greater than 0′–6″ are more efficient.
✔Coves using asymmetric luminaires specifically designed for cove applications are more efficient.

Application Key

Commercial
Gallery
Hospitality[1]
Institutional
Manufacturing
Residential
Retail
Exterior

1/includes casual commercial and institutional applications and decorative retail applications

bold = primary application
partial fade = minimal application
fade = unlikely application

Net Addresses/Luminaires
See page 11–5

CONNECT FOR MORE

DetailConcepts™

Coves Striplights
10' Ceiling Height/9' x 9' Cove Opening/1'–0" Clear Cove Height

Performance Perspectives

[All data outlined here is based on information from boldfaced manufacturer's published data at time of manuscript preparation. Cove opening located in room corner and starts 2' from each of two walls. Maintained values based on analysis grid 12'x12' centered under cove and on 0.86 light loss factor (0.95 lamp lumen depreciation; 0.95 luminaire dirt depreciation; 0.95 room surface dirt depreciation); ballast factor of 0.98 and very light reflectances: ceiling 90%; wall 50%; and floor 20%. All views put eyes at 5'–6" AFF—cove perspective varies as ceiling height changes and/or cove height changes.]

▶ **Cove height: 1'–0" clear**
Central ceiling height: 10'
Perimeter ceiling height: 8'–6"±
Lamp: F28T5/830 or F32T8/830
Floor illuminance (12'x12' area under cove)
▶ 22 fc, avg, maintained (37 fc max)

▶ **Cove height: 1'–0" clear**
Central ceiling height: 10'
Perimeter ceiling height: 8'–6"±
Lamp: F54T5/HO/830
Floor illuminance (12'x12' area under cove)
▶ 37 fc, avg, maintained (67 fc max)

▶ **Cove height: 1'–0" clear**
Central ceiling height: 10'
Perimeter ceiling height: 8'–6"±
Lamp: F40Long/830
Floor illuminance (12'x12' area under cove)
▶ 43 fc, avg, maintained (74 fc max)

Details Concepts and Effects

Coves Striplights

10′ Ceiling Height/9′ x 9′ Cove Opening/1′–0″ Clear Cove Height

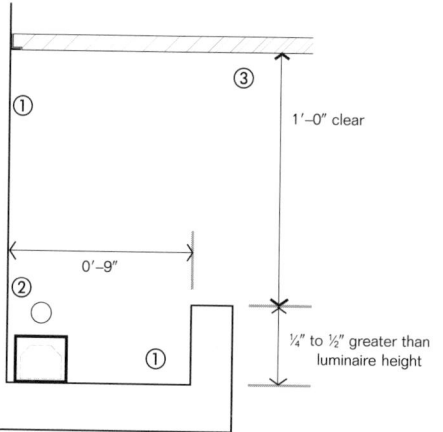

Notes
①Paint all interior cove surfaces matte white.
②Socket shadows are noticeable along back cove wall with 1–lamp striplight in this location. Single–lamp staggered strips introduce a "reverse socket shadow"—lamps closer to back cove wall create a slightly greater intensity. Butt strips end–to–end tightly.
③Ceiling reflectance significantly influences success of cove. Higher is better (preferably 90 percent).
④T8 and T5 lamps offer same light output—so striplights can be either version. Lamps must be combined with correct ballast and are not interchangeable after installation. See below for T5/HO version.

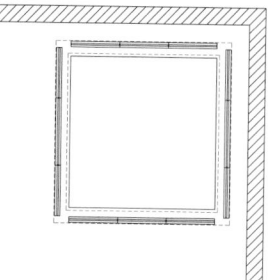

D e t a i l C o n c e p t s ™

T5 and T8 Striplights**Back of Cove**
⊃Possible vendors: Bartco, Columbia, DayBrite, Legion, Lightolier, **Lithonia**[1], Metalux
[1]T12 photometry was used and rerated for T8 lamping. This is somewhat imprecise.

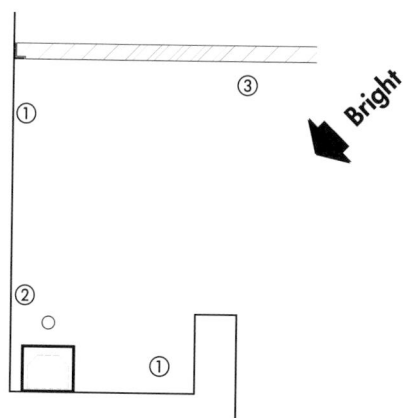

Notes
①Paint all interior cove surfaces matte white.
②Socket shadows are noticeable along back cove wall with 1–lamp striplight in this location. Single–lamp staggered strips introduce a "reverse socket shadow"—lamps closer to back cove wall create a slightly greater intensity. Butt strips end–to–end tightly.
③Ceiling reflectance significantly influences success of cove. Higher is better (preferably 90 percent).

Design Tips
✔Heavily textured ceiling material will create some shadows; as will tegular tile.
✔If ceiling tone is other than white, then consider painting interior of cove surfaces to match ceiling.
✔Corner shadowing less noticeable with strips at front of cove.
✔Clear cove heights greater than 1′–0″ are more efficient.
✔Coves using asymmetric luminaires specifically designed for cove applications are more efficient. See 11–45.

T5/HO Striplights**Back of Cove**
⊃Possible vendors: Bartco, Legion, Lightolier, **Lithonia**[1]
[1]T12 photometry was used and rerated for T5/HO lamping. This is somewhat imprecise.

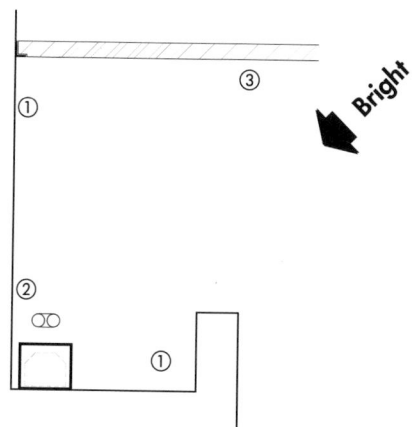

Notes
①Paint all interior cove surfaces matte white.
②Socket shadows are less noticeable along back cove wall and at ceiling with LongCompact lamps in striplight in this location.
③Ceiling reflectance significantly influences success of cove. Higher is better (preferably 90 percent).

Application Key
Commercial
Gallery
Hospitality[1]
Institutional
Manufacturing
Residential
Retail
Exterior

1/includes casual commercial and institutional applications and decorative retail applications

bold = primary application
partial fade = minimal application
fade = unlikely application

LongCompact Striplights**Back of Cove**
⊃Possible vendors: Columbia, DayBrite, Legion, Lightolier, **Lithonia**[1], Metalux
[1]T12 photometry was used and rerated for LongCompact lamping. This is somewhat imprecise.

Net Addresses/Luminaires
See page 11–5

Coves Striplights

10′ Ceiling Height/9′ x 9′ Cove Opening/1′–0″ Clear Cove Height

DetailConcepts™

Performance Perspectives

[All data outlined here is based on information from boldfaced manufacturer's published data at time of manuscript preparation. Cove opening located in room corner and starts 2′ from each of two walls. Maintained values based on analysis grid 12′x12′ centered under cove and on 0.86 light loss factor (0.95 lamp lumen depreciation; 0.95 luminaire dirt depreciation; 0.95 room surface dirt depreciation); ballast factor of 0.98 and very light reflectances: ceiling 90%; wall 50%; and floor 20%. All views put eyes at 5′–6″ AFF—cove perspective varies as ceiling height changes and/or cove height changes.]

▶ **Cove height: 1′–0″ clear**
Central ceiling height: 10′
Perimeter ceiling height: 8′–6″±
Lamp: F28T5/830 or F32T8/830
Floor illuminance (12′x12′ area under cove)
▶ 26 fc, avg, maintained (41 fc max)

▶ **Cove height: 1′–0″ clear**
Central ceiling height: 10′
Perimeter ceiling height: 8′–6″±
Lamp: F54T5/HO/830
Floor illuminance (12′x12′ area under cove)
▶ 42 fc, avg, maintained (70 fc max)

▶ **Cove height: 1′–0″ clear**
Central ceiling height: 10′
Perimeter ceiling height: 8′–6″±
Lamp: F40Long/830
Floor illuminance (12′x12′ area under cove)
▶ 50 fc, avg, maintained (83 fc max)

Details Concepts and Effects

11

Coves Striplights

10' Ceiling Height/9' x 9' Cove Opening/ 1'–0" Clear Cove Height

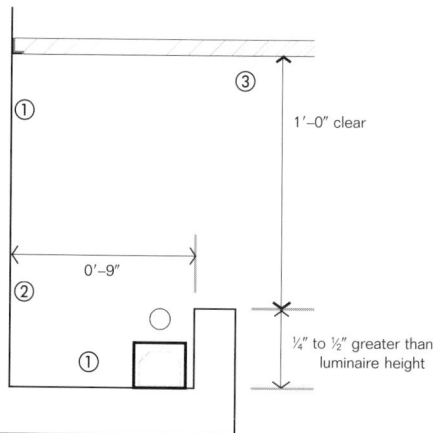

Notes
①Paint all interior cove surfaces matte white.
②Socket shadows are less noticeable with 1–lamp striplight in this location. Shadow line on ceiling may be visible a few feet or so out from front edge of cove lip, however.
③Ceiling reflectance significantly influences success of cove. Higher is better (preferably 90 percent).
④T8 and T5 lamps offer same light output—so striplights can be either version. Lamps must be combined with correct ballast and are not interchangeable after installation. See below for T5/HO version.

DetailConcepts™

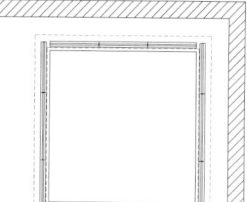

T5 and T8 Striplights **Front of Cove**
⊃Possible vendors: Bartco, Columbia, DayBrite, Legion, Lightolier, **Lithonia**[1], Metalux
[1]T12 photometry was used and rerated for T8 lamping. This is somewhat imprecise.

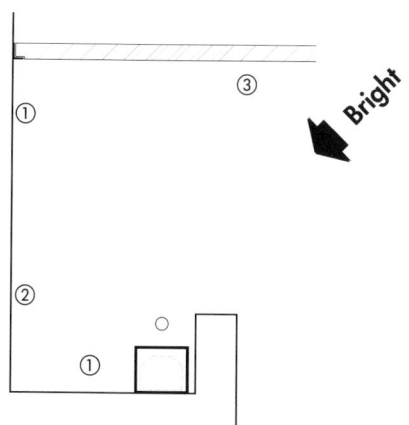

Notes
①Paint all interior cove surfaces matte white.
②Socket shadows are less noticeable with 1–lamp striplight in this location. Shadow line on ceiling may be visible a few feet or so out from front edge of cove lip, however.
③Ceiling reflectance significantly influences success of cove. Higher is better (preferably 90 percent).

Design Tips
✔Heavily textured ceiling material will create some shadows; as will tegular tile.
✔If ceiling tone is other than white, then consider painting interior of cove surfaces to match ceiling.
✔Clear cove heights greater than 0'–6" are more efficient.
✔Coves using asymmetric luminaires specifically designed for cove applications are more efficient. See 11–45.

T5/HO Striplights **Front of Cove**
⊃Possible vendors: Bartco, Legion, Lightolier, **Lithonia**[1]
[1]T12 photometry was used and rerated for T5/HO lamping. This is somewhat imprecise.

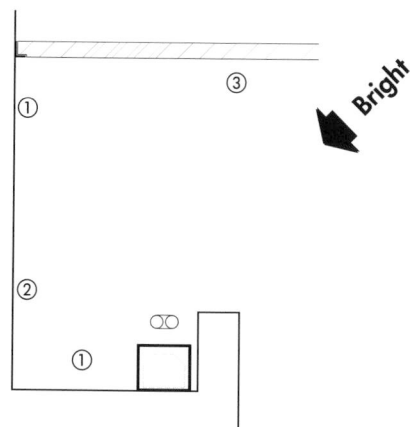

Notes
①Paint all interior cove surfaces matte white.
②Socket shadows are not noticeable with LongCompact lamps in striplight in this location. Shadow line on ceiling may be visible a few feet or so out from front edge of cove lip, however.
③Ceiling reflectance significantly influences success of cove. Higher is better (preferably 90 percent).

Application Key

Commercial
Gallery
Hospitality[1]
Institutional
Manufacturing
Residential
Retail
Exterior

1/includes casual commercial and institutional applications and decorative retail applications

bold = primary application
partial fade = minimal application
fade = unlikely application

LongCompact Striplights **Front of Cove**
⊃Possible vendors: Columbia, DayBrite, Legion, Lightolier, **Lithonia**[1], Metalux
[1]T12 photometry was used and rerated for LongCompact lamping. This is somewhat imprecise.

Net Addresses/Luminaires
See page 11–5

CONNECT FOR MORE

©2000 Howard Grey/Tony Stone Images

DetailConcepts™

Coves Asymmetric Luminaires

10' Ceiling Height/9' x 9' Cove Opening/ 1'–0" Clear Cove Height

Performance Perspectives

[All data outlined here is based on information from boldfaced manufacturer's published data at time of manuscript preparation. Cove opening located in room corner and starts 2' from each of two walls. Maintained values based on analysis grid 12'x12' centered under cove and on 0.86 light loss factor (0.95 lamp lumen depreciation; 0.95 luminaire dirt depreciation; 0.95 room surface dirt depreciation); ballast factor of 0.98 and very light reflectances: ceiling 90%; wall 50%; and floor 20%. All views put eyes at 5'–6" AFF—cove perspective varies as ceiling height changes and/or cove height changes.]

▶ **Cove height: 1'–0" clear**
Central ceiling height: 10'
Perimeter ceiling height: 8'–6"±
Lamp: F32T8/830
Floor illuminance (12'x12' area under cove)
▶ 28 fc, avg, maintained (44 fc max)

▶ **Cove height: 1'–0" clear**
Central ceiling height: 10'
Perimeter ceiling height: 8'–6"±
Lamp: F54T5/HO/830
Floor illuminance (12'x12' area under cove)
▶ 50 fc, avg, maintained (80 fc max)

▶ **Cove height: 1'–0" clear**
Central ceiling height: 10'
Perimeter ceiling height: 8'–6"±
Lamp: F40Long/830
Floor illuminance (12'x12' area under cove)
▶ 62 fc, avg, maintained (104 fc max)

Details
Concepts and Effects

11

Coves Asymmetric Luminaires

10′ Ceiling Height/9′ x 9′ Cove Opening/1′–0″ Clear Cove Height

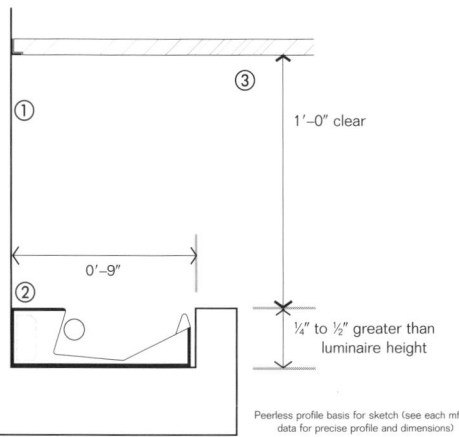

Notes
①Paint all interior cove surfaces matte white.
②Socket shadows are not noticeable with optically designed asymmetric luminaires.
③Ceiling reflectance significantly influences success of cove. Higher is better (preferably 90 percent).

1′–0″ clear

0′–9″

¼″ to ½″ greater than luminaire height

Peerless profile basis for sketch (see each mfgr's data for precise profile and dimensions)

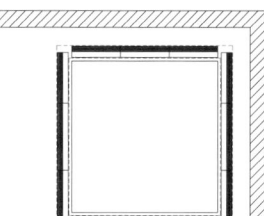

DetailConcepts™

T8 Asymmetric Luminaires
⊃Possible vendors: Columbia (PIC), Elliptipar(301, 306/T8), Ledalite (In–Cove Asymmetry™), LITECONTROL (Cove–45), Neoray (74IC), **Peerless** (ECX®)

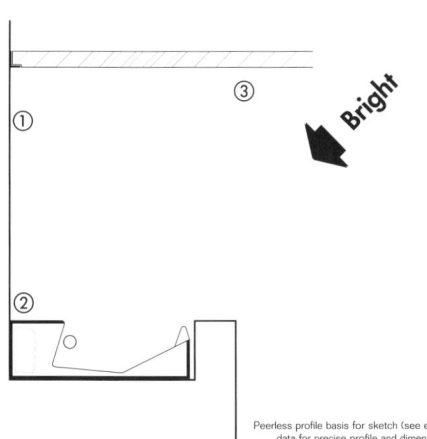

Bright

Notes
①Paint all interior cove surfaces matte white.
②Socket shadows are not noticeable with optically designed asymmetric luminaires.
③Ceiling reflectance significantly influences success of cove. Higher is better (preferably 90 percent).
④Standard T5 lamp/ballast will produce 42% less light and use 45% less energy.

Peerless profile basis for sketch (see each mfgr's data for precise profile and dimensions)

Design Tips
✔Asymmetric luminaires will perform better than striplights—better uniformity across the ceiling and no socket shadows.
✔Asymmetric luminaires are more efficient in coves than striplights.

T5/HO Asymmetric Luminaires
⊃Possible vendors: Elliptipar (303, 305), **Peerless** (ECX®)

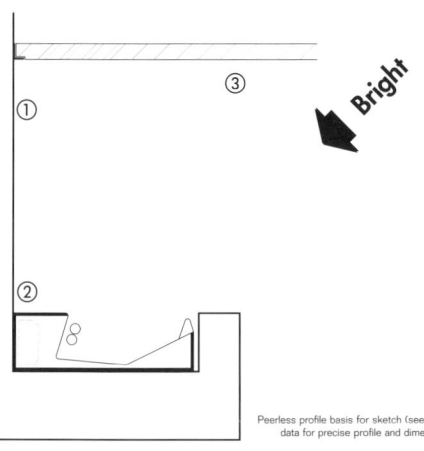

Bright

Notes
①Paint all interior cove surfaces matte white.
②Socket shadows are not noticeable with optically designed asymmetric luminaires.
③Ceiling reflectance significantly influences success of cove. Higher is better (preferably 90 percent).

Peerless profile basis for sketch (see each mfgr's data for precise profile and dimensions)

Application Key

Commercial
Gallery
Hospitality¹
Institutional
Manufacturing
Residential
Retail
Exterior

1/includes casual commercial and institutional applications and decorative retail applications

bold = primary application
partial fade = minimal application
fade = unlikely application

LongCompact Asymmetric Luminaires
⊃Possible vendors: Columbia (PIC), Elliptipar (302), ELP (CL series), Ledalite (In–Cove Asymmetry™), Lightolier (Covelite™), LITECONTROL (Cove–45), Neoray (74IC), **Peerless** (ECX®)
Trademarks, service marks and product names are owned and registered by respective manufacturers.

Net Addresses/Luminaires
See page 11–5

 CONNECT FOR MORE

Details Concepts and Effects **11**

©2000 Howard Grey/Tony Stone Images

DetailConcepts™

Coves Striplights

10' Ceiling Height/12' x 12' Cove Opening/0'–6" Clear Cove Height

Performance Perspectives

[All data outlined here is based on information from boldfaced manufacturer's published data at time of manuscript preparation. Cove opening located in room corner and starts 2' from each of two walls. Maintained values based on analysis grid 15'x15' centered under cove and on 0.86 light loss factor (0.95 lamp lumen depreciation; 0.95 luminaire dirt depreciation; 0.95 room surface dirt depreciation); ballast factor of 0.98 and very light reflectances: ceiling 90%; wall 50%; and floor 20%. All views put eyes at 5'–6" AFF—cove perspective varies as ceiling height changes and/or cove height changes.]

▶ **Cove height: 0'–6" clear**
Central ceiling height: 10'
Perimeter ceiling height: 9'±
Lamp: F28T5/830 or F32T8/830
Floor illuminance (15'x15' area under cove):
  ▶ 13 fc, avg, maintained (25 fc max)

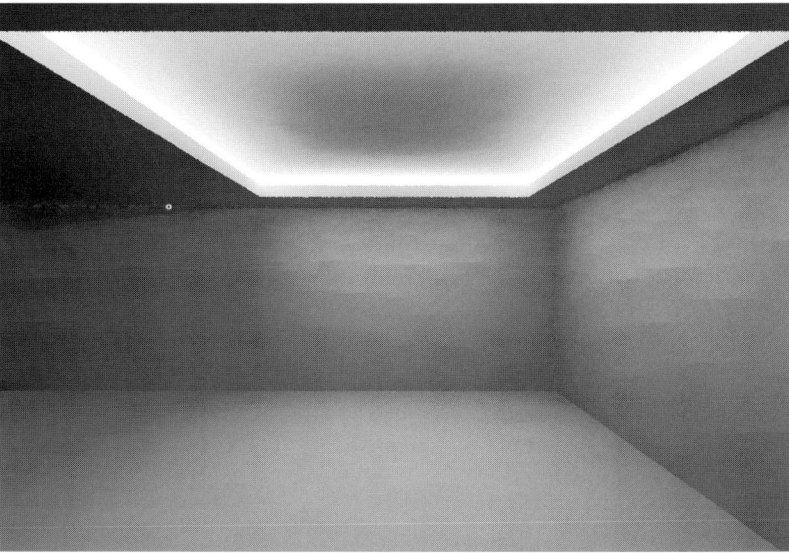

▶ **Cove height: 0'–6" clear**
Central ceiling height: 10'
Perimeter ceiling height: 9'±
Lamp: F54T5/HO/830
Floor illuminance (15'x15' area under cove):
  ▶ 23 fc, avg, maintained (42 fc max)

▶ **Cove height: 0'–6" clear**
Central ceiling height: 10'
Perimeter ceiling height: 9'±
Lamp: F40Long/830
Floor illuminance (15'x15' area under cove):
  ▶ 26 fc, avg, maintained (47 fc max)

Details Concepts and Effects

11

Coves Striplights

10' Ceiling Height/12' x 12' Cove Opening/0'–6" Clear Cove Height

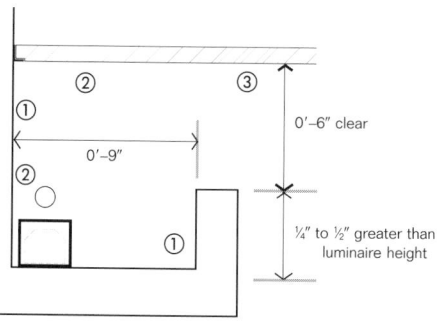

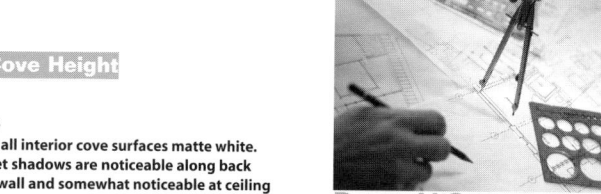

0'–6" clear

0'–9"

¼" to ½" greater than luminaire height

Notes
①Paint all interior cove surfaces matte white.
②Socket shadows are noticeable along back cove wall and somewhat noticeable at ceiling with 1–lamp striplight in this location. Single–lamp staggered strips introduce a "reverse socket shadow"—lamps closer to back cove wall create a slightly greater intensity. Butt strips end–to–end tightly.
③Ceiling reflectance significantly influences success of cove. Higher is better (preferably 90 percent).
④T8 and T5 lamps offer same light output—so striplights can be either version. Lamps must be combined with correct ballast and are not interchangeable after installation . T5 striplights are a few inches shorter than T8s. See below for T5/HO version.

T5 and T8 Striplights**Back of Cove**
⊃Possible vendors: Bartco, Columbia, DayBrite, Legion, Lightolier, **Lithonia**[1], Metalux
[1]T12 photometry was used and rerated for T8 lamping. This is somewhat imprecise.

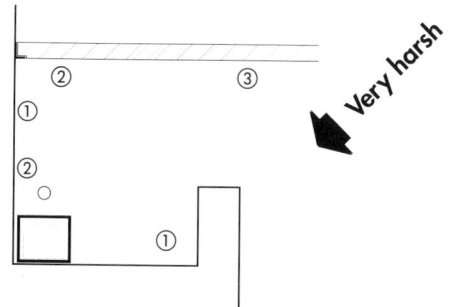

Very harsh

Notes
①Paint all interior cove surfaces matte white.
②Socket shadows are noticeable along back cove wall and somewhat noticeable at ceiling with 1–lamp striplight in this location. Single–lamp staggered strips introduce a "reverse socket shadow"—lamps closer to back cove wall create a slightly greater intensity. Butt strips end–to–end tightly.
③Ceiling reflectance significantly influences success of cove. Higher is better (preferably 90 percent).

T5/HO Striplights**Back of Cove**
⊃Possible vendors: Bartco, Legion, Lightolier, **Lithonia**[1]
[1]T12 photometry was used and rerated for T5/HO lamping. This is somewhat imprecise.

Very harsh

Notes
①Paint all interior cove surfaces matte white.
②Socket shadows are less noticeable along back cove wall and at ceiling with LongCompact lamps in striplight in this location.
③Ceiling reflectance significantly influences success of cove. Higher is better (preferably 90 percent).

LongCompact Striplights**Back of Cove**
⊃Possible vendors: Columbia, DayBrite, Legion, Lightolier, **Lithonia**[1], Metalux
[1]T12 photometry was used and rerated for LongCompact lamping. This is somewhat imprecise.

D e t a i l C o n c e p t s ™

Design Tips
✔Lamp and ballast maintenance is difficult. Lay–in ceiling can help— lift tiles for better reach/view into cove.
✔Heavily textured ceiling material will create significant shadows; as will tegular tile.
✔If ceiling is drywall, then any imperfections will telegraph with such grazing light.
✔If ceiling tone is other than white, then consider painting interior of cove surfaces to match ceiling.
✔Corner shadowing less noticeable with strips at front of cove.
✔Clear cove heights greater than 0'–6" are more efficient.
✔Coves using asymmetric luminaires specifically designed for cove applications are more efficient.

Details Concepts and Effects

11

Application Key

Commercial
Gallery
Hospitality[1]
Institutional
Manufacturing
Residential
Retail
Exterior

1/includes casual commercial and institutional applications and decorative retail applications

bold = primary application
partial fade = minimal application
fade = unlikely application

Net Addresses/Luminaires
See page 11–5

Coves Striplights

10′ Ceiling Height/12′ x 12′ Cove Opening/0′–6″ Clear Cove Height

DetailConcepts™

Performance Perspectives

[All data outlined here is based on information from boldfaced manufacturer's published data at time of manuscript preparation. Cove opening located in room corner and starts 2′ from each of two walls. Maintained values based on analysis grid 15′x15′ centered under cove and on 0.86 light loss factor (0.95 lamp lumen depreciation; 0.95 luminaire dirt depreciation; 0.95 room surface dirt depreciation); ballast factor of 0.98 and very light reflectances: ceiling 90%; wall 50%; and floor 20%. All views put eyes at 5′–6″ AFF—cove perspective varies as ceiling height changes and/or cove height changes.]

▶ **Cove height: 0′–6″ clear**
Central ceiling height: 10′
Perimeter ceiling height: 9′±
Lamp: F28T5/830 or F32T8/830
Floor illuminance (15′x15′ area under cove)
▶ 18 fc, avg, maintained (30 fc max)

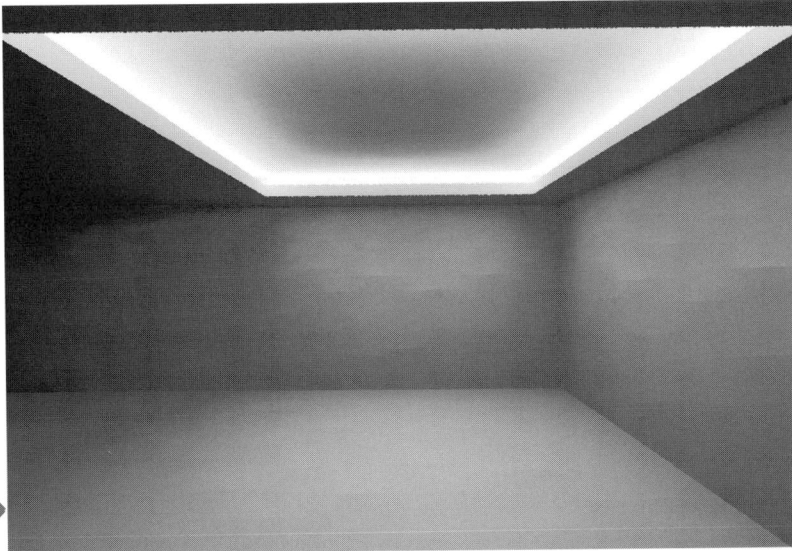

▶ **Cove height: 0′–6″ clear**
Central ceiling height: 10′
Perimeter ceiling height: 9′±
Lamp: F54T5/HO/830
Floor illuminance (15′x15′ area under cove)
▶ 31 fc, avg, maintained (53 fc max)

▶ **Cove height: 0′–6″ clear**
Central ceiling height: 10′
Perimeter ceiling height: 9′±
Lamp: F40Long/830
Floor illuminance (15′x15′ area under cove)
▶ 36 fc, avg, maintained (67 fc max)

Details Concepts and Effects

11

Coves Striplights
10' Ceiling Height/12' x 12' Cove Opening/0'–6" Clear Cove Height

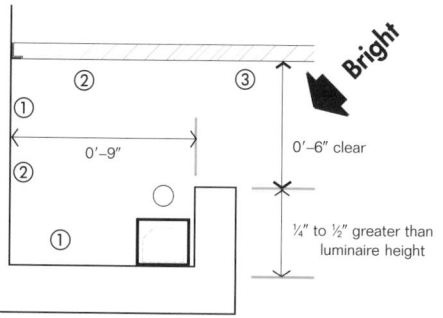

Detail Concepts™

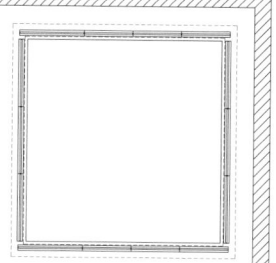

Notes
①Paint all interior cove surfaces matte white.
②Socket shadows are less noticeable with 1–lamp striplight in this location. Shadow line on ceiling may be visible a foot or so out from front edge of cove lip, however.
③Ceiling reflectance significantly influences success of cove. Higher is better (preferably 90 percent).
④T8 and T5 lamps offer same light output—so striplights can be either version. Lamps must be combined with correct ballast and are not interchangeable after installation. See below for T5/HO version.

Bright

0'–9"

0'–6" clear

¼" to ½" greater than luminaire height

T5 and T8 Striplights**Front of Cove**
⊃Possible vendors: Bartco, Columbia, DayBrite, Legion, Lightolier, **Lithonia**[1], Metalux
[1]T12 photometry was used and rerated for T8 lamping. This is somewhat imprecise.

Notes
①Paint all interior cove surfaces matte white.
②Socket shadows are less noticeable with 1–lamp striplight in this location. Shadow line on ceiling may be visible a foot or so out from front edge of cove lip, however.
③Ceiling reflectance significantly influences success of cove. Higher is better (preferably 90 percent).

Very harsh

T5/HO Striplights**Front of Cove**
⊃Possible vendors: Bartco, Legion, Lightolier, **Lithonia**[1]
[1]T12 photometry was used and rerated for T5/HO lamping. This is somewhat imprecise.

Design Tips
✔Lamp and ballast maintenance is difficult. Lay–in ceiling can help—lift tiles for better reach/view into cove.
✔Heavily textured ceiling material will create significant shadows; as will tegular tile.
✔If ceiling is drywall, then any imperfections will telegraph with such grazing light.
✔If ceiling tone is other than white, then consider painting interior of cove surfaces to match ceiling.
✔Clear cove heights greater than 0'–6" are more efficient.
✔Coves using asymmetric luminaires specifically designed for cove applications are more efficient.

Details Concepts and Effects **11**

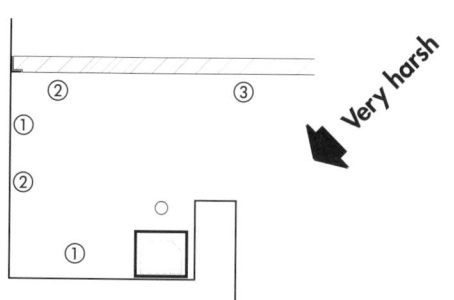

Very harsh

Notes
①Paint all interior cove surfaces matte white.
②Socket shadows are not noticeable with LongCompact lamps in striplight in this location.
③Ceiling reflectance significantly influences success of cove. Higher is better (preferably 90 percent).

LongCompact Striplights**Front of Cove**
⊃Possible vendors: Columbia, DayBrite, Legion, Lightolier, **Lithonia**[1], Metalux
[1]T12 photometry was used and rerated for LongCompact lamping. This is somewhat imprecise.

Application Key

Commercial
Gallery
Hospitality[1]
Institutional
Manufacturing
Residential
Retail
Exterior

1/includes casual commercial and institutional applications and decorative retail applications

bold = primary application
partial fade = minimal application
fade = unlikely application

Net Addresses/Luminaires
See page 11–5

DetailConcepts™

Coves Striplights

10′ Ceiling Height/12′ x 12′ Cove Opening/1′–0″ Clear Cove Height

Performance Perspectives

[All data outlined here is based on information from boldfaced manufacturer's published data at time of manuscript preparation. Cove opening located in room corner and starts 2′ from each of two walls. Maintained values based on analysis grid 15′x15′ centered under cove and on 0.86 light loss factor (0.95 lamp lumen depreciation; 0.95 luminaire dirt depreciation; 0.95 room surface dirt depreciation); ballast factor of 0.98 and very light reflectances: ceiling 90%; wall 50%; and floor 20%. All views put eyes at 5′–6″ AFF—cove perspective varies as ceiling height changes and/or cove height changes.]

▶ **Cove height: 1′–0″ clear**
Central ceiling height: 10′
Perimeter ceiling height: 8′–6″±
Lamp: F28T5/830 or F32T8/830
Floor illuminance (15′x15′ area under cove)
  ▶ 23 fc, avg, maintained (38 fc max)

▶ **Cove height: 1′–0″ clear**
Central ceiling height: 10′
Perimeter ceiling height: 8′–6″±
Lamp: F54T5/HO/830
Floor illuminance (15′x15′ area under cove)
  ▶ 35 fc, avg, maintained (61 fc max)

▶ **Cove height: 1′–0″ clear**
Central ceiling height: 10′
Perimeter ceiling height: 8′–6″±
Lamp: F40Long/830
Floor illuminance (15′x15′ area under cove)
  ▶ 44 fc, avg, maintained (76 fc max)

Details Concepts and Effects **11**

Coves Striplights

10' Ceiling Height/12' x 12' Cove Opening/1'–0" Clear Cove Height

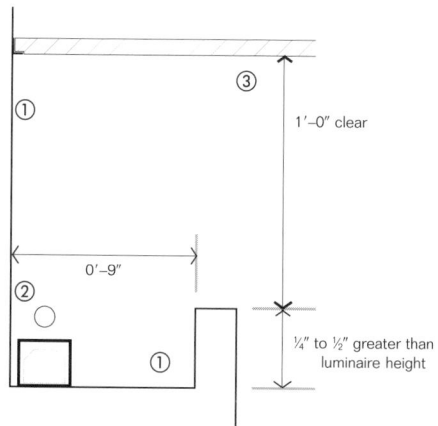

DetailConcepts™

Notes
①Paint all interior cove surfaces matte white.
②Socket shadows are noticeable along back cove wall with 1–lamp striplight in this location. Single–lamp staggered strips introduce a "reverse socket shadow"—lamps closer to back cove wall create a slightly greater intensity. Butt strips end–to–end tightly.
③Ceiling reflectance significantly influences success of cove. Higher is better (preferably 90 percent).
④T8 and T5 lamps offer same light output—so striplights can be either version. Lamps must be combined with correct ballast and are not interchangeable after installation. See below for T5/HO version.

T5 and T8 Striplights **Back of Cove**
➲Possible vendors: Bartco, Columbia, DayBrite, Legion, Lightolier, **Lithonia**[1], Metalux
[1]T12 photometry was used and rerated for T8 lamping. This is somewhat imprecise.

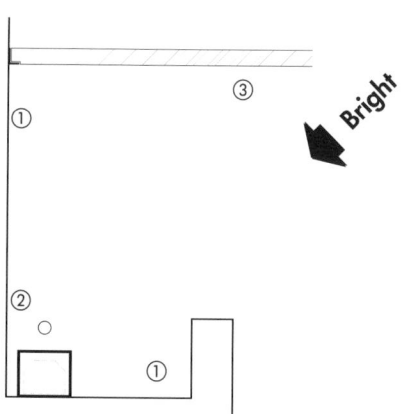

Bright

Notes
①Paint all interior cove surfaces matte white.
②Socket shadows are noticeable along back cove wall with 1–lamp striplight in this location. Single–lamp staggered strips introduce a "reverse socket shadow"—lamps closer to back cove wall create a slightly greater intensity. Butt strips end–to–end tightly.
③Ceiling reflectance significantly influences success of cove. Higher is better (preferably 90 percent).

Details
Concepts and Effects

11

Design Tips
✔Heavily textured ceiling material will create some shadows; as will tegular tile.
✔If ceiling tone is other than white, then consider painting interior of cove surfaces to match ceiling.
✔Corner shadowing less noticeable with strips at front of cove.
✔Clear cove heights greater than 1'–0" are more efficient.
✔Coves using asymmetric luminaires specifically designed for cove applications are more efficient. See 11–55.

T5/HO Striplights **Back of Cove**
➲Possible vendors: Bartco, Legion, Lightolier, **Lithonia**[1]
[1]T12 photometry was used and rerated for T5/HO lamping. This is somewhat imprecise.

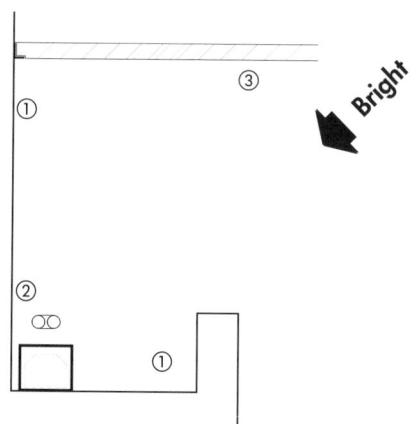

Bright

Notes
①Paint all interior cove surfaces matte white.
②Socket shadows are less noticeable along back cove wall and at ceiling with LongCompact lamps in striplight in this location.
③Ceiling reflectance significantly influences success of cove. Higher is better (preferably 90 percent).

Application Key

Commercial
Gallery
Hospitality[1]
Institutional
Manufacturing
Residential
Retail
Exterior

1/includes casual commercial and institutional applications and decorative retail applications

bold = primary application
partial fade = minimal application
fade = unlikely application

LongCompact Striplights **Back of Cove**
➲Possible vendors: Columbia, DayBrite, Legion, Lightolier, **Lithonia**[1], Metalux
[1]T12 photometry was used and rerated for LongCompact lamping. This is somewhat imprecise.

Net Addresses/Luminaires
See page 11–5

Coves Striplights

10′ Ceiling Height/12′ x 12′ Cove Opening/1′–0″ Clear Cove Height

DetailConcepts™

Performance Perspectives

[All data outlined here is based on information from boldfaced manufacturer's published data at time of manuscript preparation. Cove opening located in room corner and starts 2′ from each of two walls. Maintained values based on analysis grid 15′x15′ centered under cove and on 0.86 light loss factor (0.95 lamp lumen depreciation; 0.95 luminaire dirt depreciation; 0.95 room surface dirt depreciation); ballast factor of 0.98 and very light reflectances: ceiling 90%; wall 50%; and floor 20%. All views put eyes at 5′–6″ AFF—cove perspective varies as ceiling height changes and/or cove height changes.]

▶ **Cove height: 1′–0″ clear**
Central ceiling height: 10′
Perimeter ceiling height: 8′–6″±
Lamp: F28T5/830 or F32T8/830
Floor illuminance (15′x15′ area under cove):
▶ 26 fc, avg, maintained (42 fc max)

▶ **Cove height: 1′–0″ clear**
Central ceiling height: 10′
Perimeter ceiling height: 8′–6″±
Lamp: F54T5/HO/830
Floor illuminance (15′x15′ area under cove):
▶ 43 fc, avg, maintained (68 fc max)

▶ **Cove height: 1′–0″ clear**
Central ceiling height: 10′
Perimeter ceiling height: 8′–6″±
Lamp: F40Long/830
Floor illuminance (15′x15′ area under cove):
▶ 49 fc, avg, maintained (84 fc max)

Details Concepts and Effects

11

Coves Striplights

10' Ceiling Height/12' x 12' Cove Opening/ 1'–0" Clear Cove Height

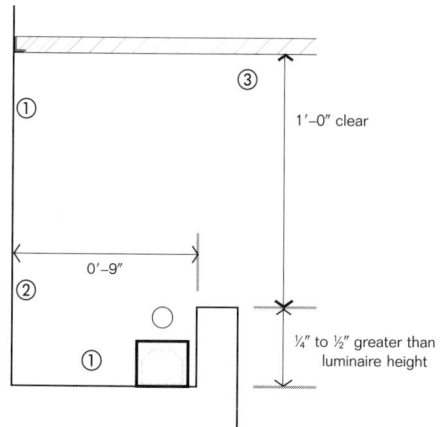

DetailConcepts™

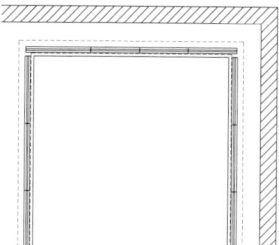

Notes
① Paint all interior cove surfaces matte white.
② Socket shadows are less noticeable with 1–lamp striplight in this location. Shadow line on ceiling may be visible a few feet or so out from front edge of cove lip, however.
③ Ceiling reflectance significantly influences success of cove. Higher is better (preferably 90 percent).
④ T8 and T5 lamps offer same light output—so striplights can be either version. Lamps must be combined with correct ballast and are not interchangeable after installation. See below for T5/HO version.

1'–0" clear

0'–9"

¼" to ½" greater than luminaire height

T5 and T8 Striplights**Front of Cove**
⊃ Possible vendors: Bartco, Columbia, DayBrite, Legion, Lightolier, **Lithonia**[1], Metalux
[1]T12 photometry was used and rerated for T8 lamping. This is somewhat imprecise.

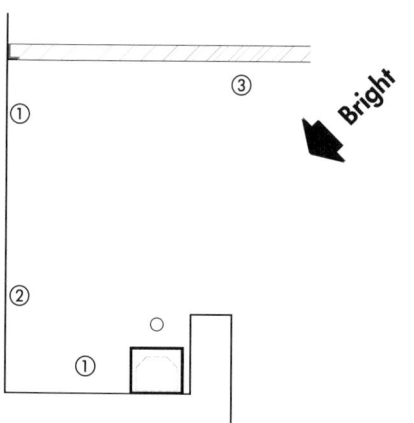

Bright

Notes
① Paint all interior cove surfaces matte white.
② Socket shadows are less noticeable with 1–lamp striplight in this location. Shadow line on ceiling may be visible a few feet or so out from front edge of cove lip, however.
③ Ceiling reflectance significantly influences success of cove. Higher is better (preferably 90 percent).

Design Tips
✔ Heavily textured ceiling material will create some shadows; as will tegular tile.
✔ If ceiling tone is other than white, then consider painting interior of cove surfaces to match ceiling.
✔ Clear cove heights greater than 0'–6" are more efficient.
✔ Coves using asymmetric luminaires specifically designed for cove applications are more efficient. See 11–55.

Details Concepts and Effects

11

T5/HO Striplights**Front of Cove**
⊃ Possible vendors: Bartco, Legion, Lightolier, **Lithonia**[1]
[1]T12 photometry was used and rerated for T5/HO lamping. This is somewhat imprecise.

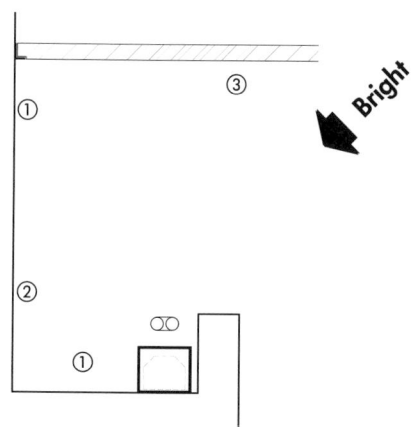

Bright

Notes
① Paint all interior cove surfaces matte white.
② Socket shadows are not noticeable with LongCompact lamps in striplight in this location. Shadow line on ceiling may be visible a few feet or so out from front edge of cove lip, however.
③ Ceiling reflectance significantly influences success of cove. Higher is better (preferably 90 percent).

Application Key

Commercial
Gallery
Hospitality[1]
Institutional
Manufacturing
Residential
Retail
Exterior

1/includes casual commercial and institutional applications and decorative retail applications

bold = primary application
partial fade = minimal application
fade = unlikely application

LongCompact Striplights**Front of Cove**
⊃ Possible vendors: Columbia, DayBrite, Legion, Lightolier, **Lithonia**[1], Metalux
[1]T12 photometry was used and rerated for LongCompact lamping. This is somewhat imprecise.

Net Addresses/Luminaires
See page 11–5

DetailConcepts™

Coves Asymmetric Luminaires

10′ Ceiling Height/12′ x 12′ Cove Opening/ 1′–0″ Clear Cove Height

Performance Perspectives

[All data outlined here is based on information from boldfaced manufacturer's published data at time of manuscript preparation. Cove opening located in room corner and starts 2′ from each of two walls. Maintained values based on analysis grid 15′x15′ centered under cove and on 0.86 light loss factor (0.95 lamp lumen depreciation; 0.95 luminaire dirt depreciation; 0.95 room surface dirt depreciation); ballast factor of 0.98 and very light reflectances: ceiling 90%; wall 50%; and floor 20%. All views put eyes at 5′–6″ AFF—cove perspective varies as ceiling height changes and/or cove height changes.]

▶ **Cove height: 1′–0″ clear**
Central ceiling height: 10′
Perimeter ceiling height: 8′–6″±
Lamp: F32T8/830
Floor illuminance (15′x15′ area under cove):
▶ 28 fc, avg, maintained (44 fc max)

▶ **Cove height: 1′–0″ clear**
Central ceiling height: 10′
Perimeter ceiling height: 8′–6″±
Lamp: F54T5/HO/830
Floor illuminance (15′x15′ area under cove):
▶ 50 fc, avg, maintained (72 fc max)

▶ **Cove height: 1′–0″ clear**
Central ceiling height: 10′
Perimeter ceiling height: 8′–6″±
Lamp: F40Long/830
Floor illuminance (15′x15′ area under cove):
▶ 61 fc, avg, maintained (93 fc max)

Details Concepts and Effects

11

Coves Asymmetric Luminaires

10' Ceiling Height/12' x 12' Cove Opening/1'–0" Clear Cove Height

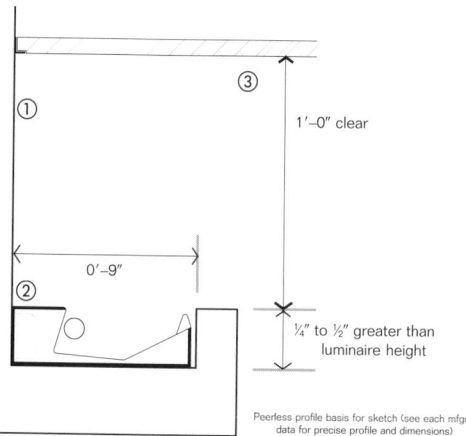

1'–0" clear

0'–9"

¼" to ½" greater than luminaire height

Peerless profile basis for sketch (see each mfgr's data for precise profile and dimensions)

Notes
①Paint all interior cove surfaces matte white.
②Socket shadows are not noticeable with optically designed asymmetric luminaires.
③Ceiling reflectance significantly influences success of cove. Higher is better (preferably 90 percent).

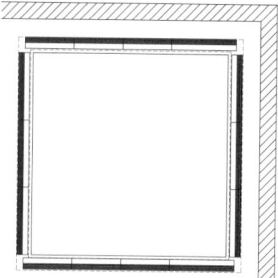

DetailConcepts™

T8 Asymmetric Luminaires
↪Possible vendors: Columbia (PIC), Elliptipar (301, 306/T8), Ledalite (In–Cove Asymmetry™), LITECONTROL (Cove–45), Neoray (74IC), **Peerless** (ECX®)

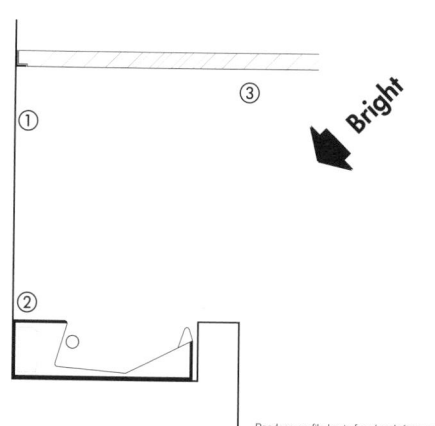

Bright

Peerless profile basis for sketch (see each mfgr's data for precise profile and dimensions)

Notes
①Paint all interior cove surfaces matte white.
②Socket shadows are not noticeable with optically designed asymmetric luminaires.
③Ceiling reflectance significantly influences success of cove. Higher is better (preferably 90 percent).
④Standard T5 lamp/ballast will produce 42% less light and use 45% less energy.

Design Tips
✔Asymmetric luminaires will perform better than striplights—better uniformity across the ceiling and no socket shadows.
✔Asymmetric luminaires are more efficient in coves than striplights.

T5/HO Asymmetric Luminaires
↪Possible vendors: Elliptipar (303, 305), **Peerless** (ECX®)

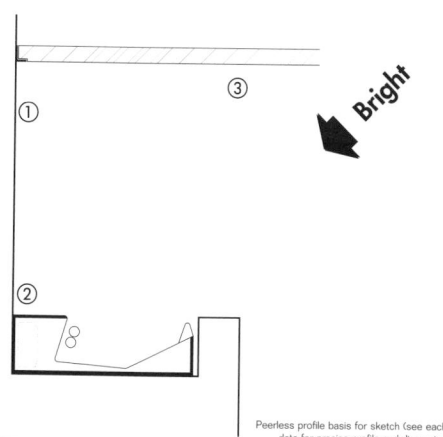

Bright

Peerless profile basis for sketch (see each mfgr's data for precise profile and dimensions)

Notes
①Paint all interior cove surfaces matte white.
②Socket shadows are not noticeable with optically designed asymmetric luminaires.
③Ceiling reflectance significantly influences success of cove. Higher is better (preferably 90 percent).

Application Key

Commercial
Gallery
Hospitality[1]
Institutional
Manufacturing
Residential
Retail
Exterior

1/includes casual commercial and institutional applications and decorative retail applications

bold = primary application
partial fade = minimal application
fade = unlikely application

LongCompact Asymmetric Luminaires
↪Possible vendors: Columbia (PIC), Elliptipar (302), ELP (CL series), Ledalite (In–Cove Asymmetry™), Lightolier (Covelite™), LITECONTROL (Cove–45), Neoray (74IC), **Peerless** (ECX®)

Net Addresses/Luminaires
See page 11–5

Details
Concepts and Effects

11

©2000 Howard Grey/Tony Stone Images

DetailConcepts ™

Coves Striplights

10′ Ceiling Height/12′ x 12′ Cove Opening/2′–0″ Clear Cove Height

Performance Perspectives

[All data outlined here is based on information from boldfaced manufacturer's published data at time of manuscript preparation. Cove opening located in room corner and starts 2′ from each of two walls. Maintained values based on analysis grid 15′x15′ centered under cove and on 0.86 light loss factor (0.95 lamp lumen depreciation; 0.95 luminaire dirt depreciation; 0.95 room surface dirt depreciation); ballast factor of 0.98 and very light reflectances: ceiling 90%; wall 50%; and floor 20%. All views put eyes at 5′–6″ AFF—cove perspective varies as ceiling height changes and/or cove height changes.]

▶ **Cove height: 2′–0″ clear**
Central ceiling height: 10′
Perimeter ceiling height: 7′–6″±
Lamp: F28T5/830 or F32T8/830
Floor illuminance (15′x15′ area under cove)
▶ 29 fc, avg, maintained (46 fc max)

▶ **Cove height: 2′–0″ clear**
Central ceiling height: 10′
Perimeter ceiling height: 7′–6″±
Lamp: F54T5/HO/830
Floor illuminance (15′x15′ area under cove)
▶ 47 fc, avg, maintained (76 fc max)

▶ **Cove height: 2′–0″ clear**
Central ceiling height: 10′
Perimeter ceiling height: 7′–6″±
Lamp: F40Long/830
Floor illuminance (15′x15′ area under cove)
▶ 55 fc, avg, maintained (89 fc max)

Details Concepts and Effects

11

Coves Striplights

10' Ceiling Height/12' x 12' Cove Opening/2'–0" Clear Cove Height

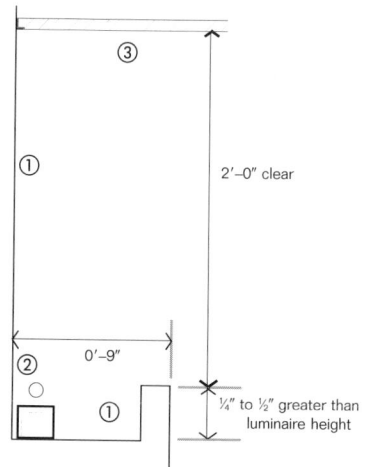

DetailConcepts™

Notes
①Paint all interior cove surfaces matte white.
②Socket shadows are noticeable along back cove wall with 1–lamp striplight in this location. Single–lamp staggered strips introduce a "reverse socket shadow"—lamps closer to back cove wall create a slightly greater intensity. Butt strips end–to–end tightly.
③Ceiling reflectance significantly influences success of cove. Higher is better (preferably 90 percent).
④T8 and T5 lamps offer same light output—so striplights can be either version. Lamps must be combined with correct ballast and are not interchangeable after installation. See below for T5/HO version.

T5 and T8 Striplights**Back of Cove**
⊃Possible vendors: Bartco, Columbia, DayBrite, Legion, Lightolier, **Lithonia**[1], Metalux
[1]T12 photometry was used and rerated for T8 lamping. This is somewhat imprecise.

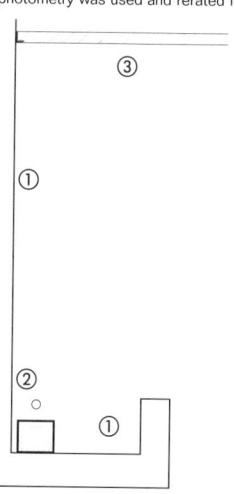

Notes
①Paint all interior cove surfaces matte white.
②Socket shadows are noticeable along back cove wall with 1–lamp striplight in this location. Single–lamp staggered strips introduce a "reverse socket shadow"—lamps closer to back cove wall create a slightly greater intensity. Butt strips end–to–end tightly.
③Ceiling reflectance significantly influences success of cove. Higher is better (preferably 90 percent).

Design Tips
✔ If ceiling tone is other than white, then consider painting interior of cove surfaces to match ceiling.
✔ Corner shadowing less noticeable with strips at front of cove.
✔ Coves using asymmetric luminaires specifically designed for cove applications are more efficient.

T5/HO Striplights**Back of Cove**
⊃Possible vendors: Bartco, Legion, Lightolier, **Lithonia**[1]
[1]T12 photometry was used and rerated for T5/HO lamping. This is somewhat imprecise.

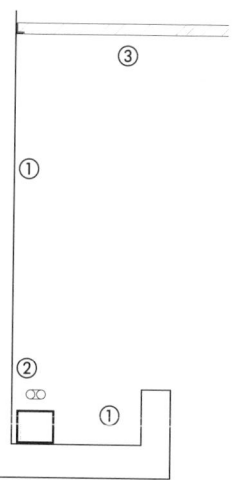

Notes
①Paint all interior cove surfaces matte white.
②Socket shadows are less noticeable along back cove wall and at ceiling with LongCompact lamps in striplight in this location.
③Ceiling reflectance significantly influences success of cove. Higher is better (preferably 90 percent).

Application Key
Commercial
Gallery
Hospitality[1]
Institutional
Manufacturing
Residential
Retail
Exterior

1/includes casual commercial and institutional applications and decorative retail applications

bold = primary application
partial fade = minimal application
fade = unlikely application

LongCompact Striplights**Back of Cove**
⊃Possible vendors: Columbia, DayBrite, Legion, Lightolier, **Lithonia**[1], Metalux
[1]T12 photometry was used and rerated for LongCompact lamping. This is somewhat imprecise.

Net Addresses/Luminaires
See page 11–5

Coves Striplights

10′ Ceiling Height/12′ x 12′ Cove Opening/2′–0″ Clear Cove Height

DetailConcepts™

Performance Perspectives

[All data outlined here is based on information from boldfaced manufacturer's published data at time of manuscript preparation. Cove opening located in room corner and starts 2′ from each of two walls. Maintained values based on analysis grid 15′x15′ centered under cove and on 0.86 light loss factor (0.95 lamp lumen depreciation; 0.95 luminaire dirt depreciation; 0.95 room surface dirt depreciation); ballast factor of 0.98 and very light reflectances: ceiling 90%; wall 50%; and floor 20%. All views put eyes at 5′–6″ AFF—cove perspective varies as ceiling height changes and/or cove height changes.]

▶ **Cove height: 2′–0″ clear**
Central ceiling height: 10′
Perimeter ceiling height: 7′–6″±
Lamp: F28T5/830 or F32T8/830
Floor illuminance (15′x15′ area under cove):
▶ 32 fc, avg, maintained (52 fc max)

▶ **Cove height: 2′–0″ clear**
Central ceiling height: 10′
Perimeter ceiling height: 7′–6″±
Lamp: F54T5/HO/830
Floor illuminance (15′x15′ area under cove):
▶ 53 fc, avg, maintained (82 fc max)

▶ **Cove height: 2′–0″ clear**
Central ceiling height: 10′
Perimeter ceiling height: 7′–6″±
Lamp: F40Long/830
Floor illuminance (15′x15′ area under cove):
▶ 61 fc, avg, maintained (99 fc max)

Details
Concepts and Effects

11

Coves Striplights
10' Ceiling Height/12' x 12' Cove Opening/2'–0" Clear Cove Height

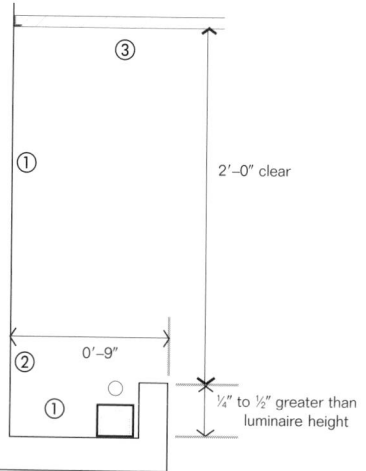

2'–0" clear

0'–9"

¼" to ½" greater than
luminaire height

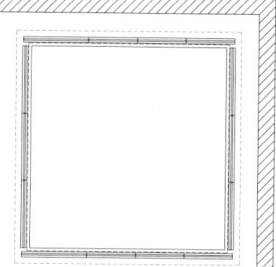

D e t a i l C o n c e p t s ™

©2000 Howard Grey/Tony Stone Images

Notes
① Paint all interior cove surfaces matte white.
② Socket shadows are less noticeable with 1–lamp striplight in this location. Shadow line on ceiling likely not noticeable.
③ Ceiling reflectance significantly influences success of cove. Higher is better (preferably 90 percent).
④ T8 and T5 lamps offer same light output—so striplights can be either version. Lamps must be combined with correct ballast and are not interchangeable after installation. See below for T5/HO version.

T5 and T8 Striplights**Front of Cove**
➲ Possible vendors: Bartco, Columbia, DayBrite, Legion, Lightolier, **Lithonia**[1], Metalux
[1]T12 photometry was used and rerated for T8 lamping. This is somewhat imprecise.

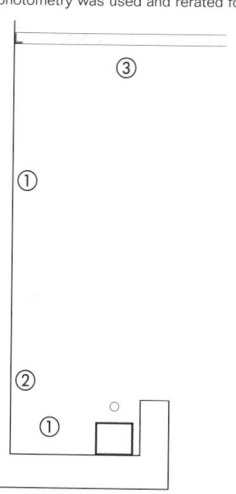

Notes
① Paint all interior cove surfaces matte white.
② Socket shadows are less noticeable with 1–lamp striplight in this location. Shadow line on ceiling likely not noticeable.
③ Ceiling reflectance significantly influences success of cove. Higher is better (preferably 90 percent).

Design Tips
✓ If ceiling tone is other than white, then consider painting interior of cove surfaces to match ceiling.
✓ Coves using asymmetric luminaires specifically designed for cove applications are more efficient.

T5/HO Striplights**Front of Cove**
➲ Possible vendors: Bartco, Legion, Lightolier, **Lithonia**[1]
[1]T12 photometry was used and rerated for T5/HO lamping. This is somewhat imprecise.

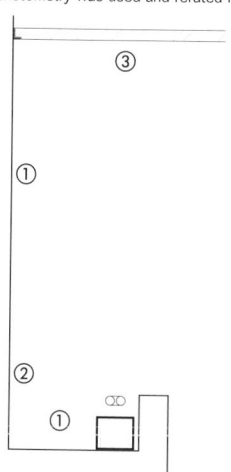

Notes
① Paint all interior cove surfaces matte white.
② Socket shadows are not noticeable along back cove wall and at ceiling with LongCompact lamps in striplight in this location. Shadow line on ceiling likely not noticeable.
③ Ceiling reflectance significantly influences success of cove. Higher is better (preferably 90 percent).

Application Key
Commercial
Gallery
Hospitality[1]
Institutional
Manufacturing
Residential
Retail
Exterior

1/includes casual commercial and institutional applications and decorative retail applications

bold = primary application
partial fade = minimal application
fade = unlikely application

LongCompact Striplights**Front of Cove**
➲ Possible vendors: Columbia, DayBrite, Legion, Lightolier, **Lithonia**[1], Metalux
[1]T12 photometry was used and rerated for LongCompact lamping. This is somewhat imprecise.

Net Addresses/Luminaires
See page 11–5

Coves Striplights

12′ Ceiling Height/4′ x 12′ Cove Opening/0′–6″ Clear Cove Height

DetailConcepts™

Performance Perspectives

[All data outlined here is based on information from boldfaced manufacturer's published data at time of manuscript preparation. Cove opening located in room corner and starts 2′ from each of two walls. Maintained values based on analysis grid 7′x15′ centered under cove and on 0.86 light loss factor (0.95 lamp lumen depreciation; 0.95 luminaire dirt depreciation; 0.95 room surface dirt depreciation); ballast factor of 0.98 and very light reflectances: ceiling 90%; wall 50%; and floor 20%. All views put eyes at 5′–6″ AFF—cove perspective varies as ceiling height changes and/or cove height changes.]

▶ **Cove height: 0′–6″ clear**
Central ceiling height: 12′
Perimeter ceiling height: 11′±
Lamp: F28T5/830 or F32T8/830
Floor illuminance (7′x15′ area under cove):
▶ 9 fc, avg, maintained (16 fc max)

▶ **Cove height: 0′–6″ clear**
Central ceiling height: 12′
Perimeter ceiling height: 11′±
Lamp: F54T5/HO/830
Floor illuminance (7′x15′ area under cove):
▶ 14 fc, avg, maintained (24 fc max)

▶ **Cove height: 0′–6″ clear**
Central ceiling height: 12′
Perimeter ceiling height: 11′±
Lamp: F40Long/830
Floor illuminance (7′x15′ area under cove):
▶ 18 fc, avg, maintained (27 fc max)

Details Concepts and Effects

11

Coves Striplights

12' Ceiling Height/4' x 12' Cove Opening/ 0'–6" Clear Cove Height

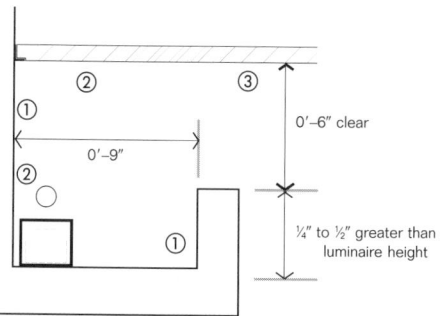

0'–6" clear

0'–9"

¼" to ½" greater than luminaire height

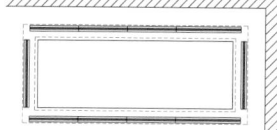

DetailConcepts™

©2000 Howard Grey/Tony Stone Images

Notes

① Paint all interior cove surfaces matte white.
② Socket shadows are noticeable along back cove wall and somewhat noticeable at ceiling with 1–lamp striplight in this location. Single–lamp staggered strips introduce a "reverse socket shadow"—lamps closer to back cove wall create a slightly greater intensity. Butt strips end–to–end tightly.
③ Ceiling reflectance significantly influences success of cove. Higher is better (preferably 90 percent).
④ T8 and T5 lamps offer same light output—so striplights can be either version. Lamps must be combined with correct ballast andare not interchangeable after installation .T5 striplights are a few inches shorter than T8s. See below for T5/HO version.

T5 and T8 Striplights Back of Cove
➲ Possible vendors: Bartco, Columbia, DayBrite, Legion, Lightolier, **Lithonia**[1], Metalux
[1] T12 photometry was used and rerated for T8 lamping. This is somewhat imprecise.

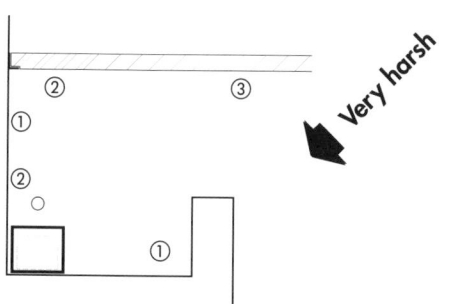

Very harsh

Notes

① Paint all interior cove surfaces matte white.
② Socket shadows are noticeable along back cove wall and somewhat noticeable at ceiling with 1–lamp striplight in this location. Single–lamp staggered strips introduce a "reverse socket shadow"—lamps closer to back cove wall create a slightly greater intensity. Butt strips end–to–end tightly.
③ Ceiling reflectance significantly influences success of cove. Higher is better (preferably 90 percent).

T5/HO Striplights Back of Cove
➲ Possible vendors: Bartco, Legion, Lightolier, **Lithonia**[1]
[1] T12 photometry was used and rerated for T5/HO lamping. This is somewhat imprecise.

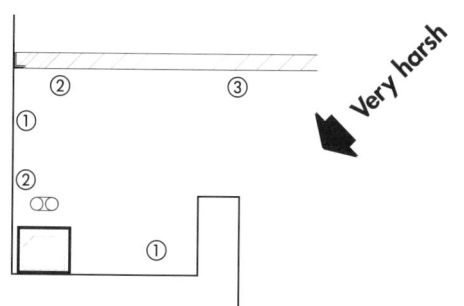

Very harsh

Notes

① Paint all interior cove surfaces matte white.
② Socket shadows are less noticeable along back cove wall and at ceiling with LongCompact lamps in striplight in this location.
③ Ceiling reflectance significantly influences success of cove. Higher is better (preferably 90 percent).

LongCompact Striplights Back of Cove
➲ Possible vendors: Columbia, DayBrite, Legion, Lightolier, **Lithonia**[1], Metalux
[1] T12 photometry was used and rerated for LongCompact lamping. This is somewhat imprecise.

Design Tips

✔ Lamp and ballast maintenance is difficult. Lay–in ceiling can help— lift tiles for better reach/view into cove.

✔ Heavily textured ceiling material will create significant shadows; as will tegular tile.

✔ If ceiling is drywall, then any imperfections will telegraph with such grazing light.

✔ If ceiling tone is other than white, then consider painting interior of cove surfaces to match ceiling.

✔ Corner shadowing less noticeable with strips at front of cove.

✔ Clear cove heights greater than 0'–6" are more efficient.

✔ Coves using asymmetric luminaires specifically designed for cove applications are more efficient. See 11–22.

Details
Concepts and Effects

11

Application Key

Commercial
Gallery
Hospitality[1]
Institutional
Manufacturing
Residential
Retail
Exterior

1/includes casual commercial and institutional applications and decorative retail applications

bold = primary application
partial fade = minimal application
fade = unlikely application

Net Addresses/Luminaires
See page 11–5

DetailConcepts™

Coves Striplights

12′ Ceiling Height/4′ x 12′ Cove Opening/0′–6″ Clear Cove Height

Performance Perspectives

[All data outlined here is based on information from boldfaced manufacturer's published data at time of manuscript preparation. Cove opening located in room corner and starts 2′ from each of two walls. Maintained values based on analysis grid 7′x15′ centered under cove and on 0.86 light loss factor (0.95 lamp lumen depreciation; 0.95 luminaire dirt depreciation; 0.95 room surface dirt depreciation); ballast factor of 0.98 and very light reflectances: ceiling 90%; wall 50%; and floor 20%. All views put eyes at 5′–6″ AFF—cove perspective varies as ceiling height changes and/or cove height changes.]

▶ **Cove height: 0′–6″ clear**
Central ceiling height: 12′
Perimeter ceiling height: 11′±
Lamp: F28T5/830 or F32T8/830
Floor illuminance (7′x15′ area under cove):
▶ 13 fc, avg, maintained (23 fc max)

▶ **Cove height: 0′–6″ clear**
Central ceiling height: 12′
Perimeter ceiling height: 11′±
Lamp: F54T5/HO/830
Floor illuminance (7′x15′ area under cove):
▶ 22 fc, avg, maintained (38 fc max)

▶ **Cove height: 0′–6″ clear**
Central ceiling height: 12′
Perimeter ceiling height: 11′±
Lamp: F40Long/830
Floor illuminance (7′x15′ area under cove):
▶ 25 fc, avg, maintained (41 fc max)

Details Concepts and Effects

11

Coves Striplights
12′ Ceiling Height/4′ x 12′ Cove Opening/0′–6″ Clear Cove Height

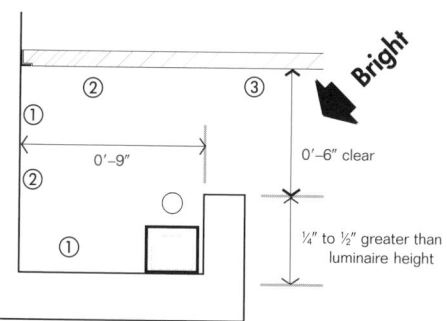

0′–9″

0′–6″ clear

¼″ to ½″ greater than luminaire height

Bright

Notes
① Paint all interior cove surfaces matte white.
② Socket shadows are less noticeable with 1-lamp striplight in this location. Shadow line on ceiling may be visible a foot or so out from front edge of cove lip, however.
③ Ceiling reflectance significantly influences success of cove. Higher is better (preferably 90 percent).
④ T8 and T5 lamps offer same light output—so striplights can be either version. Lamps must be combined with correct ballast and are not interchangeable after installation. See below for T5/HO version.

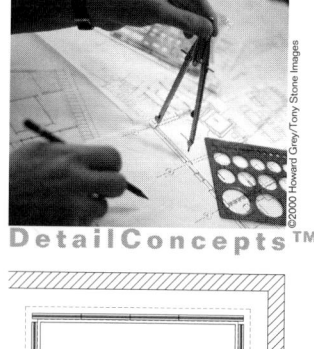

DetailConcepts™

©2000 Howard Grey/Tony Stone Images

T5 and T8 Striplights**Front of Cove**
⊃Possible vendors: Bartco, Columbia, DayBrite, Legion, Lightolier, **Lithonia**[1], Metalux
[1]T12 photometry was used and rerated for T8 lamping. This is somewhat imprecise.

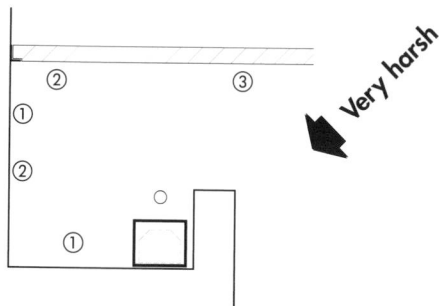

Very harsh

Notes
① Paint all interior cove surfaces matte white.
② Socket shadows are less noticeable with 1-lamp striplight in this location. Shadow line on ceiling may be visible a foot or so out from front edge of cove lip, however.
③ Ceiling reflectance significantly influences success of cove. Higher is better (preferably 90 percent).

T5/HO Striplights**Front of Cove**
⊃Possible vendors: Bartco, Legion, Lightolier, **Lithonia**[1]
[1]T12 photometry was used and rerated for T5/HO lamping. This is somewhat imprecise.

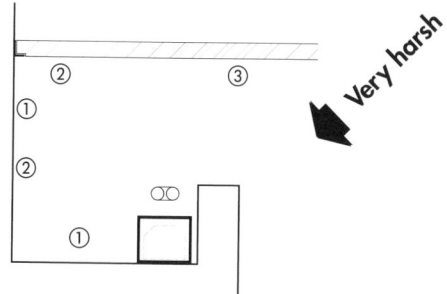

Very harsh

Notes
① Paint all interior cove surfaces matte white.
② Socket shadows are not noticeable with LongCompact lamps in striplight in this location.
③ Ceiling reflectance significantly influences success of cove. Higher is better (preferably 90 percent).

LongCompact Striplights**Front of Cove**
⊃Possible vendors: Columbia, DayBrite, Legion, Lightolier, **Lithonia**[1], Metalux
[1]T12 photometry was used and rerated for LongCompact lamping. This is somewhat imprecise.

Design Tips
✔ Lamp and ballast maintenance is difficult. Lay–in ceiling can help—lift tiles for better reach/ view into cove.
✔ Heavily textured ceiling material will create significant shadows; as will tegular tile.
✔ If ceiling is drywall, then any imperfections will telegraph with such grazing light.
✔ If ceiling tone is other than white, then consider painting interior of cove surfaces to match ceiling.
✔ Clear cove heights greater than 0′–6″ are more efficient.
✔ Coves using asymmetric luminaires specifically designed for cove applications are more efficient.

Details Concepts and Effects

11

Application Key

Commercial
Gallery
Hospitality[1]
Institutional
Manufacturing
Residential
Retail
Exterior

1/includes casual commercial and institutional applications and decorative retail applications

bold = primary application
partial fade = minimal application
fade = unlikely application

Net Addresses/Luminaires
See page 11–5

Coves striplights

12′ Ceiling Height/4′ x 12′ Cove Opening/ 1′–0″ Clear Cove Height

DetailConcepts™

Performance Perspectives

[All data outlined here is based on information from boldfaced manufacturer's published data at time of manuscript preparation. Cove opening located in room corner and starts 2′ from each of two walls. Maintained values based on analysis grid 7′x15′ centered under cove and on 0.86 light loss factor (0.95 lamp lumen depreciation; 0.95 luminaire dirt depreciation; 0.95 room surface dirt depreciation); ballast factor of 0.98 and very light reflectances: ceiling 90%; wall 50%; and floor 20%. All views put eyes at 5′–6″ AFF—cove perspective varies as ceiling height changes and/or cove height changes.]

▶ **Cove height: 1′–0″ clear**
Central ceiling height: 12′
Perimeter ceiling height: 10′–6″±
Lamp: F28T5/830 or F32T8/830
Floor illuminance (7′x15′ area under cove):
▶ 15 fc, avg, maintained (23 fc max)

▶ **Cove height: 1′–0″ clear**
Central ceiling height: 12′
Perimeter ceiling height: 10′–6″±
Lamp: F54T5/HO/830
Floor illuminance (7′x15′ area under cove):
▶ 23 fc, avg, maintained (37 fc max)

▶ **Cove height: 1′–0″ clear**
Central ceiling height: 12′
Perimeter ceiling height: 10′–6″±
Lamp: F40Long/830
Floor illuminance (7′x15′ area under cove):
▶ 29 fc, avg, maintained (45 fc max)

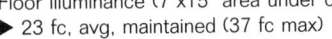

Details
Concepts and Effects

11

Coves Striplights

12′ Ceiling Height/4′ x 12′ Cove Opening/1′–0″ Clear Cove Height

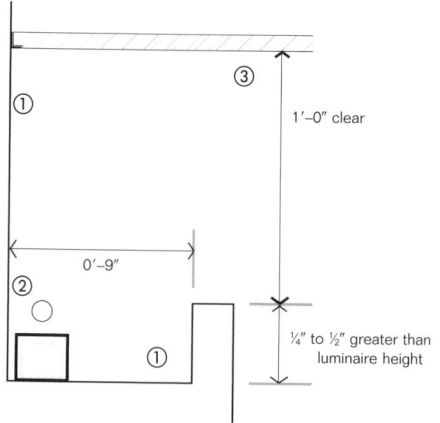

1′–0″ clear

0′–9″

¼″ to ½″ greater than luminaire height

Notes
① Paint all interior cove surfaces matte white.
② Socket shadows are noticeable along back cove wall with 1–lamp striplight in this location. Single–lamp staggered strips introduce a "reverse socket shadow"—lamps closer to back cove wall create a slightly greater intensity. Butt strips end–to–end tightly.
③ Ceiling reflectance significantly influences success of cove. Higher is better (preferably 90 percent).
④ T8 and T5 lamps offer same light output—so striplights can be either version. Lamps must be combined with correct ballast and are not interchangeable after installation. See below for T5/HO version.

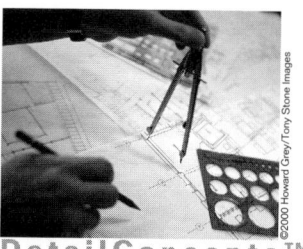
DetailConcepts™

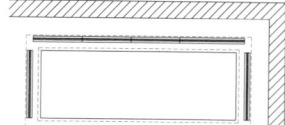

T5 and T8 Striplights **Back of Cove**
⮑ Possible vendors: Bartco, Columbia, DayBrite, Legion, Lightolier, **Lithonia**[1], Metalux
[1] T12 photometry was used and rerated for T8 lamping. This is somewhat imprecise.

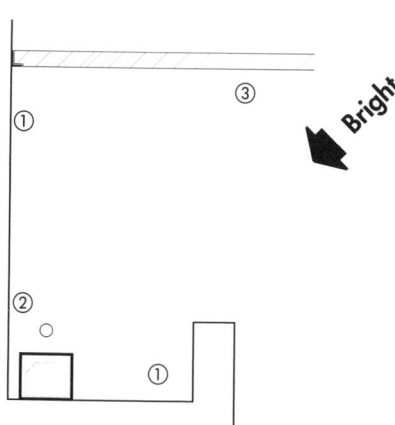

Bright

Notes
① Paint all interior cove surfaces matte white.
② Socket shadows are noticeable along back cove wall with 1–lamp striplight in this location. Single–lamp staggered strips introduce a "reverse socket shadow"—lamps closer to back cove wall create a slightly greater intensity. Butt strips end–to–end tightly.
③ Ceiling reflectance significantly influences success of cove. Higher is better (preferably 90 percent).

Design Tips
✔ Heavily textured ceiling material will create some shadows; as will tegular tile.
✔ If ceiling tone is other than white, then consider painting interior of cove surfaces to match ceiling.
✔ Corner shadowing less noticeable with strips at front of cove.
✔ Clear cove heights greater than 1′–0″ are more efficient.
✔ Coves using asymmetric luminaires specifically designed for cove applications are more efficient. See 11–69.

Details Concepts and Effects

11

T5/HO Striplights **Back of Cove**
⮑ Possible vendors: Bartco, Legion, Lightolier, **Lithonia**[1]
[1] T12 photometry was used and rerated for T5/HO lamping. This is somewhat imprecise.

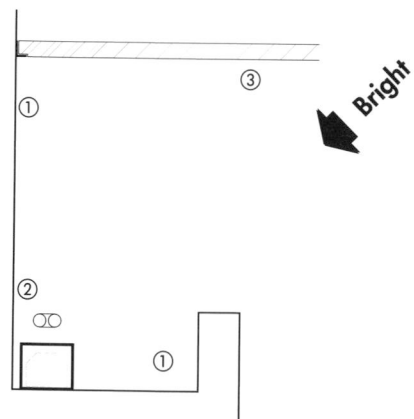

Bright

Notes
① Paint all interior cove surfaces matte white.
② Socket shadows are less noticeable along back cove wall and at ceiling with LongCompact lamps in striplight in this location.
③ Ceiling reflectance significantly influences success of cove. Higher is better (preferably 90 percent).

Application Key

Commercial
Gallery
Hospitality[1]
Institutional
Manufacturing
Residential
Retail
Exterior

1/includes casual commercial and institutional applications and decorative retail applications

bold = primary application
partial fade = minimal application
fade = unlikely application

LongCompact Striplights **Back of Cove**
⮑ Possible vendors: Columbia, DayBrite, Legion, Lightolier, **Lithonia**[1], Metalux
[1] T12 photometry was used and rerated for LongCompact lamping. This is somewhat imprecise.

Net Addresses/Luminaires
See page 11–5

DetailConcepts™

Coves Striplights

12′ Ceiling Height/4′ x 12′ Cove Opening/1′–0″ Clear Cove Height

Performance Perspectives

[All data outlined here is based on information from boldfaced manufacturer's published data at time of manuscript preparation. Cove opening located in room corner and starts 2′ from each of two walls. Maintained values based on analysis grid 7′x15′ centered under cove and on 0.86 light loss factor (0.95 lamp lumen depreciation; 0.95 luminaire dirt depreciation; 0.95 room surface dirt depreciation); ballast factor of 0.98 and very light reflectances: ceiling 90%; wall 50%; and floor 20%. All views put eyes at 5′–6″ AFF—cove perspective varies as ceiling height changes and/or cove height changes.]

▶ **Cove height: 1′–0″ clear**
Central ceiling height: 12′
Perimeter ceiling height: 10′–6″±
Lamp: F28T5/830 or F32T8/830
Floor illuminance (7′x15′ area under cove):
▶ 18 fc, avg, maintained (26 fc max)

▶ **Cove height: 1′–0″ clear**
Central ceiling height: 12′
Perimeter ceiling height: 10′–6″±
Lamp: F54T5/HO/830
Floor illuminance (7′x15′ area under cove):
▶ 27 fc, avg, maintained (40 fc max)

▶ **Cove height: 1′–0″ clear**
Central ceiling height: 12′
Perimeter ceiling height: 10′–6″±
Lamp: F40Long/830
Floor illuminance (7′x15′ area under cove):
▶ 35 fc, avg, maintained (53 fc max)

Details
Concepts and Effects

11

Coves Striplights

12′ Ceiling Height/4′ x 12′ Cove Opening/ 1′–0″ Clear Cove Height

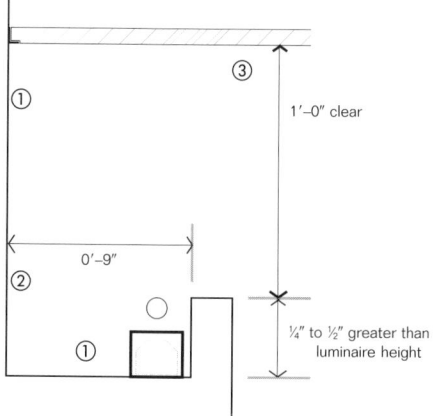

1′–0″ clear

0′–9″

¼″ to ½″ greater than
luminaire height

Notes
① Paint all interior cove surfaces matte white.
② Socket shadows are less noticeable with 1–lamp striplight in this location. Shadow line on ceiling may be visible a few feet or so out from front edge of cove lip, however.
③ Ceiling reflectance significantly influences success of cove. Higher is better (preferably 90 percent).
④ T8 and T5 lamps offer same light output—so striplights can be either version. Lamps must be combined with correct ballast and are not interchangeable after installation. See below for T5/HO version.

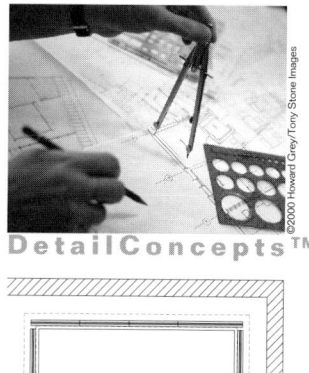

DetailConcepts™
©2000 Howard Grey/Tony Stone Images

T5 and T8 Striplights**Front of Cove**
➱Possible vendors: Bartco, Columbia, DayBrite, Legion, Lightolier, **Lithonia**[1], Metalux
[1]T12 photometry was used and rerated for T8 lamping. This is somewhat imprecise.

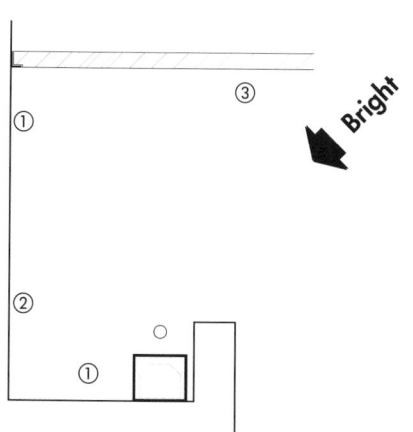

Bright

Notes
① Paint all interior cove surfaces matte white.
② Socket shadows are less noticeable with 1–lamp striplight in this location. Shadow line on ceiling may be visible a few feet or so out from front edge of cove lip, however.
③ Ceiling reflectance significantly influences success of cove. Higher is better (preferably 90 percent).

Design Tips
✔ Heavily textured ceiling material will create some shadows; as will tegular tile.
✔ If ceiling tone is other than white, then consider painting interior of cove surfaces to match ceiling.
✔ Clear cove heights greater than 0′–6″ are more efficient.
✔ Coves using asymmetric luminaires specifically designed for cove applications are more efficient. See 11–69.

Details
Concepts and Effects

11

T5/HO Striplights**Front of Cove**
➱Possible vendors: Bartco, Legion, Lightolier, **Lithonia**[1]
[1]T12 photometry was used and rerated for T5/HO lamping. This is somewhat imprecise.

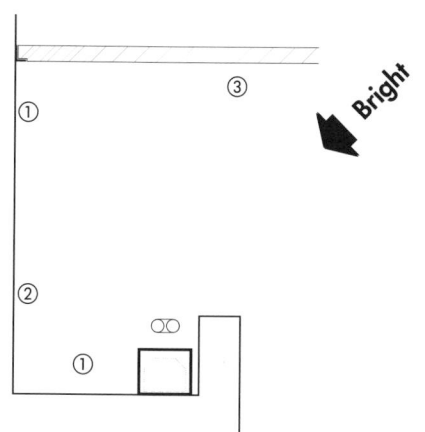

Bright

Notes
① Paint all interior cove surfaces matte white.
② Socket shadows are not noticeable with LongCompact lamps in striplight in this location. Shadow line on ceiling may be visible a few feet or so out from front edge of cove lip, however.
③ Ceiling reflectance significantly influences success of cove. Higher is better (preferably 90 percent).

Application Key

Commercial
Gallery
Hospitality[1]
Institutional
Manufacturing
Residential
Retail
Exterior

1/includes casual commercial and institutional applications and decorative retail applications

bold = primary application
partial fade = minimal application
fade = unlikely application

LongCompact Striplights**Front of Cove**
➱Possible vendors: Columbia, DayBrite, Legion, Lightolier, **Lithonia**[1], Metalux
[1]T12 photometry was used and rerated for LongCompact lamping. This is somewhat imprecise.

Net Addresses/Luminaires
See page 11–5

©2000 Howard Grey/Tony Stone Images

DetailConcepts™

Coves Asymmetric Luminaires

12′ Ceiling Height/4′ x 12′ Cove Opening/1′–0″ Clear Cove Height

Performance Perspectives

[All data outlined here is based on information from boldfaced manufacturer's published data at time of manuscript preparation. Cove opening located in room corner and starts 2′ from each of two walls. Maintained values based on analysis grid 7′x15′ centered under cove and on 0.86 light loss factor (0.95 lamp lumen depreciation; 0.95 luminaire dirt depreciation; 0.95 room surface dirt depreciation); ballast factor of 0.98 and very light reflectances: ceiling 90%; wall 50%; and floor 20%. All views put eyes at 5′–6″ AFF—cove perspective varies as ceiling height changes and/or cove height changes.]

▶ **Cove height: 1′–0″ clear**
Central ceiling height: 12′
Perimeter ceiling height: 10′–6″±
Lamp: F32T8/830
Floor illuminance (7′x15′ area under cove):
▶ 20 fc, avg, maintained (29 fc max)

▶ **Cove height: 1′–0″ clear**
Central ceiling height: 12′
Perimeter ceiling height: 10′–6″±
Lamp: F54T5/HO/830
Floor illuminance (7′x15′ area under cove):
▶ 35 fc, avg, maintained (50 fc max)

▶ **Cove height: 1′–0″ clear**
Central ceiling height: 12′
Perimeter ceiling height: 10′–6″±
Lamp: F40Long/830
Floor illuminance (7′x15′ area under cove):
▶ 42 fc, avg, maintained (67 fc max)

Details
Concepts and Effects

11

Coves Asymmetric Luminaires

12′ Ceiling Height/4′ x 12′ Cove Opening/1′–0″ Clear Cove Height

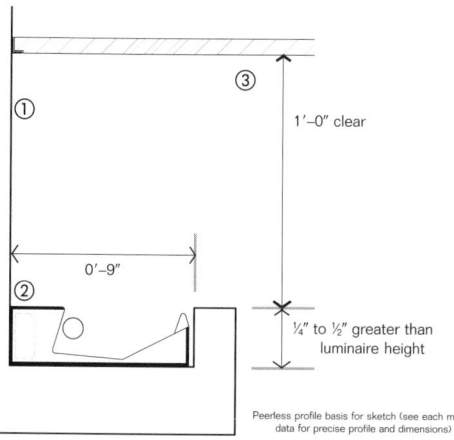

1′–0″ clear

0′–9″

¼″ to ½″ greater than luminaire height

Peerless profile basis for sketch (see each mfgr's data for precise profile and dimensions)

Notes
①Paint all interior cove surfaces matte white.
②Socket shadows are not noticeable with optically designed asymmetric luminaires.
③Ceiling reflectance significantly influences success of cove. Higher is better (preferably 90 percent).

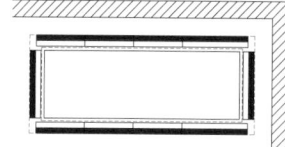

DetailConcepts™

©2000 Howard Grey/Tony Stone Images

T8 Asymmetric Luminaires

⊃Possible vendors: Columbia (PIC), Elliptipar (301, 306/T8), Ledalite (In–Cove Asymmetry™, LITECONTROL (Cove–45), Neoray (74IC), **Peerless** (ECX®)

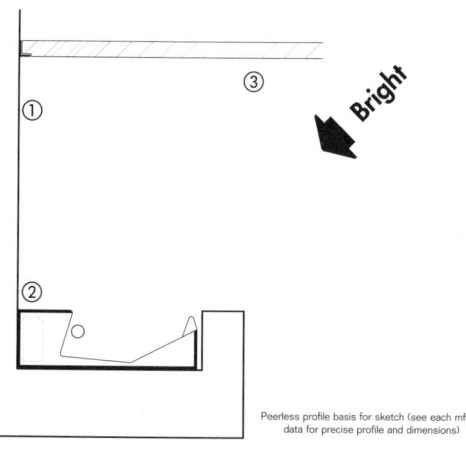

Bright

Peerless profile basis for sketch (see each mfgr's data for precise profile and dimensions)

Notes
①Paint all interior cove surfaces matte white.
②Socket shadows are not noticeable with optically designed asymmetric luminaires.
③Ceiling reflectance significantly influences success of cove. Higher is better (preferably 90 percent).
④Standard T5 lamp/ballast will produce 42% less light and use 45% less energy.

Design Tips
✔Asymmetric luminaires will perform better than striplights—better uniformity across the ceiling and no socket shadows.
✔Asymmetric luminaires are more efficient in coves than striplights.

Details Concepts and Effects

11

T5/HO Asymmetric Luminaires

⊃Possible vendors: Elliptipar (303, 305), **Peerless** (ECX®)

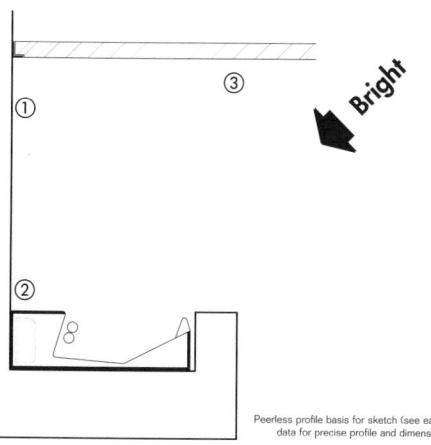

Bright

Peerless profile basis for sketch (see each mfgr's data for precise profile and dimensions)

Notes
①Paint all interior cove surfaces matte white.
②Socket shadows are not noticeable with optically designed asymmetric luminaires.
③Ceiling reflectance significantly influences success of cove. Higher is better (preferably 90 percent).

Application Key

Commercial
Gallery
Hospitality[1]
Institutional
Manufacturing
Residential
Retail
Exterior

1/includes casual commercial and institutional applications and decorative retail applications

bold = primary application
partial fade = minimal application
fade = unlikely application

LongCompact Asymmetric Luminaires

⊃Possible vendors: Columbia (PIC), Elliptipar (302), ELP (CL series), Ledalite (In–Cove Asymmetry™), Lightolier (Covelite™), LITECONTROL (Cove–45), Neoray (74IC), **Peerless** (ECX®)

Net Addresses/Luminaires
See page 11–5

Trademarks, service marks and product names are owned and registered by respective manufacturers.

Coves Asymmetric Luminaires

Examples

DetailConcepts™

Figure 11.1

Asymmetric cove luminaires can be used to provide ambient light in situations where a cove alone is too simple or institutional for the setting. Here, an architectural trellis of wood members is used to introduce warmth and intimacy. To maintain flexibility of table arrangements, minimize overhead visual clutter, and limit any direct glare exposure for most patrons, no pendent lights or accents aimed onto the tables were introduced. An asymmetric cove luminaire located at ① introduces a soft wash of light onto an upper ceiling, which reflects into the trellis. The trellis blocks more than half of the light which might otherwise be available from the cove lighting—F50Long/830 lamps were used with dimming ballasts to enable an intensity change over the course of a day. Asymmetric uplight is Lithonia/Peerless. Wallslot along mural wall is Hubbell/Sterner with Norton baffles and 100PAR38/HIR/FL lamps. Small sconces use compact fluorescent lamps and are Baldinger Architectural Custom Lighting. Also see Figure C25.

Details
Concepts and Effects

11

Image copyright 2000 C. M. Korab

Other sections in this book are illustrated in black and white. Light, however, significantly influences color perception and colored light can greatly enhance spatial and emotional experiences. Here, then, are color images of some of the projects illustrated throughout the rest of the text.

Color

and Light

Figures C.1, C.2 and C.3

On page 1 (previous page), Figures C.1 (top), C.2 and C.3 (bottom) represent time–of–day changes in the electric lighting of a hotel lobby. An astronomical timeclock slowly fades from one light intensity and/or color to another eight times in 24 hours. This is one technique to keep the visual environment fresh and exciting for guests throughout a stay. The hotel reception desk (just visible in the right background) also sees lighting changes during the course of the day. The slot above the water wall uses 100PAR38/HIR/FL lamps (p5–55). Spacing lamps on 6–inch centers with a third of the lamps dipped in blue gel, a third in amber gel and a third in red gel respectively, a range of colored light is then available depending on the intensity of the various colored lamps. The uplights are lamped with 100PAR38/HIR/NSP lamps (p5–52). Every other uplight is fitted with a blue gel.

Image copyright 2000 C. M. Korab

Figure C.4 (lower left)
Fountain

Sixty four GE Q500PAR56/NSP (quartz, 500–watt, narrow spot) lamps are used to illuminate the fountain feature. With multicolored glass gels or filters and with a programmable astronomical timeclock, lighting intensities, duration of effect and color are programmed to vary uniquely every few minutes over a 90–minute light show. The light show then repeats until the system automatically shuts down at midnight. Since the water jets are programmable, lighting can be coordinated to accentuate water jet fluctuations and overall fountain height.

Facade

Architectural details and facade features are highlighted, but with careful attention to color of light and color rendering of light. Details and features finished in earthtones are accented with warmtone lamps—F32Triple/830 fluorescent (p8–18) or 100ED17/CMH/3K/C/U ceramic metal halide (p7–10) lamps offer a warm light and also have excellent color rendering of warm finishes. Naturally cooltoned materials, like the verdigris copper chimneys and select evergreens are highlighted with 5700°K mercury lamps which accentuate blue–green finishes and provide an eerily bluetoned light.

Image copyright 2000 C. M. Korab

Figure C.5

Fountain lights are shown fading from amber to red. Although overexposed here, the trees in the background are lighted alternatively with 100ED17/CMH/3K/C/U ceramic metal halide (p7–10) and 5700°K mercury vapor lamps. Mecury vapor lamps are used primarily on coniferous trees for a year–round punch of accented bluegreen color. The warmtone ceramic metal halide lamps are used on deciduous trees for better fall color rendering of leaves as well as a more natural bark–brown tone during the winter.

Color

and Light

Images copyright 2000 C. M. Korab

Color and Light

Figure C.6 (top left)

The main entry to the hotel is accentuated much the same as the rear facade shown in Figure C.4. Warmtone light sources help set the scene for a comfortable, inviting arrival and interior atmosphere.

Figure C.7 (top right)

At the porte cochere, custom luminares are fitted with F32Triple/830 compact fluorescent lamps (p8–18) and suitably rated low temperature ballasts. To further enhance the warmth of light, the luminaire lenses are a variegated marble swirl of beige, amber and light brown acrylic (Sterling Products Natural Horn). These acrylic panels are sandblasted on the exterior to provide a soft look, belying the fact that the panels are acrylic.

Figure C.8 (lower left)

Site bollards developed for major pedestrian ways throughout the hotel site are of a style, scale and light output quality not common with standard commercial bollards. To enhance a warm, pleasant experience, the bollards use the same F32Triple/830 compact fluorescent lamp (p8–18) found in the porte cochere lanterns. This also consolidates lamp types for maintenance. The lens on the bollard also matches that of the porte cochere lantern. Finally, the bollard is just over 5 inches in diameter and stands at 36 inches in height for a nice human scale element.

Figure C.9 (lower right)

Upper facade elements are both back and front lighted. The steel "Y" is uplighted by a 100PAR38/CMH/3K/NFL ceramic metal halide lamp (p7–72) on either side. At the juncture of the legs of the "Y," a modified jelly jar luminaire is used to backlight the "Y." The jelly jar is dipped in an amber gel, so that when combined with the site–standard 100ED17/CMH/3K/C/U ceramic metal halide lamp (p7–10) a slightly warmer amber glow results. At the lower facade, the stone walls are uplighted with 100ED17/CMH/3K/C/U ceramic metal halide lamps (p7–10). Evergreen plants are lighted with 5700°K mercury lamps, while deciduous plants are lighted with 100ED17/CMH/3K/C/U ceramic metal halide lamps (p7–10). It is important to note: only ceramic metal halide lamps can create this closeness to incandescent color quality and color rendering. Standard metal halide, even warmtone versions of standard metal halide CANNOT achieve this look (color temperature) and quality (color rendering and color consistency) of light (see Figure C.41).

Color

and Light

Figure C.10 (right)

Wall sconces in the hotel lobby were custom made for several reasons. First, the style and scale of the interior architecture was not well represented with off–the–shelf solutions. Second, the articulation of the architecture required a corner–made sconce (lower left of Figure C.10). Finally, to enhance the warmth of the lobby, warmtoned variegated acrylic lens panels are used. The panels are sandblasted on the outside for a soft, nonplastic look.

Figure C.11 (lower right)

The registration desk and guest lounge (off to the left) are contiguous with the water wall feature shown in Figures C.1, C.2 and C.3. As the water wall changes in light color and intensity, the registration desk lobby also sequences through intensity changes and ceiling color changes. Neon (small diameter cold cathode) lamps are aligned three across in the faux ceiling timbers. One row of neon is triphosphor white, another row is blue and a third row is magenta. These colors are programmed to cross fade eight times in a 24–hour period.

The large pendent luminaire is fitted with sixteen F50Long/830/RS long compact fluorescent lamps (p8–38). Eight lamps sit on top of a metal pan which is regressed a few inches below the top edge of the pendent uplighting and accentuating the circular drywall reveal in the ceiling. The bottom portion of the pendent is fitted the other eight lamps and with Sterling Natural Horn acrylic panels which emphasize the warmth of the compact fluorescent lamps.

Figure C.12 (below)

The magenta neon has cycled to "full up" and the other two neon lamp colors in the faux timbers are "off," creating the intense warmth here. Such color changes can be timed with time–of–day situations like sunset or sunrise (mixing magenta and blue, for example). Additionally, the intensity can be changed over time to offer an impression of brightness which tracks daylight hours (e.g., by noon, the white neon could be dimmed up to full output and then as the early afternoon progresses, blue can be turned up for a blue/white crescendo midafternoon). Since warmtoned surfaces tend to recede in view and cooltoned surfaces tend to advance, it could be said that the magenta ceiling tone makes the ceiling appear closer (or somewhat lower) to the observer than the blue ceiling.

Image copyright 2000 C. M. Korab

Color
and Light

Color and Light

Figures C.13, C.14, C.15, CI16, C.17, C.18, C.19 and C.20 (from top right and clockwise)

The hotel ballroom has three major lighting elements along with traditional downlights. First, large, custom chandeliers house colored neon lamping (copper, magenta, blue and triphoshpor white). Second, faux timbers house colored fluorescent lamping (blue and gold). Third, vertical wall elements are backed with colored neon lamping. These various elements are controlled by preprogrammed scenes. From C.13 (all blue lamps are full on) to C.20 (all copper colored neon lamps are full on), various scenes can be set—some appropriate for dining (perhaps C.17 for breakfast; C.19 for lunch; and C.21 (below) for dinner) and some more appropriate for parties or transitions from one scene to another.

Figure C.21 (below)

A scene which may be used for dinner or as a prefunction to dinner is shown here. The "blue sky" could be ghastly on skintones and food, but the copper pendents along with the more focused 100PAR38/HIR/FL–lamped (p5–55) downlights set an appropriate color tone at the table for facial rendering and food rendering.

Color
and Light

Color and Light

Image copyright 2000 C. M. Korab

Figure C.22 (left)

The indoor pool is covered with a beautiful wood structure—a strikingly warm contrast to the coolness of the pool. To accentuate this cocoon, the ceiling was kept clear of monstrous ducts, vents and pipes. Further, downlighting was kept to a minimum and the few necessary units were tucked up into the wood trusswork. The ceiling is uplighted from a perimeter valance detail which also holds all of the mechanical ductwork. This valance, to limit its mass, was fitted with lenses and backlighted. Decorative sconces are used at the stone pilasters to emphasize the pilasters and to introduce a more humanscale lighting element.

Figure C.23 (above)

The play of acrylic lens media is important to enhancing the visual cues provided by the lighting. Sconces are lensed in Tea Stained acrylic by Sterling Products. This acrylic is sandblasted on the front side for a soft, matte look (belying the fact that the lens is plastic). The Tea Stained color, when backlighted with an F50Long/830 long compact fluorescent lamp (p8–38) provides an amber (more like dimmed incandescent) glow—effecting the more humanscale character of the light. The valance running above the sconce and around the pool house uses two kinds of acrylic. Above each sconce, a "float glass green" acrylic panel, also sandblasted on the front side, mimics backlighted float glass (standard clear glass has enough iron content to appear slightly green). The remainder of the acrylic panels are Natural Horn by Sterling Products—offering a soft, warm white glow when backlighted with F32T8/830 tubular fluorescent lamps (p8–52). This juxtaposition of these various toned backlighted acrylic elements helps introduce visual interest and visual texture to the lighting elements and complements the overall warmth of the wood ceiling enclosure.

color
and Light

Image copyright 2000 C. M. Korab

Figure C.24

The hotel prefunction lighting very much was intended to support the artwork, the architectural finishes and the tasks of the space—pre– and post–function activities as well as breakout functions associated with the large ballrooms (entries along the left wall). Asymmetric fluorescent uplights hidden in the millwork faux beam elements running the length of the Prefunction space uniformly accentuate the sepia–toned mural ceiling. 3000°K lamps, F50Long/830 (p8–38), are used to achieve maximum light output over the length of the asymmetric uplight by Lithonia/Peerless. 3000°K triphosphor neon hidden in the lower portion of the millwork faux beam elements add a layer or architectural texture by separating the beam from the wall. The sepia–toned ceiling mural adds significant warmth to the room, especially as the warm, high–color–rendering light from the deluxe triphosphor lamp interacts with it. Lightolier downlights with 250PAR38/H/SP lamps (p3–120) provide some feature downlighting onto the custom carpet inlays. Lightolier adjustable accents lamped with 39PAR20/CMH/NSP lamps (p7–36) accent the centerpiece.

Figure C.25

A trellis in the dining room is fitted with neon 3000°K triphosphor white lamps (on the top of selected wood members) and with copper neon (on top of other selected wood members). The mural is lighted with dimmed 100PAR38/HIR/FL lamps (p5–55) in a slot detail (grazing the mural wall). Decorative sconces are custom–fitted with F10Quad/830 lamps—providing an ADA–compliant sconce with very long lamp life and very low energy consumption. The apperance of an incandescent sconce was achieved by combining the fluorescent lamp with a parchment shade.

Figure C.26

The hotel bar is lighted with multicolored neon coves, 50MR16/C/CG/NFL (GE) (p4–157) accents and custom decorative animal hide pendents (lamped with 50TB19/H lamps (p3–8)). Each lighting "layer" is on a separate control zone so that during the course of the 18–hour sales period each day the lighting can be slowly changed (using long fade rates) from various intensities and colors. The neon cove consists of neo–blue, magenta, and copper colored tubing (by Voltarc and installed by Heller Signs). The interior of the cove is painted out to match the ceiling (Benjamin Moore Linen White) for maximum reflection without compromising colored surface integrity or reflected color of light.

Image copyright 2000 C. M. Korab

Figure C.27

The Elizabeth Arden boutique on Fifth Avenue in New York City has lighting expressive of the elegance of the product line and of the architectural setting. The narrow, long plan limits daylight penetration. An uplight cove along one wall continues the sense of daylight brightness into the store, but with warmtone light more favorable to skintones. Additionally, the warmtoned light has a more inviting appeal during darkened day conditions and in the evening. Finally, warmtoned light exudes a softer more casual and comfortable experience than what is typically experienced with cooltoned sources. The cove uplight uses the F28T5/830 tubular fluorescent lamps (p8–48). The small diameter lamp placed in a very small optically active luminaire by Specialty Lighting allowed the entire cove width to remain less than 100mm (4 inches). This allowed for the integration of the lighting with the millwork without sacrificing precious floor space. Finally, the show window and interior entry area are accented with 39PAR20/CMH/NSP ceramic metal halide lamps (p7–36).

Image copyright 2000 Laszlo Regos

Color

and Light

Figure C.28

The signature–red accent on the entry door (shown in Figure C.27) is repeated in the elevator lobby to the upper floor salons. Note the elevator lobby is even visible from streetside (lower middle of Figure C.27). To enhance the lacquer finish of the red walls without introducing glary reflections, 35PAR20/H/FL lamps (p3–36) are used on close spacings in a wallslot detail. Each lamp is separated by a metal baffle to limit long–run views of the lamps and sockets (bottoms of baffles are just visible in the slot running across the transome between the boutique and the elevator lobby).

Vertical displays along the wall on the right are cross washed with F28T5/830 tubular fluorescent lamps (p8–48) running vertically. 42MR16/C/NSP (p4–123) and 35MR16/C/FL (p4–121) lamps in slot luminaires overhead (see the small rectangular units in the ceiling in Figures C.27 and C.29) are then used to front light portions of the display. These MR16 lamps are also used to highlight the foreground seat and the floral arrangement in the corner.

Image copyright 2000 Laszlo Regos

Figure C.29
A view of the boutique from the elevator lobby back toward the entry shows the three major lighting layers or components used—the cove uplight using F28T5/830 lamps (p8–48) along one wall; the display millwork integrated lighting; and the MR16 rectangular slot accents by Specialty Lighting. To minimize visual clutter in the ceiling, the MR16 slot accents were used. These luminaires are about 12 inches in length and 4 inches in width. Each luminaire holds three MR16 adjustable sockets. Architectural surface and light colors are neutral warm white so that the clients' skintones and makeup are the focus of attention along with the "eye candy" appeal of the colored liquid and bottles of product on display.

The art of modern elegance involves presenting one's own personality to the greatest possible advantage.

Elizabeth Arden, 1938

" The thought of beautiful women all over the world, walking in and out of my Salons gives me a deep and satisfying happiness."

Elizabeth Arden, 19

Color and Light

Figure C.30 (right)
At the Steelcase Inc. Headquarters in Grand Rapids, Michigan, 35MR16/C/CG/FL lamps (p4–153) in low voltage cable–light luminaires by Translite provide low lighting for notetaking during AV presentations in this meeting and workroom. Blue gels on a select few of these luminaires are aimed up onto the painted–out white deck.

Figure C.31 (below)
Blue filters on 35MR16/C/CG/FL (p4–153) lamps in low voltage cable light luminaires in a skylight near the meeting/workroom help evoke the feeling of daylight in early morning and late afternoon in the winter and prevent the skylight from being a blackhole during those times. Blue, satin glass sconces by Leucos in small work enclaves (intended for temporary breakouts to make phone calls and/or check e–mail) are lighted with 3000°K compact fluorescent lamps.

Image copyright 2000 and courtesy of Steelcase Inc.

Color
and Light

Image copyright 2000 and courtesy of Steelcase Inc.

Figure C.32
For work sessions, general meetings and teleconferences at the Steelcase Headquarters meeting/workroom, ambient lighting is introduced from cove uplighting along the perimeter of the community center as well as from small profile pendent lumianaires (each about 2½ inches square in cross section). The cove uplighting uses asymmetric luminaires from Lithonia/Peerless lamped with F40Long/830 lamps (8–36). The small profile pendent lights are lamped with the same lamps. This lighting provides diffuse light for good facial recognition and task visibility while still permitting acceptable visibility of the projection screens.

Color and Light

Figure C.33

Ambient lighting in the open office area near the meeting/workroom is achieved using the same small profile extrusion indirect luminaire from Lithonia/Peerless with F32T8/830 lamping (8–52). Note the impact of furniture and flooring coloration as well as the impact of the blue gels on the MR16s in Figures C.30 and C.31 on spatial impressions. The entry to the meeting/workroom is just off the image to the right.

Color

and Light

Figure C.34

Each of the mustard piers slides into a ceiling slot into which is placed a 39PAR30L/CMH/3K/NFL (p7–38). For a fresh approach to wall lighting, and because the wall slides through a slot in the floor and onto the second floor, metal halide floor uplights are used to light the left face of the elevator core in the background. Monopoint accents lamped with 35MR16/C/CG/FL (p4–153) are used at the mustard pier near the entry to highlight artwork.

Image copyright 2000 and courtesy of Steelcase Inc.

Color and Light

Figure C.35 (left)
This closeup is taken from Figure C.34. The mustard "triple pier" detail in the entry lobby is accented with 39PAR30L/CMH/3K/NFL ceramic metal halide lamps (p7–38).

Figure C.36 (middle)
The Steelcase Corporate Services Center in Grand Rapids, Michigan, combines architectural surface color with accent lighting and colored light to provide interesting visual texture to the interiors. Note the front entry is lighted with in–pavement uplights and small scale bollards—offering a soft, more inviting entry than typically achieved with downlights, floodlights and/or postlights.

Figure C.37 (bottom)
The red interior wall is highlighted while in the distant background cove uplights using F40B (blue) fluorescent lamps offer a daylight–like crispness.

Images copyright 2000 and courtesy of Steelcase Inc.

Color and Light

Figure C.38 (below) and C.39 (right)

The Giorgio Beverly Hills Boutique uses color temperature shifts in lighting to evoke the sense of daylighted surfaces. The ceiling detail (running the length of the boutique along right) is lit with 5000°K F50Long lamps. The fluorescent–lighted wall cutouts also use 5000°K F50Long compact fluorescent lamps. These lamps offer a cooltoned appearance but maintain high color rendering necessary for richer looking merchandise and skintones. The uplighting along the wall at the left is lamped with F40Long/830 fluorescent lamps (p8–36). Halogen accenting offers warm white accenting giving the impression of visually advancing fashions against a receding background. With the 18–foot ceiling height, 50PAR36/H/VNSP5° lamps (p4–180) were used for highlighting merchandise within the wall niches.

Image copyright 2000 Stephen Graham

Color and Light

Color and Light

Figure C.41 (right)
The University of Michigan Lurie Tower (North Campus Carillon) was completed in 1997, just prior to the advent of ceramic metal halide lamps. 3200°K metal halide ED17 lamps are used in recessed spreadlens wallwash luminaires to light the lower interior brick walls. A single in–ground uplight also uses a 3200°K ED17 lamp. 3000°K metal halide ED17 spot accents are used on the upper tower. Note the variation in color of light on the central upper–tower accent. This is a characteristic of the original metal halide lamp technology. Ceramic metal halide lamps provide much better color consistency and better color rendering (as is evident in Figures C.4 and C.9).

Figure C.40
This view from a feature balcony illustrates the concentrated beam spread of the 50PAR36/H/VNSP5° lamps (p4–180) in adjustable accents in the high ceiling. The display of fragrance and skin care products off to the right is lighted by 50MR16/C/CG/FL (GE) lamps (p4–159) in adjustable accent lights recessed into the floating ceiling plane.

Color and Light

Color
and Light

Images copyright 2000 Hedrich-Blessing/Quesada

Figure C.42 (above) and C.43 (right)

St. Philip the Apostle Church, in Lewisville, Texas, is lighted primarily by suspended extruded luminaires of about 3 inches in width by about 7 inches in height. These luminaires, by Lithonia/Peerless, use F40Long/830 lamps in an upper compartment to highlight the beautiful wood cathedral ceiling and visually expand the volume of the sanctuary. In the lower compartment, the same lamps are used, with selected lamps on dimmer so that various levels can be achieved for various liturgical settings. Accent lighting in lower ceiling areas is achieved with 50PAR30S/HIR/NFL lamps (p5–28) in adjustable accents. Accenting of major focals (e.g., altar, podium, presider's chair) is achieved with 240PAR56/VNSP lamps in theatrical, pipe–mounted luminaires (see the central oculus).

Color and Light

Figure C.44
The daylighting was deemed as a necessary and very desirable part of the overall experience. However, it was recognized that uncontrolled daylighting could wreak havoc on viewers vision during many early morning services. The clerestory was thus designed with a system of large, fixed interior millwork louvers, the angles of which were determined through a series of daylighting analyses for various seasons, times–of–day and sky conditions. The tapestry is lighted with 50PAR30S/HIR/NFL lamps (p5–28) in adjustable trackheads mounted up in a skylight wall along the wall.

Color
and Light

Color
and Light

Figure C.45 (left) and C.46 (below)

The daily chapel is rendered as a warm, inviting and intimate space by the wall of vigil candles developed by the interior designers (INAI Studios). A niche to the left of the chapel entry is lighted with an architectural slot detail containing three 35PAR20/H/FL lamps (p3–36). Below, the sacred post is highlighted by 50PAR30/HIR/ NFL lamps (p5–28) in recessed luminaires. The background garden is lit by a skylight. Adjustable accent lights for focal emphasis use 39PAR20/CMH/NSP lamps (p7–36).

Images copyright 2000 Hedrich–Blessing/Quesada

Color and Light

Image copyright 1992 Balthazar Korab, Ltd.

Figure C.47

The rotunda of the Michigan State Capitol, built in 1871, is a highly decorated example of Victorian public architecture. The oculus focuses view to a silver–starred midnight blue field nearly eight stories above the main floor. This oculus is lit with blue filtered metal halide floodlights. The dome encircling the oculus is uplighted from a sixth level balcony closed to the public and therefore capable of housing adjustable accents lamped with 250PAR38/H/SP lamps (p3–120) which accentuate the gilt surfaces beautifully. The lower dome is ringed with large paintings, each of which is accented with 250PAR38/H/SP lamps (p3–120) aimed from across the opposite side of the balcony. The ceilings of the lower rotunda floor balconies are uplighted with in–floor uplights (by Lithonia/Hydrel) lamped with 75PAR30L/H/NFL lamps (p3–120). Early twentieth–century cast glass lanterns hang from cantilevers and are lit with clear extended service incandescent lamps for an authentic Edison lamp appearance.

Coves Asymmetric Luminaires

Examples

DetailConcepts™

Figure 11.2

Asymmetric cove luminaires are used here in beam–like elements to wash across the ceiling of the corridor. Single–lamp cross section units were used with F32T8/830 lamps and ballasts with 1.18 ballast factors (for an additional 18 percent light output over rated lamp output). The lighting is sufficiently diffuse to allow recognition of other folks in the corridor for the entire length of the corridor, while minimizing glare situations. This works well for security cameras—enabling the surveillance of long distances and providing the ability to recognize facial expressions, clothing color and the like—there are no dark zones where perpetrators could go unrecognized. Finally, the diffuse uplight allows for good light at the door key/handle—whereas downlighting would create severe shadowing as guests would approach the entry and attempt to find the card key access slot. Asymmetric uplight is Lithonia/ Peerless.

Details Concepts and Effects

11

DetailConcepts™

Coves Striplights

12′ Ceiling Height/9′ x 9′ Cove Opening/0′–6″ Clear Cove Height

Performance Perspectives

[All data outlined here is based on information from boldfaced manufacturer's published data at time of manuscript preparation. Cove opening located in room corner and starts 2′ from each of two walls. Maintained values based on analysis grid 12′x12′ centered under cove and on 0.86 light loss factor (0.95 lamp lumen depreciation; 0.95 luminaire dirt depreciation; 0.95 room surface dirt depreciation); ballast factor of 0.98 and very light reflectances: ceiling 90%; wall 50%; and floor 20%. All views put eyes at 5′–6″ AFF—cove perspective varies as ceiling height changes and/or cove height changes.]

▶ **Cove height: 0′–6″ clear**
Central ceiling height: 12′
Perimeter ceiling height: 11′±
Lamp: F28T5/830 or F32T8/830
Floor illuminance (12′x12′ area under cove)
▶ 10 fc, avg, maintained (16 fc max)

▶ **Cove height: 0′–6″ clear**
Central ceiling height: 12′
Perimeter ceiling height: 11′±
Lamp: F54T5/HO/830
Floor illuminance (12′x12′ area under cove)
▶ 17 fc, avg, maintained (29 fc max)

▶ **Cove height: 0′–6″ clear**
Central ceiling height: 12′
Perimeter ceiling height: 11′±
Lamp: F40Long/830
Floor illuminance (12′x12′ area under cove)
▶ 20 fc, avg, maintained (30 fc max)

Details Concepts and Effects 11

Coves Striplights

12′ Ceiling Height/9′ x 9′ Cove Opening/0′–6″ Clear Cove Height

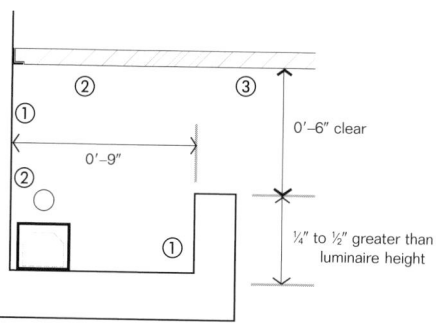

DetailConcepts™

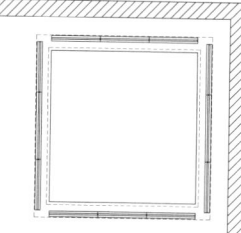

Notes
① Paint all interior cove surfaces matte white.
② Socket shadows are noticeable along back cove wall and somewhat noticeable at ceiling with 1–lamp striplight in this location. Single–lamp staggered strips introduce a "reverse socket shadow"—lamps closer to back cove wall create a slightly greater intensity. Butt strips end–to–end tightly.
③ Ceiling reflectance significantly influences success of cove. Higher is better (preferably 90 percent).
④ T8 and T5 lamps offer same light output—so striplights can be either version. Lamps must be combined with correct ballast and are not interchangeable after installation. T5 striplights are a few inches shorter than T8s. See below for T5/HO version.

Dimensions shown: 0′–6″ clear, 0′–9″, ¼″ to ½″ greater than luminaire height

T5 and T8 Striplights **Back of Cove**
➲ Possible vendors: Bartco, Columbia, DayBrite, Legion, Lightolier, **Lithonia**[1], Metalux
[1] T12 photometry was used and rerated for T8 lamping. This is somewhat imprecise.

Very harsh

Notes
① Paint all interior cove surfaces matte white.
② Socket shadows are noticeable along back cove wall and somewhat noticeable at ceiling with 1–lamp striplight in this location. Single–lamp staggered strips introduce a "reverse socket shadow"—lamps closer to back cove wall create a slightly greater intensity. Butt strips end–to–end tightly.
③ Ceiling reflectance significantly influences success of cove. Higher is better (preferably 90 percent).

T5/HO Striplights **Back of Cove**
➲ Possible vendors: Bartco, Legion, Lightolier, **Lithonia**[1]
[1] T12 photometry was used and rerated for T5/HO lamping. This is somewhat imprecise.

Very harsh

Notes
① Paint all interior cove surfaces matte white.
② Socket shadows are less noticeable along back cove wall and at ceiling with LongCompact lamps in striplight in this location.
③ Ceiling reflectance significantly influences success of cove. Higher is better (preferably 90 percent).

LongCompact Striplights **Back of Cove**
➲ Possible vendors: Columbia, DayBrite, Legion, Lightolier, **Lithonia**[1], Metalux
[1] T12 photometry was used and rerated for LongCompact lamping. This is somewhat imprecise.

Design Tips
✔ Lamp and ballast maintenance is difficult. Lay–in ceiling can help—lift tiles for better reach/view into cove.
✔ Heavily textured ceiling material will create significant shadows; as will tegular tile.
✔ If ceiling is drywall, then any imperfections will telegraph with such grazing light.
✔ If ceiling tone is other than white, then consider painting interior of cove surfaces to match ceiling.
✔ Corner shadowing less noticeable with strips at front of cove.
✔ Clear cove heights greater than 0′–6″ are more efficient.
✔ Coves using asymmetric luminaires specifically designed for cove applications are more efficient.

Details Concepts and Effects

11

Application Key
Commercial
Gallery
Hospitality[1]
Institutional
Manufacturing
Residential
Retail
Exterior

1/includes casual commercial and institutional applications and decorative retail applications

bold = primary application
partial fade = minimal application
fade = unlikely application

Net Addresses/Luminaires
See page 11–5

Coves Striplights

12' Ceiling Height/9' x 9' Cove Opening/0'–6" Clear Cove Height

DetailConcepts™

Performance Perspectives

[All data outlined here is based on information from boldfaced manufacturer's published data at time of manuscript preparation. Cove opening located in room corner and starts 2' from each of two walls. Maintained values based on analysis grid 12'x12' centered under cove and on 0.86 light loss factor (0.95 lamp lumen depreciation; 0.95 luminaire dirt depreciation; 0.95 room surface dirt depreciation); ballast factor of 0.98 and very light reflectances: ceiling 90%; wall 50%; and floor 20%. All views put eyes at 5'–6" AFF—cove perspective varies as ceiling height changes and/or cove height changes.]

▶ **Cove height: 0'–6" clear**
Central ceiling height: 12'
Perimeter ceiling height: 11'±
Lamp: F28T5/830 or F32T8/830
Floor illuminance (12'x12' area under cove)
▶ 14 fc, avg, maintained (26 fc max)

▶ **Cove height: 0'–6" clear**
Central ceiling height: 12'
Perimeter ceiling height: 11'±
Lamp: F54T5/HO/830
Floor illuminance (12'x12' area under cove)
▶ 24 fc, avg, maintained (39 fc max)

▶ **Cove height: 0'–6" clear**
Central ceiling height: 12'
Perimeter ceiling height: 11'±
Lamp: F40Long/830
Floor illuminance (12'x12' area under cove)
▶ 30 fc, avg, maintained (52 fc max)

Details
Concepts and Effects

11

Coves Striplights
12′ Ceiling Height/9′ x 9′ Cove Opening/0′–6″ Clear Cove Height

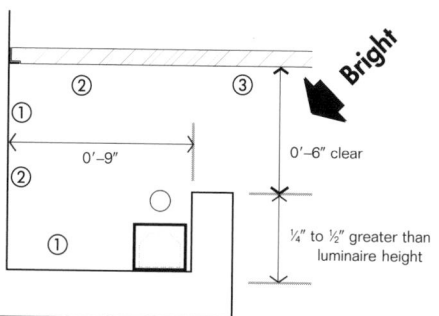

0′–9″

0′–6″ clear

¼″ to ½″ greater than luminaire height

Notes
①Paint all interior cove surfaces matte white.
②Socket shadows are less noticeable with 1–lamp striplight in this location. Shadow line on ceiling may be visible a foot or so out from front edge of cove lip, however.
③Ceiling reflectance significantly influences success of cove. Higher is better (preferably 90 percent).
④T8 and T5 lamps offer same light output—so striplights can be either version. Lamps must be combined with correct ballast and are not interchangeable after installation. See below for T5/HO version.

DetailConcepts™

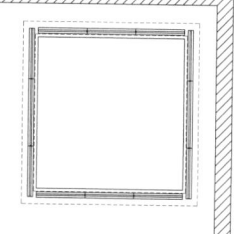

T5 and T8 Striplights**Front of Cove**
⊃Possible vendors: Bartco, Columbia, DayBrite, Legion, Lightolier, **Lithonia**[1], Metalux
[1]T12 photometry was used and rerated for T8 lamping. This is somewhat imprecise.

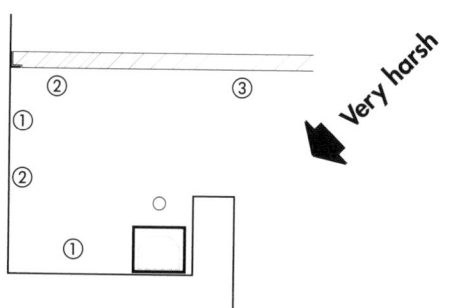

Notes
①Paint all interior cove surfaces matte white.
②Socket shadows are less noticeable with 1–lamp striplight in this location. Shadow line on ceiling may be visible a foot or so out from front edge of cove lip, however.
③Ceiling reflectance significantly influences success of cove. Higher is better (preferably 90 percent).

T5/HO Striplights**Front of Cove**
⊃Possible vendors: Bartco, Legion, Lightolier, **Lithonia**[1]
[1]T12 photometry was used and rerated for T5/HO lamping. This is somewhat imprecise.

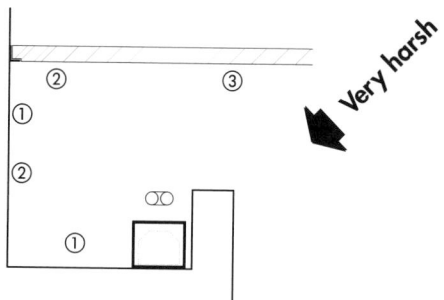

Notes
①Paint all interior cove surfaces matte white.
②Socket shadows are not noticeable with LongCompact lamps in striplight in this location.
③Ceiling reflectance significantly influences success of cove. Higher is better (preferably 90 percent).

LongCompact Striplights**Front of Cove**
⊃Possible vendors: Columbia, DayBrite, Legion, Lightolier, **Lithonia**[1], Metalux
[1]T12 photometry was used and rerated for LongCompact lamping. This is somewhat imprecise.

Design Tips
✓Lamp and ballast maintenance is difficult. Lay–in ceiling can help—lift tiles for better reach/view into cove.
✓Heavily textured ceiling material will create significant shadows; as will tegular tile.
✓If ceiling is drywall, then any imperfections will telegraph with such grazing light.
✓If ceiling tone is other than white, then consider painting interior of cove surfaces to match ceiling.
✓Clear cove heights greater than 0′–6″ are more efficient.
✓Coves using asymmetric luminaires specifically designed for cove applications are more efficient.

Details
Concepts and Effects

11

Application Key

Commercial
Gallery
Hospitality[1]
Institutional
Manufacturing
Residential
Retail
Exterior

1/includes casual commercial and institutional applications and decorative retail applications

bold = primary application
partial fade = minimal application
fade = unlikely application

Net Addresses/Luminaires
See page 11–5

©2000 Howard Grey/Tony Stone Images

DetailConcepts™

Performance Perspectives

[All data outlined here is based on information from boldfaced manufacturer's published data at time of manuscript preparation. Cove opening located in room corner and starts 2′ from each of two walls. Maintained values based on analysis grid 12′x12′ centered under cove and on 0.86 light loss factor (0.95 lamp lumen depreciation; 0.95 luminaire dirt depreciation; 0.95 room surface dirt depreciation); ballast factor of 0.98 and very light reflectances: ceiling 90%; wall 50%; and floor 20%. All views put eyes at 5′–6″ AFF—cove perspective varies as ceiling height changes and/or cove height changes.]

Coves Striplights

12′ Ceiling Height/9′ x 9′ Cove Opening/1′–0″ Clear Cove Height

▶ **Cove height: 1′–0″ clear**
Central ceiling height: 12′
Perimeter ceiling height: 10′–6″±
Lamp: F28T5/830 or F32T8/830
Floor illuminance (12′x12′ area under cove):
▶ 17 fc, avg, maintained (27 fc max)

▶ **Cove height: 1′–0″ clear**
Central ceiling height: 12′
Perimeter ceiling height: 10′–6″±
Lamp: F54T5/HO/830
Floor illuminance (12′x12′ area under cove):
▶ 28 fc, avg, maintained (43 fc max)

▶ **Cove height: 1′–0″ clear**
Central ceiling height: 12′
Perimeter ceiling height: 10′–6″±
Lamp: F40Long/830
Floor illuminance (12′x12′ area under cove):
▶ 35 fc, avg, maintained (54 fc max)

Details
Concepts and Effects

11

Coves Striplights

12′ Ceiling Height/9′ x 9′ Cove Opening/1′–0″ Clear Cove Height

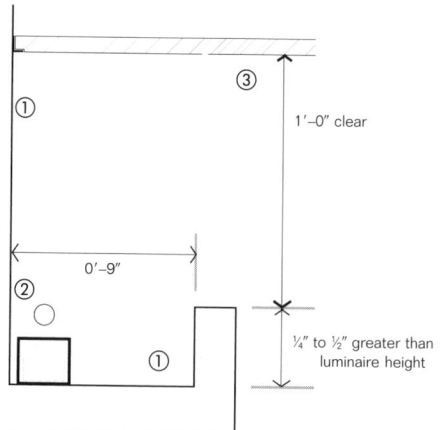

DetailConcepts™

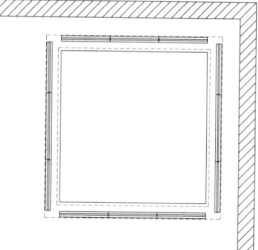

Notes
①Paint all interior cove surfaces matte white.
②Socket shadows are noticeable along back cove wall with 1–lamp striplight in this location. Single–lamp staggered strips introduce a "reverse socket shadow"—lamps closer to back cove wall create a slightly greater intensity. Butt strips end–to–end tightly.
③Ceiling reflectance significantly influences success of cove. Higher is better (preferably 90 percent).
④T8 and T5 lamps offer same light output—so striplights can be either version. Lamps must be combined with correct ballast and are not interchangeable after installation. See below for T5/HO version.

T5 and T8 Striplights **Back of Cove**
➲Possible vendors: Bartco, Columbia, DayBrite, Legion, Lightolier, **Lithonia**[1], Metalux
[1]T12 photometry was used and rerated for T8 lamping. This is somewhat imprecise.

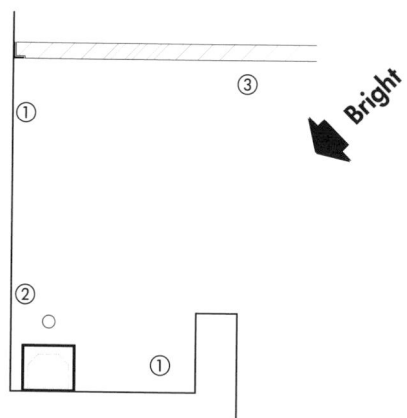

Notes
①Paint all interior cove surfaces matte white.
②Socket shadows are noticeable along back cove wall with 1–lamp striplight in this location. Single–lamp staggered strips introduce a "reverse socket shadow"—lamps closer to back cove wall create a slightly greater intensity. Butt strips end–to–end tightly.
③Ceiling reflectance significantly influences success of cove. Higher is better (preferably 90 percent).

T5/HO Striplights **Back of Cove**
➲Possible vendors: Bartco, Legion, Lightolier, **Lithonia**[1]
[1]T12 photometry was used and rerated for T5/HO lamping. This is somewhat imprecise.

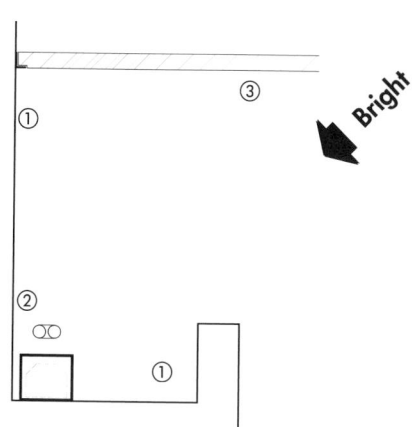

Notes
①Paint all interior cove surfaces matte white.
②Socket shadows are less noticeable along back cove wall and at ceiling with LongCompact lamps in striplight in this location.
③Ceiling reflectance significantly influences success of cove. Higher is better (preferably 90 percent).

LongCompact Striplights **Back of Cove**
➲Possible vendors: Columbia, DayBrite, Legion, Lightolier, **Lithonia**[1], Metalux
[1]T12 photometry was used and rerated for LongCompact lamping. This is somewhat imprecise.

Design Tips
✔Heavily textured ceiling material will create some shadows; as will tegular tile.
✔If ceiling tone is other than white, then consider painting interior of cove surfaces to match ceiling.
✔Corner shadowing less noticeable with strips at front of cove.
✔Clear cove heights greater than 1′–0″ are more efficient.
✔Coves using asymmetric luminaires specifically designed for cove applications are more efficient. See 11–81.

Details Concepts and Effects

11

Application Key
Commercial
Gallery
Hospitality[1]
Institutional
Manufacturing
Residential
Retail
Exterior

1/includes casual commercial and institutional applications and decorative retail applications

bold = primary application
partial fade = minimal application
fade = unlikely application

Net Addresses/Luminaires
See page 11–5

DetailConcepts™

Coves Striplights

12′ Ceiling Height/9′ x 9′ Cove Opening/ 1′–0″ Clear Cove Height

Performance Perspectives

[All data outlined here is based on information from boldfaced manufacturer's published data at time of manuscript preparation. Cove opening located in room corner and starts 2′ from each of two walls. Maintained values based on analysis grid 12′x12′ centered under cove and on 0.86 light loss factor (0.95 lamp lumen depreciation; 0.95 luminaire dirt depreciation; 0.95 room surface dirt depreciation); ballast factor of 0.98 and very light reflectances: ceiling 90%; wall 50%; and floor 20%. All views put eyes at 5′–6″ AFF—cove perspective varies as ceiling height changes and/or cove height changes.]

▶ **Cove height: 1′–0″ clear**
Central ceiling height: 12′
Perimeter ceiling height: 10′–6″±
Lamp: F28T5/830 or F32T8/830
Floor illuminance (12′x12′ area under cove)
▶ 19 fc, avg, maintained (31 fc max)

▶ **Cove height: 1′–0″ clear**
Central ceiling height: 12′
Perimeter ceiling height: 10′–6″±
Lamp: F54T5/HO/830
Floor illuminance (12′x12′ area under cove)
▶ 33 fc, avg, maintained (50 fc max)

▶ **Cove height: 1′–0″ clear**
Central ceiling height: 12′
Perimeter ceiling height: 10′–6″±
Lamp: F40Long/830
Floor illuminance (12′x12′ area under cove)
▶ 39 fc, avg, maintained (68 fc max)

Details Concepts and Effects

11

Coves Striplights
12' Ceiling Height/9' x 9' Cove Opening/1'–0" Clear Cove Height

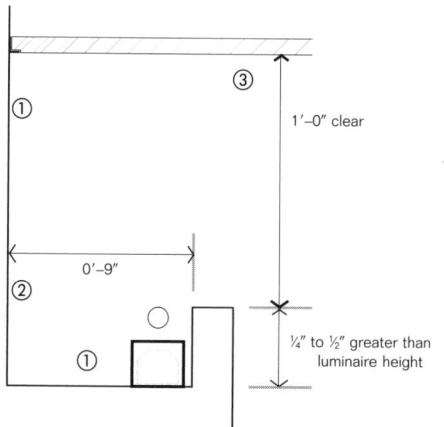

1'–0" clear

0'–9"

¼" to ½" greater than luminaire height

Notes
① Paint all interior cove surfaces matte white.
② Socket shadows are less noticeable with 1–lamp striplight in this location. Shadow line on ceiling may be visible a few feet or so out from front edge of cove lip, however.
③ Ceiling reflectance significantly influences success of cove. Higher is better (preferably 90 percent).
④ T8 and T5 lamps offer same light output—so striplights can be either version. Lamps must be combined with correct ballast and are not interchangeable after installation. See below for T5/HO version.

DetailConcepts™

©2000 Howard Grey/Tony Stone Images

T5 and T8 Striplights **Front of Cove**
⊃ Possible vendors: Bartco, Columbia, DayBrite, Legion, Lightolier, **Lithonia**[1], Metalux
[1] T12 photometry was used and rerated for T8 lamping. This is somewhat imprecise.

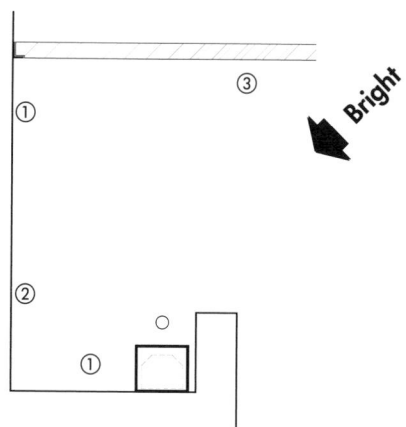

Bright

Notes
① Paint all interior cove surfaces matte white.
② Socket shadows are less noticeable with 1–lamp striplight in this location. Shadow line on ceiling may be visible a few feet or so out from front edge of cove lip, however.
③ Ceiling reflectance significantly influences success of cove. Higher is better (preferably 90 percent).

T5/HO Striplights **Front of Cove**
⊃ Possible vendors: Bartco, Legion, Lightolier, **Lithonia**[1]
[1] T12 photometry was used and rerated for T5/HO lamping. This is somewhat imprecise.

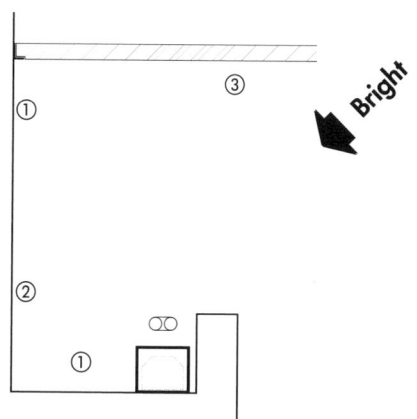

Bright

Notes
① Paint all interior cove surfaces matte white.
② Socket shadows are not noticeable with LongCompact lamps in striplight in this location. Shadow line on ceiling may be visible a few feet or so out from front edge of cove lip, however.
③ Ceiling reflectance significantly influences success of cove. Higher is better (preferably 90 percent).

LongCompact Striplights **Front of Cove**
⊃ Possible vendors: Columbia, DayBrite, Legion, Lightolier, **Lithonia**[1], Metalux
[1] T12 photometry was used and rerated for LongCompact lamping. This is somewhat imprecise.

Design Tips
✔ Heavily textured ceiling material will create some shadows; as will tegular tile.
✔ If ceiling tone is other than white, then consider painting interior of cove surfaces to match ceiling.
✔ Clear cove heights greater than 0'–6" are more efficient.
✔ Coves using asymmetric luminaires specifically designed for cove applications are more efficient. See 11–81.

Details Concepts and Effects

11

Application Key

Commercial
Gallery
Hospitality[1]
Institutional
Manufacturing
Residential
Retail
Exterior

1/includes casual commercial and institutional applications and decorative retail applications

bold = primary application
partial fade = minimal application
fade = unlikely application

Net Addresses/Luminaires
See page 11–5

Coves Asymmetric Luminaires

12′ Ceiling Height/9′ x 9′ Cove Opening/1′–0″ Clear Cove Height

DetailConcepts™

Performance Perspectives

[All data outlined here is based on information from boldfaced manufacturer's published data at time of manuscript preparation. Cove opening located in room corner and starts 2′ from each of two walls. Maintained values based on analysis grid 12′x12′ centered under cove and on 0.86 light loss factor (0.95 lamp lumen depreciation; 0.95 luminaire dirt depreciation; 0.95 room surface dirt depreciation); ballast factor of 0.98 and very light reflectances: ceiling 90%; wall 50%; and floor 20%. All views put eyes at 5′–6″ AFF—cove perspective varies as ceiling height changes and/or cove height changes.]

▶ **Cove height: 1′–0″ clear**
Central ceiling height: 12′
Perimeter ceiling height: 10′–6″±
Lamp: F32T8/830
Floor illuminance (12′x12′ area under cove)
▶ 22 fc, avg, maintained (33 fc max)

▶ **Cove height: 1′–0″ clear**
Central ceiling height: 12′
Perimeter ceiling height: 10′–6″±
Lamp: F54T5/HO/830
Floor illuminance (12′x12′ area under cove)
▶ 40 fc, avg, maintained (56 fc max)

▶ **Cove height: 1′–0″ clear**
Central ceiling height: 12′
Perimeter ceiling height: 10′–6″±
Lamp: F40Long/830
Floor illuminance (12′x12′ area under cove)
▶ 49 fc, avg, maintained (77 fc max)

Details
Concepts and Effects

11

Coves Asymmetric Luminaires

12′ Ceiling Height/9′ x 9′ Cove Opening/ 1′–0″ Clear Cove Height

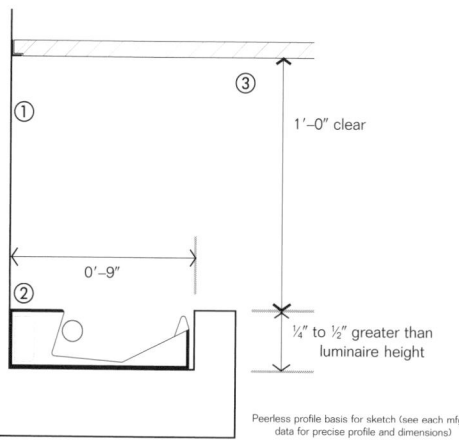

Notes
① Paint all interior cove surfaces matte white.
② Socket shadows are not noticeable with optically designed asymmetric luminaires.
③ Ceiling reflectance significantly influences success of cove. Higher is better (preferably 90 percent).

1′–0″ clear

0′–9″

¼″ to ½″ greater than luminaire height

Peerless profile basis for sketch (see each mfgr's data for precise profile and dimensions)

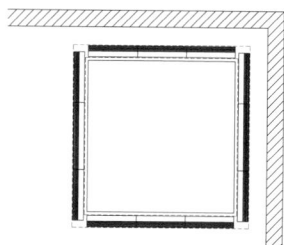

DetailConcepts™

T8 Asymmetric Luminaires

➲ Possible vendors: Columbia (PIC), Elliptipar(301, 306/T8), Ledalite (In–Cove Asymmetry™, LITECONTROL (Cove–45), Neoray (74IC), **Peerless** (ECX®)

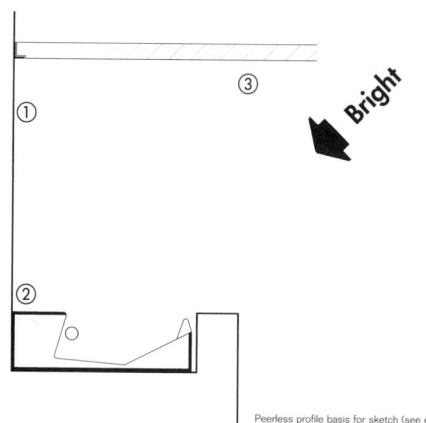

Bright

Notes
① Paint all interior cove surfaces matte white.
② Socket shadows are not noticeable with optically designed asymmetric luminaires.
③ Ceiling reflectance significantly influences success of cove. Higher is better (preferably 90 percent).
④ Standard T5 lamp/ballast will produce 42% less light and use 45% less energy.

Peerless profile basis for sketch (see each mfgr's data for precise profile and dimensions)

T5/HO Asymmetric Luminaires

➲ Possible vendors: Elliptipar (303, 305), **Peerless** (ECX®)

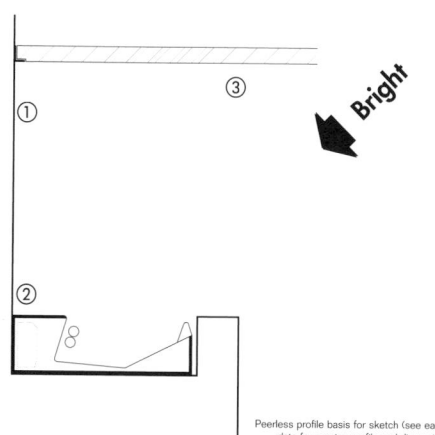

Bright

Notes
① Paint all interior cove surfaces matte white.
② Socket shadows are not noticeable with optically designed asymmetric luminaires.
③ Ceiling reflectance significantly influences success of cove. Higher is better (preferably 90 percent).

Peerless profile basis for sketch (see each mfgr's data for precise profile and dimensions)

LongCompact Asymmetric Luminaires

➲ Possible vendors: Columbia (PIC), Elliptipar (302), ELP (CL series), Ledalite (In–Cove Asymmetry™), Lightolier (Covelite™), LITECONTROL (Cove–45), Neoray (74IC), **Peerless** (ECX®)

Trademarks, service marks and product names are owned and registered by respective manufacturers.

Design Tips

✔ Asymmetric luminaires will perform better than striplights—better uniformity across the ceiling and no socket shadows.
✔ Asymmetric luminaires are more efficient in coves than striplights.

Details Concepts and Effects

11

Application Key

Commercial
Gallery
Hospitality[1]
Institutional
Manufacturing
Residential
Retail
Exterior

1/includes casual commercial and institutional applications and decorative retail applications

bold = primary application
partial fade = minimal application
fade = unlikely application

Net Addresses/Luminaires
See page 11–5

Coves Striplights

12' Ceiling Height/12' x 12' Cove Opening/0'–6" Clear Cove Height

DetailConcepts™

Performance Perspectives

[All data outlined here is based on information from boldfaced manufacturer's published data at time of manuscript preparation. Cove opening located in room corner and starts 2' from each of two walls. Maintained values based on analysis grid 15'x15' centered under cove and on 0.86 light loss factor (0.95 lamp lumen depreciation; 0.95 luminaire dirt depreciation; 0.95 room surface dirt depreciation); ballast factor of 0.98 and very light reflectances: ceiling 90%; wall 50%; and floor 20%. All views put eyes at 5'–6" AFF—cove perspective varies as ceiling height changes and/or cove height changes.]

▶ **Cove height: 0'–6" clear**
Central ceiling height: 12'
Perimeter ceiling height: 11'±
Lamp: F28T5/830 or F32T8/830
Floor illuminance (15'x15' area under cove)
▶ 11 fc, avg, maintained (19 fc max)

▶ **Cove height: 0'–6" clear**
Central ceiling height: 12'
Perimeter ceiling height: 11'±
Lamp: F54T5/HO/830
Floor illuminance (15'x15' area under cove)
▶ 17 fc, avg, maintained (29 fc max)

▶ **Cove height: 0'–6" clear**
Central ceiling height: 12'
Perimeter ceiling height: 11'±
Lamp: F40Long/830
Floor illuminance (15'x15' area under cove)
▶ 20 fc, avg, maintained (31 fc max)

Coves Striplights

12′ Ceiling Height/12′ x 12′ Cove Opening/0′–6″ Clear Cove Height

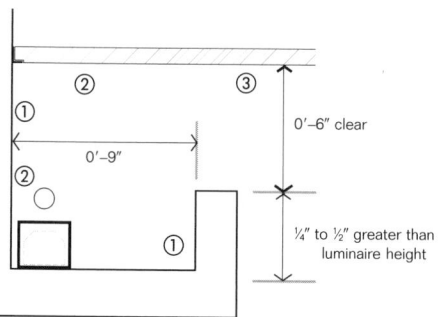

D e t a i l C o n c e p t s™

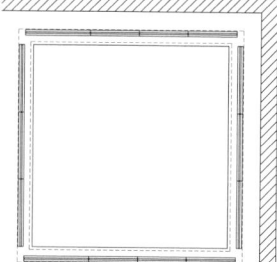

Notes
①Paint all interior cove surfaces matte white.
②Socket shadows are noticeable along back cove wall and somewhat noticeable at ceiling with 1–lamp striplight in this location. Single–lamp staggered strips introduce a "reverse socket shadow"—lamps closer to back cove wall create a slightly greater intensity. Butt strips end–to–end tightly.
③Ceiling reflectance significantly influences success of cove. Higher is better (preferably 90 percent).
④T8 and T5 lamps offer same light output—so striplights can be either version. Lamps must be combined with correct ballast and are not interchangeable after installation .T5 striplights are a few inches shorter than T8s. See below for T5/HO version.

T5 and T8 Striplights**Back of Cove**
↪Possible vendors: Bartco, Columbia, DayBrite, Legion, Lightolier, **Lithonia**[1], Metalux
[1]T12 photometry was used and rerated for T8 lamping. This is somewhat imprecise.

Very harsh

Notes
①Paint all interior cove surfaces matte white.
②Socket shadows are noticeable along back cove wall and somewhat noticeable at ceiling with 1–lamp striplight in this location. Single–lamp staggered strips introduce a "reverse socket shadow"—lamps closer to back cove wall create a slightly greater intensity. Butt strips end–to–end tightly.
③Ceiling reflectance significantly influences success of cove. Higher is better (preferably 90 percent).

T5/HO Striplights**Back of Cove**
↪Possible vendors: Bartco, Legion, Lightolier, **Lithonia**[1]
[1]T12 photometry was used and rerated for T5/HO lamping. This is somewhat imprecise.

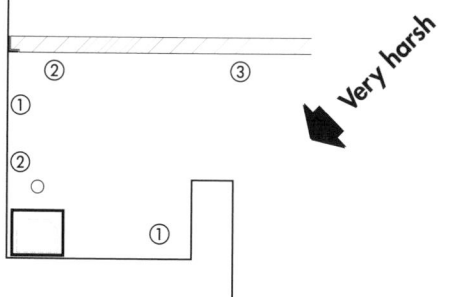

Very harsh

Notes
①Paint all interior cove surfaces matte white.
②Socket shadows are less noticeable along back cove wall and at ceiling with LongCompact lamps in striplight in this location.
③Ceiling reflectance significantly influences success of cove. Higher is better (preferably 90 percent).

LongCompact Striplights**Back of Cove**
↪Possible vendors: Columbia, DayBrite, Legion, Lightolier, **Lithonia**[1], Metalux
[1]T12 photometry was used and rerated for LongCompact lamping. This is somewhat imprecise.

Design Tips
✔Lamp and ballast maintenance is difficult. Lay–in ceiling can help—lift tiles for better reach/view into cove.
✔Heavily textured ceiling material will create significant shadows; as will tegular tile.
✔If ceiling is drywall, then any imperfections will telegraph with such grazing light.
✔If ceiling tone is other than white, then consider painting interior of cove surfaces to match ceiling.
✔Corner shadowing less noticeable with strips at front of cove.
✔Clear cove heights greater than 0′–6″ are more efficient.
✔Coves using asymmetric luminaires specifically designed for cove applications are more efficient.

Details Concepts and Effects

11

Application Key
Commercial
Gallery
Hospitality[1]
Institutional
Manufacturing
Residential
Retail
Exterior

1/includes casual commercial and institutional applications and decorative retail applications

bold = primary application
partial fade = minimal application
fade = unlikely application

Net Addresses/Luminaires
See page 11–5

Coves Striplights

12' Ceiling Height/12' x 12' Cove Opening/0'–6" Clear Cove Height

DetailConcepts™

Performance Perspectives

[All data outlined here is based on information from boldfaced manufacturer's published data at time of manuscript preparation. Cove opening located in room corner and starts 2' from each of two walls. Maintained values based on analysis grid 15'x15' centered under cove and on 0.86 light loss factor (0.95 lamp lumen depreciation; 0.95 luminaire dirt depreciation; 0.95 room surface dirt depreciation); ballast factor of 0.98 and very light reflectances: ceiling 90%; wall 50%; and floor 20%. All views put eyes at 5'–6" AFF—cove perspective varies as ceiling height changes and/or cove height changes.]

▶ **Cove height: 0'–6" clear**
Central ceiling height: 12'
Perimeter ceiling height: 11'±
Lamp: F28T5/830 or F32T8/830
Floor illuminance (15'x15' area under cove)
▶ 14 fc, avg, maintained (28 fc max)

▶ **Cove height: 0'–6" clear**
Central ceiling height: 12'
Perimeter ceiling height: 11'±
Lamp: F54T5/HO/830
Floor illuminance (15'x15' area under cove)
▶ 25 fc, avg, maintained (46 fc max)

▶ **Cove height: 0'–6" clear**
Central ceiling height: 12'
Perimeter ceiling height: 11'±
Lamp: F40Long/830
Floor illuminance (15'x15' area under cove)
▶ 30 fc, avg, maintained (50 fc max)

Details Concepts and Effects

11

Coves Striplights
12' Ceiling Height/12' x 12' Cove Opening/0'–6" Clear Cove Height

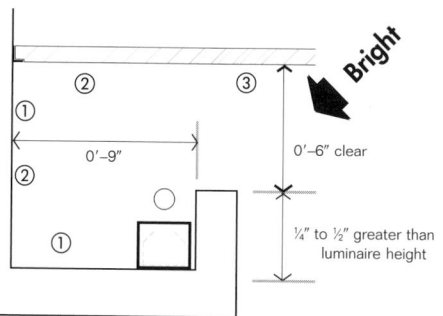

DetailConcepts™

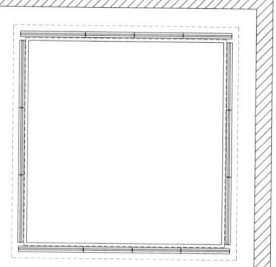

Notes
①Paint all interior cove surfaces matte white.
②Socket shadows are less noticeable with 1–lamp striplight in this location. Shadow line on ceiling may be visible a foot or so out from front edge of cove lip, however.
③Ceiling reflectance significantly influences success of cove. Higher is better (preferably 90 percent).
④T8 and T5 lamps offer same light output—so striplights can be either version. Lamps must be combined with correct ballast and are not interchangeable after installation. See below for T5/HO version.

T5 and T8 Striplights **Front of Cove**
⊃Possible vendors: Bartco, Columbia, DayBrite, Legion, Lightolier, **Lithonia**[1], Metalux
[1]T12 photometry was used and rerated for T8 lamping. This is somewhat imprecise.

Notes
①Paint all interior cove surfaces matte white.
②Socket shadows are less noticeable with 1–lamp striplight in this location. Shadow line on ceiling may be visible a foot or so out from front edge of cove lip, however.
③Ceiling reflectance significantly influences success of cove. Higher is better (preferably 90 percent).

T5/HO Striplights **Front of Cove**
⊃Possible vendors: Bartco, Legion, Lightolier, **Lithonia**[1]
[1]T12 photometry was used and rerated for T5/HO lamping. This is somewhat imprecise.

Design Tips
✔Lamp and ballast maintenance is difficult. Lay–in ceiling can help—lift tiles for better reach/view into cove.
✔Heavily textured ceiling material will create significant shadows; as will tegular tile.
✔If ceiling is drywall, then any imperfections will telegraph with such grazing light.
✔If ceiling tone is other than white, then consider painting interior of cove surfaces to match ceiling.
✔Clear cove heights greater than 0'–6" are more efficient.
✔Coves using asymmetric luminaires specifically designed for cove applications are more efficient.

Details Concepts and Effects **11**

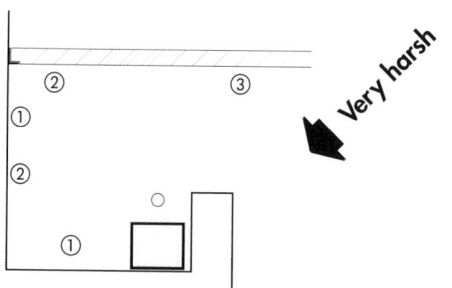

Notes
①Paint all interior cove surfaces matte white.
②Socket shadows are not noticeable with LongCompact lamps in striplight in this location.
③Ceiling reflectance significantly influences success of cove. Higher is better (preferably 90 percent).

LongCompact Striplights **Front of Cove**
⊃Possible vendors: Columbia, DayBrite, Legion, Lightolier, **Lithonia**[1], Metalux
[1]T12 photometry was used and rerated for LongCompact lamping. This is somewhat imprecise.

Application Key
Commercial
Gallery
Hospitality[1]
Institutional
Manufacturing
Residential
Retail
Exterior

1/includes casual commercial and institutional applications and decorative retail applications

bold = primary application
partial fade = minimal application
fade = unlikely application

Net Addresses/Luminaires
See page 11–5

Coves Striplights

12' Ceiling Height/12' x 12' Cove Opening/ 1'–0" Clear Cove Height

DetailConcepts™

Performance Perspectives

[All data outlined here is based on information from boldfaced manufacturer's published data at time of manuscript preparation. Cove opening located in room corner and starts 2' from each of two walls. Maintained values based on analysis grid 15'x15' centered under cove and on 0.86 light loss factor (0.95 lamp lumen depreciation; 0.95 luminaire dirt depreciation; 0.95 room surface dirt depreciation); ballast factor of 0.98 and very light reflectances: ceiling 90%; wall 50%; and floor 20%. All views put eyes at 5'–6" AFF—cove perspective varies as ceiling height changes and/or cove height changes.]

▶ **Cove height: 1'–0" clear**
Central ceiling height: 12'
Perimeter ceiling height: 10'–6"±
Lamp: F28T5/830 or F32T8/830
Floor illuminance (15'x15' area under cove)
▶ 18 fc, avg, maintained (30 fc max)

▶ **Cove height: 1'–0" clear**
Central ceiling height: 12'
Perimeter ceiling height: 10'–6"±
Lamp: F54T5/HO/830
Floor illuminance (15'x15' area under cove)
▶ 30 fc, avg, maintained (52 fc max)

▶ **Cove height: 1'–0" clear**
Central ceiling height: 12'
Perimeter ceiling height: 10'–6"±
Lamp: F40Long/830
Floor illuminance (15'x15' area under cove)
▶ 36 fc, avg, maintained (62 fc max)

Details
Concepts and Effects

11

Coves Striplights

12′ Ceiling Height/12′ x 12′ Cove Opening/1′–0″ Clear Cove Height

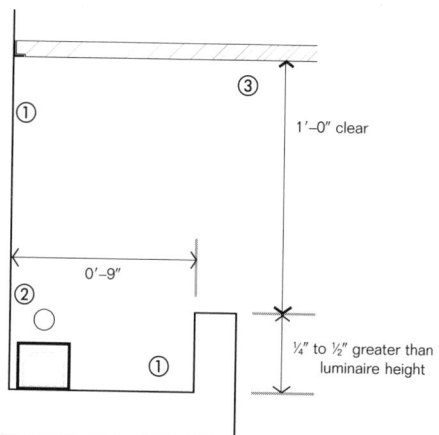

Notes

①Paint all interior cove surfaces matte white.
②Socket shadows are noticeable along back cove wall with 1–lamp striplight in this location. Single–lamp staggered strips introduce a "reverse socket shadow"—lamps closer to back cove wall create a slightly greater intensity. Butt strips end–to–end tightly.
③Ceiling reflectance significantly influences success of cove. Higher is better (preferably 90 percent).
④T8 and T5 lamps offer same light output—so striplights can be either version. Lamps must be combined with correct ballast and are not interchangeable after installation. See below for T5/HO version.

DetailConcepts™

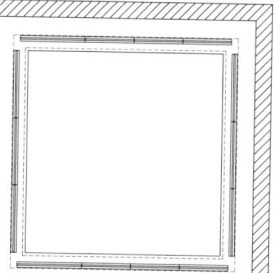

T5 and T8 Striplights**Back of Cove**

⮑Possible vendors: Bartco, Columbia, DayBrite, Legion, Lightolier, **Lithonia**[1], Metalux

[1]T12 photometry was used and rerated for T8 lamping. This is somewhat imprecise.

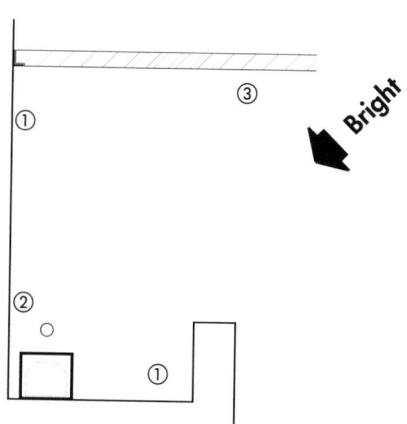

Notes

①Paint all interior cove surfaces matte white.
②Socket shadows are noticeable along back cove wall with 1–lamp striplight in this location. Single–lamp staggered strips introduce a "reverse socket shadow"—lamps closer to back cove wall create a slightly greater intensity. Butt strips end–to–end tightly.
③Ceiling reflectance significantly influences success of cove. Higher is better (preferably 90 percent).

Design Tips

✔Heavily textured ceiling material will create some shadows; as will tegular tile.
✔If ceiling tone is other than white, then consider painting interior of cove surfaces to match ceiling.
✔Corner shadowing less noticeable with strips at front of cove.
✔Clear cove heights greater than 1′–0″ are more efficient.
✔Coves using asymmetric luminaires specifically designed for cove applications are more efficient. See 11–91.

T5/HO Striplights**Back of Cove**

⮑Possible vendors: Bartco, Legion, Lightolier, **Lithonia**[1]

[1]T12 photometry was used and rerated for T5/HO lamping. This is somewhat imprecise.

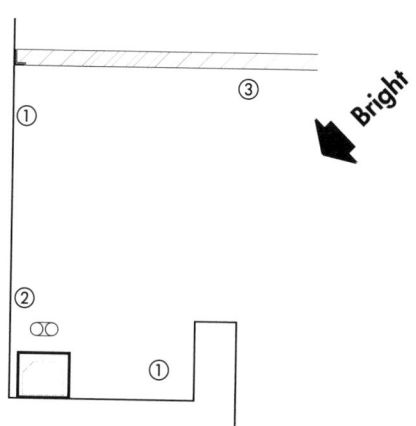

Notes

①Paint all interior cove surfaces matte white.
②Socket shadows are less noticeable along back cove wall and at ceiling with LongCompact lamps in striplight in this location.
③Ceiling reflectance significantly influences success of cove. Higher is better (preferably 90 percent).

Application Key

Commercial
Gallery
Hospitality[1]
Institutional
Manufacturing
Residential
Retail
Exterior

1/includes casual commercial and institutional applications and decorative retail applications

bold = primary application
partial fade = minimal application
fade = unlikely application

LongCompact Striplights**Back of Cove**

⮑Possible vendors: Columbia, DayBrite, Legion, Lightolier, **Lithonia**[1], Metalux

[1]T12 photometry was used and rerated for LongCompact lamping. This is somewhat imprecise.

Net Addresses/Luminaires
See page 11–5

Coves striplights

12' Ceiling Height/12' x 12' Cove Opening/1'–0" Clear Cove Height

DetailConcepts™

Performance Perspectives

[All data outlined here is based on information from boldfaced manufacturer's published data at time of manuscript preparation. Cove opening located in room corner and starts 2' from each of two walls. Maintained values based on analysis grid 15'x15' centered under cove and on 0.86 light loss factor (0.95 lamp lumen depreciation; 0.95 luminaire dirt depreciation; 0.95 room surface dirt depreciation); ballast factor of 0.98 and very light reflectances: ceiling 90%; wall 50%; and floor 20%. All views put eyes at 5'–6" AFF—cove perspective varies as ceiling height changes and/or cove height changes.]

▶ **Cove height: 1'–0" clear**
Central ceiling height: 12'
Perimeter ceiling height: 10'–6"±
Lamp: F28T5/830 or F32T8/830
Floor illuminance (15'x15' area under cove):
▶ 20 fc, avg, maintained (32 fc max)

▶ **Cove height: 1'–0" clear**
Central ceiling height: 12'
Perimeter ceiling height: 10'–6"±
Lamp: F54T5/HO/830
Floor illuminance (15'x15' area under cove):
▶ 36 fc, avg, maintained (62 fc max)

▶ **Cove height: 1'–0" clear**
Central ceiling height: 12'
Perimeter ceiling height: 10'–6"±
Lamp: F40Long/830
Floor illuminance (15'x15' area under cove):
▶ 41 fc, avg, maintained (62 fc max)

Details Concepts and Effects

11

Coves Striplights
12′ Ceiling Height/12′ x 12′ Cove Opening/1′–0″ Clear Cove Height

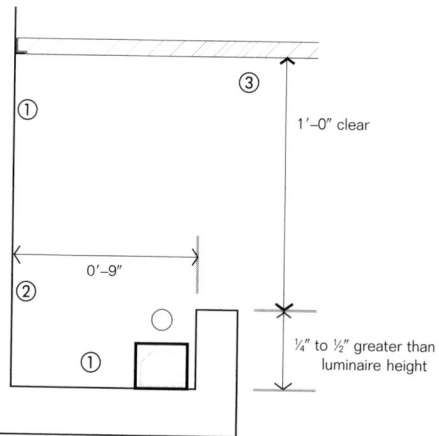

Notes
①Paint all interior cove surfaces matte white.
②Socket shadows are less noticeable with 1-lamp striplight in this location. Shadow line on ceiling may be visible a few feet or so out from front edge of cove lip, however.
③Ceiling reflectance significantly influences success of cove. Higher is better (preferably 90 percent).
④T8 and T5 lamps offer same light output—so striplights can be either version. Lamps must be combined with correct ballast and are not interchangeable after installation. See below for T5/HO version.

DetailConcepts™

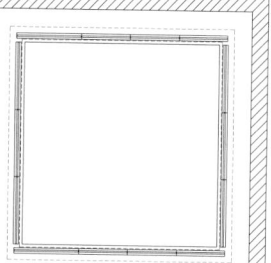

T5 and T8 Striplights**Front of Cove**
⊃Possible vendors: Bartco, Columbia, DayBrite, Legion, Lightolier, **Lithonia**[1], Metalux
[1]T12 photometry was used and rerated for T8 lamping. This is somewhat imprecise.

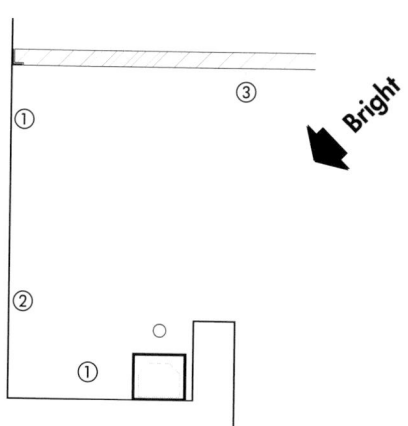

Bright

Notes
①Paint all interior cove surfaces matte white.
②Socket shadows are less noticeable with 1-lamp striplight in this location. Shadow line on ceiling may be visible a few feet or so out from front edge of cove lip, however.
③Ceiling reflectance significantly influences success of cove. Higher is better (preferably 90 percent).

Design Tips
✔Heavily textured ceiling material will create some shadows; as will tegular tile.
✔If ceiling tone is other than white, then consider painting interior of cove surfaces to match ceiling.
✔Clear cove heights greater than 0′–6″ are more efficient.
✔Coves using asymmetric luminaires specifically designed for cove applications are more efficient. See 11–91.

T5/HO Striplights**Front of Cove**
⊃Possible vendors: Bartco, Legion, Lightolier, **Lithonia**[1]
[1]T12 photometry was used and rerated for T5/HO lamping. This is somewhat imprecise.

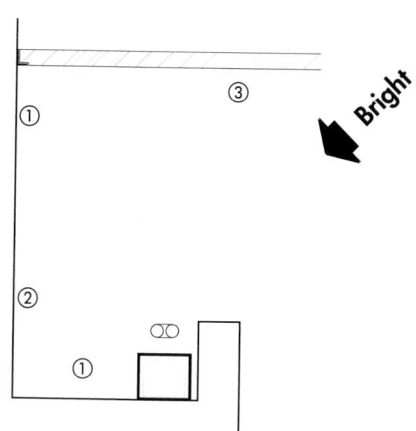

Bright

Notes
①Paint all interior cove surfaces matte white.
②Socket shadows are not noticeable with LongCompact lamps in striplight in this location. Shadow line on ceiling may be visible a few feet or so out from front edge of cove lip, however.
③Ceiling reflectance significantly influences success of cove. Higher is better (preferably 90 percent).

Application Key

Commercial
Gallery
Hospitality[1]
Institutional
Manufacturing
Residential
Retail
Exterior

1/includes casual commercial and institutional applications and decorative retail applications

bold = primary application
partial fade = minimal application
fade = unlikely application

LongCompact Striplights**Front of Cove**
⊃Possible vendors: Columbia, DayBrite, Legion, Lightolier, **Lithonia**[1], Metalux
[1]T12 photometry was used and rerated for LongCompact lamping. This is somewhat imprecise.

Net Addresses/Luminaires
See page 11–5

©2000 Howard Grey/Tony Stone Images

DetailConcepts™

Coves Asymmetric Luminaires

12′ Ceiling Height/12′ x 12′ Cove Opening/ 1′–0″ Clear Cove Height

Performance Perspectives

[All data outlined here is based on information from boldfaced manufacturer's published data at time of manuscript preparation. Cove opening located in room corner and starts 2′ from each of two walls. Maintained values based on analysis grid 15′x15′ centered under cove and on 0.86 light loss factor (0.95 lamp lumen depreciation; 0.95 luminaire dirt depreciation; 0.95 room surface dirt depreciation); ballast factor of 0.98 and very light reflectances: ceiling 90%; wall 50%; and floor 20%. All views put eyes at 5′–6″ AFF—cove perspective varies as ceiling height changes and/or cove height changes.]

▶ **Cove height: 1′–0″ clear**
Central ceiling height: 12′
Perimeter ceiling height: 10′–6″±
Lamp: F32T8/830
Floor illuminance (15′x15′ area under cove)
  ▶ 23 fc, avg, maintained (35 fc max)

▶ **Cove height: 1′–0″ clear**
Central ceiling height: 12′
Perimeter ceiling height: 10′–6″±
Lamp: F54T5/HO/830
Floor illuminance (15′x15′ area under cove)
  ▶ 42 fc, avg, maintained (64 fc max)

▶ **Cove height: 1′–0″ clear**
Central ceiling height: 12′
Perimeter ceiling height: 10′–6″±
Lamp: F40Long/830
Floor illuminance (15′x15′ area under cove)
  ▶ 51 fc, avg, maintained (80 fc max)

Details
Concepts and Effects

11

Coves Asymmetric Luminaires
12′ Ceiling Height/12′ x 12′ Cove Opening/1′–0″ Clear Cove Height

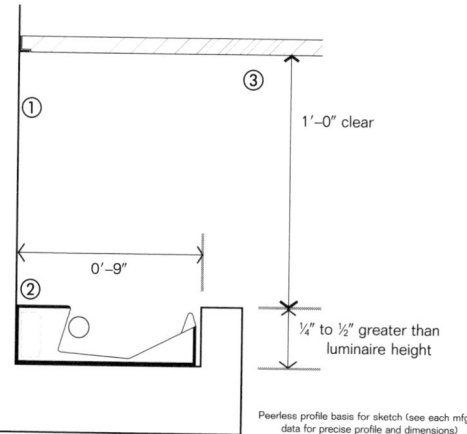

Notes
①Paint all interior cove surfaces matte white.
②Socket shadows are not noticeable with optically designed asymmetric luminaires.
③Ceiling reflectance significantly influences success of cove. Higher is better (preferably 90 percent).

Peerless profile basis for sketch (see each mfgr's data for precise profile and dimensions)

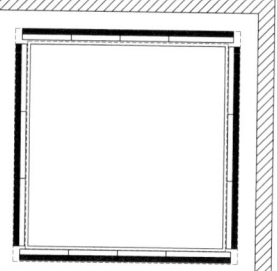

DetailConcepts™

T8 Asymmetric Luminaires
➲Possible vendors: Columbia (PIC), Elliptipar(301, 306/T8), Ledalite (In–Cove Asymmetry™, LITECONTROL (Cove–45), Neoray (74IC), **Peerless** (ECX®)

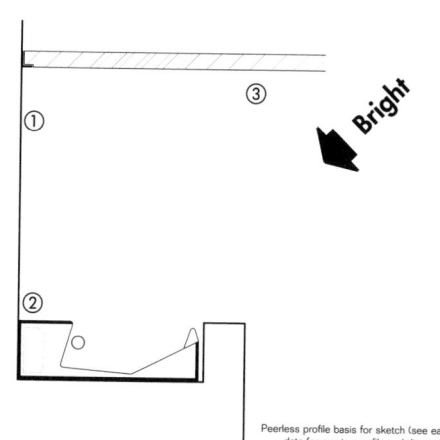

Notes
①Paint all interior cove surfaces matte white.
②Socket shadows are not noticeable with optically designed asymmetric luminaires.
③Ceiling reflectance significantly influences success of cove. Higher is better (preferably 90 percent).
④Standard T5 lamp/ballast will produce 42% less light and use 45% less energy.

Peerless profile basis for sketch (see each mfgr's data for precise profile and dimensions)

Design Tips
✔Asymmetric luminaires will perform better than striplights—better uniformity across the ceiling and no socket shadows.
✔Asymmetric luminaires are more efficient in coves than striplights.

T5/HO Asymmetric Luminaires
➲Possible vendors: Elliptipar (303, 305), **Peerless** (ECX®)

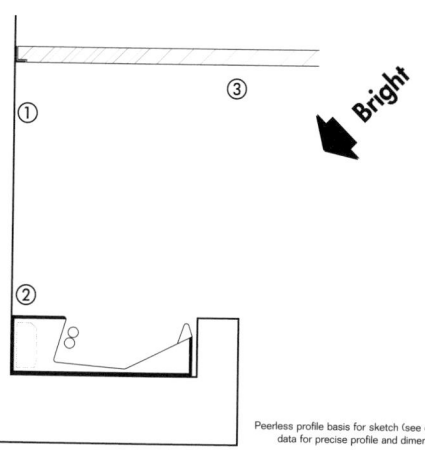

Notes
①Paint all interior cove surfaces matte white.
②Socket shadows are not noticeable with optically designed asymmetric luminaires.
③Ceiling reflectance significantly influences success of cove. Higher is better (preferably 90 percent).

Peerless profile basis for sketch (see each mfgr's data for precise profile and dimensions)

Application Key

Commercial
Gallery
Hospitality[1]
Institutional
Manufacturing
Residential
Retail
Exterior

1/includes casual commercial and institutional applications and decorative retail applications

bold = primary application
partial fade = minimal application
fade = unlikely application

LongCompact Asymmetric Luminaires
➲Possible vendors: Columbia (PIC), Elliptipar (302), ELP (CL series), Ledalite (In–Cove Asymmetry™), Lightolier (Covelite™), LITECONTROL (Cove–45), Neoray (74IC), **Peerless** (ECX®)

Trademarks, service marks and product names are owned and registered by respective manufacturers.

Net Addresses/Luminaires
See page 11–5

Coves Striplights

12' Ceiling Height/12' x 12' Cove Opening/2'–0" Clear Cove Height

DetailConcepts™

Performance Perspectives

[All data outlined here is based on information from boldfaced manufacturer's published data at time of manuscript preparation. Cove opening located in room corner and starts 2' from each of two walls. Maintained values based on analysis grid 15'x15' centered under cove and on 0.86 light loss factor (0.95 lamp lumen depreciation; 0.95 luminaire dirt depreciation; 0.95 room surface dirt depreciation); ballast factor of 0.98 and very light reflectances: ceiling 90%; wall 50%; and floor 20%. All views put eyes at 5'–6" AFF—cove perspective varies as ceiling height changes and/or cove height changes.]

▶ **Cove height: 2'–0" clear**
Central ceiling height: 12'
Perimeter ceiling height: 9'–6"±
Lamp: F28T5/830 or F32T8/830
Floor illuminance (15'x15' area under cove):
▶ 24 fc, avg, maintained (38 fc max)

▶ **Cove height: 2'–0" clear**
Central ceiling height: 12'
Perimeter ceiling height: 9'–6"±
Lamp: F54T5/HO/830
Floor illuminance (15'x15' area under cove):
▶ 37 fc, avg, maintained (58 fc max)

▶ **Cove height: 2'–0" clear**
Central ceiling height: 12'
Perimeter ceiling height: 9'–6"±
Lamp: F40Long/830
Floor illuminance (15'x15' area under cove):
▶ 45 fc, avg, maintained (69 fc max)

Coves Striplights

12′ Ceiling Height/12′ x 12′ Cove Opening/2′–0″ Clear Cove Height

2′–0″ clear

0′–9″

¼″ to ½″ greater than luminaire height

DetailConcepts™

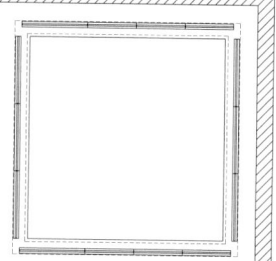

Notes
①Paint all interior cove surfaces matte white.
②Socket shadows are noticeable along back cove wall with 1–lamp striplight in this location. Single–lamp staggered strips introduce a "reverse socket shadow"—lamps closer to back cove wall create a slightly greater intensity. Butt strips end–to–end tightly.
③Ceiling reflectance significantly influences success of cove. Higher is better (preferably 90 percent).
④T8 and T5 lamps offer same light output—so striplights can be either version. Lamps must be combined with correct ballast and are not interchangeable after installation. See below for T5/HO version.

T5 and T8 Striplights **Back of Cove**
⊃Possible vendors: Bartco, Columbia, DayBrite, Legion, Lightolier, **Lithonia**[1], Metalux
[1]T12 photometry was used and rerated for T8 lamping. This is somewhat imprecise.

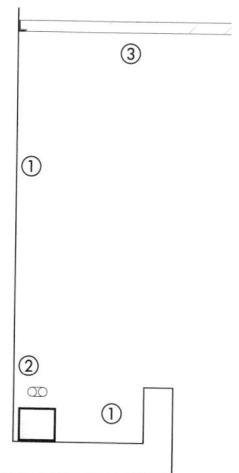

Notes
①Paint all interior cove surfaces matte white.
②Socket shadows are noticeable along back cove wall with 1–lamp striplight in this location. Single–lamp staggered strips introduce a "reverse socket shadow"—lamps closer to back cove wall create a slightly greater intensity. Butt strips end–to–end tightly.
③Ceiling reflectance significantly influences success of cove. Higher is better (preferably 90 percent).

T5/HO Striplights **Back of Cove**
⊃Possible vendors: Bartco, Legion, Lightolier, **Lithonia**[1]
[1]T12 photometry was used and rerated for T5/HO lamping. This is somewhat imprecise.

Notes
①Paint all interior cove surfaces matte white.
②Socket shadows are less noticeable along back cove wall and at ceiling with LongCompact lamps in striplight in this location.
③Ceiling reflectance significantly influences success of cove. Higher is better (preferably 90 percent).

LongCompact Striplights **Back of Cove**
⊃Possible vendors: Columbia, DayBrite, Legion, Lightolier, **Lithonia**[1], Metalux
[1]T12 photometry was used and rerated for LongCompact lamping. This is somewhat imprecise.

Design Tips
✔If ceiling tone is other than white, then consider painting interior of cove surfaces to match ceiling.
✔Corner shadowing less noticeable with strips at front of cove.
✔Coves using asymmetric luminaires specifically designed for cove applications are more efficient.

Details
Concepts and Effects

11

Application Key

Commercial
Gallery
Hospitality[1]
Institutional
Manufacturing
Residential
Retail
Exterior

1/includes casual commercial and institutional applications and decorative retail applications

bold = primary application
partial fade = minimal application
fade = unlikely application

Net Addresses/Luminaires
See page 11–5

©2000 Howard Grey/Tony Stone Images

DetailConcepts™

Coves striplights

12' Ceiling Height/12' x 12' Cove Opening/2'–0" Clear Cove Height

Performance Perspectives

[All data outlined here is based on information from boldfaced manufacturer's published data at time of manuscript preparation. Cove opening located in room corner and starts 2' from each of two walls. Maintained values based on analysis grid 15'x15' centered under cove and on 0.86 light loss factor (0.95 lamp lumen depreciation; 0.95 luminaire dirt depreciation; 0.95 room surface dirt depreciation); ballast factor of 0.98 and very light reflectances: ceiling 90%; wall 50%; and floor 20%. All views put eyes at 5'–6" AFF—cove perspective varies as ceiling height changes and/or cove height changes.]

▶ **Cove height: 2'–0" clear**
Central ceiling height: 12'
Perimeter ceiling height: 9'–6"±
Lamp: F28T5/830 or F32T8/830
Floor illuminance (15'x15' area under cove)
▶ 26 fc, avg, maintained (41 fc max)

▶ **Cove height: 2'–0" clear**
Central ceiling height: 12'
Perimeter ceiling height: 9'–6"±
Lamp: F54T5/HO/830
Floor illuminance (15'x15' area under cove)
▶ 43 fc, avg, maintained (67 fc max)

▶ **Cove height: 2'–0" clear**
Central ceiling height: 12'
Perimeter ceiling height: 9'–6"±
Lamp: F40Long/830
Floor illuminance (15'x15' area under cove)
▶ 49 fc, avg, maintained (76 fc max)

Details Concepts and Effects

11

Coves Striplights
12' Ceiling Height/12' x 12' Cove Opening/2'–0" Clear Cove Height

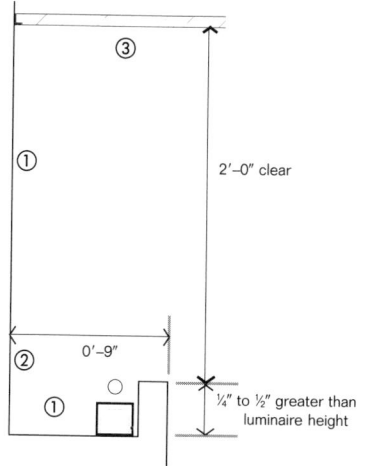

2'–0" clear

0'–9"

¼" to ½" greater than luminaire height

Notes
①Paint all interior cove surfaces matte white.
②Socket shadows are less noticeable with 1–lamp striplight in this location. Shadow line on ceiling likely not noticeable.
③Ceiling reflectance significantly influences success of cove. Higher is better (preferably 90 percent).
④T8 and T5 lamps offer same light output—so striplights can be either version. Lamps must be combined with correct ballast and are not interchangeable after installation. See below for T5/HO version.

DetailConcepts™

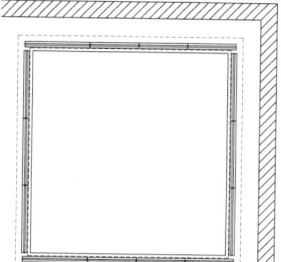

T5 and T8 Striplights**Front of Cove**
↪Possible vendors: Bartco, Columbia, DayBrite, Legion, Lightolier, **Lithonia**[1], Metalux
[1]T12 photometry was used and rerated for T8 lamping. This is somewhat imprecise.

Notes
①Paint all interior cove surfaces matte white.
②Socket shadows are less noticeable with 1–lamp striplight in this location. Shadow line on ceiling likely not noticeable.
③Ceiling reflectance significantly influences success of cove. Higher is better (preferably 90 percent).

Design Tips
✔If ceiling tone is other than white, then consider painting interior of cove surfaces to match ceiling.
✔Coves using asymmetric luminaires specifically designed for cove applications are more efficient.

T5/HO Striplights**Front of Cove**
↪Possible vendors: Bartco, Legion, Lightolier, **Lithonia**[1]
[1]T12 photometry was used and rerated for T5/HO lamping. This is somewhat imprecise.

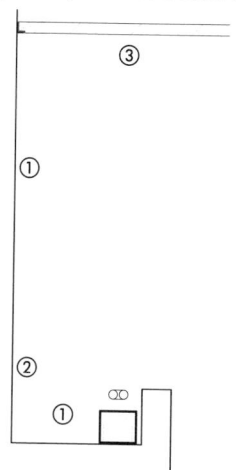

Notes
①Paint all interior cove surfaces matte white.
②Socket shadows are not noticeable along back cove wall and at ceiling with LongCompact lamps in striplight in this location. Shadow line on ceiling likely not noticeable.
③Ceiling reflectance significantly influences success of cove. Higher is better (preferably 90 percent).

Application Key

Commercial
Gallery
Hospitality[1]
Institutional
Manufacturing
Residential
Retail
Exterior

1/includes casual commercial and institutional applications and decorative retail applications

bold = primary application
partial fade = minimal application
fade = unlikely application

LongCompact Striplights**Front of Cove**
↪Possible vendors: Columbia, DayBrite, Legion, Lightolier, **Lithonia**[1], Metalux
[1]T12 photometry was used and rerated for LongCompact lamping. This is somewhat imprecise.

Net Addresses/Luminaires
See page 11–5

Details Concepts and Effects
Wallwashing

Wallwash lighting or wallwashing is an effective technique to brighten a space considerably without overlighting (and wasting energy) spaces. In the 1950's and 1960's, energy was inexpensive, greenhouse effects were ignored and lensed luminaires were sufficient for the tasks of the period. As such, by accident, so much light bounced around the room that the walls were reasonably well lighted and the environment appeared bright. Today, with more task–oriented lighting, the ambient or general lighting system cannot and should not overlight the environment. To improve the impressions of brightness, wallwashing is used. With decorative wall materials or artwork, these lighted wall areas also provide distant focal points for intermittent viewing throughout the day. This reduces eye fatigue.The advent of deluxe triphosphor compact fluorescent lamps and the small profile T5 and T8 deluxe triphosphor linear fluorescent lamps now enable the designer to use wallwash lighting which is functional, energy efficient and maintenance–reasonable.

■■■■■■■■■■■■■■■■■■■■■■■■■

Wallwashing
- ❶ Pleasant aesthetic
- ❶ Functional—use light intensities from wallwashing to conserve energy
- ❶ Helps soften harshness of downlighting
- ❶ Provides improved sense of overall brightness in environment
- ❶ Provides improved adaptation and accommodation resulting in less visual fatigue

Wallwashing for accenting
Wallwashing, like coves, can be used to help identify special areas, zones or "destinations." In retail applications, wallwashing can be used to draw attention to certain departments or areas within departments and help lead the consumer through the store. In commercial applications, wallwashing can be used to identify receptionists, waiting areas, conference centers, circulation and the like. Even if the wallwashing is intended to be primarily an accent or decorative element, the light it produces should be considered as part of the overall ambient lighting in the area. This will minimize energy consumption.

Wallwashing for function
Wallwashing can and should be considered one tool in the kit of parts to develop functional lighting. Not taking advantage of the illuminance produced by wallwashing is essentially wasting energy

Details Concepts and Effects
Wallwashing

and resources. The technique, as it's name suggests, is to rather uniformly and softly add light to a wall. Hence, this lighting is not as overt or obvious as a pinspot accent. With so many choices on wallwash aperture sizes, on the lamping, on luminaire style—optically inactive (such as a striplight in a slot) or optically active (such as a specially designed reflector slots or spreadlens wallwashers)—and on wattage, wallwashing results vary substantially. A series of Performance

Radiance (lighting rendering software)
http://www.radsite.lbl.gov/radiance/refer/
   long.html

CONNECT FOR MORE

Sketches™ are presented which offers quick visual assessment of the results of various wallwash parameters, luminaire style options and lamp type and wattage.

Performance Sketches™ are shown at a scale of ¼ inch equals 1 foot. Tracing for evaluation of various elevations is encouraged. Copying is permitted (according to the conventions outlined in "Licensing Agreement" in the Introduction of this book). Luminaire location or distance from wall is listed with each Sketch as are center beam aiming angle and average light level over the area of the wall. Maximum and minimum maintained intensities on the wall are also listed to help with assessment of uniformity. If the wallwashing is to be used for lighting artwork, then consideration should be given to the following guidelines. These vary slightly lower than criterion cited in Sections 3 through 7 since this technique—wallwashing—is intended to be more subtle than focused accents. For museum quality artwork where conservation of art is a priority, and where ambient lighting is anticipated to contribute very little light to the walls, wallwash intensities on the wall should be about 5 footcandles average maintained, with maximums of perhaps 10 fc. In other applications, total average maintained intensities on walls will depend on the general or ambient lighting within the space(s). In typical hospitality applications, wallwashing itself might contribute average maintained intensities of 10 fc, with maximum intensities of 20 fc. In commercial and institutional applications, average maintained intensities from wallwashing alone might be 15 fc or so, with maximums of 20 fc. Retail applications use wallwashing more as merchandise highlight. Here, wallwash intensities alone might be 30 fc or even greater. These Performance Sketches™ are based on

Details Concepts and Effects

11

Details Concepts and Effects

Wallwashing

the cited manufacturer's data at time of manuscript preparation. Manufacturers' catalog data may be of one date and for one set of lamps while photometric data available from manufacturers may be based on another set of lamps of another date (this may have occurred with some data in this reference). Performance Sketches™ were generated with this data in Radiance (a Unix–based lighting simulation and rendering program available free of charge at www.radsite.lbl.gov/radiance/refer/long.html).

■ ■ ■ ■ ■ ■ ■ ■ ■ ■ ■ ■ ■ ■

Performance Sketches™ and Information

ⓘ Typical application is shown
ⓘ Other applications may also be acceptable
ⓘ Other applications may depend on ceiling heights
ⓘ Maintained values based on 0.9 maintenance factor

These Performance Sketches™ illustrate specific conditions which meet the noted application criteria. As ceiling heights change, as wall finishes change, the cited lamp may be quite appropriate for other applications. Darker wall finishes will yield a dim appearance.

Selection guides

Selection guides are offered on the next several pages for various wallwashers in various applications. These guides are intended to help limit the designer's, engineer's or facility engineer's search and offer a good starting point from which design or alternative analyses can progress. While there are exceptions, wallwashers are appropriate in the following interior applications:

▶ commercial /institutional/retail
▶ gallery and residential
▶ hospitality

Recommended intensities
for wallwashing

Low (gallery, museum, residential)
5 fc average on wall
Moderate (hospitality)
10 fc average on wall
High (commercial—or higher for retail)
15 fc average on wall

These are intended to be average, maintained values WITHOUt effects of general room lighting. Remember that general room lighting contributes just a few footcandles in residential, gallery and museum settings, but contributes 5 to 10 fc in commercial settings (hence, total light on wall will be the additive effect of this ambient lighting and the wallwashing). Maximum intensities of wallwashing alone might range from 50 percent to 100 percent of the average value. Wherever artwork is highly valuable and/or sensitive to light, precautions should be taken to limit light exposure (maintaining low levels and/or switching lights off when art is not being viewed); and exposure to infrared and ultraviolet should be limited (using filters on lamps or placing art behind specially treated glass or acrylic).

Extrapolating information

There will be situations where perhaps twice as much light is desired (to provide a significant focal point) or where less light is desired. For wallwash situations where more light is desired, consider using lamps with center beam candlepower that is twice that of the lamp cited, but with same beamspread and wattage (or wattage equal to or less than rated wattage of selected luminaire (this information is available from luminaire manufacturers)). Alternatively, review other wallwash/luminaire parameters herein, as some situations will result in nearly twice as much light over some situations. Where less light is desired, consider using lamps of same beamspread but with lower center beam candlepower.

Details Concepts and Effects

Wallwashing

Net Addresses/Striplights (for wallslots)

```
http://www.bartcolighting.com/index2.html
http://www.belfer.com/prodsi.html
http://www.columbia-ltg.com/products/recessed-surface.html
http://www.thomasltg.com/
http://www.lightolier.com/
http://www.lithonia.com/
http://www.cooperlighting.com/
```

CONNECT FOR MORE

Net Addresses/Rectilinear Wallwashers

```
http://www.columbia-ltg.com/products/recessed-surface.html
http://www.elliptipar.com/home/Contents.htm
http://www.focalpointlights.com/
http://www.ledalite.com/
http://www.litecontrol.com/products/index.html
http://www.lithonia.com/
http://www.neoray.com/CATALOG/catalog.html
http://www.peerless-lighting.com/
```

CONNECT FOR MORE

Net Addresses/Spreadlens Wallwashers

```
http://www.elliptipar.com/home/Contents.htm
http://www.thomasltg.com/
http://www.cooperlighting.com/
http://www.hubbell-ltg.com/products.htm#Down&Track
http://www.danalite.com/
http://www.kramerlighting.com/
http://www.lightolier.com/
http://www.lithonia.com/
http://www.prescolite.com/
http://www.zumtobelstaff.co.at/    [Note: German language]
```

CONNECT FOR MORE

©2000 Howard Grey/Tony Stone Images

DetailConcepts™

Details Wallwash Lighting Selection Guide

Table 11.10 Wallwash Concepts/Commercial/Institutional/Retail

◀ L a m p i n g ▶

Adjustable Accents / Ceiling Height	Halogen (120V)	Halogen (12V)	HIR (120V)	HIR (12V)	CMH	Fluorescent
(page references to pages in Section 11)						
High probability of meeting intensities outlined in sidebar❶						
• 8'–6"		111				
• 9'–6"	119, 121, 123, 125	117, **119**	129			
• 10'–6"		127				
• 12'–6"						
• 15'–0"						

❶Surface reflectances won't affect intensities, but will significantly influence perceived brightness and efficiency of reflected light contributing to area lighting.
Boldfaced entry indicates commercial, institutional and retail applications may be found on cited page. All other entries reference commercial and institutional applications.

Table 11.11 Wallwash Concepts/Gallery/Residential

◀ L a m p i n g ▶

Adjustable Accents / Ceiling Height	Halogen (120V)	Halogen (12V)	HIR (120V)	HIR (12V)	CMH	Fluorescent
(page references to pages in Section 11)						
High probability of meeting intensities outlined in sidebar❶						
• 8'–6"	111, 113					
• 9'–6"	121	115				
• 10'–6"		127				
• 12'–6"						
• 15'–0"						

❶Surface reflectances won't affect intensities, but will significantly influence perceived brightness and efficiency of reflected light contributing to area lighting.

Cost magnitude[2000]
per unit for adjustable accents

Halogen/HIR (120V)
US$75. to US$100. per unit
Halogen/HIR (12V)
US$85. to US$110. per unit
CMH
US$175. to US$225. per unit
Fluorescent
not applicable

Cost magnitude for hardware only (luminaires, lamps and ballasts or transformers—as provided in the luminaire), excluding installation costs. Costs vary based on quantities, distributor and contractor markups, market conditions, manufacturing situations and annual inflation. Residential–grade products, where available as a category, will likely be at low end of range. These values are for preliminary, magnitude budgeting and do not represent quotes nor actual final pricing to client.

Details Concepts and Effects

11

Details Wallwash Lighting Selection Guide

©2000 Howard Grey/Tony Stone Images

Table 11.12 Wallwash Concepts/Hospitality

DetailConcepts™

Adjustable Accents	◀ L a m p i n g ▶					
	Halogen (120V)	Halogen (12V)	HIR (120V)	HIR (12V)	CMH	Fluorescent

(page references to pages in Section 11)

Ceiling Height	High probability of meeting intensities outlined in sidebar❶					
• 8'–6"	111, 113	109, 111				
• 9'–6"	119, 121, 123, 125	117, 119	129			
• 10'–6"		127				
• 12'–6"						
• 15'–0"						

❶Surface reflectances won't affect intensities, but will significantly influence perceived brightness and efficiency of reflected light contributing to area lighting.

Details
Concepts and Effects

11

Recommended intensities
for wallwashing

Low (gallery, museum, residential)
5 fc average on wall
Moderate (hospitality)
10 fc average on wall
High (commercial—or higher for retail)
15 fc average on wall

These are intended to be average, maintained values WITHOUT effects of general room lighting. Remember that general room lighting contributes just a few footcandles in residential, gallery and museum settings, but contributes 5 to 10 fc in commercial settings (hence, total light on wall will be the additive effect of this ambient lighting and the wallwashing). Maximum intensities of wallwashing alone might range from 50 percent to 100 percent of the average value. Wherever artwork is highly valuable and/or sensitive to light, precautions should be taken to limit light exposure (maintaining low levels and/or switching lights off when art is not being viewed); and exposure to infrared and ultraviolet should be limited (using filters on lamps or placing art behind specially treated glass or acrylic).

DetailConcepts™

Details Wallwash Lighting Selection Guide

Table 11.13 Wallwash Concepts/Commercial/Institutional/Retail

◀ L a m p i n g ▶

Spreadlens	Halogen (120V)	Halogen (12V)	HIR (120V)	HIR (12V)	CMH	Fluorescent

(page references to pages in Section 11)

Ceiling Height — High probability of meeting intensities outlined in sidebar ❶

Ceiling Height	Halogen (120V)	Halogen (12V)	HIR (120V)	HIR (12V)	CMH	Fluorescent
• 8'–6"		137		145		159
• 9'–6"	157		177		**153**	**161**
• 10'–6"		141	179, 181	149	**155**	**163**
• 12'–6"			183, 185			165
• 15'–0"						

❶Surface reflectances won't affect intensities, but will significantly influence perceived brightness and efficiency of reflected light contributing to area lighting.
Boldfaced entry indicates commercial, institutional and retail applications may be found on cited page. All other entries reference commercial and institutional applications.

Table 11.14 Wallwash Concepts/Gallery/Residential

◀ L a m p i n g ▶

Spreadlens	Halogen (120V)	Halogen (12V)	HIR (120V)	HIR (12V)	CMH	Fluorescent

(page references to pages in Section 11)

Ceiling Height — High probability of meeting intensities outlined in sidebar ❶

Ceiling Height	Halogen (120V)	Halogen (12V)	HIR (120V)	HIR (12V)	CMH	Fluorescent
• 8'–6"	167	139		147		
• 9'–6"	169	135, 157	177			
• 10'–6"	171	143	179	151		
• 12'–6"	173, 175		183			
• 15'–0"						

❶Surface reflectances won't affect intensities, but will significantly influence perceived brightness and efficiency of reflected light contributing to area lighting.

Cost magnitude[2000]
per unit for spreadlens wallwashers

Halogen/HIR (120V)
US$85. to US$115. per unit
Halogen/HIR (12V)
US$110. to US135. per unit
CMH
US$195. to US$250. per unit
Fluorescent
US$115. to US145. per unit

Cost magnitude for hardware only (luminaires, lamps and ballasts or transformers—as provided in the luminaire), excluding installation costs. Costs vary based on quantities, distributor and contractor markups, market conditions, manufacturing situations and annual inflation. Residential–grade products, where available as a category, will likely be at low end of range. These values are for preliminary, magnitude budgeting and do not represent quotes nor actual final pricing to client.

Details Concepts and Effects

11

Details Wallwash Lighting Selection Guide

©2000 Howard Grey/Tony Stone Images

Table 11.15 Wallwash Concepts/Hospitality

DetailConcepts™

Spreadlens	Halogen (120V)	Halogen (12V)	HIR (120V)	HIR (12V)	CMH	Fluorescent

◀ L a m p i n g ▶

(page references to pages in Section 11)

Ceiling Height — High probability of meeting intensities outlined in sidebar❶

Ceiling Height	Spreadlens	Halogen (120V)	Halogen (12V)	HIR (120V)	HIR (12V)	CMH	Fluorescent
• 8'–6"			137		145		
• 9'–6"		157, 169		177		153	
• 10'–6"		171	141	179, 181	149	155	
• 12'–6"		173		183, 185			
• 15'–0"				187			

❶Surface reflectances won't affect intensities, but will significantly influence perceived brightness and efficiency of reflected light contributing to area lighting.

Recommended intensities
for wallwashing

Low (gallery, museum, residential)
5 fc average on wall
Moderate (hospitality)
10 fc average on wall
High (commercial—or higher for retail)
15 fc average on wall

These are intended to be average, maintained values WITHOUT effects of general room lighting. Remember that general room lighting contributes just a few footcandles in residential, gallery and museum settings, but contributes 5 to 10 fc in commercial settings (hence, total light on wall will be the additive effect of this ambient lighting and the wallwashing). Maximum intensities of wallwashing alone might range from 50 percent to 100 percent of the average value. Wherever artwork is highly valuable and/or sensitive to light, precautions should be taken to limit light exposure (maintaining low levels and/or switching lights off when art is not being viewed); and exposure to infrared and ultraviolet should be limited (using filters on lamps or placing art behind specially treated glass or acrylic).

Details Concepts and Effects

11

Details Wallwash Lighting Selection Guide

DetailConcepts™ **Table 11.16** Wallwash Concepts/Commercial/Institutional/Retail

Rectilinear	Halogen (120V)	Halogen (12V)	HIR (120V)	HIR (12V)	CMH	Fluorescent

◄ L a m p i n g ►

(page references to pages in Section 11)

Ceiling Height	High probability of meeting intensities outlined in sidebar ❶					
• 8'–6"						189, 191
• 9'–6"						197, 199, 201
• 10'–6"						203, 205
• 12'–6"						207, 209
• 15'–0"						

❶Surface reflectances won't affect intensities, but will significantly influence perceived brightness and efficiency of reflected light contributing to area lighting.
Boldfaced entry indicates commercial, institutional and retail applications may be found on cited page. All other entries reference commercial and institutional applications.

Table 11.17 Wallwash Concepts/Gallery/Residential

Rectilinear	Halogen (120V)	Halogen (12V)	HIR (120V)	HIR (12V)	CMH	Fluorescent

◄ L a m p i n g ►

(page references to pages in Section 11)

Ceiling Height	High probability of meeting intensities outlined in sidebar ❶					
• 8'–6"						189, 191
• 9'–6"						
• 10'–6"						193, 195
• 12'–6"						
• 15'–0"						

❶Surface reflectances won't affect intensities, but will significantly influence perceived brightness and efficiency of reflected light contributing to area lighting.

Cost magnitude²⁰⁰⁰
per unit for rectilinear wallwashers

Halogen/HIR (120V)
not applicable
Halogen/HIR (12V)
not applicable
CMH
not applicable
Fluorescent
US$125. to US175. per unit

Cost magnitude for hardware only (luminaires, lamps and ballasts), excluding installation costs. Costs vary based on quantities, distributor and contractor markups, market conditions, manufacturing situations and annual inflation. Residential–grade products, where available as a category, will likely be at low end of range. These values are for preliminary, magnitude budgeting and do not represent quotes nor actual final pricing to client.

Details Wallwash Lighting Selection Guide

DetailConcepts™

Table 11.18 Wallwash Concepts/Hospitality

Rectilinear	◄ L a m p i n g ►					
	Halogen (120V)	Halogen (12V)	HIR (120V)	HIR (12V)	CMH	Fluorescent

(page references to pages in Section 11)

Ceiling Height	High probability of meeting intensities outlined in sidebar❶					
• 8'–6"						189, 191
• 9'–6"						199, 201
• 10'–6"						193, 195, 203, 205
• 12'–6"						207, 209
• 15'–0"						

❶Surface reflectances won't affect intensities, but will significantly influence perceived brightness and efficiency of reflected light contributing to area lighting.

Details
Concepts and Effects

11

Recommended intensities
for wallwashing

Low (gallery, museum, residential)
5 fc average on wall
Moderate (hospitality)
10 fc average on wall
High (commercial—or higher for retail)
15 fc average on wall

These are intended to be average, maintained values WITHOUT effects of general room lighting. Remember that general room lighting contributes just a few footcandles in residential, gallery and museum settings, but contributes 5 to 10 fc in commercial settings (hence, total light on wall will be the additive effect of this ambient lighting and the wallwashing). Maximum intensities of wallwashing alone might range from 50 percent to 100 percent of the average value. Wherever artwork is highly valuable and/or sensitive to light, precautions should be taken to limit light exposure (maintaining low levels and/or switching lights off when art is not being viewed); and exposure to infrared and ultraviolet should be limited (using filters on lamps or placing art behind specially treated glass or acrylic).

DetailConcepts™

Details Wallwash Lighting Selection Guide

Table 11.19 Wallslot Concepts/Commercial/Institutional/Retail

	◀ L a m p i n g ▶					
Wallslot	**Halogen (120V)**	**Halogen (12V)**	**HIR (120V)**	**HIR (12V)**	**CMH**	**Fluorescent**

(page references to pages in Section 11)

Ceiling Height	High probability of meeting intensities outlined in sidebar❶					
• 8'–6"		■		■		
• 9'–6"	**211**					**226**
• 10'–6"	**213**		**219**			
• 12'–6"			215, **221**		**223**	**228**
• 15'–0"			217		**225**	**230**

❶Surface reflectances won't affect intensities, but will significantly influence perceived brightness and efficiency of reflected light contributing to area lighting.
Boldfaced entry indicates commercial, institutional and retail applications may be found on cited page. All other entries reference commercial and institutional applications.

Table 11.20 Wallslot Concepts/Gallery/Residential

	◀ L a m p i n g ▶					
Wallslot	**Halogen (120V)**	**Halogen (12V)**	**HIR (120V)**	**HIR (12V)**	**CMH**	**Fluorescent**

(page references to pages in Section 11)

Ceiling Height	High probability of meeting intensities outlined in sidebar❶					
• 8'–6"		■		■		
• 9'–6"	211					
• 10'–6"	213					
• 12'–6"			215			
• 15'–0"			217			

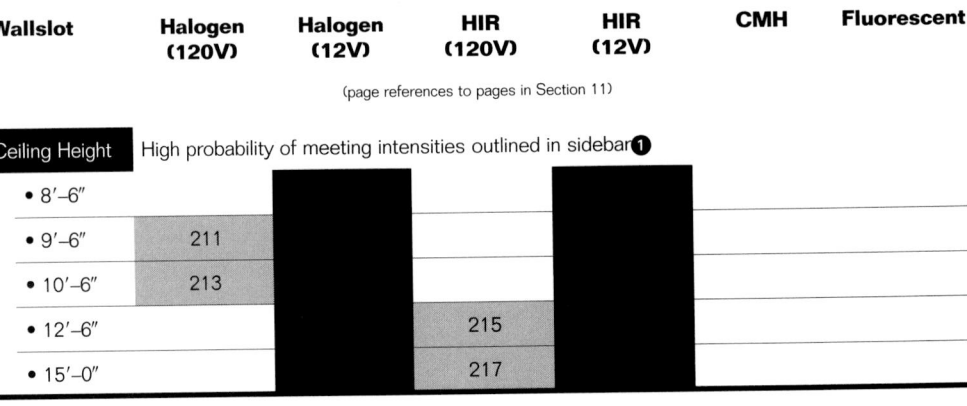

❶Surface reflectances won't affect intensities, but will significantly influence perceived brightness and efficiency of reflected light contributing to area lighting.

Cost magnitude²⁰⁰⁰
per linear foot

Halogen/HIR (120V)
US$125./ft to US$185./ft
CMH
US$200./ft to US$350./ft
Fluorescent Striplights
(T8) US$10./ft to US$12./ft
(T5) US$15./ft to US$20./ft
(F/Long) US$20./ft to US$25./ft

Cost magnitude for hardware only (luminaires, lamps and ballasts—as provided in the luminaire), excluding installation costs. Costs vary based on quantities, distributor and contractor markups, market conditions, manufacturing situations and annual inflation. Residential–grade products, where available as a category, will likely be at low end of range. These values are for preliminary, magnitude budgeting and do not represent quotes nor actual final pricing to client.

Details Wallwash Lighting Selection Guide

©2000 Howard Grey/Tony Stone Images

DetailConcepts™

Table 11.21 Wallslot Concepts/Hospitality

Wallslot	Halogen (120V)	Halogen (12V)	HIR (120V)	HIR (12V)	CMH	Fluorescent
◀ Lamping ▶						
(page references to pages in Section 11)						
Ceiling Height — High probability of meeting intensities outlined in sidebar ❶						
• 8'–6"		■		■		
• 9'–6"	211	■		■		226
• 10'–6"	213	■	219	■		
• 12'–6"		■	215, 221	■	223	228
• 15'–0"			217	■	225	

❶ Surface reflectances won't affect intensities, but will significantly influence perceived brightness and efficiency of reflected light contributing to area lighting.

Recommended intensities
for wallwashing

Low (gallery, museum, residential)
5 fc average on wall
Moderate (hospitality)
10 fc average on wall
High (commercial—or higher for retail)
15 fc average on wall

These are intended to be average, maintained values WITHOUT effects of general room lighting. Remember that general room lighting contributes just a few footcandles in residential, gallery and museum settings, but contributes 5 to 10 fc in commercial settings (hence, total light on wall will be the additive effect of this ambient lighting and the wallwashing). Maximum intensities of wallwashing alone might range from 50 percent to 100 percent of the average value. Wherever artwork is highly valuable and/or sensitive to light, precautions should be taken to limit light exposure (maintaining low levels and/or switching lights off when art is not being viewed); and exposure to infrared and ultraviolet should be limited (using filters on lamps or placing art behind specially treated glass or acrylic).

Details Concepts and Effects

11

Wallwash Adjustable Accents/Halogen

8'–6" Ceiling Height

Details Concepts and Effects

11

Design Tips

✔ IC–rated luminaires are required where contact with insulation in the construction is likely or as directed by local codes. Not all luminaire manufacturers' versions are available in IC–rated option.

✔ Use of non–IC luminaires might be accepted by local code authority for use where insulation is expected if provisions are made in construction to keep insulation at least 3 inches from luminaire housing, wiring compartment and/or ballast/transformer. Subject to local interpretation—confirm in advance of specification.

✔ Housing sizes above ceiling are relatively large—review manufacturer's datasheet to confirm fit.

For more efficient wallwashing, use fluorescent rectilinear wallwashers (see 11–199 and 201).

Wallwash Adjustable Accents/Halogen

8'–6" Ceiling Height

D e t a i l C o n c e p t s ™

Performance Sketches

[All data outlined here is based on cited lamp data at time of manuscript preparation. Maintained values based on analysis grid centered on wall and extending to 0'–6" from edges and on 0.90 light loss factor (0.95 lamp lumen depreciation; 0.95 luminaire dirt depreciation). Scale is ¼ inch equals 1 foot. Bare lamp photometry used which is somewhat imprecise.]

⊃**Possible vendors:** Cooper (Portfolio®), Edison Price, Kurt Versen, Lightolier (Calculite®), Lithonia, Prescolite, Zumtobel/Staff

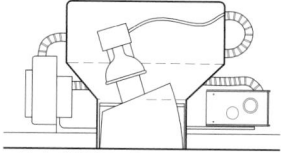

Lightolier 4½"Ø MR16 profile basis for sketch (see each mfgr's data for precise profile and dimensions)

▶ **Lamp: 20MR16/CG/FL (p4–79)**
Luminaire: Monopoint or Recessed Adjustable
Ceiling height: **8'–6"**
Distance from wall: **2'–6"**
Center beam aiming: 36° (at point 5'–0" AFF)
Spacings: **2' o.c.**
Wall illuminance (19'x7'–6" area):
▶ 5 fc, avg, maintained (23 fc max/0 fc min)
▶ Hospitality

Design Tips

✔Use where wall design is interesting and/or art potential exists but is not finalized.
✔Use luminaires with rotation lock and tilt lock to minimize misaiming on maintenance cycle.
✔For softer, more uniform wash, consider spreadlens accessory (specify luminaire with such capability) or use spreadlens wallwashers.
✔Effect is strongest with lighter colored walls. Need more light on darker walls for impact.
✔Enhances wood paneling, but consider spacing lights on center of paneling module.
✔Walls which have specular or glossy finishes will reflect lamp image back into room—less problematic the closer the lights are to the wall.

▶ **Lamp: 20MR16/C/FL (p4–117)**
Luminaire: Monopoint or Recessed Adjustable
Ceiling height: **8'–6"**
Distance from wall: **3'–0"**
Center beam aiming: 36° (at point 5'–0" AFF)
Spacings: **3' o.c.**
Wall illuminance (19'x7'–6" area):
▶ 7 fc, avg, maintained (25 fc max/1 fc min)
▶ Hospitality

Details
Concepts and Effects

11

▶ **Lamp: 20MR16/C/VWFL (p4–118)**
Luminaire: Monopoint or Recessed Adjustable
Ceiling height: **8'–6"**
Distance from wall: **3'–0"**
Center beam aiming: 29° (at point 4'–0" AFF)
Spacings: **3' o.c.**
Wall illuminance (19'x7'–6" area):
▶ 8 fc, avg, maintained (20 fc max/2 fc min)
▶ Hospitality

Net Addresses/Luminaires
See page 11–99

CONNECT FOR MORE

DetailConcepts™

Wallwash Adjustable Accents/Halogen

8'–6" Ceiling Height

Design Tips

✔ IC–rated luminaires are required where contact with insulation in the construction is likely or as directed by local codes. Not all luminaire manufacturers' versions are available in IC–rated option.

✔ Use of non–IC luminaires might be accepted by local code authority for use where insulation is expected if provisions are made in construction to keep insulation at least 3 inches from luminaire housing, wiring compartment and/or ballast/transformer. Subject to local interpretation—confirm in advance of specification.

✔ Housing sizes above ceiling are relatively large—review manufacturer's datasheet to confirm fit.

✔ Some manufacturers' luminaires have minimum spacing requirements to maintain appropriate clearance and to maintain UL listings—see specific manufacturer's data.

For more efficient wallwashing, use fluorescent rectilinear wallwashers (see 11–189 and 191) **or F32Triple spreadlens wallwashers** (see 11–159).

Wallwash Adjustable Accents/Halogen

8'–6" Ceiling Height

DetailConcepts ™

Performance Sketches

[All data outlined here is based on cited lamp data at time of manuscript preparation. Maintained values based on analysis grid centered on wall and extending to 0'–6" from edges and on 0.90 light loss factor (0.95 lamp lumen depreciation; 0.95 luminaire dirt depreciation). Scale is ¼ inch equals 1 foot. Bare lamp photometry used which is somewhat imprecise.]

↪**Possible vendors:** Cooper (Portfolio®), Edison Price, Kurt Versen, Lightolier (Calculite®), Lithonia, Prescolite, Zumtobel/Staff

▶ **Lamp: 35MR16/C/VWFL (p4–122)**
Luminaire: Monopoint or Recessed Adjustable
Ceiling height: **8'–6"**
Distance from wall: **2'–0"**
Center beam aiming: 27° (at point 4'–0" AFF)
Spacings: **3' o.c.**
Wall illuminance (19'x7'–6" area):
▶ 16 fc, avg, maintained (42 fc max/3 fc min)
▶ Commercial, Hospitality, Institutional

Lightolier 4½"Ø MR16 profile basis for sketch (see each mfgr's data for precise profile and dimensions)

Design Tips

✔Use where wall design is interesting and/or art potential exists but is not finalized.

✔Use luminaires with rotation lock and tilt lock to minimize misaiming on maintenance cycle.

✔For softer, more uniform wash, consider spreadlens accessory (specify luminaire with such capability) or use spreadlens wallwashers.

✔Effect is strongest with lighter colored walls. Need more light on darker walls for impact.

✔Enhances wood paneling, but consider spacing lights on center of paneling module.

✔Walls which have specular or glossy finishes will reflect lamp image back into room—less problematic the closer the lights are to the wall.

▶ **Lamp: 35PAR20/H/FL (p3–36)**
Luminaire: Monopoint or Recessed Adjustable
Ceiling height: **8'–6"**
Distance from wall: **2'–6"**
Center beam aiming: 32° (at point 4'–6" AFF)
Spacings: **3' o.c.**
Wall illuminance (19'x7'–6" area):
▶ 7 fc, avg, maintained (15 fc max/1 fc min)
▶ Gallery, Hospitality, Residential

▶ **Lamp: 45PAR38/H/WFL (p3–89)**
Luminaire: Monopoint or Recessed Adjustable
Ceiling height: **8'–6"**
Distance from wall: **2'–6"**
Center beam aiming: 24° (at point 3'–0" AFF)
Spacings: **4' o.c.** (Advisory: confirm allowable spacing requirements with manufacturer)
Wall illuminance (19'x7'–6" area):
▶ 8 fc, avg, maintained (19 fc max/3 fc min)
▶ Gallery, Hospitality, Residential

Details
Concepts and Effects

11

Net Addresses/Luminaires
See page 11–99

CONNECT FOR MORE

DetailConcepts™

©2000 Howard Grey/Tony Stone Images

Wallwash Adjustable Accents/Halogen

8'–6" Ceiling Height

Design Tips

✓ IC–rated luminaires are required where contact with insulation in the construction is likely or as directed by local codes. Not all luminaire manufacturers' versions are available in IC–rated option.

✓ Use of non–IC luminaires might be accepted by local code authority for use where insulation is expected if provisions are made in construction to keep insulation at least 3 inches from luminaire housing, wiring compartment and/or ballast/transformer. Subject to local interpretation—confirm in advance of specification.

✓ Housing sizes above ceiling are relatively large—review manufacturer's datasheet to confirm fit.

Wallwash Adjustable Accents/Halogen

8′–6″ Ceiling Height

DetailConcepts™

Performance Sketches

[All data outlined here is based on cited lamp data at time of manuscript preparation. Maintained values based on analysis grid centered on wall and extending to 0′–6″ from edges and on 0.90 light loss factor (0.95 lamp lumen depreciation; 0.95 luminaire dirt depreciation). Scale is ¼ inch equals 1 foot. Bare lamp photometry used which is somewhat imprecise.]

⊃Possible vendors: Cooper (Portfolio®), Edison Price, Kurt Versen, Lightolier (Calculite®), Lithonia, Prescolite, Zumtobel/Staff

▶ **Lamp: 50PAR30L/H/WFL (p3–68)**
Luminaire: Monopoint or Retrofitted Recessed Adjustable
Ceiling height: **8′–6″**
Distance from wall: **2′–6″**
Center beam aiming: 18° (at point 1′–0″ AFF)
Spacings: **4′ o.c.**
Wall illuminance (19′x7′–6″ area):
▶ 6 fc, avg, maintained (15 fc max/3 fc min)
▶ Gallery, Residential

▶ **Lamp: 50PAR30L/H/VWFL (p3–69)**
Luminaire: Monopoint or Retrofitted Recessed Adjustable
Ceiling height: 8′–6″
Distance from wall: **2′–6″**
Center beam aiming: 18° (at point 1′–0″ AFF)
Spacings: **4′ o.c.**
Wall illuminance (19′x7′–6″ area):
▶ 8 fc, avg, maintained (19 fc max/3 fc min)
▶ Gallery, Hospitality, Residential

Design Tips

✔ Use where wall design is interesting and/or art potential exists but is not finalized.
✔ Use luminaires with rotation lock and tilt lock to minimize misaiming on maintenance cycle.
✔ For softer, more uniform wash, consider spreadlens accessory (specify luminaire with such capability) or use spreadlens wallwashers.
✔ Effect is strongest with lighter colored walls. Need more light on darker walls for impact.
✔ Enhances wood paneling, but consider spacing lights on center of paneling module.
✔ Walls which have specular or glossy finishes will reflect lamp image back into room—less problematic the closer the lights are to the wall.
✔ Long neck lamps (e.g., 30L) are intended for retrofit into existing downlights or adjustable accents originally designed for R30 and R40 lamps of same or greater wattage.

Details Concepts and Effects

11

Net Addresses/Luminaires
See page 11–99

CONNECT FOR MORE

Wallwash Adjustable Accents/Halogen

9'–6" Ceiling Height

Design Tips

✔ IC–rated luminaires are required where contact with insulation in the construction is likely or as directed by local codes. Not all luminaire manufacturers' versions are available in IC–rated option.

✔ Use of non–IC luminaires might be accepted by local code authority for use where insulation is expected if provisions are made in construction to keep insulation at least 3 inches from luminaire housing, wiring compartment and/or ballast/transformer. Subject to local interpretation—confirm in advance of specification.

✔ Housing sizes above ceiling are relatively large—review manufacturer's datasheet to confirm fit.

Wallwash Adjustable Accents/Halogen

9'–6" Ceiling Height

DetailConcepts™

Performance Sketches

[All data outlined here is based on cited lamp data at time of manuscript preparation. Maintained values based on analysis grid centered on wall and extending to 0'–6" from edges and on 0.90 light loss factor (0.95 lamp lumen depreciation; 0.95 luminaire dirt depreciation). Scale is ¼ inch equals 1 foot. Bare lamp photometry used which is somewhat imprecise.]

⊃**Possible vendors:** Cooper (Portfolio®), Edison Price, Kurt Versen, Lightolier (Calculite®), Lithonia, Prescolite, Zumtobel/Staff

Lightolier 4½"Ø MR16 profile basis for sketch (see each mfgr's data for precise profile and dimensions)

▶ **Lamp: 20MR16/AL/FL (p4–89)**
Luminaire: Monopoint or Recessed Adjustable
Ceiling height: **9'–6"**
Distance from wall: **3'–0"**
Center beam aiming: 30° (at point 4'–4" AFF)
Spacings: **3' o.c.**
Wall illuminance (19'x8'–6" area):
▶ 7 fc, avg, maintained (20 fc max/2 fc min)
▶ Gallery, Residential

▶ **Lamp: 20MR16/C/CG/FL** (GE) **(p4–150)**
Luminaire: Monopoint or Recessed Adjustable
Ceiling height: **9'–6"**
Distance from wall: **3'–0"**
Center beam aiming: 27° (at point 3'–6" AFF)
Spacings: **3' o.c.**
Wall illuminance (19'x8'–6" area):
▶ 6 fc, avg, maintained (13 fc max/1 fc min)
▶ Gallery, Residential

Design Tips

✔ Use where wall design is interesting and/or art potential exists but is not finalized.
✔ Use luminaires with rotation lock and tilt lock to minimize misaiming on maintenance cycle.
✔ For softer, more uniform wash, consider spreadlens accessory (specify luminaire with such capability) or use spreadlens wallwashers.
✔ Effect is strongest with lighter colored walls. Need more light on darker walls for impact.
✔ Enhances wood paneling, but consider spacing lights on center of paneling module.
✔ Walls which have specular or glossy finishes will reflect lamp image back into room—less problematic the closer the lights are to the wall.
✔ Long neck lamps (e.g., 30L) are intended for retrofit into existing downlights or adjustable accents originally designed for R30 and R40 lamps of same or greater wattage.

▶ **Lamp: 35PAR30L/H/WFL (p3–62)**
Luminaire: Monopoint or Retrofitted Recessed Adjustable
Ceiling height: **9'–6"**
Distance from wall: **3'–0"**
Center beam aiming: 18° (at point 2'–0" AFF)
Spacings: **4' o.c.**
Wall illuminance (19'x8'–6" area):
▶ 4 fc, avg, maintained (10 fc max/2 fc min)
▶ Gallery, Residential

Details
Concepts and Effects
11

Net Addresses/Luminaires

See page 11–99

CONNECT FOR MORE

Wallwash Adjustable Accents/Halogen

9'–6" Ceiling Height

DetailConcepts™

Design Tips

✔ IC–rated luminaires are required where contact with insulation in the construction is likely or as directed by local codes. Not all luminaire manufacturers' versions are available in IC–rated option.

✔ Use of non–IC luminaires might be accepted by local code authority for use where insulation is expected if provisions are made in construction to keep insulation at least 3 inches from luminaire housing, wiring compartment and/or ballast/transformer. Subject to local interpretation—confirm in advance of specification.

✔ Housing sizes above ceiling are relatively large—review manufacturer's datasheet to confirm fit.

For more efficient wallwashing, use fluorescent rectilinear wallwashers (see 11–197, 199 and 201) **or F32Triple spreadlens wallwashers** (see 11–161).

Wallwash Adjustable Accents/Halogen

9'–6" Ceiling Height

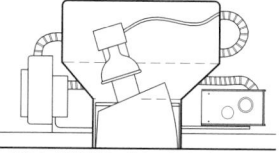

DetailConcepts™

Performance Sketches

[All data outlined here is based on cited lamp data at time of manuscript preparation. Maintained values based on analysis grid centered on wall and extending to 0'–6" from edges and on 0.90 light loss factor (0.95 lamp lumen depreciation; 0.95 luminaire dirt depreciation). Scale is ¼ inch equals 1 foot. Bare lamp photometry used which is somewhat imprecise.]

⊃**Possible vendors:** Cooper (Portfolio®), Edison Price, Kurt Versen, Lightolier (Calculite®), Lithonia, Prescolite, Zumtobel/Staff

▶ **Lamp: 50MR16/C/VWFL (p4–129)**
Luminaire: Monopoint or Recessed Adjustable
Ceiling height: **9'–6"**
Distance from wall: **3'–0"**
Center beam aiming: 31° (at point 4'–6" AFF)
Spacings: **4' o.c.**
Wall illuminance (19'x8'–6" area):
▶ 13 fc, avg, maintained (39 fc max/3 fc min)
▶ Commercial, Hospitality, Institutional

Lightolier 4½"Ø MR16 profile basis for sketch (see each mfgr's data for precise profile and dimensions)

Design Tips
✔ Use where wall design is interesting and/or art potential exists but is not finalized.
✔ Use luminaires with rotation lock and tilt lock to minimize misaiming on maintenance cycle.
✔ For softer, more uniform wash, consider spreadlens accessory (specify luminaire with such capability) or use spreadlens wallwashers.
✔ Effect is strongest with lighter colored walls. Need more light on darker walls for impact.
✔ Enhances wood paneling, but consider spacing lights on center of paneling module.
✔ Walls which have specular or glossy finishes will reflect lamp image back into room—less problematic the closer the lights are to the wall.

▶ **Lamp: 50MR16/C/CG/VWFL (p4–160)**
Luminaire: Monopoint or Recessed Adjustable
Ceiling height: **9'–6"**
Distance from wall: **3'–0"**
Center beam aiming: 22° (at point 2'–0" AFF)
Spacings: **4' o.c.**
Wall illuminance (19'x8'–6" area):
▶ 11 fc, avg, maintained (25 fc max/2 fc min)
▶ Hospitality

Details Concepts and Effects **11**

▶ **Lamp: 50MR16/VWFL (p4–65)**
Luminaire: Monopoint or Recessed Adjustable
Ceiling height: **9'–6"**
Distance from wall: **3'–0"**
Center beam aiming: 22° (at point 2'–0" AFF)
Spacings: **5' o.c.**
Wall illuminance (19'x8'–6" area):
▶ 13 fc, avg, maintained (31 fc max/2 fc min)
▶ Commercial, Hospitality, Institutional

Net Addresses/Luminaires
See page 11–99

CONNECT FOR MORE

D e t a i l C o n c e p t s ™

Wallwash Adjustable Accents/Halogen

9'–6" Ceiling Height

Details Concepts and Effects

11

Design Tips

✔ IC–rated luminaires are required where contact with insulation in the construction is likely or as directed by local codes. Not all luminaire manufacturers' versions are available in IC–rated option.

✔ Use of non–IC luminaires might be accepted by local code authority for use where insulation is expected if provisions are made in construction to keep insulation at least 3 inches from luminaire housing, wiring compartment and/or ballast/transformer. Subject to local interpretation—confirm in advance of specification.

✔ Housing sizes above ceiling are relatively large—review manufacturer's datasheet to confirm fit.

✔ Some manufacturers' luminaires have minimum spacing requirements to maintain appropriate clearance and to maintain UL listings— see specific manufacturer's data.

For more efficient wallwashing, use fluorescent rectilinear wallwashers (see 11–197, 199 and 201) **or F32Triple spreadlens wallwashers** (see 11–161).

Wallwash Adjustable Accents/Halogen

9′–6″ Ceiling Height

DetailConcepts™

Performance Sketches

[All data outlined here is based on cited lamp data at time of manuscript preparation. Maintained values based on analysis grid centered on wall and extending to 0′–6″ from edges and on 0.90 light loss factor (0.95 lamp lumen depreciation; 0.95 luminaire dirt depreciation). Scale is ¼ inch equals 1 foot. Bare lamp photometry used which is somewhat imprecise.]

⊃**Possible vendors:** Cooper (Portfolio®), Edison Price, Kurt Versen, Lightolier (Calculite®), Lithonia, Prescolite, Zumtobel/Staff

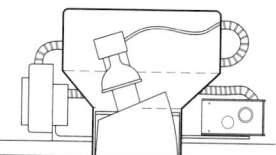

Lightolier 4½″Ø MR16 profile basis for sketch (see each mfgr's data for precise profile and dimensions)

▶ **Lamp: 60PAR38/H/VWFL (p3–96)**
Luminaire: Monopoint or Recessed Adjustable
Ceiling height: **9′–6″**
Distance from wall: **3′–0″**
Center beam aiming: 22° (at point 2′–0″ AFF)
Spacings: **4′ o.c.** (Advisory: confirm allowable spacing requirements with manufacturer)
Wall illuminance (19′x8′–6″ area):
▶ 14 fc, avg, maintained (31 fc max/6 fc min)
▶ Commercial, Hospitality, Institutional

Design Tips

✔Use where wall design is interesting and/or art potential exists but is not finalized.
✔Use luminaires with rotation lock and tilt lock to minimize misaiming on maintenance cycle.
✔For softer, more uniform wash, consider spreadlens accessory (specify luminaire with such capability) or use spreadlens wallwashers.
✔Effect is strongest with lighter colored walls. Need more light on darker walls for impact.
✔Enhances wood paneling, but consider spacing lights on center of paneling module.
✔Walls which have specular or glossy finishes will reflect lamp image back into room—less problematic the closer the lights are to the wall.

Details
Concepts and Effects
11

▶ **Lamp: 65MR16/C/VWFL (p4–134)**
Luminaire: Monopoint or Recessed Adjustable
Ceiling height: **9′–6″**
Distance from wall: **3′–0″**
Center beam aiming: 22° (at point 2′–0″ AFF)
Spacings: **4′ o.c.**
Wall illuminance (19′x8′–6″ area):
▶ 13 fc, avg, maintained (30 fc max/2 fc min)
▶ Commercial, Hospitality, Institutional

▶ **Lamp: 71MR16/C/CG/FL (p4–163)**
Luminaire: Monopoint or Recessed Adjustable
Ceiling height: **9′–6″**
Distance from wall: **2′–6″**
Center beam aiming: 24° (at point 4′–0″ AFF)
Spacings: **4′ o.c.**
Wall illuminance (19′x8′–6″ area):
▶ 20 fc, avg, maintained (67 fc max/3 fc min)
▶ Retail

Net Addresses/Luminaires
See page 11–99

CONNECT FOR MORE

DetailConcepts™

Wallwash Adjustable Accents/Halogen

9'–6" Ceiling Height

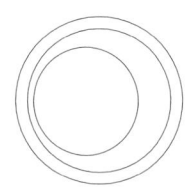

Details Concepts and Effects

11

Design Tips

✔ IC–rated luminaires are required where contact with insulation in the construction is likely or as directed by local codes. Not all luminaire manufacturers' versions are available in IC–rated option.

✔ Use of non–IC luminaires might be accepted by local code authority for use where insulation is expected if provisions are made in construction to keep insulation at least 3 inches from luminaire housing, wiring compartment and/or ballast/transformer. Subject to local interpretation—confirm in advance of specification.

✔ Housing sizes above ceiling are relatively large—review manufacturer's datasheet to confirm fit.

✔ Some manufacturers' luminaires have minimum spacing requirements to maintain appropriate clearance and to maintain UL listings—see specific manufacturer's data.

For more efficient wallwashing, use fluorescent rectilinear wallwashers (see 11–197, 199 and 201) **or F32Triple spreadlens wallwashers** (see 11–161).

Wallwash Adjustable Accents/Halogen
9'–6" Ceiling Height

Performance Sketches

[All data outlined here is based on cited lamp data at time of manuscript preparation. Maintained values based on analysis grid centered on wall and extending to 0'–6" from edges and on 0.90 light loss factor (0.95 lamp lumen depreciation; 0.95 luminaire dirt depreciation). Scale is ¼ inch equals 1 foot. Bare lamp photometry used which is somewhat imprecise.]

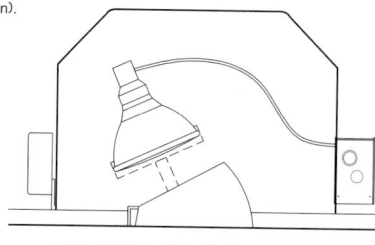

Lightolier 6"Ø PAR38 profile basis for sketch (see each mfgr's data for precise profile and dimensions)

⊃Possible vendors: Cooper (Portfolio®), Edison Price, Kurt Versen, Lightolier (Calculite®), Lithonia, Prescolite, Zumtobel/Staff

▶ **Lamp: 75PAR30L/H/WFL (p3–75)**
Luminaire: Monopoint or Retrofitted Recessed Adjustable
Ceiling height: **9'–6"**
Distance from wall: **2'–6"**
Center beam aiming: 16° (at point 1'–0" AFF)
Spacings: **4' o.c.**
Wall illuminance (19'x8'–6" area):
▶ 8 fc, avg, maintained (20 fc max/4 fc min)
▶ Gallery, Residential

▶ **Lamp: 75PAR30L/H/VWFL (p3–76)**
Luminaire: Monopoint or Retrofitted Recessed Adjustable
Ceiling height: **9'–6"**
Distance from wall: **2'–6"**
Center beam aiming: 16° (at point 1'–0" AFF)
Spacings: **4' o.c.**
Wall illuminance (19'x8'–6" area):
▶ 13 fc, avg, maintained (30 fc max/4 fc min)
▶ Commercial, Hospitality, Institutional

▶ **Lamp: 75PAR38/H/VWFL (p3–103)**
Luminaire: Monopoint or Recessed Adjustable
Ceiling height: **9'–6"**
Distance from wall: **3'–0"**
Center beam aiming: 22° (at point 2'–0" AFF)
Spacings: **4' o.c.** (Advisory: confirm allowable spacing requirements with manufacturer)
Wall illuminance (19'x8'–6" area):
▶ 9 fc, avg, maintained (20 fc max/4 fc min)
▶ Hospitality

Design Tips
✔Use where wall design is interesting and/or art potential exists but is not finalized.
✔Use luminaires with rotation lock and tilt lock to minimize misaiming on maintenance cycle.
✔For softer, more uniform wash, consider spreadlens accessory (specify luminaire with such capability) or use spreadlens wallwashers.
✔Effect is strongest with lighter colored walls. Need more light on darker walls for impact.
✔Enhances wood paneling, but consider spacing lights on center of paneling module.
✔Walls which have specular or glossy finishes will reflect lamp image back into room—less problematic the closer the lights are to the wall.
✔Long neck lamps (e.g., 30L) are intended for retrofit into existing downlights or adjustable accents originally designed for R30 and R40 lamps of same or greater wattage.

Details Concepts and Effects **11**

Net Addresses/Luminaires
See page 11–99

CONNECT FOR MORE

Wallwash Adjustable Accents/Halogen

9'–6" Ceiling Height

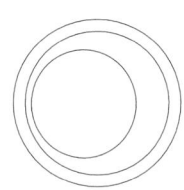

Design Tips

✔ IC–rated luminaires are required where contact with insulation in the construction is likely or as directed by local codes. Not all luminaire manufacturers' versions are available in IC–rated option.

✔ Use of non–IC luminaires might be accepted by local code authority for use where insulation is expected if provisions are made in construction to keep insulation at least 3 inches from luminaire housing, wiring compartment and/or ballast/transformer. Subject to local interpretation—confirm in advance of specification.

✔ Housing sizes above ceiling are relatively large—review manufacturer's datasheet to confirm fit.

✔ Some manufacturers' luminaires have minimum spacing requirements to maintain appropriate clearance and to maintain UL listings— see specific manufacturer's data.

For more efficient wallwashing, use fluorescent rectilinear wallwashers (see 11–197, 199 and 201) or F32Triple spreadlens wallwashers (see 11–161).

Wallwash Adjustable Accents/Halogen

9'–6" Ceiling Height

DetailConcepts™

Performance Sketches

[All data outlined here is based on cited lamp data at time of manuscript preparation. Maintained values based on analysis grid centered on wall and extending to 0'–6" from edges and on 0.90 light loss factor (0.95 lamp lumen depreciation; 0.95 luminaire dirt depreciation). Scale is ¼ inch equals 1 foot. Bare lamp photometry used which is somewhat imprecise.]

⊃ **Possible vendors:** Cooper (Portfolio®), Edison Price, Kurt Versen, Lightolier (Calculite®), Lithonia, Prescolite, Zumtobel/Staff

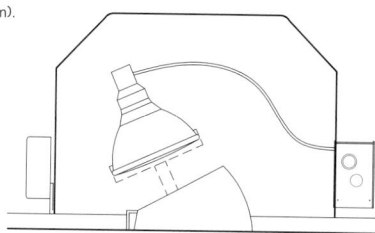

Lightolier 6"Ø PAR38 profile basis for sketch (see each mfgr's data for precise profile and dimensions)

▶ **Lamp: 90PAR38/H/WFL (p3–109)**
Luminaire: Monopoint or Recessed Adjustable
Ceiling height: **9'–6"**
Distance from wall: **3'–0"**
Center beam aiming: 22° (at point 2'–0" AFF)
Spacings: **5' o.c.** (Advisory: confirm allowable spacing requirements with manufacturer)
Wall illuminance (19'x8'–6" area):
▶ 13 fc, avg, maintained (31 fc max/0 fc min)
▶ Commercial, Hospitality, Institutional

▶ **Lamp: 90PAR38/H/VWFL (p3–110)**
Luminaire: Monopoint or Recessed Adjustable
Ceiling height: **9'–6"**
Distance from wall: **3'–0"**
Center beam aiming: 22° (at point 2'–0" AFF)
Spacings: **5' o.c.** (Advisory: confirm allowable spacing requirements with manufacturer)
Wall illuminance (19'x8'–6" area):
▶ 14 fc, avg, maintained (33 fc max/5 fc min)
▶ Commercial, Hospitality, Institutional

Design Tips

✔ Use where wall design is interesting and/or art potential exists but is not finalized.
✔ Use luminaires with rotation lock and tilt lock to minimize misaiming on maintenance cycle.
✔ For softer, more uniform wash, consider spreadlens accessory (specify luminaire with such capability) or use spreadlens wallwashers.
✔ Effect is strongest with lighter colored walls. Need more light on darker walls for impact.
✔ Enhances wood paneling, but consider spacing lights on center of paneling module.
✔ Walls which have specular or glossy finishes will reflect lamp image back into room—less problematic the closer the lights are to the wall.

Details Concepts and Effects

11

Net Addresses/Luminaires
See page 11–99

CONNECT FOR MORE

DetailConcepts™

Wallwash Adjustable Accents/Halogen

9'–6" Ceiling Height

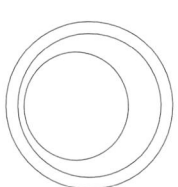

Design Tips

✔ IC–rated luminaires are required where contact with insulation in the construction is likely or as directed by local codes. Not all luminaire manufacturers' versions are available in IC–rated option.

✔ Use of non–IC luminaires might be accepted by local code authority for use where insulation is expected if provisions are made in construction to keep insulation at least 3 inches from luminaire housing, wiring compartment and/or ballast/transformer. Subject to local interpretation—confirm in advance of specification.

✔ Housing sizes above ceiling are relatively large—review manufacturer's datasheet to confirm fit.

✔ Some manufacturers' luminaires have minimum spacing requirements to maintain appropriate clearance and to maintain UL listings— see specific manufacturer's data.

For more efficient wallwashing, use fluorescent rectilinear wallwashers (see 11–197, 199 and 201) **or F32Triple spreadlens wallwashers** (see 11–161).

Details (vertical, left margin)

Concepts and Effects (vertical)

11

Wallwash Adjustable Accents/Halogen

9'–6" Ceiling Height

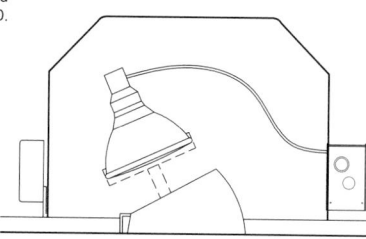

D e t a i l C o n c e p t s™

Performance Sketches

[All data outlined here is based on cited lamp data at time of manuscript preparation. Maintained values based on analysis grid centered on wall and extending to 0'–6" from edges and on 0.90 light loss factor (0.95 lamp lumen depreciation; 0.95 luminaire dirt depreciation). Scale is ¼ inch equals 1 foot. Bare lamp photometry used which is somewhat imprecise.]

⊃Possible vendors: Cooper (Portfolio®), Edison Price, Kurt Versen, Lightolier (Calculite®), Lithonia, Prescolite, Zumtobel/Staff

▶ **Lamp: 120PAR38/H/WFL (p3–119)**
Luminaire: Monopoint or Recessed Adjustable
Ceiling height: **9'–6"**
Distance from wall: **3'–0"**
Center beam aiming: 22° (at point 2'–0" AFF)
Spacings: **6' o.c.** (Advisory: confirm allowable spacing requirements with manufacturer)
Wall illuminance (19'x8'–6" area):
▶ 14 fc, avg, maintained (42 fc max/3 fc min)
▶ Commercial, Hospitality, Institutional

Lightolier 6"Ø PAR38 profile basis for sketch (see each mfgr's data for precise profile and dimensions)

Design Tips
✔Use where wall design is interesting and/or art potential exists but is not finalized.
✔Use luminaires with rotation lock and tilt lock to minimize misaiming on maintenance cycle.
✔For softer, more uniform wash, consider spreadlens accessory (specify luminaire with such capability) or use spreadlens wallwashers.
✔Effect is strongest with lighter colored walls. Need more light on darker walls for impact.
✔Enhances wood paneling, but consider spacing lights on center of paneling module.
✔Walls which have specular or glossy finishes will reflect lamp image back into room—less problematic the closer the lights are to the wall.

Details Concepts and Effects **11**

Net Addresses/Luminaires
See page 11–99

CONNECT FOR MORE

DetailConcepts™

Wallwash Adjustable Accents/Halogen

10'–6" Ceiling Height

Design Tips

✔ IC–rated luminaires are required where contact with insulation in the construction is likely or as directed by local codes. Not all luminaire manufacturers' versions are available in IC–rated option.

✔ Use of non–IC luminaires might be accepted by local code authority for use where insulation is expected if provisions are made in construction to keep insulation at least 3 inches from luminaire housing, wiring compartment and/or ballast/transformer. Subject to local interpretation—confirm in advance of specification.

✔ Housing sizes above ceiling are relatively large—review manufacturer's datasheet to confirm fit.

For more efficient wallwashing, use fluorescent rectilinear wallwashers (see 11–193, 195, 203 and 205) or F32Triple spreadlens wallwashers (see 11–163).

Wallwash Adjustable Accents/Halogen
10′–6″ Ceiling Height

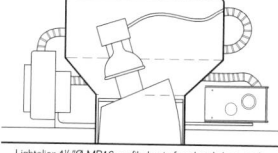

DetailConcepts™

Performance Sketches

[All data outlined here is based on cited lamp data at time of manuscript preparation. Maintained values based on analysis grid centered on wall and extending to 0′–6″ from edges and on 0.90 light loss factor (0.95 lamp lumen depreciation; 0.95 luminaire dirt depreciation). Scale is ¼ inch equals 1 foot. Bare lamp photometry used which is somewhat imprecise.]

⊃**Possible vendors:** Cooper (Portfolio®), Edison Price, Kurt Versen, Lightolier (Calculite®), Lithonia, Prescolite, Zumtobel/Staff

Lightolier 4½″Ø MR16 profile basis for sketch (see each mfgr's data for precise profile and dimensions)

▶ **Lamp: 50MR16/AL/CG/FL (p4–106)**
Luminaire: Monopoint or Recessed Adjustable
Ceiling height: **10′–6″**
Distance from wall: **5′–0″**
Center beam aiming: 42° (at point 5′–0″ AFF)
Spacings: **5′ o.c.**
Wall illuminance (19′x9′–6″ area):
▶ 4 fc, avg, maintained (13 fc max/1 fc min)
▶ Gallery, Residential

▶ **Lamp: 71MR16/C/FL (p4–137)**
Luminaire: Monopoint or Recessed Adjustable
Ceiling height: **10′–6″**
Distance from wall: **5′–0″**
Center beam aiming: 42° (at point 5′–0″ AFF)
Spacings: **5′ o.c.**
Wall illuminance (19′x9′–6″ area):
▶ 17 fc, avg, maintained (45 fc max/5 fc min)
▶ Commercial, Hospitality, Institutional

Design Tips
✔Use where wall design is interesting and/or art potential exists but is not finalized.
✔Use luminaires with rotation lock and tilt lock to minimize misaiming on maintenance cycle.
✔For softer, more uniform wash, consider spreadlens accessory (specify luminaire with such capability) or use spreadlens wallwashers.
✔Effect is strongest with lighter colored walls. Need more light on darker walls for impact.
✔Enhances wood paneling, but consider spacing lights on center of paneling module.
✔Walls which have specular or glossy finishes will reflect lamp image back into room—less problematic the closer the lights are to the wall.

Details Concepts and Effects **11**

Net Addresses/Luminaires
See page 11–99

CONNECT FOR MORE

Wallwash Adjustable Accents/HIR

9'–6" Ceiling Height

DetailConcepts™

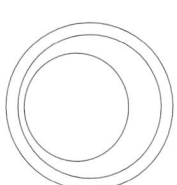

Design Tips

✔IC–rated luminaires are required where contact with insulation in the construction is likely or as directed by local codes. Not all luminaire manufacturers' versions are available in IC–rated option.

✔Use of non–IC luminaires might be accepted by local code authority for use where insulation is expected if provisions are made in construction to keep insulation at least 3 inches from luminaire housing, wiring compartment and/or ballast/transformer. Subject to local interpretation—confirm in advance of specification.

✔Housing sizes above ceiling are relatively large—review manufacturer's datasheet to confirm fit.

✔Some manufacturers' luminaires have minimum spacing requirements to maintain appropriate clearance and to maintain UL listings—see specific manufacturer's data.

Details Concepts and Effects

11

For more efficient wallwashing, use fluorescent rectilinear wallwashers (see 11–197, 199 and 201) **or F32Triple spreadlens wallwashers** (see 11–161).

Wallwash Adjustable Accents/HIR

9'–6" Ceiling Height

DetailConcepts™

Performance Sketches

[All data outlined here is based on cited lamp data at time of manuscript preparation. Maintained values based on analysis grid centered on wall and extending to 0'–6" from edges and on 0.90 light loss factor (0.95 lamp lumen depreciation; 0.95 luminaire dirt depreciation). Scale is ¼ inch equals 1 foot. Bare lamp photometry used which is somewhat imprecise.]

⊃Possible vendors: Cooper (Portfolio®), Edison Price, Kurt Versen, Lightolier (Calculite®), Lithonia, Prescolite, Zumtobel/Staff

▶ **Lamp: 60PAR38/HIR/WFL (p5–45)**
Luminaire: Monopoint or Recessed Adjustable
Ceiling height: **9'–6"**
Distance from wall: **3'–0"**
Center beam aiming: 19° (at point 1'–0" AFF)
Spacings: **4' o.c.** (Advisory: confirm allowable spacing requirements with manufacturer)
Wall illuminance (19'x8'–6" area):
▶ 14 fc, avg, maintained (26 fc max/6 fc min)
▶ Commercial, Hospitality, Institutional

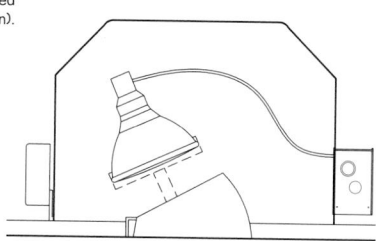

Lightolier 6"Ø PAR38 profile basis for sketch (see each mfgr's data for precise profile and dimensions)

Design Tips

✔Use where wall design is interesting and/or art potential exists but is not finalized.
✔Use luminaires with rotation lock and tilt lock to minimize misaiming on maintenance cycle.
✔For softer, more uniform wash, consider spreadlens accessory (specify luminaire with such capability) or use spreadlens wallwashers.
✔Effect is strongest with lighter colored walls. Need more light on darker walls for impact.
✔Enhances wood paneling, but consider spacing lights on center of paneling module.
✔Walls which have specular or glossy finishes will reflect lamp image back into room—less problematic the closer the lights are to the wall.

Details Concepts and Effects

11

Net Addresses/Luminaires
See page 11–99

CONNECT FOR MORE

Wallwash Adjustable Accents/CMH

12'–6" Ceiling Height

DetailConcepts™

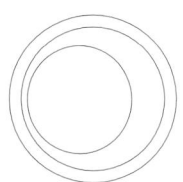

Design Tips

✔ IC–rated luminaires are required where contact with insulation in the construction is likely or as directed by local codes. Not all luminaire manufacturers' versions are available in IC–rated option.

✔ Use of non–IC luminaires might be accepted by local code authority for use where insulation is expected if provisions are made in construction to keep insulation at least 3 inches from luminaire housing, wiring compartment and/or ballast/transformer. Subject to local interpretation—confirm in advance of specification.

✔ Housing sizes above ceiling are large—review manufacturer's datasheet to confirm fit.

✔ Some manufacturers' luminaires have minimum spacing requirements to maintain appropriate clearance and to maintain UL listings—see specific manufacturer's data.

For more efficient, more uniform wallwashing, use fluorescent rectilinear wallwashers (see 11–207 and 209).

Wallwash Adjustable Accents/CMH

12'–6" Ceiling Height

Performance Sketches

[All data outlined here is based on cited lamp data at time of manuscript preparation. Maintained values based on analysis grid centered on wall and extending to 0'–6" from edges and on 0.80 light loss factor (0.85 lamp lumen depreciation; 0.95 luminaire dirt depreciation). Scale is ¼ inch equals 1 foot. Bare lamp photometry used which is somewhat imprecise.]

⊃Possible vendors: Cooper (Portfolio®), Edison Price, Kurt Versen, Lightolier (Calculite®), Lithonia, Prescolite, Zumtobel/Staff

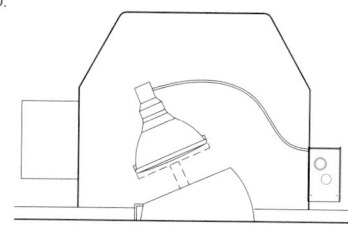

Lightolier 6"Ø PAR38/CMH profile basis for sketch (see each mfgr's data for precise profile and dimensions)

▶ **Lamp: 70PAR38/CMH/3K/VWFL (p7–66)**
Luminaire: Monopoint or Recessed Adjustable
Ceiling height: **12'–6"**
Distance from wall: **5'–0"**
Center beam aiming: 22° (at point 0'–0" AFF)
Spacings: **8' o.c.** (Advisory: confirm allowable spacing requirements with manufacturer)
Wall illuminance (19'x11'–6" area):
▶ 9 fc, avg, maintained (20 fc max/1 fc min)
▶ Hospitality

Design Tips

✔ Use where wall design is interesting and/or art potential exists but is not finalized.
✔ Use luminaires with rotation lock and tilt lock to minimize misaiming on maintenance cycle.
✔ For softer, more uniform wash, consider spreadlens accessory (specify luminaire with such capability) or use spreadlens wallwashers.
✔ Effect is strongest with lighter colored walls. Need more light on darker walls for impact.
✔ Enhances wood paneling, but consider spacing lights on center of paneling module.
✔ Walls which have specular or glossy finishes will reflect lamp image back into room—less problematic the closer the lights are to the wall.
✔ Not instant–on and requires 10– to 15–minute cycle–on period if power fails.
✔ Not dimmable in fashion meeting user expectations.

Details
Concepts and Effects
11

Net Addresses/Luminaires
See page 11–99

CONNECT FOR MORE

DetailConcepts™

Wallwash Adjustable Accents/CMH

15'–0" Ceiling Height

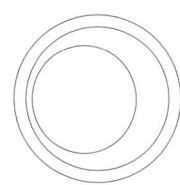

Details
Concepts and Effects

11

Design Tips

✔IC–rated luminaires are required where contact with insulation in the construction is likely or as directed by local codes. Not all luminaire manufacturers' versions are available in IC–rated option.

✔Use of non–IC luminaires might be accepted by local code authority for use where insulation is expected if provisions are made in construction to keep insulation at least 3 inches from luminaire housing, wiring compartment and/or ballast/transformer. Subject to local interpretation—confirm in advance of specification.

✔Housing sizes above ceiling are large—review manufacturer's datasheet to confirm fit.

✔Some manufacturers' luminaires have minimum spacing requirements to maintain appropriate clearance and to maintain UL listings—see specific manufacturer's data.

Wallwash Adjustable Accents/CMH

15'–0" Ceiling Height

DetailConcepts™

Performance Sketches

[All data outlined here is based on cited lamp data at time of manuscript preparation. Maintained values based on analysis grid centered on wall and extending to 0'–6" from edges and on 0.80 light loss factor (0.85 lamp lumen depreciation; 0.95 luminaire dirt depreciation). Scale is ¼ inch equals 1 foot. Bare lamp photometry used which is somewhat imprecise.]

⊃Possible vendors: Cooper (Portfolio®), Edison Price, Kurt Versen, Lightolier (Calculite®), Lithonia, Prescolite, Zumtobel/Staff

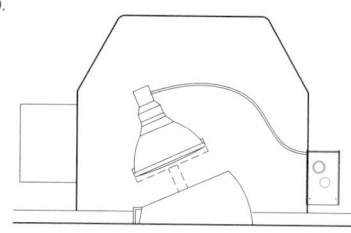

Lightolier 6"Ø PAR38/CMH profile basis for sketch (see each mfgr's data for precise profile and dimensions)

▶ **Lamp: 100PAR38/CMH/3K/VWFL (p7–76)**
Luminaire: Monopoint or Recessed Adjustable
Ceiling height: **15'–0"**
Distance from wall: **7'–0"**
Center beam aiming: 25° (at point 0'–0" AFF)
Spacings: **10' o.c.** (Advisory: confirm allowable spacing requirements with manufacturer)
Wall illuminance (19'x14' area):
▶ 10 fc, avg, maintained (17 fc max/2 fc min)
▶ Hospitality

Design Tips

✔Use where wall design is interesting and/or art potential exists but is not finalized.
✔Use luminaires with rotation lock and tilt lock to minimize misaiming on maintenance cycle.
✔For softer, more uniform wash, consider spreadlens accessory (specify luminaire with such capability) or use spreadlens wallwashers.
✔Effect is strongest with lighter colored walls. Need more light on darker walls for impact.
✔Enhances wood paneling, but consider spacing lights on center of paneling module.
✔Walls which have specular or glossy finishes will reflect lamp image back into room—less problematic the closer the lights are to the wall.
✔Not instant–on and requires 10– to 15–minute cycle–on period if power fails.
✔Not dimmable in fashion meeting user expectations.

Details Concepts and Effects

11

Net Addresses/Luminaires
See page 11–99

CONNECT FOR MORE

DetailConcepts™

Wallwash Spreadlens/4½"Ø/Halogen(120V)

9'–6" Ceiling Height

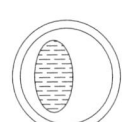

Design Tips

✔ IC–rated luminaires are required where contact with insulation in the construction is likely or as directed by local codes. Not all luminaire manufacturers' versions are available in IC–rated option.

✔ Use of non–IC luminaires might be accepted by local code authority for use where insulation is expected if provisions are made in construction to keep insulation at least 3 inches from luminaire housing, wiring compartment and/or ballast/transformer. Subject to local interpretation—confirm in advance of specification.

✔ Housing sizes above ceiling are relatively large—review manufacturer's datasheet to confirm fit.

Wallwash Spreadlens/4½"Ø/Halogen(120V)

9'–6" Ceiling Height

DetailConcepts™

Performance Sketches

[All data outlined here is based on information from boldfaced manufacturer's published data at time of manuscript preparation. Photometric data were rerated as necessary for cited lamp. Maintained values based on analysis grid centered on wall and extending to 0'–6" from edges and on 0.90 light loss factor (0.95 lamp lumen depreciation; 0.95 luminaire dirt depreciation). Scale is ¼ inch equals 1 foot.]

⊃Possible vendors: Cooper (Portfolio®), **Lightolier** (Calculite®), Lithonia, Zumtobel/Staff

Lightolier 4½"Ø PAR20 profile basis for sketch (see each mfgr's data for precise profile and dimensions)

▶ **Lamp: 50PAR20/H/NFL (p3–39)**
Luminaire: 4½"Ø Spreadlens Wallwasher
Ceiling height: **9'–6"**
Distance from wall: **2'–0"**
Spacings: **2' o.c.**
Wall illuminance (19'x8'–6" area):
▶ 4 fc, avg, maintained (6 fc max/0 fc min)
▶ Gallery, Residential

Design Tips

✔Use where wall design is interesting and/or art potential exists but is not finalized.
✔Effect is strongest with lighter colored walls. Need more light on darker walls for impact.
✔Enhances wood paneling, but consider spacing lights relative to center of paneling module.
✔Where art preservation is a necessity, specify UV–reduction spreadlens.
✔Walls which have specular or glossy finishes will reflect some lens image back into room.
✔Lens must be oriented with linear pattern perpendicular to wall (typically there is a positioning "key" in the glass to help with orientation).
✔Lens must be placed in concave orientation (lens cups upward into housing—**not** downward which is a common error).

Details
Concepts and Effects

11

Net Addresses/Luminaires
See page 11–99

CONNECT FOR MORE

Wallwash Spreadlens/4½″Ø/Halogen 12V

8′–6″ Ceiling Height

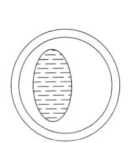

Design Tips

✔ IC–rated luminaires are required where contact with insulation in the construction is likely or as directed by local codes. Not all luminaire manufacturers' versions are available in IC–rated option.

✔ Use of non–IC luminaires might be accepted by local code authority for use where insulation is expected if provisions are made in construction to keep insulation at least 3 inches from luminaire housing, wiring compartment and/or ballast/transformer. Subject to local interpretation—confirm in advance of specification.

✔ Housing sizes above ceiling are relatively large—review manufacturer's datasheet to confirm fit.

For more efficient wallwashing, use fluorescent rectilinear wallwashers (see 11–189 and 191).

Wallwash Spreadlens/4½"Ø/Halogen (12V)

8'–6" Ceiling Height

DetailConcepts™

Performance Sketches

[All data outlined here is based on information from boldfaced manufacturer's published data at time of manuscript preparation. Photometric data were rerated as necessary for cited lamp. Maintained values based on analysis grid centered on wall and extending to 0'–6" from edges and on 0.90 light loss factor (0.95 lamp lumen depreciation; 0.95 luminaire dirt depreciation). Scale is ¼ inch equals 1 foot.]

⤴Possible vendors: Cooper (Portfolio®), **Lightolier** (Calculite®), Lithonia, Zumtobel/Staff

▶ **Lamp: 50MR16/FL (p4–64)**
Luminaire: 4½"Ø Spreadlens Wallwasher
Ceiling height: **8'–6"**
Distance from wall: **2'–0"**
Spacings: **2' o.c.**
Wall illuminance (19'x7'–6" area):
▶ 12 fc, avg, maintained (18 fc max/1 fc min)
▶ Hospitality

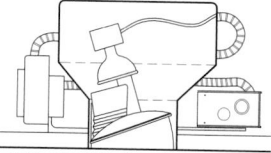

Lightolier 4½"Ø MR16 profile basis for sketch (see each mfgr's data for precise profile and dimensions)

Design Tips

✔Use where wall design is interesting and/or art potential exists but is not finalized.
✔Effect is strongest with lighter colored walls. Need more light on darker walls for impact.
✔Enhances wood paneling, but consider spacing lights relative to center of paneling module.
✔Where art preservation is a necessity, specify UV–reduction spreadlens.
✔Walls which have specular or glossy finishes will reflect some lens image back into room.
✔Lens must be oriented with linear pattern perpendicular to wall (typically there is a positioning "key" in the glass to help with orientation).
✔Lens must be placed in concave orientation (lens cups upward into housing—**not** downward which is a common error).

▶ **Lamp: 73MR16/FL (p4–72)**
Luminaire: 4½"Ø Spreadlens Wallwasher
Ceiling height: **8'–6"**
Distance from wall: **2'–0"**
Spacings: **2' o.c.**
Wall illuminance (19'x7'–6" area):
▶ 15 fc, avg, maintained (22 fc max/2 fc min)
▶ Commercial, Hospitality, Institutional

11 Details Concepts and Effects

Figure 11.3

The wall above the fireplace and the artwall in the distant background (to the left, rear) are lighted with 4½"Ø spreadlens wallwashers using 50MR16/C/CG/NFL(GE) lamps. The NFL lamp provides more intensity on the wall than the 50MR16/FL lamp outlined above on this page. Wallwash units are Lightolier.

Net Addresses/Luminaires
See page 11–99

CONNECT FOR MORE

©2000 Howard Grey/Tony Stone Images

DetailConcepts™

Wallwash Spreadlens/4½″Ø/Halogen 12V

8′–6″ Ceiling Height

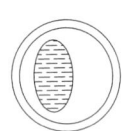

Design Tips

✔ IC–rated luminaires are required where contact with insulation in the construction is likely or as directed by local codes. Not all luminaire manufacturers' versions are available in IC–rated option.

✔ Use of non–IC luminaires might be accepted by local code authority for use where insulation is expected if provisions are made in construction to keep insulation at least 3 inches from luminaire housing, wiring compartment and/or ballast/transformer. Subject to local interpretation—confirm in advance of specification.

✔ Housing sizes above ceiling are relatively large—review manufacturer's datasheet to confirm fit.

For more efficient wallwashing, use fluorescent rectilinear wallwashers (see 11–189 and 191).

Wallwash Spreadlens/4½"Ø/Halogen 12V
8'–6" Ceiling Height

DetailConcepts™

Performance Sketches

[All data outlined here is based on information from boldfaced manufacturer's published data at time of manuscript preparation. Photometric data were rerated as necessary for cited lamp. Maintained values based on analysis grid centered on wall and extending to 0'–6" from edges and on 0.90 light loss factor (0.95 lamp lumen depreciation; 0.95 luminaire dirt depreciation). Scale is ¼ inch equals 1 foot.]

⊃Possible vendors: Cooper (Portfolio®), **Lightolier** (Calculite®), Lithonia, Zumtobel/Staff

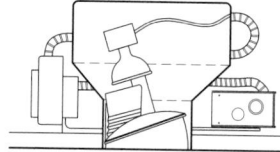

Lightolier 4½"Ø MR16 profile basis for sketch (see each mfgr's data for precise profile and dimensions)

▶ **Lamp: 50MR16/FL (p4–64)**
Luminaire: 4½"Ø Spreadlens Wallwasher
Ceiling height: **8'–6"**
Distance from wall: **4'–0"**
Spacings: **4' o.c.**
Wall illuminance (19'x7'–6" area):
▶ 3 fc, avg, maintained (5 fc max/0 fc min)
▶ Gallery, Residential

▶ **Lamp: 73MR16/FL (p4–72)**
Luminaire: 4½"Ø Spreadlens Wallwasher
Ceiling height: **8'–6"**
Distance from wall: **4'–0"**
Spacings: **4' o.c.**
Wall illuminance (19'x7'–6" area):
▶ 4 fc, avg, maintained (6 fc max/0 fc min)
▶ Gallery, Residential

Design Tips

✔ Use where wall design is interesting and/or art potential exists but is not finalized.
✔ Effect is strongest with lighter colored walls. Need more light on darker walls for impact.
✔ Enhances wood paneling, but consider spacing lights relative to center of paneling module.
✔ Where art preservation is a necessity, specify UV–reduction spreadlens.
✔ Walls which have specular or glossy finishes will reflect some lens image back into room.
✔ Lens must be oriented with linear pattern perpendicular to wall (typically there is a positioning "key" in the glass to help with orientation).
✔ Lens must be placed in concave orientation (lens cups upward into housing—**not** downward which is a common error).

Details
Concepts and Effects

11

Net Addresses/Luminaires
See page 11–99

CONNECT FOR MORE

DetailConcepts™

Wallwash Spreadlens/4½"Ø/Halogen (12V)

10'–6" Ceiling Height

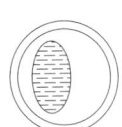

Design Tips

✔ IC–rated luminaires are required where contact with insulation in the construction is likely or as directed by local codes. Not all luminaire manufacturers' versions are available in IC–rated option.

✔ Use of non–IC luminaires might be accepted by local code authority for use where insulation is expected if provisions are made in construction to keep insulation at least 3 inches from luminaire housing, wiring compartment and/or ballast/transformer. Subject to local interpretation—confirm in advance of specification.

✔ Housing sizes above ceiling are relatively large—review manufacturer's datasheet to confirm fit.

For more efficient wallwashing, use fluorescent rectilinear wallwashers (see 11– 193, 195, 203 and 205).

Wallwash Spreadlens/4½"Ø/Halogen [12V]

10'–6" Ceiling Height

DetailConcepts™

Performance Sketches

[All data outlined here is based on information from boldfaced manufacturer's published data at time of manuscript preparation. Photometric data were rerated as necessary for cited lamp. Maintained values based on analysis grid centered on wall and extending to 0'–6" from edges and on 0.90 light loss factor (0.95 lamp lumen depreciation; 0.95 luminaire dirt depreciation). Scale is ¼ inch equals 1 foot.]

⊃Possible vendors: Cooper (Portfolio®), **Lightolier** (Calculite®), Lithonia, Zumtobel/Staff

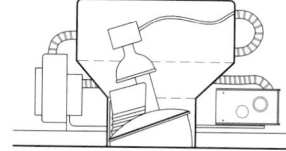

Lightolier 4½"Ø MR16 profile basis for sketch (see each mfgr's data for precise profile and dimensions)

▶ **Lamp: 50MR16/FL (p4–64)**
Luminaire: 4½"Ø Spreadlens Wallwasher
Ceiling height: **10'–6"**
Distance from wall: **2'–0"**
Spacings: **2' o.c.**
Wall illuminance (19'x9'–6" area):
▶ 10 fc, avg, maintained (17 fc max/1 fc min)
▶ Hospitality

▶ **Lamp: 73MR16/FL (p4–72)**
Luminaire: 4½"Ø Spreadlens Wallwasher
Ceiling height: **10'–6"**
Distance from wall: **2'–0"**
Spacings: **2' o.c.**
Wall illuminance (19'x9'–6" area):
▶ 13 fc, avg, maintained (22 fc max/2 fc min)
▶ Commercial, Hospitality, Institutional

Design Tips
✔Use where wall design is interesting and/or art potential exists but is not finalized.
✔Effect is strongest with lighter colored walls. Need more light on darker walls for impact.
✔Enhances wood paneling, but consider spacing lights relative to center of paneling module.
✔Where art preservation is a necessity, specify UV–reduction spreadlens.
✔Walls which have specular or glossy finishes will reflect some lens image back into room.
✔Lens must be oriented with linear pattern perpendicular to wall (typically there is a positioning "key"in the glass to help with orientation).
✔Lens must be placed in concave orientation (lens cups upward into housing—**not** downward which is a common error).

11 Details Concepts and Effects

Net Addresses/Luminaires
See page 11–99

CONNECT FOR MORE

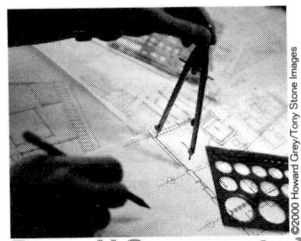

DetailConcepts™

Wallwash Spreadlens/4½"Ø/Halogen (12V)

10'–6" Ceiling Height

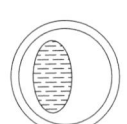

Details
Concepts and Effects

11

Design Tips

✓ IC–rated luminaires are required where contact with insulation in the construction is likely or as directed by local codes. Not all luminaire manufacturers' versions are available in IC–rated option.

✓ Use of non–IC luminaires might be accepted by local code authority for use where insulation is expected if provisions are made in construction to keep insulation at least 3 inches from luminaire housing, wiring compartment and/or ballast/transformer. Subject to local interpretation—confirm in advance of specification.

✓ Housing sizes above ceiling are relatively large—review manufacturer's datasheet to confirm fit.

For more efficient wallwashing, use fluorescent rectilinear wallwashers (see 11–193 and 195).

Wallwash Spreadlens/4½"Ø/Halogen (12V)

10'–6" Ceiling Height

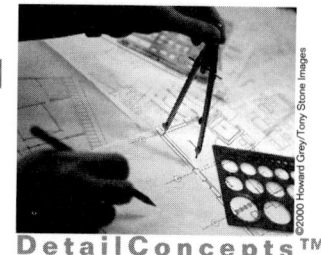

DetailConcepts™

Performance Sketches

[All data outlined here is based on information from boldfaced manufacturer's published data at time of manuscript preparation. Photometric data were rerated as necessary for cited lamp. Maintained values based on analysis grid centered on wall and extending to 0'–6" from edges and on 0.90 light loss factor (0.95 lamp lumen depreciation; 0.95 luminaire dirt depreciation). Scale is ¼ inch equals 1 foot.]

⊃Possible vendors: Cooper (Portfolio®), **Lightolier** (Calculite®), Lithonia, Zumtobel/Staff

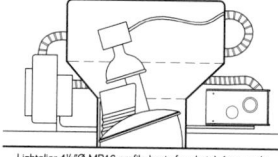

Lightolier 4½"Ø MR16 profile basis for sketch (see each mfgr's data for precise profile and dimensions)

▶ **Lamp: 50MR16/FL (p4–64)**
Luminaire: 4½"Ø Spreadlens Wallwasher
Ceiling height: **10'–6"**
Distance from wall: **4'–0"**
Spacings: **4' o.c.**
Wall illuminance (19'x9'–6" area):
▶ 3 fc, avg, maintained (4 fc max/0 fc min)
▶ Gallery, Residential

▶ **Lamp: 73MR16/FL (p4–72)**
Luminaire: 4½"Ø Spreadlens Wallwasher
Ceiling height: **10'–6"**
Distance from wall: **4'–0"**
Spacings: **4' o.c.**
Wall illuminance (19'x9'–6" area):
▶ 4 fc, avg, maintained (6 fc max/0 fc min)
▶ Gallery, Residential

Design Tips

✔ Use where wall design is interesting and/or art potential exists but is not finalized.
✔ Effect is strongest with lighter colored walls. Need more light on darker walls for impact.
✔ Enhances wood paneling, but consider spacing lights relative to center of paneling module.
✔ Where art preservation is a necessity, specify UV–reduction spreadlens.
✔ Walls which have specular or glossy finishes will reflect some lens image back into room.
✔ Lens must be oriented with linear pattern perpendicular to wall (typically there is a positioning "key" in the glass to help with orientation).
✔ Lens must be placed in concave orientation (lens cups upward into housing—**not** downward which is a common error).

Details Concepts and Effects

11

Net Addresses/Luminaires
See page 11–99

CONNECT FOR MORE

Wallwash Spreadlens/4½"Ø/HIR(12V)

8'–6" Ceiling Height

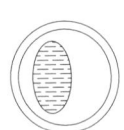

Design Tips

✓ IC–rated luminaires are required where contact with insulation in the construction is likely or as directed by local codes. Not all luminaire manufacturers' versions are available in IC–rated option.

✓ Use of non–IC luminaires might be accepted by local code authority for use where insulation is expected if provisions are made in construction to keep insulation at least 3 inches from luminaire housing, wiring compartment and/or ballast/transformer. Subject to local interpretation—confirm in advance of specification.

✓ Housing sizes above ceiling are relatively large—review manufacturer's datasheet to confirm fit.

For more efficient wallwashing use fluorescent rectilinear wallwashers (see 11–189 and 191).

Wallwash Spreadlens/4½″Ø/HIR(12V)

8′–6″ Ceiling Height

DetailConcepts™

Performance Sketches

[All data outlined here is based on information from boldfaced manufacturer's published data at time of manuscript preparation. Photometric data were rerated as necessary for cited lamp. Maintained values based on analysis grid centered on wall and extending to 0′–6″ from edges and on 0.90 light loss factor (0.95 lamp lumen depreciation; 0.95 luminaire dirt depreciation). Scale is ¼ inch equals 1 foot.]

⊃Possible vendors: Cooper (Portfolio®), **Lightolier** (Calculite®), Lithonia, Zumtobel/Staff

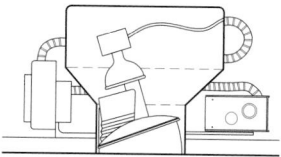

Lightolier 4½″Ø MR16 profile basis for sketch (see each mfgr's data for precise profile and dimensions)

▶ **Lamp: 37MR16/HIR/C/FL (p6–20)**
Luminaire: 4½″Ø Spreadlens Wallwasher
Ceiling height: **8′–6″**
Distance from wall: **2′–0″**
Spacings: **2′ o.c.**
Wall illuminance (19′x7′–6″ area):
▶ 12 fc, avg, maintained (18 fc max/2 fc min)
▶ Hospitality

Design Tips

✔Use where wall design is interesting and/or art potential exists but is not finalized.
✔Effect is strongest with lighter colored walls. Need more light on darker walls for impact.
✔Enhances wood paneling, but consider spacing lights relative to center of paneling module.
✔Where art preservation is a necessity, specify UV–reduction spreadlens.
✔Walls which have specular or glossy finishes will reflect some lens image back into room.
✔Lens must be oriented with linear pattern perpendicular to wall (typically there is a positioning "key"in the glass to help with orientation).
✔Lens must be placed in concave orientation (lens cups upward into housing—**not** downward which is a common error).

▶ **Lamp: 50MR16/HIR/C/FL (p6–25)**
Luminaire: 4½″Ø Spreadlens Wallwasher
Ceiling height: **8′–6″**
Distance from wall: **2′–0″**
Spacings: **2′ o.c.**
Wall illuminance (19′x7′–6″ area):
▶ 18 fc, avg, maintained (26 fc max/2 fc min)
▶ Commercial, Hospitality, Institutional

Details
Concepts and Effects

11

Net Addresses/Luminaires
See page 11–99

CONNECT FOR MORE

DetailConcepts™

Wallwash Spreadlens/4½″Ø/HIR(12V)

8′–6″ Ceiling Height

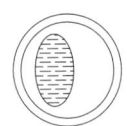

Design Tips

✔ IC–rated luminaires are required where contact with insulation in the construction is likely or as directed by local codes. Not all luminaire manufacturers' versions are available in IC–rated option.

✔ Use of non–IC luminaires might be accepted by local code authority for use where insulation is expected if provisions are made in construction to keep insulation at least 3 inches from luminaire housing, wiring compartment and/or ballast/transformer. Subject to local interpretation—confirm in advance of specification.

✔ Housing sizes above ceiling are relatively large—review manufacturer's datasheet to confirm fit.

Details Concepts and Effects **11**

For more efficient wallwashing use fluorescent rectilinear wallwashers (see 11–189 and 191).

Wallwash Spreadlens/4½"Ø/HIR(12V)
8'–6" Ceiling Height

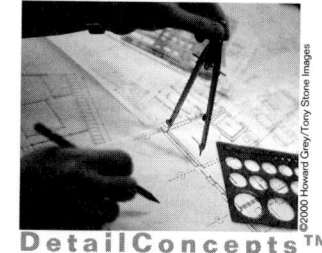

DetailConcepts™

Performance Sketches

[All data outlined here is based on information from boldfaced manufacturer's published data at time of manuscript preparation. Photometric data were rerated as necessary for cited lamp. Maintained values based on analysis grid centered on wall and extending to 0'–6" from edges and on 0.90 light loss factor (0.95 lamp lumen depreciation; 0.95 luminaire dirt depreciation). Scale is ¼ inch equals 1 foot.]

⊃Possible vendors: Cooper (Portfolio®), **Lightolier** (Calculite®), Lithonia, Zumtobel/Staff

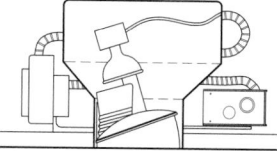

Lightolier 4½"Ø MR16 profile basis for sketch (see each mfgr's data for precise profile and dimensions)

▶ **Lamp: 37MR16/HIR/C/FL (p6–20)**
Luminaire: 4½"Ø Spreadlens Wallwasher
Ceiling height: **8'–6"**
Distance from wall: **4'–0"**
Spacings: **4' o.c.**
Wall illuminance (19'x7'–6" area):
▶ 3 fc, avg, maintained (5 fc max/0 fc min)
▶ Gallery, Residential

▶ **Lamp: 50MR16/HIR/C/FL (p6–25)**
Luminaire: 4½"Ø Spreadlens Wallwasher
Ceiling height: **8'–6"**
Distance from wall: **4'–0"**
Spacings: **4' o.c.**
Wall illuminance (19'x7'–6" area):
▶ 5 fc, avg, maintained (8 fc max/0 fc min)
▶ Gallery, Residential

Design Tips

✔Use where wall design is interesting and/or art potential exists but is not finalized.
✔Effect is strongest with lighter colored walls. Need more light on darker walls for impact.
✔Enhances wood paneling, but consider spacing lights relative to center of paneling module.
✔Where art preservation is a necessity, specify UV–reduction spreadlens.
✔Walls which have specular or glossy finishes will reflect some lens image back into room.
✔Lens must be oriented with linear pattern perpendicular to wall (typically there is a positioning "key" in the glass to help with orientation).
✔Lens must be placed in concave orientation (lens cups upward into housing—**not** downward which is a common error).

Details Concepts and Effects

11

Net Addresses/Luminaires
See page 11–99

CONNECT FOR MORE

DetailConcepts™

Wallwash Spreadlens/4½"Ø/HIR(12V)

10'–6" Ceiling Height

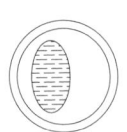

Design Tips

✔ IC–rated luminaires are required where contact with insulation in the construction is likely or as directed by local codes. Not all luminaire manufacturers' versions are available in IC–rated option.

✔ Use of non–IC luminaires might be accepted by local code authority for use where insulation is expected if provisions are made in construction to keep insulation at least 3 inches from luminaire housing, wiring compartment and/or ballast/transformer. Subject to local interpretation—confirm in advance of specification.

✔ Housing sizes above ceiling are relatively large—review manufacturer's datasheet to confirm fit.

For more efficient wallwashing use fluorescent rectilinear wallwashers (see 11–193).

Wallwash Spreadlens/4½"Ø/HIR (12V)

10′–6″ Ceiling Height

©2000 Howard Grey/Tony Stone Images

D e t a i l C o n c e p t s ™

Performance Sketches

[All data outlined here is based on information from boldfaced manufacturer's published data at time of manuscript preparation. Photometric data were rerated as necessary for cited lamp. Maintained values based on analysis grid centered on wall and extending to 0′–6″ from edges and on 0.90 light loss factor (0.95 lamp lumen depreciation; 0.95 luminaire dirt depreciation). Scale is ¼ inch equals 1 foot.)

⊃Possible vendors: Cooper (Portfolio®), **Lightolier** (Calculite®), Lithonia, Zumtobel/Staff

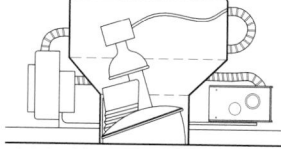

Lightolier 4½"Ø MR16 profile basis for sketch (see each mfgr's data for precise profile and dimensions)

▶ **Lamp: 37MR16/HIR/C/FL (p6–20)**
Luminaire: 4½"Ø Spreadlens Wallwasher
Ceiling height: **10′–6″**
Distance from wall: **2′–0″**
Spacings: **2′ o.c.**
Wall illuminance (19′x9′–6″ area):
▶ 11 fc, avg, maintained (18 fc max/2 fc min)
▶ Hospitality

▶ **Lamp: 50MR16/HIR/C/FL (p6–25)**
Luminaire: 4½"Ø Spreadlens Wallwasher
Ceiling height: **10′–6″**
Distance from wall: **2′–0″**
Spacings: **2′ o.c.**
Wall illuminance (19′x9′–6″ area):
▶ 16 fc, avg, maintained (26 fc max/2 fc min)
▶ Commercial, Hospitality, Institutional

Design Tips

✔Use where wall design is interesting and/or art potential exists but is not finalized.
✔Effect is strongest with lighter colored walls. Need more light on darker walls for impact.
✔Enhances wood paneling, but consider spacing lights relative to center of paneling module.
✔Where art preservation is a necessity, specify UV–reduction spreadlens.
✔Walls which have specular or glossy finishes will reflect some lens image back into room.
✔Lens must be oriented with linear pattern perpendicular to wall (typically there is a positioning "key"in the glass to help with orientation).
✔Lens must be placed in concave orientation (lens cups upward into housing—**not** downward which is a common error).

Details
Concepts and Effects

11

Net Addresses/Luminaires
See page 11–99

▷ **CONNECT FOR MORE**

Wallwash Spreadlens/4½"Ø/HIR(12V)

10'–6" Ceiling Height

DetailConcepts™

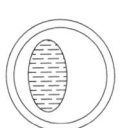

Design Tips

✔ IC–rated luminaires are required where contact with insulation in the construction is likely or as directed by local codes. Not all luminaire manufacturers' versions are available in IC–rated option.

✔ Use of non–IC luminaires might be accepted by local code authority for use where insulation is expected if provisions are made in construction to keep insulation at least 3 inches from luminaire housing, wiring compartment and/or ballast/transformer. Subject to local interpretation—confirm in advance of specification.

✔ Housing sizes above ceiling are relatively large—review manufacturer's datasheet to confirm fit.

Details
Concepts and Effects

11

For more efficient wallwashing, use fluorescent rectilinear wallwashers (see 11–195).

Wallwash Spreadlens/4½"Ø/HIR(12V)

10'–6" Ceiling Height

Performance Sketches

[All data outlined here is based on information from boldfaced manufacturer's published data at time of manuscript preparation. Photometric data were rerated as necessary for cited lamp. Maintained values based on analysis grid centered on wall and extending to 0'–6" from edges and on 0.90 light loss factor (0.95 lamp lumen depreciation; 0.95 luminaire dirt depreciation). Scale is ¼ inch equals 1 foot.]

⊃Possible vendors: Cooper (Portfolio®), **Lightolier** (Calculite®), Lithonia, Zumtobel/Staff

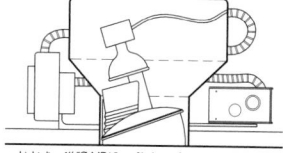

Lightolier 4½"Ø MR16 profile basis for sketch (see each mfgr's data for precise profile and dimensions)

▶ **Lamp: 37MR16/HIR/C/FL (p6–20)**
Luminaire: 4½"Ø Spreadlens Wallwasher
Ceiling height: **10'–6"**
Distance from wall: **4'–0"**
Spacings: **4' o.c.**
Wall illuminance (19'x9'–6" area):
▶ 3 fc, avg, maintained (4 fc max/0 fc min)
▶ Gallery, Residential

▶ **Lamp: 50MR16/HIR/C/FL (p6–25)**
Luminaire: 4½"Ø Spreadlens Wallwasher
Ceiling height: **10'–6"**
Distance from wall: **4'–0"**
Spacings: **4' o.c.**
Wall illuminance (19'x9'–6" area):
▶ 5 fc, avg, maintained (7 fc max/0 fc min)
▶ Gallery, Residential

Design Tips

✔Use where wall design is interesting and/or art potential exists but is not finalized.

✔Effect is strongest with lighter colored walls. Need more light on darker walls for impact.

✔Enhances wood paneling, but consider spacing lights relative to center of paneling module.

✔Where art preservation is a necessity, specify UV–reduction spreadlens.

✔Walls which have specular or glossy finishes will reflect some lens image back into room.

✔Lens must be oriented with linear pattern perpendicular to wall (typically there is a positioning "key" in the glass to help with orientation).

✔Lens must be placed in concave orientation (lens cups upward into housing—**not** downward which is a common error).

Details Concepts and Effects

11

Net Addresses/Luminaires
See page 11–99

CONNECT FOR MORE

DetailConcepts™

Wallwash Spreadlens/4½"Ø/CMH

9'–6" Ceiling Height

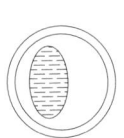

Design Tips

✔ IC–rated luminaires are required where contact with insulation in the construction is likely or as directed by local codes. Not all luminaire manufacturers' versions are available in IC–rated option.

✔ Use of non–IC luminaires might be accepted by local code authority for use where insulation is expected if provisions are made in construction to keep insulation at least 3 inches from luminaire housing, wiring compartment and/or ballast/transformer. Subject to local interpretation—confirm in advance of specification.

✔ Housing sizes above ceiling are relatively large—review manufacturer's datasheet to confirm fit.

Details Concepts and Effects

11

Wallwash Spreadlens/4½"Ø/CMH

9'–6" Ceiling Height

DetailConcepts™

Performance Sketches

[Lightolier data were unavailable. All data outlined here is based on information from a 4½" Ø 50PAR20/H/NFL Kurt Versen spreadlens luminaire. Photometric data were rerated as necessary for cited lamp. This is imprecise. Maintained values based on analysis grid centered on wall and extending to 0'–6" from edges and on 0.80 light loss factor (0.85 lamp lumen depreciation; 0.95 luminaire dirt depreciation). Scale is ¼ inch equals 1 foot.]

⊃Possible vendors: Lightolier (Calculite®)

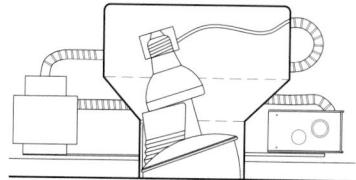

Lightolier 4½"Ø PAR20/CMH profile basis for sketch (see each mfg's data for precise profile and dimensions)

▶ **Lamp: 39PAR20/CMH/3K/NFL (p7–38)**
Luminaire: 4½"Ø Spreadlens Wallwasher
Ceiling height: **9'–6"**
Distance from wall: **2'–0"**
Spacings: **2' o.c.**
Wall illuminance (19'x8'–6" area):
▶ 24 fc, avg, maintained (48 fc max/7 fc min)
▶ Commercial, Institutional, Retail

▶ **Lamp: 39PAR20/CMH/3K/NFL (p7–38)**
Luminaire: 4½"Ø Spreadlens Wallwasher
Ceiling height: **9'–6"**
Distance from wall: **2'–0"**
Spacings: **4' o.c.**
Wall illuminance (19'x8'–6" area):
▶ 12 fc, avg, maintained (33 fc max/2 fc min)
▶ Commercial, Hospitality, Institutional

Design Tips

✔ Use where wall design is interesting and/or art potential exists but is not finalized.
✔ Effect is strongest with lighter colored walls. Need more light on darker walls for impact.
✔ Enhances wood paneling, but consider spacing lights relative to center of paneling module.
✔ Where art preservation is a necessity, specify UV–reduction spreadlens.
✔ Walls which have specular or glossy finishes will reflect some lens image back into room.
✔ Lens must be oriented with linear pattern perpendicular to wall (typically there is a positioning "key" in the glass to help with orientation).
✔ Lens must be placed in concave orientation (lens cups upward into housing—**not** downward which is a common error).
✔ Not instant–on and requires 10– to 15–minute cycle–on period if power fails.
✔ Not dimmable in fashion meeting user expectations.

Details Concepts and Effects

11

Net Addresses/Luminaires
See page 11–99

CONNECT FOR MORE

DetailConcepts™

Wallwash Spreadlens/4½″Ø/CMH

10′–6″ Ceiling Height

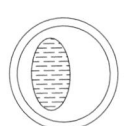

Design Tips

✔ IC–rated luminaires are required where contact with insulation in the construction is likely or as directed by local codes. Not all luminaire manufacturers' versions are available in IC–rated option.

✔ Use of non–IC luminaires might be accepted by local code authority for use where insulation is expected if provisions are made in construction to keep insulation at least 3 inches from luminaire housing, wiring compartment and/or ballast/transformer. Subject to local interpretation—confirm in advance of specification.

✔ Housing sizes above ceiling are relatively large—review manufacturer's datasheet to confirm fit.

Details
Concepts and Effects

11

Wallwash Spreadlens/4½"Ø/CMH

10'–6" Ceiling Height

DetailConcepts™

Performance Sketches

[Lightolier data were unavailable. All data outlined here is based on information from a 4½" Ø 50PAR20/H/NFL Kurt Versen spreadlens luminaire. Photometric data were rerated as necessary for cited lamp. This is imprecise. Maintained values based on analysis grid centered on wall and extending to 0'–6" from edges and on 0.80 light loss factor (0.85 lamp lumen depreciation; 0.95 luminaire dirt depreciation). Scale is ¼ inch equals 1 foot.]

⊃Possible vendors: Lightolier (Calculite®)

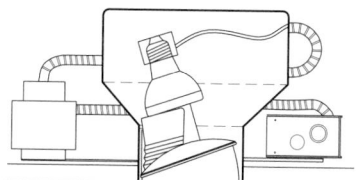

Lightolier 4½"Ø PAR20/CMH profile basis for sketch (see each mfgr's data for precise profile and dimensions)

▶ **Lamp: 39PAR20/CMH/3K/NFL (p7–38)**
Luminaire: 4½"Ø Spreadlens Wallwasher
Ceiling height: **10'–6"**
Distance from wall: **2'–0"**
Spacings: **2' o.c.**
Wall illuminance (19'x9'–6" area):
▶ 22 fc, avg, maintained (47 fc max/6 fc min)
▶ Commercial, Institutional, Retail

Design Tips

✔Use where wall design is interesting and/or art potential exists but is not finalized.
✔Effect is strongest with lighter colored walls. Need more light on darker walls for impact.
✔Enhances wood paneling, but consider spacing lights relative to center of paneling module.
✔Where art preservation is a necessity, specify UV–reduction spreadlens.
✔Walls which have specular or glossy finishes will reflect some lens image back into room.
✔Lens must be oriented with linear pattern perpendicular to wall (typically there is a positioning "key" in the glass to help with orientation).
✔Lens must be placed in concave orientation (lens cups upward into housing—**not** downward which is a common error).
✔Not instant–on and requires 10– to 15–minute cycle–on period if power fails.
✔Not dimmable in fashion meeting user expectations.

▶ **Lamp: 39PAR20/CMH/3K/NFL (p7–38)**
Luminaire: 4½"Ø Spreadlens Wallwasher
Ceiling height: **10'–6"**
Distance from wall: **2'–0"**
Spacings: **4' o.c.**
Wall illuminance (19'x9'–6" area):
▶ 11 fc, avg, maintained (33 fc max/2 fc min)
▶ Hospitality

Details
Concepts and Effects

11

Net Addresses/Luminaires
See page 11–99

CONNECT FOR MORE

Wallwash Spreadlens/6"Ø/Halogen(120V)

9'–6" Ceiling Height

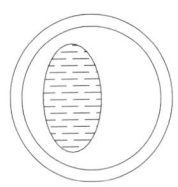

Design Tips

✔ IC–rated luminaires are required where contact with insulation in the construction is likely or as directed by local codes. Not all luminaire manufacturers' versions are available in IC–rated option.

✔ Use of non–IC luminaires might be accepted by local code authority for use where insulation is expected if provisions are made in construction to keep insulation at least 3 inches from luminaire housing, wiring compartment and/or ballast/transformer. Subject to local interpretation—confirm in advance of specification.

✔ Housing sizes above ceiling are relatively large—review manufacturer's datasheet to confirm fit.

✔ Some manufacturers' luminaires have minimum spacing requirements to maintain appropriate clearance and to maintain UL listings—see specific manufacturer's data.

For more efficient wallwashing, use fluorescent rectilinear wallwashers (see 11–197, 199 and 201) .

Wallwash Spreadlens/6"Ø/Halogen(120V)

9'–6" Ceiling Height

Performance Sketches

[All data outlined here is based on information from boldfaced manufacturer's published data at time of manuscript preparation. Photometric data were rerated as necessary for cited lamp. Maintained values based on analysis grid centered on wall and extending to 0'–6" from edges and on 0.90 light loss factor (0.95 lamp lumen depreciation; 0.95 luminaire dirt depreciation). Scale is ¼ inch equals 1 foot.]

⊃Possible vendors: Cooper (Portfolio®), Kurt Versen, **Lightolier** (Calculite®), Prescolite, Zumtobel/Staff

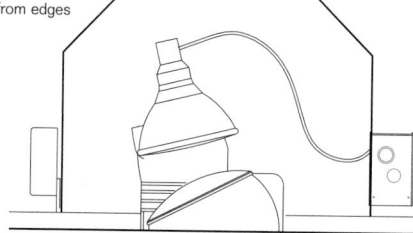

Lightolier 6"Ø PAR38 profile basis for sketch (see each mfgr's data for precise profile and dimensions)

▶ **Lamp: 75PAR30S/H/FL (p3–52)**
Luminaire: 6"Ø Spreadlens Wallwasher
Ceiling height: **9'–6"**
Distance from wall: **2'–0"**
Spacings: **2' o.c.** (Advisory: confirm allowable spacing requirements with manufacturer)
Wall illuminance (19'x8'–6" area):
▶ 17 fc, avg, maintained (43 fc max/2 fc min)
▶ Commercial, Hospitality, Institutional

Design Tips
✔Use where wall design is interesting and/or art potential exists but is not finalized.
✔Effect is strongest with lighter colored walls. Need more light on darker walls for impact.
✔Enhances wood paneling, but consider spacing lights relative to center of paneling module.
✔Where art preservation is a necessity, specify UV–reduction spreadlens.
✔Walls which have specular or glossy finishes will reflect some lens image back into room.
✔Lens must be oriented with linear pattern perpendicular to wall (typically there is a positioning "key" in the glass to help with orientation).
✔Lens must be placed in concave orientation (lens cups upward into housing—**not** downward which is a common error).

▶ **Lamp: 75PAR30S/H/FL (p3–52)**
Luminaire: 6"Ø Spreadlens Wallwasher
Ceiling height: **9'–6"**
Distance from wall: **2'–6"**
Spacings: **3' o.c.** (Advisory: confirm allowable spacing requirements with manufacturer)
Wall illuminance (19'x8'–6" area):
▶ 10 fc, avg, maintained (23 fc max/0 fc min)
▶ Gallery, Hospitality, Residential

Details Concepts and Effects

11

Net Addresses/Luminaires
See page 11–99

CONNECT FOR MORE

DetailConcepts™

Wallwash Spreadlens/6"Ø/Fluorescent

8'–6" Ceiling Height

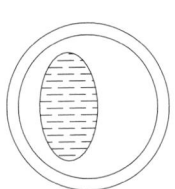

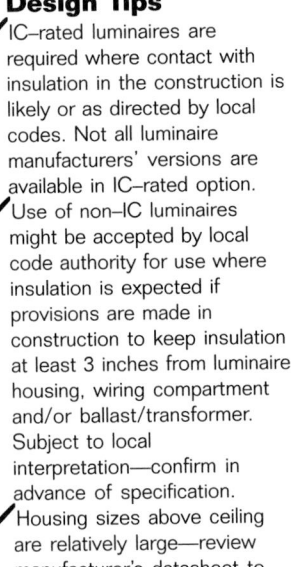

Design Tips

✔ IC–rated luminaires are required where contact with insulation in the construction is likely or as directed by local codes. Not all luminaire manufacturers' versions are available in IC–rated option.

✔ Use of non–IC luminaires might be accepted by local code authority for use where insulation is expected if provisions are made in construction to keep insulation at least 3 inches from luminaire housing, wiring compartment and/or ballast/transformer. Subject to local interpretation—confirm in advance of specification.

✔ Housing sizes above ceiling are relatively large—review manufacturer's datasheet to confirm fit.

Wallwash Spreadlens/6″Ø/Fluorescent

8′–6″ Ceiling Height

DetailConcepts™

Performance Sketches

[All data outlined here is based on information from boldfaced manufacturer's published data at time of manuscript preparation. Photometric data were rerated as necessary for cited lamp. Maintained values based on analysis grid centered on wall and extending to 0′–6″ from edges and on 0.90 light loss factor (0.95 lamp lumen depreciation; 0.95 luminaire dirt depreciation). Scale is ¼ inch equals 1 foot.]

⊃Possible vendors: Cooper (Portfolio®), **Kurt Versen**, Lightolier (Calculite®), Prescolite, Zumtobel/Staff

▶ **Lamp: F32Triple/830 (p8–18)**
Luminaire: 6″Ø Spreadlens Wallwasher
Ceiling height: **8′–6″**
Distance from wall: **2′–6″**
Spacings: **2′ o.c.**
Wall illuminance (19′x7′–6″ area):
▶ 25 fc, avg, maintained (42 fc max/6 fc min)
▶ Commercial, Institutional

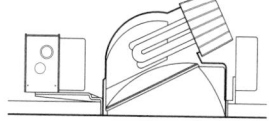

Lightolier 6″Ø profile basis for sketch (see each mfgr's data for precise profile and dimensions)

Design Tips

✔Use where wall design is interesting and/or art potential exists but is not finalized.
✔Effect is strongest with lighter colored walls. Need more light on darker walls for impact.
✔Enhances wood paneling, but consider spacing lights relative to center of paneling module.
✔Where art preservation is a necessity, specify UV–reduction spreadlens.
✔Walls which have specular or glossy finishes will reflect some lens image back into room.
✔Lens must be oriented with linear pattern perpendicular to wall (typically there is a positioning "key"in the glass to help with orientation).
✔Lens must be placed in concave orientation (lens cups upward into housing—**not** downward which is a common error).

Details
Concepts and Effects

11

Net Addresses/Luminaires
See page 11–99

CONNECT FOR MORE

DetailConcepts™

Wallwash Spreadlens/6"Ø/Fluorescent

9'–6" Ceiling Height

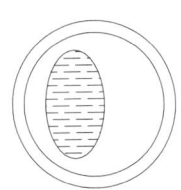

Design Tips

✔IC–rated luminaires are required where contact with insulation in the construction is likely or as directed by local codes. Not all luminaire manufacturers' versions are available in IC–rated option.

✔Use of non–IC luminaires might be accepted by local code authority for use where insulation is expected if provisions are made in construction to keep insulation at least 3 inches from luminaire housing, wiring compartment and/or ballast/transformer. Subject to local interpretation—confirm in advance of specification.

✔Housing sizes above ceiling are relatively large—review manufacturer's datasheet to confirm fit.

Wallwash Spreadlens/6"Ø/Fluorescent
9'–6" Ceiling Height

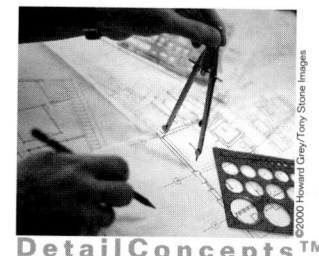

DetailConcepts™

Performance Sketches

[All data outlined here is based on information from boldfaced manufacturer's published data at time of manuscript preparation. Photometric data were rerated as necessary for cited lamp. Maintained values based on analysis grid centered on wall and extending to 0'–6" from edges and on 0.90 light loss factor (0.95 lamp lumen depreciation; 0.95 luminaire dirt depreciation). Scale is ¼ inch equals 1 foot.]

⊃**Possible vendors:** Cooper (Portfolio®), **Kurt Versen**, Lightolier (Calculite®), Prescolite, Zumtobel/Staff

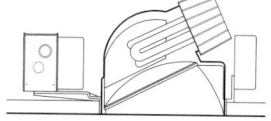

Lightolier 6"Ø profile basis for sketch (see each mfgr's data for precise profile and dimensions)

▶ **Lamp: F32Triple/830 (p8–18)**
Luminaire: 6"Ø Spreadlens Wallwasher
Ceiling height: **9'–6"**
Distance from wall: **2'–6"**
Spacings: **2' o.c.**
Wall illuminance (19'x8'–6" area):
▶ 23 fc, avg, maintained (42 fc max/6 fc min)
▶ Commercial, Institutional

▶ **Lamp: F42Triple/830 (p8–22)**
Luminaire: 6"Ø Spreadlens Wallwasher
Ceiling height: **9'–6"**
Distance from wall: **2'–6"**
Spacings: **2' o.c.**
Wall illuminance (19'x8'–6" area):
▶ 28 fc, avg, maintained (56 fc max/10 fc min)
▶ Commercial, Institutional, Retail

Design Tips
✔Use where wall design is interesting and/or art potential exists but is not finalized.
✔Effect is strongest with lighter colored walls. Need more light on darker walls for impact.
✔Enhances wood paneling, but consider spacing lights relative to center of paneling module.
✔Where art preservation is a necessity, specify UV–reduction spreadlens.
✔Walls which have specular or glossy finishes will reflect some lens image back into room.
✔Lens must be oriented with linear pattern perpendicular to wall (typically there is a positioning "key"in the glass to help with orientation).
✔Lens must be placed in concave orientation (lens cups upward into housing—**not** downward which is a common error).

Details Concepts and Effects **11**

Net Addresses/Luminaires
See page 11–99

 CONNECT FOR MORE

Wallwash Spreadlens/6″Ø/Fluorescent

10′–6″ Ceiling Height

DetailConcepts™

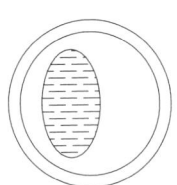

Design Tips

✔ IC–rated luminaires are required where contact with insulation in the construction is likely or as directed by local codes. Not all luminaire manufacturers' versions are available in IC–rated option.

✔ Use of non–IC luminaires might be accepted by local code authority for use where insulation is expected if provisions are made in construction to keep insulation at least 3 inches from luminaire housing, wiring compartment and/or ballast/transformer. Subject to local interpretation—confirm in advance of specification.

✔ Housing sizes above ceiling are relatively large—review manufacturer's datasheet to confirm fit.

Details
Concepts and Effects

11

Wallwash Spreadlens/6″Ø/Fluorescent

10′–6″ Ceiling Height

DetailConcepts™

Performance Sketches

[All data outlined here is based on information from boldfaced manufacturer's published data at time of manuscript preparation. Photometric data were rerated as necessary for cited lamp. Maintained values based on analysis grid centered on wall and extending to 0′–6″ from edges and on 0.90 light loss factor (0.95 lamp lumen depreciation; 0.95 luminaire dirt depreciation). Scale is ¼ inch equals 1 foot.]

⊃Possible vendors: Cooper (Portfolio®), **Kurt Versen**, Lightolier (Calculite®), Prescolite, Zumtobel/Staff

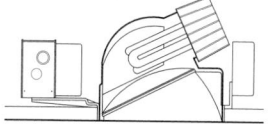

Lightolier 6″Ø profile basis for sketch (see each mfgr's data for precise profile and dimensions)

▶ **Lamp: F32Triple/830 (p8–18)**
Luminaire: **6″Ø Spreadlens Wallwasher**
Ceiling height: **10′–6″**
Distance from wall: **2′–6″**
Spacings: **2′ o.c.**
Wall illuminance (19′x9′–6″ area):
▶ 21 fc, avg, maintained (42 fc max/6 fc min)
▶ Commercial, Institutional

▶ **Lamp: F42Triple/830 (p8–22)**
Luminaire: **6″Ø Spreadlens Wallwasher**
Ceiling height: **10′–6″**
Distance from wall: **2′–6″**
Spacings: **2′ o.c.**
Wall illuminance (19′x9′–6″ area):
▶ 26 fc, avg, maintained (56 fc max/8 fc min)
▶ Commercial, Institutional, Retail

Design Tips

✔Use where wall design is interesting and/or art potential exists but is not finalized.
✔Effect is strongest with lighter colored walls. Need more light on darker walls for impact.
✔Enhances wood paneling, but consider spacing lights relative to center of paneling module.
✔Where art preservation is a necessity, specify UV–reduction spreadlens.
✔Walls which have specular or glossy finishes will reflect some lens image back into room.
✔Lens must be oriented with linear pattern perpendicular to wall (typically there is a positioning "key" in the glass to help with orientation).
✔Lens must be placed in concave orientation (lens cups upward into housing—**not** downward which is a common error).

Details
Concepts and Effects

11

Net Addresses/Luminaires
See page 11–99

CONNECT FOR MORE

DetailConcepts™

Wallwash Spreadlens/6″Ø/Fluorescent

12′–6″ Ceiling Height

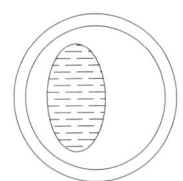

Design Tips

✔ IC–rated luminaires are required where contact with insulation in the construction is likely or as directed by local codes. Not all luminaire manufacturers' versions are available in IC–rated option.

✔ Use of non–IC luminaires might be accepted by local code authority for use where insulation is expected if provisions are made in construction to keep insulation at least 3 inches from luminaire housing, wiring compartment and/or ballast/transformer. Subject to local interpretation—confirm in advance of specification.

✔ Housing sizes above ceiling are relatively large—review manufacturer's datasheet to confirm fit.

Details
Concepts and Effects

11

Wallwash Spreadlens/6"Ø/Fluorescent

12'–6" Ceiling Height

DetailConcepts™

Performance Sketches

[All data outlined here is based on information from boldfaced manufacturer's published data at time of manuscript preparation. Photometric data were rerated as necessary for cited lamp. Maintained values based on analysis grid centered on wall and extending to 0'–6" from edges and on 0.90 light loss factor (0.95 lamp lumen depreciation; 0.95 luminaire dirt depreciation). Scale is ¼ inch equals 1 foot.]

⊃Possible vendors: Cooper (Portfolio®), **Kurt Versen**, Lightolier (Calculite®), Prescolite, Zumtobel/Staff

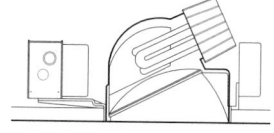

Lightolier 6"Ø profile basis for sketch (see each mfgr's data for precise profile and dimensions)

▶ **Lamp: F42Triple/830 (p8–22)**
Luminaire: 6"Ø Spreadlens Wallwasher
Ceiling height: **12'–6"**
Distance from wall: **2'–6"**
Spacings: **2' o.c.**
Wall illuminance (19'x11'–6" area):
▶ 23 fc, avg, maintained (55 fc max/6 fc min)
▶ Commercial, Institutional

Design Tips

✔ Use where wall design is interesting and/or art potential exists but is not finalized.

✔ Effect is strongest with lighter colored walls. Need more light on darker walls for impact.

✔ Enhances wood paneling, but consider spacing lights relative to center of paneling module.

✔ Where art preservation is a necessity, specify UV–reduction spreadlens.

✔ Walls which have specular or glossy finishes will reflect some lens image back into room.

✔ Lens must be oriented with linear pattern perpendicular to wall (typically there is a positioning "key" in the glass to help with orientation).

✔ Lens must be placed in concave orientation (lens cups upward into housing—**not** downward which is a common error).

Details Concepts and Effects

11

Net Addresses/Luminaires
See page 11–99

CONNECT FOR MORE

DetailConcepts™

Wallwash Spreadlens/6½"Ø/Halogen(120V)

8′–6″ Ceiling Height

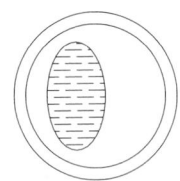

Details Concepts and Effects

11

Design Tips

✔ IC–rated luminaires are required where contact with insulation in the construction is likely or as directed by local codes. Not all luminaire manufacturers' versions are available in IC–rated option.

✔ Use of non–IC luminaires might be accepted by local code authority for use where insulation is expected if provisions are made in construction to keep insulation at least 3 inches from luminaire housing, wiring compartment and/or ballast/transformer. Subject to local interpretation—confirm in advance of specification.

✔ Housing sizes above ceiling are relatively large—review manufacturer's datasheet to confirm fit.

✔ Some manufacturers' luminaires have minimum spacing requirements to maintain appropriate clearance and to maintain UL listings—see specific manufacturer's data.

For more efficient wallwashing, use fluorescent rectilinear wallwashers (see 11–189 and 191).

Wallwash Spreadlens/6½"Ø/Halogen(120V)

8'–6" Ceiling Height

DetailConcepts™

Performance Sketches

[All data outlined here is based on information from boldfaced manufacturer's published data at time of manuscript preparation. Photometric data were rerated as necessary for cited lamp. Maintained values based on analysis grid centered on wall and extending to 0'–6" from edges and on 0.90 light loss factor (0.95 lamp lumen depreciation; 0.95 luminaire dirt depreciation). Scale is ¼ inch equals 1 foot.]

⟳Possible vendors: Cooper (Portfolio®), Edison Price, Kurt Versen, **Lightolier** (Calculite®), Prescolite, Zumtobel/Staff

▶ **Lamp: 45PAR38/H/NFL (p3–88)**
Luminaire: 6½"Ø Spreadlens Wallwasher (some versions are 6"Ø)
Ceiling height: **8'–6"**
Distance from wall: **2'–0"**
Spacings: **2' o.c.** (Advisory: confirm allowable spacing requirements with manufacturer)
Wall illuminance (19'x7'–6" area):
▶ 7 fc, avg, maintained (11 fc max/1 fc min)
▶ Gallery, Residential

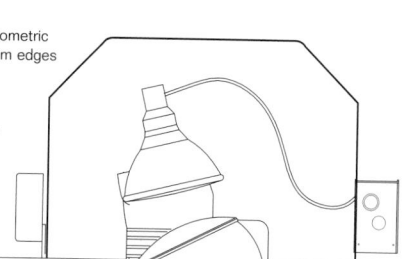

Lightolier 6"Ø PAR38 profile basis for sketch (see each mfgr's data for precise profile and dimensions)

Design Tips

✔Use where wall design is interesting and/or art potential exists but is not finalized.
✔Effect is strongest with lighter colored walls. Need more light on darker walls for impact.
✔Enhances wood paneling, but consider spacing lights relative to center of paneling module.
✔Where art preservation is a necessity, specify UV–reduction spreadlens.
✔Walls which have specular or glossy finishes will reflect some lens image back into room.
✔Lens must be oriented with linear pattern perpendicular to wall (typically there is a positioning "key" in the glass to help with orientation).
✔Lens must be placed in concave orientation (lens cups upward into housing—**not** downward which is a common error).

Details Concepts and Effects

11

Net Addresses/Luminaires
See page 11–99

CONNECT FOR MORE

DetailConcepts™

Wallwash Spreadlens/6½"Ø/Halogen(120V)

9'–6" Ceiling Height

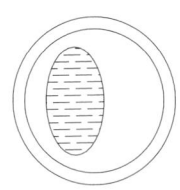

Design Tips

✔ IC–rated luminaires are required where contact with insulation in the construction is likely or as directed by local codes. Not all luminaire manufacturers' versions are available in IC–rated option.

✔ Use of non–IC luminaires might be accepted by local code authority for use where insulation is expected if provisions are made in construction to keep insulation at least 3 inches from luminaire housing, wiring compartment and/or ballast/transformer. Subject to local interpretation—confirm in advance of specification.

✔ Housing sizes above ceiling are relatively large—review manufacturer's datasheet to confirm fit.

✔ Some manufacturers' luminaires have minimum spacing requirements to maintain appropriate clearance and to maintain UL listings— see specific manufacturer's data.

Details Concepts and Effects

11

For more efficient wallwashing, use fluorescent rectilinear wallwashers (see 11–199 and 201) **.**

Wallwash Spreadlens/6½"Ø/Halogen(120V)
9'–6" Ceiling Height

Lightolier 6"Ø PAR38 profile basis for sketch (see each mfgr's data for precise profile and dimensions)

DetailConcepts™

Performance Sketches

[All data outlined here is based on information from boldfaced manufacturer's published data at time of manuscript preparation. Photometric data were rerated as necessary for cited lamp. Maintained values based on analysis grid centered on wall and extending to 0'–6" from edges and on 0.90 light loss factor (0.95 lamp lumen depreciation; 0.95 luminaire dirt depreciation). Scale is ¼ inch equals 1 foot.]

⊃Possible vendors: Cooper (Portfolio®), Edison Price, Kurt Versen, **Lightolier** (Calculite®), Prescolite, Zumtobel/Staff

▶ **Lamp: 45PAR38/H/NFL (p3–88)**
Luminaire: 6½"Ø Spreadlens Wallwasher(some versions are 6"Ø)
Ceiling height: **9'–6"**
Distance from wall: **2'–0"** (Advisory: confirm allowable spacing requirements with manufacturer)
Spacings: **2' o.c.**
Wall illuminance (19'x8'–6" area):
▶ 6 fc, avg, maintained (11 fc max/1 fc min)
▶ Gallery, Residential

▶ **Lamp: 60PAR38/H/NFL (p3–95)**
Luminaire: 6½"Ø Spreadlens Wallwasher(some versions are 6"Ø)
Ceiling height: **9'–6"**
Distance from wall: **2'–0"** (Advisory: confirm allowable spacing requirements with manufacturer)
Spacings: **2' o.c.**
Wall illuminance (19'x8'–6" area):
▶ 10 fc, avg, maintained (17 fc max/2 fc min)
▶ Hospitality

Design Tips
✔Use where wall design is interesting and/or art potential exists but is not finalized.
✔Effect is strongest with lighter colored walls. Need more light on darker walls for impact.
✔Enhances wood paneling, but consider spacing lights relative to center of paneling module.
✔Where art preservation is a necessity, specify UV–reduction spreadlens.
✔Walls which have specular or glossy finishes will reflect some lens image back into room.
✔Lens must be oriented with linear pattern perpendicular to wall (typically there is a positioning "key"in the glass to help with orientation).
✔Lens must be placed in concave orientation (lens cups upward into housing—**not** downward which is a common error).

Details
Concepts and Effects
11

Net Addresses/Luminaires
See page 11–99

CONNECT FOR MORE

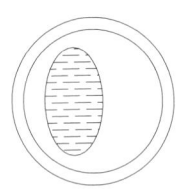

Wallwash Spreadlens/6½″Ø/Halogen(120V)

10′–6″ Ceiling Height

DetailConcepts™

Design Tips

✔ IC–rated luminaires are required where contact with insulation in the construction is likely or as directed by local codes. Not all luminaire manufacturers' versions are available in IC–rated option.

✔ Use of non–IC luminaires might be accepted by local code authority for use where insulation is expected if provisions are made in construction to keep insulation at least 3 inches from luminaire housing, wiring compartment and/or ballast/transformer. Subject to local interpretation—confirm in advance of specification.

✔ Housing sizes above ceiling are relatively large—review manufacturer's datasheet to confirm fit.

✔ Some manufacturers' luminaires have minimum spacing requirements to maintain appropriate clearance and to maintain UL listings—see specific manufacturer's data.

For more efficient wallwashing, use fluorescent rectilinear wallwashers (see 11–193, 195, 203 and 205).

Wallwash Spreadlens/6½"Ø/Halogen(120V)
10'–6" Ceiling Height

DetailConcepts™

Performance Sketches

[All data outlined here is based on information from boldfaced manufacturer's published data at time of manuscript preparation. Photometric data were rerated as necessary for cited lamp. Maintained values based on analysis grid centered on wall and extending to 0'–6" from edges and on 0.90 light loss factor (0.95 lamp lumen depreciation; 0.95 luminaire dirt depreciation). Scale is ¼ inch equals 1 foot.]

⊃Possible vendors: Cooper (Portfolio®), Edison Price, Kurt Versen, **Lightolier** (Calculite®), Prescolite, Zumtobel/Staff

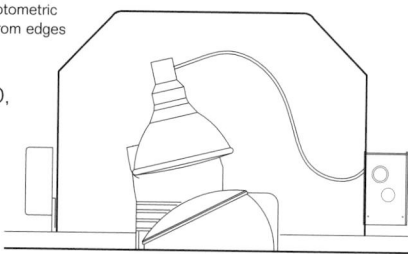

Lightolier 6"Ø PAR38 profile basis for sketch (see each mfgr's data for precise profile and dimensions)

▶ **Lamp: 60PAR38/H/NFL (p3–95)**
Luminaire: 6½"Ø Spreadlens Wallwasher (some versions are 6"Ø)
Ceiling height: **10'–6"**
Distance from wall: **2'–0"**
Spacings: **2' o.c.** (Advisory: confirm allowable spacing requirements with manufacturer)
Wall illuminance (19'x9'–6" area):
▶ 9 fc, avg, maintained (17 fc max/2 fc min)
▶ Hospitality

Design Tips
✔ Use where wall design is interesting and/or art potential exists but is not finalized.
✔ Effect is strongest with lighter colored walls. Need more light on darker walls for impact.
✔ Enhances wood paneling, but consider spacing lights relative to center of paneling module.
✔ Where art preservation is a necessity, specify UV–reduction spreadlens.
✔ Walls which have specular or glossy finishes will reflect some lens image back into room.
✔ Lens must be oriented with linear pattern perpendicular to wall (typically there is a positioning "key" in the glass to help with orientation).
✔ Lens must be placed in concave orientation (lens cups upward into housing—**not** downward which is a common error).

Details Concepts and Effects

11

▶ **Lamp: 75PAR38/H/NFL30° (p3–102)**
Luminaire: 6½"Ø Spreadlens Wallwasher (some versions are 6"Ø)
Ceiling height: **10'–6"**
Distance from wall: **2'–0"**
Spacings: **2' o.c.** (Advisory: confirm allowable spacing requirements with manufacturer)
Wall illuminance (19'x9'–6" area):
▶ 11 fc, avg, maintained (21 fc max/3 fc min)
▶ Hospitality

▶ **Lamp: 75PAR38/H/NFL30° (p3–102)**
Luminaire: 6½"Ø Spreadlens Wallwasher (some versions are 6"Ø)
Ceiling height: **10'–6"**
Distance from wall: **3'–0"**
Spacings: **3' o.c.** (Advisory: confirm allowable spacing requirements with manufacturer)
Wall illuminance (19'x9'–6" area):
▶ 6 fc, avg, maintained (9 fc max/0 fc min)
▶ Gallery, Residential

Net Addresses/Luminaires
See page 11–99

CONNECT FOR MORE

DetailConcepts™

Wallwash Spreadlens/6½″Ø/Halogen(120V)

12′–6″ Ceiling Height

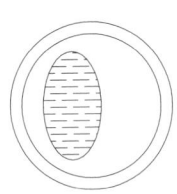

Details Concepts and Effects

11

Design Tips

✔ IC–rated luminaires are required where contact with insulation in the construction is likely or as directed by local codes. Not all luminaire manufacturers' versions are available in IC–rated option.

✔ Use of non–IC luminaires might be accepted by local code authority for use where insulation is expected if provisions are made in construction to keep insulation at least 3 inches from luminaire housing, wiring compartment and/or ballast/transformer. Subject to local interpretation—confirm in advance of specification.

✔ Housing sizes above ceiling are relatively large—review manufacturer's datasheet to confirm fit.

✔ Some manufacturers' luminaires have minimum spacing requirements to maintain appropriate clearance and to maintain UL listings—see specific manufacturer's data.

For more efficient wallwashing, use HIR spreadlens wallwashers (see 11–183) .

Wallwash Spreadlens/6½"Ø/Halogen(120V)

12'–6" Ceiling Height

Performance Sketches

[All data outlined here is based on information from boldfaced manufacturer's published data at time of manuscript preparation. Photometric data were rerated as necessary for cited lamp. Maintained values based on analysis grid centered on wall and extending to 0'–6" from edges and on 0.90 light loss factor (0.95 lamp lumen depreciation; 0.95 luminaire dirt depreciation). Scale is ¼ inch equals 1 foot.]

⊃Possible vendors: Cooper (Portfolio®), Edison Price, Kurt Versen, **Lightolier** (Calculite®), Prescolite, Zumtobel/Staff

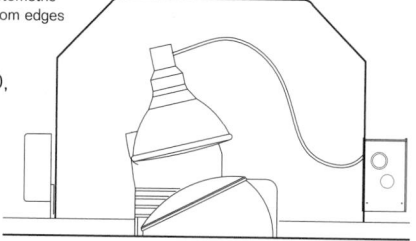

Lightolier 6"Ø PAR38 profile basis for sketch (see each mfgr's data for precise profile and dimensions)

▶ **Lamp: 75PAR38/H/NFL30° (p3–102)**

Luminaire: 6½"Ø Spreadlens Wallwasher (some versions are 6"Ø)
Ceiling height: **12'–6"**
Distance from wall: **3'–0"**
Spacings: **3' o.c.** (Advisory: confirm allowable spacing requirements with manufacturer)
Wall illuminance (19'x11'–6" area):
▶ 5 fc, avg, maintained (9 fc max/0 fc min)
▶ Gallery, Residential

▶ **Lamp: 120PAR38/H/NFL (p3–118)**

Luminaire: 6½"Ø Spreadlens Wallwasher (some versions are 6"Ø)
Ceiling height: **12'–6"**
Distance from wall: **3'–0"**
Spacings: **3' o.c.** (Advisory: confirm allowable spacing requirements with manufacturer)
Wall illuminance (19'x11'–6" area):
▶ 8 fc, avg, maintained (15 fc max/0 fc min)
▶ Hospitality

Design Tips

✔Use where wall design is interesting and/or art potential exists but is not finalized.
✔Effect is strongest with lighter colored walls. Need more light on darker walls for impact.
✔Enhances wood paneling, but consider spacing lights relative to center of paneling module.
✔Where art preservation is a necessity, specify UV–reduction spreadlens.
✔Walls which have specular or glossy finishes will reflect some lens image back into room.
✔Lens must be oriented with linear pattern perpendicular to wall (typically there is a positioning "key" in the glass to help with orientation).
✔Lens must be placed in concave orientation (lens cups upward into housing—**not** downward which is a common error).

Details
Concepts and Effects
11

Net Addresses/Luminaires
See page 11–99

DetailConcepts™

Wallwash Spreadlens/6½″Ø/Halogen(120V)

12′–6″ Ceiling Height

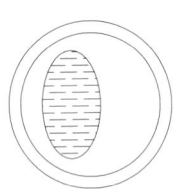

Details Concepts and Effects

11

Design Tips

✔ IC–rated luminaires are required where contact with insulation in the construction is likely or as directed by local codes. Not all luminaire manufacturers' versions are available in IC–rated option.

✔ Use of non–IC luminaires might be accepted by local code authority for use where insulation is expected if provisions are made in construction to keep insulation at least 3 inches from luminaire housing, wiring compartment and/or ballast/transformer. Subject to local interpretation—confirm in advance of specification.

✔ Housing sizes above ceiling are relatively large—review manufacturer's datasheet to confirm fit.

✔ Some manufacturers' luminaires have minimum spacing requirements to maintain appropriate clearance and to maintain UL listings— see specific manufacturer's data.

For more efficient wallwashing, use HIR spreadlens wallwashers (see 11–183).

Wallwash Spreadlens/6½″Ø/Halogen(120V)

12′–6″ Ceiling Height

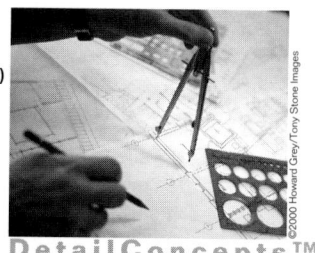

Performance Sketches

[All data outlined here is based on information from boldfaced manufacturer's published data at time of manuscript preparation. Photometric data were rerated as necessary for cited lamp. Maintained values based on analysis grid centered on wall and extending to 0′–6″ from edges and on 0.90 light loss factor (0.95 lamp lumen depreciation; 0.95 luminaire dirt depreciation). Scale is ¼ inch equals 1 foot.]

⊃Possible vendors: Cooper (Portfolio®), Edison Price, Kurt Versen, **Lightolier** (Calculite®), Prescolite, Zumtobel/Staff

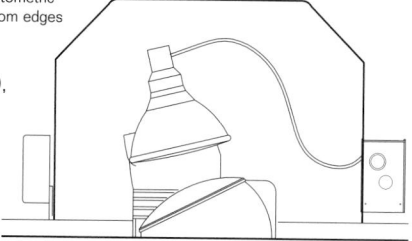

Lightolier 6″Ø PAR38 profile basis for sketch (see each mfgr's data for precise profile and dimensions)

▶ **Lamp: 120PAR38/H/NFL (p3–118)**

Luminaire: 6½″Ø Spreadlens Wallwasher (some versions are 6″Ø)

Ceiling height: **12′–6″**

Distance from wall: **4′–0″**

Spacings: **4′ o.c.** (Advisory: confirm allowable spacing requirements with manufacturer)

Wall illuminance (19′x11′–6″ area):

▶ 5 fc, avg, maintained (9 fc max/0 fc min)

▶ Gallery, Residential

Design Tips

✔Use where wall design is interesting and/or art potential exists but is not finalized.

✔Effect is strongest with lighter colored walls. Need more light on darker walls for impact.

✔Enhances wood paneling, but consider spacing lights relative to center of paneling module.

✔Where art preservation is a necessity, specify UV–reduction spreadlens.

✔Walls which have specular or glossy finishes will reflect some lens image back into room.

✔Lens must be oriented with linear pattern perpendicular to wall (typically there is a positioning "key" in the glass to help with orientation).

✔Lens must be placed in concave orientation (lens cups upward into housing—**not** downward which is a common error).

Details
Concepts and Effects

11

Net Addresses/Luminaires
See page 11–99

CONNECT FOR MORE

DetailConcepts™

Wallwash Spreadlens/6½"Ø/HIR(120V)

9'–6" Ceiling Height

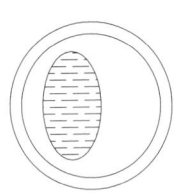

Design Tips

✔ IC–rated luminaires are required where contact with insulation in the construction is likely or as directed by local codes. Not all luminaire manufacturers' versions are available in IC–rated option.

✔ Use of non–IC luminaires might be accepted by local code authority for use where insulation is expected if provisions are made in construction to keep insulation at least 3 inches from luminaire housing, wiring compartment and/or ballast/transformer. Subject to local interpretation—confirm in advance of specification.

✔ Housing sizes above ceiling are relatively large—review manufacturer's datasheet to confirm fit.

✔ Some manufacturers' luminaires have minimum spacing requirements to maintain appropriate clearance and to maintain UL listings—see specific manufacturer's data.

Details
Concepts and Effects

11

For more efficient wallwashing, use fluorescent rectilinear wallwashers (see 11–197, 199 and 201).

Wallwash Spreadlens/6½"Ø/HIR(120V)

9'–6" Ceiling Height

Performance Sketches

[All data outlined here is based on information from boldfaced manufacturer's published data at time of manuscript preparation. Photometric data were rerated as necessary for cited lamp. Maintained values based on analysis grid centered on wall and extending to 0'–6" from edges and on 0.90 light loss factor (0.95 lamp lumen depreciation; 0.95 luminaire dirt depreciation). Scale is ¼ inch equals 1 foot.]

⊃Possible vendors: Cooper (Portfolio®), Edison Price, Kurt Versen, **Lightolier** (Calculite®), Prescolite, Zumtobel/Staff

Lightolier 6"Ø PAR38 profile basis for sketch (see each mfgr's data for precise profile and dimensions)

▶ **Lamp: 60PAR38/HIR/NFL30° (p5–43)**
Luminaire: 6½"Ø Spreadlens Wallwasher (some versions are 6"Ø)
Ceiling height: **9'–6"**
Distance from wall: **2'–0"**
Spacings: **2' o.c.** (Advisory: confirm allowable spacing requirements with manufacturer)
Wall illuminance (19'x8'–6" area):
▶ 14 fc, avg, maintained (24 fc max/3 fc min)
▶ Commercial, Hospitality, Institutional

▶ **Lamp: 60PAR38/HIR/NFL30° (p5–43)**
Luminaire: 6½"Ø Spreadlens Wallwasher (some versions are 6"Ø)
Ceiling height: **9'–6"**
Distance from wall: **3'–0"**
Spacings: **3' o.c.** (Advisory: confirm allowable spacing requirements with manufacturer)
Wall illuminance (19'x8'–6" area):
▶ 7 fc, avg, maintained (11 fc max/0 fc min)
▶ Gallery, Residential

Design Tips
✔Use where wall design is interesting and/or art potential exists but is not finalized.
✔Effect is strongest with lighter colored walls. Need more light on darker walls for impact.
✔Enhances wood paneling, but consider spacing lights relative to center of paneling module.
✔Where art preservation is a necessity, specify UV–reduction spreadlens.
✔Walls which have specular or glossy finishes will reflect some lens image back into room.
✔Lens must be oriented with linear pattern perpendicular to wall (typically there is a positioning "key" in the glass to help with orientation).
✔Lens must be placed in concave orientation (lens cups upward into housing—**not** downward which is a common error).

Details
Concepts and Effects

11

Net Addresses/Luminaires
See page 11–99

CONNECT FOR MORE

DetailConcepts™

Wallwash Spreadlens/6½"Ø/HIR(120V)

10′–6″ Ceiling Height

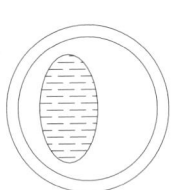

Details
Concepts and Effects

11

Design Tips

✔ IC–rated luminaires are required where contact with insulation in the construction is likely or as directed by local codes. Not all luminaire manufacturers' versions are available in IC–rated option.

✔ Use of non–IC luminaires might be accepted by local code authority for use where insulation is expected if provisions are made in construction to keep insulation at least 3 inches from luminaire housing, wiring compartment and/or ballast/transformer. Subject to local interpretation—confirm in advance of specification.

✔ Housing sizes above ceiling are relatively large—review manufacturer's datasheet to confirm fit.

✔ Some manufacturers' luminaires have minimum spacing requirements to maintain appropriate clearance and to maintain UL listings—see specific manufacturer's data.

For more efficient wallwashing, use fluorescent rectilinear wallwashers (see 11– 193, 195, 203 and 205).

Wallwash Spreadlens/6½″Ø/HIR(120V)

10′–6″ Ceiling Height

D e t a i l C o n c e p t s ™

Performance Sketches

[All data outlined here is based on information from boldfaced manufacturer's published data at time of manuscript preparation. Photometric data were rerated as necessary for cited lamp. Maintained values based on analysis grid centered on wall and extending to 0′–6″ from edges and on 0.90 light loss factor (0.95 lamp lumen depreciation; 0.95 luminaire dirt depreciation). Scale is ¼ inch equals 1 foot.]

⊃Possible vendors: Cooper (Portfolio®), Edison Price, Kurt Versen, **Lightolier** (Calculite®), Prescolite, Zumtobel/Staff

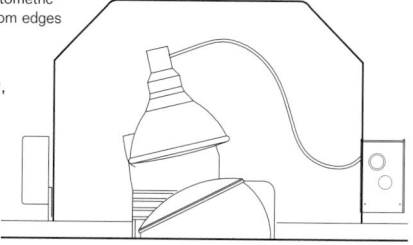

Lightolier 6″Ø PAR38 profile basis for sketch (see each mfgr's data for precise profile and dimensions)

▶ **Lamp: 60PAR38/HIR/NFL30° (p5–43)**

Luminaire: 6½″Ø Spreadlens Wallwasher (some versions are 6″Ø)
Ceiling height: **10′–6″**
Distance from wall: **2′–0″**
Spacings: **2′ o.c.** (Advisory: confirm allowable spacing requirements with manufacturer)
Wall illuminance (19′x9′–6″ area):
▶ 13 fc, avg, maintained (24 fc max/3 fc min)
▶ Commercial, Hospitality, Institutional

▶ **Lamp: 60PAR38/HIR/NFL30° (p5–43)**

Luminaire: 6½″Ø Spreadlens Wallwasher (some versions are 6″Ø)
Ceiling height: **10′–6″**
Distance from wall: **3′–0″**
Spacings: **3′ o.c.** (Advisory: confirm allowable spacing requirements with manufacturer)
Wall illuminance (19′x9′–6″ area):
▶ 6 fc, avg, maintained (11 fc max/0 fc min)
▶ Gallery, Residential

Design Tips

✔Use where wall design is interesting and/or art potential exists but is not finalized.
✔Effect is strongest with lighter colored walls. Need more light on darker walls for impact.
✔Enhances wood paneling, but consider spacing lights relative to center of paneling module.
✔Where art preservation is a necessity, specify UV–reduction spreadlens.
✔Walls which have specular or glossy finishes will reflect some lens image back into room.
✔Lens must be oriented with linear pattern perpendicular to wall (typically there is a positioning "key"in the glass to help with orientation).
✔Lens must be placed in concave orientation (lens cups upward into housing—**not** downward which is a common error).

Details
Concepts and Effects
11

Net Addresses/Luminaires
See page 11–99

CONNECT FOR MORE

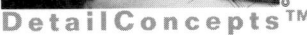

Wallwash Spreadlens/6½"Ø/HIR(120V)

10'–6" Ceiling Height

DetailConcepts™

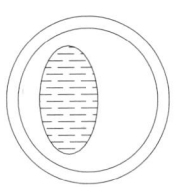

Design Tips

✔ IC–rated luminaires are required where contact with insulation in the construction is likely or as directed by local codes. Not all luminaire manufacturers' versions are available in IC–rated option.

✔ Use of non–IC luminaires might be accepted by local code authority for use where insulation is expected if provisions are made in construction to keep insulation at least 3 inches from luminaire housing, wiring compartment and/or ballast/transformer. Subject to local interpretation—confirm in advance of specification.

✔ Housing sizes above ceiling are relatively large—review manufacturer's datasheet to confirm fit.

✔ Some manufacturers' luminaires have minimum spacing requirements to maintain appropriate clearance and to maintain UL listings— see specific manufacturer's data.

Details Concepts and Effects

11

For more efficient wallwashing, use fluorescent rectilinear wallwashers (see 11–193, 195, 203 and 205).

Wallwash Spreadlens/6½"Ø/HIR(120V)

10'–6" Ceiling Height

Performance Sketches

[All data outlined here is based on information from boldfaced manufacturer's published data at time of manuscript preparation. Photometric data were rerated as necessary for cited lamp. Maintained values based on analysis grid centered on wall and extending to 0'–6" from edges and on 0.90 light loss factor (0.95 lamp lumen depreciation; 0.95 luminaire dirt depreciation). Scale is ¼ inch equals 1 foot.]

⊃**Possible vendors:** Cooper (Portfolio®), Edison Price, Kurt Versen, **Lightolier** (Calculite®), Prescolite, Zumtobel/Staff

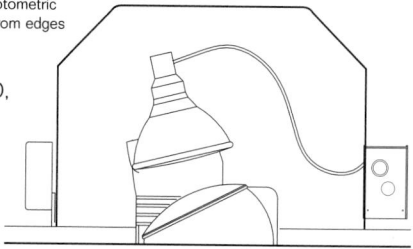

Lightolier 6"Ø PAR38 profile basis for sketch (see each mfgr's data for precise profile and dimensions)

▶ **Lamp: 100PAR38/HIR/NFL (p5–54)**
Luminaire: 6½"Ø Spreadlens Wallwasher (some versions are 6"Ø)
Ceiling height: **10'–6"**
Distance from wall: **3'–0"**
Spacings: **3' o.c.** (Advisory: confirm allowable spacing requirements with manufacturer)
Wall illuminance (19'x9'–6" area):
▶ 14 fc, avg, maintained (22 fc max/1 fc min)
▶ Commercial, Institutional

▶ **Lamp: 100PAR38/HIR/NFL (p5–54)**
Luminaire: 6½"Ø Spreadlens Wallwasher (some versions are 6"Ø)
Ceiling height: **10'–6"**
Distance from wall: **4'–0"**
Spacings: **4' o.c.** (Advisory: confirm allowable spacing requirements with manufacturer)
Wall illuminance (19'x9'–6" area):
▶ 9 fc, avg, maintained (14 fc max/0 fc min)
▶ Hospitality

Design Tips
✔Use where wall design is interesting and/or art potential exists but is not finalized.
✔Effect is strongest with lighter colored walls. Need more light on darker walls for impact.
✔Enhances wood paneling, but consider spacing lights relative to center of paneling module.
✔Where art preservation is a necessity, specify UV–reduction spreadlens.
✔Walls which have specular or glossy finishes will reflect some lens image back into room.
✔Lens must be oriented with linear pattern perpendicular to wall (typically there is a positioning "key"in the glass to help with orientation).
✔Lens must be placed in concave orientation (lens cups upward into housing—**not** downward which is a common error).

Details
Concepts and Effects
11

Net Addresses/Luminaires
See page 11–99

CONNECT FOR MORE

©2000 Howard Grey/Tony Stone Images

DetailConcepts™

Wallwash Spreadlens/6½"Ø/HIR(120V)

12'–6" Ceiling Height

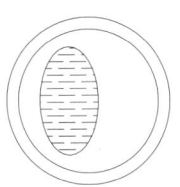

Design Tips

✔ IC–rated luminaires are required where contact with insulation in the construction is likely or as directed by local codes. Not all luminaire manufacturers' versions are available in IC–rated option.

✔ Use of non–IC luminaires might be accepted by local code authority for use where insulation is expected if provisions are made in construction to keep insulation at least 3 inches from luminaire housing, wiring compartment and/or ballast/transformer. Subject to local interpretation—confirm in advance of specification.

✔ Housing sizes above ceiling are relatively large—review manufacturer's datasheet to confirm fit.

✔ Some manufacturers' luminaires have minimum spacing requirements to maintain appropriate clearance and to maintain UL listings— see specific manufacturer's data.

For more efficient wallwashing, use fluorescent rectilinear wallwashers (see 11–207 and 209).

Wallwash Spreadlens/6½"Ø/HIR(120V)

12'–6" Ceiling Height

DetailConcepts™

Performance Sketches

[All data outlined here is based on information from boldfaced manufacturer's published data at time of manuscript preparation. Photometric data were rerated as necessary for cited lamp. Maintained values based on analysis grid centered on wall and extending to 0'–6" from edges and on 0.90 light loss factor (0.95 lamp lumen depreciation; 0.95 luminaire dirt depreciation). Scale is ¼ inch equals 1 foot.]

⊃Possible vendors: Cooper (Portfolio®), Edison Price, Kurt Versen, **Lightolier** (Calculite®), Prescolite, Zumtobel/Staff

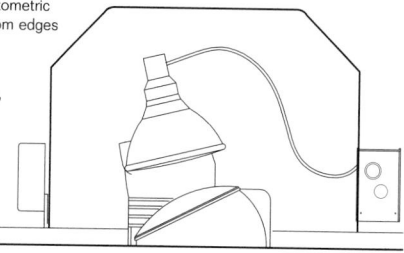

Lightolier 6"Ø PAR38 profile basis for sketch (see each mfgr's data for precise profile and dimensions)

▶ **Lamp: 60PAR38/HIR/NFL30° (p5–43)**
Luminaire: 6½"Ø Spreadlens Wallwasher (some versions are 6"Ø)
Ceiling height: **12'–6"**
Distance from wall: **2'–0"**
Spacings: **2' o.c.** (Advisory: confirm allowable spacing requirements with manufacturer)
Wall illuminance (19'x11'–6" area):
▶ 13 fc, avg, maintained (24 fc max/3 fc min)
▶ Commercial, Hospitality, Institutional

Design Tips

✔Use where wall design is interesting and/or art potential exists but is not finalized.
✔Effect is strongest with lighter colored walls. Need more light on darker walls for impact.
✔Enhances wood paneling, but consider spacing lights relative to center of paneling module.
✔Where art preservation is a necessity, specify UV–reduction spreadlens.
✔Walls which have specular or glossy finishes will reflect some lens image back into room.
✔Lens must be oriented with linear pattern perpendicular to wall (typically there is a positioning "key" in the glass to help with orientation).
✔Lens must be placed in concave orientation (lens cups upward into housing—**not** downward which is a common error).

Details Concepts and Effects **11**

▶ **Lamp: 60PAR38/HIR/NFL30° (p5–43)**
Luminaire: 6½"Ø Spreadlens Wallwasher (some versions are 6"Ø)
Ceiling height: **12'–6"**
Distance from wall: **3'–0"**
Spacings: **3' o.c.** (Advisory: confirm allowable spacing requirements with manufacturer)
Wall illuminance (19'x11'–6" area):
▶ 6 fc, avg, maintained (11 fc max/0 fc min)
▶ Gallery, Residential

Net Addresses/Luminaires
See page 11–99

CONNECT FOR MORE

DetailConcepts™

Wallwash Spreadlens/6½"Ø/HIR(120V)

12'–6" Ceiling Height

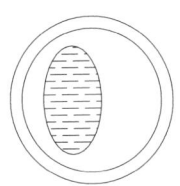

Design Tips

✔ IC–rated luminaires are required where contact with insulation in the construction is likely or as directed by local codes. Not all luminaire manufacturers' versions are available in IC–rated option.

✔ Use of non–IC luminaires might be accepted by local code authority for use where insulation is expected if provisions are made in construction to keep insulation at least 3 inches from luminaire housing, wiring compartment and/or ballast/transformer. Subject to local interpretation—confirm in advance of specification.

✔ Housing sizes above ceiling are relatively large—review manufacturer's datasheet to confirm fit.

✔ Some manufacturers' luminaires have minimum spacing requirements to maintain appropriate clearance and to maintain UL listings— see specific manufacturer's data.

For more efficient wallwashing, use fluorescent rectilinear wallwashers (see 11–207 and 209).

Wallwash Spreadlens/6½"Ø/HIR(120V)
12'–6" Ceiling Height

DetailConcepts™

Performance Sketches

[All data outlined here is based on information from boldfaced manufacturer's published data at time of manuscript preparation. Photometric data were rerated as necessary for cited lamp. Maintained values based on analysis grid centered on wall and extending to 0'–6" from edges and on 0.90 light loss factor (0.95 lamp lumen depreciation; 0.95 luminaire dirt depreciation). Scale is ¼ inch equals 1 foot.]

⊃Possible vendors: Cooper (Portfolio®), Edison Price, Kurt Versen, **Lightolier** (Calculite®), Prescolite, Zumtobel/Staff

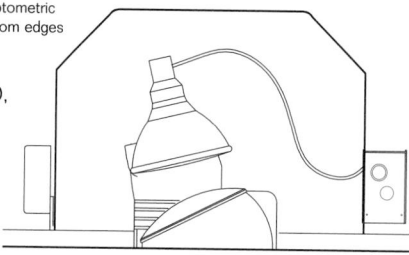

Lightolier 6"Ø PAR38 profile basis for sketch (see each mfgr's data for precise profile and dimensions)

▶ **Lamp: 100PAR38/HIR/NFL (p5–54)**
Luminaire: 6½"Ø Spreadlens Wallwasher (some versions are 6"Ø)
Ceiling height: **12'–6"**
Distance from wall: **3'–0"**
Spacings: **3' o.c.** (Advisory: confirm allowable spacing requirements with manufacturer)
Wall illuminance (19'x11'–6" area):
▶ 12 fc, avg, maintained (22 fc max/1 fc min)
▶ Commercial, Hospitality, Institutional

Design Tips
✔Use where wall design is interesting and/or art potential exists but is not finalized.
✔Effect is strongest with lighter colored walls. Need more light on darker walls for impact.
✔Enhances wood paneling, but consider spacing lights relative to center of paneling module.
✔Where art preservation is a necessity, specify UV–reduction spreadlens.
✔Walls which have specular or glossy finishes will reflect some lens image back into room.
✔Lens must be oriented with linear pattern perpendicular to wall (typically there is a positioning "key"in the glass to help with orientation).
✔Lens must be placed in concave orientation (lens cups upward into housing—**not** downward which is a common error).

Details
Concepts and Effects

11

▶ **Lamp: 100PAR38/HIR/NFL (p5–54)**
Luminaire: 6½"Ø Spreadlens Wallwasher (some versions are 6"Ø)
Ceiling height: **12'–6"**
Distance from wall: **4'–0"**
Spacings: **4' o.c.** (Advisory: confirm allowable spacing requirements with manufacturer)
Wall illuminance (19'x11'–6" area):
▶ 8 fc, avg, maintained (13 fc max/0 fc min)
▶ Hospitality

Net Addresses/Luminaires
See page 11–99

CONNECT FOR MORE

DetailConcepts™

Wallwash Spreadlens/6½″Ø/HIR(120V)

15'–0″ Ceiling Height

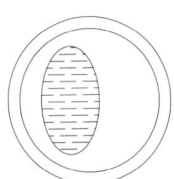

Details Concepts and Effects

11

Design Tips

✔ IC–rated luminaires are required where contact with insulation in the construction is likely or as directed by local codes. Not all luminaire manufacturers' versions are available in IC–rated option.

✔ Use of non–IC luminaires might be accepted by local code authority for use where insulation is expected if provisions are made in construction to keep insulation at least 3 inches from luminaire housing, wiring compartment and/or ballast/transformer. Subject to local interpretation—confirm in advance of specification.

✔ Housing sizes above ceiling are relatively large—review manufacturer's datasheet to confirm fit.

✔ Some manufacturers' luminaires have minimum spacing requirements to maintain appropriate clearance and to maintain UL listings— see specific manufacturer's data.

Wallwash Spreadlens/6½"Ø/HIR(120V)

15'–0" Ceiling Height

DetailConcepts™

Performance Sketches

[All data outlined here is based on information from boldfaced manufacturer's published data at time of manuscript preparation. Photometric data were rerated as necessary for cited lamp. Maintained values based on analysis grid centered on wall and extending to 0'–6" from edges and on 0.90 light loss factor (0.95 lamp lumen depreciation; 0.95 luminaire dirt depreciation). Scale is ¼ inch equals 1 foot.]

⊃Possible vendors: Cooper (Portfolio®), Edison Price, Kurt Versen, **Lightolier** (Calculite®), Prescolite, Zumtobel/Staff

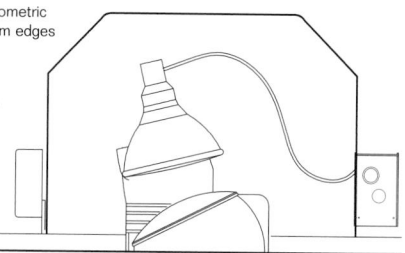

Lightolier 6"Ø PAR38 profile basis for sketch (see each mfgr's data for precise profile and dimensions)

Design Tips

✔Use where wall design is interesting and/or art potential exists but is not finalized.

✔Effect is strongest with lighter colored walls. Need more light on darker walls for impact.

✔Enhances wood paneling, but consider spacing lights relative to center of paneling module.

✔Where art preservation is a necessity, specify UV–reduction spreadlens.

✔Walls which have specular or glossy finishes will reflect some lens image back into room.

✔Lens must be oriented with linear pattern perpendicular to wall (typically there is a positioning "key" in the glass to help with orientation).

✔Lens must be placed in concave orientation (lens cups upward into housing—**not** downward which is a common error).

Details
Concepts and Effects
11

▶ **Lamp: 100PAR38/HIR/NFL (p5–54)**
Luminaire: 6½"Ø Spreadlens Wallwasher (some versions are 6"Ø)
Ceiling height: **15'–0"**
Distance from wall: **3'–0"**
Spacings: **3' o.c.** (Advisory: confirm allowable spacing requirements with manufacturer)
Wall illuminance (19'x14' area):
▶ 11 fc, avg, maintained (22 fc max/2 fc min)
▶ Hospitality

▶ **Lamp: 100PAR38/HIR/NFL (p5–54)**
Luminaire: 6½"Ø Spreadlens Wallwasher (some versions are 6"Ø)
Ceiling height: **15'–0"**
Distance from wall: **4'–0"**
Spacings: **4' o.c.** (Advisory: confirm allowable spacing requirements with manufacturer)
Wall illuminance (19'x14' area):
▶ 8 fc, avg, maintained (13 fc max/1 fc min)
▶ Hospitality

Net Addresses/Luminaires
See page 11–99

CONNECT FOR MORE

©2000 Howard Grey/Tony Stone Images

DetailConcepts™

Wallwash Linear/8"x12"/Fluorescent

8'–6" Ceiling Height

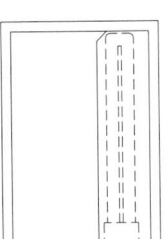

Design Tips

✔ IC–rated luminaires are required where contact with insulation in the construction is likely or as directed by local codes. Not all luminaire manufacturers' versions are available in IC–rated option.

✔ Use of non–IC luminaires might be accepted by local code authority for use where insulation is expected if provisions are made in construction to keep insulation at least 3 inches from luminaire housing, wiring compartment and/or ballast/transformer. Subject to local interpretation—confirm in advance of specification.

Wallwash Linear/8"x12"/Fluorescent

8'–6" Ceiling Height

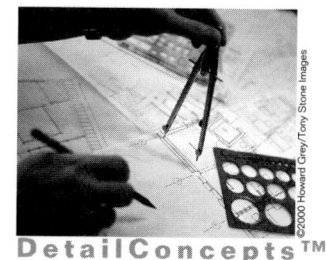

DetailConcepts™

Performance Sketches

[All data outlined here is based on information from boldfaced manufacturer's published data at time of manuscript preparation. Photometric data were rerated as necessary for cited lamp. Maintained values based on analysis grid centered on wall and extending to 0'–6" from edges and on 0.90 light loss factor (0.95 lamp lumen depreciation; 0.95 luminaire dirt depreciation). Scale is ¼ inch equals 1 foot.]

⊃Possible vendors: **Columbia** (Parawash), ELP, Indy, Lithonia

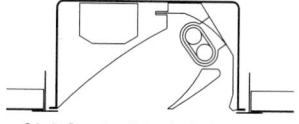

Columbia Parawash profile basis for sketch (see each mfgr's data for precise profile and dimensions)

▶ **Lamp: F18Long/830/RS (p8–32)**
Luminaire: 8"x12" Linear Wallwasher
Ceiling height: **8'–6"**
Distance from wall: **3'–0"**
Spacings: **4' o.c.**
Wall illuminance (19'x8'–6" area):
  ▶ 12 fc, avg, maintained (26 fc max/3 fc min)
  ▶ Commercial (see example installation on p9–6 and 9–7), Hospitality, Institutional

▶ **Lamp: F18Long/830/RS (p8–32)**
Luminaire: 8"x12" Linear Wallwasher
Ceiling height: **8'–6"**
Distance from wall: **3'–0"**
Spacings: **6' o.c.**
Wall illuminance (19'x8'–6" area):
  ▶ 8 fc, avg, maintained (20 fc max/2 fc min)
  ▶ Gallery, Hospitality

Design Tips

✔Use where wall design is interesting and/or art potential exists but is not finalized.

✔Effect is strongest with lighter colored walls. Need more light on darker walls for impact.

✔Enhances wood paneling, but consider spacing lights relative to center of paneling module.

✔Where art preservation is a necessity, specify UV–sleeve for fluorescent lamp.

✔These are the most efficient method of wallwashing.

✔Walls which have specular or glossy finishes will reflect the linear lamp image back into room.

Details Concepts and Effects

11

Net Addresses/Luminaires
See page 11–99

CONNECT FOR MORE

Wallwash Linear/8"x12"/Fluorescent

8'–6" Ceiling Height

DetailConcepts™

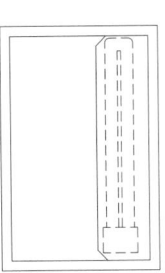

Design Tips

✔ IC–rated luminaires are required where contact with insulation in the construction is likely or as directed by local codes. Not all luminaire manufacturers' versions are available in IC–rated option.

✔ Use of non–IC luminaires might be accepted by local code authority for use where insulation is expected if provisions are made in construction to keep insulation at least 3 inches from luminaire housing, wiring compartment and/or ballast/transformer. Subject to local interpretation—confirm in advance of specification.

Details
Concepts and Effects

11

Wallwash Linear/8"x12"/Fluorescent

8'–6" Ceiling Height

D e t a i l C o n c e p t s ™

Performance Sketches

[All data outlined here is based on information from boldfaced manufacturer's published data at time of manuscript preparation. Photometric data were rerated as necessary for cited lamp. Maintained values based on analysis grid centered on wall and extending to 0'–6" from edges and on 0.90 light loss factor (0.95 lamp lumen depreciation; 0.95 luminaire dirt depreciation). Scale is ¼ inch equals 1 foot.]

⊃Possible vendors: **Columbia** (Parawash), ELP, Indy, Lithonia

▶ **Lamp: F18Long/830/RS (p8–32)**
Luminaire: 8"x12" Linear Wallwasher
Ceiling height: **8'–6"**
Distance from wall: **4'–0"**
Spacings: **4' o.c.**
Wall illuminance (19'x8'–6" area):
▶ 11 fc, avg, maintained (20 fc max/4 fc min)
▶ Commercial, Hospitality, Institutional

▶ **Lamp: F18Long/830/RS (p8–32)**
Luminaire: 8"x12" Linear Wallwasher
Ceiling height: **8'–6"**
Distance from wall: **4'–0"**
Spacings: **6' o.c.**
Wall illuminance (19'x8'–6" area):
▶ 7 fc, avg, maintained (14 fc max/1 fc min)
▶ Gallery, Hospitality

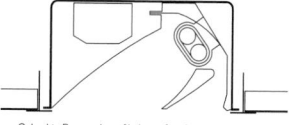

Columbia Parawash profile basis for sketch (see each
mfgr's data for precise profile and dimensions)

Design Tips

✔Use where wall design is interesting and/or art potential exists but is not finalized.
✔Effect is strongest with lighter colored walls. Need more light on darker walls for impact.
✔Enhances wood paneling, but consider spacing lights relative to center of paneling module.
✔Where art preservation is a necessity, specify UV–sleeve for fluorescent lamp.
✔These are the most efficient method of wallwashing.
✔Walls which have specular or glossy finishes will reflect the linear lamp image back into room.

Details
Concepts and Effects

11

Net Addresses/Luminaires
See page 11–99

CONNECT FOR MORE

Wallwash Linear/8"x12"/Fluorescent

10'–6" Ceiling Height

DetailConcepts™

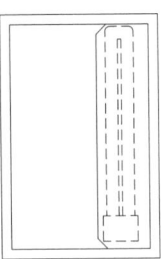

Design Tips

✔ IC–rated luminaires are required where contact with insulation in the construction is likely or as directed by local codes. Not all luminaire manufacturers' versions are available in IC–rated option.

✔ Use of non–IC luminaires might be accepted by local code authority for use where insulation is expected if provisions are made in construction to keep insulation at least 3 inches from luminaire housing, wiring compartment and/or ballast/transformer. Subject to local interpretation—confirm in advance of specification.

Details Concepts and Effects

11

Wallwash Linear/8″x12″/Fluorescent

10′–6″ Ceiling Height

DetailConcepts™

©2000 Howard Grey/Tony Stone Images

Performance Sketches

[All data outlined here is based on information from boldfaced manufacturer's published data at time of manuscript preparation. Photometric data were rerated as necessary for cited lamp. Maintained values based on analysis grid centered on wall and extending to 0′–6″ from edges and on 0.90 light loss factor (0.95 lamp lumen depreciation; 0.95 luminaire dirt depreciation). Scale is ¼ inch equals 1 foot.]

⊃Possible vendors: **Columbia** (Parawash), ELP, Indy, Lithonia

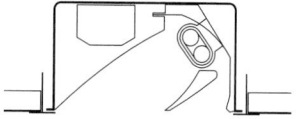

Columbia Parawash profile basis for sketch (see each mfgr's data for precise profile and dimensions)

▶ **Lamp: F18Long/830/RS (p8–32)**
Luminaire: 8″x12″ Linear Wallwasher
Ceiling height: **10′–6″**
Distance from wall: **3′–0″**
Spacings: **4′ o.c.**
Wall illuminance (19′x9′–6″ area):
▶ 10 fc, avg, maintained (26 fc max/2 fc min)
▶ Hospitality

Design Tips

✔Use where wall design is interesting and/or art potential exists but is not finalized.
✔Effect is strongest with lighter colored walls. Need more light on darker walls for impact.
✔Enhances wood paneling, but consider spacing lights relative to center of paneling module.
✔Where art preservation is a necessity, specify UV–sleeve for fluorescent lamp.
✔These are the most efficient method of wallwashing.
✔Walls which have specular or glossy finishes will reflect the linear lamp image back into room.

▶ **Lamp: F18Long/830/RS (p8–32)**
Luminaire: 8″x12″ Linear Wallwasher
Ceiling height: **10′–6″**
Distance from wall: **3′–0″**
Spacings: **6′ o.c.**
Wall illuminance (19′x9′–6″ area):
▶ 7 fc, avg, maintained (20 fc max/1 fc min)
▶ Gallery, Hospitality

Details
Concepts and Effects

11

Net Addresses/Luminaires
See page 11–99

CONNECT FOR MORE

DetailConcepts™

Wallwash Linear/8″x12″/Fluorescent

10′–6″ Ceiling Height

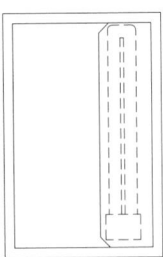

Design Tips

✔ IC–rated luminaires are required where contact with insulation in the construction is likely or as directed by local codes. Not all luminaire manufacturers' versions are available in IC–rated option.

✔ Use of non–IC luminaires might be accepted by local code authority for use where insulation is expected if provisions are made in construction to keep insulation at least 3 inches from luminaire housing, wiring compartment and/or ballast/transformer. Subject to local interpretation—confirm in advance of specification.

(sidebar) Details — Concepts and Effects — 11

Wallwash Linear/8″x12″/Fluorescent

10′–6″ Ceiling Height

©2000 Howard Grey/Tony Stone Images

D e t a i l C o n c e p t s ™

Performance Sketches

[All data outlined here is based on information from boldfaced manufacturer's published data at time of manuscript preparation. Photometric data were rerated as necessary for cited lamp. Maintained values based on analysis grid centered on wall and extending to 0′–6″ from edges and on 0.90 light loss factor (0.95 lamp lumen depreciation; 0.95 luminaire dirt depreciation). Scale is ¼ inch equals 1 foot.]

⊃Possible vendors: **Columbia** (Parawash), ELP, Indy, Lithonia

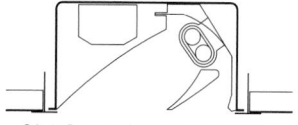

Columbia Parawash profile basis for sketch (see each mfgr's data for precise profile and dimensions)

▶ **Lamp: F18Long/830/RS (p8–32)**
Luminaire: 8″x12″ Linear Wallwasher
Ceiling height: **10′–6″**
Distance from wall: **4′–0″**
Spacings: **4′ o.c.**
Wall illuminance (19′x9′–6″ area):
▶ 9 fc, avg, maintained (19 fc max/2 fc min)
▶ Gallery, Hospitality

Design Tips

✔Use where wall design is interesting and/or art potential exists but is not finalized.
✔Effect is strongest with lighter colored walls. Need more light on darker walls for impact.
✔Enhances wood paneling, but consider spacing lights relative to center of paneling module.
✔Where art preservation is a necessity, specify UV–sleeve for fluorescent lamp.
✔These are the most efficient method of wallwashing.
✔Walls which have specular or glossy finishes will reflect the linear lamp image back into room.

Details Concepts and Effects

11

▶ **Lamp: F18Long/830/RS (p8–32)**
Luminaire: 8″x12″ Linear Wallwasher
Ceiling height: **10′–6″**
Distance from wall: **4′–0″**
Spacings: **6′ o.c.**
Wall illuminance (19′x9′–6″ area):
▶ 6 fc, avg, maintained (14 fc max/1 fc min)
▶ Gallery

Net Addresses/Luminaires
See page 11–99

CONNECT FOR MORE

etailConcepts™

Wallwash Linear/8″x24″/Fluorescent

9′–6″ Ceiling Height

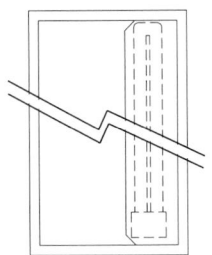

Design Tips

✔ IC–rated luminaires are required where contact with insulation in the construction is likely or as directed by local codes. Not all luminaire manufacturers' versions are available in IC–rated option.

✔ Use of non–IC luminaires might be accepted by local code authority for use where insulation is expected if provisions are made in construction to keep insulation at least 3 inches from luminaire housing, wiring compartment and/or ballast/transformer. Subject to local interpretation—confirm in advance of specification.

Details Concepts and Effects

11

Wallwash Linear/8"x24"/Fluorescent

9'–6" Ceiling Height

DetailConcepts™

Performance Sketches

[All data outlined here is based on information from boldfaced manufacturer's published data at time of manuscript preparation. Photometric data were rerated as necessary for cited lamp. Maintained values based on analysis grid centered on wall and extending to 0'–6" from edges and on 0.90 light loss factor (0.95 lamp lumen depreciation; 0.95 luminaire dirt depreciation). Scale is ¼ inch equals 1 foot.]

⟳Possible vendors: **Columbia** (Parawash), Cooper (Horizon™), ELP, Indy, Lithonia

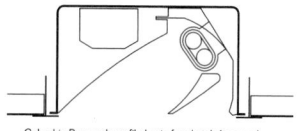

Columbia Parawash profile basis for sketch (see each mfgr's data for precise profile and dimensions)

▶ **Lamp: F40Long/830/RS (p8–36)**
Luminaire: 8"x24" Linear Wallwasher
Ceiling height: **9'–6"**
Distance from wall: **3'–0"**
Spacings: **4' o.c.**
Wall illuminance (19'x8'–6" area):
▶ 28 fc, avg, maintained (63 fc max/5 fc min)
▶ Commercial, Institutional, Retail

Design Tips

✔Use where wall design is interesting and/or art potential exists but is not finalized.
✔Effect is strongest with lighter colored walls. Need more light on darker walls for impact.
✔Enhances wood paneling, but consider spacing lights relative to center of paneling module.
✔Where art preservation is a necessity, specify UV–sleeve for fluorescent lamp.
✔These are the most efficient method of wallwashing.
✔Walls which have specular or glossy finishes will reflect the linear lamp image back into room.

Details Concepts and Effects

11

▶ **Lamp: F40Long/830/RS (p8–36)**
Luminaire: 8"x24" Linear Wallwasher
Ceiling height: **9'–6"**
Distance from wall: **3'–0"**
Spacings: **6' o.c.**
Wall illuminance (19'x8'–6" area):
▶ 18 fc, avg, maintained (49 fc max/3 fc min)
▶ Commercial, Institutional, Retail

Net Addresses/Luminaires

See page 11–99

CONNECT FOR MORE

Trademarks, service marks and product names are owned and registered by respective manufacturers.

Wallwash Linear/8"x24"/Fluorescent

9'–6" Ceiling Height

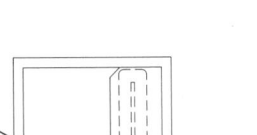

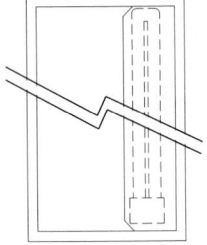

Design Tips

✓ IC–rated luminaires are required where contact with insulation in the construction is likely or as directed by local codes. Not all luminaire manufacturers' versions are available in IC–rated option.

✓ Use of non–IC luminaires might be accepted by local code authority for use where insulation is expected if provisions are made in construction to keep insulation at least 3 inches from luminaire housing, wiring compartment and/or ballast/transformer. Subject to local interpretation—confirm in advance of specification.

Details Concepts and Effects

11

Wallwash Linear/8"x24"/Fluorescent
9'–6" Ceiling Height

DetailConcepts™

Performance Sketches

[All data outlined here is based on information from boldfaced manufacturer's published data at time of manuscript preparation. Photometric data were rerated as necessary for cited lamp. Maintained values based on analysis grid centered on wall and extending to 0'–6" from edges and on 0.90 light loss factor (0.95 lamp lumen depreciation; 0.95 luminaire dirt depreciation). Scale is ¼ inch equals 1 foot.]

⊃Possible vendors: **Columbia** (Parawash), Cooper (Horizon™), ELP, Indy, Lithonia

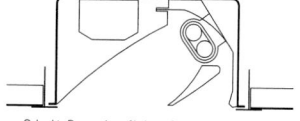

Columbia Parawash profile basis for sketch (see each mfgr's data for precise profile and dimensions)

▶ **Lamp: F40Long/830/RS (p8–36)**
Luminaire: 8"x24" Linear Wallwasher
Ceiling height: **9'–6"**
Distance from wall: **4'–0"**
Spacings: **4' o.c.**
Wall illuminance (19'x8'–6" area):
▶ 25 fc, avg, maintained (48 fc max/8 fc min)
▶ Commercial, Institutional, Retail

▶ **Lamp: F40Long/830/RS (p8–36)**
Luminaire: 8"x24" Linear Wallwasher
Ceiling height: **9'–6"**
Distance from wall: **4'–0"**
Spacings: **6' o.c.**
Wall illuminance (19'x8'–6" area):
▶ 16 fc, avg, maintained (35 fc max/4 fc min)
▶ Commercial, Hospitality Institutional

Design Tips
✔Use where wall design is interesting and/or art potential exists but is not finalized.
✔Effect is strongest with lighter colored walls. Need more light on darker walls for impact.
✔Enhances wood paneling, but consider spacing lights relative to center of paneling module.
✔Where art preservation is a necessity, specify UV–sleeve for fluorescent lamp.
✔These are the most efficient method of wallwashing.
✔Walls which have specular or glossy finishes will reflect the linear lamp image back into room.

Details Concepts and Effects

11

Net Addresses/Luminaires
See page 11–99

CONNECT FOR MORE

Wallwash Linear/8"x24"/Fluorescent

9'–6" Ceiling Height

DetailConcepts™

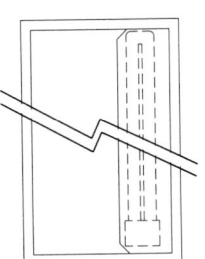

Design Tips

✔ IC–rated luminaires are required where contact with insulation in the construction is likely or as directed by local codes. Not all luminaire manufacturers' versions are available in IC–rated option.

✔ Use of non–IC luminaires might be accepted by local code authority for use where insulation is expected if provisions are made in construction to keep insulation at least 3 inches from luminaire housing, wiring compartment and/or ballast/transformer. Subject to local interpretation—confirm in advance of specification.

Details Concepts and Effects

11

Wallwash Linear/8"x24"/Fluorescent

9'–6" Ceiling Height

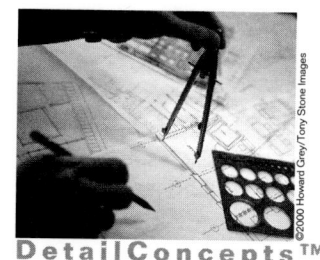

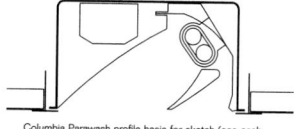

DetailConcepts™

Performance Sketches

[All data outlined here is based on information from boldfaced manufacturer's published data at time of manuscript preparation. Photometric data were rerated as necessary for cited lamp. Maintained values based on analysis grid centered on wall and extending to 0'–6" from edges and on 0.90 light loss factor (0.95 lamp lumen depreciation; 0.95 luminaire dirt depreciation). Scale is ¼ inch equals 1 foot.]

⊃Possible vendors: **Columbia** (Parawash), Cooper (Horizon™), ELP, Indy, Lithonia

Columbia Parawash profile basis for sketch (see each mfgr's data for precise profile and dimensions)

▶ **Lamp: F40Long/830/RS (p8–36)**
Luminaire: 8"x24" Linear Wallwasher
Ceiling height: **9'–6"**
Distance from wall: **5'–0"**
Spacings: **4' o.c.**
Wall illuminance (19'x8'–6" area):
▶ 21 fc, avg, maintained (38 fc max/6 fc min)
▶ Commercial, Institutional, Retail

▶ **Lamp: F40Long/830/RS (p8–36)**
Luminaire: 8"x24" Linear Wallwasher
Ceiling height: **9'–6"**
Distance from wall: **5'–0"**
Spacings: **6' o.c.**
Wall illuminance (19'x8'–6" area):
▶ 14 fc, avg, maintained (26 fc max/3 fc min)
▶ Commercial, Hospitality Institutional

Design Tips

✔ Use where wall design is interesting and/or art potential exists but is not finalized.
✔ Effect is strongest with lighter colored walls. Need more light on darker walls for impact.
✔ Enhances wood paneling, but consider spacing lights relative to center of paneling module.
✔ Where art preservation is a necessity, specify UV–sleeve for fluorescent lamp.
✔ These are the most efficient method of wallwashing.
✔ Walls which have specular or glossy finishes will reflect the linear lamp image back into room.

Details Concepts and Effects

11

Net Addresses/Luminaires
See page 11–99

CONNECT FOR MORE

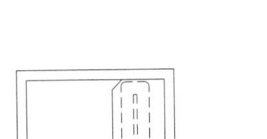

Wallwash Linear/8"x24"/Fluorescent

10'–6" Ceiling Height

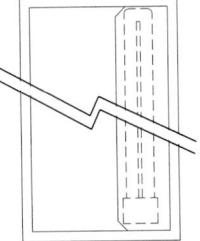

Design Tips

✓ IC–rated luminaires are required where contact with insulation in the construction is likely or as directed by local codes. Not all luminaire manufacturers' versions are available in IC–rated option.

✓ Use of non–IC luminaires might be accepted by local code authority for use where insulation is expected if provisions are made in construction to keep insulation at least 3 inches from luminaire housing, wiring compartment and/or ballast/transformer. Subject to local interpretation—confirm in advance of specification.

Wallwash Linear/8"x24"/Fluorescent

10'–6" Ceiling Height

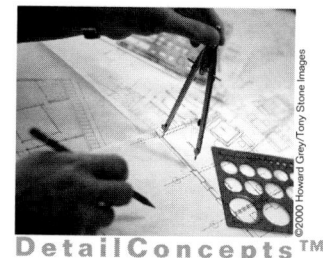

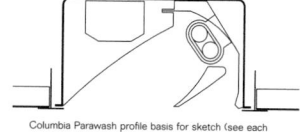

D e t a i l C o n c e p t s ™

Performance Sketches

[All data outlined here is based on information from boldfaced manufacturer's published data at time of manuscript preparation. Photometric data were rerated as necessary for cited lamp. Maintained values based on analysis grid centered on wall and extending to 0'–6" from edges and on 0.90 light loss factor (0.95 lamp lumen depreciation; 0.95 luminaire dirt depreciation). Scale is ¼ inch equals 1 foot.]

↺Possible vendors: **Columbia** (Parawash), Cooper (Horizon™), ELP, Indy, Lithonia

Columbia Parawash profile basis for sketch (see each
mfgr's data for precise profile and dimensions)

▶ **Lamp: F40Long/830/RS (p8–36)**
Luminaire: 8"x24" Linear Wallwasher
Ceiling height: **10'–6"**
Distance from wall: **4'–0"**
Spacings: **4' o.c.**
Wall illuminance (19'x9'–6" area):
▶ 23 fc, avg, maintained (48 fc max/6 fc min)
▶ Commercial, Institutional, Retail

Design Tips

✔Use where wall design is interesting and/or art potential exists but is not finalized.
✔Effect is strongest with lighter colored walls. Need more light on darker walls for impact.
✔Enhances wood paneling, but consider spacing lights relative to center of paneling module.
✔Where art preservation is a necessity, specify UV–sleeve for fluorescent lamp.
✔These are the most efficient method of wallwashing.
✔Walls which have specular or glossy finishes will reflect the linear lamp image back into room.

▶ **Lamp: F40Long/830/RS (p8–36)**
Luminaire: 8"x24" Linear Wallwasher
Ceiling height: **10'–6"**
Distance from wall: **4'–0"**
Spacings: **6' o.c.**
Wall illuminance (19'x9'–6" area):
▶ 15 fc, avg, maintained (35 fc max/4 fc min)
▶ Commercial, Hospitality, Institutional

Details Concepts and Effects

11

Net Addresses/Luminaires
See page 11–99

CONNECT FOR MORE

Wallwash Linear/8″x24″/Fluorescent

10′–6″ Ceiling Height

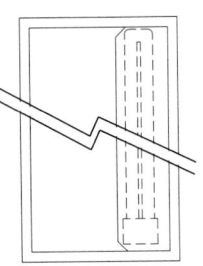

Design Tips

✔ IC–rated luminaires are required where contact with insulation in the construction is likely or as directed by local codes. Not all luminaire manufacturers' versions are available in IC–rated option.

✔ Use of non–IC luminaires might be accepted by local code authority for use where insulation is expected if provisions are made in construction to keep insulation at least 3 inches from luminaire housing, wiring compartment and/or ballast/transformer. Subject to local interpretation—confirm in advance of specification.

Details Concepts and Effects

11

Wallwash Linear/8"x24"/Fluorescent

10'–6" Ceiling Height

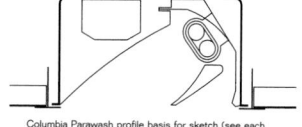

DetailConcepts™

Performance Sketches

[All data outlined here is based on information from boldfaced manufacturer's published data at time of manuscript preparation. Photometric data were rerated as necessary for cited lamp. Maintained values based on analysis grid centered on wall and extending to 0'–6" from edges and on 0.90 light loss factor (0.95 lamp lumen depreciation; 0.95 luminaire dirt depreciation). Scale is ¼ inch equals 1 foot.]

⊃Possible vendors: **Columbia** (Parawash), Cooper (Horizon™), ELP, Indy, Lithonia

Columbia Parawash profile basis for sketch (see each mfgr's data for precise profile and dimensions)

▶ **Lamp: F40Long/830/RS (p8–36)**
Luminaire: 8"x24" Linear Wallwasher
Ceiling height: **10'–6"**
Distance from wall: **5'–0"**
Spacings: **4' o.c.**
Wall illuminance (19'x9'–6" area):
▶ 20 fc, avg, maintained (37 fc max/6 fc min)
▶ Commercial, Institutional, Retail

Design Tips
✔Use where wall design is interesting and/or art potential exists but is not finalized.
✔Effect is strongest with lighter colored walls. Need more light on darker walls for impact.
✔Enhances wood paneling, but consider spacing lights relative to center of paneling module.
✔Where art preservation is a necessity, specify UV–sleeve for fluorescent lamp.
✔These are the most efficient method of wallwashing.
✔Walls which have specular or glossy finishes will reflect the linear lamp image back into room.

Details Concepts and Effects

11

▶ **Lamp: F40Long/830/RS (p8–36)**
Luminaire: 8"x24" Linear Wallwasher
Ceiling height: **10'–6"**
Distance from wall: **5'–0"**
Spacings: **6' o.c.**
Wall illuminance (19'x9'–6" area):
▶ 13 fc, avg, maintained (26 fc max/3 fc min)
▶ Commercial, Hospitality, Institutional

Net Addresses/Luminaires
See page 11–99

CONNECT FOR MORE

Trademarks, service marks and product names are owned and registered by respective manufacturers.

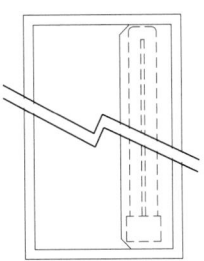

Wallwash Linear/8"x24"/Fluorescent

12'–6" Ceiling Height

DetailConcepts™

Design Tips

✔ IC–rated luminaires are required where contact with insulation in the construction is likely or as directed by local codes. Not all luminaire manufacturers' versions are available in IC–rated option.

✔ Use of non–IC luminaires might be accepted by local code authority for use where insulation is expected if provisions are made in construction to keep insulation at least 3 inches from luminaire housing, wiring compartment and/or ballast/transformer. Subject to local interpretation—confirm in advance of specification.

Details Concepts and Effects **11**

Wallwash Linear/8"x24"/Fluorescent

12'–6" Ceiling Height

DetailConcepts™

Performance Sketches

[All data outlined here is based on information from boldfaced manufacturer's published data at time of manuscript preparation. Photometric data were rerated as necessary for cited lamp. Maintained values based on analysis grid centered on wall and extending to 0'–6" from edges and on 0.90 light loss factor (0.95 lamp lumen depreciation; 0.95 luminaire dirt depreciation). Scale is ¼ inch equals 1 foot.]

↻Possible vendors: **Columbia** (Parawash), Cooper (Horizon™), ELP, Indy, Lithonia

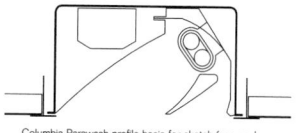

Columbia Parawash profile basis for sketch (see each mfgr's data for precise profile and dimensions)

▶ **Lamp: F40Long/830/RS (p8–36)**
Luminaire: 8"x24" Linear Wallwasher
Ceiling height: **12'–6"**
Distance from wall: **4'–0"**
Spacings: **4' o.c.**
Wall illuminance (19'x11'–6" area):
▶ 20 fc, avg, maintained (48 fc max/4 fc min)
▶ Commercial, Institutional, Retail

Design Tips
✔Use where wall design is interesting and/or art potential exists but is not finalized.
✔Effect is strongest with lighter colored walls. Need more light on darker walls for impact.
✔Enhances wood paneling, but consider spacing lights relative to center of paneling module.
✔Where art preservation is a necessity, specify UV–sleeve for fluorescent lamp.
✔These are the most efficient method of wallwashing.
✔Walls which have specular or glossy finishes will reflect the linear lamp image back into room.

Details
Concepts and Effects

11

▶ **Lamp: F40Long/830/RS (p8–36)**
Luminaire: 8"x24" Linear Wallwasher
Ceiling height: **12'–6"**
Distance from wall: **4'–0"**
Spacings: **6' o.c.**
Wall illuminance (19'x11'–6" area):
▶ 13 fc, avg, maintained (35 fc max/2 fc min)
▶ Commercial, Hospitality, Institutional

Net Addresses/Luminaires
See page 11–99

CONNECT FOR MORE

Trademarks, service marks and product names are owned and registered by respective manufacturers.

DetailConcepts™

Wallwash Linear/8"x24"/Fluorescent

12'–6" Ceiling Height

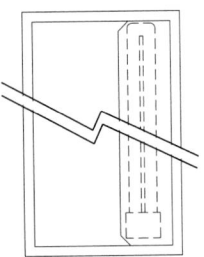

Design Tips

✔ IC–rated luminaires are required where contact with insulation in the construction is likely or as directed by local codes. Not all luminaire manufacturers' versions are available in IC–rated option.

✔ Use of non–IC luminaires might be accepted by local code authority for use where insulation is expected if provisions are made in construction to keep insulation at least 3 inches from luminaire housing, wiring compartment and/or ballast/transformer. Subject to local interpretation—confirm in advance of specification.

Wallwash Linear/8"x24"/Fluorescent

12'–6" Ceiling Height

DetailConcepts™

©2000 Howard Grey/Tony Stone Images

Performance Sketches

[All data outlined here is based on information from boldfaced manufacturer's published data at time of manuscript preparation. Photometric data were rerated as necessary for cited lamp. Maintained values based on analysis grid centered on wall and extending to 0'–6" from edges and on 0.90 light loss factor (0.95 lamp lumen depreciation; 0.95 luminaire dirt depreciation). Scale is ¼ inch equals 1 foot.]

⊃Possible vendors: **Columbia** (Parawash), Cooper (Horizon™), ELP, Indy, Lithonia

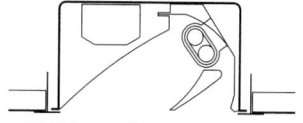

Columbia Parawash profile basis for sketch (see each mfg's data for precise profile and dimensions)

▶ **Lamp: F40Long/830/RS (p8–36)**
Luminaire: 8"x24" Linear Wallwasher
Ceiling height: **12'–6"**
Distance from wall: **5'–0"**
Spacings: **4' o.c.**
Wall illuminance (19'x11'–6" area):
▶ 18 fc, avg, maintained (37 fc max/5 fc min)
▶ Commercial, Institutional

Design Tips

✔Use where wall design is interesting and/or art potential exists but is not finalized.
✔Effect is strongest with lighter colored walls. Need more light on darker walls for impact.
✔Enhances wood paneling, but consider spacing lights relative to center of paneling module.
✔Where art preservation is a necessity, specify UV–sleeve for fluorescent lamp.
✔These are the most efficient method of wallwashing.
✔Walls which have specular or glossy finishes will reflect the linear lamp image back into room.

Details
Concepts and Effects

11

▶ **Lamp: F40Long/830/RS (p8–36)**
Luminaire: 8"x24" Linear Wallwasher
Ceiling height: **12'–6"**
Distance from wall: **5'–0"**
Spacings: **6' o.c.**
Wall illuminance (19'x11'–6" area):
▶ 11 fc, avg, maintained (26 fc max/3 fc min)
▶ Commercial, Hospitality, Institutional

Net Addresses/Luminaires
See page 11–99

CONNECT FOR MORE

DetailConcepts™

Wallslots Spreadlens/Baffle/Halogen(120V)

9'–6" Ceiling Height/0'–9" Lamp Spacing/0'–9" Slot Width

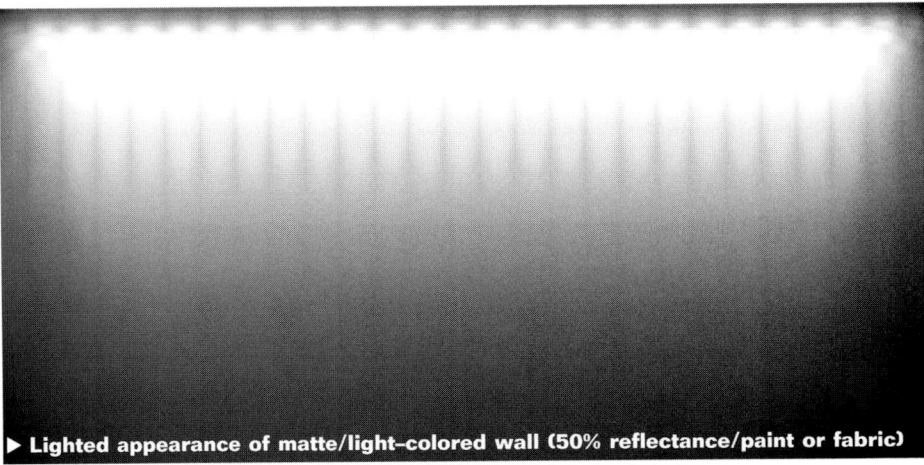

▶ **Lighted appearance of matte/light–colored wall (50% reflectance/paint or fabric)**

Design Tips

✔ Excellent technique (the only successful technique) for lighting specular (polished) materials without serious glare reflections back into room.

✔ This is an energy intensive technique and should be limited to lighting very detailed/ornate/expensive walls in highly visible/public areas.

▶ **Lighted appearance of specular/light–colored wall (50% reflectance/marble)**

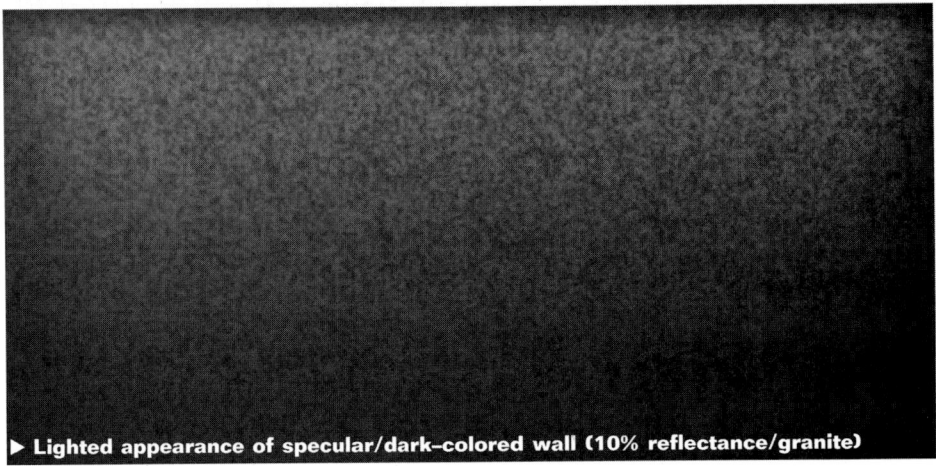

▶ **Lighted appearance of specular/dark–colored wall (10% reflectance/granite)**

Details
Concepts and Effects

11

Wallslots Spreadlens/Baffle/Halogen(120V)

9'–6" Ceiling Height/0'–9" Lamp Spacing/0'–9" Slot Width

DetailConcepts™

Performance Sketches

[All data outlined here is based on information from boldfaced manufacturer's published data at time of manuscript preparation. Photometric data were rerated as necessary for cited lamp. Maintained values based on analysis grid centered on wall and extending to 0'–6" from edges and on 0.90 light loss factor (0.95 lamp lumen depreciation; 0.95 luminaire dirt depreciation). Scale is ¼ inch equals 1 foot.]

⊃Possible vendors: **Edison Price** (Spredlite®)

Spredlite is a registered trademark of Edison Price.

▶ **Lamp: 50PAR20/H/NSP (p3–37)**
Luminaire: 5⅝"W by 7⁵⁄₁₆"H Spreadlens/Baffle Assembly in Drywall Slot Detail
Ceiling height: **9'–6"**
Drywall slot opening width: **0'–9"**
Lamp spacings: **0'–9" o.c.**
Wall illuminance (19'x8'–6" area):
▶ 30 fc, avg, maintained (51 fc max/9 fc min)
▶ Commercial, Hospitality, Residential,Retail

Design Tips

✔Use where wall design is interesting.
✔Effect is strongest with lighter colored walls.
✔Enhances wood paneling and stone or metal wall materials.
✔Where art preservation is a necessity, specify UV–reduction spreadlens.
✔Grazing light will telegraph wall imperfections.
✔If artwork is located on wall, select frameless or minimally framed pieces as grazing light on frames creates significant shadow lines.
✔Maintenance intensive (may want to reserve use for installations with few hours of daily operation).

Details Concepts and Effects

11

Notes

①Paint slot "ceiling" to match wall tone.
②Continue finished wall material up into slot to slot "ceiling."
③Manufacturing and photometric characteristics of luminaire may change from time to time—reconfirm detail dimensioning with luminaire manufacturer prior to final specification.

0'–7⁵⁄₁₆"
③

5" or greater
③

0'–9"
③

0'–2"
③

Edison Price profile basis for sketch (see each mfgr's data for precise profile, dimensions and detail requirements)

Toward wall

① ②

Net Addresses/Luminaires
See page 11–99

CONNECT FOR MORE

Trademarks, service marks and product names are owned and registered by respective manufacturers.

DetailConcepts™

Wallslots Spreadlens/Baffle/Halogen(120V)

10'–6" Ceiling Height/0'–9" Lamp Spacing/0'–9" Slot Width

Design Tips

✔Excellent technique (the only successful technique) for lighting specular (polished) materials without serious glare reflections back into room.

✔This is an energy intensive technique and should be limited to lighting very detailed/ornate/expensive walls in highly visible/public areas.

▶ **Lighted appearance of matte/light–colored wall (50% reflectance/paint or fabric)**

▶ **Lighted appearance of specular/light–colored wall (50% reflectance/marble)**

▶ **Lighted appearance of specular/dark–colored wall (10% reflectance/granite)**

Details
Concepts and Effects

11

Wallslots Spreadlens/Baffle/Halogen(120V)

10'–6" Ceiling Height/0'–9" Lamp Spacing/0'–9" Slot Width

DetailConcepts™

Performance Sketches

[All data outlined here is based on information from boldfaced manufacturer's published data at time of manuscript preparation. Photometric data were rerated as necessary for cited lamp. Maintained values based on analysis grid centered on wall and extending to 0'–6" from edges and on 0.90 light loss factor (0.95 lamp lumen depreciation; 0.95 luminaire dirt depreciation). Scale is ¼ inch equals 1 foot.]

⊃Possible vendors: **Edison Price** (Spredlite®)

▶ **Lamp: 50PAR20/H/NSP (p3–37)**
Luminaire: 5⅝"W by 7⁵⁄₁₆"H Spreadlens/Baffle Assembly in Drywall Slot Detail
Ceiling height: **10'–6"**
Drywall slot opening width: **0'–9"**
Lamp spacings: **0'–9" o.c.**
Wall illuminance (19'x9'–6" area):
▶ 28 fc, avg, maintained (51 fc max/7 fc min)
▶ Commercial, Hospitality, Residential, Retail

Design Tips
✔Use where wall design is interesting.
✔Effect is strongest with lighter colored walls.
✔Enhances wood paneling and stone or metal wall materials.
✔Where art preservation is a necessity, specify UV–reduction spreadlens.
✔Grazing light will telegraph wall imperfections.
✔If artwork is located on wall, select frameless or minimally framed pieces as grazing light on frames creates significant shadow lines.
✔Maintenance intensive (may want to reserve use for installations with few hours of daily operation).

Details
Concepts and Effects

11

Notes
①Paint slot "ceiling" to match wall tone.
②Continue finished wall material up into slot to slot "ceiling."
③Manufacturing and photometric characteristics of luminaire may change from time to time—reconfirm detail dimensioning with luminaire manufacturer prior to final specification.

0'–7⁵⁄₁₆"
③

5" or greater
③

0'–9"
③

0'–2"
③

Edison Price profile basis for sketch (see each mfgr's data for precise profile, dimensions and detail requirements)

Toward wall

Net Addresses/Luminaires
See page 11–99

CONNECT FOR MORE

DetailConcepts™

Wallslots Spreadlens/Baffle/HIR(120V)

12'–6" Ceiling Height/1'–0" Lamp Spacing/0'–10" Slot Width

Design Tips

✔ Excellent technique (the only successful technique) for lighting specular (polished) materials without serious glare reflections back into room.

✔ This is an energy intensive technique and should be limited to lighting very detailed/ornate/expensive walls in highly visible/public areas.

▶ **Lighted appearance of specular/light–colored wall (50% reflectance/marble)**

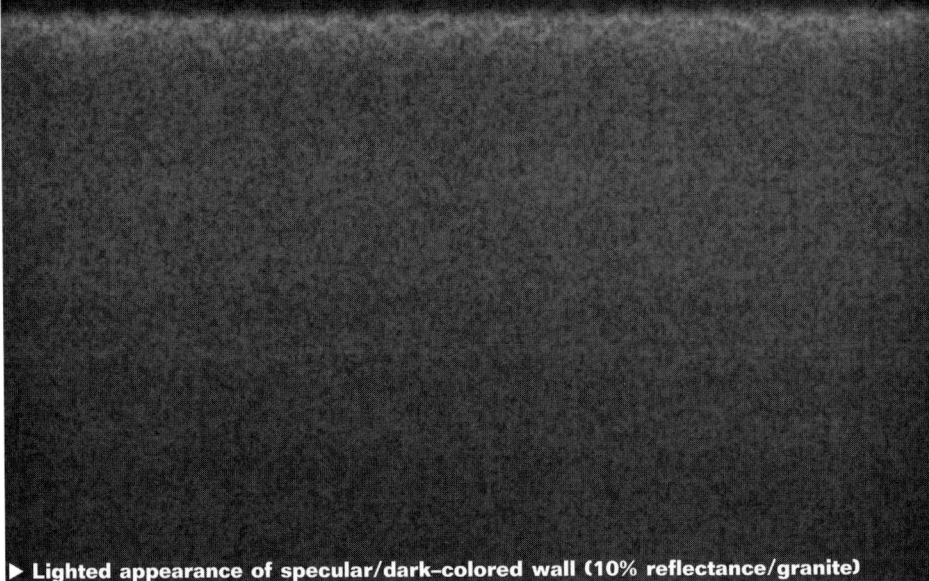

▶ **Lighted appearance of specular/dark–colored wall (10% reflectance/granite)**

Wallslots Spreadlens/Baffle/HIR(120V)

12'–6" Ceiling Height/1'–0" Lamp Spacing/0'–10" Slot Width

DetailConcepts™

Performance Sketches

[All data outlined here is based on information from boldfaced manufacturer's published data at time of manuscript preparation. Photometric data were rerated as necessary for cited lamp. Maintained values based on analysis grid centered on wall and extending to 0'–6" from edges and on 0.90 light loss factor (0.95 lamp lumen depreciation; 0.95 luminaire dirt depreciation). Scale is ¼ inch equals 1 foot.]

⊃Possible vendors: **Edison Price** (Spredlite®)

▶ **Lamp: 50PAR38/HIR/NSP (p5–38)**
Luminaire: 8⅝"W by 1'–0¹¹/₁₆"H Spreadlens/Baffle Assembly in Drywall Slot Detail
Ceiling height: **12'–6"**
Drywall slot opening width: **0'–10"**
Lamp spacings: **1'–0" o.c.**
Wall illuminance (19'x11'–6" area):
▶ 20 fc, avg, maintained (29 fc max/7 fc min)
▶ Commercial, Hospitality, Residential

Design Tips

✔Use where wall design is interesting.
✔Effect is strongest with lighter colored walls.
✔Enhances wood paneling and stone or metal wall materials.
✔Where art preservation is a necessity, specify UV–reduction spreadlens.
✔Grazing light will telegraph wall imperfections.
✔If artwork is located on wall, select frameless or minimally framed pieces as grazing light on frames creates significant shadow lines.
✔Maintenance intensive (may want to reserve use for installations with few hours of daily operation).

Details
Concepts and Effects
11

Notes
①**Paint slot "ceiling" to match wall tone.**
②**Continue finished wall material up into slot to slot "ceiling."**
③**Manufacturing and photometric characteristics of luminaire may change from time to time—reconfirm detail dimensioning with luminaire manufacturer prior to final specification.**

①
②
③ 1'–0¹¹/₁₆"
③ 4" or greater

Edison Price profile basis for sketch (see each mfgr's data for precise profile, dimensions and detail requirements)

③ 0'–10"   ③ 0'–3"

◄ Toward wall

Net Addresses/Luminaires
See page 11–99

CONNECT FOR MORE

DetailConcepts™

Wallslots Spreadlens/Baffle/HIR(120V)

15'–0" Ceiling Height/1'–0" Lamp Spacing/0'–10" Slot Width

Design Tips
✔Excellent technique (the only successful technique) for lighting specular (polished) materials without serious glare reflections back into room.
✔This is an energy intensive technique and should be limited to lighting very detailed/ornate/expensive walls in highly visible/public areas.

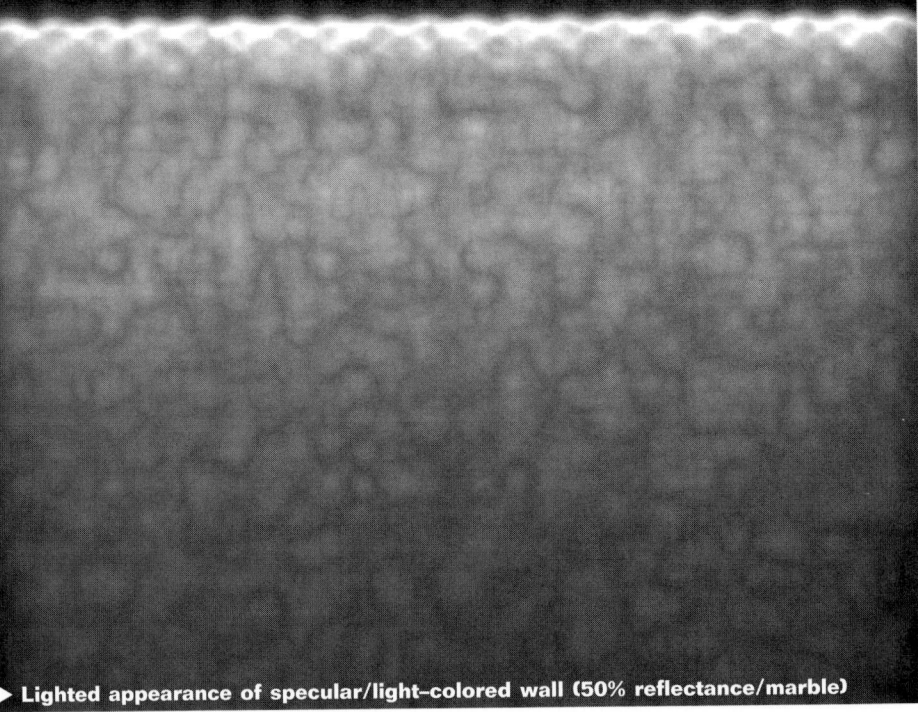

▶ Lighted appearance of specular/light–colored wall (50% reflectance/marble)

▶ Lighted appearance of specular/dark–colored wall (10% reflectance/granite)

Details Concepts and Effects

11

Wallslots Spreadlens/Baffle/HIR(120V)

15'–0" Ceiling Height/1'–0" Lamp Spacing/0'–10" Slot Width

DetailConcepts™

©2000 Howard Grey/Tony Stone Images

Performance Sketches

[All data outlined here is based on information from boldfaced manufacturer's published data at time of manuscript preparation. Photometric data were rerated as necessary for cited lamp. Maintained values based on analysis grid centered on wall and extending to 0'–6" from edges and on 0.90 light loss factor (0.95 lamp lumen depreciation; 0.95 luminaire dirt depreciation). Scale is ¼ inch equals 1 foot.]

⊃Possible vendors: **Edison Price** (Spredlite®)

▶ **Lamp: 50PAR38/HIR/NSP (p5–38)**

Luminaire: 8⅝"W by 1'–0¹¹⁄₁₆"H Spreadlens/Baffle Assembly in Drywall Slot Detail
Ceiling height: **15'–6"**
Drywall slot opening width: **0'–10"**
Lamp spacings: **1'–0" o.c.**
Wall illuminance (19'x14'–0" area):
▶ 20 fc, avg, maintained (48 fc max/11 fc min)
▶ Commercial, Hospitality, Residential

Design Tips

✔Use where wall design is interesting.
✔Effect is strongest with lighter colored walls.
✔Enhances wood paneling and stone or metal wall materials.
✔Where art preservation is a necessity, specify UV–reduction spreadlens.
✔Grazing light will telegraph wall imperfections.
✔If artwork is located on wall, select frameless or minimally framed pieces as grazing light on frames creates significant shadow lines.
✔Maintenance intensive (may want to reserve use for installations with few hours of daily operation).

Details Concepts and Effects

11

Notes

① Paint slot "ceiling" to match wall tone.
② Continue finished wall material up into slot to slot "ceiling."
③ Manufacturing and photometric characteristics of luminaire may change from time to time—reconfirm detail dimensioning with luminaire manufacturer prior to final specification.

① ② 1'–0¹¹⁄₁₆" ③ 4" or greater ③

Edison Price profile basis for sketch (see each mfgr's data for precise profile, dimensions and detail requirements)

0'–10" ③ 0'–3" ③

← Toward wall

Net Addresses/Luminaires
See page 11–99

CONNECT FOR MORE

DetailConcepts™

©2000 Howard Grey/Tony Stone Images

Wallslots Bare Lamp/Baffle/HIR(120V)

10'–6" Ceiling Height/1'–0" Lamp Spacing/0'–8" Slot Width

▶ **Lighted appearance of specular/light–colored wall (50% reflectance/marble)**

Design Tips

✔ Excellent technique (the only successful technique) for lighting specular (polished) materials without serious glare reflections back into room.

✔ This is an energy intensive technique and should be limited to lighting very detailed/ ornate/expensive walls in highly visible/public areas.

▶ **Lighted appearance of specular/dark–colored wall (10% reflectance/granite)**

Details

Concepts and Effects

11

Wallslots Bare Lamp/Baffle/HIR(120V)

10'–6" Ceiling Height/1'–0" Lamp Spacing/0'–8" Slot Width

DetailConcepts™

Performance Sketches

[All data outlined here is based on information from boldfaced manufacturer's published data at time of manuscript preparation. Photometric data were rerated as necessary for cited lamp. Maintained values based on analysis grid centered on wall and extending to 0'–6" from edges and on 0.90 light loss factor (0.95 lamp lumen depreciation; 0.95 luminaire dirt depreciation). Scale is ¼ inch equals 1 foot.]

⊃Possible vendors: **Hubbell/Sterner and Norton Industries** (socket strip, spreadlens and baffle), Litelab and Norton Industries

▶ **Lamp: 100PAR38/HIR/FL (p5–55)**
Luminaire: 5⅝"W by 7⁵⁄₁₆"H Socket strip/Spreadlens/Baffle Components in Drywall Slot Detail
Ceiling height: **10'–6"**
Drywall slot opening width: **0'–8"**
Lamp spacings: **1'–0" o.c.**
Wall illuminance (19'x9'–6" area):
▶ 77 fc, avg, maintained (159 fc max/20 fc min)
▶ Commercial, Hospitality, Retail

Design Tips

✔Use where wall design is interesting.
✔Effect is strongest with lighter colored walls.
✔Enhances wood paneling and stone or metal wall materials.
✔Where art preservation is a necessity, specify UV–reduction spreadlens.
✔Grazing light will telegraph wall imperfections.
✔If artwork is located on wall, select frameless or minimally framed pieces as grazing light on frames creates significant shadow lines.
✔Maintenance intensive (may want to reserve use for installations with few hours of daily operation).
✔Greater chance for viewing lamp/socket/baffle assembly directly (no view shield). This is minimized by deepening pocket.

Details Concepts and Effects

11

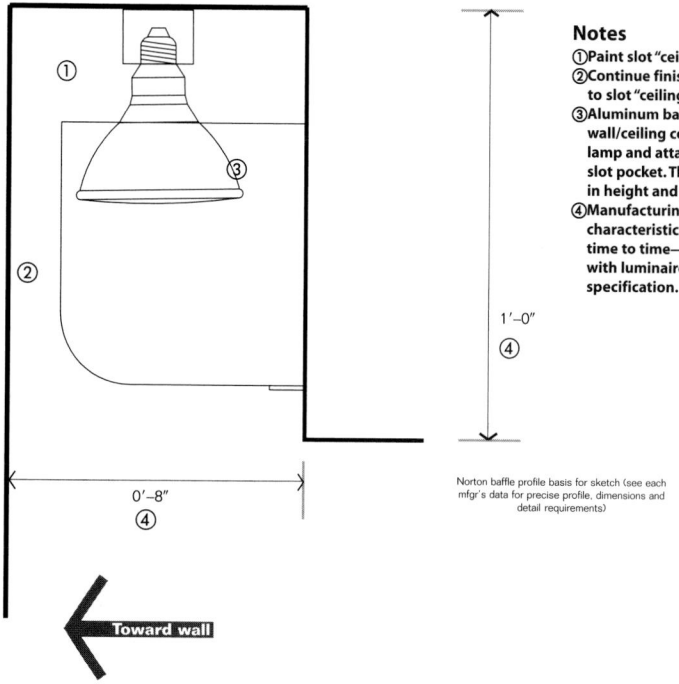

0'–8"
④

Toward wall

1'–0"
④

Notes
①Paint slot "ceiling" to match wall tone.
②Continue finished wall material up into slot to slot "ceiling."
③Aluminum baffle painted black or to match wall/ceiling color is centered between each lamp and attached to "inside" soffit wall of slot pocket. This is labor intensive. Baffle is 7" in height and projects 7".
④Manufacturing and photometric characteristics of luminaire may change from time to time—reconfirm detail dimensioning with luminaire manufacturer prior to final specification.

Norton baffle profile basis for sketch (see each mfgr's data for precise profile, dimensions and detail requirements)

Net Addresses/Luminaires
See page 11–99

CONNECT FOR MORE

DetailConcepts™

Wallslots Bare Lamp/Baffle/HIR(120V)

12'–6" Ceiling Height/1'–0" Lamp Spacing/0'–8" Slot Width

▶ **Lighted appearance of specular/light–colored wall (50% reflectance/marble)**

Design Tips

✔ Excellent technique (the only successful technique) for lighting specular (polished) materials without serious glare reflections back into room.

✔ This is an energy intensive technique and should be limited to lighting very detailed/ornate/expensive walls in highly visible/public areas.

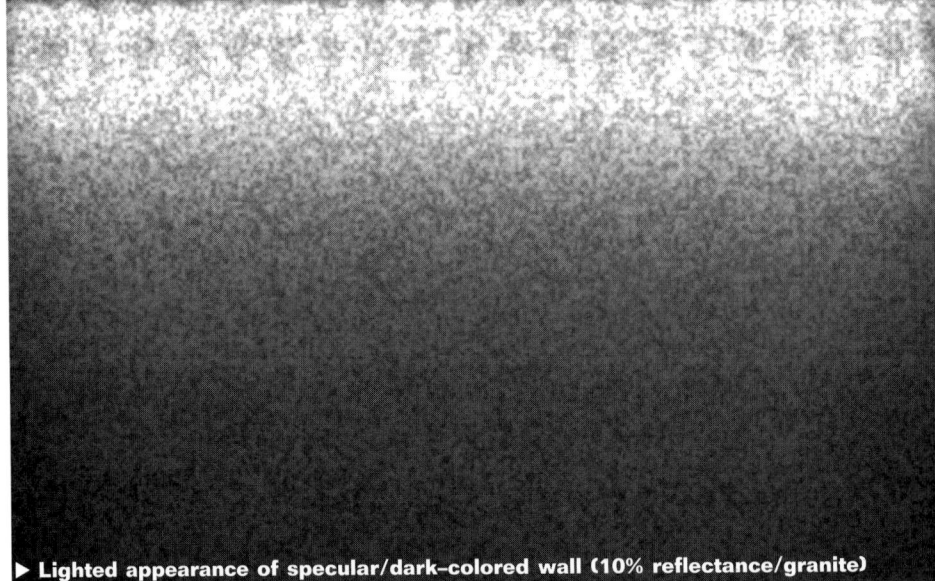

▶ **Lighted appearance of specular/dark–colored wall (10% reflectance/granite)**

Wallslots Bare Lamp/Baffle/HIR(120V)

12'–6" Ceiling Height/1'–0" Lamp Spacing/0'–8" Slot Width

DetailConcepts™

Performance Sketches

[All data outlined here is based on information from boldfaced manufacturer's published data at time of manuscript preparation. Photometric data were rerated as necessary for cited lamp. Maintained values based on analysis grid centered on wall and extending to 0'–6" from edges and on 0.90 light loss factor (0.95 lamp lumen depreciation; 0.95 luminaire dirt depreciation). Scale is ¼ inch equals 1 foot.]

⮌Possible vendors: **Hubbell/Sterner and Norton Industries** (socket strip, spreadlens and baffle), Litelab and Norton Industries

▶ **Lamp: 100PAR38/HIR/FL (p5–55)**
Luminaire: 5⅝"W by 7⁵⁄₁₆"H Socket strip/Spreadlens/Baffle Components in Drywall Slot Detail
Ceiling height: **12'–6"**
Drywall slot opening width: **0'–8"**
Lamp spacings: **1'–0" o.c.**
Wall illuminance (19'x9'–6" area):
▶ 66 fc, avg, maintained (159 fc max/14 fc min)
▶ Commercial, Hospitality, Retail

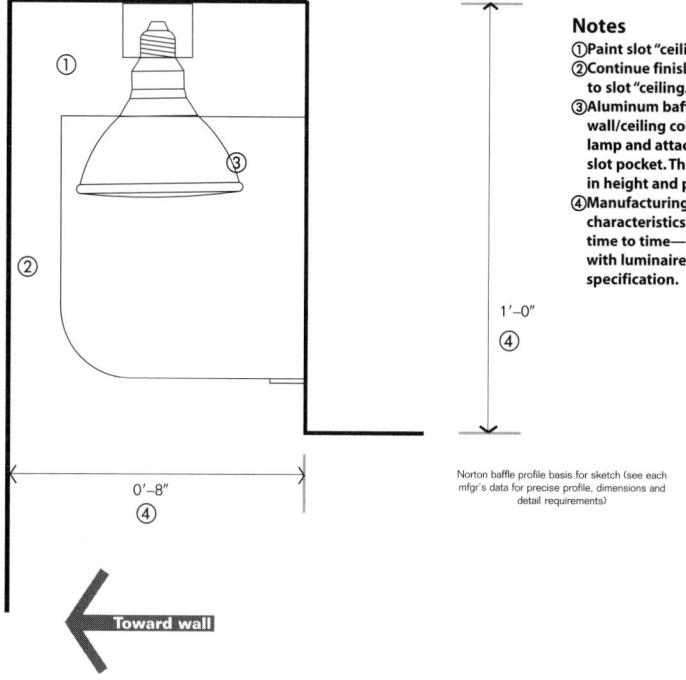

Notes
①Paint slot "ceiling" to match wall tone.
②Continue finished wall material up into slot to slot "ceiling."
③Aluminum baffle painted black or to match wall/ceiling color is centered between each lamp and attached to "inside" soffit wall of slot pocket. This is labor intensive. Baffle is 7" in height and projects 7".
④Manufacturing and photometric characteristics of luminaire may change from time to time—reconfirm detail dimensioning with luminaire manufacturer prior to final specification.

1'–0"
④

0'–8"
④

Norton baffle profile basis for sketch (see each mfgr's data for precise profile, dimensions and detail requirements)

Toward wall

Design Tips

✔Use where wall design is interesting.
✔Effect is strongest with lighter colored walls.
✔Enhances wood paneling and stone or metal wall materials.
✔Where art preservation is a necessity, specify UV–reduction spreadlens.
✔Grazing light will telegraph wall imperfections.
✔If artwork is located on wall, select frameless or minimally framed pieces as grazing light on frames creates significant shadow lines.
✔Maintenance intensive (may want to reserve use for installations with few hours of daily operation).
✔Greater chance for viewing lamp/socket/baffle assembly directly (no view shield). This is minimized by deepening pocket.

Details
Concepts and Effects

11

Net Addresses/Luminaires
See page 11–99

CONNECT FOR MORE

DetailConcepts™

Wallslots Bare Lamp/Baffle/CMH

12'–6" Ceiling Height/1'–0" Lamp Spacing/0'–6" Slot Width

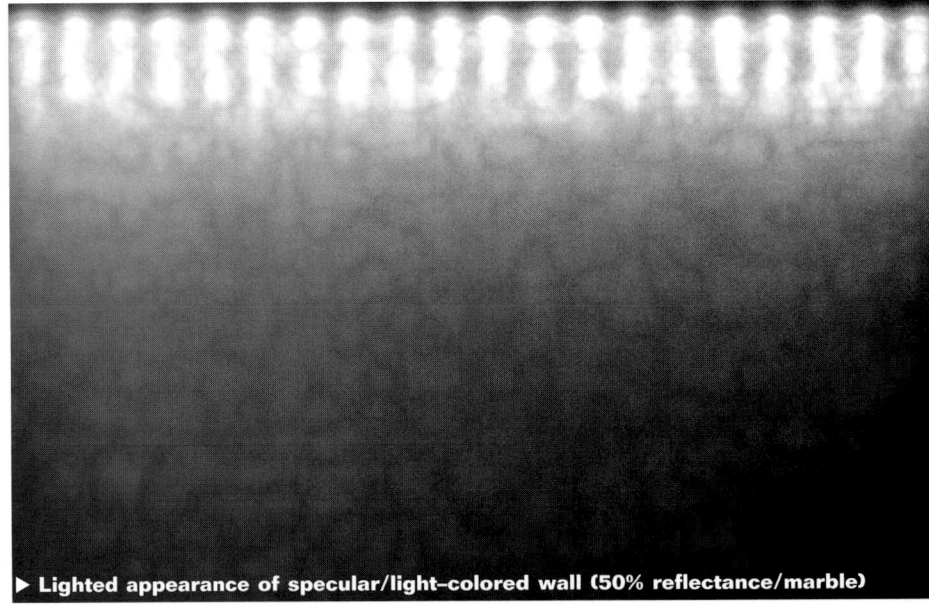

▶ **Lighted appearance of specular/light–colored wall (50% reflectance/marble)**

Design Tips

✔ Excellent technique (the only successful technique) for lighting specular (polished) materials without serious glare reflections back into room.

✔ This is an energy intensive technique and should be limited to lighting very detailed/ ornate/expensive walls in highly visible/public areas.

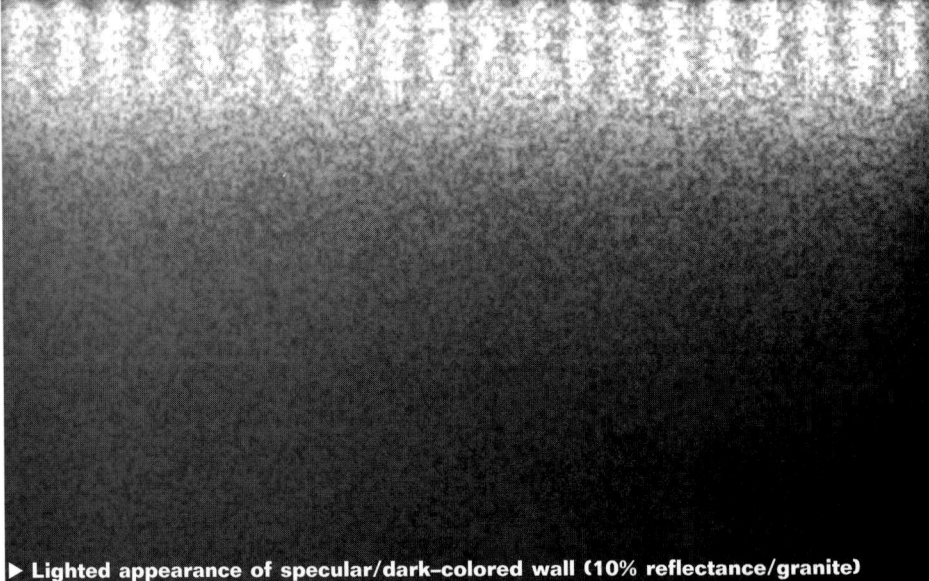

▶ **Lighted appearance of specular/dark–colored wall (10% reflectance/granite)**

Details Concepts and Effects

11

Wallslots Bare Lamp/Baffle/CMH
12'–6" Ceiling Height/1'–0" Lamp Spacing/0'–6" Slot Width

DetailConcepts™

Performance Sketches

[All data outlined here is based on information from boldfaced manufacturer's published data at time of manuscript preparation. Photometric data were rerated as necessary for cited lamp. Maintained values based on analysis grid centered on wall and extending to 0'–6" from edges and on 0.90 light loss factor (0.95 lamp lumen depreciation; 0.95 luminaire dirt depreciation). Scale is ¼ inch equals 1 foot.]

⊃Possible vendors: **Hubbell/Sterner and Norton Industries** (socket strip, spreadlens and baffle), Litelab and Norton Industries

▶ **Lamp: 39PAR20/CMH/FL (p5–55)**
Luminaire: 5⅝"W by 7⁵⁄₁₆"H Socket strip/Spreadlens/Baffle Components in Drywall Slot Detail
Ceiling height: **12'–6"**
Drywall slot opening width: **0'–6"**
Lamp spacings: **1'–0" o.c.**
Wall illuminance (19'x9'–6" area):
▶ 57 fc, avg, maintained (133 fc max/13 fc min)
▶ Commercial, Hospitality, Retail

Design Tips
✔Use where wall design is interesting.
✔Effect is strongest with lighter colored walls.
✔Enhances wood paneling and stone or metal wall materials.
✔Where art preservation is a necessity, specify UV–reduction spreadlens.
✔Grazing light will telegraph wall imperfections.
✔If artwork is located on wall, select frameless or minimally framed pieces as grazing light on frames creates significant shadow lines.
✔Maintenance intensive (may want to reserve use for installations with few hours of daily operation).
✔Greater chance for viewing lamp/socket/baffle assembly directly (no view shield). This is minimized by deepening pocket.

Details
Concepts and Effects

11

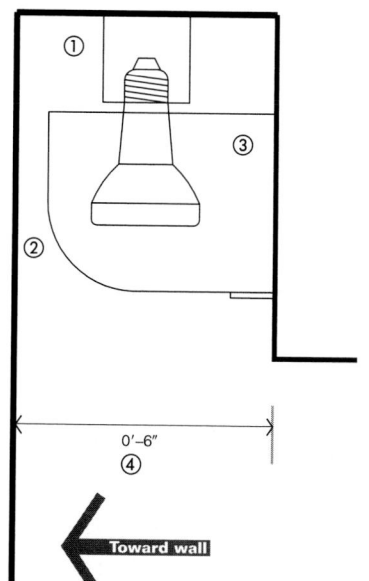

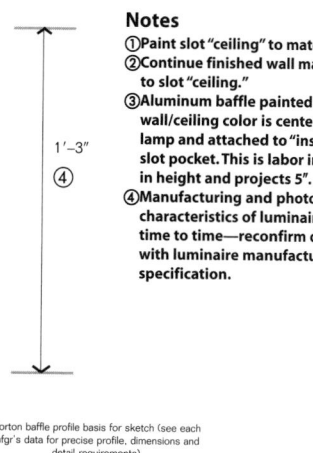

1'–3"
④

0'–6"
④

Toward wall

Norton baffle profile basis for sketch (see each mfgr's data for precise profile, dimensions and detail requirements)

Notes
①Paint slot "ceiling" to match wall tone.
②Continue finished wall material up into slot to slot "ceiling."
③Aluminum baffle painted black or to match wall/ceiling color is centered between each lamp and attached to "inside" soffit wall of slot pocket. This is labor intensive. Baffle is 4" in height and projects 5".
④Manufacturing and photometric characteristics of luminaire may change from time to time—reconfirm detail dimensioning with luminaire manufacturer prior to final specification.

Net Addresses/Luminaires
See page 11–99

CONNECT FOR MORE

DetailConcepts™ ©2000 Howard Grey/Tony Stone Images

Wallslots Bare Lamp/Baffle/CMH

15'–0" Ceiling Height/1'–0" Lamp Spacing/0'–6" Slot Width

Design Tips
✔ Excellent technique (the only successful technique) for lighting specular (polished) materials without serious glare reflections back into room.
✔ This is an energy intensive technique and should be limited to lighting very detailed/ornate/expensive walls in highly visible/public areas.

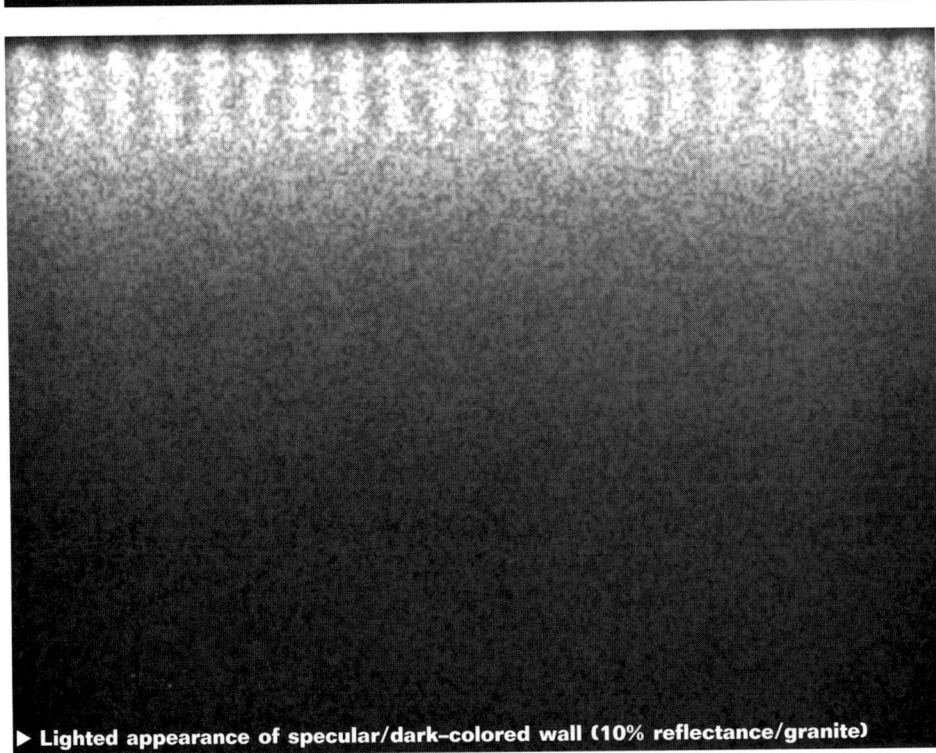

▶ Lighted appearance of specular/light–colored wall (50% reflectance/marble)

▶ Lighted appearance of specular/dark–colored wall (10% reflectance/granite)

Details Concepts and Effects

11

Wallslots Bare Lamp/Baffle/CMH

15'–0" Ceiling Height/1'–0" Lamp Spacing/0'–6" Slot Width

D e t a i l C o n c e p t s ™

Performance Sketches

[All data outlined here is based on information from boldfaced manufacturer's published data at time of manuscript preparation. Photometric data were rerated as necessary for cited lamp. Maintained values based on analysis grid centered on wall and extending to 0'–6" from edges and on 0.90 light loss factor (0.95 lamp lumen depreciation; 0.95 luminaire dirt depreciation). Scale is ¼ inch equals 1 foot.]

⊃Possible vendors: **Hubbell/Sterner and Norton Industries** (socket strip, spreadlens and baffle), Litelab and Norton Industries

▶ **Lamp: 39PAR20/CMH/FL (p5–55)**
Luminaire: 5⅝"W by 7⁵⁄₁₆"H Socket strip/Spreadlens/Baffle Components in Drywall Slot Detail
Ceiling height: **15'–0"**
Drywall slot opening width: **0'–6"**
Lamp spacings: **1'–0" o.c.**
Wall illuminance (19'x9'–6" area):
▶ 47 fc, avg, maintained (142 fc max/9 fc min)
▶ Commercial, Hospitality, Retail

Design Tips
✔Use where wall design is interesting.
✔Effect is strongest with lighter colored walls.
✔Enhances wood paneling and stone or metal wall materials.
✔Where art preservation is a necessity, specify UV–reduction spreadlens.
✔Grazing light will telegraph wall imperfections.
✔If artwork is located on wall, select frameless or minimally framed pieces as grazing light on frames creates significant shadow lines.
✔Maintenance intensive (may want to reserve use for installations with few hours of daily operation).
✔Greater chance for viewing lamp/socket/baffle assembly directly (no view shield). This is minimized by deepening pocket.

Details Concepts and Effects

11

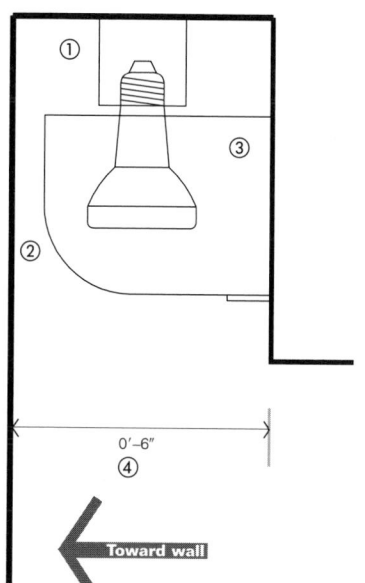

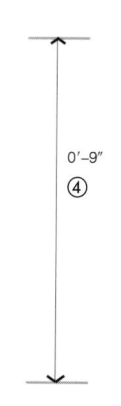

0'–9"
④

Notes
①Paint slot "ceiling" to match wall tone.
②Continue finished wall material up into slot to slot "ceiling."
③Aluminum baffle painted black or to match wall/ceiling color is centered between each lamp and attached to "inside" soffit wall of slot pocket. This is labor intensive. Baffle is 4" in height and projects 5".
④Manufacturing and photometric characteristics of luminaire may change from time to time—reconfirm detail dimensioning with luminaire manufacturer prior to final specification.

0'–6"
④

Toward wall

Norton baffle profile basis for sketch (see each mfgr's data for precise profile, dimensions and detail requirements)

Net Addresses/Luminaires
See page 11–99

CONNECT FOR MORE

Wallslots Drywall/Fluorescent Striplights

9′–6″ Ceiling Height/1–lamp Cross–section/0′–8″ Slot Width

DetailConcepts™

▶ **Lamp: F32T8/830 (p8–52)**
Luminaire: Striplight/Drywall Slot
Ceiling height: **9′–6″**
Wall illuminance (19′x8′–6″ area):
▶ 15 fc, avg, maintained (94 fc max/1 fc min)
▶ Commercial, Hospitality, Institutional

▶ Lighted appearance of matte/light–colored wall (50% reflectance/paint or fabric)

▶ **Lamp: F54T5/HO/830 (p8–50)**
Luminaire: Striplight/Drywall Slot
Ceiling height: **9′–6″**
Wall illuminance (19′x8′–6″ area):
▶ 26 fc, avg, maintained (168 fc max/2 fc min)
▶ Commercial, Institutional, Retail

▶ Lighted appearance of matte/light–colored wall (50% reflectance/paint or fabric)

▶ **Lamp: F40Long/830 (p8–36)**
Luminaire: Striplight/Drywall Slot
Ceiling height: **9′–6″**
Wall illuminance (19′x8′–6″ area):
▶ 30 fc, avg, maintained (187 fc max/2 fc min)
▶ Commercial, Institutional, Retail

▶ Lighted appearance of matte/light–colored wall (50% reflectance/paint or fabric)

Wallslots Drywall/Fluorescent Striplights

9'–6" Ceiling Height/1–lamp Cross–section/0'–8" Slot Width

Performance Sketches

[All data outlined here is based on information from boldfaced manufacturer's published data at time of manuscript preparation. Photometric data were rerated as necessary for cited lamp. Maintained values based on analysis grid centered on wall and extending to 0'–6" from edges and on 0.90 light loss factor (0.95 lamp lumen depreciation; 0.95 luminaire dirt depreciation). Scale is ¼ inch equals 1 foot.]

DetailConcepts™

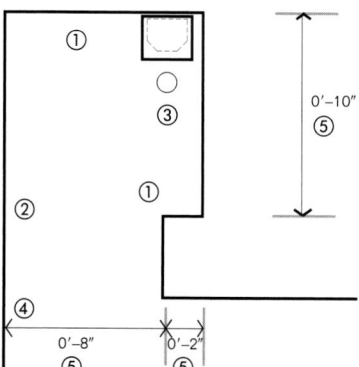

Notes
①Paint all interior cove surfaces matte white.
②Continue finished wall material up into slot.
③Center lamp over lower lip for optimal light coverage down the wall while minimizing direct view from room. Advisory: without baffling, lamps/luminaires are visible from end viewing situations.
④Socket shadows are noticeable along back slot wall with 1–lamp striplight in this location. Single–lamp staggered strips introduce a "reverse socket shadow"—lamps closer to back cove wall create a slightly greater intensity. Butt strips end–to–end tightly.
⑤Manufacturing and photometric characteristics of luminaire may change from time to time—reconfirm detail dimensioning with luminaire manufacturer prior to final specification.
⑥T8 and T5 lamps offer same light output—so striplights can be either version. Lamps must be combined with correct ballast and are not interchangeable after installation. See below for T5/HO version.

T5 and T8 Striplights

⊃Possible vendors: Bartco, Columbia, DayBrite, Legion, Lightolier, **Lithonia**[1], Metalux

[1]T12 photometry was used and rerated for T8 lamping. This is somewhat imprecise.

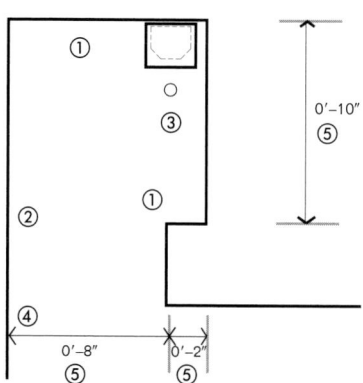

Notes
①Paint all interior cove surfaces matte white.
②Continue finished wall material up into slot.
③Center lamp over lower lip for optimal light coverage down the wall while minimizing direct view from room. Advisory: without baffling, lamps/luminaires are visible from end viewing situations.
④Socket shadows are noticeable along back slot wall with 1–lamp striplight in this location. Single–lamp staggered strips introduce a "reverse socket shadow"—lamps closer to back cove wall create a slightly greater intensity. Butt strips end–to–end tightly.
⑤Manufacturing and photometric characteristics of luminaire may change from time to time—reconfirm detail dimensioning with luminaire manufacturer prior to final specification.

T5/HO Striplights

⊃Possible vendors: Bartco, Legion, Lightolier, **Lithonia**[1]

[1]T12 photometry was used and rerated for T5/HO lamping. This is somewhat imprecise.

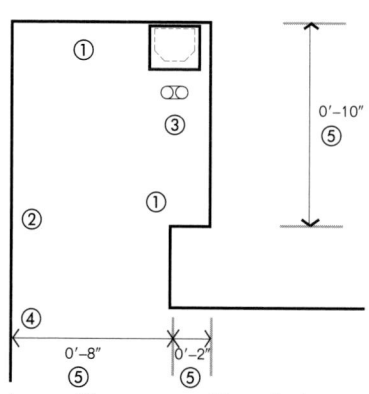

Notes
①Paint all interior cove surfaces matte white.
②Continue finished wall material up into slot.
③Center lamp over lower lip for optimal light coverage down the wall while minimizing direct view from room. Advisory: without baffling, lamps/luminaires are visible from end viewing situations.
④Socket shadows are less noticeable with the LongCompact lamp. Butt strips end–to–end tightly.
⑤Manufacturing and photometric characteristics of luminaire may change from time to time—reconfirm detail dimensioning with luminaire manufacturer prior to final specification.

LongCompact Striplights

⊃Possible vendors: Columbia, DayBrite, Legion, Lightolier, **Lithonia**[1], Metalux

[1]T12 photometry was used and rerated for LongCompact lamping. This is somewhat imprecise.

Details
Concepts and Effects

11

Design Tips

✔Effect is strongest with lighter colored walls.
✔Does not enhance wood or stone materials—good on plain painted drywall construction and fabric covered walls.
✔Grazing light will telegraph wall imperfections—particularly problematic on drywall construction and pillowed fabric.
✔Not suggested for artwork walls unless an additional layer of accent lighting is used to highlight artwork.
✔Energy–effective solution.
✔Baffling may be necessary if end viewing situations will be encountered.
✔Manufactured wallslots are available from the likes of Columbia, Neoray, Linear Lighting and Lithonia providing better optical control of the light (better uniformity) and better glare control.

Net Addresses/Luminaires
See page 11–99

CONNECT FOR MORE

DetailConcepts™

Wallslots Drywall/Fluorescent Striplights

12'–6" Ceiling Height/1–lamp Cross–section/0'–8" Slot Width

Details Concepts and Effects

11

▶ **Lamp: F32T8/830 (p8–52)**
Luminaire: Striplight/Drywall Slot
Ceiling height: **12'–6"**
Wall illuminance (19'x11'–6" area):
▶ 12 fc, avg, maintained (101 fc max/1 fc min)
▶ Commercial, Hospitality, Institutional

▶ **Lighted appearance of matte/light–colored wall (50% reflectance/paint or fabric)**

▶ **Lamp: F54T5/HO/830 (p8–50)**
Luminaire: Striplight/Drywall Slot
Ceiling height: **12'–6"**
Wall illuminance (19'x11'–6" area):
▶ 20 fc, avg, maintained (153 fc max/1 fc min)
▶ Commercial, Institutional, Retail

▶ **Lighted appearance of matte/light–colored wall (50% reflectance/paint or fabric)**

Wallslots Drywall/Fluorescent Striplights

12'–6" Ceiling Height/1–lamp Cross–section/ 0'–8" Slot Width

Performance Sketches

[All data outlined here is based on information from boldfaced manufacturer's published data at time of manuscript preparation. Photometric data were rerated as necessary for cited lamp. Maintained values based on analysis grid centered on wall and extending to 0'–6" from edges and on 0.90 light loss factor (0.95 lamp lumen depreciation; 0.95 luminaire dirt depreciation). Scale is ¼ inch equals 1 foot.]

Detail Concepts™

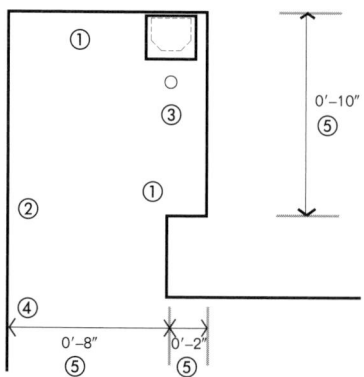

Notes
①Paint all interior cove surfaces matte white.
②Continue finished wall material up into slot.
③Center lamp over lower lip for optimal light coverage down the wall while minimizing direct view from room. Advisory: without baffling, lamps/luminaires are visible from end viewing situations.
④Socket shadows are noticeable along back slot wall with 1–lamp striplight in this location. Single–lamp staggered strips introduce a "reverse socket shadow"—lamps closer to back cove wall create a slightly greater intensity. Butt strips end–to–end tightly.
⑤Manufacturing and photometric characteristics of luminaire may change from time to time—reconfirm detail dimensioning with luminaire manufacturer prior to final specification.
⑥T8 and T5 lamps offer same light output—so striplights can be either version. Lamps must be combined with correct ballast and are not interchangeable after installation. See below for T5/HO version.

T5 and T8 Striplights
⮕Possible vendors: Bartco, Columbia, DayBrite, Legion, Lightolier, **Lithonia**[1], Metalux
[1]T12 photometry was used and rerated for T8 lamping. This is somewhat imprecise.

Notes
①Paint all interior cove surfaces matte white.
②Continue finished wall material up into slot.
③Center lamp over lower lip for optimal light coverage down the wall while minimizing direct view from room. Advisory: without baffling, lamps/luminaires are visible from end viewing situations.
④Socket shadows are noticeable along back slot wall with 1–lamp striplight in this location. Single–lamp staggered strips introduce a "reverse socket shadow"—lamps closer to back cove wall create a slightly greater intensity. Butt strips end–to–end tightly.
⑤Manufacturing and photometric characteristics of luminaire may change from time to time—reconfirm detail dimensioning with luminaire manufacturer prior to final specification.

T5/HO Striplights
⮕Possible vendors: Bartco, Legion, Lightolier, **Lithonia**[1]
[1]T12 photometry was used and rerated for T5/HO lamping. This is somewhat imprecise.

Design Tips
✔Effect is strongest with lighter colored walls.
✔Does not enhance wood or stone materials—good on plain painted drywall construction and fabric covered walls.
✔Grazing light will telegraph wall imperfections—particularly problematic on drywall construction and pillowed fabric.
✔Not suggested for artwork walls unless an additional layer of accent lighting is used to highlight artwork.
✔Energy–effective solution.
✔Baffling may be necessary if end viewing situations will be encountered.
✔Manufactured wallslots are available from the likes of Columbia, Neoray, Linear Lighting and Lithonia providing better optical control of the light (better uniformity) and better glare control.

Details Concepts and Effects **11**

Net Addresses/Luminaires
See page 11–99

DetailConcepts™

Wallslots Drywall/Fluorescent Striplights

12'–6" Ceiling Height/1–lamp Cross–section/0'–8" Slot Width

▶ **Lamp: F40Long/830 (p8–36)**

Luminaire: Striplight/Drywall Slot
Ceiling height: **12'–6"**
Wall illuminance (19'x11'–6" area):
▶ 23 fc, avg, maintained (185 fc max/2 fc min)
▶ Commercial, Institutional, Retail

▶ **Lighted appearance of matte/light–colored wall (50% reflectance/paint or fabric)**

Details
Concepts and Effects

11

Wallslots Drywall/Fluorescent Striplights

12'–6" Ceiling Height/1–lamp Cross–section/ 0'–8" Slot Width

Performance Sketches

[All data outlined here is based on information from boldfaced manufacturer's published data at time of manuscript preparation. Photometric data were rerated as necessary for cited lamp. Maintained values based on analysis grid centered on wall and extending to 0'–6" from edges and on 0.90 light loss factor (0.95 lamp lumen depreciation; 0.95 luminaire dirt depreciation). Scale is ¼ inch equals 1 foot.]

DetailConcepts™

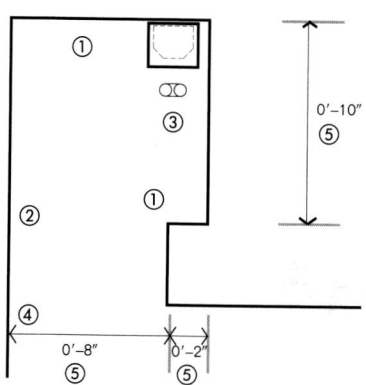

Notes
① Paint all interior cove surfaces matte white.
② Continue finished wall material up into slot.
③ Center lamp over lower lip for optimal light coverage down the wall while minimizing direct view from room. **Advisory: without baffling, lamps/luminaires are visible from end viewing situations.**
④ Socket shadows are less noticeable with the LongCompact lamp. Butt strips end–to–end tightly.
⑤ Manufacturing and photometric characteristics of luminaire may change from time to time—reconfirm detail dimensioning with luminaire manufacturer prior to final specification.

LongCompact Striplights

⊃ Possible vendors: Columbia, DayBrite, Legion, Lightolier, **Lithonia**[1], Metalux

[1] T12 photometry was used and rerated for LongCompact lamping. This is somewhat imprecise.

Design Tips

✔ Effect is strongest with lighter colored walls.
✔ Does not enhance wood or stone materials—good on plain painted drywall construction and fabric covered walls.
✔ Grazing light will telegraph wall imperfections—particularly problematic on drywall construction and pillowed fabric.
✔ Not suggested for artwork walls unless an additional layer of accent lighting is used to highlight artwork.
✔ Energy–effective solution.
✔ Baffling may be necessary if end viewing situations will be encountered.
✔ Manufactured wallslots are available from the likes of Columbia, Neoray, Linear Lighting and Lithonia providing better optical control of the light (better uniformity) and better glare control.
✔ Using asymmetric luminaires in the slot (from the likes of Elliptipar) in place of striplights will provide better optical control of the light (better uniformity).

Details Concepts and Effects

11

Net Addresses/Luminaires
See page 11–99

CONNECT FOR MORE

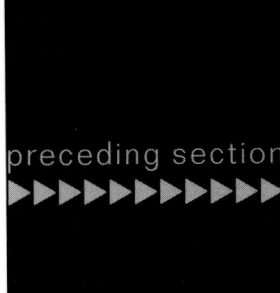

preceding section
▶▶▶▶▶▶▶▶▶▶▶

next section
▶▶▶▶▶▶▶▶▶▶▶

11

Details

Index

Index

Index

Note: all lamp entries are cited in order of lowest wattage to highest wattage. Within each wattage family, lamps are cited in order of narrowest (spottiest) distribution to widest (floodiest) distribution.

Index

Index

Index

Index

Index

Index

Index

Index

Index

Index

Index

Index

Index

Index

Index

About the Author

Gary Steffy is the principal of Gary Steffy Lighting Design Inc. and is one of the world's preeminent lighting designers. He is a fellow and past president of the International Association of Lighting Designers (IALD), a member of the U.S. National Committee of the International Commission on Illumination (CIE), a member of the Illuminating Engineering Society of North America (IESNA), and a founding director of the National Council on Qualifications for the Lighting Professions (NCQLP). His activities in the IESNA include appointment to the Technical Review Council. He is the author of *Lighting the Electronic Office* and *Architectural Lighting Design*. He has written many articles for the lighting magazines *Architectural Lighting* and *Lighting Design + Application*. He has taught lighting design at Michigan State University, The Pennsylvania State University, University of Michigan and Wayne State University. His firm's work has been featured in *Architectural Record* and *Interior Design*.